T0418742

Antarctic Climate Evolution

Antarctic Climate Evolution

Second Edition

Edited by

Fabio Florindo
National Institute of Geophysics and Volcanology, Rome, Italy

Martin Siegert
Grantham Institute and Department of Earth Science and Engineering, Imperial College London, London, United Kingdom

Laura De Santis
National Institute of Oceanography and Applied Geophysics—OGS, Sgonico, Trieste, Italy

Tim Naish
Antarctic Research Centre, Victoria University of Wellington, Wellington, New Zealand

Elsevier
Radarweg 29, PO Box 211, 1000 AE Amsterdam, Netherlands
The Boulevard, Langford Lane, Kidlington, Oxford OX5 1GB, United Kingdom
50 Hampshire Street, 5th Floor, Cambridge, MA 02139, United States

British Library Cataloguing-in-Publication Data
A catalogue record for this book is available from the British Library

Library of Congress Cataloging-in-Publication Data
A catalog record for this book is available from the Library of Congress

ISBN: 978-0-12-819109-5

For Information on all Elsevier publications
visit our website at https://www.elsevier.com/books-and-journals

Publisher: Candice Janco
Acquisitions Editor: Marisa LaFleur
Editorial Project Manager: Andrea Dulberger
Production Project Manager: Paul Prasad Chandramohan
Cover Designer: Miles Hitchen

Typeset by MPS Limited, Chennai, India

Contents

List of contributors

Michael J. Bentley Department of Geography, Durham University, Durham, United Kingdom

Peter Bijl Laboratory of Palaeobotany and Palynology, Department of Earth Sciences, Marine Palynology and Paleoceanography, Utrecht University, Utrecht, the Netherlands

Helen Bostock-Lyman School of Earth and Environmental Sciences, University of Queensland, Brisbane, QLD, Australia

Melissa Bowen School of Environment, University of Auckland, Auckland, New Zealand

Henk Brinkuis Laboratory of Palaeobotany and Palynology, Department of Earth Sciences, Marine Palynology and Paleoceanography, Utrecht University, Utrecht, the Netherlands; Coastal Systems Department, Royal Netherlands Institute for Sea Research, Utrecht University, Den Burg, the Netherlands

Lionel Carter Antarctic Research Centre, Victoria University of Wellington, Wellington, New Zealand

Hannah K. Chorley Antarctic Research Centre, Victoria University of Wellington, Wellington New Zealand

Florence Colleoni National Institute of Oceanography and Applied Geophysics – OGS, Sgonico, Italy

Laura De Santis National Institute of Oceanography and Applied Geophysics – OGS, Sgonico, Italy

Robert M. DeConto Department of Geosciences, University of Massachusetts, Amherst, Amherst, MA, United States; Institute for Climate Change Solutions, Frontone, Italy

Warren Dickinson Antarctic Research Centre, Victoria University of Wellington, Wellington, New Zealand

Aisling M. Dolan School of Earth and Environment, University of Leeds, Leeds, United Kingdom

Federica Donda National Institute of Oceanography and Applied Geophysics – OGS, Sgonico, Italy

Bella Duncan Antarctic Research Centre, Victoria University of Wellington, Wellington, New Zealand

Carlota Escutia Andalusian Institute of Earth Sciences, CSIC and Universidad de Granada, Armilla, Spain

Tina van de Flierdt Department of Earth Science and Engineering, Imperial College London, London, United Kingdom

Fabio Florindo National Institute of Geophysics and Volcanology, Rome, Italy; Institute for Climate Change Solutions, Frontone, Italy

Jane Francis British Antarctic Survey, Cambridge, United Kingdom

Simone Galeotti Department of Pure and Applied Sciences, University of Urbino Carlo Bo, Urbino, Italy; Institute for Climate Change Solutions, Frontone, Italy

Edward G.W. Gasson School of Geographical Sciences, University of Bristol, Bristol, United Kingdom; Centre for Geography and Environmental Science, University of Exeter, Cornwall Campus, United Kingdom

Claudio Ghezzo Department of Physical, Earth and Environmental Sciences, University of Siena, Siena, Italy

Karsten Gohl Alfred Wegener Institute, Helmholtz-Center for Polar and Marine Science, Bremerhaven, Germany; School of Geography, Geology and the Environment, University of Leicester, Leicester, United Kingdom

Nicholas R. Golledge Antarctic Research Centre, Victoria University of Wellington, Wellington, New Zealand

Damian B. Gore Department of Earth and Environmental Sciences, Macquarie University, Sydney, NSW, Australia

Georgia R. Grant GNS Science, Avalon, Lower Hutt, New Zealand

Sean Gulick Institute for Geophysics & Deptartment of Geological Sciences, Jackson School of Geosciences, University of Texas at Austin, Austin, TX, United States

Richard H. Levy Antarctic Research Centre, Victoria University of Wellington, Wellington, New Zealand; GNS Science, Lower Hutt, New Zealand

Anna Ruth W. Halberstadt Climate System Research Center, University of Massachusetts, Amherst, MA, United States

David M. Harwood Department of Earth and Atmospheric Sciences, University of Nebraska, Lincoln, NE, United States

Andrew S. Hein School of GeoSciences, University of Edinburgh, Edinburgh, United Kingdom

Javier Hernández-Molina Department of Earth Sciences, Royal Holloway University of London, Egham, Surrey, United Kingdom

Claus-Dieter Hillenbrand British Antarctic Survey, Cambridge, United Kingdom

Katharina Hochmuth School of Geography, Geology and the Environment, University of Leicester, Leicester, United Kingdom; Alfred Wegener Institute, Helmholtz-Center for Polar and Marine Science, Bremerhaven, Germany

David Hutchinson Department of Geological Sciences and Bolin Centre for Climate Research, Stockholm University, Stockholm, Sweden

Stewart Jamieson Department of Geography, Durham University, Durham, United Kingdom

Alan Kennedy-Asser BRIDGE, School of Geographical Sciences, University of Bristol, Bristol, United Kingdom

Sookwan Kim Korea Polar Research Institute, Incheon, Republic of Korea

Georg Kleinschmidt Institute for Geosciences, University of Frankfurt, Frankfurt, Germany

Douglas E. Kowalewski Department of Earth, Environment, and Physics, Worcester State University, Worcester, MA, United States

Gerhard Kuhn Alfred Wegener Institute, Helmholtz-Center for Polar and Marine Science, Bremerhaven, Germany; School of Geography, Geology and the Environment, University of Leicester, Leicester, United Kingdom

Luca Lanci Department of Pure and Applied Sciences, University of Urbino Carlo Bo, Urbino, Italy; Institute for Climate Change Solutions, Frontone, Italy

Robert Larter British Antarctic Survey, Cambridge, United Kingdom

German Leitchenkov Institute for Geology and Mineral Resources of the World Ocean, St. Petersburg, Russia; Institute of Earth Sciences, St. Petersburg State University, St. Petersburg, Russia

Richard H. Levy GNS Science, Avalon, Lower Hutt, New Zealand; Antarctic Research Centre, Victoria University of Wellington, Wellington, New Zealand

Adam R. Lewis Department of Geosciences, North Dakota State University, Fargo, ND, United States

Robert M. McKay Antarctic Research Centre, Victoria University of Wellington, Wellington, New Zealand; School of Earth and Environment, University of Leeds, Leeds, United Kingdom

Antonio Meloni National Institute of Geophysics and Volcanology, Rome, Italy

Stephen R. Meyers Department of Geoscience, University of Wisconsin-Madison, Madison, WI, United States

Tim R. Naish Antarctic Research Centre, Victoria University of Wellington, Wellington, New Zealand

Christian Ohneiser Department of Geology, University of Otago, Dunedin, New Zealand

Phil O'Brien Department of Environment and Geography, Macquarie University, Sydney, NSW, Australia

Molly O. Patterson Department of Geological Sciences and Environmental Studies, Binghamton University, Binghamton, NY, United States

Lara F. Pérez British Antarctic Survey, Cambridge, United Kingdom

Ross Powell Department of Geology and Environmental Geosciences, Northern Illinois University, DeKalb, IL, United States

Francesca Sangiorgi Laboratory of Palaeobotany and Palynology, Department of Earth Sciences, Marine Palynology and Paleoceanography, Utrecht University, Utrecht, The Netherlands

Laura De Santis National Institute of Oceanography and Applied Geophysics, Trieste, Italy

Isabel Sauermilch Laboratory of Palaeobotany and Palynology, Department of Earth Sciences, Marine Palynology and Paleoceanography, Utrecht University, Utrecht, the Netherlands; Institute for Marine and Antarctic Studies, University of Tasmania, Hobart, TAS, Australia

Amelia E. Shevenell College of Marine Science, University of South Florida, St. Petersburg, FL, United States

Martin Siegert Grantham Institute and Department of Earth Science and Engineering, Imperial College London, London, United Kingdom

Appy Sluijs Laboratory of Palaeobotany and Palynology, Department of Earth Sciences, Marine Palynology and Paleoceanography, Utrecht University, Utrecht, the Netherlands; Institute for Climate Change Solutions, Frontone, Italy

Paolo Stocchi Laboratory of Palaeobotany and Palynology, Department of Earth Sciences, Marine Palynology and Paleoceanography, Utrecht University, Utrecht, The Netherlands; Coastal Systems Department, Royal Netherlands Institute for Sea Research, Utrecht University, Den Burg, the Netherlands; Institute for Climate Change Solutions, Frontone, Italy

Franco Talarico Department of Physical, Earth and Environmental Sciences, University of Siena, Siena, Italy; National Museum for Antarctica, University of Siena, Siena, Italy

Gabriele Uenzelmann-Neben Alfred Wegener Institute Helmholtz Center for Polar and Marine Research, Bremerhaven, Germany

Tina van de Flierdt Department of Earth Science and Engineering, Imperial College London, London, United Kingdom

Marjolaine Verret Antarctic Research Centre, Victoria University of Wellington, Wellington, New Zealand

Duanne A. White Institute for Applied Ecology, University of Canberra, Canberra, ACT, Australia

Trevor Williams International Ocean Discovery Program, Texas A&M University, College Station, TX, United States

David J. Wilson Institute of Earth and Planetary Sciences, University College London and Birkbeck, University of London, London, United Kingdom; Department of Earth Sciences, University College London, London, United Kingdom

Gary Wilson GNS Science, Lower Hutt, New Zealand

Preface

In July 2008, we published the first edition of *Antarctic Climate Evolution*; the first book dedicated to understanding the origin and development of the world's largest ice sheet and, in particular, how it responded to and influenced climate change during the Cenozoic. The book's content largely mirrored the structure of the Antarctic Climate Evolution (ACE) program, which was an international initiative of the Scientific Committee on Antarctic Research (SCAR), to investigate past changes in Antarctica by linking climate and ice-sheet modelling studies with terrestrial and marine geological and geophysical records. ACE was succeeded by another SCAR programme named Past Antarctic Ice Sheet dynamics (PAIS), which existed between 2013 and 2020. By building on the ACE legacy, and because of significant improvements in ice-sheet modelling and the acquisition of palaeoclimate records in key regions, PAIS led to new insights into Antarctica's contribution to former global sea-level change over timescales from centuries to multiple millennia. PAIS also helped to understand better the interconnections between ice-sheet mass loss and atmospheric and oceanic processes at local, regional and global levels.

The second edition of *Antarctic Climate Evolution* is a result of both SCAR programmes, and serves to document the 'state of knowledge' concerning ice and climate evolution of the Antarctic continent and its surrounding seas from the beginning of the Cenozoic era to the present day. We hope the book will continue to be of interest to research scientists from a wide range of disciplines including glaciology, palaeoclimatology, sedimentology, climate change, environmental science, oceanography and palaeoentology. We also anticipate that it can serve as a guide to those wishing to understand how Antarctica has changed in the past, and how past change can inform our future.

Fabio Florindo[1], Martin J. Siegert[2], Laura De Santis[3] and Tim R. Naish[4]
[1]National Institute of Geophysics and Volcanology, Rome, Italy, [2]Grantham Institute and Department of Earth Science and Engineering, Imperial College London, London, United Kingdom, [3]National Institute of Oceanography and Applied Geophysics—OGS, Sgonico, Trieste, Italy, [4]Antarctic Research Centre, Victoria University of Wellington, Wellington, New Zealand

Chapter 1

Antarctic Climate Evolution – second edition

Fabio Florindo[1], Martin Siegert[2], Laura De Santis[3] and Tim R. Naish[4]

[1]National Institute of Geophysics and Volcanology, Rome, Italy, [2]Grantham Institute and Department of Earth Science and Engineering, Imperial College London, London, United Kingdom, [3]National Institute of Oceanography and Applied Geophysics – OGS, Sgonico, Italy, [4]Antarctic Research Centre, Victoria University of Wellington, Wellington, New Zealand

1.1 Introduction

The Antarctic continent and the Southern Ocean are influential components of the Earth System. Central to the understanding of global climate change (including increases in temperature, precipitation and ocean pH) is an appreciation of how the Antarctic Ice Sheet interacts with climate, especially during times of rapid change. To comprehend the rates, mechanisms and impact of the processes involved, one must look into the geological record for evidence of past changes, on time scales from centuries and up to millions of years. For several decades, international efforts have been made to determine the glacial, tectonic and climate history of Antarctica and the Southern Ocean. Much of this information derives from studies of sedimentary sequences, drilled and correlated via seismic reflection data in and around the continent (e.g., Cooper et al., 2009). In addition, there have been numerous terrestrial geological expeditions to the mountains exposed above the ice, usually close to the margin of the ice sheet (e.g., GANOVEX expeditions). Holistic interpretation of these data is now being made, and new challenging hypotheses on the size and timing of past changes in Antarctica are being developed.

In 2004 the Scientific Committee on Antarctic Research (SCAR) commissioned a scientific research programme on Antarctic Climate Evolution (ACE) to quantify the glacial and climate history of Antarctica by linking climate and ice sheet modelling studies with terrestrial and marine geological and geophysical evidence of past changes. ACE grew out of the ANTOSTRAT (ANTarctic Offshore STRATigraphy) project, which was

Antarctic Climate Evolution. DOI: https://doi.org/10.1016/B978-0-12-819109-5.00007-4

sanctioned by SCAR in 1990 to reconstruct the Cenozoic palaeoclimatic and glacial history of the Antarctic region from the study of the sedimentary record surrounding the continent. The main achievements of ANTOSTRAT and ACE were published in a set of special issues (Barrett et al., 2006; Escutia et al., 2012; Florindo et al., 2003, 2005, 2008, 2009) and summarised in the first edition of this book published in December 2008 (Florindo and Siegert, 2008). ACE was followed from 2013 to 2020 by the Past Antarctic Ice Sheet (PAIS) dynamics programme, which continued to work on constraining Antarctica's contribution to sea level resulting from past changes in ice sheet mass loss, and understanding its impacts on global environments through changes to atmospheric and oceanic circulation. Based on paleo analysis, PAIS aimed to bound the estimates of future ice loss in key areas of the Antarctic margin with a multidisciplinary geoscientific approach and, importantly, by integrating observations and records with numerical models.

The PAIS research philosophy was based on data–data and data–model integration and intercomparison, and the development of 'ice-to-abyss' data transects, extending from the ice-sheet interior to the deep sea (Fig. 1.1). The 'data transect' concept links ice cores, ice sheet-proximal information, offshore sediments and far-field records of past ice sheet behaviour and sea level change, allowing reconstructions of former ice sheet geometries, and ice sheet–ocean processes and their interactions. Different sectors respond differently to external forcing due to a variety of constraints including bed topography and geology, proximity to warm water masses and ice accumulation rates (to list a few), so results from one sector are not necessarily representative of the whole of Antarctica. Therefore PAIS aimed to develop several transects across numerous regions. These integrated datasets enable robust testing of a new generation of ice-sheet models (Siegert and Gollege, 2021), which are beginning to be coupled with glacial isostatic, atmosphere and ocean models.

ANTOSTRAT, ACE and PAIS stratigraphic studies were based on a huge compilation of multichannel seismic profiles, collected by many nations and made freely available via the Antarctic Seismic Data Library System, established and endorsed in 1991 by SCAR and by the Antarctic Treaty for scientific cooperation and research purposes (see McKay et al., 2021).

Extensive PAIS-facilitated fieldwork on land and at sea has been planned and undertaken within a framework of national and multinational projects, including International Ocean Discovery Program (IODP) expeditions 374 (in 2018) (McKay et al., 2019), 379 (Gohl et al., 2021) and 382 (Weber et al., 2021) (in 2019). PAIS research addressed some of the key questions formulated by the 20-year Scientific Horizon Scan for understanding Antarctic and Southern Ocean processes (Kennicutt et al., 2014, 2015, 2016, 2019), was influential in the Intergovernmental Panel on Climate Change (IPCC) Fifth Assessment Report (AR5) (IPCC, 2014), the IPCC Special

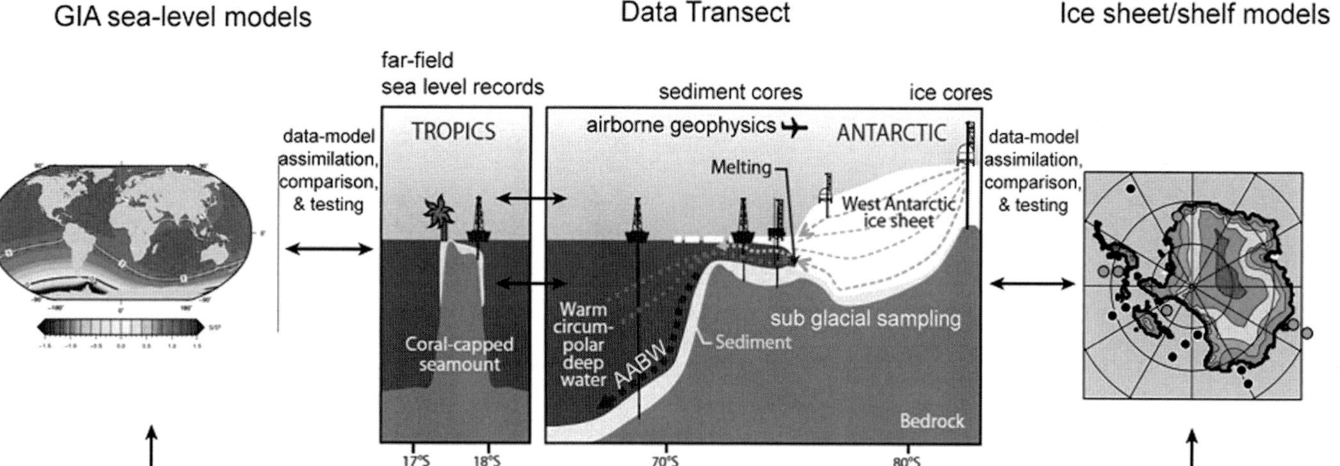

FIGURE 1.1 PAIS research approach. *Left panel*: Change of relative sea level caused by a deglaciation episode in West Antarctica normalised by the ocean-averaged value, obtained solving the Sea Level Equation for an elastic Earth. *Central panel*: Proposed drilling strategy using International Ocean Discovery Program drilling platforms to collect records linking climate, ice sheet and sea level histories on geologic time scales. *Right panel*: Ice sheet volume reconstructions for extreme interglacials considering four forcing mechanisms (sub-ice-shelf oceanic melting, sea level changes, annual precipitation changes and temperature changes from present). Black dots in the right panel indicate the ANDRILL, the International Ocean Discovery Program (IODP) deep and shallow (SHALDRILL and MeBO) drill sites recovered from 2010 to 2019. Grey dots are pending/approved IODP proposals developed during PAIS. *Left panel: Adapted from Spada, 2017, (Spada and Melini, 2019), Central panel: Adapted from figure 2.6 of (Bickle et al., 2011); Right panel: Adapted from Pollard and DeConto (2009).*

Report on the Ocean and Cryosphere in a Changing Climate (IPCC, 2019) and the IPCC Sixth Assessment Report (AR6), (IPCC, 2021), which will be published in full during 2022.

The deep-sea oxygen isotope and atmospheric CO_2 records (Fig. 1.2) show an overall climate cooling combined with ice volume increase and CO_2 decline during the Cenozoic, punctuated by events recognised in both seismic and sedimentary records. The reconstructions of environmental conditions, and ice-sheet response, during such events have been a focus of PAIS, with the aim to provide the IPCC with such knowledge to better understand future changes we are locked into and those we can still avoid.

This book is a culmination of findings from both ACE and PAIS, and still benefiting from the legacy of ANTOSTRAT. It documents the state of knowledge concerning the ice and climate evolution of the Antarctic continent and its surrounding seas through the Cenozoic era.

1.2 Structure and content of the book

The book opens with a chapter (Chapter 2) (Florindo et al., 2021a) that provides a background to the role of SCAR in over 60 years of coordination and support of high-quality scientific research in the Antarctic and Southern Ocean. Chapter 3 (McKay et al., 2021) summarises the current state of knowledge of Cenozoic climate history in Antarctica in the context of near- and far-field records, followed by a detailed discussion of the seismic stratigraphy, and drill core constraints and palaeoclimatic records preserved within this stratigraphy, from around the continental margin. Chapter 4 (Carter et al., 2021) focuses on

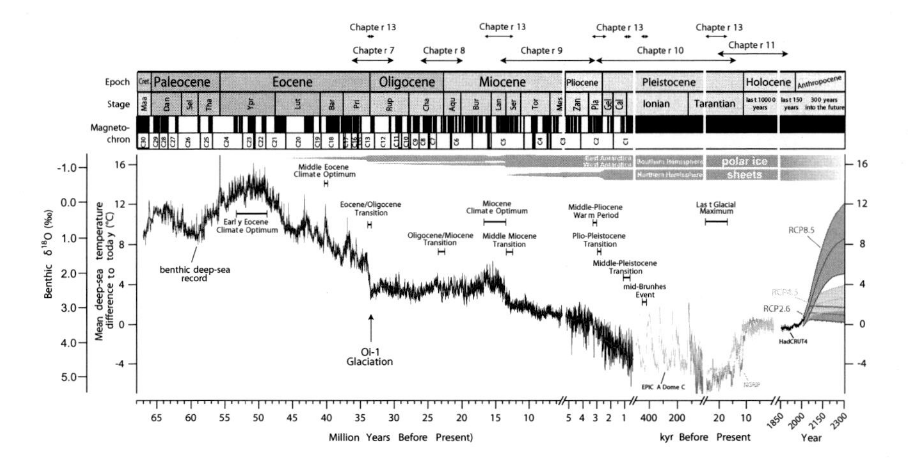

FIGURE 1.2 Global climate spanning the last 67 million years and future projections of climate change. Specific time periods in Antarctic history that are discussed in this book (Chapters 7–11 and 13) are indicated at the top. *Published with permission by Thomas Westerhold.*

the modern oceanography to provide a physical basis for realistic reconstructions of past environments in Antarctica. Chapter 5 (Siegert and Golledge, 2021) synthesises developments in ice sheet modelling over the last decade, and how they have helped to understand the growth and decay of ice sheets during the glaciated history of Antarctica. Chapter 6 (Talarico et al., 2021) provides an overview of the Antarctic continent evolution from its inclusion as part of the Gondwana supercontinent to the break-up of this landmass and the repositioning of Antarctica at southern polar latitudes since the Early Cretaceous.

From Chapters 7 to 11, the book presents a series of reviews dealing with specific time periods in Antarctic history: Eocene/Oligocene (Chapter 7) (Galeotti et al., 2021), Oligocene/Miocene (Chapter 8) (Naish et al., 2021), Miocene to Pliocene (Chapter 9) (Levy et al., 2021), Pleistocene (Chapter 10) (Wilson et al., 2021) and the Last Glacial Maximum and Holocene (Chapter 11) (Siegert et al., 2021a). Chapter 12 (Colleoni et al., 2021) focuses on how PAIS research has improved our understanding about the ice and climate evolution of the Antarctic continent and its surrounding seas through the last 65 million years. The final Chapter 13 (Siegert et al., 2021b) briefly summarises the PAIS legacy and highlights research priorities needed for over the next decade to answer key scientific questions on the role of the Antarctic continent in global climate change.

Acknowledgements

We are grateful to our many friends and colleagues for advice and encouragement through the production of this volume. We thank Andrea Dulberger, Editorial Project Manager of Elsevier Science, for the support in the production of this book. During the preparation of this book, our colleague and friend Franco Talarico, lead author of Chapter 6 (The Antarctic Continent in Gondwana: A perspective from the Ross Embayment and Potential Research Targets for Future Investigations) died unexpectedly on 15th December 2020. This book is dedicated to his memory.

References

Bamber, J.L., Riva, R.E.M., Vermeersen, B.L.A., LeBrocq, A.M., 2009. Reassessment of the potential sea-level rise from a collapse of the West Antarctic ice sheet. Science 324 (5929), 901–903. Available from: https://doi.org/10.1126/science.1169335.

Barrett, P.J., Florindo, F., Cooper, A.K. (Eds.), 2006. Antarctic climate evolution: geological records from the margin and modelling. Palaegeography, Palaeoclimatology, Palaeoecology, 231, 1–252.

Bickle, M., Arculus, R., Barrett, P., DeConto, R., Camoin, G., Edwards, K., Fisher, F., Inagaki, F., Kodaira, S., Ohkouchi, N., Palike, H., Ravelo, C., Saffer, D., Teagle, D., 2011. Illuminating Earth's Past, Present and Future. The Science Plan for the International Ocean Discovery Program 2013 – 2023. IODP: Integrated ocean drilling program, Washington, D.C.: Integrated Ocean Drilling Program.

Carter, L., et al., 2021. Circulation and water masses. In: Florindo, F., et al., (Eds.), Antarctic Climate Evolution, Second ed. Elsevier.

Colleoni, F., et al.,PAIS community 2021. Past Antarctic ice sheet dynamics (PAIS) and implications for future sea-level change. In: Florindo, F., et al., (Eds.), Antarctic Climate Evolution, Second ed. Elsevier.

Cooper, A.K., Brancolini, G., Escutia, C., Kristoffersen, Y., Larter, R., Leitchenkov, G., et al., 2009. Cenozoic climate history from seismic reflection and drilling studies on the Antarctic continental margin. In: Florindo, F., Siegert, M. (Eds.), Developments in Earth and Environmental Sciences, Vol. 8: Antarctic Climate Evolution. Elsevier, Amsterdam, pp. 115–228.

Escutia, C., DeConto, R., Florindo, F., Bentley, M. (Eds.), 2012. Cenozoic evolution of Antarctic climates, Ocean and Ice Sheets. Palaegeography, Palaeoclimatology, Palaeoecology 335–336, 1–94.

Florindo, F., et al., 2021a. Sixty-years of Coordination and Support for Antarctic-Science - The role of SCAR. In: Florindo, F., et al., (Eds.), Second ed. Elsevier.

Florindo, F., et al., 2021. Antarctic Climate and Ice Sheet Evolution. In: Second ed. Elsevier.

Florindo, F., Cooper, A.K., O'Brien, P. (Eds.), 2003. Cenozoic paleoenvironments: geologic record and models. Palaegeography, Palaeoclimatology, Palaeoecology 198, 1–278.

Florindo, F., Harwood, D.M., Wilson, G.S. (Eds.), 2005. Long-term changes in Southern high-latitude ice sheets and climate, the Cenozoic history. Global and Planetary Change 45, 1–264.

Florindo, F., Harwood, D.M., Levy, R. (Eds.), 2009. Cenozoic Antarctic glacial history. Global and Planetary Change 69 (3), 1–184.

Florindo, F., Haywood, A.M., Nelson, A.E. (Eds.), 2008. Antarctic cryosphere and Southern Ocean climate evolution (Cenozoic-Holocene). Palaegeography, Palaeoclimatology, Palaeoecology 260 (1–2), 1–298.

Florindo, F., Siegert, M. (Eds.), 2008. Antarctic Climate Evolution. Developments in Earth and Environmental Sciences Series, vol. 8, Elsevier, pp. 1–593.

Galeotti, S., et al., 2021. The Eocene/Oligocene boundary climate transition: an Antarctic perspective. In: Florindo, F., et al., (Eds.), Antarctic Climate Evolution, Second ed. Elsevier.

Gohl, K., Wellner S., J., Klaus, A., Expedition 379 Scientists, 2021. *Amundsen Sea West Antarctic Ice Sheet History*. Proceedings of the International Ocean Discovery Program, 379: College Station, TX (International Ocean Discovery Program). Available from: https://doi.org/10.14379/iodp.proc.379.2021.

IPCC, 2014. Climate change 2014: Synthesis report. In: Core Writing Team, Pachauri, R.K., Meyer, L.A. (Eds.), Contribution of Working Groups I, II and III to the Fifth Assessment Report of the Intergovernmental Panel on Climate Change. IPCC, Geneva, Switzerland, p. 151.

IPCC, 2019. IPCC Special Report on the Ocean and Cryosphere in a Changing Climate. Pörtner, H.-O., Roberts, D.C., Masson-Delmotte, V., Zhai, P., Tignor, M., Poloczanska, E., et al. (Eds.). Available from: https://www.ipcc.ch/srocc/.

IPCC, 2021. Climate Change 2021: The Physical Science Basis. Contribution of Working Group I to the Sixth Assessment Report of the Intergovernmental Panel on Climate Change. Masson-Delmotte, V., Zhai, P., Pirani, A., Connors, S.L., Péan, C., Berger, S., et al. (Eds.), Cambridge University Press. Available from: https://www.ipcc.ch/report/sixth-assessment-report-working-group-i/.

Kennicutt, M., et al., 2014. Six priorities for Antarctic science. Nature 512, 23–25.

Kennicutt, M., et al., 2015. A roadmap for Antarctic and Southern Ocean science for the next two decades and beyond. Antarctic Science 27, 3–18. Available from: https://doi.org/10.1017/S0954102014000674.

Kennicutt, M.C., Bromwich, D., Liggett, D., Njåstad, B., Peck, L., Rintoul, S.R., et al., 2019. Sustained Antarctic Research: A 21st Century Imperative. One Earth. 1, 95–113. Available from: https://doi.org/10.1016/j.oneear.2019.08.014.

Kennicutt, M.C., Kim, Y., Finnemore-Rogan, M., Anandakrishnan, S., Chown, S.L., Colwell, S., et al., 2016. Delivering 21st century Antarctic and Southern Ocean science. Antarctic Science 28, 407–423. Available from: https://doi.org/10.1017/S0954102016000481.

Levy, R.H., et al., 2021. Antarctic Environmental Change and Ice Sheet Evolution through the Miocene to Pliocene – A perspective from the Ross Sea, George V Coast, and Wilkes Land. In: Florindo, F., et al., (Eds.), Antarctic Climate Evolution, Second ed. Elsevier.

McKay, R., et al., 2021. Cenozoic History of Antarctic Glaciation and Climate from Onshore and Offshore Studies. In: Florindo, F., et al., (Eds.), Antarctic Climate Evolution, Second ed. Elsevier.

McKay, R.M., De Santis, L., Kulhanek, D.K., the Expedition 374 Scientists, 2019. Ross sea West Antarctic Ice Sheet history. Proceedings of the International Ocean Discovery Program 374. Available from: https://doi.org/10.14379/iodp.proc.374.2019. College Station, TX (International Ocean Discovery Program).

Naish, T., et al., 2021. Antarctic Ice Sheet dynamics during the Late Oligocene and Early Miocene: Climatic conundrums revisited. In: Florindo, F., et al., (Eds.), Antarctic Climate Evolution, Second ed. Elsevier.

Pollard, D., De Conto, R., 2009. Modelling West Antarctic ice sheet growth and collapse through the past five million years. Nature 458, 329–332. Available from: https://doi.org/10.1038/nature07809.

Siegert, M.J., et al., 2021a. Antarctic Ice Sheet Changes since the Last Glacial Maximum. In: Florindo, F. (Ed.), Antarctic Climate Evolution, Second ed. Elsevier.

Siegert, M.J., et al., 2021b. The Future Evolution of Antarctic Climate: Conclusions and Upcoming Programmes. In: Florindo, F., et al., (Eds.), Antarctic Climate Evolution, Second ed. Elsevier.

Siegert, M.J., Golledge, N.R., 2021. Advances in Numerical Modelling of the Antarctic Ice Sheet. In: Florindo, F., et al., (Eds.), Antarctic Climate Evolution, Second ed. Elsevier.

Spada, G., 2017. Glacial Isostatic Adjustment and Contemporary Sea Level Rise: An Overview. Surveys in Geophysics 38 (1), 153–185. Available from: https://doi.org/10.14379/iodp.proc.379.2021.

Spada, G., Melini, D., 2019. SELEN 4 (SELEN version 4.0): A Fortran program for solving the gravitationally and topographically self-consistent sea-level equation in glacial isostatic adjustment modeling. Geoscientific Model Development 12 (12), 5055–5075. Available from: https://doi.org/10.5194/gmd-12-5055-2019.

Talarico, F., et al., 2021. The Antarctic Continent in Gondwana: A perspective from the Ross Embayment and Potential Research Targets for Future Investigations. In: Florindo, F., et al., (Eds.), Antarctic Climate Evolution, Second ed. Elsevier.

Weber, M.E., Raymo, M.E., William, T., Expedition 382 Scientists, 2021. Iceberg Alley and Subantarctic Ice and Ocean Dynamics. Proceedings of the International Ocean Discovery Program, 382: College Station, TX (International Ocean Discovery Program). Available from: https://doi.org/10.14379/iodp.proc.382.2021.

Wilson, D.J., et al., 2021. Pleistocene Antarctic Climate Variability: Ice Sheet – Ocean – Climate Interactions. In: Florindo, F., et al., (Eds.), Antarctic Climate Evolution, Second ed. Elsevier.

Chapter 2

Sixty years of coordination and support for Antarctic science – the role of SCAR

Fabio Florindo[1], Antonio Meloni[1] and Martin Siegert[2]
[1]National Institute of Geophysics and Volcanology, Rome, Italy, [2]Grantham Institute and Department of Earth Science and Engineering, Imperial College London, London, United Kingdom

2.1 Introduction

International scientific collaboration in Antarctica and the Southern Ocean is essential if we are to answer key questions on how the region is changing in relation to anthropogenic warming and on the global impacts that will result, as they can only be tackled seriously by sharing resources, logistics, skills, experience and infrastructure across the region (e.g., Kennicutt et al., 2016). Aside from its contribution to scientific research per se, the Scientific Committee on Antarctic Research (SCAR) is widely accepted as the most influential non-governmental organisation contributing to international Antarctic governance through the provision of impartial expert scientific advice (Walton, 2009). SCAR aims to aid international scientific collaboration, while facilitating the science that occurs in the Antarctic region through a variety of initiatives. One scheme that SCAR operates is its formal Scientific Research Programmes (SRPs), which are bold, thematic, interdisciplinary, multi-year efforts to advance fundamental knowledge on specific issues. Two SRPs that relate to past changes in Antarctica have been Antarctic Climate Evolution (ACE) and its successor Past Antarctic Ice Sheet dynamics (PAIS). The main achievements from ACE were summarised in the first edition of this book (Florindo and Siegert, 2008). In 2013 SCAR awarded funding to PAIS, which has led to a number of research achievements relating to constraining Antarctica's contribution to sea level that resulted from past changes in ice sheet mass loss, and its general impacts on the environment, and atmospheric and oceanic circulation. In the following

Antarctic Climate Evolution. DOI: https://doi.org/10.1016/B978-0-12-819109-5.00011-6

sections, we note how SCAR has achieved its goals, especially with regard to understanding past climate, ocean and ice-sheet change.

2.2 Scientific value of research in Antarctica and the Southern Ocean

Antarctica is the least explored continent on our planet. Similarly, its surrounding ocean is poorly charted and understood. Because of its geographical position, and extreme physical characteristics, Antarctica is home to a unique flora and fauna, whose study enlightens understanding of the complex relationships between living organisms and the environment. Due to its distance from the main sources of pollution and, on land at least, the almost total absence of anthropogenic disturbance, Antarctica provides us with the opportunity to obtain knowledge about the function of the planet from a remote, largely unpolluted observation point.

The thick ice sheet of Antarctica (over 4.5 km in some places) contains records of snow precipitation over several hundred thousand years. Air bubbles trapped in the ice are time capsules of previous atmospheric composition, providing unique insights on past climate change and its drivers (e.g., Brook and Buizert, 2018). Because of its location with respect to the geomagnetic field, Antarctica (like the Arctic) enables us to study phenomena ensuing from the interactions between the Sun and the Earth (Weller et al., 1987). The transparency of the high-altitude Antarctic atmosphere makes it ideal for astronomic observations and, consequently, for cosmological research (e.g., Kim et al., 2018).

The harsh conditions and isolation of Antarctic research stations also make the region ideal as a training field for future space missions. The uniqueness of the Antarctic continent has driven scientists to become used to dealing with specific scientific problems with bespoke technological solutions and, often through trial and error, perfecting the technologies necessary to complete and repeat such science, like those necessary for deep-ice drilling (Talalay, 2020).

Antarctic scientists have been providing information about the state of the continent and its surrounding seas since polar exploration began back in the early 19th century and, increasingly, as that exploration became more scientific and sophisticated in the latter part of the last century. Antarctic research endeavour was galvanised in the International Geophysical Year (IGY), also known as the third International Polar Year (IPY), which spanned an 18-month period from 1 July 1957 to 31 December 1958 and represented the first coordinated study to measure the continent of Antarctica (Florindo et al., 2008; Summerhayes, 2008; Walton, 2009) (Fig. 2.1).

The IGY was one of the largest organised international scientific endeavours of the 20th century and led to significant advances in meteorology, atmospheric sciences and glaciology. World Data Centers were established in order for the new measurements to be stored and shared. During the IGY, 12 nations

FIGURE 2.1 One of the many postage stamps issued for the International Geophysical Year 1957–1958. *By Bureau of Engraving and Printing. Designed by Ervine Metzl. (Public domain), via Wikimedia Commons.*

FIGURE 2.2 Signature of the Antarctic Treaty on 1 December 1959 in Washington, DC, by Ambassador Herman Phleger from the United States, who chaired the Conference on Antarctica from 15 October to 1 December 1959 (Department of State, 1960). *Courtesy of the Carleton College Archives.*

established Antarctic research stations including those at the South Pole (Amundsen-Scott, USA), temporarily at the Pole of Inaccessibility (Polyus Nedostupnosti, USSR) and, importantly for past climate studies, at Vostok Station.

The IGY paved the way for an international agreement like no other – the Antarctic Treaty – which reserves the entire continent for peace and science. In Washington DC on the 1st of December 1959 (Fig. 2.2), government representatives of Argentina, Australia, Belgium, Chile, France, Japan, New Zealand, Norway, South Africa, the then USSR, the UK and the USA became the first signatories of the Antarctic Treaty (Berkman, 2011) (see Appendix).

The Treaty entered into force on 23 June 1961. Today, 54 nations have signed the Treaty – 29 of which have voting rights (see http://www.scar.org/policy/antarctic-treaty-system/). The Antarctic Treaty contains 14 articles (see Appendix), which enshrine the following three principles:

- Antarctica (meaning the entire region south of latitude 60° South) is to be used for peaceful purposes only and military bases, manoeuvres and weapons testing are prohibited. The prohibition also extends to nuclear explosions and the disposal of nuclear waste.
- The promotion of scientific investigation and cooperation, with the exchange of information, plans, results and personnel to be actively encouraged. This also includes freedom of access for the purpose of scientific investigation.
- Territorial claims are not recognised, disputed or established by the Treaty, and no new claims are to be asserted.

Realising the importance of continuing international Antarctic collaboration at the end of the IGY, it was decided that there was a need for further international organisation of scientific activity in Antarctica and that a committee should be set up for this purpose. The 4th Special Committee for the International Geophysical Year (CSAGI) Antarctic Conference in Paris in June 1957 passed a resolution recommending that the International Council of Scientific Unions (ICSU) should appoint a committee *ad hoc* with Professor C.-G. Rossby as convener (Fig. 2.3) to

FIGURE 2.3 Professor Carl-Gustaf Arvid Rossby (Stockholm, 28 December 1898 – Stockholm, 19 August 1957). *Harris & Ewing Collection (Public domain), via Wikimedia Commons.*

examine the merits of further investigation in the Antarctic after the end of the IGY, covering the entire field of science. The committee met at the ICSU Antarctic meeting held in Stockholm between 9 and 11 September 1957, with members from Argentina, Chile, France, Japan, Norway, UK, USA and USSR being present.

Later in September a Special Committee on Antarctic Research (SCAR) was established as an inter-disciplinary committee of ICSU to facilitate and coordinate activities; twelve nations and four Unions (International Union of Geodesy and Geophysics – IUGG; International Geographical Union – IGU; International Union of Biological Sciences – IUBS; and Union Radio Scientifique International – URSI) were invited to nominate delegates.

SCAR held its first meeting in the Administrative Office of ICSU in The Hague (the Netherlands) on 3–6 February 1958; the 12 participating nations of the IGY were invited to attend, as well as representatives from five scientific unions (Fig. 2.4). Subsequently SCAR was renamed the 'Scientific' Committee on Antarctic Research. In 1987 SCAR was appointed as an observer to the Antarctic Treaty Consultative Meeting (ATCM) to ensure political and governance decisions are informed by the best available science.

FIGURE 2.4 Participants at the first SCAR meeting in The Hague (the Netherlands) on 3–6 February 1958. (1) Dr. L.M. Gould, USA; (2) Dr. Ronald Fraser, ICSU; (3) Dr. N. Herlofson, Convenor; (4) Col. E. Herbays, ICSU; (5) Prof. T. Rikitake, Japan; (6) Prof. Leiv Harang, Norway; (7) Dr. Valter Schytt, IGU; (8) Dr. Anton F. Bruun, IUBS; (9) Mr. J.J. Taljaard, South Africa; (10) Capt. F. Bastin, Belgium; (11) Capt. Luis de la Canal, Argentina; (12) Sir James Wordie, UK; (13) Prof. K. E. Bullen, Australia; (14) Dr. H. Wexler, USA; (15) Ing. Gén. Georges Laclavère, IUGG; (16) Ing. Gén. M.A. Gougenheim, France; (17) Mr. Luis Renard, Chile; (18) Dr. M.M. Somov, USSR; (19) Prof. J. van Mieghen, Belgium. *From Wolff, T., 2010. The Birth and First Years of the Scientific Committee on Oceanic Research, SCOR History Report #1. Scientific Committee on Oceanic Research, Newark, DE, photograph courtesy of the Scientific Committee on Oceanic Research.*

SCAR is responsible for initiating, developing and coordinating high-quality international research in the Antarctic region within three scientific standing groups: physical sciences, geosciences and life sciences. Its scientific business is conducted in over 30 Science Groups including SRPs, standing committees, and action and expert groups. SCAR not only provides objective, independent scientific advice to the ATCM but also to other organisations such as the UNFCCC (United Nations Framework Convention on Climate Change) and the IPCC (Intergovernmental Panel on Climate Change) on matters relating to the science and conservation affecting the management of Antarctica and the Southern Ocean, and on the role of the Antarctic region in the wider connected and multi-process Earth system. SCAR's Standing Committee on the Antarctic Treaty System (SCATS) is responsible for coordinating the advice presented to the ATCMs. This is mainly done through: (1) the presentation of Information Papers and Working Papers, most commonly involving contributions from scientists from around the world helping to convey the up-to-date status of research in any particular area; and (2) the Antarctic Environments Portal, which provides an important link between Antarctic science and Antarctic policy. The portal makes science-based information available to the Antarctic Treaty System's Committee for Environmental Protection (CEP) and all the Antarctic Treaty nations.

A modern view of 'Antarctic Science' comes not only from the knowledge of the continent's life, structure and history but also from an understanding of the wide-ranging regional and global changes taking place in Antarctica and the Southern Ocean. SCAR's scientific work is achieved through the engagement and support of thousands of researchers from around the world who together comprise the SCAR community, supported by SCAR's 44 national committees, which report to their respective academies of science or equivalent bodies. SCAR adds value to national scientific activities by addressing topics covering the whole of Antarctica and/or the surrounding Southern Ocean in ways impossible for any one nation to achieve alone. SCAR's governing body, ICSU, recently merged with the International Social Science Council to form the International Science Council (ISC). For this reason, amendments to the SCAR organisation and website are in progress and the SCAR logo has been updated to reflect this change of name (Fig. 2.5).

SCAR's mission remains to be engaged, active and forward-looking in an organisation that promotes, facilitates and delivers scientific excellence and evidence-based policy advice on globally significant issues in and about Antarctica. SCAR has also taken a leading role in supporting early career scientists and in recognising the importance and value of inclusion and diversity in fulfilling its mission. The 32 full members in SCAR's family are as follows: Argentina, Australia, Belgium, Brazil, Bulgaria, Canada, Chile, China, Ecuador, Finland, France, Germany, India, Italy, Japan, Korea (Rep.

FIGURE 2.5 The SCAR logo. *With permission from SCAR.*

of), Malaysia, the Netherlands, New Zealand, Norway, Peru, Poland, Portugal, Russia, South Africa, Spain, Sweden, Switzerland, Ukraine, the UK, Uruguay and the USA. Twelve associate members are: Austria, Belarus, Colombia, Czech Republic, Denmark, Iran, Monaco, Pakistan, Romania, Thailand, Turkey and Venezuela. Nine ISC union members participate in the work of SCAR: IUGG; IGU; IUBS; URSI; the International Astronomical Union – IAU; International Union for Quaternary Research – INQUA; International Union of Geological Sciences – IUGS; International Union of Physiological Sciences – IUPS; and the International Union of Pure and Applied Chemistry – IUPAC.

SCAR is governed by its Memorandum of Association (the legal statement agreed when the organisation became a registered company and charity) and its Articles of Association (the legal rules about how the organisation is run), and these two documents form SCAR's Constitution. More detailed rules about the duties and responsibilities of SCAR's members are laid out in SCAR's Rules of Procedure, describing how SCAR's working groups are established and governed.

2.3 The international framework in which SCAR operates

Although SCAR is primarily focused on science, it has close connections to the Antarctic Treaty System (ATS), which incorporates a whole complex of arrangements made for the purpose of regulating relations among States working in the Antarctic. The primary purpose of the Antarctic Treaty is to ensure "in the interests of all mankind, that Antarctica shall continue forever to be used exclusively for peaceful purposes and shall not become the scene or object of international discord". To this end, the ATS prohibits military activity (except in direct support of science), prohibits nuclear explosions and the disposal of nuclear waste, promotes scientific research and the exchange of data, and holds all territorial claims in abeyance. The Treaty applies to the area south of 60° South latitude, including all floating ice shelves and islands.

The Treaty is augmented by recommendations adopted at Consultative Meetings, by the Protocol on Environmental Protection to the Antarctic Treaty (Madrid, 1991), and by two separate conventions dealing with the Conservation of Antarctic Seals (London, 1972) and the Conservation of Antarctic Marine Living Resources (CCAMLR, Canberra, 1980). The Convention on the Regulation of Antarctic Mineral Resource Activities (Wellington, 1988), negotiated between 1982 and 1988, has so far not entered into force.

In October 2016 the world's largest marine reserve was created in the Ross Sea by a unanimous decision of CCAMLR's 24 member states. The Ross Sea Marine Protected Area (MPA) came into force in December 2017. The 598,000 square-mile MPA (more than twice the size of Texas) consists of:

- A 'no take' General Protection Zone (a fully protected area where no commercial fishing is permitted) split into three separate areas;
- A Special Research Zone which allows for limited research fishing for krill and toothfish – see below; and
- A Krill Research Zone which allows for controlled research fishing for krill, in accordance with the objectives of the MPA.

The Antarctic's nutrient-rich waters are highly productive, leading to huge plankton and krill blooms that support vast numbers of fish, seals, penguins and whales. Importantly, and inspite of whaling that nearly drove extinction of some species, the Ross Sea is 'the least altered marine ecosystem on Earth', containing intact communities of emperor and Adelie penguins, crabeater seals, orcas and minke whales (that are recovering in numbers).

2.4 The organisation of SCAR

SCAR's three scientific groups (discussed earlier) are responsible for sharing information on disciplinary scientific research, identifying research areas or fields where current research is lacking, coordinating and stimulating proposals for future research, and establishing scientific Programme Planning Groups (PPGs) to develop formal proposals on future work to the Delegates, who meet every 2 years. At the heart of this coordination are SCAR's SRPs, which address major cutting-edge research questions. To develop an SRP requires first a 2-year PPG. Proposals are fully peer-reviewed and exist in the first instance for a 4-year term. Data management policies and outreach plans are required to ensure there is a lasting legacy of dissemination and knowledge exchange. In 2004 SCAR Delegates approved the following SRPs:

FIGURE 2.6 Antarctic Climate Evolution (ACE) logo.

- Antarctica and the Global Climate System (AGCS);
- Antarctic Climate Evolution (ACE) (Fig. 2.6);
- Evolution and Biodiversity in the Antarctic (EBA);
- Subglacial Antarctic Lake Exploration (SALE); and
- Interhemispheric Conjugacy Effects in Solar-Terrestrial and Aeronomy Research (ICESTAR).

The subsequent generation of SCAR SRPs (2013–2020) included (Fig. 2.7):

- State of the Antarctic Ecosystem (AntEco), which aimed to increase the scientific knowledge of biodiversity, from genes to ecosystems that, coupled with increased knowledge of species biology, can be used for the conservation and management of Antarctic ecosystems;
- Antarctic Ecosystems: Adaptations, Thresholds and Resilience (AntERA), which aimed to provide a platform for the exchange of knowledge and for the support of research on biological processes at ecological time scales especially related to environmental change;
- Solid Earth Responses and Influences on Cryospheric Evolution (SERCE), which aimed to advance understanding of the interactions between the solid earth and the cryosphere to better constrain ice mass balance, ice dynamics and sea level change in a warming world;
- Antarctic Climate Change in the 21st Century (AntClim21), which aimed to deliver improved regional projections of key elements of the Antarctic atmosphere, ocean and cryosphere for the next 20–200 years and to understand the responses of the physical and biological systems (through multidisciplinary collaboration) to natural and anthropogenic climate drivers;
- Astronomy and Astrophysics from Antarctica (AAA), which aimed to coordinate astronomical activities in Antarctica in a way that ensures the

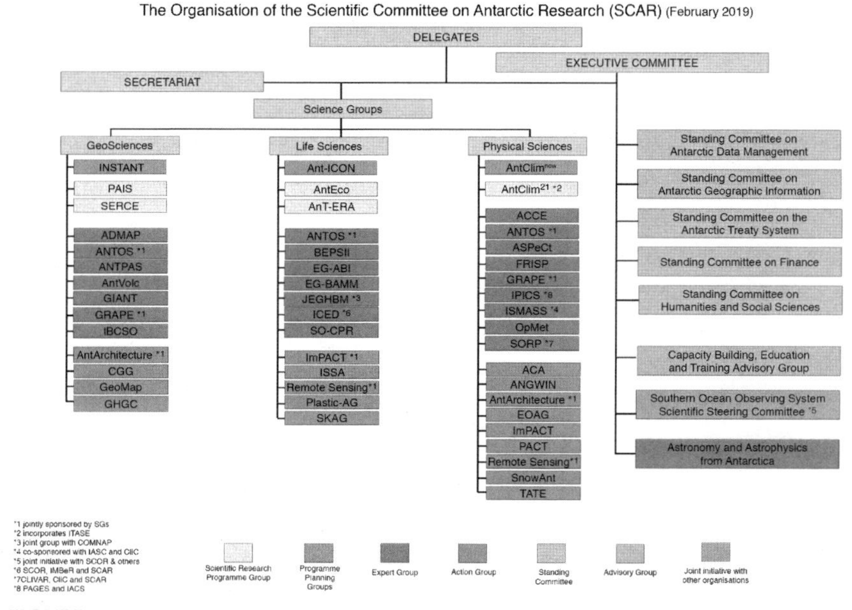

FIGURE 2.7 Organisation of SCAR. *http://www.scar.org (retrieved February 2019).*

FIGURE 2.8 Past Antarctic Ice Sheet Dynamics (PAIS) logo.

best possible outcomes from international investment in Antarctic astronomy and maximises the opportunities for productive interaction with other disciplines; and

- Past Antarctic Ice Sheet dynamics (PAIS), which built on the ACE legacy in constraining Antarctica's contribution to sea level, resulting from past changes in ice sheet mass loss and its impacts on the environment, and atmospheric and oceanic circulation (Fig. 2.8).

Based on the analysis of Antarctica's past, PAIS aimed to bound estimates of future ice loss in key areas of the Antarctic margin with a multidisciplinary approach and by integrating geological data with computer models. PAIS research has been influential in the IPCC's fifth assessment report (AR5) and its current sixth cycle climate assessments (AR6). Extensive PAIS-facilitated fieldwork on land and at sea has been carried out within the framework of national and multi-national projects, including the International Ocean Discovery Program (IODP) expeditions 374 in 2018, and 389 and 382 in 2019.

The IODP is among the main organisations external to SCAR, in addition to National Antarctic Programs, providing enormous support for the PAIS drilling expeditions in Antarctica, both in terms of offshore and shore-based science, education and communication-outreach programmes and for pre-cruise work and meetings. A recent paper highlighting progress made in the past 45 years between the SCAR geoscience paleoclimate projects and IODP has been published in the special issue on Scientific Ocean Drilling – 'Keeping an Eye on Antarctic Ice Sheet Stability' (Escutia et al., 2019).

Importantly, numerical ice sheet modelling has developed significantly since the first edition of the ACE book (Siegert and Golledge, 2021), and these developments have led to important advances in our understanding of how ice sheets have changed with past climate, and how they will likely change in future under global warming.

In line with its predecessors ANTOSTRAT and ACE, PAIS fulfilled an important role in informing and coordinating the scientific community by organising scientific conferences, workshops, schools (Fig. 2.9), facilitating the planning of new data-acquisition missions using emerging technologies, encouraging data sharing (e.g., the update and use of

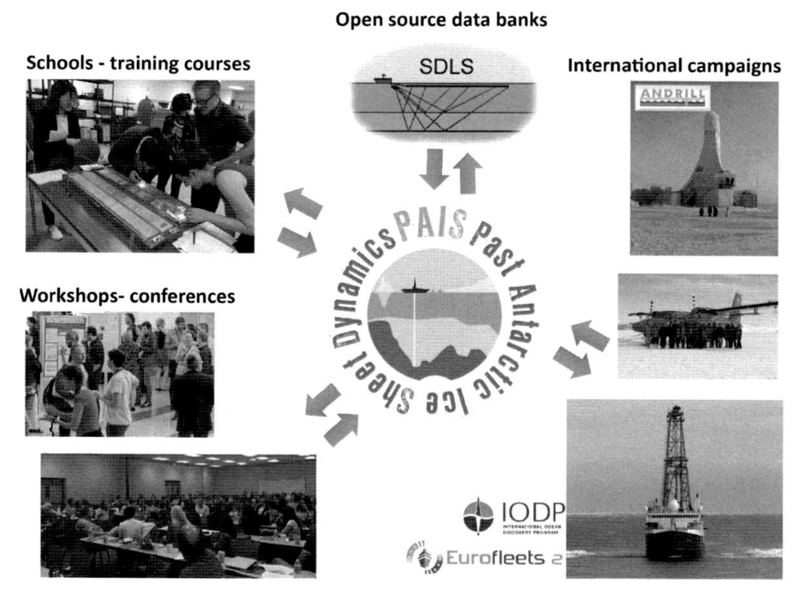

FIGURE 2.9 The different activities carried out by the PAIS programme in support of scientific advances, training, collaboration, knowledge exchange and data sharing.

the Antarctic Seismic Data Library System, see McKay et al., 2021) and initiating/expanding cross-linkages among Antarctic research communities.

All the SRPs ended in December 2020 and a new set of SCAR programmes have come into play for the next 8 years (https://www.scar.org/science/srp/):

- Integrated Science to Inform Antarctic and Southern Ocean Conservation (Ant-ICON), which aims to answer fundamental science questions (as identified by the SCAR Horizon Scan, Kennicutt et al., 2015) relating to the conservation and management of Antarctica and the Southern Ocean with a focus on research to drive and inform international decision-making and policy change;
- Near-term Variability and Prediction of the Antarctic Climate System (AntClim^{now}) that aims to investigate the prediction of near-term conditions in the Antarctic climate system on timescales of years to multiple decades;
- INStabilities and Thresholds in ANTarctica (INSTANT) will address the first-order question about Antarctica's contribution to sea level. It encompasses geoscience, physical sciences and biological sciences, to investigate the ways in which interactions between the ocean, atmosphere and cryosphere have influenced ice sheets in the past, and how such interplay may occur in the future, with a special focus on quantifying the contributions to global sea level change. INSTANT builds on a white paper developed during the PAIS conference held in Italy in 2017, involving over 200 scientists from 18 nations and spanning different disciplines and with representatives from the other SCAR SRPs (see details in the PAIS web site http://www.scar-pais.org/index.php/highlights/past-antarctic-ice-sheet-dynamics-pais-conference-2017-trieste-italy). The white paper recognised the importance of a transdisciplinary approach in understanding and quantifying the Antarctic ice sheet contribution to past and future global sea-level change, from improved understanding of climate, ocean and solid Earth interactions and feedbacks with the ice. The white paper also acknowledged the importance of understanding the global consequences and impacts of Antarctic change so that decision-makers can better anticipate and assess the risk of sea level rise in order to evaluate adaptation and mitigation pathways.

2.5 Sixty years of significant Antarctic science discoveries

Antarctic scientists working with SCAR have been involved in many leading scientific discoveries, such as:

- The discovery of the ozone hole and elucidation of its chemistry;
- The history of the ice sheet and its implications for changing sea level;

- The circulation of the Southern Ocean and its role in the storage and emission of CO_2 and heat;
- The fossilised flora of Antarctica, which was covered by flourishing vegetation 100 million years ago, and of Antarctic dinosaurs;
- The 600 million years journey of Antarctica from North Pole to South Pole, under the influence of plate tectonics;
- The sub-ice topography, including the existence of subglacial lakes and rivers;
- The amazing circum-Antarctic land-free travel of albatrosses;
- The extraordinary diversity of marine life;
- The detection of neutrinos originating in outer space; and
- Antarctica as an analogue for extra-terrestrial life and other aspects of planetary exploration.

Assemblies of Delegates and Open Science Conferences are key events in the life of SCAR. The last two were POLAR2018 and SCAR 2020 Online. The former took place in Davos, Switzerland (15–26 June 2018), and was also home to the XXXV SCAR Biennial Meeting, the Arctic Science Summit Week (ASSW) and a joint SCAR/IASC Open Science Conference. The SCAR meetings, the ASSW and the Open Science Conference were hosted by the Swiss Federal Institute for Forest, Snow and Landscape Research (WSL) under the patronage of the Swiss Committee on Polar and High-Altitude Research.

Over 2500 attendees presented ~1600 posters and ~1000 oral papers as well as plenary sessions, panel discussions, side meetings and social events. At POLAR2018 SCAR agreed to support the following action groups: Earth Observation (Physical Sciences), ANGWIN – ANtarctic Gravity Wave Instrument Network (Physical Sciences), AntArchitecture (Geosciences and Physical Sciences), IMPACT – Input Pathways of Persistent Organic Pollutants to Antarctica (Life Sciences), SKAG – SCAR Krill action group (Life Sciences) and Plastic in Polar Environment (Life Sciences). At the Delegates Meeting, three new PPGs were proposed in order to develop future SRPs: Integrated science to support Antarctic and Southern Ocean conservation (ANT-ICON); Near-term Variability and Prediction of the Antarctic Climate System (AntClim^{now}); and INSTANT (initially called AISSL, e.g., Fig. 2.7).

Due to the COVID-19 pandemic, the SCAR-OSC 2020 was held online and remotely (03–07 August 2020). The registration was free, and there were over 2700 people in attendance. The SCAR-OSC 2020 Portal will remain open indefinitely, so everyone who registers will be able to view sessions and also plenaries, workshops and mini-symposia recordings at any time and browse the Contributing Authors Gallery (almost 600 virtual displays). All videos uploaded to the event portal are available on the SCAR 2020 Online YouTube channel and event website scar2020.org, which that will also remain open indefinitely.

Excellence in research and leadership has been recognised by prestigious awards to ACE/PAIS leaders. Dr. Carlota Escutia, who successfully led the PAIS programme, was awarded the SCAR 2020 Medal for International Scientific Coordination, during the week of SCAR 2020 Online. Past ACE/PAIS Co-Chief Officers Dr. Martin Seigert, Dr. Tim Naish, and Dr. Rob DeConto have been awarded the prestigious Tinker-Muse Prize for their contribution to Antarctic Science and Policy. In 2016, the SCAR Research Medal was awarded to Dr Rob Dunbar who was an inaugural Co-Chief Officer of ACE.

2.6 Scientific Horizon Scan

In 2014 SCAR organised a formal Scientific Horizon Scanning exercise to determine the most pressing eighty scientific questions that require answers within the next two decades. The Horizon Scan was conducted through a formal online process involving the entire Antarctic research community, followed up by a residential retreat in Queenstown, New Zealand from 20 to 23 April 2014 (Kennicutt et al., 2014, 2015). The questions were agreed under six main themes: (1) define the global reach of the Antarctic atmosphere and Southern Ocean; (2) understand how, where and why ice sheets lose mass; (3) reveal Antarctica's history; (4) learn how Antarctic life evolved and survived; (5) observe space and the Universe; and (6) recognise and mitigate human influences. At least 17 of the 80 scientific questions relate to past changes in Antarctica and the Southern Ocean (see Table 2.1). In 2016 COMNAP (the Council of Managers of National Antarctic Programs) responded to the results of the horizon scan in its Antarctic Roadmap Challenge (22–24 August, Tromsø, Norway), by ascertaining the logistics and equipment necessary to provide answers to the scientific questions (Kennicutt et al., 2016). Again, a formal process was used involving scientists, managers and logistics agencies from around the world. Most recently, 5 years since the Horizon Scan, Kennicutt et al. (2019) explored the progress the SCAR community has made in answering the scientific questions. While some key questions have received attention and progress, others – especially those requiring new equipment or fieldwork in remote regions – require significant attention if the promise of the Horizon Scan is to be fulfilled within its 20-year vision (Table 2.1). For research into past conditions in the Antarctic, most questions score '2', meaning that progress has been limited. However, for a few questions, a score of 1 was awarded, indicating that very little if any progress has been made in 5 years. Questions such as: *How did the climate and atmospheric composition vary prior to the oldest ice records?*; *How will the sedimentary record beneath the ice sheet inform our knowledge of the presence or absence of continental ice?*; *Do variations in geothermal heat flux in Antarctica provide a diagnostic signature of sub-ice geology?*; and *How have ecosystems in the Antarctic and the Southern Ocean responded to warmer climate conditions in the past?* make up some of the most important scientific questions concerning past changes in Antarctica. These are now internationally-agreed as essential scientific imperatives to which SCAR and its community hope to focus

TABLE 2.1 Table of SCAR horizon scan questions related to past conditions in Antarctica.

Question	Progress made since 2014
#3 How have teleconnections, feedbacks and thresholds in decadal and longer term climate variability affected ice sheet response since the Last Glacial Maximum, and how can this inform future climate projections?	The absence of annually resolved records limits progress on this question. Rating = 1 – 2
#5 How did the climate and atmospheric composition vary prior to the oldest ice records?	More studies of the Pliocene in the Antarctic are needed, and collection of sediment and rock records is required. Rating = 1 – 2
#8 Does past amplified warming of Antarctica provide insight into the effects of future warming on climate and ice sheets?	Research on this question is at its beginnings. Incorporating ozone variability could improve seasonal predictions. Rating = 2
# 9 Are there CO_2 equivalent thresholds that foretell collapse of all or part of the Antarctic ice sheet?	It has been 3 million years Before Present since CO_2 values reached those of the present day. Additional geological records are needed. Rating = 2
#17 How has Antarctic sea ice extent and volume varied over decadal to millennial timescales?	A multi-decadal trend of slow expansion of Antarctic sea ice was interrupted by a sharp decline to record low levels in 2016 and has remained low, while regional trends in sea ice extent have been linked to well-known modes of climate variability. Rating = 2
#21 How did the Antarctic cryosphere and the Southern Ocean contribute to glacial/inter-glacial cycles?	There are new insights into the contribution of changes in Southern Ocean overturning circulation to glacial/inter-glacial cycles. Rating = 3
#28 What are the thresholds that lead to irreversible loss of all or part of the Antarctic ice sheet?	The existence of various tipping points has been proposed. Rating = 2
#32 How fast has the Antarctic ice sheet changed in the past and what does that tell us about the future?	Improved understanding of how the Antarctic Ice Sheet responded to forcings in the past is crucial to improving the reliability of forecasts. Most studies to date are for the last deglaciation. Rating = 2

(*Continued*)

TABLE 2.1 (Continued)

Question	Progress made since 2014
#33 How did marine-based Antarctic ice sheets change during previous interglacial periods?	Information on past interglacial periods is mainly based on sea-level records in both the north and south. In these records it is difficult to disentangle contributions from the Greenland and Antarctic ice sheets. Rating = 2
#34 How will the sedimentary record beneath the ice sheet inform our knowledge of the presence or absence of continental ice?	Even if robust indicators of ice-free conditions can be extracted from sedimentary records, the difficulties of subglacial access means that only sparse point data could be recovered, making large-scale inferences about ice sheet presence/absence challenging. Rating = 1
#35 How does the bedrock geology under the Antarctic ice sheet inform our understanding of supercontinent assembly and break-up through Earth's history?	Machine learning approaches have not been adapted. Samples from the bed are key – hence fast drill/Rapid Access Ice Drill are important technologies, but little progress has been made in perfecting these tools. Rating = 2
#36 Do variations in geothermal heat flux in Antarctica provide a diagnostic signature of subice geology?	Available geothermal maps differ from each other making it difficult to resolve geology. In situ samples are key. Rating = 1 – 2
#37 How does volcanism affect the evolution of the Antarctic lithosphere, ice sheet dynamics, and global climate?	Significant progress on this question but gaps in knowledge persist. Major advances have been accomplished by the Polar Observing Network (POLENET). Detailed models exist for lithospheric thickness and other variables, but data are needed for validations. Rating = 2 – 3
#39 What are and have been the rates of geomorphic change in different Antarctic regions, and what are the ages of preserved landscapes?	There are ideas on how to do this, but they need to be expanded continent-wide. Links to modelling need improvements. Rating = 2
#40 How do tectonics, dynamic topography, ice loading and isostatic adjustment affect the spatial pattern of sea level change on all timescales?	Dynamic topography may be important. Pore pressures may contribute to a better understanding of aseismicity. Rating = 2

(*Continued*)

TABLE 2.1 (Continued)

Question	Progress made since 2014
#45 How have ecosystems in the Antarctic and the Southern Ocean responded to warmer climate conditions in the past?	There is progress regarding the last Mya (glacial cycles of warming and cooling) with identification of refugia (both marine and terrestrial habitats). Knowledge is starting to be developed for some marine and terrestrial systems, but ecosystem responses remain largely unknown. Rating = 1
#46 How has life evolved in the Antarctic in response to dramatic events in the Earth's history?	1, 2 or 3 depending on the adaptation and event. Significant adaptations are known in relation to cooling (antifreeze, haemoglobin loss etc.) and in other areas (e.g., refugia) good progress has been made. Rating = 1 − 3

The ratings are between 1 and 4; 1 being unaddressed, 4 being answered.
Source: Question numbers as in Kennicutt et al. (2014); progress report from Kennicutt et al. (2019).

on for at least the next decade. Answers will increasingly depend on drilling on land for the oldest ice (expected to be about 1.5 million years old), and deep ocean drilling to explore the sedimentary record of the discharge of ice from the continent and the thermal history of the Southern Ocean (from the physical and chemical remains of fossil plankton).

2.7 Summary

Antarctic research is conducted with the assistance and support of SCAR, the most important non-governmental body overseeing coordination and collaboration of science in Antarctica and the Southern Ocean. SCAR has been in existence since 1958 and has offered a unique forum through which international dialogue, exchange and cooperation can take place regarding forming answers to scientific questions relevant to understanding processes in the most remote and extreme place on Earth. This approach was demonstrated in 1957–1958 as part of the IGY (which in effect was the third International Polar Year or IPY), and in 2007–2009, when SCAR played a leading role in the fourth IPY. SCAR has been instrumental in recent years in focusing resources on major research priorities, and research into past environments has benefitted significantly from this approach – first through the ACE SRP and in the last few years in the PAIS SRP. In 2014, a formal international Horizon Scan of Antarctic research was undertaken, which led to the emergence of an agreed set of the most important 80 scientific questions for understanding Antarctic and Southern Ocean processes. At least 17 of these questions are directly germane to understanding past changes. A recent inspection of the progress made on each question tells us that much work still remains to be done on most of the Horizon Scan questions, and especially on those relating to the past. Future drilling offshore and through the ice sheet is essential for further progress. Through SCAR, the work of ACE and PAIS has led to significant improvements in our appreciation of past conditions in Antarctica and their global significance, but much work remains to be done.

References

Berkman, P.A., 2011. President Eisenhower, the Antarctic Treaty and the origin of international spaces. In: Berkman, P.A., Lang, M.A., Walton, D.W.H., Young, O.R. (Eds.), Science Diplomacy. Smithsonian Institution Scholarly Press, Washington, DC, pp. 17–27.

Brook, E.J., Buizert, C., 2018. Antarctic and global climate history viewed from ice cores. Nature 558, 200–208. Available from: https://doi.org/10.1038/s41586-018-0172-5.

Escutia, C., DeConto, R.M., Dunbar, R., De Santis, L., Shevenell, A., Naish, T., 2019. Keeping an eye on Antarctic ice sheet stability. Oceanography 32 (1), 32–46.

Florindo, F., Siegert, M.J. (Eds.), 2008. Developments in Earth and Environmental Sciences, Vol. 8: Antarctic Climate Evolution. Elsevier, Amsterdam.

Florindo, F., Meloni, A., Siegert, M., 2008. The international polar years: a history of developments in Antarctic climate evolution. In: Florindo, F., Siegert, M. (Eds.), Developments in Earth and Environmental Sciences, Vol. 8: Antarctic Climate Evolution. Elsevier, Amsterdam, pp. 13–31.

Kennicutt, M.C., Chown, S.L., Cassano, J.J., Liggett, D., Massom, R., Peck, L.S., et al., 2014. Polar research: six priorities for Antarctic science. Nature 512, 23–25. Available from: https://doi.org/10.1038/512023a.

Kennicutt, M.C., et al., 2015. A roadmap for Antarctic and Southern Ocean science for the next two decades and beyond. Antarctic Science 27, 318. Available from: https://doi.org/10.1017/S0954102014000674.

Kennicutt, M.C., Bromwich, D., Liggett, D., Njåstad, B., Peck, L., Rintoul, S.R., et al., 2019. Sustained Antarctic research – a 21st century imperative. One Earth 1, 95–113. Available from: https://doi.org/10.1016/j.oneear.2019.08.014.

Kennicutt, M.C., Kim, Y., Finnemore-Rogan, M., Anandakrishnan, S., Chown, S.L., Colwell, S., et al., 2016. Delivering 21st century Antarctic and Southern Ocean science. Antarctic Science 28, 407–423. Available from: https://doi.org/10.1017/S0954102016000481.

Kim, J., Marrone, D.P., Beaudoin, C., Carlstrom, J.E., Doeleman, S.S., Folkers, T.W., et al., 2018. A VLBI receiving system for the south pole telescope. Proceedings of the SPIE, vol. 10708, Millimeter, Submillimeter, and Far-Infrared Detectors and Instrumentation for Astronomy IX. p. 107082S. Available from: https://doi.org/10.1117/12.2301005.

McKay, R., et al., 2021. Cenozoic History of Antarctic Glaciation and Climate from Onshore and Offshore Studies. In: Florindo, F. (Ed.), et al., Antarctic Climate Evolution, second ed. Elsevier.

Siegert, M.J., Golledge, N.R., 2021. Advances in Numerical Modelling of the Antarctic Ice Sheet. In: Florindo, F., et al. (Eds.), Antarctic Climate Evolution, Second ed. Elsevier.

Summerhayes, C.P., 2008. International collaboration in Antarctica: the international polar years, the international geophysical year, and the scientific committee on Antarctic research. Polar Record 44, 321–334.

Talalay, P.G., 2020. Thermal Ice Drilling Technology. Springer, Singapore. Available from: https://doi.org/10.1007/978-981-13-8848-4.

Walton, D.W.H., 2009. The Scientific Committee on Antarctic Research and the Antarctic Treaty, Science Diplomacy: Antarctic Treaty Summit, first ed. Smithsonian Institute, Washington, DC.

Weller, G., Bentley, C.R., Elliot, D.H., Lanzerotti, L.J., Webber, P.J., 1987. Laboratory Antarctica: contributions to global problems. Science 238, 1361–1368. Available from: https://doi.org/10.1126/science.238.4832.1361.

Wolff, T., 2010. The Birth and First Years of the Scientific Committee on Oceanic Research (SCOR) History Report #1. Scientific Committee on Oceanic Research, Newark, DE.

Appendix

The Antarctic Treaty signed at Washington on 1 December 1959, by plenipotentiaries of the United States of America and 11 other countries.

DOC. FILE

86TH CONGRESS 2d Session	SENATE	EXECUTIVE B

THE ANTARCTIC TREATY

MESSAGE

FROM

THE PRESIDENT OF THE UNITED STATES

TRANSMITTING

A CERTIFIED COPY OF THE ANTARCTIC TREATY, SIGNED AT WASHINGTON ON DECEMBER 1, 1959, BY THE PLENIPOTENTIARIES OF THE UNITED STATES OF AMERICA AND ELEVEN OTHER COUNTRIES

FEBRUARY 15, 1960.—Treaty was read the first time and the injunction of secrecy was removed therefrom. The treaty, the President's message of transmittal, and all accompanying papers were referred to the Committee on Foreign Relations and ordered to be printed for the use of the Senate

THE WHITE HOUSE,
February 15, 1960.

To the Senate of the United States:

With a view to receiving the advice and consent of the Senate to ratification, I transmit herewith a certified copy of the Antarctic Treaty, signed at Washington on December 1, 1959, by plenipotentiaries of the United States of America and 11 other countries.

This is a unique and historic treaty. It provides that a large area of the world—an area equal in size to Europe and the United States combined—will be used for peaceful purposes only. It contains a broad, unrestricted inspection system to ensure that the nonmilitarization provisions will be carried out.

The purposes and provisions of the treaty are explained in the report of the Secretary of State, which is transmitted herewith.

I transmit also, for the information of the Senate, a certified copy of the final act of the Conference on Antarctica, held at Washington October 15 to December 1, 1959, at which the treaty was formulated. The final act does not require ratification.

49118—60——1

I am gratified to recall that it was at the initiative of the United States that the Conference on Antarctica was convened. On May 2, 1958, the United States extended to the 11 other countries which participated in the Antarctic program of the International Geophysical Year an invitation to participate in a conference to consider the conclusion of a treaty on Antarctica for certain stated purposes. The invitation was accepted by all 11 countries: Argentina, Australia, Belgium, Chile, the French Republic, Japan, New Zealand, Norway, the Union of South Africa, the Union of Soviet Socialist Republics, and the United Kingdom of Great Britain and Northern Ireland.

The spirit of cooperation and mutual understanding with which representatives of the 12 countries drafted the Antarctic Treaty and signed it for their respective governments is an inspiring example of what can be accomplished by international cooperation in the field of science and in the pursuit of peace.

I believe that the Antarctic Treaty is a significant advance toward the goal of a peaceful world with justice. In the hope that the United States, which initiated the idea of the Antarctic Treaty, may be one of the first to ratify it, I recommend that the Senate give it early and favorable consideration.

DWIGHT D. EISENHOWER.

THE WHITE HOUSE, *February 15, 1960.*

(Enclosures: (1) Report of the Secretary of State; (2) the Antarctic Treaty (certified copy); (3) U.S. Government note of May 2, 1958; (4) final act of the Conference on Antarctica (certified copy).)

DEPARTMENT OF STATE,
Washington, February 4, 1960.

The PRESIDENT,
The White House:

I have the honor to submit to you, with a view to its transmission to the Senate for advice and consent to ratification, a certified copy of the Antarctic Treaty, signed at Washington on December 1, 1959, on behalf of the United States of America and 11 other countries. Those countries are Argentina, Australia, Belgium, Chile, the French Republic, Japan, New Zealand, Norway, the Union of South Africa, the Union of Soviet Socialist Republics, and the United Kingdom of Great Britain and Northern Ireland.

The treaty was formulated at the Conference on Antarctica held at Washington from October 15 to December 1, 1959. The idea for the Conference was initiated by the Government of the United States which, on May 2, 1958, extended an invitation to take part in such a conference to the 11 other countries which participated in the Antarctic program of the International Geophysical Year. A copy of the U.S. note of invitation, dated May 2, 1958, is enclosed for transmittal to the Senate for its information.

Acceptances were received from all 11 governments, and subsequently informal preparatory talks were held in Washington among representatives of the 12 countries. When the Conference convened on October 15, 1959, it used as a basis for discussion working papers considered in the course of the preparatory talks.

The treaty formulated at the Conference and signed on behalf of all 12 countries incorporates the basic purposes of the U.S. proposal and provides practical means for their fulfillment.

The treaty consists of a preamble and 14 articles. It was drafted, as stated in the preamble in recognition:

> * * * that it is in the interest of all mankind that Antarctica shall continue forever to be used exclusively for peaceful purposes and shall not become the scene or object of international discord.

The preamble also makes clear that the treaty is designed to further the purposes and principles embodied in the Charter of the United Nations.

Article I dedicates Antarctica to peaceful purposes only. It outlaws measures of a military nature, such as the establishment of military bases and fortifications, the carrying out of maneuvers, and the testing of weapons. It specifies that military personnel or equipment may be used there for scientific research or any other peaceful purpose. As the United States and a few other countries have conducted their Antarctic programs with logistic support provided by their military establishments the latter provision was considered appropriate to dispel any doubt that peaceful programs could continue to be carried out in this way.

Article II provides that freedom of scientific investigation in Antarctica and cooperation toward that end, as applied during the International Geophysical Year, shall continue subject to the provisions of the treaty.

Article III contains provisions for the promotion of such international scientific cooperation. Under its terms the parties agree that scientists may be exchanged between expeditions and stations in Antarctica. The parties also shall keep one another informed of their plans for scientific programs in Antarctica and shall make freely available scientific observations. Such exchanges would be made to the extent that the parties consider it feasible and practicable. The article also encourages the establishment of cooperative working relations with those specialized agencies of the United Nations and other international organizations having a scientific or technical interest in Antarctica.

Article IV specifies that nothing in the treaty will be interpreted as a renunciation of any party's claim to sovereignty, as a renunciation or diminution of any party's basis of claim, or as prejudicing the position of any party regarding recognition or nonrecognition of another party's claim or basis of claim. The article also specifies that while the treaty is in force no acts or activities will constitute a basis for asserting, supporting or denying a claim or create any rights of sovereignty in Antarctica. It is finally provided that no new claims can be made and no existing claims enlarged while the treaty is in force.

It is believed that the manner in which the treaty deals with the sensitive problem of territorial claims is one of its most significant aspects. Seven of the twelve countries which signed the treaty have for many years asserted claims of sovereignty to portions of Antarctica, some of which overlap and have given rise to occasional frictions. The claimants are Argentina, Australia, Chile, France, New Zealand, Norway, and the United Kingdom. Neither the United States nor the Soviet Union has made any territorial claims, nor do they recognize the claims of others. Other nonclaimant countries which are signatories to the treaty are Belgium, Japan, and the Union of South Africa. In essence, article IV minimizes the possibility that disputes over claims to sovereignty will erupt and interfere with constructive scientific work in Antarctica.

Article V bans all nuclear explosions in Antarctica and the dumping there of radioactive waste material pending the conclusion of international agreements on nuclear uses. In effect this provision prevents Antarctica from being used as a nuclear proving ground or as a dumping ground for radioactive wastes. It prevents the possibility that harmful fallout will be carried to neighboring regions. However, this article does not prevent the use of nuclear energy in atomic powerplants.

Article VI establishes the zone of application of the treaty. By its terms the treaty applies to the area south of 60° south latitude, including ice shelves, that is, thick portions of ice attached to the land and extending seaward; but the rights of any state under international law with regard to the high seas within the area are not affected.

Article VII contains provisions designed to ensure that the peaceful intent of the treaty is being carried out. It permits the signatory parties, and any acceding parties qualified to participate in the consultative meetings, to send observers anywhere in Antarctica at any time. The observers must be nationals of the sending party and their designation must be made known to every other party having the right to send observers. They are to have complete freedom of access to all areas of Antarctica and must be permitted to inspect stations, installations, and equipment as well as ships and aircraft at points of discharging or embarking cargoes or personnel in Antarctica. Aerial observation by any country having the right to send observers is also permitted. Parties are required to furnish advance notice of all Antarctic expeditions by their ships or nationals and all Antarctic expeditions organized in or proceeding from their territories. They must also report all stations in Antarctica occupied by their nationals and all military personnel or equipment intended to be sent to Antarctica for peaceful purposes. The requirement to give advance information would not, of course, prevent previously notified plans from being modified or revised, upon further notice, whenever advisable because of unforeseen events such as budgetary limitations, weather, or damages to ships or equipment.

Under article VIII each party has exclusive jurisdiction over its own nationals who are observers designated under the treaty for inspection purposes or scientific personnel exchanged between expeditions or stations in Antarctica, in respect of all acts or omissions, occurring while such persons are in Antarctica for the purpose of exercising their functions. Members of the staffs accompanying such persons are also covered. The positions of the parties relating to jurisdiction over all other persons in Antarctica are not affected. The parties agree to consult together immediately should any dispute arise concerning the exercise of jurisdiction in Antarctica.

Under article IX, representatives of the 12 signatory states are to meet in Canberra, Australia, within 2 months after the treaty enters into force and thereafter at times and places which they deem suitable. Their functions will be to exchange information, to consult together on matters of common interest pertaining to Antarctica, and to recommend to their governments measures in furtherance of the principles and objectives of the treaty. These measures are to become effective when approved by all of the parties who were entitled to participate in the meetings held to consider those measures.

A country which has become party to the treaty by accession may qualify to participate in the meetings during such time as it demonstrates its interest in Antarctica by conducting substantial scientific research there.

Representatives participating in the meetings will receive the reports of observers carrying out inspection under the treaty.

Regardless of whether measures to facilitate the exercise of treaty rights are adopted, all rights established by the treaty, including, of course, rights to conduct inspection, may be exercised from the date the treaty enters into force.

By article X the parties are obliged to exert appropriate efforts, consistent with the Charter of the United Nations, to see that no one engages in any activity in Antarctica contrary to the principles of the treaty. Its aim is not only to prevent such activity by nationals and organizations under the jurisdiction of the parties but to deter countries which are not parties to the treaty and their nationals and organizations, from engaging in nonpeaceful activities in Antarctica. In effect it pledges the parties not only to refrain from giving assistance to persons or countries which might engage in nonpeaceful activities or atomic tests in Antarctica, but to take active steps to discourage any such activity.

Under article XI disputes among the parties arising under the treaty are to be resolved by peaceful means of their own choice, such as arbitration, conciliation, or the like. If this proves unsuccessful, the dispute is to be referred to the International Court of Justice, with the consent of all parties to the dispute.

Article XII provides a method for modifying or amending the treaty at any time by unanimous consent of the parties entitled to participate in the consultative meetings. An amendment will enter into force when all such parties give notice that they have ratified it. Parties not entitled to participate in the meetings may accept the amendment within 2 years. If they fail to do so within that time they will be deemed to have withdrawn from the treaty.

The treaty has no specified duration, but article XII provides that after 30 years any of the parties participating in consultative meetings may ask for a Conference to review the operation of the treaty. Amendments approved at such a Conference by a majority of those represented, including a majority of the consultative parties, will enter into force when ratified by all of the consultative parties. If a modification or amendment approved at such a Conference does not enter into force within 2 years, any country may withdraw from the treaty, effective 2 years from its notification to that effect.

Article XIII provides that the treaty will enter into force when ratified by all 12 signatory states. It contains an accession clause by which countries other than the original 12 may acquire the rights and assume the obligations embodied in the treaty. All states members of the United Nations and any other state which is unanimously invited by the consultative parties may accede.

Article XIII also names the United States as depositary government and contains other provisions of a formal nature relating to ratification, accession, and registration with the United Nations.

By article XIV the English, French, Russian, and Spanish language versions of the treaty are declared to be equally authentic.

On the occasion of the signing of the Antarctic Treaty the Governments of the United States, Argentina, and Chile declared that the treaty does not affect their obligations under the Inter-American Treaty of Reciprocal Assistance, signed at Rio de Janeiro on September 2, 1947 (62 Stat. 1681).

The U.S. representative to the Conference on Antarctica was Ambassador Herman Phleger, former Legal Adviser of the Department of State. Alternate Representatives were Ambassador Paul C. Daniels and Mr. George H. Owen. The U.S. delegation included, in addition to officers of the Department of State, a representative of the Department of Defense. Congressional advisers were the Honorable Frank Carlson and the Honorable Gale W. McGee, U.S. Senators. The delegation received advice directly from the National Science Foundation, the agency responsible for coordinating the planning and management of the U.S. scientific program in Antarctica. It was also advised by a committee of six distinguished scientists appointed by the National Academy of Sciences because of their active interest in scientific investigations in Antarctica.

There is transmitted for your information, and for that of the Senate, the final act of the Conference on Antarctica, signed at Washington on December 1, 1959, by plenipotentiaries of the 12 participating nations. The final act does not require ratification.

I believe that the signing of the Antarctic Treaty is a substantial achievement. Its ratification by all of the signatory states would further, in an entire continent, peaceful cooperation in the attainment of scientific progress. It is based on the will to maintain peace in an important area of the world. The United States, which has engaged in extensive exploratory and scientific activities in Antarctica, initiated the idea of the Antarctic Treaty, which is believed to be in the best interests of this country and of all mankind. It is hoped therefore that the United States may be among the first to ratify it.

Respectfully submitted.

CHRISTIAN A. HERTER.

Enclosures: (1) the Antarctic Treaty (certified copy); (2) U.S. Government note of May 2, 1958 (copy); (3) final act of the Conference on Antarctica (certified copy).

THE ANTARCTIC TREATY

The Governments of Argentina, Australia, Belgium, Chile, the French Republic, Japan, New Zealand, Norway, the Union of South Africa, the Union of Soviet Socialist Republics, the United Kingdom of Great Britain and Northern Ireland, and the United States of America,

Recognizing that it is in the interest of all mankind that Antarctica shall continue forever to be used exclusively for peaceful purposes and shall not become the scene or object of international discord;

Acknowledging the substantial contributions to scientific knowledge resulting from international cooperation in scientific investigation in Antarctica;

Convinced that the establishment of a firm foundation for the continuation and development of such cooperation on the basis of freedom of scientific investigation in Antarctica as applied during the International Geophysical Year accords with the interests of science and the progress of all mankind;

Convinced also that a treaty ensuring the use of Antarctica for peaceful purposes only and the continuance of international harmony in Antarctica will further the purposes and principles embodied in the Charter of the United Nations;

Have agreed as follows:

ARTICLE I

1. Antarctica shall be used for peaceful purposes only. There shall be prohibited, *inter alia*, any measures of a military nature, such as the establishment of military bases and fortifications, the carrying out of military maneuvers, as well as the testing of any type of weapons.

2. The present Treaty shall not prevent the use of military personnel or equipment for scientific research or for any other peaceful purpose.

ARTICLE II

Freedom of scientific investigation in Antarctica and cooperation toward that end, as applied during the International Geophysical Year, shall continue, subject to the provisions of the present Treaty.

ARTICLE III

1. In order to promote international cooperation in scientific investigation in Antarctica, as provided for in Article II of the present Treaty, the Contracting Parties agree that, to the greatest extent feasible and practicable:

(a) information regarding plans for scientific programs in Antarctica shall be exchanged to permit maximum economy and efficiency of operations;

(b) scientific personnel shall be exchanged in Antarctica between expeditions and stations;

(c) scientific observations and results from Antarctica shall be exchanged and made freely available.

2. In implementing this Article, every encouragement shall be given to the establishment of cooperative working relations with those Specialized Agencies of the United Nations and other international organizations having a scientific or technical interest in Antarctica.

ARTICLE IV

1. Nothing contained in the present Treaty shall be interpreted as:

(a) a renunciation by any Contracting Party of previously asserted rights of or claims to territorial sovereignty in Antarctica;

(b) a renunciation or diminution by any Contracting Party of any basis of claim to territorial sovereignty in Antarctica which it may have whether as a result of its activities or those of its nationals in Antarctica, or otherwise;

(c) prejudicing the position of any Contracting Party as regards its recognition or non-recognition of any other State's right of or claim or basis of claim to territorial sovereignty in Antarctica.

2. No acts or activities taking place while the present Treaty is in force shall constitute a basis for asserting, supporting or denying a claim to territorial sovereignty in Antarctica or create any rights of sovereignty in Antarctica. No new claim, or enlargement of an existing claim, to territorial sovereignty in Antarctica shall be asserted while the present Treaty is in force.

ARTICLE V

1. Any nuclear explosions in Antarctica and the disposal there of radioactive waste material shall be prohibited.

2. In the event of the conclusion of international agreements concerning the use of nuclear energy, including nuclear explosions and the disposal of radioactive waste material, to which all of the Contracting Parties whose representatives are entitled to participate in the meetings provided for under Article IX are parties, the rules established under such agreements shall apply in Antarctica.

ARTICLE VI

The provisions of the present Treaty shall apply to the area south of 60° South Latitude, including all ice shelves, but nothing in the present Treaty shall prejudice or in any way affect the rights, or the exercise of the rights, of any State under international law with regard to the high seas within that area.

ARTICLE VII

1. In order to promote the objectives and ensure the observance of the provisions of the present Treaty, each Contracting Party whose representatives are entitled to participate in the meetings referred to in Article IX of the Treaty shall have the right to designate observers to carry out any inspection provided for by the present Article. Observers shall be nationals of the Contracting Parties which designate them. The names of observers shall be communicated to every other Contracting Party having the right to designate observers, and like notice shall be given of the termination of their appointment.

2. Each observer designated in accordance with the provisions of paragraph 1 of this Article shall have complete freedom of access at any time to any or all areas of Antarctica.

3. All areas of Antarctica, including all stations, installations and equipment within those areas, and all ships and aircraft at points of discharging or embarking cargoes or personnel in Antarctica, shall be open at all times to inspection by any observers designated in accordance with paragraph 1 of this Article.

4. Aerial observation may be carried out at any time over any or all areas of Antarctica by any of the Contracting Parties having the right to designate observers.

5. Each Contracting Party shall, at the time when the present Treaty enters into force for it, inform the other Contracting Parties, and thereafter shall give them notice in advance, of

(a) all expeditions to and within Antarctica, on the part of its ships or nationals, and all expeditions to Antarctica organized in or proceeding from its territory;

(b) all stations in Antarctica occupied by its nationals; and

(c) any military personnel or equipment intended to be introduced by it into Antarctica subject to the conditions prescribed in paragraph 2 of Article I of the present Treaty.

ARTICLE VIII

1. In order to facilitate the exercise of their functions under the present Treaty, and without prejudice to the respective positions of the Contracting Parties relating to jurisdiction over all other persons in Antarctica, observers designated under paragraph 1 of Article VII and scientific personnel exchanged under subparagraph 1(b) of Article III of the Treaty, and members of the staffs accompanying any such persons, shall be subject only to the jurisdiction of the Contracting Party of which they are nationals in respect of all acts or omissions occurring while they are in Antarctica for the purpose of exercising their functions.

2. Without prejudice to the provisions of paragraph 1 of this Article, and pending the adoption of measures in pursuance of subparagraph 1(e) of Article IX, the Contracting Parties concerned in any case of dispute with regard to the exercise of jurisdiction in Antarctica shall immediately consult together with a view to reaching a mutually acceptable solution.

ARTICLE IX

1. Representatives of the Contracting Parties named in the preamble to the present Treaty shall meet at the City of Canberra within two months after the date of entry into force of the Treaty, and thereafter at suitable intervals and places, for the purpose of exchanging information, consulting together on matters of common interest pertaining to Antarctica, and formulating and considering, and recommending to their Governments, measures in furtherance of the principles and objectives of the Treaty, including measures regarding:

(a) use of Antarctica for peaceful purposes only;

(b) facilitation of scientific research in Antarctica;

(c) facilitation of international scientific cooperation in Antarctica;

(d) facilitation of the exercise of the rights of inspection provided for in Article VII of the Treaty;

(e) questions relating to the exercise of jurisdiction in Antarctica;

(f) preservation and conservation of living resources in Antarctica.

2. Each Contracting Party which has become a party to the present Treaty by accession under Article XIII shall be entitled to appoint representatives to participate in the meetings referred to in paragraph 1 of the present Article, during such time as that Contracting Party demonstrates its interest in Antarctica by conducting substantial

49118—60——2

scientific research activity there, such as the establishment of a scientific station or the despatch of a scientific expedition.

3. Reports from the observers referred to in Article VII of the present Treaty shall be transmitted to the representatives of the Contracting Parties participating in the meetings referred to in paragraph 1 of the present Article.

4. The measures referred to in paragraph 1 of this Article shall become effective when approved by all the Contracting Parties whose representatives were entitled to participate in the meetings held to consider those measures.

5. Any or all of the rights established in the present Treaty may be exercised as from the date of entry into force of the Treaty whether or not any measures facilitating the exercise of such rights have been proposed, considered or approved as provided in this Article.

Article X

Each of the Contracting Parties undertakes to exert appropriate efforts, consistent with the Charter of the United Nations, to the end that no one engages in any activity in Antarctica contrary to the principles or purposes of the present Treaty.

Article XI

1. If any dispute arises between two or more of the Contracting Parties concerning the interpretation or application of the present Treaty, those Contracting Parties shall consult among themselves with a view to having the dispute resolved by negotiation, inquiry, mediation, conciliation, arbitration, judicial settlement or other peaceful means of their own choice.

2. Any dispute of this character not so resolved shall, with the consent, in each case, of all parties to the dispute, be referred to the International Court of Justice for settlement; but failure to reach agreement on reference to the International Court shall not absolve parties to the dispute from the responsibility of continuing to seek to resolve it by any of the various peaceful means referred to in paragraph 1 of this Article.

Article XII

1. (a) The present Treaty may be modified or amended at any time by unanimous agreement of the Contracting Parties whose representatives are entitled to participate in the meetings provided for under Article IX. Any such modification or amendment shall enter into force when the depositary Government has received notice from all such Contracting Parties that they have ratified it.

(b) Such modification or amendment shall thereafter enter into force as to any other Contracting Party when notice of ratification by it has been received by the depositary Government. Any such Contracting Party from which no notice of ratification is received within a period of two years from the date of entry into force of the modification or amendment in accordance with the provisions of subparagraph 1(a) of this Article shall be deemed to have withdrawn from the present Treaty on the date of the expiration of such period.

2. (a) If after the expiration of thirty years from the date of entry into force of the present Treaty, any of the Contracting Parties whose

representatives are entitled to participate in the meetings provided for under Article IX so requests by a communication addressed to the depositary Government, a Conference of all the Contracting Parties shall be held as soon as practicable to review the operation of the Treaty.

(b) Any modification or amendment to the present Treaty which is approved at such a Conference by a majority of the Contracting Parties there represented, including a majority of those whose representatives are entitled to participate in the meetings provided for under Article IX, shall be communicated by the depositary Government to all the Contracting Parties immediately after the termination of the Conference and shall enter into force in accordance with the provisions of paragraph 1 of the present Article.

(c) If any such modification or amendment has not entered into force in accordance with the provisions of subparagraph 1(a) of this Article within a period of two years after the date of its communication to all the Contracting Parties, any Contracting Party may at any time after the expiration of that period give notice to the depositary Government of its withdrawal from the present Treaty; and such withdrawal shall take effect two years after the receipt of the notice by the depositary Government.

ARTICLE XIII

1. The present Treaty shall be subject to ratification by the signatory States. It shall be open for accession by any State which is a Member of the United Nations, or by any other State which may be invited to accede to the Treaty with the consent of all the Contracting Parties whose representatives are entitled to participate in the meetings provided for under Article IX of the Treaty.
2. Ratification of or accession to the present Treaty shall be effected by each State in accordance with its constitutional processes.
3. Instruments of ratification and instruments of accession shall be deposited with the Government of the United States of America, hereby designated as the depositary Government.
4. The depositary Government shall inform all signatory and acceding States of the date of each deposit of an instrument of ratification or accession, and the date of entry into force of the Treaty and of any modification or amendment thereto.
5. Upon the deposit of instruments of ratification by all the signatory States, the present Treaty shall enter into force for those States and for States which have deposited instruments of accession. Thereafter the Treaty shall enter into force for any acceding State upon the deposit of its instrument of accession.
6. The present Treaty shall be registered by the depositary Government pursuant to Article 102 of the Charter of the United Nations.

ARTICLE XIV

The present Treaty, done in the English, French, Russian and Spanish languages, each version being equally authentic, shall be deposited in the archives of the Government of the United States of

America, which shall transmit duly certified copies thereof to the Governments of the signatory and acceding States.

IN WITNESS WHEREOF, the undersigned Plenipotentiaries, duly authorized, have signed the present Treaty.

DONE at Washington this first day of December, one thousand nine hundred and fity-nine.

For Argentina:
ADOLFO SCILINGO
F BELLO

For Australia:
HOWARD BEALE

For Belgium:
OBERT DE THIEUSIES

For Chile:
MARCIAL MORA M
E GAJARDO V
JULIO ESCUDERO

For the French Republic:
PIERRE CHARPENTIER

For Japan:
KOICHIRO ASAKAI
T. SHIMODA

For New Zealand:
G D L WHITE

For NORWAY:
PAUL KOHT

For the Union of South Africa:
WENTZEL C. DU PLESSIS

For the Union of Soviet Socialist Republics:
V. KUZNETSOV [Romanization]

For the United Kingdom of Great Britain and Northern Ireland:
HAROLD CACCIA

For the United States of America:
HERMAN PHLEGER
PAUL C. DANIELS

I CERTIFY THAT the foregoing is a true copy of the Antarctic Treaty signed at Washington on December 1, 1959 in the English, French, Russian, and Spanish languages, the signed original of which is deposited in the archives of the Government of the United States of America.

IN TESTIMONY WHEREOF, I, CHRISTIAN A. HERTER, Secretary of State of the United States of America, have hereunto caused the seal of the Department of State to be affixed and my name subscribed by the Authentication Officer of the said Department, at the city of

Washington, in the District of Columbia, this second day of December, 1959.

[SEAL]

CHRISTIAN A. HERTER,
Secretary of State.

By BARBARA HARTMAN,
Authentication Officer, Department of State.

TEXT OF UNITED STATES NOTE OF MAY 2, 1958 [1]

EXCELLENCY:

I have the honor to refer to the splendid example of international cooperation which can now be observed in many parts of the world because of the coordinated efforts of scientists of many countries in seeking a better understanding of geophysical phenomena during the current International Geophysical Year. These coordinated efforts of the scientists of many lands have as their objective a greatly increased knowledge of the planet on which we live and will no doubt contribute directly and indirectly to the welfare of the human race for many generations to come.

Among the various portions of the globe where these cooperative scientific endeavors are being carried on with singular success and with a sincere consciousness of the high ideals of mankind to which they are dedicated is the vast and relatively remote continent of Antarctica. The scientific research being conducted in that continent by the cooperative efforts of distinguished scientists from many countries is producing information of practical as well as theoretical value for all mankind.

The International Geophysical Year comes to a close at the end of 1958. The need for coordinated scientific research in Antarctica, however, will continue for many more years into the future. Accordingly it would appear desirable for those countries participating in the Antarctic program of the International Geophysical Year to reach agreement among themselves on a program to assure the continuation of the fruitful scientific cooperation referred to above. Such an arrangement could have the additional advantage of preventing unnecessary and undesirable political rivalries in that continent, the uneconomic expenditure of funds to defend individual national interests, and the recurrent possibility of international misunderstanding. It would appear that if harmonious agreement can be reached among the countries directly concerned in regard to friendly cooperation in Antarctica, there would be advantages not only to those countries but to all other countries as well.

The present situation in Antarctica is characterized by diverse legal, political, and administrative concepts which render friendly cooperation difficult in the absence of an understanding among the countries involved. Seven countries have asserted claims of sovereignty to portions of Antarctica, some of which overlap and give rise to occasional frictions. Other countries have a direct interest in that

[1] Addressed to the Foreign Ministers of each of the 11 other countries participating in the International Geophysical Year activities in Antarctica: Argentina, Australia, Belgium, Chile, France, Japan, New Zealand, Norway, the Union of South Africa, the U.S.S.R., and the United Kingdom. Each note was signed and delivered by the American ambassador to that country.

Chapter 3

Cenozoic history of Antarctic glaciation and climate from onshore and offshore studies

Robert M. McKay[1], Carlota Escutia[2], Laura De Santis[3], Federica Donda[3], Bella Duncan[1], Karsten Gohl[4], Sean Gulick[5], Javier Hernández-Molina[6], Claus-Dieter Hillenbrand[7], Katharina Hochmuth[4], Sookwan Kim[8], Gerhard Kuhn[4], Robert Larter[7], German Leitchenkov[9,10], Richard H. Levy[1,11], Tim R. Naish[1], Phil O'Brien[12], Lara F. Pérez[7], Amelia E. Shevenell[13] and Trevor Williams[14]

[1]Antarctic Research Centre, Victoria University of Wellington, Wellington, New Zealand, [2]Andalusian Institute of Earth Sciences, CSIC and Universidad de Granada, Armilla, Spain, [3]National Institute of Oceanography and Applied Geophysics – OGS, Sgonico, Italy, [4]Alfred Wegener Institute, Helmholtz-Center for Polar and Marine Science, Bremerhaven, Germany, School of Geography, Geology and the Environment, University of Leicester, Leicester, United Kingdom, [5]Institute for Geophysics & Deptartment of Geological Sciences, Jackson School of Geosciences, University of Texas at Austin, Austin, TX, United States, [6]Department of Earth Sciences, Royal Holloway University of London, Egham, Surrey, United Kingdom, [7]British Antarctic Survey, Cambridge, United Kingdom, [8]Korea Polar Research Institute, Incheon, Republic of Korea, [9]Institute for Geology and Mineral Resources of the World Ocean, St. Petersburg, Russia, [10]Institute of Earth Sciences, St. Petersburg State University, St. Petersburg, Russia, [11]GNS Science, Lower Hutt, New Zealand, [12]Department of Environment and Geography, Macquarie University, Sydney, NSW, Australia, [13]College of Marine Science, University of South Florida, St. Petersburg, FL, United States, [14]International Ocean Discovery Program, Texas A&M University, College Station, TX, United States

3.1 Introduction

The influence of Antarctica's ice sheets and fringing sea ice belt on global climate is profound, and processes relating to changes in Antarctica's cryosphere are central to many of the largest feedbacks in the Earth System. Variations in polar ice cover and ice sheet volume directly affect Earth's absorption of solar energy through changes in planetary albedo, while shifts in atmospheric and oceanic currents associated with high latitude processes influence the way heat and marine nutrients are redistributed around the

Antarctic Climate Evolution. DOI: https://doi.org/10.1016/B978-0-12-819109-5.00008-6

planet, and in turn form a significant control on global carbon cycling through time. Indeed, ice core records provide unequivocal evidence of a strong linear relationship between global temperatures, ice volume and atmospheric CO_2 concentrations through the glacial-interglacial cycles of the past 800,000 years (EPICA community Members, 2004; Lüthi et al., 2008).

For records older than 800 ka, the relationship between these three variables remains less certain. In part, this is due to significant uncertainties in measuring atmospheric CO_2 from geological proxies, such as boron isotopes from marine foraminifera, the carbon isotopic composition of marine algal biomarkers and stomata on terrestrial leaf fossils (Foster et al., 2017). Adding further uncertainty to understanding the relationship between CO_2, global temperature and ice sheet volumes is the history of the Antarctic ice sheets from geological data, which is limited by the lack of rock exposures and the small number of drill cores from the Antarctic region. In this context, understanding of long-term geological processes are essential to constrain, as factors such as the tectonic subsidence of the parts of the Antarctic continent through time invalidate the assumption of a uniform relationship between global temperature change, global ice volume and atmospheric CO_2 (Gasson et al., 2016b; Naish et al., 2021 (from this volume); Wilson et al., 2013).

Numerous long-term Cenozoic global climate and polar ice volume records have been inferred from far-field datasets, either from deep sea oxygen isotope records or sea level estimates derived from the continental margins in low- to mid-latitudes (Holbourn et al., 2013; Liebrand et al., 2016; Miller et al., 2020; Vleeschouwer et al., 2017; Zachos et al., 2001). While these records are fundamental to our understanding of Earth's climate and ice sheet history, they remain complicated by a range of uncertainties and assumptions. For example, deep sea foraminiferal oxygen isotopes record a signal of both temperature and global ice volume, which may not covary in a linear manner (Cramer et al., 2011; Elderfield et al., 2012; Miller et al., 2020), and cannot identify the hemisphere in which polar ice sheet volume change was occurring.

Identifying locations of ice volume change is also a significant issue for sea level reconstructions derived from low- to mid-latitude continental margins, but to obtain a true eustatic signal these records also require significant corrections for sediment supply and compaction, tectonic subsidence, and an understanding of the influence of glacio-hydrostatic adjustment and mantle dynamics (Grant et al., 2019; Kominz et al., 2008; Miller et al., 2020; Moucha et al., 2008). Consequently, these far-field proxy records remain subject to uncertainties and assumptions when assessing high-latitude climatic change. Examples include determining the magnitude of polar amplification of temperatures in past high CO_2 worlds and identifying the synchronicity of ice sheet behaviour in either hemisphere. Direct records from the Antarctic margin are needed to best understand the direct response of the Antarctic ice sheets to global climate and oceanographic shifts, as well as the influence of high latitude processes on global climate feedbacks and sea level change. Only these can provide the critical and unique datasets to refine interpretations derived from far-field records.

Direct records of Antarctic Cenozoic glacial history come from rare, disparate rock outcrops, and from more than 60 drill holes (of variable recovery and quality) that are mostly located around the periphery of the East Antarctic Ice Sheet (EAIS) or the deeper offshore waters of the Southern Ocean (see recent reviews by Barrett, 2008; Escutia et al., 2019; McKay et al., 2016) (Fig. 3.1). Here, we outline the current state of knowledge of Cenozoic climate history in Antarctica from near- and far-field records,

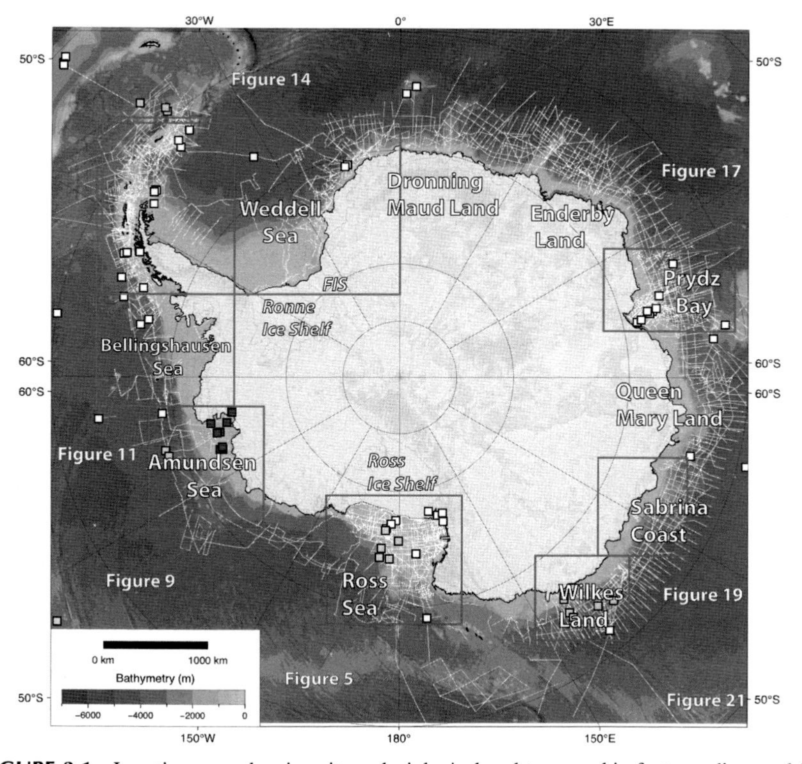

FIGURE 3.1 Location map showing sites, glaciological and topographic features discussed in text. Present-day ice sheet cover is slightly transparent to highlight the location of subglacial basins. Seismic tracks are from the SCAR Antarctic Seismic Data Library System (SDLS). Bathymetry is from Bathymetric Chart of the Southern Ocean (Arndt et al., 2013) and subglacial topography from BedMachine Antarctica dataset (Morlighem et al., 2020). Boxes show location of detailed map figures in the chapter. Small white filled squares represent deep drill sites collected prior to 2008, and yellow squares represent sites drilled since 2008 by the IODP program (see detailed map figures). Blue squares are sites collected by the PS104/MeBo expedition in the inner Amundsen Sea. *EAIS*, East Antarctic Ice Sheet; *FIS*, Filchner Ice Shelf; *WAIS*, West Antarctic Ice Sheet.

followed by a discussion of the seismic stratigraphy and drill core constraints and paleoclimatic records preserved within this stratigraphy from around the margin. In this sector-by-sector summary of the stratigraphy, we also provide a direction for future drilling projects in the Antarctic region that will address the outstanding challenge that extracting this history presents.

3.2 Long-term tectonic drivers and ice sheet evolution

Tectonic events are one of two fundamental controls on the long-term Antarctic climate trend over geological timescales, the other being atmospheric CO_2 levels. The most significant of these events was the break-up of the Gondwana supercontinent, which resulted in the formation of the Southern Ocean and allowed the modern global overturning circulation system to develop (Talarico et al., 2021, this volume). The opening of ocean gateways led to cooling and ultimately to the development of the Antarctic Circumpolar Current (ACC). This contributed to the Cenozoic cooling and expansion of Antarctica's continental ice sheets (Bijl et al., 2013; DeConto and Pollard, 2003; Hochmuth et al., 2020a,b; Kennett, 1977; Kennett et al., 1974), alongside concomitant shifts in global carbon cycling. Feedbacks associated with ice sheet initiation in Antarctica in turn acted to amplify the Earth system response to orbital forcing, creating large shifts in Antarctic ice volumes and eustatic sea level, atmospheric and oceanic circulation further altering global carbon cycling processes and heat transport around the planet.

Associated with post-Gondwana break-up is the evolution of the West Antarctic Rift System (WARS) (Figs. 3.1 and 3.2), a major control on ice sheet evolution through the Cenozoic, particularly for the West Antarctic Ice Sheet (WAIS). This rift system has influenced the timing and amount of West Antarctic subsidence, Transantarctic Mountain (TAM) uplift, sediment redistribution, mantle viscosity and crustal heat flux – all of which would have influenced the distribution of land and ocean areas, and in turn EAIS and WAIS evolution (Colleoni et al., 2018; Gasson et al., 2016b; Hochmuth and Gohl, 2019; LeMasurier et al., 1982; Levy et al., 2019; Paxman et al., 2019; Wilson et al., 2013). Rifting in West Antarctica likely initiated in Marie Byrd Land (between the Amundsen and Ross Sea sectors) at ~104 Ma.

Rifting continued through the Cenozoic, moving into the central then western Ross Sea Embayment, where it still persists today in the Terror Rift and is associated with the McMurdo Volcanic Group (Decesari et al., 2007; Eagles et al., 2004; Fielding et al., 2008; Wenman et al., 2020; Wilson and Luyendyk, 2009; Wobbe et al., 2012). Neogene to Quaternary rifting and active volcanism is also thought to have been widespread across much of Marie Byrd Land and the Ross Sea (Jordan and Siddoway, 2020; Kalberg and Gohl, 2014; Sauli et al., 2021; Wobbe et al., 2012).

Critically, the WARS has created large sedimentary basins in the Ross and Amundsen Sea region. These contain accessible archives of Antarctic

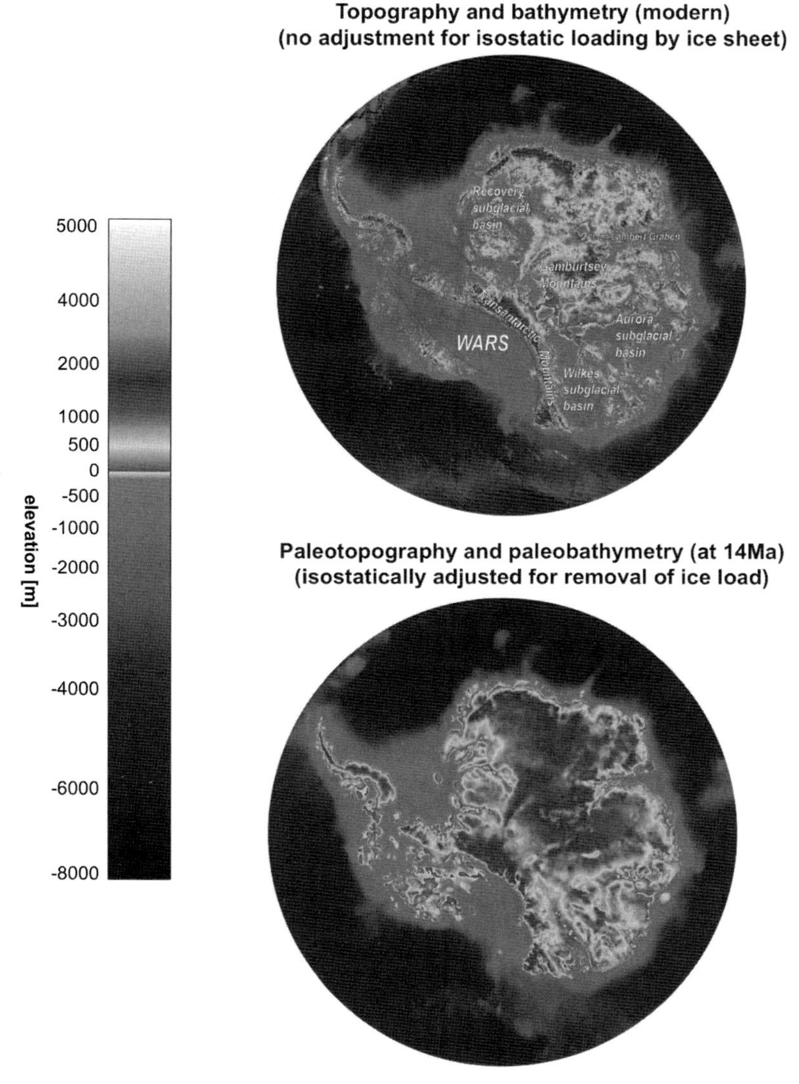

FIGURE 3.2 (Top) Map showing modern subglacial topography and bathymetry, and major geological and topographic features discussed in text. Base map above is plotted from MEaSUREs BedMachine Antarctica, version 2 (Morlighem, 2020). The approximate boundary of the West Antarctic Rift System (WARS) is shown as red lines. No isostatic adjustments have been made for removal of ice load. (Bottom) Middle Miocene (14 Ma) merged palaeotopographic and paleobathymetric models of the Southern Ocean and the Antarctic continent reconstructed based on all available geophysical and geological data after Paxman et al. (2019) and Hochmuth et al. (2020a,b). This map has been fully isostatically relaxed under ice-free conditions. The contrast in topography between the upper and lower panels highlights the importance of tectonic subsidence, erosion and isostatic responses to ice loading in influencing the sensitivity of the ice sheet to marine vs terrestrial mass balance controls.

environmental change since the Late Cretaceous, shedding light on the early history of the Cenozoic Antarctic ice sheets (Bart and De Santis, 2012; Brancolini et al., 1995; Decesari et al., 2007; Fielding et al., 2008; Gohl et al., 2013; Sorlien et al., 2007). In contrast, the rift basins in the Weddell Sea are much older, associated with back-arc rifting relating to subduction along the Pacific margin of the Gondwanan continent in the Jurassic (Huang et al., 2014; Jordan and Siddoway, 2020; Jordan et al., 2017; Riley et al., 2020). Drivers of topographic change and sedimentation in this region are directly related to erosion associated with the Cenozoic evolution of the EAIS and WAIS rather than active rifting processes (Hochmuth and Gohl, 2019; Paxman et al., 2019; Thomson et al., 2013).

Compared to the thinner crust of West Antarctica, the thicker, more tectonically stable cratonic crust of East Antarctica led to fundamentally different ice sheet dynamics and history there. Ancient high elevation orogenic belts in Dronning Maud Land and the Gamburtsev Mountains region were the most likely sites for the continent's first glaciers in the distant past, growing from alpine to ice caps with timing unknown but predating the first continental ice sheets 34 Ma ago (Bo et al., 2009; Ferraccioli et al., 2011; Rose et al., 2013). They also provided nuclei of the early glacial systems in East Antarctica (DeConto and Pollard, 2003; Galeotti et al., 2016; Stocchi et al., 2013). The Transantarctic Mountains, which began rising as a flank of the WARS from around 55 Ma ago (Gleadow and Fitzgerald, 1987), provided a growing buttress for East Antarctic ice flowing towards West Antarctica from the time of the first continental ice sheets. (DeConto and Pollard, 2003).

Today, approximately one-third of the EAIS is grounded below sea level. If this marine-based ice sheet were to melt it would contribute ~19.2 m to global sea level (Fretwell et al., 2013). The largest EAIS basins with marine-based ice are the Recovery Subglacial Basin (Weddell Sea sector) and the Wilkes and Aurora Subglacial Basins, with a smaller basin existing in the Prydz Bay region (Lambert Trough). These basins resulted from a range of predominately Mesozoic tectonic rifting processes, but have evolved through a range of subsequent erosion and sedimentary processes through the Cenozoic (Aitken et al., 2014, 2016; Ferraccioli et al., 2009, 2001; Harrowfield et al., 2005).

3.3 Global climate variability and direct evidence for Antarctic ice sheet variability in the Cenozoic

Several reviews on the long-term history of Antarctica's ice sheets have recently been compiled (Escutia et al., 2019; McKay et al., 2016; Noble et al., 2020; Shevenell and Bohaty, 2012). Chapters 3 (Barrett, 2008) and 5 (Cooper et al., 2008) of the first edition of this book (Florindo and Siegert, 2008) recorded discoveries made and strategies adopted to reveal Antarctic ice sheet age and evolution since the first International Geophysical Year (1957–1958).

These were largely achieved by over-snow traverses, the employment of post-World War II geophysical techniques in offshore surveying, and by ship-based and land-based deep drilling. Much progress has been made since that first edition. Areas like the Ross Sea and Prydz Bay, already relatively known, and less known areas (like the Amundsen Sea and most of the East Antarctic margin), have been revisited with new geophysical techniques, deep drilling and other multidisciplinary approaches. In this review, we provide an update of the current knowledge with a focus on regions seismically surveyed and sampled over the past two decades by geological drilling on the Antarctic margin, e.g., the ANDRILL and SHALDRIL projects, Ocean Drilling Program (ODP) and the Integrated Ocean Drilling Program/International Ocean Discovery Program (IODP). For detailed assessments of Antarctic ice sheet histories of specific time periods, or from more distal paleoceanographic records in the Southern Ocean and beyond, the reader is referred to Colleoni et al. (2021), Galeotti et al. (2021), Levy et al. (2021), Naish et al. (2021), Siegert et al. (2021) and Wilson et al. (2021).

3.3.1 Late Cretaceous to early Oligocene evidence of Antarctic ice sheets and climate variability

Low- and mid-latitude passive margins record moderate- to high-amplitude sea level variations (40–100 m) since Cretaceous times (Fig. 3.3), implying substantial ice sheets on Antarctica long before the assumed first glaciation in the earliest Oligocene (Hollis et al., 2014; Kominz et al., 2008; Miller et al., 2020, 2008). However, mantle dynamics may exert a significant influence that overwhelms a eustatic signal in the far-field sequence stratigraphic record (Moucha et al., 2008; Raymo et al., 2011), providing an alternative to ice sheets as the cause of pre-Oligocene sea level variations.

The only in situ Antarctic continuous continental shelf records obtained for the Late Cretaceous to earliest Paleogene are exposed on Seymour Island in the Antarctic Peninsula (Ivany et al., 2008; Mohr et al., 2020; Scasso et al., 2020), which at the time of deposition was an emergent volcanic arc at a paleolatitude of ~65°S, rather than part of continental Antarctica (Elliot, 1988; Lawver et al., 1992). Pollen assemblages indicate a cool to warm temperate coastal vegetation in the Late Cretaceous, with Mean Annual Temperatures (MAT) of ~10°C–15°C and dinoflagellate assemblages suggesting the possibility of winter sea ice (Bowman et al., 2014, 2013). ODP Leg 113 also recovered late Cretaceous (Maastrichtian) on the Maud Rise, with calcareous chalks and oozes that were interpreted to record warm temperate to cool subtropical climates with high seasonality in the Weddell Sea, with no clear evidence of cryospheric development (Kennett and Barker, 1990). This cooling trend in surface and intermediate Weddell Sea waters during the Maastrichtian is similar to that reported from Seymour Island.

Drilling in Prydz Bay recovered a short interval of terrestrial mid-Cretaceous sediment, with conifer-dominated woodland pollen suggesting a cool, humid climate (Macphail and Truswell, 2004a,b). As rifting in the eastern Ross Sea and Amundsen Sea sectors initiated during the Late Cretaceous, it is likely there are significant thicknesses of contemporaneous strata in these basins (Luyendyk et al., 2001; Wilson and Luyendyk, 2009). Indeed, reworked terrestrial Cretaceous palynomorphs are observed in surface sediments from the eastern Ross Sea, suggesting Cretaceous sediments are present in the rift basins (Kemp and Barrett, 1975; Truswell and Drewry, 1984). A recent short seafloor drill core from the Amundsen Sea region of West Antarctica recovered a mid-Cretaceous (~92–84 Ma) paleosol formed at a paleolatitude of ~82°S during the early phases of rifting with Zealandia (Gohl et al., 2017; Klages et al., 2020), when atmospheric CO_2 was estimated be ~1000 ppm (Foster et al., 2017). This paleosol contained in situ fossil tree roots, and fossil pollen and spores that indicate a diverse temperate lowland rainforest biome flourishing in about 4 months of complete polar night darkness and at mean annual temperatures of 13°C with precipitation of 1120 mm/yr. The reconstructed temperate climate at this high latitude

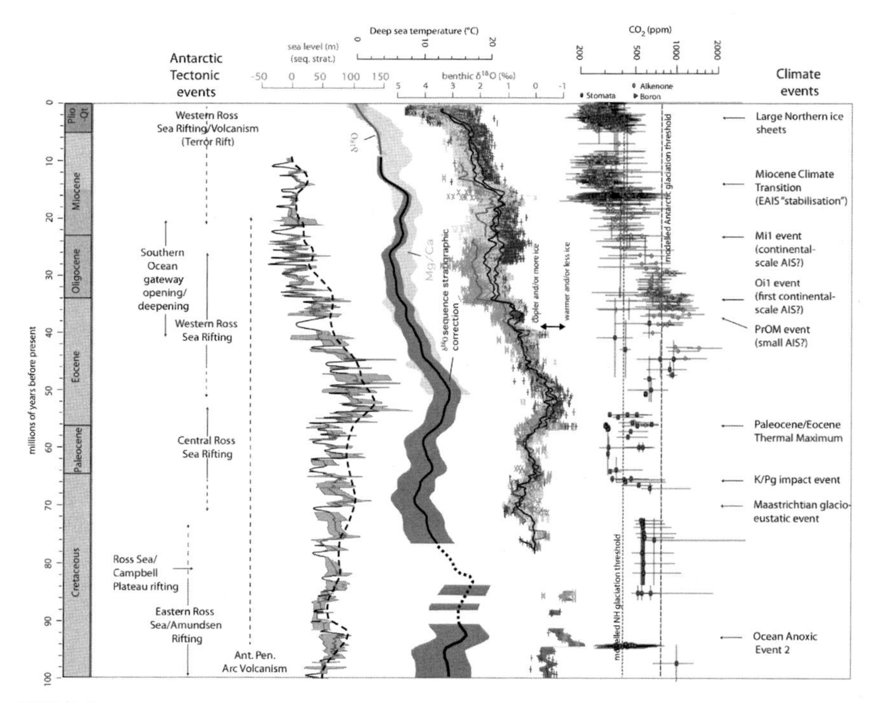

FIGURE 3.3 Chronostratigraphic summary of far-field sea level records (Kominz et al., 2008), climate proxies (deep sea temperature and $\delta^{18}O$; Cramer et al., 2009), atmospheric CO_2 (Foster et al., 2017) and relevant Cenozoic climate and tectonic events discussed in text also shown.

requires elevated atmospheric carbon dioxide levels and a vegetated land surface without major Antarctic glaciation (Klages et al., 2020).

The Eocene epoch was the last period when atmospheric CO_2 exceeded three to five times preindustrial concentrations (Anagnostou et al., 2016; Foster et al., 2017; Jagniecki et al., 2015; Liu et al., 2009; Pagani et al., 2011, 2005) (Fig. 3.3). Ocean modelling studies indicate that poleward heat transport in the Eocene was similar to today, but polar sea-surface temperatures (SSTs) were up to ~14°C warmer, appearing to preclude significant ice sheets in Antarctica (Hollis et al., 2019, 2012; Huber et al., 2004). An ice-free and vegetated Antarctica may have also exerted a significant control on the global carbon cycle by thawing of organic-rich permafrost soils during rapid and extreme warming events (hyperthermals), including the Paleocene-Eocene Thermal Maximum (PETM) (DeConto et al., 2012). Isotope studies on molluscs in the Antarctic Peninsula region suggest SSTs reached ~15°C in the early Eocene, cooling to ~5°C by the Late Eocene (Ivany et al., 2008). Similar SSTs are estimated from ODP Leg 113 stable isotope proxies (Kennett and Stott, 1990). Proxies from early Eocene erratics in the Ross Sea and offshore strata in Wilkes Land indicate much warmer summer temperatures (~25°C), frost-free winters (~10°C) and SSTs >15°C warmer than present, despite polar winter darkness (Askin, 2000; Bijl et al., 2013; Levy and Harwood, 2000; Pross et al., 2012).

Most modelling studies fail to produce extreme Eocene temperatures at high latitudes, unless CO_2 levels were significantly higher than current proxy records suggest (Caballero and Huber, 2013; Hollis et al., 2019, 2012, 2009). This implies that models might be undersensitive and lack some processes critical for simulation of polar climates. In contrast to the apparent warmth of the Eocene, sea level records from low-latitude passive margins in the Paleocene to Eocene record large amplitude variations (~40 m), which could relate to episodic growth of substantial ice sheets on Antarctica and/or mantle dynamics affecting these relative sea level records (Kominz et al., 2008; Miller et al., 2020). If eustatic in origin, these far-field sea level records contradict established models of Cenozoic cryosphere evolution and expansion of the Antarctic ice sheets at 34 Ma due to the development of the ACC (Kennett, 1977). However, they have never been fully tested against semicontinuous direct records proximal to the Antarctic continent.

Short cores containing ice transported clasts collected in the Sabrina Coast and Prydz Bay regions of East Antarctica and the Weddell Sea suggest the presence of some marine-terminating glaciers in the early Eocene to late Eocene (Carter et al., 2017; Gulick et al., 2017; Passchier et al., 2017; Shevenell et al., 2017). This evidence of earlier glacial advances is supported by radiolarian assemblages indicative of surface cooling at high latitudes in the Southern Ocean that also coincide with a positive oxygen isotopic excursion at ~37 Ma (the Priabonian Oxygen Isotope Maximum) (Pascher et al.,

2015; Scher et al., 2014). Whether these were local mountain glaciers or ice sheets, prior to the Eocene/Oligocene boundary, it is clear from a range of proxy records around the margin that the late Eocene was significantly cooler than the mid Eocene. Temperate coastal pollen assemblages have been recorded from drill cores in the Weddell Sea, Ross Sea, Wilkes Land and Prydz Bay regions (Anderson et al., 2011; Cooper and O'Brien, 2004; Pross et al., 2012; Raine and Askin, 2001; Warny and Askin, 2013).

3.3.2 The Eocene-Oligocene transition and continental-scale glaciation of Antarctica

High-resolution deep sea $\delta^{18}O$ records, palaeontological studies and modelling experiments suggest the first continental-scale ice sheets were triggered at the Eocene/Oligocene boundary by an optimally cold orbital configuration (Coxall et al., 2005), following cooling due to the tectonic opening of Southern Ocean gateways (Bijl et al., 2013; Kennett, 1977) and decreasing atmospheric CO_2 below a threshold value (DeConto et al., 2008; Galeotti et al., 2016). This cooling also led to significant shifts in Southern Ocean planktic ecosystems (Houben et al., 2013). Once initiated, models suggest that ice sheet hysteresis behaviour, relating largely to height-mass balance and albedo feedbacks, ensured that the first ice persisted as a nucleus, allowing the ice sheet to expand and contract in response to orbital forcing throughout the Oligocene (DeConto and Pollard, 2003).

In the Antarctic Peninsula region, shallow sediment cores dated at $\sim 36 \pm 1$ Ma contain pollen and leaf wax biomarkers indicative of a significant cooling and drying prior to the Eocene/Oligocene boundary (Feakins et al., 2014). A similar cooling signal between the Eocene and Oligocene is identified in the Ross Sea, with a shift to shorter chain length leaf wax biomarkers reflecting a vegetation response to cooling climate (Duncan et al., 2019). Late Eocene cooling and drying is further supported by geochemical and clay mineral evidence from drill cores in Prydz Bay indicating reduced chemical weathering and enhanced physical weathering by glacial activity after 34.4 Ma, but with periods of ephemeral glaciation interpreted between 36 and 34.4 Ma (Forsberg et al., 2008; Hambrey et al., 1991; Passchier et al., 2017). Late Eocene palynological assemblages at Site 1166 in Prydz Bay indicate stunted *Nothofagus* rainforest, reflecting a cool to cold temperate environment at sea level (Macphail and Truswell, 2004a,b). Thermochronological studies from Prydz Bay also indicate greatly enhanced erosion by the EAIS in that region since 34 Ma (Thomson et al., 2013).

Paleotopographic reconstructions derived from tectonic, seismic and drilling studies reveal that West Antarctica and the East Antarctic marine basins were largely above sea level at 34 Ma (Paxman et al., 2019; Wilson et al., 2012). Therefore Antarctica could hold more terrestrial ice in the Oligocene than it can today, even though the climate was warmer than present (Wilson et al., 2013).

This is because the cooling threshold required for the development of a high elevation terrestrial-based ice sheet is lower than that of a marine-based ice sheet, which is highly sensitive to changes in oceanic heat flux. Consequently, models indicate a largely terrestrial West Antarctica could potentially accommodate an extra ~13 million km^2 of grounded ice (i.e. ~30 m SLE) during the early Oligocene, while the increased buttressing provided by a larger WAIS also leads to a larger and higher EAIS (Bart et al., 2016; Colleoni et al., 2018; Wilson et al., 2013) (Fig. 3.4). In addition, marine ice sheets also displace some of their mass in the ocean, so even if the volumes of a marine-based vs terrestrial-based ice sheet are the same, the resulting sea level changes are less for marine-based ice sheets (Gasson et al., 2016a).

However, these reconstructions present maximum and minimum estimates for the amount of subaerial land at the Eocene/Oligocene boundary, and Coenen et al. (2020) suggested that microfossils and biomarkers indicate marine embayments and lowlands in West Antarctica in the late Eocene, which favour the minimum topographic reconstructions of Paxman et al. (2019) and Wilson et al. (2012). However, an important caveat with this is that the modelled reconstructions have a very coarse (and smoothed) resolution. A more complex topography that included narrow marine embayments formed during rifting could have existed. These topographic reconstructions are also critical in assessing the role of glacial isostatic adjustment (GIA) on regional vs global sea level changes (e.g., Stocchi et al., 2013; Whitehouse et al., 2019).

Constraining Antarctica's paleotopography is critical for determining Antarctic Ice Sheet (AIS) contributions to eustatic sea level variability and ice volume throughout the Cenozoic era. Recent syntheses of seismic datasets constrained by geological data obtained from on-land outcrops and drill cores have provided significant advances in understanding the role that paleotopography has played on AIS history and offshore sediment deposition (Colleoni et al., 2018; Hochmuth et al., 2020a,b; Levy et al., 2019; Paxman et al., 2019). Furthermore, the accumulation of sediment through erosion and sedimentation in offshore basins is a fundamental control on ice sheet dynamics in its own right, and modellers are increasingly recognising the importance of this process on the long-term evolution of the AIS (Pollard and DeConto, 2020).

3.3.3 Transient glaciations of the Oligocene and Miocene

During the earliest Oligocene (and potentially late Eocene), marine-terminating glaciers episodically extended beyond the present-day coastline around much of the margin, with sediment core evidence of this in the Prydz Bay, Sabrina Coast, Wilkes Land and Ross Sea regions (Barrett, 1989; Escutia and Brinkhuis, 2014; Galeotti et al., 2016; Hambrey et al., 1991; Kulhanek et al., 2019; Levy et al., 2019). A high-resolution assessment of the sedimentology in the Cape Roberts Project drill cores indicated a stepped evolution of the EAIS expansion between 34 and 31 million years ago, with smaller, oscillatory terrestrial ice sheets varying at orbital timescales when atmospheric CO_2 levels were >600 ppm. However, at 32.8 Ma, a large, more stable continental-scale ice sheet formed at a time when proxies suggest CO_2 fell below 600 ppm (Galeotti et al., 2016).

In Prydz Bay and Wilkes Land, late Eocene to early Oligocene diamictites were recovered from the continental shelf, but are interpreted as glaciomarine deposits indicative of marine-terminating glaciers, rather than grounding of ice sheets on the outer continental shelf (Escutia et al., 2011a; O'Brien et al., 2001). However, determining whether or not Antarctic ice sheet expansion resulted in a ice advancing to the coastline in all sectors of Antarctica remains equivocal

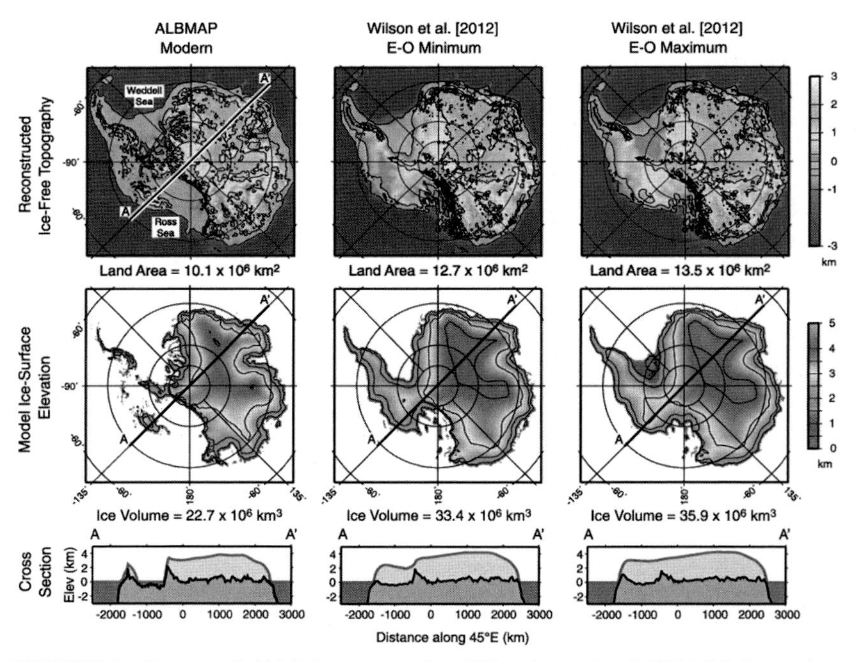

FIGURE 3.4 Top row: (left) Modern topography of West Antarctica; (middle) Minimum elevation of reconstructed E/O topography of West Antarctica, based tectonic history, seismic stratigraphic interpretations and drill core observation. (right) Maximum elevation of reconstructed E/O topography of West Antarctica using the same methodology. Maximum and minimum reconstruction accounting for uncertainties in the various datasets used to compile the paleotopographies. Middle and Bottom rows: (left) modelled ice sheet volume on topographies show in top row under an E/O boundary climate state (700 ppm CO_2) indicating a lack of ice in West Antarctica as marine ice sheet does not form in this warmer-than-present climate state. (middle) Reconstructed E/O topography of West Antarctica, with modelled ice sheet volume run in same climate as shown in top middle panel. (right) Reconstructed E/O topography of West Antarctica, with modelled ice sheet volume run in same climate as shown in top right panel.

from geological datasets. Evidence of a marine-terminating ice sheet grounding on the continental shelf comes from: (1) sedimentological data from the deep-sea margin of Antarctica which indicates only presence (rather than size) of marine-terminating glaciers; (2) sites in close proximity to the TAM [e.g., CIROS-1 and Cape Roberts Project (CRP)]; and (3) continental shelf regions in Wilkes Land and Prydz Bay. However, these records are highly discontinuous and potentially record smaller-scale alpine glacial advances or are hard to distinguish between glaciomarine and subglacial sedimentation (Cooper and O'Brien, 2004; Escutia et al., 2011a; Galeotti et al., 2016).

Consequently, further drilling campaigns in targeted regions are required to test the hypothesis of continental-wide expansion of the Antarctic ice sheets near the Eocene/Oligocene boundary, as implied by deep sea records and model-based experiments (Fig. 3.3). Drill core records from the Antarctic margin can also provide valuable constraints on relative sea level records in order to test sea-level models relating to EAIS growth. Such data-model comparisons suggest a near-field increase in relative sea level occurred, as a consequence of glacial isostatic adjustment and gravitational attraction as the EAIS grew (Stocchi et al., 2013). Increased spatial resolution in proximal marginal areas of drill cores is required to fully test these models and the full extent of Antarctic ice sheet growth across the continent.

Orbital pacing of Oligocene ice sheet variability is recorded from numerous locations around the East Antarctic margin, including the Ross Sea (Barrett, 2007, 1989; Galeotti et al., 2016; Naish et al., 2001), and offshore from Wilkes Land and Dronning Maud Land (Hartman et al., 2018; Hauptvogel et al., 2017; Salabarnada et al., 2018). The Oligocene-Miocene boundary (23 Ma) is characterised by an isotopic excursion of ~1‰ attributed to an abrupt increase in Antarctic ice volume driven by changes in Earth's orbital parameters (Barrett, 1989; Flower et al., 2006; Mawbey and Lear, 2013; Naish et al., 2001; Zachos et al., 1997) and a decrease in atmospheric CO_2 (Pagani et al., 2005). Deep Sea Drilling Project (DSDP) Site 270 records a phase of proximal glaciomarine sedimentation on the outer Ross Sea continental shelf at ~24.5 Ma, inferred to be the consequence local advances of ice caps nucleating on nearby subaerial basement (De Santis et al., 1995; Kulhanek et al., 2019) (for details see Naish et al., 2021, this volume).

A comprehensive synthesis of geological drill core data and the seismic stratigraphy in the Ross Sea was conducted by Levy et al. (2019), who compared these data to orbital scale timeseries analysis on deep sea oxygen isotope records. In their study, the authors noted a large increase in the sensitivity of benthic $\delta^{18}O$ records to obliquity pacing (40-kyr orbital cycles relating to the tilt of the Earth's axis) during the latest Oligocene. This increased sensitivity coincided with direct evidence of AIS advance onto the Ross Sea continental shelf – as determined by seismic unconformities and sedimentary facies in drill cores, as well as evidence from global proxies of lower atmospheric CO_2 conditions following the Oligocene/Miocene boundary (Foster and Rohling, 2013; Foster et al., 2017). As obliquity is an important regulator of latitudinal temperature gradients, and therefore the wind-driven ocean currents that strongly influence marine ice sheet mass balance, these results suggest increased frequency of orbitally paced marine-terminating ice sheet advances and retreats in this region following the Oligocene/Miocene boundary (Levy et al., 2019).

Following the periodic transient glaciations during the Oligocene to early Miocene, both near- and far-field proxy data indicate a period of sustained warmth, referred to as the Miocene Climatic Optimum (MCO; ~17–15 Ma; Flower and Kennett, 1994; Shevenell et al., 2008) (for more detail see Levy et al 2021), this volume. At this time, changes in global carbon cycling occurred (e.g., Holbourn et al., 2015; Vincent and Berger, 1985), average global temperatures were ~3°C warmer than present, and while polar amplification is suggested from proxy data, models require atmospheric CO_2 in the range of 460–580 ppmv (Feakins et al., 2012; Lewis et al., 2008; Shevenell and Kennett, 2004; Warny et al., 2009; You et al., 2009).

The ANDRILL site AND-2A was recovered from the Western Ross Sea and was interpreted to record TAM tidewater outlet glaciers overriding and/or calving near the site (Fielding et al., 2011; Passchier et al., 2011). At 15.7 Ma, a diatomite with abundant pollen, algae and other biomarkers suggests a warmer than present (Mean Summer Temperature of ~10°C) climate during the MCO (Feakins et al., 2012; Warny et al., 2009). A compilation of sedimentological and geochemical evidence from the AND-2A drill core combined with modelling experiments suggest retreat of the EAIS to its terrestrial margin during peak warm interglacials of the MCO (Gasson et al., 2016a; Levy et al., 2016). Further offshore, IODP Site U1521 recovered a thick interval of diatom-rich mudstone deposited during the MCO (McKay et al., 2019), while offshore of Wilkes Land marine sediments demonstrates a relative lack of ice rafted debris, with temperate pollen assemblages of *Nothofagidites* and *Podocarps* indicative of mean summer on-land temperatures of >10°C (Sangiorgi et al., 2018). On-land outcrops in the Dry Valleys provide evidence of voluminous outbursts of subglacial meltwater derived from the EAIS margin prior to 14 Ma (Lewis et al., 2006; Sugden and Denton, 2004), while a tundra vegetation and wet-based alpine glaciation persisted at high elevations in the TAM until at least 14.07 Ma (Lewis et al., 2008).

The Middle Miocene Climate Transition (MMCT) is identified by a ~1% enrichment in the $\delta^{18}O$ record of deep sea benthic foraminifera at 13.8 Ma (Flower and Kennett, 1994; Holbourn et al., 2014; Kennett, 1977; Shevenell and Kennett, 2004), with sea level and temperature reconstructions implying the majority of this $\delta^{18}O$ enrichment is related to ice sheet expansion (John et al., 2004; Shevenell et al., 2008; Wright et al., 1991) (for more detail see Levy et al. 2021, this volume). This isotopic excursion is associated with rapid cooling to dry-based glaciation and aridification of the high elevation regions of the TAM, alongside inferred extinction of alpine tundra vegetation (Lewis et al., 2008, 2007). Consequently, this expansion has commonly been inferred to represent the development of a permanent and relatively stable EAIS (Sugden, 1996).

This idea of a stable EAIS since ~14 Ma was intensely debated throughout the 1980s and 1990s on the basis of Pliocene-age marine diatoms observed in the Transantarctic Mountain Sirius Group deposit. These diatoms were inferred to have been sourced from Pliocene-age marine strata deposited in the Wilkes Land subglacial basin and transported into the TAM via a subsequent readvance of the EAIS (Webb et al., 1984), implying this subglacial basin was deglaciated in the Pliocene (see review by Barrett, 2013 for details on this debate). Subsequent studies suggest the diatoms were windblown from isostatically uplifted Pliocene marine strata in the coastal regions of the Ross Sea and incorporated into the surface layers of the Sirius Group after deposition and are therefore more indicative of a Pliocene WAIS retreat (McKay et al., 2008). This hypothesis was recently expanded by model experiments indicating potential windblown pathways may have been more widespread and Pliocene-aged marine diatoms could have been sourced from widespread uplifted marine sediments along the margin of the EAIS, as well as the WAIS (Scherer et al., 2016).

A glacial unconformity observed in seismic records from the Ross Sea, RSU4, is constrained to have formed between 16 and 14 Ma and suggests substantial marine-based ice advance of the EAIS into the Ross Sea associated with the MMCT (Anderson and Bartek, 1992; Bart, 2003; De Santis et al., 1995). Resolving the timing of this unconformity was a primary objective of the recent IODP Expedition 374 (McKay et al., 2019). In Prydz Bay and Wilkes Land (Sites 1165 and U1356), greatly increased deposition of gravel and pebble-sized ice rafted debris from 14.1 to 13.7 Ma provides direct evidence for ice expansion of the EAIS at the MMCT (Pierce et al., 2017).

The ANDRILL sites AND-1B and AND-2A in the Western Ross Sea recovered several diamictite-rich deposits with a maximum age of 13.57 Ma, interpreted to represent a phase of Antarctic ice sheet expansion in the Ross Embayment, potentially during the MMCT (Fielding et al., 2011; Levy et al., 2016; McKay et al., 2009; Passchier et al., 2011). A depositional change observed in seismic profiles from the Ross Sea around this time has been attributed to EAIS and WAIS expansion (De Santis et al., 1999). In the Wilkes Land region, a depositional shift in the channel levee systems on the continental margin is interpreted to have been associated with expansion of the EAIS to the shelf edge at the MMCT (Escutia et al., 2011a; Sangiorgi et al., 2018). Decreasing sedimentation rates and increased ice rafted debris deposition after 14.5 Ma on the East Antarctic margin in Prydz Bay potentially signifies reduced erosion by the EAIS under the influence of a cooler glacial regime (Florindo et al., 2003), although ODP Site 1166 recovered sediments of this age containing benthic foraminifers recycled from Eocene shallow water sediments, suggesting physical erosion of the continental shelf continued through this period (Cooper and O'Brien, 2004). At this time, a hiatus or condensed sequence that spans most of the middle Miocene (<6 m/Ma, 15–8 Ma) has been reported along the Pacific Margin of the Antarctic Peninsula (Hernández-Molina et al., 2017).

Following the MMCT, several other Miocene glaciations inferred from $\delta^{18}O$ excursions probably occurred between 13 and 8 Ma (Holbourn et al., 2013; Miller et al., 1991), but geological records of this time period are sparse in Antarctica. In the Antarctic Peninsula and Wilkes Land region, pollen contained in continental shelf drill cores suggest tundra vegetation may have persisted until 12.8–11 Ma (Anderson et al., 2011; Sangiorgi et al., 2018). This implies that at lower latitudes and/or low elevations, climate conditions may have allowed vegetation to persist well into the late Miocene. Biomarker evidence in the Dry Valleys and sparse pollen data from DSDP Site 274 in the northwestern Ross Sea hints at coastal refugia of vegetation persisting into the early Pliocene, although unequivocal evidence of this occurring is still lacking (Fleming and Barron, 1996; Ohneiser et al., 2020).

An increased abundance of hemipelagic sediments and ice rafted debris in the Weddell Sea was originally interpreted as representing the development of the first major 'marine-based' WAIS advance occurring at ~8 Ma (Kennett and Barker, 1990). Late Miocene interglacials (11–6 Ma) in AND-1B are represented by terrigenous facies that are similar to those deposited widely in proximal glaciomarine setting in Greenland today, such as outwash gravels and coarse inclined sand beds interpreted as glaciomarine fan systems, and mudstones that contain intervals of rhythmically laminated sand/silts formed by large volumes of turbid subglacial meltwater discharge (McKay et al., 2009). Although such facies exist in some Late Pleistocene and modern sediments, they are generally thin (<1 m) and sparsely distributed (Mckay et al., 2009; Prothro et al., 2018). In contrast, these facies in late Miocene interglacial intervals in AND-1B are >10 m thick, indicating elevated levels of subglacial meltwater discharge relative to Plio-Pleistocene times, which contain diatom ooze that are relatively starved of terrigenous sediment supply (McKay et al., 2009; Rosenblume and Powell, 2019). Similar sedimentary facies are observed in the middle to late Miocene Fisher Bench Formation from the Pagodroma Group in the Prydz Bay region of East Antarctica, while seismic sequences and short cores near Totten Glacier also provide clear evidence that elevated subglacial meltwater discharge was widespread during the late Miocene around East Antarctica, with a warmer subpolar style glacial regime relative to the Pliocene (Donda et al., 2020; Gulick et al., 2017; Hambrey and McKelvey, 2000).

Recent global temperature reconstructions indicate a significant late Miocene cooling event between 7 and 5.4 Ma (Herbert et al., 2016), broadly corresponding to a reduction in the presence of sedimentary facies associated with large turbid subglacial meltwater discharge around the East Antarctica (Gulick et al., 2017; Hambrey and McKelvey, 2000; McKay et al., 2009;

Rosenblume and Powell, 2019). Despite this inferred warmth and surface melt in East Antarctica in the late Miocene, cosmogenic nuclides in sand grains from AND-1B suggest minimal surface exposure of these grains prior to subglacial transport to the offshore drillsite, indicating the EAIS had not experienced significant on-land retreat of its margin within the past 8 Myr (Shakun et al., 2018). During the latest Miocene, and coveal to this cooling event, a shift in the development of glacial margin sequences along the Pacific Margin of the Antarctic Peninsula marks the time that regular ice advances began to reach the shelf break (Larter and Barker, 1989, 1991b; Larter et al. 1997; Hernández-Molina et al., 2017).

3.3.4 Pliocene to Pleistocene

The Pliocene and Pleistocene history of the AIS has been an area of intense study over the past decade, and here we only provide a first-order summary of recent developments in understanding, with a focus on how this history may manifest itself in the sedimentary basins of the Antarctic margin. Levy et al. (2021) and Wilson et al. (2021) provide a detailed review of the Pliocene and Pleistocene history of the AIS and linkages to global records. Even more extensively studied is the Last Glacial Maximum (LGM) deglaciation, which has recently been reviewed by the Reconstruction of Antarctic Ice Sheet Deglaciation (RAISED) consortium (The RAISED Consortium, 2014), and is covered in detail by Siegert et al. (2021). For conciseness, the LGM and last deglaciation histories are not repeated in this chapter, except where relevant context for the deeper time records is provided.

During the early to mid-Pliocene warmth, global sea levels are thought to have been 20 ± 10 m above present-day levels, indicating a reduction/collapse of both the Greenland Ice Sheet and the WAIS, and potentially large parts of the marine-based sectors of the EAIS (Dumitru et al., 2019; Grant et al., 2019; Miller et al., 2012; Naish and Wilson, 2009). Although the exact timing of the subsidence of West Antarctica remains equivocal, seismic stratigraphic arguments suggest that between the late Miocene and Pliocene, there was a major shift in the style of ice sheet erosion. In West Antarctica, and parts of East Antarctica, it is inferred that larger, cold polar, marine ice sheets led to increased erosion and overdeepening of the continental shelf (Bart and De Santis, 2012; Bart and Iwai, 2012; Cooper and O'Brien, 2004; De Santis et al., 1995; Hernández-Molina et al., 2017). This eroded continental shelf sediment was subsequently redeposited to the continental slope to rise, resulting in the development of large trough mouth fan and mass transport systems (O'Brien et al., 2007; Rebesco et al., 2006). This overdeepening would have increased the sensitivity of the ice sheets margin to Marine Ice Sheet Instability (MISI) processes (Bart et al., 2016; Colleoni et al., 2018). Currently, it remains uncertain how diachronous overdeepening of the continental shelf was across the late Miocene to Pliocene, as tectonic drivers of sedimentary basin development played a role alongside the climatic or glaciological processes. However, as discussed in later region-specific sections of this chapter, it is clear there was a significant shift in stratigraphic architecture around the Antarctic margin during the late Neogene that appears to be associated with more frequent occurrences of marine ice sheet advances.

Sedimentary facies from the AND-1B drill core indicate numerous orbitally-paced advances of the WAIS to the mid-continental shelf during the early Pliocene (Naish et al., 2009). This is supported by seismic stratigraphic evidence of truncated glacially eroded surfaces that extended to continental shelf edge (Alonso et al., 1992; Anderson et al., 2018; Bart, 2001). Coastal diatom oozes indicate significant sediment starvation during Pliocene interglacials, implying reduced surface and subglacial meltwater from the adjacent coastline relative to the late Miocene (McKay et al., 2009), an observation that is mirrored in outcrops from the Prydz Bay region, implying a shift towards cold polar glaciation in the Pliocene around the EAIS margin (Hambrey and McKelvey, 2000; Whitehead et al., 2004). AND-1B also records periodic retreats of the WAIS that continued through the Plio-Pleistocene until at least 1.0 Ma, and although other proximal ice sheet and distal sea level/ice core records suggest WAIS may have collapsed since this time (as recently as the last interglacial), there is a lack of direct Antarctic records to confirm this (Hillenbrand et al., 2002; McKay et al., 2012b; Scherer et al., 1998). Surface waters during the warmest interglacial of the Pliocene in AND-1B indicate greatly reduced sea ice in the Ross Sea, and diatom species that today live north of the Polar Front, implying sea surface temperatures were 4°C–5°C warmer than present (McKay et al., 2012a; Riesselman and Dunbar, 2013). In the Wilkes Land and Prydz Bay region, similar microfossil-derived sea surface temperatures and open water conditions suggest migration of the Polar Front southward during the Pliocene (Bohaty and Harwood, 1998; Escutia et al., 2009; Taylor-Silva and Riesselman, 2018; Whitehead and Bohaty, 2003). Warmer than present waters were also reconstructed during Marine Isotope Stage 31 at ~1 Ma (Beltran et al., 2016; Scherer et al., 2008; Villa et al., 2008), when a significant deglaciation of the WAIS was proposed (Naish et al., 2009; Scherer et al., 2008). Offshore provenance and ice rafted debris data suggests the marine-terminating margin of the EAIS fluctuated greatly at orbital timescales and periodically retreated significantly inland during the Pliocene (Bertram et al., 2018; Cook et al., 2013; Hansen et al., 2015; Passchier, 2011; Patterson et al., 2014; Williams et al., 2010) while continental shelf deposits in Wilkes Land (IODP Site 1358) show advances and retreats of the ice sheet from the continental shelf edge at these times (Reinardy et al., 2015).

Global cooling and the onset of Northern Hemisphere glaciation at ~2.7 Ma corresponded with a continued cooling trend in Antarctica and increasing seasonal sea ice persistence/extent around the continental margin (Armbrecht et al., 2018; Cortese and Gersonde, 2008; Escutia et al., 2009;

Kennett and Barker, 1990; McKay et al., 2012a; Riesselman and Dunbar, 2013; Taylor-Silva and Riesselman, 2018). Cosmogenic isotope dating studies around the EAIS margin suggest a thicker ice sheet existed during glacials in the late Pliocene and early Pleistocene, compared to late Pleistocene glacials. This is interpreted to represent a shift to cooler, more arid conditions (i.e., less precipitation of snow) in the late Pleistocene (Jones et al., 2017; O'Brien et al., 2007; Suganuma et al., 2014; Yamane et al., 2015). Despite this cooling, provenance indicators from Site 1361 offshore of Wilkes Land provide evidence that some inland retreat of the marine-based EAIS margin occurred during Marine Isotope Stages 11, 9 and 5, (~400, 320 and 120 ka, respectively), although the exact extent of EAIS retreat remains difficult to constrain from offshore provenance studies (Wilson et al., 2018). A novel method to constrain the extent of the last major inland retreat of the EAIS has recently been presented using the uranium isotopic composition of carbonate precipitates in morainal deposits and indicates a ~700 km retreat inland of the Wilkes Land margin occurred during Marine Isotope Stage 11 (Blackburn et al., 2020). During the last interglaciation of Marine Isotope Stage 5, ice core evidence combined with far-field sea level fingerprinting and modelling experiments suggests substantial ice loss driven by oceanic warming in the Weddell Sea sector of the WAIS, although it remains equivocal if this corresponded to full marine-based WAIS collapse (Clark et al., 2020; Turney et al., 2020).

The most accessible geological records in Antarctica capture the retreat of the ice sheets since the Last Glacial Maximum and consequently has been heavily studied by a range of offshore (e.g., shallow sediment cores, multibeam bathymetric surveys) and onshore studies (e.g., mapping and cosmogenic nuclide dating of moraines). Given the breadth of studies in this time period and that such studies are generally not captured by geological drilling methods, we do not discuss post-LGM glacial retreat in detail in this chapter (unless it is relevant to the deeper time records) and the reader is referred to Siegert et al. (2021).

3.4 Regional seismic stratigraphies and drill core correlations, and future priorities to reconstruct Antarctica's Cenozoic ice sheet history

Extensive seismic reflection data have been acquired by many nations around Antarctica, and a compilation of almost all multichannel profiles are available in stack version at the SCAR Seismic Data Library System for Cooperative Research (SDLS) (https://sdls.ogs.trieste.it), established and endorsed in 1991 by SCAR (Report 9) and the Antarctic Treaty (ATCM Recommendation XVI-12). This cooperative library has sparked many successful collaborations and the exchange of data between scientists of all SCAR countries and beyond. The existence and structure of SDLS allowed to maximise reuse of existing data and strategically collect new data where necessary, often coordinating data collection between countries. This has reduced costs and logistic effort for data collection and also minimises unnecessary exposure of possible environmental impacts through seismic data acquisition. There are currently data from 153 surveys with over 336,000 km on seismic lines from 16 countries included in the SDLS (Fig. 3.1). This represents ~87% of the known multichannel seismic reflection data collected in Antarctica since the late 1970s.

The seismic stratigraphic framework above the acoustic basement was defined regionally by the ANTOSTRAT (Antarctic Offshore Acoustic Stratigraphy) project for the Ross Sea, Antarctic Peninsula, Weddell Sea, Prydz Bay and Wilkes Land in the mid-1990s (Brancolini et al., 1995; Cooper et al., 2011, 2008) and represented one of the precursor programs for ACE, PAIS, PRAMSO and for the circum-Antarctic paleobathymetric reconstructions. However, the ages of the ANTOSTRAT unconformities on the continental shelf, rise and abyssal plain have been only partially constrained by drilling, mainly in the Ross Sea and Prydz Bay (see synthesis by Bart and De Santis 2012; Cooper et al., 2008), the Antarctic Peninsula (Barker and Camerlenghi, 2002) and Wilkes Land (Escutia et al., 2011a). In addition to continental shelf drilling, partial age control of continental rise unconformities has allowed several region-to-region correlations to be made (Close, 2010; Donda et al., 2007; Hernández-Molina et al., 2017; Hochmuth et al., 2020a,b; Hochmuth and Gohl, 2019; Leitchenkov et al., 2007, 2015; Lindeque et al., 2016, 2013; Rebesco et al., 2006; Sauermilch et al., 2019). In the frame of the Palaeotopographic-Palaeobathymetric Reconstructions PAIS subcommittee, these have allowed for updated paleodepth contour maps of the main seismic horizons across all the Antarctic margin and Southern Ocean from the Eocene-Oligocene Boundary to modern times to be presented (Hochmuth et al., 2020a,b). Such data helps inform models restoring the Antarctic margin paleotopography since the Eocene, by measuring the volume of sediment that had been transported offshore by erosive ice sheets. However, such models also need to consider the compaction of the sediments, as well as the lithosphere isostatic subsidence, in addition to the tectonic and thermal history that can only be verified by deep geological drilling (Paxman et al., 2019; Wilson et al., 2012). All these reconstructions represent a significant step forward, but gaps still remain and areas where maps are extrapolated need to be verified by future seismic survey and drilling campaigns. Despite these remaining uncertainties, over the past decade we have gained an increasingly detailed view of the morphological and tectonic evolution of the Southern Ocean, including the opening and deepening of the Tasmanian and Drake gateways that allowed for the establishment of the ACC.

In this section, we summarise the seismic stratigraphy of major sedimentary basins around the Antarctic margin, how the integration of this stratigraphic architecture with geological drilling may inform on the long-term climate and ice sheet evolution described above, and how future campaigns could be conducted to resolve outstanding aspects of this history.

3.4.1 Ross Sea

The Ross Sea contains the most well-defined seismic stratigraphic framework for Antarctica's continental shelves, with a dense network of seismic profiles that have been constrained by numerous geological drilling projects since the 1970s (Fig. 3.5). A significant number of regional, large-scale geophysical exploration surveys of the Ross Sea continental margin architecture were conducted in the 1980s and 1990s, but during the last two decades, a series of cruises collecting closely spaced grids were conducted, in order to reconstruct the WAIS and EAIS dynamics across the Late Cenozoic (Anderson et al., 2018; Bart, 2003; Bart and De Santis, 2012; Bart et al., 2011; Böhm et al., 2009; Chow and Bart, 2003; Kim et al., 2018; Mosola and Anderson, 2006; Sauli et al., 2014). Recent work focused on understanding the relationship between the ocean circulation, the morphology of the sea floor, the processes affecting the sub-ice shelf cavities and the ice sheet dynamics, in the present day and since the LGM (Ashley et al., 2020; Gales et al., 2021; Tinto et al., 2019).

Although numerous different frameworks have been proposed over this time, for simplification purposes, the synthesis below largely adopts the ANTOSTRAT nomenclature (e.g., ANTOSTRAT Ross Sea atlas of Brancolini et al., 1995), but discusses these with caveats relating to basin-wide correlations of these sedimentary packages and erosive surfaces. In this framework, eight major Ross Sea Sequence (RSS) units bounded by seven major Ross Sea Unconformities (RSU) are mapped across the Ross Sea (Brancolini, et al., 1995), although subdivision of these sequences has been conducted in subsequent studies (Fig. 3.6). These sequences are largely defined by studies in the Eastern and Central basins of the Ross Sea, and represent major steps in Antarctica's tectonic and climatic evolution, whereas in the active rift zones of the Western Ross Sea, different nomenclatures exist (Fielding et al., 2008; Levy et al., 2012, 2019; Pekar et al., 2013, Sauli et al., 2021). These differences arise due to the difficulties of correlation between basins, as well as the potential for significant diachronism relating to the complex history of Cenozoic rifting, subsidence, sedimentation and ice sheet expansion across the Ross Sea (Fig. 3.7). Work combining seismic, magnetic and gravity data highlight the complex structural evolution of the western Ross Sea during the late Cenozoic, with the diachronous propagation of rifting southwards between 46 and ~11 Ma, with a prominent pulse of rifting in the early Miocene (~17 Ma) (Davey et al., 2016; Ferraccioli et al., 2009; Fielding et al., 2008; Granot and Dyment, 2018; Granot et al., 2013). Late Miocene and Pliocene subsidence was associated with intense volcanism (Lisker et al., 2014; Wenman et al., 2020), and recent extensional tectonics

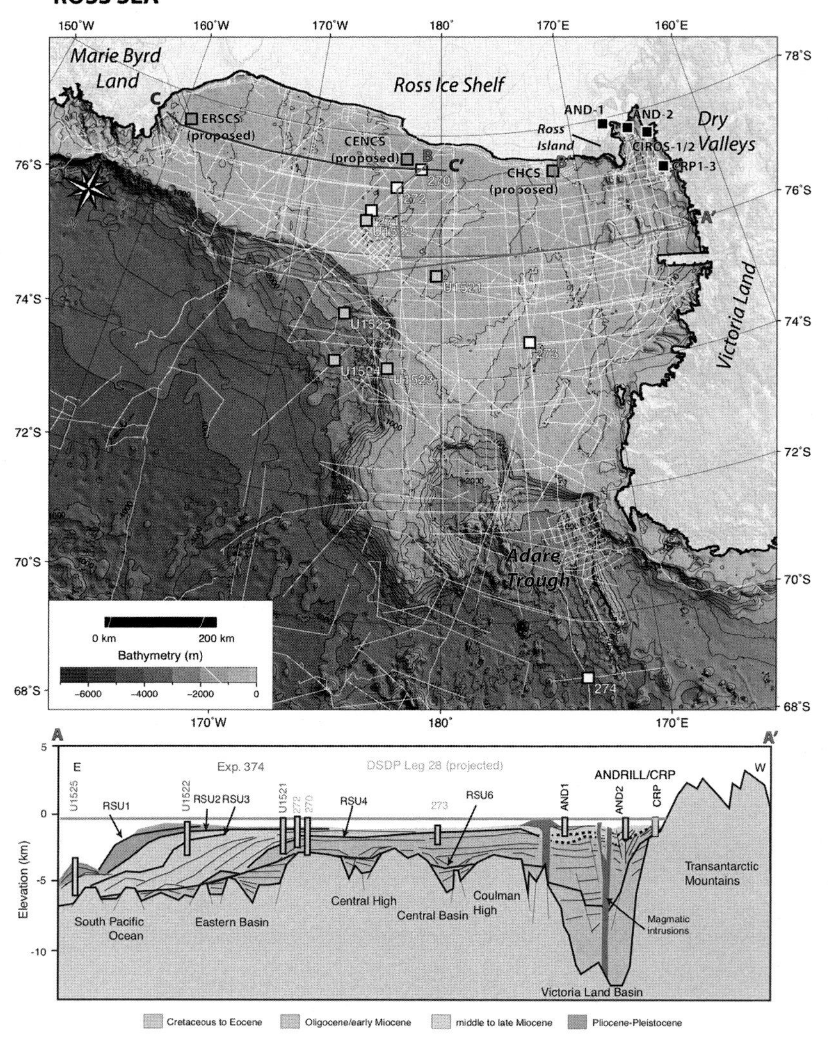

FIGURE 3.5 (Top) Overview map of Ross Sea sector with existing seismic lines (white lines) and drill sites. DSDP Leg 28 sites are shown as white squares, ice-platform drilling projects as black squares (CRP, CIROS, ANDRILL) and IODP Expedition 374 sites as yellow squares. Future proposed sites discussed in text are shown as green squares. Seismic tracks are from the SCAR Antarctic Seismic Data Library System (SDLS) and additional survey information. Bathymetry is from Bathymetric Chart of the Southern Ocean (Arndt et al., 2013) and subglacial topography from BedMachine Antarctica dataset (Morlighem et al., 2020). Track lines of the seismic profiles shown in bottom panel and Fig. 3.6 are marked with bold red (A–A′), green (B–B′) and blue (C–C′) lines, respectively. (Bottom) Ross Sea seismic stratigraphy showing major unconformities (RSU6,4-1) and associated shifts in the geometry of sedimentary packages on the continental shelf, and previous drilling (see upper panel for transect line). DSDP Leg 28, ANDRILL, CRP and Expedition 374 sites. *Adapted from McKay, R.M., De Santis, L., Kulhanek, D.K., IODP Expedition 374 Science Team, 2019. Proceedings of the International Ocean Discovery Program, vol. 374, https://doi.org/10.14379/iodp.proc.374.2019.*

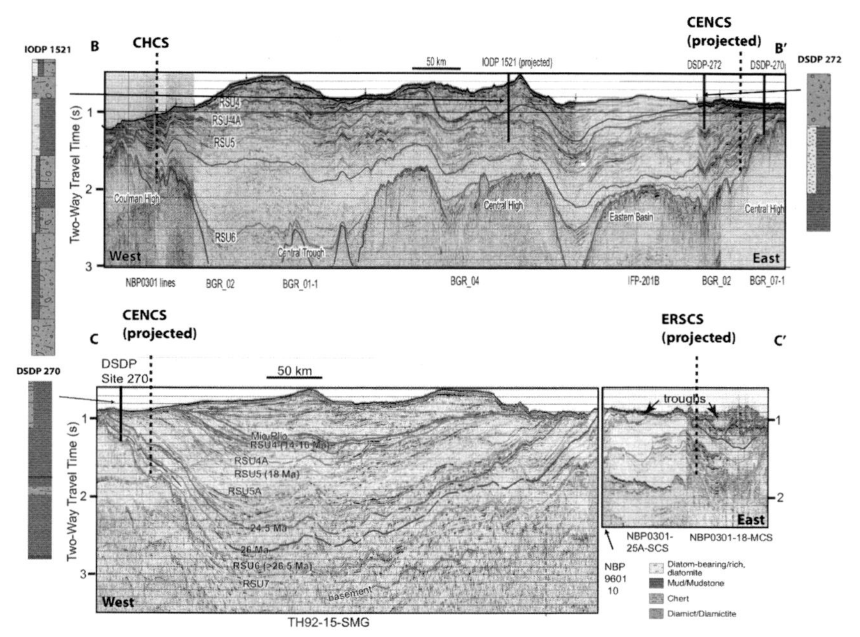

FIGURE 3.6 Detailed seismic cross sections and correlations of major unconformities and ties to DSDP Leg 28 sites (Sites 270, 272) compiled for IODP pre-proposal 998 (Table 3.1) from: (Top) the Central Trough to western flank of the Eastern Basin (Line B-B' in Fig. 3.5); (Bottom) the western flank to Eastern Flank of the Eastern Basin (Line C-C' in Fig. 3.5), based on correlations presented by Sorlien et al. (2007). Approximate location of proposed (projected) future drill sites discussed in text is shown. Seismic line numbers used to construct profile and are available in SDLS are labelled for each section and are available in SDLS. Vertical exaggeration in both lines is large (> 80x). *Note*: Labels are at base of upper section and top of lower section. *Drill core lithologies are based on data from Hayes, D.E., Frakes, L.A., et al., 1975. Initial Reports of the Deep Sea Drilling Project, vol. 28. US Government Printing Office. Available from: http://deepseadrilling.org/28/dsdp_toc.htm, Kulhanek, D.K., Levy, R.H., Clowes, C.D., Prebble, J.G., Rodelli, D., Jovane, L., et al., 2019. Revised chronostratigraphy of DSDP Site 270 and late Oligocene to early Miocene paleoecology of the Ross Sea sector of Antarctica. Global and Planetary Change 178, 46–64. Available from: https://doi.org/10.1016/j.glopla-cha.2019.04.002, McKay, R.M., De Santis, L., Kulhanek, D.K., IODP Expedition 374 Science Team, 2019. Proceedings of the International Ocean Discovery Program, vol. 374, https://doi.org/10.14379/iodp.proc.374.2019.*

affecting the sea floor has been reported in the western Ross Sea (Geletti and Busetti, 2011; Hall et al., 2007; Sauli et al., 2021). This tectonic system appears to be associated with the presence of mud volcanoes and pockmarks fed by fluids/gas seeping migrating upward along faults. Similar morphological mounds of possibly carbonatic and/or volcanic origin also occur within the Terror Rift, close to volcanic centres including Franklin Island (Lawver et al., 2012).

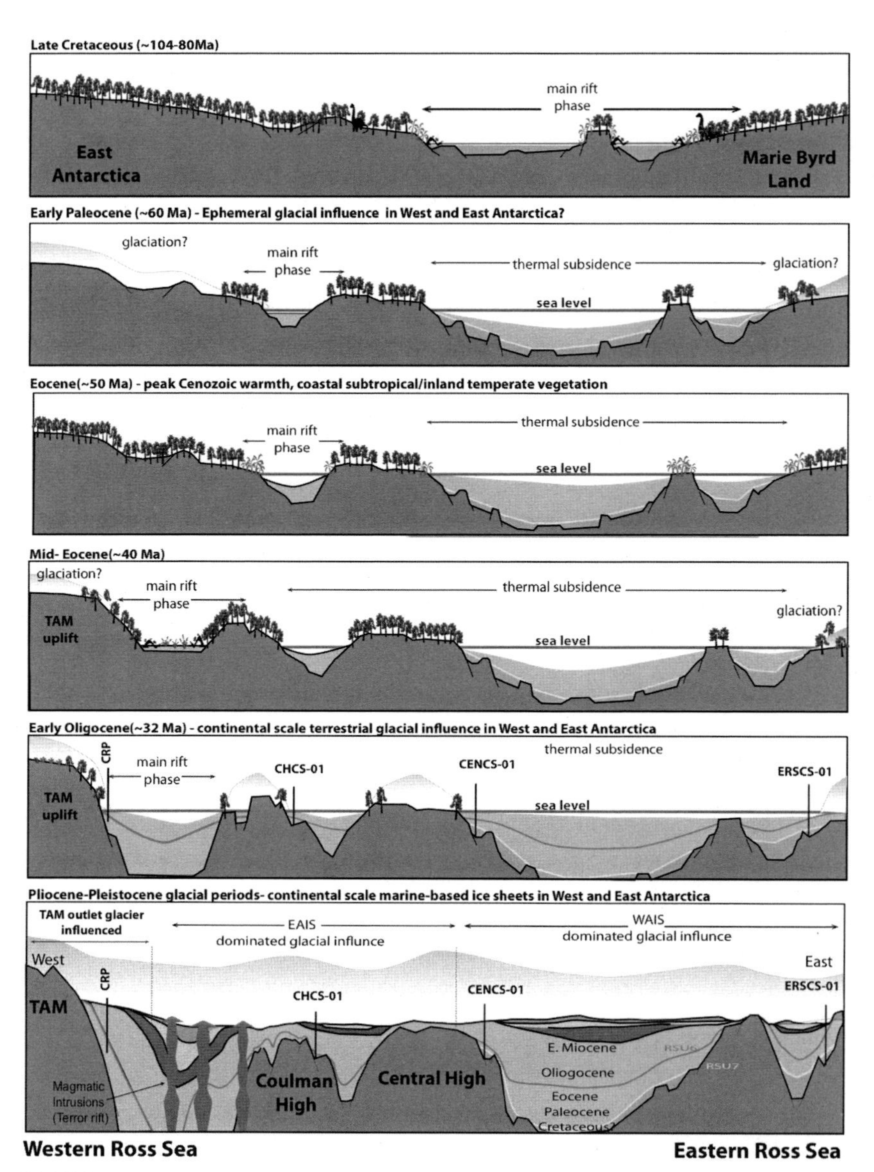

FIGURE 3.7 West Antarctic Rift System (WARS) history in the Ross Sea and relationship to basin infill and ice sheet/TAM outlet glacier influences through time. The history is based on the rift model presented by Wilson and Luyendyk (2009), and the profile shown in Fig. 3.6. Targeted drilling in a proposed E–W transect across the continental shelf is intended to assess the timing of active rift termination and will obtain high-resolution climate records back to the onset of rifting (~Late Cretaceous).

The geological basement depth in the Ross Sea is reconstructed by combining seismic profile interpretation with inversion of gravity data (Brancolini et al., 1995; Decesari et al., 2007; Ji et al., 2018; Luyendyk et al., 2001; Tinto et al., 2019). The oldest and deepest ANTOSTRAT unit in the Ross Sea is RSS-1, which consists of graben-bound rift-fill strata and is separated into two units, representing early- (RSS-1 Lower) and late-rift (RSS-1 Upper) sequences bounded by unconformity RSU7. These units are likely to be diachronous across basins, a consequence of rifting propagating toward the west (Luyendyk et al., 2001; Wilson and Luyendyk, 2009). Unit RSS-1-lower is likely Late Cretaceous in age in the Eastern Basin and is characterised by dipping and disrupted reflectors. RSS-1-lower may represent a fluvial terrestrial setting and potentially provides a unique opportunity to sample terrestrial Antarctic environments during the Cretaceous (Luyendyk et al., 2001). RSU7 is inferred to have formed during regional extension caused by the break-up of the Gondwana continent (Decesari et al., 2007; Luyendyk et al., 2001), although glacial erosion or relative sea level changes could also be responsible. Unit RSS-1-upper is a thicker sequence of flat lying, faulted strata onlapping onto RSU7, and inferred to contain marine strata deposited during marine transgression across thermally subsiding crust (Decesari et al., 2007).

RSU6 is a prominent seismic unconformity across the Ross Sea, but its origin and age is enigmatic. It is possible that RSU6 was cored at CRP and CIROS-1 (Figs. 3.5 and 3.7), but its age at these sites is highly ambiguous due to extensive unconformities truncated by alpine TAM glaciation. In addition, the occurrence of faults bounding the western Ross Sea basins (Sauli et al., 2021) makes correlations to the broader Ross Sea seismic stratigraphy tenuous (Davey et al., 2000). DSDP Site 270 provides a minimum constraint for RSU6 of >26.5 Ma, but this site cored into a basement high and likely postdates RSU6. However, correlation of this onlapping surface onto the basement high suggests DSDP Site 270 provides a constraint for the onset of sedimentation following RSU6 at 26.5 Ma in this locality (Kulhanek et al., 2019) (Fig. 3.6). An earlier Oligocene age (>28 Ma) is consistent with the hypothesis that RSU6 represents the transition from the Eocene 'greenhouse' to early Oligocene 'icehouse' and has been proposed to be related to early Oligocene sea level falls associated with ice sheet growth (Anderson and Bartek, 1992; Bartek et al., 1991; De Santis et al., 1995), but drilling will help constrain whether climatic, tectonic or glacial driver are responsible for the origin of RSU6. However, GIA theory and experiments indicate initial ice sheet growth would result in rising sea levels in the near-field Antarctic and falling sea levels away from the ice grounding zone, in the peripheral bulge (Stocchi et al., 2013). Depending on the ice sheet proximity and mass, the depositional response and resulting stratigraphic architecture could lead to widespread disconformities and stratigraphic truncations, or maximum flooding surfaces containing highly condensed sections. In the Western Ross

Sea, RSU6 is possibly amalgamated over some of the structural highs with the older Coulman High Major Unconformity (CHMU), which separates rift fill strata (RSS-1) from overlying post-rift, glaciomarine strata of RSS-2 and was the proposed target for the ANDRILL-Coulman High project (Rack et al., 2012). The ages of pre-CHMU strata are likely younger than the pre-RSU6 strata (RSS-1) in the Central and Eastern Ross Sea. This truncated surface is likely diachronous across the Ross Sea and also likely to be younger in the Western Ross Sea than in the Eastern Basin based on the inferred propagation of the rifting and subsidence histories from east to west (Wilson et al., 2012). Consequently, the origin of RSU6 remains ambiguous and needs to be assessed by future geological drilling.

Overlying RSU6, glacially-influenced marine sediments dominate RSS-2. This unit provides a record of the early Oligocene to early Miocene history of the Antarctic ice sheets, although it has only been sparsely sampled. The CRP and CIROS-1 drill cores obtained discontinuous records of RSS-2 that indicate the Victoria Land basin was influenced by deposition from local TAM alpine glaciers and trunk glaciers directly connected to teh EAIS and incised through the TAM since the earliest Oligocene (Barrett, 2007; Galeotti et al., 2016). DSDP Site 270 only sampled the latest Oligocene in the upper part of RSS-2, but included a discontinuity near the O/M boundary (Kulhanek et al., 2019; Levy et al., 2019). Much of the early history of WAIS expansion in the central to eastern Ross Sea is currently unsampled, despite seismic evidence of large WAIS expansion sometime during the early to late Oligocene (Sorlien et al., 2007). This provides a compelling target for future drilling campaigns, either via ice shelf drilling platforms such as the ANDRILL project (Levy et al., 2016), ship-based systems such as the JOIDES Resolution, or through IODP Mission Specific Platforms. As noted earlier, expansion of the WAIS during the relatively warmer climates (compared to today) of the early Oligocene is possible if West Antarctica was more elevated at that time (Wilson et al., 2013) (Fig. 3.4). Consequently, the ice sheet evolution in the Ross Sea is hypothesised to be strongly-coupled to the tectonic and subsidence history of West Antarctica, rather than climate forcing alone (Colleoni et al., 2018; Hochmuth and Gohl, 2019; Paxman et al., 2019; Wilson et al., 2013). Several large truncations are noted across the Ross Sea during the Oligocene to early Miocene (RSU5 and RSU4a) and are interpreted as phases of localised advances of marine-terminating ice caps nucleating on basements highs that remained subaerially exposed at this time (De Santis et al., 1995). Similar erosional surfaces are noted in the ANDRILL and Cape Roberts Project cores during this time period, suggesting advances of marine-terminating EAIS outlet glaciers extended into the Western Ross Sea during the late Oligocene and early Miocene (Levy et al., 2019). A recent proposal (998-pre) was submitted to IODP to drill three continental shelf sites in the Ross Sea (sites CHCS, CENCS, ERSCS in Fig. 3.6). The three sites are located along an E–W longitudinal-transect designed to capture the integrated history of tectonic, climate and glacial influences from both East and

West Antarctica (Fig. 3.7). By drilling sequences between RSU7 and RSU4, the specific objectives are: (1) to obtain direct evidence of the earliest ice sheets in East and West Antarctica expanding into the Ross Sea; (2) to obtain geological reconstructions of 'pre-icehouse' climates at high latitudes in Antarctica during the Late Cretaceous to Eocene; and (3) to constrain the timing of late rift phases in the Ross Sea to resolve mechanisms of crustal extension in the Ross Sea, in order to test hypotheses of global plate tectonic models and understand tectonic controls on ice sheet evolution. The proposed sites are intended to constrain the pre-RSU5 stratigraphy in the Ross Sea.

The recent IODP Expedition 374 obtained excellent records post-dating 18 Ma (RSS-3 and younger) in the Central Ross Sea and was successful in refining the post-RSU5 stratigraphy (McKay et al., 2019) (Figs. 3.5 and 3.6) and a revision of the depth and age of the RSU5 is ongoing, but is constrained to be ~18 Ma at the base of U1521 (Mckay et al., 2019; Pérez et al., 2021; Sauli et al., 2021). These cores will document the evolution of the marine-based ice sheets in the Ross Sea, as glacial erosion resulted in progradation of the continental shelf (older strata are generally aggradational), followed by a transition from a seaward dipping shallow continental shelf to an overdeepened (i.e. landward deepening) continental shelf. Site U1521 drilled through a thick seismic sequence of progradational foresets above RSU5, after ~18 Ma, recovering a ~300 m interval of diamictites interbedded with thin mudstone layers suggestive of the input of a large volume of glacially eroded material by marine-terminating glaciers or ice sheets. This new evidence of marine-terminating glaciers discharging large volumes of sediment in the early Miocene is consistent with sedimentary evidence from coastal drill cores from the Cape Roberts and AND-2A drilling projects. However, the Early Miocene diamictites in U1521 are overlain by ~120 m of diatom-rich muds, provisionally dated at 16.7 to 15.8 Ma and inferred to relate to ice sheet retreat during the MCO and a sustained period of seasonally ice-free surface waters that allowed for diatom production and accumulation (McKay et al., 2019). Facies sequences at these sites are indicative of glacial depositional systems characterised by sediment-laden melt water plumes originating from marine-terminating glaciers of the EAIS, or outflow of glacifluvial sediments during periods when the EAIS had retreated on-land (Fielding et al., 2011; Levy et al., 2019; Naish et al., 2001; Passchier et al., 2011). Seismic profiles indicate the presence of deep channels that are associated with these strata and are indicative of the discharge of large volumes of subglacial outwash in warmer-than-present glacial regimes (Bart and De Santis, 2012; Chow and Bart, 2003; De Santis et al., 1995; McKay et al., 2019). On land, geologic and geomorphic studies indicate that this period was associated with a wet-based style of glaciations experiencing significant surface melt (Lewis et al., 2007, 2006; Smellie et al., 2011).

Truncation of glaciomarine foresets defines RSU4, which provides the first seismic evidence of an expanded grounded ice sheet advancing across much of the outer western Ross Sea. This surface is interpreted as an expansion of a grounded ice stream originating from the west (i.e., East Antarctica) in the central Ross Sea between 16 and 14 Ma, and work is ongoing on IODP Exp374 Site U1521 to refine the exact timing and magnitude of this event and the provenance of the grounded ice sheet (Anderson and Bartek, 1992; Colleoni et al., 2018; De Santis et al., 1995; Kim et al., 2018; McKay et al., 2019; Pérez et al., 2021; Ten Brink et al., 1995).

Unconformity RSU3 provides the first clear evidence of major cross-shelf paleotroughs, associated with enhanced progradation of the continental shelf further in the Eastern Ross Sea (Bart and De Santis, 2012; Chow and Bart, 2003; De Santis et al., 1995). This is interpreted as an expansion of the marine-based WAIS, which likely coalesced with ice derived from the EAIS to create a continental shelf wide advance of ice across the Ross Sea (De Santis et al., 1995, 1999). Resolving the age of RSU3 was a primary focus of IODP Expedition 374 Site U1522, with preliminary age models indicating a late Miocene hiatus associated with RSU3 between ~9 and 5.5 Ma (McKay et al., 2019). Meltwater and outwash features are rare in strata younger than RSU3, and, where present, are greatly reduced in scale when compared to Pleistocene examples (c.f., Simkins et al., 2017). Above RSU3, progradational seismic facies are progressively thinner and less common (Fig. 3.5), suggesting gradual sediment starvation as the ice sheets transited to a colder glacial regime, and marine-based ice sheet expansions eroded and truncated preexisting sediment infill on the outer shelf.

Paleobathymetric reconstructions suggest the RSU2 surface (Pliocene age) is associated with the establishment of an overdeepened continental shelf in the Ross Sea (De Santis et al., 1999). This overdeepening likely occurred as a result of an increasingly marine setting over much of West Antarctica due to extensive Plio-Pleistocene glacial erosion, following Cretaceous to Neogene rifting. This must have also occurred on the background of a cooling climate that allowed expansion of marine-based ice sheets. Such expansions across the Ross Sea acted to erode continental shelf sediments in units below RSU3. Most of this sediment was transported to the continental shelf break and redeposited in the form of large trough-mouth fans on the upper slope (Bart, 2003; Colleoni et al., 2018; Cooper et al., 2008; De Santis et al., 1999; Kim et al., 2018).

Several large canyons exist on the continental rise of the Ross Sea, acting as conduits funnelling cascading High Salinity Shelf Water from the continental shelf into the abyssal ocean to form Antarctic Bottom Water. On the flanks of these canyon systems are thick levee deposits that contain high-resolution archives of oceanographic change. In the central Ross Sea, the late Pliocene to Pleistocene intervals of these levees were cored at sites U1524 and U1525 by IODP Expedition 374 (McKay et al., 2019). In addition to this, the upper slope and outermost continental rise display mounded drift deposits emplaced during the Neogene to Pleistocene (Kim et al.,

2018). These drifts are associated with easterly currents of the Antarctic Slope Current, and one of these drifts was the target of IODP Expedition 374 Site U1523 (McKay et al., 2019). Analysis of this site and nearby gravity cores covering the Plio-Pleistocene will allow changes in the strength of this current to be reconstructed, which is important as this current is thought to regulate water mass (and therefore heat and nutrient) exchange between the continental shelf and offshore waters (Kim et al., 2020; McKay et al., 2019).

On the continental shelf, seismic facies above RSU2 consist of massive tabular sheets generally displaying an aggradational pattern, interpreted as till sheets deposited by the increased frequency of marine-based ice sheet advances (Accaino et al., 2005; Alonso et al., 1992; Böhm et al., 2009). These facies are also associated with widespread deposition of low-relief, asymmetrical Grounding Zones Wedges deposited at the margin of a marine-based ice sheet over the Plio-Pleistocene (Bart et al., 2011; Bart and De Santis, 2012; Bart and Owolana, 2012; Batchelor and Dowdeswell, 2015). Coastal tidewater glaciers built morainal features in the Victoria Land offshore (Sauli et al., 2014). Above RSU1 (0.8 Ma?), sediment packages on the shelf edge show distinctive aggrading or backstepping (rather than prograding), indicating that most sediment delivered from land by marine-based ice sheets was reworked and sequestered on the outer shelf or continental slope/rise (Bart and Tulaczyk, 2020; De Santis et al., 1999; Shipp et al., 1999). Anderson et al. (2018) highlight a disconnect over the past 0.8 Ma between the frequency of advances noted in outer shelf seismic facies (two advances) and those recorded in the mid-shelf site AND-1B core (eight advances). A critical point of difference in these locations is that AND-1B was recovered from an actively forming flexural moat around the volcanic centre of Ross Island, providing continued development of accommodation space to protect deposited strata from glacial erosion (Horgan et al., 2005; Naish et al., 2009). Consequently, this disconnect may indicate, that either, not all glacial periods of the late Pleistocene represent shelf wide expansion, or that the outer shelf sequences in RSS-8 represent amalgamations of numerous glacial advances, with the largest advances eroding evidence of previous expansion.

A high priority ambition for the drilling community is to obtain records of ice sheet retreat during Plio-Pleistocene interglacials near the present-day grounding line of the WAIS, with the primary goal of acquiring records of WAIS extent during late Quaternary 'super-interglacials' of the past 400 kyrs, which were characterised by warmer than usual conditions, both in Antarctica and globally (McKay et al., 2016). Drilling in this region is critical to ground-truth the full extent of WAIS retreat during these warmer interglacials and allows for a complete transect from the outer to inner continental shelf to be developed (Fig. 3.8). Such a transect would enable a better understanding of ocean–ice interactions that may have influenced past retreat events. Although several sediment cores have been collected from hot water drill access holes in the Siple Coast region of the Ross Sea, they have

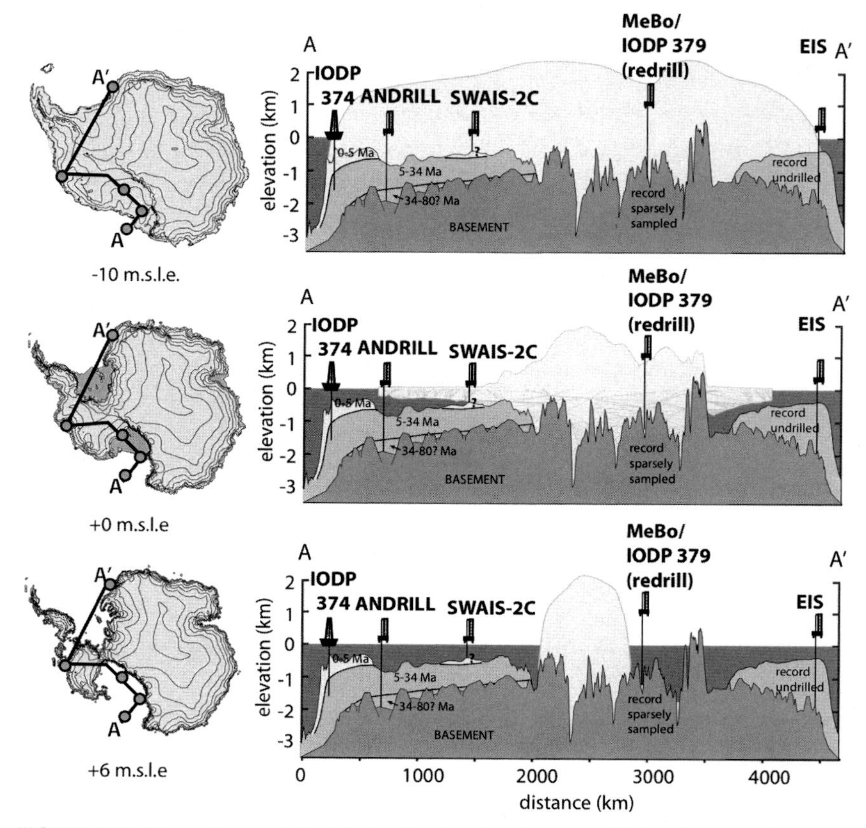

FIGURE 3.8 Transect across West Antarctica showing three possible ice sheet configurations, and the southward extension of transect-based drilling strategy (circles in maps) to obtain direct records of its past climatic history. Past drill sites from ANDRILL and Expedition 374 are shown, alongside proposed SWAIS 2C (International Continental Drilling Program; ICDP) drill sites and Ekström Ice Shelf sites, as well as undrilled IODP Expedition 379 continental shelf sites that remain viable future targets. *Modified from McKay, R.M., Barrett, P.J., Levy, R.S., Naish, T.R., Golledge, N.R., Pyne, A., 2016. Antarctic Cenozoic climate history from sedimentary records: ANDRILL and beyond. Philosophical Transactions of the Royal Society A 374 (2059), 20140301. Available from: https://doi.org/10.1098/rsta.2014.0301.*

yet to penetrate the LGM diamict (Kingslake et al., 2018; Priscu et al., 2019; Scherer et al., 1998). To overcome this problem, New Zealand researchers (Antarctic Research Centre at Victoria University of Wellington, and GNS Science) are currently building an integrated hot water/rock drilling system capable of recovering ~200 m of sediment in places where the combined depth of the ice shelf (or sea ice) and water column is <1000 m thick. The sediment/rock drill is similar in functionality to the ANDRILL system, but is lighter weight and its smaller footprint is designed to be capable of deep field deployment within the constraints of existing Antarctic science support programmes. The combined rig, drill pipe and casing package weighs ~30 tonnes and can undertake soft sediment coring using hydraulic piston and punch corers. For hard rock drilling, it will use off-the-shelf diamond-bit rotary coring technology capable of rates of recovery approaching 100%, as was the case for ANDRILL. It is anticipated that this system will drill at the grounding line of the Kamb Ice Stream (KIS) and Crary Ice Rise (CIR), to obtain Neogene and Quaternary sediments – as part of a proposed Intercontinental Drilling Program (ICDP) project (SWAIS 2C) project from 2021 onward.

3.4.2 Amundsen Sea

The Amundsen Sea continental shelf and rise (Fig. 3.9) developed after the Cretaceous break-up of Zealandia from West Antarctica (see Section 3.2), but the paleoenvironment from the Cretaceous to the Neogene in the Amundsen Sea sector remains poorly sampled. The first drill cores from the shelf were collected during a MeBo seabed drilling expedition in 2017 (Gohl et al., 2017). A drill core from the inner/middle shelf showed evidence for a temperate rainforest and swamp environment formed in a rift basin in the middle Cretaceous, with a ~40-Myr hiatus separating these sediments from an overlying sandstone formation of late Eocene age (Klages et al., 2020). More analyses of core samples from the other MeBo drill sites, which according to preliminary age estimates span various time slices from the Oligocene to Holocene (Gohl et al., 2017), are in progress. In terms of paleoclimate-related and glacially-driven sedimentary processes in the Neogene and Quaternary, this region is dominated by sediment erosion, transport and deposition driven mainly by the large outlet ice streams of the Pine Island, Thwaites, Haynes, Pope and Smith glaciers of the eastern Amundsen Sea Embayment. As well as these, many smaller glaciers drain ice from the elevated central Marie Byrd Land and feed into the Dotson Ice Shelf and the various segments of the Getz Ice Shelf of the central and western Amundsen Sea.

The analysis of seismic lines crossing the slope and shelf, most of them collected since 2000 (Dowdeswell et al., 2006; Graham et al., 2009; Hochmuth and Gohl, 2013; Klages et al., 2014, 2015; Lowe and Anderson, 2002; Uenzelmann-Neben et al., 2007; Weigelt et al., 2012, 2009), resulted in a seismic stratigraphic model (Gohl et al., 2013) very much analogous to the dated Ross Sea shelf stratigraphy with a similar seismic reflection signature (Fig. 3.9). However, deep drill sites currently do not exist on the Amundsen Sea shelf and this was a key focus of IODP Expedition 379, but a near 'worst-case scenario' sea ice season during the drilling window in 2018/19 precluded access of the drill ship to the continental shelf (Gohl et al., 2021). However, the proposal targets remain viable and important drilling targets for the future and will assist in validating these correlations to the

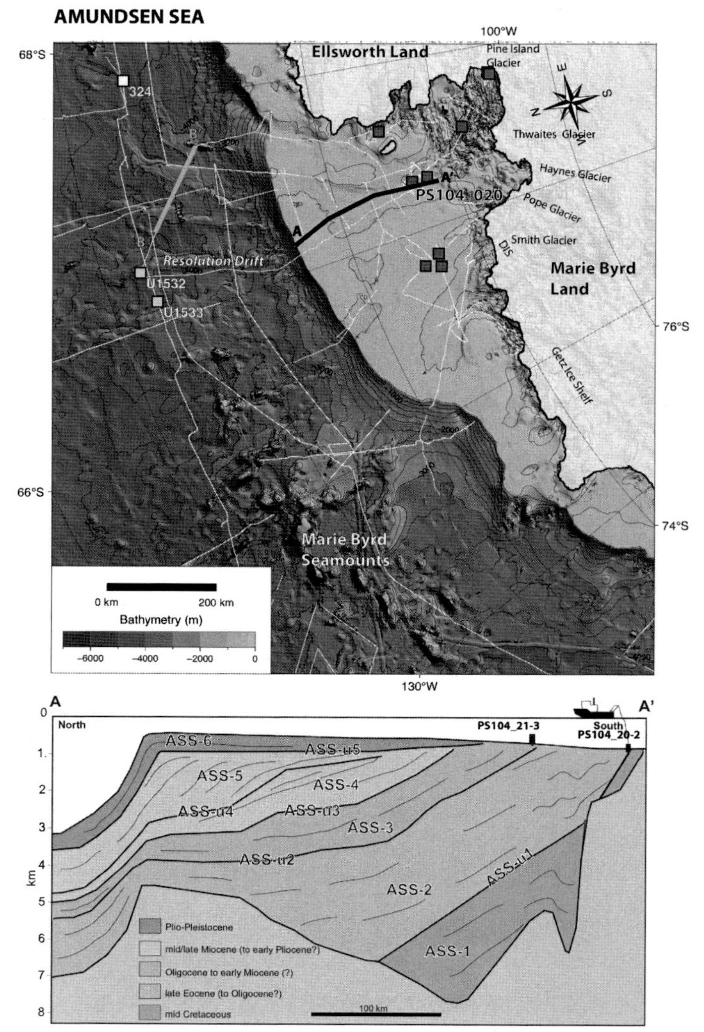

FIGURE 3.9 (Top) Overview map of Amundsen Sea sector with existing seismic lines (white lines) and drill sites, including DSDP Leg 35 Site 324 (white square), IODP Expedition 379 sites U1532 and U1533 (yellow squares) and MeBo seabed drill sites (purple squares), including Site PS104_20−2 (Gohl et al., 2017; Klages et al., 2020). Seismic tracks are from the SCAR Antarctic Seismic Data Library System (SDLS) and additional survey information Bathymetry is from Bathymetric Chart of the Southern Ocean (Arndt et al., 2013) and subglacial topography from BedMachine Antarctica dataset (Morlighem et al., 2020). Segments of seismic profiles shown in Figs. 3.10 are marked with bold green (B−B′) and red (C−C′) lines, respectively. (Bottom) Seismic stratigraphy and major unconformities (discussed in text) across the Amundsen Sea Embayment shelf (track-line shown as a black bold line (A−A′) in top panel), using updated age constraints from Klages et al. (2020). Most of the seismic stratigraphy is still undated as deep drill holes do not exist on the shelf. *(Bottom) Modified from Gohl, K., Uenzelmann-Neben, G., Larter, R.D., Hillenbrand, C.-D., Hochmuth, K., Kalberg, T.,et al., 2013. Seismic stratigraphic record of the Amundsen Sea Embayment shelf from preglacial to recent times: evidence for a dynamic West Antarctic ice sheet. Marine Geology 344, 115−131. Available from: https://doi.org/10.1016/j.margeo.2013.06.011.*

Ross Sea (Fig. 3.8). This will determine whether the ice sheet histories in these regions are fundamentally different, with local influences by tectonic processes potentially playing as an important role as climate and oceanographic processes. Existing seismic coverage indicates the total pre-glacial to glacial sediment cover on the shelf is up to 7 km thick in places. Based on the observation of glacially-driven truncational unconformities (surfaces ASS-u4 and above; Fig. 3.9), an early advance of grounded ice onto the continental inner to middle shelf is interpreted to have not occurred prior to the Miocene. A large proportion of the present outer shelf consists of a 70 km broad zone of prograding sequences that were deposited after transport by advancing grounded ice (Gohl et al., 2013; Hochmuth and Gohl, 2019, 2013).

Studies of the large number of samples from conventional coring systems, and geomorphological studies, have yielded a reasonably detailed record of the ice retreat in the Amundsen Sea since the LGM. These works were comprehensively synthesised by Larter et al. (2014) as part of the RAISED consortium project, although numerous new studies have provided new insights (Klages et al., 2017, 2014; Kuhn et al., 2017; Smith et al., 2017). Combined, these works indicate that the WAIS retreated rapidly from the outer shelf at the LGM (about 18 ka) to the inner shelf at about 10 ka, where it halted until the current retreat from coastal locations started in the mid-20th century (Hillenbrand et al., 2013; Klages et al., 2017; Larter et al., 2014; Smith et al., 2017, 2014). Sedimentary, geochemical and microfossil proxies from post-LGM sediment cores on the continental shelf suggest incursions of relatively warm Circumpolar Deep Water onto the shelf forced deglaciation during the early Holocene, as well as ice shelf thinning since the mid-20th century (Hillenbrand et al., 2017; Minzoni et al., 2017). Circumpolar Deep Water incursions have been identified as the main driver for present ice shelf cavity melting in the Amundsen Sea embayment (e.g., Nakayama et al., 2013; Scambos et al., 2017). Present research and future drilling in this region are consequently focusing on testing the hypothesis that Circumpolar Deep Water incursions onto the shelf were the main driver of WAIS retreat during past warm periods of the Quaternary and Neogene (Gohl et al., 2021).

Similar to the continental shelf, the coverage of seismic lines on the continental rise and deep sea of the Amundsen Sea has increased in recent years (Fig. 3.9), although large unsurveyed areas still exist, in particular in the central and western Amundsen Sea. A single transect along the rise from the Ross Sea to the Amundsen Sea was used to establish the first seismic stratigraphic record on the full sedimentary cover for the western Amundsen Sea with the identification of distinct pre-glacial, transitional and full glacial sequences (Fig. 3.10) (Lindeque et al., 2016). The seismic stratigraphy was derived by long-distance correlation to the Ross Sea chronostratigraphic record based on the shelf drill sites of DSDP Leg 28 (e.g., De Santis et al.,

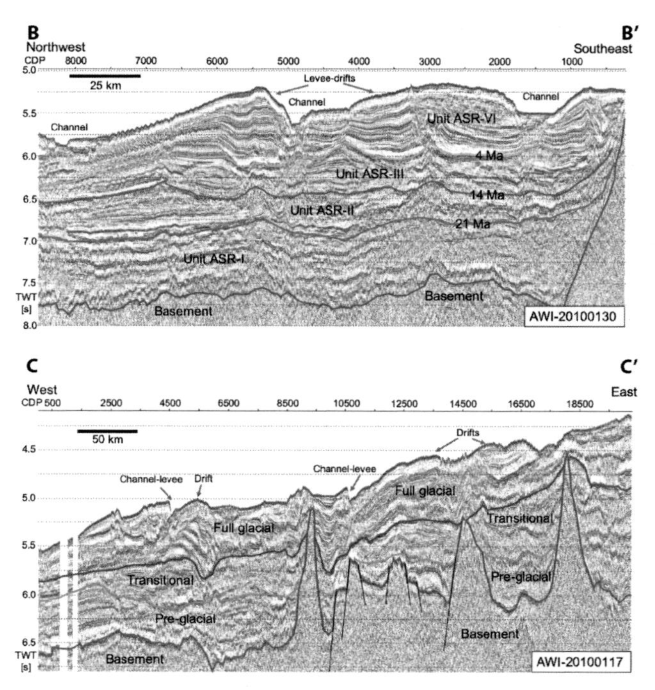

FIGURE 3.10 (Top) Section of seismic profile AWI-20100130 from the eastern Amundsen Sea rise crossing major sediment drifts and adjacent deep-sea channels. The long-distance correlation and interpretation of key horizons and units is slightly modified from Uenzelmann-Neben and Gohl (2014). Section location is marked in Fig. 3.9 as a bold green line (B—B′). (Bottom) Section of seismic profile AWI-20100117 from the western Amundsen Sea rise with interpretation of three distinct pre-glacial (Late Cretaceous to late Eocene), transitional (late Eocene to mid-Miocene) and full glacial (mid-Miocene to present) units, slightly modified from Lindeque et al. (2016). Section location is marked in Fig. 3.9 as a bold red line (C—C′).

1999; Hayes et al., 1975). The transitional period includes the late Eocene to mid-Miocene when grounded ice first expanded onto the continental shelves, while the full glacial period describes the interval from the mid-Miocene to Quaternary, with intensified ice advances onto the outer shelves. No apparent difference in the deep-sea sedimentation transport processes or temporal shift in deposition between the Amundsen Sea and Ross Sea is observed. Additional new seismic data were collected on a Russian expedition in 2019 and will help quantify the extent of the depositional sequences in the western Amundsen Sea.

The continental rise of the eastern Amundsen Sea is dominated by large contourite drifts, some of them rising several hundred metres above the surrounding seafloor (Nitsche et al., 2000; Scheuer et al., 2006a,b; Uenzelmann-Neben, 2018; Uenzelmann-Neben and Gohl, 2014, 2012; Yamaguchi et al., 1988) (Fig. 3.10). Most drift systems are elongated in a north–south direction and flanked by deep-sea channels eroded by turbidity currents carrying suspended detritus supplied by downslope transport processes from the slope and shelf. The suspended particles were entrained in strong bottom currents and were subsequently deposited on the flanks of the channels to form the drifts. Seismic analyses with first estimates of a chronostratigraphy by long-distance and jump correlation of seismic horizons to DSDP and ODP drill sites in the Bellingshausen Sea and Ross Sea indicate that early drift formation by enhanced bottom-current activity began in the Eocene/Oligocene. Drift formation intensified in the Miocene, likely caused by expansion of the WAIS during global cooling, increased sea-ice cover and, as a result, the formation of Antarctic Bottom Water and enhanced bottom-current flow (Uenzelmann-Neben and Gohl, 2012).

Although only two sites on the continental rise and none on the shelf could be drilled during IODP Expedition 379 in 2019 (Fig. 3.1), almost continuous sequences spanning the latest Miocene to Pleistocene, deposited with high sedimentation rates, were recovered from the lower and upper western flank of the Resolution Drift (Gohl et al., 2021). The cores from both sites recovered predominantly glaciomarine, fine-grained terrigenous sediments intercalated with pelagic and hemipelagic deposits that, in the younger parts of the cores, contain more biogenic material. The records show an interplay of glacially transported shelf sediments that were transported downslope to the continental rise by gravitational processes and redistributed across the rise by turbidity and bottom currents. Although IODP Expedition 379 was unable to retrieve cores from the shelf, the drill records from the continental rise reveal the timing of glacial advances across the shelf and, thus, the existence of a large ice sheet in West Antarctica for prolonged time periods since the late Miocene. Detailed analyses of the IODP Expedition 379 core samples and data are in progress (Gohl et al., 2021).

In contrast to the Amundsen Sea continental shelf, information from conventional sediment cores from the continental slope and rise is sparse. Dowdeswell et al. (2006) reported debris and grain flow deposits presumably of LGM age from three short cores collected on the continental slope. In a long gravity core from the continental rise (Hillenbrand et al., 2002), thick beds of turbidites and contourites deposited during late Pleistocene glacial periods alternate with thin beds of foraminifera-bearing, bioturbated muds deposited during interglacials. A condensed, but well dated core from a seamount location on the continental slope retrieved sediments mainly consisting of pelagic material, i.e. planktic foraminifera and iceberg-rafted debris. Notably, this record did not provide evidence for a WAIS collapse during the last 800 ka (Hillenbrand et al., 2002). By comparing palaeoceanographic data from a core recovered from the same seamount location to other Southern Ocean cores, Williams et al. (2019) found evidence that deep- and bottom-water mass formation varied between the different sectors of the Antarctic margin during glacial periods of the last 800 ka. Hillenbrand et al.

(2009) analysed Pleistocene to Holocene glacial-interglacial cycles in a sediment core recovered from the Resolution Drift (site located to the south of the two IODP Expedition edition 379 drill core sites) and found, based on proxies for biological productivity and lithogenic sediment supply, that in the Amundsen Sea the interval from Marine Isotope Stage (MIS) 15 to MIS 13 (621–478 ka) was a single prolonged interglacial period, during which the WAIS may have collapsed. A palaeoceanographic study on the same core by Konfirst et al. (2012) concluded that after the end of the Mid-Pleistocene Transition (MPT) at ca. 620 ka the Amundsen Sea low pressure system shifted farther south during interglacial periods than during earlier interglacials, thereby increasing Circumpolar Deep Water upwelling onto the shelf. Recent provenance studies provide information about the sub-ice geology and drainage systems in the hinterland that may allow past ice sheet retreat events to be identified in offshore sediments (Simões Pereira et al., 2018, 2020).

3.4.3 Bellingshausen Sea and Pacific coastline of Antarctic Peninsula

The continental margin in the Bellingshausen Sea and along the west side of the Antarctic Peninsula is a former active margin where subduction of successive ridge crest segments proceeded from southwest to northeast (Eagles et al., 2009; Larter and Barker, 1991a) (Fig. 3.11). The arrival of each ridge crest segment at the margin was followed by 1–4 million years of uplift and then long-term subsidence (Bart and Anderson, 1995, 2000; Hernández-Molina et al., 2017; Larter and Barker, 1991b; Larter et al., 1997). Along the Pacific margin northeast of Alexander Island, multichannel seismic profiles show that outer shelf sequences are separated from NE-SW trending mid-shelf basins by the Mid-shelf High (Gambôa et al., 1990; Kimura, 1982; Larter et al., 1997). On the outer shelf aggrading sequences without a distinct paleo-shelf edge are unconformably overlain by prograding sequences with an abrupt paleo-shelf edge (Fig. 3.12). This change is observed all along the Pacific margin (Bart and Anderson, 1995; Jin et al., 2002; Larter and Barker, 1989, 1991b; Larter et al., 1997) and west along the margin of the Bellingshausen Sea (Nitsche et al., 1997). Above a widespread unconformity within the prograding sequences there is less progradation than in the earlier units. Four lobes along the margin where the extent of progradation is greater represent late Quaternary and earlier termini of paleo-ice streams (Bart and Anderson, 1995; Larter et al., 1997). Marguerite Trough, the pathway of the largest paleo-ice stream on the Pacific margin northeast of Alexander Island during the late Quaternary, reaches the margin between two of these lobes as a consequence of lateral migration of the outer part of the trough (Bart and Anderson, 1995; Hernández-Molina et al., 2017; Ó Cofaigh et al., 2005a).

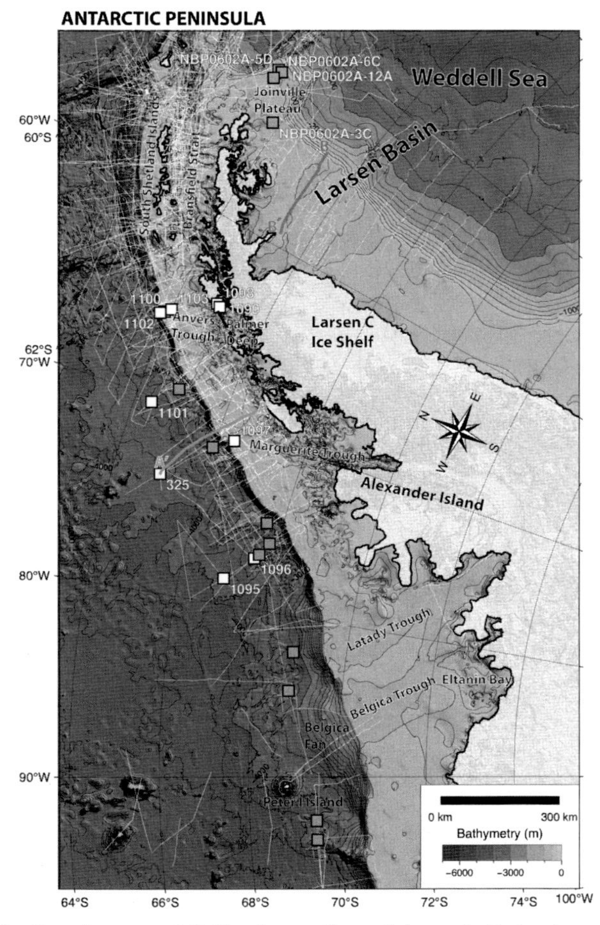

FIGURE 3.11 Overview map of Bellingshausen Sea and Antarctic Peninsula sector with existing seismic lines (white lines), drill sites and topographic features mentioned in text. Bathymetry is from Bathymetric Chart of the Southern Ocean (Arndt et al., 2013) and subglacial topography from BedMachine Antarctica dataset (Morlighem et al., 2020). Drill sites, including DSDP Leg 35, ODP Leg 178 sites (white square), and SHALDRIL drill sites (orange squares; NBP0602A). IODP Proposal 732-FULL2 sites are shown as dark green squares.

A relatively steep continental slope along the Pacific margin northeast of Alexander Island, with typical maximum gradients of 13°–17° and a zone of rugged topography near its base, make it difficult to trace seismic units between the continental shelf and rise (Hernández-Molina et al., 2017; Larter and Cunningham, 1993). The upper continental rise hosts large sediment drifts that are up to 250 km long (Rebesco et al., 1996, 1997, 2002). These are separated by large turbidity current channels, with up to 1 km of relief

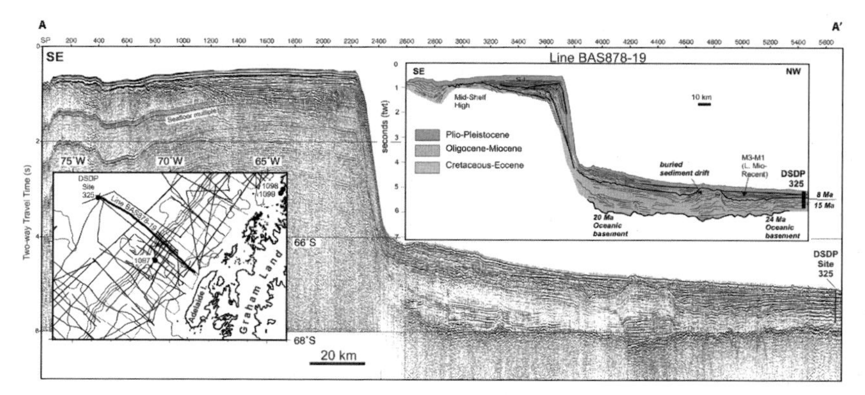

FIGURE 3.12 Seismic profile BAS878-19, and interpretation (insert) through correlation to DSDP Site 325 across the Bellingshausen Sea continental shelf and rise. Profile line as green line A–A′ is shown in Fig. 3.11. *Modified after Cooper, A.K., Brancolini, G., Escutia, C., Kristoffersen, Y., Larter, R., Leitchenkov, G., et al., 2008. Cenozoic climate history from seismic reflection and drilling studies on the Antarctic continental margin. In: Florindo, F., Siegert, M. (Eds.), Antarctic Climate Evolution, vol. 8. Elsevier, pp. 115–234.*

between the drift crests and the channel thalwegs (Larter et al., 2016a; Rebesco et al., 2002). The channels originate from networks of tributaries at the base of the continental slope, which have generated the rugged topography there (Amblas et al., 2006; Rebesco et al., 2002). The drifts developed after an earlier period of sediment accumulation in a lower energy regime that infilled depressions in basement and aggraded over most of the continental rise, resulting in a low relief paleo-seabed (Larter and Cunningham, 1993; Rebesco et al., 1997, 2002). However, Hernández-Molina et al. (2006) identified a buried sediment drift of early Miocene age that formed to the northeast of a group of seamounts near DSDP Site 325 (Fig. 3.12).

The dominant features on the broad southern Bellingshausen Sea continental shelf west of Alexander Island are the Belgica and Latady troughs, the former being >100 km wide. These troughs were the pathways of two very large paleo-ice streams that deposited extensive prograded sequences on the outer shelf and constructed large trough mouth fans on the continental slope (Dowdeswell et al., 2008; Larter et al., 2016b; Nitsche et al., 1997, 2000; Ó Cofaigh et al., 2005b; Scheuer et al., 2006). The slope on the fans is much less steep than along the part of the margin to the northeast, mostly between 1° and 2°. The slope is steeper again to the west of the Belgica Fan, where the shelf is narrower. One more large sediment drift extends from this western part of the margin towards Peter I Island, near the boundary between the Amundsen and Bellingshausen seas.

Scientific drilling in this subregion was conducted on DSDP Leg 35 in 1974 and on ODP Leg 178 in 1998 (Barker et al., 1999; Hollister et al., 1976) (Fig. 3.1). Of the sites drilled during Leg 35, Site 325 was the only one that provides useful chronostratigraphic constraints on the long-term evolution of the margin (Fig. 3.12). Drilling penetrated close to latest Oligocene oceanic basement on the central continental rise, but the single hole was only spot cored. Later correlations along seismic lines connecting the site to the margin showed that prograded sequences on the outer shelf are all younger than a condensed unit or hiatus from 15 to 8 Ma at Site 325 (Hernández-Molina et al., 2017; Larter and Barker, 1991b). Correlations to the sediment drifts on the upper continental rise showed that they started to develop during the condensed interval or hiatus at Site 325, but their growth accelerated in the late Miocene (Hernández-Molina et al., 2017; Rebesco et al., 1997, 2002).

During ODP Leg 178, two sites were drilled on the continental shelf (albeit with poor core recovery) and three sites on the continental rise, which provided detailed constraints on the late Miocene to recent evolution of the margin. Results from sites 1097 and 1103, on the outer shelf in the axes of Marguerite Trough and Anvers Trough, respectively, showed that progradation started at around 8 Ma, but also confirmed that the earlier late Miocene aggradational units were glacially influenced (Bart et al., 2005, 2007; Eyles et al., 2001). Two further sites, 1098 and 1099, were drilled on the inner shelf in the 1400-m deep Palmer Deep basin, recovering postglacial and Holocene successions of centennial to millennial scale resolution, ~47 and ~108 m thick, respectively (Barker et al., 1999; Domack et al., 2001; Ishman and Sperling, 2002; Leventer et al., 2002; Shevenell and Kennett, 2002). Continental rise sites 1095, 1097 and Site 1101 drilled into sediment drifts that contained continuous, expanded sedimentary records extending back to the earliest late Miocene. At Site 1095, on the distal part of one of these drifts (Drift 7), sedimentation rates were highest in the late Miocene and decreased progressively through the Pliocene and Quaternary. No sites have been drilled in the mid-shelf basins west of the Antarctic Peninsula or in the Bellingshausen Sea. Configuration of seismic reflectors on the northwest flank of the Antarctic Peninsula mid-shelf basins suggests deposition in them pre-dates uplift of the mid-shelf high (Larter et al., 1997). Therefore, they probably contain a record of early Neogene to Paleogene, possibly extending to Cretaceous, climate history of the Antarctic Peninsula.

DSDP Site 324, at 69°03.21′S, 98°47.20′W, where Pliocene and Quaternary sediments were recovered by spot coring to 199 m below sea floor (mbsf), has sometimes been described as being in the southern Bellingshausen Sea (Hollister et al., 1976). However, its location is >300 km west of Peter I Island, and so actually in the Amundsen Sea. The earliest indication of ice rafted debris in cores recovered by scientific drilling in this subregion is in latest early Miocene sediments at Site 325, with none having been found in three cores from earlier intervals in the early Miocene at the same site. This suggests that any marine-terminating glaciers existing before this time were rare, although there is local evidence of marine-terminating

Oligocene glaciers on the South Shetland Islands (Birkenmajer, 1991; Dingle and Lavelle, 1998; Troedson and Smellie, 2002). The start of deposition of prograded units on the outer shelf at around 8 Ma suggests that this was the time of onset of frequent grounded ice advances to the shelf edge, i.e. this was the time of transition from a glacially-influenced to a fully glacial regime on the continental marine margin (Hernández-Molina et al., 2017). The widespread unconformity within the outer shelf prograding sequences above which there is less progradation may have resulted from deeper erosion during glacial periods with lower sea levels after the Late Pliocene increase in the volume of Northern Hemisphere ice sheets (Larter and Barker, 1989, 1991b). The limited age control available from the outer shelf drill sites is consistent with this interpretation.

The large sediment drifts on the continental rise have formed beneath a bottom current that flows southwest along the margin (Giorgetti et al., 2003). However, the position of the buried early Miocene drift identified by Hernández-Molina et al. (2006) in relation to a group of seamounts implies a bottom current that flowed in the opposite direction at that time, and therefore suggests a reversal of bottom current flow along the continental rise during the middle or late Miocene. The observation from Site 1095 that accumulation rates of dominantly terrigenous sediments on the continental rise were highest in the late Miocene and decreased progressively through the Pliocene and Quaternary suggests that most erosional overdeepening of the inner shelf and cross-shelf troughs occurred during the late Miocene. However, from a synthesis of ODP Leg 178 diatom assemblage data (Bart and Iwai, 2012) proposed that such overdeepening occurred later, in the earliest Pliocene. Seismic correlations from Site 1095 and 1096 to the large trough mouth fans in the Bellingshausen Sea suggest the highest rates of sediment accumulation in them occurred during the Pliocene (Scheuer et al., 2006). The observation that fairly high rates of terrigenous sedimentation on the continental rise in this subregion were sustained through the Pliocene before declining during the Quaternary suggests that the early Pliocene warm period did not result in a long interruption between grounded ice advances across the continental shelf. In relation to this observation it is interesting to note that most model simulations for the warmest part of the Pliocene show an ice sheet or ice caps remaining on the spine of the Antarctic Peninsula (e.g., Pollard et al., 2015; Smellie et al., 2009). This is a consequence of the fact that the spine of the Peninsula has an average elevation of around 2000 m, maintaining low temperatures over its summit plateau and leading to high snow accumulation rates from maritime air masses moving across it (Turner et al., 2002). The high accumulation rates enable rapid and extensive ice sheet growth during cold periods, even if they are of relatively short duration.

The evidence of sustained terrigenous sediment supply to the continental rise through the Pliocene is just one striking example of the potential of the records in the sediment drifts. Complete recovery of the continuous, expanded records they contain will provide new insights into details of the

Miocene to Recent glacial history of the Antarctic Peninsula and the Bellingshausen Sea sector of the WAIS. The results of Leg 178 combined with new seismic profiles, have guided the development of a new IODP proposal that was approved for drilling but is awaiting scheduling (732-Full2). The aim of this new proposal is to obtain ultra-high precision chronostratigraphic control that can be established using palaeomagnetic methods (Channell et al., 2019), to evaluate the past history and stability of the WAIS and Antarctic Peninsula Ice Sheet. This will enable, for example, determination of the conditions in, the duration of, and rate of change through, glacial periods and deglacial transitions into Pliocene and Quaternary interglacials.

3.4.4 The Northern Antarctic Peninsula and South Shetland Islands

This subregion includes the Bransfield Strait and the continental margin around the South Shetland Islands (Fig. 3.11). Bransfield Strait is a 2000 m deep rift basin that is actively extending at 7 mm/yr (Dietrich et al., 2004). The time of initial extension and the oldest age of basin sediments are uncertain, but most studies suggest a probable onset between ~6 Ma (Larter and Barker, 1991a) and 3.3–4 Ma (Barker and Dalziel, 1983; Lodolo and Pérez, 2015). Gambôa et al. (1990) speculated that Bransfield Strait may have opened earlier, during the early Miocene and have been continuous with mid-shelf basins to the southwest. A new investigation of the crustal structure of the Bransfield rift and associated volcanic edifices has been conducted recently involving combined use of a large number of land and ocean bottom passive seismometers and marine surveys (Almendros et al., 2020).

From multichannel seismic data, Gambôa et al. (1990) identified ‘rift’ and ‘drift’ sequences in Bransfield Strait. ‘Drift’ sequences prograde the shelf and are about 1 km thick. In single channel seismic data, Jeffers et al. (1990) defined four glacio-eustatic units within the ‘drift’ sequences, and interpreted all four as being younger than 3 Ma. Prieto et al. (1999) used different single channel seismic data to define eight seismic units that comprise interfingering slope and basinal deposits within the ‘drift’ sequences. They interpret the slope units as having been deposited directly from grounded ice during glacial periods, and the basinal units as having been deposited by mass flow processes during deglaciations and interglacial periods. García et al. (2008) interpreted data from a multichannel seismic survey in the context of comprehensive multibeam bathymetry coverage of Bransfield Strait (Gràcia et al., 1997; Lawver et al., 1996). They highlighted changes through time in the post-rift succession to increased aggradation over progradation and to more localised sediment delivery to the basin. From single channel seismic data, Banfield and Anderson (1995) identified sediment mound features that they infer to be glacial grounding moraines in up to 1000 m water depth on the southeastern flank of Bransfield Strait. They speculated that mounds at ~700 m depth mark the maximum ice advance during the LGM. Deep troughs with mega-scale lineations are incised into the shelf and are interpreted as the paths of paleo-ice streams (Banfield and Anderson, 1995; Canals et al., 2002), while the Bransfield Basin contains an expanded Plio-Pleistocene section that to date has not been targeted for drilling.

Multichannel seismic profiles across the continental slope NW of the South Shetland Islands reveal a forearc basin, with more than 1.5 km sediments, that is bounded to the NW by a small accretionary prism (Maldonado et al., 1994). The prism overthrusts trench-fill sediments that may have been deposited rapidly and are up to 1 km thick (Kim et al., 1995; Maldonado et al., 1994). The only scientific drilling in the South Shetland Islands region was done at SHALDRIL Site 1 in Maxwell Bay, where an expanded succession of Holocene diatomaceous muds, ~105 m thick, overlying a clay-rich diamicton was recovered (Milliken et al., 2009).

3.4.5 The Eastern Margin of the Antarctic Peninsula

This subregion, which includes the Weddell Sea margin of the Antarctic Peninsula and Larsen Basin, is a passive margin that originally formed in the Jurassic during initial opening of the Weddell Sea (Fig. 3.11). Multichannel seismic reflection data show at least 8 km of sediment at the base of the northernmost part of the continental slope, overlying likely Early Cretaceous age basement (Barker and Lonsdale, 1991). Seismic data farther south are sparse as there is persistent sea ice cover throughout most austral summers (Parkinson, 2019).

Four main seismic stratigraphic units are identified from single channel seismic reflection data over a wide area of the shelf and upper slope (Anderson et al., 1992; Sloan et al., 1995), (Fig. 3.13): Unit 4, acoustic basement, interpreted as Jurassic and younger volcanic rocks; Unit 3, parallel to sub-parallel seaward-dipping reflections; Unit 2, prograding sequences with truncated foresets that downlap onto Unit 3; Unit 1, aggrading reflections. The northern continental slope has plastered contourite drift deposits up to 900 m thick, thought to have been deposited by bottom currents flowing northward along the western limb of the Weddell Gyre (Pudsey, 2002). Reflectors within the older part of Unit 3 are observed close to Seymour Island where Late Cretaceous to Oligocene sediments with similar dips crop out (Anderson et al., 1992, 2011). The only scientific drilling in this subregion was carried out through the SHALDRIL project, which used a lightweight drill rig installed over a moonpool on RV *Nathaniel B Palmer* during two cruises in early 2005 (cruise NBP0502) and 2006 (cruise NBP0602A; sites shown in Fig. 3.11). The holes drilled in this subregion were all relatively shallow, with the maximum depth below sea floor from which cores were recovered being 31.4 m at site NBP0602A-5. Anderson et al. (2011) presented a synthesis of the SHALDRIL

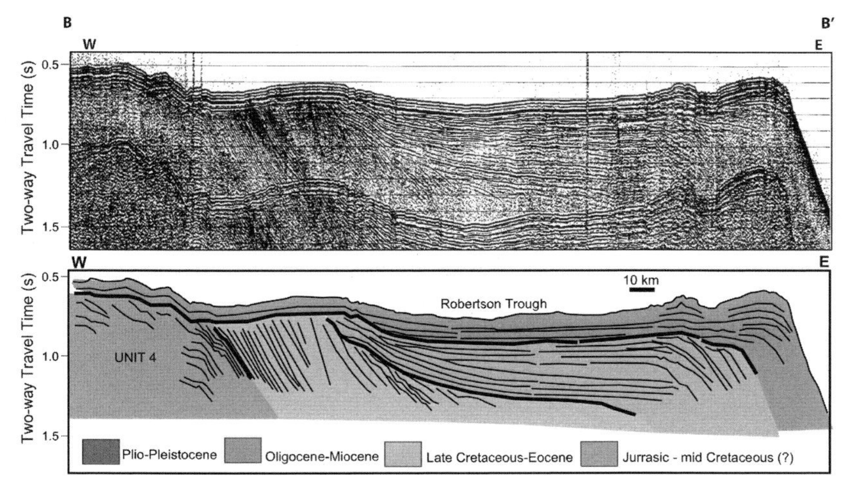

FIGURE 3.13 Seismic profile and interpretation across the Larsen Basin continental shelf. Profile line as red line B–B′ is shown in Fig. 3.11. *Modified from Sloan, B.J., Lawver, L.A., Anderson, J.B., 1995. Seismic stratigraphy of the Larsen Basin, Eastern Antarctic Peninsula. In: Cooper, A.K., Barker, P.F., Brancolini, G. (Eds.), Geology and Seismic Stratigraphy of the Antarctic Margin. American Geophysical Union, pp. 59–74.*

results, which provide the only offshore chronostratigraphic constraints in this region. Sediments of late Eocene age were recovered at SHALDRIL Site 3, above a prominent unconformity. The unconformity is probably within sediments that correlate with seismic Unit 3 described above. The other SHALDRIL sites in the subregion are located near the northeastern corner of the Joinville Plateau on seismic profiles in which stratigraphic units have not been correlated with those identified further south. The late Eocene age (37–34 Ma) of the sediments recovered at SHALDRIL Site NBP0602A-3 is consistent with previous interpretations of seismic Unit 3 as consisting of Late Cretaceous to Oligocene marine shelf deposits (Sloan et al., 1995; Smith and Anderson, 2010). Seismic Unit 2 has been interpreted as Miocene to Pliocene in age and seismic Unit 1 as Pliocene to recent.

The seismic units on the profiles through the SHALDRIL sites near the northeastern corner of the Joinville Plateau exhibit rather uniform seismic facies and contain reflectors that are truncated near the sea floor and dip fairly consistently at a low-angle towards the continental margin. NBP0602A sites 12, 5 and 6 were at locations progressively nearer to the continental shelf break and recovered sediments of late Oligocene (28.4–23.3 Ma), middle Miocene (12.8–11.7 Ma) and early Pliocene age, respectively. Seismic profiles suggest the Joinville Plateau is characterised by a late Oligocene to Pliocene sediment wedge containing no large unconformities, with contour currents and glacimarine sedimentation dominating deposition in this region since at least the Late Oligocene, and a lack of a grounded Antarctic Peninsula Ice Sheet until early

Pliocene (Smith and Anderson, 2010). SHALDRIL site NBP0602A-5 penetrated an unconformity between middle Miocene sediments below and early Pliocene sediments above, similar to those recovered at site NBP0602A-6. Smith and Anderson (2010) used regional seismic stratigraphic ties to the SHALDRIL cores to constrain an expanded Plio-Pleistocene section in James Ross Basin. They identified 10 unconformity bounded units, suggesting this area could yield an excellent Pliocene to Pleistocene record of climate-ice sheet conditions if it was to be drilled in the future.

Results from analysis of the SHALDRIL cores indicate progressive cooling and increasing glacial influence in the northern Antarctic Peninsula region since the late Eocene (Anderson et al., 2011). In the late Eocene, despite palynological evidence of diverse flora similar to that in forested parts of southern Chile and New Zealand today, rare ice-rafted pebbles and sand grain surface textures suggest some tidewater glaciers were also present. The late Oligocene cores had a markedly greater content of ice-rafted sand grains and indications that seasonal sea ice may have been present. The recovered middle Miocene sediments (~12.8 Ma) contain diverse ice-rafted pebbles from local and distant sources and provide clear indications of sea ice presence. However, palynological analyses show that tundra vegetation still persisted in this subregion at this time, and therefore persisted through the Middle Miocene Climate Transition, but evidence of this vegetation is absent in the cores by early Pliocene time, by which time there had certainly been grounded ice advances across the northernmost continental shelf. Sand grains in Pleistocene sediments recovered in SHALDRIL NBP0602A-5 exhibit greater grain roughness and more abundant glacial surface textures relative to the middle Miocene and early Pliocene sediments, suggesting a further intensification of glaciation.

The SHALDRIL results show that the sedimentary units on the continental shelf in this subregion contain a rich record of Cenozoic climate change and cryospheric evolution that is accessible to shallow drilling. However, the existing cores are from condensed sections spanning only a limited number of time windows, and therefore only provide a tantalising glimpse of the long-term climate in this region. These cores, alongside extensive seismic data in this region, show there are accessible strata that clearly present opportunities for further shallow drilling expeditions, perhaps using remotely-operated sea-floor drilling systems. However, any such expeditions will be vulnerable to the variable sea ice conditions in this region, so a wide range of alternate sites would be required to mitigate against this risk.

3.4.6 The South Orkney Microcontinent and adjacent deep-water basins

This subregion includes the South Orkney Microcontinent (SOM) and the adjacent deep-water Jane and Powell basins (Fig. 3.14). The SOM extends about 350 km from east to west and 250 km from north to south and is

WEDDELL SEA

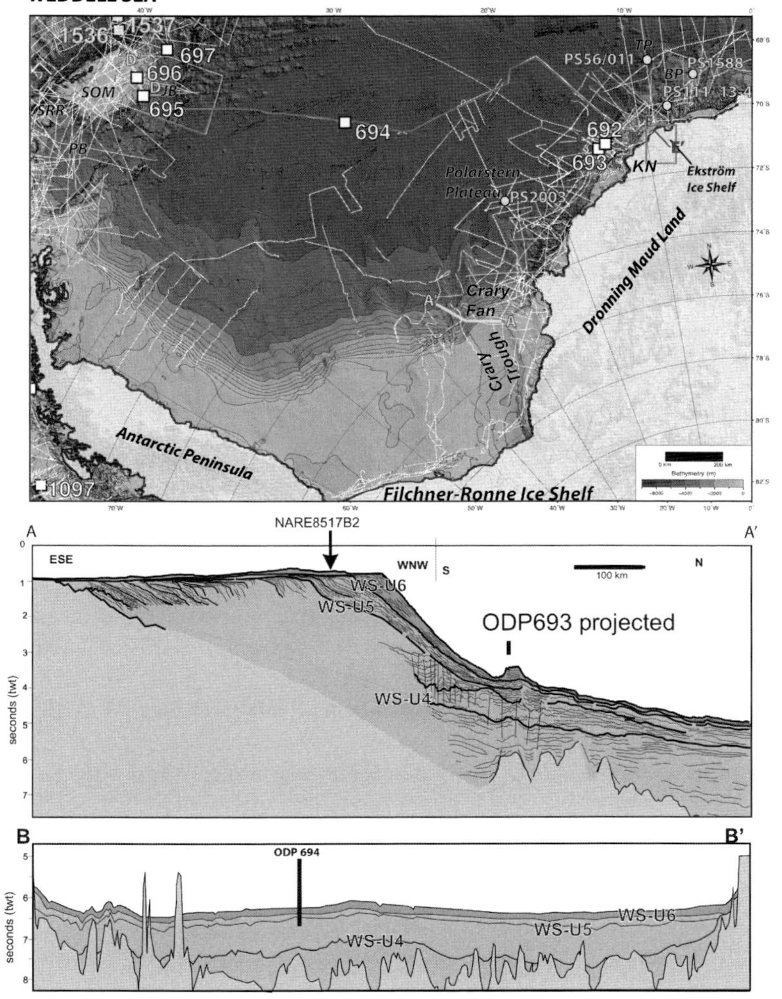

FIGURE 3.14 (Top) Location map of Weddell Sea sector with existing seismic lines (white lines). Yellow sites indicate sediment cores and the Red box shows the Ekström Ice Shelf area for pre-site drilling investigations (*BP*, Bungenstock Plateau; *KN*, Kapp Norvegia; *SSR*, South Shetland Rise; *TP*, Torge Plateau; *SOM*; South Orkney Microcontinent; *JB*, Jane Basin; *PB*, Powell Basin). Seismic tracks are from the SCAR Antarctic Seismic Data Library System (SDLS) and additional survey information. Bathymetry is from Bathymetric Chart of the Southern Ocean (Arndt et al., 2013) and subglacial topography from BedMachine Antarctica dataset (Morlighem et al., 2020). Bold green line (B–B′) shows seismic line for figure in bottom panel. Segments of seismic profiles shown in Figs. 3.15 and 3.16 are marked with red line (C–C′) and blue line (D–D′), respectively. (Middle) Interpretation of seismic lines BGR86-11 and BGR86-12 (available on the SDLS), shown as profile A–A′ in upper panel. Crossing line NARE8517B2 was used to check the depth of the unconformities WS4 and WS5. (Bottom) Interpretative line drawing of the seismic image and the sequences identified by Lindeque et al. (2013). Unconformities WS-U4 and WS-U6 were correlated to drill sites around Antarctic to about 34 and 5 Ma, respectively, although these units may be diachronous across the basin (Hochmuth et al., 2020a,b; Lindeque et al., 2013). *(Middle) From Cooper, A.K., Brancolini, G., Escutia, C., Kristoffersen, Y., Larter, R., Leitchenkov, G., et al., 2008. Cenozoic climate history from seismic reflection and drilling studies on the Antarctic continental margin. In: Florindo, F., Siegert, M. (Eds.), Antarctic Climate Evolution, vol. 8. Elsevier, pp. 115–234. Lower panel modified from Lindeque, A., Martos Martin, Y., Gohl, K., Maldonado, A., 2013. Deep-sea pre-glacial to glacial sedimentation in the Weddell Sea and southern Scotia Sea from a cross-basin seismic transect. Marine Geology 336 (0), 61–83. Available from: https://doi.org/10.1016/j.margeo.2012.11.004.*

underlain by Mesozoic metamorphic and sedimentary rocks (Dalziel, 1984; Thomson, 1981). Offshore, the SOM includes four Cenozoic sedimentary basins (King and Barker, 1988) with up to 5 km of sediment (Busetti et al., 2001, 2002; Harrington et al., 1972). Powell Basin (up to 3600 m deep) formed as the SOM rifted and drifted away from the tip of the Antarctic Peninsula in late Eocene to early Miocene time (Catalán et al., 2020; Coren et al., 1997; Eagles and Livermore, 2002; King and Barker, 1988; Lawver et al., 1994; Rodríguez-Fernández et al., 1997). Opening of Jane Basin (up to 3300 m deep) probably began later (Lawver et al., 1991, 1994) and may have continued until the middle Miocene (Bohoyo et al., 2002; Eagles and Jokat, 2014; Maldonado et al., 1998).

From single channel seismic data on the SOM, King and Barker (1988) defined pre-rift, syn-rift and post-rift units (S3, S2 and S1, respectively) (Fig. 3.15). The post-rift sediments are less than 1 km thick (e.g., Busetti et al., 2001, 2002) and were drilled at ODP Site 695 (1300 m water depth) and ODP Site 696 (600 m water depth) (Fig. 3.15). They comprise Oligocene or early Miocene to Quaternary terrigenous sediments, with rare coarse-grained ice rafted debris until the late Miocene (~ 8.7 Ma) and common ice rafted debris thereafter. Middle Miocene to Quaternary sediments are hemipelagic and diatomaceous muds and oozes (Barker and Kennett, 1988; Barker et al., 1988). ODP Site 696 also sampled syn-rift late Eocene sandy mudstones (S2) that have nannofossil assemblages and clay minerals suggesting a relatively warm climate, and palynoflora indicating temperate beech forests and ferns on the northern Antarctic Peninsula. However, the same late Eocene sediments also contain ice-rafted sand grains of the southern Weddell Sea provenance, indicating that tidewater glaciers were present in that area even before the Eocene-Oligocene climate transition (Carter et al., 2017). The occurrence of autochthonous glaucony in these sediments indicates that they were deposited during a transgressional event just before the Eocene-Oligocene climate transition (López-Quirós et al., 2019). Other results from ODP sites 695 and 696 suggest intermittent glaciation with little sea ice during most of the Miocene and a persistent ice cap to sea level on the Antarctic Peninsula since the late Miocene. Herron and Anderson (1990) place the maximum late Quaternary grounding line advance at the 300 m isobath, and consider open-marine conditions to have existed over the SOM since 6000 years BP based on single-channel seismic and seafloor core data. A detailed bathymetric compilation confirmed a laterally extensive break of slope across the middle of the SOM shelf at this depth, and reexamination of a seismic profile crossing it suggests this represents deposition at a former grounding zone (Dickens et al., 2014).

In Powell Basin, post-early-rift sediments are up to 3 km thick. King et al. (1997) identified two seismic units with low reflectivity below and high reflectivity above. They interpreted the change as recording the onset of glacial-interglacial cyclicity in the supply of coarse detritus to the basin in

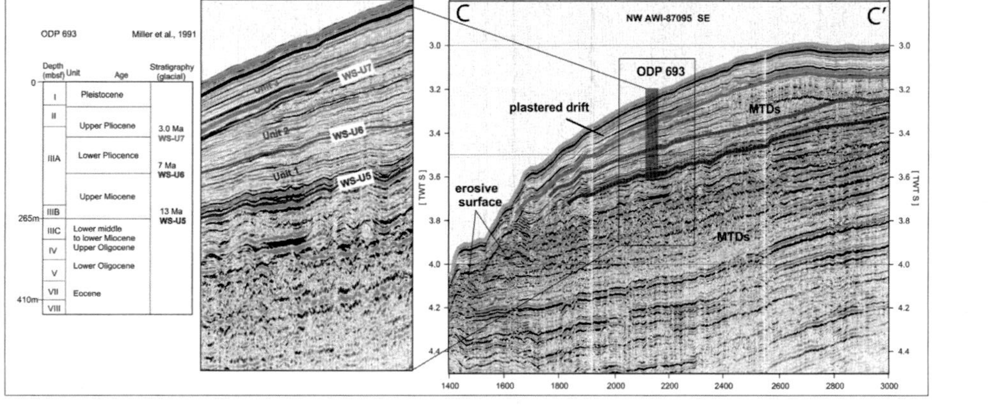

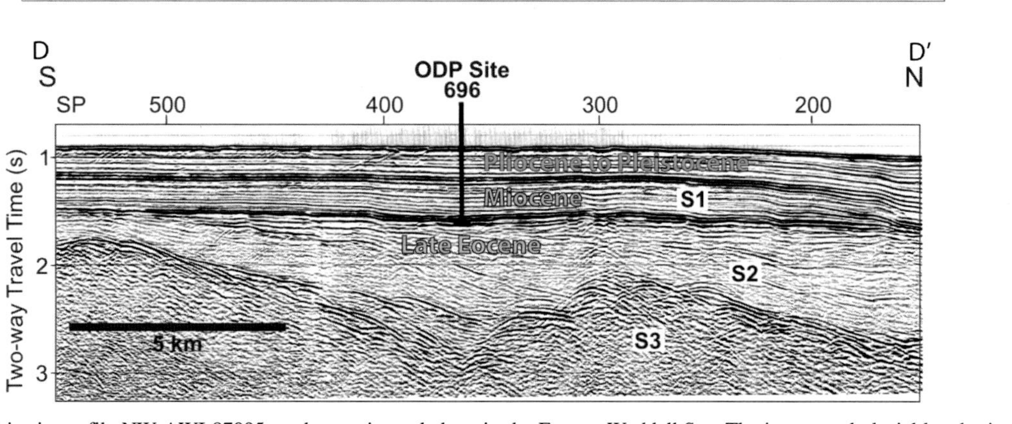

FIGURE 3.15 (Top) Seismic profile NW AWI-87095 on the continental slope in the Eastern Weddell Sea. The interpreted glacial key horizons constrained by ODP site 693. Seismic line (B–B′) and drill site location shown in Fig. 3.14; *MTDs*, Mass Transport Deposits. (Bottom) Seismic profile AMG845-18 over the South Orkney Microcontinent in the Western Weddell Sea, with location of ODP site 696. *(Top) Modified from Huang, X., Jokat, W., 2016. Middle Miocene to present sediment transport and deposits in the Southeastern Weddell Sea, Antarctica. Global and Planetary Change 139, 211–225. Available from: https://doi.org/10.1016/j.gloplacha.2016.03.002. (Bottom) Modified from Barker, P.F., Kennett, J.P., Scientific, P., 1988. Weddell sea palaeoceanography: preliminary results of ODP Leg 113. Palaeogeography, Palaeoclimatology, Palaeoecology 67 (12), 75–102.*

the late Miocene. A similar upward change in reflectivity is observed in Jane Basin (Maldonado et al., 1998). The reflectivity change may also be due to silica diagenesis (e.g., Lonsdale, 1990; Volpi et al., 2003). Rodríguez-Fernández et al. (1997) interpreted four seismic units in Powell Basin, which they characterised as pre-rift, syn-rift, syn-drift and post-drift. Maldonado et al. (2006) defined five seismic units in Jane Basin and neighbouring ocean basins, and related changes in seismic characteristics since the middle Miocene to variations in bottom water flow. ODP Site 697 was drilled in Jane Basin (Fig. 3.14) to ~323 mbsf, and recovered mainly early Pliocene and younger hemipelagic sediments with ice rafted debris throughout; however, ice rafted debris is abundant only near the base of the sequence (Barker and Kennett, 1988; Barker et al., 1988). Other seismic studies of Powell Basin (e.g., Coren et al., 1997; Kavoun and Vinnikovskaya, 1994; Viseras and Maldonado, 1999) focus on the post-early Oligocene rift history of the basin, but are limited in paleoclimate interpretations by the lack of drilling data.

Results from the ODP Leg 113 sites in this subregion show that the sedimentary units on the SOM and in the adjacent basins hold valuable long-term records of Cenozoic climate change and development of glaciation in this key area near the outflow of the Weddell Gyre and the Southern Boundary of the ACC. In particular, the succession holds good records of the conditions prior to the Eocene-Oligocene climate transition that are accessible by drilling, and records of the transition itself are probably present in some locations. The ice rafted debris in these sediments also provides evidence of the development of glaciation in the southern Weddell Sea as icebergs transported around the Weddell Gyre are carried across and around the SOM (Weber et al., 2019, 2014).

In 2019 IODP Expedition 382 drilled three sites in basins to the immediate north of the SOM, in the Pirie and Dove Basin. These sites are situated in the primary pathway of icebergs discharged from the Weddell Sea margin of the WAIS and EAIS and will allow researchers working on those cores to reconstruct ice sheet and oceanographic dynamics in this sector of Antarctic back to ~6 Ma (Weber et al., 2019). This will complement similar resolution records influenced by WAIS ice dynamics recently captured by IODP Expeditions 374 and 379, in the Ross Sea and Amundsen Sea, as well as the marine-based margin of Wilkes Land (Patterson et al., 2014). Future scientific drilling in this region could target earlier Eocene sedimentary records and more expanded records of some later periods.

3.4.7 East Antarctic Margin

The EAIS holds the world's largest freshwater reservoir with an equivalent of ~52 m sea level rise (Fretwell et al., 2013). It was long argued that the EAIS remained stable on million-year timescales (e.g., Sugden et al., 1993), supported by ice-sheet models requiring unlikely warming ($>10°C$) before exhibiting ice mass loss (e.g., Huybrechts, 1993). The last decade, however, has seen a significant increase in ice mass discharge in east Antarctica (Rignot et al., 2019; Shen et al., 2018), with the Wilkes Land loosing 51 ± 13 Gt/yr of ice between 2009 and 2017, which accounts for 20% of the total mass loss of Antarctica during this period (Rignot et al., 2019). Furthermore, for the last four decades, the cumulative contribution to sea level from East Antarctica is not far behind that of West Antarctica. These observations point to East Antarctica as a major participant in the ice mass loss from Antarctica in addition to the rapid mass loss reported from West Antarctica (Rignot et al., 2019).

The most recent Antarctic bed topography reconstruction by Morlighem et al. (2020) (Fig. 3.2) shows the glaciers with the most potential of becoming unstable draining through deep throughs with inland-sloping bedrock topography. In East Antarctica, these include glaciers draining through the Wilkes Subglacial Basin, Queen Mary Land and glaciers feeding the eastern Filchner Ice Shelf. Conversely, the Totten Glacier and Moscow University glaciers draining through the Aurora Subglacial Basin, and glaciers feeding the Amery Ice Shelf (except along the East Lambert Rift) are shown to have stabilising slopes beneath them and therefore would have to retreat several km inland before reaching a zone of retrograde bed. In Enderby and Queen Maud Land, the drainage basins are mostly above sea level and glaciers flow over prograde bed slopes, except along a narrow coastal margin, and are hence more protected from MISI processes. These findings signal the relevance of understanding past East Antarctica's marine ice sheet dynamics for assessing their response to ongoing climate change and to better determine Antarctica's future contribution to sea-level change.

3.4.7.1 Weddell Sea

The Weddell Sea comprises a wide, deeply-incised continental shelf, that is currently characterised by heavy sea ice cover on the outer continental shelf, and the Filchner-Ronne Ice Shelf (FRIS) on the inner shelf, which is the largest ice shelf on Earth by volume (Akhoudas et al., 2020; Arndt et al., 2013). This extensive ice cover has precluded detailed surveying by seismic vessels and drilling on the continental shelf, so most seismic studies and drill cores are restricted to offshore regions (Figs. 3.1 and 3.14). In the southeastern Weddell Sea, the Filchner Depression/Crary Trough represents a glacially carved seafloor depression that drained a large portion of the marine-based sector of the EAIS in the Recovery Ice Stream catchment. Model-based experiments suggest this sector of the EAIS is the most sensitive to oceanic warming, and potentially susceptible to threshold behaviour under moderate levels of oceanic warming ($\sim 1°C-2°C$) (Golledge et al., 2017). Such warming could occur rapidly in this region, if there is a shift in wind-driven advection of modified Circumpolar Deep Water into Filchner Depression/Crary Trough on the continental shelf, via the southern limb of the Weddell Gyre (Hattermann, 2018; Hellmer et al., 2012).

Kristoffersen and Jokat (2008) previously described the regional seismic stratigraphy of the Weddell Sea continental shelf/slope/rise in the first edition of Antarctic Climate Evolution (Florindo and Siegert, 2008). Broadly, the principal stratigraphic features in the Weddell Sea sedimentary basins are prograding wedges of glacigenic sediments along the entire continental shelf margin. A large trough-mouth fan (Crary Fan) north of the Filchner Depression/Crary Trough extends into much of the continental rise to abyssal plain of the southeastern Weddell Sea and is associated with a complex channel-levee and gully system (Gales et al., 2016, 2014, 2012). The present-day sedimentary depositional system in the Weddell Sea can be split into three geomorphological regions (Jerosch et al., 2016): (1) the continental margin in the southern and eastern region is heavily influenced by downslope transport of sediments near the glaciated continental shelf edge, with offshore channel-levee and contourite systems (Michels et al., 2002); (2) slowly accumulating hemipelagic sediments of mostly terrigenous origin further offshore in the central and northern Weddell Gyre; and (3) hemipelagites and contouritic sediment drifts in the north-western Weddell Sea (Vernet et al., 2019).

Very few multichannel reflection seismic surveys have explored the southern Weddell Sea since the review by Kristoffersen and Jokat (2008), partly because of increasing difficulties to fulfil environmental protection restrictions, but also due to the pervasive ice conditions in this region. However, ongoing data interpretation and review papers have provided new results for paleoenvironmental changes in the Weddell Sea area, with implications for understanding the evolution of Antarctica's cryosphere and global impacts of climate change. Here, we provide an update on scientific knowledge about (1) Gondwana break-up, Weddell Sea opening and the pre-ice-sheet depositional environment from mainly geophysical research and (2) the Eocene-Oligocene transition and younger geological units that have been sampled by drilling. A wealth of studies have been undertaken on deglaciation processes (from LGM, to recent), and these are discussed in more detail by Siegert et al. (2021) and by Hillenbrand et al. (2014). We also discuss some outstanding scientific questions that could be addressed by future geoscientific drilling based on recent seismic surveys beneath the Ekström Ice Shelf (EIS).

3.4.7.1.1 Gondwana break-up, Weddell Sea opening and pre-ice-sheet depositional environment

U-Pb zircon studies revealed that the southern side of Dronning Maud Land has a protracted late Neoproterozoic/early Palaeozoic metamorphic overprint, accompanied by igneous activity, most likely related to the East African-Antarctic Orogen (Jacobs et al., 2017). The lack of preserved Mesozoic–Cenozoic sediments and structures in central Dronning Maud Land has limited our understanding of the post-Pan-African evolution of this important part of East Antarctica (Sirevaag et al., 2018). Vibroseismic site surveys conducted for planned drilling studies to improve this understanding have been conducted below the Ekström Ice Shelf (Kuhn and Gaedicke, 2015; Smith et al., 2020) and show a possible outcrop of the syn-rift volcanic Explora Wedge unit at the seafloor (Fig. 3.16). A recent synthesis of ship-based and aeromagnetic surveys has provided new insights into the reconstruction of the spreading history of the Weddell Sea area (Eagles and Jokat, 2014). The separation of Antarctica from former South Africa in the Dronning Maud Land area has been assumed to have occurred between 82 and 159 Ma (Kristoffersen et al., 2014). In the southern Weddell Sea in front of the FRIS, a failed rift system is observed, and oceanic crust in the centre of this failed rift has a minimum age of 145/148 Ma (magnetic chron M19/M20) and maximum age of 160 Ma (Jokat and Herter, 2016).

The Weddell Sea formed as a consequence of the break-up of the Gondwana super-continent in Jurassic times. This break-up was associated with massive, but short-lived volcanic activity, that possibly formed the Explora Wedge (Kristoffersen et al., 2014). Dredges have recovered organic rich marine sediments, indicating temperate climate conditions, and a belemnite of likely early Cretaceous age from the slopes of the Wegener Canyon (from 2700 to 3250 m water depth). This is very close to ODP Site 693, where upper Cretaceous marine sediments were drilled at the borehole base

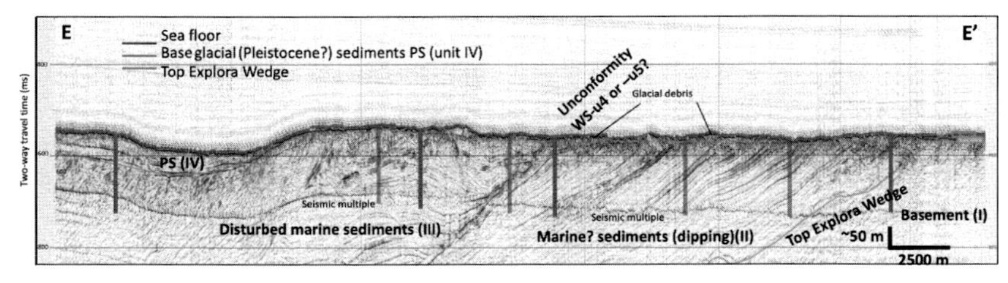

FIGURE 3.16 Vibroseis seismic profile from below the Ekström Ice Shelf. Four seismic units were detected: continental basement (I) with the Explora Wedge and overlying unconformity (red line). Seismic unit II could include the oldest post-rift sediments (Jurassic–Cretaceous) in the Dronning Maud Land area and is interpreted as a shallow marine (or terrestrial?) environment. An angular unconformity separates this from seismic unit III that indicates more turbulent deposition (drifts?, canyon infill?), interpreted as being deposited after initiation of the EAIS (younger than 34 Ma?). (Plio-?) Pleistocene sediments (Unit IV) stratified parallel to the seafloor occur in the most northern part of the profile. Whole seismic units could be sampled at four (up to 200 m deep) drill-coring sites (vertical red lines) for bedrock and covering sediment characterisation. Four additional drill cores (transparent red) would allow coverage of the entire sedimentary sequences. Seismic line (E-E′) location shown in Fig. 3.14.

(Kennett and Barker, 1990; Fütterer et al., 1990), suggesting significant potential for obtaining longer-term Mesozoic records in this region. Continued seafloor spreading led to expansion of the Weddell Sea between the Cretaceous and the Eocene, but a lack of geological sampling of this stratigraphy do not allow any further reliable statements about the sedimentary and oceanographic environment during this period. However, some inferences can be made from intraregional seismic correlations (Lindeque et al., 2013; Maldonado et al., 2006; Pérez et al., 2021), geologic data (Carter et al., 2017; Mackensen and Ehrmann, 1992) and modelling results (Douglas et al., 2014; Ladant et al., 2014; Wilson et al., 2012) that suggest a proto-Weddell Gyre existed throughout much of this period. Model experiments indicate that deep convection was more intense prior to the opening of the Drake Passage, but changed to a more stagnant stage when the Drake Passage opened (Ladant et al., 2018).

3.4.7.1.2 The Eocene-Oligocene transition and paleoenvironment during increasing glacial conditions

After the opening of Drake Passage at ~34–35 Ma, the Weddell Sea became tectonically passive with approximately the same topography as it has today (Vernet et al., 2019). The opening of Drake Passage allowed the ACC to develop and has long been thought to be a key factor in ice sheet expansion across the Antarctic continent (Kennett, 1977). On the continental rise and abyssal plain of the Weddell Sea, a prominent unconformity (WS-U4) marks these changes, inferred to coincide with the Eocene/Oligocene boundary but has yet to be constrained by drilling (Lindeque et al., 2013) (Fig. 3.14). A correlation of the deep water Weddell Sea seismic stratigraphy to ODP/IODP/DSDP drill cores in the Southern Ocean and Antarctic margin was conducted by Huang et al. (2014) and Hochmuth et al. (2020a,b). This work presented a state-of-the-art reconstruction of sediment accumulation in the Southern Ocean with time slice presentations of back-stripped sediment accumulation maps. It highlights that the failed rift basin in the Weddell Sea contains the thickest sequence of pre-Oligocene sediments on the Antarctic margin, with thicknesses reaching more than 7 km (Jokat and Herter, 2016; Leitchenkov and Kudryavtsev, 2000), whereas extensive fluvial erosion is inferred for the southeastern Weddell Sea (Hochmuth et al., 2020a,b). Between 34 and 14 Ma, sediment accumulation rates increased slowly and were most focussed in regions where there was enhanced erosion by expanding glacial systems on the continental margin of the eastern Weddell Sea (34–24 Ma), Ronne (24–21 Ma) and Filchner (21–14 Ma) shelf areas.

Unconformity WS-U5 is constrained by ODP Site 693 to between 15 and 13 Ma (Huang and Jokat, 2016; Kennett and Barker, 1990), correlating with the MMCT (Fig. 3.15). After this time, sedimentation rates increased significantly in the Weddell Sea, and stronger bottom-water currents formed expansive contouritic drifts (Hochmuth et al., 2020a,b).

Particularly high sedimentation rates from ODP Leg 113 cores are reconstructed for the period between 14 and 3 Ma in the Crary Fan area, and between 5 Ma and recent on the continental slope and rise offshore from Dronning Maud Land (Barker et al., 1988; Hochmuth et al., 2020a,b; Huang and Jokat, 2016; Huang et al., 2014; Kennett and Barker, 1990). This suggests increased expansion of marine-terminating ice sheets delivering sediment to the continental shelf edge at these times. Highest last glacial sedimentation rates of up to 300 cm/kyr were recovered in sediment cores from the southern Weddell Sea on the Crary Fan rise (Sprenk et al., 2014; Weber et al., 2011). These data suggest that if sea ice-related risks for drilling vessels could be mitigated, there is excellent potential to obtain Neogene to Quaternary archives in this high-latitude, polar region for decadal and perhaps annual resolution of sedimentation relating to ice dynamics, oceanic process and depositional history. This is in contrast to very low sedimentation rates in the northern abyssal plain (Geibert et al., 2005; Honjo et al., 2008; Howe et al., 2007). Seafloor plateaus on the abyssal plain, such as the Polarstern or Torge Plateaus (Arndt et al., 2013; Bart et al., 1999; Kuhn et al., 2002), and sediment starved slopes like the Bungenstock Plateau provide low but continuous pelagic sedimentation that are covered with biogenic sediments (Abelmann et al., 1990). Gravity cores from the Torge, Bungenstock and Polarstern Plateaus collected sediment dating back to the lower Pliocene (ca. 4.5 Ma) and show a shift from biogenic opal rich to carbonate-rich sediments near the Pliocene/Pleistocene boundary (Hillenbrand and Ehrmann, 2005). This shift is suggested to be related to oceanic cooling and freshening by enhanced meltwater input (higher ice volume dynamics), increased sea-ice coverage, less general overturning circulation due to more Antarctic Bottom Water formation, or a deeper carbonate compensation depth and higher alkalinities as observed during late Pleistocene glacial periods (Rickaby et al., 2010).

3.4.7.1.3 Recent geophysical survey beneath the Ekström Ice Shelf and future directions for drilling

Satellite gravimetry data comparisons to vibroseis measurements conducted on the Ekström Ice Shelf (EIS) have enabled an improved bathymetry to be developed below the Dronning Maud Land ice shelves. These interdisciplinary observations have been conducted as part of a site survey for future potential geological drilling below the EIS, and 615 km of high-resolution vibroseismic lines were collected on the ice shelf between 2010 and 2018 (Eisermann et al., 2020; Kuhn and Gaedicke, 2015; Smith et al., 2018, 2020) (Fig. 3.16). Vibroseismic profiles near the calving front on the ice shelf indicate onlapping strata in a $\sim$1000 m thick dipping sedimentary sequence covering the continental basement (Kristoffersen et al., 2014). Although the seismic data are currently unpublished, preliminary analysis indicates three distinct sedimentary units above the continental basement. The volcanic Explora Wedge is interpreted to likely form the top of this basement and dips ~20° to the north (Kristoffersen et al., 2014) (Fig. 3.16).

A well stratified seismic interval (unit II) overlies the basement and dips in the same direction and is hypothesised to have been deposited in a quiet, shallow marine environment after the Weddell Sea opening during the Jurassic. A clear unconformity with a steeper dipping angle cross cuts unit II and is overlain by irregularly folded, but clearly stratified sediments (unit III) that outcrop on the seafloor in the northern part of the profiled area. This unconformity is tentatively interpreted to represent the WS-U4 (~Eocene/Oligocene boundary) or WS-U5 (~14 Ma) surface discussed above. The uppermost seismic unit (IV) shows deposition and bedding parallel to the seafloor and could have been deposited during interglacials (or glacials?) through the Plio-Pleistocene. Glacial debris and diamicton covers the older sequences in an up to 10-m-thick layer on the seafloor. Depending on the utilised drilling technology, a drilling plan that revolved around five cores penetrating 200 m below the seafloor would enable sampling of all the identified seismic units, allowing a tectonic and climate history to be reconstructed in this region for key 'snapshot' time periods since the Jurassic. This would provide a history of early sedimentation after the rifting and EAIS build up, alongside a reconstruction of associated variability of productivity and oceanographic changes related to past climates. Accompanied drilling further to the north would enable more comprehensive changes of Weddell Gyre dynamics and Antarctic Bottom Water formation during Cenozoic glacial/interglacial cycles.

High-quality Plio-Pleistocene records from this sector of the margin would also build on the excellent high-resolution long sediment cores collected further north in the Scotia Sea during 2019 by IODP Expedition 382. This recent IODP expedition aimed to capture millennial scale resolution records of ice rafted debris deposition from cores in the 'Iceberg Alley' region of the Weddell Gyre/ACC confluence (Weber et al., 2019). Sedimentary records deposited at locations along the pathways of the returning Weddell Gyre surface currents that pass over Dronning Maud Land and the Weddell Sea Deep Water overflow to the Scan Basin in the Scotia Sea (Pérez et al., 2014) are connected to the IODP Expedition 382 sites and offer unique opportunities for reconstructing past changes in bottom water flow and iceberg discharge in these areas. With hot-water drilling required to access the seafloor below the EIS, this drilling should be complimented by a deployment of instrumentation to obtain an oceanographic time series observing upwelling of modified Warm Deep Water onto the shelf and in the deepest part of the glacially eroded trough below the EIS. This will help constrain processes associated with ocean-cryosphere interaction in the cavity of this ice shelf, supplementing similar measurements below the Fimbul Ice Shelf to the east (Hattermann, 2018; Hattermann et al., 2012, 2014), as well as enabling a better understanding of the relevance of the paleoclimate records and model experiments in this region.

3.4.7.2 Prydz Bay

Prydz Bay overlies a rift structure that extends about 500 km into the interior of the continent and has channelled drainage at least since the early Cretaceous (Arne, 1994). It presently contains the Lambert Glacier-Amery Ice Shelf drainage system that drains more than 16% of the EAIS. As with the Aurora Basin, this drainage basin extends to the Gamburtsev Mountains, a subglacial range in which the Cenozoic ice sheets may have nucleated (Bo et al., 2009). An extensive grid of seismic surveys exists in this region, with age constraints provided by the two ODP Legs 119 and 188 (Fig. 3.17). The offshore paleoclimate records and seismic stratigraphic framework in this region are

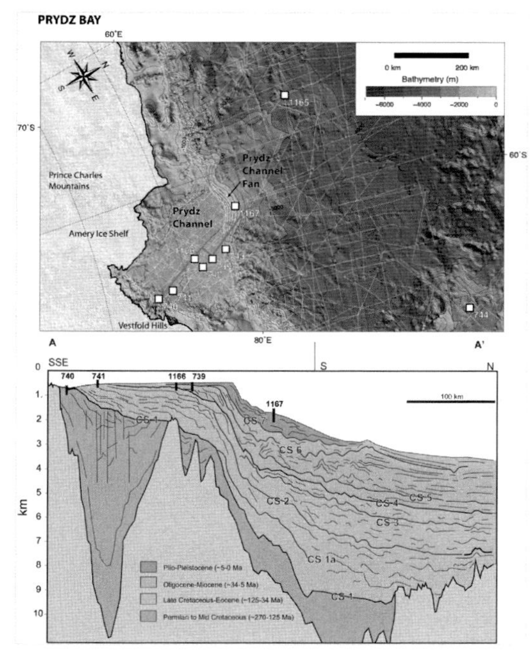

FIGURE 3.17 (Top) Location map of Prydz Bay sector with existing seismic lines (white lines) and drill sites (white boxes). Seismic tracks are from the SCAR Antarctic Seismic Data Library System (SDLS) and additional surveys. Bathymetry is from Bathymetric Chart of the Southern Ocean (Arndt et al., 2013) and subglacial topography from BedMachine Antarctica dataset (Morlighem et al., 2020). (Bottom) Interpretive seismic section across Prydz Bay shelf, continental slope and continental rise based on Russian seismic lines RAE 5702, 5701 and 5206. See bold green line A–A′ in upper panel for section trackline. *Adapted from Cooper, A.K., Brancolini, G., Escutia, C., Kristoffersen, Y., Larter, R., Leitchenkov, G., et al., 2008. Cenozoic climate history from seismic reflection and drilling studies on the Antarctic continental margin. In: Florindo, F., Siegert, M. (Eds.), Antarctic Climate Evolution, vol. 8. Elsevier, pp. 115–234. with annotated seismic horizons from Leitchenkov, G., Galushkin, Y., Guseva, Y., Dubinin, E., 2020. Evolution of the Sedimentary Basin of the Continental Margin of Antarctica in the cooperation sea (from results of numerical modeling). Russian Geology and Geophysics 61 (1), 6878. Available from: https://doi.org/10.15372/RGG2019079.*

further constrained by a wide range of on-land field studies and exposure dating. Two Phanerozoic phases of enhanced denudation are inferred for the Prince Charles Mountains, with an initial phase of 1.6–5.0 km of focussed erosion during the initial rift phase in the early Palaeozoic, followed by 1–4.5 km of erosion during rifting reactivation in the early Cretaceous (Lisker et al., 2003). Zircons and hornblendes recovered from Eocene fluvial sediments draining from the Gamburtsev mountains in ODP Leg 188 Site 1166 provide $^{40}Ar/^{39}Ar$ dates of ~519–530 Ma, indicative of a continental origin for the Gamburtsev mountains as opposed to a younger volcanic origin (van de Flierdt, 2008). Veevers (2008) examined zircons for U-Pb ages and Hf isotopes from Permian to Cenozoic sediments from across the Prince Charles Mountains and Prydz Bay and identified clusters of ages between 700–460 Ma, 1200–900 Ma and smaller clusters as old as 3350 Ma, suggesting the region comprises a core complex of Grenville age (1200–900 Ma) rocks surrounded by fold belts formed by the assembly of Gondwana (700–460 Ma).

3.4.7.2.1 Early Cenozoic greenhouse and earliest glacial phase in late Eocene

The formation of the Lambert Graben and Prydz Bay basin during the Carboniferous to early Cretaceous provided sedimentary depocentres prior to the inferred onset of Cenozoic glaciation near the Eocene/Oligocene boundary, and both basins contain up to 9.0 km of rift-fill sediment (Arne, 1994; Leitchenkov et al., 2020, 2018, 2014; Lisker et al., 2005). Seismic reflection data show Mesozoic sequences of parallel, moderately continuous reflectors (Figs 3.17 and 3.18), which were penetrated at ODP Sites 740, 741 and 1166.

ODP Site 740 intersected interbedded sandstone, siltstone and mudstone with reddish coloration (Barron and Larsen, 1989), interpreted as fluvial flood plain deposits (Turner, 1991). The age of this unit remains unknown, but could be as old as Triassic (Mcloughlin and Drinnan, 1997a,b; Truswell, 1991). Multichannel seismic data highlights a thick (up to 5 km), faulted and high-velocity (up to 5.2 km/s) unit underlying these red beds (Leitchenkov et al., 2020, 2015). This sequence predates the main phase of break-up-related crustal extension and is thought to correlate with the Permian-Triassic sediments of the northern Prince Charles Mountains. This indicates the reddish beds in ODP Site 740 are likely to be early Cretaceous or Late Jurassic in age. The base of ODP sites 741 and 1166 recovered Turonian-Santonian(?) (94–84 Ma) and middle Aptian age (125–113 Ma) sediments, respectively, and contain siltstone and sandstone interbeds with minor coal. These strata are interpreted as probable delta plain to lagoonal in origin, and palynomorph assemblages from ODP Site 1166 indicate a conifer-dominated woodland vegetation, consistent with a cool, humid climate (Macphail and Truswell, 2004a,b).

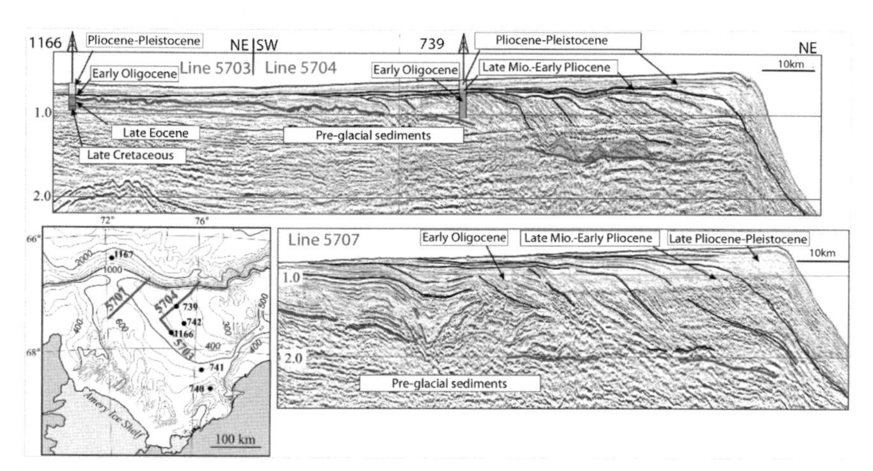

FIGURE 3.18 Seismic lines showing shelf and upper slope development during the Cenozoic in Prydz Bay. Prograding clinoforms developed in the Oligocene with episodes of shelf and slope erosion. *From Leitchenkov, G., Galushkin, Y., Guseva, Y., Dubinin, E., 2020. Evolution of the Sedimentary Basin of the Continental Margin of Antarctica in the cooperation sea (from results of numerical modeling). Russian Geology and Geophysics 61 (1), 6878. Available from: https://doi.org/10.15372/RGG2019079.*

The continental rise offshore of Prydz Bay contains up to 7 km of post-rift sediments (Fig. 3.17) (Mizukoshi et al., 1986; Stagg et al., 2005, 2004). The lowermost seismic stratigraphic unit has parallel, mostly continuous, reflectors typical of deep-ocean deposition that probably occurred during pre-ice-sheet times (Kuvaas et al., 2005; Kuvaas and Leitchenkov, 1992; Leitchenkov et al., 2015, 2007; Mizukoshi et al., 1986). Nine seismic horizons were identified from the continental slope and rise (Fig. 3.17) starting from horizon CS1, which corresponds to the top of rift fill unit. Horizons CS1a, CS2 and CS3 bounding early post-rift units likely formed in response to tectonic, ocean circulation and climate changes in the Southern Ocean prior to the onset of major glaciation at about 34 Ma, which is marked by Horizon CS4 (Leitchenkov et al., 2015, 2007) (Fig. 3.17).

Although syn-glacial strata (post-Eocene) are nearly absent on much of the inner shelf, they are up to 1.5 km thick on the outer shelf, and up to 2.8 km thick on the continental rise (Leitchenkov et al., 2015). In Prydz Bay, ODP Sites 739, 742 and 1166 recovered sediments deposited immediately before major glaciation (Barron et al., 1991; Cooper and O'Brien, 2004). Seismic correlations to these cores indicate that poorly sorted sands overlie an undulating erosion surface, suggesting a period of erosion possibly related to a relative low stand of sea level (Erohina et al., 2004) (Fig. 3.18). These sands fine upsection into mudstone containing lonestones, marine diatoms and dynocysts (O'Brien et al., 2001). An integrated bio-(diatoms, palynomorphs) and magneto-stratigraphic age model developed for ODP Sites 739,

742 and 1166 indicates the late Eocene to early Oligocene intervals of these sites form a relatively continuous succession from 36 Ma (C16.1n) to 33 Ma (C13n) (Macphail and Truswell, 2004a,b; O'Brien et al., 2001; Passchier et al., 2017). Sand-grain surface textures in fluvial sands from Hole 1166A suggest erosion and breakage by glaciers, implying the presence of at least valley glaciers in the hinterland of Prydz Bay prior to the arrival of ice-rafted debris offshore (Strand et al., 2003).

3.4.7.2.2 Oligocene–Miocene ice-sheet development

The early glacial section of the shelf comprises tabular units that pinch out shoreward due to inner-shelf erosion, and extend seaward into prograding slope deposits (Cooper et al., 1991). The paleo-continental shelf edges for these units are better defined upsection, as the foreset strata steepen seaward (Fig. 3.18). On the outer shelf, the early glacial time is characterised by crosscutting mounded seismic facies, indicative of migrating grounding line fans formed at the front of meltwater rich glaciers (Leitchenkov et al., 2015). Several separate foreset packages interpreted as representing ice advances are identified on the shelf, the oldest of which occurs 50–100 km landward of the present-day shelf break. The foresets range in age from late Oligocene to late Miocene (early Pliocene?) based on correlations to ODP Site 739 (Leitchenkov et al., 2015). Shelf-edge clinoforms developed during the Oligocene, with foresets interrupted by numerous erosion surfaces (Fig. 3.18).

ODP sites 739, 740, 741, 742 and 1166 on the continental shelf recovered probable subglacial and glaciomarine diamicts, with thin interbedded diatomaceous mudstones deposited during warm episodes (Erohina et al., 2004; Hambrey et al., 1991). The drilling and seismic evidence indicates glacial advance across the Prydz Bay continental shelf during cold episodes, probably reaching the shelf edge. Over-compacted horizons indicate periods of glacial erosion and ice loading during the early Oligocene, Miocene and Plio-Pleistocene (O'Brien et al., 2001; Solheim, 1991). Prior to the late Miocene, the Prydz Bay shelf prograded uniformly across its width, with the bulk of the ice and entrained sediment coming from the southern end of the bay (i.e. from the Lambert Graben). The Prydz Bay continental slope became progressively steeper from the early phase of glaciation in the early Oligocene, to reach angles of as much as 8° on the present slope (Figs. 3.17 and 3.18).

On the continental rise, a pre-ice-sheet unit is overlain by another unit exhibiting channel-levee geometries. The nature of the change in geometry and the tracing of reflectors to the shelf drill sites suggest that this change originated from the glacial expansion and increased sediment supply in the early Oligocene (Kuvaas and Leitchenkov, 1992). Overlying the channel-levee complexes are thick mounds and sediment waves that show evidence of modifications by bottom currents although sediment waves may also form under influence overspill of turbidite flows and flow stripping from the main channels (Huang et al., 2020) (Fig. 3.17).

ODP Site 1165 drilled a thick mound of lower Miocene and younger contourite sediments with turbidites only in the upper 5 m (Cooper and O'Brien, 2004). The surface marking the base of the mounded sequences was dated as being early Miocene and was interpreted as representing a shift from low relief submarine fans to highly mounded deposits. However, no obviously lithological change is noted across this boundary, and interbedded bioturbated claystone with ice rafted debris and claystones with silt laminae are described above and below this surface. The facies associations suggest that sediment was delivered by turbidity currents and subsequently entrained and redeposited by deep sea contour currents, while the interbeds represent Milankovich-scale forcing of paleoenvironmental processes relating to ice rafting, biogenic sediment supply, oxygenation of bottom waters and current speed (Williams and Handwerger, 2005). At the middle Miocene (14–16 Ma), a shift from laminated contourites to alternating hemipelagic and pelagic facies occurs, with the alternation interpreted to represent Milankovitch cycle modulation of biological productivity relating to sea ice state and ice sheet extent (Cooper and O'Brien, 2004; Florindo et al., 2003; Grutzner et al., 2003). At this time, recycled microfossils increase significantly, suggesting the start of intense erosion by ice sheets of sediments on the continental shelf.

3.4.7.2.3 The Polar Ice Sheet (late Miocene(?)–Pleistocene)

In the early Pliocene, ice flow regimes changed and ice was focused into an ice stream on the western side of Prydz Bay, cutting a cross-shelf trough, the Prydz Channel (Taylor et al., 2004). The ice stream delivered basal debris to the shelf edge, where the debris built a trough mouth fan on the upper continental slope (O'Brien and Harris, 1996; O'Brien and Leitchenkov, 1997; O'Brien et al., 2004). On the banks adjacent to Prydz Channel, vertical aggradation of subglacial debris produced tabular units while glacial erosion overdeepened the inner shelf. The inferred early Pliocene base of the Prydz Channel Fan is the prominent unconformity mapped as reflector A by Mizukoshi et al. (1986) and reflector PP-12 by O'Brien et al. (2004).

While drilling and seismic evidence indicate that glaciers periodically advanced to the continental shelf edge in Prydz Bay during cold episodes of the Pliocene and early Pleistocene, evidence of warmer-than-present episodes and reduced ice extent also exists. Sediments in the Prince Charles Mountains indicate open-water fjordal environments in the Miocene to Pliocene, and an ice sheet margin that retreated several hundred kilometres inland from its present position (Hambrey and McKelvey, 2000; Whitehead, 2003; Whitehead et al., 2004). Lower Pliocene marine diatomite in the

Vestfold Hills, on the eastern side of Prydz Bay, contains evidence of temperatures 4°C warmer than today (Quilty et al., 2000; Whitehead et al., 2001). Diatoms in ODP sites 1166 and 1165 suggest reduced sea ice during the Pliocene (Whitehead et al., 2005), with silicoflagellates indicating higher surface water temperatures (Escutia et al., 2005; Whitehead and Bohaty, 2003). Calcareous nannoplankton at ~1.1 Ma indicates warmer conditions at that time (Pospichal, 2003).

ODP Site 743 was drilled into the eastern, steep part of the slope, and recovered diamict, while ODP Site 1167 was drilled into the Prydz Fan, and recovered muddy, pebbly sands and diamicts, interpreted to be deposited by slumping of subglacial debris delivered to the shelf edge by past expansion of the EAIS (Huang et al., 2020; O'Brien et al., 2001; Passchier, 2003). ODP Site 1167 also recovered thin mudstone units deposited during periods of reduced ice extent. More than 90% of the fan was deposited before the mid-Pleistocene, and only three advances of the Amery Ice Shelf to the shelf edge are constrained for the late Pleistocene (O'Brien et al., 2004). The seismic stratigraphy of the margin reflects the decrease of sediment supply from the continental shelf in the late Pliocene and the increase of influence of bottom currents in reworking fine grained material, leading to the formation of sediment drifts and sediment waves on the rise (Huang et al., 2020). Clay mineralogy, magnetic properties and clast composition at ODP Site 1167 show changes suggesting that the Pleistocene peak of erosion and ice volume in the Lambert-Amery drainage system occurred in the early Pleistocene (O'Brien et al., 2004). Oxygen isotope measurements on foraminifera from ODP Site 1167 also suggest that sedimentation was reduced after the mid-Pleistocene, with the last ice advance to the shelf edge at about Marine Isotope Stage 16 (612–698 ka; Lisiecki and Raymo, 2005). However, the stratigraphic record is fragmentary because hiatuses are common, which leads to a tentative identification of isotope stages (Theissen et al., 2003).

The late Pleistocene saw the Amery Ice Shelf grounding zone moving seaward to a position in mid-shelf during glaciations with grounded ice likely on the adjacent banks and an ice shelf in Prydz Channel (Domack et al., 1998). This landward shift in ice sheet margin extent during glacials is inferred to be the consequence of several factors including overdeepening of the inner shelf that acted to restrict marine-based ice sheet advance after the mid-Pleistocene; shifts in orbital influences on ice sheet mass balance at the Mid-Pleistocene Transition; and increased aridity due to late Pleistocene cooling may have also contributed to reduced ice extent (O'Brien et al., 2007; Suganuma et al., 2014).

3.4.7.3 East Antarctic Margin – Sabrina Coast

The low-lying, glacially sculpted Aurora Subglacial Basin (ASB) contains ~3–5 m sea level equivalent of marine-based ice (Fig. 3.19), and its

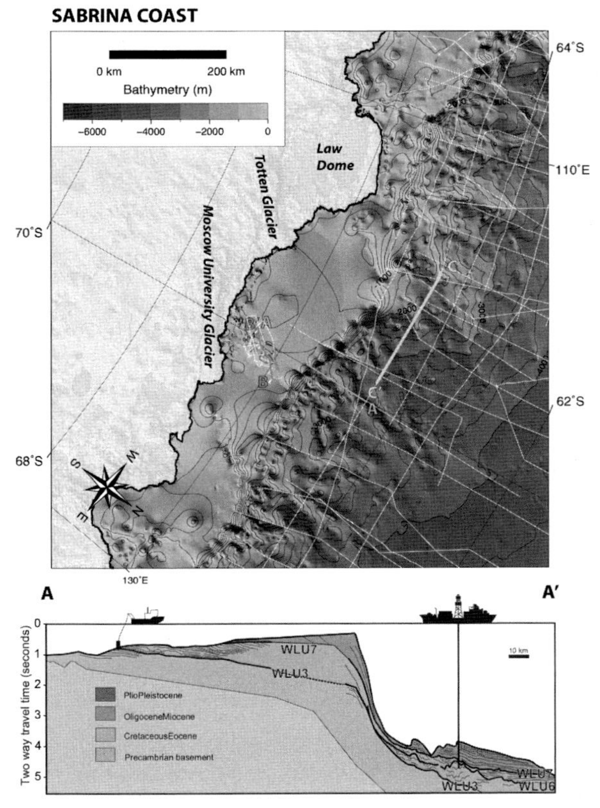

FIGURE 3.19 (Top) Overview map of Sabrina coast region of East Antarctica with existing seismic lines (white lines). No drill sites exist in the region, which is the focus of two planned drilling projects on the continental shelf and rise (see text for details). Seismic tracks are from the SCAR Antarctic Seismic Data Library System (SDLS) and additional surveys. Seismic track-lines in lower panel (orange line B′−B and green line A−A′) and Fig. 3.20 (yellow line C−C′) are shown. Bathymetry is from Bathymetric Chart of the Southern Ocean (Arndt et al., 2013) and subglacial topography from BedMachine Antarctica dataset (Morlighem et al., 2020). (Bottom) Sabrina Coast seismic stratigraphy showing major unconformities using the nomenclature for the Wilkes Land margin (WLU3 to WLU7) and associated shifts in the geometry of sedimentary packages on the continental shelf, and previous drilling (see upper panel for transect line). Profile is constructed from seismic lines 10 and 21 presented by Gulick et al. (2017) and RAE5103, with a tentative correlation between these lines based on Donda et al. (2020).

broader catchment drains ice from the Gamburtsev Mountains to the Sabrina Coast (Aitken et al., 2016; Fretwell et al., 2013; Greenbaum et al., 2015; Young et al., 2011a). The Totten Glacier is one of the catchment's largest marine-terminating outlet glaciers (Khazendar, 2013). Its grounding line is presently influenced by warm subsurface (<400 m) ocean waters and it is experiencing East Antarctica's largest mass loss (Li et al., 2016; Rintoul et al., 2016; Roberts et al., 2017). The ASB catchment consists of several overdeepened basins (Aitken et al., 2016; Young et al., 2011a) and hosts an active subglacial hydrological system that drains meltwater to the ocean (Wright et al., 2012), suggesting that regional ice may be susceptible to progressive retreat and changing subglacial hydrology. Models suggest significant ASB catchment mass loss is possible under some Intergovernmental Panel on Climate Change (IPCC) Representative Concentration Pathways, but there are large differences in the scale of this change between models (DeConto and Pollard, 2016; Golledge et al., 2015). Furthermore, model experiments indicate that both the Wilkes Subglacial Basin and ASB are more sensitive to atmospheric warming than other sectors of the EAIS (Golledge et al., 2017). Consequently, identifying regional climate threshold responses to oceanic and/or atmospheric warming from geological archives remains a high priority to provide important constraints on model-based interpretations and to allow assessment of the potential future sensitivity of this region to atmospheric vs oceanic influences on ice sheet mass balance.

On the Sabrina Coast continental shelf, three seismic megasequences were defined by Gulick et al. (2017). Chronostratigraphy of these megasequences was constrained by short piston cores targeting locations where seismic reflectors outcroped at the seafloor (Fernandez et al., 2018; Gulick et al., 2017; Montelli et al., 2020; Smith et al., 2019). Megasequence I (Ms-1) consists of low to moderate amplitude seaward-dipping discontinuous to continuous reflectors. Clinoforms imaged on the middle shelf indicate enhanced sediment supply to the margin, and terrestrial palynomorphs and benthic foraminifers from sediment cores that penetrated outcropping reflectors stratigraphically below and above the uppermost clinoform (Fig. 3.19) suggest a latest Paleocene and early to middle Eocene age range for the uppermost feature (Gulick et al., 2017; Smith et al., 2019). A series of discontinuous moderate- to high-amplitude reflectors at the top of Ms-I and the presence of lonestones in recovered sediments indicate that marine-terminating tidewater glaciers were present at the coastline by the early-to-middle Eocene, well before continental-scale ice sheets were thought to have been established. The top of Ms-1 is defined by a sharp erosive surface in the seismic data and is interpreted as an initial ice advance across the continental shelf, but is currently undated (Fig. 3.19; Gulick et al., 2017). Above this surface, Ms-II consists of strata with parallel high-amplitude reflectivity and prograding strata that are interrupted by ten undated erosive surfaces (Fig. 3.20; Gulick et al., 2017). Five of these surfaces display prominent tunnel valleys, providing evidence of a grounded ice sheet with a substantial meltwater channel system, that in turn implies climates warm enough for surface-derived subglacial meltwater (Fig. 3.20; Gulick et al., 2017; Ó Cofaigh, 1996). Thus, seismic and sedimentologic data from Ms-II suggest that meltwater-rich glaciers advanced and retreated across the shelf at least 10 times between the early to middle Eocene and the late Miocene. Ms-II is truncated by a regional

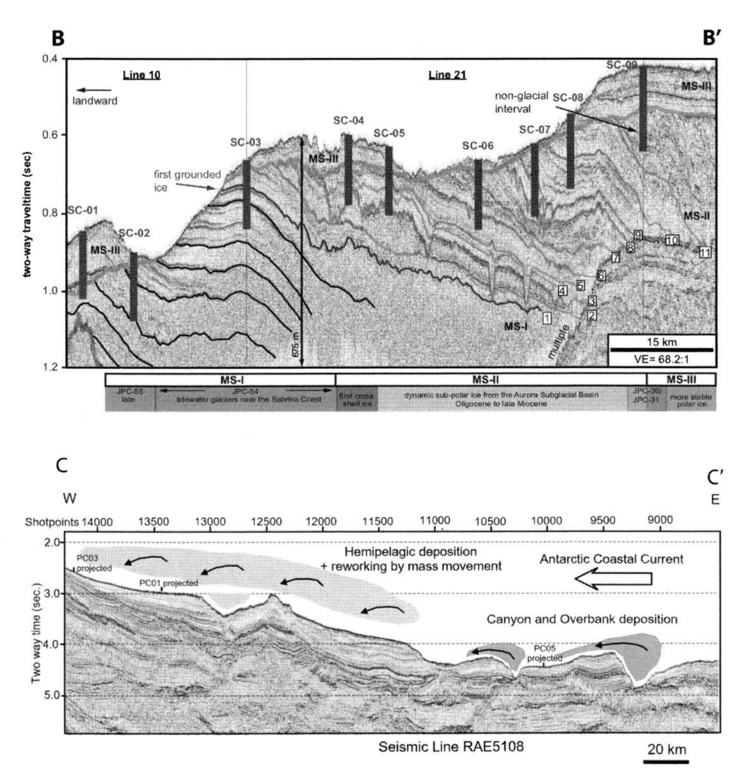

FIGURE 3.20 (Top) Composite seismic line with pre- to pro-glacial Megasequence I (Ms-I; black), meltwater-rich glacial Ms-II with grounded ice-eroded surfaces (initial, dark blue horizon; subsequent, grey) overlain by a non-glacial interval and regional unconformity (light blue horizon), and polar glacial Ms-III. Proposed drill sites are indicated in red (sites SC-01 to SC-09). (Bottom) Seismic line RAE5108 [profile shown in Fig. 3.19 as orange line (B–B′)]. Suspended sediment supplied by turbidity currents passing down continental rise canyon are entrained by alongslope currents building sediment ridges adjacent to the canyons. These channel levee and drifts provide ideal drilling target to obtain continuous offshore oceanographic records adjacent to the Sabrina Coast, complimenting the less continuous direct ice sheet records from proposed drilling on the continental shelf (see text and Fig. 3.19 for details). *(Top) Modified from Gulick, S.P.S., Shevenell, A.E., Montelli, A., Fernandez, R., Smith, C., Warny, S., et al., 2017. Initiation and long-term instability of the East Antarctic Ice Sheet. Nature 552, 225. (Bottom) From O'Brien, P.E., Post, A.L., Edwards, S., Martin, T., Caburlotto, A., Donda, F., et al., 2020. Continental slope and rise geomorphology seaward of the Totten Glacier, East Antarctica (112°E-122°E). Marine Geology 427, 106–221. Available from: https://doi.org/10.1016/j.margeo.2020.106221.*

landward-dipping unconformity. Diatom data from short sediment cores that penetrated the unconformity indicate diatomites of late Miocene to earliest Pliocene age overly the regional unconformity (Gulick et al., 2017; Montelli et al., 2020). Above the regional unconformity, late Miocene to recent glacial to glaciomarine sediments of variable (≤110 m) thickness are preserved in

Ms-III. Inter-till reflectors suggest glacial variability in the catchment since the late Miocene (Fernandez et al., 2018). However, subglacial meltwater features are absent, suggesting a late Miocene change in ice sheet thermal regime (sub-polar to polar) (Fig. 3.20; Gulick et al., 2017).

A tentative correlation between the shelf and rise seismic sequences (Fig. 3.19; schematic section of the margin, modified from Donda et al., 2020) suggests that the boundary between the pre-glacial and ice sheet growth phases is the seismic horizon WL3. This major erosional unconformity is inferred to be related to the onset of glaciation, thus marking the transition from the greenhouse to the ice-house era. There is evidence that WL3 is Late Eocene in age (Leitchenkov et al., 2015). Based on the tentative shelf-rise correlation, WL3 correlates to the boundary between Ms-I and Ms-II megasequences of Gulick et al. (2017), which has been interpreted as providing the first preserved evidence of grounded ice on the Sabrina Coast shelf in the Late Eocene (ca. 38 Ma).

The progressive increase of the sediment input from the continent as recorded on the continental rise deposits is consistent with the existence of a fluvial-like drainage system possibly related to the nucleation of the EAIS in the Gamburtsev Mountains (DeConto and Pollard, 2003) and the arrival of marine-terminating glaciers near the Sabrina Coast by the Mid-Eocene (Donda et al., 2020; Gulick et al., 2017). Since then, sediments on the continental rise record the widespread occurrence of gravity-driven processes and, in general, the dominance of downslope processes. This change should be related to the growth of a continental-scale ice sheet, similarly to Prydz Bay and western Wilkes Land (Donda et al., 2020). The occurrence of highly fluctuating, warm-based outlet glaciers dynamically responding to climate change (Levy et al., 2019) would have led to sediment delivery from the continent interior to the shelf edge, leading to the formation of turbidity flows down to the slope and rise area. This statement is supported by the findings from Gulick et al. (2017), based upon which Oligocene–Miocene sequences record 11 glacial advances and retreats from the Aurora Subglacial Basin. A remarkable change in the depositional style above unconformity WL7 is represented by a steep prograding wedge on the continental shelf and slope, produced during maximum advances of the ice sheet (Donda et al., 2007 and references therein). Based on a tentative slope-to-rise correlation, WL7 would correspond to the boundary between the Ms-II and Ms-III megasequences of Gulick et al. (2017), the latter being considered the expression of an expanded polar ice sheet that has occupied the Sabrina Coast continental shelf since the late Miocene. Mass-wasting deposits on the continental rise are interpreted to represent the distal record of glacial advance and collapse and slope failures, suggesting the occurrence of a dynamic ice sheet with a well-organised subglacial drainage system. Recurrent outburst flooding events may have occurred here even under polar conditions (Donda et al., 2020 and references therein).

Seismic surveys of the Sabrina Coast shelf were collected by the United States Antarctic Program cruise NBP14-02 (2014) and from the continental slope and rise by the international collaborative project IN2017-V01 (led by Australia, and involving the United States, Italy, Spain; 2017). Other regional seismic profiles collected in previous cruises include Russian, Australian and Japanese multichannel seismic profiles, available through the SDLS.

The Sabrina Coast shelf sequences are proposed as a future IODP drilling target (Proposal 931-Pre) that aims to investigate high-latitude greenhouse warmth, the timing of initial EAIS development, and its subsequent history and sensitivity to past climatic and oceanographic boundary conditions. Drilling targets are anticipated to enable scientists to date the arrival of the first ice to the shelf, constrain the timing of 10 unconformities present in Ms-II, as well as to refine the dating of the regional unconformity. The scientific rationale of proposed future drilling in this frontier region are broad ranging and include:

Investigate how Antarctic terrestrial and proximal marine environments contributed and/or responded to Cretaceous Paleogene greenhouse climate. Eocene strata will provide comparison with other sectors of East Antarctica already obtained for this time period (c.f. Contreras et al., 2014; Pross et al., 2012), but may also capture new records of hyperthermal events, such as the Paleocene/Eocene Thermal Maximum (PETM), which are proposed to have resulted from high-latitude melting of permafrost (DeConto et al., 2012).

Identify the nature of pre-Eocene glaciation. This region is ideally suited to obtain Cretaceous to Paleogene coastal plain sediment derived from the catchment of East Antarctica's Gamburtsev Mountains. This is the region where East Antarctica's ice caps were first thought to nucleate and flow towards the coast, eventually coalescing as continental-scale ice sheets (Bo et al., 2009; DeConto and Pollard, 2003), but the timing of initial terrestrial alpine glaciation remains unknown (Miller et al., 2005, 2020).

Characterise the Paleogene greenhouse to Neogene icehouse transition in East Antarctica. A key objective of future drilling would be to constrain the timing of the first regional marine-terminating glaciers and subsequent ice advance across the shelf (Gulick et al., 2017). Such data, when combined with modelling experiments, can help elucidate if the first marine ice advances out of the ASB were synchronous to, or pre/post-dated, similar advances in the other sectors of Antarctica.

Constrain the frequency and style of Oligocene to late Miocene marine ice sheet and climate variability. This would focus on dating the 10 imaged unconformities in Ms-II to assess their relationship with isotopic or sea level excursions noted in far-field geologic records. Drilling of these sequences would enable assessment of how regional glacial systems responded during warmer-than-present Eocene to Miocene climates (e.g., Middle Eocene Climate Optimum, MECO). Drilling would also help constrain the paleoenvironmental conditions during Miocene shelf progradation, and how this relates to similar shifts in sedimentation around the continental margin (e.g., Prydz Bay and the Ross Sea; Cooper and O'Brien, 2004; Hambrey and McKelvey, 2000; McKay et al., 2009, 2019). This interval is of interest as seismic lines indicate large meltwater drainage systems existed, helping to identify thresholds for surface melt processes around the EAIS margin (Levy et al., 2019; Lewis et al., 2006; Liebrand et al., 2016; Shevenell et al., 2008).

Characterise the Aurora Subglacial Basin (ASB) catchment response to Pliocene warmth. This would complement significantly the records of substantial glacial retreat that occurred in the Wilkes Basin, Lambert and Ross Sea catchments during the warm Pliocene. Notably, the ASB is potentially the last marine-based catchment to deglaciate under scenarios of oceanic warming (Golledge et al., 2017). Therefore geologic evidence for past regional retreat will help identify climatic and oceanic thresholds for marine-based ice sustainability.

Seismic surveys of the Sabrina Coast slope and rise (Fig. 3.20) highlight the potential to obtain Miocene to Pleistocene geologic records to assess oceanic drivers and responses to EAIS variability (Donda et al., 2020; Leitchenkov et al., 2015; O'Brien et al., 2020; Post et al., 2020). Levee deposits on the banks of large channel systems offshore of the Sabrina Coast are proposed as drilling targets. These types of settings were demonstrated during both IODP Expeditions 318 (Escutia et al., 2011a; see Wilkes Land margin discussion in this chapter) and 374 (McKay et al., 2019) to provide high-resolution archives of glacially-influenced sedimentation (McKay et al., 2019; O'Brien et al., 2020; Post et al., 2020) (Fig. 3.20). IODP pre-proposal 1002 was submitted in 2021 to drill sites on the continental rise in this sector, based on a recent Australian RV Investigator cruise which collected piston cores, entitled 'Totten Glacier Climate Vulnerability under varying Neogene climate conditions: lessons for East Antarctica Ice Sheet climate sensitivity' (Table 3.1).

While drilling on the continental shelf requires a mission-specific style drilling platform to capture the critically important direct record of ice sheet variability in this sector of the EAIS, a standalone, but complimentary proposal would use a ship-based drilling system, such as the JOIDES resolution to obtain records from the continental slope/rise. The integration of data from these two continental shelf and slope to rise drilling proposals is essential to fully capture the range of complex ocean-ice-biogeochemical processes that operated in this climatically sensitive region of East Antarctica over the past 66 million years (Adusumilli et al., 2020; Gulick et al., 2017; Rintoul et al., 2016).

3.4.7.4 Wilkes Land margin and Georges V Land

The termini of large Antarctic outlet glaciers draining the Wilkes Subglacial Basin are located along the George V Land and the eastern sector of the Wilkes Land coasts, including Adélie Land (Fig. 3.21). Topographic

TABLE 3.1 List of past drilling projects on the Antarctic continental margin (excluding expeditions to the north of the Antarctic polar front), and proposed expeditions discussed by PRAMSO under various stages of development.

Drilling project	Region	Sites	Year(s) drilled	Age range of primary targets	Initial results reference
Dry Valley Drilling Project (DVDP)	Ross Sea	DVDP 1–15	1972–1975	Late Miocene-Quaternary	McGinnis (1981)
DSDP Leg 28	Ross Sea/Wilkes Land	DSDP 264–274	1973	Oligocene-Quaternary	Hayes et al. (1975)
DSDP Leg 35	Bellinghausen/ Amundsen Sea	DSDP 322–325	1974	Late Oligocene-Quaternary (thin Cretaceous)	Hollister et al. (1976)
MSSTS	Ross Sea	MSSTS-1	1979	Late Oligocene to Early Miocene	Barrett (1986)
CIROS	Ross Sea	CIROS-1,2	1984–1986	Eocene to Quaternary	Barrett (1989)
ODP 113	Weddell Sea	ODP 689–697	1987	Cretaceous to Quaternary	Barker and Kennett (1988)
ODP119	Prydz Bay	ODP 736–746	1987–1988	Cretaceous to Quaternary	Barron and Larsen (1989)
Cape Roberts Project	Ross Sea	CRP-1, CRP-2/2A, CRP-3	1997–1999	Eocene to Miocene, (thin Quaternary)	Barrett et al. (1998, 2000), Fielding et al. (1999)
ODP 178	Bellingshausen Sea/Antarctic Peninsula	1095–1103	1998	Late Miocene-Quaternary (including high-res. Holocene)	Barker et al. (1999)
ODP 188	Prydz Bay	ODP 1165–1167	2000	Late Eocene-Quaternary (snapshot Cretaceous)	O'Brien et al. (2001)

(*Continued*)

TABLE 3.1 (Continued)

Drilling project	Region	Sites	Year(s) drilled	Age range of primary targets	Initial results reference
ANDRILL	Ross Sea	AND-1, AND-2	2006–2007	Early Miocene to Quaternary	Naish et al. (2007), Florindo et al. (2008)
SHALDRIL	Weddell Sea, Antarctic Peninsula	NBP0602A-01 to NBP0602A-12	2005–2006	Late Eocene to Holocene (including high-res. Holocene)	Anderson (2006)
IODP Expedition 318	Wilkes Land	IODP U1355-U1361	2010	Middle Eocene to Holocene (including high-res. Holocene)	Escutia et al. (2011a)
PS104/MeBo	Amundsen Sea	PS104-006, -009, -020, -021, -024, -038, -040, -041, -042	2017	Cretaceous to Holocene	Gohl et al. (2017)
IODP Expedition 374	Ross Sea	IODP U1521–U1525	2018	Early Miocene-Quaternary	McKay et al. (2019)
IODP Expedition 379	Amundsen Sea	IODP U1532–U1533	2019	Late Miocene-Quaternary	Gohl et al. (2021)
IODP Expedition 382	Scotia Sea	IODP U1534–U1538	2019	Late Miocene-Quaternary	Weber et al. (2019)

PROPOSED EXPEDITIONS	Region	Sites	Proposal status	Age Range	Lead proponent(s)
IODP Proposal 732 Full	Bellinghausen Sea/Antarctic Peninsula	Eight primary piston core sites (PEN1–5, BEL1–3)	Approved by IODP – awaiting scheduling	Miocene-Pleistocene	James Channell, Rob Larter
IODP 813 Full (Expedition 373)	Wilkes Land	Sixteen sites (seabed drill) to 80 m penetration	Approved by IODP – awaiting scheduling	Eocene to Pliocene	Trevor Williams, Carlota Escutia
IODP 931-Pre (EAIS evolution)	Sabrina Coast	Six sites (mission-specific platform) to 80 m penetration	Assessed by IODP Science Evaluation Panel – invited to submit full proposal	Paleocene to Quaternary	Amelia Shevenell, Sean Gulick
IODP 1002-Pre-(Totten Glacier Climate Vulnerability)	Sabrina Coast	Five sites using standard ship-based	Submitted to IODP – April 2021	Miocene-Pleistocene	Bradley Opdyke, Federica Donda
IODP Proposal 953-Pre	Australian-Antarctic Rift-Drift	Four sites using standard ship-based	Assessed by IODP Science Evaluation Panel – invited to submit full proposal	Cretaceous to Quaternary	Peter Bijl, Isabel Sauermilch
IODP 998-pre (Ross Sea)	Ross Sea	Four sites using standard ship-based	Submitted to IODP – October 2020	Cretaceous to Early Miocene	Robert Mckay, Laura De Santis
SWAIS-2C	Ross Sea (Siple Coast)	2(+) Sites using custom-designed sub-ice shelf drill	Approved – to be drilled 2021/22	Pliocene to Quaternary	Richard Levy, Molly Patterson
Ekström Ice Shelf	Dronning Maud Land/Weddell Sea	3(+) sites using custom-designed sub-ice shelf drill	Proposal in development	Cretaceous to Early Miocene	Gerhard Kuhn

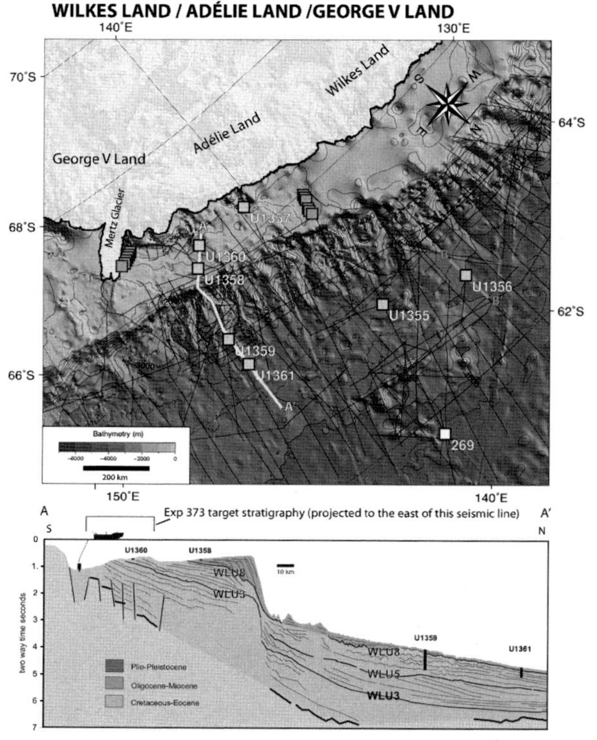

FIGURE 3.21 Overview map of Eastern Wilkes Land sector with existing seismic lines (white lines), drill sites and topographic features mentioned in text.Bathymetry is from Bathymetric Chart of the Southern Ocean (Arndt et al., 2013) and subglacial topography from BedMachine Antarctica dataset (Morlighem et al., 2020). DSDP Leg 28 sites are shown as white squares (site 269), and IODP Expedition 318 sites as yellow squares (sites U1355 to U1361). Future sites (IODP Proposal Full-813/Exp 373) system (discussed in text) are shown as green squares (unlabelled site numbers). These future sites are proposed to be collected via a seafloor drilling system with shallow (~80 m) penetration, and a transect of sites is intended to recover key snapshots of this stratigraphic framework. Seismic tracks are from the SCAR Antarctic Seismic Data Library System (SDLS) and additional survey information. Track lines of the seismic profiles shown in bottom panel and Fig. 3.22 are marked with bold yellow (A–A′), green (B–B′) and red (C–C′) lines, respectively. (Bottom) Wilkes Land seismic stratigraphy showing major unconformities (WLU3–WLU8) and associated shifts in the geometry of sedimentary packages on the continental shelf, location of key IODP Exp 318 sites. Line profile in shown in upper panel, and constructed from seismic lines ATC82-107, GA2904 and GA4201 presented by Escutia et al. (2011a).

reconstructions in this region using satellite radar altimetry and airborne radio-echo sounding surveys show fast flowing glaciers within the Wilkes Subglacial Basin are located at the mouth of deep submarine basins (Ferraccioli et al., 2009; Fretwell et al., 2013; Morlighem et al., 2020; Paxman et al., 2019, 2018) (Fig. 3.2). In these basins, the bed from the periphery of the ice sheet is inland-sloping (Fretwell et al., 2013; Morlighem et al., 2020) suggesting their susceptibility to MISI if warmer Circumpolar Deep Water access the base of the glaciers. Moreover, ice sheet dynamic simulations show the existence of a small ice plug on a ridge at the margin of the Wilkes Subglacial Basin (Mengel and Levermann, 2014). If the ice plug was removed by melting it would contribute less than 80 mm of global sea-level rise but most importantly, it could destabilise the regional ice flow leading to a self-sustained discharge of the entire basin and a global sea-level rise of 3–4 m (Mengel and Levermann, 2014). However, a recent ice sheet model sensitivity test investigating EAIS ice drainage basin responses to atmospheric and oceanic warming, show the Wilkes Land sector is likely to be more sensitive to atmospheric warming than other EAIS basins, and relatively insensitive to ocean warming alone (Golledge et al., 2017). Consequently, determining the past response of this sector to both atmospheric and oceanic change is critical to ground-truth these models and help project the future response of the Wilkes Subglacial Basin drainage basin to future change.

Until 2010, our understanding of past EAIS dynamics in this sector of East Antarctica was based on a regional seismic stratigraphic framework based on the available network of seismic reflection profiles in the area (e.g., see Cooper et al., 2008, and references therein for a review). The Deep Sea Drilling Project (DSDP) Leg 28, intermittently cored Site 269 (4285 m water depth) to a depth of 958 m below sea floor (mbsf) and obtained sediments spanning the Oligocene to recent (Hayes et al., 1975). The cores documented extensive Antarctic glaciation beginning at least by Oligocene to early Miocene times and indicated that water temperatures were cool to temperate in the late Oligocene and early Miocene and then cooled during the Neogene, presumably as glaciation intensified (Hayes et al., 1975). Unfortunately, spot-coring, low recoveries (42%), poor age control and basement highs surrounding the basin where Site 269 lies, causing thinning and/or onlapping of the seismic units, prevented confidently extending the results from coring regionally. It was not until 2010, when IODP Expedition 318 drilled seven sites in a latitudinal transect from the continental shelf to the lower continental rise providing the first robust geological constrains for Cenozoic strata offshore the Wilkes Subglacial Basin (Escutia et al., 2011a). Continental shelf sites targeted strata across seismic unconformities interpreted to contain the record of the transition to a glaciated EAIS, and the grounding line advances and retreats across the continental shelf that followed (e.g., De Santis et al., 2003; Eittreim et al., 1995; Escutia et al., 2005; Tanahashi et al., 1994; Wannesson et al., 1985), in addition to the oceanographic conditions that prevailed with each ice configuration. Continental rise sites targeted levee systems developed on the flank of deep-sea channels, which serve as conduits for turbidity flows and for cascading Adélie Land Bottom Waters, respectively (Bindoff et al., 2000; Close, 2010; Eittreim et al., 1972; Escutia et al., 2000; Fukamachi et al., 2000). Furthermore, along-slope bottom currents had been reported to interact with gravity flows forming contourite ridges or mixed turbidite-contourite levee deposits (e.g., Caburlotto et al., 2010, 2006; Close et al., 2009; Donda et al., 2003; Escutia et al., 2003, 2002). Thus, these levee deposits are ideal targets for high-resolution archives of distal ice sheet dynamics and oceanographic change.

To provide for an account of the stratigraphic control achieved by Expedition 318, we adopt the nomenclature of unconformities and seismic units in the region of drilling by De Santis et al. (2003) and Escutia et al. (2005), with eight major regional unconformities (WL-U1-WL-U8) bounding seismic units WL-S2 to WL-S9 (Fig. 3.22). IODP data were also used for improvement of regional seismic stratigraphy (Leitchenkov et al., 2015). In addition, we refer the reader to previous papers that provide for a detailed description of seismic attributes of the seismic sequences, the processes

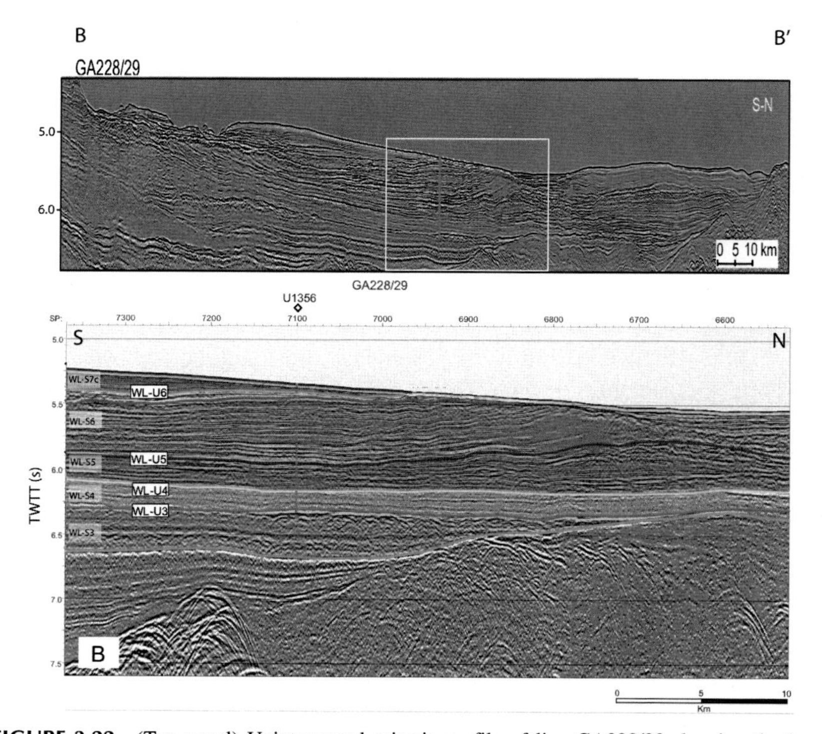

FIGURE 3.22 (Top panel) Uninterpreted seismic profile of line GA228/29 showing the location and the depth reached during drilling of Site U1356. White box shows a detailed seismic line in (Bottom panel) Interpreted seismic profile showing unconformities and seismic units defined in the region of the eastern Wilkes Land and Georges V margins. Seismic line locations are shown in Fig. 3.21. *(B) Modified from Escutia, C., Brinkhuis, H., Klaus, A., 2011a. Cenozoic East Antarctic ice sheet evolution from Wilkes Land margin sediments. Proceedings of the Integrated Ocean Drilling Program 318.*

involved in their development and the inferred record of glaciation before Expedition 318 drilling (e.g., Close et al., 2009; De Santis et al., 2003; Donda et al., 2003; Eittreim et al., 1995; Escutia et al., 2005, 2002, 2000; Tanahashi et al., 1994; Wannesson et al., 1985). Furthermore, a recent paper by Sauermilch et al. (2019) provides a regional comparison between horizons and sedimentary units, with inferred and assigned ages ranging from the Cretaceous to the Miocene, along the whole Wilkes Land and Australian margins.

The oldest sediments drilled during Expedition 318 were recovered from Site U1356 located in the lowermost continental rise (3992 m water depth). The site drilled on the flank of a levee deposit to a depth of 1006.4 mbsf, and recovered sediments below and above unconformity WL-U3 that was reached at 895 mbsf (Escutia et al., 2011a; Tauxe et al., 2012) (Figs. 3.21 and 3.22). Sediments below 1006 mbsf, and their bonding unconformities remain unsampled to date, although drilling at U1356 suggests a late Cretaceous-Paleocene age for much of these strata (Escutia et al., 2014; Leitchenkov et al., 2015). Sauermilch et al. (2019) provide a summary of previous interpretations for these units and, based on correlations between the Antarctic and the Australian margins, infer ages spanning from 83 to 65 Ma in the eastern Wilkes Land and George V Land margins.

The ~100 m of sediments below unconformity WL-U3 recovered from Site U1356 were dated early to middle Eocene (54–46 Ma), providing age and paleoenvironmental constrains for the upper part of seismic unit WL-S3 (Escutia et al., 2011a; Tauxe et al., 2012) (Fig. 3.22). Paleotopographic reconstructions show the Wilkes Subglacial Basin at that time to be occupied by low-lying lands (Paxman et al., 2019, 2018; Wilson and Luyendyk, 2009; Wilson et al., 2012), which based on palynomorph data from Site U1356 were covered by diverse near-tropical forests at least from 54 to 51.9 Ma (Contreras et al., 2014, 2013; Pross et al., 2012). SSTs are estimated to have ranged between 31°C (±2.5°C) and 24°C (±4.0°C), depending on whether a TEX^{H}_{86} or TEX^{L}_{86} calibration is used, respectively (Kim et al., 2010; Schouten et al., 2013). Continental temperatures indicate mean summer air temperatures in the Antarctic coastal regions to be 20°C–23°C (±5.0°C) (Contreras et al., 2013; Peterse et al., 2012; Pross et al., 2012). This record is interrupted by a hiatus spanning from ~52 to 51 Ma (Bijl et al., 2013; Tauxe et al., 2012). Middle Eocene sediments overlying the hiatus record the dominance of a temperate rainforest biome extending into the coastal regions and a decline of around 2°C–4°C in temperature since the early Eocene (Contreras et al., 2013; Pross et al., 2012). This cooling has been attributed to the onset of westward throughflow of the Antarctic Counter Current from the southwest Pacific Ocean into the southern Australo-Antarctic Gulf at ~49–50 Ma, based on the appearance of dominant endemic Antarctic dinocysts on the Wilkes Land Margin from 50 Ma onwards (Bijl et al., 2013). It has been suggested that this throughflow could have been favoured by a gradual drowning of continental blocks in the southern part of the Tasmanian Gateway that resulted in accelerated rifting during the early-to-middle Eocene transition (52–48 Ma) (Bijl et al., 2013; Close et al., 2009; Stagg et al., 2004).

The upper middle Eocene and the late Eocene are conspicuously missing at Site U1356 in a hiatus spanning from ~46 to 33.6 Ma, which is associated with unconformity WL-U3, (Escutia et al., 2011a; Tauxe et al., 2012). Unconformity WL-U3 was previously interpreted as an erosional surface based on reflector truncation and downlap surfaces on the continental shelf (De Santis et al., 2003; Eittreim et al., 1995). The erosion was inferred to be related to the development of the first continental ice sheet at ~34 Ma (e.g., De Santis et al., 2003; Escutia et al., 2005), at ~40 Ma (Eittreim et al., 1995), or to erosion by strong bottom currents related to an increase in seafloor spreading rates and margin subsidence at ~45 Ma (Close et al., 2007). More recently, non-deposition rather than erosion has been proposed as the dominant driver for the formation of WL-U3 based on stratigraphic similarities between the Antarctic and Australian conjugate slopes (Sauermilch et al., 2019).

Despite the distal setting of Site U1356, earliest Oligocene (33.6 Ma) sediments within seismic unit WL-S4 immediately above unconformity WL-U3 unequivocally reflect icehouse environments (Escutia et al., 2011a; Escutia and Brinkhuis, 2014). Among the evidence, the presence of dropstones (up to boulder-sized) in early Oligocene sediments indicates iceberg activity on the margin. A regime shift in zooplankton–phytoplankton interactions and community structure is interpreted as the appearance of eutrophic and seasonally productive environments on the Antarctic margin (Houben et al., 2013). In addition, sediments record a distinct shift from to smectite- and kaolinite-dominated clays below WL-U3, suggestive of chemical weathering under warm and humid climates to illite- and chlorite-dominated assemblages above the unconformity pointing to much colder/drier physical weathering regimes (Escutia et al., 2011a; Passchier et al., 2013). Coeval earliest Oligocene sediments were also recovered from Site U1360, drilled on the continental shelf to a total depth of 70 mbsf (Escutia et al., 2011a; Houben et al., 2013). Although recovery was poor, partly a function of the shallow penetration depth that prevented sufficient weight being placed on the drill bit, sediments recovered from Site U1360 were interpreted as ice-proximal to ice-distal glaciomarine deposits, similar to those from the Oligocene and Miocene strata of the Ross Sea (Kulhanek et al., 2019; Naish et al., 2001; Powell and Cooper, 2002).

These results point to processes related with ice sheet growth, including crustal deformation (i.e., glacial isostatic adjustment, GIA) and gravitational perturbations in addition to tectonic subsidence resulting in a complex spatial pattern of relative sea-level change around the Antarctic margin, as the principal mechanisms in the development of unconformity WL-U3 (Escutia and Brinkhuis, 2014; Stocchi et al., 2013). The continuous presence of reworked middle-late Eocene dinocyst species in the overlying Oligocene sediments (Bijl et al., 2018, 2013; Houben et al., 2013) suggests unabated submarine erosion of the Antarctic shelf as reported by Eittreim et al. (1995), rather than non-deposition (Sauermilch et al., 2019). Eocene material in dredge samples offshore of the Mertz Glacier (Truswell, 1982) support this interpretation.

Sediments recovered from seismic units WL-S4 and WL-S5 are dated to the Oligocene and earliest Miocene (Escutia et al., 2011a; Tauxe et al., 2012) (Fig. 3.22). Dinoflagellate cyst records from Sites U1356 and U1360, combined with those from other locations across the Antarctic margin, suggest an abrupt shift to high seasonal primary productivity associated with the development of seasonal sea ice at Oi-1 times (33.6 Ma, Houben et al., 2013). This is in agreement with numerical climate models simulations indicating that sea-ice formation along Antarctic margins may have followed full-scale Antarctic glaciation (DeConto et al., 2007). However, sea ice—related dinocyst species suggest the occurrence of sea ice near-site U1356 only during the first 1.5 million years (33.6—32.1 Ma) of the Oligocene (Bijl et al., 2013). For the remainder of the Oligocene dinocyst assemblages resemble present-day open-ocean north of the sea-ice edge, with episodic dominance of temperate species similar to those found in the present-day subtropical front (Bijl et al., 2013). The prevalence of oligotrophic and temperate surface waters over the site notably during interglacial times is supported by the presence of carbonate-rich intervals at Site U1356 and Site 269, along with sedimentological and geochemical evidence for considerable latitudinal frontal migrations over the course of Oligocene glacial-interglacial cycles (Bijl et al., 2013; Evangelinos et al., 2020; Salabarnada et al., 2018) that at least during the late Oligocene (within WL-S5) are paced by obliquity (Salabarnada et al., 2018). The oligotrophic, temperate dinocysts and nannofossil remains suggest fundamentally warmer surface water conditions than today. This is supported by SST reconstructions at Site U1356 averaging 17°C albeit highly variable and with estimated temperature differences of 1.5°C—3.1°C between glacials and interglacials cycles (Hartman et al., 2018). The records show a long-term temperature increase towards 30.5 Ma, followed by a minimum around 27 Ma and an optimum around 25 Ma (Hartman et al., 2018).

A thick succession of Mass Transport Deposits (MTDs) with interbedded calcareous, bioturbated and laminated claystones characterise the latest late Oligocene record (24.76—23.23 Ma) at Site U1356 (Escutia et al., 2011a,b). The stacked MTDs succession has been interpreted to record repeated ice sheet advances across the continental shelf during the cooling trend leading to the Mi-1 event glaciation (Escutia and Brinkhuis, 2014; Escutia et al., 2011a). In situ deposits interbedded between the U1356 MTDs indeed record a cooling trend in SST towards the end of the Oligocene, following the climatic optimum at 25 Ma (Hartman et al., 2018). DSDP Site 269 sediments between ~24 and 23.6 Ma provide a window to oceanic conditions at this time characterised by upwelling, high-nutrient waters and TEX_{86}-derived SST between 9°C and 13.5°C (Evangelinos et al., 2020). In addition, two expansions of proto-Circumpolar Deep Water closer to the Wilkes Land margin at ~23.6 and ~23.23 Ma are recorded by higher Ca XRF values, better preservation of calcareous microfossils, higher Br/Ti ratios, high amounts of dinocyst that characterise temperate and oligotrophic waters, and high SSTs (from 11.5°C to 12.9°C) (Evangelinos et al., 2020).

Site U1356 sediments recovered above unconformity WL-U5 and within seismic unit WL-S6 record three important events in the evolution of Earth climates: the Oligocene-Miocene transition at 23.03 Ma, the MCO (17–15 Ma) and the MMCT (14.2–13.8 Ma) (Fig. 3.3). Early Miocene deposits record similar conditions to those described for the end of the Oligocene characterised by a relative lack of ice rafted debris, and temperate pollen assemblages indicative of mean summer on-land temperatures of >10°C (Sangiorgi et al., 2018). The MCO is characterised by inland retreat of the ice sheet and temperate vegetation (Sangiorgi et al., 2018). SST records show that waters were warm (i.e., 11.2°C–16.6°C ±2.8°C calibration error), lacked a strong sea ice component and were relatively low in nutrients compared to today (Sangiorgi et al., 2018). Specifically, dinocyst assemblages resemble those found today in the Pacific sector of the Southern Ocean at around or north of the Subtropical Fronts, where sea-surface

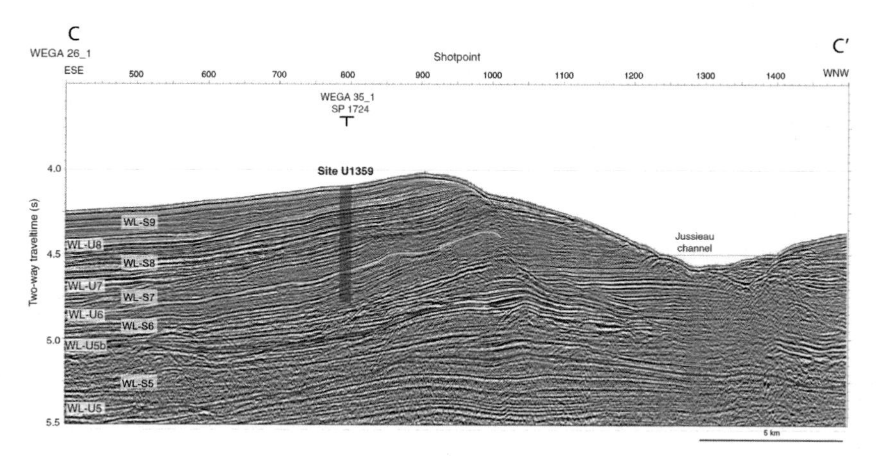

FIGURE 3.23 Seismic line WEGA 26_1 across IODP Site U1359 showing regional unconformities and seismic units defined in the region of the eastern Wilkes Land and Georges V margins. Seismic line locations are shown in Fig. 3.21. *Modified from Escutia, C., Brinkhuis, H., Klaus, A., 2011a. Cenozoic East Antarctic ice sheet evolution from Wilkes Land margin sediments. Proceedings of the Integrated Ocean Drilling Program 318.*

temperatures vary between 8°C and 17°C (Prebble et al., 2013). After the MCO, the MMCT expansion of marine-based ice sheets occurred, recorded by an interval of increased gravel and pebble-sized ice rafted debris after 14.0 to 13.8 Ma (Escutia et al., 2011a; Pierce et al., 2017; Sangiorgi et al., 2018), and dinocyst taxa (i.e., Selenopemphix Antarctica) exclusively known from the present-day seasonal sea ice zone south of the Antarctic Polar Front (Prebble et al., 2013).

Sediments from seismic unit WL-S9 are dated latest-late Miocene to recent (Escutia et al., 2011a; Tauxe et al., 2012) (Fig. 3.23), encompassing the Pliocene and the Pleistocene epochs when the Earth transitioned from having only ice sheets in Antarctica to having ice sheets occupying both polar regions. Sediments recovered from the continental rise at Sites U1359 and U1361 contain a distal record of ice sheet dynamics during Pliocene-Pleistocene glacial and interglacial cycles (Escutia and Brinkhuis, 2014; Escutia et al., 2011a). Iceberg debris accumulation in sediments from Sites U1359 and 1361 provide a record of ice sheet growth and decay that is orbitally paced (Hansen et al., 2015; Patterson et al., 2014). High fidelity spectral analyses in sediments from Site U1361 show that during the warm Pliocene period (4.3 to 3.5 Ma), glacial-interglacial cycles occur with a periodicity of about 40,000 years (i.e., paced by obliquity) with maximum iceberg debris accumulation associated with enhanced calving of icebergs during ice-sheet margin retreat (Patterson et al., 2014). Evidence for repeated retreat of the marine-based EAIS into the Wilkes Subglacial Basin between 5 and 3 Ma is provided by the geochemical provenance of fine-grained detrital material recovered from this deep-water site (Bertram et al., 2018; Cook et al., 2013). Shifts in sediment provenance were paralleled by increases in marine productivity, while the onset of such changes was marked by peaks in ice-rafted debris mass accumulation rates (Bertram et al., 2018). This argues against a switch in sediment delivery between ice streams and instead suggests that deglacial warming triggered increased rates of iceberg calving, followed by inland retreat of the ice margin (Bertram et al., 2018). Early Pliocene ice sheet retreats from the outer continental shelf are also recorded in coeval sediments recovered from Site U1358, indicating repeated times of grounding line migration across the outer continental shelf during past glacials, with interglacial times characterised by open marine conditions with reduced summer sea ice (Orejola and Passchier, 2014; Reinardy et al., 2015).

After 3.3 Ma, the pacing of the glacial-interglacial cycles is in the 100,000-year and 20,000-year frequencies (i.e., eccentricity and precession, respectively) (Patterson et al., 2014). These authors suggest that the Southern Ocean cooled after 3.5 Ma, which is supported by a shift in surface water productivity recorded by a decrease in biogenic opal content at Site U1359 after 3.7 Ma (Khim et al., 2017). As cooling progressed, the development of a perennial sea-ice field limited the oceanic forcing of the EAIS driven by obliquity, leaving atmospheric forcing driven by precession to regulate both sea ice and ice sheet dynamics in this region (Patterson et al., 2014). Despite this shift in orbital pacing, the marine margin of EAIS of Wilkes Land retreated inland from its present-day margin throughout Late Pliocene and Pleistocene interglacials. For example, a temporary southward migration of the Antarctic Polar Front is recorded by the presence of subantarctic diatoms at Site U1361 during marine isotope stage KM3 (3.17–3.15 Ma) (Taylor-Silva and Riesselman, 2018). In addition, marine sedimentological and geochemical records from Site U1361 provide evidence for ice margin retreat in the vicinity of the Wilkes Subglacial Basin Basin during MIS 5 (130–82 ka), MIS 9 (337 ka) and MIS 11 (424 ka) (Jimenez-Espejo et al., 2020; Wilson et al., 2018). At Site U1359, sediments sourced from the east Antarctic Craton and the Ross Orogen are interpreted as an indication for retreats of the East Antarctic ice sheet into the Wilkes Subglacial Basin (Pant et al., 2013). Furthermore, based on studies on the accumulation of ^{234}U in carbonates from inland glacial erratics, Blackburn et al. (2020) suggest that during the warmer and longer MIS11 interglacial, the EAIS grounding line may have retreated about 700 km into the Wilkes Subglacial Basin from the current coastline, which assuming current ice volumes would have contributed about 3–4 m to global sea levels from the Wilkes Subglacial Basin alone. However, inferred higher-than-modern temperatures during the shorter MIS5 and MIS9 interglacial periods did not result in a comparable retreat.

Despite the wealth of information resulting from IODP Expedition 318, it did not recover the 12-Myr interval between 46 Ma, when temperate rainforests covered the coastal areas of Wilkes Land, and 34 Ma, the start of major Antarctic glaciation at the Eocene-Oligocene boundary (e.g., Escutia et al., 2019), an interval when CO_2 fell from approx. 1000 to 600 ppm (Fig. 3.3). The evolution of environments and vegetation during the intervening 12 million years is unknown for mainland Antarctica. This interval is likely to be present in strata on the continental shelf, accessible at shallow drilling depths in tilted and truncated strata, based on the presence of reworked late Eocene sediments at Site U1356, of middle and late Eocene dinocysts in dredged sedimentary rocks from the continental shelf, seismic stratigraphy and lowermost Oligocene sediments cored at Site U1360 (Escutia and Brinkhuis, 2014). Therefore, IODP Expedition 373 (Proposal 813) aims to investigate the development of Antarctic climate over the Eocene, using shallow marine sediment cores from the George V Land shelf. Currently, the most southerly Antarctic Eocene record is from Seymour Island at the northern end of the Antarctic Peninsula; while Eocene data from the Ross Sea are extremely limited (e.g., Galeotti et al., 2016). This mission-specific platform expedition is awaiting rescheduling after initially being planned for early 2018 and intends to use seafloor drill system to obtain a series of shallow drill cores ($\sim$80 m deep) (Table 3.1, Fig. 3.21).

IODP Expedition 373 aims to sample a wide range of Antarctic climates, vegetation and paleoceanography between the sub-tropical and glaciated extremes to provide tests for climate models and a better understanding of how greenhouse climates operate (e.g., Baatsen et al., 2020). High-latitude paleotemperatures are warmer than the current generation of climate models can reproduce, implying that models are missing a key mechanism that enhances climate sensitivity and polar amplification across a range of geographies and greenhouse gas concentrations (Huber and Caballero, 2011). In addition to investigating the basic operation of high-latitude greenhouse climates, the proposed drill sites will allow us to examine the following four questions:

1. *What is the relation of Antarctic climate to tectonic opening of the Drake and Tasman ocean gateways?* The possible dates for full Drake passage opening range from 49 Ma to as young as 12 Ma (Dalziel et al., 2013; Scher and Martin, 2006), and Tasman opening has been proposed to be between 50 Ma (Bijl et al., 2013) and 30 Ma (Scher et al., 2015), or close to the E/O boundary itself, 34 Ma.
2. *Were there short-lived glaciations in the late Eocene prior to the main initiation of continent-wide glaciation* (e.g., Scher et al., 2014)? So far, no direct evidence has been found. Such insights are needed to understand the ocean–atmosphere–cryosphere interactions that set the stability threshold for ice on Antarctica.
3. *Do early Eocene sediments hold evidence for the polar permafrost hypothesis for Eocene hyperthermal events* (DeConto et al., 2012), *where vast amounts of carbon are proposed to have been released from Antarctic permafrost?*
4. *What is the nature of the E/O boundary on the George V Land shelf?* Depending on whether or not the sites were situated in a foreland basin caused by glacio-isostatic adjustment, the boundary may be erosive or there may be continuous sediment accumulation. This provides a test of glacio-isostatic adjustment models (e.g., Stocchi et al., 2013) that are critical to understanding solid earth–ice sheet interactions and therefore the potential contribution to sea level change (Whitehouse et al., 2019).

Another high priority for this region is to obtain direct records of early Oligocene to the Pleistocene ice sheet dynamics, paleoclimate and palaeoceanography from the continental shelf region and on-land (i.e., subglacial drilling, cosmogenic nuclide dating of exposed strata). Some sites in Expedition 373 also target Oligocene ice sheet–proximal sediments to understand past ice sheet extent and climate during this time. They can be integrated into coupled ice sheet-climate models for improved projections of sea level change. Alternate sites target younger strata. Younger Neogene records are also critical to relate ice behaviour and ice-proximal ocean configuration to open ocean and paleoclimate conditions recorded in IODP Expedition 318 sediments and other Southern Ocean and lower latitude records. In terms of deeper water sites across the Australian-Antarctic basin,

IODP Proposal Pre-953 (Table 3.1) targets the Australian-Antarctic rift-drift transition and development of the ACC. While this pre-proposal aims to reconstruct the early Cenozoic history of the ACC onset and the tectonic separation of Australia and Antarctica, and is therefore largely outside the scope of this chapter on continental margin stratigraphy, one site is proposed on the Antarctic continental rise. This site is intended to reveal the subsidence history conjugate to the Australian margin and will provide a record of sea surface temperature gradients, polar front migrations, and the glacial and terrestrial history of the Wilkes Land margin over the Oligocene to Neogene.

3.5 Summary, future directions and challenges

Since the first edition of this book (Florindo and Siegert, 2008), the SCAR *Past Antarctic Ice Sheet Dynamics* (PAIS) research programme and its two key subcommittees: *Paleoclimate Records from the Antarctic Margin and Southern Ocean* (PRAMSO) and *Palaeotopographic-Palaeobathymetric Reconstructions* have made significant advances in constraining the Antarctic margin stratigraphy and obtaining new paleoclimatic records in order to better understand Antarctica's role in the Earth system since the Cretaceous. Highlights include a total of four completed IODP expeditions (Expeditions 318, 374, 379 and 382) that were explicitly designed to capture direct records of past ice sheet dynamics and ice sheet–ocean interactions. In addition, the scientific results from DSDP, ODP, IODP, MeBo, ANDRILL and SHALDRIL projects published over the past 15 years have revolutionised our understanding of the sensitivity of Antarctica's ice sheets to the warming events and have provided critical benchmarks against which models used to project the future response of the AIS to anthropogenic forcing have been tested (DeConto and Pollard, 2016; Golledge et al., 2015, 2019; Pollard and DeConto, 2009). Three of these IODP expeditions had the explicit aim to extensively sample continental shelf strata, which would have significantly advanced our ability to understand if major shifts in offshore depositional patterns observed in seismic survey data were directly related to continent wide climatic shifts and ice sheet behaviour, or whether these shifts were diachronous and related to regional tectonic and erosional events.

Drilling successes since the initiation of the PRAMSO initiative in 2009 have been abundant, even in the face of difficulties associated with riserless ship-based drilling on Antarctica's continental shelves, which can result in variable core recovery. These difficulties include: (1) poor recovery of rotary coring of unconsolidated Plio-Pleistocene sediments; (2) poor recovery in the upper 50 to 100 m of each hole, until weight on the bit stabilises drilling; and (3) adverse weather and ice conditions resulting in ship heave and site abandonment in the upper 100 mbsf. These three challenges are often associated with each other, as experience from drilling the past 12 years has indicated the greatest challenge relates to weather and ice, which in turn significantly shortens drilling windows and restricts the ability to penetrate to enough depth to drill past into consolidated strata and place sufficient weight on the drill-bit to improve recovery. While IODP Expedition 318 to Wilkes Land drilled three sites on the shelf, recovery was low in the cores targeting sequences to date Cenozoic variability of the ice sheet (although valuable snapshot of the early Oligocene and late Miocene-early Pliocene were recovered in addition to a $\sim$180 m thick Holocene diatom ooze section), due to narrow drilling windows between regular poor weather excursions and ice conditions. More recently IODP Expedition 379 planned to drill numerous sites on the continental shelf, but a poor sea ice season restricted access to this region and only drifts on the lowermost continental rise were cored. However, there remains a need to return to these regions for drilling and, while similar risks will exist in the future, these areas remain high priorities to understand the sensitivity of Antarctica's marine ice sheet in warmer than present climate. Some of these challenges could be overcome by the use of riser drilling (e.g., Mission Specific Platforms), as well as continued development of novel drilling seafloor and sub-ice sheet drilling systems (Escutia et al., 2019; Kennicutt et al., 2016; Mckay et al., 2016). More success was obtained with ship-based drilling in the Ross Sea with IODP Expedition 374, where regular open water conditions and predictable ice conditions allowed for high recovery of Late Miocene and older continental shelf strata. Although diamict lithologies are commonly perceived as the largest issue for drilling recovery in Antarctic, experience from DSDP Leg 28, ODP Leg 119 and IODP Expedition 374 demonstrate this is incorrect. While IODP Expedition 374 indicates that unconsolidated diamict remains difficult to recover (e.g., Plio-Pleistocene strata) from ship-based drill systems, drilling lithified glacial diamicts with a hard mud matrix is much easier as it is homogenous and cohesive, and recovery rates from rotary drilling actually exceed those of many other regions globally.

The challenges of continental shelf drilling of Plio-Pleistocene strata are also offset by the ability to piston core sequences on the continental rise and abyssal plain. In this depositional setting, IODP expeditions 318, 374, 379 and 382, following the pioneering ODP 178 and 188 expeditions, all obtained high-recovery sequences of orbital scale to millennial scale variability of Plio-Pleistocene ice sheet dynamics and oceanic change adjacent to the Antarctica margin. Even in scenarios of low recovery rates, 'snapshot' records from shallow and deep coring the Wilkes Land, Amundsen Sea and Sabrina Coast have provided unique glimpses into Cretaceous to early Cenozoic Antarctic environment from the Antarctic margin.

This review highlights the complexities of tectonic vs climate controls on the climate, ice sheet evolution and stratigraphic architecture of the Antarctic margin. Each sector of Antarctica discussed in the chapter has fundamentally different tectonic histories, including rifting on the Gondwana margin during the Mesozoic in East Antarctica and the Weddell Sea; through to rifting associated with the break-up of Gondwana during the Early Cenozoic in the Amundsen and Ross Sea; and continued rifting into the Miocene and Quaternary persisting in several focussed regions along the Pacific margin of West Antarctica. Despite these differences, first-order similarities in the stratigraphic architecture of the continental shelf are evident, with basin/rift-fill strata (of variable ages) overlain by Early Cenozoic strata that are predominately aggradational, resulting from run-off of fluvial systems into the marine margin that capture depositional records of inland temperate to subtropical coastal climates. By latest Eocene to Early Miocene times, an increased glacial influence is apparent in many offshore records, through the presence of iceberg rafted debris. This glacial influence resulted in a shift towards progradational strata becoming more predominant by the early Miocene in most of the basins around the margin, and is suggested to be the consequence of glacial expansions of highly-erosive polythermal glacial systems characterised by large volumes of turbid meltwater and glaciofluvial discharge.

By the middle to late Miocene times, a shift in glacial thermal regime led to reduced sediment supply from turbid meltwater input and glaciofluvial run-off, but erosion of basin fill sediments continued due to expansion of marine-based ice sheets leading to overdeepening of the continental shelf in many sectors. This resulted in an aggradational deposition pattern on the inner- to middle-continental shelf around much of the margin, accompanied by the development of large prograding fan systems on the continental slope and rise, fed by ice stream sediment transport and deposition processes. During periods of ice sheet retreat, thick hemipelagic drapes on the continental shelf provide unique archives of continental margin climate and oceanographic changes, most notably in the form of Pliocene and Miocene diatom oozes and diatom rich muds from ANDRILL and IODP cores.

New sub-ice shelf drilling campaigns, similar to that used by ANDRILL, are being planned near the modern-day grounding line in the Siple Coast region to assess if complete collapse of the marine-based WAIS occurred during Late Pleistocene interglacials. It is also capable of drilling semicontinuous stratigraphic marine archives of Early Pleistocene, Pliocene (and older climates) from the highest latitudes yet obtained in Antarctica. This will use the next generation of light weight drill systems capable of being deployed in the deep field. The use of rapid-access ice drills to recover shallow cores from beneath grounded ice sheets will likely also be utilised more widely in the future, since airborne geophysical surveys have highlighted several spots where to penetrate thick sediment basins near the Antarctic coast. These sub-ice shelf/ice sheet drill records to constrain past marine-based ice sheet collapse in Plio-Pleistocene signature should be complemented by further drilling of sediment drifts growing on the continental slope/rise near the main route for mass wasting discharge and sediment delivery from the inner shelf to the ocean.

Further offshore, continental rise to abyssal plain record continue to provide continuous records of dynamic ice sheet behaviour through the development of iceberg rafted debris and provenance proxies, while geochemical, sedimentological and palaeontological proxies also provide indications of oceanographic and biological system shifts in the Southern Ocean. These offshore records also provide age control for the continental shelf stratigraphy discussed above, whereby dated shifts in the continental slope/rise stratigraphic architecture and offshore depositional systems (e.g., levee/drift building, channel incision) are tied to changes in the seismic facies and continental shelf stratigraphic architecture. There are still large areas of the continental slope and rise that remain unexplored despite the discovery of the potential high resolution and relatively continuous records that can be obtained from drilling sedimentary mounds accumulating on the continental slope and rise, allowing ice sheet dynamics to be linked with Southern Ocean circulation and global changes. The huge and comprehensive work done by the PAIS Paleo-sea bed and paleotopographic subcommittee represents a milestone in understanding the changes affecting the pan-Antarctic continent across past climatic thresholds. However, this work needs to be refined with further seismic surveys to provide a better constrained and detailed picture of processes, causes and feedbacks, and estimates on the rates of change. Even though grids of seismic data were collected locally (and successfully drilled), most of the WAIS margin has been surveyed only with a few long, seismic profiles, connecting different regions, and some regions of East Antarctica are completely unexplored. It is therefore only indirectly possible in most cases to chronstratigraphically correlate sedimentary sequences and compare processes acting in the different sectors of the Antarctic margin. In addition, only a few seismic transects go across the continental shelf edge to the deep ocean, both in West and East Antarctica. We understand that trough-mouth fans were built in front of major ice streams, but only some of these have been partially surveyed. Slope canyons are rare in Antarctica, but they are the main transfer conduits for Antarctic dense water to the global thermohaline circulation. However, despite their significance, we still know very little about the formation and evolution of these canyons. Moreover, an improved understanding of how water masses are exchanged between the Southern Ocean and the continental shelf is fundamental to assessing the processes the lead to increases in southward ocean heat flux, which may have triggered past ice sheet collapses and warming of terrestrial climates. Therefore, more data are needed to fully understand the processes acting near the shelf edge, to help constrain past changes in ocean–ice sheet interactions.

Most of the East Antarctic continental shelf is still unexplored due to severe sea ice coverage and long distance from scientific stations. A large effort in collecting geophysical and geomorphological data is key to identifying high-resolution paleoclimate and paleoceanographic records that, in turn, will provide valuable boundary conditions and benchmarks for testing climate and ice sheet models used to project future change. As highlighted in the recent 30-year framework for the future of scientific ocean drilling (Koppers and Coggon, 2020), there remains a need to continue drilling on the Antarctic margin in order to ground-truth future climate change in the polar regions, and to fully succeed in this vision we need to continue to improve seismic coverage and drilling from sectors of the ice sheet margin that are understudied. At the same time, improvement of existing, and development of new, technologies for seabed and sub-ice shelf/sheet drilling in remote regions remains an imperative for the community (Kennicutt et al., 2016; Mckay et al., 2016).

The international coordination and multidisciplinary approaches nurtured by the SCAR PAIS program and PRAMSO/Palaeotopographic-Palaeobathymetric Reconstructions subcommittees over the last 10 years has led to a wealth of new data and environmental proxies, allowing a fundamental shift in our understanding of the functioning of natural mechanisms affecting the Antarctic's climate and ice sheets over the Cenozoic. Most of the results collected during completed expeditions, especially the recent deep drilling campaigns in 2018 and 2019, will be published in the next ten years and will be integrated with increasingly more sophisticated and complex modelling experiments. In turn, these will lead to new hypotheses that can tested by future drilling and geological studies on the margin. The success of the coordinated PRAMSO community approach will continue in the frame of the new SCAR program INStabilities and Thresholds in ANTarctica (INSTANT). There are still gaps to be filled in the least accessible sectors of the West and East Antarctic margin, as well as in previously drilled regions of the Ross Sea, Amundsen, Wilkes Land and Antarctic Peninsula. The regional differences in the tectonic, ice sheet and climate histories discussed in this chapter highlight that wide spatial and temporal coverage will to be key to understanding the full complexity of past ice sheet dynamics and their relevance for estimates of future rates of change driven by anthropogenic climate change. The collection of these new geomorphological, seismic and drilling data will be challenging and will require innovative approaches but will open a new and exciting phase of Antarctic research.

Acknowledgements

We thank many people who collaborated, by sharing data and ideas, on geoscience research projects under the umbrella of the highly successful Paleoclimate Records from the Antarctic Margin and Southern Ocean (PRAMSO) and Palaeotopographic-Palaeobathymetric Reconstructions subcommittees of the Scientific Committee on

Antarctic Research (SCAR) Past Antarctic Ice Sheet scientific program. This synthesis, which reflects our views, would not have been possible without the efforts of these many investigators, most of whom continue their collaborative Antarctic studies, now under the successor SCAR INSTANT programme. Chris Sorlien is thanked for drafting Fig. 3.6. We thank John Anderson, Peter Barrett, Giuliano Brancolini and Alan Cooper for their useful comments and for their continuous dedication to the past Antarctic Ice Sheet evolution reconstructions. We thank Nigel Wardell, Frank Nitsche and Paolo Diviacco for maintaining the Seismic Data Library System and the National Antarctic funding agencies of many countries (Australia, China, Germany, Italy, Japan, Korea, New Zealand, Russia, Spain, the UK, the United States) for supporting geophysical and geological surveys essential for Paleotopographic and Paleobathymetric reconstructions. We thank the International Ocean Discovery Program (IODP) for its support of recent expeditions that arose out of PRAMSO discussions. R.M. was funded by the Royal Society Te Apārangi NZ Marsden Fund (grant 18-VUW-089). C.E. acknowledges funding by the Spanish Ministry of Economy, Industry and Competitivity (grants CTM2017-89711-C2-1/2-P), cofunded by the European Union through FEDER funds. L.D.S. and F.D. were funded by the Programma Nazionale delle Ricerche in Antartide (PNRA16_00016 project and PNRA 14_00119). R.Larter and C.D.H. were funded by the BAS Polar Science for Planet Earth Programme and NERC UK IODP grant NE/J006548/1. S.K. was supported by the KOPRI Grant (PE21050). L.P. was funded by the European Union's Horizon 2020 research and innovation programme under the Marie Sklodowska-Curie grant agreement No. 792773 WAMSISE. A.S. and S.G. were funded by NSF Office of Polar Programs (Grants OPP-1744970 (A.S.), -1143836 (A.S.), and -1143843 (S.G.). This is University of Texas Institute for Geophysics Contribution #3784. B.D. acknowledges funding from a Rutherford Foundation Postdoctoral Fellowship (RFT-VUW1804-PD). K.G. and G.K. were funded by AWI research programme Polar Regions and Coasts in the changing Earth System (PACES II) and the Sub-EIS-Obs programme by the Bundesanstalt für Geowissenschaften und Rohstoffe (BGR). RL, RM, TN acknowledge support from MBIE Antarctic Science Platform contract ANTA1801.

References

Abelmann, A., Gersonde, R., Spiess, V., 1990. Pliocene—Pleistocene paleoceanography in the Weddell Sea—Siliceous microfossil evidence. Geological History of the Polar Oceans: Arctic vs Antarctic. Springer, pp. 729–759.

Accaino, F., Böhm, G., Brancolini, G., 2005. Analysis of Antarctic glaciations by seismic reflection and refraction tomography. Marine Geology 216 (3), 145–154. Available from: https://doi.org/10.1016/j.margeo.2005.02.008.

Adusumilli, S., Fricker, H.A., Medley, B., Padman, L., Siegfried, M.R., 2020. Interannual variations in meltwater input to the Southern Ocean from Antarctic ice shelves. Nature Geoscience 1–5. Available from: https://doi.org/10.1038/s41561-020-0616-z.

Aitken, A.R.A., Roberts, J.L., Ommen, T.D., van, Young, D.A., Golledge, N.R., Greenbaum, J.S., et al., 2016. Repeated large-scale retreat and advance of Totten Glacier indicated by inland bed erosion. Nature 533, 385.

Aitken, A.R.A., Young, D.A., Ferraccioli, F., Betts, P.G., Greenbaum, J.S., Richter, T.G., et al., 2014. The subglacial geology of Wilkes Land, East Antarctica. Geophysical Research Letters 41 (7). Available from: https://doi.org/10.1002/2014GL059405, 2014GL059405.

Akhoudas, C., Sallée, J.-B., Reverdin, G., Aloisi, G., Benetti, M., Vignes, L., et al., 2020. Ice Shelf Basal Melt and influence on dense water outflow in the Southern Weddell Sea. Journal of Geophysical Research: Oceans 125 (2). Available from: https://doi.org/10.1029/2019jc015710, e2019JC015710.

Almendros, J., Wilcock, W., Soule, D., Teixidó, T., Vizcaíno, L., Ardanaz, O., et al., 2020. BRAVOSEIS: Geophysical investigation of rifting and volcanism in the Bransfield strait, Antarctica. Journal of South American Earth Sciences 104, 102834. Available from: https://doi.org/10.1016/j.jsames.2020.102834.

Alonso, B., Anderson, J.B., Díaz, J.I., Bartek, L.R., 1992. Pliocene-Pleistocene seismic stratigraphy of the Ross Sea: evidence for multiple ice sheet grounding episodes. In: Elliot, D.H. (Ed.), Contributions to Antarctic Research III. American Geophysical Union, pp. 93–103. Available from: https://doi.org/10.1029/AR057p0093.

Amblas, D., Urgeles, R., Canals, M., Calafat, A.M., Rebesco, M., Camerlenghi, A., et al., 2006. Relationship between continental rise development and palaeo-ice sheet dynamics, Northern Antarctic Peninsula Pacific margin. Quaternary Science Reviews 25 (9), 933–944. Available from: https://doi.org/10.1016/j.quascirev.2005.07.012.

Anagnostou, E., John, E.H., Edgar, K.M., Foster, G.L., Ridgwell, A., Inglis, G.N., et al., 2016. Changing atmospheric CO_2 concentration was the primary driver of early Cenozoic climate. Nature 533 (7603), 380–384. Available from: https://doi.org/10.1038/nature17423.

Anderson, J.B., 2006. SHALDRIL II 2006 NBP0602A Cruise Report. Rice University, Department of Earth Science, Houston, TX.

Anderson, J.B., Bartek, L.R., 1992. Cenozoic glacial history of the Ross Sea revealed by intermediate resolution seismic reflection data combined with drill site information. In: James, P., Kennett, Warkne, D.A. (Eds.), The Antarctic Paleoenvironment: A Perspective on Global Change: Part One. American Geophysical Union, pp. 231–264. Available from: https://doi.org/10.1029/AR056p0231.

Anderson, J.B., Shipp, S.S., Siringan, F.P., 1992. Preliminary seismic stratigraphy of the northwestern Weddell Sea continental shelf. In: Yoshida, Y., Kaminuma, K., Shiraishi, K. (Eds.), Recent Progress in Antarctic Earth Science. Terra Scientific Publishing, pp. 603–612.

Anderson, J.B., Simkins, L.M., Bart, P.J., Santis, L.D., Halberstadt, A.R.W., Olivo, E., et al., 2018. Seismic and geomorphic records of Antarctic Ice Sheet evolution in the Ross Sea and controlling factors in its behaviour. Geological Society, London, Special Publications 475, SP475.5. Available from: https://doi.org/10.1144/SP475.5.

Anderson, J.B., Warny, S., Askin, R.A., Wellner, J.S., Bohaty, S.M., Kirshner, A.E., et al., 2011. Progressive Cenozoic cooling and the demise of Antarctica's last refugium. Proceedings of the National Academy of Sciences 108 (28), 11356–11360. Available from: https://doi.org/10.1073/pnas.1014885108.

Armbrecht, L.H., Lowe, V., Escutia, C., Iwai, M., McKay, R., Armand, L.K., 2018. Variability in diatom and silicoflagellate assemblages during mid-Pliocene glacial-interglacial cycles determined in Hole U1361A of IODP Expedition 318, Antarctic Wilkes Land Margin. Marine Micropaleontology 139, 28–41. Available from: https://doi.org/10.1016/j.marmicro.2017.10.008.

Arndt, J.E., Schenke, H.W., Jakobsson, M., Nitsche, F.-O., Buys, G., Goleby, B., Rebesco, M., et al., 2013. The International Bathymetric Chart of the Southern Ocean (IBCSO) Version 1.0—a new bathymetric compilation covering circum-Antarctic waters. Geophysical Research Letters, 40 (12), 3111–3117. Available from: https://doi.org/10.1002/grl.50413.

Arne, D.C., 1994. Phanerozoic exhumation history of northern Prince Charles Mountains (East Antarctica). Antarctic Science 6 (1), 69–84. Available from: https://doi.org/10.1017/S0954102094000106.

Ashley, K.E., Bendle, J.A., McKay, R., Etourneau, J., Jimenez-Espejo, F.J., Condron, A., et al., 2020. Mid-Holocene Antarctic sea-ice increase driven by marine ice sheet retreat. Climate of the Past Discussions 1–36. Available from: https://doi.org/10.5194/cp-2020-3.

Askin, R.A., 2000. Spores and pollen from the McMurdo Sound erratics, Antarctica. In: Stillwell, J.D., Feldmann, R.M. (Eds.), Paleobiology and Paleoenvironments of Eocene Rocks, McMurdo Sound, East Antarctica. Antarctic Research Series 76, 161–181.

Baatsen, M., von der Heydt, A.S., Huber, M., Kliphuis, M.A., Bijl, P.K., Sluijs, A., et al., 2020. The middle to late Eocene greenhouse climate modelled using the CESM 1.0.5. Climate of the Past 16 (6), 2573–2597. Available from: https://doi.org/10.5194/cp-16-2573-2020.

Banfield, L., Anderson, J., 1995. Seismic facies investigation of the late Quaternary Glacial history of Bransfield Basin, Antarctica. In: Cooper, A.K., Barker, P.F., Brancolini, G. (Eds.), Antarctic Research Series, vol. 68. American Geophysical Union, pp. 123–140.

Barker, P.F., Camerlenghi, A., 2002. Glacial history of the Antarctic Peninsula from Pacific margin sediments. Proceedings of the Ocean Drilling Program, Scientific Results 178, 1–40.

Barker, P.F., Camerlenghi, A., Acton, G.D., et al., 1999. Proceedings of the Ocean Drilling Program, Initial Reports, vol. 178: vol. CD-ROM. Ocean Drilling Program, Texas A&M University.

Barker, P.F., Dalziel, I.W.D., 1983. Progress in geodynamics in the Scotia arc region. In: Cabre, R. (Ed.), Geodynamics of the Eastern Pacific Region, Caribbean and Scotia Arcs, vol. 9. American Geophysical Union, pp. 137–170.

Barker, P.F., Kennett, J.P., 1988. Proceedings, initial reports, Ocean Drilling Program, Leg 113, Weddell Sea, Antarctica.

Barker, P.F., Lonsdale, M.J., 1991. A multichannel seismic profile across the Weddell Sea margin of the Antarctic Peninsula: regional tectonic implications. In: Thomson, M.R.A., Crame, J.A., Thomson, J.W. (Eds.), Geological Evolution of Antarctica. Cambridge University Press, pp. 237–241.

Barker, P.F., Kennett, J.P., Scientific, P., 1988. Weddell sea palaeoceanography: preliminary results of ODP Leg 113. Palaeogeography, Palaeoclimatology, Palaeoecology 67 (1–2), 75–102.

Barrett, P., 2008. Chapter 3: A history of Antarctic Cenozoic Glaciation – view from the margin. In: Florindo, F., Siegert, M. (Eds.), Developments in Earth and Environmental Sciences, Vol. 8. Elsevier, pp. 33–83. Available from: http://www.sciencedirect.com/science/article/pii/S1571919708000037.

Barrett, P.J., 1986. Antarctic Cenozoic History from the MSSTS-1 Drillhole McMurdo Sound. DSIR Bulletin 235, Wellington, 174 p.

Barrett, P.J., 1989. Antarctic Cenozoic History from the CIROS-1 Drillhole, McMurdo Sound. DRIS Bulletin 245, NZ, 254 p.

Barrett, P.J., 2007. Cenozoic climate and sea level history from Glacimarine Strata off the Victoria Land Coast, Cape Roberts Project, Antarctica. In: Hambrey, Michael J., Christoffersen, P., Glasser, N.F., Hubbard, B. (Eds.), Glacial Sedimentary Processes and Products. Blackwell Publishing Ltd, pp. 259–287. Available from: http://onlinelibrary.wiley.com/doi/10.1002/9781444304435.ch15/summary.

Barrett, P.J., 2013. Resolving views on Antarctic Neogene glacial history – the Sirius debate. Earth and Environmental Science Transactions of the Royal Society of Edinburgh 104 (01), 31–53. Available from: https://doi.org/10.1017/S175569101300008X.

Barrett, P.J., Fielding, C.R., Wise, S.W., Cape Roberts Project Science Team, 1998. Initial report on CRP-1, Cape Roberts Project, Antarctica. Terra Antartica 5 (1).

Barrett, P.J., Sarti, M., Wise, S.W., Cape Roberts Project science team, 2000. Studies from the Cape Roberts project, Ross Sea, Antarctica, initial reports on CRP-3. Terra Antartica 7 (1).

Barron, J., Larsen, B., 1989. Proceedings ODP, Init. Repts., 119.

Barron, J., Larsen, B., et al., (Eds.), 1991. Proceedings of the Ocean Drilling Program, 119 Scientific Results, vol. 119. Ocean Drilling Program. Available from: http://www-odp.tamu.edu/publications/119_SR/119TOC.HTM.

Bart, P.J., 2001. Did the Antarctic ice sheets expand during the early Pliocene? Geology 29 (1), 67–70. Available from: https://doi.org/10.1130/0091-7613(2001)029 < 0067:DTAISE > 2.0.CO;2.

Bart, P.J., 2003. Were West Antarctic Ice Sheet grounding events in the Ross Sea a consequence of East Antarctic Ice Sheet expansion during the middle Miocene? Earth and Planetary Science Letters 216 (1–2), 93–107. Available from: https://doi.org/10.1016/S0012-821X(03)00509-0.

Bart, P.J., Anderson, J.B., 1995. Seismic record of glacial events affecting the Pacific margin of the Northwestern Antarctic Peninsula. In: Cooper, A.K., Barker, P.F., Brancolini, G. (Eds.), Geology and Seismic Stratigraphy of the Antarctic Margin, Vol. 68. American Geophysical Union, pp. 75–95.

Bart, P.J., Anderson, J.B., 2000. Relative temporal stability of the Antarctic ice sheets during the late Neogene based on the minimum frequency of outer shelf grounding events. Earth and Planetary Science Letters 182 (3–4), 259–272. Available from: https://doi.org/10.1016/S0012-821X(00)00257-0.

Bart, P.J., De Santis, L., 2012. Glacial intensification during the neogene: a review of seismic stratigraphic evidence from the Ross Sea, Antarctica, Continental Shelf. Oceanography . Available from: http://agris.fao.org/agris-search/search.do?recordID = DJ2012092517.

Bart, P.J., De Batist, M., Jokat, W., 1999. Interglacial collapse of Crary Trough-mouth fan, Weddell Sea, Antarctica; implications for Antarctic glacial history. Journal of Sedimentary Research 69 (6), 1276–1289.

Bart, P.J., Egan, D., Warny, S.A., 2005. Direct constraints on Antarctic Peninsula Ice Sheet grounding events between 5.12 and 7.94 Ma. Journal of Geophysical Research: Earth Surface 110 (F4). Available from: https://doi.org/10.1029/2004JF000254.

Bart, P.J., Hillenbrand, C.D., Ehrmann, W., Iwai, M., Winter, D., Warny, S.A., 2007. Are Antarctic Peninsula Ice Sheet grounding events manifest in sedimentary cycles on the adjacent continental rise? Marine Geology 236 (1), 1–13. Available from: https://doi.org/10.1016/j.margeo.2006.09.008.

Bart, P.J., Iwai, M., 2012. The over-deepening hypothesis: how erosional modification of the marine-scape during the early Pliocene altered glacial dynamics on the Antarctic Peninsula's Pacific margin. Palaeogeography, Palaeoclimatology, Palaeoecology 335–336, 42–51. Available from: https://doi.org/10.1016/j.palaeo.2011.06.010.

Bart, P.J., Mullally, D., Golledge, N.R., 2016. The influence of continental shelf bathymetry on Antarctic Ice Sheet response to climate forcing. Global and Planetary Change 142, 87–95. Available from: https://doi.org/10.1016/j.gloplacha.2016.04.009.

Bart, P.J., Owolana, B., 2012. On the duration of West Antarctic Ice Sheet grounding events in Ross Sea during the Quaternary. Quaternary Science Reviews 47 (0), 101–115. Available from: https://doi.org/10.1016/j.quascirev.2012.04.023.

Bart, P.J., Sjunneskog, C., Chow, J.M., 2011. Piston-core based biostratigraphic constraints on Pleistocene oscillations of the West Antarctic Ice Sheet in western Ross Sea between North Basin and AND-1B drill site. Marine Geology 289 (1–4), 86–99. Available from: https://doi.org/10.1016/j.margeo.2011.09.005.

Bart, P.J., Tulaczyk, S., 2020. A significant acceleration of ice volume discharge preceded a major retreat of a West Antarctic paleo–ice stream. Geology 48 (4), 313–317. Available from: https://doi.org/10.1130/G46916.1.

Bartek, L.R., Vail, P.R., Anderson, J.B., Emmet, P.A., Wu, S., 1991. Effect of Cenozoic ice sheet fluctuations in Antarctica on the stratigraphic signature of the neogene. Journal of Geophysical Research 96, 6753–6778. Available from: https://doi.org/10.1029/90JB02528.

Batchelor, C.L., Dowdeswell, J.A., 2015. Ice-sheet grounding-zone wedges (GZWs) on high-latitude continental margins. Marine Geology 363, 65–92. Available from: https://doi.org/10.1016/j.margeo.2015.02.001.

Beltran, C., Ohneiser, C., Hageman, K.J., Scanlan, E., 2016. Evolution of the southwestern Pacific surface waters during the early Pleistocene. New Zealand Journal of Geology and Geophysics 59 (4), 514–521. Available from: https://doi.org/10.1080/00288306.2016.1195756.

Bertram, R.A., Wilson, D.J., van de Flierdt, T., McKay, R.M., Patterson, M.O., Jimenez-Espejo, F.J., et al., 2018. Pliocene deglacial event timelines and the biogeochemical response offshore Wilkes Subglacial Basin, East Antarctica. Earth and Planetary Science Letters 494, 109–116. Available from: https://doi.org/10.1016/j.epsl.2018.04.054.

Bijl, P.K., Bendle, J.A.P., Bohaty, S.M., Pross, J., Schouten, S., Tauxe, L., et al., 2013. Eocene cooling linked to early flow across the Tasmanian Gateway. Proceedings of the National Academy of Sciences . Available from: https://doi.org/10.1073/pnas.1220872110.

Bijl, P.K., Houben, A.J.P., Hartman, J.D., Pross, J., Salabarnada, A., Escutia, C., et al., 2018. Paleoceanography and ice sheet variability offshore Wilkes Land, Antarctica – part 2: insights from Oligocene–Miocene dinoflagellate cyst assemblages. Climate of the Past 14 (7), 1015–1033. Available from: https://doi.org/10.5194/cp-14-1015-2018.

Bindoff, N.L., Rosenberg, M.A., Warner, M.J., 2000. On the circulation and water masses over the Antarctic continental slope and rise between 80 and 150°E. Deep Sea Research Part II: Topical Studies in Oceanography 47 (12–13), 2299–2326. Available from: https://doi.org/10.1016/S0967-0645(00)00038-2.

Birkenmajer, K., 1991. Tertiary glaciation in the South Shetland Islands, West Antarctica: evaluation of data. In: Thomson, M.R.A., Crame, J.A., Thomson, J.W. (Eds.), Geological Evolution of Antarctica. Cambridge University Press, pp. 629–632.

Blackburn, T., Edwards, G.H., Tulaczyk, S., Scudder, M., Piccione, G., Hallet, B., et al., 2020. Ice retreat in Wilkes Basin of East Antarctica during a warm interglacial. Nature 583 (7817), 554–559. Available from: https://doi.org/10.1038/s41586-020-2484-5.

Bo, S., Siegert, M.J., Mudd, S.M., Sugden, D., Fujita, S., Xiangbin, C., et al., 2009. The Gamburtsev mountains and the origin and early evolution of the Antarctic Ice Sheet. Nature 459 (7247), 690–693. Available from: https://doi.org/10.1038/nature08024.

Bohaty, S.M., Harwood, D.M., 1998. Southern Ocean Pliocene paleotemperature variation from high-resolution silicoflagellate biostratigraphy. Marine Micropaleontology 33 (3–4), 241–272. Available from: https://doi.org/10.1016/S0377-8398(97)00037-6.

Böhm, G., Ocakoğlu, N., Picotti, S., De Santis, L., 2009. West Antarctic Ice Sheet evolution: new insights from a seismic tomographic 3D depth model in the Eastern Ross Sea (Antarctica). Marine Geology 266 (1–4), 109–128. Available from: https://doi.org/10.1016/j.margeo.2009.07.016.

Bohoyo, F., Galindo-Zaldivar, J., Maldonado, A., Schreider, A.A., Suriñach, E., 2002. Basin development subsequent to ridge-trench collision: The Jane Basin, Antarctica. Marine Geophysical Research 23 (5–6), 413–421.

Bowman, V.C., Francis, J.E., Riding, J.B., 2013. Late Cretaceous winter sea ice in Antarctica? Geology 41 (12), 1227–1230. Available from: https://doi.org/10.1130/G34891.1.

Bowman, V.C., Francis, J.E., Askin, R.A., Riding, J.B., Swindles, G.T., 2014. Latest Cretaceous–earliest Paleogene vegetation and climate change at the high southern latitudes: palynological evidence from Seymour Island, Antarctic Peninsula. Palaeogeography, Palaeoclimatology, Palaeoecology 408, 26–47. Available from: https://doi.org/10.1016/j.palaeo.2014.04.018.

Brancolini, G., Busetti, M., Coren, F., De Cillia, C., Marchetti, M., De Santis, L., et al., 1995. ANTOSTRAT Project, seismic stratigraphic atlas of the Ross Sea, Antarctica. Geology and Seismic Stratigraphy of the Antarctic Margin. Antarctic Research Series. AGU, Washington, DC, p. 68.

Busetti, M., Marchetti, A., Zanolla, C., De Cillia, C., 2002. Tectonic history of the South Orkney Microcontinent from seismic structure and stratigraphy. Royal Society of New Zealand Bulletin 35, 507–513.

Busetti, M., Zanolla, C., Marchetti, A., 2001. Geological structure of the South Orkney Microcontinent. Terra Antartica 8, 1–8.

Caballero, R., Huber, M., 2013. State-dependent climate sensitivity in past warm climates and its implications for future climate projections. Proceedings of the National Academy of Sciences 110 (35), 14162–14167. Available from: https://doi.org/10.1073/pnas.1303365110.

Caburlotto, A., De Santis, L., Zanolla, C., Camerlenghi, A., Dix, J.K., 2006. New insights into Quaternary glacial dynamic changes on the George V Land continental margin (East Antarctica). Quaternary Science Reviews 25 (21), 3029–3049. Available from: https://doi.org/10.1016/j.quascirev.2006.06.012.

Caburlotto, A., Lucchi, R.G., Santis, L.D., Macrì, P., Tolotti, R., 2010. Sedimentary processes on the Wilkes Land continental rise reflect changes in glacial dynamic and bottom water flow. International Journal of Earth Sciences 99 (4), 909–926. Available from: https://doi.org/10.1007/s00531-009-0422-8.

Canals, M., Casamor, J.L., Urgeles, R., Calafat, A., Domack, E., Baraza, J., et al., 2002. Seafloor evidence of a subglacial sedimentary system off the northern Antarctic Peninsula. Geology 30. Available from: https://doi.org/10.1130/0091-7613(2002)030 < 0603:SEOASS > 2.0.CO;2.

Carter, A., Riley, T.R., Hillenbrand, C.-D., Rittner, M., 2017. Widespread Antarctic glaciation during the late Eocene. Earth and Planetary Science Letters 458, 49–57. Available from: https://doi.org/10.1016/j.epsl.2016.10.045.

Catalán, M., Martos, Y.M., Galindo-Zaldivar, J., Perez, L.F., Bohoyo, F., 2020. Unveiling Powell Basin's Tectonic domains and understanding its abnormal magnetic anomaly signature. Is heat the key? Frontiers in Earth Science 8. Available from: https://doi.org/10.3389/feart.2020.580675.

Channell, J.E.T., Xuan, C., Hodell, D.A., Crowhurst, S.J., Larter, R.D., 2019. Relative paleointensity (RPI) and age control in Quaternary sediment drifts off the Antarctic Peninsula. Quaternary Science Reviews 211, 17–33. Available from: https://doi.org/10.1016/j.quascirev.2019.03.006.

Chow, J.M., Bart, P.J., 2003. West Antarctic Ice Sheet grounding events on the Ross Sea outer continental shelf during the middle Miocene. Palaeogeography, Palaeoclimatology, Palaeoecology 198 (1–2), 169–186. Available from: https://doi.org/10.1016/S0031-0182(03)00400-0.

Clark, P.U., He, F., Golledge, N.R., Mitrovica, J.X., Dutton, A., Hoffman, J.S., et al., 2020. Oceanic forcing of penultimate deglacial and last interglacial sea-level rise. Nature 577 (7792), 660–664. Available from: https://doi.org/10.1038/s41586-020-1931-7.

Close, D.I., 2010. Slope and fan deposition in deep-water turbidite systems, East Antarctica. Marine Geology 274 (1), 21–31. Available from: https://doi.org/10.1016/j.margeo.2010.03.002.

Close, D.I., Stagg, H.M.J., O'Brien, P.E., 2007. Seismic stratigraphy and sediment distribution on the Wilkes Land and Terre Adélie margins, East Antarctica. Marine Geology 239 (1–2), 33–57. Available from: https://doi.org/10.1016/j.margeo.2006.12.010. Scopus.

Close, D.I., Watts, A.B., Stagg, H.M.J., 2009. A marine geophysical study of the Wilkes Land rifted continental margin, Antarctica. Geophysical Journal International 177 (2), 430–450. Available from: https://doi.org/10.1111/j.1365-246X.2008.04066.x.

Coenen, J.J., Scherer, R.P., Baudoin, P., Warny, S., Castañeda, I.S., Askin, R., 2020. Paleogene marine and terrestrial development of the West Antarctic Rift System. Geophysical Research Letters 47 (3). Available from: https://doi.org/10.1029/2019GL085281, e2019GL085281.

Colleoni, F., De Santis, L., Montoli, E., Olivo, E., Sorlien, C.C., Bart, P.J., et al., 2018. Past continental shelf evolution increased Antarctic ice sheet sensitivity to climatic conditions. Scientific Reports 8. Available from: https://doi.org/10.1038/s41598-018-29718-7.

Colleoni, F., et al., 2021. Past Antarctic ice sheet dynamics and implications for future sea-level change. In: Florindo, F., et al. (Eds.), Antarctic Climate Evolution, second ed. Elsevier (this volume).

Contreras, L., Pross, J., Bijl, P.K., Koutsodendris, A., Raine, J.I., van de Schootbrugge, B., et al., 2013. Early to Middle Eocene vegetation dynamics at the Wilkes Land Margin (Antarctica). Review of Palaeobotany and Palynology 197, 119–142. Available from: https://doi.org/10.1016/j.revpalbo.2013.05.009.

Contreras, L., Pross, J., Bijl, P.K., O'Hara, R.B., Raine, J.I., Sluijs, A., et al., 2014. Southern high-latitude terrestrial climate change during the Palaeocene–Eocene derived from a marine pollen record (ODP Site 1172, East Tasman Plateau). Climate of the Past 10 (4), 1401–1420. Available from: https://doi.org/10.5194/cp-10-1401-2014.

Cook, C.P., van de Flierdt, T., Williams, T., Hemming, S.R., Iwai, M., Kobayashi, M., et al., 2013. Dynamic behaviour of the East Antarctic ice sheet during Pliocene warmth. Nature Geoscience 6 (9), 765–769. Available from: https://doi.org/10.1038/ngeo1889.

Cooper, A., Barker, P., Barrett, P., Behrendt, J., Brancolini, G., Childs, J., et al., 2011. Proceedings of the ANTOSTRAT Legacy: Science Collaboration and International Transparency in Potential Marine Mineral Resource Exploitation of Antarctica. <http://repository.si.edu/xmlui/handle/10088/16178>.

Cooper, A., Stagg, H., Geist, E., 1991. Seismic stratigraphy and structure of Prydz Bay, Antarctica: implications from Leg 119 drilling. Proceedings, of ODP, Scientific Results, Leg 119, Kerguelen Plateau-Prydz Bay 5–25. Available from: https://doi.org/10.2973/odp.proc.sr.119.181.1991.

Cooper, A.K., Brancolini, G., Escutia, C., Kristoffersen, Y., Larter, R., Leitchenkov, G., et al., 2008. Cenozoic climate history from seismic reflection and drilling studies on the Antarctic continental margin. In: Florindo, F., Siegert, M. (Eds.), Antarctic Climate Evolution, vol. 8. Elsevier, pp. 115–234.

Cooper, A.K., O'Brien, P.E., 2004. Leg 188 synthesis: transitions in the glacial history of the Prydz Bay region, East Antarctica, from ODP drilling, Proceedings of the ODP, Science Results, 188. pp. 1–42.

Coren, F., Ceccone, G., Lodolo, E., Zanolla, C., Zitellini, N., Bonazzi, C., et al., 1997. Morphology, seismic structure and tectonic development of the Powell Basin, Antarctica. Journal of the Geological Society, London 154 (5), 849–862.

Cortese, G., Gersonde, R., 2008. Plio/Pleistocene changes in the main biogenic silica carrier in the Southern Ocean, Atlantic Sector. Marine Geology 252 (3–4), 100–110. Available from: https://doi.org/10.1016/j.margeo.2008.03.015.

Coxall, H.K., Wilson, P.A., Palike, H., Lear, C.H., Backman, J., 2005. Rapid stepwise onset of Antarctic glaciation and deeper calcite compensation in the Pacific Ocean. Nature 433 (7021), 53–57. Available from: https://doi.org/10.1038/nature03135.

Cramer, B.S., Toggweiler, J.R., Wright, J.D., Katz, M.E., Miller, K.G. (2009). Ocean overturning since the Late Cretaceous: Inferences from a new benthic foraminiferal isotope compilation. Paleoceanography, 24, 14 PP. https://doi.org/10.1029/2008PA001683.

Cramer, B.S., Miller, K.G., Barrett, P.J., Wright, J.D., 2011. Late Cretaceous–Neogene trends in deep ocean temperature and continental ice volume: reconciling records of benthic foraminiferal geochemistry ($\delta^{18}O$ and Mg/Ca) with sea level history. Journal of Geophysical Research 116, 23. Available from: https://doi.org/10.1029/2011JC007255.

Dalziel, I.W.D., 1984. The scotia arc: an international geological laboratory. Episodes 7 (3), 8–13.

Dalziel, I.W.D., Lawver, L.A., Pearce, J.A., Barker, P.F., Hastie, A.R., Barfod, D.N., Schenke, H.W., Davis, M.B., 2013. A potential barrier to deep Antarctic circumpolar flow until the late Miocene? Geology. Available from: https://doi.org/10.1130/G34352.

Davey, F., Brancolini, G., Hamilton, R., Henrys, S., Sorlien, C., Bartek, L., 2000. A revised correlation of the seismic stratigraphy at the Cape Roberts drill sites with the seismic stratigraphy of the Victoria Land Basin, Antarctica. Terra Antartica 7 (3), 215–220.

Davey, F.J., Granot, R., Cande, S.C., Stock, J.M., Selvans, M., Ferraccioli, F., 2016. Synchronous oceanic spreading and continental rifting in West Antarctica. Geophysical Research Letters 43 (12), 6162–6169. Available from: https://doi.org/10.1002/2016GL069087.

Decesari, R.C., Sorlien, C.C., Luyendyk, B.P., Wilson, D.S., Bartek, L., Diebold, J., et al., 2007. Regional seismic stratigraphic correlations of the Ross Sea: implications for the tectonic history of the West Antarctic Rift System. In: Antarctica: A Keystone in a Changing World – Online Proceedings of the Tenth International Symposium on Antarctic Earth Sciences, August 26 to September 1, 2007, Santa Barbara, CA. Available from: https://doi.org/10.17226/12168.

De Santis, L., Anderson, J.B., Brancolini, G., Zayatz, I., 1995. Seismic record of late Oligocene through Miocene Glaciation on the central and eastern continental shelf of the Ross Sea. In: Cooper, A.K., Barker, P.F., Brancolini, G. (Eds.), Geology and Seismic Stratigraphy of the Antarctic Margin. American Geophysical Union, pp. 235–260. Available from: http://onlinelibrary.wiley.com/doi/10.1029/AR068p0235/summary.

De Santis, L., Brancolini, G., Donda, F., 2003. Seismo-stratigraphic analysis of the Wilkes Land continental margin (East Antarctica): influence of glacially driven processes on the Cenozoic deposition. Deep Sea Research Part II: Topical Studies in Oceanography 50 (8–9), 1563–1594. Available from: https://doi.org/10.1016/S0967-0645(03)00079-1.

De Santis, L., Prato, S., Brancolini, G., Lovo, M., Torelli, L., 1999. The Eastern Ross Sea continental shelf during the Cenozoic: implications for the West Antarctic ice sheet development. Global and Planetary Change 23 (1–4), 173–196. Available from: https://doi.org/10.1016/S0921-8181(99)00056-9.

DeConto, R.M., Pollard, D., 2003. Rapid Cenozoic glaciation of Antarctica induced by declining atmospheric CO_2. Nature 421 (6920), 245–249. Available from: https://doi.org/10.1038/nature01290.

DeConto, R., Pollard, D., Harwood, D., 2007. Sea ice feedback and Cenozoic evolution of Antarctic climate and ice sheets. Paleoceanography 22, PA3214. Available from: https://doi.org/10.1029/2006PA001350.

DeConto, R.M., Galeotti, S., Pagani, M., Tracy, D., Schaefer, K., Zhang, T., et al., 2012. Past extreme warming events linked to massive carbon release from thawing permafrost. Nature 484 (7392), 87–91. Available from: https://doi.org/10.1038/nature10929.

DeConto, R.M., Pollard, D., 2016. Contribution of Antarctica to past and future sea-level rise. Nature 531 (7596), 591–597. Available from: https://doi.org/10.1038/nature17145.

DeConto, R.M., Pollard, D., Wilson, P.A., Palike, H., Lear, C.H., Pagani, M., 2008. Thresholds for Cenozoic bipolar glaciation. Nature 455 (7213), 652–656. Available from: https://doi.org/10.1038/nature07337.

Dickens, W.A., Graham, A.G.C., Smith, J.A., Dowdeswell, J.A., Larter, R.D., Hillenbrand, C.-D., et al., 2014. A new bathymetric compilation for the South Orkney Islands region, Antarctic Peninsula (49°–39°W to 64°–59°S): insights into the glacial development of the continental shelf. Geochemistry, Geophysics, Geosystems 15, 2494–2514. Available from: https://doi.org/10.1002/2014GC005323.

Dietrich, R., Rülke, A., Ihde, J., Lindner, K., Miller, H., Niemeier, W., et al., 2004. Plate kinematics and deformation status of the Antarctic Peninsula based on GPS. Global and Planetary Change 42 (1–4), 313–321. Available from: https://doi.org/10.1016/j.gloplacha.2003.12.003.

Dingle, R.V., Lavelle, M., 1998. Antarctic Peninsular cryosphere: early Oligocene (c.30 Ma) initiation and a revised glacial chronology. Journal of the Geological Society 155 (3), 433–437. Available from: https://doi.org/10.1144/gsjgs.155.3.0433.

Domack, E., Leventer, A., Dunbar, R., Taylor, F., Brachfeld, S., Sjunneskog, C., 2001. Chronology of the Palmer Deep site, Antarctic Peninsula: a Holocene palaeoenvironmental reference for the circum-Antarctic. Holocene 11, 1–9. Available from: https://doi.org/10.1191/095968301673881493.

Domack, E., O'Brien, P., Harris, P., Taylor, F., Quilty, P.G., De Santis, L., et al., 1998. Late Quaternary sediment facies in Prydz Bay, East Antarctica and their relationship to glacial advance onto the continental shelf. Antarctic Science 10 (3), 236–246. Available from: https://doi.org/10.1017/s0954102098000339.

Donda, F., Brancolini, G., Santis, L.D., Trincardi, F., 2003. Seismic facies and sedimentary processes on the continental rise off Wilkes Land (East Antarctica): evidence of bottom current activity. Deep Sea Research Part II: Topical Studies in Oceanography 50 (8–9), 1509–1527. Available from: https://doi.org/10.1016/S0967-0645(03)00075-4.

Donda, F., Brancolini, G., O'Brien, P.E., De Santis, L., Escutia, C., 2007. Sedimentary processes in the Wilkes Land margin: a record of the Cenozoic East Antarctic Ice Sheet evolution. Journal of the Geological Society 164 (1), 243–256. Available from: https://doi.org/10.1144/0016-76492004-159.

Donda, F., Leitchenkov, G., Brancolini, G., Romeo, R., De Santis, L., Escutia, C., et al., 2020. The influence of Totten Glacier on the late Cenozoic sedimentary record. Antarctic Science. Available from: https://doi.org/10.1017/S0954102020000188.

Douglas, P.M.J., Affek, H.P., Ivany, L.C., Houben, A.J.P., Sijp, W.P., Sluijs, A., et al., 2014. Pronounced zonal heterogeneity in Eocene southern high-latitude sea surface temperatures. Proceedings of the National Academy of Sciences 111 (18), 6582–6587. Available from: https://doi.org/10.1073/pnas.1321441111.

Dowdeswell, J.A., Cofaigh, C.Ó., Noormets, R., Larter, R.D., Hillenbrand, C.D., Benetti, S., et al., 2008. A major trough-mouth fan on the continental margin of the Bellingshausen Sea, West Antarctica: The Belgica Fan. Marine Geology 252 (3–4), 129–140. Available from: https://doi.org/10.1016/j.margeo.2008.03.017.

Dowdeswell, J.A., Evans, J., Cofaigh, C., Anderson, J., 2006. Morphology and sedimentary processes on the continental slope off Pine Island Bay, Amundsen Sea, West Antarctica. Geological Society of America Bulletin 118, 606–619. Available from: https://doi.org/10.1130/B25791.1.

Dumitru, O.A., Austermann, J., Polyak, V.J., Fornós, J.J., Asmerom, Y., Ginés, J., et al., 2019. Constraints on global mean sea level during Pliocene warmth. Nature 574 (7777), 233–236. Available from: https://doi.org/10.1038/s41586-019-1543-2.

Duncan, B., McKay, R., Bendle, J., Naish, T., Inglis, G.N., Moossen, H., et al., 2019. Lipid biomarker distributions in Oligocene and Miocene sediments from the Ross Sea region, Antarctica: implications for use of biomarker proxies in glacially-influenced settings. Palaeogeography, Palaeoclimatology, Palaeoecology 516, 71–89. Available from: https://doi.org/10.1016/j.palaeo.2018.11.028.

Eagles, G., Jokat, W., 2014. Tectonic reconstructions for paleobathymetry in Drake Passage. Tectonophysics 611, 28–50. Available from: https://doi.org/10.1016/j.tecto.2013.11.021.

Eagles, G., Gohl, K., Larter, R.D., 2004. High-resolution animated tectonic reconstruction of the South Pacific and West Antarctic Margin. Geochemistry, Geophysics, Geosystems 5 (7). Available from: https://doi.org/10.1029/2003GC000657.

Eagles, G., Gohl, K., Larter, R.D., 2009. Animated tectonic reconstruction of the Southern Pacific and alkaline volcanism at its convergent margins since Eocene times. Tectonophysics 464 (1–4), 21–29. Available from: https://doi.org/10.1016/j.tecto.2007.10.005.

Eagles, G., Livermore, R.A., 2002. Opening history of Powell Basin, Antarctic Peninsula. Marine Geology 185 (3–4), 195–205.

Eisermann, H., Eagles, G., Ruppel, A., Smith, E.C., Jokat, W., 2020. Bathymetry beneath Ice Shelves of Western Dronning Maud Land, East Antarctica, and Implications on Ice Shelf Stability. Geophysical Research Letters 47 (12). Available from: https://doi.org/10.1029/2019gl086724, e2019GL086724.

Eittreim, S., Gordon, A.L., Ewing, M., Thorndike, E.M., Bruchhausen, P., 1972. The nepheloid layer and observed bottom currents in the Indian-Pacific Antarctic Sea. Studies in Physical Oceanography 2, 19–35.

Eittreim, S.L., Cooper, A.K., Wannesson, J., 1995. Seismic stratigraphic evidence of ice-sheet advances on the Wilkes Land margin of Antarctica. Sedimentary Geology 96 (1–2), 131–156. Available from: https://doi.org/10.1016/0037-0738(94)00130-M.

Elderfield, H., Ferretti, P., Greaves, M., Crowhurst, S., McCave, I.N., Hodell, D., et al., 2012. Evolution of Ocean Temperature and Ice Volume Through the Mid-Pleistocene Climate Transition. Science 337 (6095), 704–709. Available from: https://doi.org/10.1126/science.1221294.

Elliot, D.H., 1988. Tectonic setting and evolution of the James Ross Basin, northern Antarctic Peninsula. In: Geological Society of America, Boulder, CO, Memoirs, pp. 541–555.

EPICA Community Members, 2004. Eight glacial cycles from an Antarctic ice core. Nature 429 (6992), 623–628. Available from: https://doi.org/10.1038/nature02599.

Erohina, T., Cooper, A.K., Handwerger, D.A., Dunbar, R.B., Cooper, A.K., 2004. Seismic stratigraphic correlations between ODP Sites 742 and 1166; implications for depositional paleoenvironments in Prydz Bay, Antarctica. Proceedings of the Ocean Drilling Program, Scientific Results (CD ROM) 188, 21.

Escutia, C., Brinkhuis, H., 2014. Chapter 3.3 – From greenhouse to icehouse at the Wilkes Land Antarctic Margin: IODP Expedition 318 synthesis of results. In: Blackman, D.K., Stein, R., Inagaki, F., Larsen, H.-C. (Eds.), Developments in Marine Geology, vol. 7.

Elsevier, pp. 295–328. Available from: http://www.sciencedirect.com/science/article/pii/B9780444626172000128.

Escutia, C., Brinkhuis, H., Expedition 318 Scientists, 2014. From Greenhouse to Icehouse at the Wilkes Land Antarctic Margin: IODP Expedition 318 Synthesis of Results. In: Stein, R., Blackman, D.K., Inagaki, F., Larsen, H.-C. (Eds.), Developments in Marine Geology, Vol. 7. Elsevier, pp. 295–328. Available from: https://doi.org/10.1016/B978-0-444-62617-2.00012-8.

Escutia, C., Bárcena, M.A., Lucchi, R.G., Romero, O., Ballegeer, A.M., Gonzalez, J.J., et al., 2009. Circum-Antarctic warming events between 4 and 3.5 Ma recorded in marine sediments from the Prydz Bay (ODP Leg 188) and the Antarctic Peninsula (ODP Leg 178) margins. Global and Planetary Change 69 (3), 170–184. Available from: https://doi.org/10.1016/j.gloplacha.2009.09.003.

Escutia, C., Eittreim, S.L., Cooper, A.K., Nelson, C.H., 2000. Morphology and Acoustic Character of the Antarctic Wilkes Land Turbidite Systems: Ice-Sheet-Sourced vs River-Sourced Fans. Journal of Sedimentary Research 70 (1), 84–93. Available from: https://doi.org/10.1306/2DC40900-0E47-11D7-8643000102C1865D.

Escutia, C., Nelson, C.H., Acton, G.D., Eittreim, S.L., Cooper, A.K., Warnke, D.A., et al., 2002. Current controlled deposition on the Wilkes Land continental rise, Antarctica. Geological Society, London, Memoirs 22 (1), 373–384. Available from: https://doi.org/10.1144/GSL.MEM.2002.022.01.26.

Escutia, C., Warnke, D., Acton, G.D., Barcena, A., Burckle, L., Canals, M., et al., 2003. Sediment distribution and sedimentary processes across the Antarctic Wilkes Land margin during the Quaternary. Deep Sea Research Part II: Topical Studies in Oceanography 50 (8–9), 1481–1508. Available from: https://doi.org/10.1016/S0967-0645(03)00073-0.

Escutia, C., De Santis, L., Donda, F., Dunbar, R.B., Cooper, A.K., Brancolini, G., et al., 2005. Cenozoic ice sheet history from East Antarctic Wilkes Land continental margin sediments. Global and Planetary Change 45 (1–3), 51–81. Available from: https://doi.org/10.1016/j.gloplacha.2004.09.010.

Escutia, C., Brinkhuis, H., Klaus, A., 2011a. Cenozoic East Antarctic ice sheet evolution from Wilkes Land margin sediments. Proceedings of the Integrated Ocean Drilling Program 318.

Escutia, C., Brinkhuis, H., Klaus, A., the IODP Expedition 318 Scientists, 2011b. IODP Expedition 318: from Greenhouse to Icehouse at the Wilkes Land Antarctic Margin. Scientific Drilling 12, 15–23. Available from: https://doi.org/10.2204/iodp.sd.12.02.2011.

Escutia, C., DeConto, R.M., Dunbar, R., Santis, L.D., Shevenell, A., Naish, T., 2019. Keeping an eye on Antarctic ice sheet stability. Oceanography 32 (1), 32–46.

Evangelinos, D., Escutia, C., Etourneau, J., Hoem, F., Bijl, P., Boterblom, W., et al., 2020. Late Oligocene-Miocene proto-Antarctic Circumpolar Current dynamics off the Wilkes Land margin, East Antarctica. Global and Planetary Change 191, 103221. Available from: https://doi.org/10.1016/j.gloplacha.2020.103221.

Eyles, N., Daniels, J., Osterman, L., Januszczak, N., 2001. Ocean Drilling Program Leg 178 (Antarctic Peninsula): sedimentology of glacially influenced continental margin topsets and foresets. Marine Geology 178, 135–156. Available from: https://doi.org/10.1016/S0025-3227(01)00184-0.

Feakins, S.J., Warny, S., DeConto, R.M., 2014. Snapshot of cooling and drying before onset of Antarctic Glaciation. Earth and Planetary Science Letters 404, 154–166. Available from: https://doi.org/10.1016/j.epsl.2014.07.032.

Feakins, S.J., Warny, S., Lee, J.-E., 2012. Hydrologic cycling over Antarctica during the middle Miocene warming. Nature Geoscience 5 (8), 557–560. Available from: https://doi.org/10.1038/ngeo1498.

Fernandez, R.L., Gulick, S.P.S., Domack, E.W., Montelli, A., Leventer, A.R., Shevenell, A.E., et al., 2018. Past ice stream and ice sheet changes on the continental shelf off the Sabrina Coast, East Antarctica. Geomorphology 317, 10–22. Available from: https://doi.org/10.1016/J.GEOMORPH.2018.05.020.

Ferraccioli, F., Armadillo, E., Jordan, T., Bozzo, E., Corr, H., 2009. Aeromagnetic exploration over the East Antarctic Ice Sheet: a new view of the Wilkes Subglacial Basin. Tectonophysics 478 (1–2), 62–77. Available from: https://doi.org/10.1016/j.tecto.2009.03.013.

Ferraccioli, F., Coren, F., Bozzo, E., Zanolla, C., Gandolfi, S., Tabacco, I., et al., 2001. Rifted(?) crust at the East Antarctic Craton margin: gravity and magnetic interpretation along a traverse across the Wilkes Subglacial Basin region. Earth and Planetary Science Letters 192 (3), 407–421. Available from: https://doi.org/10.1016/S0012-821X(01)00459-9.

Ferraccioli, F., Finn, C.A., Jordan, T.A., Bell, R.E., Anderson, L.M., Damaske, D., 2011. East Antarctic rifting triggers uplift of the Gamburtsev Mountains. Nature 479 (7373), 388–392. Available from: https://doi.org/10.1038/nature10566.

Fielding, C.R., Browne, G.H., Field, B., Florindo, F., Harwood, D.M., Krissek, L.A., et al., 2011. Sequence stratigraphy of the ANDRILL AND-2A drillcore, Antarctica: a long-term, ice-proximal record of Early to Mid-Miocene climate, sea-level and glacial dynamism. Palaeogeography, Palaeoclimatology, Palaeoecology 305 (1–4), 337–351. Available from: https://doi.org/10.1016/j.palaeo.2011.03.026.

Fielding, C.R., Thomson, M.R., Cape Roberts Project science team, 1999. Studies from the Cape Roberts Project, Ross Sea Antarctica, Initial report on CRP-2/2A. Terra Antartica 6 (1).

Fielding, C.R., Whittaker, J., Henrys, S.A., Wilson, T.J., Naish, T.R., 2008. Seismic facies and stratigraphy of the Cenozoic succession in McMurdo Sound, Antarctica: implications for tectonic, climatic and glacial history. Palaeogeography, Palaeoclimatology, Palaeoecology 260 (1–2), 8–29. Available from: https://doi.org/10.1016/j.palaeo.2007.08.016.

Fleming, R.F., Barron, J.A., 1996. Evidence of Pliocene Nothofagus in Antarctica from Pliocene marine sedimentary deposits (DSDP Site 274). Marine Micropaleontology 27 (1), 227–236. Available from: https://doi.org/10.1016/0377-8398(95)00062-3.

Florindo, F., Bohaty, S.M., Erwin, P.S., Richter, C., Roberts, A.P., Whalen, P.A., et al., 2003. Magnetobiostratigraphic chronology and palaeoenvironmental history of Cenozoic sequences from ODP sites 1165 and 1166, Prydz Bay, Antarctica. Palaeogeography, Palaeoclimatology, Palaeoecology 198 (1–2), 69–100. Available from: https://doi.org/10.1016/S0031-0182(03)00395-X.

Florindo, F., Harwood, D.M., Talarico, F., Levy, R., the ANDRILL-SMS Science Team, 2008. Studies from the ANDRILL Southern McMurdo Sound Project, Antarctica – Initial Science Report on AND-2A. Terra Antartica 15 (1), 235.

Florindo, F., Siegert, M. (Eds.), 2008. Antarctic Climate Evolution, vol. 8. Elsevier.

Flower, B.P., Kennett, J.P., 1994. The middle Miocene climatic transition: East Antarctic ice sheet development, deep ocean circulation and global carbon cycling. Palaeogeography, Palaeoclimatology, Palaeoecology 108 (3–4), 537–555. Available from: https://doi.org/10.1016/0031-0182(94)90251-8.

Flower, B.P., Zachos, J.C., Pearson, P.N., 2006. Astronomic and oceanographic influences on global carbon cycling across the Oligocene/Miocene boundary. In: Sinha, D.K. (Ed.), Micropaleontology: Application in Stratigraphy and Paleoceanography. Narosa Publishing House. Available from: http://orca.cf.ac.uk/15523/.

Forsberg, C.F., Florindo, F., Grützner, J., Venuti, A., Solheim, A., 2008. Sedimentation and aspects of glacial dynamics from physical properties, mineralogy and magnetic properties at ODP Sites 1166 and 1167, Prydz Bay, Antarctica. Palaeogeography, Palaeoclimatology, Palaeoecology 260 (1), 184–201. Available from: https://doi.org/10.1016/j.palaeo.2007.08.022.

Foster, G.L., Rohling, E.J., 2013. Relationship between sea level and climate forcing by CO2 on geological timescales. Proceedings of the National Academy of Sciences 110 (4), 1209–1214. Available from: https://doi.org/10.1073/pnas.1216073110.

Foster, G.L., Royer, D.L., Lunt, D.J., 2017. Future climate forcing potentially without precedent in the last 420 million years. Nature Communications 8, 14845. Available from: https://doi.org/10.1038/ncomms14845.

Fretwell, P., Pritchard, H.D., Vaughan, D.G., Bamber, J.L., Barrand, N.E., Bell, R., et al., 2013. Bedmap2: improved ice bed, surface and thickness datasets for Antarctica. The Cryosphere 7 (1), 375–393. Available from: https://doi.org/10.5194/tc-7-375-2013.

Fukamachi, Y., Wakatsuchi, M., Taira, K., Kitagawa, S., Ushio, S., Takahashi, A., et al., 2000. Seasonal variability of bottom water properties off Adélie Land, Antarctica. Journal of Geophysical Research 105, 6531–6540. Available from: https://doi.org/10.1029/1999JC900292.

Fütterer, D.K., Kuhn, G., Schenke, H.W., 1990. Wegener Canyon bathymetry and results from rock dredging near ODP Sites 691–693, eastern Weddell Sea, Antarctica. In: Proceedings, Scientific Results, ODP, Leg 113, Weddell Sea, Antarctica, pp. 39–48.

Galeotti, S., et al., 2021. The Eocene/Oligocene boundary climate transition: an Antarctic perspective. In: Florindo, F., et al. (Eds.), Antarctic Climate Evolution, second ed. Elsevier (this volume).

Galeotti, S., DeConto, R., Naish, T., Stocchi, P., Florindo, F., Pagani, M., et al., 2016. Antarctic Ice Sheet variability across the Eocene-Oligocene boundary climate transition. Science 352 (6281), 76–80. Available from: https://doi.org/10.1126/science.aab0669.

Gales, J.A., Larter, R.D., Mitchell, N.C., Hillenbrand, C.D., Østerhus, S., Shoosmith, D.R., 2012. Southern Weddell Sea shelf edge geomorphology: implications for gully formation by the overflow of high-salinity water. Journal of Geophysical Research: Earth Surface 117 (F4), F04021. Available from: https://doi.org/10.1029/2012JF002357.

Gales, J.A., Larter, R.D., Leat, P.T., Jokat, W., 2016. Components of an Antarctic trough-mouth fan: examples from the Crary Fan, Weddell Sea. Geological Society, London, Memoirs 46 (1), 377–378. Available from: https://doi.org/10.1144/M46.82.

Gales, J.A., Leat, P.T., Larter, R.D., Kuhn, G., Hillenbrand, C.D., Graham, A.G.C., et al., 2014. Large-scale submarine landslides, channel and gully systems on the southern Weddell Sea margin, Antarctica. Marine Geology 348, 73–87. Available from: https://doi.org/10.1016/j.margeo.2013.12.002.

Gales, J.A., Rebesco, M., De Santis, L., Bergamasco, A., Colleoni, F., Kim, S., et al., 2021. Role of dense shelf water in the development of Antarctic submarine canyon morphology. Geomorphology 372, 107453. Available from: https://doi.org/10.1016/j.geomorph.2020.107453.

Gambôa, L.A.P., Maldonado, P.R., John, B.S., 1990. Geophysical Investigations in the Bransfield Strait and in the Bellingshausen Sea—Antarctica, Antarctica as an Exploration Frontier—Hydrocarbon Potential, Geology, and Hazards, vol. 31. American Association of Petroleum Geologists.

García, M., Ercilla, G., Anderson, J.B., Alonso, B., 2008. New insights on the post-rift seismic stratigraphic architecture and sedimentary evolution of the Antarctic Peninsula margin (Central Bransfield Basin). Marine Geology 251 (3–4), 167–182. Available from: https://doi.org/10.1016/j.margeo.2008.02.006.

Gasson, E., DeConto, R.M., Pollard, D., 2016a. Modeling the oxygen isotope composition of the Antarctic ice sheet and its significance to Pliocene sea level. Geology 44 (10), 827–830. Available from: https://doi.org/10.1130/G38104.1.

Gasson, E., DeConto, R.M., Pollard, D., Levy, R.H., 2016b. Dynamic Antarctic ice sheet during the early to mid-Miocene. Proceedings of the National Academy of Sciences 113 (13), 3459–3464. Available from: https://doi.org/10.1073/pnas.1516130113.

Geibert, W., Rutgers van der Loeff, M.M., Usbeck, R., Gersonde, R., Kuhn, G., Seeberg-Elverfeldt, J., 2005. Quantifying the opal belt in the Atlantic and southeast Pacific sector of the Southern Ocean by means of ^{230}Th normalization. Global Biogeochemical Cycles 19, GB4001. Available from: https://doi.org/10.1029/2005GB002465.

Geletti, R., Busetti, M., 2011. A double bottom simulating reflector in the western Ross Sea, Antarctica. Journal of Geophysical Research: Solid Earth 116 (B4), B04101. Available from: https://doi.org/10.1029/2010JB007864.

Giorgetti, A., Crise, A., Laterza, R., Perini, L., Rebesco, M., Camerlenghi, A., 2003. Water masses and bottom boundary layer dynamics above a sediment drift of the Antarctic Peninsula Pacific Margin. Antarctic Science 15 (4), 537–546. Available from: https://doi.org/10.1017/S0954102003001652.

Gleadow, A.J.W., Fitzgerald, P.G., 1987. Uplift history and structure of the Transantarctic Mountains: new evidence from fission track dating of basement apatites in the Dry Valleys area, southern Victoria Land. Earth and Planetary Science Letters 82 (1), 1–14. Available from: https://doi.org/10.1016/0012-821X(87)90102-6.

Gohl, K., Freudenthal, T., Hillenbrand, C.-D., Klages, J., Larter, R., Bickert, T., et al., 2017. MeBo70 seabed drilling on a Polar Continental Shelf: operational report and lessons from drilling in the Amundsen Sea Embayment of West Antarctica. Geochemistry, Geophysics, Geosystems 18 (11), 4235–4250. Available from: https://doi.org/10.1002/2017GC007081.

Gohl, K., Uenzelmann-Neben, G., Larter, R.D., Hillenbrand, C.-D., Hochmuth, K., Kalberg, T., et al., 2013. Seismic stratigraphic record of the Amundsen Sea Embayment shelf from pre-glacial to recent times: evidence for a dynamic West Antarctic ice sheet. Marine Geology 344, 115–131. Available from: https://doi.org/10.1016/j.margeo.2013.06.011.

Gohl, K., Wellner, J.S., Klaus, A., the Expedition 379 Scientists, 2021. Amundsen Sea West Antarctic Ice Sheet History. In: Proceedings of the International Ocean Discovery Program, 379, International Ocean Discovery Program, College Station, TX. <https://doi.org/10.14379/iodp.proc.379.2021>.

Golledge, N.R., Keller, E.D., Gomez, N., Naughten, K.A., Bernales, J., Trusel, L.D., et al., 2019. Global environmental consequences of twenty-first-century ice-sheet melt. Nature 566 (7742), 65. Available from: https://doi.org/10.1038/s41586-019-0889-9.

Golledge, N.R., Kowalewski, D.E., Naish, T.R., Levy, R.H., Fogwill, C.J., Gasson, E.G.W., 2015. The multi-millennial Antarctic commitment to future sea-level rise. Nature 526 (7573), 421–425. Available from: https://doi.org/10.1038/nature15706.

Golledge, N.R., Levy, R.H., McKay, R.M., Naish, T.R., 2017. East Antarctic ice sheet most vulnerable to Weddell Sea warming. Geophysical Research Letters 44 (5), 2343–2351.

Gràcia, E., Canals, M., Farràn, M., Sorribas, J., Pallàs, R., 1997. Central and eastern Bransfield basins (Antarctica) from high-resolution swath-bathymetry data. Antarctic Science 9 (2), 168–180.

Graham, A.G.C., Larter, R.D., Gohl, K., Hillenbrand, C.-D., Smith, J.A., Kuhn, G., 2009. Bedform signature of a West Antarctic palaeo-ice stream reveals a multi-temporal record of flow and substrate control. Quaternary Science Reviews 28 (25), 2774–2793. Available from: https://doi.org/10.1016/j.quascirev.2009.07.003.

Granot, R., Dyment, J., 2018. Late Cenozoic unification of East and West Antarctica. Nature Communications 9 (1), 3189. Available from: https://doi.org/10.1038/s41467-018-05270-w.

Granot, R., Cande, S.C., Stock, J.M., Damaske, D., 2013. Revised Eocene-Oligocene kinematics for the West Antarctic rift system. Geophysical Research Letters 40 (2), 279–284. Available from: https://doi.org/10.1029/2012GL054181.

Grant, G.R., Naish, T.R., Dunbar, G.B., Stocchi, P., Kominz, M.A., Kamp, P.J.J., et al., 2019. The amplitude and origin of sea-level variability during the Pliocene epoch. Nature 574 (7777), 237–241. Available from: https://doi.org/10.1038/s41586-019-1619-z.

Greenbaum, J.S., Blankenship, D.D., Young, D.A., Richter, T.G., Roberts, J.L., Aitken, A.R.A., et al., 2015. Ocean access to a cavity beneath Totten Glacier in East Antarctica. Nature Geoscience. Available from: https://doi.org/10.1038/ngeo2388, *advance online publication*.

Grutzner, J., Rebesco, M.A., Cooper, A.K., Forsberg, C.F., Kryc, K.A., Wefer, G., 2003. Evidence for orbitally controlled size variations of the East Antarctic Ice Sheet during the late Miocene. Geology 31 (9), 777–780. Available from: https://doi.org/10.1130/G19574.1.

Gulick, S.P.S., Shevenell, A.E., Montelli, A., Fernandez, R., Smith, C., Warny, S., et al., 2017. Initiation and long-term instability of the East Antarctic Ice Sheet. Nature 552, 225.

Hall, J., Wilson, T., Henrys, S., 2007. Structure of the central Terror Rift, western Ross Sea, Antarctica. United States Geological Survey and The National Academies; USGS OFR-2007-1047, Short Research Paper 108. <https://doi.org/10.3133/of2007-1047.srp108>.

Hambrey, M.J., Ehrmann, W.U., Larsen, B., 1991. Cenozoic glacial record of the Prydz Bay continental shelf, East Antarctica. In: Proceedings Scientific Results, ODP, Leg 119, Kerguelen Plateau-Prydz Bay, pp. 77–132.

Hambrey, M.J., McKelvey, B., 2000. Neogene fjordal sedimentation on the western margin of the Lambert Graben, East Antarctica. Sedimentology 47 (3), 577–607. Available from: https://doi.org/10.1046/j.1365-3091.2000.00308.x.

Hansen, M.A., Passchier, S., Khim, B.-K., Song, B., Williams, T., 2015. Threshold behavior of a marine-based sector of the East Antarctic Ice Sheet in response to early Pliocene ocean warming. Paleoceanography 30 (6). Available from: https://doi.org/10.1002/2014PA002704, 2014PA002704.

Harrington, P.K., Barker, P.F., Griffiths, D.H., 1972. Continental structure of the South Orkney Islands area from seismic refraction and magnetic measurements. In: Adie, J. (Ed.), Antarctic Geology and Geophysics. Universitets Forlaget, pp. 27–32.

Harrowfield, M., Holdgate, G.R., Wilson, C.J.L., McLoughlin, S., 2005. Tectonic significance of the Lambert graben, East Antarctica: reconstructing the Gondwanan rift. Geology 33 (3), 197–200. Available from: https://doi.org/10.1130/G21081.1.

Hartman, J.D., Sangiorgi, F., Salabarnada, A., Peterse, F., Houben, A.J.P., Schouten, S., et al., 2018. Paleoceanography and ice sheet variability offshore Wilkes Land, Antarctica – part 3: insights from Oligocene–Miocene TEX_{86}-based sea surface temperature reconstructions. Climate of the Past 14 (9), 1275–1297. Available from: https://doi.org/10.5194/cp-14-1275-2018.

Hattermann, T., 2018. Antarctic thermocline dynamics along a narrow shelf with easterly winds. Journal of Physical Oceanography 48 (10), 2419–2443. Available from: https://doi.org/10.1175/jpo-d-18-0064.1.

Hattermann, T., Nøst, O.A., Lilly, J.M., Smedsrud, L.H., 2012. Two years of oceanic observations below the Fimbul Ice Shelf, Antarctica. Geophysical Research Letters 39 (12). Available from: https://doi.org/10.1029/2012gl051012.

Hattermann, T., Smedsrud, L.H., Nøst, O.A., Lilly, J.M., Galton-Fenzi, B.K., 2014. Eddy-resolving simulations of the Fimbul Ice Shelf cavity circulation: basal melting and exchange with open ocean. Ocean Modelling 82, 28–44. Available from: https://doi.org/10.1016/j.ocemod.2014.07.004.

Hauptvogel, D.W., Pekar, S.F., Pincay, V., 2017. Evidence for a heavily glaciated Antarctica during the late Oligocene "warming" (27.8–24.5 Ma): stable isotope records from ODP Site 690. Paleoceanography 32 (4), 384–396. Available from: https://doi.org/10.1002/2016PA002972.

Hayes, D.E., Frakes, L.A., et al., 1975. Initial Reports of the Deep Sea Drilling Project, vol. 28. US Government Printing Office. Available from: http://deepseadrilling.org/28/dsdp_toc.htm.

Hellmer, H.H., Kauker, F., Timmermann, R., Determann, J., Rae, J., 2012. Twenty-first-century warming of a large Antarctic ice-shelf cavity by a redirected coastal current. Nature 485 (7397), 225–228. Available from: https://doi.org/10.1038/nature11064.

Herbert, T.D., Lawrence, K.T., Tzanova, A., Peterson, L.C., Caballero-Gill, R., Kelly, C.S., 2016. Late Miocene global cooling and the rise of modern ecosystems. Nature Geoscience 9 (11), 843–847. Available from: https://doi.org/10.1038/ngeo2813.

Hernández-Molina, F.J., Larter, R.D., Maldonado, A., 2017. Neogene to Quaternary stratigraphic evolution of the Antarctic Peninsula, Pacific Margin offshore of Adelaide Island: transitions from a non-glacial, through glacially-influenced to a fully glacial state. Global and Planetary Change 156, 80–111. Available from: https://doi.org/10.1016/j.gloplacha.2017.07.002.

Hernández-Molina, F.J., Larter, R.D., Rebesco, M., Maldonado, A., 2006. Miocene reversal of bottom water flow along the Pacific Margin of the Antarctic Peninsula: stratigraphic evidence from a contourite sedimentary tail. Marine Geology 228 (1–4), 93–116.

Herron, M.J., Anderson, J.B., 1990. Late Quaternary glacial history of the South Orkney Plateau, Antarctica. Quaternary Research 33 (3), 265–275. Available from: https://doi.org/10.1016/0033-5894(90)90055-P.

Hillenbrand, C.-D., Kuhn, G., Smith, J.A., Gohl, K., Graham, A.G.C., Larter, R.D., Klages, J.P., Downey, R., Moreton, S.G., Forwick, M., Vaughan, D.G. 2013. Grounding-line retreat of the West Antarctic Ice Sheet from inner Pine Island Bay. Geology 41, 35–38, Available from: https://doi.org/10.1130/G33469.1

Hillenbrand, C.-D., Bentley, M.J., Stolldorf, T.D., Hein, A.S., Kuhn, G., Graham, A.G.C., et al., 2014. Reconstruction of changes in the Weddell Sea sector of the Antarctic Ice Sheet since the Last Glacial Maximum. Quaternary Science Reviews 100, 111–136. Available from: https://doi.org/10.1016/j.quascirev.2013.07.020.

Hillenbrand, C.-D., Ehrmann, W., 2005. Late Neogene to Quaternary environmental changes in the Antarctic Peninsula region: evidence from drift sediments. Global and Planetary Change 45 (1–3), 165.

Hillenbrand, C.-D., Fütterer, D., Grobe, H., Frederichs, T., 2002. No evidence for a Pleistocene collapse of the West Antarctic Ice Sheet from continental margin sediments recovered in the Amundsen Sea. Geo-Marine Letters 22 (2), 51–59. Available from: https://doi.org/10.1007/s00367-002-0097-7.

Hillenbrand, C.-D., Kuhn, G., Frederichs, T., 2009. Record of a Mid-Pleistocene depositional anomaly in West Antarctic continental margin sediments: an indicator for ice-sheet collapse. Quaternary Science Reviews 28 (13–14), 1147–1159. Available from: https://doi.org/10.1016/j.quascirev.2008.12.010.

Hillenbrand, C.-D., Smith, J.A., Hodell, D.A., Greaves, M., Poole, C.R., Kender, S., et al., 2017. West Antarctic Ice Sheet retreat driven by Holocene warm water incursions. Nature 547 (7661), 43. Available from: https://doi.org/10.1038/nature22995.

Hochmuth, K., Gohl, K., 2013. Glaciomarine sedimentation dynamics of the Abbot glacial trough of the Amundsen Sea Embayment shelf, West Antarctica. Geological Society,

London, Special Publications 381 (1), 233–244. Available from: https://doi.org/10.1144/SP381.21.

Hochmuth, K., Gohl, K., 2019. Seaward growth of Antarctic continental shelves since establishment of a continent-wide ice sheet: patterns and mechanisms. Palaeogeography, Palaeoclimatology, Palaeoecology 520, 44–54. Available from: https://doi.org/10.1016/j.palaeo.2019.01.025.

Hochmuth, K., Gohl, K., Leitchenkov, G., Sauermilch, I., Whittaker, J.M., Uenzelmann-Neben, G., et al., 2020a. The evolving paleobathymetry of the circum-Antarctic Southern Ocean since 34 Ma – a key to understanding past cryosphere-ocean developments. Geochemistry, Geophysics, Geosystems. Available from: https://doi.org/10.1029/2020gc009122, e2020GC009122.

Hochmuth, K., Paxman, G., Gohl, K., Jamieson, S., Leitchenkov, G., Bentley, M., et al., 2020b. Combined palaeotopography and palaeobathymetry of the Antarctic continent and the Southern Ocean since 34 Ma. PANGAEA. Available from: https://doi.org/10.1594/PANGAEA.923109.

Holbourn, A., Kuhnt, W., Clemens, S., Prell, W., Andersen, N., 2013. Middle to late Miocene stepwise climate cooling: evidence from a high-resolution deep water isotope curve spanning 8 million years. Paleoceanography 28 (4), 688–699. Available from: https://doi.org/10.1002/2013PA002538.

Holbourn, A., Kuhnt, W., Kochhann, K.G.D., Andersen, N., Meier, K.J.S., 2015. Global perturbation of the carbon cycle at the onset of the Miocene Climatic Optimum. Geology 43 (2), 123–126. Available from: https://doi.org/10.1130/G36317.1.

Holbourn, A., Kuhnt, W., Lyle, M., Schneider, L., Romero, O., Andersen, N., 2014. Middle Miocene climate cooling linked to intensification of eastern equatorial Pacific upwelling. Geology 42 (1), 19–22. Available from: https://doi.org/10.1130/G34890.1.

Hollis, C.J., Dunkley Jones, T., Anagnostou, E., Bijl, P.K., Cramwinckel, M.J., Cui, Y., et al., 2019. The DeepMIP contribution to PMIP4: methodologies for selection, compilation and analysis of latest Paleocene and early Eocene climate proxy data, incorporating version 0.1 of the DeepMIP database. Geoscientific Model Development Discussions 1–98. Available from: https://doi.org/10.5194/gmd-2018-309.

Hollis, C.J., Handley, L., Crouch, E.M., Morgans, H.E.G., Baker, J.A., Creech, J., et al., 2009. Tropical Sea Temperatures in the High-Latitude South Pacific During the Eocene. Geology 37 (2), 99–102. Available from: https://doi.org/10.1130/G25200A.1.

Hollis, C.J., Tayler, M.J.S., Andrew, B., Taylor, K.W., Lurcock, P., Bijl, P.K., et al., 2014. Organic-rich sedimentation in the South Pacific Ocean associated with late Paleocene climatic cooling. Earth-Science Reviews 134, 81–97. Available from: https://doi.org/10.1016/j.earscirev.2014.03.006.

Hollis, C.J., Taylor, K.W.R., Handley, L., Pancost, R.D., Huber, M., Creech, J.B., et al., 2012. Early Paleogene temperature history of the Southwest Pacific Ocean: reconciling proxies and models. Earth and Planetary Science Letters 349–350 (0), 53–66. Available from: https://doi.org/10.1016/j.epsl.2012.06.024.

Hollister, C.D., Craddock, C., et al., 1976. Initial Reports of the Deep Sea Drilling Project, vol. 35. US Government Printing Office, p. 35. Available from: https://doi.org/10.2973/dsdp.proc.35.1976.

Honjo, S., Manganini, S.J., Krishfield, R.A., Francois, R., 2008. Particulate organic carbon fluxes to the ocean interior and factors controlling the biological pump: a synthesis of global sediment trap programs since 1983. Progress in Oceanography 76 (3), 217–285. Available from: https://doi.org/10.1016/j.pocean.2007.11.003.

Horgan, H., Naish, T., Bannister, S., Balfour, N., Wilson, G., 2005. Seismic stratigraphy of the Plio-Pleistocene Ross Island flexural moat-fill: a prognosis for ANDRILL Program drilling beneath McMurdo-Ross Ice Shelf. Global and Planetary Change 45 (1–3), 83–97. Available from: https://doi.org/10.1016/j.gloplacha.2004.09.014.

Houben, A.J.P., Bijl, P.K., Pross, J., Bohaty, S.M., Passchier, S., Stickley, C.E., et al., 2013. Reorganization of Southern Ocean plankton ecosystem at the onset of Antarctic glaciation. Science 340 (6130), 341–344.

Howe, J.A., Wilson, C.R., Shimmield, T.M., Diaz, R.J., Carpenter, L.W., 2007. Recent deep-water sedimentation, trace metal and radioisotope geochemistry across the Southern Ocean and Northern Weddell Sea, Antarctica. Deep Sea Research Part II: Topical Studies in Oceanography 54 (16), 1652–1681. Available from: https://doi.org/10.1016/j.dsr2.2007.07.007.

Huang, X., Bernhardt, A., De Santis, L., Wu, S., Leitchenkov, G., Harris, P., et al., 2020. Depositional and erosional signatures in sedimentary successions on the continental slope and rise off Prydz Bay, East Antarctica– implications for Pliocene paleoclimate. Marine Geology 430, 106339. Available from: https://doi.org/10.1016/j.margeo.2020.106339.

Huang, X., Gohl, K., Jokat, W., 2014. Variability in Cenozoic sedimentation and paleo-water depths of the Weddell Sea basin related to pre-glacial and glacial conditions of Antarctica. Global and Planetary Change 118, 25–41. Available from: https://doi.org/10.1016/j.gloplacha.2014.03.010.

Huang, X., Jokat, W., 2016. Middle Miocene to present sediment transport and deposits in the Southeastern Weddell Sea, Antarctica. Global and Planetary Change 139, 211–225. Available from: https://doi.org/10.1016/j.gloplacha.2016.03.002.

Huber, M., Brinkhuis, H., Stickley, C.E., Döös, K., Sluijs, A., Warnaar, J., et al., 2004. Eocene circulation of the Southern Ocean: was Antarctica kept warm by subtropical waters? Paleoceanography 19 (4), PA4026. Available from: https://doi.org/10.1029/2004PA001014.

Huber, M., Caballero, R., 2011. The early Eocene equable climate problem revisited. Climate of the Past 7 (2), 603–633.

Huybrechts, P., 1993. Glaciological modelling of the late Cenozoic East Antarctic ice sheet: stability or dynamism? Geografiska Annaler, Series A 75A (4), 221–238. Available from: https://doi.org/10.1080/04353676.1993.11880395.

Ishman, S.E., Sperling, M.R., 2002. Benthic foraminiferal record of Holocene deep-water evolution in the Palmer Deep, western Antarctic Peninsula. Geology 30 (5), 435–438. Available from: https://doi.org/10.1130/0091-7613(2002)030 < 0435:Bfrohd > 2.0.Co;2.

Ivany, L.C., Lohmann, K.C., Hasiuk, F., Blake, D.B., Glass, A., Aronson, R.B., et al., 2008. Eocene Climate Record of a High Southern Latitude Continental Shelf: Seymour Island, Antarctica. Geological Society of America Bulletin 120 (5–6), 659–678. Available from: https://doi.org/10.1130/B26269.1.

Jacobs, J., Opås, B., Elburg, M.A., Läufer, A., Estrada, S., Ksienzyk, A.K., et al., 2017. Cryptic sub-ice geology revealed by a U-Pb zircon study of glacial till in Dronning Maud Land, East Antarctica. Precambrian Research 294, 1–14. Available from: https://doi.org/10.1016/j.precamres.2017.03.012.

Jagniecki, E.A., Lowenstein, T.K., Jenkins, D.M., Demicco, R.V., 2015. Eocene atmospheric CO_2 from the nahcolite proxy. Geology G36886.1. Available from: https://doi.org/10.1130/G36886.1.

Jeffers, J.D., Anderson, J.B., John, B.S., 1990. Sequence stratigraphy of the bransfield basin, Antarctica: implications for tectonic history and hydrocarbon potential, Antarctica as an Exploration Frontier—Hydrocarbon Potential, Geology, and Hazards, vol. 31. American Association of Petroleum Geologists.

Jerosch, K., Kuhn, G., Krajnik, I., Scharf, F.K., Dorschel, B., 2016. A geomorphological seabed classification for the Weddell Sea, Antarctica. Marine Geophysical Research 37 (2), 127–141. Available from: https://doi.org/10.1007/s11001-015-9256-x.

Ji, F., Li, F., Gao, J.-Y., Zhang, Q., Hao, W.-F., 2018. 3-D density structure of the Ross Sea basins, West Antarctica from constrained gravity inversion and their tectonic implications. Geophysical Journal International 215 (2), 1241–1256. Available from: https://doi.org/10.1093/gji/ggy343.

Jimenez-Espejo, F.J., Presti, M., Kuhn, G., Mckay, R., Crosta, X., Escutia, C., et al., 2020. Late Pleistocene oceanographic and depositional variations along the Wilkes Land margin (East Antarctica) reconstructed with geochemical proxies in deep-sea sediments. Global and Planetary Change 184, 103045. Available from: https://doi.org/10.1016/j.gloplacha.2019.103045.

Jin, Y.K., Larter, R.D., Kim, Y., Nam, S.H., Kim, K.J., 2002. Post-subduction margin structures along Boyd Strait, Antarctic Peninsula. Tectonophysics 346 (3), 187–200. Available from: https://doi.org/10.1016/S0040-1951(01)00281-5.

John, C.M., Karner, G.D., Mutti, M., 2004. $\delta^{18}O$ and Marion Plateau backstripping: combining two approaches to constrain late middle Miocene eustatic amplitude. Geology 32 (9), 829–832. Available from: https://doi.org/10.1130/G20580.1.

Jokat, W., Herter, U., 2016. Jurassic failed rift system below the Filchner-Ronne-Shelf, Antarctica: new evidence from geophysical data. Tectonophysics 688, 65–83. Available from: https://doi.org/10.1016/j.tecto.2016.09.018.

Jones, R.S., Norton, K.P., Mackintosh, A.N., Anderson, J.T.H., Kubik, P., Vockenhuber, C., et al., 2017. Cosmogenic nuclides constrain surface fluctuations of an East Antarctic outlet glacier since the Pliocene. Earth and Planetary Science Letters 480 (Suppl. C), 75–86. Available from: https://doi.org/10.1016/j.epsl.2017.09.014.

Jordan, T.A., Ferraccioli, F., Leat, P.T., 2017. New geophysical compilations link crustal block motion to Jurassic extension and strike-slip faulting in the Weddell Sea Rift System of West Antarctica. Gondwana Research 42, 29–48. Available from: https://doi.org/10.1016/j.gr.2016.09.009.

Jordan, T.A., Siddoway, C.S., 2020. The geological history and evolution of West Antarctica. Nature Reviews Earth & Environment 1 (2), 117–133. Available from: https://doi.org/10.1038/s43017-019-0013-6.

Kalberg, T., Gohl, K., 2014. The crustal structure and tectonic development of the continental margin of the Amundsen Sea Embayment, West Antarctica: implications from geophysical data. Geophysical Journal International 198 (1), 327–341. Available from: https://doi.org/10.1093/GJI/GGU118.

Kavoun, M., Vinnikovskaya, O., 1994. Seismic stratigraphy and tectonics of the northwestern Weddell Sea (Antarctica) inferred from marine geophysical surveys. Tectonophysics 240, 299–341.

Kemp, E.M., Barrett, P.J., 1975. Antarctic glaciation and early Tertiary vegetation. Nature 258 (5535), 507. Available from: https://doi.org/10.1038/258507a0.

Kennett, J.P., 1977. Cenozoic Evolution of Antarctic Glaciation, the Circum-Antarctic Ocean, and Their Impact on Global Paleoceanography. Journal of Geophysical Research 82 (27), 3843–3860. Available from: https://doi.org/10.1029/JC082i027p03843.

Kennett, J.P., Barker, P.F., 1990. Latest Cretaceous to Cenozoic climate and oceanographic developments in the Weddell Sea, Antarctica: an ocean-drilling perspective. Proceedings of the Ocean Drilling Program, Scientific Results 113, 937–960.

Kennett, J., Stott, L., 1990. Proteus and Proto-Oceanus: Ancestral Paleogene Oceans as Revealed from Antarctic Stable Isotopic Results; ODP Leg 113. <https://doi.org/10.2973/ODP.PROC.SR.113.188.1990>.

Kennett, J.P., Houtz, R.E., Andrews, P.B., Edwards, A.R., Gostin, V.A., Hajos, M., et al., 1974. Development of the circum-antarctic current. Science 186 (4159), 144–147. Available from: https://doi.org/10.1126/science.186.4159.144.

Kennicutt, M., Kim, Y., Rogan-Finnemore, M., Anandakrishnan, S., Chown, S., Colwell, S., et al., 2016. Delivering 21st century Antarctic and Southern Ocean science. Antarctic Science 28 (6), 407–423. Available from: https://doi.org/10.1017/S0954102016000481.

Khazendar, A., Schodlok, M.P., Fenty, I., Ligtenberg, S.R.M., Rignot, E., Van Den Broeke, M. R., 2013. Observed thinning of Totten Glacier is linked to coastal polynya variability. Nature Communications 4. Available from: https://doi.org/10.1038/ncomms3857.

Khim, B.-K., Song, B., Cho, H.G., Williams, T., Escutia, C., 2017. Late Neogene sediment properties in the Wilkes Land continental rise (IODP Exp. 318 Hole U1359A), East Antarctica. Geosciences Journal 21 (1), 21–32. Available from: https://doi.org/10.1007/s12303-016-0037-6.

Kim, Y., Kim, H.-S., Larter, R.D., Camerlenghi, A., Gambôa, L.A.P., Rudowski, S., 1995. Tectonic deformation in the upper crust and sediments at the South Shetland Trench. In: Cooper, A.K., Barker, P.F., Brancolini, G. (Eds.), Geology and Seismic Stratigraphy of the Antarctic Margin, vol. 68. American Geophysical Union, pp. 157–166.

Kim, S., Lee, J.I., McKay, R.M., Yoo, K.-C., Bak, Y.-S., Lee, M.K., et al., 2020. Late Pleistocene paleoceanographic changes in the Ross Sea – Glacial-interglacial variations in paleoproductivity, nutrient utilization, and deep-water formation. Quaternary Science Reviews 239, 106356. Available from: https://doi.org/10.1016/j.quascirev.2020.106356.

Kim, J.-H., van der Meer, J., Schouten, S., Helmke, P., Willmott, V., Sangiorgi, F., et al., 2010. New indices and calibrations derived from the distribution of crenarchaeal isoprenoid tetraether lipids: implications for past sea surface temperature reconstructions. Geochimica et Cosmochimica Acta 74 (16), 4639–4654. Available from: https://doi.org/10.1016/j.gca.2010.05.027.

Kim, S., De Santis, L., Hong, J.K., Cottlerle, D., Petronio, L., Colizza, E., et al., 2018. Seismic stratigraphy of the Central Basin in northwestern Ross Sea slope and rise, Antarctica: clues to the late Cenozoic ice-sheet dynamics and bottom-current activity. Marine Geology 395, 363–379. Available from: https://doi.org/10.1016/j.margeo.2017.10.013.

Kimura, K., 1982. Geological and geophysical survey in the Bellingshausen Basin, off Antarctica. Antarctic Record 75, 12–24.

King, E.C., Barker, P.F., 1988. The margins of the South Orkney microcontinent. Journal of Geological Society (London) 145 (2), 317–331.

King, E., Leitchenkov, G., Galindo Zaldivar, J., Maldonado, A., Lodolo, E., 1997. Crustal structure and sedimentation in Powell Basin. In: Barker, P.F., Cooper, A. (Eds.), Geology and Seismic Stratigraphy of the Antarctic Margin, Part 2, vol. 71. American Geophysical Union, pp. 75–93.

Kingslake, J., Scherer, R.P., Albrecht, T., Coenen, J., Powell, R.D., Reese, R., et al., 2018. Extensive retreat and re-advance of the West Antarctic Ice Sheet during the Holocene. Nature 558 (7710), 430–434. Available from: https://doi.org/10.1038/s41586-018-0208-x.

Klages, J.P., Kuhn, G., Graham, A.G.C., Hillenbrand, C.-D., Smith, J.A., Nitsche, F.O., et al., 2015. Palaeo-ice stream pathways and retreat style in the easternmost Amundsen Sea Embayment, West Antarctica, revealed by combined multibeam bathymetric and seismic data. Geomorphology 245, 207–222. Available from: https://doi.org/10.1016/j.geomorph.2015.05.020.

Klages, J.P., Kuhn, G., Hillenbrand, C.-D., Graham, A.G.C., Smith, J.A., Larter, R.D., et al., 2014. Retreat of the West Antarctic Ice Sheet from the western Amundsen Sea shelf at a pre- or early LGM stage. Quaternary Science Reviews 91, 1–15. Available from: https://doi.org/10.1016/j.quascirev.2014.02.017.

Klages, J.P., Kuhn, G., Hillenbrand, C.-D., Smith, J.A., Graham, A.G.C., Nitsche, F.O., et al., 2017. Limited grounding-line advance onto the West Antarctic continental shelf in the easternmost Amundsen Sea Embayment during the last glacial period. PLoS One 12 (7), e0181593. Available from: https://doi.org/10.1371/journal.pone.0181593.

Klages, J.P., Salzmann, U., Bickert, T., Hillenbrand, C.-D., Gohl, K., Kuhn, G., et al., 2020. Temperate rainforests near the South Pole during peak Cretaceous warmth. Nature 580 (7801), 81–86. Available from: https://doi.org/10.1038/s41586-020-2148-5.

Kominz, M.A., Browning, J.V., Miller, K.G., Sugarman, P.J., Mizintseva, S., Scotese, C.R., 2008. Late Cretaceous to Miocene sea-level estimates from the New Jersey and Delaware coastal plain coreholes: an error analysis. Basin Research 20 (2), 211–226. Available from: https://doi.org/10.1111/j.1365-2117.2008.00354.x.

Konfirst, M.A., Scherer, R.P., Hillenbrand, C.-D., Kuhn, G., 2012. A marine diatom record from the Amundsen Sea—insights into oceanographic and climatic response to the Mid-Pleistocene transition in the West Antarctic sector of the Southern Ocean. Marine Micropaleontology 92–93, 40–51. Available from: https://doi.org/10.1016/j.marmicro.2012.05.001.

Koppers, A.A.P., R. Coggon (Eds.), 2020. Exploring Earth by Scientific Ocean Drilling: 2050 Science Framework. 124 pp. <https://doi.org/10.6075/J0W66J9H>.

Kristoffersen, Y., Hofstede, C., Diez, A., Blenkner, R., Lambrecht, A., Mayer, C., et al., 2014. Reassembling Gondwana: a new high quality constraint from vibroseis exploration of the sub-ice shelf geology of the East Antarctic continental margin. Journal of Geophysical Research: Solid Earth 119 (12), 9171–9182. Available from: https://doi.org/10.1002/2014JB011479.

Kristoffersen, Y., Jokat, W., 2008. The Weddell Sea. In: Florindo, Fabio, Siegert, M. (Eds.), Antarctic Climate Evolution, vol. 8. Elsevier, pp. 144–152. In: Cooper et al., Cenozoic Climate History from Seismic-Reflection and Drilling Studies on the Antarctic Continental Margin.

Kuhn, G., Gaedicke, C., 2015. A plan for interdisciplinary process-studies and geoscientific observations beneath the Ekstroem Ice Shelf (Sub-EIS-Obs). Polarforschung 84 (2), 99–102.

Kuhn, G., Hass, H.C., Censarek, B., Rudalph, M., Forwick, M., Quirós-Alperta, S., 2002. Meeresgeologie, vol. 404. AWI.

Kuhn, G., Hillenbrand, C.-D., Kasten, S., Smith, J.A., Nitsche, F.O., Frederichs, T., et al., 2017. Evidence for a palaeo-subglacial lake on the Antarctic continental shelf. Nature Communications 8 (1), 15591. Available from: https://doi.org/10.1038/ncomms15591.

Kulhanek, D.K., Levy, R.H., Clowes, C.D., Prebble, J.G., Rodelli, D., Jovane, L., et al., 2019. Revised chronostratigraphy of DSDP Site 270 and late Oligocene to early Miocene paleoecology of the Ross Sea sector of Antarctica. Global and Planetary Change 178, 46–64. Available from: https://doi.org/10.1016/j.gloplacha.2019.04.002.

Kuvaas, B., Kristoffersen, Y., Guseva, J., Leitchenkov, G., Gandjukhin, V., Løvås, O., et al., 2005. Interplay of turbidite and contourite deposition along the Cosmonaut Sea/Enderby Land margin, East Antarctica. Marine Geology 217 (1–2), 143–159. Available from: https://doi.org/10.1016/j.margeo.2005.02.025.

Kuvaas, B., Leitchenkov, G., 1992. Glaciomarine turbidite and current controlled deposits in Prydz Bay, Antarctica. Marine Geology 108 (3–4), 365–381. Available from: https://doi.org/10.1016/0025-3227(92)90205-V.

Ladant, J.B., Donnadieu, Y., Dumas, C., 2014. Links between CO_2, glaciation and water flow: reconciling the Cenozoic history of the Antarctic Circumpolar Current. Climate of the Past 10 (6), 1957–1966. Available from: https://doi.org/10.5194/cp-10-1957-2014.

Ladant, J.-B., Donnadieu, Y., Bopp, L., Lear, C.H., Wilson, P.A., 2018. Meridional contrasts in productivity changes driven by the opening of Drake Passage. Paleoceanography and Paleoclimatology 33 (3), 302–317. Available from: https://doi.org/10.1002/2017pa003211.

Larter, R.D., Barker, P.F., 1989. Seismic stratigraphy of the Antarctic Peninsula Pacific margin: a record of Pliocene-Pleistocene ice volume and paleoclimate. Geology 17 (8), 731–734. Available from: https://doi.org/10.1130/0091-7613(1989)017%3C0731:Ssotap%3E2.3.Co;2.

Larter, R.D., Barker, P.F., 1991a. Effects of ridge crest-trench interaction on Antarctic-Phoenix spreading: forces on a young subducting plate. Journal of Geophysical Research 96 (B12), 19,583–19,607.

Larter, R.D., Barker, P.F., 1991b. Neogene interaction of Tectonic and glacial processes at the Pacific Margin of the Antarctic Peninsula. In: Macdonald, D.I.M. (Ed.), Sedimentation, Tectonics and Eustasy, vol. 12. International Association of Sedimentologists, pp. 165–186.

Larter, R.D., Cunningham, A.P., 1993. The depositional pattern and distribution of glacial-interglacial sequences on the Antarctic Peninsula Pacific margin. Marine Geology 109 (3–4), 203–219. Available from: https://doi.org/10.1016/0025-3227(93)90061-Y.

Larter, R.D., Anderson, J.B., Graham, A.G.C., Gohl, K., Hillenbrand, C.-D., Jakobsson, M., et al., 2014. Reconstruction of changes in the Amundsen Sea and Bellingshausen Sea sector of the West Antarctic Ice Sheet since the Last Glacial Maximum. Quaternary Science Reviews 100, 55–86. Available from: https://doi.org/10.1016/j.quascirev.2013.10.016.

Larter, R.D., Hogan, K.A., Dowdeswell, J., 2016a. Large sediment drifts on the upper continental rise west of the Antarctic Peninsula. Geological Society, London, Memoirs 46, 401–402. Available from: https://doi.org/10.1144/M46.132.

Larter, R.D., Hogan, K.A., Hillenbrand, C., Benetti, S., 2016b. Debris-flow deposits on the West Antarctic continental slope. Geological Society, London, Memoirs 46, 375–376. Available from: https://doi.org/10.1144/M46.155.

Larter, R.D., Rebesco, M., Vanneste, L.E., Gambôa, L.A.P., Barker, P.F., 1997. Cenozoic tectonic, sedimentary and glacial history of the continental Shelf West of Graham Land, Antarctic Peninsula. In: Barker, P.F., Cooper, A.K. (Eds.), Geology and Seismic Stratigraphy of the Antarctic Margin, 2. American Geophysical Union, pp. 1–27.

Lawver, L., Lee, J., Kim, Y., Davey, F., 2012. Flat-topped mounds in western Ross Sea: carbonate mounds or subglacial volcanic features? Geosphere 8 (3), 645–653. Available from: https://doi.org/10.1130/GES00766.1.

Lawver, L.A., Della Vedova, B., Von Herzen, R.P., 1991. Heat flow in Jane Basin, northwest Weddell Sea. Journal of Geophysical Research 96 (B2), 2019–2038.

Lawver, L.A., Gahagan, L.M., Coffin, M.F., 1992. The development of Paleoseaways around Antarctica. In: James, P., Kennett, Warkne, D.A. (Eds.), The Antarctic Paleoenvironment: A Perspective on Global Change: Part One. American Geophysical Union, pp. 7–30. Available from: http://onlinelibrary.wiley.com/doi/10.1029/AR056p0007/summary.

Lawver, L.A., Williams, T., Sloan, B., 1994. Seismic stratigraphy and heat flow of Powell Basin. Terra Antarctica 1.

Lawver, L.A., Sloan, B.J., Barker, D.H.N., Ghidella, M., Von Herzen, R.P., Keller, R.A., et al., 1996. Distributed, active extension in Bransfield basin Antarctic Peninsula: evidence from multibeam bathymetry. GSA Today 6 (11), 1–6.

Leitchenkov, G.L., Belyatsky, B.V., Kaminsky, V.D., 2018. The age of rift-related basalts in East Antarctica. Doklady Earth Sciences 478 (1), 11–14. Available from: https://doi.org/10.1134/S1028334X18010051.

Leitchenkov, G.L., Guseva, Y.B., Gandyukhin, V.V., 2007. Cenozoic environmental changes along the East Antarctic continental margin inferred from regional seismic stratigraphy. In: Cenozoic Environmental Changes along the East Antarctic Continental Margin Inferred from Regional Seismic Stratigraphy (USGS Numbered Series No. 2007-1047-SRP-005; Open-File Report, Vols. 2007-1047-SRP-005). United States Geological Survey. <https://doi.org/10.3133/ofr20071047SRP005>.

Leitchenkov, G., Guseva, K., Gandyukhin, V., Ivanov, S., 2015. Crustal Structure, Tectonic Evolution and Seismic Stratigraphy of the Southern Indian Ocean. VNIIOkeangeologia, St. Petersburg, p. 200.

Leitchenkov, G.L., Guseva, Y.B., Gandyukhin, V.V., Ivanov, S.V., Safonova, L.V., 2014. Structure of the Earth's crust and tectonic evolution history of the Southern Indian Ocean (Antarctica). Geotectonics 48 (1), 5–23. Available from: https://doi.org/10.1134/S001685211401004X.

Leitchenkov, G., Galushkin, Y., Guseva, Y., Dubinin, E., 2020. Evolution of the Sedimentary Basin of the Continental Margin of Antarctica in the cooperation sea (from results of numerical modeling). Russian Geology and Geophysics 61 (1), 68–78. Available from: https://doi.org/10.15372/RGG2019079.

Leitchenkov, G.L., Kudryavtsev, G.A., 2000. Structure and origin of the Earth's crust in the Weddell Sea Embayment (beneath the front of the Filchner and Ronne Ice Shelves) from the Deep Seismic Soundings data. Polarforschung. 67 (3), 143–154.

LeMasurier, W.E., Rex, D.C., Craddock, C., 1982. Antarctic geoscience. Volcanic Record of Cenozoic Glacial History in Marie Byrd Land, and Western Ellsworth Land: Revised Chronology and Evaluation of Tectonic Factors. The University of Wisconsin Press, Madison, pp. 725–734.

Leventer, A., Domack, E., Barkoukis, A., McAndrews, B., Murray, J., 2002. Laminations from the Palmer Deep: a diatom-based interpretation. Paleoceanography 17 (3), PAL3-1–PAL3-15. Available from: https://doi.org/10.1029/2001PA000624.

Levy, R., Cody, R., Crampton, J., Fielding, C., Golledge, N., Harwood, D., et al., 2012. Late Neogene climate and glacial history of the Southern Victoria Land coast from integrated drill core, seismic and outcrop data. Global and Planetary Change 80–81, 61–84. Available from: https://doi.org/10.1016/j.gloplacha.2011.10.002.

Levy, R., Harwood, D., Florindo, F., Sangiorgi, F., Tripati, R., Von Eynatten, H., et al., 2016. Antarctic ice sheet sensitivity to atmospheric CO_2 variations in the early to mid-Miocene. Proceedings of the National Academy of Sciences of the United States of America 113 (13), 3453–3458.

Levy, R.H., Harwood, D.M., 2000. Tertiary marine palynomorphs from the McMurdo Sound erratics, Antarctica. Paleobiology and Paleoenvironments of Eocene Rocks: McMurdo Sound, East Antarctica 76, 183–242.

Levy, R.H., Meyers, S.R., Naish, T.R., Golledge, N.R., McKay, R.M., Crampton, J.S., et al., 2019. Antarctic ice-sheet sensitivity to obliquity forcing enhanced through ocean connections. Nature Geoscience, 1 . Available from: https://doi.org/10.1038/s41561-018-0284-4.

Levy, R.H., et al., 2021. Antarctic environmental change and ice sheet evolution through the Miocene to Pliocene: a perspective from the Ross Sea, George V Coast, and Wilkes Land. In: Florindo, F., et al. (Eds.), Antarctic Climate Evolution, second ed. Elsevier (this volume).

Lewis, A.R., Marchant, D.R., Ashworth, A.C., Hedenäs, L., Hemming, S.R., Johnson, J.V., et al., 2008. Mid-Miocene cooling and the extinction of tundra in continental Antarctica. Proceedings of the National Academy of Sciences . Available from: https://doi.org/10.1073/pnas.0802501105.

Lewis, A.R., Marchant, D.R., Ashworth, A.C., Hemming, S.R., Machlus, M.L., 2007. Major middle Miocene global climate change: evidence from East Antarctica and the Transantarctic Mountains. Geological Society of America Bulletin 119 (11–12), 1449–1461. Available from: https://doi.org/10.1130/0016-7606(2007)119[1449:MMMGCC]2.0.CO;2.

Lewis, A.R., Marchant, D.R., Kowalewski, D.E., Baldwin, S.L., Webb, L.E., 2006. The age and origin of the Labyrinth, western Dry Valleys, Antarctica: evidence for extensive middle Miocene subglacial floods and freshwater discharge to the Southern Ocean. Geology 34 (7), 513–516. Available from: https://doi.org/10.1130/G22145.1.

Li, X., Rignot, E., Mouginot, J., Scheuchl, B., 2016. Ice flow dynamics and mass loss of Totten Glacier, East Antarctica, from 1989 to 2015. Geophysical Research Letters 43 (12), 6366–6373. Available from: https://doi.org/10.1002/2016GL069173.

Liebrand, D., Beddow, H.M., Lourens, L.J., Pälike, H., Raffi, I., Bohaty, S.M., et al., 2016. Cyclostratigraphy and eccentricity tuning of the early Oligocene through early Miocene (30.1–17.1 Ma): Cibicides mundulus stable oxygen and carbon isotope records from Walvis Ridge Site 1264. Earth and Planetary Science Letters 450, 392–405. Available from: https://doi.org/10.1016/j.epsl.2016.06.007.

Lindeque, A., Gohl, K., Henrys, S., Wobbe, F., Davy, B., 2016. Seismic stratigraphy along the Amundsen Sea to Ross Sea continental rise: a cross-regional record of pre-glacial to glacial processes of the West Antarctic margin. Palaeogeography, Palaeoclimatology, Palaeoecology 443, 183–202. Available from: https://doi.org/10.1016/j.palaeo.2015.11.017.

Lindeque, A., Martos Martin, Y., Gohl, K., Maldonado, A., 2013. Deep-sea pre-glacial to glacial sedimentation in the Weddell Sea and southern Scotia Sea from a cross-basin seismic transect. Marine Geology 336 (0), 61–83. Available from: https://doi.org/10.1016/j.margeo.2012.11.004.

Lisiecki, L.E., Raymo, M.E., 2005. A Pliocene-Pleistocene stack of 57 globally distributed benthic $\delta^{18}O$ records. Paleoceanography 20, 17. Available from: https://doi.org/10.1029/2004PA001071.

Lisker, F., Belton, D., Kroner, U., 2005. Thermochronological investigation around the Lambert Graben: review of Pre-existing data and field work during PCMEGA. Terra Antartica 12, 45–50.

Lisker, F., Brown, R., Fabel, D., 2003. Denudational and thermal history along a transect across the Lambert Graben, northern Prince Charles Mountains, Antarctica, derived from apatite fission track thermochronology. Tectonics 22 (5). Available from: https://doi.org/10.1029/2002TC001477.

Lisker, F., Prenzel, J., Läufer, A., Spiegel, C., 2014. Recent thermochronological research in northern Victoria Land, Antarctica. Polarforschung 84, 59–66.

Liu, Z., Pagani, M., Zinniker, D., DeConto, R., Huber, M., Brinkhuis, H., et al., 2009. Global cooling during the Eocene-Oligocene climate transition. Science 323 (5918), 1187–1190. Available from: https://doi.org/10.1126/science.1166368.

Lodolo, E., Pérez, L.F., 2015. An abandoned rift in the southwestern part of the South Scotia Ridge (Antarctica): implications for the genesis of the Bransfield Strait. Tectonics 34 (12), 2451–2464. Available from: https://doi.org/10.1002/2015tc004041.

Lonsdale, M.J., 1990. The relationship between silica diagenesis, methane, and seismic reflections on the South Orkney microcontinent. In: Proceedings, Scientific Results, ODP, Leg 113, Weddell Sea, Antarctica, pp. 27–37.

López-Quirós, A., Escutia, C., Sánchez-Navas, A., Nieto, F., Garcia-Casco, A., Martín-Algarra, A., et al., 2019. Glaucony authigenesis, maturity and alteration in the Weddell Sea: an indicator of paleoenvironmental conditions before the onset of Antarctic glaciation. Scientific Reports 9 (1), 13580. Available from: https://doi.org/10.1038/s41598-019-50107-1.

Lowe, A.L., Anderson, J.B., 2002. Reconstruction of the West Antarctic ice sheet in Pine Island Bay during the Last Glacial Maximum and its subsequent retreat history. Quaternary Science Reviews 21 (16–17), 1879–1897. Available from: https://doi.org/10.1016/S0277-3791(02)00006-9.

Lüthi, D., Le Floch, M., Bereiter, B., Blunier, T., Barnola, J.-M., Siegenthaler, U., et al., 2008. High-resolution carbon dioxide concentration record 650,000–800,000 years before present. Nature 453 (7193), 379–382. Available from: https://doi.org/10.1038/nature06949.

Luyendyk, B.P., Sorlien, C.C., Wilson, D.S., Bartek, L.R., Siddoway, C.S., 2001. Structural and tectonic evolution of the Ross Sea rift in the Cape Colbeck region, Eastern Ross Sea, Antarctica. Tectonics 20 (6), 933–958. Available from: https://doi.org/10.1029/2000TC001260.

Mackensen, A., Ehrmann, W.U., 1992. Middle Eocene through Early Oligocene climate history and paleoceanography in the Southern Ocean: stable oxygen and carbon isotopes from ODP Sites on Maud Rise and Kerguelen Plateau. Marine Geology 108 (1), 1–27. Available from: https://doi.org/10.1016/0025-3227(92)90210-9.

Macphail, M.K., Truswell, E.M., 2004a. Palynology of Neogene slope and rise deposits from ODP sites 1165 and 1167, East Antarctica. Proceedings of the Ocean Drilling Program: Scientific Results 188. Available from: https://doi.org/10.2973/odp.proc.sr.188.012.2004.

Macphail, M.K., Truswell, E.M., 2004b. Palynology of Site 1166, Prydz Bay, East Antarctica. Proceedings of the Ocean Drilling Program, Scientific Results (CD ROM) 188, 43. Available from: https://doi.org/10.2973/odp.proc.sr.188.013.2004.

Maldonado, A., Bohoyo, F., Galindo-Zaldívar, J., Hernández-Molina, F.J., Jabaloy, A., Lobo, F.J., et al., 2006. Ocean basins near the Scotia—Antarctic plate boundary: influence of tectonics and paleoceanography on the Cenozoic deposits. Marine Geophysics Research 27 (2), 83–107.

Maldonado, A., Larter, R.D., Aldaya, F., 1994. Forearc tectonic evolution of the South Shetland margin, Antarctic Peninsula. Tectonics 13 (6), 1345–1370.

Maldonado, A., Zitellini, N., Leitchenkov, G., Balanyá, J.C., Coren, F., Galindo-Zaldivar, J., et al., 1998. Small ocean basin development along the Scotia-Antarctica plate boundary and in the northern Weddell Sea. Tectonophysics 296 (3–4), 371–402.

Mawbey, E.M., Lear, C.H., 2013. Carbon cycle feedbacks during the Oligocene-Miocene transient glaciation. Geology 41 (9), 963–966. Available from: https://doi.org/10.1130/G34422.1.

McGinnis, L.D., 1981. Dry Valley Drilling Project, vol. 33. American Geophysical Union, Washington, DC, p. 465.

McKay, R., Browne, G., Carter, L., Cowan, E., Dunbar, G., Krissek, L., et al., 2009. The stratigraphic signature of the late Cenozoic Antarctic Ice Sheets in the Ross Embayment. Geological Society of America Bulletin 121 (11–12), 1537–1561. Available from: https://doi.org/10.1130/B26540.1.

McKay, R., Naish, T., Carter, L., Riesselman, C., Dunbar, R., Sjunneskog, C., et al., 2012a. Antarctic and Southern Ocean influences on late Pliocene global cooling. Proceedings of the

National Academy of Sciences 109 (17), 6423–6428. Available from: https://doi.org/10.1073/pnas.1112248109.

McKay, R., Naish, T., Powell, R., Barrett, P., Scherer, R., Talarico, F., et al., 2012b. Pleistocene variability of Antarctic Ice Sheet extent in the Ross Embayment. Quaternary Science Reviews 34, 93–112. Available from: https://doi.org/10.1016/j.quascirev.2011.12.012.

McKay, R.M., Barrett, P.J., Harper, M.A., Hannah, M.J., 2008. Atmospheric transport and concentration of diatoms in surficial and glacial sediments of the Allan Hills, Transantarctic Mountains. Palaeogeography, Palaeoclimatology, Palaeoecology 260 (1–2), 168–183. Available from: https://doi.org/10.1016/j.palaeo.2007.08.014.

McKay, R.M., Barrett, P.J., Levy, R.S., Naish, T.R., Golledge, N.R., Pyne, A., 2016. Antarctic Cenozoic climate history from sedimentary records: ANDRILL and beyond. Philosophical Transactions of the Royal Society A 374 (2059), 20140301. Available from: https://doi.org/10.1098/rsta.2014.0301.

McKay, R.M., De Santis, L., Kulhanek, D.K., IODP Expedition 374 Science Team, 2019. Proceedings of the International Ocean Discovery Program, vol. 374. <https://doi.org/10.14379/iodp.proc.374.2019>.

McLoughlin, S., Drinnan, A.N., 1997a. Fluvial sedimentology and revised stratigraphy of the Triassic Flagstone Bench Formation, northern Prince Charles Mountains, East Antarctica. Geological Magazine 134 (6), 781–806. Available from: https://doi.org/10.1017/S0016756897007528.

McLoughlin, S., Drinnan, A.N., 1997b. Revised stratigraphy of the Permian Bainmedart Coal Measures, northern Prince Charles Mountains, East Antarctica. Geological Magazine 134 (3), 335–353. Available from: https://doi.org/10.1017/S0016756897006870.

Mengel, M., Levermann, A., 2014. Ice plug prevents irreversible discharge from East Antarctica. Nature Climate Change 4 (6), 451–455. Available from: https://doi.org/10.1038/nclimate2226.

Michels, K.H., Kuhn, G., Hillenbrand, C.-D., Diekmann, B., Fütterer, D.K., Grobe, H., et al., 2002. The southern Weddell Sea: combined contourite-turbidite sedimentation at the southeastern margin of the Weddell Gyre. In: Stow, D.A.V., Pudsey, C., Howe, J.C., Faugères, J.-C., Viana, A.R. (Eds.), Deep-Water Contourite Systems: Modern Drifts and Ancient Series, Seismic and Sedimentary Characteristics, vol. 22. Geological Society, pp. 305–323.

Miller, K.G., Browning, J.V., Aubry, M.-P., Wade, B.S., Katz, M.E., Kulpecz, A.A., et al., 2008. Eocene–Oligocene global climate and sea-level changes: St. Stephens Quarry, Alabama. GSA Bulletin 120 (1–2), 34–53. Available from: https://doi.org/10.1130/B26105.1.

Miller, K.G., Browning, J.V., Schmelz, W.J., Kopp, R.E., Mountain, G.S., Wright, J.D., 2020. Cenozoic sea-level and cryospheric evolution from deep-sea geochemical and continental margin records. Science Advances 6 (20), eaaz1346. Available from: https://doi.org/10.1126/sciadv.aaz1346.

Miller, K.G., Kominz, M.A., Browning, J.V., Wright, J.D., Mountain, G.S., Katz, M.E., Sugarman, P.J., Cramer, B.S., Christie-Blick, N., Pekar, S.F., 2005. The Phanerozoic Record of Global Sea-Level Change. Science 310 (5752), 1293–1298. Available from: https://doi.org/10.1126/science.1116412.

Miller, K.G., Wright, J.D., Browning, J.V., Kulpecz, A., Kominz, M., Naish, T.R., et al., 2012. High tide of the warm Pliocene: implications of global sea level for Antarctic deglaciation. Geology . Available from: https://doi.org/10.1130/G32869.1.

Miller, K.G., Wright, J.D., Fairbanks, R.G., 1991. Unlocking the Ice House: Oligocene-Miocene oxygen isotopes, eustasy, and margin erosion. Journal of Geophysical Research: Solid Earth 96 (B4), 6829–6848. Available from: https://doi.org/10.1029/90JB02015.

Milliken, K.T., Anderson, J.B., Wellner, J.S., Bohaty, S.M., Manley, P.L., 2009. High-resolution Holocene climate record from Maxwell Bay, South Shetland Islands, Antarctica. GSA Bulletin 121 (11–12), 1711–1725. Available from: https://doi.org/10.1130/b26478.1.

Minzoni, R.T., Majewski, W., Anderson, J.B., Yokoyama, Y., Fernandez, R., Jakobsson, M., 2017. Oceanographic influences on the stability of the Cosgrove Ice Shelf, Antarctica. The Holocene 27 (11), 1645–1658. Available from: https://doi.org/10.1177/0959683617702226.

Mizukoshi, I., Sunouchi, H., Saki, T., Sato, S., Tanahashi, M., 1986. Preliminary report of geological and geophysical surveys off Amery Ice Shelf, East Antarctica. Memoirs of the National Institute of Polar Research, Japan, Special Issue 43 (Spec. Issue), 48–61.

Mohr, R.C., Tobin, T.S., Petersen, S.V., Dutton, A., Oliphant, E., 2020. Subannual stable isotope records reveal climate warming and seasonal anoxia associated with two extinction intervals across the Cretaceous-Paleogene boundary on Seymour Island, Antarctica. Geology 48. Available from: https://doi.org/10.1130/g47758.1.

Montelli, A., Gulick, S.P.S., Fernandez, R., Frederick, B.C., Shevenell, A.E., Leventer, A., et al., 2020. Seismic stratigraphy of the Sabrina Coast shelf, East Antarctica: early history of dynamic meltwater-rich glaciations. GSA Bulletin 132 (3–4), 545–561. Available from: https://doi.org/10.1130/B35100.1.

Morlighem, M., 2020. MEaSUREs BedMachine Antarctica, Version 2 [bed subset]. NASA National Snow and Ice Data Center Distributed Active Archive Center, Boulder, CO. Available from: https://doi.org/10.5067/E1QL9HFQ7A8M.

Morlighem, M., Rignot, E., Binder, T., Blankenship, D., Drews, R., Eagles, G., et al., 2020. Deep glacial troughs and stabilizing ridges unveiled beneath the margins of the Antarctic ice sheet. Nature Geoscience 13 (2), 132–137. Available from: https://doi.org/10.1038/s41561-019-0510-8.

Mosola, A.B., Anderson, J.B., 2006. Expansion and rapid retreat of the West Antarctic Ice Sheet in eastern Ross Sea: possible consequence of over-extended ice streams? Quaternary Science Reviews 25 (17–18), 2177–2196. Available from: https://doi.org/10.1016/j.quascirev.2005.12.013.

Moucha, R., Forte, A.M., Mitrovica, J.X., Rowley, D.B., Quéré, S., Simmons, N.A., et al., 2008. Dynamic topography and long-term sea-level variations: there is no such thing as a stable continental platform. Earth and Planetary Science Letters 271 (1–4), 101–108. Available from: https://doi.org/10.1016/j.epsl.2008.03.056.

Naish, T., et al., 2021. Antarctic Ice Sheet dynamics during the Late Oligocene and Early Miocene: climatic conundrums revisited. In: Florindo, F., et al. (Eds.), Antarctic Climate Evolution, second ed. Elsevier (in press).

Naish, T., Powell, R., Levy, R.H., the ANDRILL-MIS Science Team, 2007. Studies from the ANDRILL, McMurdo Ice Shelf Project, Antarctica – initial science report on AND-1B. Terra Antartica 14 (3), 328.

Naish, T., Powell, R., Levy, R., Wilson, G., Scherer, R., Talarico, F., et al., 2009. Obliquity-paced Pliocene West Antarctic ice sheet oscillations. Nature 458 (7236), 322–328. Available from: https://doi.org/10.1038/nature07867.

Naish, T.R., Wilson, G.S., January 13, 2009. Constraints on the amplitude of Mid-Pliocene (3.6–2.4 Ma) eustatic sea-level fluctuations from the New Zealand shallow-marine sediment record. <http://rsta.royalsocietypublishing.org/content/367/1886/169.full>.

Naish, T.R., Woolfe, K.J., Barrett, P.J., Wilson, G.S., Atkins, C., Bohaty, S.M., et al., 2001. Orbitally induced oscillations in the East Antarctic ice sheet at the Oligocene/Miocene boundary. Nature 413 (6857), 719–723. Available from: https://doi.org/10.1038/35099534.

Nakayama, Y., Schröder, M., Hellmer, H.H., 2013. From circumpolar deep water to the glacial meltwater plume on the eastern Amundsen Shelf. Deep Sea Research Part I: Oceanographic Research Papers 77, 50–62. Available from: https://doi.org/10.1016/j.dsr.2013.04.001.

Nitsche, F.O., Cunningham, A.P., Larter, R.D., Gohl, K., 2000. Geometry and development of glacial continental margin depositional systems in the Bellingshausen Sea. Marine Geology 162 (2–4), 277–302. Available from: https://doi.org/10.1016/S0025-3227(99)00074-2.

Nitsche, F.O., Gohl, K., Vanneste, K., Miller, H., 1997. Seismic expression of glacially deposited sequences in the Bellingshausen and Amundsen Seas, West Antarctica. In: Barker, P.F., Cooper, A.K. (Eds.), Geology and Seismic Stratigraphy of the Antarctic Margin, 2, vol. 71. American Geophysical Union, pp. 95–108.

Noble, T.L., Rohling, E.J., Aitken, A.R.A., Bostock, H.C., Chase, Z., Gomez, N., et al., 2020. The Sensitivity of the Antarctic Ice Sheet to a Changing Climate: past, present, and future. Reviews of Geophysics 58 (4). Available from: https://doi.org/10.1029/2019RG000663, e2019RG000663.

O'Brien, P.E., Harris, P.T., 1996. Patterns of glacial erosion and deposition in Prydz Bay and the past behaviour of the Lambert Glacier. Papers and Proceedings – Royal Society of Tasmania 130 (2), 79–85. Available from: https://doi.org/10.26749/rstpp.130.2.79.

O'Brien, P.E., Leitchenkov, G., 1997. Deglaciation of Prydz Bay, East Antarctica, based on echo sounder and topographic features. Antarctic Research Series 71, 109–125.

O'Brien, P.E., Cooper, A.K., Florindo, F., Handwerger, D.A., Lavelle, M., 2004. Prydz channel fan and the history of extreme ice advances in Prydz Bay. Proceedings of the Ocean Drilling Program, Scientific Results (CD ROM) 188, 32.

O'Brien, P.E., Cooper, A.K., Richter, C., et al., (Eds.), 2001. Proceedings of the Ocean Drilling Program, 188 Initial Reports, vol. 188. Ocean Drilling Program. Available from: http://www-odp.tamu.edu/publications/188_IR/188TOC.HTM.

O'Brien, P.E., Goodwin, I., Forsberg, C.-F., Cooper, A.K., Whitehead, J., 2007. Late Neogene ice drainage changes in Prydz Bay, East Antarctica and the interaction of Antarctic ice sheet evolution and climate. Palaeogeography, Palaeoclimatology, Palaeoecology 245 (3–4), 390–410. Available from: https://doi.org/10.1016/j.palaeo.2006.09.002.

O'Brien, P.E., Post, A.L., Edwards, S., Martin, T., Caburlotto, A., Donda, F., et al., 2020. Continental slope and rise geomorphology seaward of the Totten Glacier, East Antarctica (112°E–122°E). Marine Geology 427, 106221. Available from: https://doi.org/10.1016/j.margeo.2020.106221.

Ó Cofaigh, C., 1996. Tunnel valley genesis. Progress in Physical Geography: Earth and Environment 20 (1), 1–19. Available from: https://doi.org/10.1177/030913339602000101.

Ó Cofaigh, C., Dowdeswell, J.A., Allen, C.S., Hiemstra, J.F., Pudsey, C.J., Evans, J., et al., 2005a. Flow dynamics and till genesis associated with a marine-based Antarctic palaeo-ice stream. Quaternary Science Reviews 24 (5–6), 709–740. Available from: https://doi.org/10.1016/j.quascirev.2004.10.006.

Ó Cofaigh, C., Larter, R.D., Dowdeswell, J.A., Hillenbrand, C.-D., Pudsey, C.J., Evans, J., et al., 2005b. Flow of the West Antarctic Ice Sheet on the continental margin of the Bellingshausen Sea at the Last Glacial Maximum. Journal of Geophysical Research: Solid Earth 110 (B11). Available from: https://doi.org/10.1029/2005jb003619.

Ohneiser, C., Wilson, G.S., Beltran, C., Dolan, A.M., Hill, D.J., Prebble, J.G., 2020. Warm fjords and vegetated landscapes in early Pliocene East Antarctica. Earth and Planetary Science Letters 534, 116045. Available from: https://doi.org/10.1016/j.epsl.2019.116045.

Orejola, N., Passchier, S., 2014. Sedimentology of lower Pliocene to Upper Pleistocene diamictons from IODP Site U1358, Wilkes Land margin, and implications for East Antarctic Ice

Sheet dynamics. Antarctic Science 26 (02), 183–192. Available from: https://doi.org/10.1017/S0954102013000527.

Pagani, M., Huber, M., Liu, Z., Bohaty, S.M., Henderiks, J., Sijp, W., et al., 2011. The Role of Carbon Dioxide During the Onset of Antarctic Glaciation. Science 334 (6060), 1261–1264. Available from: https://doi.org/10.1126/science.1203909.

Pagani, M., Zachos, J.C., Freeman, K.H., Tipple, B., Bohaty, S., 2005. Marked Decline in Atmospheric Carbon Dioxide Concentrations During the Paleogene. Science 309 (5734), 600–603. Available from: https://doi.org/10.1126/science.1110063.

Pant, N.C., Biswas, P., Shrivastava, P.K., Bhattacharya, S., Verma, K., Pandey, M., et al., 2013. Provenance of Pleistocene sediments from Site U1359 of the Wilkes Land IODP Leg 318 – evidence for multiple sourcing from the East Antarctic Craton and Ross Orogen. Geological Society, London, Special Publications 381 (1), 277–297. Available from: https://doi.org/10.1144/SP381.11.

Parkinson, C.L., 2019. A 40-y record reveals gradual Antarctic sea ice increases followed by decreases at rates far exceeding the rates seen in the Arctic. Proceedings of the National Academy of Sciences 116 (29), 14414–14423. Available from: https://doi.org/10.1073/pnas.1906556116.

Pascher, K.M., Hollis, C.J., Bohaty, S.M., Cortese, G., McKay, R.M., Seebeck, H., et al., 2015. Expansion and diversification of high-latitude radiolarian assemblages in the late Eocene linked to a cooling event in the southwest Pacific. Climate of the Past 11 (12), 1599–1620. Available from: https://doi.org/10.5194/cp-11-1599-2015.

Passchier, S., 2003. Pliocene-Pleistocene glaciomarine sedimentation in eastern Prydz Bay and development of the Prydz trough-mouth fan, ODP Sites 1166 and 1167, East Antarctica. Marine Geology 199(3–4), 279–305.

Passchier, S., 2011. Linkages between East Antarctic Ice Sheet extent and Southern Ocean temperatures based on a Pliocene high-resolution record of ice-rafted debris off Prydz Bay, East Antarctica. Paleoceanography 26, 13. Available from: https://doi.org/10.1029/2010PA002061.

Passchier, S., Bohaty, S.M., Jiménez-Espejo, F., Pross, J., Röhl, U., Flierdt, T., et al., 2013. Early Eocene to middle Miocene cooling and aridification of East Antarctica. Geochemistry, Geophysics, Geosystems 14 (5), 1399–1410. Available from: https://doi.org/10.1002/ggge.20106.

Passchier, S., Browne, G., Field, B., Fielding, C.R., Krissek, L.A., Panter, K., et al., 2011. Early and middle Miocene Antarctic glacial history from the sedimentary facies distribution in the AND-2A drill hole, Ross Sea, Antarctica. Geological Society of America Bulletin B30334.1. Available from: https://doi.org/10.1130/B30334.1.

Passchier, S., Ciarletta, D.J., Miriagos, T.E., Bijl, P.K., Bohaty, S.M., 2017. An Antarctic stratigraphic record of stepwise ice growth through the Eocene-Oligocene transition. GSA Bulletin 129 (3–4), 318–330. Available from: https://doi.org/10.1130/B31482.1.

Patterson, M.O., McKay, R., Naish, T., Escutia, C., Jimenez-Espejo, F.J., Raymo, M.E., et al., 2014. Orbital forcing of the East Antarctic ice sheet during the Pliocene and Early Pleistocene. Nature Geoscience 7 (11), 841–847.

Paxman, G.J.G., Jamieson, S.S.R., Ferraccioli, F., Bentley, M.J., Ross, N., Armadillo, E., et al., 2018. Bedrock Erosion Surfaces Record former East Antarctic Ice Sheet Extent. Geophysical Research Letters 45 (9), 4114–4123. Available from: https://doi.org/10.1029/2018GL077268.

Paxman, G.J.G., Jamieson, S.S.R., Hochmuth, K., Gohl, K., Bentley, M.J., Leitchenkov, G., et al., 2019. Reconstructions of Antarctic topography since the Eocene–Oligocene boundary. Palaeogeography, Palaeoclimatology, Palaeoecology 535, 109346. Available from: https://doi.org/10.1016/j.palaeo.2019.109346.

Pekar, S.F., Speece, M.A., Wilson, G.S., Sunwall, D.S., Tinto, K.J., 2013. The Offshore New Harbour Project: deciphering the Middle Miocene through Late Eocene seismic stratigraphy of Offshore New Harbour, western Ross Sea, Antarctica. Geological Society, London, Special Publications 381 (1), 199–213. Available from: https://doi.org/10.1144/SP381.2.

Pérez, L.F., Maldonado, A., Bohoyo, F., Hernández-Molina, F.J., Vázquez, J.T., Lobo, F.J., et al., 2014. Depositional processes and growth patterns of isolated oceanic basins: the protector and Pirie basins of the Southern Scotia Sea (Antarctica). Marine Geology 357, 163–181. Available from: https://doi.org/10.1016/j.margeo.2014.08.001.

Pérez, L. F., De Santis, L., McKay, R. M., Larter, R. D., Ash, J., Bart, P. J., Böhm, G., Brancatelli, G., Browne, I., Colleoni, F., Dodd, J. P., Geletti, R., Harwood, D. M., Kuhn, G., Sverre Laberg, J., Leckie, R. M., Levy, R. H., Marschalek, J., Mateo, Z., ... 374 Scientists, I. O. D. P. E. (2021). Early and middle Miocene ice sheet dynamics in the Ross Sea: Results from integrated core-log-seismic interpretation. GSA Bulletin. Available from: https://doi.org/10.1130/B35814.1

Pérez, L.F., Martos, Y.M., García, M., Weber, M.E., Raymo, M.E., Williams, T., et al., 2021. Miocene to present oceanographic variability in the Scotia Sea and Antarctic ice sheets dynamics: insight from revised seismic-stratigraphy following IODP Expedition 382. Earth and Planetary Science Letters 553, 116657. Available from: https://doi.org/10.1016/j.epsl.2020.116657.

Peterse, F., van der Meer, J., Schouten, S., Weijers, J.W.H., Fierer, N., Jackson, R.B., et al., 2012. Revised calibration of the MBT–CBT paleotemperature proxy based on branched tetraether membrane lipids in surface soils. Geochimica et Cosmochimica Acta 96, 215–229. Available from: https://doi.org/10.1016/j.gca.2012.08.011.

Pierce, E.L., van de Flierdt, T., Williams, T., Hemming, S.R., Cook, C.P., Passchier, S., 2017. Evidence for a dynamic East Antarctic ice sheet during the mid-Miocene climate transition. Earth and Planetary Science Letters 478, 1–13. Available from: https://doi.org/10.1016/j.epsl.2017.08.011.

Pollard, D., DeConto, R.M., 2009. Modelling West Antarctic ice sheet growth and collapse through the past five million years. Nature 458 (7236), 329–332. Available from: https://doi.org/10.1038/nature07809.

Pollard, D., DeConto, R.M., 2020. Continuous simulations over the last 40 million years with a coupled Antarctic ice sheet-sediment model. Palaeogeography, Palaeoclimatology, Palaeoecology 537, 109374. Available from: https://doi.org/10.1016/j.palaeo.2019.109374.

Pollard, D., DeConto, R.M., Alley, R.B., 2015. Potential Antarctic Ice Sheet retreat driven by hydrofracturing and ice cliff failure. Earth and Planetary Science Letters 412, 112–121. Available from: https://doi.org/10.1016/j.epsl.2014.12.035.

Pospichal, J.J., 2003. Calcareous nannofossils from continental rise Site 1165, ODP Leg 188, Prydz Bay, Antarctica. Proceedings of ODP, Science Results 188, 1–14.

Post, A.L., O'Brien, P.E., Edwards, S., Carroll, A.G., Malakoff, K., Armand, L.K., 2020. Upper slope processes and seafloor ecosystems on the Sabrina continental slope, East Antarctica. Marine Geology 422, 106091. Available from: https://doi.org/10.1016/j.margeo.2019.106091.

Powell, R.D., Cooper, J.M., 2002. A glacial sequence stratigraphic model for temperate, glaciated continental shelves. Geological Society, London, Special Publications 203 (1), 215–244. Available from: https://doi.org/10.1144/GSL.SP.2002.203.01.12.

Prebble, J.G., Crouch, E.M., Carter, L., Cortese, G., Bostock, H., Neil, H., 2013. An expanded modern dinoflagellate cyst dataset for the Southwest Pacific and Southern Hemisphere with environmental associations. Marine Micropaleontology 101, 33–48. Available from: https://doi.org/10.1016/j.marmicro.2013.04.004.

Prieto, M.J., Ercilla, G., Canals, M., de Batist, M., 1999. Seismic stratigraphy of the Central Bransfield Basin (NW Antarctic Peninsula): interpretation of deposits and sedimentary processes in a glacio-marine environment. Marine Geology 157 (1), 47–68. Available from: https://doi.org/10.1016/S0025-3227(98)00149-2.

Priscu, J.C., Barker, J.D., Campbell, T., Christner, B.C., Davis, C., Dore, J.E., et al., 2019. SALSA: an integrated program focusing on carbon transformations in Mercer Subglacial Lake located 1100 m beneath the West Antarctic Ice Sheet. AGU Fall Meeting Abstracts 53. Available from: http://adsabs.harvard.edu/abs/2019AGUFM.P53B.03P.

Pross, J., Contreras, L., Bijl, P.K., Greenwood, D.R., Bohaty, S.M., Schouten, S., et al., 2012. Persistent near-tropical warmth on the Antarctic continent during the early Eocene epoch. Nature 487 (7409), 73–77.

Prothro, L.O., Simkins, L.M., Majewski, W., Anderson, J.B., 2018. Glacial retreat patterns and processes determined from integrated sedimentology and geomorphology records. Marine Geology 395, 104–119. Available from: https://doi.org/10.1016/j.margeo.2017.09.012.

Pudsey, C.J., 2002. The Weddell Sea: contourites and hemipelagites at the northern margin of the Weddell Gyre. Geological Society, London, Memoirs 22 (1), 289–303.

Quilty, P.G., Lirio, J.M., Jillett, D., 2000. Stratigraphy of the Pliocene Sørsdal Formation, Marine Plain, Vestfold Hills, East Antarctica. Antarctic Science 12 (02), 205–216. Available from: https://doi.org/10.1017/S0954102000000262.

Rack, F., Zook, R., Levy, R., Limeburner, R., Stewart, C., Williams, M., et al., 2012. What Lies Beneath? Interdisciplinary Outcomes of the ANDRILL Coulman High Project Site Surveys on the Ross Ice Shelf. Oceanography 25 (3), 84–89. Available from: https://doi.org/10.5670/oceanog.2012.79.

Raine, J., Askin, R., 2001. Terrestrial palynology of Cape Roberts Project Drillhole CRP-3, Victoria Land Basin, Antarctica. Terra Antartica 8 (4), 389–400.

Raymo, M.E., Mitrovica, J.X., O'Leary, M.J., DeConto, R.M., Hearty, P.J., 2011. Departures from eustasy in Pliocene sea-level records. Nature Geoscience 4 (5), 328–332. Available from: https://doi.org/10.1038/ngeo1118.

Rebesco, M., Camerlenghi, A., Geletti, R., Canals, M., 2006. Margin architecture reveals the transition to the modern Antarctic ice sheet ca. 3 Ma. Geology 34 (4), 301–304. Available from: https://doi.org/10.1130/G22000.1.

Rebesco, M., Larter, R.D., Barker, P.F., Camerlenghi, A., Vanneste, L.E., 1997. The history of sedimentation on the continental rise west of the Antarctic Peninsula. In: Barker, P.F., Cooper, A.K. (Eds.), Geology and Seismic Stratigraphy of the Antarctic Margin, Part 2, vol. 71. American Geophysical Union, pp. 29–49.

Rebesco, M., Larter, R.D., Camerlenghi, A., Barker, P.F., 1996. Giant sediment drifts on the continental rise west of the Antarctic Peninsula. Geo-Marine Letters 16, 65–75.

Rebesco, M., Pudsey, C.J., Canals, M., Camerlenghi, A., Barker, P.F., Estrada, F., et al., 2002. Sediment drifts and deep-sea channel systems, Antarctic Peninsula Pacific Margin. Geological Society, London, Memoirs 22, 353–371. Available from: https://doi.org/10.1144/GSL.MEM.2002.022.01.25.

Reinardy, B.T.I., Escutia, C., Iwai, M., Jimenez-Espejo, F.J., Cook, C., van de Flierdt, T., et al., 2015. Repeated advance and retreat of the East Antarctic Ice Sheet on the continental shelf during the early Pliocene warm period. Palaeogeography, Palaeoclimatology, Palaeoecology 422, 65–84. Available from: https://doi.org/10.1016/j.palaeo.2015.01.009.

Rickaby, R.E.M., Elderfield, H., Roberts, N., Hillenbrand, C.D., Mackensen, A., 2010. Evidence for elevated alkalinity in the glacial Southern Ocean. Paleoceanography 25 (1), PA1209. Available from: https://doi.org/10.1029/2009pa001762.

Riesselman, C.R., Dunbar, R.B., 2013. Diatom evidence for the onset of Pliocene cooling from AND-1B, McMurdo Sound, Antarctica. Palaeogeography, Palaeoclimatology, Palaeoecology 369, 136–153. Available from: https://doi.org/10.1016/j.palaeo.2012.10.014.

Rignot, E., Mouginot, J., Scheuchl, B., van den Broeke, M., van Wessem, M.J., Morlighem, M., 2019. Four decades of Antarctic Ice Sheet mass balance from 1979–2017. Proceedings of the National Academy of Sciences 116 (4), 1095. Available from: https://doi.org/10.1073/pnas.1812883116.

Riley, T.R., Jordan, T.A., Leat, P.T., Curtis, M.L., Millar, I.L., 2020. Magmatism of the Weddell Sea rift system in Antarctica: implications for the age and mechanism of rifting and early stage Gondwana breakup. Gondwana Research 79, 185–196. Available from: https://doi.org/10.1016/j.gr.2019.09.014.

Rintoul, S.R., Silvano, A., Pena-Molino, B., Wijk, E., van, Rosenberg, M., Greenbaum, J.S., et al., 2016. Ocean heat drives rapid basal melt of the Totten Ice Shelf. Science Advances 2 (12), e1601610. Available from: https://doi.org/10.1126/sciadv.1601610.

Roberts, J., Galton-Fenzi, B.K., Paolo, F.S., Donnelly, C., Gwyther, D.E., Padman, L., et al., 2017. Ocean forced variability of Totten Glacier mass loss. Geological Society, London, Special Publications 461 (1), 175–186. Available from: https://doi.org/10.1144/SP461.6.

Rodríguez-Fernández, J., Balanyá, J.C., Galindo-Zaldivar, J., Maldonado, A., 1997. Tectonic evolution of a restricted ocean basin: The Powell Basin (northeastern Antarctic Peninsula). Geodinamica Acta 10 (4), 159–174.

Rose, K.C., Ferraccioli, F., Jamieson, S.S.R., Bell, R.E., Corr, H., Creyts, T.T., et al., 2013. Early East Antarctic Ice Sheet growth recorded in the landscape of the Gamburtsev Subglacial Mountains. Earth and Planetary Science Letters 375, 1–12. Available from: https://doi.org/10.1016/j.epsl.2013.03.053.

Rosenblume, J.A., Powell, R.D., 2019. Glacial sequence stratigraphy of ANDRILL-1B core reveals a dynamic subpolar Antarctic Ice Sheet in Ross Sea during the late Miocene. Sedimentology 66 (6), 2072–2097. Available from: https://doi.org/10.1111/sed.12592.

Salabarnada, A., Escutia, C., Röhl, U., Nelson, C.H., McKay, R., Jiménez-Espejo, F.J., et al., 2018. Paleoceanography and ice sheet variability offshore Wilkes Land, Antarctica—part 1: insights from late Oligocene astronomically paced contourite sedimentation. Climate of the Past 14 (7), 991–1014. Available from: https://doi.org/10.5194/cp-14-991-2018.

Sangiorgi, F., Bijl, P.K., Passchier, S., Salzmann, U., Schouten, S., McKay, R., et al., 2018. Southern Ocean warming and Wilkes Land ice sheet retreat during the mid-Miocene. Nature Communications 9 (1), 317. Available from: https://doi.org/10.1038/s41467-017-02609-7.

Sauermilch, I., Whittaker, J.M., Bijl, P.K., Totterdell, J.M., Jokat, W., 2019. Tectonic, Oceanographic, and Climatic Controls on the Cretaceous-Cenozoic Sedimentary Record of the Australian-Antarctic Basin. Journal of Geophysical Research: Solid Earth 124 (8), 7699–7724. Available from: https://doi.org/10.1029/2018JB016683.

Sauli, C., Busetti, M., De Santis, L., Wardell, N., 2014. Late Neogene geomorphological and glacial reconstruction of the northern Victoria Land coast, western Ross Sea (Antarctica). Marine Geology 355, 297–309. Available from: https://doi.org/10.1016/j.margeo.2014.06.008.

Sauli, C., Sorlien, C., Busetti, M., De Santis, L., Geletti, R., Wardell, N., et al., 2021. Neogene development of the Terror Rift, western Ross Sea, Antarctica. Geochemistry, Geophysics, Geosystems. Available from: https://doi.org/10.1029/2020GC009076.

Scambos, T.A., Bell, R.E., Alley, R.B., Anandakrishnan, S., Bromwich, D.H., Brunt, K., et al., 2017. How much, how fast?: a science review and outlook for research on the instability of Antarctica's Thwaites Glacier in the 21st century. Global and Planetary Change 153, 16–34. Available from: https://doi.org/10.1016/j.gloplacha.2017.04.008.

Scasso, R.A., Prámparo, M.B., Vellekoop, J., Franzosi, C., Castro, L.N., Sinninghe Damsté, J.S., 2020. A high-resolution record of environmental changes from a Cretaceous-Paleogene section of Seymour Island, Antarctica. Palaeogeography, Palaeoclimatology, Palaeoecology 555, 109844. Available from: https://doi.org/10.1016/j.palaeo.2020.109844.

Scher, H.D., Bohaty, S.M., Smith, B.W., Munn, G.H., 2014. Isotopic interrogation of a suspected late Eocene glaciation. Paleoceanography 29 (6). Available from: https://doi.org/10.1002/2014PA002648, 2014PA002648.

Scher, H.D., Martin, E.E., 2006. Timing and Climatic Consequences of the Opening of Drake Passage. Science 312 (5772), 428–430. Available from: https://doi.org/10.1126/science.1120044.

Scher, H.D., Whittaker, J.M., Williams, S.E., Latimer, J.C., Kordesch, W.E.C., Delaney, M.L., 2015. Onset of Antarctic Circumpolar Current 30 million years ago as Tasmanian Gateway aligned with westerlies. Nature 523 (7562), 580–583. Available from: https://doi.org/10.1038/nature14598.

Scherer, R.P., Aldahan, A., Tulaczyk, S., Possnert, G., Engelhardt, H., Kamb, B., 1998. Pleistocene Collapse of the West Antarctic Ice Sheet. Science 281 (5373), 82–85. Available from: https://doi.org/10.1126/science.281.5373.82.

Scherer, R.P., Bohaty, S.M., Dunbar, R.B., Esper, O., Flores, J.-A., Gersonde, R., et al., 2008. Antarctic records of precession-paced insolation-driven warming during early Pleistocene Marine Isotope Stage 31. Geophysical Research Letters 35, 5. Available from: https://doi.org/10.1029/2007GL032254.

Scherer, R.P., DeConto, R.M., Pollard, D., Alley, R.B., 2016. Windblown Pliocene diatoms and East Antarctic Ice Sheet retreat. Nature Communications 7. Available from: https://doi.org/10.1038/ncomms12957, ncomms12957.

Scheuer, C., Gohl, K., Larter, R.D., Rebesco, M., Udintsev, G., 2006a. Variability in Cenozoic sedimentation along the continental rise of the Bellingshausen Sea, West Antarctica. Marine Geology 227 (3–4), 279–298. Available from: https://doi.org/10.1016/j.margeo.2005.12.007.

Scheuer, C., Gohl, K., Udintsev, G., 2006b. Bottom-current control on sedimentation in the western Bellingshausen Sea, West Antarctica. Geo-Marine Letters 26 (2), 90–101.

Schouten, S., Hopmans, E.C., Sinninghe Damsté, J.S., 2013. The organic geochemistry of glycerol dialkyl glycerol tetraether lipids: a review. Organic Geochemistry 54, 19–61. Available from: https://doi.org/10.1016/j.orggeochem.2012.09.006.

Shakun, J.D., Corbett, L.B., Bierman, P.R., Underwood, K., Rizzo, D.M., Zimmerman, S.R., et al., 2018. Minimal East Antarctic Ice Sheet retreat onto land during the past eight million years. Nature 558 (7709), 284–287. Available from: https://doi.org/10.1038/s41586-018-0155-6.

Shen, Q., Wang, H., Shum, C.K., Jiang, L., Hsu, H.T., Dong, J., 2018. Recent high-resolution Antarctic ice velocity maps reveal increased mass loss in Wilkes Land, East Antarctica. Scientific Reports 8 (1), 4477. Available from: https://doi.org/10.1038/s41598-018-22765-0.

Shevenell, A.E., Kennett, J.P., 2002. Antarctic Holocene climate change: a benthic foraminiferal stable isotope record from Palmer Deep. Paleoceanography 17 (2), PAL9-1–PAL9-12. Available from: https://doi.org/10.1029/2000PA000596.

Shevenell, A.E., Kennett, J.P., 2004. Paleoceanographic change during the middle miocene climate revolution: an antarctic stable isotope perspective. Geophysical monograph 151. Available from: https://doi.org/10.1029/151GM14.

Shevenell, A.E., Kennett, J.P., Lea, D.W., 2008. Middle Miocene ice sheet dynamics, deep-sea temperatures, and carbon cycling: a Southern Ocean perspective. Geochemistry, Geophysics, Geosystems 9 (2). Available from: https://doi.org/10.1029/2007GC001736.

Shevenell, A.E., Bohaty, S.M., 2012. Southern exposure new paleoclimate insights from southern ocean and Antarctic margin sediments. Oceanography 25 (3), 106–117. Available from: https://doi.org/10.5670/oceanog.2012.82. Scopus.

Shipp, S., Anderson, J., Domack, E., 1999. Late Pleistocene-Holocene retreat of the West Antarctic Ice-Sheet system in the Ross Sea: part 1–geophysical results. Geological Society of America Bulletin 111 (10), 1486–1516. Available from: https://doi.org/10.1130/0016-7606(1999)111 < 1486:LPHROT > 2.3.CO;2.

Siegert, M.J., et al., 2021. Antarctic Ice Sheet Changes since the Last Glacial Maximum. In: Florindo, F., et al. (Eds.), Antarctic Climate Evolution, Second ed. Elsevier.

Simkins, L.M., Anderson, J.B., Greenwood, S.L., Gonnermann, H.M., Prothro, L.O., Halberstadt, A.R.W., et al., 2017. Anatomy of a meltwater drainage system beneath the ancestral East Antarctic ice sheet. Nature Geoscience 10 (9), 691. Available from: https://doi.org/10.1038/ngeo3012.

Simões Pereira, P., van de Flierdt, T., Hemming, S.R., Frederichs, T., Hammond, S.J., Brachfeld, S., et al., 2020. The geochemical and mineralogical fingerprint of West Antarctica's weak underbelly: Pine Island and Thwaites glaciers. Chemical Geology 550, 119649. Available from: https://doi.org/10.1016/j.chemgeo.2020.119649.

Simões Pereira, P., van de Flierdt, T., Hemming, S.R., Hammond, S.J., Kuhn, G., Brachfeld, S., et al., 2018. Geochemical fingerprints of glacially eroded bedrock from West Antarctica: detrital thermochronology, radiogenic isotope systematics and trace element geochemistry in late Holocene glacial-marine sediments. Earth-Science Reviews 182, 204–232. Available from: https://doi.org/10.1016/j.earscirev.2018.04.011.

Sirevaag, H., Ksienzyk, A.K., Jacobs, J., Dunkl, I., Läufer, A., 2018. Tectono-thermal evolution and morphodynamics of the Central Dronning Maud Land Mountains, East Antarctica, based on new thermochronological data. Geosciences 8 (11), 390.

Sloan, B.J., Lawver, L.A., Anderson, J.B., 1995. Seismic stratigraphy of the Larsen Basin, Eastern Antarctic Peninsula. In: Cooper, A.K., Barker, P.F., Brancolini, G. (Eds.), Geology and Seismic Stratigraphy of the Antarctic Margin. American Geophysical Union, pp. 59–74.

Smellie, J.L., Haywood, A.M., Hillenbrand, C.-D., Lunt, D.J., Valdes, P.J., 2009. Nature of the Antarctic Peninsula Ice Sheet during the Pliocene: geological evidence and modelling results compared. Earth-Science Reviews 94 (1), 79–94. Available from: https://doi.org/10.1016/j.earscirev.2009.03.005.

Smellie, J.L., Rocchi, S., Armienti, P., 2011. Late Miocene volcanic sequences in northern Victoria Land, Antarctica: products of glaciovolcanic eruptions under different thermal regimes. Bulletin of Volcanology 73 (1), 1–25. Available from: https://doi.org/10.1007/s00445-010-0399-y.

Smith, R.T., Anderson, J.B., 2010. Ice-sheet evolution in James Ross Basin, Weddell Sea margin of the Antarctic Peninsula: the seismic stratigraphic record. GSA Bulletin 122 (5–6), 830–842. Available from: https://doi.org/10.1130/B26486.1.

Smith, J.A., Hillenbrand, C.-D., Kuhn, G., Klages, J.P., Graham, A.G.C., Larter, R.D., et al., 2014. New constraints on the timing of West Antarctic Ice Sheet retreat in the eastern Amundsen Sea since the Last Glacial Maximum. Global and Planetary Change 122, 224–237. Available from: https://doi.org/10.1016/j.gloplacha.2014.07.015.

Smith, J.A., Andersen, T.J., Shortt, M., Gaffney, A.M., Truffer, M., Stanton, T.P., et al., 2017. Sub-ice-shelf sediments record history of twentieth-century retreat of Pine Island Glacier. Nature 541 (7635), 77–80. Available from: https://doi.org/10.1038/nature20136.

Smith, E., Drews, R., Ehlers, T., Franke, D., Gaedicke, C., Hofstede, C., et al., 2018. On-ice Vibroseis: Sediment Features Below Ekström Ice Shelf, East Antarctica. In: POLAR 2018 (SCAR and IASC Conference), 19 June 2018–23 June 2018, Davos, Switzerland.

Smith, C., Warny, S., Shevenell, A.E., Gulick, S.P.S., Leventer, A., 2019. New species from the Sabrina Flora: an early Paleogene pollen and spore assemblage from the Sabrina Coast, East Antarctica. Palynology 43 (4), 650–659. Available from: https://doi.org/10.1080/01916122.2018.1471422.

Smith, E.C., Hattermann, T., Kuhn, G., Gaedicke, C., Berger, S., Drews, R., et al., 2020. Detailed seismic bathymetry beneath Ekström Ice Shelf, Antarctica: implications for glacial history and ice-ocean interaction. Geophysical Research Letters 47 (10). Available from: https://doi.org/10.1029/2019gl086187, e2019GL086187.

Solheim, A., 1991. Stepwise consolidation of glacigenic sediments related to the glacial history of Prydz Bay, East Antarctica. Proceedings of the Ocean Drilling Program, Scientific Results 119, 169.

Sorlien, C.C., Luyendyk, B.P., Wilson, D.S., Decesari, R.C., Bartek, L.R., Diebold, J.B., 2007. Oligocene development of the West Antarctic Ice Sheet recorded in eastern Ross Sea strata. Geology 35 (5), 467–470. Available from: https://doi.org/10.1130/G23387A.1.

Sprenk, D., Weber, M.E., Kuhn, G., Wennrich, V., Hartmann, T., Seelos, K., 2014. Seasonal changes in glacial polynya activity inferred from Weddell Sea varves. Climate of the Past 10 (3), 1239–1251. Available from: https://doi.org/10.5194/cp-10-1239-2014.

Stagg, H.M.J., Colwel, J.B., Direen, N.G., O'Brien, P.E., Bernardel, G., Borissova, I., et al., 2004. Geology of the continental margin of Enderby and Mac. Robertson Lands, East Antarctica: insights from a regional data set. Marine Geophysical Research 25 (3–4), 183–219. Available from: https://doi.org/10.1007/s11001-005-1316-1.

Stagg, H.M.J., Colwell, J.B., Direen, N.G., O'Brien, P.E., Browning, B.J., Bernardel, G., et al., 2005. Geological framework of the continental margin in the region of the Australian Antarctic Territory. Geoscience Australia Record 2004/2, 1–373.

Stocchi, P., Escutia, C., Houben, A.J.P., Vermeersen, B.L.A., Bijl, P.K., Brinkhuis, H., et al., 2013. Relative sea-level rise around East Antarctica during Oligocene glaciation. Nature Geoscience 6 (5), 380–384.

Strand, K., Passchier, S., Näsi, J., 2003. Implications of quartz grain microtextures for onset Eocene/Oligocene Glaciation in Prydz Bay, ODP Site 1166, Antarctica. Palaeogeography, Palaeoclimatology, Palaeoecology 198 (1–2), 101–111. Available from: https://doi.org/10.1016/S0031-0182(03)00396-1.

Suganuma, Y., Miura, H., Zondervan, A., Okuno, J., 2014. East Antarctic deglaciation and the link to global cooling during the Quaternary: evidence from glacial geomorphology and 10Be surface exposure dating of the Sør Rondane Mountains, Dronning Maud Land. Quaternary Science Reviews 97, 102–120. Available from: https://doi.org/10.1016/j.quascirev.2014.05.007.

Sugden, D.E., 1996. The East Antarctic Ice Sheet: unstable ice or unstable ideas? Transactions of the Institute of British Geographers 21 (3), 443–454. Available from: https://doi.org/10.2307/622590.

Sugden, D., Denton, G., 2004. Cenozoic landscape evolution of the Convoy Range to Mackay Glacier area, Transantarctic Mountains: onshore to offshore synthesis. Geological Society of America Bulletin 116 (7–8), 840–857. Available from: https://doi.org/10.1130/B25356.1.

Sugden, D.E., Marchant, D.R., Denton, G.H., 1993. The case for a stable East Antarctic Ice Sheet: the background. Geografiska Annaler. Series A, Physical Geography 75 (4), 151–154. Available from: https://doi.org/10.2307/521199.

Talarico, F., et al., 2021. The Antarctic Continent in Gondwana: A perspective from the Ross Embayment and Potential Research Targets for Future Investigations. In: Florindo, F., et al., (Eds.), Antarctic Climate Evolution, Second ed. Elsevier.

Tanahashi, M., Eittreim, S., Wanneson, J., 1994. Seismic stratigraphic sequences of the Wilkes Land margin. Terra Antarctica 1 (2), 391–393.

Tauxe, L., Stickley, C.E., Sugisaki, S., Bijl, P.K., Bohaty, S.M., Brinkhuis, H., et al., 2012. Chronostratigraphic framework for the IODP Expedition 318 cores from the Wilkes Land Margin: constraints for paleoceanographic reconstruction. Paleoceanography 27 (2). Available from: https://doi.org/10.1029/2012PA002308.

Taylor, J., Siegert, M.J., Payne, A.J., Hambrey, M.J., O'Brien, P.E., Leitchenkov, G., et al., 2004. Late Miocene/early Pliocene changes in sedimentation paths, Prydz Bay, Antarctica: changes in ice-sheet dynamics? Geology, 32 (3), 197–200. Available from: https://doi.org/10.1130/G20275.1.

Taylor-Silva, B.I., Riesselman, C.R., 2018. Polar frontal migration in the warm late Pliocene: diatom evidence from the Wilkes Land Margin, East Antarctica. Paleoceanography and Paleoclimatology 33 (1), 76–92. Available from: https://doi.org/10.1002/2017PA003225.

Ten Brink, U.S., Schneider, C., Johnson, A.H., 1995. Morphology and stratal geometry of the Antarctic Continental Shelf: insights from models. In: Cooper, Alan K., Barker, P.F., Brancolini, G. (Eds.), Geology and Seismic Stratigraphy of the Antarctic Margin. American Geophysical Union, pp. 1–24. Available from: https://doi.org/10.1029/AR068p0001.

The RAISED Consortium, 2014. A community-based geological reconstruction of Antarctic Ice Sheet deglaciation since the Last Glacial Maximum. Quaternary Science Reviews 100, 1–9. Available from: https://doi.org/10.1016/j.quascirev.2014.06.025.

Theissen, K.M., Dunbar, R.B., Cooper, A.K., Mucciarone, D.A., Hoffmann, D., 2003. The Pleistocene evolution of the East Antarctic Ice Sheet in the Prydz bay region: Stable isotopic evidence from ODP Site 1167. Global and Planetary Change 39 (3–4), 227–256. Available from: https://doi.org/10.1016/S0921-8181(03)00118-8.

Thomson, M.R.A., 1981. Late Mesozoic stratigraphy and invertebrate palaeontology of the South Orkney Islands. British Antarctic Survey Bulletin 54, 65–83.

Thomson, S.N., Reiners, P.W., Hemming, S.R., Gehrels, G.E., 2013. The contribution of glacial erosion to shaping the hidden landscape of East Antarctica. Nature Geoscience 6 (3), 203–207. Available from: https://doi.org/10.1038/ngeo1722.

Tinto, K.J., Padman, L., Siddoway, C.S., Springer, S.R., Fricker, H.A., Das, I., et al., 2019. Ross Ice Shelf response to climate driven by the tectonic imprint on seafloor bathymetry. Nature Geoscience 12 (6), 441–449. Available from: https://doi.org/10.1038/s41561-019-0370-2.

Troedson, A.L., Smellie, J.L., 2002. The Polonez Cove Formation of King George Island, Antarctica: stratigraphy, facies and implications for mid-Cenozoic cryosphere development. Sedimentology 49 (2), 277–301. Available from: https://doi.org/10.1046/j.1365-3091.2002.00441.x.

Truswell, E.M., 1982. Palynology of seafloor samples collected by the 1911–14 Australasian Antarctic expedition: implications for the geology of coastal East Antarctica. Journal of the Geological Society of Australia 29 (3–4). Available from: https://doi.org/10.1080/00167618208729218.

Truswell, E.M., 1991. Data report: palynology of sediments from Leg 119 drill sites in Prydz Bay, East Antarctica. Proceedings of Scientific Results, ODP, Leg 119, Kerguelen Plateau-Prydz Bay 941–945. Available from: https://doi.org/10.2973/odp.proc.sr.119.162.1991.

Truswell, E.M., Drewry, D.J., 1984. Distribution and provenance of recycled palynomorphs in surficial sediments of the Ross Sea, Antarctica. Marine Geology 59 (1), 187–214. Available from: https://doi.org/10.1016/0025-3227(84)90093-8.

Turner, B.R., 1991. Depositional environment and petrography of preglacial continental sediments from Hole 740A, Prydz Bay, Antarctica. In: Barron, J., Larsen, B., et al.,Proceedings of the Ocean Drilling Program, Scientific Results, 119. Ocean Drilling Program, College Station, TX, pp. 45–56. Available from: https://doi.org/10.2973/odp.proc.sr.119.133.1991.

Turner, J., Lachlan-Cope, T.A., Marshall, G.J., Morris, E.M., Mulvaney, R., Winter, W., 2002. Spatial variability of Antarctic Peninsula net surface mass balance. Journal of Geophysical Research: Atmospheres 107 (D13), AAC4-1–AAC4-18. Available from: https://doi.org/10.1029/2001JD000755.

Turney, C.S.M., Fogwill, C.J., Golledge, N.R., McKay, N.P., Sebille, E., van, Jones, R.T., et al., 2020. Early Last Interglacial ocean warming drove substantial ice mass loss from Antarctica. Proceedings of the National Academy of Sciences 117 (8), 3996–4006. Available from: https://doi.org/10.1073/pnas.1902469117.

Uenzelmann-Neben, G., 2018. Variations in ice-sheet dynamics along the Amundsen Sea and Bellingshausen Sea West Antarctic Ice Sheet margin. Bulletin of the Geological Society of America 131 (3–4), 479–498. Available from: https://doi.org/10.1130/B31744.1.

Uenzelmann-Neben, G., Gohl, K., 2012. Amundsen Sea sediment drifts: archives of modifications in oceanographic and climatic conditions. Marine Geology 299–302, 51–62. Available from: https://doi.org/10.1016/j.margeo.2011.12.007.

Uenzelmann-Neben, G., Gohl, K., 2014. Early glaciation already during the Early Miocene in the Amundsen Sea, Southern Pacific: indications from the distribution of sedimentary sequences. Global and Planetary Change 120, 92–104. Available from: https://doi.org/10.1016/j.gloplacha.2014.06.004.

Uenzelmann-Neben, G., Gohl, K., Larter, R.D., Schlüter, P., 2007. Differences in ice retreat across Pine Island Bay, West Antarctica, since the Last Glacial Maximum: indications from multichannel seismic reflection data. In: Differences in Ice Retreat Across Pine Island Bay, West Antarctica, Since the Last Glacial Maximum: indications from Multichannel Seismic Reflection Data (USGS Numbered Series No. 2007-1047-SRP-084; Open-File Report, Vols. 2007-1047-SRP-084). United States Geological Survey. <https://doi.org/10.3133/ofr20071047SRP084>.

van de Flierdt, T., 2008. Evidence against a young volcanic origin of the Gamburtsev subglacial mountains, Antarctica. Geophysical Research Letters 35 (21). Available from: https://doi.org/10.1029/2008GL035564.

van de Flierdt, T., Hemming, S.R., Goldstein, S.L., Gehrels, G.E., Cox, S.E., 2008. Evidence against a young volcanic origin of the Gamburtsev Subglacial Mountains, Antarctica. Geophysical Research Letters 35 (21).

Veevers, J.J., 2008. Provenance of the Gamburtsev subglacial mountains from U-Pb and Hf analysis of detrital zircons in Cretaceous to Quaternary sediments in Prydz Bay and beneath the Amery Ice Shelf. Sedimentary Geology 211 (1–2), 12–32. Available from: https://doi.org/10.1016/j.sedgeo.2008.08.003.

Vernet, M., Geibert, W., Hoppema, M., Brown, P.J., Haas, C., Hellmer, H.H., et al., 2019. The Weddell Gyre, Southern Ocean: present knowledge and future challenges. Reviews of Geophysics 57 (3), 623–708. Available from: https://doi.org/10.1029/2018rg000604.

Villa, G., Lupi, C., Cobianchi, M., Florindo, F., Pekar, S.F., 2008. A Pleistocene warming event at 1 Ma in Prydz Bay, East Antarctica: evidence from ODP Site 1165. Palaeogeography, Palaeoclimatology, Palaeoecology 260 (1–2), 230–244. Available from: https://doi.org/10.1016/j.palaeo.2007.08.017.

Vincent, E., Berger, W.H., 1985. Carbon Dioxide and Polar Cooling in the Miocene: the Monterey hypothesis. In: Sundquist, E.T., Broecker, W.S. (Eds.), The Carbon Cycle and

Atmospheric CO_2: Natural Variations Archean to Present. American Geophysical Union, pp. 455–468. Available from: https://doi.org/10.1029/GM032p0455.

Viseras, C., Maldonado, A., 1999. Facies architecture, seismic stratigraphy and development of a high-latitude basin: The Powell Basin (Antarctica). Marine Geology 157 (1–2), 69–87.

Vleeschouwer, D.D., Vahlenkamp, M., Crucifix, M., Pälike, H., 2017. Alternating Southern and Northern Hemisphere climate response to astronomical forcing during the past 35 m.y. Geology 45 (4), 375–378. Available from: https://doi.org/10.1130/G38663.1.

Volpi, V., Camerlenghi, A., Hillenbrand, C.-D., Rebesco, M., Ivaldi, R., 2003. Effects of biogenic silica on sediment compaction and slope stability on the Pacific margin of the Antarctic Peninsula. Basin Research 15 (3), 339–363. Available from: https://doi.org/10.1046/j.1365-2117.2003.00210.x.

Wannesson, J., Pelras, M., Petitperrin, B., Perret, M., Segoufin, J., 1985. A geophysical transect of the Adélie Margin, East Antarctica. Marine and Petroleum Geology 2 (3), 192–200. Available from: https://doi.org/10.1016/0264-8172(85)90009-1.

Warny, S., Askin, R.A., Hannah, M.J., Mohr, B.A.R., Raine, J.I., Harwood, D.M., et al., 2009. Palynomorphs from a sediment core reveal a sudden remarkably warm Antarctica during the middle Miocene. Geology 37 (10), 955–958. Available from: https://doi.org/10.1130/G30139A.1.

Warny, S.S., Askin, R.R., 2013. Vegetation and organic-walled phytoplankton at the end of the Antarctic greenhouse world: latest Eocene cooling events. Tectonic, Climatic, and Cryospheric Evolution of the Antarctic Peninsula. American Geophysical Union (AGU), pp. 193–210. Available from: https://doi.org/10.1029/2010SP000965.

Webb, P.N., Harwood, D.M., McKelvey, B.C., Mercer, J.H., Stott, L.D., 1984. Cenozoic marine sedimentation and ice-volume variation on the East Antarctic craton. Geology 12 (5), 287–291. Available from: https://doi.org/10.1130/0091-7613(1984)12 < 287:CMSAIV > 2.0.CO;2.

Weber, M.E., Clark, P.U., Kuhn, G., Timmermann, A., Sprenk, D., Gladstone, R., et al., 2014. Millennial-scale variability in Antarctic ice-sheet discharge during the last deglaciation. Nature 510 (7503), 134–138. Available from: https://doi.org/10.1038/nature13397.

Weber, M.E., Clark, P.U., Ricken, W., Mitrovica, J.X., Hostetler, S.W., Kuhn, G., 2011. Interhemispheric ice-sheet synchronicity during the last glacial maximum. Science 334 (6060), 1265–1269. Available from: https://doi.org/10.1126/science.1209299.

Weber, M.E., Raymo, M.E., Peck, V.L., Williams, T., 2019. Expedition 382 Preliminary Report: Iceberg Alley and subAntarctic ice and ocean dynamics, 20 March–20 May 2019. International Ocean Discovery Program, 39 pp.

Weigelt, E., Gohl, K., Uenzelmann-Neben, G., Larter, R.D., 2009. Late Cenozoic ice sheet cyclicity in the western Amundsen Sea Embayment—evidence from seismic records. Global and Planetary Change 69 (3), 162–169. Available from: https://doi.org/10.1016/j.gloplacha.2009.07.004.

Weigelt, E., Uenzelmann-Neben, G., Gohl, K., Larter, R.D., 2012. Did massive glacial dewatering modify sedimentary structures on the Amundsen Sea Embayment shelf, West Antarctica? Global and Planetary Change 92–93, 8–16. Available from: https://doi.org/10.1016/j.gloplacha.2012.04.006.

Wenman, C.P., Harry, D.L., Jha, S., 2020. Post Middle Miocene Tectonomagmatic and Stratigraphic Evolution of the Victoria Land Basin, West Antarctica. Geochemistry, Geophysics, Geosystems 21 (3). Available from: https://doi.org/10.1029/2019GC008568, e2019GC008568.

Whitehead, J.M., 2003. Ice-distal upper Miocene marine strata from inland Antarctica. Sedimentology 50 (3), 531–552. Available from: https://doi.org/10.1046/j.1365-3091.2003.00563.x.

Whitehead, J.M., Bohaty, S.M., 2003. Pliocene summer sea surface temperature reconstruction using silicoflagellates from Southern Ocean ODP Site 1165. Paleoceanography 18 (3), 20-1–20-11. Available from: https://doi.org/10.1029/2002PA000829.

Whitehead, J.M., Harwood, D.M., McKelvey, B.C., Hambrey, M.J., McMinn, A., 2004. Diatom biostratigraphy of the Cenozoic glaciomarine Pagodroma Group, northern Prince Charles Mountains, East Antarctica. Australian Journal of Earth Sciences 51 (4), 521–547. Available from: https://doi.org/10.1111/j.1400-0952.2004.01072.x.

Whitehead, J.M., Quilty, P.G., Harwood, D.M., McMinn, A., 2001. Early Pliocene paleoenvironment of the Sørsdal Formation, Vestfold Hills, based on diatom data. Marine Micropaleontology 41 (3–4), 125–152. Available from: https://doi.org/10.1016/S0377-8398(00)00060-8.

Whitehead, J.M., Wotherspoon, S., Bohaty, S.M., 2005. Minimal Antarctic sea ice during the Pliocene. Geology 33 (2), 137–140. Available from: https://doi.org/10.1130/G21013.1.

Whitehouse, P.L., Gomez, N., King, M.A., Wiens, D.A., 2019. Solid Earth change and the evolution of the Antarctic Ice Sheet. Nature Communications 10 (1), 503. Available from: https://doi.org/10.1038/s41467-018-08068-y.

Williams, T., Handwerger, D., 2005. A high-resolution record of early Miocene Antarctic glacial history from ODP Site 1165, Prydz Bay. Paleoceanography 20 (2), PA2017. Available from: https://doi.org/10.1029/2004PA001067.

Williams, T., van de Flierdt, T., Hemming, S.R., Chung, E., Roy, M., Goldstein, S.L., 2010. Evidence for iceberg armadas from East Antarctica in the Southern Ocean during the late Miocene and early Pliocene. Earth and Planetary Science Letters 290 (3–4), 351–361. Available from: https://doi.org/10.1016/j.epsl.2009.12.031.

Williams, T.J., Hillenbrand, C.-D., Piotrowski, A.M., Hodell, D., Allen, C.S., Frederichs, T., et al., 2019. Paleocirculation and ventilation history of Southern Ocean sourced deep water masses during the last 800,000 years. Paleoceanography and Paleoclimatology 34 (5), 833–852. Available from: https://doi.org/10.1029/2018PA003472.

Wilson, D.J., Bertram, R.A., Needham, E.F., Flierdt, T., van de, Welsh, K.J., McKay, R.M., et al., 2018. Ice loss from the East Antarctic Ice Sheet during late Pleistocene interglacials. Nature 561 (7723), 383–386. Available from: https://doi.org/10.1038/s41586-018-0501-8.

Wilson, D.J., et al., 2021. Pleistocene Antarctic climate variability: ice sheet–ocean–climate interactions. In: Florindo, F., et al. (Eds.), Antarctic Climate Evolution, second ed. Elsevier (this volume).

Wilson, D.S., Jamieson, S.S.R., Barrett, P.J., Leitchenkov, G., Gohl, K., Larter, R.D., 2012. Antarctic topography at the Eocene–Oligocene boundary. Palaeogeography, Palaeoclimatology, Palaeoecology 335–336, 24–34. Available from: https://doi.org/10.1016/j.palaeo.2011.05.028.

Wilson, D.S., Luyendyk, B.P., 2009. West Antarctic paleotopography estimated at the Eocene-Oligocene climate transition. Geophysical Research Letters 36, 4. Available from: https://doi.org/10.1029/2009GL039297.

Wilson, D.S., Pollard, D., DeConto, R.M., Jamieson, S.S.R., Luyendyk, B.P., 2013. Initiation of the West Antarctic Ice Sheet and estimates of total Antarctic ice volume in the earliest Oligocene. Geophysical Research Letters 40 (16), 4305–4309. Available from: https://doi.org/10.1002/grl.50797.

Wobbe, F., Gohl, K., Chambord, A., Sutherland, R., 2012. Structure and breakup history of the rifted margin of West Antarctica in relation to Cretaceous separation from Zealandia and Bellingshausen plate motion. Geochemistry, Geophysics, Geosystems 13 (4). Available from: https://doi.org/10.1029/2011GC003742.

Wright, J.D., Miller, K.G., Fairbanks, R.G., 1991. Evolution of modern deepwater circulation: evidence from the late Miocene Southern Ocean. Paleoceanography 6 (2), 275–290. Available from: https://doi.org/10.1029/90PA02498.

Wright, A.P., Young, D.A., Roberts, J.L., Schroeder, D.M., Bamber, J.L., Dowdeswell, J.A., et al., 2012. Evidence of a hydrological connection between the ice divide and ice sheet margin in the Aurora Subglacial Basin, East Antarctica. Journal of Geophysical Research: Earth Surface 117 (F1). Available from: https://doi.org/10.1029/2011JF002066.

Yamaguchi, K., Tamura, Y., Mizukoshi, I., Tsuru, T., 1988. Preliminary report of geophysical and geological surveys in the Amundsen Sea, West Antarctica. Proceedings of NIPR Symposium on Antarctic Geosciences 2, 55–67.

Yamane, M., Yokoyama, Y., Abe-Ouchi, A., Obrochta, S., Saito, F., Moriwaki, K., et al., 2015. Exposure age and ice-sheet model constraints on Pliocene East Antarctic ice sheet dynamics. Nature Communications 6 (1), 7016. Available from: https://doi.org/10.1038/ncomms8016.

You, Y., Huber, M., Müller, R.D., Poulsen, C.J., Ribbe, J., 2009. Simulation of the Middle Miocene Climate Optimum. Geophysical Research Letters 36 (4). Available from: https://doi.org/10.1029/2008GL036571.

Young, D.A., Wright, A.P., Roberts, J.L., Warner, R.C., Young, N.W., Greenbaum, J.S., et al., 2011a. A dynamic early East Antarctic Ice Sheet suggested by ice-covered fjord landscapes. Nature 474 (7349), 72–75. Available from: https://doi.org/10.1038/nature10114.

Zachos, J.C., Flower, B.P., Paul, H., 1997. Orbitally paced climate oscillations across the Oligocene/Miocene boundary. Nature 388 (6642), 567–570. Available from: https://doi.org/10.1038/41528.

Zachos, J., Pagani, M., Sloan, L., Thomas, E., Billups, K., 2001. Trends, rhythms, and aberrations in global climate 65 Ma to present. Science 292 (5517), 686–693. Available from: https://doi.org/10.1126/science.1059412.

Chapter 4

Water masses, circulation and change in the modern Southern Ocean

Lionel Carter[1], Helen Bostock-Lyman[2] and Melissa Bowen[3]
[1]Antarctic Research Centre, Victoria University of Wellington, Wellington, New Zealand, [2]School of Earth and Environmental Sciences, University of Queensland, Brisbane, QLD, Australia, [3]School of Environment, University of Auckland, Auckland, New Zealand

4.1 Introduction

The break-up of the southern supercontinent of Gondwana was a major event in Earth's evolution. As continental plates drifted towards their present sites, a proto-Southern Ocean developed that eventually isolated continental Antarctica. That isolation depended on the opening of the Drake Passage and Tasman Gateway – two pathways with debatable histories (e.g., Scher and Martin, 2006). A predecessor of the Antarctic Circumpolar Current (ACC) may have commenced in the Drake Passage as an intermediate depth flow around 50 Ma with a deep flow established through the topographically complex passage around 30 Ma (Eagles and Jokat, 2014; Galeotti et al., 2021). The opening of the Tasman Gateway in the Australasian region was similarly complex. A west-bound flow developed around 50 Ma (Bijl et al., 2013), but it was not until ~33.5 Ma that an eastbound ACC was established (Carter et al., 2004; Exon et al., 2004), with the ACC fully formed after 23.95 Ma (Pfuhl and McCave, 2005). This development occurred at a time of declining atmospheric CO_2 and, together with the thermal isolation of Antarctica, contributed to a marked cooling of Earth (e.g., Kennett and von der Borch, 1985; McKay et al., 2016; Zachos et al., 2008). Geographically, the evolving Southern Ocean linked the Pacific, Indian and Atlantic Oceans to become a major influence on the global distribution of heat, salt, carbon, nutrients and dissolved gases. Accordingly, Southern Ocean water masses and currents became strongly interwoven with global climate/ocean change. In that context, this chapter focuses on the modern oceanography to provide a physical basis for realistic reconstructions of past environments – a theme of this book, *Antarctic Climate Evolution*. This

Antarctic Climate Evolution. DOI: https://doi.org/10.1016/B978-0-12-819109-5.00003-7

approach is made in recognition of an issue, eloquently highlighted by Wunsch (2010) that '*Paleoclimate reconstruction and understanding present some of the most intriguing data and problems in all of science. Progress clearly requires combining the remarkable achievements in producing proxy data with similar achievements in understanding dynamics*'.

Mindful of the range of disciplines and subdisciplines presented in this book, this chapter is, where possible, written for the non-specialist in order to improve accessibility and understanding of physical oceanographic processes. Three facets of the Southern Ocean are covered; water masses, their dynamics and recent changes especially, post-1950. That period welcomed regular ship-borne hydrographic surveys of the entire water column exemplified by the World Ocean Circulation Experiment (WOCE) of 1990–2002 (Orsi and Whitworth, 2005). WOCE surveyed key transects of the global ocean, which serve as reference sections for repeat oceanographic observations that continue today with the GO-SHIP program (Sloyan et al., 2019). Other significant years are 1979 when satellite observation of sea surface temperatures started and 1992 when satellite-borne altimeters provided sea surface height information with an accuracy that resolved ocean dynamics. Recent insights into the Southern Ocean come from instrumented floats, primarily those in the Argo program (Argo, 2020). That initiative began in the early 2000s and currently has almost 4000 floats monitoring the ocean's interior to ~2000 m water depth. New innovations in the Argo program involving biogeochemical systems, ice-avoiding algorithms, ice-strengthened floats and floats capable of operating to 6000 m water depth have allowed deployments south of latitude 60°S to improve coverage of this sparsely-sampled region. Instrumented fauna, such as elephant seals, have also been providing measurements of water properties through the winter, in sea ice and close to shore (Roquet et al., 2014). Much of the data from the aforementioned endeavours can be found through the Southern Ocean Observing System (SOOS, 2020).

4.1.1 Defining the Southern Ocean

The ocean's southern limit is acknowledged as the coast line of Antarctica. However, its northern limit is less well defined, reflecting the difficulty in placing a boundary on such a dynamic feature as the ocean. The International Hydrographic Organisation chose latitude 60°S – a land-free parallel that encircles Antarctica – as the northern boundary (IHO, 2002). However, this demarcation has yet to be officially adopted. A more realistic boundary is the Subtropical Front (STF) that separates subtropical surface water from subantarctic surface water (e.g., Bostock et al., 2012; Gille, 2010; Orsi et al., 1995). This front follows a circumpolar path that fluctuates about 40°S, with exception of the interruption by South America (Fig. 4.1). In that configuration, the STF separates the subtropical gyres from the ACC – the main flow in the Southern Ocean (Orsi et al., 1995; Rintoul et al., 2001; Sokolov and Rintoul, 2009a,b). Studies related to global climate change (Bindoff et al., 2007; Shi et al., 2018)

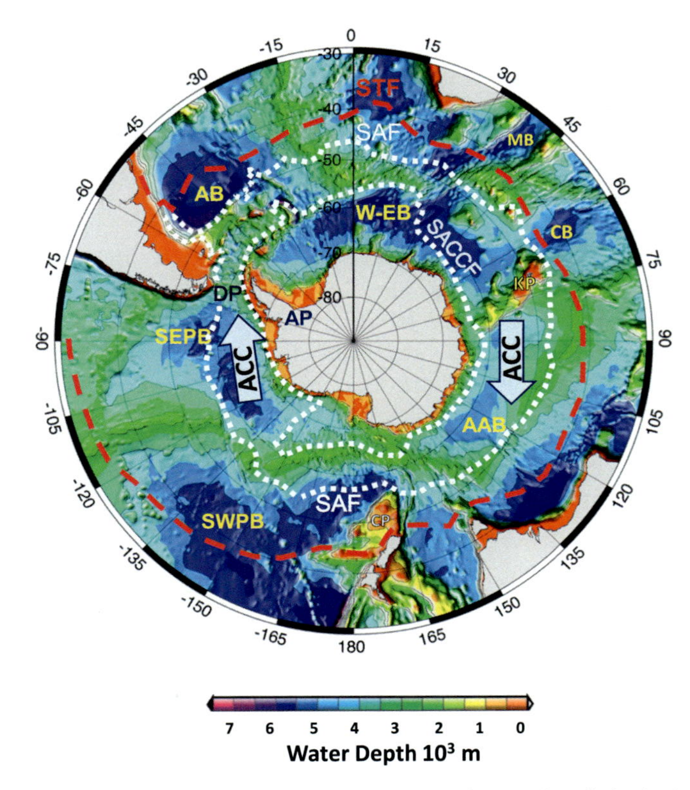

FIGURE 4.1 Southern Ocean bathymetry together with its various northern limits that include latitude 30°S, which is latitude limit of the chart (e.g., Bindoff et al., 2007), latitude 60°S (IHO, 2002) and the Subtropical Front = red dashed line. ACC is the Antarctic Circumpolar Current outlined by the Subantarctic Front (SAF) and Southern ACC Front (SACCF) = white dotted lines. Ocean ridges outline the Weddell–Enderby Basin (W-EB), Crozet Basin (CB), Mozambique Basin (MB), Australian-Antarctic Basin (AAB), SW Pacific Basin (SWPB), SE Pacific Basin (SEPB), Argentine Basin (AB), Campbell Plateau (CP), Kerguelen Plateau (KP) and Drake Passage (DP). Seabed relief model adapted from Hayes et al. (2009).

extend the Southern Ocean's boundary to latitude 30°S on the basis that this wider latitudinal zone stores more anthropogenic CO_2 and heat than any other ocean region. In this chapter the northern limit of the Southern Ocean is taken as the STF (Fig. 4.1).

4.2 Water masses – characteristics and distribution

4.2.1 Upper ocean

The uppermost 100–200 m of the Southern Ocean is separated into zones defined by a series of ocean fronts. These were originally identified by marked changes in water mass properties that gave rise to strong currents or *jets*. Early observations identified four to five frontal systems that included, from north to south, the STF, Subantarctic Front (SAF), Polar Front (PF), Southern ACC Front (SACCF) and Southern Boundary Front (SBF) (e.g., Deacon, 1982; Nowlin and Klinck, 1986; Orsi et al., 1995). Far from simple linear features, remote sensing, hydrographic time-series and instrumented floats now show fronts to be more dynamic and complex than suggested previously by limited ship-based hydrographic data. Instead of single jets, the SAF and PF are now shown to each have three jets labelled *north*, *mid* and *south*, whereas the SACCF comprises two jets, *mid* and *south*, and SBF a single *mid* jet (Fig. 4.2; British Oceanographic Data Centre BODC, 2020; Sokolov and Rintoul, 2007, 2009a). For the purpose of this general overview, we use the *mid* or mean position to delimit the fronts (Figs. 4.2 to 4.4).

SACCF to SBF: The STF is the southern limit of warm, saline Subtropical Surface Water (STSW: potential temperature $\theta \geq 12°C$ and salinity $S \geq 34.7$). Its transition to Subantarctic Surface Water (SASW) involves marked cooling and freshening with $\theta = 7°C–4°C$ and $S = 34.0$. SASW is

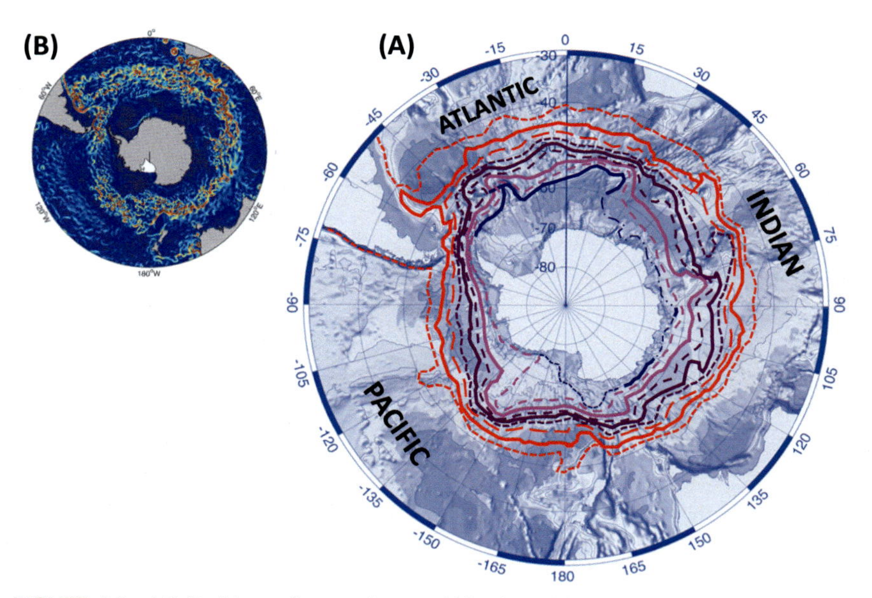

FIGURE 4.2 (A) Positions of ocean fronts within the ACC as identified by Sokolov and Rintoul (2009a) from satellite-based sea surface height, validated by hydrographic data. From north to south: the Subantarctic Front (red with short dash = **North** front; solid line = **Mid** front; and long dash = **South** front); Polar Front (purple = N:Mid:S), South ACCF (pink = Mid:S) and Southern Boundary Front (blue = Mid), which is incomplete due to masking by sea ice. Base chart is from Hayes et al. (2009). (B) Inset outlines the eddy-rich ACC circulation as simulated by NOAA/Goddard Fluid Dynamic Model.

also enriched with nutrients including nitrate and phosphate that when mixed with nutrient-poor, but iron-rich STSW enhances primary productivity to provide another identifier of the zone, namely, satellite-sensed chlorophyll *a* (e.g., Bostock et al., 2012; Chiswell et al., 2013; Jasmine et al., 2009; Murphy et al., 2001).

SAF to PF: This zone accommodates the transition of SASW to Antarctic Surface Water (AASW). While temperature and salinity continue to decline southward ($\theta = \sim 4°C$ and $S = 34.0–33.8$), the water structure is complex and probably reflects significant eddy activity in this region (Rintoul, 2018).

PF to SACCF: This zone is characterised by a thin (~100 m thick) surface layer of AASW that extends to the Antarctic continental margin (Fig. 4.4). Temperatures are typically $\theta \leq 4°C$ and $S = \sim 34.0$ with elevated nitrate, phosphate and silicate (Bostock et al., 2012).

SACCF to SBF and Antarctica: Cooling of the AASW layer continues south ($\theta = 0$ to $1.5°C$ and $S = \sim 34.0$) for much of eastern Antarctica, but salinity freshens to $S \leq 33.0$ off western Antarctica to the Ross Sea. This change is attributed to a combination of factors including increased meltwater and precipitation plus reduced sea ice (e.g., Jacobs et al., 2002) (Fig. 4.3). However, locally within the Weddell and Ross Seas, salinity rises to $S \geq 34.4$, which together with elevated nutrients suggest upwelling of Circumpolar Deep Water (CDW) (Deacon, 1982; Gordon, 1975; Smith et al., 2012).

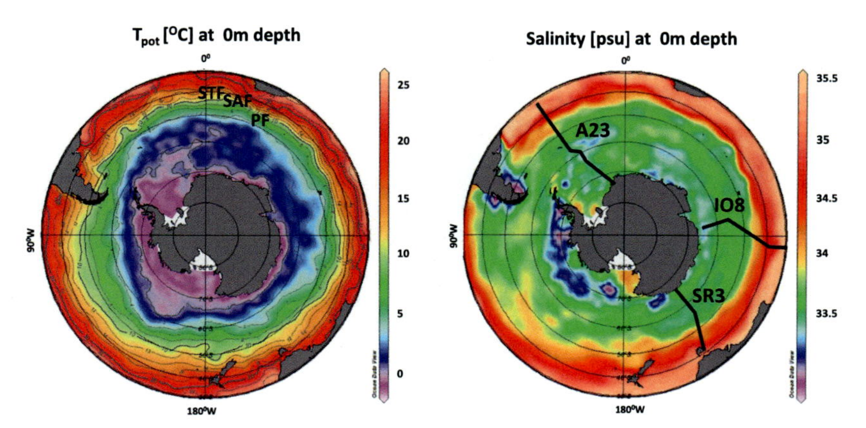

FIGURE 4.3 Potential surface temperature and surface salinity highlighting their distribution and gradients from the Subtropical Front (STF), Subantarctic Front (SAF), Polar Front (PF) to coastal Antarctica. Transects A23, IO8 and SR3 mark positions of hydrographic profiles of Fig. 4.4. Localised freshening off west Antarctica to the Ross Sea region may reflect increased precipitation and meltwater and/or reduced sea ice, whereas coastal zones of increased salinity in the Weddell and Ross Seas are probably due to upwelling of CDW. Based on data from WOCE/GLODAP (Orsi and Whitworth, 2005) and plotted in Ocean Data View (Schlitzer, 2009).

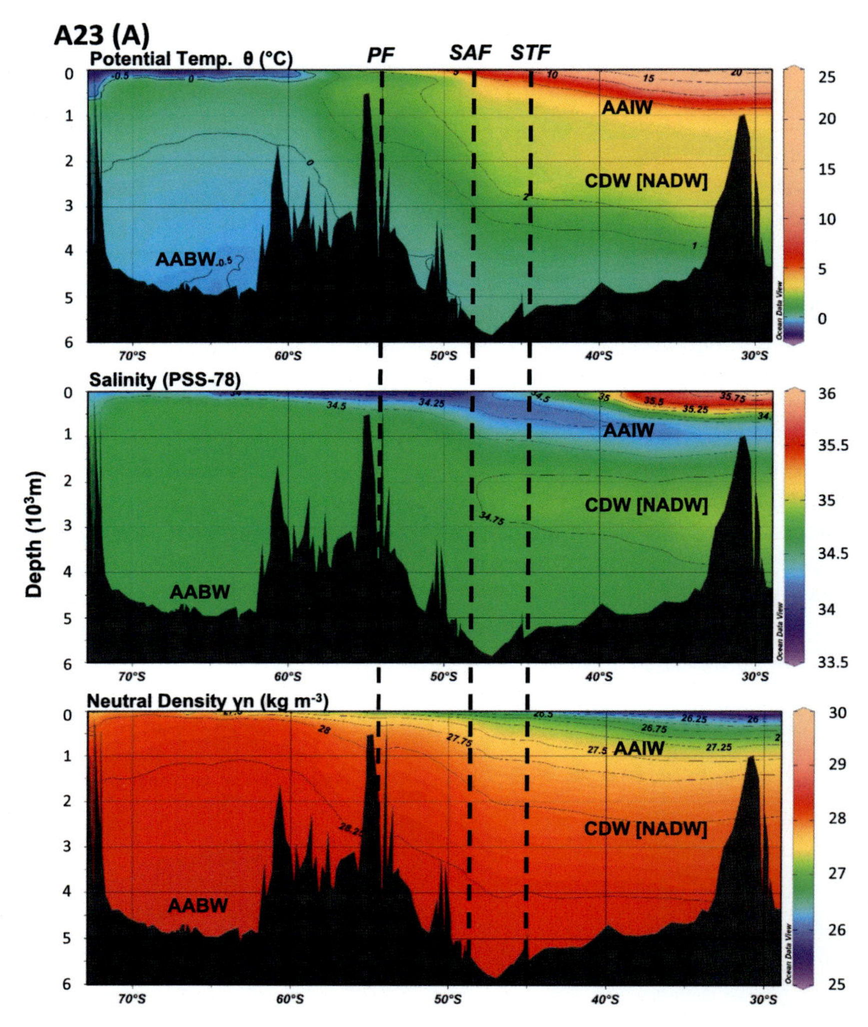

FIGURE 4.4 Hydrographic profiles of potential temperature, salinity and neutral density along lines A23 (A), IO8 (B) and SR3 (C), which are located in Fig. 4.3. *AABW*, Antarctic Bottom Water; *AAIW*, Antarctic Intermediate Water; *CDW*, Circumpolar Deep Water together with its prominent regional components that include North Atlantic Deep Water (NADW) and Indian Deep Water (IDW). Data are from WOCE/GLODAP (Orsi and Whitworth, 2005) and plotted in Ocean Data View (Schlitzer, 2009).

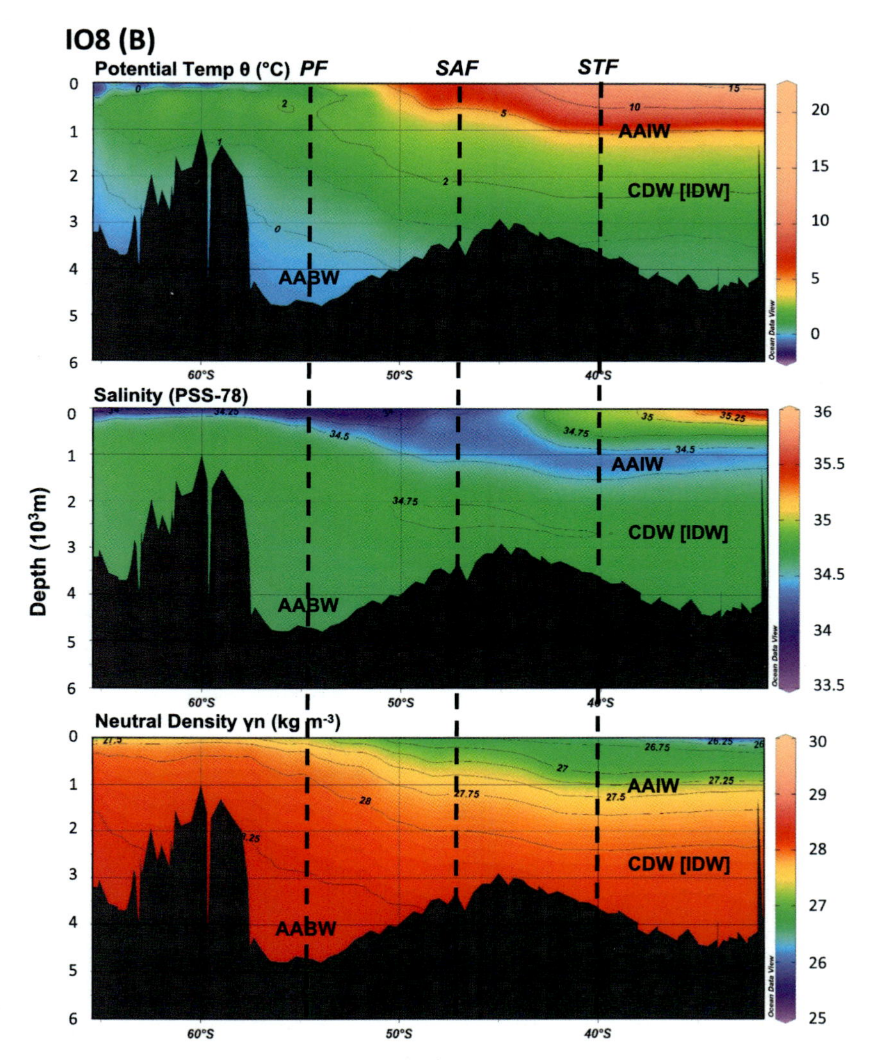

FIGURE 4.4 Continued

4.2.2 Intermediate depth waters

Surface waters that subduct to depth and ventilate the ocean interior include Subantarctic Mode Water (SAMW) and Antarctic Intermediate Water (AAIW) (Fig. 4.4). SAMW forms within the subantarctic zone between the SAF and STF. There, waters are subject to intense convection during winter to form a layer of uniform density or *pycnostad* occupying water depths of ~450–700 m (Aoki et al., 2007; McCartney, 1977; Rintoul and England,

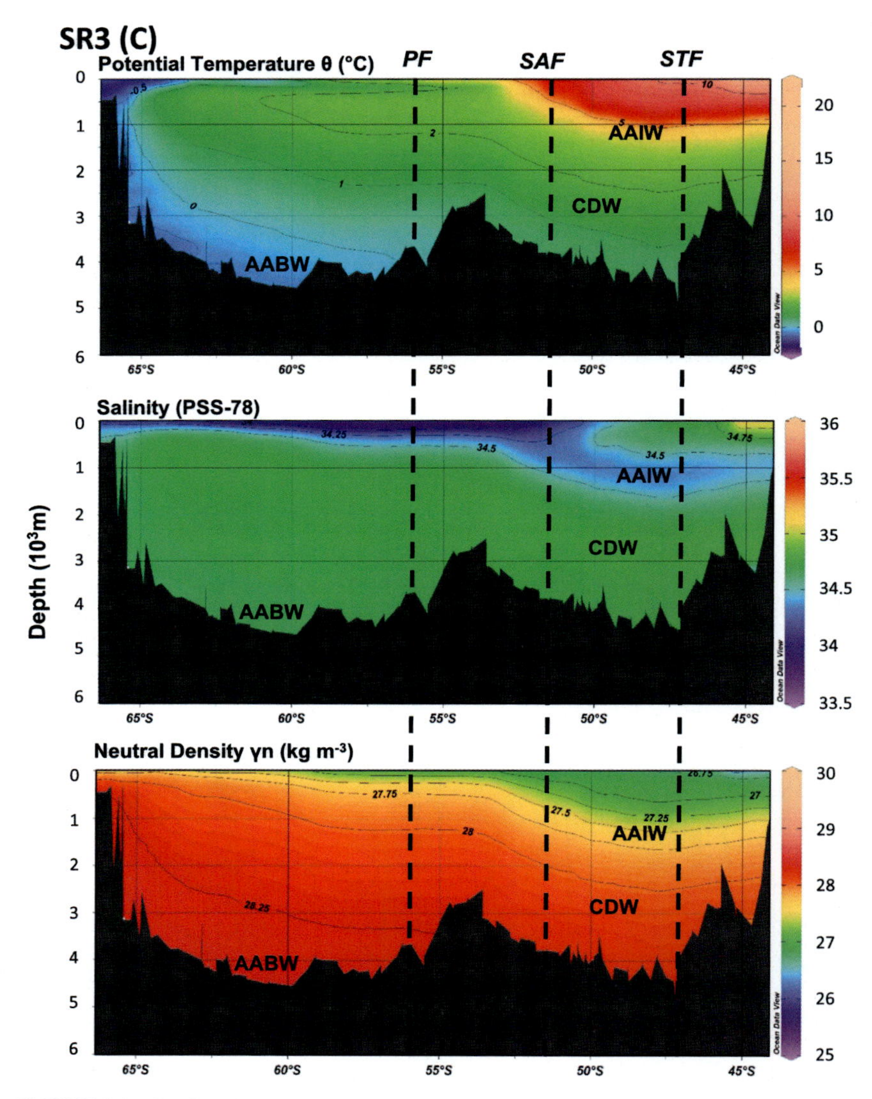

FIGURE 4.4 Continued

2002). SAMW is also distinguished by its high oxygen content and minimum in potential vorticity (Herraiz-Borreguero and Rintoul, 2011). SAMW forms in several 'hotspots' with variations in properties across the major oceans (Sallée et al., 2006). Temperature ($\theta = 7°C-9°C$) and salinity ($S = 34.35-34.6$) vary regionally in part reflecting differences in AASW transported by the northward Ekman transport associated with westerly wind-driven ACC. Subduction and export of lightest SAMW ($2.68\sigma_\theta$) is focused in the eastern Indian and westernmost Pacific Oceans, a denser mode (2.69 to 2.70σ_θ) in the central Pacific Ocean and the densest mode (2.71 to 2.72σ_θ) in the Southeast Pacific and Drake Passage (Herraiz-Borreguero and Rintoul, 2011; Sallée et al., 2010b).

SAMW is underlain by AAIW that subducts at the SAF (Fig. 4.4) ~500–1500 m water depths (Fig. 4.3; Sloyan et al., 2010). Production of AAIW is focused in the Southeast Pacific and Southwest Atlantic sectors of the Southern Ocean plus smaller areas such as the Kerguelen region of the Indian sector (Fig. 4.1; Bostock et al., 2013; Sallée et al., 2010b; Talley, 1996). AAIW is fresh ($S = 34.3–34.6$, typically including the salinity minimum), oxygen-rich water and has temperatures of $\theta = 3.5°C–10°C$. These and other tracers (nutrients, radiocarbon, carbon isotopes) enabled Bostock et al. (2013) to identify at least four different types of AAIW in the South Pacific that reflect their varied formation and subsequent mixing with other water masses.

4.2.3 Deep water

The most prominent water mass in the Southern Ocean is CDW that extends from ~1500 to >3500 m water depth (Fig. 4.4). It is basically a hybrid formed from deep waters present in all the major oceans. Nevertheless, two basic types are evident: (1) upper Circumpolar Deep Water (UCDW) is identified by its oxygen minimum and high nutrient content, and commonly occupies around 1500–2500 m water depth; and (2) lower Circumpolar Deep Water (LCDW), which occurs down to ~3500 m and is characterised by the salinity maximum ($S = 34.70–34.75$) and elevated nutrients (Gordon, 1975; Orsi et al., 1995; Schmidtko et al., 2014). The high salinity signature of LCDW reflects the input of North Atlantic Deep Water via the Atlantic Ocean. That signal is subsequently redistributed northward to the Pacific and Indian Oceans via deep western boundary currents (Rintoul et al., 2001; Sloyan, 2006; Warren, 1981).

South of the PF, CDW upwells to the ocean surface. This is the result of wind-induced upwelling at the Antarctic Divergence where westerly winds transition to polar easterly winds (Deacon 1982; Rintoul et al., 2001). This is one of the few regions of the world where deep waters can exchange directly with the atmosphere. As a result, CO_2 stored in the deep waters can be exchanged with the atmosphere. Such upwelling also bring nutrients to the ocean surface.

CDW is present around much of the Antarctic continental shelf edge except in the Weddell Sea (Dinniman et al., 2010). Mixing of CDW with locally generated waters such as AASW and high salinity shelf water creates modified CDW (MCDW) whose properties vary depending upon the mixing regime (Whitworth et al., 1998).

Model simulations suggest that half of the total CDW crosses the 1000 m isobath on to the Antarctic shelf in the Ross Sea where the CDW is cooled through vigorous wind-induced mixing with cold shelf waters (Dinniman et al., 2010; Morrison et al., 2020). This contrasts with the West Antarctic Peninsula where CDW intrusions are warmer, subject to less vertical mixing and have a shorter passage across the shelf. Recent observations in East Antarctica also show relatively warm MCDW can traverse the narrow continental shelf via a deep channel to affect basal melting of the Totten Ice Shelf (Greenbaum et al., 2015; Rintoul et al., 2016).

4.2.4 Bottom water

Warren (1971) encapsulated the significance of abyssal flows emanating from Antarctica, namely, '*water from the Antarctic is largely responsible for keeping the rest of the deep sea cold*'. A key component of that influence is the production and circulation of Antarctic Bottom Water (AABW), e.g., Johnson (2008). AABW is a generic term that includes all circumpolar waters with the neutral density of $\gamma^n \geq 28.27\ \mathrm{kg/m^3}$ generated over Antarctica's continental margin (Orsi et al., 1999). However, this density should not be regarded as a minimum. Williams et al. (2010) show that the threshold neutral density of shelf water to form AABW in the Australian-Antarctic Basin can be as low as $\gamma^n = 27.80\ \mathrm{kg/m^3}$. Thus, there are site-specific bottom waters that reflect local oceanographic and bathymetric conditions (Bindoff et al., 1999; Johnson, 2008; Orsi et al., 1999; Williams et al., 2010). Some examples from Ohshima et al. (2013) and Orsi et al.

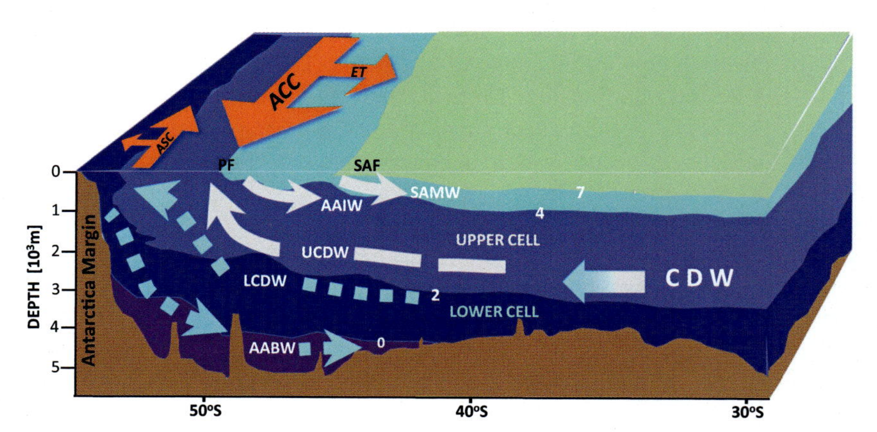

FIGURE 4.5 Outline of the Southern Ocean meridional overturning circulation (SOMOC). The lower cell has LCDW ascending along southward rising isopycnals to the Antarctic margin where cooling and addition of salt lower its buoyancy to form AABW that descends and spreads northwards. In the upper cell, UCDW upwells north of Antarctica and becomes more buoyant through freshening and warming. These modified waters move north in the Ekman transport (ET) layer to eventually subduct as SAMW and AAIW. White numerals are indicative potential temperatures.

(1999) are as follows: the warmest, most saline bottom water is from the NW Ross Sea (RSBW: $-0.6°C < \theta < -0.3°C$; $S \geq 34.72$), the coldest, freshest water is in the SW Weddell Sea (WSBW: $\theta \leq -1°C$; $S \leq 34.64$), and intermediate varieties such as Adélie Land Bottom Water (ALBW: $0.8°C < \theta < -0.4°C$; $<34.62 < S < 34.68$) and Cape Darnley Bottom Water (CDBW: $\theta < -0.4°C$ and $S < 34.64$).

Several processes affect the density of continental shelf and deeper waters to form AABW. These include: (1) the supply of CDW, which increases where the ACC nears the continental margin (Fig. 4.4; Orsi et al., 1995); (2) brine rejection during the formation of seasonal sea ice (Foster and Carmack, 1976); and (3) the addition of ultra-cold Ice Shelf Water (ISW: $\theta \geq -1.9°C$) where the circulation extends beneath ice shelves (Foldvik et al., 2004; Jacobs, 2004). AABW is often associated with large active coastal polynyas, which are regions of high sea-ice production where seawater is cooled and salt expelled (e.g., Fusco et al., 2009; Ohshima et al., 2013; Williams et al., 2010). Estimates of total AABW transport rates are highly variable but there appears to be some consensus on an Antarctic-wide rate of $\sim 16-20$ Sverdrups (Sv; 1Sv = 10^6 m^3/s) (Jacobs, 2004).

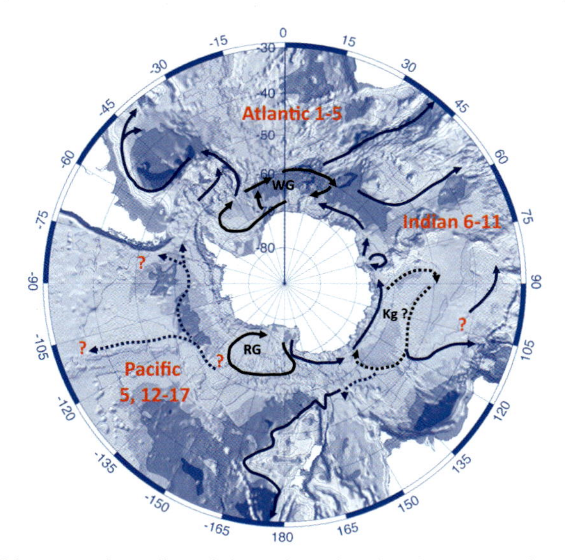

FIGURE 4.6 Diagrammatic outline of the main paths of various types of AABW and dense LCDW in the Southern Ocean as depicted by solid black lines and by dashed lines plus '*?*' where paths are suggested or poorly defined. Also included are the Weddell Gyre (WG) and Ross Gyre (RG) plus the informally termed Kerguelen gyre (Kg). For the ***Atlantic sector*:** information comes from: (1) Garabato et al. (2002), (2) Hogg (2001), (3) Locarnini et al. (1993), (4) Klatt et al. (2005) and (5) Orsi et al. (1999). For the ***Indian sector***: (6) Fukamachi et al. (2010), (7) Heywood et al. (2004), (8) McCartney and Donohue (2007), (9) Ohshima et al. (2013), (10) Sloyan (2006) and (11) Warren and Johnson (2002). For the ***Pacific sector***: (12) Budillon et al. (2002), (13) Carter and McCave (1997), (14) Jacobs et al. (2002), (15) McCave et al. (2008), (16) Orsi et al. (1999) and (17) Smith et al. (2012).

4.3 Southern Ocean circulation

The ocean circulation is dominated by the westerly wind-driven ACC, which has an eastward, circumpolar path covering most of the Southern Ocean (Figs 4.1 and 4.2). Perpendicular to this flow is the north-south Southern Ocean meridional overturning circulation (SOMOC) that links all the oceans to distribute heat, salt, carbon, dissolved gases and nutrients globally (Fig. 4.5). Other circulatory features include the Weddell and Ross Sea Gyres and potentially another large scale gyre, informally named here as the Kerguelen Gyre (Fig. 4.6). These deep-reaching cyclonic flows are located between the ACC and Antarctica's continental margin where they interact with westward Antarctic slope and shelf currents formed under the polar easterly winds (Deacon, 1982; Dotto et al., 2018; Vernet et al., 2019).

4.3.1 Antarctic Circumpolar Current (ACC)

The ACC is a current of superlatives. Its course at latitude 55°S exceeds 23,000 km to make it the world's longest current and the only one to link all the oceans (Klinck and Nowlin, 2001; Rintoul, 2018). The eastward path of the ACC is strongly influenced by bathymetry. The current is narrowest in the Drake Passage (~660 km) with other prominent constrictions caused by Kerguelen Plateau (Fukamachi et al., 2010), Macquarie Ridge (Gordon, 1972) and Campbell Plateau (Carter and Wilkin, 1999). Where unconstrained, the ACC path widens to ~2000 km (Klinck and Nowlin, 2001). It is also a full depth flow, affecting surface to bottom waters and the seabed. Such interaction with the seabed has a critical role: wind stress is partly balanced by bottom stress and linked to eddy activity (e.g., Marshall et al., 2017; Rintoul et al., 2001; Whitworth, 1988). Measurements of >0.2 m/s have been made at depths >3000 m by moorings positioned in gaps on Macquarie Ridge (Rintoul et al., 2014). Indeed, observations of the seabed beneath the ACC along the Campbell Plateau margin in water depths >3500 m reveal active benthic sediment transport in a zone subject to ~9 eddies annually (Carter and McCave, 1997; Stanton and Morris, 2004). While direct observation of eddy-driven transport along Campbell Plateau has yet to be monitored, Hollister and McCave (1984) show it is a viable process as shown by active benthic sediment plumes at 4800–4900 m water depth beneath the eddy-rich Gulf Stream.

As noted in Section 2.1, the ACC is a complex system of fronts (Fig. 4.2; Moore et al., 1999; Rintoul et al., 2001; Sokolov and Rintoul, 2007, 2009a, b). These are eddy-rich, meandering features with multiple jets (Fig. 4.2). Although jets may vary in location and intensity due to seasonality, interactions with bathymetry and baroclinic instabilities – Sokolov and Rintoul (2009a,b) show from 15 years of SSH observations that the fronts align with specific regional streamlines. With the exception of the northern jet of the

SAF, which intercepts South America, the remaining jets are circumpolar although any continuity of the SBF is partially masked by sea ice (Fig. 4.2).

Given the ACC's exposure to strong westerly winds over a fetch that is largely unimpeded by land, it is not surprising that it has the largest volume transport. Off southern Australia, the mean baroclinic transport is 147 ± 10 Sv (Rintoul and Sokolov, 2001), whereas Drake Passage registers a mean of 136.7 ± 7.8 Sv (Cunningham et al., 2003). Those transports are two to four times larger than any other major current, e.g., the Agulhas Current (~73 Sv; Beal and Bryden, 1999), East Australian Current (25−37 Sv; Ridgway and Hill, 2009) and Gulf Stream (~31 Sv; Lund et al., 2006). Most of the ACC transport is concentrated in its frontal jets especially the SAF, which can accommodate over two-thirds of the total ACC transport, e.g., 105.7 Sv off Australia (Rintoul and Sokolov, 2001) and 90 Sv off southern New Zealand (Morris et al., 2001).

4.3.2 Southern Ocean meridional overturning circulation (SOMOC)

Despite its dominant eastward transport, the ACC is intimately associated with a north-south global overturning system that extends into each of the major ocean basins (Marshall and Speer, 2012; Rintoul, 2018). The SOMOC is comprised of two cells that centre on the southward migration of CDW following upward tilting isopycnals and the CDW's subsequent transformation into northward moving (1) intermediate depth AAIW and SAMW and (2) deep AABW. Those transformations involve changes in CDW buoyancy, wind forcing and the locations where CDW comes close to the surface (Fig. 4.5).

In the anticlockwise lower cell, LCDW typically upwells close to the Antarctic continental margin under eddy transport and wind-induced upwelling associated with the Antarctic Divergence. Depending upon the location, upwelled water may interact with eddies, and currents occupying the slope and shelf. En route, LCDW mixes with various dense waters generated in polynyas, under sea ice and/or ice shelves. The resultant AABW descends northwards to circulate into the deep ocean basins mainly via deep western boundary currents (Fig. 4.6).

In the upper cell, UCDW upwells further offshore than the lower cell. Upwelling occurs under the influence of winds and increased buoyancy caused by mixing with fresh AASW. This modified UCDW is exported northwards in the Ekman to eventually subduct to intermediate depths as SAMW and AAIW (Hogg, 2001) (Figs 4.4 and 4.5). This equatorward transport of surface-intermediate waters and dense LCDW/AABW balances the poleward migration of CDW (Rintoul, 2018).

4.3.3 Deep western boundary currents

Export of dense shelf water to form AABW in the lower SOMOC cell commonly occurs along submarine canyons, channels and troughs incised into the continental margin (e.g., Ohshima et al., 2013; Williams et al., 2010), although other delivery mechanisms such as unchanneled dense plumes may also play a role (Baines and Condie, 1998).

Upon leaving the continental margin, AABW is initially steered and contained by the marked ridge and basin topography surrounding Antarctica (Figs 4.1 and 4.6) before entering the global ocean primarily via equatorward deep western boundary currents (DWBCs) located along the eastern and western Indian, western Atlantic and western Pacific Oceans (Schmitz, 1995; Stommel, 1958; Warren, 1981). Since those studies, considerable advances have been made in charting and quantifying the volume transports of those flows (Aoki et al., 2008; Fukamachi et al., 2010; Garabato et al., 2002; Johnson, 2008; Rintoul, 1998; Whitworth et al., 1999). However, as Fukamachi et al. (2010) point out difficulties remain in quantifying abyssal transport associated with the SOMOC. These include: (1) a paucity of long-term data; (2) change in AABW properties due to mixing or entrainment along its flow path; (3) interaction with topography and other current systems; and (4) recirculation of flows as occurs in gyres (McCartney and Donohue, 2007). With those caveats in mind, the following section briefly summarises the main DWBC pathways.

4.3.3.1 Pacific deep western boundary current

The formation and export of dense water into the Pacific sector of the Southern Ocean is focused on the western to central Ross Sea (Smith et al., 2012). There, AABW passes westward as well as recirculating within the Ross Gyre (Fig. 4.6). Orsi et al. (1999) hint at an eastward transport of AABW into the Southeast Pacific Basin where McCave et al. (2008) suggest a possible DWBC along the Pacific-Antarctic Ridge. In contrast, westward plumes of dense water are observed moving with the Antarctic Slope Current into the Kerguelen Gyre and associated boundary currents (Fukamachi et al., 2010; Gordon et al., 2009, 2015). Part of that circulation extends north into the Indian Ocean via a powerful DWBC, whereas the remainder turns eastward into the western Pacific Ocean. Because the Tasman Sea is a *cul de sac* for abyssal currents, the main DWBC inflow is steered through and around Macquarie Ridge and then northeast along the 3000–3500 m high flanks of Campbell Plateau (Boyer and Guala, 1972; Carter and Wilkin, 1999; Gordon, 1972; Warren, 1973). The Plateau also steers the Subantarctic Front of the ACC, which overrides and dominates the DWBC (Carter and Wilkin, 1999) especially during the passage of deep-reaching eddies (Carter and McCave, 1997; Stanton and Morris 2004). The combined ACC/DWBC flow south of New Zealand extends to around latitude 49°S where the ACC turns east leaving the underlying DWBC to continue north and enter the subtropical Pacific via the 2600-km-long Tonga-Kermadec Ridge. A 22-month time-series of the DWBC at latitude 32°30′S reveals a highly variable northward transport of 16 ± 11.9 Sv (Whitworth et al., 1999). Overall, the volume transport of these DWBCs appears to reduce northwards (Hogg, 2001), a trend that is consistent with the capture and recirculation of bottom waters within the deep ocean basins (Stommel, 1958).

4.3.3.2 Indian deep western boundary currents

The Indian Ocean accommodates three DWBCs that primarily transport waters from the Atlantic and Pacific sectors of the Southern Ocean. In the eastern Indian Ocean, the DWBC follows a circuitous route steered by Kerguelen Plateau, Southeast Indian Ridge, Ninety-east Ridge and intervening basins (Fukamachi et al., 2010; Toole and Warren, 1993). AABW is retained in the Australian-Antarctic Basin, whereas overlying LCDW moves equatorward (Orsi et al., 1999). Studies across the eastern margin of Kerguelen Plateau provide new insights into DWBCs (Aoki et al., 2008; Fukamachi et al., 2010). Due to its proximity to the ACC and associated gyres, the local DWBC is exceptionally powerful (Fig. 4.6). Its speed at 3500 m water depth has a maximum 2-yr mean of >20 cm/s, which is the fastest mean for a DWBC at that water depth. At the base of the continental slope, AABW ($\theta \leq 0$°C) forms a ~1500 m-thick layer with an average equatorward transport of 12.3 ± 1.2 Sv thus identifying the Kerguelen DWBC as a major exporter of Antarctic water.

In the west, dense LCDW and AABW from the Weddell Sea also encounter complex basin-and-ridge topography (Warren, 1981). Initially, waters enter Crozet and Mozambique Basins, which trap AABW while the overlying LCDW passes northwards via fractures in the Southwest Indian Ridge (Johnson et al., 1991; Mantyla and Reid, 1995; Warren, 1974). There the DWBC intensifies against the western sides of Madagascar and Mascarene Basins. Like the Pacific Basin, there is a northward reduction in volume transport ranging from 7 to 11 Sv at latitude 32°S to 1–5 Sv at latitude 10°S (Johnson et al., 1998). This general decrease likely reflects recirculation and mixing influenced by the marked basin and ridge topography (Fig. 4.1).

4.3.3.3 Atlantic deep western boundary current

The prime source for the Atlantic DWBC is the Weddell Sea. Because of its high density, most Weddell Sea Bottom Water (WSBW) is retained in the Weddell Basin. The lighter WSDW fraction is exported north through gaps in the South Scotia Ridge into the Georgia Basin via the South Sandwich Trench (Fig. 4.6; Fahrbach et al., 2001; Garabato et al., 2002; Whitworth et al., 1991). Estimated transport of the WSDW-bearing AABW from the Weddell Sea is ~9.7 Sv.

From a broad perspective, the resultant Atlantic DWBC flows along the continental margin off South America commonly within a water depth range of 3500–4000 m. Again, the current pathway is complex due to the presence of the deep Georgia, Argentine and Brazil Basins and the inter-basin ridges (Fig. 4.1; Schmitz, 1995; Warren, 1981). These basins trap and recirculate AABW while ridge gaps allow LCDW to continue north (Whitworth et al., 1991). For example, ~6.9 Sv flow from Argentine Basin into Brazil Basin through Vema and Hunter Channels. However, only ~3.2 Sv exits Brazil Basin, suggesting a recirculation of ~3.7 Sv (Hogg, 2001). These observations were made against a backdrop of strong variability in abyssal transport, recently recorded in a 3.8-year-long time-series from the South Atlantic (Kersalé et al., 2020). Temporal variations ranged from days to weeks and involved both strength and direction of transport. Nevertheless, estimates based on this recent data are consistent with the historic measurements from Vema and Hunter Channels (Hogg, 2001).

4.3.4 Subpolar circulation – gyres, slope and coastal currents

The circulation over Antarctica's continental margin consists of a westward current associated with the Antarctic Slope Front (Heywood et al., 2004). This flow locally interacts with several cyclonic gyres positioned in the abyssal Southwest Pacific, Australian-Antarctic and Weddell–Enderby Basins (Fig. 4.2; Garabato et al., 2002; McCartney and Donohue, 2007; Orsi et al., 1999). South of the continental slope, a system of coastal currents occupies the shelf to connect CDW with land-based ice (Heywood et al., 2004; Holland et al., 2020).

4.3.4.1 Gyres

The Antarctic gyres are major features dominated by the Weddell and Ross Gyres (Fig. 4.2) (Jacobs et al., 2002; Orsi et al., 1993). There is some evidence suggesting the presence of a third, basin-wide recirculation, informally named here as the Kerguelen Gyre (e.g., McCartney and Donohue, 2007). However, its structure is not as well established as the Weddell and Ross Gyres. Rather than single feature, the Kerguelen circulation appears to be several smaller sub-gyres (Yamazaki et al., 2020).

In general, the main gyres are deep-reaching features whose location and oceanographic properties are influenced by the regional bathymetry, namely, the Antarctic continental margin and adjacent >4000-m-deep circumpolar basins (Fig. 4.1). Cyclonic circulations prevail with the northern gyre limbs flowing east in concert with the ACC. The southern limbs flow west and connect with the Antarctic Slope Current (ASC). Thus, gyres transfer heat and salt from the ACC to the Antarctic margin, which usually involves the export of relatively warm CDW ($\theta \geq 1°C-2°C$) on to the shelf to be transformed into dense deep and bottom waters (Jullion et al., 2014; Smith et al., 2012). These can be gyre-specific with freshest, coldest and densest AABW associated with the Weddell Gyre, warmest and most saline with the Ross Gyre and potentially intermediate varieties with the Kerguelen Gyre (Orsi et al., 1999).

4.3.4.2 Antarctic slope and coastal currents

The Antarctic Slope Front extends along most of the continental slope immediately seaward of the shelf edge in water depths down to ~1500 m (Heywood et al., 1998, 2004; Holland et al., 2020; Whitworth et al., 1998). As noted previously, the front separates the relatively warm, saline CDW that has upwelled at the slope from cold, fresh shelf waters. As such, the front has a strong but variable density field as evinced by local temperature gradients as steep as 3°C over 25 km (Jacobs, 1991). Thus, it supports the fast-flowing ASC, which passes westward at average speeds ranging over 10–50 cm/s depending upon the location and water depth (Jacobs, 1991). Observations from the eastern Weddell Sea found the ASC flowed southwest with a surface speed of ~30 cm/s but was accompanied by an undercurrent moving northeast at 6–9 cm/s (Chavanne et al., 2010).

As a dynamic barrier, the Antarctic Slope Front affects the transfer of CDW-borne heat, salt and nutrients to the Antarctic shelf (Chavanne et al., 2010; Holland et al., 2020) and hence influences ice shelf stability, the buoyancy of SOMOC water masses, and primary productivity that underpins the polar food chain (Heywood et al., 2004). Such transfer to the shelf varies according to ASC dynamics (Thompson et al., 2018), water mass properties and continental margin topography. In general terms, where the ASF is weak – perhaps reflecting a modest temperature gradient or effects of eddy overturning (Nøst et al., 2011) – relatively warm CDW may be readily transferred to the continental shelf (Holland et al., 2020).

Once on the shelf, CDW and MCDW are distributed by coastal currents, for example the narrow, rapid flows observed on the Southwest Weddell Sea shelf (Heywood et al., 1998, 2004). This current has a well-defined path along the 400–500 m isobaths and has a modest transport of <1 Sv. Another example occurs on the ~130-km-wide, western shelf off the Antarctic Peninsula. Locally known as the Antarctic Peninsular coastal current (Moffat et al., 2008), this buoyant flow travels counter to the local ACC, along the coastline at 30–40 cm/s and with a transport of 0.32 Sv. It may be a seasonal feature being absent during winter-early spring but reappearing in ice-free months when precipitation and freshwater run-off increase to drive the current, probably assisted by local winds and eddies (Moffat et al., 2008).

In contrast to wide continental shelves, prominent coastal currents on narrow shelves may either be absent because of the strongly embayed bathymetry or are difficult to distinguish from the ASC. This may be the case off Wilkes Land – Adélie Land where the ASC, locally augmented by gyres, appears to dominate the upper reaches of the continental slope/shelf (McCartney and Donohue, 2007).

4.4 Modern Southern Ocean change

Given the Southern Ocean's roles in affecting the worldwide distribution of heat, salt, gases and nutrients as well as influencing global sea level rise and the uptake of anthropogenic heat and CO_2, its response to modern climate change is highly relevant (e.g., Bindoff et al., 2011). This is the clear, science-based message from Assessment and Special Reports of the Intergovernmental Panel on Climate Change (IPCC, 2013, 2019, 2021), as well as the science literature, especially a recent review of Southern Ocean dynamics by Rintoul (2018).

4.4.1 Climate change

With regard to changes in near-surface air temperatures since the 1950s, Turner et al. (2005) reveal strong regional and temporal variability on the basis of meteorological records from 19 terrestrial Antarctic sites. The Antarctic Peninsula warmed at 0.56°C/decade up to the late 1990s followed by the western Ross Sea at 0.29°C/decade, although this latter rate is not statistically significant. The remaining sites highlight Antarctica's variability with eastern coasts and the continental interior subject to varying degrees of cooling per annum whereas western coasts warmed. With exception of the South Pole, all sites had warmer winters.

More recent data from the northern Antarctic Peninsula highlight a reversal in the warming trend (Turner et al., 2016). For 1979–1999, a statistically significant warming trend of 0.32°C/decade was recorded. In contrast, 1999–2014 witnessed a cooling of −0.47°C/decade. That change is attributed to more frequent cold, east to southeast winds and the accumulation of sea ice off the eastern Peninsula. This cooling coincides with a slowdown of global warming in the early 2000s (Fyfe et al., 2016). However, the size and frequency of the Peninsula events are within the bounds of a marked decadal natural variability extending back to at least year 1702 as recorded in an ice core from coastal West Antarctica (Thomas et al., 2013). Ice cores from other parts of Antarctica confirm the strong variability of air temperatures that reflect infuences of climate modes such as ENSO and SAM as well as anthropogenic forcing (e.g., Bertler et al., 2011; Clem et al., 2016; Maasch et al., 2005; Wintrup et al., 2019).

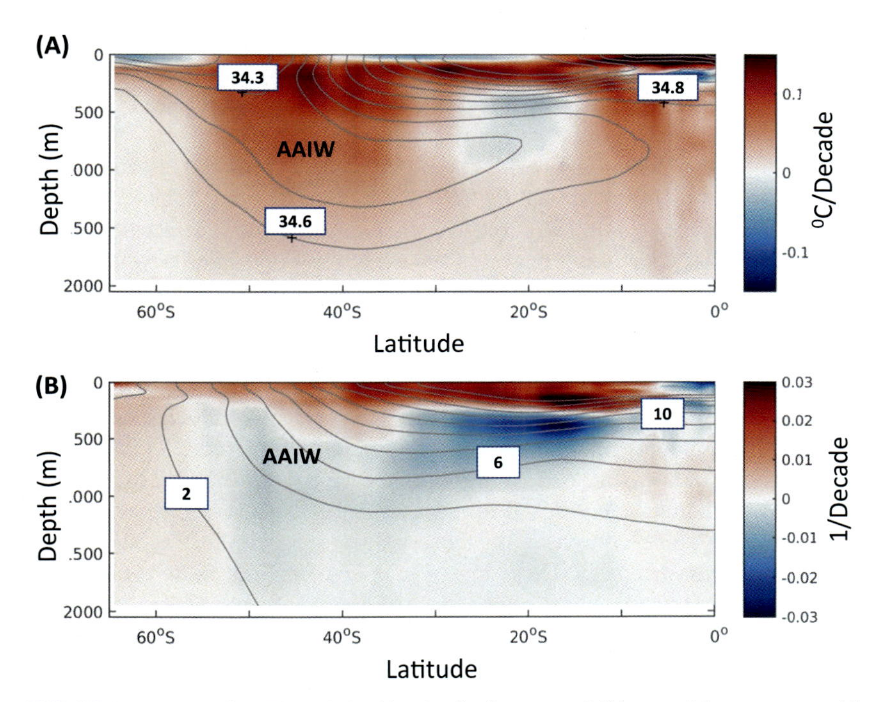

FIGURE 4.7 Trends for 2004–2019 of longitudinally-averaged (A) potential temperature with mean salinities depicted by grey contours spaced 0.15 apart and (B) salinity with mean temperatures in grey contours spaced 1°C apart highlighting freshening and warming of subducted AAIW. Derived for this paper from gridded Argo observations (Roemmich and Gilson, 2009).

4.4.2 Ocean change

Historical oceanographic data and measurements from Argo and Alace floats show the Southern Ocean down to ~2000 m water depth has become warmer and fresher since the 1950s (IPCC, 2019; Jacobs, 2006; Jacobs and Giulivi, 2010). For example, Böning et al. (2008) show the ocean interior at 300 to 500 m water depth warmed at $\theta \geq 0.1$°C/decade and freshened at $S \geq -0.01$/decade. Likewise, Gille (2002) recorded an overall warming of 0.17°C at 300–500 m water depth between the 1950s and 1980s. The upper ocean has also warmed with SST rates of 0.1°C–0.4°C/decade for 1982–2012 (Fig. 4.7; Armour et al., 2016). These changes centred on the northern ACC, which has the largest increase in global ocean heat content (Meijers et al., 2011). In contrast, SSTs to the south have mainly cooled at $\theta = -0.05$°C to -0.3°C/decade apart from a modest warming of $\theta = \sim 0.1$°C/decade off West Antarctica including the Antarctic Peninsula (Fig. 4.7). This apparent delay in Southern Ocean warming is attributed to the upper cell of the SOMOC as follows: (1) upwelling of relatively cold water to ameliorate any warming; (2) northward transport of warmed surface water; and (3) its subduction to intermediate depths and transport into adjacent subtropical gyres. Eventually, warmer CDW from the North Atlantic may enter the upwelling system to accelerate ocean warming, but this change is likely to be several centuries away (Armour et al., 2016).

Warming continues below 2000 m water depth especially in the abyssal basins surrounding Antarctica (IPCC, 2019; Purkey and Johnson, 2010; Rintoul et al., 2014). For 1981 to 2019, the warming trend at 2000 m water depth was 0.18°C/century, but at 4400 m water depth the trend rose to 0.3°C/century, considerably more than the global mean of 0.05°C/century for that depth. Enhanced warming of the circumpolar abyss coincides with the reduced production of AABW, which may reflect freshening of source waters (van Wijk and Rintoul, 2014).

With respect to dissolved gases, the Southern Ocean continues to take up CO_2. CMIP5 model simulations for 1861–2005, for example, show the ocean accommodated up to 43% ± 3% of the CO_2 and 75% ± 22% of heat generated by human activities (Frölicher et al., 2015; Gruber et al., 2019). While there is some consensus on those values, e.g., Bindoff et al. (2007), the mechanisms that affect heat and gas uptake are not well established. Rintoul (2018) notes that just over a decade ago some studies suggested the Southern Ocean was saturated with respect to CO_2 (Le Quéré et al., 2007), and it was reasoned that it may not cope with projected increases in anthropogenic and natural CO_2 – the latter released by an anticipated strengthened upwelling of CDW. However, as the regional coverage of marine carbon cycle research expands, it is clear that the Southern Ocean's role varies in time and space in response to the seasons, wind-driven vertical movement of the SOMOC and warmer ocean temperatures (Landschützer et al., 2015; Lovenduski et al., 2007; McNeil and Matear, 2008).

Concomitant with the uptake of CO_2 is a lowering of pH, i.e. ocean acidification (e.g., Fabry et al., 2009). Evidence of this change in the Southern Ocean comes from the *Munida* CO_2 time-series collected from dominantly SASW near the STF off the South Island, New Zealand (Bates et al., 2014). Still active, the 22-yr *Munida* record shows seawater pCO_2 anomalies are increasing at 1.28 μatm/ yr, whereas pH is reducing at −0.0013/yr.

In contrast to dissolved CO_2, concentrations of dissolved oxygen have generally declined over the past four to five decades with some of the largest changes occurring in the Southern Ocean (Andrews et al., 2017; IPCC, 2019; Jacobs, 2006; Keeling et al., 2010). While the accuracy of some old oceanographic data may be insufficient to identify trends, observations and model simulations cited by IPCC (2019) suggest dissolved O_2 losses of −0.5 to −3.0 μmol/kg/decade in upper 1200 m of the ocean and −0.5 to −2.0 μmol / kg/decade from 1200 m to the seabed (Frölicher et al., 2020; Talley et al., 2016). Attribution of these trends in O_2 to anthropogenic climate change is equivocal, but the O_2 decline is consistent with warmer temperatures and enhanced stratification of the upper ocean (see Section 4.3).

4.4.3 Change in dynamics and circulation

Since the 1950s, the Southern Ocean and atmosphere have been subject to a strong interannual to decadal variability that reflects the inter-relationships between westerly winds, sea ice distribution, ozone depletion and interacting climate modes, in particular, the Southern Annular Mode (SAM) and El Niño Southern Oscillation (ENSO) (Gille, 2008; Jones et al., 2016; Lovenduski and Gruber, 2005; Sallée et al., 2008, 2010; Turner et al., 2016). A high-resolution subantarctic climate/ocean record, extending back to the 1870s, suggests the present level of strong variability in terrestrial temperatures and SSTs was reached in the 1940s. That timing coincided with warming SSTs in the equatorial Pacific and subsequent generation of a Rossby wave train across much of the Southern Hemisphere (Turney et al., 2017). Mindful of that variability, several trends are still evident for the post-1950s ocean. From the 1970s to early 2000s, the westerly wind storm belt shifted south, forced by stratospheric depletion of ozone concomitant with a prolonged positive phase of the SAM (Thompson et al., 2011; Toggweiler et al., 2006). The effect on the ACC flow is not clear cut. Using SSH data for 1992–2007, Sokolov and Rintoul (2009b) show ACC fronts shifted south over a mean distance of 60 km. That shift, however, is locally constrained by bathymetry as proposed by Sallée et al. (2008). Gille (2014) suggests that changes in SSH may be associated with large-scale changes in ocean height rather than local shifts in the position of fronts. An increase in eddy kinetic energy has been observed over the last few decades in the ACC from satellite altimeters consistent with the increasing winds over the region (Martínez-Moreno et al., 2021).

Winds associated with an increasingly more positive SAM can affect the mixed layer depth (Sallée et al., 2010; Sallée et al., 2021). This layer connects the air/ocean interface with the deep ocean to provide a pathway for the uptake of heat and CO_2. Drawing on direct measurements of the mixed layer depth by Argo floats, Sallée et al. (2010) show that, during positive SAM, the mixed layer deepens by ~100 m in the eastern Indian and central Pacific Oceans but shallows in the western Pacific and central Indian Oceans. This pattern reflects anomalous meridional winds that introduce cold air to the eastern Indian and central Pacific. Surface waters are cooled and descend to increase the depth of the mixed layer (Gille, 2010).

Strengthening westerly winds also enhance components of the SOMOC including: (1) circumpolar upwelling south of the westerly wind maximum presently positioned about an average latitude of 52°S; (2) northward migration of surface waters by Ekman transport; and (3) subduction of SAMW/AAIW to the north. However, as noted previously, winds are only one forcing mechanism. Buoyancy forcing is a key if not the main driver of the SOMOC (Marshall and Speer, 2012; Rintoul, 2018). In that context, upwelling in the upper cell can also be attributed to increased buoyancy caused by the addition of freshwater and/or heat. Indeed, Haumann et al. (2016) suggest that freshwater from Antarctic sea ice contributes to the transformation of deep water to more buoyant SAMW and AAIW. Northward transport of sea ice increased 20% ± 10% between 1982 and 2008 and was accompanied by reduced salinities (Jacobs, 2006). Sea ice may also contribute to the lower SOMOC cell through brine rejection and formation of dense shelf water in continental margin settings. However, this reduced buoyancy may be offset by an overall freshening of shelf and deep waters due to ice sheet melting (Aoki et al., 2005; Jacobs et al., 2002; Rintoul, 2007). In the Australian-Antarctic Basin, the volume of AABW with neutral density $\gamma^n \geq 28.30$ kg/m^3, decreased by ≥ 50% between 1969–1971 and 2008–2012 and was replaced by less dense water (Sallée et al., 2021; Silvano et al., 2020; van Wijk and Rintoul, 2014).

Freshening and warming of the Southern Ocean have increased stratification of the uppermost 200 m since 1970 (Meredith and King, 2005). This has the potential to restrict the vertical exchange of heat, salt, nutrients and gases between surface, intermediate and deep waters (Talley et al., 2016) as well as favouring dynamical processes such as propagation of internal waves (IPCC, 2019).

The build-up of heat in the northern ACC is reinforced by heat moving poleward from the subtropics. Over the past seven decades, the subtropical gyres have intensified in response to an increased curl of the wind stress associated with a more positive SAM (e.g., Cai et al., 2010; Hill et al., 2008; Wu et al., 2012). Stronger winds have increased heat content in the upper 2000 m in the South Pacific subtropical gyre (Roemmich et al., 2016) and in the thickness and heat content of the SAMW (Gao et al., 2018). The strengthened winds have also intensified the western boundary currents of the subtropical gyres, such as the East Australian, Agulhas and Brazil Currents causing then to extend poleward (Wu et al., 2012). The East Australian Current, for example, has moved 350 km south since the 1940s (Hill et al., 2008). As a result, the marine environment off Tasmania, where the time-series were recorded, changed from a subantarctic to subtropical marine ecosystem (Johnson et al., 2011). Extension of the subtropical western boundary currents may contribute warm water directly into the ACC especially where subtropical and ACC frontal systems converge, as occurs off eastern New Zealand (Fernandez et al., 2014).

4.5 Concluding remarks

In the decade following publication of the first ACE volume (Florindo and Siegert, 2009), our knowledge of the Southern Ocean has advanced considerably. Now, there is an improved appreciation of the intimate links between the ACC and overturning circulations; the importance of eddies in affecting the mixing and transport of water masses; the importance of buoyancy in driving the SOMOC and new insights into the formation and distribution of

AABW especially in the Indian Ocean sector of the Southern Ocean. These advances are made possible by the continuation and expansion of high-quality global databases relating to remote sensing of the upper ocean and monitoring of the ocean interior via Argo floats and ocean observatories, e.g., SOOS (2020).

So how can these advances improve reconstructions of past ocean change - a major theme of this book? Essentially, physical oceanography identifies the dynamic processes that affect change over time scales of hours to decades. As such they present a mechanistic understanding of modern change, which provides insights into the processes affecting longer-term changes such as those captured by palaeo-environmental reconstructions. In return, such reconstructions measure the outcomes of prolonged change at time scales ranging from 3 to 7-yr periodicity of ENSO cycles to 100-kyr glacial/interglacial cycles and longer. In doing so, such reconstructions may identify other influences on the modern record that were previously undetected because of its brief time span.

Of immediate relevance is the provision of continuity between past reconstructions and modern records of ocean change. To some extent this continuity is achieved for the Southern Ocean by Antarctic ice cores, which carry a range of atmospheric and ocean proxies with annual to decadal resolution and hence improves correlation with the modern ocean/climate records e.g., Sinclair et al. (2014). This is well shown by Abram et al. (2014) whom, on the basis of the James Ross Island ice core and other temperature-sensitive proxies from Antarctica and South America, provide a record of the SAM for the past millennium. It shows the march towards a positive SAM began in the 15th century and was accompanied by a cooling of continental Antarctica with exception of the Antarctica Peninsula, which became warmer. Before the 20th Century, SAM trends may have reflected a teleconnection with the tropical Pacific climate, whereas after 1940, the positive trend was stimulated by increased anthropogenic greenhouse emissions and ozone depletion. Thus, the integration of modern observations, past-reconstructions and numerical modelling place the modern positive trend of SAM into a temporal context, i.e. the anthropogenic signal is superimposed on a long-term, positive trend that may be related to a strong natural signal from the central Pacific.

Such an example leads us to suggest that knowledge and understanding of environmental change would be better served by improved integration of modern and past observations. The aim is to provide long-term records of change that are well founded by physical dynamics. Such records would enhance validation of model simulations of projected change.

References

Abram, N., Mulvaney, R., Vimeux, F., et al., 2014. Evolution of the Southern annular mode during the past millennium. Nature Climate Change 4, 564–569. Available from: https://doi.org/10.1038/nclimate2235.

Andrews, O., Buitenhuis, E., Le Quéré, C., et al., 2017. Biogeochemical modelling of dissolved oxygen in a changing ocean. Philosophical Transactions of the Royal Society A 375.

Aoki, S., Fujii, N., Ushio, S., et al., 2008. Deep western boundary current and southern frontal systems of the Antarctic circumpolar current southeast of the Kerguelen Plateau. Journal of Geophysical Research 113, C08038. Available from: https://doi.org/10.1029/2007JC004627.

Aoki, S., Hariyama, M., Mitsudera, H., et al., 2007. Formation regions of Subantarctic mode water detected by OFES and Argo profiling floats. Geophysical Research Letters 34, L10606. Available from: https://doi.org/10.1029/2007GL029828.

Aoki, S., Rintoul, S.R., Ushio, S., et al., 2005. Freshening of the Adelie Land bottom water near 140°E. Geophysical Research Letters 32, L23601. Available from: https://doi.org/10.1029/2005GL024246.

Argo, 2020. <http://www.argo.ucsd.edu/>.

Armour, K.C., Marshall, J., Scott, J.R., Donohoe, A., Newsom, E.R., 2016. Southern Ocean warming delayed by circumpolar upwelling and equatorward transport. Nature Geoscience 9, 549–555.

Baines, P.G., Condie, S., 1998. Observations and modelling of Antarctic downslope flows: a review. In: Jacobs, S.S., Weiss, R. (Eds.), Ocean, Ice, and Atmosphere: Interactions at the Antarctic Continental Margin. Antarctic Research Series, 75. American Geophysical Union, pp. 29–49.

Bates, N.R., Astor, Y.M., Church, M.J., et al., 2014. A time-series view of changing surface ocean chemistry due to ocean uptake of anthropogenic CO_2 and ocean acidification. Oceanography 27, 126–141.

Beal, L.M., Bryden, H.L., 1999. The velocity and vorticity structure of the Agulhas Current at 32°S. Journal of Geophysical Research 104, 5151–5176.

Bertler, N., Mayewski, P., Carter, L., 2011. Cold conditions in Antarctica during the Little Ice Age: implications for abrupt climate change mechanisms. Earth and Planetary Science Letters 308, 41–51.

Bijl, P.K., Bendle, J.A.P., Bohaty, S.M., et al., 2013. Eocene cooling and Tasmanian gateway. Proceedings of the National Academy of Sciences 1109645–1109650. Available from: https://doi.org/10.1073/pnas.1220872110.

Bindoff, N.L., Rintoul, S., Haward, M., 2011. Position Analysis: Climate Change and the Southern Ocean. Antarctic Climate Ecosystems. 28 pp. ISBN 978-0-9871939-0-2.

Bindoff, N.L., Rosenberg, M.A., Warner, M.J., 2000. On the circulation and water masses over the Antarctic continental slope and rise between 80 and 150°E. Deep Sea Research 47, 2299–2326.

Bindoff, N.L., Willebrand, V., Artale, A., et al., 2007. Observations: oceanic climate change and sea level. In: Solomon, S., Qin, D., Manning, M., Chen, Z., Marquis, M., Averyt, K.B., Tignor, M., Miller, H.L. (Eds.), Climate Change 2007: The Physical Science Basis. Cambridge University Press, pp. 385–432. Working Group I Contribution to the Intergovernmental Panel on Climate Change Fourth Assessment Report.

Böning, C.W., Dispert, A., Visbeck, M., et al., 2008. Response of the Antarctic circumpolar current to recent climate change. Nature Geoscience 1, 864–869.

Bostock, H.C., Barrows, T.T., Carter, L., et al., 2012. A review of the Australian-New Zealand sector of the Southern Ocean over the last 30 ka (Aus-INTIMATE project). Quaternary Science Reviews 74, 35–57.

Bostock, H.C., Sutton, P.J., Williams, M.J.M., et al., 2013. Reviewing the circulation and mixing of Antarctic intermediate water in the South Pacific using evidence from geochemical tracers and Argo float trajectories. Deep Sea Research Part I 73, 84–98.

Boyer, D.L., Guala, J.R., 1972. Model of the Antarctic circumpolar current in the vicinity of the Macquarie Ridge. In: Hayes, D.E. (Ed.), Antarctic Oceanology II – The Australian-New Zealand Sector. Antarctic Research Series, 19. American Geophysical Union, pp. 79–94.

British Oceanographic Data Centre (BODC), 2020. Southern Ocean Argo Regional Data Centre. National Oceanographic Centre. Available from: https://www.bodc.ac.uk/projects/data_management/international/argo/southern_ocean/.

Budillon, G., Cordero, S.G., Salusti, E., 2002. On the dense water spreading off the Ross Sea shelf (Southern Ocean). Journal of Marine System 35, 207–227.

Cai, W., Cowan, T., Stuart, G., et al., 2010. Simulations of processes associated with the fast warming rate of the Southern midlatitude Ocean. Journal of Climate 23, 197–206.

Carter, L., McCave, I.N., 1997. The sedimentary regime beneath the deep western boundary current inflow to the Southwest Pacific Ocean. Journal of Sedimentary Research 67, 1005–1017.

Carter, L., Wilkin, J., 1999. Abyssal circulation around New Zealand – a comparison between observations and a global circulation model. Marine Geology 159, 221–239.

Carter, R.M., McCave, I.N., Carter, L., 2004. Fronts, flows, drifts, volcanoes, and the evolution of the southwestern gateway to the Pacific Ocean: A review of Ocean Drilling Program Leg 181, Eastern New Zealand. Scientific Results Volume, Leg 181 Ocean Drilling Program, College Station, TX. Available from: https://doi.org/10.2973/odp.proc.sr.181.210.2004.

Chavanne, C.P., Heywood, K.J., Nicholls, K.W., et al., 2010. Observations of the Antarctic slope undercurrent in the Southeastern Weddell Sea. Geophysical Research Letters 37, L13601. Available from: https://doi.org/10.1029/2010GL043603.

Chiswell, S.M., Bradford-Grieve, J., Hadfield, M.G., et al., 2013. Climatology of Surface chlorophyll a, autumn-winter and spring blooms in the southwest Pacific Ocean. Journal of Geophysical Research Oceans 118, 1003–1018. Available from: https://doi.org/10.1002/jgrc.20088.

Clem, K.R., Renwick, J.A., McGregor, J., Fogt, R.L., 2016. The relative influence of ENSO and SAM on Antarctic Peninsula climate. Journal of Geophysical Research: Atmospheres 121, 9324–9341.

Cunningham, S.A., Alderman, S.G., King, B.A., et al., 2003. Transport and variability of the Antarctic circumpolar current in drake passage. Journal of Geophysical Research 108. Available from: https://doi.org/10.1029/2001JC001147.

Deacon, G.E.R., 1982. Physical and biological zonation in the Southern Ocean. Deep Sea Research 29, 1–15.

Dinniman, M.S., Klinck, J.M., Smith Jr, W.O., 2010. A model study of circumpolar deep water on the West Antarctic Peninsula and Ross Sea continental shelves. Deep Sea Research Part II 58, 1508–1523.

Dotto, T.S., Naveira Garabato, A., Bacon, S., et al., 2018. Variability of the Ross Gyre, Southern Ocean: Drivers and responses revealed by satellite altimetry. Geophysical Research Letters 45, 6195–6204.

Eagles, G., Jokat, W., 2014. Tectonic reconstructions for paleobathymetry in Drake passage. Tectonophysics 611, 28–50.

Exon, N.F., Kennett, J.P., Malone, M.J., 2004. Leg 189 synthesis: Cretaceous–Holocene history of the Tasmanian gateway. In: Exon, N.F., Kennett, J.P., Malone, M.J. (Eds.), Proceeding of ODP, Science Results, vol. 189. pp. 1–37. College Station, TX (Ocean Drilling Program).

Fabry, V., McClintock, J., Mathis, J., et al., 2009. Ocean acidification at high latitudes: the Bellwether. Oceanography 22, 160–171.

Fahrbach, E., Harms, H.S., Rohardt, G., et al., 2001. Flow of bottom water in the northwestern Weddell Sea. Journal of Geophysical Research 106 (C2), 2761–2778.

Fernandez, D., Bowen, M., Carter, L., 2014. Intensification and variability of the confluence of subtropical and subantarctic boundary currents east of New Zealand. Journal of Geophysical Research Oceans 119, 1146–1160.

Florindo, F., Siegert, M. (Eds.), 2009. Developments in Earth & Environmental Sciences, Volume 8: Antarctic Climate Evolution. Elsevier, Amsterdam, 606 pp. ISBN: 978-0-444-52847-6.

Foldvik, A., Gammelsrod, T., Osterhus, S., et al., 2004. Ice shelf water overflow and bottom water formation in the southern Weddell Sea. Journal of Geophysical Research 109, C02015. Available from: https://doi.org/10.1029/2003JC002008.

Foster, T.D., Carmack, E.C., 1976. Frontal zone mixing and Antarctic bottom water formation in the Southern Weddell Sea. Deep Sea Research 23, 301–317.

Frölicher, T.L., Aschwanden, M.T., Gruber, N., et al., 2020. Contrasting upper and deep ocean oxygen response to protracted global warming. Global Biogeochemical Cycles 34. Available from: https://doi.org/10.1029/2020GB006601e2020GB006601.

Frölicher, T.L., Sarmiento, J.L., Paynter, D.J., et al., 2015. Dominance of the Southern Ocean in anthropogenic carbon and heat uptake in CMIP5 models. Journal of Climate 28, 862–886.

Fukamachi, Y., Rintoul, S., Church, J., et al., 2010. Strong export of Antarctic bottom water east of the Kerguelen Plateau. Nature Geoscience 3, 327–331. Available from: https://doi.org/10.1038/ngeo842.

Fusco, G., Budillon, G., Spezie, G., 2009. Surface heat fluxes and thermohaline variability in the Ross Sea and in TerraNova Bay polynya. Continental Shelf Research 29, 1887–1895.

Fyfe, J.C., Meehl, G.A., England, M.H., et al., 2016. Making sense of the early-2000s warming slowdown. Nature Climate Change 6, 224–228.

Galeotti, S., et al., 2021. The Eocene-Oligocene boundary climate transition: an Antarctic perspective. In: Florindo, F., et al. (Eds.), Antarctic Climate Evolution, second ed. Elsevier.

Gao, L., Rintoul, S.R., Yu, W., 2018. Recent wind-driven change in sub-Antarctic mode eater and its impact on ocean heat storage. Nature Climate Change 8, 58–63.

Garabato, A.C., Mcdonagh, E.L., Stevens, D.P., et al., 2002. On the export of Antarctic bottom water from the Weddell Sea. Deep Sea Research Part II 49, 4715–4742.

Gille, S.T., 2002. Warming of the Southern Ocean since the 1950s. Science (New York, N.Y.) 295, 1275–1277.

Gille, S.T., 2008. Decadal-scale temperature trends in the Southern Hemisphere ocean. Journal of Climate 21, 4749–4765.

Gille, S.T., 2010. Asymmetric response. Nature Geoscience 3, 227–228.

Gille, S.T., 2014. Meridional displacement of the Antarctic circumpolar current. Philosophical Transactions of the Royal Society A 372.

Gordon, A.L., 1972. On the interaction of the Antarctic circumpolar current and the Macquarie Ridge. In: Hayes, D.E. (Ed.), Antarctic Oceanology II – The Australian-New Zealand Sector. Antarctic Research Series, vol. 19, pp. 71–78.

Gordon, A.L., 1975. An Antarctic oceanographic section along 170°E. Deep Sea Research 22, 357–377.

Gordon, A.L., Huber, B.A., Busecke, J., 2015. Bottom water export from the western Ross Sea, 2007 through 2010. Geophysical Research Letters 42, 5387–5394. Available from: https://doi.org/10.1002/2015GL064457.

Gordon, A.L., Orsi, A.H., Muench, R., et al., 2009. Western Ross Sea continental slope gravity currents. Deep Sea Research Part II 56, 796–817. Available from: https://doi.org/10.1016/j.dsr2.2008.10.037.

Greenbaum, J.S., Blankenship, D.D., Young, D.A., et al., 2015. Ocean access to a cavity beneath Totten Glacier in East Antarctica. Nature Geoscience 8, 294–298. Available from: https://doi.org/10.1038/NGEO2388.

Gruber, N., Clement, D., Carter, B.R., et al., 2019. The oceanic sink for anthropogenic CO_2 from 1994 to 2007. Science (New York, N.Y.) 363, 1193–1199. Available from: https://doi.org/10.1126/science.aau5153.

Haumann, F.A., Gruber, N., Münnich, M., et al., 2016. Sea-ice transport driving Southern Ocean salinity and its recent trends. Nature 537, 89–92.

Hayes, D.E., Zhang, C., Weissel, R.A., 2009. Modelling paleobathymetry in the Southern Ocean. Eos, Transactions, American Geophysical Union 90, 165–166. Available from: https://doi.org/10.1029/2009EO190001.

Herraiz-Borreguero, L., Rintoul, S.R., 2011. Subantarctic mode water: distribution and circulation. Ocean Dynamics 61, 103–126. Available from: https://doi.org/10.1007/s10236-010-0352-9.

Heywood, K.J., Garabato, A.C.N., Stevens, D.P., et al., 2004. On the fate of the Antarctic slope front and the origin of the Weddell Front. Journal of Geophysical Research 109. Available from: https://doi.org/10.1029/2003JC002053.

Heywood, K.J., Locarnini, R.A., Frew, R.D., et al., 1998. Transport and water masses of the Antarctic slope front system in the Eastern Weddell Sea. In: Jacobs, S.S., Weiss, R.F. (Eds.), Ocean, Ice, and Atmosphere: Interactions at the Antarctic Continental Margin, Antarctic Research Series., 75. AGU, Washington, D.C, pp. 203–214.

Hill, K.L., Rintoul, S.R., Coleman, R., et al., 2008. Wind forced low frequency variability of the East Australian Current. Geophysical Research Letters 35 (L08602). Available from: https://doi.org/10.1029/2007/GL032912.

Hogg, N.G., 2001. Quantification of the deep circulation. In: Siedler, G., Church, J., Gould, J. (Eds.), Ocean Circulation and Climate. Academic Press, London, pp. 259–270.

Holland, D.M., Nicholls, K.W., Basinski, A., 2020. The Southern Ocean and its interaction with the Antarctic Ice Sheet. Science (New York, N.Y.) 367, 1326–1330. Available from: https://doi.org/10.1126/science.aaz5491.

Hollister, C.D., McCave, I.N., 1984. Sedimentation under deep sea storms. Nature 309, 220–225.

IHO (International Hydrographic Organisation), 2002. Names and Limits of Oceans and Seas, Special Publication, fourth ed. IHO Publication S-23, 249 pp. Available from: http://wiki.geosys.ru/lib/exe/fetch.php/ru/portal/lib/iho/s23.los.ed4draft.2002.pdf.

IPCC, 2013. Climate change 2013: the physical science basis. In: Stocker, T.F., Qin, D., Plattner, G.-K., Tignor, M., Allen, S.K., Boschung, J., Nauels, A., Xia, Y., Bex, V., Midgley, P.M. (Eds.), Contribution of Working Group I to the Fifth Assessment Report of the Intergovernmental Panel on Climate Change. Cambridge University Press, Cambridge, UK and New York, NY, p. 1535.

IPCC, 2019. Summary for policymakers. In: Pörtner, H. -O., Roberts, D. C., Masson-Delmotte, V., Zhai, P., Tignor, M., Poloczanska, E., et al. (Eds.), IPCC Special Report on the Ocean and Cryosphere in a Changing Climate.

IPCC, 2021. Climate Change 2021: The Physical Science Basis. In: Masson-Delmotte, V., Zhai, P., Pirani, A., Connors, S.L., Péan, C., Berger, S., et al. (Eds.), Contribution of Working Group I to the Sixth Assessment Report of the Intergovernmental Panel on Climate Change. Cambridge University Press. Available from: https://www.ipcc.ch/report/sixth-assessment-report-working-group-i/.

Jacobs, S.S., 1991. On the nature and significance of the Antarctic slope front. Marine Chemistry 35, 9–24.

Jacobs, S.S., 2004. Bottom water production and its links with the thermohaline circulation. Antarctic Science 16, 427–437.

Jacobs, S.S., 2006. Observations of change in the Southern Ocean. Philosophical Transactions of the Royal Society A 364, 1657–1681.

Jacobs, S.S., Giulivi, C.F., 2010. Large multidecadal salinity trends near the Pacific–Antarctic continental margin. Journal of Climate 23, 4508–4524. Available from: https://doi.org/10.1175/2010JCLI3284.1.

Jacobs, S.S., Giulivi, C.F., Mele, P.A., 2002. Freshening of the Ross Sea during the late 20th Century. Nature 297, 386–389.

Jasmine, P., Muraleedharan, K.R., Madhu, N.V., et al., 2009. Hydrographic and productivity characteristics along 45°E longitude in the Southwestern Indian Ocean and Southern Ocean during austral summer 2004. Marine Ecology Progress Series 389, 97–116. Available from: https://doi.org/10.3354/meps08126.

Johnson, C.R., Banks, S.C., Barrett, N.S., et al., 2011. Climate change cascades: shifts in oceanography, species' ranges and subtidal marine community dynamics in eastern Tasmania. Journal of Experimental Marine Biology and Ecology 400, 17–32.

Johnson, G.C., 2008. Quantifying Antarctic bottom water and North Atlantic deep water volumes. Journal of Geophysical Research 113, C05027. Available from: https://doi.org/10.1029/2007JC004477.

Johnson, G.C., Musgrave, D.L., Warren, B.A., et al., 1998. Flow of bottom and deep water in the Amirante passage and Mascarene Basin. Journal of Geophysical Research 103, 30,973–30,984.

Johnson, G.C., Warren, B.A., Olson, D.B., 1991. Flow of bottom water in the Somali Basin. Deep Sea Research 38, 637–652.

Jones, J.M., Gille, S.T., Goosse, H., et al., 2016. Assessing recent trends in high-latitude Southern Hemisphere surface climate. Nature Climate Change 6, 917–926. Available from: https://doi.org/10.1038/nclimate3103.

Jullion, L., Garabato, A.C.N., Bacon, S., et al., 2014. The contribution of the Weddell Gyre to the lower limb of the global overturning circulation. Journal of Geophysical Research Oceans 119, 3357–3377. Available from: https://doi.org/10.1002/2013JC009725.

Keeling, R.F., Körtzinger, A., Gruber, N., 2010. Ocean deoxygenation in a warming world. Annual Review of Marine Science 2, 199–229. Available from: https://doi.org/10.1146/annurev.marine.010908.163855.

Kennett, J.P., von der Borch, C.C., 1985. Southwest Pacific Cenozoic paleoceanography. In: Kennett, J.P., von der Borch, C.C. (Eds.), Initial Reports of the Deep Sea Drilling Project, vol. 90. United States Government Printing Office, Washington, DC, pp. 1493–1517.

Kersalé, M., Meinen, C.S., Perez, R.C., et al., 2020. Highly variable upper and abyssal overturning cells in the South Atlantic. Science Advances 6, eaba7573. Available from: https://doi.org/10.1126/sciadv.aba7573.

Klatt, O., Fahrbach, E., Hoppema, M., et al., 2005. The transport of the Weddell Gyre across the prime meridian. Deep Sea Research 52, 513–528.

Klinck, J.M., Nowlin Jr, W.D., 2001. The Southern Ocean – Antarctic Circumpolar Current. Academic Press, pp. 1–11. Available from: https://doi.org/10.1006/rwos.2001.0370.

Landschützer, P., Gruber, N., Haumann, F.A., et al., 2015. The reinvigoration of the Southern Ocean carbon sink. Science (New York, N.Y.) 349, 1221–1224. Available from: https://doi.org/10.1126/science.aab2620.

Le Quéré, C., Rödenbeck, C., Buitenhuis, E.T., et al., 2007. Saturation of the Southern Ocean CO_2 sink due to recent climate change. Science (New York, N.Y.) 316, 1735–1738. Available from: https://doi.org/10.1126/science.1136188.

Locarnini, R., Whitworth III, T., Nowlin Jr, W.D., 1993. The importance of the Scotia Sea on the outflow of Weddell Sea Deep Water. Journal of Marine Research 51, 135–153.

Lovenduski, N.S., Gruber, N., 2005. The impact of the Southern annular mode on Southern Ocean circulation and biology. Geophysical Research Letters 32, L11603. Available from: https://doi.org/10.1029/2005GL022727.

Lovenduski, N.S., Gruber, N., Doney, S.C., et al., 2007. Enhanced CO_2 outgassing in the Southern Ocean from a positive phase of the Southern annular mode. Global Biogeochemical Cycles 21, GB2026. Available from: https://doi.org/10.1029/2006GB002900.

Lund, D., Lynch-Stieglitz, J., Curry, W., 2006. Gulf Stream density structure and transport during the past millennium. Nature 444, 601–604. Available from: https://doi.org/10.1038/nature05277.

Maasch, K., Mayewski, P.A., Rohling, E., et al., 2005. Climate of the past 2000 years. Geografiska Annaler 87A, 7–15.

Mantyla, A.W., Reid, J.L., 1995. On the origins of deep and bottom waters of the Indian Ocean. Journal of Geophysical Research 100, 2417–2439.

Marshall, D.P., Ambaum, M.H.P., Maddison, J.R., et al., 2017. Eddy saturation and frictional control of the Antarctic Circumpolar Current. Geophysical Research Letters 44, 286–292.

Marshall, J., Speer, K., 2012. Closure of the meridional overturning circulation through Southern Ocean upwelling. Nature Geoscience 5, 171–180.

Martínez-Moreno, J., Hogg, A.M., England, M.H., et al., 2021. Global changes in oceanic mesoscale currents over the satellite altimetry record. Nature Climate Change 11, 397–403.

McCartney, M.S., 1977. Subantarctic mode water. In: Angel, M.V. (Ed.), A Voyage of Discovery: George Deacon 70th Anniversary Volume. Pergamon Press, Oxford, pp. 103–119. Supplement to Deep-Sea Research.

McCartney, M.S., Donohue, K.A., 2007. A deep cyclonic gyre in the Australian–Antarctic Basin. Progress in Oceanography 75, 675–750.

McCave, I.N., Carter, L., Hall, I.R., 2008. Glacial-interglacial changes in water mass structure and flow in the SW Pacific Ocean. Quaternary Science Reviews 27, 1886–1908.

McKay, R.M., Barrett, P.J., Levy, R.S., et al., 2016. Antarctic Cenozoic climate history from sedimentary records: ANDRILL and beyond. Philosophical Transactions of the Royal Society A 374. Available from: https://doi.org/10.1098/rsta.2014.0301.

McNeil, B.I., Matear, R.J., 2008. Southern Ocean acidification: a tipping point at 450-ppm atmospheric CO_2. Proceedings of the National Academy of Sciences 105, 18860–18864. Available from: https://doi.org/10.1073/pnas.0806318105.

Meijers, A.J.S., Bindoff, N.L., Rintoul, S.R., 2011. Frontal movements and property fluxes: contributions to heat and freshwater trends in the Southern Ocean. Journal of Geophysical Research Oceans 116, C08024.

Meredith, M.P., King, J., 2005. Rapid climate change in the ocean west of the Antarctic Peninsula during the second half of the 20th century. Geophysical Research Letters 32, L19604. Available from: https://doi.org/10.1029/2005GL024042.

Moffat, C., Beardsley, R.C., Owens, B., et al., 2008. A first description of the Antarctic Peninsula Coastal current. Deep Sea Research Part II 55, 277–293.

Moore, J.K., Abbott, M.R., Richman, J.G., 1999. Location and dynamics of the Antarctic circumpolar front from satellite sea surface temperature data. Journal of Geophysical Research 104, 3059–3073.

Morris, M., Stanton, B.R., Neil, H.L., 2001. Subantarctic oceanography around New Zealand: preliminary results from an ongoing survey. New Zealand Journal of Marine and Freshwater Research 35, 499–519.

Morrison, A.K., Hogg, A.M., England, M.H., et al., 2020. Warm circumpolar deep water transport towards Antarctica driven by local dense water export in canyons. Science Advances 6. Available from: https://doi.org/10.1126/sciadv.aav2516.

Murphy, R.J., Pinkerton, M.H., Richardson, K.M., et al., 2001. Phytoplankton distributions around New Zealand derived from SeaWiFs remotely sensed ocean colour data. New Zealand Journal of Marine and Freshwater Research 35, 343–362.

Nøst, O.A., Biuw, M., Tverberg, V., et al., 2011. Eddy overturning of the Antarctic slope front controls glacial melting in the Eastern Weddell Sea. Journal of Geophysical Research 116, C11014. Available from: https://doi.org/10.1029/2011JC006965.

Nowlin, W.D., Klinck, J.M., 1986. The physics of the Antarctic circumpolar current. Reviews of Geophysics 24, 469–491.

Ohshima, K., Fukamachi, Y., Williams, G.D., et al., 2013. Antarctic bottom water production by intense sea-ice formation in the Cape Darnley polynya. Nature Geoscience 6, 235–240.

Orsi, A.H., Johnson, G.C., Bullister, J.L., 1999. Circulation, mixing and production of Antarctic bottom water. Progress in Oceanography 43, 55–109.

Orsi, A.H., Nowlin Jr, W.D., Whitworth III, T., 1993. On the circulation and stratification of the Weddell Gyre. Deep Sea Research 40, 169–203.

Orsi, A.H., Whitworth III, T., 2005. Hydrographic atlas of the world ocean circulation experiment (WOCE). In: Sparrow, M., Chapman, P., Gould, J. (Eds.), Southern Ocean, Volume 1. International WOCE Project Office, Southampton, ISBN 0-904175-49-9.

Orsi, A.H., Whitworth III, T., Nowlin Jr, W.D., 1995. On the meridional extent and fronts of the Antarctic circumpolar current. Deep Sea Research 42, 641–673.

Pfuhl, H.A., McCave, I.N., 2005. Evidence for late Oligocene establishment of the Antarctic circumpolar current. Earth and Planetary Science Letters 235, 715–728. Available from: https://doi.org/10.1016/j.epsl.2005.04.025.

Purkey, S.G., Johnson, G.C., 2010. Warming of global abyssal and deep Southern Ocean waters between the1990s and 2000s: contributions to global heat and sea level rise budgets. J. Clim. 23, 6336–6351.

Ridgway, K., Hill, K., 2009. The East Australian current. In: Poloczanska, E.S., Hobday, A.J., Richardson, A.J. (Eds.), A Marine Climate Change Impacts and Adaptation Report Card for Australia 2009. NCCARF Publication 05/09, ISBN 978-1-921609-03-9.

Rintoul, S.R., 1998. On the origin and influence of Adelie land bottom water, Ocean, Ice, and Atmosphere – Interactions at the Antarctic Continental Margin. Antarctic Research Series, vol. 75. pp. 151–171.

Rintoul, S.R., 2007. Rapid freshening of Antarctic bottom water formed in the Indian and Pacific Oceans. Geophysical Research Letters 34, L06606. Available from: https://doi.org/10.1029/2006GL028550.

Rintoul, S.R., 2018. The global influence of localized dynamics in the Southern Ocean. Nature 558, 209–218. Available from: https://doi.org/10.1038/s41586-018-0182-3.

Rintoul, S., Sokolov, S., 2001. Baroclinic transport variability of the Antarctic circumpolar current south of Australia (WOCE repeat section SR3). Journal of Geophysical Research 106 (C2). Available from: https://doi.org/10.1029/2000JC900107.

Rintoul, S.R., England, M.H., 2002. Ekman transport dominates local air–sea fluxes in driving variability of subantarctic mode water. Journal of Physical Oceanography 32, 1308–1321.

Rintoul, S.R., Hughes, C.W., Olbers, D., 2001. The Antarctic circumpolar current system. In: Siedler, G., Church, J., Gould, J. (Eds.), Ocean Circulation and Climate: Observing and Modelling the Global Ocean. Academic Press, London, pp. 271–302.

Rintoul, S.R., Sokolov, S., Williams, M.J.M., et al., 2014. Antarctic Circumpolar Current transport and barotropic transition at Macquarie Ridge. Geophysical Research Letters 41, 7254–7261. Available from: https://doi.org/10.1002/2014GL061880.

Rintoul, S.R., Silvano, A., Pena-Molino, B., 2016. Ocean heat drives rapid basal melt of the Totten Ice Shelf. Science Advances 2, e1601610. Available from: https://doi.org/10.1126/sciadv.1601610.

Rintoul, S.R., Sokolov, S., Williams, M.J.M., et al., 2014. Antarctic Circumpolar Current transport and barotropic transition at Macquarie Ridge. Geophysical Research Letters 41, 7254–7261. Available from: https://doi.org/10.1002/2014GL061880.

Roemmich, D., Gilson, J., 2009. The 2004–2008 mean and annual cycle of temperature, salinity, and steric height in the global ocean from the Argo Programme. Progress in Oceanography 82, 81–100.

Roemmich, D., Gilson, J., Sutton, P., et al., 2016. Multidecadal change of the South Pacific Gyre circulation. Journal of Physical Oceanography 46, 1871–1883. Available from: https://doi.org/10.1175/JPO-D-15-0237.1.

Roquet, F., Williams, G., Hindell, M., et al., 2014. A Southern Indian Ocean database of hydrographic profiles obtained with instrumented elephant seals. Scientific Data 1, 140028. Available from: https://doi.org/10.1038/sdata.2014.28.

Sallée, J., Speer, K., Morrow, R., 2008. Response of the Antarctic circumpolar current to atmospheric variability. Journal of Climate 21, 3020–3039.

Sallée, J., Speer, K., Rintoul, S., 2010. Zonally asymmetric response of the Southern Ocean mixed-layer depth to the Southern Annular Mode. Nature Geoscience 3, 273–279. Available from: https://doi.org/10.1038/ngeo812.

Sallée, J.B., Pellichero, V., Akhoudas, C., et al., 2021. Summertime increases in upper-ocean stratification and mixed-layer depth. Nature 591, 592–598.

Sallée, J.-B., Speer, K., Rintoul, S., et al., 2010. Southern Ocean thermocline ventilation. Journal of Physical Oceanography 40, 509–529.

Sallée, J.B., Wienders, N., Speer, K., et al., 2006. Formation of subantarctic mode water in the southeastern Indian Ocean. Ocean Dynamics 56, 525–542. Available from: https://doi.org/10.1007/s10236-005-0054-x.

Scher, H.D., Martin, E.E., 2006. Timing and climatic consequences of the opening of drake passage. Science (New York, N.Y.) 312 (5772), 428–430. Available from: https://doi.org/10.1126/science.1120044.

Schlitzer, R., 2009. Ocean Data View Software. <http://odv.awi.de/>.

Schmidtko, S., Heywood, K.J., Thompson, A.F., et al., 2014. Multidecadal warming of Antarctic waters. Science (New York, N.Y.) 346, 1227–1231. Available from: https://doi.org/10.1126/science.1256117.

Schmitz, W.J., 1995. On the interbasin-scale thermohaline circulation. Reviews of Geophysics 33, 151–173.

Shi, J., Xie, S., Talley, L.D., 2018. Evolving relative importance of the Southern Ocean and North Atlantic in Anthropogenic Ocean heat uptake. Journal of Climate 31, 7459–7479. Available from: https://doi.org/10.1175/JCLI-D-18-0170.1.

Silvano, A., Foppert, A., Rintoul, S.R., et al., 2020. Recent recovery of Antarctic Bottom Water formation in the Ross Sea driven by climate anomalies. Nature Geoscience 13, 780–786.

Sinclair, K.E., Bertler, N.A.N., Bowen, M.M., et al., 2014. Twentieth century sea-ice trends in the Ross Sea from a high-resolution, coastal ice-core record. Geophysical Research Letters 41, 3510–3516.

Sloyan, B.M., 2006. Antarctic bottom and lower circumpolar deep water circulation in the eastern Indian Ocean. Journal of Geophysical Research 111, C02006. Available from: https://doi.org/10.1029/2005JC003011.

Sloyan, B.M., Talley, L.D., Chereskin, T.K., et al., 2010. Antarctic intermediate water and subantarctic mode water formation in the Southeast Pacific: the role of turbulent mixing. Journal of Physical Oceanography 40, 1558–1574.

Sloyan, B.M., Wanninkhof, R., Kramp, M., et al., 2019. The Global Ocean Ship-Based Hydrographic Investigations Program (GO-SHIP): A Platform for Integrated Multidisciplinary Ocean Science. Frontiers in Marine Science 6, 445. Available from: https://doi.org/10.3389/fmars.2019.00445.

Smith Jr., W.O., Sedwick, P.N., Arrigo, K.R., et al., 2012. The Ross Sea in a sea of change. Oceanography 25, 90–103. Available from: https://doi.org/10.5670/oceanog.2012.80.

Sokolov, S., Rintoul, S.R., 2007. On the relationship between fronts of the Antarctic circumpolar current and surface chlorophyll concentrations in the Southern Ocean. Journal of Geophysical Research 112, C07030. Available from: https://doi.org/10.1029/2006JC004072.

Sokolov, S., Rintoul, S.R., 2009a. Circumpolar structure and distribution of the Antarctic circumpolar current fronts: 1. Mean circumpolar paths. Journal of Geophysical Research 114, C11018. Available from: https://doi.org/10.1029/2008JC005108. 19 pp.

Sokolov, S., Rintoul, S.R., 2009b. The circumpolar structure and distribution of the Antarctic circumpolar current fronts part 2: variability and relationship to sea surface height. Journal of Geophysical Research 114, C1101915 pp. Available from: https://doi.org/10.1029/2008JC005248.

SOOS, 2020. The Southern Ocean Observing System. <http://www.soos.aq/>.

Stanton, B.R., Morris, M., 2004. Direct velocity measurements in the Subantarctic Front and over Campbell Plateau, south-east of New Zealand. Journal of Geophysical Research 109, C01028. Available from: https://doi.org/10.1029/2002JC001339.

Stommel, H., 1958. The abyssal circulation. Deep Sea Research 5, 80–82.

Talley, L.D., 1996. Antarctic intermediate water in the South Atlantic. The South Atlantic. Springer, Berlin, Heidelberg.

Talley, L.D., Feely, R.A., Sloyan, B.M., et al., 2016. Changes in Ocean Heat, carbon content, and ventilation: a review of the first decade of GO-SHIP global repeat hydrography. Annual Review of Marine Science 8, 185–215.

Thomas, E.R., Bracegirdle, T.J., Turner, J., et al., 2013. A 308 year record of climate variability in West Antarctica. Geophysical Research Letters 40, 5492–5496. Available from: https://doi.org/10.1002/2013GL057782.

Thompson, A.F., Stewart, A.L., Spence, P., et al., 2018. The Antarctic slope current in a changing climate. Reviews of Geophysics 56, 741–770. Available from: https://doi.org/10.1029/2018RG000624.

Thompson, D., Solomon, S., Kushner, P., et al., 2011. Signatures of the Antarctic ozone hole in Southern Hemisphere surface climate change. Nature Geoscience 4, 741–749. Available from: https://doi.org/10.1038/ngeo1296.

Toggweiler, J.R., Russell, J.L., Carson, S.R., 2006. Midlatitude westerlies, atmospheric CO_2, and climate change. Paleoceanography 21, PA2005. Available from: https://doi.org/10.1029/2005PA001154.

Toole, J.M., Warren, B.A., 1993. A hydrographic section across the subtropical Indian Ocean. Deep Sea Research 40, 1973–2019.

Turner, J., Colwell, S.R., Marshall, G.J., et al., 2005. Antarctic climate change during the last 50 years. International Journal of Climatology 25, 279–294.

Turner, J., Lu, H., White, I., et al., 2016. Absence of 21st century warming on Antarctic Peninsula consistent with natural variability. Nature 535, 411–415. Available from: https://doi.org/10.1038/nature18645.

Turney, C.S.M., Fogwill, C.J., Palmer, J.G., et al., 2017. Tropical forcing of increased Southern Ocean climate variability revealed by a 140-year subantarctic temperature reconstruction. Climate of the Past 13, 231–248. Available from: https://doi.org/10.5194/cp-13-231-2017.

van Wijk, E.M., Rintoul, S.R., 2014. Freshening drives contraction of Antarctic bottom water in the Australian Antarctic Basin. Geophysical Research Letters 41, 1657–1664.

Vernet, M., Geibert, W., Hoppema, M., Brown, P.J., Haas, C., Hellmer, H.H., et al., 2019. The Weddell Gyre, Southern Ocean: Present knowledge and future challenges. Reviews of Geophysics 57 (3), 623–708.

Warren, B.A., 1971. Antarctic deep water contribution to the world ocean. Research in the Antarctic. American Association for the Advancement of Science, Washington, D.C, pp. 631–643.

Warren, B.A., 1973. TransPacific hydrographic sections at latitudes 43°S and 28°S: the SCORPIO Expedition—II. Deep water. Deep Sea Research 20, 9–38.

Warren, B.A., 1974. Deep flow in the Madagascar and Mascarene basins. Deep Sea Research 21, 1–21.

Warren, B.A., 1981. Deep circulation of the world ocean. In: Warren, B.A., Wunsch, C. (Eds.), Evolution of Physical Oceanography. Massachusetts Institute of Technology Press, Cambridge, pp. 6–41.

Warren, B.A., Johnson, G.C., 2002. The overflows across Ninetyeast Ridge. Deep Sea Research Part II 49, 1423–1429.

Whitworth III, T., 1988. The Antarctic Circumpolar Current. Oceanus 32, 53–58.

Whitworth III, T., Nowlin Jr., W.D., Pillsbury, R.D., et al., 1991. Observations of the Antarctic circumpolar current and deep boundary current in the Southwest Atlantic. Journal of Geophysical Research 96, 15,105–15,118.

Whitworth III, T., Orsi, A.H., Kim, S.-J., et al., 1998. Water masses and mixing near the Antarctic slope front. Antarctic Research Series 75, 1–27.

Whitworth, T., Warren, B.A., Nowlin Jr, W.D., et al., 1999. On the deep western-boundary current in the Southwest Pacific Basin. Progress in Oceanography 43, 1–54.

Williams, G.D., Aoki, S., Jacobs, S.S., et al., 2010. Antarctic bottom water from the Adélie and George V Land coast, East Antarctica (140–149°E). Journal of Geophysical Research 115, C04027. Available from: https://doi.org/10.1029/2009JC005812.

Wintrup, M., Vallelonga, P., Kjær, H.A., et al., 2019. A 2700-year annual timescale and accumulation history for an ice core from Roosevelt Island, West Antarctica. Climate of the Past 15, 751–779. Available from: https://doi.org/10.5194/cp-15-751-2019.

Wu, L., Cai, W., Zhang, L., et al., 2012. Enhanced warming over the global subtropical western boundary currents. Nature Climate Change 2, 161–166.

Wunsch, C., 2010. Towards understanding the Paleocean. Quaternary Science Reviews 29, 1960–1967.

Yamazaki, K., Aoki, S., Shimada, K., et al., 2020. Structure of the subpolar gyre in the Australian-Antarctic Basin derived from Argo floats. Journal of Geophysical Research: Oceans 125, e2019JC015406. Available from: https://doi.org/10.1029/2019JC015406.

Zachos, J., Dickens, G., Zeebe, R., 2008. An early Cenozoic perspective on greenhouse warming and carbon-cycle dynamics. Nature 451, 279–283. Available from: https://doi.org/10.1038/nature06588.

Chapter 5

Advances in numerical modelling of the Antarctic ice sheet

Martin Siegert[1] and Nicholas R. Golledge[2]
[1]Grantham Institute and Department of Earth Science and Engineering, Imperial College London, London, United Kingdom, [2]Antarctic Research Centre, Victoria University of Wellington, Wellington, New Zealand

5.1 Introduction and aims

In the first edition of Antarctic Climate Evolution, Siegert (2008a) provided a basic assessment of (1) the function of numerical ice sheet modelling, (2) the data needed as input, (3) inter-comparison and calibration exercises and (4) examples of how models have been used to constrain the dynamics and form of past ice sheets. In this chapter, we examine progress made in the last 10 years in each of these areas.

Ice sheet modelling has advanced considerably in the last decade, with substantial improvements in performance and resolution. A critical region of the ice sheet to predicting glacial change is the grounding zone. Our ability to model this part of the ice sheet has improved markedly through better knowledge of the physical processes at play here, and also with anisotropic or adaptive grids that offer high resolution where it is needed most.

In 2008 the bed elevation of Antarctica was poorly constrained in several places, which hindered the usefulness of ice sheet models. In 2013 a revised bed elevation, Bedmap2, was introduced as a consequence of imaginative interpolation procedures and new bed data. Bedmap3 is in the process of being configured, which would be the first full appreciation of continent-wide Antarctic topography without major data gaps. In addition, there have also been attempts to determine the paleotopography of Antarctica, which would allow better appreciation of former ice sheets through numerical modelling (Siegert, 2008b).

Ice sheet model intercomparison has also improved in recent years, with it being acknowledged as a critical way to advance coupled systems. The ISMIP6 (Ice Sheet Modelling Intercomparison Project - Phase 6)

Antarctic Climate Evolution. DOI: https://doi.org/10.1016/B978-0-12-819109-5.00006-2

programme, used in the CMIP6 (Coupled Model Intercomparison Project - Phase 6) project (Nowicki et al., 2016), allows coupled and stand-alone ice sheet models to be tested with a set of climate scenarios, adding confidence to model results and better quantification of their uncertainties.

Given the improvements in models, their inputs and intercomparisons, it is hardly surprising that they have been used well in the last few years to examine former Antarctic ice sheets. They have provided useful information to determine the last glacial maximum (LGM), Eemian and Pliocene ice sheet configurations, as well as further back in time. An important idea is the degree to which 'marine ice sheet instability', which takes effect as the grounding line retreats across deepening topography, and 'marine ice cliff instability', which would act following the break-up of large ice shelves, might lead to wide-spread deglaciation in Antarctica.

Ice sheet models from 2008 look primitive compared to today's. This is a reflection of an area of glaciology that is developing quite quickly. That said, ice sheet models have some way to go before they can offer ideas of former and future ice sheets with a high degree of certainty. It is reasonable to conclude that in 10 years time ice sheet models will be better connected to geophysical data, as well as benefiting from other advances, and will make today's models appear similarly old fashioned.

While geological evidence of former ice sheet changes is paramount to evaluating how ice sheets have behaved in the past, ice-sheet models provide the only means by which quantitative knowledge of the flow and form of past ice sheets can be assessed. In the first volume of Antarctic Climate Evolution, Siegert (2008a) presented a summary of ice sheet modelling activities and how they have been used to define past ice sheets. That chapter discussed how ice sheet models work and offered some examples of their use. In this short chapter, rather than replicate the background to ice sheet modelling, we will instead focus on developments in ice sheet modelling over the last decade, and how they have helped us to understand the growth and decay of ice sheets during the glaciated history of Antarctica. The chapter has five sections. The first summarises recent advances in ice sheet modelling. In the second, improvements in the inputs to numerical models are discussed. The third section looks at new model inter-comparison exercises, as a means to better evaluate model performance, and the fourth provides an assessment of several case studies where modelling has been used to define former ice sheets. In the fifth section, the future of numerical ice sheet modelling is discussed, focusing on areas that would make the greatest positive difference to determining the dynamics and growth cycles of former ice sheets.

5.2 Advances in ice sheet modelling

5.2.1 Grounding line physics

A critical zone of the ice sheet is at its marine margin, where grounded ice either terminates in the ocean or transitions to form a floating ice shelf. The physics under which this transition occurs is non-trivial and has been the subject of much concern and research in glaciology. This concern is supported by satellite observations of ice-surface elevation change over the past 30 years, which show the regions where mass is being lost at the greatest rates correspond to locations with submarine grounding lines (Shepherd et al., 2018). Further, particularly in West Antarctica, if upstream migration of the grounding line occurs over a retro-grade sloping bed, a positive feedback may lead to further recession until the bed slope becomes 'normal' to flow (Joughin et al., 2014), so-called 'marine ice-sheet instability' or MISI.

The relevance of MISI to the potential collapse of the West Antarctic Ice Sheet (WAIS) under anthropogenic global warming has been understood for over 40 years (Mercer, 1978) – the key issue being that ocean warmth could melt marine sections of the ice sheet, triggering ice-sheet collapse through MISI. Clearly, this is a problem that numerical ice-sheet models would be well-suited to investigate, yet until quite recently models treated the grounding line casually and without appreciation of the physics and processes operating in this key location.

A mathematical simplification of grounding line physics was described in a seminal work by Schoof (2007), who demonstrated how it may be parameterised using a simple flux condition that could be incorporated into continental scale ice sheet models. Adoption of this approach has been widespread in the ice-sheet modelling community, particularly among more traditional fixed-grid models (e.g., Pollard and DeConto, 2009). This advance has allowed such models to provide a better appreciation of the evolution of marine sections of the Antarctic ice sheet in the past, and projections of its change over the coming centuries. Models that do not incorporate this Schoof-based flux adjustment have instead chosen to implement other methods, such as sub-grid calculations of the grounding line stress field (Feldmann et al., 2014).

While these advances have been important, they have not led to full resolution of the problem around grounding lines. Aside from the challenges in oceanographic modelling to describe how ocean heat is supplied to the ice-sheet margin, grounding lines represent one of the most challenging places to undertake glaciological research, due to ice-surface crevassing and inaccessibility of the ocean–ice interface. Hence, there is a dearth of measurements from even the most vulnerable marine ice-sheet margins with which to guide ice-sheet models. Moreover, far from being a simple transition between grounded and floating ice, we now appreciate the grounding line resembles more of a 'zone' in many places, heavily influenced by tides that lift and lower significant sections of ice streams and ice rises (Jeofry et al., 2018a) allowing episodic incursions of warm ocean water into subglacial cavities (Milillo et al., 2019). As a consequence, while advances in our ability to model grounding lines have been made, there is much that is needed to be done before we can model their function and change in the detail necessary for reliable forecasting.

5.2.2 Adaptive grids

The first Antarctic ice sheet models ran using a regular sized (fixed resolution) grid with a cell width of 50 km (Budd and Smith, 1982). While slow flow at the centre of the ice sheet can be approximated reasonably well by such a grid – i.e. averaging little change over long distances – as gradients in speed, surface slope, basal conditions and ice dynamics increase coarse grids are unable to simulate realistic ice flow. One solution is to decrease the cell width uniformly across the ice sheet but with a downside of significantly increasing computer running time. By the early 1990s, the Antarctic ice sheet was being modelled with a cell width of 20 km (Huybrechts, 1990), but while this increased resolution led to more plausible ice sheets, the same limitations existed across grounding zones and shear margins. This problem was compounded by the discovery that ice streams are fed by tributaries that stretch well into the ice sheet interior (Bamber et al., 2000), meaning that slow interior ice flow averaged across large distances results in missing key ice stream processes. The numerical solution is to design an approach where grid-cell size can 'adapt' in recognition of where gradients are steep – such as at grounding zones and ice-stream shear margins – allowing ice processes to be accounted for appropriately, while minimising computer running time. This is non-trivial compared to the numerical treatment of a fixed grid, however.

It is fairly straight forward to consider a fixed 'nested' grid of varying cell widths, where resolution is applied to the ice-sheet regions that require it. The more challenging issue is how to allow the grid to 'adapt' as glaciological circumstances change – such as grounding line or shear margin migration.

The first person to consider adaptive grids in ice sheet models was Andrew Starr – a PhD student from Aberystwyth University in the UK (Starr, 2001). He used a 'quad-tree' grid approach as it led to simple modification of the grid. He based his work around a simple shallow-ice approximation model, using rule-based adaptation criteria. Studies included using the EISMINT (European Ice Sheet Modelling Initiative) inter-comparison standards (Huybrechts et al., 1996) to ensure the adaptation was performing appropriately and then application of the validated model to the case of the Loch Lomond Stadial ice sheet of Scotland (Starr, 2001). The adaptive grid model delivered similar accuracy as the fixed grid model but with reductions in computational load of between 40% and 90%. Although this work demonstrated the utility of adaptive grids in glaciology, it only considered the numerical treatment of the adaptation and not how to apply specific glaciological processes associated with the change in resolution. Hence, while the model ran faster than a fixed-grid model, it was not necessarily any better in terms of glaciological processes.

The next step was to incorporate improved ice flow processes into numerical ice sheet models, using adaptive grids to determine where to employ them. The optimal outcome is a model that runs efficiently, applies processes where needed and avoids unnecessary calculations. Cornford et al. (2013) produced the first such model, BISICLES (Berkeley Ice Sheet Initiative for Climate at Extreme Scales), and applied it to the West Antarctic ice sheet. In their model, a 'block-structured finite volume method with adaptive mesh refinement' was used to apply enhanced resolution at the grounding line within a three-dimensional ice sheet model. As in Starr (2001), the model was compared against fixed-grid alternatives revealing a similar overall function but with reduced computation time. However, the major advantage was that grounding line physics was included, and with adaptation of the grid around the grounding line, its migration was able to be modelled appropriately and in relatively small increments, rather than in jumps dictated by a fixed grid size.

5.2.3 Parallel ice sheet model – PISM

Adding functionality to ice sheet models such as three-dimensional flow, longitudinal stresses, thermodynamics and grounding line physics adds to computational time. This is a major issue if ice sheet evolution is being investigated over long time periods, such as glacial cycles. In addition to adaptive grids, a further improvement in numerical methods is exemplified by the Parallel Ice Sheet Model (PISM). This model uses an approach whereby the 'shallow ice approximation', which calculates the flow of ice frozen to the bed well, and the 'shallow shelf approximation', which determines flow driven by longitudinal stresses such as in floating ice shelves, are calculated across the ice sheet – meaning a realistic ice flow field can be derived without boundaries between flow systems (Winkelmann et al., 2011).

Since it was launched, PISM has been used in over 100 published studies of both the Greenland and Antarctic ice sheets in a variety of settings from deep geological time to predictions of future change. One recent application of PISM was to model Antarctica over several glacial cycles (Albrecht et al., 2020a), to determine the influence of boundary conditions and parameter choices in model output – the issue being that modelling is being significantly hampered by a lack of reliable input data in some key areas that make the use of sophisticated ice sheet models unsuitable.

5.2.4 Coupled models

Having established more sophisticated numerical methods in ice sheet models, a final improvement has been in the coupling of the ice sheet system to other models (atmosphere, ocean and solid earth). Most numerical models now couple ice sheet and ice shelf components to achieve realistic ice dynamic simulations across the grounding zone (e.g., Bueler and Brown, 2009; Huybrechts, 1990; Martin et al., 2019; Pollard and DeConto, 2009;

Sato and Greve, 2012). However, major advances have also been made in terms of coupling either models, or model outputs, from atmosphere–ocean general circulation models (GCMs) into ice sheet models run over long time periods. Pollard and DeConto (2009), for example, did this by forming a library of GCM solutions for a variety of ice sheet scenarios and were able to then select the most appropriate to drive ice sheet change through the model run. While this is clearly a simplification, the outcome was that the ice sheet model could be accompanied for the last 5 million years by relatively detailed inputs of air temperature and precipitation.

In the last few years, other components of the earth system have been coupled to ice sheet models, such as interactions with the solid earth and with spatially varying sea level (Gomez et al., 2010, 2015, 2018), improving the robustness of the simulations. However, full integrated coupling of the entire Earth system has yet to be achieved in a manner that allows reliable prediction of future ice sheet change over centennial or longer timescales (Siegert et al., 2020). Parameterisation remains a more appropriate choice in many circumstances, to avoid computational time burdens and unrealistic scenarios (where attempts to incorporate basal processes lead to ice-sheet outcomes that are unlikely).

5.3 Model input – bed data

Regardless of their sophistication, and improvements in their capabilities, numerical ice-sheet models require reliable and sufficiently detailed boundary conditions and input data in order to yield plausible results. The most essential of these is the subglacial topography of Antarctica, which has been the subject of substantial research in the last decade. Bed topography is a major control on ice dynamics, and so unless it is known well ice-sheet models run the risk of significant mistakes regarding ice-sheet flow and change.

At the time of the first Antarctic Climate Evolution volume (Florindo and Siegert, 2008), the most complete assembly of bed data was known as Bedmap (Lythe and the BEDMAP Consortium, 2000) – the first compilation of bed data since the Antarctic Geophysical Folio in the 1980s (Drewry, 1983). While Bedmap was an advance on the 'Folio', and contained detailed bed information from several key places such as the Siple Coast in West Antarctica, it still left several large (100s km wide) regions without any bed data to constrain the subglacial landscape.

In the last two decades, a number of teams of glaciologists have been acquiring new data from, primarily, airborne radio-echo sounding missions. The data-free regions in Bedmap have been a major focus of geophysical exploration, allowing discovery of uncharted mountains and lowlands, lakes and sedimentary basins – each of which would be influential on ice flow and so critical as modelling boundary conditions. For example, the Chinese Antarctic programme and the AGAP (Antarctica's GAmburtsev Province) programme mapped the enigmatic Gamburtsev Subglacial Mountains (Rose et al., 2013; Sun et al., 2009), revealing upland glacial landforms that show it as the birthplace of the modern Antarctic ice sheet. In addition, the ICECAP (Ice and Crustal Evolution of the Central Antarctic Plate) collaborative undertook several seasons of fieldwork to characterise the landscape of the Aurora and Wilkes subglacial basins (Wright et al., 2012), revealing how water could run from the ice sheet interior to its margin, and discovering ice-cut fjords deep within the ice sheet interior as evidence of a former, smaller, dynamic ice sheet (Young et al., 2011).

These, and many other smaller datasets, led to an updated compilation of bed topography, named Bedmap2 (Fretwell et al., 2013). Indeed so much new data was included in Bedmap2 that the difference between it and Bedmap was far greater than between Bedmap and the 'Folio'. This was a major boost to ice-sheet modelling, and hundreds of projects and papers have subsequently taken advantage of it. However, several data issues remain in Bedmap2, in three forms. The first is substantial geographical regions devoid of basal measurements, including upstream of Recovery Ice Stream, Princess Elizabeth Land and South Pole. The second is data-free zones between existing geophysical transects, some of which are spaced 50 km apart. The third is a reliance on 1970s analogue data, recorded every ~2 km with navigation inaccuracies of up to 5 km.

On the first issue, since 2013 there have been several new airborne geophysical surveys targeted at filling the remaining data gaps (e.g., Paxman et al., 2019a,b) for the vicinity around South Pole; Humbert et al. (2018) for the Recovery system; and Cui et al. (2020) under the second phase of ICECAP for Princess Elizabeth Land. On the second issue, data-free zones between transects is a continuing problem. To create a bed surface between irregular sets of data, an interpolation procedure is needed. Bedmap2 and many other bed surface products use 'krigging'. However, more sophisticated approaches are starting to be used, including 'mass conservation' techniques, where the bed is determined by considering the flux of ice necessary to yield measured surface velocities. Morlighem et al. (2017, 2019) used this approach for the most up-to-date subglacial bed surfaces in Greenland and Antarctica; and Jeofry et al. (2018b) used it for the Weddell Sea sector in West Antarctica. In addition, Graham et al. (2017) calculated a 'synthetic' bed surface at 100 m resolution by utilising roughness measurements within Bedmap and new radio-echo sounding (RES) data and assimilating the results into Bedmap2; the result is a bed product that contains morphological information in fine detail as well as large-scale topography. The third issue is being addressed by the recent digitisation of historical analogue records (Schroeder et al., 2019), allowing along-track data between the officially logged recordings to be introduced, navigation to be improved at cross-overs with new data and, importantly, a times series of >40 years in RES measurements to be established for ~40% of the continent. As a consequence of these recent improvements, a new international project – Bedmap3 – is being set up by the Scientific Committee on Antarctic Research (SCAR) to provide the first complete observation-based bed surface of subglacial Antarctica.

Interestingly, the geophysical data that have been acquired to form the Bedmap products allow closer inspection of subglacial morphology, which has in places revealed planation surfaces that are likely relics from preglacial times (Rose et al., 2015). Further, using Bedmap2, new data and sedimentary back-filling, it is possible to provide assessments of former Antarctic landscapes that are crucial to our ability to model ancient ice-sheet evolution (as we can be certain that the modern bed topography reduces in relevance with past time). Sugden and Jamieson (2018) provide an overview of the preglacial landscape of Antarctica, while Paxman et al. (2019a,b) have revealed the role of lithospheric flexure and erosion in determining landscape evolution; hence, it is now becoming possible to run numerical ice-sheet models over beds more representative of ancient periods than the Bedmap products.

Naturally, ice sheet models require more inputs than merely bed topography, and a good example of how they can be set up to model past ice sheets is provided by Albrecht et al. (2020b) for their assessment of Antarctica at the LGM using PISM.

5.4 Advances in knowledge of bed processes

Most ice sheet models are validated by comparing outputs against the modern Antarctic ice sheet. Commonly, model initialisation is optimised in such a way as to yield the best match between simulated and satellite-observed surface ice velocities, either by modifying basal drag or other ice flow parameters. While ice sheet models are capable of calculating basal processes (melting, sliding, for example) to dictate the flow of ice, such models are rarely capable of mimicking the real flow. Hence, basal conditions (such as the rheology and saturation state of sediments, or the plasticity of substrate deformation) often need to be parameterised. This approach is valid if the modern ice sheet is being assessed, but where the ice sheet evolves significantly in modelling experiments, the approach is less appropriate.

Geophysical data used to obtain ice thickness and bed topography often contain information that can resolve basal conditions and processes, but these are rarely assessed in numerical modelling investigations (Jeofry et al., 2019). The problem is twofold; first the modelling community often fails to see the benefit of geophysical data in refining basal processes and, second, the geophysical community often fails to provide the information needed to improve models. Furthermore, it is not always clear how local-scale or point geophysical interpretations can be meaningfully integrated into much larger-scale (for example, whole continent) ice sheet simulations.

In the last 10 years, there have been several key papers that have advanced our knowledge of bed processes in Antarctica, both now and in the past, and such information should be assimilated into numerical modelling procedures if they are to advance to become process-oriented, capable of predicting glaciological change. The most obvious areas in which ice sheet modelling can improve are in its consideration of basal hydrology and bed conditions, and their combined role in modulating the flow of ice above. Here there have been some notable geophysical advances.

Glacier hydrology has been an active area in glaciology for over 50 years (e.g., Nye, 1969). We know that basal water lubricates the bed, especially in the presence of weak unconsolidated sediments, and that fast flow within ice streams is controlled by that water (among other things, such as topography and geology). Clearly, as the ice sheet changes, so too does the subglacial hydrological regime. Hence process-based ice sheet models must advance by learning, from geophysical measurements, of the basal environment. To that end, some ice sheet models now include basal hydrological components. For example, in PISM, basal resistance evolves from an initialisation state depending on the availability and transport of subglacial water (Bueler and Brown, 2009; Bueler and van Pelt, 2015). This time-evolving scheme allows ice streams to switch on and off entirely naturally within the model, without specific prescription of where or when rapid sliding should occur (Golledge et al., 2015; van Pelt and Oerlemans, 2012).

Across Antarctica, the last decade has seen significant ideas for how water flows beneath the ice. This is critical in glaciology, as work on valley glaciers over many years has shown how specific hydraulics affect ice flow. For Thwaites Glacier in West Antarctica, which is a priority for investigation given it is losing mass and susceptible to MISI, Schroeder et al. (2013) used a novel data processing procedure, involving the 'angular distribution of energy in radar bed echoes' to decipher precisely where basal water exists. They used this technique to reveal basal canals within which water can flow and showed that different hydrological systems exist, which are related to gross basal topography. In one location, the canals are distributed and thus conducive to high water pressures, ponding and reduced basal friction, but across a bedrock ridge this transitions to a system of more concentrated well-organised channels in which water flow to the margin is more efficient. This pioneering study was supported by Le Brocq et al. (2013), who showed how channels carved upwards into floating ice shelves at the grounding line were caused by channelised flow of basal water upstream. Later, similar channels were explained by subglacial eskers (Drews et al., 2017) upstream of the Roi Baudouin Ice Shelf in East Antarctica, and water flow alongside large flow-parallel bedrock landforms in Foundation Ice Stream in West Antarctica (Jeofry et al., 2018a). Thus, there is emerging but convincing geophysical evidence for widespread channelised flow of basal water from several sites in Antarctica. Uptake of this knowledge within ice sheet models, and 'training them' to match the data, is possible in coming years. We know this would be useful, as measured discharges from subglacial lakes have been shown to influence the flow speed of outlet glaciers (Stearns et al., 2008). Water flow into and out of hydraulically 'active' subglacial lakes may be more complex than thought initially (e.g., Wingham et al., 2006), but geophysical evidence is now able to measure the topographic conditions that bound them (Siegert et al., 2014), offering the possibility of being able to model them. Basal water channels in ancient glaciological settings (i.e. not possible to form them in today's ice sheet configuration) have also been identified (Rose et al., 2014), though the timing of when they formed is not yet known.

A new frontier in subglacial hydrology lies beneath the ice–water interface as groundwater (Siegert et al., 2018), especially within large sedimentary basins (Siegert et al., 2016), and has the potential to contribute significantly to the water supply to ice stream beds (Christoffersen et al., 2014) and the distribution and exchange of basal heat (Gooch et al., 2016) – both having the potential to significantly influence the flow of ice above. While our ability to model such systems is not yet possible due to a lack of data, several plans exist to identify, measure and assess Antarctic groundwater making it a potential new element for ice sheet models of the future.

5.5 Model intercomparison

For many years in the early history of ice-sheet modelling, a number of models were built and run independently. While these used the same glaciological knowledge, differences in numerical schemes, resolution and process-prioritisation led to differences in past ice-sheet reconstructions depending on the model used. To help resolve this, formal model intercomparison exercises were established, to show where consistencies exist and how differences occur. The first, as reported in Siegert and Payne (2001), was the EISMINT (European Ice Sheet Modelling Intercomparison) exercise. In the last decade, new intercomparison exercises have accompanied enhanced numerical modelling sophistication, focusing on key components of the ice-sheet system. For example, Pattyn et al. (2012) formed the MISMIP (Marine Ice Sheet Model Intercomparison) project, to investigate how different models can reproduce marine ice sheet dynamics and change. This is especially relevant to knowledge of how the WAIS has changed in the past, and will change in the future, and underlined the importance of the appropriate adoption of grounding line physics. More recently, ISMIP6 has allowed ice-sheet models to be intercompared with respect to IPCC scenarios, in order to improve estimates of future ice sheet contributions to sea level. Importantly, this allows ice-sheet models that are coupled into Earth System models, as well as standalone models, to be inter-compared and assessed. The first output from this exercise identified how models of Antarctica should be set up and initialised (Seroussi et al., 2019), with subsequent work presenting the scenario-based projections (Edwards et al., 2021; Seroussi et al., 2020). Additional projects have been established to inter-compare models from specific periods in the past, such as the Pliocene (Haywood et al., 2013), or to investigate the way in which particular processes, such as MISI, are captured (Sun et al., 2020).

5.6 Brief case studies

Now that we have covered how the last decade has seen advances in ice sheet modelling, input data, geophysical evidence and model intercomparison exercises, we can examine how models have been used to reconstruct former Antarctic ice sheets. The discussion will be brief, as it will be provided in detail in other chapters, but key outcomes from such numerical modelling exercises are worth stating for the LGM (Siegert et al., 2021), the mid-Pleistocene (Wilson et al., 2021) and the Pliocene (Levy et al., 2021).

For the LGM, Kusahara et al. (2015) produced a simulation of the LGM Antarctic climate system to understand relationships between the atmosphere, ocean, sea-ice and ice shelves. They showed that ice shelf cavities were shaped very differently from today – being much larger for one thing, but also with thicker floating ice in many places. Because the ice shelf margins were closer to the continental shelf edge, warmth from Circumpolar Deep Water could readily flow into the cavity, causing far more ice-shelf melting than today. Further, the modelling revealed that the winter sea ice limit was around 7° further north than today, and that in summer the Atlantic sector retained sea ice cover with limited amounts elsewhere. Golledge et al. (2012) produced a numerical model of the LGM ice sheet, focusing on how ocean heat led to its subsequent retreat. They examined whether ocean heat, or eustatic sea level rise, could be responsible for the retreat measured in the geological record. They found that while these drivers led to some change to the ice margin, once this had been triggered an impact was felt on ice sheet dynamics. Modification to the flow of ice, and in particular increased flow in ice streams, led to drawdown of ice from the ice-sheet interior and, hence, further ice loss to the ocean. They suggest that this finding has obvious consequences for today's ice sheet that is presently losing mass in several places due to ocean warming (Shepherd et al., 2018).

The last interglacial has also been the subject of recent ice-sheet modelling, with several studies concluding that the Antarctic Ice Sheet was smaller than today at that time, probably leading to a sea-level contribution of 2.5–3 m (Clark et al., 2020; DeConto and Pollard, 2016; Sutter et al., 2020).

An interesting use of PISM was made by Sutter et al. (2019) who focused on the mid-Pleistocene transition (1.2–0.9 Ma), which saw a change from glacial cycles paced at a frequency of ~100 ka to ~40 ka. Their simulations covered the last 2 Ma and revealed how the ice sheet form has changed over this time, especially in West Antarctica as this period may have seen it transition to a large and almost entirely marine-based ice sheet. Importantly, the model also revealed sites where little glaciological change had occurred, predominantly in East Antarctica, which has relevance to the preservation of very old ice at the base of the modern ice sheet that can potentially be sampled by coring. The conditions for such ice are quite stringent – it must be thick enough to contain a record, but thin enough so as not to melt at the base. This requirement means that there are relatively few sites in East Antarctica that appear suitable, but some exist around major ice domes.

Further back in time, the Pliocene has also been the subject of considerable ice-sheet modelling activities, not least because the period was influenced by an atmospheric CO_2 concentration of over 400 ppm, which is similar to today's. Thus, the period likely represents an example of how the Antarctic ice sheet may react ultimately to the warming that occurs under such a scenario. Modelling can help reveal the end-state conditions – the maximum effect of the warming on the ice sheet – and the processes responsible. If the rate of climate forcing is also known, simulations can also help constrain rates of ice sheet loss. In terms of forcing, it is essential to recognise that the Antarctic continent was an integral part of Pliocene Earth, not a passive component influenced by external forcing (Dolan et al., 2018; Haywood et al., 2013). One important ice sheet modelling investigation was undertaken by Pollard and DeConto (2009), which revealed that the Pliocene interglacial ice sheet was far smaller than today's; possibly contributing around 5–7 m to a global sea level rise of approximately 20 m, which is consistent with geological records of major eustatic change at this time (Dumitru et al., 2019; Grant et al., 2019; Raymo et al., 2011). More recently, Golledge et al. (2017) specifically focused on a Pliocene interglaciation at 4.23 Ma, when Southern Hemisphere insolation was at its maximum. They used a one-way coupled ice-sheet/climate model to produce a range of ice-sheet scenarios and concluded that Antarctica contributed ~8.6 m to sea level. Given that the WAIS presently contains enough ice to raise sea level by ~3.5 m, sea level contributions above this point to major ice loss from East Antarctica and, specifically through modelling, sedimentary geology (e.g., Aitken et al., 2016; Cook et al., 2013) and glaciological theory from within its deep marine settings such as the Wilkes and Aurora basins. One controversy in recent years has focused on the rates of past ice-sheet change experienced in Antarctica. The motivation for this is to understand how quickly the present ice sheet can respond to greenhouse-gas driven warming. In a well-document study, Pollard et al. (2015) put the case for Antarctic loss driven by 'hydrofracturing and cliff failure'; the marine ice cliff instability (MICI) hypothesis. This, they argue, is possible if ice-shelves disintegrate rapidly under extreme ocean and atmospheric heating and the resulting hydrofacturing of floating ice, which then produces vertical ice cliffs at grounding lines. If the thickness of the vertical wall of ice is over ~100 m above the ocean surface, it was suggested to be unable to support itself mechanically and, thus, collapses. Pollard et al. (2015) suggest through ice-sheet modelling that this was a mechanism responsible for ice loss in the Pliocene and, potentially, for the future Antarctic ice sheet. This is a controversial proposal because the rate of change is far greater (10s–100s vs 1000s years) than shown by other ice sheet models. DeConto and Pollard (2016) argue that the rate of future sea level rise is significantly underpredicted if this process is not taken into account. However, this theory has been rebutted recently by Edwards et al. (2019) who used a statistical reassessment of the original DeConto and Pollard (2016) simulations to show that Pliocene variations were as possible without invoking cliff failure as they were with the mechanism switched on. Furthermore, recently published process-based (rather than parameterised) models have shown that ice cliffs of the height necessary to trigger cliff collapse in the Pollard et al. (2015) scheme are unlikely to develop unless the buttressing ice shelf can be entirely removed in less than an hour (Clerc et al., 2019). Even though MICI is thought (by some) to be unlikely to occur in this century, if it did, there would be significantly higher rates of rise in the sea level than what we have experienced in the last few decades (Siegert et al., 2020) and predicted by models that do not account for it. Hence, under a precautionary principle it should not be discounted. Knowledge of the conditions that may permit MICI is a major area of research in glaciology, and a priority for the INSTANT programme is to evaluate evidence for its influence on past ice sheet changes (Colleoni et al., 2021).

5.7 Future work

While ice sheet modelling, and its use, has developed considerably in the last decade, several areas where improvements could be made remain (see Siegert et al., 2020). Colleoni et al. (2018) point to where attention is needed at the interfaces between well-studied systems such as at ice—ocean and ice—bed boundaries. Such locations are often very challenging to access physically; hence, we rely on remotely-sensed information to guide models. However, in critical locations such as grounding lines and ice shelf cavities, we are only now beginning to understand the complexities that exist there and how simplistic our approach to modelling them has been (Jeofry et al., 2018a; Siegert et al., 2020). Colleoni et al. (2018) argue that greater attention is needed by the modelling community in these places, with targeted geophysical and oceanographic missions to acquire remote and in situ data, if we are to advance ice-sheet modelling still further to meet the ultimate ambition of reliably predicting ice sheet change and reducing uncertainty in future sea-level estimates.

Another area where more work is needed is in the use of geophysical data by numerical ice-sheet models. In other modelling communities, data assimilation and machine learning utilise geophysical measurements to train and improve model performance. An obvious example is in atmospheric models for weather predictions. While this is now beginning in ice-sheet modelling (for example, the ISMIP6 Antarctic Ice Sheet assessment incorporates six data assimilation ice sheet models and 10 that use more traditional spin-up approaches – matching geometry and velocities with remote sensing measurements), there remains much more to be done. Issues with ISMIP6 include the use of ensemble members that cannot simulate the present-day ice sheet, and the need for more consistency and understanding of the important rate limiting processes.

The most obvious geophysical measurement required by ice-sheet models is the bed topography from ice-penetrating radar (Fretwell et al., 2013). However, such data contain a wealth of information about conditions at the ice sheet beds that are rarely utilised, if at all, in modelling schemes. Jeofry et al. (2019) analysed the performance of the BISICLES ice sheet model, which was optimised for surface flow by reducing basal drag, against radar evidence for basal water, sediment and morphology. The results revealed that while the model performs reasonably well in a gross sense, several discrepancies were revealed. Obviously, if the model optimised for present-day conditions cannot replicate actual measurements as well as it might, how can we have confidence in models when they predict ice sheet evolution? Hence, assimilating geophysical evidence, such as hydrological conditions in ice-sheet models, appears to be an obvious next step.

Finally, and most obviously, Bedmap2 – which is used as a basic boundary condition for most Antarctic ice sheet models – contains several large data gaps, and areas where data coverage is relatively sparse. Graham et al. (2017) showed that getting basal topography right in ice sheet models was critical to their performance and provided a 'synthetic' bed elevation grid for models to be tested. Morlighem et al. (2019), using the 'mass conservation' technique, have recently calculated the most accurate depiction of Antarctic bed elevation (Bedmachine Antarctica). The technique works by measuring ice surface velocities from satellite data and, assuming the vertical profile of ice flow, a bed profile that allows ice flux along flowlines to be preserved is established. The technique works best where velocities are greatest, as the bulk of ice flow is from basal processes, and allows bed morphology to be determined in the key area of ice sheet grounding lines, where they have revealed unmeasured deep troughs bounding the ice sheet. However, Bedmachine Antarctica is not a data base and, as discussed above, there may be additional benefits to the modelling community to fully, systematically and accurately measure the base of the Antarctic ice sheet. Plans are in place to update Bedmap2 with new data that cover most of its data gaps (i.e., Bedmap3), and this should appear in the coming years.

References

Aitken, A.R.A., Roberts, J.L., Van Ommen, T., Young, D.A., Golledge, N.R., Greenbaum, J.S., et al., 2016. Repeated large-scale retreat and advance of Totten glacier indicated by inland bed erosion. Nature 533, 385–389. Available from: https://doi.org/10.1038/nature17447.

Albrecht, T., Winkelmann, R., Levermann, A., 2020a. Glacial-cycle simulations of the Antarctic Ice Sheet with the Parallel Ice Sheet Model (PISM) – Part 1: Boundary conditions and climatic forcing. The Cryosphere 14, 599–632. Available from: https://doi.org/10.5194/tc-14-599-2020.

Albrecht, T., Winkelmann, R., Levermann, A., 2020b. Glacial-cycle simulations of the Antarctic Ice Sheet with the Parallel Ice Sheet Model (PISM) – Part 2: Parameter ensemble analysis. The Cryosphere 14, 633–656.

Bamber, J.L., Vaughan, D.G., Joughin, I., 2000. Widespread complex flow in the interior of the Antarctic Ice Sheet. Science (New York, N.Y.) 287, 1248–1250.

Budd, W.F., Smith, I.N., 1982. Large-scale numerical modelling of the Antarctic Ice Sheet. Annals of Glaciology 3, 42–49.

Bueler, E., Brown, J., 2009. Shallow shelf approximation as a "sliding law" in a thermomechanically coupled icesheet model. Journal of Geophysical Research 114, F03008. Available from: https://doi.org/10.1029/2008JF001179.

Bueler, E., van Pelt, W., 2015. Mass-conserving subglacial hydrology in the Parallel Ice Sheet Model version 0.6. Geoscientific Model Development 8, 1613–1635. Available from: https://doi.org/10.5194/gmd-8-1613-2015.

Christoffersen, P., Bougamont, M., Carter, S.P., Fricker, H.A., Tulaczyk, S., 2014. Significant groundwater contribution to Antarctic ice streams hydrologic budget. Geophysical Research Letters 41, 2003–2010. Available from: https://doi.org/10.1002/2014GL059250.

Clark, P.U., He, F., Golledge, N.R., et al., 2020. Oceanic forcing of penultimate deglacial and last interglacial sea-level rise. Nature 577, 660–664. Available from: https://doi.org/10.1038/s41586-020-1931-7.

Clerc, F., Minchew, B.M., Behn, M.D., 2019. Marine Ice Cliff Instability Mitigated by Slow Removal of Ice Shelves. Geophysical Research Letters 46, 12,108–12,116. Available from: https://doi.org/10.1029/2019GL084183.

Colleoni, F., De Santis, L., Siddoway, C.S., Bergamasco, A., Golledge, N., Lohmann, G., et al., 2018. Spatio-temporal variability of processes across Antarctic ice-bed-ocean interfaces. Nature Communications 8, 2289. Available from: https://doi.org/10.1038/s41467-018-04583-0.

Colleoni, F., et al., 2021. Past Antarctic ice sheet dynamics (PAIS) and implications for future sea-level change. In: Florindo, F., et al., (Eds.), Antarctic Climate Evolution, second ed. Elsevier (this volume).

Cook, C.P., et al., 2013. Dynamic behaviour of the East Antarctic Ice Sheet during Pliocene warmth. Nature Geoscience 6, 765–769.

Cornford, S.L., et al., 2013. Adaptive mesh, finite volume modeling of marine ice sheets. Journal of Computational Physics 232, 529–549.

Cui, X., Jeofry, H., Greenbaum, J.S., Guo, J., Li, L., Lindzey, L.E., Habbal, F.A., Wei, W., Young, D.A., Ross, N., Morlighem, M., Jong, L.M., Roberts, J.L., Blankenship, D.D., Bo, S., Siegert, M.J., 2020. Bed topography of Princess Elizabeth Land in East Antarctica. Earth System Science Data 12, 2765–2774. Available from: https://doi.org/10.5194/essd-12-2765-2020.

DeConto, R.M., Pollard, D., 2016. Contribution of Antarctica to past and future sea-level rise. Nature 531, 591–597. Available from: https://doi.org/10.1038/nature17145.

Dolan, A.M., de Boer, B., Bernales, J., Hill, D.J., Haywood, A.M., 2018. High climate model dependency of Pliocene Antarctic Ice-Sheet predictions. Nature Communications 9, 2799.

Drewry, D.J., 1983. Antarctica: Glaciological and Geophysical Folio. Scott Polar Research Institute, University of Cambridge.

Drews, R., Pattyn, F., Hewitt, I.J., Ng, F.S.L., Berger, S., Matsuoka, K., et al., 2017. Actively evolving subglacial conduits and eskers initiate ice shelf channels at an Antarctic grounding line. Nature Communications 8, 15228.

Dumitru, O.A., Austermann, J., Polyak, V.J., Fornós, J.J., Asmerom, Y., Ginés, J., et al., 2019. Constraints on global mean sea level during Pliocene warmth. Nature 574, 233–236. Available from: https://doi.org/10.1038/s41586-019-1543-2.

Edwards, T.L., Brandon, M.A., Durand, G., et al., 2019. Revisiting Antarctic ice loss due to marine ice-cliff instability. Nature 566, 58–64. Available from: https://doi.org/10.1038/s41586-019-0901-4.

Edwards, T.L., Nowicki, S., Marzeion, B., et al., 2021. Projected land ice contributions to twenty-first-century sea level rise. Nature 593, 74–82. Available from: https://doi.org/10.1038/s41586-021-03302-y.

Feldmann, J., Albrecht, T., Khroulev, C., Pattyn, F., Levermann, A., 2014. Resolution-dependent performance of grounding line motion in a shallow model compared with a full-Stokes model according to the MISMIP3d intercomparison. Journal of Glaciology 60, 353–360. Available from: https://doi.org/10.3189/2014JoG13J093.

Florindo, F., Siegert, M.J. (Eds.). Antarctic Climate Evolution. Developments in Earth & Environmental Sciences, vol. 8. Elsevier, Amsterdam. 606 pp. ISBN: 9780444528476 (2008).

Fretwell, P., et al., 2013. Bedmap2: improved ice bed, surface and thickness datasets for Antarctica. The Cryosphere 7, 375–393. Available from: https://doi.org/10.5194/tc-7-375-2013. Available from: http://www.the-cryosphere.net/7/375/2013/.

Golledge, N.R., Fogwill, C.J., Mackintosh, A.N., Buckley, K.M., 2012. Dynamics of the last glacial maximum Antarctic ice-sheet and its response to ocean forcing. Proceedings of the National Academy of Sciences 109, 16052–16056. Available from: https://doi.org/10.1073/pnas.1205385109.

Golledge, N.R., Kowalewski, D.E., Naish, T.R., Levy, R.H., Fogwill, C.J., Gasson, E.G.W., 2015. The multi-millennial Antarctic commitment to future sea-level rise. Nature 526, 421–425. Available from: https://doi.org/10.1038/nature15706.

Golledge, N.R., et al., 2017. Antarctic climate and ice-sheet configuration during the early Pliocene interglacial at 4.23 Ma. Climate of the Past 13, 959–975.

Gomez, N., Mitrovica, J.X., Huybers, P., Clark, P.U., 2010. Sea level as a stabilizing factor for marine-ice-sheet grounding lines. Nature Geoscience 3, 850–853. Available from: https://doi.org/10.1038/ngeo1012.

Gomez, N., Pollard, D., Holland, D., 2015. Sea-level feedback lowers projections of future Antarctic Ice-Sheet mass loss. Nature Communications 6, 8798. Available from: https://doi.org/10.1038/ncomms9798.

Gomez, N., Latychev, K., Pollard, D., 2018. A Coupled Ice Sheet–Sea Level Model Incorporating 3D Earth Structure: Variations in Antarctica during the Last Deglacial Retreat. Journal of Climate 31, 4041–4054. Available from: https://doi.org/10.1175/JCLI-D-17-0352.1.

Gooch, B.T., Young, D.A., Blankenship, D.D., 2016. Potential groundwater and heterogeneous heat source contributions to ice sheet dynamics in critical submarine basins of East Antarctica. Geochemistry, Geophysics, Geosystems 17, 395–409. Available from: https://doi.org/10.1002/2015GC006117.

Graham, F.S., Roberts, J.L., Galton-Fenzi, B., Young, D., Blankenship, D., Siegert, M.J., 2017. A high-resolution synthetic bed elevation grid of the Antarctic continent. Earth System Science Data 9, 267–279. Available from: https://doi.org/10.5194/essd-9-267-2017. Available from: http://www.earth-syst-sci-data.net/9/267/2017/.

Grant, G.R., Naish, T.R., Dunbar, G.B., Stocchi, P., Kominz, A.A., Kamp, P.J.J., et al., 2019. The amplitude and origin of sea-level variability during the Pliocene epoch. Nature 574, 237–3240. Available from: https://doi.org/10.1038/s41586-019-1619-z.

Haywood, A.M., et al., 2013. Large-scale features of Pliocene climate: Results from the Pliocene Model Intercomparison Project. Climate of the Past 9, 191–209.

Humbert, A., Steinhage, D., Helm, V., Beyer, S., Kleiner, T., 2018. Missing evidence of widespread subglacial lakes at Recovery Glacier, Antarctica. Journal of Geophysical Research: Earth Surface 123, 2802–2826. Available from: https://doi.org/10.1029/2017JF004591.

Huybrechts, P., 1990. A 3-D model for the Antarctic Ice Sheet: A sensitivity study on the glacial-interglacial contrast. Climate Dynamics 5, 79–92.

Huybrechts, P., Payne, T., et al., 1996. The EISMINT benchmarks for testing ice sheet models. Annals of Glaciology 23, 1–12.

Jeofry, H., Ross, N., Le Brocq, A., Graham, A., Li, J., Gogineni, P., Morlighem, M., Jordan, T., Siegert, M.J., 2018a. Hard rock landforms generate 130 km ice shelf channels through water focusing in basal corrugations. Nature Communications 9, 4576. Available from: https://doi.org/10.1038/s41467-018-06679-z.

Jeofry, H., Ross, N., Corr, H.F.J., Li, J., Morlighem, M., Gogineni, P., et al., 2018b. A new digital elevation model for the Weddell Sea sector of the West Antarctic Ice Sheet. Earth System Science Data 10, 711–725. Available from: https://doi.org/10.5194/essd-10-711-2018.

Jeofry, H., Ross, N., Siegert, M.J., 2019. Comparing numerical ice-sheet model output with radio-echo sounding measurements in the Weddell Sea sector of West Antarctica. Annals of Glaciology 81, 1–10. Available from: https://doi.org/10.1017/aog.2019.39.

Joughin, I., Smith, B.E., Medley, B., 2014. Marine Ice Sheet collapse potentially under way for the Thwaites Glacier Basin, West Antarctica. Science (New York, N.Y.) 244, 735–738. Available from: https://doi.org/10.1126/science.1249055.

Kusahara, K., Sato, T., Oka, A., Obase, T., Greve, R., Abe-Ouchi, A., et al., 2015. Modelling the Antarctic marine cryosphere at the Last Glacial Maximum Kazuya. Annals of Glaciology 56, 425–435. Available from: https://doi.org/10.3189/2015AoG69A792.

Le Brocq, A., Ross, N., Griggs, J., Bingham, R., Corr, H., Ferroccioli, F., Jenkins, A., Jordan, T., Payne, A., Rippin, D., Siegert, M.J., 2013. Ice shelves record the history of channelised flow beneath the Antarctic ice sheet. Nature Geoscience 6, 945–948. Available from: https://doi.org/10.1038/ngeo1977.

Levy, R.H., et al., 2021. Antarctic environmental change and ice sheet evolution through the Miocene to Pliocene – a perspective from the Ross Sea and George V to Wilkes Land Coasts. In: Florindo, F., et al., (Eds.), Antarctic Climate Evolution, second edition Elsevier (this volume).

Lythe, M.B., the BEDMAP Consortium, 2000. BEDMAP – Bed Topography of the Antarctic. 1:10,000,000 map. BAS (Misc.) 9. British Antarctic Survey, Cambridge.

Martin, D.F., Cornford, S.L., Payne, A.J., 2019. Millennial- scalevulnerability of the Antarctic IceSheet to regional ice shelf collapse. Geophysical Research Letters 46, 1467–1475. Available from: https://doi.org/10.1029/2018GL081229.

Mercer, J.H., 1978. West Antarctic ice sheet and CO_2 greenhouse effect: A threat of disaster. Nature 271, 321–325.

Milillo, P., Rignot, E., Rizzoli, P., Scheuchl, B., Bueso-Bello, J., Prats-Iraola, P., 2019. Heterogeneous retreat and ice melt of Thwaites Glacier, West Antarctica. Science Advances 5, eaau3433 . Available from: https://doi.org/10.1126/sciadv.aau3433.

Morlighem, M., Williams, C., Rignot, E., An, L., Bamber, J., Chauche, N., et al., 2017. BedMachine v3: Complete bed topography and ocean bathymetry mapping of Greenland from multi-beam radar sounding combined with mass conservation. Geophysical Research Letters 44. Available from: https://doi.org/10.1002/2017GL074954.

Morlighem, M., Rignot, E., Binder, T., et al., 2019. Deep glacial troughs and stabilizing ridges unveiled beneath the margins of the Antarctic ice sheet. Nature Geoscience. Available from: https://doi.org/10.1038/s41561-019-0510-8.

Nowicki, S.M.J., Payne, T., Larour, E., Seroussi, H., Goelzer, H., Lipscomb, W., Gregory, J., Abe-Ouchi, A., Shepherd, A., 2016. Ice Sheet Model Intercomparison Project (ISMIP6) contribution to CMIP6. Geoscientific Model Development Discussions. Available from: https://doi.org/10.5194/gmd-9-4521-2016.

Nye, J.F., 1969. A calculation on the sliding of ice over a wavy surface using a Newtonian viscous approximation. Proceedings of the Royal Society A 311, 445–467.

Pattyn, F., Schoof, C., Perichon, L., Hindmarsh, R.C.A., Bueler, E., de Fleurian, B., et al., 2012. Results of the Marine Ice Sheet Model Intercomparison project, MISMIP. The Cryosphere 6, 573–588. Available from: https://doi.org/10.5194/tc-6-573-2012.

Paxman, G.J.G., Jamieson, S.S.R., Ferraccioli, F., Jordan, T.A., Bentley, M.J., Ross, N., et al., 2019a. Subglacial geology and geomorphology of the Pensacola-Pole Basin, East Antarctica. Geochemistry, Geophysics, Geosystems 20, 2786–2807. Available from: https://doi.org/10.1029/2018GC008126.

Paxman, G.J.G., Jamieson, S.S.R., Ferraccioli, F., Bentley, M.J., Ross, N., Watts, A.B., et al., 2019b. The role of lithospheric flexure in the landscape evolution of the Wilkes Subglacial Basin and Transantarctic Mountains, East Antarctica. Journal of Geophysical Research: Earth Surface 124, 812–829. Available from: https://doi.org/10.1029/2018JF004705.

Pollard, D., DeConto, R.M., 2009. Modelling West Antarctic ice sheet growth and collapse through the past five million years. Nature 458, 329–332.

Pollard, D., DeConto, R.M., Alley, R.B., 2015. Potential Antarctic Ice Sheet retreat driven by hydrofracturing and ice cliff failure. Earth and Planetary Science Letters 412, 112–121.

Raymo, M.E., Mitrovica, J.X., O'Leary, M.J., DeConto, R.M., Hearty, P.J., 2011. Departures from eustasy in Pliocene sea-level records. Nature Geoscience 4, 328–332.

Rose, K.C., Ferraccioli, F., Jamieson, S.S.R., Bell, R.E., Corr, H., Creyts, T.T., et al., 2013. Early East Antarctic Ice Sheet growth recorded in the landscape of the Gamburtsev Subglacial Mountains. Earth and Planetary Science Letters 375, 1–12. Available from: https://doi.org/10.1016/j.epsl.2013.03.053.

Rose, K.C., Ross, N., Bingham, R.G., Corr, H.F.J., Ferraccioli, F., Jordan, T.A., et al., 2014. A temperate former West Antarctic ice sheet suggested by an extensive zone of bed channels. Geology 42, 971–974. Available from: https://doi.org/10.1130/G35980.1.

Rose, K.C., Ross, N., Jordan, T.A., Bingham, R.G., Corr, H., Ferraccioli, F., et al., 2015. Ancient pre-glacial erosion surfaces preserved beneath the West Antarctic ice sheet. Earth Surface Dynamics 3, 139–152. Available from: https://doi.org/10.5194/esurf-3-139-2015.

Sato, T., Greve, R., 2012. Sensitivity experiments for the Antarctic ice sheet with varied sub-ice-shelf melting rates. Annals of Glaciology 53, 221–228. Available from: https://doi.org/10.3189/2012AoG60A042.

Schoof, C., 2007. Ice sheet grounding line dynamics: Steady states, stability, and hysteresis. Journal of Geophysical Research 112, F03S28. Available from: https://doi.org/10.1029/2006JF000664.

Schroeder, D.M., Blankenship, D.D., Young, D.A., 2013. Water system transition beneath Thwaites Glacier. Proceedings of the National Academy of Sciences 110, 12225–12228. Available from: https://doi.org/10.1073/pnas.1302828110.

Schroeder, D.M., Dowdeswell, J.A., Siegert, M.J., Bingham, R.G., Chu, W., MacKie, E.J., et al., 2019. Multi-decadal observations of the Antarctic Ice Sheet from restored analogue radar records. Proceedings of the National Academy of Sciences . Available from: https://doi.org/10.1073/pnas.1821646116.

Seroussi, H., Nowicki, S., Simon, E., Abe-Ouchi, A., Albrecht, T., Brondex, J., et al., 2019. initMIP-Antarctica: An ice sheet model initialization experiment of ISMIP6. The Cryosphere 13, 1441–1471. Available from: https://doi.org/10.5194/tc-13-1441-2019.

Seroussi, H., Nowicki, S., Payne, A.J., Goelzer, H., Lipscomb, W.H., et al., 2020. ISMIP6 Antarctica: a multi-model ensemble of the Antarctic ice sheet evolution over the 21st century. The Cryosphere 14, 3033–3070. Available from: https://doi.org/10.5194/tc-14-3033-2020.

Shepherd, A., Ivins, E., Rignot, E., et al., 2018. Mass balance of the Antarctic Ice Sheet from 1992 to 2017. Nature 558, 219–222. Available from: https://doi.org/10.1038/s41586-018-0179-y.

Siegert, M.J., 2008a. Numerical modelling of the Antarctic Ice Sheet. In: Florindo, F., Siegert, M.J. (Eds.), Antarctic Climate Evolution. Developments in Earth & Environmental Sciences, vol. 8, 235–256. Available from: https://doi.org/10.1016/S1571-9197(08)00006-2.

Siegert, M.J., 2008b. Antarctic subglacial topography and ice sheet evolution. Earth Surface Processes and Landforms 33, 646–660. Available from: https://doi.org/10.1002/esp.1670.

Siegert, M.J., Kulessa, B., Bougamont, M., Christoffersen, P., Key, K., Andersen, K.R., et al., 2018. Antarctic subglacial groundwater: A concept paper on its measurement and potential influence on ice flow. In: Siegert, M.J., Jamieson, S.S.R., White, D.A. (Eds.), Exploration of Subsurface Antarctica: Uncovering Past Changes and Modern Processes, vol. 461. Geological Society, London, Special Publications, pp. 197–214. Available from: https://doi.org/10.1144/SP461.8.

Siegert, M.J., Hein, A.S., White, D.A., Gore, D.B., De Santis, L., Hillenbrand, C.D., 2021. Antarctic ice sheet changes since the Last Glacial Maximum. In: Florindo, F., et al., (Eds.), Antarctic Climate Evolution, second edition Elsevier (this volume).

Siegert, M.J., Ross, N., Corr, H., Smith, B., Jordan, T., Bingham, R., et al., 2014. Boundary conditions of an active West Antarctic subglacial lake: Implications for storage of water beneath the ice sheet. The Cryosphere 8, 15–24. Available from: https://doi.org/10.5194/tc-8-15-2014.

Siegert, M.J., Ross, N., Li, J., Schroeder, D., Rippin, D., Ashmore, D., et al., 2016. Controls on the onset and flow of Institute Ice Stream, West Antarctica. Annals of Glaciology 57, 19–24. Available from: https://doi.org/10.1017/aog.2016.17.

Siegert, M.J., Alley, R.B., Rignot, E., Englander, J., Corell, R., 2020. 21st Century sea-level rise could exceed IPCC predictions for strong-warming futures. One Earth 3, 691–703. Available from: https://doi.org/10.1016/j.oneear.2020.11.002(2020).

Siegert, M.J., Payne, T., 2001. Validation of ice sheet models. In: Anderson, M.G., Bates, P.D. (Eds.), Model validation in hydrological science. John Wiley and Sons Ltd, Chichester, England, pp. 439–460.

Starr, A., 2001. Numerical Modelling of Ice Sheets Using Adaptive Grids. Department of Computer Science, University of Wales, Aberystwyth. Available from: http://users.aber.ac.uk/aos/thesis_final.pdf.

Stearns, L.A., Smith, B.E., Hamilton, G.S., 2008. Increased flow speed on a large East Antarctic outlet glacier caused by subglacial floods. Nature Geoscience 1. Available from: https://doi.org/10.1038/ngeo356.

Sugden, D.E., Jamieson, S.S.R., 2018. The pre-glacial landscape of Antarctica. Scottish Geographical Journal 134 (3–4), 203–223. Available from: https://doi.org/10.1080/14702541.2018.1535090.

Sun, B., Siegert, M.J., et al., 2009. The Gamburtsev Mountains and the origin and early evolution of the Antarctic Ice Sheet. Nature 459, 690–693. Available from: https://doi.org/10.1038/nature08024.

Sun, S., Pattyn, F., et al., 2020. Antarctic ice sheet response to sudden and sustained ice-shelf collapse (ABUMIP). Journal of Glaciology 66, 891–904. Available from: https://doi.org/10.1017/jog.2020.67.

Sutter, J., Gierz, P., Grosfeld, K., Thoma, M., Lohmann, G., 2016. Ocean temperature thresholds for Last Interglacial West Antarctic Ice Sheet collapse. Geophysical Research Letters 43, 2675–2682. Available from: https://doi.org/10.1002/2016gl067818.

Sutter, J., Eisen, O., Werner, M., Grosfeld, K., Kleiner, T., Fischer, H., 2020. Limited retreat of the Wilkes Basin ice sheet during the Last Interglacial. Geophysical Research Letters 47, Available from: https://doi.org/10.1029/2020GL088131. e2020GL088131.

van Pelt, W.J.J., Oerlemans, J., 2012. Numerical simulations of cyclic behaviour in the Parallel Ice Sheet Model (PISM). Journal of Glaciology 58, 347–360. Available from: https://doi.org/10.3189/2012JoG11J217.

Wilson, D.J., et al., 2021. Pleistocene Antarctic climate variability: ice sheet ocean climate interactions. In: Florindo, F., et al., (Eds.), Antarctic Climate Evolution, second ed. Elsevier (this volume).

Wingham, D.J., Siegert, M.J., Shepherd, A.P., Muir, A.S., 2006. Rapid discharge connects Antarctic subglacial lakes. Nature 440, 1033–1036.

Winkelmann, R., Martin, M.A., Haseloff, M., Albrecht, T., Bueler, E., Khroulev, C., et al., 2011. The Potsdam Parallel Ice Sheet Model (PISM-PIK) – Part 1: Model description. The Cryosphere 5, 715–726. Available from: https://doi.org/10.5194/tc-5-715-2011.

Wright, A.P., Young, D.A., Roberts, J.L., Dowdeswell, J.A., Bamber, J.L., Young, N., et al., 2012. Evidence for a hydrological connection between the ice divide and ice sheet margin in the Aurora Subglacial Basin sector of East Antarctica. Journal of Geophysical Research: Earth Surface 117, F01033. Available from: https://doi.org/10.1029/2011JF002066. 15 pp.

Young, D.A., Wright, A.P., Roberts, J.L., Warner, R.C., Young, N., Greenbaum, J.S., et al., 2011. A dynamic early East Antarctic Ice Sheet suggested by ice covered fjord landscapes. Nature 474, 72–75.

Chapter 6

The Antarctic Continent in Gondwana: a perspective from the Ross Embayment and Potential Research Targets for Future Investigations

Franco Talarico[1,2,*], Claudio Ghezzo[1] and Georg Kleinschmidt[3]
[1]*Department of Physical, Earth and Environmental Sciences, University of Siena, Siena, Italy,* [2]*National Museum for Antarctica, University of Siena, Siena, Italy,* [3]*Institute for Geosciences, University of Frankfurt, Frankfurt, Germany*

6.1 Introduction

The geological record has, for a long time, demonstrated that Antarctica holds a key role in deep-time supercontinent reconstructions. More recently, because of the well-documented relevance of the polar regions' processes in influencing the global changes of both ocean circulation and climate patterns, Antarctica has increasingly shown to be of similar importance in the context of paleoenvironmental and paleoclimatic investigations, particularly in those focused on the Cenozoic greenhouse to icehouse evolution.

Geological data in the Ross Embayment, the large region between the Transantarctic Mountains and West Antarctica including the Ross Sea and Ross Ice shelf, have been collected since the very early phase of heroic exploration (Anderson, 1965). Samples of the fossilised Glossopteris flora were found in the last field camp of the last expedition of Captain Scott's party, on their way back to Ross Island. Du Toit's (1937) Gondwana reconstruction was proven on the basis of the results of these first expeditions. The Mesoproterozoic Rodinia supercontinent (Dalziel, 1991; Li et al., 2008; Merdith et al., 2017a,b; Moores, 1991; Wingate et al., 2002) provides another example of where fundamental pieces of evidence revealed the

* Deceased in December 2020.

Antarctic Climate Evolution. DOI: https://doi.org/10.1016/B978-0-12-819109-5.00004-9

central position held by Antarctica with respect to the other continental blocks, even though the details of some of the proposed configurations differ from each other.

Several air images were collected during an early reconnaissance period in the 1930s–40s, and the first detailed mapping started in the late 1950s–60s. From the 1960s to the early years of the 21st century, geological expeditions have faced the challenging conditions of remote field work necessary to evaluate the geological record held within the Antarctic rocks. Progressively revealed by the increased research activities of national and multinational expeditions following the 1957–1958 International Geophysical Year, to the activities planned during the International Polar Year 2007/2008, the geological record of Antarctica has now acquired a valuable role in comprehending paleoenvironmental and paleoclimatic investigations, particularly those focused on the initiation of Cenozoic glaciations, the stability of the polar ice sheets, and the complex interactions among tectonic, sedimentary and climatic processes. The last century also included some challenging drilling projects both onshore or very close to the coast (e.g., DVDP, CIROS, CRP and ANDRILL) and offshore (e.g., DSDP leg 28; IODP Expedition 318, IODP Expedition 374) (see McKay et al., 2021, figure 5). The former were benchmarks for the first proximal glaciological reconstructions of the entire continent. The latter played a critical role in the first set up of modern paleoclimate research, providing the first record of Cenozoic oxygen isotope data in the Southern Ocean. In a mostly ice-covered continent as Antarctica, remarkable results on the bedrock geology were also obtained from geophysical (e.g., Golynsky et al., 2018 and ref. therein), geochronological and geochemical studies on rocks and mineral detrital fragments included in metasedimentary and sedimentary rocks of various ages in circum Antarctic outcrops (Cooper et al., 2011; Elliot, 2013; Elliot and Fanning, 2008; Elliot et al., 2015, 2016, 2017; Elsner et al., 2013; Estrada et al., 2016; Goodge et al., 2001, 2002, 2004a,b, 2010, 2012, 2017; Henjes-Kunst et al., 2004; Paulsen et al., 2015, 2016; Paulsen et al., 2017; Rocchi et al., 2015; Schulz and Schüssler, 2013; Stump et al., 2006; Veevers and Saeed, 2008, 2011, 2013; Wisoczanski and Allibone, 2004). Recent detailed reviews of the Antarctic geological features discussed here are reported in detail within Kleinschmidt (2021a).

As a result of many international collaborative projects, and several national programs supporting geoscientific research, the big picture of the Antarctic plate, its relations to the surrounding ones and its most prominent tectonic features, are increasingly becoming clearer. Knowledge of basement geology, and cover sedimentary rocks, determined by conventional and more recent analytical methods have opened up future field and laboratory-based investigations. With the switch in many areas from the initial phase of reconnaissance work to more detailed high-resolution mapping and sampling, geophysical surveys, the intense use of new facilities providing isotopic data and remote sensing data, the anatomy of the basement is now revealed as much more variegated and complex than expected from the preliminary data. Nevertheless, areas where there is insufficient information, or lack of appropriate level of resolution, still persist and require new basic research efforts before a robust reconstruction of the basement deformational and petrological evolution, as well of the sedimentary processes in spatially associated basin set-ups, can be considered as fully documented.

After a brief description of the present-day geodynamic setting and a summary of the tectonostratigraphic framework, we focus on three main topics: (1) the important phase of Proterozoic to Paleozoic crustal evolution, which marked the transition between the two supercontinents of Rodinia and Gondwana; (2) the complex interplay between tectonic, sedimentary and paleoclimatic processes recorded in the Paleozoic to Mesozoic cover rocks; and (3) the relationships between compressional and extensional events in West Antarctica during Phanerozoic time after the end of the Ross Orogeny, and the close links between onshore geological data and offshore (sedimentary cores) information to decipher the younger, Cenozoic to Holocene, events.

The chapter describes the evolution of the Antarctic continent from its inclusion as part of the Gondwana supercontinent to the breakup of this landmass and the repositioning of Antarctica at southern polar latitudes since the Early Cretaceous (c. 120 Ma). The chapter also highlights some of the most interesting paleoclimatic issues, which are considered essential to improve our understanding of the polar climate and ice ages and their influences on Earth's climate system in the Cenozoic to present time. Our intention is to give a general overview, which should be complementary to the more detailed information included in the accompanying chapters devoted to specific aspects of the geological record. The chapter's conclusions highlight some of the persisting open problems in Antarctica's tectonic evolution, and areas where major research themes are needed both on- and off-shore.

6.2 The Antarctic plate and the present-day geological setting of the Ross Embayment

The Antarctic continent comprises two primary tectonic regions: (1) East Antarctica and (2) West Antarctica, with the associated West Antarctic Rift System (WARS) (Fig. 6.1B). East Antarctica is thought to feature Precambrian continental crust generally c.35–45 km thick (Bentley, 1991), and up to c.58 km in the Gamburtsev Mountains (Ferraccioli et al. (2011) and ref. therein) that are stable, coherent and often topographically high (Cogley, 1984) material that held a central position in the Paleozoic supercontinent of Gondwana (Tingley, 1991), as it did in the Mesoproterozoic supercontinent Rodinia (Dalziel, 1991; Moores, 1991). In contrast, West Antarctica is an amalgamation of low-lying, 20–35 km thick, mainly younger crustal blocks (Dalziel and Elliot, 1982; Janowski and Drewry, 1981; Jordan et al., 2020).

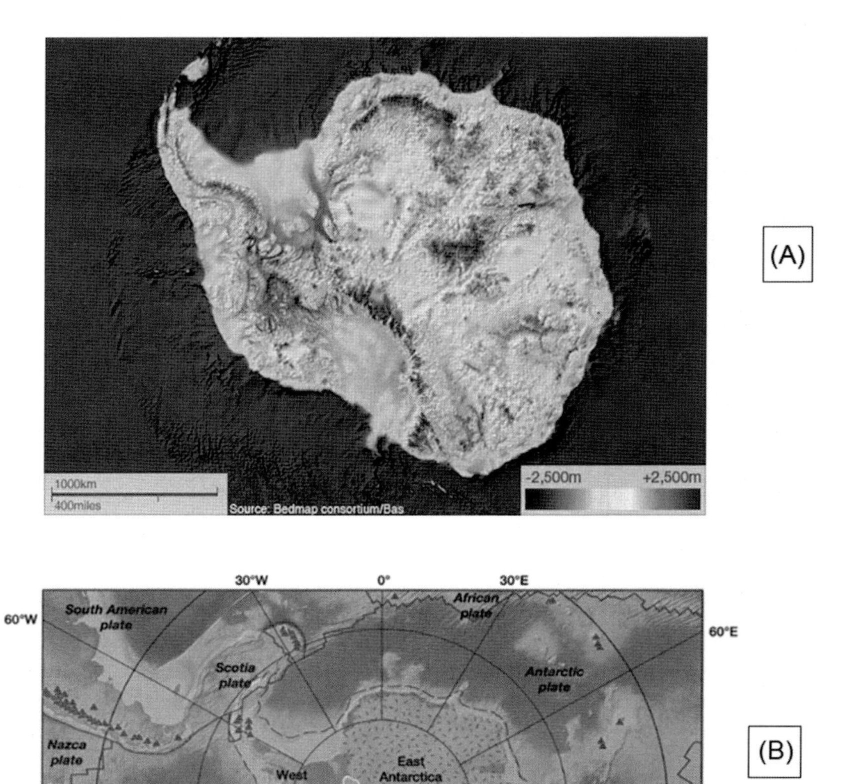

FIGURE 6.1 (A) Map showing the main Antarctic bed topography, based on BEDMAP 2 (Fretwell et al., 2013). (B) Schematic tectonic map of the Antarctic plate, showing bounding spreading ridges and subduction zones (after Goodge, 2020). Continental Antarctica extends to the shelf edge. Thick cratonic crust and lithosphere of East Antarctica (dashed pattern) and extensional West Antarctic rift system (WARS, diagonal ruling) are separated by the intraplate Transantarctic Mountains. Base from GeoMapApp (http://www.geomapapp.org).

The West Antarctic Rift lithosphere (WARS) compares well to other major Cenozoic continental rift systems (Behrendt, 1999). In the Ross Sea, the WARS borders the Transantarctic Mountains on their eastern side as a broad region of thinned continental crust associated with Cretaceous and episodic Cenozoic extension (Behrendt et al., 1991a,b; Behrendt, 1999). The WARS has high heat flow (83–126 mWm^2) (Berg et al., 1989; Blackman et al., 1987a,b) and thick (5–14 km) sedimentary basins with recent faulting (Cande and Leslie, 1986; Cooper et al., 1987; Hamilton et al., 2001). The Ross Archipelago (LeMasurier and Thomson, 1990) is currently active with fumarolic activity associated with alkaline volcanism at Mt. Erebus and Mt. Melbourne. The crust in the WARS is currently 20–25 km thick (Cooper et al., 1997; Jordan et al., 2020; Trehu, 1989).

The Transantarctic Mountains are approximately 2500 km long and 200 km wide, dividing East Antarctica from West Antarctica with peaks that rise over 4 km above sea level. Crustal thickness under the Transantarctic Mountains varies between 20 and 45 km (Bannister et al., 2003; Busetti et al., 1999; Cooper et al., 1997; Kanao et al., 2002; ten Brink et al., 1993, 1997). These mountains differ sharply from most mountain ranges of similar size and lateral extent because their formation did not reflect any compressional orogenic phases, but rather a process thought to have been directly-linked and causally-related to the development of the WARS (e.g., Studinger et al., 2004; ten Brink et al., 1997, and ref. therein). The Transantarctic Mountains were uplifted 6–10 km in an asymmetric tilt block formation and underwent denudation from the Cenozoic to the Cretaceous (Fitzgerald, 1992, 1995; Studinger et al., 2004). The later part of the uplift and denudation phases occurred under persisting spreading/extension within the western Ross Sea (Cande et al., 2000) and has been concomitant with voluminous sediment infilling from the Late Eocene/Oligocene to present (Barrett et al., 2000, 2001; Hamilton et al., 2001).

Detailed information about the different reconstructions and tectonic scenarios for Antarctica in Gondwana can be found in a range of previous work, including in Lawver et al. (1998), Larter et al. (2002), Boger and Miller (2004), Fitzsimons (2000a,b, 2003), Cawood (2005), Collins and Pisarevsky (2005), Payne et al. (2009), Boger (2011), and Merdith et al. (2017a,b).

Here, we will focus on the major geological constraints, key regions and datasets available in Antarctica, which represent a basis for reconstructing the southern continents in Gondwana. In this context, our review will also face some of the most controversial aspects that are presently under debate concerning the reconstruction of the main phases before and during the amalgamation of the Gondwana supercontinent. The geological evolution of the Antarctic continent is reviewed in two main periods: (1) before c. 450 Ma, covering the processes that were active during the amalgamation of Rodinia and Gondwana; and (2) c. 450 Ma to present day, including all the major events that occurred after the final stage of Gondwana amalgamation along its paleo-Pacific margin.

6.3 East Antarctica

6.3.1 The Main Geological Units during the Paleoproterozoic–Early Neoproterozoic Rodinia Assemblage

As with all other continents, Antarctica comprises a number of Archaean/Early Proterozoic cratons (older than c.1.5 Ga) surrounded by successively younger belts that formed, and/or accreted to the continental margins, as products of convergent plate tectonic events such as subduction of oceanic crust underneath continental crust and/or collision of two former separated continents (Fig. 6.2). The Precambrian geological evolution of the East Antarctica is still highly debated, but some key points are now well defined by a large number of recent papers (see Aitken et al., 2016; Bauer et al., 2003a,b, 2009; Bogdanova et al., 2009; Boger, 2011; Boger et al., 2015; Cawood and Buchan, 2007; Collins and Pisarevsky, 2005; Daczko et al., 2018; Evans, 2009; Fitzsimons, 2000a,b, 2003; Godard and Palmeri, 2013; Golynsky and Jacobs, 2001; Golynsky et al., 2018; Goodge, 2002, 2020, 2021; Goodge and Severinghaus, 2016; Goodge et al., 2002, 2008, 2010, 2012, 2017; Harley et al., 2013; Jacobs, 2009; Jacobs and Lisker, 1999; Jacobs and Thomas, 2002, 2004; Jacobs et al., 1998, 2003a,b, 2008a,b, 2015, 2017, 2020; Kleinschmidt, 2021a,b; Kleinschmidt and Boger, 2008; Läufer, 2021; Läufer et al., 2021; Li et al., 1995, 2008; Liu et al., 2006, 2007, 2009, 2013, 2014, 2016, 2017, 2018; Meert and Torsvik, 2003; Ménot, 2021; Merdith et al., 2017a,b; Mieth and Jokat, 2014; Mikhalsky and Kamenev, 2013; Mikhalsky et al., 2013, 2015; Morrissey et al., 2017; Osanai et al., 2013; Paech et al., 2005; Palmeri et al., 2018; Payne et al., 2009; Pisarevsky et al., 2003; Riley et al., 2020; Roland, 2021; Ruppel et al., 2015, 2018, 2020; Siddoway, 2021; Smellie, 2021; Thomson, 2021; Tucker et al., 2017; Veevers et al., 2016; Wang et al., 2020; Yoshida et al., 2003). In Fig. 6.2 a schematic highly-speculative geological map is proposed with the hypothesised sub-ice extent of the main crustal domains. Archaean-Proterozoic Cratons are confined to East Antarctica and several are generally accepted as follows:

- the small 'Grunehogna Craton' (c. 3.1 Ga basement – Marschall et al., 2010, 2013) covered by flat-lying and undeformed sediments older than 1 Ga, interpreted as a fragment of the African 'Proto-Kalahari Craton';
- the 'Napier Craton' (considered a fragment of the 'Dwarhai Craton' in India) (Kelly and Harley, 2005);
- the 'Mawson Craton' (Fitzsimons, 2000a,b, 2003; Naumenko-Dèzez et al., 2020 and ref. therein) exposed only in limited areas near the continental margin in Terre Adélie and in George V Land, and their Australian counterpart the 'Gawler Craton', extending into the Miller and Geologists Ranges of the central Transantarctic Mountains;
- the 'Ruker Craton', probably extending to south-west and the cryptic 'Valkyrie Craton' (part of the 'Crohn Craton') (Boger, 2011; Boger et al., 2006);

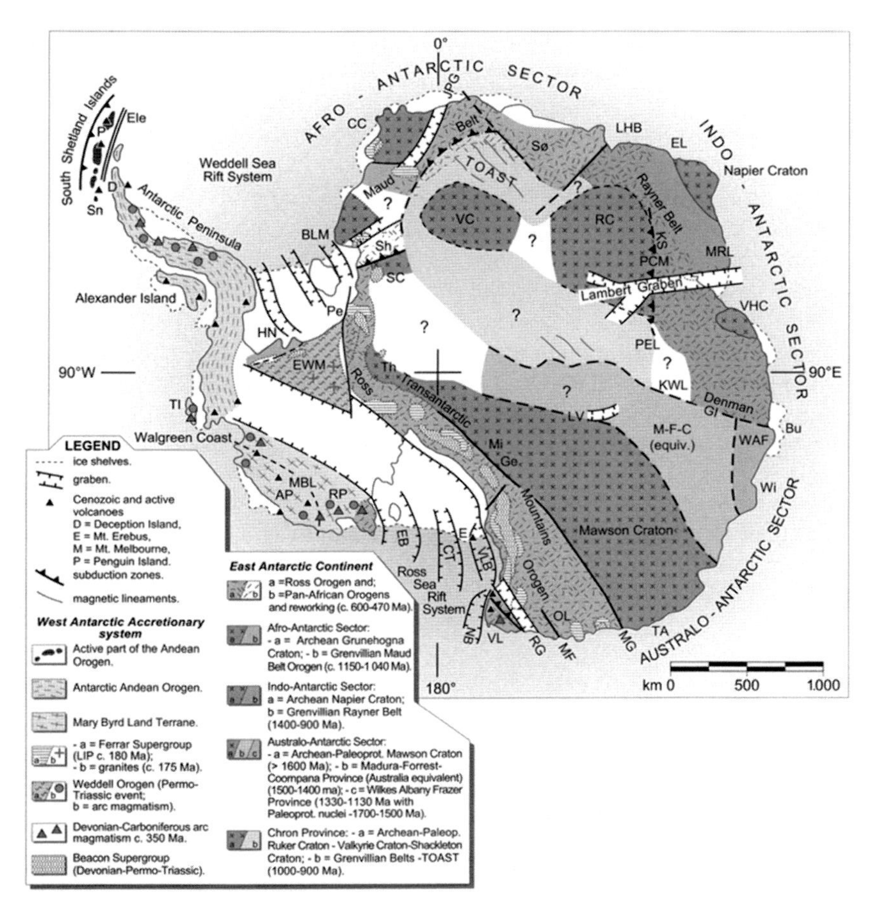

FIGURE 6.2 Schematic geological map of Antarctica (modified after Kleinschmidt, 2014). The map shows the possible sub-ice extent of the cratonic areas, the four 'Grenvillian' belts, the Pan-African orogenic belts and the Ross Orogen. The map also shows the distribution of Beacon Supergroup and Ferrar Supergroup outcrops and of the two main Mesozoic and Cenozoic orogens (Ellsworth or Weddell Orogeny and Antarctic Andean Orogen) which formed along the paleo-Pacific margin of Gondwana in West Antarctica, and the major intra-plate fracture zones in East Antarctica. Abbreviations: *AP*, Amundsen Province; *BLM*, Bertrab, Littlewood, Moltke Nunataks; *Bu*, Bunger Hills; *CT*, Central Trough; *EB*, Eastern Basin; *EL*, Enderby Land; *Ele*, Elephant Island; *EWM*, Ellsworth-Whitmore Mountains; *GC*, Grunehogna Craton; *Ge*, Geologists Range; *HN*, Haag Nunatak; *JPG*, Jutul Penck Graben; *KL*, Kemp Land; *KWL*, Kaiser-Wilhelm-II-Land; *KS*, Kuunga Suture; *LHB*, Lützow-Holmbukta; *LV*, Lake Vostok; *MBL*, Marie Byrd Land; *MF*, Matusevich strike–slip fault; *MG*, Metz Glacier; *Mi*, Miller Range; *MRL*, Mac.Robertson Land; *NB*, Northern Basin; *OL*, Oates Land; PCM, Prince Charles Mts.; *Pe*, Pensacola Mts.; *PEL*, Princess Elizabeth Land; *RG*, Rennick Graben; *RC*, Ruker Craton; *RP*, Ross Province; *SC*, Shackleton Cratonic domain; *Sh*, Shackleton Range; *Sn*, Snow Island; *Sø*, Sør Rondane; *TA*, Terre Adélie; *Th*, Thiel Mts.; *TI*, Thurston Island; *TOAST*, Tonian Oceanic Arc Super Terrane; *VC*, Valkyrie Craton; *VHC*, Vestfold Hills Craton; *VL*, Victoria Land; *VLB*, Victoria Land Basin; *WAF*, Wilkes-Albany-Frazer Orogen; *Wi*, Windmill Islands. The Precambrian to Cambrian geological structure of East Antarctica, which is ice covered for the most part, is still highly speculative.

- the southern Shackleton Range (Riley et al., 2020; Will et al., 2009 and ref. therein) and perhaps the eastern Thiel Mountains (Ford, 1963) (part of the 'Mawson Craton'?);
- the 'Vestfold Hills' cratonic fragment in Prydz Bay region (Clark et al., 2012; Zulbati and Harley, 2007); and
- the BLM cratonic block in the Coats Land, which is an enigmatic crustal domain, hypothesised to be of pre-Grenvillian age (Kleinschmidt and Boger, 2008) or alleged Paleoproterozoic age (Jacobs et al., 2015), where only an undeformed c. 1100 Ma rhyolites and granophyres cover emerges from the ice. In a recent paper, Riley et al. (2020) dated a granite pegmatite cobble ice-transported to and recovered from the Brunt Ice Shelf: its U-Pb zircon age (c. 1100 Ma) is similar to that of the felsic volcanic cover, but also to the widespread Grenvillian age of the Maud Belt.

The exposed basement of these cratonic blocks, whose internal evolutionary history is not detailed here, only occasionally covered by almost flat-lying platform sediments (e.g., Thiel Mountains), consists of high-grade gneisses and granulite facies metamorphic complexes showing radiometric ages older than c. 1500 Ma up to just over 3800 Ma. Since the relationships between the different cratons are still poorly known, a tectonic reconstruction is mainly speculative. How and when did these cratonic nuclei assemble together? Until about the late 2010s the prevailing hypothesis was that in a long-lasting period, from the Mesoproterozoic to early Neoproterozoic (c. 1300–800 Ma), their docking was related to the Grenvillian age orogens well known in the Laurentian paleoplate (e.g., Bogdanova et al., 2009; Li et al., 2008; Meert and Torsvik, 2003 and ref. therein).

Along the margins of these cratons, large accretionary belts are found. These orogenic belts, separating the cratons as internal belts in East Antarctica, or as elongated peripheral belts at the paleo-Pacific margin of Gondwana, developed during two major orogenic cycles spanning in time from c.1.3–0.9 Ga (Grenvillian-aged orogens), through 500–600 Ma (Ross and the Pan-African orogens). As in eastern North America and other continents, the Grenvillian-aged orogens in Antarctica are considered to record the fundamental orogenic event marking the amalgamation of the Rodinia supercontinent in Meso-Neoproterozoic times. Until the early 1990s, only one very long Antarctic Grenvillian orogen was assumed, following the Antarctic coast as a 250 km wide strip from Coats Land in the west up to George V Land in the east, occasionally specially named the 'Circum East Antarctic Mobile Belt' (Yoshida, 1992). Such an extension has been demonstrated to be incorrect for Terre Adélie and George V Land and unproven for Coats Land, where the tiny outcrops represented by three little groups of nunataks (Littlewood, Bertrab and Moltke) consist of c.1100 Ma rhyolites and granophyres that are absolutely undeformed (Kleinschmidt and Boger, 2008; Storey et al., 1994). The existence of a Circum East Antarctic Mobile

Belt is no more supported by new geological evidence, which instead indicates the presence of at least four distinct Grenvillian-aged domains (Fitzsimons, 2003; Jacobs et al., 2015, 2020; Ruppel et al., 2015, 2020 and ref. therein) (Fig. 6.2) described as follows:

1. the 'Maud Belt', extended from western to eastern Dronning Maud Land to parts of Sør Rondane, interpreted as the trace of an active Grenvillian continental margin at the 'proto-Kalahari Craton' (Jacobs et al., 2008a,b, 2015, 2020 and ref. therein);
2. the 'Rayner Belt' (Yoshida and Kizaki, 1983) that in Enderby, Kemp and MacRobertson Lands separates the 'Napier Craton' from the southern Prince Charles Mountains cratonic area ('Ruker Craton') and which was verified in the northern Prince Charles Mountains (e.g., Boger et al., 2002; Morrissey et al., 2016);
3. the 'Wilkes Province Belt' (Fitzsimons, 2000a; the 'Wilkes-Albany-Frazer Orogen' in Fig. 6.2) of Wilkes Land that is exposed in the Bunger Hills and Windmill Islands, and merges and divides the Vestfold Hills domain from the Mawson continent; and
4. the 'TOAST' ('Tonian Oceanic Antarctic Super Terrane') domain interpreted as an oceanic arc belt of Tonian age (Jacobs et al., 2008a,b, 2015, 2020).

At the same time, some detailed research proposed that these Grenvillian belts are active continental arcs accreted to the proto-Kalahari and proto-Indian cratons, far from the Mawson-Gauler cratons (Jacobs, 2009; Jacobs et al., 1998, 2003a,b, 2008a,b; Liu et al., 2009; Mikhalsky et al., 2009); a proposal now widely accepted (see Boger, 2011; Jacobs et al., 2020; Merdith et al., 2017a,b and ref. therein). Below a short summary is given of these events, subdivided in the four different crustal domains according the proposed paleoplate affinity (Fig. 6.2).

1. ***Afro-Antarctic Sector***. In this wide region the basement is composed by high-grade ortho and paragneisses (the Grenvillian 'Maud Belt') dated at 1170 to 1050 Ma followed by high-grade metamorphism at c. 1080–1030 Ma as part of an active continental margin on the Kalahari craton at that time (Arndt et al., 1991; Bauer et al., 2003a,b; Board et al., 2005; Grantham et al., 2013; Groenewald et al., 1995; Jacobs, 2009; Jacobs et al., 1998, 2003a,b, 2008b, 2009, 2015, 2020; Kamei et al., 2013; Paulsson and Austrheim, 2003; Ruppel et al., 2020; Wang et al., 2020; Will et al., 2009). The sector extends from the Coats Land, through the Dronning Maud Land to the Sør Rondane region, north of the 'Forster Magnetic Anomaly'. Evidence of a late stage (c.785-760 Ma) magmatic event in the Schirmacher Oasis suggests a reactivation of this Grenvillian continental margin (Jacobs et al., 2020).
2. ***Indo-Antarctic Sector***. Here, covering circum-Antarctica coastal regions from Lützow-Holmbukta Bay to the Denman Glacier, the wide and composite 'Rayner Belt' (or 'Rayner Complex') is characterised by magmatic and metamorphic events dated c. 1400–800 Ma (Kelly and Harley, 2004;

Kelly et al., 2002; Kelsey et al., 2007; Liu et al., 2009, 2016, 2017, 2018; Mikhalsky and Leitchenkov, 2018; Mikhalsky et al., 2009, 2013, 2015; Phillips et al., 2006) testifying a long-lived Grenvillian active continental arc. Magmatic rocks of 'charnockite type' and granulites are common. The composite character of this orogenic belt is strengthened by the presence of crustal domains with Paleoproterozoic protoliths ages of felsic to mafic orthogneisses (e.g., in the Lützow-Holm Bay area, Dunkley et al., 2020; Takahashi et al., 2018 and ref. therein). The western boundary with the 'Maud Belt' in the Lützow-Holmbukta region is still highly debated and inferred mainly from geophysical data (see Ruppel et al., 2018). In the oceanward side two Archaean cratonic fragments are present: the Napier Craton (Kròl et al., 2020 and ref. therein), and the Vestfold Craton with Indian affinity (Clark et al., 2012; Zulbati et al., 2007).

3. ***Australo-Antarctic Sector***. Contrary to the sectors described above, the Australo-Antarctic sector, extending between the Denman Glacier region to the northern coast of northern Victoria Land, was in Precambrian-Paleozoic time always welded to the Australian units. Along the coastal rocks exposure a long-lasting research activity defined different lithotectonic units from the ('Lachlan Orogen' equivalent) Robertson Bay Terrane, to the Bowers Terrane and the Wilson Mobile Belt ('Ross Orogen'; Tessensohn and Henjes-Kunst, 2005 and ref. therein), to the Mertz Glacier Shear Zone where the 'Mawson Craton' is transected (Di Vincenzo et al., 2007; Lamarque et al., 2018; Naumenko-Dèzes et al., 2020; Talarico and Kleinschmidt, 2003 and ref. therein). To the west of the 'Mawson Craton', according to Payne et al. (2009), Morrissey et al. (2017), and Liu et al. (2018), a younger, poorly exposed terrane remains hypothetical: the Australian 'Madura-Forrest-Coompana Province' (M-F-C) equivalent province (Fig. 6.3). This domain is proposed also on the basis of a magnetic grain different from that of the Mawson Craton (Golynsky et al., 2018) and as a source area for the glacially-derived granitoid clasts dated c. 1500−1400 Ma by Goodge et al. (2008, 2010, 2017). The Bunger Hills and Windmill Islands regions are located and interpreted as the north-western border zone of the Mawson Continent correlable to the Grenvillian 'Albany-Frazer Orogen' in Australia developed at the margin of the Yilgarn West Australian Craton (Liu et al., 2018; Morrissey et al., 2017; Tucker et al., 2020). The paleo-plate boundary between the Australo-Antarctic sector and the Indo-Antarctic sector is not exposed. Its position is generally inferred to be located within the wide (c. 250 km) region between Mirny Station and the Denman Glaciers (Aitken et al., 2014, 2016; Daczko et al., 2018; Fitzsimons, 2000a,b, 2003; Tucker et al., 2017, 2020). In Fig. 6.2 this boundary is located along Mirny Fault according to the proposal of Daczko et al. (2018). This is a key region (see discussion in Tucker et al., 2020) where both the 'Kuunga Orogen' and the 'Pinjarra Orogen' overlapped over a Paleo-Meso-Proterozoic basement (Daczko et al., 2018; Mikhalsky et al., 2015).

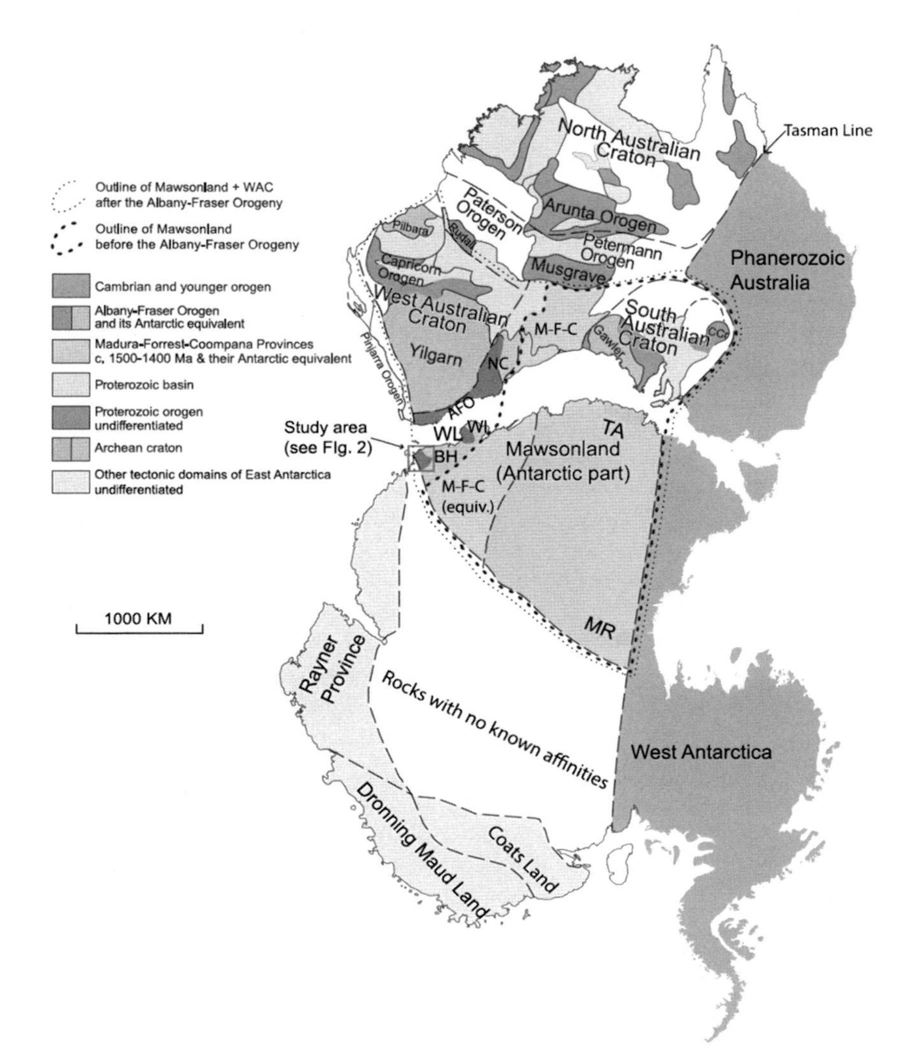

FIGURE 6.3 Tectonic map of Australia and Antarctica in a Gondwana configuration (after Liu et al., 2018). Abbreviations: *AFO*, Albany-Fraser Orogen; *BH*, Bunger Hills; *CCr*, Curnamona Craton; *M-F-C*, Madura-Forrest-Coompana Provinces; *Mr*, Miller Range; *NC*, Nornalup Complex; *TA*, Terre Adélie craton; *WI*, Windmill Islands; *WL*, Wilkes Land.

In the Wilkes–Albany-Frazer Orogen, in spite of the mainly ice covered field situation, detailed geochronological and petrological studies on Obruchev Hills, Bunger Hills and Windmill Islands (Morrissey et al., 2017; Tucker et al., 2017; Tucker et al., 2020; Zhang et al., 2012 and ref. therein), aided by aerogeophysical surveys on the subglacial tectonic structure (Aitken et al., 2014, 2016; Maritati et al., 2016), has led to better knowledge of the geological evolution of this province. Evidence for felsic to mafic magmatism and high-grade metamorphism are widespread and dated between c.1345 and c.1130 Ma, coeval with Stages I and II of the Albany-Frazer Orogen in Australia (Spaggiari et al., 2015). For some lithological domains an Archaean to Paleoproterozoic protoliths age is proposed (Black et al., 1992; Sheraton et al., 1992, 1993; Tucker et al., 2017).

4. **The '*Crohn Province*'**. This wide domain of East Antarctica, mainly covered by ice, matches what Boger (2011) defined the 'Crohn Craton'. Due to the extremely poor conditions of exposure, all statements about this province are highly speculative. Available field and geophysical data suggest that it is a composite crustal block made by cratonic nuclei (the 'Ruker Craton' and the cryptic 'Valkyrie Craton') and enveloping mobile belts at least in part of Grenvillian age (c.1000–900 Ma) as proposed for the TOAST in SE Dronning Maud Land (Elburg et al., 2014, 2016; Jacobs et al., 2015, 2018; Kamei et al., 2013; Ruppel et al., 2018, 2020 and ref. therein). These age data (SHRIMP or LA-MC-ICPMS on zircon) are mainly obtained on a gabbro-tonalite-trondhjemite-granodiorite (meta)intrusive suite (GTTS, Kamei et al., 2013) cropping out in the Sør Rondane Mountains region (a Pan-African sliver of a Tonian juvenile island arc according to Ruppel et al., 2020), and from samples selected from moraine deposits in the SW Terrane of this region (Jacobs et al., 2017). A characteristic structural feature in some areas of these belts are an alternating parallel, elongated, magnetic anomalies trending NW-SE (Mieth and Jokat, 2014; Mieth et al., 2014; Ruppel et al., 2018). According to Ruppel et al. (2020) the TOAST first accreted to the cryptic Valkyrie Craton in early Neoproterozoic times, then during the Ediacaran-Early Cambrian was sandwiched against the Grenvillian Maud Belt. Geophysical data provide the best constraints on subglacial geological structures but there is little information on their age. Fitzsimons (2003) suggested three possible paths for extension in East Antarctica of the composite Pinijarra Orogen (assuming a Pan-African age for these tectonothermal events) and one of these, the PD3 path, transects East Antarctica along the boundary between the Crohn Province and the Mawson Continent including the Gamburtsev Mountains subglacial region. There are no robust data supporting this hypothesis, however. Instead, as the Ruker domain has Pb-isotopic composition similar to basement terranes in Indo-Antarctica (Flowerdew et al., 2013) a possible alternative model is that proposed by Boger (2011): first the composite oceanic Crohn Province accreted in Grenvillian time to the Mawson Continent, then the Indo-Antarctic Sector collided along the 'Kuunga Suture' during the Pan-African events.

6.3.2 From Rodinia breakup to Gondwana (c. 800–650 Ma)

The amalgamation phase of Gondwana necessarily involved the aggregation of various continental fragments, which derived from the fragmentation and dispersal of a former supercontinent – variously named Ur-Gondwana (Hartnady, 1991), Katania (Young, 1995), Paleopangea (Piper, 2000) or Rodinia (McMenamin and McMenamin, 1990). However, the precise configuration and modality of breakup of this supercontinent are yet not completely known and these uncertainties obviously propagate to the formulation of tectonic models for the constructive phase of Gondwana. Since the focus of this paper is on the Cretaceous–Cenozoic record, we will briefly summarise the main recent results in the reconstruction of the Late Precambrian–Early Paleozoic tectonic evolution of Antarctica in Gondwana avoiding a detailed review of Rodinia models: SWEAT (Dalziel, 1991; Hoffman, 1991; Moores, 1991); AUSWUS (Karlstrom et al., 1999); AUSMEX (Wingate et al., 2002); and reviews by Dalziel (1997), Meert and Torsvik (2003), Evans (2013), Merdith et al. (2017a,b) and alternative models (e.g., Paleopangea; Piper, 1982, 2000). As recently summarised by Merdith et al. (2017a), and shown in Fig. 6.4, in the SWEAT fit, eastern Antarctica is pushed against the southwest United States, while Australia lies further north near the United States-Canadian border (Dalziel, 1991; Hoffman, 1991; Moores, 1991; Wingate et al., 2002); in the AUSWUS fit, eastern Australia is matched against the southwest United States of America (Karlstrom et al., 1999); in the AUSMEX, Australia has only a small connection with Laurentia, where the north tip of Queensland fits against Mexico (Wingate et al., 2002), Kalahari Craton is shifted further south to accommodate Mawson; in the 'Missing-Link' model (Li et al., 2008), South China is considered as a continental sliver between Australia and Laurentia.

In spite of significant uncertainties about the precise reconstruction of the global paleogeography in Neoproterozoic time, and the still incomplete geochronological framework, most authors agree that the breakup of Rodinia led to the development of extensive passive continental margins which are documented in the late Neoproterozic (c. 750–600 Ma) record of most present-day continents (Cawood, 2005; Dalziel, 1991, 1992, 1997; Goodge, 2020; Goodge et al., 2002, 2004a,b; Meert and Van der Voo, 1997; Powell et al., 1993). In Antarctica, following Dalziel's hypothesis (Dalziel, 1991), the first stage of Rodinia breakup involved a rifting phase which started at c. 800–750 Ma leading to the separation of Laurentia from the East Antarctica-Australia block and the formation of the intervening proto-Pacific Ocean. This process was accompanied by the drift of many cratonic blocks, presently exposed in South America and Africa (Amazonian, Rio de la Plata, Western African and India + South China cratons) (i.e. West Gondwana). While Laurentia remained thereafter always separated from Mawson-Australian cratons, the Kalahari and India + South China moved against proto-East Antarctica and collided with it. Most workers agree that key evidence of this evolution is stored in the 'Mozambique Belt' (Holmes, 1951), 'East African Orogen' (Stern, 1994), or 'East African Antarctica Orogen' (Jacobs et al., 2003a, 2015), and that

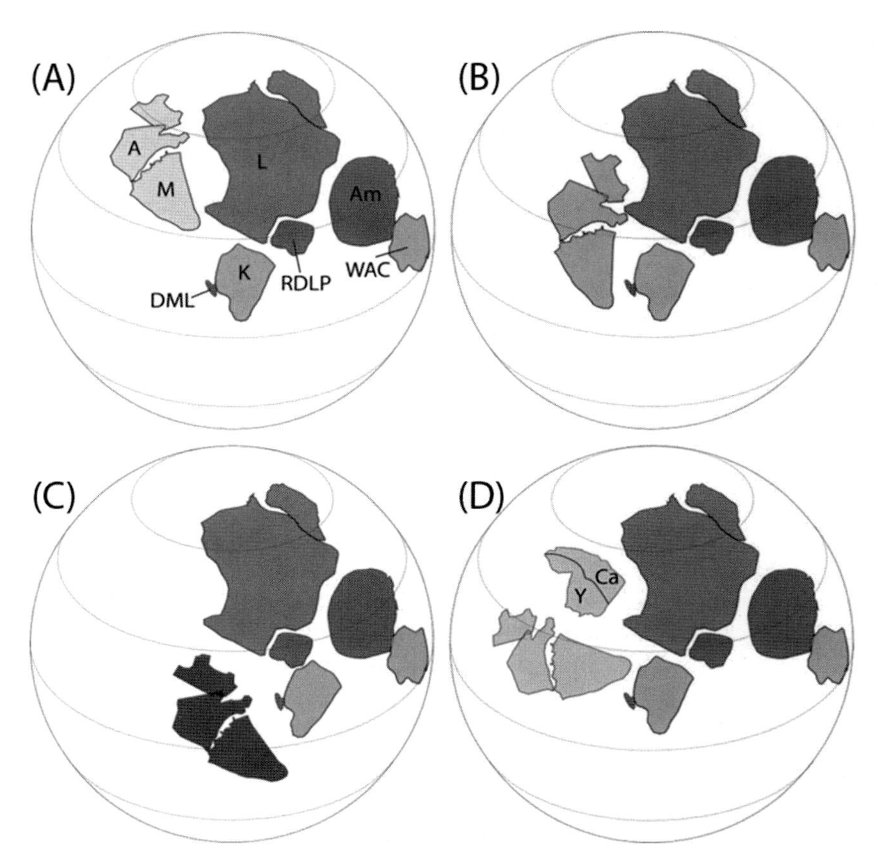

FIGURE 6.4 The different configurations of Laurentia with Australia-East Antarctica and South China with pre-Neoproterozoic geology overlain. Laurentia is fixed in its present-day position with all other blocks rotated relative to it at 800 Ma (from Merdith et al., 2017a). (A) SWEAT fit (green); (B) AUSWUS fit (blue); (C) AUSMEX (black); (D) Missing-Link model (tan). *A*, Australia; *Am*, Amazonia; *Ca*, Cathaysia (part of South China); *DML*, Dronning Maud Land; *K*, Kalahari; *L*, Laurentia; *M*, Mawson (East Antarctica); *RDLP*, Rio de la Plata; *WAC*, West Africa Craton; *Y*, Yangtze (part of South China). The colour of the other blocks represents their present-day geographical position. South America – dark blue; Africa – orange; Antarctica – purple; North America – red; China – yellow.

this extensive orogenic belt formed as a result of the closure of a 'Mozambique Ocean' and subsequent collision and amalgamation of East and West Gondwana during the Pan-African events (Dalziel, 1992; Jacobs et al., 2015; Shackleton, 1996; Stern, 1994 and ref. therein).

The Neoproterozoic-Cambrian process of amalgamation in Antarctica involved the following three regions: (1) the Afro-Antarctic Sector; (2) the Indo-Antarctica Sector; and (3) Wilkes Land in the Australo-Antarctic Sector. All are currently the subject of much debate.

1. **Afro-Antarctic Sector**. A review of structural and geochronological data from Dronning Maud Land-Lützow Holm Bay region and comparison with the adjacent (in Gondwana) Falkland Microplate and south-eastern Africa, led Jacobs and Thomas (2002) to corroborate the proposal by Jacobs et al. (1998) of a southward continuation of the Mozambique Belt into Dronning Maud Land in Antarctica. The Grenvillian 'Maud Belt' suffered a pervasive high-grade tectono-thermal overprinting and reworking at c. 650–500 Ma testifying the Pan-African collision of the West Gondwana with the East Antarctic proto-continent. The Pan-African events are characterised by strong variable metamorphic events up to the granulitic grade. In the Northern Shackleton Range only Pan-African ages have been detected: here the metamorphic basement is characterised by extensive thrust and nappe tectonics, widespread and distinct late orogenic collapse structures, thick molasse formations ('Blaiklock Glacier Group') (see Kleinschmidt, 2021a, 2021b). In most of Dronning Maud Land, and the eastern Sør Rondane region (Ruppel et al., 2020), metamorphic reworking up to HP-HT granulite grade is pervasive and also syn- and post-orogenic magmatism (Baba et al., 2010, 2013; Board et al., 2005; Elburg et al., 2016; Grantham et al., 2013; Jacobs et al., 2008a; Kleinhans et al., 2013; Osanai et al., 2013; Paech, 2004; Pauly et al., 2016; Roland, 2004; Roland and Olesch, 2004; Ruppel et al., 2020; Shiraishi et al., 2008). Many alternative locations have been proposed for the Pan-African suture in the Dronning Maud Land (e.g., figure 1 in Mieth and Jokat, 2014, and ref. therein). If the TOAST region is unbundled from the Afro-Antarctic Grenvillian Belt, a plausible suture-boundary between the Grenvillian Afro-Antarctic sector and the 'Crohn Province' is represented by the Forster Magnetic Anomaly (Riedel et al., 2013). A Pan-African suture zone in the Shackleton Range was suggested by Grunow et al. (1996) and conclusive evidence was found by Talarico et al. (1999) who described relics of ophiolites consisting of serpentinites and amphibolites with N-type MORB to OIB geochemistry and a maximum Sm–Nd age of c. 900 Ma. Metamorphic reworking of these rocks occurred under variable high P to medium P conditions in the eastern Shackleton Range; a stage of eclogite-facies metamorphism at c. 510 Ma has been recently reported by Schmädicke and Will (2006) in the central Shackleton Range. A detailed review of the Shackleton Range geology was recently published by Kleinschmidt (2021a, 2021b). On the basis of these discoveries, and thrust patterns in both the Shackleton Range and Western Dronning Maud Land, Kleinschmidt et al. (2002) proposed that the ophiolites may have formed part of a connection between the Paleo-Pacific and the Mozambique oceans, in the way the Drake Passage is linking the present Pacific and the Atlantic Oceans. The Mozambique Belt or EAAO ('East African Antarctic Orogen', as renamed by Jacobs and Thomas, 2002) and its continuation in Antarctica show a general N–S trending (Jacobs and Thomas, 2002).

The 'Maud Belt' belt has therefore provided sound evidence in contrast to the classical assumption that the whole Eastern Gondwana formed during the consolidation of Rodinia in the Mesoproterozoic time and supported the proposal that the 'Maud Belt' collided only in Neoproterozoic time with the 'Crohn Provence'. It then remained tectonically stable until the modern continents rifted from Gondwana in the Mesozoic. The Lützow Holm–Prydz Bay–Pan-African structures were initially considered to be part of a wider Pan-African belt termed the 'Kuunga Orogen' by Meert et al. (1995) or 'Kuunga Suture' by Boger and Miller (2004), and interpreted as the result of the collision between East Antarctica (+Australia) and India (+Madagascar + Sri Lanka) at c. 535–520 Ma, after the amalgamation of India with the rest of Gondwana along the Mozambique suture (Fig. 6.5). More recent data suggest that there are actually three orogenic belts involved in the orogenic collisions at about the same Pan-African period of c. 650–500 Ma in East Antarctica, highlighted by a often high-grade metamorphic reworking and intrusive bodies of Pan-African age (Fig. 6.2): (1) the belt in the Shackleton Range–Dronning Maud Land–southern Sør Rondane–Lützow-Holmbukta region ('East Antarctic Orogen' or 'East Antarctic Belt' of Jacobs et al., 1998, renamed 'Lützow Holm Belt' by Fitzsimons, 2000b); (2) a wide elongated portion of the Grenvillian 'Rayner Belt' from the Enderby Land to the Wilhelm II Land through the southern Prince Charles Mountains–Grove Mountains ('Kuunga Suture' according to Boger et al., 2002); and (3) the belt in the Denman Glacier region, interpreted as prolongation of the Leeuwin Complex of Australia's Pinjarra Orogen (e.g., Fitzsimons, 2000b). However, the exact extent of these orogens is doubtful because of the extensive ice cover, and this applies especially to the Denman Glacier belt.

As proposed by Boger et al. (2001, 2002), Gondwana's amalgamation in Antarctica may have taken place in two steps: the first before 550 Ma and the second after 550 Ma. The first step involved the amalgamation of West Gondwana with 'Indo-Antarctica' (i.e. India and the 'Rayner Belt' – 'Napier Craton') documented in Dronning Maud Land. The second step led to the aggregation of these terranes to the rest of East Gondwana, i.e. the rest of East Antarctica thus producing the 'Kuunga Suture'. Geophysical data from Mieth and Jokat (2014) suggest that the Lützow Holm Belt (=East Antarctic Orogen) is not a continuous belt from the Shackleton Range to the Lützow-Holmbukta, but interrupted by a crustal domain characterised by NW–SE trending aeromagnetic anomalies. They indicate a NW-directed Tonian indenter pushing apart the Shackleton Range region to the west and the Sør Rondane–Lützow-Holmbukta region to the east during Pan-African times (Jacobs et al., 2015; Ruppel et al., 2015) (Fig. 6.6). The indenter has been recently interpreted as an Oceanic Arc Super Terrane (Tonian Oceanic Arc Super Terrane – TOAST; Jacobs et al., 2015) that is sandwiched in between Kalahari margin and Indo-Antarctica – Rukerland. In the Sør Rondane

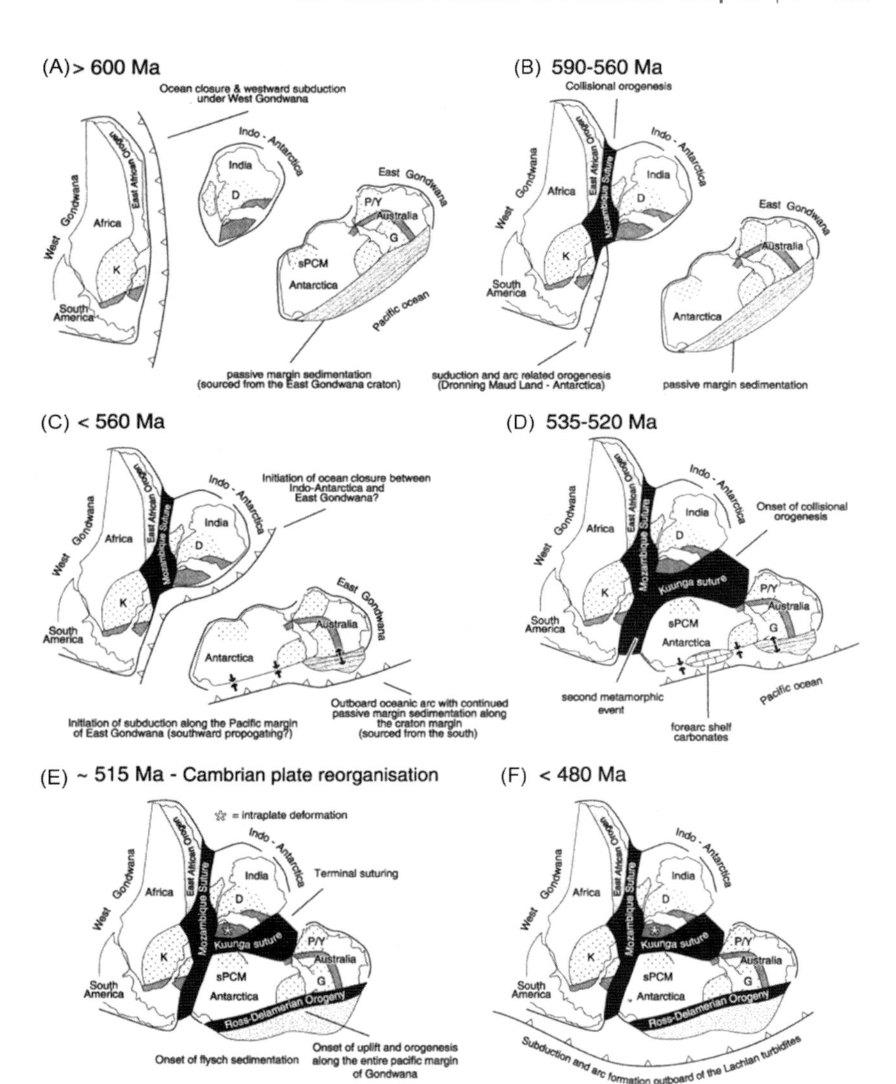

FIGURE 6.5 Main tectonic stages of the amalgamation of Gondwana (after Boger and Miller, 2004). Abbreviated Cratons – *D*, Dwarhai Craton; *G*, Gawler Craton; *K*, Kalahary Craton; *P/Y*, Pilbara/Yilgarn Craton; *sPCM*, Southern Prince Charles Mountains.

region, according to Ruppel et al., 2020 (and ref. therein), a polyphase tectono-metamorphic stage up to HP-HT grade and a coeval sequence of magmatic pulses from c. 650 to 500 Ma define a wide collisional margin of Pan-African age characterized by a fold and thrust tectonics and a protracted cooling pattern. Geophysical data (Mieth and Jokat., 2014) might indicate that the southern continuation of the TOAST terminates against an as yet unidentified craton further inland (the cryptic 'Valkyrie Craton'). Jacobs et al. (2015)

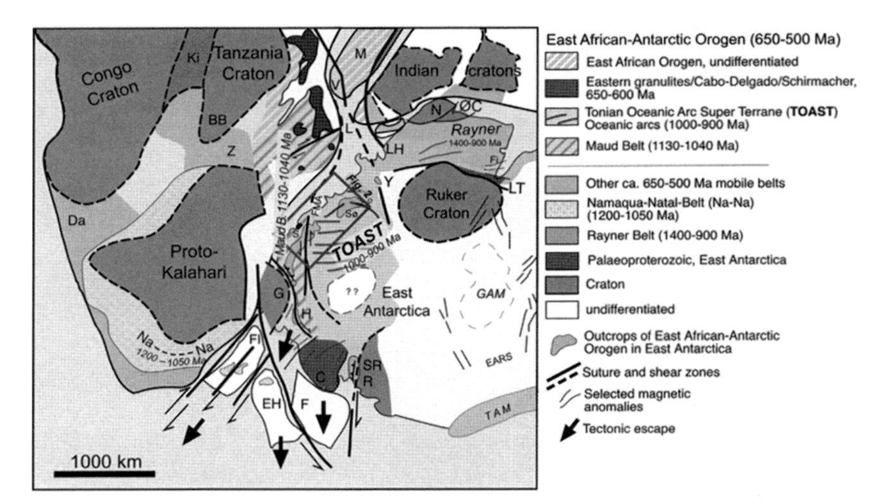

FIGURE 6.6 Geology of the African-Antarctica-India sector of Gondwana (after Jacobs et al., 2020). Abbreviations: *BB*, Bangweulu Block; *C*, Coats Land, *Da*, Damara belt; *EARS*, East Antarctic Rift System; *EH*, Ellsworth-Haag; *F*, Filchner block; *FI*, Falkland Islands, *Fi*, Fisher Terrane; *FMA*, Forster Magnetic Anomaly; G, Grunehogna; *GAM*, Gamburtsev Mts.; *H*, Heimefrontfjella; *Ki*, Kibaran; *L*, Lurio Belt; *LH*, Lützow-Holm Bay; *LT*, Lambert Terrane; *N*, Napier Complex; *Na-Na*, Namaqua-Natal; *M*, Madagascar; *ØC*, Øygarden Complex; *R*, Read Block; *S*, Schirmacher Oasis; *Sø*, Sør Rondane; *SR*, Shackleton Range; *TAM*, Transantarctic Mountains.

included the TOAST intraoceanic juvenile arc in the EAAO, but some features raise doubts on this interpretation: for example, the structural grain of this belt is orthogonal to that of the Dronning Maud Land and similar to other domains of interior Antarctica continent as observed by Mieth and Jokat (2014). A different model appears plausible. The composite 'Crohn Province' (that is the 'Crohn Craton' of Boger, 2011) could be involved in the Neoproterozoic drifting of the Indian block against the Mawson Continent as also proposed by Mulder et al. (2019) along the 'Pinjarra Orogen' in a way similar to that proposed by Markwitz et al. (2016) for western Australia. Alternatively, could the 'India-Napier-Rayner Belt' block have collided with the 'Crohn Province' when already joined to the 'Mawson Continent' in Mesoproterozoic time? This remains an open question.

2. **Indo-Antarctic Sector**. Almost all the authors involved in studies in this sector describe metamorphic events overprinting the Grenvillian basement, often grouped under the term 'reworking'. Inboard in the continent, along the 'Rayner Belt' and also in the Kaiser-Wilhelm Land (KWL) and Princess Elisabeth Land (PEL), a sequence of reworked rocks of high variable metamorphic grade dated c. 620−500 Ma, including many intrusive bodies, testifies a Pan-African event (the 'Kuunga Orogen' s.l.; Arora et al., 2020; Boger et al., 2001; Corvino et al., 2016; Fitzsimons, 1997; Fitzsimons, 2003; Grew et al., 2012; Kelsey et al., 2007;

Liu et al., 2013, 2014, 2020; Morrissey et al., 2016; Phillips et al., 2009; Wang et al., 2008) related to the final collision of Indian Cratons and 'Rayner Belt' with a composite block including 'Ruker Craton', and here named the 'Crohn Province' including the TOAST. A hypothetical Pan-African suture has been introduced along the border of the Archaean Ruker Craton (Jacobs et al., 2015; Mulder et al., 2019). Some authors exclude the existence of a Pan-African suture in the Lambert Rift region, suggesting that the Kuunga Orogen represents a belt of intraplate reactivation (Phillips et al., 2006; Wilson et al., 2007). Arora et al. (2020) proposed that these Pan-African events in PEL are correlated to the Eastern Ghat Mobile Belt of India. A sequence of high-grade metamorphic reworking events related to the final assembly of Gondwana in the time range 650–500 Ma are also reported from different localities in the Lützow-Holm Bay region (Dunkley et al., 2020; Takahashi et al., 2018; Tsunogae et al., 2016 and ref. therein).

3. **Wilkes –Albany-Frazer Orogen in the Australo-Antarctic Sector**. In this province few evidence of Pan-African events are described. Sheraton et al. (1992) reported an age of c.515 Ma for outcrops of granites and syenites from David Island and c.500 Ma for alkaline mafic dykes at Bunger Hills. Daczko et al. (2018) reported U-Pb SHRIMP age (lower intercept) of c.533 Ma for a granite orthogneiss from Cape Harrison, interpreted as the age of metamorphism.

6.3.3 The 'Ross Orogen' in the Transantarctic Mountains during the late Precambrian–early Paleozoic evolution of the paleo-Pacific margin of Gondwana (c. 600–450 Ma)

The Transantarctic Mountains (Fig. 6.7), at the margin of the East Antarctica, represent a key element in providing an important but still cryptic record of Proterozoic and Early Paleozoic supercontinent history in the period from c.800 to 450 Ma. The existing dataset, and data that can be provided by future research, can be gathered from sedimentary, plutonic and volcanic assemblages that potentially reflect different tectonic events integral to the Rodinia–Gondwana transformation: from the breakup of Rodinia to the development of a transitional margin and plate-margin activity during Gondwana assembly. In the Transantarctic Mountains, constraints on the timing of Rodinia breakup are provided by siliciclastic sedimentation of Neoproterozoic age from two areas.

In the Skelton Glacier area, basalts interlayered with Skelton Group sediments yielded a Sm–Nd model age of 700–800 Ma (Rowell et al., 1993) and were dated (U–Pb zircon age) at c. 650 Ma by Cooper et al. (2011). In the Nimrod Glacier area (Cotton Plateau), gabbros and basalts interlayered with sediments of the Beardmore Group previously dated at 762 Ma (Sm–Nd isochron age, Borg et al., 1990) are now considered to have been

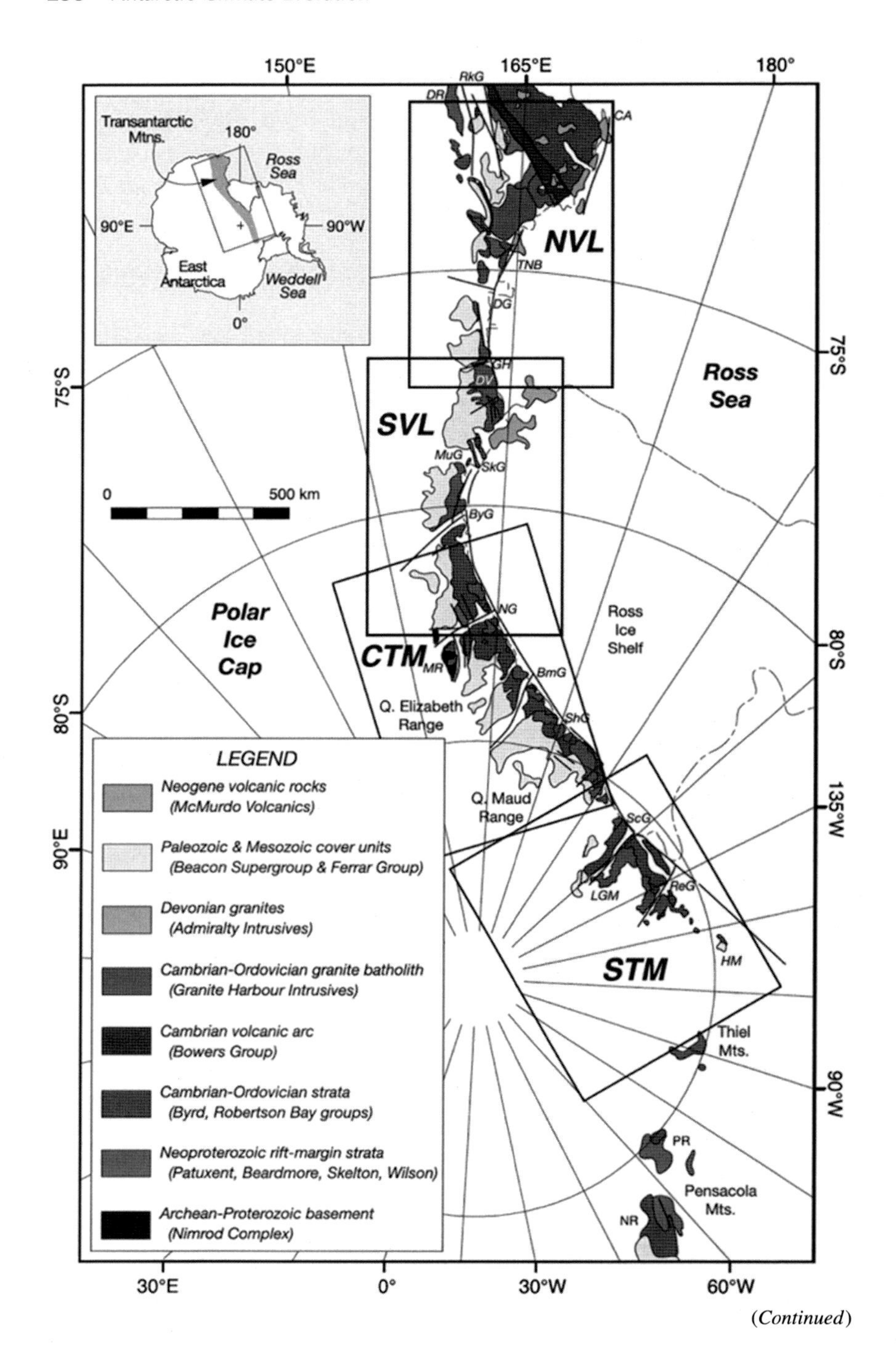

150°E
165°E
180°
Transantarctic Mtns.
180°
Ross Sea
90°E
90°W
East Antarctica
Weddell Sea
0°
NVL
SVL
CTM
STM
Ross Sea
Polar Ice Cap
Ross Ice Shelf
Q. Elizabeth Range
Q. Maud Range
Thiel Mts.
Pensacola Mts.
0
500 km
75°S
80°S
135°W
90°W
30°E
0°
30°W
60°W
LEGEND
Neogene volcanic rocks (McMurdo Volcanics)
Paleozoic & Mesozoic cover units (Beacon Supergroup & Ferrar Group)
Devonian granites (Admiralty Intrusives)
Cambrian-Ordovician granite batholith (Granite Harbour Intrusives)
Cambrian volcanic arc (Bowers Group)
Cambrian-Ordovician strata (Byrd, Robertson Bay groups)
Neoproterozoic rift-margin strata (Patuxent, Beardmore, Skelton, Wilson)
Archean-Proterozoic basement (Nimrod Complex)

(Continued)

emplaced at 668 Ma (U–Pb zircon age, Goodge et al., 2002). Thick siliciclastic successions were long interpreted as deep-water turbidities deposited in Proterozoic time along a rifted margin. Although the older parts of some successions may relate depositionally to the rifting process, recent investigations have demonstrated that some units are nearshore deposits and the depositional age of several major successions must be revised upward. For example, in the central Transantarctic Mountains, sandstones formerly included in the Beardmore Group and considered Neoproterozoic are now assigned to the Middle Cambrian or younger (Goodge et al., 2002). A transformation from drifting to active subduction mode is inferred in the late Neoproterozoic, starting at c. 615–620 Ma and continuing as a protracted contractional tectonic cycle including several discrete tectonic events during the Ross Orogenic cycle, until the Ordovician (Allibone et al., 1993; Armienti et al., 1990; Borg et al., 1987; Goodge, 2020; Goodge et al., 2002, 2010; Talarico et al., 2004 and ref. therein; Hagen-Peter and Cottle, 2016, 2018; Paulsen et al., 2007, 2013; Rocchi et al., 2004, 2011).

As result of these processes the Ross orogenic belt is exposed from Northern Victoria Land at the Pacific end up to the Pensacola Mountains at the Atlantic end. Westernmost Marie Byrd Land (Edward VII Peninsula) has to be considered as part of the same orogenic belt, from which it became isolated only much later, during a major phase of the evolution of the WARS around the end of the Cretaceous. The 'Ross Orogen' is characterised by folds, thrusts, very low- to high-grade metamorphism, granitoids, and flysch- and molasse-type sediments. Remarkable thrusts have been reported from Oates Land, which could be traced into the Australian continuation of the 'Ross Orogen' and the 'Delamerian Orogen' (Flöttmann et al., 1993). The systematic distribution of high- and low-pressure types of metamorphism (e.g., Talarico et al., 2004) and of S- and I-type granitoids (e.g., Vetter and Tessensohn, 1987) led to the model of subduction of the paleo-Pacific beneath East Antarctica. In southern Victoria Land, the older plutons also include a peculiar suite of highly alkaline rocks (nepheline syenites and carbonatites in the 'Koettlitz Glacier Alkaline Province', Cooper et al., 1997) and a stage of Precambrian rift-related magmatism and sedimentation has been documented by Cook (2007) in the Skelton Glacier area and dated at c. 650 Ma by Cooper et al. (2011). In northern Victoria Land, the Ross Orogen is made up

◀ **FIGURE 6.7** Simplified map of the major geologic units underlying the Transantarctic Mountains (after Goodge, 2020). Boxes indicate the main regions: *NVL*, North Victoria Land; *SVL*, South Victoria Land; *CTM*, Central Transantarctic Mountains; *STM*, Southern Transantarctic Mountains. *BmG*, Beardmore Glacier; *ByG*, Byrd Glacier; *CA*, Cape Adare; *DG*, Davis Glacier; *Dr*, Daniels Range; *DV*, Dry Valleys; *GH*, Granite Harbour; *HM*, Horlick Mountains; *LGM*, La Gorce Mountains; *Mr*, Miller Range; *MuG*, Mulock Glacier; *NG*, Nimrod Glacier; *NR*, Neptune Range; *PR*, Patuxent Range; *ReG*, Reedy Glacier; *ScG*, Scott Glacier; *ShG*, Shackleton Glacier; *SkG*, Skelton Glacier; *TNB*, Terra Nova Bay.

of three so-called terranes, as follows: the high- to medium-grade and granite-dominated Wilson Terrane to the west; the low-grade turbiditic Robertson Bay Terrane to the east; and the low-grade and volcanic-rich Bowers Terrane in between. These terranes are considered to be allochthonous or just adjacent paleogeographic domains including an intra-oceanic island arc (Bowers Terrane) and an accretionary wedge (Robertson Bay Terrane) (Tessensohn and Henjes-Kunst, 2005). In northern Victoria Land, a unique occurrence of ultra-high-pressure rocks, including well-preserved mafic eclogites as lenses and pods within metasedimentary gneisses and quartzites, decorates the tectonic boundary between the inboard Wilson Terrane and the Bower Terrane in the Lanterman Range. Geological, petrological and geochronological studies indicate that mafic, ultramafic and felsic host rocks in this region underwent a common metamorphic evolution with an eclogite facies stage about 500 Ma ago at temperatures of up to about 850°C and pressures greater than 2.6 Gpa (Di Vincenzo et al., 1997; Palmeri et al., 2003, 2007). This stage is probably preceded by an eclogitic event at about 530 Ma (Di Vincenzo et al., 2016).

As shown in the timeline reported in Fig. 6.8, a Cryogenian to early Ediacaran stage of extension and subsidence during a pre-orogenic passive-margin phase is marked by bimodal and mafic volcanism and silicoclastic sedimentation between about 670–650 Ma. The onset of the Ross orogenic cycle was marked by c. 600–615 Ma high-grade metamorphism (Hagen-Peter et al., 2016b) and by magmatism at c. 590–565 Ma (Hagen-Peter and Cottle, 2016a) and U-Pb data from glacial granitic clasts (Goodge et al., 2010, 2012). Continuity of magmatism between about 620 and 475 Ma is shown by major peaks in detrital-zircon age spectra from syn-orogenic silicoclastic rocks (Goodge et al., 2002, 2004; Paulsen et al., 2016). Periods of alternating convergent-margin upper plate contraction and extension are inferred from regional structural data and alternating changes in syn-tectonic magma compositions in the Granite Harbour Intrusives (Fig. 6.9), with structural data from northern Victoria Land, Central Transantarctic Mountains, Southern Transantarctic Mountains, and Pensacola Mountains also indicating strain partitioning during sinistral-oblique underflow. The Ordovician upper bound of the Ross orogenic cycle, and establishment of stable post-orogenic crust, is marked at about 470 Ma by an end of magmatism and low-T metamorphic cooling ages in all areas.

The total duration of the Ross cycle is therefore in the range of about 145 Ma and a typical contractional period in the long-lived evolution of the orogenic belt (Fig. 6.10) is commonly modelled based on data from the central Transantarctic Mountains, as the result of relative plate motions and partitioned into both contractional and strike-slip components of deformation. According to this model (Goodge et al., 1993a, 2004a) oceanward retreat of the subduction zone to the east (present-day coordinates) is inferred in order to explain offshore migration of the magmatic axis. The late Neoproterozoic and early Paleozoic tectonic evolution of the East Antarctic margin relate to the long-lived subduction between the East Antarctic shield and proto-Pacific oceanic lithosphere and occurred in four

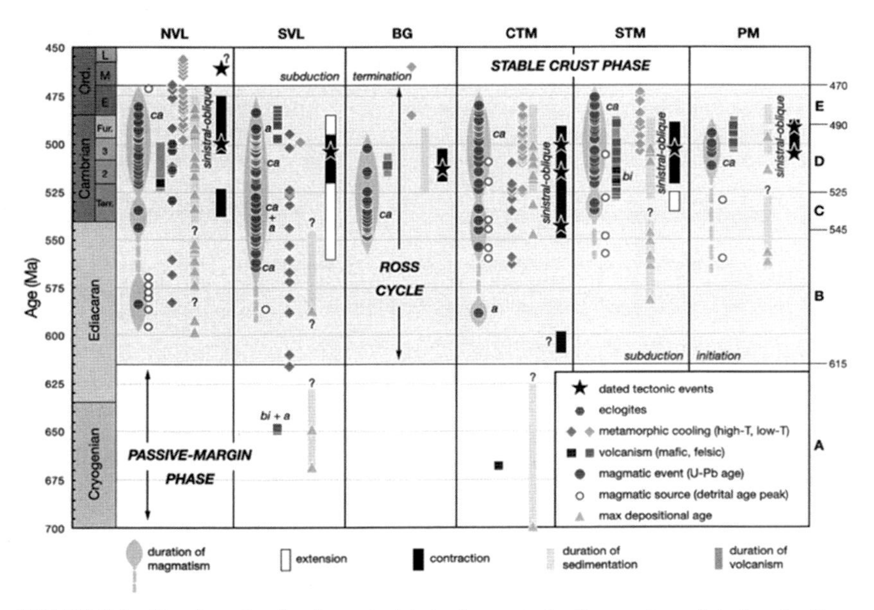

FIGURE 6.8 Timeline showing important tectonic events leading up to and during the Ross Orogenic cycle (after Goodge, 2020). Events with geochronological control are shown by symbols in the legend, and duration of events inferred from stratigraphic, magmatic, and structural relations are shown by colored vertical bars indicated in the diagram. *BG*, Byrd Glacier area; *CNVL*, northern Victoria Land; *CTM*, central Transantarctic Mountains; *PM*, Pensacola Mountains; *STM*, southern Transantarctic Mountains; SVL, southern Victoria Land.

(Goodge et al., 2004a, Fig. 6.11) or five main subsequent stages (synoptic diagram of Fig. 6.8 from Goodge, 2020): 700–615 Ma, a passive-margin stage (A); 615–545 Ma, a platform and incipient arc stage (B); 545-525 Ma, a synorogenic stage (C); 525–490 Ma, a late-orogenic stage (D); 490–470 Ma, a post-orogenic stage (E). The whole tectonomagmatic evolution is confirmed as a complex long-lived active continental margin in an accretionary-type orogenic system.

6.4 West Antarctic Accretionary System

Many papers have been published on the geological events that have modified the original Gondwanian paleo-Pacific margin of the 'Ross Orogeny' up to the present. Detailed recent reviews are those of Goodge (2020), Jordan et al. (2020) and Siddoway (2021). A highly dynamic history of convergence, continental growth and magmatism, alternating with extensional phases and fragmentation-rifting–rotation of microplates, built a complex accretionary system grossly starting from the inboard areas up to the most external ones where a subduction zone is still active in the South Shetland Islands. Such a tectonic evolution is common within accretionary orogens developed at active plate

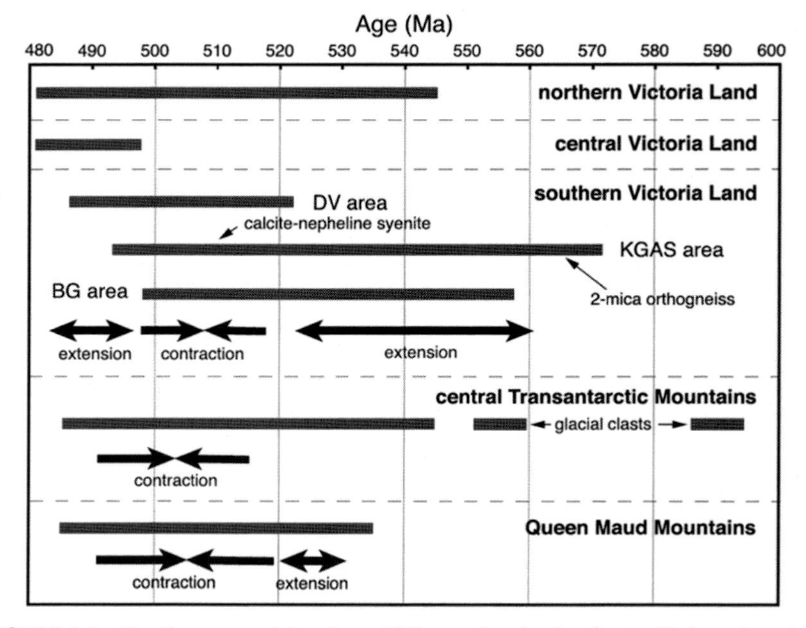

FIGURE 6.9 Timeline summarising zircon U\Pb age data for the Granite Harbour Intrusives along the Ross Orogen and related tectonic events as proposed by Hagen-Peter et al. (2016).

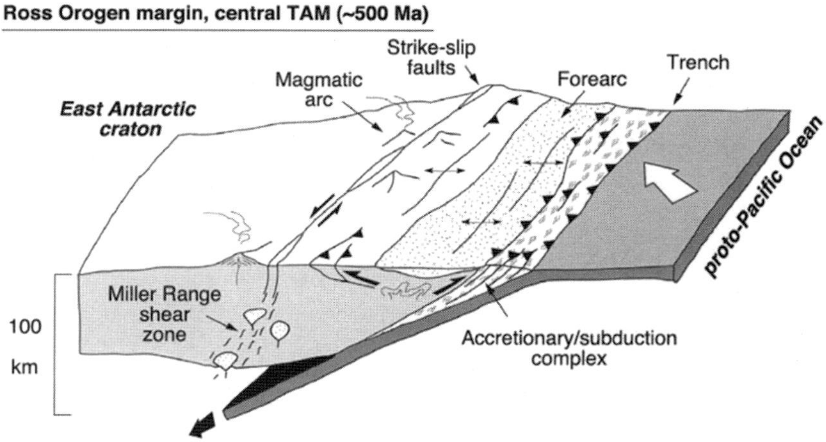

FIGURE 6.10 Schematic diagram of Transantarctic Mountains plate-boundary regime at ~500 Ma (after Goodge, 2020).

margins such as the American Cordillera or the Australian Tasmanides (Cawood and Buchan, 2007; Cawood et al., 2016; Collins, 2002). Convergence along the continental margins is often oblique, so continental fragments of the margin can easily be translated along strike and rotated.

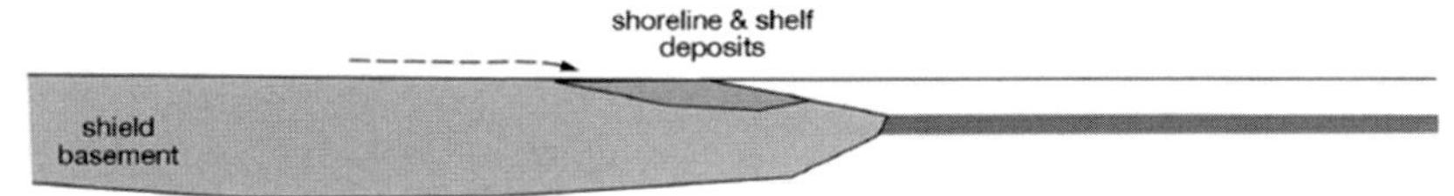

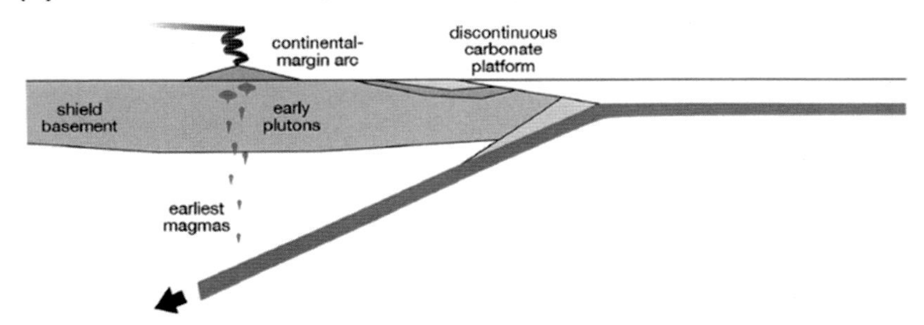

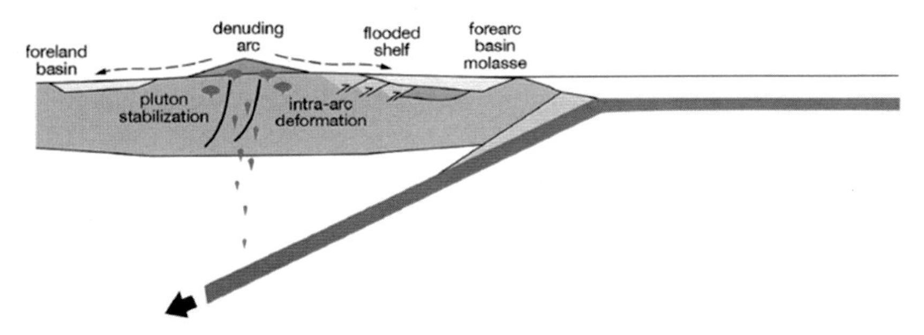

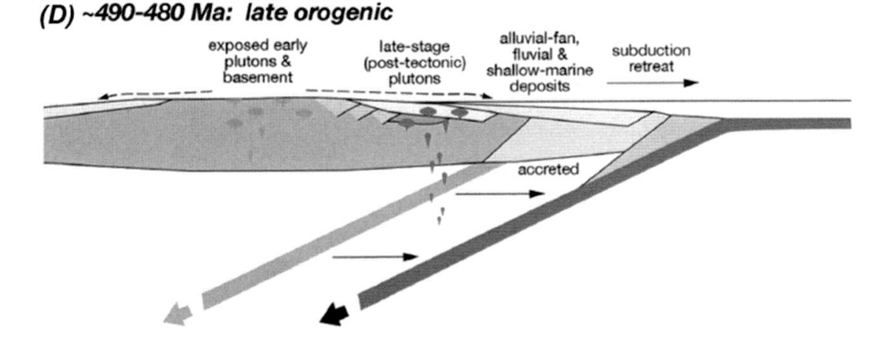

FIGURE 6.11 Model of the late Neoproterozoic and early Paleozoic tectonic evolution of the East Antarctic margin in the central Transantarctic Mountains (after Goodge, 2020). Crustal thicknesses approximately to scale, but sedimentary basins exaggerated in thickness for clarity.

6.4.1 West Antarctica in the Precambrian to Mesozoic (c. 180 Ma) evolution of Gondwana until the middle Jurassic breakup

In this time window, the main geological events so far known in the Antarctic geological record include the formation of a long-lived continental basin where a dominantly fluvial succession, the Beacon Supergroup, was deposited in Devonian to early Jurassic times, and the development of two main magmatic and deformational belts generally known as the middle Paleozoic 'Borchgrevink' and the Permo-Triassic 'Weddell' orogens.

6.4.1.1 Precambrian to Cambrian metamorphic basement

A wide domain of the West Antarctic Rift System, in inboard position adjacent to the Transantarctic Mountains, exposes a lower continental crust composed by crystalline rocks of Cambrian or older Mesoproterozoic age.

The Haag Nunataks crustal block (Jordan et al., 2017) is composed by felsic orthogneisses dated in the range of c. 1240–1060 Ma (emplacement) and deformed at c. 1060 Ma, that is during the Grenvillian Orogeny (Millar and Pankhurst, 1987; Riley et al., 2020). A similar Grenvillian age is proposed for the unexposed basement in the Ellsworth-Mountain crustal block (Castillo et al., 2017).

In the Ross Sea Rift System a thinned continental crustal of Cambro-Ordovician age is commonly assumed but recovered only from the DSDP site 270 and Iselin Bank (Mortimer et al., 2011). In the Ross Province of Mary Byrd Land several outcrops of Cambrian rocks are found. The Swanson Formation is the oldest exposed metasedimentary turbidite sequence related to the late – Ross forearc sedimentation (c. 514–490 Ma) and deformed by the subsequent tectonic events (Siddoway and Fanning, 2009 and ref. therein). At Mount Murphy a granodiorite orthogneiss has been dated at c. 505 Ma and correlated to the late magmatic pulses of the Ross Orogeny (Pankhurst et al., 1998).

6.4.1.2 Devono-Carboniferous arc magmatism ('Borchgrevink Event') (c. 370–350 Ma)

The occurrence in northern Victoria Land of a calc-alkaline magmatic suite of middle Paleozoic age (c. 370–350 Ma; Borg et al., 1986, 1987; GANOVEX Team, 1987; Vetter et al., 1983) is attributed to an orogenic event separated from the waning phase of the Ross Orogeny (Craddock, 1970) by about 80 Ma. Often related to the middle phase of the 'Lachlan fold and thrust Belt' in eastern Australia (Forster and Goscombe, 2013 and ref. therein) the plutonic suite includes dominant granitoids (Admiralty Intrusives) and minor felsic volcanic (Gallipoli Volcanics, Fioretti et al., 2001). The granitoids form several plutons which mainly occur in the Robertson Bay Terrane, although some stitch the tectonic boundary between the Bowers and Wilson Terrane, where the coeval volcanics are also concentrated. All the intrusions are characterised by isotropic fabrics and discordant contacts with respect to the surrounding metasediments suggesting a post-tectonic emplacement. These features, and meagre evidence from radiometric data for concomitant metamorphism and deformation, have so far prevented a comprehensive reconstruction of the tectonic setting and development of an orogenic event called the 'Borchgrevink Orogeny' (Craddock, 1970) in northern Victoria Land. Nevertheless, the continental arc geochemical signature of the Devonian–Carboniferous magmatic suite indicates that the plutonism most likely occurred as the result of a renewed period of subduction activity along the-paleo-Pacific margin (Borg and DePaolo, 1991; Kleinschmidt and Tessensohn, 1987). Similar suites are known in Tasmania, New Zealand and the Campbell Plateau (Bradshaw et al., 1997; Gibson and Ireland, 1996), and in the Ford Ranges of Marie Byrd Land (emplacement age c. 375–345 Ma, Nelson and Cottle, 2018; Yakymchuk et al., 2015 and ref. therein) and in the Antarctic Peninsula (Riley et al., 2016). In Thurston Island granodiorite orthogneisses in the basement are dated at c. 349 Ma (Riley et al., 2016) and 347 Ma (Nelson and Cottle, 2018) suggesting the presence of a rifted crustal micro-plate traslated in to West Antarctica from an inboard position (see Fig. 6.14 from Paulsen et al., 2017).

A correlation between the 'Borchgrevink Event' of northern Victoria Land and the Tasmanian Orogeny was put forward by Findlay (1987). Elsewhere in Antarctica, marine sediments from the Ellsworth Mountains and Pensacola Mountains are considered to have been deposited in the same time window, within an intracontinental basin that, may have extended from the Weddell Sea to the Ross Sea (Elliot, 1975).

6.4.1.3 Beacon Supergroup (Devonian-Permo-Triassic-earliest Jurassic)

The Beacon Supergroup consists of dominantly continental sedimentary deposits forming a generally flat-lying, 0.5–3.5 km thick cover developed over a marked unconformity (Kukri Peneplain) above the Ross orogenic belt throughout most of the Transantarctic Mountains (Elliot, 2013 and ref. therein) (Fig. 6.12). Similar sequences are known in limited outcrops in East Antarctica (e.g., Prince Charles Mountains, Dronning Maud Land and the Ellsworth Mountains) (Barrett, 1991) and as bedrock of the Cenozoic glaciomarine sediments in the Victoria Land Basin (VLB) in the Ross Sea as documented by the CRP-3 drill-hole (Cape Roberts Science Team, 2000). Outside Antarctica, similar sedimentary rocks, collectively called Gondwanian Sequences, occur in South Africa, Australia and South America. The deposition of these sediments in Antarctica started in Devonian time with the Taylor Group, consisting of dominantly quartz-arenites and conglomerates. A product of erosion and fluvial processes under arid and semiarid conditions (Campbell and Claridge, 1987), the Taylor Group accumulated in a series of basins along the paleo-Pacific margin of Gondwana. The deposition of fossiliferous

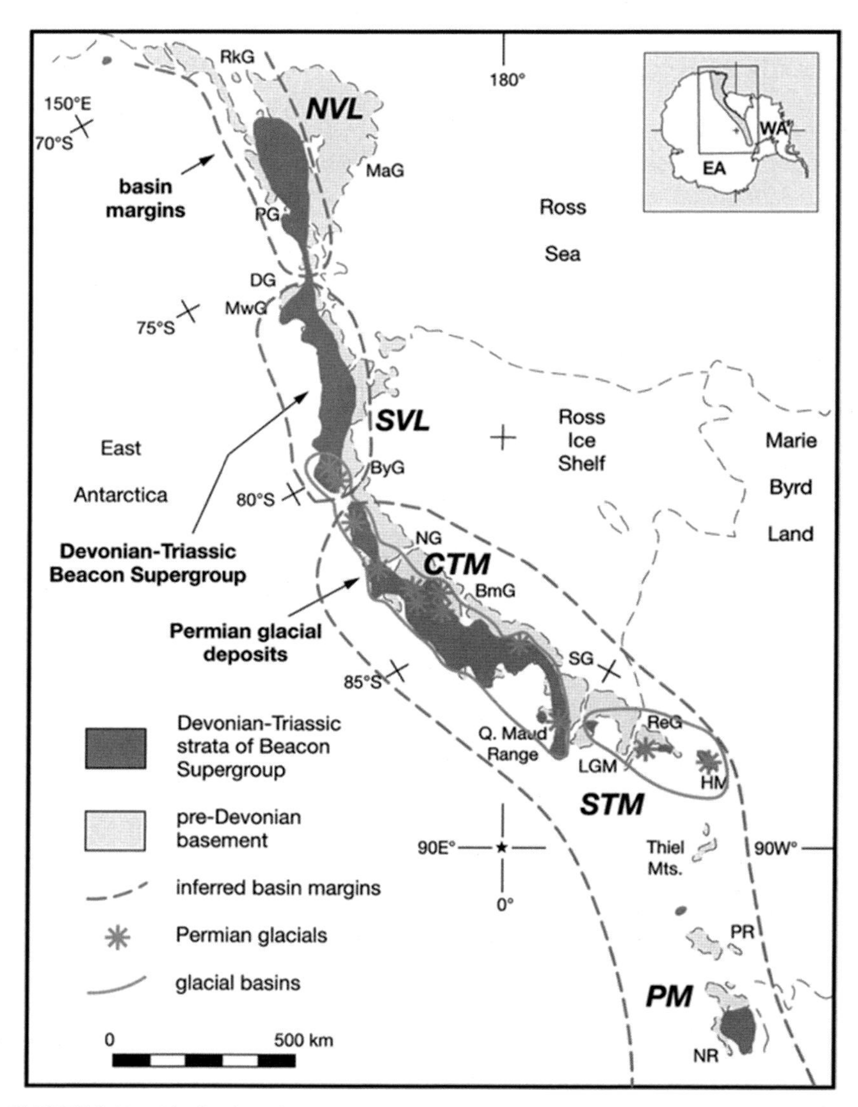

FIGURE 6.12 Distribution of Beacon Supergroup strata in the Transantarctic Mountains (after Elliot, 2013, as modified by Goodge, 2020).

siltites was followed by an erosional phase, probably related to a glacial event which led to the deposition of glaciogenic sediments in late Carboniferous–Early Permian time. In the Transantarctic Mountains, the Carboniferous to Triassic sediments form the Victoria Group, which includes carbonaceous layers and feldspathic sandstones. A summary of the main lithostratigraphic features in the central Transantarctic Mountains, southern Victoria Land and northern Victoria Land is displayed in Fig. 6.13. The low sedimentation rate (c.12.5 m/My) and absence of

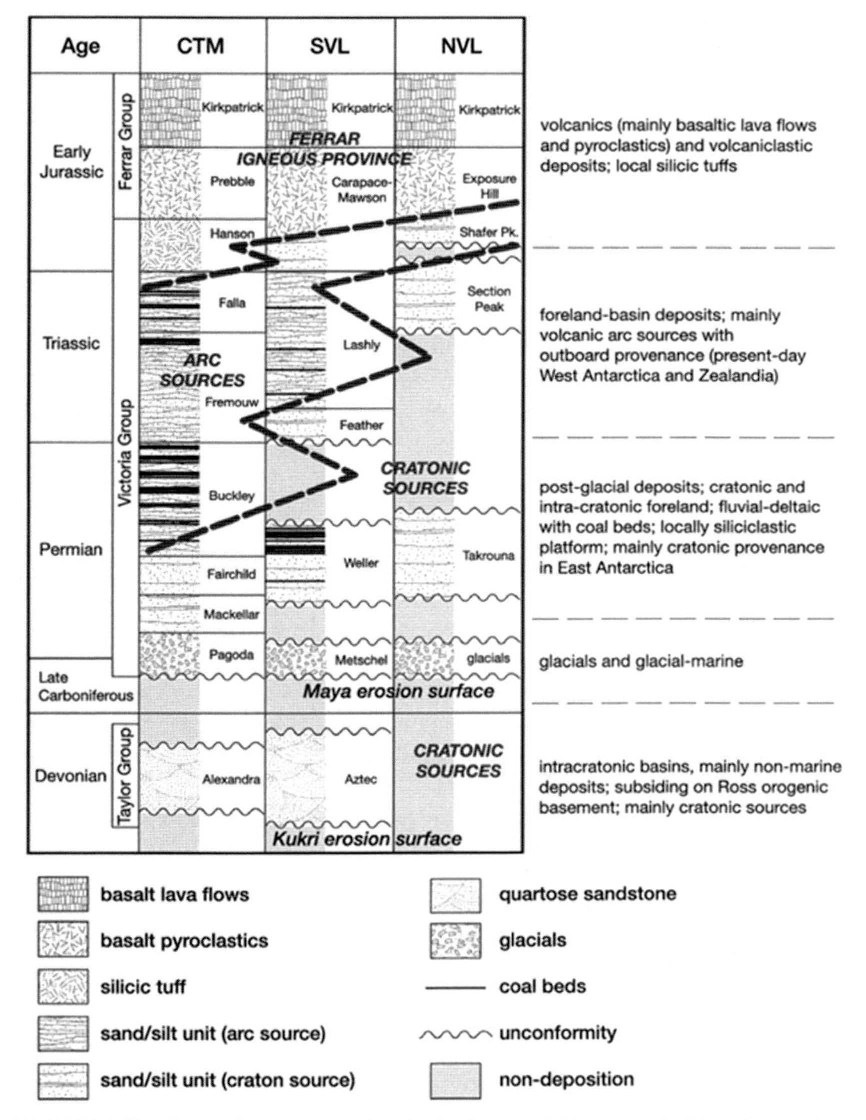

FIGURE 6.13 Beacon Supergroup stratigraphy in the central Transantarctic Mountains, southern Victoria Land and northern Victoria Land, showing major stratigraphic units of the Beacon Supergroup and Ferrar Group. Trends in depositional environments and sediment sources are generalised (after Elliot, 2013 as modified by Goodge, 2020).

concomitant compressional deformation suggest that deposition occurred over a thick continental crust (Barrett, 1991) but different tectonic models have been proposed, including a passive margin (Isbell, 1999), intra-cratonic (Barrett, 1991; Woolfe and Barrett, 1995) or marginal/back-arc (Bradshaw and Webers, 1988).

6.4.1.4 The Ellsworth-Whitmore Mountains Terrane and the Permo-Triassic arc magmatism

The Ellsworth or Weddell Orogeny (or – in a larger context – also known as the Gondwanide Orogeny which spanned parts of southern Africa – Cape Fold Belt – and South America – Sierra de la Ventana Fold Belt) occurred in Permo-Triassic time c. 250–200 Ma. As noted by Cawood (2005), this orogeny overlaps with the end Paleozoic assembly of Pangea (Li and Powell, 2001), through ocean closure and accretion of Gondwana, Laurussia (Laurentia + Baltica) and Siberia, as well as completion of terrane accretion in the Altaids. Stratigraphic and geochronological data (Dalziel, 1982; Dalziel and Elliot, 1982; Johnston, 2000; Storey et al., 1987; Trouw and De Wit, 1999) indicate that Permo-Triassic deformation of variable intensity and distribution occurs throughout West Antarctica and the adjoining Cape Fold Belt of southern Africa. In Antarctica, this orogenic event is well documented in the Ellsworth–Whitmore Mountains and in the Pensacola Mountains, where upright to inclined folds with axial planar cleavage are inferred to have formed in a dextral transpressive environment (Curtis, 1998). The Ellsworth–Pensacola Mountains chain represents the fold belt. It merges with the Ross Orogen in the Pensacola Mountains (Fig. 6.2), where the tectonism partly overprinted the older Ross-aged structures. Elsewhere, deformation is heterogeneously distributed, with Storey et al. (1987) noting that in the Antarctic Peninsula, unconformities previously ascribed to the Gondwanide Orogeny are younger and that the only event related to Gondwanide deformation is the regional metamorphism at 245 Ma. Parts of the Trinity Peninsula Group where affected. The trend of the Ellsworth Orogen is at a high angle to the paleo-Pacific margin of Antarctica and the Ross Orogen. This obliqueness is due to secondary rotation, determined from paleomagnetic investigations by Funaki et al. (1991) and confirmed by Randall and Mac Niocaill (2004).

Triassic magmatism is widespread in the Antarctic Peninsula (Figs. 6.2 and 6.14) and dated c.200–230 Ma (Rb/Sr and U/Pb methods) in Graham Land, eastern Palmer Land and Thurston Island (Bastias et al., 2020; Flowerdew et al., 2006; Pankhurst, 1982; Riley et al., 2012, 2017; Vaughan and Storey, 2000; Weaver et al., 1994) and eastern Mary Byrd Land (Nelson and Cottle, 2018). These magmatic rocks are generally orthogneisses derived by protoliths with a calcalkaline to alkali-calcic affinity and a variable peraluminous character. They testify to the presence of an active Andean-style continental magmatic arc all along the Antarctic margin to the Patagonian belts. Metamorphic age of these intrusives is debated, and referred mainly to a middle Cretaceous contractional event (Vaughan et al., 2012).

6.4.1.5 Ferrar Supergroup and the Gondwana breakup (c. 180 Ma)

The first major tectonic stage in the breakup of Gondwana corresponds to an initial rifting phase that started in the Weddell Sea, initially as a back-arc

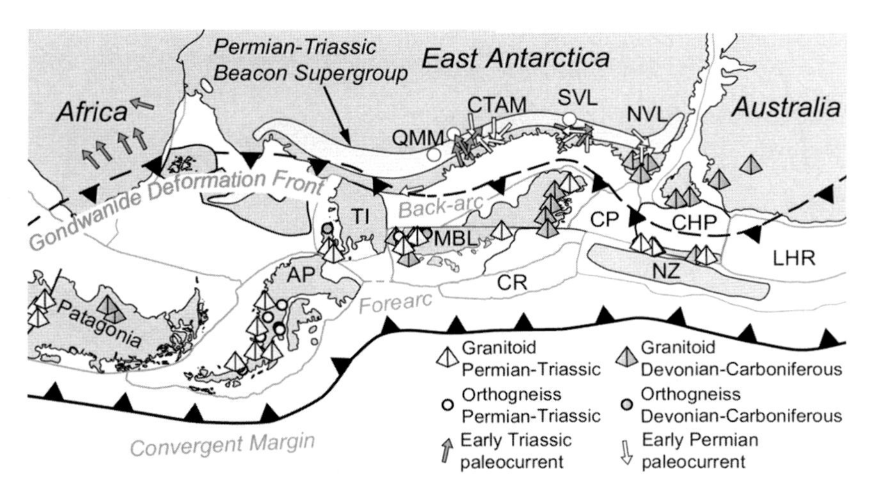

FIGURE 6.14 Permian–Triassic Gondwana reconstruction showing the distribution of Permian– Triassic strata of the Beacon Supergroup with respect to the trace of the Gondwanide deformation front and Permian–Triassic plutonic and metamorphic rocks located in the orogenic hinterland. *AP*, Antarctic Peninsula; *CR*, Chatham Rise; *CP*, Campbell Plateau; *CHP*, Challenger Plateau; *CTAM*, central Transantarctic Mountains; *LHR*, Lord Howe Rise; *MBL*, Marie Byrd Land; *NVL*, north Victoria Land; *NZ*, New Zealand; *QMM*, Queen Maud Mountains; *SVL*, south Victoria Land; *TI*, Thurston Island (after Paulsen et al., 2017).

basin, in the Late Jurassic (Fig. 6.15) (Heimann et al., 1994; Jordan et al., 2017, 2020; Lawver et al., 1991; Riley et al., 2020 and ref. therein). This stage involved right-lateral transtension as East Gondwana (Antarctica, Australia, India and New Zealand) and West Gondwana (South America and Africa) moved apart with stretching beginning in the north and propagating southward (Lawver et al., 1992). Initial breakup involved a complex geodynamic evolution characterised by the rotation and translation of several microplates such as the Ellsworth Mountains block; a displaced part of the Gondwanide fold belt. The original position of the various microplates is still controversial, but the Ellsworth–Whitmore Mountains crustal block most likely originated from the paleopacific margin of the Ross Orogen (Elliot et al., 2016). Rotation of West Antarctic microplates must have been accomplished by c. 165 Ma (the time of opening of the Weddell Sea), and rotation of the Ellsworth–Whitmore Mountains crustal block was finished before translation into its present position by 175 Ma (Grunow et al., 1987).

Rotation of microplates did not apparently involve the production of oceanic crust (Marshall, 1994) but rather a crustal block rotation with controlling faults concealed beneath Mesozoic sedimentary basins in the Weddell Sea (Storey, 1996; Storey et al., 1988). The breakup has been explained by several authors as the result of a hot mega-plume, which, according to Storey and Kyle (1999), could have promoted domal uplift and formation of a triple junction. According to Dalziel et al. (1999), the Gondwana plume

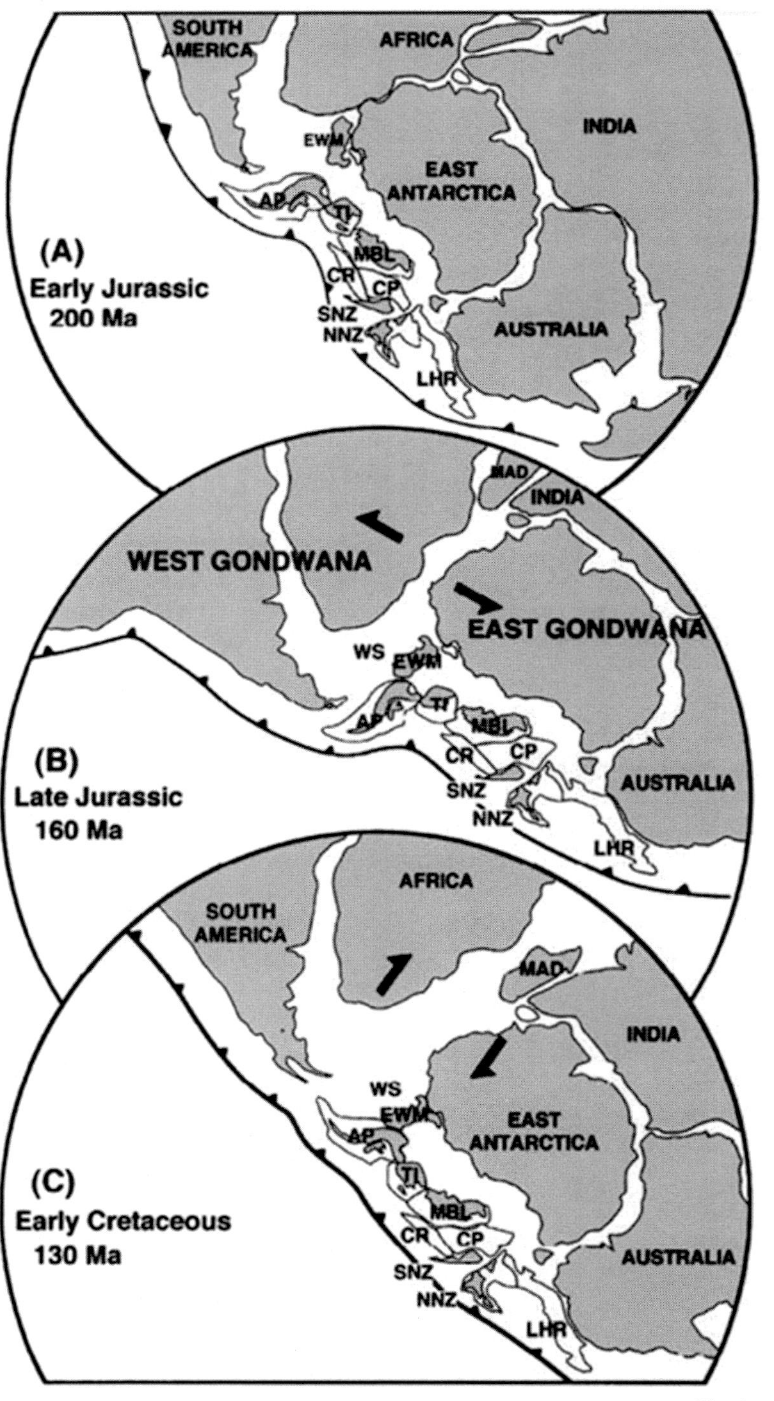
SOUTH AMERICA
AFRICA
INDIA
EWM
EAST ANTARCTICA
AP
TI
(A)
Early Jurassic
200 Ma
MBL
CR
CP
SNZ
NNZ
AUSTRALIA
LHR
MAD
INDIA
WEST GONDWANA
EAST GONDWANA
WS
EWM
AP
TI
MBL
(B)
Late Jurassic
160 Ma
CR
CP
SNZ
NNZ
AUSTRALIA
LHR
AFRICA
SOUTH AMERICA
MAD
INDIA
WS
EWM
EAST ANTARCTICA
AP
(C)
Early Cretaceous
130 Ma
TI
MBL
CR
CP
SNZ
NNZ
AUSTRALIA
LHR

(Continued)

may have also caused or expedited formation of the early Mesozoic Gondwanide fold belt, due to the buoyancy of a hot plume acting on the downgoing slab and causing it to flatten. In both cases, plume activity led to the production of magma batches reflecting different degrees of plume–lithosphere interaction, which migrated along crustal shear zones to ultimately form the various large igneous provinces. The plume-related magmatic products are represented by huge within-plate mafic and felsic magmatic provinces in many Gondwana continents as well as Antarctica (Cox, 1988; Storey, 1996; White and MacKenzie, 1989). In the Transantarctic Mountains (LeMasurier and Thomson, 1990), Jurassic mafic rocks are generally known as the Ferrar Supergroup or the Jurassic Ferrar Large Igneous Province (FLIP) (Fig. 6.16). They are divided into a volcanic component, the Kirkpatrick Basalt – preceded by extensive phreatomagmatic volcanoclastic rocks (Elliot et al., 2006; Viereck-Götte et al., 2007) – and the intrusive dolerite (diabase) sills and dikes of the Ferrar Dolerite. Geochemically, the FLIP is unusual with upper crustal-like characteristics such as high-large ion lithophile element concentrations and enriched isotopic signatures ($^{87}Sr/^{86}Sr > 0.709$). The crust-like signature suggests derivation from an enriched lithosphere, possibly connected to subduction along the Pacific margin of Gondwana during the Paleozoic and Mesozoic. The mafic rocks are concentrated within a long linear belt exposed in Tasmania, Antarctica and South Africa (Ivanov et al., 2017). Cox (1988) considered that the linear pattern of the Ferrar and Tasman provinces could not be compatible with classic circular plumes and proposed a hot line rather than a hot spot. A number of rifts that intersected at the Dufek Massif (Elliot, 1992) could have favored the development of zones of weakened lithosphere, which acted as pathways for lateral migration of magmas derived from lithospheric sources. In spite of the still not completely understood tectonic setting, it is important to note that, similarly to other continental flood basalt provinces, all the mafic products formed during a short period of eruption (Tasman province: 175 ± 18 Ma; Ferrar Supergroup: 180–183 Ma; Dronning Maud Land province: 177 ± 2 Ma; Karoo province: 182 ± 2 Ma) (Heimann et al., 1994; Hergt et al., 1989). Coeval felsic intrusions considered to be rift-related are known in West Antarctica (Storey et al., 1988) and southern South America (Chon-Aike or Tobıfera province) (Gust et al., 1985). The

◀ **FIGURE 6.15** (A) Gondwana tight fit reconstruction (Early Jurassic). (B and C) Major initial stage in the break-up of Gondwana: (A) Initial rifting stage (Late Jurassic); (B) Change in the Gondwana break-up stress regime from dominantly North–South between East and West Gondwana to dominantly East–West with the two-plate system being replaced by a multiple-plate system (Early Cretaceous) (after Fitzgerald, 2002; Lawver et al., 1992, 1998). *AP*, Antarctic Peninsula; *TI*, Thurston Island; *MBL*, Marie Byrd Land; *CR*, Chatham Rise; *CP*, Campbell Plateau; *SNZ*, Southern New Zealand; *NNZ*, Northern New Zealand; *LHR*, Lord Howe Rise; *WS*, Weddell Sea. Continent and microplate positions are from Lawver et al. (1992, 1998) with other information from Storey (1996).

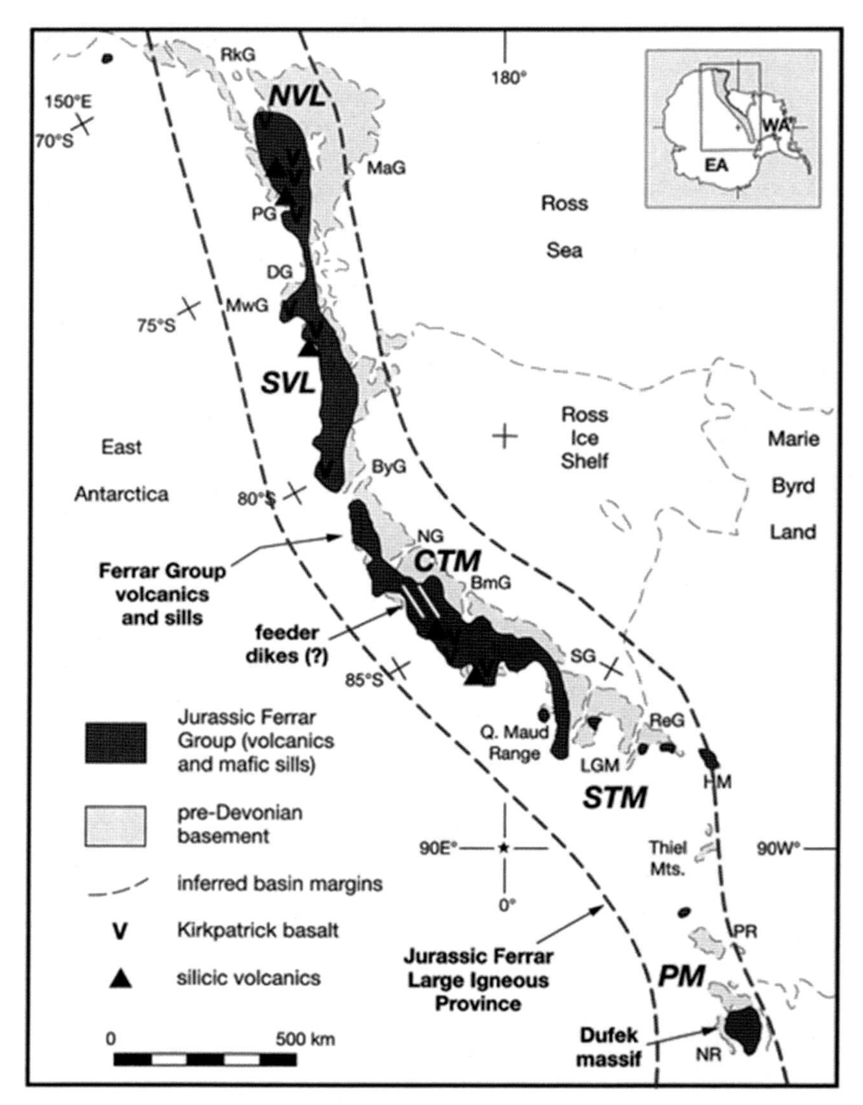

FIGURE 6.16 Distribution of the Ferrar Large Igneous Province in the Transantarctic Mountains (after Elliot, 2013; Elliot and Fanning, 2008 as modified by Goodge, 2020). Location of possible feeder dikes in the Nimrod Glacier area imaged in aeromagnetic data (Goodge and Finn, 2010).

second major geodynamic stage occurred in the Early Cretaceous (Fig. 6.15), as a consequence of the change of the Gondwana breakup stress regime from dominantly north–south between East and West Gondwana to dominantly east–west with the two-plate system being replaced by a multiple-plate system (e.g., Lawver et al., 1992).

Felsic intrusives outcrop in the Whitmore Mountains region (Craddock et al., 2017; Leat et al., 2018) at Pagano Nunatak, Pirrit Hills, Nash–Martin Hills and Linch Nunataks. These granites, emplaced at c. 174–178 Ma, post-date the Ferrar effusive event. They are mildly peraluminous, two-mica granites with chemical composition suggesting a within-plate affinity. While the origin is debated, Craddock et al. (2017) considered that they formed from hybridised magmas between mantle melts and a lower continental crustal contribution.

This modification in stress regime is thought to have induced large-scale ductile deformation concentrated along shear zones in the Antarctic Peninsula (Storey et al., 1996) and thin-skinned deformation and inversion of existing sedimentary basins such as the Latady Basin (Kellogg and Rowley, 1989). By ca. 110 Ma, the microplates of West Antarctica had nearly reached their present location with respect to East Antarctica. In the same time period, separation also began between India and Antarctica (Lawver et al., 1991). Initial stretching between Australia and Antarctica began as early as 125 Ma (Stagg and Willcox, 1992), but sea-floor spreading was delayed to c. 95 Ma (Cande and Mütter, 1982; Royer and Rollet, 1997; Veevers et al., 1990) in the Ross embayment (Elliot, 1992), as well as extension between the Lord Howe Rise and northern New Zealand (Lawver et al., 1992). By the Late Cretaceous, Antarctica had reached its final polar location and configuration, and the final stage of break-up was completed when Zealandia (New Zealand plus Campbell Plateau; Mortimer, 2017) rifted from Marie Byrd Land at 84 Ma (e.g., Lawver et al., 1991; Mortimer et al., 2019; Stock and Molnar, 1987). In this geodynamic context, many particularly large and conspicuous intraplate fracture zones have been investigated. All related to extensive and prolonged extensional regimes spanning in time from Mesozoic to present (Fig. 6.2). These major extensional zones include:

- the Lambert Graben or Lambert Rift (East Antarctica);
- the Graben of Jutulstraumen and Penckmulde (occasionally called Jutul Penck Graben East Antarctica);
- the WARS (West Antarctic Rift System) including the Ross Sea Rift System (RSRS) and the Weddel Sea Rift System (WSRS); and
- the Rennick Graben as the main element of a strike–slip fault system in Victoria and Oates Lands.

The **Lambert Graben**, developed in the East Antarctic Craton, is filled mainly by sediments of the Permo-Triassic Beacon Supergroup and was interpreted by Harrowfield et al. (2015) as an accommodation zone of a wide intracontinental rift that extended from Australia's North West Shelf, between India and Antarctica, to southern Africa. Faulting started during the early Palaeozoic, reached its peak in the Permian and continued to the Early Cretaceous (Hofmann, 1996; Mikhalsky and Leitchenkov, 2018). Possibly, the trench in which subglacial Lake Vostok (Kapitisa et al., 1996; Siegert et al., 2011) sits belongs to the same rift system, but somewhat offset. The continuation of the Lambert Graben in the Indian landmass is the Mahanadi Rift south-west of Calcutta in the state of Orissa (Hofmann, 1996), filled with sediments of the same type and age as the Lambert Graben. The reconstruction of the Gondwanian India–Antarctica fit using these graben systems coincides with reconstructions by Archaean to Early Proterozoic elements for Gondwana.

Many other subglacial topographic basins similar to rift structures have been identified by geophysical methods beneath the thick Antarctic Ice Sheet. These basins, whose surface is often located below sea level, record a multi-phase history of rifting and subsidence and are the depocentres of thick sedimentary sequences (Aitken et al., 2014; Maritati et al., 2016 and ref. therein). Several hundred subglacial lakes have been identified in the topographic depressions (Siegert et al., 2005; Wright and Siegert, 2012).

The **Ross Sea Rift System** is extremely wide (about 1000 km). Extension and subsidence started during the late Mesozoic (about 140 Ma ago) and reached its main activity in the Cretaceous and Early Tertiary. An event at 55-50 Ma produced an enormous relief at its western shoulder, the Transantarctic Mountains. The crustal extension, combined with alkaline intra-continental volcanism, is still active at Mt. Erebus (3794 m) and at Mt. Melbourne (2732 m), both located in Victoria Land (Kyle, 1990a,b; Kyle and Cole, 1974; Tessensohn and Wörner, 1991; Wörner et al., 1989).

The Weddel Sea Rift System.

The **Jutul Penck Graben** of western Dronning Maud Land originated probably around 140 Ma ago or a little bit later (Jacobs and Lisker, 1999). The graben marks the boundary of the Grunehogna Craton towards the south-east which follows an ancestral geological structure.

The Rennick Graben. The strike–slip fault system of Victoria and Oates Lands runs obliquely to the Ross Sea Rift and is cut by it. The principal element of the system is the Rennick Graben, which is presently active, as demonstrated by earthquakes in 1952, 1974 and 1998. The graben contains downfaulted Ferrar volcanics and sediments of the Beacon Supergroup, which have been spectacularly folded and squeezed along the graben shoulders (Rossetti et al., 2003). These structures demonstrate alternating dextral transpression and transtension (with formation of pull-apart basins), within the setting of a complicated strike–slip system (Rossetti et al., 2003). The 1974 earth quake occurred some 120 km to the west at the parallel structure of the Matusevich Glacier. There is an ongoing discussion of the interpretation that the strike–slip fault system represents the continuation of oceanic fracture zones between Australia and Antarctica (e.g., the Tasman Fracture Zone) into the continental crust of Antarctica (Kleinschmidt and Läufer, 2006; Salvini and Storti, 2003; Salvini et al., 1997).

6.4.1.6 The Antarctic Andean Orogen

The orogen of the Antarctic Andes occupies the entire Antarctic Peninsula down to the Walgreen Coast (Fig. 6.2). It formed mainly in three episodes partly involving an older Permo-Triassic basement and may be an even older Paleozoic lower crust (Bastias et al., 2020 and ref. therein): (1) Late Jurassic through Early Cretaceous (150–140 Ma); (2) mid-Cretaceous (c.105 Ma); and (3) Tertiary (c.50 Ma to recent), and is partly still active (e.g., Birkenmajer, 1994; Burton-Johnson and Riley, 2015; Vaughan and Storey, 1997). Thus, it represents the youngest growth zone of the continent. The Antarctic Andes are a typical subduction orogen accompanied by orogenic magmatism in the form of granitic plutons and volcanic rocks. In detail, the deformation and metamorphism are very complicated, because they are polyphase. Folding and thrust faulting is reported mainly from the southern portion of the Antarctic Peninsula (Palmer Land and Alexander Island) and from the extreme north (Trinity Peninsula and eastern South Shetland Islands). The distribution of related metamorphism is also heterogeneous, including high-pressure metamorphism with blueschists characteristic of subduction complexes, e.g., on Elephant Island (Trouw et al., 1991).

The only plate-tectonically more or less active in Antarctica is situated north-west of the Antarctic Peninsula, in the South Shetland Islands (from Snow Island in the south-west up to Elephant Island in the north-east) and the Bransfield Strait (Fig. 6.2). North-west of the South Shetland Islands, a small section of the ocean floor, called the Drake Plate (i.e. the remnant of the older, but largely subducted Phoenix Plate), is being subducted at the South Shetland Trench beneath the Antarctic Plate. Related, mainly andesitic, volcanism forms the island arc of the South Shetland Islands. Parts of the South Shetland Islands (part of Livingston Island and Elephant Island) belong – as does the Peninsula itself – to earlier stages of the Antarctic Andean orogeny and consist of strongly deformed Jurassic trench sediments. The Bransfield Strait is located south of the subduction-related volcanic island arc and forms an active extensional basin accompanied by tholeiitic volcanism, partly submarine, partly as active island volcanoes (e.g., Penguin Island and Deception Island; Smellie and Lopez-Martınez, 2002). The Bransfield Strait often is regarded as a classic example of a back-arc basin, but recently this has been disputed (Gonzales-Casado et al., 2000).

6.5 Mesozoic to Cenozoic Tectonic Evolution of the Transantarctic Mountains

The Transantarctic Mountains, one of the major young mountain chains on Earth, separate East Antarctica from the West Antarctica Rift System and the largely land-based East Antarctic Ice Sheet from the marine-based West Antarctic Ice Sheet over a substantial portion of the Antarctic interior along the margin of the Ross embayment (Fig. 6.1B). The regional structural architecture of the Transantarctic Mountains remains poorly known in most regions because of the extensive ice cover. The East Antarctic Ice Sheet hides the structure of the mountains along the Polar Plateau, preventing the identification of the extent of mountain structures into East Antarctica. In the McMurdo Sound coastal area of the Ross embayment, thin but extensive piedmont glaciers obscure the structural boundary ('Transantarctic Mountains Front') with the off-shore Victoria Land rift basin. The main drainage for the East Antarctic Ice Sheet into the Ross embayment of West Antarctica is through outlet glaciers carved through the mountains. It has long been inferred that these outlet glaciers developed where faults cut transverse to the mountain trend (e.g., Cooper et al., 1991; Davey, 1981; Fitzgerald, 1992; Gould, 1935; Grindley and Laird, 1969; Gunn and Warren, 1962; Tessensohn and Wörner, 1991; Wrenn and Webb, 1982), and the occurrence of pseudotachylites has been proved a valuable tool to constrain the age of some faults (Di Vincenzo et al., 2004).

In most cases, however, there is little direct evidence either for the existence of a fault, or of its age. This is a key issue to address if the structures responsible for localized differential uplift of discrete mountain blocks are to be identified (e.g., Van der Wateren et al., 1999). The results will aid in understanding of localized valley incisions and how much erosion has contributed to mountain uplift. This could provide constraints on the development of the pathways for drainage of the East Antarctic Ice Sheet. Unlike other young mountain belts formed at the convergence plate boundaries (Alpine–Himalayan system and North American and Andean Cordilleras), the Transantarctic Mountains display a tight genetic link between mountain uplift and intra-plate rifting processes within the Antarctic plate.

The Transantarctic Mountains are generally interpreted as a high-relief rift flank uplift (Van der Wateren et al., 1999), formed during the Mesozoic–Cenozoic breakup of the Gondwana supercontinent (Cooper et al., 1987, 1991; Davey and Brancolini, 1995; Fitzgerald and Stump, 1997; Tessensohn and Wörner, 1991). The position of the Transantarctic Mountains in an extensional, rather than a contractional, tectonic regime had been recognised by pioneering Antarctic geologists, who interpreted the mountain chain as a fault-bounded horst block (David and Priestley, 1914; Gould, 1935). More recent structural investigations indicate that the Transantarctic Mountains consist of a linear to curvilinear chain of asymmetric tilt blocks bounded on the West Antarctic edge by a major normal fault zone and subdivided by transverse faults (Fitzgerald, 1992; Fitzgerald and Baldwin, 1997; Fitzgerald et al., 1986; Tessensohn and Wörner, 1991; Tessensohn, 1994a,b). Active rift tectonics and mountain uplift have been inferred from the presence of active volcanism and Neogene–Quaternary age faulting in the western portion of the rift and the Transantarctic Mountains (Behrendt and Cooper, 1991; Davey and Brancolini, 1995; Jones, 1997). The Cenozoic–Cretaceous asymmetric uplift and subsequent erosion exposed basement rock and older sediments along the coastward side of the Transantarctic Mountains, leaving younger sediments only on the inland side. Apatite fission track analysis in the McMurdo Sound area indicates an uplift of c. 6 km since c. 55 Ma. Other sectors of the Transantarctic Mountains record denudation events in the Late Cretaceous (Fitzgerald, 1992, 1995, 2002; Studinger et al., 2004) (Fig. 6.17) and in the Early Cretaceous (e.g., Scott Glacier area: Fitzgerald and Stump, 1997; Stump and Fitzgerald, 1992; Wannamaker et al., 2017 and ref. therein). Based on new apatite fission track data, Lisker and Läufer (2013) postulate a totally vanished Mesozoic basin spanning the Transantarctic Mountains in Victoria Land and the Wilkes Basin.

The cause of Transantarctic Mountains uplift and denudation is the subject of continuing debate, the reconstruction of Cenozoic tectonic processes being complicated by the complex interplay between a number of factors including the regional plate geodynamics, rifting style, erosion rates, subsidence and formation of thick sedimentary layers, the volcanic activity and the glacial processes. The possible mechanisms for the uplift (Fig. 6.18) include thermal buoyancy due to conductive or advective heating from the extended upper mantle of the hotter West Antarctic lithosphere (Stern and ten Brink, 1989; ten Brink and Stern, 1992), simple shear extension (Fitzgerald et al., 1986), isostatic rebound due to stretching of the lithosphere

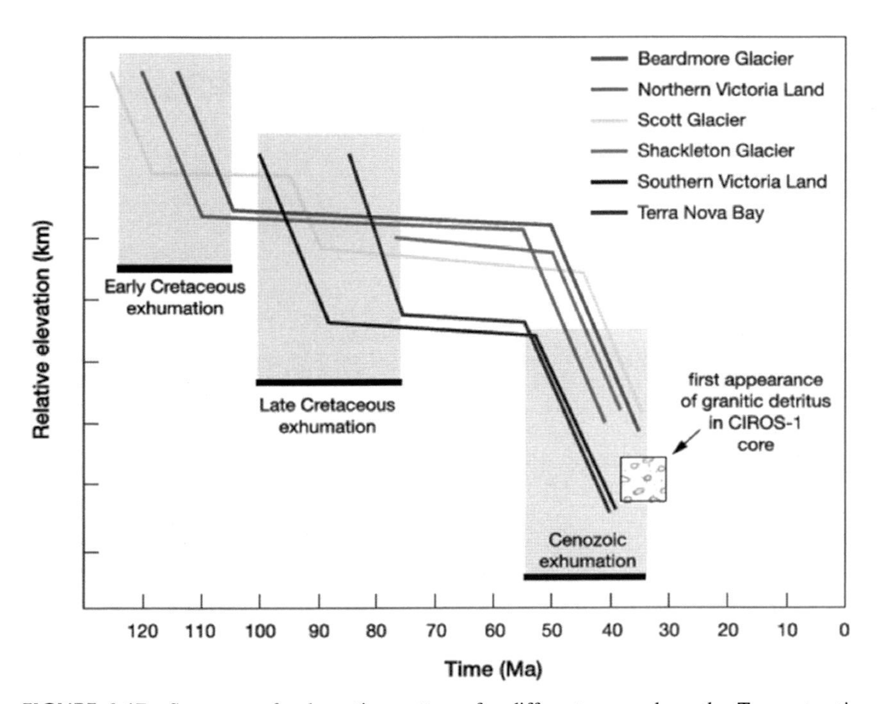

FIGURE 6.17 Summary of exhumation patterns for different areas along the Transantarctic Mountains, obtained from apatite fission track cooling ages (after Fitzgerald, 2002). *BDM*, Beardmore Glacier area; *NVL*, northern Victoria Land; *SCG*, Scott Glacier area; *SHG*, Shackleton Glacier area; *SVL*, southern Victoria Land; *TNB*, Terra Nova Bay.

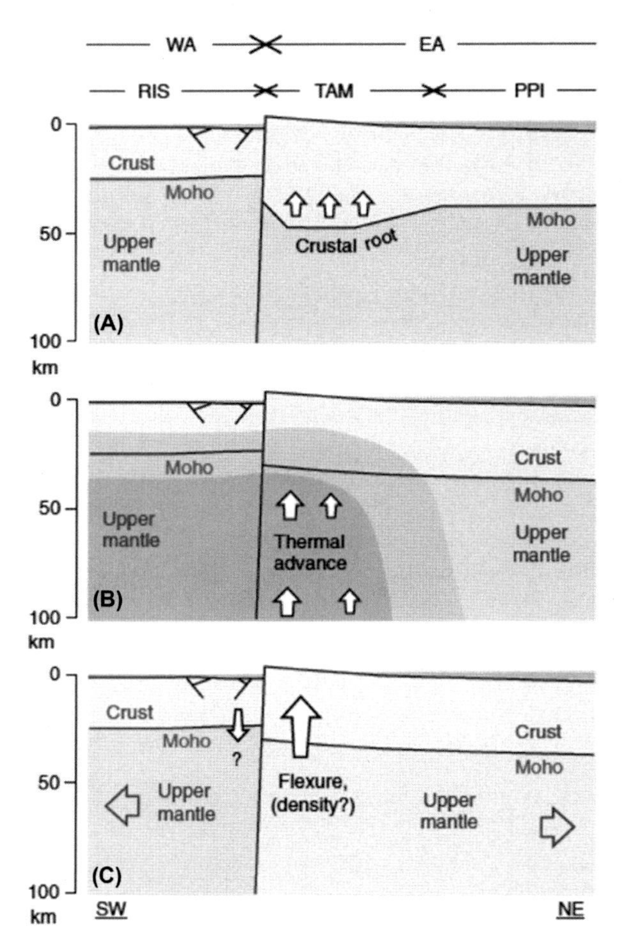

FIGURE 6.18 Hypothetical uplift mechanisms for Transantarctic Mountains rift shoulder (after Wannamaker et al., 2017). These are: (A) buoyant uplift via low-density crustal root; (B) uplift via lateral heating, thermal expansion, and possible melting; and (C) uplift via other mechanisms such as lithospheric cantilevered flexure with or without regional density contrasts. Physiographic regions include West Antarctica (WA), East Antarctica (EA), Ross Ice Shelf (RIS), Transantarctic Mountains (TAM), and Polar Plateau (PPl). Diagram not to scale.

through normal faulting (Bott and Stern, 1992), plastic necking (Chery et al., 1992), elastic necking (Van der Beek et al., 1994) and rebound in response to erosion (Stern and ten Brink, 1989). In the McMurdo Sound area, some of these mechanisms are based on specific assumptions about the crustal and upper mantle structure beneath the 'Transantarctic Mountains Front', as well as about the timing of the rift-related processes in the nearby VLB. Some constraints have already been provided by gravity studies (e.g., Davey and Cooper, 1987; Reitmayr, 1997), and by seismic reflection data (Della Vedova et al., 1997;

O'Connell and Stepp, 1993). Data from the ACRUP seismic experiment indicate thickening crust beneath the Transantarctic Mountains to a depth of 38 km, and quite low P-wave velocities (7.6–7.7 km/s) in the mantle beneath the VLB (Della Vedova et al., 1997), while low S-wave velocities are inferred at 60–160 km depth from surface wave analysis (Bannister et al., 1999), suggesting that the upper mantle is anomalously warm at that depth. More recently, the findings of drilling projects in the McMurdo Sound area (CIROS, Cape Roberts Project) have resolutely constrained the age of the onset of subsidence at the westernmost margin of the Victoria Land Basin. The first direct geological evidence of a major pre-Oligocene uplift phase of the Transantarctic Mountains comes from the oldest strata cored in the CIROS-1 and CRP-3 drill-holes (Barrett et al., 1989, 2001). These include granitic clasts eroded from exposed basement to the west, implying that the Transantarctic Mountains were at least half of their present height by then, as erosion had cut through more than 2000 m of Devonian–Jurassic Gondwana cover beds to basement (Barrett et al., 1989, 2001). In the Cape Roberts drill core, the presence of the Devonian Arena Formation (Beacon Supergroup) as bedrock beneath the Cenozoic sediments indicates that significant uplift and unroofing of the Transantarctic Mountains must have occurred prior to the Oligocene (Barrett et al., 2001).

6.6 Tectonic evolution in the Ross Sea Sector during the Cenozoic

The Ross embayment is one of the most striking morphological expressions of the WARS; a region of thin and, by inference, extended continental crust whose regional boundaries are difficult to define precisely and may have been different for different episodes of extension (Fig. 6.5). Three major episodes of extension have been proposed, with rifting starting in the Middle Jurassic, coincident with the onset of Gondwana breakup and the associated Middle Jurassic magmatism (Ferrar Group), and with subsequent episodes in the Cretaceous and late Cenozoic. The active rifting may continue to the present as suggested by the active volcanoes along the western margin of the Ross Sea. However, most extension and thinning of the Ross Sea crust is considered to have taken place during the Mesozoic rift period (Siddoway, 2008) and the tectonic connection between the Ross and Weddell embayments is largely unclear (Behrendt et al., 1992; LeMasurier, 2007, and ref. therein). In particular, our knowledge of the Jurassic episode is problematic since, although the location of rifting is well constrained in the South Africa – Queen Maud Land region, neither the location nor amount of extension is well known in the Transantarctic Mountains and Ross Sea. By comparison, the Cretaceous episode is better documented (LeMasurier, 2007, and ref. therein) and is considered to be closely related to the stretching and rifting events in the Weddell Sea region, where disruption of the plate margin involved the rotation and translation of the several crustal blocks forming the present West

Antarctica, including the Ellsworth/Whitmore Block, which was translated to the boundary between the Ross and Weddell embayments (Jordan et al., 2013, 2014, 2017, 2020).

Since movement of the West Antarctic crustal blocks was largely completed by 110 Ma, later tectonic activity in the WARS has been restricted to the Ross sector of the Transantarctic Mountains and the margin of the Ellsworth–Whitmore Mountains, with the continuation of the Transantarctic Mountains into the Weddell Sea region considered a remnant Jurassic rift (LeMasurier, 2007; Schmidt and Rowley, 1986). The first direct evidence of rifting in the Ross embayment is documented in Marie Byrd Land where mafic dike arrays (Siddoway et al., 2005; Storey et al., 1999) and A-sub-type granites ('Byrd Coast Granite'; Wever et al., 1994) were emplaced as early as 115,5 ± 3,7 Ma, leading to increased magmatic activity until 97 ± 2 Ma (Brown et al., 2016; Siddoway et al., 2004). Considering Bradshaw's (1989) suggestion that a sea-floor spreading centre intersected the subduction margin of Mesozoic Zealandia (within Gondwana) around 105 Ma, the production of the mafic dikes likely reflect regional extension in the overriding plate, followed by backarc felsic magmatism at 102–97 Ma. On the basis of the age of magnetic anomaly 33 (83 Ma) identified off Campbell Plateau, sea-floor spreading commenced between Campbell Plateau and Marie Byrd Land at c. 85 Ma. The reconstruction of the Campbell Plateau against Marie Byrd Land and the remarkable match of the deduced ocean–continent boundary, as well as the alignment of the eastern edge of the Eastern Ross Basin with the northern edge of the Campbell Basin on the Campbell Plateau, are evidence that the entire Eastern Ross Basin experienced extension prior to initiation of the Pacific–Antarctic Ridge at c. 85 Ma (Lawver et al., 1992).

The plate tectonic reconstruction presented by Jordan et al. (2020) (Fig. 6.19), shows the motion of West Antarctic blocks (Marie Byrd Land, Thurston Island, the Antarctic Peninsula and the Ellsworth–Whitmore Mountains) from the Middle Jurassic initiation of Gondwana breakup (Elliot et al., 2015; Jordan et al., 2017), through the separation of Southern Africa (König and Jokat, 2006), the initiation of Marie Byrd Land, Ross Sea and West Antarctic rift system extension (Lawver et al., 1992), and the separation of Zealandia (Mortimer et al., 2019), to the final stages of West Antarctic rift development linked to extension in the Adare Trough (Davey et al., 2016).

As evident in Fig. 6.20 the Pacific–Antarctic spreading centre developed in late Cenozoic time. With its initiation, a triple junction developed with three extensional arms (Australian–Antarctic spreading centre, Australian–Pacific arm and the Pacific–Antarctic spreading). Early seafloor spreading anomalies between Tasmania and the South Tasman Rise provide age control. Key tectonic events include: (1) the Australia–Antarctica spreading centre about magnetic anomaly 30 (c. 65 Ma) jumped to a position between South Tasman Rise and

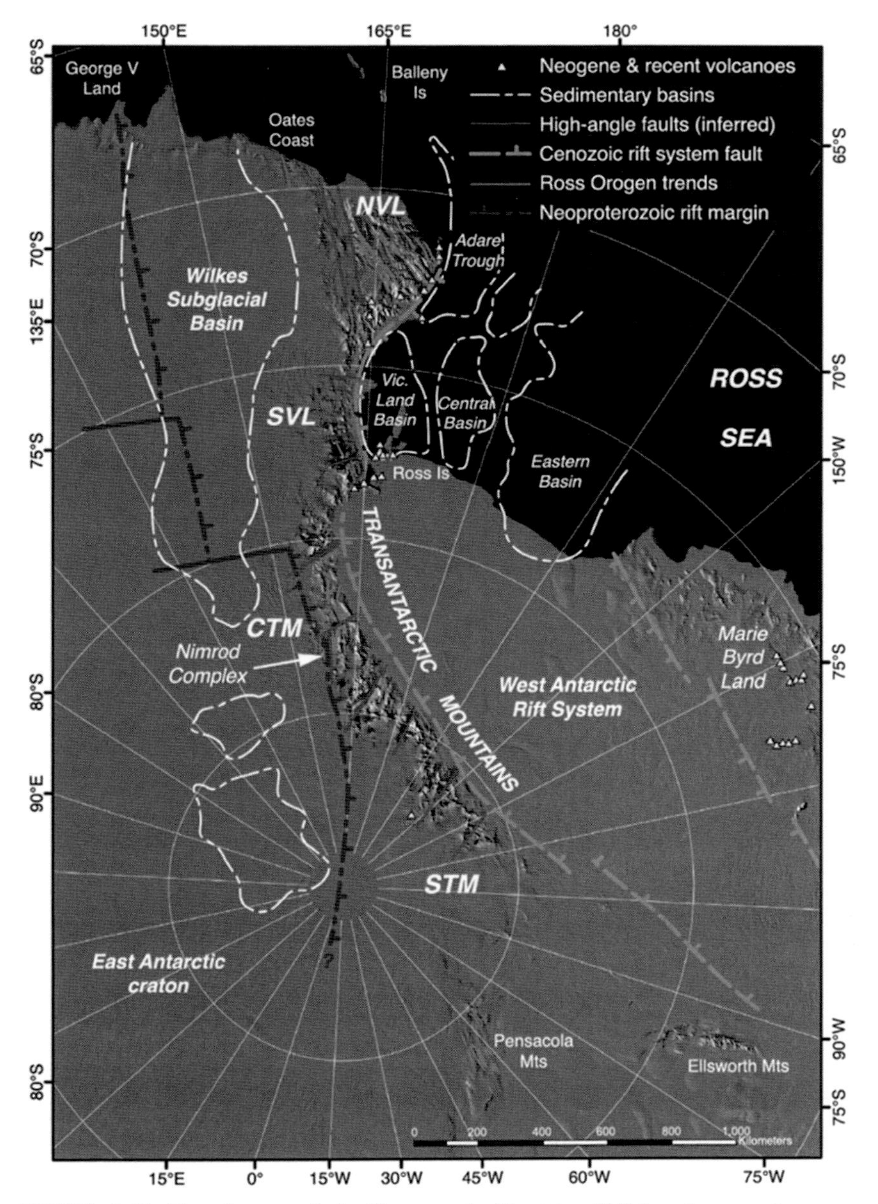

FIGURE 6.19 Tectonic map of the Transantarctic Mountains (TAM) and surrounding areas (after Goodge 2020). Neoproterozoic rift margin boundary is inferred from magnetic, gravity and seismic geophysical data (Goodge and Finn, 2010), and marks the Ross margin of the Precambrian East Antarctic shield. Neogene faults bounding the West Antarctic Rift System (WARS; from Wilson, 1999) also form the TAM front. High-angle faults inferred to underlie the major TAM outlet glaciers are related to movement on the TAM frontal fault system, but they may also be locally reactivated from Neoproterozoic transfer faults. Sedimentary basins in the Ross Sea area are related to opening of the WARS, but the origin of the interior basins such as the Wilkes Subglacial Basin is enigmatic. Neogene and recent volcanoes in the TAM and Marie Byrd Land are related to extension in the WARS, including the active systems on Ross Island.

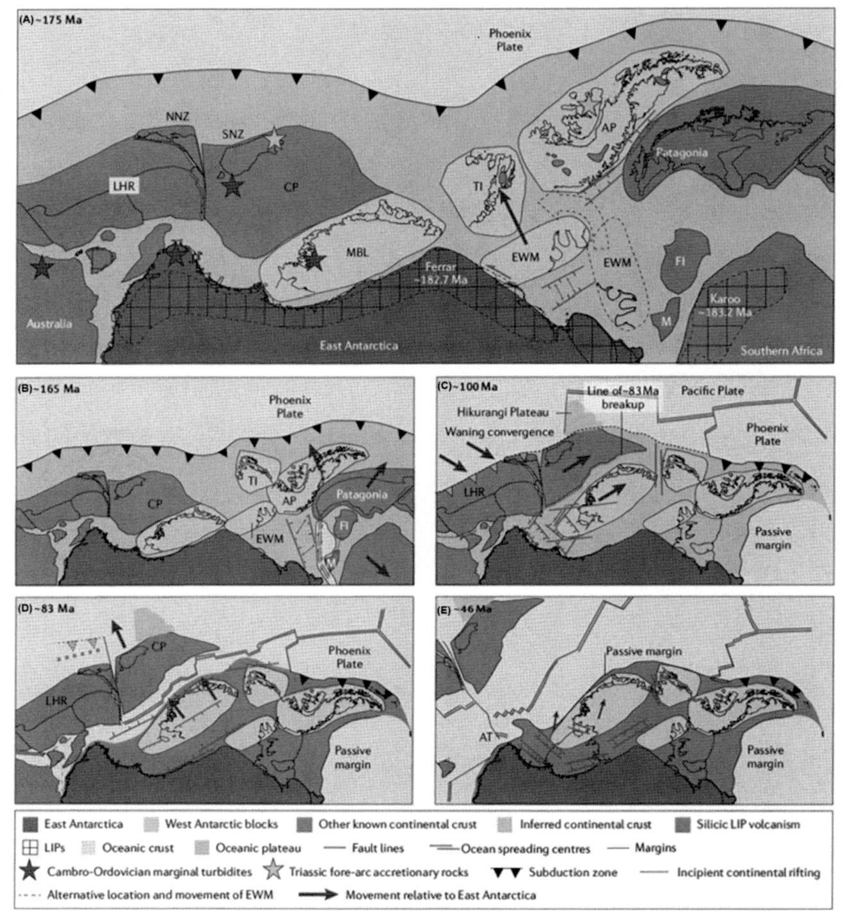

FIGURE 6.20 Western Antarctic plate tectonic reconstruction from 175 to 45 Ma (after Jordan et al., 2020), showing motion in an East Antarctica fixed reference frame. West Antarctic blocks, including Marie Byrd Land (MBL), Thurston Island (TI), the Antarctic Peninsula (AP) and the Ellsworth–Whitmore Mountains (EWM), are depicted as pale pink regions and modern coastlines are shown for reference. Other known present-day continental regions are shown in dark grey. Inferred intervening areas of continental crust are shown in light grey; such regions have been deformed and distorted to such an extent that they cannot be meaningfully traced through time. (A) Middle Jurassic initiation of Gondwana breakup is shown; the hashed regions mark Jurassic mafic large igneous provinces (LIPs) and the dark-pink regions mark silicic LIP volcanism. Cambro-Ordovician marginal turbidites (blue stars) were widespread, whereas Triassic fore-arc accretionary rocks (orange stars) were rare. The dashed outline marks alternative location and rotation of the EWM. (B) Separation of Southern Africa. (C) Initiation of MBL, Ross Sea and West Antarctic rift system extension. (D) Separation of Zealandia. (E) Final stages of West Antarctic rift development linked to extension in the Adare Trough (AT). *CP*, Campbell Plateau; *FI*, Falkland Island Plateau; *LHR*, Lord Howe Rise; *M*, Maurice Ewing Bank; *NNZ*, North Island New Zealand; *SNZ*, South Island New Zealand.

northern Victoria Land; (2) the cessation of sea-floor spreading in the Tasman Sea at magnetic anomaly 24 time (c. 54 Ma) as a consequence of the reformation of the ridge–ridge–ridge triple junction between Campbell Plateau and the Tasman Sea, and (3) the c. 40 km wide Adare Rift opened within 'older' sea-floor north-west of the Ross Sea (after magnetic anomaly 13; 33 Ma) (Cande and Stock, 2006; Cande et al., 2000) and up to 11 Ma (Granot et al., 2018). The Ross Sea rift basins include from E to W (Cooper et al., 1987; Davey and De Santis, 2006; Fielding et al., 2006; Jordan et al., 2020): the Eastern Basin – underlying most of eastern Ross-Sea; the Central Trough – running north–south discontinuously through central Ross Sea; the Northern Basin – underlying the north-eastern Ross Sea margin; and the Victoria Land Basin (VLB) underlying south-western Ross Sea, adjacent to the Transantarctic Mountains. The VLB is one of these four sedimentary basins developed due to rifting of previously thinned continental crust, along pre-existing crustal faults. Localized extension reduced the crustal thickness in some sectors to less than 10 Km. The VLB (the westernmost basin) is marked by high heat flow and alkaline volcanism (Behrendt et al., 1993; Blackman et al., 1987a,b; Cooper et al., 1991; LeMasurier, 2007; White, 1989). Extensive late Cenozoic volcanism is also inferred from aeromagnetic data. Terror Rift is located along the western edge of the VLB in the western Ross Sea (Sauli et al., 2021). The Ross Embayment has been extensively investigated through numerous airborne and marine-based geophysical surveys, highlighting the complex structural setting and tectonic evolution of the western Ross Sea (Davey et al., 2016; Ferraccioli et al., 2009; Fielding et al., 2008; Granot and Dyment, 2018; Granot et al., 2013). There is a lack of appropriate age data for the main seismo-stratigraphical units, inferences on the timing of Cenozoic rifting in the entire Ross Sea remain somewhat speculative. The results of the Cape Roberts Drilling Project (CRP) (Barrett et al., 2001) show that the VLB is mostly late Eocene or younger in age. Records recovered by drilling on the western margin of this basin indicated the onset of subsidence at about 34 Ma, significantly younger than the onset of uplift of the adjacent Transantarctic Mountains (about 55 Ma) (Fitzgerald, 1992) or before (Goodge, 2020). Hence there is a discrepancy in the relationship between uplift of the rift margin mountains and the subsidence of the adjacent rift basin. The fact that the VLB basin extends only from Ross Island to Terra Nova Bay, whereas the Transantarctic Mountains continue further north and south, would indicate that the basin may have originated from some processes other than simple extension. A transtension or 'pull-apart' process has been suggested, with the igneous activity within the 'Polar Three' anomaly (near Coulman Island) providing the transfer mechanism in the north and magnetic anomalies south of the Ross Sea Fault (Bosum et al., 1989).

Another significant geotectonic component in the Ross Sea area is represented by the extensive alkalic McMurdo magmatic province; one of the largest in the world and including two active volcanoes (Mt. Erebus at Ross

Island and Mt. Melbourne on the Ross Sea coast in northern Victoria Land; LeMasurier and Thomson, 1990). Volcanic rocks, either exposed or suggested by aeromagnetic studies in the area under the Ross Sea and West Antarctic Ice Sheet (Behrendt et al., 2002; Bell et al., 2006), occur on either side of the WARS in Marie Byrd Land (West Antarctica) and in the western Ross Sea. Earliest volcanic rocks in the western Ross Sea are Eocene to Oligocene alkali intrusive rocks ('Meander Intrusive'), interpreted to be the eroded remnants of subvolcanic magmatic complexes (Müller et al., 1991; Rocchi et al., 1999, 2002). Elsewhere, the alkaline magmatism is predominantly volcanic and has been subdivided into the informal, but geographic and petrologic distinct, Hallett, Melbourne and Erebus volcanic provinces (Kyle, 1990b). Two small isolated occurrences of basalts dated at 16–20 Ma occur around the head of the Scott Glacier (Stump et al., 1980). In the Erebus volcanic province (Kyle, 1990a), there is a continuous eruptive sequence from 19 Ma to the present day, with the Mt. Erebus crater the site of a persistent anorthoclase phonolite lava lake. The oldest exposed rocks of this province occur on the northern slopes of Mt. Morning where 19 Ma trachyandesite lavas are intruded by 16–18 Ma trachyte dikes. Volcanic ash (tephra) layers within CRP drill cores provide evidence of older Miocene trachytic eruptions. The tectonic setting suggests that the southern extent of the volcanism may be controlled by a major transfer fault, which coincides with the southern boundary of the Terror Rift (Wilson, 1999). Many of the larger volcanic centres have a radial distribution around Mt. Discovery or Mt. Erebus. The radial distribution is interpreted as a result of upwelling of a mantle plume (the Erebus plume), which was located under Mt. Discovery prior to 4 Ma and then migrated to its present position under Mt. Erebus (Kyle, 1990a).

6.7 Concluding remarks, open problems and potential research themes for future geoscience investigations in Antarctica

6.7.1 Persistent challenges for onshore geoscience investigations

In contrast to all other continents, Antarctica and its rocks, and thus its geological structures, are covered by the gigantic Antarctic Ice Sheet, which in places is over 4700 m thick. Less than 2% of the continent is uncovered, and provides the base from which geological knowledge is established, and even these exposed rocks have not been investigated thoroughly due to their remoteness and the environmental challenges faced in deep-field research. On the other hand, more than half of the ice-covered area has been surveyed geophysically, mainly aeromagnetically and gravimetrically. Further investigations are obviously needed, and the image of the Antarctic geological structure and its history will likely change, improve and be completed in the future.

Several mountain ranges in Antarctica have been seldom visited. However, their study would decidedly improve the understanding of geologic and geotectonic connections. It is evident that there are still conspicuous gaps in our knowledge of Marie Byrd Land, the Pensacola Mountains, eastern Dronning Maud Land and East Antarctica between 60° and 120°E. Moreover, the unknown ice-covered interior needs additional studies starting from the more well-known areas. For example, tracing known geological rock complexes from their exposed areas under the ice with the help of suitable geophysical methods. The confidence we can have in geophysical interpretations declines with increasing distance from directly accessible rock complexes. Even a few spot checks of rock samples from isolated deep drill-holes would provide a better reliability of interpretations (i.e. calibration of airborne geophysical data). New data from the RAID (Rapid Access Ice Drill) on key regions of the subglacial geology will be decisive for paleogeodynamic reconstructions (Goodge and Severinghaus, 2016). Another example, as clearly reported in a recent geological review of West Antarctica (Jordan et al., 2020), is to decide whether the region is best conceived as an accreted collection of rigid microcontinental blocks (as commonly depicted) or as a plastically deforming and constantly growing melange of continental fragments and juvenile magmatic regions. Jordan et al. (2020) highlight the importance of new techniques, such as finite-element modelling, and the need to couple models with more detailed geophysical and geological studies. Geophysical data can provide new constraints on the extent of magmatism and the areal extent and geometry of the underlying provinces, and geological observations and dating can provide information about how and when the different components of the system were active.

6.7.2 Antarctica and the Ross Orogen in the Transantarctic Mountains

Other important geological research problems relate to the evolution of Rodinia and Gondwana supercontinents. Since Antarctica, and particularly East Antarctica, had a central position in both Rodinia, between 1300 and 800 million years ago, and Gondwana, between 600 and 200 million years ago (today's southern continents), geological information archived in this 'keystone' region are important not only for a well-founded analysis of local conditions, but also in relation to our understanding of Earth system processes in deep time. Abundant evidence exists for the former connection between Antarctic cratonic areas and geologically similar provinces on neighboring continents.

The study of East Antarctica is a prerequisite for the reconstruction of the assembly and breakup processes of both supercontinents. Geological and paleomagnetic data show that East Antarctica comprises older cratonic fragments and Grenvillian and Pan-African structures in coastal exposures support this notion. The larger part of the so-called 'East Antarctic Craton' continues into India and Australia, while a small fragment (Grunehogna Craton) connects to the Kalahari Craton of Africa. The 'East Antarctic Craton' consists of a number of cratonic nuclei (Boger and Miller, 2004; Fitzsimons, 2000a), confirming the importance of Antarctica for the reconstruction of continental distribution in early Earth history. Although numerous alternative scenarios have been proposed for Rodinia, refined global reconstructions based on paleomagnetic data, supported by geological correlation, show the Transantarctic Mountains margin of East Antarctica, in continuity with the Neoproterozoic margin of eastern Australia, as conjugate to western Laurentia from c.1080 to c.750 Ma (so-called SWEAT model; Dalziel, 1997; Meert and Torsvik 2003; Torsvik, 2003). Recent geological studies, as reviewed by Goodge (2020), appear to confirm the cratonic linkages and Neoproterozoic rift-margin associations proposed earlier by Moores (1991) and Dalziel (1991). As stated by Goodge (2020), despite a more solid understanding of the Neoproterozoic history along the Transantarctic Mountains margin of the Mawson Craton, additional geophysical observations and research are required to resolve uncertain features, including the position of the rift margin, the geometry of rifting, the extent of crustal thinning, the extent of rift margin sedimentation, the location of possible transform offsets and the influence of these structures on younger orogens (Goodge and Finn, 2010).

Significant contributions to the understanding of global plate tectonic and geodynamic processes are also stored in the orogenic belts of Pan-African age (600–500 million years ago), which contain important information about the juxtaposition of West and East Gondwana. In Antarctica, the Shackleton Range and parts of East Antarctica between Dronning Maud Land, Lützow HolmBukta and Prydz Bay belong to these belts (Boger et al., 2002; Buggisch and Kleinschmidt, 2007; Buggisch et al., 1990; Fitzsimons, 2000b; Jacobs et al., 1998; Paech, 2005; Tessensohn et al., 1999). Equivalent rocks have been found in the African Mozambique Belt (Jacobs et al., 1998; Paech, 1985; Paech et al., 2005). So far, little is known about how these fragments continue under the ice and how they can be connected with the better researched mountains in Dronning Maud Land and the Transantarctic Mountains. Equally unknown is the formation, age and relation of subglacial mountains in the East Antarctic interior (e.g., the Gamburtsev Subglacial Mountains).

At approximately the same time as the development of internal Pan-African orogenic belts, the geographic domain corresponding to today's Transantarctic Mountains experienced the Ross Orogeny; the result of dominant accretionary tectonic processes produced by the evolution from a passive, drifting margin to an active, subducting margin. This Pan-African activity was characterised in the Transantarctic Mountains by an accretionary-type convergent plate margin along the southern outer perimeter of Gondwana. Subduction of paleo-Pacific oceanic lithosphere beneath the active Gondwana margin of continental East Antarctica spanned as much as 145 million years between about 615–470 Ma, reflecting a prolonged period of sustained underflow.

There is widespread consensus in considering the end of the Ross orogeny as a nearly synchronous event along the different sectors of the belt (e.g., Encarnacion and Grunow, 1996; Stump, 1995). Voluminous granitoids (e.g., Granite Harbour Intrusive Complex) intruded as dated batholiths at c. 560–470 Ma (Encarnacion and Grunow, 1996; Hagen-Peter and Cottle, 2016) represent a unifying feature throughout the length of the Transantarctic Mountains (Borg et al., 1990; Stump, 1995). But although the general tectonic history of the Ross Orogeny is fairly well known within each of the major segments of the Transantarctic Mountains, significant variations in lithostratigraphic, structural and metamorphic patterns, as well as in granitoid geochemical affinity, are evident between the different segments. Considerable uncertainty remains about the onset of the subduction, the tectonic setting of the early granitoids with variable chemical affinities (from calc-alkaline to alkaline and carbonatite) of southern Victoria Land, the nature of the contact between the orogenic belt and the Mawson Craton, and the relations with the Pan-African structures of the Shackleton Range.

Many other broad first-order questions still wait for future research focused on the Ross Orogeny, as acknowledged by Goodge (2020). Until our knowledge of the relationships of the tectono-metamorphic histories, and of the detailed chronology of the magmatic, tectonic, sedimentary and metamorphic episodes between the various segments of the orogenic belt is understood, a comprehensive tectonic model of the development of the Ross Orogeny remains to be formulated.

6.7.3 Antarctica after Gondwana fragmentation

The importance of Antarctica with regard to the destructive processes of plate tectonics, including the fragmentation of supercontinents is also well recognised. The young continental margins and rift structures of Antarctica, as well as their development, document the breakup of Gondwana. Aside from a small section along the Antarctic Peninsula, the Antarctic continental margins depict the fault structures of Gondwana breakup, leading ultimately to the formation of the present southern continents and oceans.

After fragmentation started and initial drifting began, Antarctica reached and remained at its south polar position since at least the Late Cretaceous (approximately 130 Ma). Its isolation from the neighbouring continents, through the early opening of the Southern Ocean and the formation of the Antarctic Circumpolar Current, also began at this time. The further breakup process led to the full formation of the Southern Ocean. The exact opening processes, and their effects on paleoceanography, have not yet been satisfactorily reconstructed because high-quality magnetic data are lacking in many key areas of the oceans. This is especially the case for the South Pacific and areas between Antarctica and Africa. For an understanding of the plate tectonic development of the Antarctic continent several events must be considered: (1) 130–100 million years ago, the opening of the Weddell, Lazarev and Riiser-Larsen Seas, and of the southern Atlantic and Indian Oceans; (2) 110–80 million years ago, the genesis of the southern Kerguelen Plateau; (3) 80–40 million years ago, the separation of Tasmania/Australia/Zealandia from Antarctica; (4) 90–14 million years ago, the development of the Ross Sea Rift; and (5) 30–20 million years ago when South America separated from Antarctica, and the Drake Passage and Scotia Sea opened, leading to the formation of the Circumpolar Current. In this time context critical data for the Ross Embayment are lacking (Goodge, 2020; Jordan et al., 2020). There is particular need to add details to the bounding structure between the Transantarctic Mountains and WARS provinces, in respect to their geometry and variability with depth. Cryptic evidence suggests the presence of fragments of the Transantarctic Mountains in Marie Byrd Land (Bradshaw, 2007) and/or beneath the Ross Ice Shelf (Tinto et al., 2019). The relationships can be better understood through integration between surface geology data and subsurface geophysical data (see also McKay et al., 2021).

Despite the many questions about the mechanisms and the consequences of these global processes (e.g., of climatic nature) it is certain that the breakup of Gondwana led to the current configuration of continental plates and the isolation of continental Antarctica. Breakup provided the starting point for many current processes and patterns, including the glaciation of the polar regions, the present-day oceanic and atmospheric circulation, climate patterns and the distribution of biota, and the global environmental conditions for human existence. To improve the present state of knowledge, the distinct progression of Gondwana breakup and global ocean circulation should be determined more precisely. The prominent mountains chains represented by the Transantarctic Mountains await further research addressing mechanisms of uplift, and regional variability of paleothermal histories that indicate diverse exhumation mechanisms, and the connection to Cenozoic denudation through tectonic and glacial process.

An increased research effort in these areas is essential to allow a better timing of the Antarctic Circumpolar Current onset (Barker and Thomas, 2004; Barker et al., 1998, 2007), to understand the links between glacial history and Transantarctic Mountains uplift, the feedbacks between glacial erosion and uplift rates, and how surface processes connect to both tectonic and climatic influences. These aspects underlie the central influence of the polar regions in general circulation models (Oglesby, 1999; Sloan et al., 1996). The thick continental ice sheets drive latitudinal gradients and sea-ice formation, leading to the formation of cold Antarctic bottom water, which through deep ocean currents reaches the lower latitudes (Carter et al., 2021), and indeed the other hemisphere. The oxygen isotope record from deep-sea cores, and eustatic changes inferred from sequence stratigraphic records on passive continental margins, have been leading paradigms for the interpretations regarding Antarctic ice sheet history. However, these global proxy records of glacio-eustasy are contradicted by geologic records in Antarctica (Harwood et al., 1991, 1993; Miller and Mabin, 1998; Moriwaki et al., 1992; Wilson, 1995). Because direct data from the Antarctic region are necessary for deep-time climate model-validation, a series of drilling projects has targeted the Antarctic continental margin to retrieve high-resolution stratigraphic records of Antarctica's glacial and climatic history (Barker et al., 1998; Barrett et al., 2000, 2001; Colleoni et al., 2021; Cooper and Webb, 1994; Gohl et al., 2021; Hambrey, 2002; Hambrey et al., 1998; Harwood et al., 2002; McKay et al., 2019; McKay et al., 2021). Paleoclimatic reconstructions based on the results of the CRP in the Ross Sea region for the early Oligocene to early Miocene time show the occurrence of ice but also a warmer climate in the Antarctic than today (Barrett et al., 2000, 2001; Hambrey et al., 1998; Naish et al., 2001). Unresolved, however, is the question of how stable the Antarctic ice sheets were during the last 20 My and the timing of the onset of glaciation. Building upon knowledge from ANDRILL 2006 and 2007 (Harwood et al., 2002), a primary objective of ongoing investigations is to decipher the responses of past ice sheets to climate forcing, including variability at a range of time-scales (e.g., Colleoni et al., 2018 and Colleoni et al., 2021). Both CIROS and CRP pursued this elusive target. Indeed, in CRP-3, Oligocene strata passed via an unconformity into the pre-glacial Devonian–Triassic Beacon Supergroup rocks, proving that sites are to be found that contain Eocene and earlier Cenozoic, or even Cretaceous records. These opportunities, along with others, will likely be identified through future drilling programs that complement Ross Sea data with data from other coastal regions of Antarctica, and will provide a deeper insight into the tectonic and climatic relevance of the Ross Embayment region.

Acknowledgements

This chapter is a contribution to the SCAR program Antarctic Climate Evolution (ACE) and Past Antarctic Ice Sheets (PAIS). Acknowledgements for funding and logistical and technical support with the field campaigns and drilling projects in the McMurdo Sound area are made to the United States Office of Polar Programs (NSFOPP), Antarctica New Zealand, the Italian National Antarctic Program (PNRA), the Alfred Wegener Institute for Polar and Marine Research (AWI) and the Deutsche Forschungsgemeinschaft (DFG). Valuable comments by F. Florindo and L. De Santis, and careful corrections and reviews by Christine Siddoway and Martin Siegert that greatly improved the manuscript are also much appreciated.

References

Aitken, A.R.A., Young, D.A., Ferraccioli, F., Betts, P.G., Greenbaum, J.S., Richter, T.G., et al., 2014. The subglacial geology of Wilkes Land, East Antarctica. Geophysical Research Letters 41, 2390–2400.

Aitken, A.R.A., Betts, P.G., Young, D.A., Blankenship, D.D., Roberts, J.L., Siegert, M.J., 2016. The Australo-Antarctic Columbia to Gondwana transition. Gondwana Research 29 (1), 136–152.

Allibone, A.H., Cox, S.C., Smillie, R.W., 1993. Granitoids of the Dry Valleys area, Southern Victoria Land: geochemistry and evolution along the early Palaeozoic Antarctic craton margin. New Zealand Journal of Geology and Geophysics 36, 299–316.

Anderson, J.J., 1965. Bedrock geology of Antarctica—a summary of exploration, 1831–1962. American Geophysical Union Antarctic Research Series 6, 1–70.

Armienti, P., Ghezzo, C., Innocenti, F., Manetti, P., Rocchi, S., Tonarini, S., 1990. Isotope geochemistry and petrology of granitoid suites from Granite Harbour Intrusives of the Wilson Terrane, North Victoria Land, Antarctica. European Journal of Mineralogy 2, 103–123.

Arndt, N.T., Todt, W., Chauvel, C., Tapfer, M., Weber, K., 1991. U-Pb zircon age and Nd isotopic composition of granitoids, charnockites and supracrustal rocks from Heimefrontfjella, Antarctica. Geologische Rundschau 80 (3), 759–777.

Arora, D., Pant, N., Pandey, M., Chattopadhyay, A., Greenbaum, J., Siegert, M., et al., 2020. Insights into geological evolution of Princess Elisabeth Land, East Antarctica-clues for continental suturing and breakup since Rodinian time. Gondwana Research 84, 260–283.

Baba, S., Hokada, T., Dunkley, D.J., Owada, M., Shiraishi, K., 2010. SHRIMP Zircon U-Pb Dating of Sapphirine-Bearing Granulite and Biotite-Hornblende Gneiss in the Schirmacher Hills, East Antarctica: implications for neoproterozoic ultrahigh-temperature metamorphism predating the assembly of Gondwana, The Journal of Geology, 118. p. 621.

Baba, S., Osanai, Y., Nakano, N., Owada, M., Hokada, T., Horie, K., et al., 2013. Counterclockwise P-T path and isobaric cooling of metapelites from Brattnipene, Sør Rondane Mountains, East Antarctica: implications for a tectonothermal event at the proto-Gondwana margin. Precambrian Research 234, 210–228.

Bannister, S., Snieder, R.K., Passier, M.L., 1999. Shear-wave velocities under the Transantarctic Mountains and terror rift from surface wave inversion. Geophysical Research Letters 27, 281–284.

Bannister, S., Yu, J., Leitner, B., Kennett, B.L.N., 2003. Variations in crustal structure across the transition from west to east Antarctica, southern Victoria Land. Geophysical Journal International 155, 870–884.

Barker, P.F., Thomas, E., 2004. Origin, signature and palaeoclimatic influence of the Antarctic circumpolar current. Earth-Science Reviews 66, 143–162.

Barker, P.F., Barrett, P.J., Camerlenghi, A., Cooper, A.K., Davey, F.J., Domack, E.W., et al., 1998. Ice sheet history from Antarctic continental margin sediments: The ANTOSTRAT approach. Terra Antartica 5 (4), 737–760.

Barker, P.F., Filippelli, G.M., Florindo, F., Martin, E.E., Scher, H.D., 2007. Onset and role of the Antarctic circumpolar current. Deep-Sea Research Part II 54, 2388–2398.

Barrett, P.J., 1991. The Devonian to Triassic Beacon Supergroup of the Transantarctic Mountains and correlatives in other parts of Antarctica. In: Tingey, R.J. (Ed.), The Geology of Antarctica. Oxford Monographs on Geology and Geophysics. Clarendon Press, Oxford, pp. 120–152.

Barrett, P.J., Hambrey, M.J., Harwood, D.M., Pyne, A.R., Webb, P.N., 1989. Synthesis. In: Barrett, P. J. (Ed.), Antarctic Cenozoic History from the CIROS-1 Drillhole, McMurdo Sound, vol. 245, DSIR Bulletin, pp. 241–252.

Barrett, P.J., Davey, F.J., Ehrmann, W.U., Hambrey, M.J., Jarrard, R., van der Meer, J.J.M., et al., 2000. Studies from the Cape Roberts Project, Ross Sea, Antarctica, Scientific Results of CRP-2/2A, Parts I and II. Terra Antartica 7 (4/5), 665 pp.

Barrett, P.J., Ricci, C.A., Bucker, C.J., Davey, F.J., Ehrmann, W.U., Laird, M.G., et al., 2001. Studies from the Cape Roberts Project, Ross Sea, Antarctica, scientific report of CRP-3, Parts I and II. Terra Antartica 8 (3–4), 620.

Bastias, J., Spikings, R., Ulianov, A., Riley, T., Burston-Johnson, A., Chiaradia, M., et al., 2020. The Gondwana margin in West Antarctica: insights from late Triassic magmatism of the Antarctic Peninsula. Gondwana Research 81, 1–20.

Bauer, W., Jacobs, J., Fanning, M., Schmidt, R., 2003a. Geochemical constraints for Late Mesoproterozoic arc and back-arc volcanism in the Heimefrontfjella (East Antarctica) and implications for the palaeography at the Southeastern Margin of the Kaapvaal-Grunehogna Craton. Gondwana Research 6, 449–465.

Bauer, W., Thomas, R.J., Jacobs, J., 2003b. Proterozoic-Cambrian history of Dronning Maud Land in the context of Gondwana assembly. Geological Society of London, Special Publications 206 (1), 247–269.

Bauer, W., Jacobs, J., Thomas, R.J., Spaeth, G., Weber, K., 2009. Geology of the Vardeklettane Terrane, Heimefrontfjella (East Antarctica). Polarforschung 79, 29–32.

Behrendt, J.C., 1999. Crustal and lithospheric structure of the west Antarctic rift system from geophysical investigations—a review. Global and Planetary Change 23, 25–44.

Behrendt, J.C., Cooper, A.K., 1991. Evidence of rapid Cenozoic uplift of the shoulder escarpment of the Cenozoic west Antarctic rift system and a speculation on possible climate forcing. Geology 19, 315–319.

Behrendt, J.C., Duerbaum, H.J., Damaske, D., Saltus, R., Bosum, W., Cooper, A., 1991a. Extensive volcanism and related tectonism beneath the Ross Sea continental shelf, Antarctica: Interpretation of an aeromagnetic survey. In: Thompson, M.R.A., Crame, J.A., Thomson, J.W. (Eds.), Geological Evolution of Antarctica. Cambridge University Press, Cambridge, pp. 299–304.

Behrendt, J.C., LeMasurier, W.E., Cooper, A.K., Tessensohn, F., Trehu, A., Damaske, D., 1991b. Geophysical studies of the west Antarctic rift system. Tectonics 10, 1257–1273.

Behrendt, J.C., LeMasurier, W., Cooper, A.K., 1992. The west Antarctic rift system – a propagating rift ''captured'' by a mantle plume? In: Yoshida, Y. (Ed.), Recent Progress in Antarctic Earth Science. Terrapub, Tokyo, pp. 315–322.

Behrendt, J.C., Damaske, D., Fritsch, J., 1993. Geophysical characteristics of the west Antarctic rift system. Geologisches Jahrbuch E47, 49–101.

Behrendt, J.C., Blankenship, D.D., Morse, D.L., Finn, C.A., Bell, R.E., 2002. Removal of subglacially erupted volcanic edifices beneath the divide of the west Antarctic ice sheet interpreted from aeromagnetic and radar ice-sounding surveys. In: Gamble, J.A., Skinner, D.N. B., Henrys, S. (Eds.), Antarctica at the Close of a Millennium, vol. 35. Royal Society of New Zealand Bulletin, pp. 579–587.

Bell, R.E., Studinger, M., Karner, G., Finn, C.A., Blankenship, D.D., 2006. Identifying major sedimentary basins beneath the west Antarctic ice sheet from aeromagnetic data analysis. In: Fütterer, D.K., Damaske, D., Kleinschmidt, G., Miller, H., Tessensohn, F. (Eds.), Antarctica: Contributions to Global Earth Sciences. Springer, Berlin, pp. 117–122.

Bentley, C.R., 1991. Configuration and structure of the subglacial crust. In: Tingey, R.J. (Ed.), The Geology of Antarctica. Clarendon, Oxford, pp. 335–364.

Berg, J.H., Moscati, R.J., Herz, D.L., 1989. A petrologic geotherm from a continental rift in Antarctica. Earth and Planetary Science Letters 93, 98–108.

Birkenmajer, K., 1994. Evolution of the Pacific margin of the northern Antarctic peninsula: An overview. Geologische Rundschau 83, 309–321.

Black, L.P., Sheraton, J.W., Tingey, R.J., McCulloch, M.T., 1992. New U–Pb zircon ages from the Denman Glacier area, East Antarctica, and their significance for Gondwana reconstruction. Antarctic Science 4, 447–460.

Blackman, D.K., Herzen, R.P., Lawver, L.A., 1987a. Heat flow and tectonics in the western Ross Sea, Antarctica. In: Cooper, A., Davey, F.J. (Eds.), The Antarctic Continental Margin, Geology and Geophysics of the Western Ross Sea, CPCEMR Earth Sciences Series, vol. 5b, pp. 179–189.

Blackman, D.K., Von Herzen, R.P., Lawver, L.A., 1987b. Heat flow and tectonics in the western Ross Sea, Antarctica. In: Cooper, A.K., Davey, F.J. (Eds.), The Antarctic Continental Margin: Geology and Geophysics of the Western Ross Sea. Earth Science Series, vol. 5B. Circum-Pacific Council for Energy and Natural Resources, Houston, TX, pp. 179–189.

Board, W.S., Frimmel, H.E., Armstrong, R.A., 2005. Pan-African Tectonism in the Western Maud Belt: P–T–t Path for High-grade Gneisses in the H.U. Sverdrupfjella, East Antarctica. Journal of Petrology 46 (4), 671–699.

Boger, S.D., 2011. Antarctica – before and after Gondwana. Gondwana Research 19, 335–371.

Boger, S.D., Miller, J.M., 2004. Terminal suturing of Gondwana and the onset of the Ross–Delamerian orogeny: the cause and effect of an early Cambrian reconstruction of plate motions. Earth and Planetary Science Letters 219, 35–48.

Boger, S.D., Wilson, C.J.L., Fanning, C.M., 2001. Early Paleozoic tectonism within the East Antarctic Craton: the final suture between East and West Gondwana? Geology 29, 463–466.

Boger, S.D., Carson, C.J., Fanning, C.M., Hergt, J.M., Wilson, C.J.L., Woodhead, J.D., 2002. Pan-African intraplate deformation in the northern Prince Charles Mountains, east Antarctica. Earth and Planetary Science Letters 195, 195–210.

Boger, S.D., Wilson, C.J.L., Fanning, C.M., 2006. An Archaean province in the southern Prince Charles Mountains, East Antarctica: U–Pb zircon evidence for c. 3170 Ma granite plutonism and c. 2780 Ma partial melting and orogenesis. Precambrian Research 145, 207–228.

Boger, S.D., Hirdes, W., Ferreira, C.A.M., Jenett, T., Dallwig, R., Fanning, C.M., 2015. The 580–520 Ma Gondwana suture of Madagascar and its continuation into Antarctica and Africa. Gondwana Research 28, 1048–1060.

Bogdanova, S.V., Pisarevsky, S.A., Li, Z.X., 2009. Assembly and breakup of Rodinia (some results of IGCP project 440). Stratigraphy and Geological Correlation 17 (3), 259–274. Pleiades Publishing, Ltd.

Borg, S.G., DePaolo, D.J., 1991. A tectonic model of the Antarctic Gondwana margin with implications for southeastern Australia: isotopic and geochemical evidence. Tectonophysics 196, 339–358.

Borg, S.G., Stump, E., Chappell, B.W., McCulloch, M.T., Wyborn, D., Armstrong, R.L., et al., 1987. Granitoids of northern Victoria Land, Antarctica: implications of chemical and isotopic variations to regional crustal structure and tectonics. American Journal of Science 287, 127–169.

Borg, S.G., DePaolo, D.J., Smith, B.M., 1990. Isotopic structure and tectonics of the central Transantarctic Mountains. Journal of Geophysical Research 95, 6647–6669.

Bosum, W., Damaske, D., Roland, N.W., Behrendt, J., Saltus, R., 1989. The GANOVEX IV Victoria Land/Ross Sea aeromagnetic survey: interpretation of anomalies. Geologisches. Jahrbuch. E38, 153–230.

Bott, M.H.P., Stern, T.A., 1992. Finite element analysis of Transantarctic Mountain uplift and coeval subsidence in the Ross embayment. Tectonophysics 201, 341–356.

Bradshaw, J.D., 1989. Cretaceous geotectonic patterns in the New Zealand region. Tectonics 8, 803–820.

Bradshaw, J.D., 2007. The Ross Orogen and Lachlan Fold Belt in Marie Byrd Land, Northern Victoria Land and New Zealand: Implication for the tectonic setting of the Lachland Fold Belt in Antarctica. In: Cooper, A.K., Raymond, C., et al. (Eds.), Antarctica: A Keystone in a Changing World-Online Proceedings of the 10th ISAES. U. S. Geological Survey and The National Academies. Available from: http://pubs.usgs.gov/of/2007/1047/srp/srp059.

Bradshaw, M.A., Webers, G.F., 1988. The Devonian rocks of Antarctica. In: McMillan, N.J., Embry, A.F., Glass, D.J. (Eds.), Devonian of the World. Canadian Society of Petroleum Geologists, Calgary, pp. 783–795.

Bradshaw, J.D., Weaver, S.D., Pankurst, R.J., Storey, B.C., Muir, R.J., 1997. New Zealand superterranes recognized in Marie Byrd Land and Thurston Island. Terra Antartica 3, 429–443.

Brown, C., Yakymchuk, C., Brown, M., Fanning, C.M., Korhonen, F.J., Siddoway, C.S., 2016. From source to sink: petrogenesis of cretaceous anatectic granites from the Fosdick migmatite–granite complex, West Antarctica. Journal of Petrology. Available from: https://doi.org/10.1093/petrology/egw039.

Buggisch, W., Kleinschmidt, G., 2007. The Pan-African nappe-tectonics in the Shackleton Range, in: Cooper, A., et al. (Eds.), 10th International Symposium on Antarctic Earth Sciences Program Book, Santa Barbara, p. 63.

Buggisch, W., Kleinschmidt, G., Kreuzer, H., Krumm, S., 1990. Stratigraphy, metamorphism and nappe-tectonics in the Shackleton Range. Geodätische Geophysikalische Veröff. Reihe I 15, 64–86.

Burton-Johnson, A., Riley, T.R., 2015. Autochthonous v. accreted terrane development of continental margins: a revised in situ tectonic history of the Antarctic Peninsula. Journal of Geological Society, London. Available from: https://doi.org/10.1144/jgs2014-1110.

Busetti, M., Spadini, G., Van der Wateren, F.M., Cloetingh, S., Zanolla, C., 1999. Kinematic modelling of the west Antarctic rift system, Ross Sea, Antarctica. Global and Planetary Change 23, 79–103.

Campbell, I.B., Claridge, G.G.C., 1987. Antarctica: Soils, Weathering Processes and Environment. Elsevier, Amsterdam, p. 368.

Cande, S.C., Leslie, R.B., 1986. Late Cenozoic tectonics of the southern Chile trench. Journal of Geophysical Research 91, 451–496.

Cande, S.C., Mütter, J.C., 1982. A revised identification of the oldest sea-floorspreading anomalies between Australia and Antarctica. Earth and Planetary Science Letters 58, 151–160.

Cande, S.C., Stock, J.M., 2006. Constraints on the timing of extension in the northern basin, Ross Sea. In: Fütterer, D.K., Damaske, D., Kleinschmidt, G., Miller, H., Tessensohn, F. (Eds.), Antarctica: Contributions to Global Earth Sciences. Springer, Berlin, pp. 19–326.

Cande, S.C., Stock, J.M., Müller, R.D., Ishihara, T., 2000. Cenozoic motion between east and west Antarctica. Nature 404, 145–150.

Cape Roberts Science Team, 2000. Studies from the Cape Roberts Project, Ross Sea, Antarctica, Initial report on CRP-3. Terra Antartica 7, 1–209.

Carter, L., et al., 2021. Water masses, circulation and change in the modern Southern Ocean. In: Florindo, F., et al. (Eds.), Antarctic Climate Evolution, second ed. Elsevier (this volume).

Castillo, P., Fanning, C.M., Fernandez, R., Poblete, F., Hervé, F., 2017. Provenance and constrains of Paleozoic siliciclastic rocks from the Ellsworth Mountains in West Antarctica, as determined by detrital zircon geochronology. Geological Society of American Bulletin 129 (11/12), 1568–1584.

Cawood, P.A., 2005. Terra Australis orogen: Rodinia breakup and development of the Pacific and Iapetus margins of Gondwana during the Neoproterozoic and Paleozoic. Earth-Science Reviews 69, 249–279.

Cawood, P.A., Buchan, C., 2007. Linking accretionary orogenesis with supercontinent assembly. Earth-Science Reviews 82, 217–256.

Cawood, P.A., Strachan, R.A., Pisarevsky, S.A., Gladkochub, D.P., Murphy, J.B., 2016. Linking collisional and accretionary orogens during Rodinia assembly and breaup: implications for models of supercontinent cycles. Earth and Planetary Science Letters 449, 118–126.

Chery, J., Lucazeau, F., Daignières, M., Vilotte, J.P., 1992. Large uplift of rift flanks: a genetic link with lithospheric rigidity? Earth and Planetary Science Letters 112, 195–211.

Clark, C., Kinny, P.D., Harley, S.L., 2012. Sedimentary provenance and age of metamorphism of the Vestfold Hills, East Antarctica: evidence for a oiece of Chinese Antarctica? Precambrian Research, 196–197, pp. 23–45.

Cogley, J.G., 1984. Deglacial hypsometry of Antarctica. Earth and Planetary Science Letters 67, 151–177.

Collins, W.J., 2002. Nature of extensional accretionary orogens. Tectonics 21 (4), 1–12.

Colleoni, F., De Santis, L., Siddoway, C.S., Bergamasco, A., Golledge, N., Lohmann, G., et al., 2018. Spatio-temporal variability of processes across Antarctic ice-bed-ocean interfaces. Nature Communications 9, 2289.

Colleoni, F., et al., 2021. Past Antarctic ice sheet dynamics and implications for future sea-level change. In: Florindo, F., et al. (Eds.), Antarctic Climate Evolution, second ed. Elsevier (this volume).

Collins, A.S., Pisarevsky, S.A., 2005. Amalgamating eastern Gondwana: the evolution of the Circum-Indian Orogens. Earth-Science Reviews 71, 229–270.

Cook, Y.A., 2007. Precambrian rift-related magmatism and sedimentation, south Victoria Land, Antarctica. Antarctic Science 8, 1–14.

Cooper, A.K., Webb, P.N., 1994. The ANTASTRAT Project-an international effort to investigate Cenozoic Antarctic Glacial History, climates, and sea-level changes. Terra Antartica 2, 239–242.

Cooper, A.K., Davey, F.J., Behrendt, J.C., 1987. Seismic stratigraphy and structure of the Victoria Land Basin, Western Ross Sea, Antarctica. In: Cooper, A. K., Davey, F. J. (Eds.), The Antarctic Continental Margin. Geology and Geophysics of the Western Ross Sea, Earth Science Series, vol. 5B. Circum-Pacific Council for Energy and Mineral Resources, Houston, TX, pp. 27–76.

Cooper, A.K., Davey, F.J., Hinz, K., 1991. Crustal extension and origin of sedimentary basins beneath the Ross Sea and Ross ice shelf, Antarctica. In: Thomson, M.R.A., Crame, J.A., Thomson, J.W. (Eds.), Geological Evolution of Antarctica. Cambridge University Press, Cambridge, pp. 285–292.

Cooper, A.K., Trey, H., Pellis, G., Cochrane, G., Egloff, F., Busetti, M., et al., 1997. Crustal structure of the southern central trough, western Ross Sea. In: Ricci, C.A. (Ed.), The Antarctic Region: Geological Evolution and Processes. Terra Antartica Publication, Siena, pp. 637–642.

Cooper, A.F., Maas, R., Scott, J.M., Barber, A.J.W., 2011. Dating of volcanism and sedimentation in the Skelton Group, Transantarctic Mountains: implications for the Rodinia-Gondwana transition in southern Victoria Land, Antarctica. Geological Society of America Bulletin 123, 681–702.

Corvino, A.F., Boger, Clement, S.D., Fay, C., 2016. Constriction structures related to viscous collision, southern Prince Charles Mountains. Antarctica Journal of Structural Geology 90, 128–143.

Cox, K.G., 1988. The Karoo Province. In: MacDougall, J.D. (Ed.), Continental Flood Basalts. Kluwer Academic Publishers, Dordrecht, pp. 239–271.

Craddock, C., 1970. Tectonic map of Gondwana. In: Bushnell, V.C., Craddock, C. (Eds.), Geologic Maps of Antarctica. American Geographic Society, New York, Antarctic Map Folio Series, Folio 12, Plate XXIII.

Craddock, J.P., Schmitz, M.D., Crowley, J.L., Larocque, J., Pankhurst, R.J., Juda, N., et al., 2017. Precise U-Pb zircon ages and geochemistry of Jurassic granites, Ellsworth-Whitmore terrane, central Antarctica. Geological Society of America Bulletin 129, 118–136.

Curtis, M.L., 1998. Development of kinematic partitioning within a pure-shear dominated dextral transpression zone: The southern Ellsworth Mountains, Antarctica. In: Strachan, R.A., Dewey, J.F. (Eds.), Continental Transpressional and Transtensional Tectonics, 135. Geological Society, London, pp. 289–306. Special Publication.

Daczko, N.R., Halpin, J.A., Fitzsimons, I.C.W., Whittaker, J.M., 2018. A cryptic Gondwana-forming orogen located in Antarctica. Scientific Reports 8, 8371. Available from: https://doi.org/10.1038/s41598-018-26530-1.

Dalziel, I.W.D., 1982. The early (pre-middle Jurassic) history of the Scotia Arc region: A review and progress report. In: Craddock, C. (Ed.), Antarctic Geoscience. University of Wisconsin Press, Madison, pp. 111–126.

Dalziel, I.W.D., 1991. Pacific margins of Laurentia and east Antarctica–Australia as a conjugate rift pair: Evidence and implications for an Eocambrian supercontinent. Geology 19, 598–601.

Dalziel, I.W.D., 1992. Antarctica: a tale of two supercontinents. Annual Review of Earth and Planetary Science 20, 501–526.

Dalziel, I.W.D., 1997. Neoproterozoic–Paleozoic geography and tectonics: review, hypothesis, environmental speculation. Geological Society of America Bulletin 109, 16–42.

Dalziel, I.W.D., Elliot, D.H., 1982. West Antarctica: Problem child of Gondwanaland. Tectonics 1, 3–19.

Dalziel, I.W.D., Lawver, L.A., Murphy, J.B., 1999. Plume, orogenesis, supercontinental fragmentation and ice-sheets. In: Skinner, D.N.B. (Ed.), 8th Symposium on Antarctic Earth Sciences, 5–9 July, Programme and Abstracts. Victoria University of Wellington, Wellington, New Zealand, p. 78.

Davey, F.J., 1981. Geophysical studies in the Ross Sea region. Journal of Royal Society of New Zealand 11, 465–479.

Davey, F.J., Brancolini, G., 1995. The late Mesozoic and Cenozoic structural setting of the Ross Sea region, Geology and Stratigraphy of the Antarctic Margin, vol. 68. American Geophysical Union, Washington, DC, pp. 167–182, AU: 14, Antarctic Research Series.

Davey, F.J., Cooper, A.K., 1987. Gravity studies of the Victoria Land Basin and Iselin Bank. In: Cooper, A.K., Davey, F.J. (Eds.), The Antarctic Continental Margin: Geology and Geophysics of the Western Ross Sea, vol. 5B. Circum-Pacific Council for Energy and Natural Resources, Houston, TX, pp. 119–137. Earth Science Series.

Davey, R.J., De Santis, L., 2006. A multi-phase rifting model for the Victoria Land Basin, Western Ross Sea. In: Fütterer, D.K., Damaske, D., Kleinschmidt, G., Miller, H., Tessensohn, F. (Eds.), Antarctica – Contributions to Global Earth Sciences. Springer, Berlin/Heidelberg, pp. 303–308.

Davey, F.J., Granot, R., Cande, S.C., Stock, J.M., Selvans, M., Ferraccioli, F., 2016. Synchronous oceanic spreading and continental rifting in West Antarctica. Geophysical Research Letters 43 (12), 6162–6169.

David, T.E., Priestley, R.E., 1914. British Antarctic Expedition 1907–09, AU Geology, 1, London.

Della Vedova, B., Pellis, G., Trey, H., Zhang, J., Cooper, A.K., Makris, J., et al., 1997. Crustal structure of the Transantarctic Mountains, western Ross Sea. In: Ricci, C.A. (Ed.), The Antarctic Region: Geological Evolution and Processes. Terra Antartica Publication, Siena, pp. 609–618.

Di Vincenzo, G., Palmeri, R., Talarico, F., Andriessen, P.A.M., Ricci, C.A., 1997. Petrology and geochronology of eclogites from the Lanterman Range, Antarctica. Journal of Petrology 38, 1391–1417.

Di Vincenzo, G., Rocchi, S., Rossetti, F., Storti, F., 2004. 40Ar–39Ar dating of pseudotachylytes: The effect of clast-hosted extraneous argon in Cenozoic fault generated friction melts from the west Antarctic rift system. Earth and Planetary Science Letters 223, 349–364.

Di Vincenzo, G., Talarico, F., Kleinschmidt, G., 2007. An ^{40}Ar-^{39}Ar investigation of the Mertz Glacier area (Georg V Land, Antarctica): Implications for the Ross Orogen-East Antarctica Craton relationship and Gondwana reconstructions. Precambrian Research 152 (3–4), 93–118.

Di Vincenzo, G., Horton, F., Palmeri, R., 2016. Protracted (~30 Ma) eclogite-facies metamorphism in northern Victoria Land (Antarctica): Implications for the geodynamics of the Ross/Delamerian Orogen. Gondwana Research 40, 91–106.

Dunkley, D.J., Hokada, T., Shiraishi, K., Hiroi, Y., Nogi, Y., Motoyoshi, Y., 2020. Geological subdivision of the Lützow–Holm Complex in East Antarctica: From the Neoarchean to the Neoproterozoic. Polar Science 26, 100606.

Du Toit, A.L., 1937. Our Wandering Continents. Oliver and Boyd, Edinburgh, 366 pp.

Elburg, M., Jacobs, J., Andersen, T., Clark, C., Läufer, A., Ruppel, A., et al., 2014. Early Neoproterozoic metagabbro-tonalite-trondhjemite of Sor Rondane (East Antarctica): implications for supercontinent assembly. Precambrian Research 259, 189–206.

Elburg, M., Andersen, T., Jacobs, J., Läufer, A., Ruppel, A., Krohne, N., et al., 2016. One hundred and fifty million years of Pan-African magmatism in the Sør Rondane Mountains (East Antarctica): implications for Gondwana assembly. The Journal of Geology 124 (1), 1–26.

Elliot, D.H., 1975. Tectonics of Antarctica: a review. American Journal of Science 275, 45–106.

Elliot, D.H., 1992. Jurassic magmatism and tectonism associated with Gondwanaland breakup: an Antarctic perspective. In: Storey, B.C., Alabaster, T., Pankhurst, R.J. (Eds.), Magmatism and the Causes of Continental Breakup. Geological Society, London, pp. 165–184. Special Publication 68.

Elliot, D.H., 2013. The Geological and Tectonic Evolution of the Transantarctic Mountains: a review. Geological Society, London, pp. 7–35, Special Publication 381.

Elliot, D.H., Fanning, C.M., 2008. Detrital zircons from upper Permian and lower Triassic Victoria Group sandstones, Shackleton Glacier region, Antarctica: evidence for multiple sources along the Gondwana plate margin. Gondwana Research 259–274.

Elliot, D.H., Fortner, E.H., Grimes, C.B., 2006. Mawson Breccias intrude Beacon strata at Allan Hills, South Victoria Land: regional implications. In: Fütterer, D.K., et al., (Eds.), Antarctica – Contributions to Global Earth Sciences. Springer, Berlin/Heidelberg, pp. 291–298.

Elliot, D.H., Fanning, C.M., Hulett, S.R.W., 2015. Age provinces in the Antarctic craton: evidence from detrital zircons in Permian strata from the Beardmore Glacier region, Antarctica. Gondwana Research 28, 152–164.

Elliot, D.H., Fanning, C.M., Laudon, T.S., 2016. The Gondwana Plate margin in the Weddell Sea sector: Zircon geochronology of Upper Paleozoic (mainly Permian) strata from the Ellsworth Mountains and eastern Ellsworth Land, Antarctica. Gondwana Research 29, 234–247.

Elliot, D.H., Fanning, C.M., Isbell, J.L., Hulett, S.R.W., 2017. The Permo-Triassic Gondwana sequence, central Transantarctic Mountains, Antarctica: Zircon geochronology, provenance, and basin evolution. Geosphere 13, 155–178.

Elsner, M., Schöner, R., Gerdes, A., Gaupp, R., 2013. Reconstruction of the early Mesozoic plate margin of Gondwana by U–Pb ages of detrital zircons from northern Victoria Land, Antarctica. In: Harley, S.L., Fitzsimons, I.C.W., Zhao, Y. (Eds.), Antarctica and Supercontinent Evolution. Geological Society, London, pp. 211–232. Special Publications 383.

Encarnacion, J., Grunow, A., 1996. Changing magmatic and tectonic styles along the paleo-Pacific margin of Gondwana and the onset of early Paleozoic magmatism in Antarctica. Tectonics 15, 1325–1341.

Estrada, S., Läufer, A., Eckelmann, K., Hofmann, M., Gärtner, A., Linnemann, U., 2016. Continuous Neoproterozoic to Ordovician sedimentation at the East Gondwana margin –implications from detrital zircons of the Ross Orogen in northern Victoria Land, Antarctica. Gondwana Research 37, 426–448.

Evans, D.A.D., 2009. The palaeomagnetically viable, longlived and all-inclusive Rodinia supercontinent reconstruction, in Murphy, J.B., Keppie, J.D., Hynes, A.J. (Eds.), Ancient Orogens and Modern Analogues. Geological Society of London Special Publication 327, pp. 371–404.

Evans, D.A.D., 2013. Reconstructing pre-Pangean supercontinents. GSA Bulletin 125 (11/12), 1735–1751.

Ferraccioli, F., Armadillo, E., Jordan, T., Bozzo, E., Corr, H., 2009. Aeromagnetic exploration over the East Antarctic Ice Sheet: a new view of the Wilkes Subglacial Basin. Tectonophysics 478, 62–77.

Ferraccioli, F., Finn, C.A., Jordan, T.A., Bell, R.E., Anderson, L.M., Damaske, D., 2011. East Antarctic rifting triggers uplift of the Gamburtsev Mountains. Nature 479, 388–393.

Fielding, C.R., Henrys, S.A., Wilson, T.J., 2006. Rift history of the Western Victoria Land Basin: a new perspective based on integration of cores with seismic reflection data. In: Fütterer, D.K., et al., (Eds.), Antarctica – Contributions to Global Earth Sciences. Springer, Berlin/Heidelberg, pp. 309–318.

Fielding, C.R., Whittaker, J., Henrys, S.A., Wilson, T.J., Naish, T.R., 2008. Seismic facies and stratigraphy of the Cenozoic succession in McMurdo Sound, Antarctica: implications for tectonic, climatic and glacial history. Palaeogeography, Palaeoclimatology, Palaeoecology 260 (1–2), 8–29.

Findlay, R.H., 1987. A review of the problems important for the interpretation of the Cambro-Ordovician paleogeography of northern Victoria Land (Antarctica), Tasmania, and New Zealand. In: McKenzie, G.D. (Ed.), Gondwana Six: Structure, Tectonics, and Geophysics. American Geophysical Union Monographs, vol. 40, pp. 49–66.

Fioretti, A.M., Black, P., Varne, R., 2001. U-Pb Zircon SHRIMP dating of the Gallipoli Volcanics, Northern Victoria Land (Antarctica): EUG XI, Strasbourg, LS08: age growth and evolution of Antarctica (AGEANT), 08.04.–12.04.02.

Fitzgerald, P.G., 1992. The Transantarctic Mountains of southern Victoria Land: the application of apatite fission track analysis to a rift shoulder uplift. Tectonics 11, 634–662.

Fitzgerald, P.G., 1995. Cretaceous and Cenozoic exhumation of the Transantarctic Mountains: Evidence from the Kukri Hills of southern Victoria Land compared to fission track data from Gneiss at DSDP site 270. In: Ricci, C.A. (Ed.), 7th International Symposium on Antarctic Earth Sciences, Siena, Italy, p. 133.

Fitzgerald, P.G., 2002. Tectonics and landscape evolution of the Antarctic plate since the breakup of Gondwana, with an emphasis on the west Antarctic rift system and the Transantarctic Mountains. In: Gamble, J., Skinner, D.A. (Eds.), Proceedings of the 8th International Symposium on Antarctic Earth Science, vol. 35. Royal Society of New Zealand Bulletin, pp. 435–469.

Fitzgerald, P.G., Baldwin, S.L., 1997. Detachment fault model for the evolution of the Ross embayment. In: Ricci, C.A. (Ed.), The Antarctic Region: Geological Evolution and Processes. Terra Antartica Publication, Siena, pp. 555–564.

Fitzgerald, P.G., Stump, E., 1997. Cretaceous and Cenozoic episodic denudation of the Transantarctic Mountains, Antarctica: new constraints from apatite fission track thermochronology in the Scott Glacier region, Journal of Geophysical Research, 102. pp. 7747–7765.

Fitzgerald, P.G., Sandiford, M., Barrett, P.J., Gleadow, A.J.W., 1986. Asymmetric extension associated with uplift and subsidence of the Transantarctic Mountains and Ross embayment. Earth and Planetary Science Letters 81, 67–78.

Fitzsimons, I.C.W., 1997. The Brattstrand paragneiss and the Sostrene orthogneiss: a review of Pan-African metamorphism and Grenvillian relics in southern Prydz Bay. In: Ricci, C.A. (Ed.), The Antarctic Region: Geological Evolution and Processes. Terra Antartica, Siena, pp. 121–130.

Fitzsimons, I.C.W., 2000a. Grenville-age basement provinces in East Antarctica: evidence for three separate collisional orogens. Geology 28, 879–882.

Fitzsimons, I.C.W., 2000b. A review of tectonic events in the east Antarctic shield and their implications for Gondwana and earlier supercontinents. Journal of African Earth Sciences, 31, 3–23.

Fitzsimons, I.C.W., 2003. Proterozoic basement provinces of southern and south-western Australia, and their correlation with Antarctica. In: Yoshida, M., Windley, B.F., Dasgupta, S. (Eds.), Proterozoic East Gondwana: Supercontinent Assembly and Breakup. Geological Society, London, pp. 93–129.

Flöttmann, T., Gibson, G.M., Kleinschmidt, G., 1993. Structural continuity of the Ross and Delamerian orogens of Antarctica and Australia along the margin of the paleo-Pacific. Geology 21, 319–322.

Flowerdew, M.J., Millar, Vaughan, A.P.M., Horstwood, M.S.A., Fanning, C.M., 2006. The source of granitic gneisses and migmatites in the Antarctic Peninsula: a combined U–Pb SHRIMP and laser ablation Hf isotope study of complex zircons. Contributions to Mineralogy and Petrology 151, 751–768.

Flowerdew, M.J., Tyrrel, S., Boger, S.D., Fitzsimons, I.C.W., Harley, S.L., Mikhalsky, E.V., Vaughan, A.P.M., 2013. Pb isotopic domains from the Indian Ocean sector of Antarctica: implications for past Antarctica-India connections. Geological Society, London, Special Publications 383 (1), 59–72.

Ford, A.B., 1963. Cordierite-bearing hypersthene–quartz–monzonite–porphyry in the Thiel Mountains and its regional importance. In: Adie, R.J. (Ed.), Antarctic Geology. North-Holland Publishing Company, Amsterdam, pp. 429–441.

Forster, D.A., Goscombe, B.D., 2013. Continental Grouth and recycling in convergent orogens with large Turbidite Fans on Oceanic Crust. Geosciences 3, 354–388.

Fretwell, P., et al., 2013. Bedmap2: improved ice bed, surface and thickness datasets for Antarctica. The Cryosphere 7, 375–393.

Funaki, M., Yoshida, M., Matsueda, H., 1991. Palaeomagnetic studies of palaeozoic rocks from the Ellsworth Mountains. In: Thomson, M.R.A., Crame, J.A., Thomson, J.W. (Eds.), Geological Evolution of Antarctica. Cambridge University Press, Cambridge, pp. 257–260.

GANOVEX Team, 1987. Geological map of North Victoria Land, Antarctica, 1:500000 explanation notes. Geologisches Jahrbuch Reihe B 66, 7–79.

Gibson, G.M., Ireland, T.R., 1996. Extension of Delamerian (Ross) orogen into western New Zealand: evidence from Zircon ages and implications for crustal growth along the Pacific margin of Gondwana. Geology 24, 1087–1090.

Godard, G., Palmeri, R., 2013. High-pressure metamorphism in Antarctica from the Proterozoic to the Cenozoic: a review and geodynamic implications. Gondwana Research 23, 844–864.

Gohl, K., Wellner, J.S., Klaus, A., the Expedition 379 Scientists, 2021. Amundsen Sea West Antarctic Ice Sheet history. In: Proceedings of the International Ocean Discovery Program, vol. 379. International Ocean Discovery Program, College Station, TX. <https://doi.org/10.14379/iodp.proc.379.2021>.

Golynsky, A., Jacobs, J., 2001. Grenville-age vs pan-african magnetic anomaly imprints in Western Dronning Maud Land, East Antarctica. The Journal of Geology 109, 136–142.

Golynsky, A.V., Ferraccioli, F., Hong, J.K., Golynsky, D.A., von Frese, R.R.B., Young, D.A., et al., 2018. New magnetic anomaly map of the Antarctic. Geophysical Research Letters 45, 6437–6449.

Gonzales-Casado, J.M., Giner-Robles, J.L., Lopez-Martinez, J., 2000. Bransfield Basin, Antarctic Peninsula: Not a normal Backarc Basin. Geology 28, 1043–1046.

Goodge, J.W., 2002. From Rodinia to Gondwana: supercontinent evolution in the Transantarctic Mountains. In: Gamble, J., Skinner, D. (Eds.), Proceedings of the 8th International Symposium on Antarctic Earth Science, vol. 35. Royal Society of New Zealand Bulletin, pp. 61–74.

Goodge, J.W., 2020. Geological and tectonic evolution of the Transantarctic Mountains, from ancient craton to recent enigma. Gondwana Research 80, 50–122.

Goodge, J.W., 2021. The Geology of the Transantarctic Mountains. In: Kleinschmidt G. (Ed.), The Geology of the Antarctic Continent. Beiträge zur regionalen Geologie der Erde, Band 33, pp. 132–217.

Goodge, J.W., Finn, C.A., 2010. Glimpses of East Antarctica: aeromagnetic and satellite magnetic view from the central Transantarctic Mountains of East Antarctica. Journal of Geophysical Research 115 (B09103), 1–22.

Goodge, J.W., Severinghaus, J.P., 2016. Rapid Access Ice Drill: a new tool for exploration of the deep Antarctic ice sheets and subglacial geology. Journal of Glaciology 62 (236), 1049–1064.

Goodge, J.W., Hansen, V.L., Peacock, S.M., Smith, B.K., Walker, N.W., 1993a. Kinematic evolution of the Miller Range shear zone, central Transantarctic Mountains, Antarctica, and implications for Neoproterozoic to early Paleozoic tectonics of the East Antarctic margin of Gondwana. Tectonics 12, 1460–1478.

Goodge, J.W., Fanning, C.M., Bennett, V.C., 2001. U-Pb evidence of ~1.7 Ga crustal tectonism during the Nimrod Orogeny in the Transantarctic Mountains, Antarctica: implications for Proterozoic plate reconstructions. Precambrian Research 112, 261–288.

Goodge, J.W., Myrow, P., Williams, I.S., Bowring, S.A., 2002. Age and provenance of the Beardmore Group, Antarctica: Constraints on Rodinia supercontinent breakup. Journal of Geology 110, 393–406.

Goodge, J.W., Myrow, P., Phillips, D., Fanning, C.M., Williams, I.S., 2004a. Siliciclastic record of rapid denudation in response to convergent-margin orogenesis, Ross Orogen, Antarctica. In: Bernet, M., Spiegel, C. (Eds.), Detrital Thermochronology—Provenance Analysis, Exhumation, and Landscape Evolution of Mountain Belts: Boulder, vol. 378, Colorado, Geological Society of America Special Paper, pp. 101–122.

Goodge, J.W., Williams, I.S., Myrow, P., 2004b. Provenance of Neoproterozoic and lower Paleozoic siliciclastic rocks of the Central Ross Orogen, Antarctica: detrital record of rift-, passive- and active-margin sedimentation, Geological Society of America Bulletin, 116. pp. 1253–1279.

Goodge, J.W., Fanning, C.M., Brecke, D.M., Licht, K.J., Palmer, E.F., 2010. Continuation of the Laurentian Grenville province across the Ross Sea margin of East Antarctica. Journal of Geology 118, 601–619.

Goodge, J.W., Fanning, C.M., Norman, M., Bennett, V.C., 2012. Temporal, isotopic and spatial relations of early Paleozoic Gondwana-margin arc magmatism, central Transantarctic Mountains, Antarctica. Journal of Petrology 53, 2027–2065.

Goodge, J.W., Fanning, C.M., Fisher, C.M., Vervoort, J.D., 2017. Proterozoic crustal evolution of central East Antarctica: Age and isotopic evidence from glacial igneous clasts, and links with Australia and Laurentia. Precambrian Research 299, 151–176.

Goodge, J.W., Vervoort, J.D., Fanning, C.M., Brecke, D.M., Farmer, G.L., Williams, I.S., et al., 2008. A positive test of East Antarctica-Laurentia juxtaposition within the Rodinia supercontinent. Science 321, 235–240.

Gould, L.M., 1935. Structure of the Queen Maud Mountains, Antarctica. Geological Society of America Bulletin 46, 973–984.

Granot, R., Cande, S.C., Stock, J.M., Damaske, D., 2013. Revised Eocene-Oligocene kinematics for the West Antarctic rift system. Geophysical Research Letters, 40 (2), 279–284.

Granot, R., Dyment, J., 2018. Late Cenozoic unification of East andWest Antarctica. Nature Communications 9, 3189. Available from: https://doi.org/10.1038/s41467-018-05270-w.

Grantham, G.H., Macey, P.H., Horie, K., Kawakami, T., Ishikawa, M., Satish-Kumar, M., et al., 2013. Comparison of the metamorphic history of the Monapo Complex, northern Mozambique and Balchenfjella and Austhameren areas, Sør Rondane, Antarctica: implications for the Kuunga Orogeny and the amalgamation of N and S Gondwana. Precambrian Research 234, 85–135.

Groenewald, P.B., Moyes, A.B., Grantham, G.H., Krynauw, J.R., 1995. East Antarctic crustal evolution: geological constraints and modeling in western Dronning Maud Land. Precambrian Research 75 (3–4), 231–250.

Grew, E.S., Carson, C.J., Christy, A.G., Maas, R., Yaxley, G.M., Boger, S.D., et al., 2012. New constraints from U–Pb, Lu–Hf and Sm–Nd isotopic data on the timing of sedimentation and felsic magmatism in the Larsemann Hills, Prydz Bay, East Antarctica. Precambrian Research 206–207, 87–108.

Grindley, G.W. Laird, M.G., 1969. Geology of the Shackleton Coast. Antarctic Map Folio Series, v. Folio 12, XIV.

Grunow, A., Hanson, R., Wilson, T., 1996. Were aspects of Pan-African deformation linked to Iapetus opening? Geology 24, 1063–1066.

Grunow, A.M., Kent, D.V., Dalziel, I.W.D., 1987. Evolution of the Weddell Sea Basin, new paleomagnetic constraints. Earth and Planetary Science Letters 86, 16–26.

Gunn, B.M., Warren, G., 1962. Geology of Victoria Land between the Mawson and Mulock Glaciers, Antartica. New Zealand Geological Survey Bulletin 71, 157.

Gust, D.A., Biddle, K.T., Phelps, D.W., Uliana, M.A., 1985. Associated middle to late Jurassic volcanism and extension in southern South America. Tectonophysics 116, 223–253.

Hagen-Peter, G., Cottle, C., 2016. Synchronous alkaline and subalkaline magmatism during the late Neoproterozoic–early Paleozoic Ross orogeny, Antarctica: insights into magmatic sources and processes within a continental arc, Lithos, 262. pp. 677–698.

Hagen-Peter, G., Cottle, C., 2018. Evaluating the relative roles of crustal growth vs reworking through continental arc magmatism: a case study from the Ross orogen, Antarctica. Gondwana Research 55, 153–166.

Hagen-Peter, G., Cottle, C., Smit, M., Cooper, A.F., 2016. Coupled garnet Lu–Hf and monazite U–Pb geochronology constrain early convergent margin dynamics in the Ross orogen, Antarctica. Journal of Metamorphic Geology 34, 293–319.

Hambrey, M.J., 2002. Historical review of drilling in the western Ross Sea. In: Harwood, D.M., Lacy, L., Levy, R.H. (Eds.), Future Antarctic Margin Drilling: Developing a Science Plan for McMurdo Sound. ANDRILL Science Management Office (SMO) Contribution 1, University of Nebraska-Lincoln, Lincoln, NE.

Hambrey, M.J., Wise Jr., S.W., Barrett, P.J., Davey, F.J., Ehrmann, W.U., Smellie, J.L. (Eds.), 1998. Scientific Report on CRP-1, Cape Roberts Project, Antarctica. Terra Antartica 5, 713.

Hamilton, R.J., Luyendyk, B.P., Sorlien, C.C., Bartek, L.R., 2001. Cenozoic tectonics of the Cape Roberts Rift Basin and Transantarctic Mountains front, southwestern Ross Sea, Antarctica. Tectonics 20, 325–342.

Harley, S.I., Fitzsimons, I.C.W., Zhao, Y., 2013. Antarctica and supercontinent evolution: historical perspectives, recent advances and unresolved issues. Geological Society of London, Special Publications 383, 1–34.

Harrowfield, M., Holdgate, G.R., Wilson, C.J.L., McLoughlin, S., 2015. Tectonic significance of the Lambert graben, East Antarctica: Reconstructing the Gondwanan rift. Geology 33 (3), 197.

Hartnady, C.J.H., 1991. About turn for supercontinents. Nature 352, 476–478.

Harwood, D.M., Lacy, L., Levy, R.H. (Eds.), 2002. Future Antarctic Margin Drilling: Developing a Science Plan for McMurdo Sound. ANDRILL Science Management Office (SMO) Contribution 1. University of Nebraska-Lincoln, Lincoln, NE, p. 301.

Harwood, D.M., Webb, P.-N., Barrett, P.J., McKelvey, B.C., August 1991. The changing style of Antarctic glaciations. In: 6th International Antarctic Earth Science Symposium, Tokyo, p. 20.

Harwood, D.M., Webb, P.-N., Barrett, P.J., 1993. The search for consistency between indices of Antarctic Cenozoic glaciation: Cenozoic glaciations and deglaciations. Geological Society of London, Stratigraphic Committee and the Quaternary Research Association, Quaternary Newsletter, 69. pp. 61–63.

Heimann, A., Fleming, T.H., Elliot, D.H., Foland, K.A., 1994. A short interval of Jurassic continental flood basalt volcanism in Antarctica as demonstrated by $^{40}Ar/^{39}Ar$ geochronology. Earth and Planetary Science Letters 121, 19–41.

Henjes-Kunst, F., Roland, N.W., Dumphy, J.M., Fletcher, I.R., 2004. SHRIMP U-Pb dating of high-grade migmatites and related magmatites from Northwestern Oates Land (East Antarctica): evidence for a single high-grade event of Ross-Orogenic age. Terra Antartica 11 (1), 67–84.

Hergt, J.M., Chappell, M.T., McCullock, M., MacDougall, I., Chivas, A.R., 1989. Geochemical and isotopic constraints on the origin of the Jurassic dolerites of Tasmania. Journal of Petroleum Science 30, 841–883.

Hoffman, P.F., 1991. Did the breakout of Laurentia turn Gondwanaland insideout? Science 252, 1409–1412.

Hofmann, J., 1996. Fragmente Intragondwanischer Rifte als Werkzeug der Gondwana-Rekonstruktion-das Beispiel des Lambert-Mahanadi-Riftes (Ostantarktika-Peninsular Indien). Neues Jahrbuch. Geol. Palaeontol. Abhandlungen 199, 33–48.

Holmes, A., 1951. The sequence of pre-Cambrian orogenic belts in south and central Africa. In: Proceedings of the 18th International Geological Congress, vol. 24, London, pp. 254–269.

Isbell, J.L., 1999. The Kukri erosion surface: a reassessment of its relationship to rocks of the Beacon Supergroup in the central Transantarctic Mountains, Antarctica. Antarctic Science 11, 228–238.

Jacobs, J., (Ed.), 2009. Antarctica and Supercontinent Evolution. Geological Society, London (Special Publication 283. Polarforschung, 79 (1), 47–57).

Jacobs, J., Lisker, F., 1999. Post Permian tectono-thermal evolution of western Dronning Maud Land, east Antarctica: an apatite fission-track approach. Antarctic Science 11, 451–460.

Jacobs, J., Thomas, R.J., 2002. The Mozambique belt from an east Antarctic perspective. In: Gamble, J.A., Skinner, D.N.B., Henrys, S. (Eds.), Antarctica at the Close of a Millennium. Proceedings of the 8th International Symposium on Antarctic Earth Sciences, vol. 35. Royal Society of New Zealand Bulletin, pp. 3–18.

Jacobs, J., Fanning, C.M., Henjes-Kunst, F., Olesch, M., Paech, H.J., 1998. Continuation of the Mozambique belt into east Antarctica: Grenville-age metamorphism and polyphase Pan-African high-grade events in central Dronning Maud Land. Journal of Geology 106, 385–406.

Jacobs, J., Bauer, W., Fanning, C.M., 2003a. New age constraints for Grenville-age metamorphism in western central Dronning Maud Land (East Antarctica), and implications for the palaeogeography of Kalahari in Rodinia. International Journal of Earth Sciences 92, 3031–3315.

Jacobs, J., Thomas, R.J., 2004. Himalayan-type indenter-escape tectonics model for the southern part of the late Neoproterozoic-early Paleozoic East African-Antarctic orogen. Geology 32, 721–724.

Jacobs, J., Fanning, C.M., Bauer, W., 2003b. Timing of Grenville-age vs. Pan-African medium- to high-grade metamorphism in western Dronning Maud Land (East Antarctica) and significance for correlations in Rodinia and Gondwana. Precambrian Research 125, 1–20.

Jacobs, J., Bingen, B., Thomas, R.J., Bauer, W., Wingate, M.T.D., Feito, P., 2008a. Early Palaezoic orogenic collapse and voluminous late-tectonic magmatism in Dronning Maud Land and Mozambique: insights into the partially delaminated orogenic root of the East African Antarctic Orogen? Geological Society, London, Special Publications, 308. pp. 69–90.

Jacobs, J., Pisarevsky, S., Thomas, R.J., Beckere, T., 2008b. The Kalahari Craton during the assembly and dispersal of Rodinia. Precambrian Research 160, 142–158.

Jacobs, J., Elburg, M., Läufer, A., Kleinhanns, I.C., Henjes-Kunst, F., Estrada, S., et al., 2015. Two distinct Late Mesoproterozoic/Early Neoproterozoic basement provinces in central/eastern Dronning Maud Land, East Antarctica: the missing link, 15–21E. Precambrian Research 265, 249–272.

Jacobs, J., Opas, B., Elburg, M.A., Läufer, A., Estrada, S., Ksienzyk, A.K., et al., 2017. Cryptic sub-ice geology revealed by U-Pb zircon study of glacial till in Dronning Maud Land, East Antarctica. Precambrian Research 294, 1–14.

Jacobs, J., Mikhalsky, E., Henjes-Kunstc, F., Läufer, A., Thomas, R.J., Elburg, M.A., et al., 2020. Neoproterozoic geodynamic evolution of easternmost Kalahari: Constraints from U-Pb-Hf-O zircon, Sm-Nd isotope and geochemical data from the Schirmacher Oasis, East Antarctica. Precambrian Research 342, 239–252.

Janowski, E.J., Drewry, D.J., 1981. The structure of west Antarctic from geophysical studies. Nature 291, 17–21.

Johnston, S.T., 2000. The Cape Fold Belt and syntaxis and the rotated Falklands Islands: dextral transpressional tectonics along the southwest margin of Gondwana. Journal of African Earth Sciences 31, 51–63.

Jones, S., 1997. Late quaternary faulting and neotectonics, south Victoria Land, Antarctica. Journal of Geological Society, London 154 (4), 645–652.

Jordan, T.A., Ferraccioli, F., Ross, N., Corr, H.F.J., Leat, P.T., Bingham, R.G., et al., 2013. Inland extent of the Weddell Sea Rift imaged by new aerogeophysical data. Tectonophysics 585, 137–160. Available from: https://doi.org/10.1016/j.tecto.2012.09.010.

Jordan, T.A., Neale, R.F., Leat, P.T., Vaghan, A.P.M., Flowerdew, M.J., Riley, T.R., et al., 2014. Structure and evolution of Cenozoic arc magmatism on the Antarctic Peninsula: a high resolution aeromagnetic perspective. Geophysical Journal International 198, 1758–1774.

Jordan, T., Ferraccioli, F., Leat, P., 2017. New geophysical compilations links crustal block motion to Jurassic extension and strike-slip faulting in the Weddell Sea Rift System of West Antarctica. Gondwana Research 42, 29–48.

Jordan, T.A., Riley, T.R., Siddoway, C.S., 2020. The geological history and evolution of West Antarctica. Nature Reviews. Earth & Environment 1, 117–133.

Kamei, A., Horie, K., Owada, M., Yukara, M., Nakano, N., Osanai, Y., et al., 2013. Late Proterozoic juvenile arc metatonalite and adakitic intrusions in the Sør Rondane Mountains, eastern Dronning Maud Land, Antarctica. Precambrian Research 234 (0), 47–62.

Kanao, M., Shibutani, T., Negishi, H., Tono, H., 2002. Crustal structure around the Antarctic margin by teleseismic receiver function analyses. In: Gamble, J.A., Skinner, D.N.B., Henrys, S. (Eds.), Antarctica at the Close of a Millennium, vol. 35. Royal Society of New Zealand, Wellington, pp. 485–491.

Kapitsa, A., Ridley, J.K., Robin, G. de Q., Siegert, M.J., Zotikov, I., 1996. Large deep freshwater lake beneath the ice of central East Antarctica. Nature 381, 684–686.

Karlstrom, K.E., Williams, M.L., McLelland, J., Geissman, J.W., Ahall, K.-I., 1999. Refining Rodinia: geologic evidence for the Australia–western United States connection in the Proterozoic. GSA Today 9 (10), 1–7.

Kellogg, K.S., Rowley, P.D., 1989. Structural Geology and Tectonics of the Orville Coast Region, Southern Antarctic Peninsula, Antarctica. United States Geological Survey, p. 25.

Kelly, N.M., Harley, S.L., 2004. Orthopyroxene–corundum in Mg–Al-rich granulites from the Oygarden Islands, east Antarctica. Journal of Petrology 45, 1481–1512.

Kelly, N.M., Harley, S.L., 2005. An integrated microtextural and chemical approach to zircon geochronology: refining the Archean history of the Napier Complex, East Antarctica. Contributions to Mineralogy and Petrology 149, 57–84.

Kelly, N.M., Clarke, G.L., Fanning, C.M., 2002. A two-stage evolution of the Neoproterozoic Rayner structural episode: new U–Pb sensitive high resolution ion microprobe constraints from the Oygarden Group, Kemp Land, East Antarctica. Precambrian Research 116, 307–330.

Kelsey, D.E., Hand, M., Clark, C., Wilson, C.J.L., 2007. On the application of in situ monazite chemical geochronology to constraining P–T–t histories in high-temperature (>850C) polymetamorphic granulites from Prydz Bay, East Antarctica. Journal of the Geological Society, London 164, 667–683.

Kleinhans, I.C., Jacobs, J., Engvik, A., Bingen, B., Roland, N.W., Läufer, A., et al., 2013. Tracing old SCLM in Pan-African granitoids from Dronning Maud Land (East Antarctica) with Sr-Nd isotope signatures. Mineralogical Magazine 77 (5), 1476.

Kleinschmidt, G., 2014. Geologische Entwicklung und tektonischer Bau der Antarktis. In: Lozan, J.L., Grassel, H., Piepenburg, D., Notz, D. (Eds.), Warnsignal Klima. Die Polarregionen, Hamburg, pp. 18–28.

Kleinschmidt, G., Boger, S.D., 2008. The Bertrab, Littlewood and Moltke Nunataks of Prinz-Regent-Luitpold-Land (Coats Land) –Enigma of East Antarctic Geology. Polarforschung 78 (3), 95–104.

Kleinschmidt, G., Tessensohn, F., 1987. Early Paleozoic westward directed subduction at the Pacific margin of Antarctica. In: McKenzie, G.D. (Ed.), Gondwana Six: Structure, Tectonics, and Geophysics, vol. 40, American Geophysical Union Monographs, pp. 89–105.

Kleinschmidt, G., Buggisch, W., Läufer, A.L., Helferich, S., Tessensohn, F., 2002. The ''Ross orogenic'' structures in the Shackleton Range and their meaning for Antarctica. In: Gamble, J.A., Skinner, D.N.B., Henrys, S. (Eds.), Antarctica at the Close of a Millennium, vol. 35. Royal Society of New Zealand, Wellington, pp. 75–83.

Kleinschmidt, G., Läufer, A.L., 2006. The Matusevich Fracture Zone, in Oates Land, east Antarctica. In: Fütterer, D.K., Damaske, D., Kleinschmidt, G., Miller, H., Tessensohn, F. (Eds.), Antarctica: Contributions to Global Earth Sciences, Berlin/Heidelberg, pp. 175–180.

Kleinschmidt, G., 2021a. The Geology of the Antarctic Continent. Beiträge zur regionalen Geologie der Erde, Band 33, 613 pp.

Kleinschmidt, G., 2021b. The Shackleton Range and Surroundings. In: Kleinschmidt, G. (Ed.), The Geology of the Antarctic Continent. Beiträge zur regionalen Geologie der Erde, Band 33, pp. 218–253.

König, M., Jokat, W., 2006. The Mesozoic breakup of the Weddell Sea. Journal of Geophysical Research, Solid Earth 111, B12102.

Kròl, P., Kusiak, M.A., Dunkley, D.J., Wilde, S.A., Yi, K., Lee, S., et al., 2020. Diversity of Archean crust in the eastern Tula Mountains, Napier complex, East Antarctica. Gondwana Research 82, 151–170.

Kyle, P.R., 1990a. Erebus Volcanic Province. In: LeMasurier, W.E., Thompson, J.W. (Eds.), Volcanoes of the Antarctic Plate and Southern Oceans, Antarctic Research Series, vol. 48. American Geophysical Union, Washington, DC, pp. 81–135.

Kyle, P.R., 1990b. McMurdo Volcanic Group, western Ross embayment. Introduction. In: LeMasurier, W.E., Thompson, J.W. (Eds.), Volcanoes of the Antarctic Plate and Southern Oceans: Antarctic Research Series, vol. 48. American Geophysical Union, Washington, DC, pp. 19–25.

Kyle, P.R., Cole, J.W., 1974. Structural control of volcanism in the McMurdo Volcanic Group, Antarctica. Bulletin Volcanologique 38, 16–25.

Ivanov, A.V., Meffre, S., Thompson, J., Corfu, F., Kamenetsky, V.S., Kamenetsky, M.B., et al., 2017. Timing and genesis of the Karoo-Ferrar large igneous province: new high precision U-Pb data for Tasmania confirm short duration of the major magmatic pulse. Chemical Geology 455, 32–43.

Lamarque, G., Bascou, J., Ménot, R.P., Paquette, J.L., Couzinié, S., Rolland, Y., et al., 2018. Ediacaran to lower Cambrian basement in eastern George V Land (Antarctica): evidence from UPb dating of gneiss xenoliths and implications for the South Australia-East Antarctica connection. Lithos 318, 219–229.

Larter, R.D., Cunningham, A.P., Barker, P.F., Gohl, K., Nitsche, F.O., 2002. Tectonic evolution of the Pacific margin of Antarctica 1. Late Cretaceous tectonic reconstructions. Journal of Geophysical Research 107 (B12), 2345. Available from: https://doi.org/10.1029/2000JB000052.

Läufer, A., 2021. Geology and Geodynamic Evolution of the Dronning Maud Land, East Antarctica. In: Kleinschmidt, G. (Ed.), The Geology of the Antarctic Continent, Beiträge zur regionalen Geologie der Erde, Band 33, pp. 254–295.

Läufer, A., Lisker,F., Roland,N.W., Kleinschmidt, G., 2021. Lambert Glacier Area (East Antarctica between 45°E and 85°E). In: Kleinschmidt, G. (Ed.), The Geology of the Antarctic Continent. Beiträge zur regionalen Geologie der Erde, Band 33, pp. 296–319.

Lawver, L.A., Royer, J.-Y., Sandwell, D.A., Scotese, C.T., 1991. Evolution of the Antarctic continental margins. In: Thomson, M.R.A., Crame, J.A., Thomson, J.W. (Eds.), Geological Evolution of Antarctica. Cambridge University Press, Cambridge, pp. 533–539.

Lawver, L.A., Gahagan, L.M., Coffin, M.F., 1992. The development of Paleoseaways around Antarctica, The Antarctic Paleoenvironment: A Perspective on Global Change, vol. 56. American Geophysical Union, Washington, DC, pp. 7–30, Antarctic Research Series.

Lawver, L.A., Gahagan, L.M., Dalziel, I.W.D., 1998. A tight fit – Early Mesozoic Gondwana: a plate tectonic perspective. In: Origin and Evolution of Continents. Memoirs of the National Institute of Polar Research (Special issue), National Institute of Polar Research, Tokyo, pp. 214–229.

Leat, P.T., Jordan, T.A., Flowerdew, M.J., Riley, T.R., Ferraccioli, F., Whitehouse, M.J., 2018. Jurassic high heat production granites associated with the Weddell Rift System. Antarctica. Tectonophysics 722, 249–264.

LeMasurier, W.E., 2007. West Antarctic rift system. In: Riffenburgh, B. (Ed.), Encyclopedia of the Antarctic, Vol. 2. Routledge, New York, pp. 1060–1066.

LeMasurier, W.E., Thomson, J.W., 1990. Volcanoes of the Antarctic Plate and Southern Oceans, vol. 48. American Geophysical Union, Washington, DC, Antarctic Research Series.

Li, Z.X., Zhang, L., Powell, C.M., 1995. South China in Rodinia: part of the missing linkbetween Australia–East Antarctica and Laurentia? Geology 23, 407–410.

Li, Z.X., Powell, C.M., 2001. An outline of the palaeogeographic evolution of the Australasian region since the beginning of the Neoproterozoic. Earth-Science Reviews 53, 237–277.

Li, Z.X., Bogdanova, S.V., Collins, A.S., Davidson, A., De Waele, B., Ernst, R.E., et al., 2008. Assembly, configuration, and break-up history of Rodinia: a synthesis. Precambrian Research 160 (1), 179–210.

Lisker, F., Läufer, A., 2013. The Mesozoic Victoria Basin: vanished link between Antarctica and Australia. Geology 41, 1043–1046.

Liu, X., Jahn, B.-m., Zhao, Y., Li, M., Li, H., Liu, X., 2006. Late Pan-African granitoids from the Grove Mountains, East Antarctica: age, origin and tectonic implications. Precambrian Research 145, 131–154.

Liu, X., Jahn, B.-M., Zhao, Y., Zhao, G., Liu, X., 2007. Geochemistry and geochronology of high-grade rocks from the Grove Mountains, East Antarctica: evidence for an early Neoproterozoic basement metamorphosed during a single Late Neoproterozoic/Cambrian tectonic cycle. Precambrian Research 158, 93–118.

Liu, X.C., Hu, J.M., Zhao, Y., Lou, Y.X., Wei, C.J., Liu, X.H., 2009. Late Neoproterozoic/Cambrian high-pressure mafic granulites from the Grove Mountains, East Antarctica: P–T–t path, collisional orogeny and implications for assembly of East Gondwana. Precambrian Research 174, 181–199.

Liu, X., Zhao, Y., Hu, J., 2013. The c. 1000–900 Ma and c. 550–500 Ma tectonothermal events in the Prince Charles Mountains–Prydz Bay region, East Antarctica, and their relations to supercontinent evolution. In: Harley, S.L., Fitzsimons, I.C.W., Zhao, Y. (Eds.), Antarctica and Supercontinent Evolution. Geological Society, London, pp. 95–112. Special Publication 383.

Liu, X., Jahn, B.-m, Zhao, Y., Liu, J., Ren, L., 2014. Geochemistry and geochronology of Mesoproterozoic basement rocks from the Eastern Amery Ice Shelf and southwestern Prydz Bay, East Antarctica: Implications for a longlived magmatic accretion in a continental arc. American Journal of Science 314, 508–547.

Liu, X., Wang Wei, R.Z., Zhao, Y., Liu, J., Chen, H., Cui, Y., et al., 2016. Early Mesoproterozoic arc magmatism followed by early Neoproterozoic granulite facies metamorphism with a near-isobaric cooling path at Mount Brown, Princess Elizabeth Land, East Antarctica. Precambrian Research 284, 30–48.

Liu, Z., Zhao, Y., Chen, H., Song, B.S., 2017. New zircon U–Pb and Hf–Nd isotopic constraints on the timing of magmatism, sedimentation and metamorphism in the northern Prince Charles Mountains, East Antarctica. Precambrian Research 299, 15–33.

Liu, Y., Li, Z.-X., Pisarevsky, S.A., Kirscher, U., Mitchell, R.N., Stark, J.C., et al., 2018. First Precambrian palaeomagnetic data from the Mawson Craton (East Antarctica) and tectonic implications. Scientific Reports 8, 16403.

Liu, X., Fu, B., Ki, Q., Zhao, Y., Kiu, J., Chen, H., 2020. The impact of Pan-African-aged tectonothermal event on high-grade rocks at Mount Brown, East Antarctica. Antarctic Science 32 (1), 45–57.

Maritati, A., Aitken, A.R.A., Young, D.A., Roberts, J.L., Blankenship, D.D., Sieghert, M.J., 2016. The tectonic development and erosion of the Knox Subglacial Sedimentary Basin, East Antarctica. Geophysical Research Letters 43. Available from: https://doi.org/10.1002/2016gl071063.

Markwitz, V., Kirkland, C.L., Evans, N.J., 2016. Early Cambrian metamorphic zircon in the northern Pinjarra Orogen: implications for the structure of the West Australian Craton margin. Lithosphere 9 (1), 3–13.

Marschall, H.R., Hawkesworth, C.J., Storey, C.D., Dhuime, B., Leat, P.T., Meyer, H.-P., et al., 2010. The Annandagstoppane Granite, East Antarctica: Evidence for the Archaean intracrustal recycling in the Kaapvaal-Grunehogna Craton from zircon O and Hf isotopes. Journal of Petrology 51, 2277–2301.

Marschall, H.R., Hawkesworth, C.J., Leat, P.T., 2013. Mesoproterozoic subduction under the eastern edge of the Kalahari-Grunehogna Craton preceding Rodinia assembly: the Ritscherflya detrital zircon record, Ahlmannryggen (Dronning Maud Land, Antarctica). Precambrian Research. Available from: http://doi.org/10.1016/j.precamres.2013.07.006.

Marshall, J.E.A., 1994. The Falkland Islands: a key element in Gondwana paleogeography. Tectonics 13, 449–514.

McKay, R.M., De Santis, L., Kulhanek, D.K., the Expedition 374 Scientists, 2019. Ross Sea West Antarctic Ice Sheet History. In: Proceedings of the International Ocean Discovery Program, vol. 374. International Ocean Discovery Program, College Station, TX. Available from: https://doi.org/10.14379/iodp.proc.374.2019.

McKay, R., et al., 2021. Cenozoic history of Antarctic glaciation and climate from onshore and offshore studies. In: Florindo, F., et al. (Eds.), Antarctic Climate Evolution, second ed. Elsevier (this volume).

McMenamin, M.A.S., McMenamin, D.L.S., 1990. The Emergence of Animals: The Cambrian Break Through. Columbia University Press, New York, p. 217.

Meert, J.G., Torsvik, T.H., 2003. The making and unmaking of a supercontinent: Rodinia revisited. Tectonophysics 375, 261–288.

Meert, J.G., Van der Voo, R., 1997. The assembly of Gondwana 800–550 Ma. Journal of Geodynamics 23, 223–235.

Meert, J.G., Van der Voo, R., Ayub, S., 1995. Paleomagnetic investigation of the Neoproterozoic Gagwe Lavas and Mbozi complex, Tanzania and the assembly of Gondwana. Precambrian Research 74, 225–244.

Ménot, R.P., 2021. The Geology of East Antarctica (between 5° and 145°E). In: Kleinschmidt, G. (Ed.), The Geology of the Antarctic Continent. Beiträge zur regionalen Geologie der Erde, Band 33, pp. 322–392.

Merdith, A.S., Collins, A.S., Williams, S.E., Pisarevsky, S., Foden, J.D., Archibald, D.B., et al., 2017a. A full-plate global reconstruction of the Neoproterozoic. Gondwana Research 50, 84–134.

Merdith, A.S., Williams, S.E., Müller, R.D., Collins, A.S., 2017b. Kinematic constraints on the Rodinia to Gondwana transition. Precambrian Research 299, 132–150.

Mieth, M., Jokat, W., 2014. New aeromagnetic view of the geological fabric of southern Dronning Maud Land and Coats Land, East Antarctica. Gondwana Research 25, 358–367.

Mieth, M., Jacobs, J., Ruppel, A., Damaske, D., Laufer, A., Jokat, W.W., 2014. New detailed aeromagnetic and geological data of eastern Dronning Maud Land: implications for refining the tectonic and structural framework of Sor Rondane, East Antarctica. Precambrian Research 245, 174–185.

Millar, I.L., Pankhurst, R.J., 1987. Rb-Sr geochronology of the region between the Antarctic Peninsula and the Transantarctic Mountains: Haag Nunataks and Mesozoic granitoids. In: McKenzie, G.D. (Ed.), Gondwana Six: Structure, Tectonics, and Geophysics. American Geophysical Union, Washington, DC, pp. 151–160. Geophysical Monograph 40.

Miller, M.F., Mabin, M.C.G., 1998. Antarctic Neogene landscapes – in the refrigerator or in the deep freeze? GSA Today 8, 1–3.

Mikhalsky, E.V., Kamenev, I.A., 2013. Recurrent transitional group charnockites in the east Amery Ice Shelf coast (East Antarctica): Petrogenesis and implications on tectonic evolution. Lithos 175–176, 230–243.

Mikhalsky, E.V., Leitchenkov, G.L. (Eds.), 2018. Geological map of Mac. Robertson Land, Princess Elizabeth Land, and Prydz Bay (East Antarctica) in scale 1:1 000 000 (Map Sheet and Explanatory Notes). St.-Petersburg. VNII Okeangeologia.

Mikhalsky, E.V., Belyatskii, B.V., Sergeev, S.A., 2009. New data on the age of rocks inthe Mirny Station area, East Antarctica. Doklady Earth Sciences 426A, 527–531.

Mikhalsky, E.V., Sheraton, J.W., Kudriavtsev, I.V., Sergeev, S.A., Kovach, V.P., Kamenev, I.A., et al., 2013. The Mesoproterozoic Rayner Province in the Lambert Glacier area: its age, origin, isotopic structure and implications for Australia-Antarctica correlations. In: Harley, S. L., Fitzsimons, I.C.W., Zhao, Y. (Eds.), Antarctica and Supercontinent Evolution. Geological Society, London, pp. 35–57. Special Publication 283.

Mikhalsky, E.V., Belyatsky, B.V., Presnyakov, S.L., Skublov, S.G., Kovach, V.P., Rodionov, N. V., et al., 2015. The geological composition of the hidden Wilhelm II Land in East Antarctica: SHRIMP zircon, Nd isotopic and geochemical studies with implications for Proterozoic supercontinent reconstructions. Precambrian Research 258, 171–185.

Moores, E.M., 1991. Southwest United States–east Antarctica (SWEAT) connection: a hypothesis. Geology 19, 425–428.

Moriwaki, K., Yoshida, Y., Harwood, D.M., 1992. Cenozoic glacial history of Antarctica – a correlative synthesis. In: Yoshida, Y., Kaminuma, K., Shiraishi, K. (Eds.), Recent Progress in Antarctic Earth Science. Terra Scientific Publishing Company, Tokyo, pp. 773–780.

Morrissey, L.J., Martin, H., Kelsey, D.E., Wade, B.P., 2016. Cambrian high-temperature reworking of the Rayner–Eastern Ghats Terrane: constraints from the Northern Prince Charles Mountains Region, East Antarctica. Journal of Petrology 57 (1), 53–92.

Morrissey, L.-J., Payne, J.P., Hand, M., Clark, C., Taylor, R., Kirkland, C.L., et al., 2017. Linking the Windmill Islands, east Antarctica and the Albany–Fraser Orogen: insights from U–Pb zircon geochronology and Hf isotopes. Precambrian Research 293 (2017), 131–149.

Mortimer, N., Palin, J.M., Dunlap, W.J., Hauff, F., 2011. Extent of the Ross Orogen in Antarctica: new data from DSDP 270 and Iselin Bank. Antarctic Science 23, 297–306. Available from: https://doi.org/10.1017/S0954102010000969.

Mortimer, N., et al., 2017. Zealandia: Earth's hidden continent. GSA Today 27. Available from: https://doi.org/10.1130/GSATG321A.1.

Mortimer, N., van den Bogaard, P., Hoernle, K., Timm, C., Gans, P.B., Werner, R., et al., 2019. Late Cretaceous oceanic plate reorganization and the breakup of Zealandia and Gondwana. Gondwana Research 65, 31–42.

Mulder, J.A., Halpin, J.A., Daczko, N.R., Orth, K., Meffre, S., Thompson, J.M., et al., 2019. A multiproxy provenance approach to uncovering the assembly of East Gondwana in Antarctica. Geology 47, 645–649.

Müller, P., Schmidt-Thome, M., Kreuzer, H., Tessensohn, F., Vetter, U., 1991. Cenozoic peralkaline magmatism at the western margin of the Ross Sea, Antarctica. Memorie della Società Geologica Italiana 46, 315–336.

Naish, T.R., Woolfe, K.J., Barrett, P.J., Wilson, G.S., Atkins, C., Bohaty, S.M., et al., 2001. Orbitally induced oscillations in the east Antarctic ice sheet at the Oligocene/Miocene boundary: direct evidence from Antarctic margin drilling. Nature 413, 719–723.

Naumenko-Dèzes, M.O., Rolland, Y., Lamarque, G., et al., 2020. Petrochronology of the Terre Adélie Craton (East Antarctica) evidences a long-lasting Proterozoic (1.7–1.5 Ga) tectono-metamorphic evolution—insights for the connections with the Gawler Craton and Laurentia. Gondwana Research 81, 21–57.

Nelson, D.A., Cottle, J.M., 2018. The secular development of accretionary orogens: linking the Gondwana magmatic arc record of West Antarctica, Australia and South America. Gondwana Research 63, 15–33.

O'Connell, D.R.H., Stepp, T.M., 1993. Structure and evolution of the crust at the Transantarctic Mountains–Ross Sea crustal transition: results from the Tourmaline Plateau seismic array of the GANOVEX V ship-to-shore seismic refraction experiment. Geologisches. Jahrbuch. 47, 229–276.

Oglesby, R., 1999. Use of climate models to extend paleoclimatic data. In: Barrett, P.J., Orombelli, G. (Eds.), Geological Records of Global and Planetary Change, vol. 3, Terra Antartica Report, pp. 131–149.

Osanai, Y., Nogi, Y., Baba, S., Nakano, N., Adachia, T., Hokada, T., et al., 2013. Geologic evolution of the Sør Rondane Mountains, East Antarctica: collision tectonics proposed based on metamorphic processes and magnetic anomalies. Precambrian Research 234, 8–29.

Paech, H.-J., 2004. GEOMAUD, Part 1, Geologisches Jahrbuch, B96.

Paech, H.-J., Bauer, W., Piazolo, S., Jacobs, J., Markl, G., 2005. Comparison of the geology of central Droning Maud Land, east Antarctica, with other areas of the Gondwana supercontinent. Geol. Jb. B97, 309–340.

Palmeri, R., Ghiribelli, B., Talarico, F., Ricci, C.A., 2003. Ultra-high-pressure metamorphism in felsic rocks: the garnet-phengite gneisses and quartzites from the Lanterman Range, Antarctica. European Journal of Mineralogy 15, 513–525.

Palmeri, R., Ghiribelli, B., Ranalli, G., Talarico, F., Ricci, C.A., 2007. Ultrahigh-pressure metamorphism and exhumation of garnet-bearing ultramafic rocks from the Lanterman Range (northern Victoria Land, Antarctica). Journal of Metamorphic Geology 25 (2), 225–243.

Palmeri, R., Godard, G., Di Vincenzo, G., Sandroni, S., Talarico, F.M., 2018. High-pressure granulite-facies metamorphism in central Dronning Maud Land (East Antarctica): implications for Gondwana assembly. Lithos 300, 361–377.

Pankhurst, R.J., 1982. Rb-Sr geochronology of Graham Land. Journal of Geological Society, London 139, 701–711.

Pankhurst, R.J., Weaver, S.D., Bradshaw, J.D., Storey, B.C., Ireland, T.R., 1998. Geochronology and geochemistry of pre-Jurassic superterranes in Marie Byrd Land, Antarctica. Journal of Geophysical Research 103, 2529–2547.

Paulsen, T.S., Encarnación, J., Grunow, A.M., Layer, P.W., Watkeys, M., 2007. New age constraints for a short pulse in Ross orogen deformation triggered by East-West Gondwana suturing. Gondwana Research 12, 417–427.

Paulsen, T., Encarnación, J., Grunow, A.M., Valencia, V.A., Pecha, M., Layer, P.W., et al., 2013. Age and significance of 'outboard' high-grade metamorphics and intrusives of the Ross orogen, Antarctica. Gondwana Research 24, 349–358.

Paulsen, T.S., Encarnación, J., Grunow, A.M., Valencia, V.A., Layer, P.W., Pecha, M., et al., 2015. Detrital mineral ages from the Ross Supergroup, Antarctica: implications for the Queen Maud terrane and outboard sediment provenance on the Gondwana margin. Gondwana Research 27 (1), 377–391.

Paulsen, T.S., Deering, C., Sliwinski, J., Bachmann, O., Guillong, M., 2016. Detrital zircon ages from the Ross Supergroup, north Victoria Land, Antarctica: implications for the tectonostratigraphic evolution of the Pacific Gondwana margin. Gondwana Research 35, 79–96.

Paulsen, T., Deering, C., Sliwinski, J., Bachmann, O., Gulillong, M., 2017. Detrital zircon ages and trace element compositions of Permian–Triassic foreland basin strata of the Gondwanide orogen, Antarctica. Geosphere 13, 6.

Paulsson, O., Austrheim, H., 2003. A geochronologcial and geochemical study of rocks from Gjelsvikfjella, Dronning Maud Land, Antarctica – implications for Mesoproterozoic correlations and assembly of Gondwana. Precambrian Research 125, 113–138.

Pauly, J., Marschall, H.R., Meyer, H.P., Chatterjee, N., Monteleone, B., 2016. Prolonged Ediacaran-Cambrian metamorphic history and short-lived high-pressure granulite-facies metamorphism in the H.U.Sverdrupfjella, Dronning Maud Land (East Antarctica): evidence for continental collision during Gondwana Assembly. Journal of Petrology 57 (1), 185–228.

Payne, J.L., Hand, M., Barovich, K.M., Reid, A., Evans, D.A.D., 2009. Correlations and Reconstruction Models for the 2500–1500 Ma Evolution of the Mawson Continent. Geological Society, London, pp. 319–355, Special Publication 323.

Phillips, G., Wilson, C.J.L., Campbell, I.H., Allen, C.M., 2006. U-Th-Pb detrital zircon geochronology from the southern Prince Charles Mountains, East Antarctica-defining the Archaean to Neoproterozoic Ruker Province. Precambrian Research 148, 292–306.

Phillips, G., Kelsey, D.E., Corvino, A.F., Dutch, R.A., 2009. Continental reworking during overprinting orogenic events, southern Prince Charles Mountains, East Antarctica. Journal of Petrology 50, 2017–2041.

Piper, J.D.A., 1982. The Precambrian palaeomagnetic record: The case for the Proterozoic supercontinent. Earth and Planetary Science Letters 59, 61–89.

Piper, J.D.A., 2000. The Neoproterozoic supercontinent: Rodinia or palaeopangea? Earth and Planetary Science Letters 176, 131–146.

Pisarevsky, S.A., Wingate, M.T., Powell, C.M., Johnson, S., Evans, D.A., 2003. Models of Rodinia Assembly and Fragmentation. Geological Society, London, pp. 35–55, Special Publication 206.

Powell, C.M., Li, Z.X., McElhinny, M.W., Meert, J.G., Park, J.K., 1993. Paleomagnetic constraints on timing of the Neoproterozoic breakup of Rodinia and the Cambrian formation of Gondwana. Geology 21, 889–892.

Randall, D.E., Mac Niocaill, C., 2004. Cambrian palaeomagnetic data confirm a natal embayment location for the Ellsworth–Withmore Mountains, Antarctica, in Gondwana reconstructions. Geophysical Journal International 157, 105–116.

Reitmayr, G., 1997. Gravity studies of Victoria Land and adjacent oceans, Antarctica. In: Ricci, C.A. (Ed.), The Antarctic Region: Geological Evolution and Processes. Terra Antartica, Siena, pp. 597–602.

Riedel, S., Jacobs, J., Jokat, W., 2013. Interpretation of new regional aeromagnetic data over Dronning Maud Land (East Antarctica). Tectonophysics 585, 161–171.

Riley, T.R., Flowerdew, M.J., Whitehouse, M.J., 2012. U-Pb ion-microprobe zircon geochronology from the basement inliers of eastern Graham Land, Antarctic Peninsula. Journal of Geological Society, London. 169, 381–393.

Riley, T.R., Curtis, M.L., Flowerdew, M.J., Whitehouse, M.J., 2016. Evolution of the Antarctic Peninsula lithosphere: evidence from Mesozoic mafic rocks. Lithos 244, 59–73.

Riley, T.R., Flowerdew, M.J., Pankhurst, R.J., Curtis, M.L., Millar, I.L., Fanning, C.M., Whitehouse, M.J., 2017. Early Jurassic subduction-related magmatism on the Antarctic Peninsula and potential correlation with the Subcordilleran plutonic belt of Patagonia. Journal Geological Society. London 174, 365–376.

Riley, T.R., Flowerdew, M.J., Pankhurst, R.J., Millar, I.L., Whitehouse, M.J., 2020. U-Pb zircon geochronology from Haag Nunataks, Coats Land and Shackleton Range (Antarctica): constraining the extent of juvenile Late Mesoproterozoic arc terranes. Precambrian Research 340, 105646.

Rocchi, S., Armienti, P., D'Orazio, M., Tonarini, S., Wijbrans, J., Di Vincenzo, G., 1999. Eocene–Oligocene rift related plutonic to subvolcanic magmatism in the western Ross embayment, Antarctica. In: Skinner, D.N.B. (Ed.), 8th International Symposium on Antarctic Earth Sciences, 5–9 July, Wellington, Programme and Abstracts. Royal Society of New Zealand, Wellington, p. 269.

Rocchi, S., Fioretti, A.M., Cavazzini, G., 2002. Petrography, geochemistry, and geochronology of the Cenozoic Cape crossfire, Cape King, and no ridge igneous complexes (northern Victoria Land, Antarctica). In: Gamble, J.A., Skinner, D.N.B., Henrys, S. (Eds.), Antarctica at the Close of a Millennium, vol. 35. Royal Society of New Zealand Bulletin, pp. 215–225.

Rocchi, S., Di Vincenzo, G., Ghezzo, C., 2004. The Terra Nova Intrusive Complex (Victoria Land, Antarctica). Terra Antartica Reports 10, 1–51.

Rocchi, S., Bracciali, L., Di Vincenzo, G., Gemelli, M., Ghezzo, C., 2011. Arc accretion to the early Paleozoic Antarctic margin of Gondwana in Victoria Land. Gondwana Research 19, 594–607.

Rocchi, S., Di Vincenzo, G., Dini, A., Petrelli, M., Vezzoni, S., 2015. Time–space focused intrusion of genetically unrelated arc magmas in the early Paleozoic Ross–Delamerian Orogen (Morozumi Range, Antarctica). Lithos 232, 84–99.

Roland, N., 2004. Pan-African granite–charnockite magmatism in central Dronning Maud Land, East Antarctica: petrography, geochemistry and plate tectonic implications. Geologisches Jahrbuch Reihe B 96, 187–231.

Roland, N.W., 2021. The mineral resources of the Antarctic: facts and problems. In: Kleinschmidt, G. (Ed.), The Geology of the Antarctic Continent. Beitrage zur regionalen Geologie der Erde, Band 33, pp. 393–425.

Roland, N.W., Olesch, M., 2004. Pan-African granites in central Dronning Maud Land, East Antarctica: product of collisionor intra-plate event? Zeitschrift der Deutschen Geologischen Gesellschaft Band 154 (4), 469–479.

Rossetti, F., Lisker, F., Storti, F., Läufer, A.L., 2003. Tectonic and denudation history of the Rennick Graben (north Victoria Land): implications for the evolution of rifting between east and west Antarctica. Tectonics 22, 11–18.

Rowell, A.J., Rees, M.N., Duebendorfer, E.M., Wallin, E.T., Van Schmus, W.R., Smith, E.I., 1993. An active Neoproterozoic margin: evidence from the Skelton Glacier area, Transantarctic Mountains. Journal of Geological Society, London 150, 677–682.

Royer, J.Y., Rollet, N., 1997. Plate-tectonic setting of the Tasmanian region. In: Exon, N.F., Crawford, A.J. (Eds.), West Tasmanian Margin and Offshore Plateaus: Geology, Tectonic and Climatic History, and Resource Potential. Australian Journal of Earth Sciences 44, 543–560.

Ruppel, A.S., Läufer, A., Jacobs, J., Elburg, M., Krohne, N., Damaske, D., et al., 2015. The Main Shear Zone in Sør Rondane, East Antarctica: implications for the late-Pan-African tectonic evolution of Dronning Maud Land. Tectonics 34, 1290–1305.

Ruppel, A., Jacobs, J., Eagles, G., Läufer, A., Jokat, W., 2018. New geophysical data from a key region in East Antarctica: estimates for the spatial extent of the Tonian Oceanic Arc Super Terrane (TOAST). Gondwana Research 59, 97–107.

Ruppel, A.S., Jacobs, J., Läufer, A., Ratschbacher, L., Pfänder, J.A., Sonntag, B.-L., et al., 2020. Protracted late Neoproterozic – early Palaeozoic deformation and cooling history of Sør Rondane, East Antarctica, from 40Ar/39Ar and U–Pb geochronology. Geological Magazine 1–21. First View.

Salvini, S., Storti, F., 2003. Do transform faults propagate and terminate in east Antarctica continental lithosphere? Terra Nostra 4, 283–284.

Salvini, F., Brancolini, G., Busetti, M., Storti, F., Mazzarini, F., Coren, F., 1997. Cenozoic geodynamics of the Ross Sea region, Antarctica: crustal extension, intraplate strike–slip faulting, and tectonic inheritance. Journal of Geophysical Research 102, 24669–24696.

Sauli, C., Sorlien, C., Busetti, M., et al., 2021. Neogene development of the Terror Rift, western Ross Sea, Antarctica. Geochemistry, Geophysics, Geosystems. Available from: https://doi.org/10.1029/2020GC009076.

Schmädicke, E., Will, T.M., 2006. First evidence of eclogite facies metamorphism in the Shackleton Range, Antarctica: trace of a suture between east and west Gondwana? Geology 34 (3), 133–136. Available from: https://doi.org/10.1130/G22170.1.

Schmidt, D.L., Rowley, P.D., 1986. Continental rifting and transform faulting along the Jurassic Transantarctic Rift, Antarctica. Tectonics 5, 279–291.

Schulz, B., Schüssler, U., 2013. Electron-microprobe Th-U-Pb monazite dating in Early-Palaeozoic high-grade gneisses as a completion of U-Pb isotopic ages (Wilson Terrane, Antarctica). Lithos 175, 178–192.

Shackleton, R.M., 1996. The final collision zone between east and west Gondwana: where is it? Journal of African Earth Sciences 23 (3), 271–287.

Sheraton, J.W., Black, L.P., Tindle, A.G., 1992. Petrogenesis of plutonic rocks in a Proterozoic granulite-facies terrane—the Bunger Hills, East Antarctica. Chemical Geology 97, 163–198.

Sheraton, J.W., Tingey, R.J., Black, L.P., Oliver, R.L., 1993. Geology of the Bunger Hills area, Antarctica: implications for Gondwana correlations. Antarctic Science 5, 85–102.

Shiraishi, K., Dunkley, D.J., Hokada, T., Fanning, C.M., Kagami, H., Hamamoto, T., 2008. Geochronological constraints on the Late Proterozoic to Cambrian crustal evolution of eastern Dronning Maud Land, East Antarctica: a synthesis of SHRIMP U-Pb age and Nd model age data. Geological Society, London, Special Publications 308, 21–67.

Siddoway, C.S., 2021. The Geology of west Antarctica. In: Kleinschmidt, G. (Ed.), The Geology of the Antarctic Continent. Beitrage zur regionalen Geologie der Erde, Band 33, pp. 87–131.

Siddoway, C.S., Baldwin, S., Fitzgerald, P.G., Fanning, C.M., Luyendyk, B.P., 2004. Ross Sea mylonites and the timing of intracontinental extension within the West Antarctic rift system. Geology 32 (1), 57–60.

Siddoway, C.S., Sass III, L.C., Esser, R.P., 2005. Kinematic history of the Marie Byrd Land terrane, West Antarctica: direct evidence from cretaceous mafic dykes. In: Vaughan, A., Leat, P., Pankhurst, R.J. (Eds.), Terrane Processes at the Margin of Gondwana. Geological Society of London, pp. 417–438. Special Publication 246.

Siddoway, C.S., 2008. Tectonics of the West Antarctic Rift System: newlight on the history and dynamics of distributed intracontinental extension. In: Cooper, A.K., Raymond, C.R., et al., Antarctica: A Keystone in a Changing World. National Academies Press, Washington, DC, pp. 91–114.

Siddoway, C.S., Fanning, C.M., 2009. Paleozoic tectonism on the East Gondwana margin: Evidence from SHRIMP U–Pb zircon geochronology of a migmatite– granite complex in West Antarctica. Tectonophysics 477, 262–277.

Siegert, M.J., Carter, S., Tabacco, I., Popov, S., Blankenship, D.D., 2005. A revised inventory of Antarctic subglacial lakes. Antarctic Science 17 (03), 453–460.

Siegert, M., Popov, S., Studinger, M., 2011. Subglacial Lake Vostok: a review of geophysical data regarding its physiographical setting. In: Siegert, M., Kennicutt, C., Bindschadler, B. (Eds.), Subglacial Antarctic Aquatic Environments. AGU Geophysical Monograph 192. Washington DC, pp. 45–60.

Sloan, L.C., Crowley, T.J., Pollard, D., 1996. Modeling of middle Pliocene climate with the NCAR GENESIS general circulation model. Marine Micropaleontology 27 (1–4), 51–61.

Smellie, J.L., 2021. Antarctic Peninsula-geology and dynamic development. In: Kleinschmidt, G. (Ed.), The Geology of the Antarctic Continent. Beiträge zur regionalen Geologie der Erde, Band 33, pp. 18–86.

Smellie, J.L., Lopez-Martınez, J., 2002. Geology and geomorphology of Deception Island. In: BAS GEOMAP Series Sheets 6-A and 6-B, 1:25.000. British Antarctic Survey, Cambridge.

Spaggiari, C.V., Kirkland, C.L., Smithies, R.H., Wingate, M.T., Belousova, E.A., 2015. Transformation of an Archean craton margin during Proterozoic basin formation and magmatism: The Albany-Frazer Orogen, Western Australia. Precambrian Research 266, 440–466.

Stagg, H.M.J., Willcox, J.B., 1992. A case for Australia–Antarctica separation in the Neocomian (ca. 125 Ma). Tectonophysics 210, 21–32.

Stern, R.J., 1994. Arc assembly and continental collision in the Neoproterozoic east African Orogeny – implications for the consolidation of Gondwana. Annual. Review of Earth and Planetary Sciences 22, 319–351.

Stern, T., ten Brink, U.S., 1989. Flexural uplift of the Transantarctic Mountains. Journal of Geophysical Research 94, 10315–10330.

Stock, J., Molnar, P., 1987. Revised history of early tertiary plate motion in the southwest Pacific. Nature 325, 495–499.

Storey, B.C., 1996. Microplates and mantle plumes in Antarctica. Terra Antartica 3, 91–102.

Storey, B.C., Kyle, P.R., 1999. The Weddell Sea region: A plume impact site and source for Gondwana magmas during the initial stages of Gondwana breakup. In: Skinner, D.N.B. (Ed.), 8th International Symposium on Antarctic Earth Sciences, 5–9 July, Programme and Abstracts. Victoria University of Wellington, Wellington, New Zealand, p. 292.

Storey, B.C., Thompson, M.R.A., Meneilly, A.W., 1987. The Gondwanian orogeny within the Antarctic Peninsula: a discussion. In: McKenzie, G.D. (Ed.), Gondwana Six: Structure, Tectonics, and Geophysics. Geophysical Monograph, vol. 40. American Geophysical Union, Washington, DC, 191–198.

Storey, B.C., Hole, M.J., Pankhurst, R.J., Millar, I.L., Vennum, W.R., 1988. Middle Jurassic within-plate granites in west Antarctica and their bearing on the breakup of Gondwanaland. Journal of Geological Society, London 145, 999–1007.

Storey, B.C., Pankhurst, R.J., Johnson, A.C., 1994. The Grenville Province within Antarctica: a test of the SWEAT hypothesis. Journal of the Geological Society 151, 1–4.

Storey, B.C., Vaughan, A.P.M., Millar, I.L., 1996. Geodynamic evolution of the Antarctic Peninsula during Mesozoic times and its bearing on Weddell Sea history. In: Storey, B.C., King, E.C., Livermore, R.A. (Eds.), Weddell Sea Tectonics and Gondwana Breakup. Geological Society, London, pp. 87–103. Special Publication 108.

Storey, B.C., Dalziel, I.W.D., Garrett, S.W., Grunow, A.M., Pankhurst, R.J., Vennum, W.R., 1998. West Antarctica in Gondwanaland: crustal blocks, reconstruction and breakup processes. Tectonophysics 155, 381–390.

Storey, B.C., Leat, P.T., Weaver, S.D., Pankhurst, R.J., 1999. Mantle plumes and Antarctica–New Zealand rifting: evidence from mid-Cretaceous mafic dykes. Journal of Geological Society, London 156, 659–671.

Studinger, M., Bell, R.E., Buck, W.R., Karner, G.D., Blankenship, D.D., 2004. Subglacial geology inland of the Transantarctic Mountains in light of new aerogeophysical data. Earth and Planetary Science Letters 220, 391–408.

Stump, E., 1995. The Ross Orogen of the Transantarctic Mountains. Cambridge University Press, Cambridge, p. 284.

Stump, E., Fitzgerald, P.G., 1992. Episodic uplift of the Transantarctic Mountains. Geology 20, 161–164.

Stump, E., Sheridan, M.F., Borg, S.G., Sutter, J.E., 1980. Early Miocene subglacial basalts, the east Antarctic ice sheet and uplift of the Transantarctic Mountains. Science 207, 757–759.

Stump, E., Gootee, B., Talarico, F., 2006. Tectonic model for development of the Byrd Glacier discontinuity and surrounding regions of the Transantarctic Mountains during the Neoproterozoic-Early Paleozoic. In: Fütterer, D.K., Kleinschmidt, G., Miller, H., Tessensohn, F. (Eds.), Antarctica: Contributions to Global Earth Sciences. Springer-Verlag, Berlin, pp. 181–190.

Takahashi, K., Tsunogae, T., Santosh, M., Takamura, Y., 2018. Paleoproterozoic (ca. 1.8 Ga) arc magmatism in the Lützow-Halm Complex, East Amtarctica: implications for crustal growth and terrane assembly in erstwhile Gondwana fragments. Journal of Asian Earth Sciences 157, 245–268.

Talarico, F., Kleinschmidt, G., 2003. Structural and Metamorphic Evolution of the Mertz Shear Zone (East Antarctic Craton, George V Land): implications for Australia/Antarctica Correlations and East Antarctic Craton/Ross Orogen Relationships. Terra Antartica 10 (3), 229–248.

Talarico, F., Kleinschmidt, G., Henjes-Kunst, F., 1999. An ophiolitic complex in the northern Shackleton Range, Antarctica. Terra Antartica 6, 293–315.

Talarico, F., Palmeri, R., Ricci, C.A., 2004. Regional metamorphism and P-T evolution of the Ross orogen in northern Victoria Land (Antarctica): a review. Periodico di Mineralogia 73, 185–196.

ten Brink, U.T., Stern, T., 1992. Rift flank uplifts and hinterland basins: comparison of the Transantarctic Mountains with the Great Escarpment of southern Africa. Journal of Geophysical Research 97, 569–585.

ten Brink, U.S., Bannister, S., Beaudoin, B.C., Stern, T.A., 1993. Geophysical investigations of the tectonic boundary between east and west Antarctica. Science 261, 45–50.

ten Brink, U.S., Hackney, R.I., Bannister, S., Stern, T.A., Makovsky, Y., 1997. Uplift of the Transantarctic Mountains and the bedrock beneath the east Antarctic ice sheet. Journal of Geophysical Research 102, 27603–27621.

Tessensohn, F., 1994a. Structural evolution of the northern end of the Transantarctic Mountains. In: van der Wateren, F.M., Verbers, A.L.L.M., Tessensohn, F. (Eds.), Landscape Evolution in the Ross Sea Area, Antarctica. Rijks Geologische Dienst, Haarlem, pp. 57–61.

Tessensohn, F., 1994b. The Ross Sea region, Antarctica: Structural interpretation in relation to the evolution of the Southern Ocean, Terra Antartica, 1. pp. 553–558.

Tessensohn, F., Henjes-Kunst, F., 2005. Northern Victoria Land terranes, Antarctica: Far-travelled or local product? In: Vaughan, A.P.M., Leat, P.T., Pankhurst, R.J. (Eds.), Terrane Processes at the Margins of Gondwana, vol. 246, Geologial Society of London, Special publication, pp. 275–291.

Tessensohn, F., Wörner, G., 1991. The Ross Sea rift system, Antarctica: structure, evolution, and analogues. In: Thompson, M.R.A., Crame, J.A., Thomson, J.W. (Eds.), Geological Evolution of Antarctica. Cambridge University Press, Cambridge, pp. 273–277.

Tessensohn, F., Kleinschmidt, G., Talarico, F., Buggisch, W., Brommer, A., Henjes-Kunst, H., et al., 1999. Ross-age amalgamation of east and west Gondwana: evidence from the Shackleton Range, east Antarctica. Terra Antartica 6, 317–325.

Thomson, M.R.A., 2021. The palaeontological Record of Antarctica. In: Kleinschmidt, G. (Ed.), The Geology of the Antarctic Continent. Beiträge zur regionalen Geologie der Erde, Band 33, pp. 426–484.

Tingley, R.J., 1991. The Geology of Antarctica. Oxford University Press, Oxford.

Tinto, K.J., Padman, L., Siddoway, C.S., et al., 2019. Ross Ice Shelf response to climate driven by the tectonic imprint on seafloor bathymetry. Nature Geoscience 12, 441–449.

Torsvik, T.H., 2003. The Rodinia jigsaw puzzle. Science 300, 1379–1381.

Trehu, A.M., 1989. Crustal structure in the Ross Sea, Antarctica: preliminary results from GANOVEX V. Eos Transactions, American Geophysical Union 70, 1344.

Trouw, R.A.J., De Wit, M.J., 1999. Relation between Gondwanide orogen and contemporaneous intracratonic deformation. Journal of African Earth Sciences 28, 203–213.

Trouw, R.A.J., Ribeiro, A., Paciullo, F.V.P., 1991. Structural and metamorphic evolution of the Elephant Island Group and Smith Island, south Shetland Islands. In: Thomson, M.R.A., Crame, J.A., Thomson, J.W. (Eds.), Geological Evolution of Antarctica. Cambridge University Press, Cambridge.

Tsunogae, T., Yang, Q.Y., Santosh, M., 2016. Neoarchean-Early Paleoproterozoic and Early Neoproterozoic arc magmatism in the Lützow-Holm Complex, East Antarctica: insights from petrology, geochemistry, zircon U-Pb geochronology and Lu-Hf isotopes. Lithos 263, 239–256.

Tucker, N.M., Payne, J.L., Clark, C., Hand, M., Taylor, R.J.M., Andrew, R.C., et al., 2017. Proterozoic reworking of Archean (Yilgarn) basement in the Bunger Hills, East Antarctica. Precambrian Research 298, 16–38.

Tucker, N.M., Hand, M., Clark, C., 2020. The Bunger Hills: 60 years of geological and geophysical research. Antarctic Science 32 (2), 85–106.

Van der Beek, P., Cloetingh, S., Andriessen, P., 1994. Mechanisms of extensionalbasin formation and vertical motions at rift flanks: constraints from tectonic modeling and fission track thermochronology. Earth and Planetary Science Letters 121, 417–433.

Van der Wateren, F.M., Dunaia, T.J., Van Balena, R.T., Klasc, W., Verbersd, A.L.L.M., Passchiere, S., et al., 1999. Contrasting neogene denudation histories of different structural regions in the Transantarctic Mountains rift flank constrained by cosmogenic isotope measurements. Global and Planetary Change 23 (1–4), 145–172.

Vaughan, A.P.M., Storey, B.C., 1997. Mesozoic geodynamic evolution of the Antarctic Peninsula. In: Ricci, C.A. (Ed.), The Antarctic Region: Geological Evolution and Processes. Terra Antartica Publication, Siena, pp. 373–382.

Vaughan, A.P.M., Storey, B.C., 2000. The eastern Palmer Land shear zone: a new terrane accretion model for the Mesozoic development of the Antarctic Peninsula. Journal of Geological Society, London 157, 1243–1256. Available from: https://doi.org/10.1144/jgs.157.6.1243.

Vaughan, A.P.M., Storey, C., Kelley, S.P., Barry, T.L., Curtis, M.L., 2012. Synkinematic emplacement of Lassiter Coast Intrusive Suite plutons during the Palmer Land Event: evidence for mid-Cretaceous sinistral transpression at the Beaumont Glacier in eastern Palmer Land. Journal of the Geological Society of London 169, 759–771.

Veevers, J.J., Saeed, A., 2008. Gamburtsev Subglacial Mountains provenance of Permian– Triassic sandstones in the Prince CharlesMountains and offshore Prydz Bay: integrated U–Pb and TDM ages and host-rock affinity from detrital zircons. Gondwana Research 14, 316–342.

Veevers, J.J., Saeed, A., 2011. Age and composition of Antarctic bedrock reflected by detrital zircons, erratics, and recycled microfossils in the Wilkes Land–Ross Sea–Marie Byrd Land sector (100°–240°E). Gondwana Research 20, 710–738.

Veevers, J.J., Saeed, A., 2013. Age and composition of Antarctic sub-glacial bedrock reflected by detrital zircons, erratics, and recycled microfossils in the Ellsworth Land–Antarctic Peninsula–Weddell Sea–Dronning Maud Land sector (240°E–0°–015°E). Gondwana Research 23, 296–332.

Veevers, J.J., Powell, C.M., Rotts, S.R., 1990. Review of seafloor spreading around Australia. I. Synthesis of the patterns of spreading. Australian Journal of Earth Sciences 38, 373–389.

Veevers, J.J., Belousova, E.A., Saeed, A., 2016. Zircons traced from the 700–500 Ma Transgondwanan Supermountains and the Gamburtsev Subglacial Mountains to the Ordovician Lachlan Orogen, Cretaceous Ceduna Delta, and modern Channel Country, central-southern Australia. Sedimentary Geology 334, 115–141.

Vetter, U., Tessensohn, F., 1987. S- and I-type granitoids of north Victoria Land, Antarctica, and their inferred geotectonic setting. Geologische Rundschau 76, 233–243.

Vetter, U., Roland, N.W., Kreuzer, H., Höhndorf, A., Lenz, C., Besang, H., 1983. Geochemistry, petrography and geochronology of the Cambro-Ordovician and Devonian-Carboniferous granitoids of Northern Victoria Land, Antarctica. In: Oliver, R.L., James, P.R., Jago, J.B. (Eds.), Antarctic Earth Science. Australian Academy of Science, Canberra, pp. 140–143.

Viereck-Götte, L., Schöner, R., Bomfleur, B., Schneider, J., 2007. Multiple shallow level sill intrusions coupled with hydromagmatic explosive eruptions mark the initial phase of ferrar magmatism in north Victoria Land, Antarctica. In: Cooper, A.K., Raymond, A.C., ISAES Editorial Team (Eds.), 10th International Symposium on Antarctic Earth Sciences Program Book, Santa Barbara, p. 60.

Wever, H.E., Millar, I.L., Pankhurst, R.J., 1994. Geochronology and radiogenic isotope geology of Mesozoic rocks from eastern Palmer Land, Antarctic Peninsula: Crustal anatexis in arc-related granitoid genesis. Journal of South American Earth Sciences 7, 69–83.

Wang, Y., Liu, D., Chung, S.-L., Tong, L., Ren, L., 2008. SHRIMP zircon age constraints from the Larsemann Hills region, Prydz Bay, for a late Mesoproterozoic to early Neoproterozoic tectono-thermal event in East Antarctica. American Journal of Science 308, 573–617.

Wang, C.-C., Jacobs, J., Elburg, M.A., Läufer, A., Thomas, R.J., Elvevold, S., 2020. Grenville-age continental arc magmatism and crustal evolution in central Dronning Maud Land (East Antarctica): Zircon geochronological and Hf-O isotopic evidence. Gondwana Research 82.

Wannamaker, P., Hill, G., Stodt, J., Maris, V., Ogawa, Y., Selway, K., et al., 2017. Uplift of the central transantarctic mountains. Nature Communications 8, 1588.

White, P., 1989. Downhole logging. In: Barrett, P.J. (Ed.), Antarctic Cenozoic History from the CIROS-1 Drillhole, vol. 245, McMurdo Sound, DSIR Bulletin, pp. 7–14.

White, R.S., MacKenzie, D., 1989. Magmatism at rift zones: the generation of volcanic continental margins and flood basalts. Journal of Geophysical Research 94, 7685–7729.

Will, T.M., Zeh, A., Gerdes, A., Frimmel, H.E., Millar, I.L., Schmadicke, E., 2009. Palaeoproterozoic to Palaeozoic magmatic and metamorphic events in the Shackleton Range, East Antarctica: constraints from zircon and monazite dating, and implications for the amalgamation of Gondwana. Precambrian Research 172 (2009), 25–45.

Wilson, G.S., 1995. The Neogene Antarctic ice sheet. A dynamic or stable feature? Quaternary Science Review 14, 101–123.

Wilson, T.J., 1999. Cenozoic structural segmentation of the Transantarctic Mountains rift flank in southern Victoria Land. Global and Planetary Change 23, 105–127.

Wilson, C.J.L., Quinn, C., Tong, L., Phillips, D., 2007. Early Palaeozoic intracratonic shears and post-tectonic cooling in the Rauer Group, Prydz Bay, East Antarctica constrained by $^{40}Ar/^{39}Ar$ thermochronology. Antarctic Science 19, 339–353.

Wingate, M.T., Pisarevsky, S.A., Evans, D.A., 2002. Rodinia connections between Australia and Laurentia: no SWEAT, no AUSWUS? Terra Nova 14 (2), 121–128.

Wisoczanski, R.J., Allibone, A.H., 2004. Age, correlation, and provenance of the Neoproterozoic Skelton Group, Antarctica: Grenville age detritus on the margin of East Antarctica. Journal of Geology 112, 401–416.

Woolfe, K.J., Barrett, P.J., 1995. Constraining the Devonian to Triassic tectonic evolution of the Ross Sea sector. Terra Antartica 2, 7–21.

Wörner, G., Viereck, L., Hertogen, J., Niephaus, H., 1989. The Mt. Melbourne volcanic field (Victoria Land, Antarctica) II. Geochemistry and magma genesis. Geologisches. Jahrbuch. E38, 395–433.

Wrenn, J.H., Webb, P.-N., 1982. Physiographic analysis and interpretation of the Ferrar Glacier–Victoria Valley area, Antarctica. In: Craddock, C. (Ed.), Antarctic Geoscience. University of Wisconsin Press, Madison, WI, pp. 1091–1099.

Wright, A., Siegert, M., 2012. A fourth inventory of Antarctic subglacial lakes. Antarctic Science 24, 659–664.

Yakymchuk, C., Siddoway, C.S., Fanning, C.M., McFadden, R., Korhonen, F.J., Brown, M., 2013. Anatectic reworking and differentiation of continental crust along the active margin of Gondwana; a zircon Hf-O perspective from West Antarctica: Special Publication 383. Geological Society of London. Available from: https://doi.org/10.1144/sp383.7.

Yakymchuk, C., Brown, C.R., Brown, M., Siddoway, C.S., Fanning, C.M., Korhonen, F.J., 2015. Paleozoic evolution of West Marie Byrd Land, West Antarctica. Geological Society of America Bulletin 127, 1464–1484.

Yoshida, M., 1992. Precambrian tectonothermal events in east Gondwanian crustal fragments and their correlation (IGCP-288). In: Japan Contribution to the IGCP. IGCP National Committee of Japan, Tokyo, pp. 51–62.

Yoshida, M., Kizaki, K., 1983. Tectonic situation of Lützow-Holm Bay in east Antarctica and its significance in Gondwanaland. In: Oliver, R.L., James, P.R., Jago, L.B. (Eds.), Antarctic Earth Science. Australian Academy of Science, Canberra, pp. 36–39.

Yoshida, M., Windley, B.F., Dasgupta, S. (Eds.), 2003. Proterozoic East Gondwana: Supercontinent Assembly and Breakup. Geological Society of London, Special Publication 206, 472 pp.

Young, G.M., 1995. Are Neoproterozoic glacial deposits preserved on the margins of Laurentia related to the fragmentation of two supercontinents? Geology 23, 153–156.

Zhang, S., Zhao, Y., Liu, X., Liu, Y., Hou, K., Li, C., et al., 2012. U–Pb geochronology and geochemistry of the bedrocks and moraine sediments from the Windmill Islands: implications for Proterozoic evolution of East Antarctica. Precambrian Research 206–207, 52–71.

Zulbati, F., Harley, S.L., 2007. Late archean granulite facies metamorphism in the Vestfold Hills, East Antarctica. Lithos 93 (1–2), 39–67.

Chapter 7

The Eocene-Oligocene boundary climate transition: an Antarctic perspective

Simone Galeotti[1,12], Peter Bijl[2], Henk Brinkuis[2,11], Robert M. DeConto[3,12], Carlota Escutia[4], Fabio Florindo[5,12], Edward G.W. Gasson[6], Jane Francis[7], David Hutchinson[8], Alan Kennedy-Asser[9], Luca Lanci[1,12], Isabel Sauermilch[2,10], Appy Sluijs[2,12] and Paolo Stocchi[11,12]
[1]Department of Pure and Applied Sciences, University of Urbino Carlo Bo, Urbino, Italy, [2]Laboratory of Palaeobotany and Palynology, Department of Earth Sciences, Marine Palynology and Paleoceanography, Utrecht University, Utrecht, the Netherlands, [3]Department of Geosciences, University of Massachusetts, Amherst, Amherst, MA, United States, [4]Andalusian Institute of Earth Sciences, CSIC and Universidad de Granada, Armilla, Spain, [5]National Institute of Geophysics and Volcanology, Rome, Italy, [6]Centre for Geography and Environmental Science, University of Exeter, Cornwall Campus, Penryn, United Kingdom, [7]British Antarctic Survey, Cambridge, United Kingdom, [8]Department of Geological Sciences and Bolin Centre for Climate Research, Stockholm University, Stockholm, Sweden, [9]BRIDGE, School of Geographical Sciences, University of Bristol, Bristol, United Kingdom, [10]Institute for Marine and Antarctic Studies, University of Tasmania, Hobart, TAS, Australia, [11]Coastal Systems Department, Royal Netherlands Institute for Sea Research, Utrecht University, Den Burg, the Netherlands, [12]Institute for Climate Change Solutions, Frontone, Italy

7.1 Introduction

The Eocene-Oligocene transition (EOT), defined as a 500-kyr long phase of accelerated climatic and biotic change that began before and ended after the Eocene-Oligocene boundary (EOB) about 34 million years ago (Ma), marks a fundamental change in the climate of our planet from a warm, ice-free greenhouse state to an icehouse state (Westerhold et al., 2020). This climatic transition is associated with an $\sim 1.5‰$ increase in benthic foraminiferal oxygen isotopic ($\delta^{18}O$) values corresponding to the 'earliest Oligocene oxygen isotope shift' (EOIS) (Hutchinson et al., 2021). Mass-balance equations, deconvolving temperature and ice volume effects, demonstrate that the EOIS cannot be caused by either deep-sea temperature or ice volume change alone; both must have contributed (Bohaty et al., 2012; Coxall et al., 2005; Lear et al., 2008;

Antarctic Climate Evolution. DOI: https://doi.org/10.1016/B978-0-12-819109-5.00009-8

Shackleton and Kennett, 1975; Zachos et al., 1996). From various high-resolution isotope records, it was shown that the $\delta^{18}O$ shift associated with the Eocene-Oligocene transition (EOT) is a stepwise event, involving several inflections in the $\delta^{18}O$ records which each represents varying contributions of deep-sea temperature and ice volume change (e.g., Bohaty et al., 2012; Lear et al., 2008). These different phases in the EOT have been given different names by a number of authors. For the sake of consistency, we use the terminology of Hutchinson et al. (2021) as presented in Fig. 7.1.

High-resolution proxy records and coupled climate–ice sheet modelling suggest that the ultimate driver for initiating large-scale Antarctic glaciation came from radiative forcing: the Antarctic Ice Sheet (AIS) formed as the

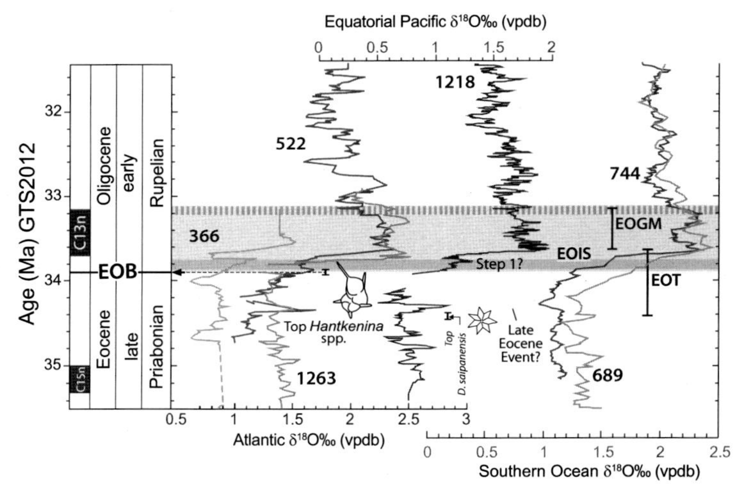

FIGURE 7.1 Oxygen stable isotope and chronostratigraphic characteristics of the Eocene-Oligocene transition (EOT) from deep marine records and EOT terminology on the Geological Time Scale 2012 (Vandenberghe et al., 2012) Benthic foraminiferal $\delta^{18}O$ from six deep sea drill holes are shown: Atlantic Sites DSDP 366, 145 522 and ODP 1263 (Langton et al., 2016; Zachos et al., 1996); Southern Ocean Sites ODP 744 and 689 (Diester-Haass and Zahn, 1996; Zachos et al., 1996) and the Equatorial Pacific Site ODP 1218 (Coxall and Wilson, 2011). Due to different sample resolutions, running means are applied using a 3-point filter for Sites DSDP 522 and ODP 689 and 1263, 5-point filter for Sites DSDP 366 and ODP 744 and a 7-point filter for Site ODP 1218. Time scale conversions were made by aligning common magneto-stratigraphic tie points. The EOT is defined as a 500-kyr long phase of accelerated climatic and biotic change that began before and ended after the Eocene-Oligocene boundary (EOB) (after Coxall and Pearson, 2007). Benthic data are 150 all *Cibicidoides* spp. or 'Cibs. Equivalent' and have not been adjusted to sea water equilibrium values. 'Step-1' comprises a modest $\delta^{18}O$ increase linked to ocean cooling (Bohaty et al., 2012; Lear et al., 2008). 'Step-2' corresponds to the Early Oligocene oxygen Isotope Step (EOIS), defined herein. The 'Top (T) *Hantkenina* spp.' marker corresponds to the position of this extinction event at DSDP Site 522 (including sampling bracket) with respect to the corresponding Site DSDP 522 $\delta^{18}O$ curve, which coincides with the published calibrated age of this event (33.9 Ma). The 'Late Eocene Event' $\delta^{18}O$ maximum (after Katz et al., 2008) may represent a failed glaciation. *From Hutchinson et al. (2021).*

system crossed a climatic threshold following a gradual decrease of atmospheric CO_2 levels under favourable orbital conditions for ice sheet expansion (DeConto and Pollard, 2003; DeConto et al., 2008; Pearson et al., 2009). However, many uncertainties remain. Ice sheet model simulations are sensitive to poorly-constrained boundary conditions, such as paleotopography, paleoceanographic conditions and indirect coupling to climate models, all of which affect the simulated timing, amplitude and volume of ice formation across the EOT (Paxman et al., 2019). Also, uncertainty exists as to the volume of ice already present prior to the EOT (Carter et al., 2017; Scher et al., 2014). Although the prevailing view is that tectonic opening of Southern Ocean gateways was not the direct cause of Antarctic glaciation (Huber et al., 2004), questions remain about what the impact of evolving oceanographic changes was on the inception of the ice sheet and its subsequent history (Houben, 2019; Sijp et al., 2011). Further questions remain about the response of regional climate and oceanography to the ice sheet's early growth (e.g., DeConto et al., 2007; Goldner et al., 2013). Reconstruction of the Cenozoic continental environmental history of Antarctica is, therefore, crucial to a thorough understanding of the role of high-latitude physical and biogeochemical processes in the global ocean and climate system and across the EOT. Precise stratigraphic correlation between the Antarctic continental margin and global marginal and open ocean records provides clues on the evolution of global climate. This global perspective is key to constraining the sensitivity of the AIS to changing global boundary conditions. Recently, major advancements have been made in reconstructing the paleogeographic, oceanographic and climatologic framework under which Antarctica became fully glaciated, as well as the response of regional climate, ocean and sea level to that glaciation.

This chapter reviews our current understanding of the greenhouse-icehouse transition from an Antarctic perspective. Before we present a summary of the evidence for the environmental evolution on the continent and marginal marine settings derived from land-based sections and drill-holes, we first lay out some crucial concepts necessary to interpret or reinterpret these observations. Although many of these concepts are not new, their importance has become appreciated in climate and ice sheet reconstructions across the EOT.

7.2 Background

7.2.1 Plate tectonic setting

Crucial boundary conditions for EOT glaciation relate to the plate tectonic setting, as it determines the solar energy on, and oceanographic conditions around, Antarctica. We distinguish here two important aspects: the relative and the absolute position of Antarctica and surrounding tectonic plates.

1. Relative plate position: This involves plate motions relative to one another (the plate circuit) and is reconstructed through stripping back ocean crust based on its age, and restoring intraplate and plate-boundary kinematics. Complex but critical areas involve the relative plate motions between the East Antarctic, West Antarctic and Antarctic Peninsula plates, which may have influenced the paleotopography of Antarctica at the EOT. Relative plate motions within Antarctica are poorly constrained, but Dalziel (2006) suggests that the West Antarctic Rift System (from the Ross Sea to the Ferrigno Rift) may have continued until the Miocene, with a total displacement of $\sim$200 km. The opening of the Drake Passage/Scotia Sea and the Australian-Antarctic separation have implications for the paleoceanographic setting around Antarctica, the presence of circum-Antarctic ocean circulation and heat transport. Of these, the opening of the Tasman Gateway is stratigraphically more tightly correlated with the EOT. Slow seafloor spreading between Antarctica and Australia started in the Late Cretaceous ($\sim$94 Ma along the western part; Cande and Mutter, 1982), the breakup of the final continental connection (between the South Tasman Rise and Terre Adelie) occurred no earlier than 33.5 $\pm$ 1.5 Ma (ref; e.g., Cande and Stock, 2004; Williams et al., 2019), quasi-synchronous with the EOT and concomitant to strong subsidence of continental margins in the Tasman region (Stickley et al., 2004). The shallowly open gateway allows initial westward countercurrent flow from 50 Ma onwards (Bijl et al., 2013), but with continuous gateway deepening, intensified atmospheric circulation (Houben, 2019) and northward tectonic drift into the strong westerlies wind system. The eastward flowing proto-Antarctic circumpolar current (ACC) establishes slightly after the EOT, around 30 Ma (Scher et al., 2015). In addition to the direct effect of gateway opening on the Antarctic climate, an indirect role has also been proposed through the ocean circulation reorganisation causing atmospheric CO_2 drawdown (Scher et al., 2015). The plate tectonic reconstruction of Drake Passage is extremely complex due to the abundance of sub-continents, various rift basins and complex absolute motions (e.g., Pérez et al., 2019). The timing of deepening of the tectonically complex Drake Passage is strongly debated, with proposed opening ranging from 49 Ma (Eagles et al., 2006; Livermore et al., 2007) to $\sim$17 Ma (Barker, 2001). Particularly, the motions of continental blocks within the Drake Passage are believed to restrict the gateway from being deeply open (Barker, 2001; Pérez et al., 2019) and uplift processes may have closed the DP region again around 21 Ma, after its initial opening (Lagabrielle et al., 2009). Recent palaeogeography reconstructions characterise the Drake Passage as narrow and shallow in both the late Eocene and early Oligocene, with minimal widening or deepening in a $\sim$5–10 Ma window around the EOT (e.g., Scotese and Wright, 2018). Therefore, even if opening Drake Passage has a long-term cooling effect

on Antarctic climate, there is no clear evidence that a significant opening occurred at the EOT.

2. Absolute plate position: The absolute position of continents involves the position of the spin axis of the Earth on the plate circuit. This changes through time as a result of true polar wander. It determines the absolute latitudinal coordinates, and thereby determines the amount of incoming sunlight and height of the snow line on Antarctica. The south polar position of Antarctica makes it extremely sensitive to changes in true polar wander. It makes true polar wander changes act as an antagonist: a change in true polar wander yields opposing latitude changes on either side of a longitudinal line. This becomes important when one compares paleoclimate reconstructions from East and West Antarctica. It is also important for the absolute position of Southern Ocean gateways. With their paleolatitude around 60°S, these gateways are at the boundary of prevailing easterly winds in the South and westerly winds in the North. A gateway position change of a couple of degrees relative to the spin axis of the Earth thus may have consequences for its throughflow. Conventional paleogeographic grids (e.g., for modelling studies) use (moving) hotspot reference frames (see discussion in Baatsen et al., 2016), that do not account for true polar wander. Using a paleomagnetic reference frame to determine the absolute position of the place circuit relative to the spin axis of the Earth does take true polar wander into account. This changes the paleo-position of Antarctica by ~2° towards the Pacific side in the Eocene (van Hinsbergen et al., 2015), when true polar wander was particularly strong. While in hotspot reference frames, Wilkes Land and Seymour Island appear at ~equal paleolatitudes in the Eocene, in paleomagnetic reference frames their latitudinal difference is up to 4°. Although absolute paleolatitude reconstructions using paleomagnetic reference frames come with error margins, they illustrate the importance of accurate absolute plate positions and motions for paleoclimate reconstructions.

7.2.2 Antarctic paleotopography

The Antarctic topography on which the ice sheet formed at the EOT was different from the modern-day topography. The onset of Antarctic glaciation and subsequent ice flow caused extensive glacial weathering and sediment transport towards the continental margin. Thermal subsidence, tectonic processes and isostatic loading also altered the landscape, and each needs to be considered when reconstructing the past Antarctic topography. Earlier reconstructions of Antarctic paleotopography (Wilson and Luyendyk, 2009) resulted in vast areas of west Antarctica being restored to above sea level. Subsequent reconstructions by Wilson et al. (2012) restored sediment packages back onto the whole continent, which resulted in higher paleotopographies and a vastly larger landmass than earlier reconstructions and than today. Ice sheet model simulations using this reconstruction suggested that the continent could have supported the growth of ice sheet with a volume 1.4 times larger at the EOT than when using a modern-day topography (Wilson et al., 2013). The growth of an ice sheet of this size would have caused an increase in benthic $\delta^{18}O$ of up to 1‰ (Wilson et al., 2013).

Continued improvements to the reconstruction technique, and more data constraining the modern-day Antarctic topography, have led to a new suite of reconstructions of the past Antarctic topography for the Oligocene-Miocene transition, mid-Miocene and mid-Pliocene, in addition to the EOT (Paxman et al., 2018, 2019). All this work has shown that the evolution of the Antarctic landscape since the EOT, and the erosion of deep subglacial basins, has continually increased the sensitivity of the ice sheet to ocean warming (Colleoni et al., 2018; Gasson and Keisling, 2020; Paxman et al., 2021).

7.2.3 Paleoceanographic setting

The Tasmanian Gateway was deeply open by the EOT (Cande and Stock, 2004; Stickley et al., 2004), and the Drake Passage allowed throughflow from the Pacific into the Atlantic (Scher and Martin, 2008). Although this must have initiated the onset of circumpolar flow, both numerical modelling (Hill et al., 2013) and field data (e.g., Bijl et al., 2018; Salabarnada et al., 2018) suggest that this proto-ACC was far from present-day flow strengths. How the proto-ACC oceanographic setting influenced poleward ocean heat transport remains under debate (Baatsen et al., 2020; Huber and Nof, 2006). Instead of a strong ACC, and particularly with restricted or closed gateways in the Eocene, Southern Ocean circulation featured strong, clockwise circulating gyres: one in the Pacific Ocean and another in the Atlantic-Indian ocean (Huber et al., 2004; Sloan and Rea, 1995). Arguably the onset of a Proto-ACC influenced this gyral circulation (Houben, 2019), but to what extent is as yet unknown.

7.2.4 Global average and regional sea level response

Over the past few years, the paleoclimate community has started to appreciate the complexity of sea level reconstructions from a geodynamic perspective. Previously, a conceptual model of glacio-eustatic sea level change was applied (e.g., Miller et al., 2008), which assumes that extraction of water from ocean basins and its storage in continental ice sheets causes a globally quasi-uniform sea level change. Connections with the geodynamic community since ~2010 led to significant reinterpretation, however. Geodynamic principles prescribe that any relocation of mass (in the case of the EOT relocation of water from the ocean basins onto the Antarctic continent) creates a redistribution of both gravitational forces and loading of the lithosphere. The first induces a displacement of the leftover, deformable ocean water, towards the more rigid ice mass. The overall relative sea level (RSL) pattern, as a result of ice formation, is consistent with the typical wave-like deformation that is induced by glacial- and hydro-isostatic adjustment (hereafter GIA). GIA consists of the response of solid Earth and gravity potential to redistributions in surface mass (from ocean water to continental ice). GIA can be modelled by solving the so-called sea level equation (SLE, Farrell and Clark, 1976; Spada and Stocchi, 2007). The SLE is a linear integral equation that yields the regionally varying RSL change for a prescribed ice sheet thickness chronology and solid Earth rheological model (the latter prescribes the deformation of the lithosphere). The area that deforms and subsides as a result of ice sheet growth is not limited to the ice-covered area, but extends further outside the glaciated area, thanks to the flexure of the underlying lithosphere. Lithospheric deformation can be rheologically described as purely elastic, thus providing an instantaneous response to any incremental surface loading. The areal extent and depth of the subsidence directly depends on the thickness of the lithosphere. Underneath the deformed lithosphere, the viscous upper mantle starts to flow, causing a further subsidence that is delayed by the viscosity. The upper mantle material flows outwards and upwards, forming the so-called uplifting peripheral forebulge that surrounds the ice sheet. As the ice sheet grows thicker and extends further offshore across the inner shelf, the forebulge is pushed outward. As a result, the offshore areas that were initially shoaling as upper mantle material upwelled, and a peripheral forebulge developed, now subside rapidly in response to the lithospheric flexure extending outwards from the ice-sheet margins. Both act to restore equilibrium. As a result, the magnitude and even the sign of sea level change as a result of ice sheet installation might be different per place on Earth. These geodynamical principles have most profound effects for sea level reconstructions close to the locus of ice sheet formation. Further away from the ice sheet sea level change will approach the eustatic value. We will further discuss the implications of these concepts below.

7.2.5 Proxies to reconstruct past Antarctic climatic and environmental evolution

Over the past years, proxies to reconstruct past climatic, oceanographic and environmental conditions around Antarctica have become more quantitative.

The most reliable atmospheric CO_2 reconstructions for the EOT interval are based on stable carbon isotopic composition of organic phytoplankton biomarkers (e.g., Pagani et al., 2005; Zhang et al., 2013) and boron isotopes in calcium carbonate microfossils (Anagnostou et al., 2020; Pearson et al., 2009). Absolute, quantitative proxies for sea surface temperature (SST) come from organic biomarkers (TEX_{86} and $U_{37}^{K'}$) (Douglas et al., 2014; Liu et al., 2009) and clumped isotopes on bivalves and planktonic foraminifera (Judd et al., 2019; Petersen and Schrag, 2015).

Mean annual air temperature (MAT) reconstructions come from a suite of proxies, based on vegetation (pollen or leaf assemblages) nearest living relative approaches (e.g., Francis et al., 2009), soil-derived organic biomarkers (MBT) (Pross et al., 2012) and the chemical index of alteration measured on detrital material (e.g., Passchier et al., 2013, 2017). The last of these is specifically a quantitative proxy for air temperature at the site of chemical weathering. The vegetation composition, as well as plant-derived organic biomarkers (n-alkanes) and weathering indices provide mostly qualitative information on hydrological conditions on the continent.

Qualitative reconstructions come from calcareous nannofossil, dinoflagellate cyst and diatom assemblages (Houben et al., 2013; Villa et al., 2013).

Independent reconstructions of deep-sea temperature come from Mg/Ca ratios in benthic foraminifera (e.g., Bohaty et al., 2012). These are then used to deconvolve temperature and ice volume effects on $\delta^{18}O$ for ice volume reconstructions. This exercise requires assumptions on the isotopic composition of the ice sheet.

Qualitative evidence for marine-terminating glaciers come from the presence of iceberg-rafted debris (IRD) in marine sediments (e.g., Ehrmann and Mackensen, 1992; Zachos et al., 1992).

All of these proxies come with underlying assumptions, uncertainties, error margins and quantified confidence intervals. The relationship between the measured parameter in the proxy carrier and the environmental parameter of interest is sometimes poorly understood mechanistically. These proxies are still very much in development while being applied to reconstruct past Antarctic climate and environmental conditions.

7.2.6 Far-field proxies

In marine geochemical records, the primary evidence for climate change leading to the establishment of the AIS is a relatively rapid ($\sim$200 kyr) $\sim$1.5‰ increase in $\delta^{18}O$ values of deep-sea benthic foraminifera (Coxall et al., 2005; Diester-Haass and Zahn, 1996, 2001; Miller et al., 1987; Zachos et al., 1992, 1996). This EOIS captured by marine carbonates has been well documented from deep-sea cores (Kennett and Shackleton, 1976; Savin, 1977; Savin et al., 1975; Shackleton and Kennett, 1975) as well as outcrop sections (e.g., Devereux, 1967). The isotope shift was initially attributed to 4°C–6°C of marine cooling. Later studies confirmed the occurrence of the climate shift and interpreted it to partly reflect the emplacement of a minor-to-moderate (e.g., Keigwin, 1980; Keigwin and Corliss, 1986; Miller and Curry, 1982; Murphy and Kennett, 1986; Shackleton et al., 1984) or major (Matthews and Poore, 1980; Poore and Matthews, 1984; Prentice and Matthews, 1988) ice volume in the earliest Oligocene. The discovery of Oligocene proximal glaciomarine sediments from the Ross Sea and Prydz Bay (Hambrey and Barrett, 1993; Hambrey et al., 1991) and IRD associated with the oxygen isotope shift at Ocean Drilling Programme (ODP) Site 748 near Antarctica confirmed this interpretation (Zachos et al., 1994).

The acquisition of high-resolution records has later shown that the EOIS and the preceding cooling trend were not monotonic. A detailed record from the equatorial Pacific ODP Site 1218 revealed that the EOIS is actually made of two steps separated by ~200 kyr (Coxall et al., 2005). The first step, 'Step 1' of Pearson et al. (2008) or EOT-1 of Coxall and Wilson (2011), which occurs between 34.2 and 33.8 Ma, is not readily identified at all sites. The use of independent temperature proxies such as Mg/Ca and TEX_{86} has improved estimates of ice volume from oxygen isotope records. The first step, starting at ca. 34.2 Ma, is thought to primarily reflect a temperature decrease while the second, which unravels between 33.7 and 33.6 Ma, has been interpreted as the onset of a prolonged interval of maximum ice extent often referred to as the Earliest Oligocene Glacial Maximum (EOGM) between 33.6 and 33.2 Ma (Bohaty et al., 2012; Katz et al., 2008; Lear et al., 2008). Other oxygen isotope maxima in the late Eocene have been interpreted to reflect ephemeral precursor glaciations during the late Eocene (Houben et al., 2012; Katz et al., 2008; Scher et al., 2011, 2014). Among these, the oldest and most prominent one is reported to have occurred at 37.3 Ma (Scher et al., 2014). Another event of hypothesised transient glaciation occurred at about 34.15 Ma (Katz et al., 2008). As recently reviewed by Hutchinson et al. (2021), inter-site correlation for some of these features in the oxygen isotope record is not straightforward and the usage of certain terms has changed through time, which make it difficult to settle a consistent terminology for all of them.

However, benthic foraminiferal $\delta^{18}O$ values are influenced by both deep-sea temperature and the volume of continental ice, and the exact magnitude of deep-sea temperature decrease across the EOT is far from clear. Progress in paleotemperature proxies, including Mg/Ca, clumped isotopes and TEX_{86}, have led to independent means of quantifying changes in bottom and surface waters across the EOT (Bohaty et al., 2012; Evans et al., 2016; Lear et al., 2008; Liu et al., 2009).

These records suggest that deep-water temperature cooled by 3°C–5°C across the EOT (Liu et al., 2009). Global average sea-surface temperature dropped by approximately 2.9°C although with large inter-site differences (Hutchinson et al., 2021). The associated estimated volume of Antarctic ice depends upon the assumed isotopic composition of the ice sheet but was likely between 70% and 110% of the size of the modern-day AIS (Bohaty et al., 2012; Lear et al., 2008). Accordingly, a ca. 70 m sea-level fall is estimated from low-latitude shallow marine sequences (Katz et al., 2008; Miller et al., 2008).

7.3 Antarctic Sedimentary Archives

7.3.1 Land-based outcrops

Documenting the evolution of continental environmental conditions across the EOT is crucial to our understanding of the transition from greenhouse to icehouse. Extensive ice cover and erosion hinders access to land-based sedimentary archives, which are confined to locations along the continental margin. Sedimentary successions outcropping on the continental margin are known from the Antarctic Peninsula and King George Island (Fig. 7.2). Information is also obtained from glacial erratics, known as the McMurdo erratics, from the Ross Sea margin. Although not capturing the EOT, these sedimentary archives record the longer-term evolution of climate leading up to and following this transition. A summary of important outcrops and the environmental signal obtained from them is presented below.

7.3.1.1 Antarctic Peninsula Region

Eocene sediments comprising the La Meseta Formation (Elliot and Trautman, 1982) are exposed on Seymour Island, Antarctica (~64°S, 54°W), located approximately 100 km east of the Antarctic Peninsula (Fig. 7.2). Based on magnetostratigraphic studies (Beamud et al., 2015) and Sr isotopes (Ivany et al., 2008), the La Meseta Formation was attributed an age of ~56.2 to ~58.8 Ma. However, dinoflagellate cyst biostratigraphy (Amenábar et al., 2020; Douglas et al., 2014) correlated to the zonation described in Bijl et al. (2013) provides a younger age (Lutetian-Priabonian; ~45 to ~34 Ma) for these strata, which we follow here.

The La Meseta Formation has been originally subdivided into seven informal mapping units 'Tertiary Eocene La Meseta' (Telms 1 to 7) (Sadler, 1988) and later refined into six erosional-based units, or allomembers (Marenssi et al., 1998). The formation consists of mudstones and sandstones with interbedded conglomerates with horizons characterised by an accumulation of molluscs, which were deposited in deltaic, estuarine and shallow

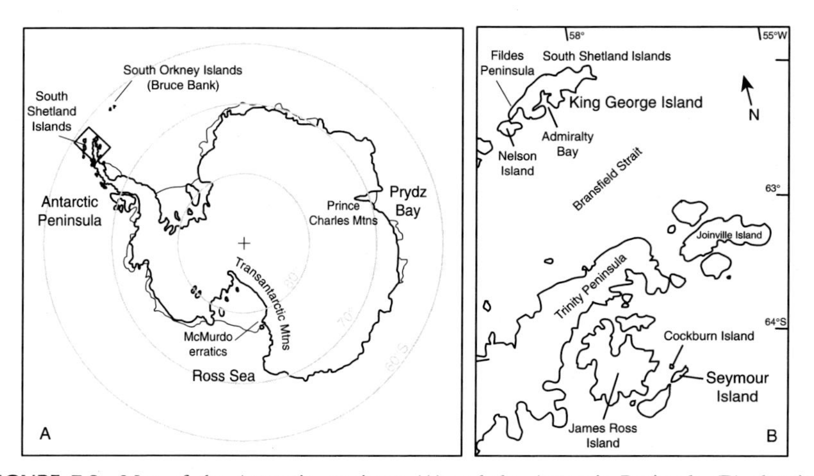

FIGURE 7.2 Map of the Antarctic continent (A) and the Antarctic Peninsula (B) showing some of the locations mentioned in the text. *From Francis et al., 2009.*

marine environments as part of a tectonically-controlled incised valley system (Marenssi et al., 1998; Porebski, 1995). Provenance studies on sandstones of the La Meseta Formation demonstrate that the source rock was located west-northwest along the present-day Antarctic Peninsula. The formation provides high diversity and abundance of fossil remains (Marenssi, 2006), including the only land mammal fossil findings of the entire continent of Antarctica (Davis et al., 2020; Gelfo et al., 2019; Reguero et al., 2002) along with marine vertebrates (including giant penguins) and invertebrates, fossil wood, fossil leaves and a rare flower (Acosta Hospitaleche, 2014; Acosta Hospitaleche and Reguero, 2014; Davis et al., 2020; Francis et al., 2004; Gandolfo et al., 1998a,b,c; Jadwiszczak and Mörs, 2019; Stilwell and Zinsmeister, 1992; Tambussi et al., 2006). While detailed correlation of the Eocene open ocean paleoclimate record with that of the Seymour Island continental margin sequence has always proved problematic, sedimentological, palaeontological and geochemical proxy records from the latter provide significant insights into the Antarctic environmental and climate evolution of this area across the middle to latest Eocene.

The Ypresian *Cucullaea* shell bed of La Meseta Formation (Telm 4 of Sadler, 1988; ~42 Ma) contains a remarkably diverse assemblage of plants, invertebrates and vertebrates that have provided a valuable record of both continental and marginal settings, which indicate an environment completely different to the present-day Antarctica (Reguero et al., 2012). Seasonally-resolved proxy data derived from bivalves estimate winter SSTs near 9°C and summer SSTs around 17°C, indicating that the Antarctic Peninsula was far too warm to support significant ice accumulation, at least at low elevation (Telm 5 ~41–38 Ma; Judd et al., 2019). Isotope-derived mean-annual temperature (MAT) is in good agreement with previous estimates from clumped isotopes (12.6°C; Douglas et al., 2014) TEX_{86} (15.4°C; Douglas et al., 2014) and bulk oxygen isotope analyses of Telm 5 carbonates (~10°C–14.5°C; Dutton et al., 2002; Ivany et al., 2008, and references therein). A stepwise increase in the oxygen isotopic composition ($\delta^{18}O$) of bivalve shells from Telm 6 (~38–36 Ma) suggests ~10°C of cooling (Ivany et al., 2008). Much of the recorded increase of oxygen isotope values took place across a short interval centred at ~37 Ma (Douglas et al., 2014). TEX_{86} and clumped isotope data from Douglas et al. (2014) do not show such an abrupt temperature decrease although they indicate a generally warmer temperature from Telms 2, 3 and 4 (45–41 Ma) compared to Telms 5, 6 and 7 (41–34 Ma). Yet, the occurrence of ice-rafted debris has been reported from the upper part of the La Meseta Formation (Telm 6 or 7) (Doktor et al., 1988) and paleoclimatic evidence of a severe climatic deterioration towards the end of the Eocene. It is possible that by the end of the Eocene, limited ice, perhaps as valley glaciers, was already present in the area. The presence of local glaciers and ice caps at the northern Antarctic Peninsula during the middle Eocene-early Oligocene (49–32 Ma) was also argued by Anderson et al. (2011). Accordingly, high diversity Neogastropoda populations indicative of warm conditions in the lower Telms terminate abruptly in the upper La Meseta Formation with an extinction event that most likely heralds the onset of global cooling (Crame et al., 2014). Other signals of cooling are provided by a swift decrease of the Chemical Alteration Index (CIA) values and a concomitant inception of illite-dominated clay mineral associations at the top of the La Meseta Formation, suggestive of a transition to cold, frost-prone and relatively dry conditions during the late Eocene.

Further evidence for a climatic deterioration on Seymour Island during the late Eocene comes from floral records. Leaves and wood of both angiosperm and conifer affinity occur with fern fossils and a flower (Case, 1988; Francis et al., 2004; Reguero et al., 2002). *Nothofagus* leaves were found to be notophyllous (a leaf-size category 7.5–12.5 cm long). Angiosperm fossils affiliated to families including Nothofagaceae, Dilleniaceae, Myricaceae, Myrtaceae, Elaeocarpaceae, Moraceae, Cunoniaceae, Winteraceae and Lauraceae have been described (Francis et al., 2004; Gandolfo et al., 1998a,b; Reguero et al., 2002). Doktor et al. (1996) also described leaves affiliated with Podocarpaceae, Araucariaceae, Nothofagaceae and Proteaceae. Gothan (1908) was the first to describe fossil wood from Seymour Island, which has subsequently been reexamined by several authors and identified as having both angiosperm and coniferous affinities (Brea, 1996, 1998; Francis, 1991; Francis and Poole, 2002; Reguero et al., 2002; Torres et al., 1994).

A decrease in leaf sizes in the upper Telms suggests that the climate deteriorated towards the end of the Eocene, as observed in studies of the La Meseta Formation by Case (1988) and Reguero et al. (2002). Gandolfo et al. (1998a,b) suggested a temperate MAT of 11°C–13°C for the Cucullaea I Allomember during the early late Eocene. Further climate data were provided by leaf margin analyses of a late Palaeocene flora (Cross Valley Formation) and of the early late Eocene Cucullaea 1 flora, which indicate a decrease in floral diversity and a change from MATs of 14°C during the late Palaeocene to 11°C during the early late Eocene, with signs of freezing winters in the late Eocene (Francis et al., 2004). The shift from cool temperate, humid Valdivian-type forest to a more depauperate vegetation was accompanied by a decrease in typical Eocene dinoflagellate cysts, an increase in sea ice–indicative marine phytoplankton, and an increase in reworked palynomorphs, suggesting the onset of periglacial conditions and a subpolar climate just before the EOT boundary in the back-arc James Ross Basin (Warny et al., 2018).

All in all, the persistence of vegetation and geochemical evidence suggest that if some ice developed during the latest Eocene its distribution was limited to elevated areas.

7.3.1.2 King George (25 de Mayo) Island, South Shetland Islands

King George Island and neighbouring Nelson Island consist of several tectonic blocks bounded by two systems of strike-slip faults of Paleogene-early Neogene (54–21 Ma) age (Birkenmajer, 1989). Thus, considerable differences in stratigraphic succession, age and character of the rocks occur between particular blocks. The stratigraphic sequence includes mainly Late Cretaceous to early Miocene island-arc extrusive and intrusive rocks comprising mainly terrestrial lavas, pyroclastic and volcaniclastic sediments often with terrestrial plant fossils. Fossiliferous marine and glaciomarine sediments are also represented, and provide clues to paleoenvironmental conditions during the early-middle Oligocene.

Several sequences of tillites crop out within these complicated sequences, representing glacial and interglacial events. Initial reports of supposed Eocene-age tillites at Magda Nunatak (Birkenmajer, 1980a,b), named the Krakow Glaciation and dated at 49 Ma, have been disproved by Sr dating (Dingle and Lavelle, 1998). Tillites on King George Island have been reported from the Point Thomas Formation in Admiralty Bay, which would provide evidence for the presence of alpine glaciers during the middle Eocene (Birkenmajer et al., 2005). However, age control is poor and the potential extent of glaciers and their drainage pattern on the eastern side of the peninsula is unconstrained.

The earliest clear evidence for glacial activity from the King George Island are diamictites and ice-rafted deposits from the Krakowiak Glacial Member of the Polonez Cove Formation, which are dated to a mid-Oligocene age (30–26 Ma) (Troedson and Smellie, 2002). At its maximum extent, ice was grounded on a shallow marine shelf. Interestingly, exotic clasts within this sequence may represent ice-rafted debris that was derived from as far away as the Transantarctic Mountains, suggesting marine-based glaciation in the Weddell Sea region. As reported by Warny et al. (2018), however, unpublished strontium isotope results point to older geological ages (early Oligocene, ca. 32–30 Ma) for parts of the glacio-marine deposits of the Krakowiak Glacier Member. The only other terrestrial evidence for Oligocene ice, possibly representing local alpine glaciation, is from Mount Petras in Marie Byrd Land, West Antarctica, where deposits indicate volcanic eruptions beneath ice (Wilch and McIntosh, 2000).

Many macrofloras have been discovered on King George Island in the South Shetland Island group, north of the Bransfield Strait in the Antarctic Peninsula region, currently dated between late Palaeocene and late Eocene. The floras may have lived at a paleolatitude of ~67°S (Van Hinsbergen et al., 2015). The stratigraphy is complex: Birkenmajer (1981, 1989, 1990) and Birkenmajer et al. (1986) erected many local formations, but a simpler scheme was created by Smellie et al. (1984). No single stratigraphic scheme exists and so the stratigraphic context of the floras remains unclear. The stratigraphic framework used here includes both schemes (also reviewed by Hunt, 2001).

Leaf macrofloras, currently understood to be of late Palaeocene to middle Eocene age, have been described in varying completeness from the Admiralty Bay and Fildes Peninsula areas of the island. In the Admiralty Bay area, the Mt. Wawel Formation (Point Hennequin Group), attributed to the middle Eocene based on U-Pb and ^{40}Ar-^{39}Ar analyses (Nawrocki et al., 2011), contains the macroflora deposits collectively known as the Point Hennequin Flora with individual localities named Mount Wawel and Dragon Glacier Moraine floras (Askin, 1992; Birkenmajer and Zastawniak, 1989a; Hunt, 2001; Hunt and Poole, 2003; Zastawniak et al., 1985). The Mount Wawel flora comprises macrofossils of *Equisetum* (horsetail), ferns and several *Nothofagus* species as microphyllous leaves (a leaf-size category of 2.5–7.5 cm long), in addition to a few other angiosperms and Podocarpaceae. The Dragon Glacier Moraine flora is similar; the angiosperm leaves being dominated by *Nothofagus* and the conifers including Araucariaceae and Cupressaceae, in addition to Podocarpaceae. The middle Eocene Petrified Forest Creek flora from the Arctowski Cove Formation and the late Eocene *Cytadela* flora from the Point Thomas Formation are both within the Ezcurra Inlet Group. The middle Eocene Petrified Forest Creek flora is a wood flora requiring revision, but intermediate *Fagus–Nothofagus*-type species are recorded. The *Cytadela* leaf flora includes ferns (including a *Blechnum*-affinity species), mostly small *Nothofagus*-type leaves with pinnately veined leaves of other dicotyledonous types and possible Podocarpaceae (Askin, 1992; Birkenmajer, 1997; Birkenmajer and Zastawniak, 1989a). Birkenmajer and Zastawniak (1989a) considered this flora of an age close to the E/O transition.

In this region, therefore, *Nothofagus*-dominated forests were the norm in the middle to late Eocene with ferns and tree ferns becoming increasingly important. Estimated MATs of 5°C–8°C are slightly cooler than those on Seymour Island to the east during the middle Eocene, and the vegetation was similar to the southernmost Patagonian–Magellanic forests of southern Chile (Askin, 1992; Birkenmajer and Zastawniak, 1989a; Hunt, 2001; Zastawniak et al., 1985). By the late Eocene, vegetation was more comparable to the recent fern-bush communities of southern oceanic islands (e.g., the Auckland Islands), interpreted from the *Cytadela* and Petrified Forest Creek floras (Askin, 1992; Birkenmajer, 1997; Birkenmajer and Zastawniak, 1989a). However, this observation entails MAT estimates of 11.7°C–15°C, which appear too high especially considering the small size of the leaves (Francis, 1999).

In the Fildes Peninsula area in the southwest of King George Island, the Fildes Peninsula Group contains the contemporary middle Eocene Collins Glacier and Rocky Cove floras within the Fildes Formation, and the diverse late Palaeocene–middle Eocene Fossil Hill flora (Fossil Hill Formation). The latter is a leaf flora containing 40 recognised taxa, including mixed broadleaf angiosperms (with large-leaved *Nothofagus* species), conifers (podocarp, araucarian and cupressacean) and ferns (Birkenmajer and Zastawniak, 1989a,b; Francis, 1999; Haomin, 1994; Li, 1992; Reguero et al., 2002). Neotropical and sub-Antarctic elements appear to be mixed perhaps indicating a collection derived from communities at different altitudes (Li, 1992), although this mixed signature may be a feature of Paleogene polar biomes (Francis et al., 2004). The *Nothofagus* leaves are much larger than their modern relatives, suggesting warm and humid climate conditions during the early part of the Eocene. Estimates of MAT suggest >10°C (from 40% entire-margined leaves) and a small annual temperature range (Li, 1992).

Fossil leaves remain undescribed from the Rocky Cove flora; however, wood from this locality has been identified as *Nothofagoxylon antarcticus* (Hunt, 2001; Shen, 1994). The Collins Glacier deposit is primarily a wood flora that includes wood of both coniferous (*Cupressinoxylon* sp. and *Podocarpoxylon fildesense*) and angiospermous (*Nothofagoxylon* spp., *Weinmannioxylon eucryphioides* (Cunoniaceae) and *Myceugenelloxylon antarcticus* (Myrtaceae) affinity (Hunt, 2001; Poole et al., 2001, 2005). Changes in the vegetation suggest that MAT had dropped to ~9°C by the middle Eocene, which compares well with an estimate of ca. 8°C derived from leaf analysis of the same flora. This temperature drop was accompanied by a concomitant increase in precipitation. However, the absence of a change from a semiring to ring porous condition in angiosperm wood indicates that seasonality had not greatly intensified (Poole et al., 2005).

7.3.1.3 The Ross Sea Region

Several hundred glacial erratic boulders and cobbles recovered from coastal moraines around the shores of Mount Discovery, Brown Peninsula and Minna Bluff provide a window on Eocene environmental conditions in the Ross Region. This ensemble, collectively known as the McMurdo Erratics, is most likely derived from sub-glacial basins, such as Discovery Deep, that lie along the coast of the Transantarctic Mountains or basement highs situated to the east of the discovery accommodation zone (Wilson, 1999; Wilson et al., 2006). The erratics were distributed into their distinctive pattern of terminal and lateral retreat moraines during relatively recent advance and retreat of grounded ice into southern McMurdo Sound (Wilson, 2000). Subsequent basal adfreezing and surface ablation has transported the erratics to the surface of the McMurdo Ice Shelf. Although currently out of their original stratigraphic position, this suite of erratics provides a means to obtain geologic data that are otherwise buried beneath the AIS and fringing ice shelves.

The McMurdo Erratics comprise a range of lithotypes and ages. Eocene rocks contain a rich suite of fossil flora and fauna including marine and terrestrial palynomorphs, diatoms, ebridians, marine vertebrates and invertebrates, terrestrial plant remains and a bird humerus. Biostratigraphic data from dinoflagellate cyst (dinocyst), ebridian and mollusc assemblages recovered from many of the erratics indicate that the majority of fossiliferous rocks range from middle to late Eocene, 43–34 Ma. Erratics collected between 1993 and 1996 (Stilwell and Feldmann, 2000) include several hundred samples of Oligocene, Miocene and Pliocene sediment. Although relatively rich dinoflagellate cyst (dinocyst) assemblages have been described from Oligocene-Miocene Sequences from the Cape Roberts Cores, assemblages in post-Eocene erratics comprise few taxa (typically of five species), which may be because of the proximal paleo-depositional setting of the sediments that the erratics represent.

The majority of the Eocene erratics record a suite of lithofacies that were deposited in coastal–terrestrial to inner shelf marine environments (Levy and Harwood, 2000a,b). These sediments were probably deposited within fan deltas that formed along the rugged coastline of the rapidly rising Transantarctic Mountains. Abundant macroinvertebrate faunas, including bivalves, gastropods, scaphopods, cirripeds, bryzoans, decapods and brachiopods, indicate that many of these sediments were deposited in a spectrum of predominantly shallow marine environments. The presence of terrestrial plant material and palynomorphs also suggests that the majority of the rocks were formed in nearshore environments. However, the occurrence of outer shelf dinocyst species and the absence of benthic diatom taxa in many of the fine-grained lithofacies indicate that outer shelf environments were also present in the source region. The Eocene erratics contain no direct or unequivocal sedimentological evidence for the presence of ice close to the basins in which the sediments were originally deposited. It is notable that erratics composed of diamictites recovered from the coastal moraines are all Oligocene and younger in age.

Although rare, fossil leaves, wood and pollen recovered from several erratics provide a glimpse of the Eocene climate for the region. One erratic contains wood and leaves from *Araucaria* and *Nothofagus* trees, which suggests cool temperate conditions with some winter snow, but temperatures were probably not cold enough to allow extensive ice at sea level (Francis, 2000; Pole et al., 2000). Spore and pollen assemblages recovered from the erratics reflect *Nothofagus*-podocarpaceous conifer-Proteaceae vegetation with other angiosperms growing in temperate climate conditions (Askin, 2000). Oligocene and younger erratics show a major drop in species richness, which is also noted in sequences recovered in CIROS-1 and the Cape Roberts Project (CRP) cores (Mildenhall, 1989; Raine and Askin, 2001).

Fossil invertebrate remains recovered from the erratics include a humerus shaft from a pseudodontorn (giant bony-toothed sea bird) (Jones, 2000), a probable crocodile tooth (Willis and Stilwell, 2000) and teeth from two species of shark (Long and Stilwell, 2000). The small but significant record of East Antarctic invertebrate fauna indicates a temperate to cool-temperate marine environment.

The n-alkane distributions from the Eocene McMurdo erratics are different from those from the Oligocene and Miocene, specifically in the prominence of the n-C29 as opposed to the n-C27 (Duncan et al., 2019). According to these authors, this observation can be explained by a combination of climate cooling as the AIS developed, and a shift in plant community towards low diversity tundra: *Nothofagus*, podocarpidites and bryophytes (Askin, 2000; Askin and Raine, 2000; Lewis et al., 2008; Prebble et al., 2006).

7.3.2 Sedimentary archives from drilling on the Antarctic Margin

Over the last decades, much effort has been put into the search for continuous sedimentary records of the EOT through drilling on the Antarctic margin. Yet, there are very few ice sheet–proximal records through this interval mainly because marine sedimentary successions on the Antarctic margin are exposed to glacial erosion, similarly to land-based successions and because the age models developed for high latitude settings might be difficult to correlate to lower latitude chronologies.

7.3.2.1 Drill cores in the western Ross Sea

Two drill holes have recovered sediment sequences that approach or span the E/O boundary with exceptionally high recovery in the western Ross Sea area: the CIROS-1 and CRP-3 drillholes (Fig. 7.3). In the 702-m-deep CIROS-1 hole (Barrett, 1989; Barrett et al., 1991), the lower part of the core was originally regarded as late Eocene, with a breccia passing up into mudstone and sandstone. The boundary with the Oligocene was originally placed at about 570 m (Barrett et al., 1989), but magnetobiostratigraphic data (Wilson et al., 1998) suggest that the EOB is much higher, at about 410–420 m. In either case, there is no obvious lithological transition, and finer grained facies include alternations of weakly stratified sand and mud, with intraformational conglomerate and occasional diamictite. Moving up core, a major hiatus exists at 366 m, which coincides with the early/late Oligocene boundary (around 28.1 Ma).

The CIROS-1 core was originally interpreted in terms of depositional setting, ice proximity and water depth (Barrett, 1996; Hambrey and Barrett, 1993; Hambrey et al., 1989). The breccia at the base of the hole is interpreted as a fault-brecciated conglomerate. The overlying sandstone/mudstone/diamictite succession is marine, influenced to varying degrees by resedimentation and iceberg-rafting. Above the major hiatus at 366 m, the

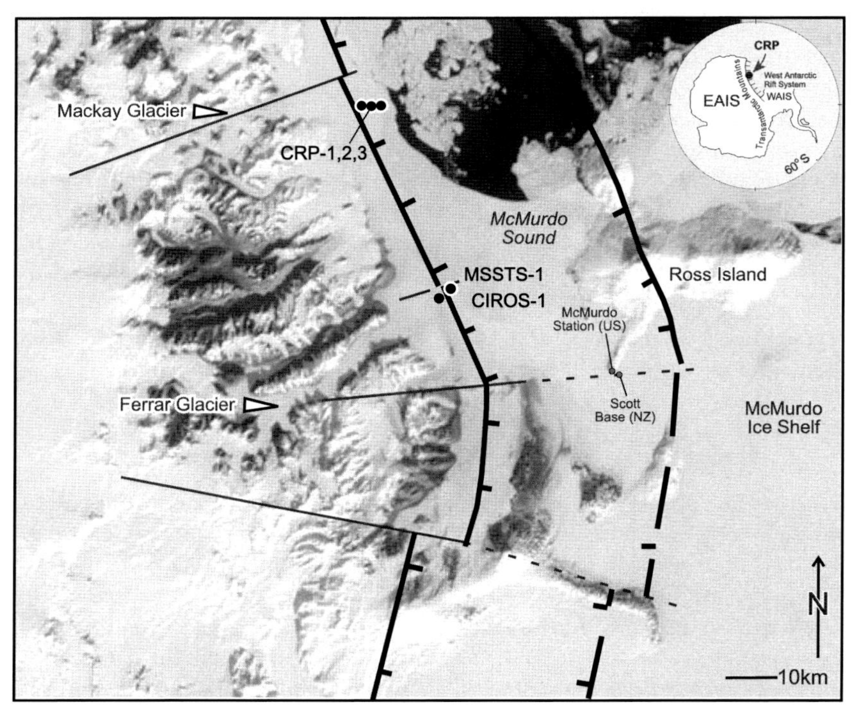

FIGURE 7.3 Satellite image with the location of the CRP, MSSTS-1 and CIROS-1 drill sites in the Southern McMurdo Sound. Key geographical and geological features of the area, including the boundary faults of the southern extension of Terror Rift, are shown. *From Galeotti et al. (2016).*

sandstones were regarded as fluvial, and the diamictites as basal glacial deposits, indicating ice overriding the site. However, Fielding et al. (1997) argued, based on a sequence stratigraphic analysis, that the late Oligocene diamictite was also glaciomarine. In contrast, Hiemstra (1999) reverted in part to the original view of grounded ice on the basis of microstructural studies. Whichever solution is the correct one, there is no clear evidence for a major environmental shift at the EOB, but there is evidence for more ice-proximal conditions at the early/late Oligocene transition.

A record of climate change through the EOT has also been determined from the environmental magnetic record in the CIROS-1 core (Sagnotti et al., 1998). Variations in magnetite were related to the concentration of detrital material transported into the Victoria Land Basin, influenced by climate and weathering rates on the Antarctic continent (especially of the Ferrar Group). Sagnotti et al. (1998) determined, from changes in the abundance of magnetite, that although there were some cold dry intervals (35–36 Ma) alternating with warm humid climates during the late Eocene, a stable cold dry climate was not established in Antarctica until the EOB, with major ice-sheet growth occurring at the early/late Oligocene boundary. This pattern matches the clay mineral history, which shows a shift from smectite-rich to smectite-poor assemblages in Antarctica at the EOB (Ehrmann, 1997; Ehrmann and Mackensen, 1992). A single *Nothofagus* leaf was found in the CIROS-1 core, originally thought to be Oligocene in age (Hill, 1989), but after recent refinement of the age model, it is now considered early Miocene (Roberts et al., 2003). Terrestrial temperature estimates for the CIROS-1 core suggest the temperature in the late Eocene was ca. 9°C (Passchier et al., 2013), broadly in agreement with fossil wood from a McMurdo glacial erratic that puts the maximum seasonal temperature at 13°C (Francis, 2000).

The second core that contains the EOT is the Cape Roberts Project Core CRP-3. Drilling at CRP-3, ~12 km east of Cape Roberts provided an almost continuous core through 823 m of Cenozoic sedimentary strata on the western edge of the Victoria Land Basin (VLB), at the Western margin of the Ross Sea continental shelf (CRP Science Team, 2000). At 823 m below the sea floor (mbsf), the basement of the VLB was penetrated and a further 133 m of basement rocks was recovered and correlated with the Devonian-age, Arena Sandstone of the Beacon Supergroup.

The CRP-3 succession contains an array of lithofacies comprising fine-grained mudrocks, interlaminated and interbedded mudrocks/sandstones, mud-rich and mud-poor sandstones, conglomerates and diamictites that are together interpreted as the products of shallow marine to possibly non-marine environments of deposition, affected by the periodic advance and retreat of a land-terminating and tidewater glacier (Fielding et al., 2001; Powell et al., 2001).

The uppermost 330 mbsf shows a cyclical arrangement of lithofacies and is interpreted to reflect cyclical variations in relative sea level (>20 m) in concert with variations in the proximity of a marine-terminating glacier, ultimately regulated by fluctuations in the volume of the AIS (Naish et al., 2001). A conceptual depositional model for the late Oligocene-Miocene interval was developed by Powell et al. (2000), based on facies associations and comparison with modern glaciomarine environments, such as those in Alaska and Greenland. This shows that during an advance and still-stand, a grounding-line fan develops, and this is followed by rapid recession until another fan develops. The sequence becomes even more complex when the glacier overrides previously formed fans. Fig. 7.4 is a simplified version of this model.

Between 330 and 780 mbsf, cyclical units generally characterised by fining-upward successions from conglomerate above a sharp boundary passing into sandstone facies were interpreted as fluvial to shallow-marine deltaic depositional sequences recording cyclical variations of relative sea level (<20 m) in concert with a more 'indirect' record of the proximity of an advancing and retreating land-terminating ice margin (Fielding et al., 2001). The early Oligocene landscape was characterised by temperate glaciers

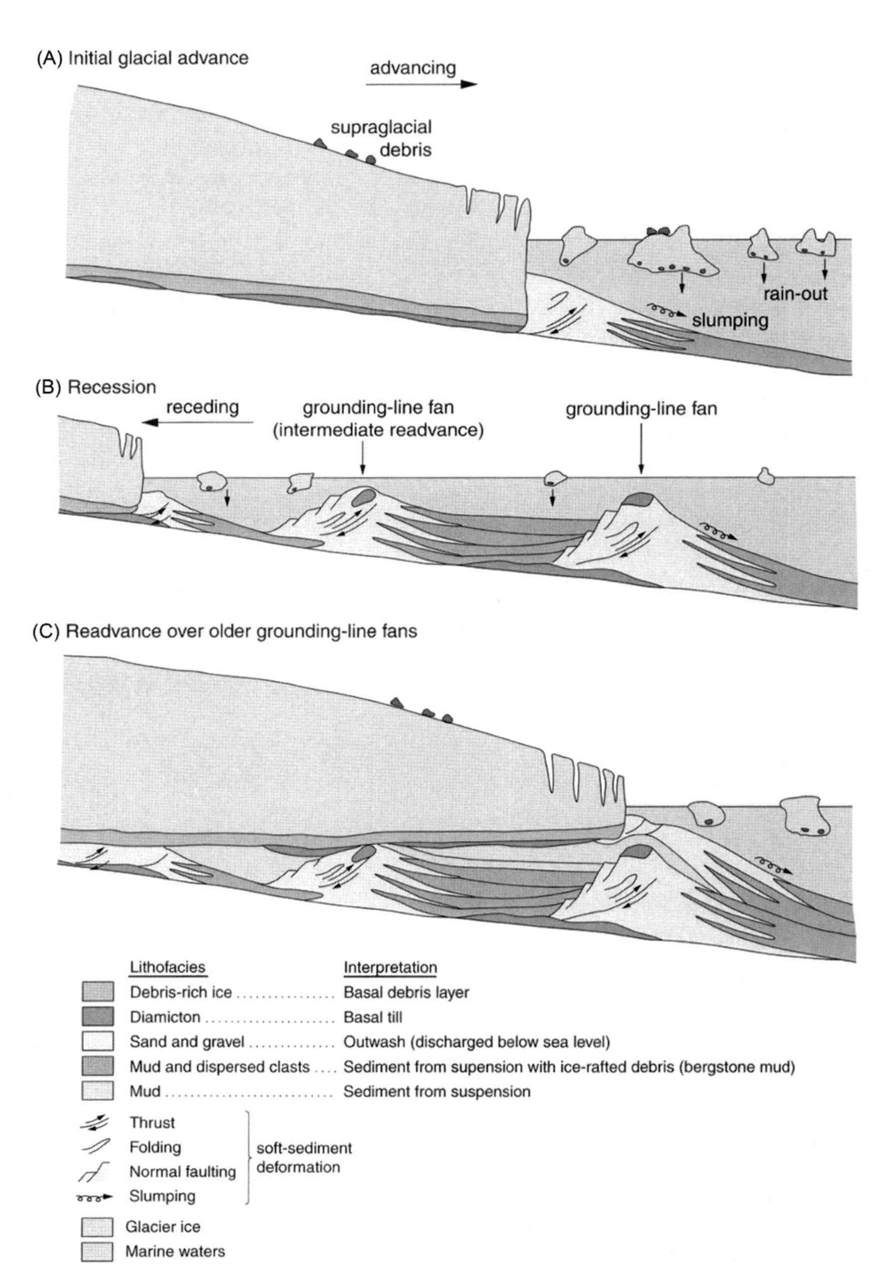

FIGURE 7.4 Grounding-line fan model of glaciomarine sedimentation for late Oligocene/early Miocene time. *From Hambrey et al. (2002); Simplified from Powell et al. (2000). Reproduced with permission of The Geological Society Publishing House, Bath, UK.*

flowing from the early East Antarctic Ice Sheet (EAIS), some terminating in the sea, and others on braided outwash plains (Fig. 7.5).

A total of thirty-seven sedimentary cycles occur below 300 mbsf and are associated with <20 m oscillations in relative sea level (Galeotti et al., 2012; Naish et al., 2001). At about 300 mbsf, the first diamictites occur recording the development of a more expansive marine-terminating ice sheet on the western Ross Sea continental shelf for the first time. Above this level, eleven glaciomarine sedimentary cycles associated with larger sea-level fluctuations of >20 m are bounded by glacial surfaces of erosion, which implies the loss of part of the sedimentary record (Fielding et al., 2001). Water depth changes across these cycles represent oscillations between innermost shelf (~5 m) and the offshore low-energy shelf below wave base (up to ~50 m) (Dunbar et al., 2008). Accommodation space for the preservation of such a remarkably complete sedimentary record was provided by high rates of tectonic subsidence during the initial stages of rift extension beginning in the latest Eocene (Fielding et al., 2008; Henrys et al., 2007).

Based on spectral analysis of the clast abundance in the lower part of the CRP-3 drillhole and using the available magnetostratigraphic correlation (Florindo et al., 2005), Galeotti et al. (2012) showed that the facies fluctuations in the lower part of CRP-3 reflect astronomically forced cycles of

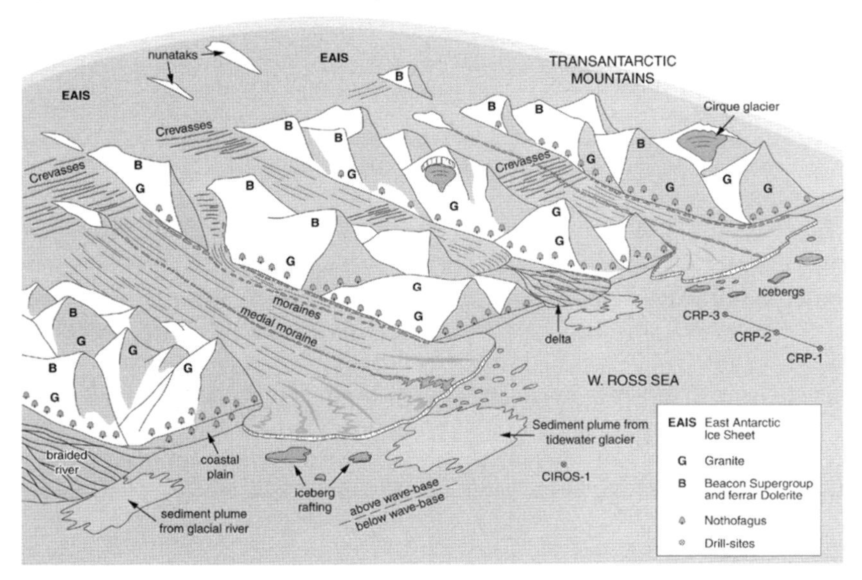

FIGURE 7.5 Cartoon depicting the Victoria Land coast in early/late Oligocene time, with glacier- and river-influenced coast, and vegetated mountainsides and lowlands. This scenario is based on a combination of sedimentological evidence and floral data from CIROS-1 and Cape Roberts cores. *From Hambrey et al. (2002). Reproduced with permission of The Geological Society Publishing House, Bath, UK.*

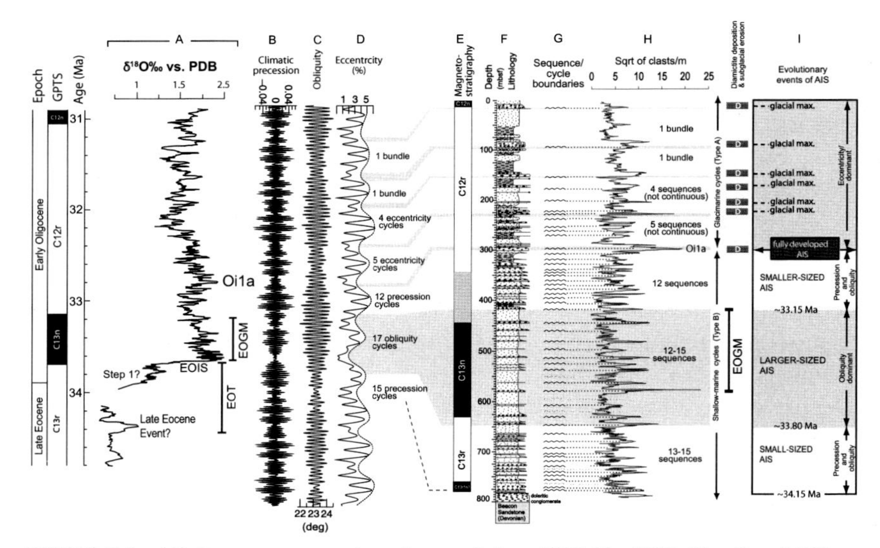

FIGURE 7.6 (A) Deep-sea oxygen isotopic record from ODP Site 1218 (Coxall and Wilson, 2011; Coxall et al., 2005), and time series for (B) climatic precession, (C) obliquity and (D) eccentricity correlated with the (E, F, and G) magnetostratigraphy, lithostratigraphy and sequence stratigraphy (Fielding et al., 2001; Florindo et al., 2005), and (H) square root of clast abundance (Sandroni and Talarico, 2001) for the late Eocene—early Oligocene CRP-3 drill core. (G) Thirty-seven shallow-marine sedimentary cycles (sequences; Type B) occur in the lower 500 m of the core record, controlled by advances and retreats of land-terminating glaciers associated with <20 m sea-level oscillations. Eleven overlying glaciomarine sedimentary cycles (sequences; Type A), each bounded by glacial surfaces of erosion, occur in the upper 300 m of the CRP-3 core, and record oscillations in the extent of a more expansive marine-terminating ice sheet in Ross Embayment. (I) Inferred stages and events in the development of the AIS across the E-O boundary and the relationship to orbital forcing are summarised. *From Galeotti et al. (2016), modified.*

glacial advancement and retreat. The resulting astrochronology provides an independently-derived age model for CRP-3, which enabled a precise definition of the E/O boundary and the first one-to-one correlation of direct physical evidence of astronomically-controlled glaciation from the Antarctic margin to the highly-resolved, orbitally tuned $\delta^{18}O$ record of paleoclimate changes across the E/O boundary climate from the deep sea (Coxall and Wilson, 2011; Coxall et al., 2005) (Fig. 7.6).

Extension of the astrochronological interpretation to the upper part of the drillhole (Galeotti et al., 2016) further constrained the age of the first marine calving ice at 32.8 Ma, coinciding with the Oi1a glacial episode of Miller et al. (1991) (Fig. 7.6).

7.3.2.2 The Prydz Bay Region

Drilling in Prydz Bay was undertaken by two ODP legs (Fig. 7.7). In contrast to the western Ross Sea cores, core-recovery rates here were much

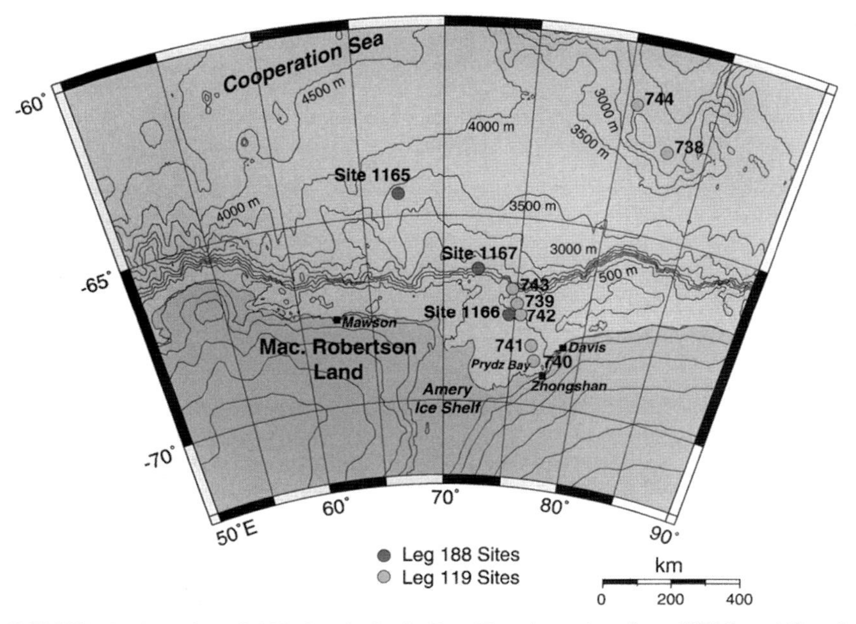

FIGURE 7.7 Location of drill sites in Prydz Bay, East Antarctica, from ODP legs 119 and 188. Continental shelf sites 742 and 1166 include strata that cross the E-O transition. *From Shipboard Scientific Party (2001a). Reproduced with permission of the Ocean Drilling Program, College Station, Texas.*

less satisfactory; hence, interpreting depositional processes is more questionable. Nevertheless, plausible paleoenvironmental scenarios have been derived, albeit lacking precise constraints owing to core loss. Prydz Bay represents the continuation of the Lambert Graben, which contains the Lambert Glacier–Amery Ice Shelf System, an ice drainage basin covering approximately 1 M km^2, draining 13% of the EAIS by surface area. Thus, the record in Prydz Bay provides a signal of the interior of the ice sheet since its inception and complements the Oligocene to Pliocene uplifted glaciomarine record in the Prince Charles Mountains (see Haywood et al., 2008). Prydz Bay itself is dominated by a trough-mouth fan that prograded during phases of glacier advance to the shelf break. Like the western Ross Sea, large data sets are available covering all aspects of core analysis from ODP Legs 119 and 188 (Barron et al., 1991; Cooper and O'Brien, 2004; Cooper et al., 2004; O'Brien et al., 2001) and a convenient summary has been provided by Whitehead et al. (2006) and McKay et al. (2021).

ODP Leg 119 drilled at two sites, 739 (480 mbsf) and 742 (316 mbsf), the lower parts of both were loosely dated shipboard as middle Eocene to early Oligocene. The dominant facies recovered was massive diamictite, with minor stratified diamictite and sand (Hambrey et al., 1991, 1992). Poorly consolidated fine-grained facies may well have been washed away during the drilling, since core-recovery rates were less than 50%. A few broken shell fragments are present, but there was a dearth of material suitable for precise dating. The base of Site 742 is represented by a zone of soft-sediment deformation. The Oligocene succession forms part of a prograding unit as defined in seismic profiles but is truncated by a regional unconformity. Above lies a flat-lying sequence of more diamictite, some with preferred clast orientation and overcompaction, of late Miocene to Pliocene age (Cooper et al., 1991).

The interpretation of the Leg 119 facies is as follows: the deformed bed at the base of Site 742 may represent the first stages of glaciation, with the ancestral Lambert Glacier extending across the continental shelf for the first time. Then the bulk of the recovered facies in Sites 739 and 742 (diamictite) records deposition from the grounding-line of a tidewater glacier margin, by debris rain-out and submarine sediment gravity flow beyond the shelf break, conditions which characterise much of early Oligocene time.

Leg 188 drilled Site 1166 on the continental shelf near Sites 739 and 742 in order to obtain a more complete record of the EOT, but again core recovery was poor (19%). From the base upwards, late Eocene matrix-supported sand was followed by a transgressive surface and the late Eocene to early Oligocene graded sand and diatom-bearing claystone with dispersed granules. These facies are capped by an unconformity and 'clast-rich clayey sandy silt' (diamicton/ite) of Neogene age (Shipboard Scientific Party, 2001a,b,c,d).

Site 1166 begins at the base with late Eocene fluvio-deltaic sands, which are inferred to be pre-glacial. The overlying late Eocene to early Oligocene sand and claystone represent shallow marine and open marine conditions, respectively, but in a proglacial setting as indicated by ice-rafted granule-sized material. The Neogene strata that lie unconformably above represent full glacial conditions.

A revised biostratigraphic age model based on dinocysts, calcareous nannofossils and diatoms (Houben et al., 2013) allowed the EOT to be identified within Site 739. The first common occurrence of the calcareous nannofossil *Reticulofenestra daviesii* in Core 739C-38R at ~300 mbsf (Wei and Thierstein, 1991) likely correlates with the onset of an early Oligocene acme documented for this species, which has been dated to 33.7 Ma (Persico et al., 2012). At nearby Site 744, the onset of this acme is calibrated to the base of subchron C13n in the earliest Oligocene (~33.7 Ma) (Fioroni et al., 2012). The first occurrence of the dinoflagellate *Malvinia escutiana* in Core 739C-39R at 310.73 mbsf is correlated to the Oi-1 excursion (Houben et al., 2011). These datums constrain the age for the upper 310 m of strata drilled in Hole 739 C to early Oligocene and younger. A late Eocene dinoflagellate assemblage, which includes *Deflandrea* sp. A sensu Brinkhuis et al. (2003) and *Vozzhennikovia* sp., is present from the base of the section up to Core 739C-41R

(~330 mbsf) (Houben et al., 2013; Truswell, 1997). At ODP Site 1172 on the East Tasman Plateau *Vozzhennikovia* sp. has a last occurrence in Chron C13r (Brinkhuis et al., 2003; Houben, 2019). Houben et al. (2013) documented the onset of typical sea ice—associated dinocyst assemblages at the Oi-1 equivalent stratigraphic level at Site 739.

The revised stratigraphies of ODP Sites 739, 742 and 1166 were put into context of the seismostratigraphy in Passchier et al. (2016), and geochemical weathering data provide a well-resolved near-field record of Antarctic continental ice growth, temperature and weathering changes from Prydz Bay. Variations in CIA and the S-index suggest the presence of ephemeral mountain glaciers on East Antarctica during the late Eocene between 35.9 and 34.4 Ma. The onset of diamict deposition, the prograding clinoforms in seismic data, the declining values of the CIA, and enhanced erosion and glacial weathering rates (Scher et al., 2011; Tochilin et al., 2012) all suggest that high-latitude climate deterioration and ice growth in the hinterland of Prydz Bay intensified 0.5 Myr prior to the EOIS with major ice growth coincident with Eocene-Oligocene precursor glaciation. Glacial microtextures are present on sand grains at ~240 mbsf in Core 1166A-26R (Strand et al., 2003), which date the onset of some glaciation to ~35.9 Ma. However, in contrast to the glaciation episodes associated with the EOIS, evidence from Holes 742A and 1166A points to ephemeral, partial, glaciation prior to EOT. A stepwise climate cooling of the hinterland is documented at Prydz Bay starting from 34.4 Ma as the ice sheet advanced towards the edges of the continent during the Eocene-Oligocene precursor interval (Passchier et al., 2016). Different than records in the Ross Sea, sedimentary archives from Prydz Bay provide evidence for full glacial development at the EOT when the ice sheet reached the continental margin.

7.3.2.3 Weddell Sea

ODP Leg 113 recovered the EOB at Sites 689 on the Maud Rise and 696 on the South Orkney Microcontinent (SOM) (Fig. 7.8). Maud Rise lies 100 km south of the present-day Polar Front, 100 km north of the Antarctic Divergence and ~700 km from the Antarctic continent (Barker et al., 1988, 2007). Although at a more distal position, the open-ocean pelagic sedimentary succession that includes the EOB cored at ODP Site 689 provides evidence for iceberg calving from both West and East Antarctica from ca. 36.5 Ma (Carter et al., 2017; Ehrmann and Mackensen, 1992).

Site 696 was drilled at 650 m water depth in the southeastern margin of the SOM, northern Weddell Sea (Barker et al., 1988). A revised age model suggests that the EOT was well recovered at this site (Houben, 2019; Houben et al., 2013). The late Eocene section is characterised by a change from organic-rich sandy mudstone facies to glaucony-bearing packstone facies (Barker et al., 1988). Mature glauconitic sediments formed in open

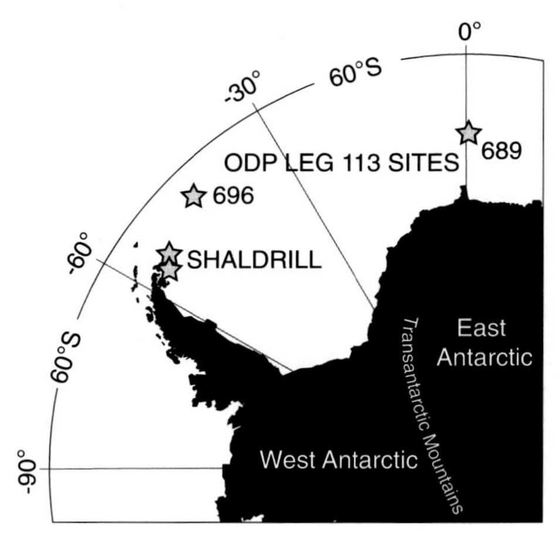

FIGURE 7.8 Location of drill sites in the Weddell Sea, East Antarctica, from ODP legs 113.

marine, shelf-slope transition under sub-oxic conditions are interpreted to record a transgressive condensed sequence deposited at the SOM prior to the EOT (López-Quirós et al., 2019). A change in environmental conditions is also recorded by a relatively diverse dinocyst Eocene assemblage, interpreted to result from temperate or warmer conditions in the northern Antarctic Peninsula area, followed by cooling shown by a decrease in diversity (Mohr, 1990). Crucial was the finding of dinocysts closely resembling the modern species *Selenopemphix antarctica* at ODP Site 696 and other circum-Antarctic sites corresponding closely to the Oi-1 event (Houben et al., 2013). *S. antarctica* is the dominant dinocyst in the modern seasonal Southern Ocean sea-ice zone and derives from a heterotrophic dinoflagellate that primarily feeds on diatoms (e.g., Houben et al., 2013). Its first occurrence and moderate abundance immediately following Antarctic glacial expansion was therefore interpreted to reflect the inception of seasonal sea ice (Houben et al., 2013). Following modelling simulations, this is likely associated with glacial expansion to the coastline (DeConto et al., 2007), consistent with time-equivalent ice rafted debris (Salamy and Zachos, 1999).

Evidence for cool but not polar conditions in the late Eocene are derived from the analysis of marine and terrestrial taxa from the SHALDRIL site 3 Hole C off James Ross Island, which also provides evidence for increased physical weathering towards the end of the Eocene (Anderson et al., 2011). In line with observations at Ste 689, the presence of algae species associated with modern-day Arctic sea ice at the SHALDRIL 3-C site provides evidence for further cooling across the EOT (Anderson et al., 2011).

Seismic reflection profiles in the Weddell Sea region, constrained by sediment ages (ODP Sites 693, 694 and 697; and SHALDRIL 3C), and magnetic anomalies indicate that after the EOIS there is an approximate doubling in sedimentation rates suggesting increased erosion, but not enough to suggest the West Antarctic Ice Sheet (WAIS) was expanded to modern proportions (Huang et al., 2014).

7.3.2.4 Wilkes Land

The Integrated Ocean Drilling Program (IODP) Expedition 318 drilled the eastern segment of the Wilkes Land margin and recovered sediments from the early-middle Eocene and Oligocene at Site U1356 located on the lower continental rise, and earliest Oligocene sediments at Site U1360 located on the continental shelf (Escutia et al., 2011, Tauxe et al., 2012) (Fig. 7.9).

Today, Site U1356 lies offshore the Wilkes Subglacial Basin (WSB), where the EAIS is marine-based (i.e., grounded on a bed that is below sea level), and is close to the southern boundary of the ACC, near the Antarctic Divergence at 63°S (Bindoff et al., 2000p; Orsi et al., 1995). During the Eocene and early Oligocene, topographic reconstructions show the WSB to be mostly emerged and occupied by low-lying lands (Paxman et al., 2018, 2019; Wilson and Luyendyk, 2009). Reconstructions of the position of the Antarctic Divergence during the early Oligocene, based on a bed the distribution of terrigenous and biogenic (calcareous and siliceous microfossils) sedimentation, Nd isotopes and Al/Ti ratios through a core transect across the Australian–Antarctic basin in the Southern Ocean, place it north of Site U1356 at around 60°S, (Scher and Martin, 2008; Scher et al., 2015). Paleogeographic reconstructions place Site U1356 during the early and middle Eocene to the south of the Antarctic Divergence at paleolatitudes of 70 and 65°S, respectively (Bijl et al., 2013; Pross et al., 2012).

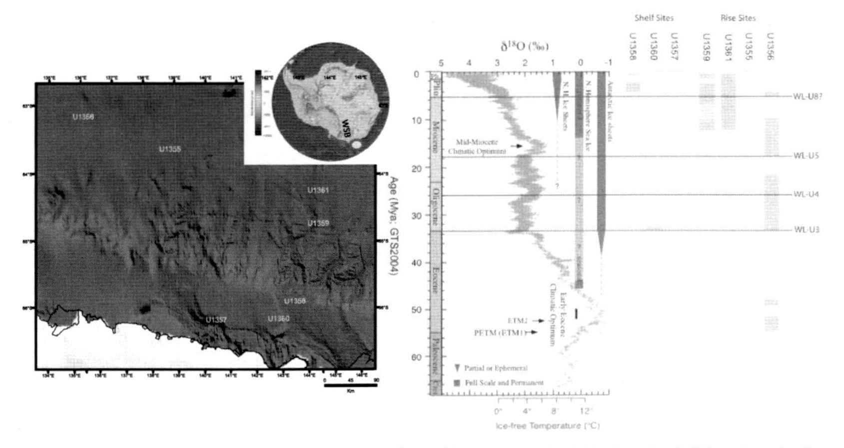

FIGURE 7.9 Location map and chronostratigraphic extent of drill sites in Wilkes Land, from Integrated Ocean Drilling Program (IODP) Expedition 318. *Adapted from Escutia and Brinkhuis (2014).*

Pollen and spores preserved in early Eocene sediments (~55 Ma) reveal that climates along the Wilkes Land coast during peak greenhouse conditions supported the growth of highly diverse, near-tropical (including palms and Bombacoideae) and temperate forests (Contreras et al., 2013; Pross et al., 2012). Using the nearest living relative approach on pollen and spore assemblages to reconstruct paleotemperatures, the paratropical rainforest biome yields winters that were frost-free and mild, with cold monthly mean winter temperatures (MWT) above 10°C, despite polar darkness during winter. Most samples indicated MATs of 16°C ± 5°C with mean summer temperatures (MST) of 21°C ± 5°C and MWT of 11°C ± 5°C. The temperate rainforest biome yields cooler climates with values for MAT and MWT of 9°C ± 3°C and 5°C ± 2°C, respectively, and MST between 14°C ± 1°C and 18°C ± 3°C (Pross et al., 2012). For both biomes, the mean annual precipitation was more than 100 cm/yr. The temperatures indicated by the palynomorphs are similar to those obtained with independent organic geochemical paleotemperature proxies [i.e., branched tetraether lipids, TEX86 and methylation of branched tetraether (MBT)−cyclisation of branched tetraether (CBT) ratios]. MBT/CBT data yield soil temperatures of 24°C−27°C similar to the MSTs derived for the paratropical forest biome, suggesting a common source.

Paratropical conditions persisted in the lowlands of this segment of the Wilkes Land margin at least until 53.9−51.9 Myr ago (Contreras et al., 2013; Pross et al., 2012). The palynological content of the middle Eocene section (49−47 Ma), above a 2 m.y. hiatus, reflects 2°C−3°C cooler temperatures for both winter and summer. MBT/CBT data also record cooling in the middle Eocene with soil temperatures of 17°C−20°C (Pross et al., 2012). Pollen and spore assemblages lacking the megathermal flora were dominated by *Nothofagus*.

In addition to the insights into continental climates, the Eocene succession from Site U1356 allows assessment of the biogeographic patterns resulting from throughflow of surface waters as the Tasmanian Gateway opened (Bijl et al., 2013). Dinocyst analyses from Site U1356, compared to ODP Leg 189 sites around Tasmania, the Australian Bight (ODP Leg 182) and sections in the southeast Australian Margin suggest the earliest throughflow of South Pacific Antarctic waters through the Tasmanian Gateway to be coeval with the shift in rifting direction from SE−NW to S−N (Cande and Stock, 2004) and the onset of a gradual deepening of the South Tasman Rise (Hill and Exon, 2004). Moreover, the onset of throughflow coincides with the earliest signs of cooling following the Early Eocene Climatic Optimum (EECO) (Bijl et al., 2013; Pross et al., 2012). The tectonic opening of the Tasmanian Gateway provides a plausible explanation of southern high latitude cooling following the EECO in the absence of significant equatorial cooling in the middle Eocene (Bijl et al., 2013, 2009). However, more recent work has indicated that tropical cooling did occur in correspondence with

Southern Ocean cooling, pointing to greenhouse gas forcing as the main cause (Cramwinckel et al., 2018).

At Site U1356, the upper middle Eocene to the late Eocene is conspicuously missing in a ~13 m.y. hiatus (~47–33.6 Ma) associated with the unconformity WL-U3 (Escutia et al., 2011; Tauxe et al., 2012). Sediments directly overlying WL-U3 at Site U1356 are dated earliest Oligocene (33.6 Ma) based on the presence of the dinocyst species *Malvinia escutiana* (Houben et al., 2011), the calcareous nannofossil assemblage suggesting an age older than 31.5 Ma (Escutia et al., 2011; Tauxe et al., 2012) and a normal polarity interval interpreted to correspond to Chron C13n (Tauxe et al., 2012). Across the hiatus, environments changed from non-glacial shallow-water depositional environments below to glaciomarine deep-water settings above WL-U3 (Escutia et al., 2011).

These findings imply significant crustal stretching, subsidence of the outer margin and deepening of the Tasman Rise and the Adélie Rift Block some time between 47 and 33.6 Ma (Escutia et al., 2011). Indeed, final loss of a continental connection between Tasmania and Antarctica occurred around 34 Ma, between the South Tasman Rise and Adélie Land (Cande and Stock, 2004). Despite ongoing subsidence, it has been interpreted that the erosive nature of unconformity WL-U3 is related to the early stages of EAIS development. The impact of ice sheet growth, including crustal and sea level response is proposed as the principal mechanism that formed unconformity WL-U3 (Escutia et al., 2011; Stocchi et al., 2013). Microfossils, sedimentology and geochemistry of the early Oligocene sediments immediately above unconformity WL-U3 unequivocally reflect ice-house environments with evidence of iceberg activity (dropstones) and seasonal sea ice cover (Escutia et al., 2011; Houben et al., 2013). In addition, clay mineralogy suggests a shift from chemical weathering to physical weathering, suggestive of much colder/arid weathering regimes (Escutia et al., 2011; Passchier et al., 2013).

In addition to the ice-distal record, the stratigraphic marker for the Oi-1, *M. escutiana* allowed an earliest Oligocene age to be assigned to lowermost sediments recovered from Wilkes Land continental shelf Site U1360 (Escutia et al., 2011; Houben et al., 2013). Glacial diamicton characterises earliest Oligocene sedimentation at Site U1360, indicating an ice proximal to ice distal environment (Escutia et al., 2011). At Site U1360, however, earliest Oligocene glacimarine sediments lie at around 90 m above unconformity WL-U3. This suggests that progressive tectonic subsidence, the large accommodation space created by erosion in the margin (300–600 m of missing strata on the shelf; Eittreim et al., 1995), and partial eustatic recovery (Stocchi et al., 2013) allowed Eocene sediments, in addition to those from the early Oligocene, to accumulate above unconformity WL-U3 on the continental shelf while a hiatus formed at the distal U1356 (Escutia and Brinkhuis, 2014). The continuous presence of reworked middle-late

Eocene dinocyst species in Oligocene sediments from Site U1356 supports unabated submarine erosion of the Antarctic shelf late Eocene strata (Houben et al., 2013).

Early Oligocene dinoflagellate cyst records from U1356 and U1360, combined with those from other locations across the Antarctic Margin, suggest a major restructuring of the Southern Ocean plankton ecosystem, which occurred abruptly and concomitant with the first major Antarctic glaciation in the earliest Oligocene (Houben et al., 2013; Salamy and Zachos, 1999). An abrupt regional increase in siliceous sedimentation at Southern Ocean sites across the EOT glaciation of Antarctica indicates the initiation of seasonal blooms of phytoplankton in circum-Antarctic seas around this time, analogous to modern ecosystems (Houben et al., 2013). The proposed scenario involves an abrupt shift to high seasonal primary productivity associated with the development of seasonal sea ice. It provides the most parsimonious explanation for the abundant appearance of protoperidiniacean dinocysts at EOIS times (Houben et al., 2013). This is in agreement with numerical climate models simulations indicating that sea-ice formation along Antarctic margins may have followed full-scale Antarctic glaciation (DeConto et al., 2007). However, sea ice–related dinocyst species, *Selenopemphix antarctica*, occurs only for the first 1.5 Myr of the early Oligocene, following the onset of full continental glaciation on Antarctica, and after the mid-Miocene Transition (Bijl et al., 2018). For the remainder of the Oligocene and Miocene, less extensive sea ice season is suggested by dinocyst assemblages generally bearing strong similarity to present-day open-ocean, high-nutrient settings north of the sea-ice edge, with episodic dominance of temperate species similar to those found in the present-day subtropical front (Bijl et al., 2018). This agrees with repetitive incursions of north component waters to Site U1356, bathed today by cold Antarctic Bottom Waters (AABW) (Salabarnada et al., 2018), and with temperate surface waters that prevailed over the site notably during interglacial times (Hartman et al., 2018). All evidence points to the existence of a weaker-than present ACC (Bijl et al., 2018, Hartman et al., 2018; Salabarnada et al., 2018). This interpretation is in line with numerical ocean modelling (Herold et al., 2012; Hill et al., 2013) and argues against the formation of a vigorous ACC at 30 Ma as inferred by Scher et al. (2015).

7.4 Summary of climate signals from Antarctic sedimentary archives

7.4.1 Longer-term changes

For tens of millions of years preceding the EOT global climate remained in a greenhouse state, characterised by atmospheric CO_2 levels up to 2000 ppmv and global average temperature much higher than present day (e.g., Anagnostou et al., 2020; Foster et al., 2017). Under these conditions, the Antarctic continent remained largely ice free, in spite of its polar position (e.g., Francis and Poole, 2002; Klages et al., 2020; Pross et al., 2012). During the early Eocene when mean global temperature was as high as 29°C (Cramwinckel et al., 2018), near-tropical forests could develop over the Antarctic continent (Contreras et al., 2013; Pross et al., 2012;). Organic biomarker paleotemperature reconstructions also reveal extremely warm Southern Ocean SSTs of 28°C (calibration error ± 4.0°C) based on TEX_{86}, in the Wilkes Land margin at that time (Bijl et al., 2013). While the temperature estimates remain uncertain, both lines of evidence indicate remarkably warm climate conditions for the highest southern latitudes during peak Cenozoic greenhouse conditions, although the Wilkes Land region was likely the warmest Antarctic sector due to its northern paleolatitude (van Hinsbergen et al., 2015) and oceanographic setting (Bijl et al., 2013).

Mean global temperatures remained high through the early middle Eocene (26°C) and late middle Eocene (23°C) (Cramwinckel et al., 2018). Importantly, Eocene climates were characterised by a reduced equator-to-pole temperature gradient compared to present day (Baatsen et al., 2018, 2020; Bijl et al., 2009; Huber and Caballero, 2011; Greenwood and Wing, 1995). This characteristic of the global climate system has proven difficult to adequately simulate with climate models (e.g., Caballero and Huber, 2013; Cramwinckel et al., 2018; Huber and Caballero, 2011; Klages et al., 2020; Lunt et al., 2012; Spicer et al., 2008), although recent modelling efforts come to closer agreement (Baatsen et al., 2018, 2020).

Significant information on the evolution of climate on the Antarctica continent have been derived from fossil plants and palynomorphs, mineralogy and geochemistry through the analysis of outcrops, marine sedimentary cores and glacial erratics. Paleogene fossil plants are indicative of a high southern latitude flora of variable diversity but dominated by fossils comparable to modern Nothofagus and conifer trees. The fossil plant record suggests that during the late Palaeocene to early Eocene moist, cool temperate rainforests existed in Antarctica, similar to modern low to mid-latitude Valdivian rainforests in southern Chile. These forests were dominated by Nothofagus and conifer trees, with ferns, horsetails and some less-prominent angiosperm groups. TEX_{86} temperature proxies show cooling of around 4°C in the Wilkes Land margin during the middle Eocene. Multiproxy data from Seymour Island provide well-constrained evidence for annual SST of 10°C–17°C during the middle and late Eocene with distinctly higher temperatures recorded between ca. 45 and 41 Ma (Douglas et al., 2014). The difference was originally attributed to zonal heterogeneity (Douglas et al., 2014), but the application of the proper paleomagnetic reference frame in tectonic reconstructions (Van Hinsbergen et al., 2015) implies a much higher latitude for the Antarctic Peninsula relative to East Antarctica at that time.

From the early middle to late Eocene (38–35 Ma), global temperatures dropped to as low as 19°C (Cramwinckel et al., 2018). Evidence for limited glaciation during this time interval comes from different areas around Antarctica. Gulick et al. (2017) found evidence that marine-terminating glaciers existed in the Wilkes Land area at some point during the early to middle Eocene, the precise age of which is poorly constrained. This was based on ice rafted debris found in piston cores and seismostratigraphic interpretations from offshore of the Sabrina Coast in Wilkes Land. There is additional information from airborne radar data of a preserved alpine landscape in the Gamburtsev Mountains, suggesting mountain glaciers existed in central Antarctica prior to the EOT (Bo et al., 2009; Rose et al., 2013). It is possible that ice located in the Gamburtsevs during the Eocene fed glaciers terminating at the Sabrina Coast (Gulick et al., 2017) supported by high precipitation (Baatsen et al., 2018, 2020). Further evidence for partial glaciation prior to the EOT is seen in records from the South Orkney Islands, which show the delivery of distal ice rafted sediments 2.5 Myr prior to this chronohorizon. The sediment provenance signature supports rafting of sediments from glaciers terminating at the coast in the Weddell Sea sector (Carter et al., 2017).

Cooling during the late Eocene is apparent at several sites, including the Weddell Sea based on terrestrial (Carter et al., 2017) and marine (Houben, 2019) records, Prydz Bay (Passchier et al., 2017), Maud Rise SST records (Petersen and Schrag, 2015), and the vegetation records from South Australia and the South Atlantic (Pound and Salzmann, 2017). Evidence of pronounced cooling and ice preceding the EOT have been found in terrestrial records from the Weddell Sea (Carter et al., 2017; Ehrmann and Mackensen, 1992), Maud Rise SST records (Petersen and Schrag, 2015), Seymour Island (Anderson et al., 2011), and vegetation from the most distal site in South Australia and the South Atlantic (Pound and Salzmann, 2017). The results indicate that from about 35.7 million years ago onward, enhanced surface ocean circulation led to sediment winnowing, higher biological productivity and cooling of surface waters around Antarctica (Houben, 2019). Similarly, organic biomarker paleothermometry and dinocyst distribution from different sites located in the southwest Atlantic (DSDP Site 511), Weddell Sea (ODP Site 696) and the south margin of Australia (ODP Sites 1128, 1168 and 1172) indicate a major oceanographic invigoration including the onset of an Antarctic Countercurrent, starting at 35.7 Ma (Houben, 2019). At localities affected by the Antarctic Countercurrent, sea surface productivity increased and simultaneously circum-Antarctic surface waters cooled, which might have preconditioned the Antarctic continent for glaciation. Significant regional differences in terms of absolute temperature values might reflect a large-scale zonal SST gradient between the South Atlantic and the southwest Pacific sectors throughout the middle-late Eocene, which has been interpreted to reflect water mass organisation (Douglas et al., 2014; Houben et al., 2013).

Sedimentary archives from various Antarctic localities, hence, provide a consistent but broad picture of climate evolution that is in line with the Cenozoic climate history derived from various proxies in lower latitude settings. Although there is evidence for glaciation and for marine-terminating glaciers, much uncertainty remains about the nature and geographical extent of ice on Antarctica before the EOT.

7.4.2 The climate of the Eocene-Oligocene transition

While a consistent picture on the long-term environmental and climatic evolution during the middle-late Eocene and the Oligocene emerges from a wealth of data at different localities on the Antarctic margin, detailed insights into the timing and paleoenvironmental implications of ice growth close to the locus of ice formation on Antarctica are sparse. A comprehensive summary of temperature proxy records across the EOT in Antarctic and peri-Antarctic settings has been recently compiled by Kennedy-Asser et al. (2020) including 14 sites (10 marine and four terrestrial) ranging in paleolatitude from 53 to 77°S (Fig. 7.10). Comparison of late Eocene and early Oligocene temperatures provide a coherent picture of climate deterioration over the Antarctic margin with a mean temperature drop of around 3.4°C (Kennedy-Asser et al., 2020) (Fig. 7.11).

However, only a few Antarctic sites provide a continuous record across the EOT, which is crucial to determine the tempo and mode of the evolution of the nascent ice sheet in comparison with data from far-field proxies. These include the CRP-3 (Florindo et al., 2005; Galeotti et al., 2012; Passchier et al., 2013) and, possibly, CIROS-1 (Roberts et al., 2003) drillholes in the Ross Sea, ODP Site 696 in the Weddell Sea (Carter et al., 2017; Ehrmann and Mackensen, 1992) and ODP Site 739 in the Prydz Bay (O'Brien et al., 2001; Passchier et al., 2013, 2017). In addition, sediments recovered from the eastern Wilkes Land margin at Site U1360 attest to ice sheet advance to the coast/continental shelf by the earliest Oligocene (33.6 Ma; Escutia and Brinkhuis, 2014; Escutia et al., 2011). All these records provide evidence for a major glacial expansion across the EOT. However, there is no unequivocal evidence for a fully developed ice sheet at all sites.

At Prydz Bay three ODP drill holes were drilled at Sites 739, 742 and 1166 (O'Brien et al., 2001). The position of the EOT has been constrained by dinoflagellate cyst biostratigraphy in Hole 739C (Houben et al., 2013), which, using seismostratigraphy, allowed the age model for the three sites to be refined (Passchier et al., 2017) with a strong age control. Analysis of the Chemical Index of Alteration and the S-index (Sheldon et al., 2002) allows reconstruction of near-field Antarctic continental ice growth, temperature and weathering changes during the EOT on the East Antarctic margin. These geochemical data suggest the presence of ephemeral

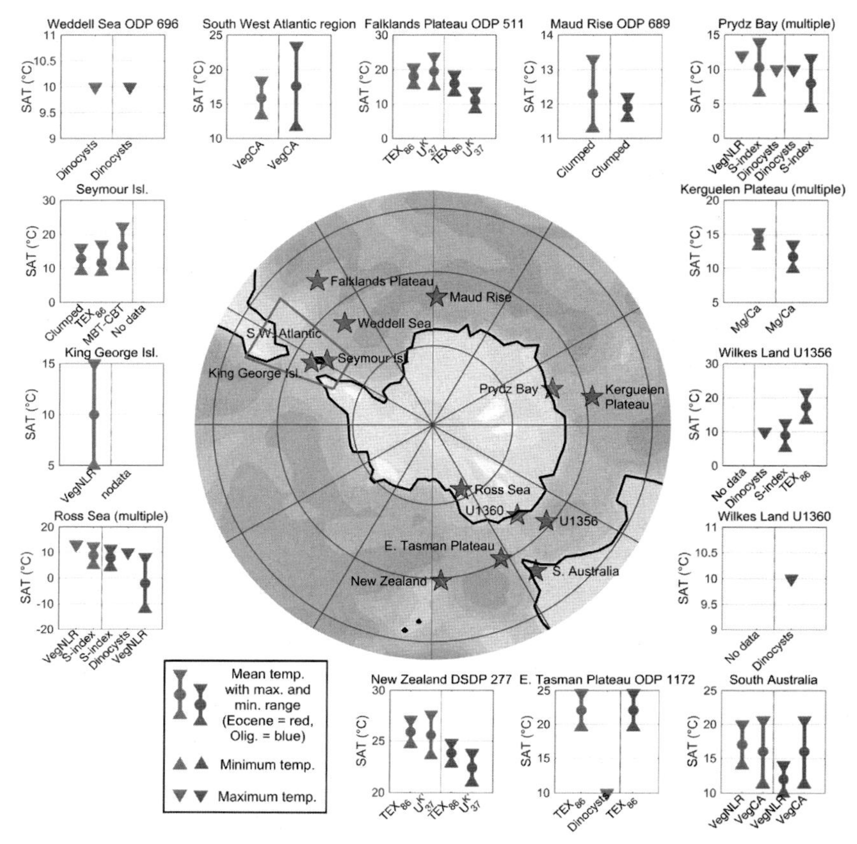

FIGURE 7.10 Mean annual temperature (°C) from proxy records for all sites during the late Eocene and early Oligocene. The mean values (circles) are shown with maximum and minimum values (error bars), while ordinal limits are shown by upward-pointing (greater than) or downward-pointing (less than) triangles. Late Eocene records are in red and early Oligocene records in blue. *From Kennedy-Asser et al. (2020).*

mountain glaciers on East Antarctica during the late Eocene between 35.9 and 34.4 Ma. High latitude warming events are recognised at ~35.8 and 34.8 Ma. A stepwise climate cooling of the Antarctic hinterland occurred from 34.4 Ma as the ice sheet advanced towards the edges of the continent slightly before the EOB. A temperature decline of >3°C is captured by S-index data between 34.4 Ma and the Oi-1 when, finally, the ice sheet extended to the outer shelf.

A different history comes from other sites where the EOT did not lead to fully glaciated conditions. Seismic reflection profiles show a relatively limited extent of glacial deposits in the southern Weddell Sea, which is indicative of a WAIS less expanded than present day (Huang et al., 2014).

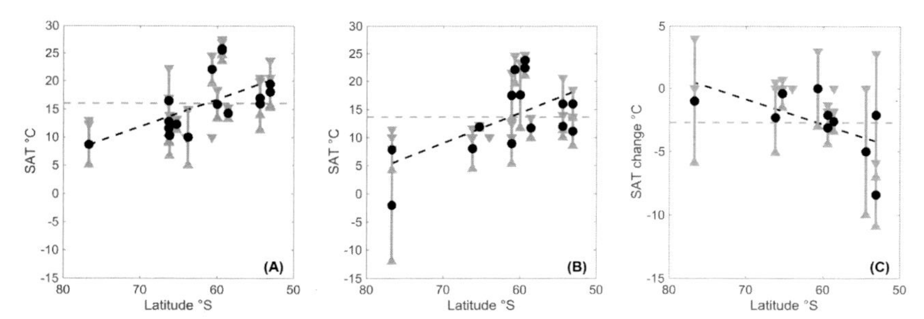

FIGURE 7.11 Latitudinal profiles of (A) late Eocene absolute temperature, (B) early Oligocene absolute temperature and (C) EOT temperature change from proxy records. The mean values are plotted in grey dotted lines and latitudinal gradients (calculated using ordinary least squares) in black dotted lines. Circles show mean values, uncertainty ranges and maximum/minimum limits are shown by the bars and triangles. *Adapted from Kennedy-Asser et al. (2020).*

Fully glaciated conditions were reached only during the Miocene in this area, according to Huang et al. (2014). In line with this observation, relatively diverse vegetation biomes persisted around most of the coastal regions of the continent well in into the early Oligocene (Francis et al., 2008). Yet, significant ice must have been present across Antarctica in the late Eocene to enable it to calve icebergs along the coastlines of the southern Weddell Sea by ca. 36.5 Ma or slightly earlier (Scher et al., 2014), as also revealed by the occurrence of iceberg-rafted sediment grains with both East and West Antarctic provenance at Weddell Sea ODP Site 696 (Barker et al., 1988; Carter et al., 2017).

The orbitally-resolved record of the shallow-water glaciomarine sedimentary succession recovered with the CRP-3 drillhole offers a detailed history of glaciation in the Ross Sea (Galeotti et al., 2012, 2016). The well-dated CRP-3 drill core suggests the development of a stable continental-scale ice sheet calving at the coastline only at 32.8 Ma (Galeotti et al., 2016). There is evidence for orbital pacing of glacial advance and retreat cycles between 34 and 31 Ma, indicating that the nascent Antarctic ice responded to local insolation forcing. The stabilisation of a continental scale ice sheet at 32.8 Ma appears to have been related to the crossing of a CO_2 threshold, although the precise CO_2 level triggering ice expansion into the marine realm remains uncertain (Anagnostou et al., 2016; Gasson et al., 2014).

In the eastern sector of the Wilkes Land margin, the EOT was not fully recovered. Instead at Site U1356, a ~13 m.y. hiatus separates middle Eocene from earliest Oligocene (33.6 Ma) sediments (Escutia et al., 2011; Tauxe et al., 2012). Coeval early Oligocene sediments were also recovered from continental shelf Site U1360. Both sites provide evidence for the existence of an ice sheet advanced to the continental shelf depositing ice-proximal sediments (diamictons) at Site U1360 and dropstones resulting from calved icebergs at distal Site U1356 (Escutia et al., 2011).

Despite the development of a large AIS at the EOT, the presence of tundra paleo-biome localities on Antarctica demonstrates the survival of vegetation in the early Oligocene on the continent (Francis, 1991; Francis and Poole, 2002; Francis et al., 2004, 2008; Poole et al., 2005; Prebble et al., 2006; Strother et al., 2017). Interpretations based on fossil plants are supported by the analysis of n-alkane from various localities on the Antarctic margin, which shows a shift from n-C29 to n-C27 dominated chain length n-alkane between the late Eocene and the Oligocene (Duncan et al., 2019). Such a change is inferred to result from both a shift in plant community, as well as a response to significant climate cooling.

The evidence for the presence of vegetation in the early Oligocene is at odds with geochemical proxy records from marine sedimentary archives suggesting that the EOT ice sheet had a volume of 60%–130% of the modern day, depending on the oxygen isotopic composition of the ice sheet (Bohaty et al., 2012). On the one hand, sub-aerial continental areas were larger than present day (Paxman et al., 2019; Wilson et al., 2012), thus leaving more land available to sustain vegetation. On the other hand, however, the warm Southern Ocean (Hartman et al., 2018; Houben, 2019) provided a local, isotopically heavier than previously assumed, source of precipitation entailing a larger ice sheet to explain the geochemical record. Fossil records indicating the survival of vegetation in the early Oligocene and geochemical proxy records from marine sedimentary successions remain, therefore, difficult to reconcile.

7.5 The global context of Earth and climate system changes across the EOT

The E-O transition marks a momentous shift in the state of the Earth system when a long-term decline in global temperatures and atmospheric CO_2 concentration that began in the early Eocene at ~50 Ma (Anagnostou et al., 2020) culminated in the rapid expansion of a permanent, albeit dynamic, ice sheet over the Antarctic continent at ~34 Ma (DeConto and Pollard, 2003; Katz et al., 2008; Lear et al., 2008; Zachos et al., 2001). Proposed causes for this fundamental change in Earth's climate state include changes in ocean circulation (and associated ocean heat transport) due to the opening of Southern Ocean gateways (Kennett, 1977), a decrease in atmospheric CO_2 (DeConto and Pollard, 2003) and a minimum in solar insolation (Coxall et al., 2005). Also, there is evidence for a long-term climate transition that lasted up to several millions of years before the inception of an AIS (e.g., Bohaty et al., 2012; Coxall and Pearson, 2007; Coxall et al., 2005, 2018; Scher et al., 2011). Associated changes in the global ecosystem were influenced by ocean circulation and overturning (e.g., Coxall et al., 2018; Katz et al., 2011), ocean biogeochemistry (e.g., Lear et al., 2008; Pälike et al., 2012), global temperatures (e.g., Liu et al., 2009), marine biology (e.g., Houben et al., 2013; Villa et al., 2013) and terrestrial fauna (e.g., Costa et al., 2011; Kraatz and Geisler, 2010; Prothero, 1994). Here, we summarise existing model experiments that test the mechanisms proposed to have been driving the EOT.

7.5.1 Climate modelling

The Eocene-Oligocene transition has been the target of climate and ice sheet modelling studies for over two decades. An array of computer models have been used, including ocean-only models (Sauermilch et al., 2021), coupled atmosphere-ocean models (Baatsen et al., 2018, 2020; Goldner et al., 2014; Hutchinson et al., 2019; Kennedy-Asser et al., 2015; Sijp et al., 2011) and coupled climate–ice sheet models (DeConto and Pollard, 2003; Gasson et al., 2014; Ladant et al., 2014), with the common goal of finding a mechanistic explanation for the global cooling that culminated in Antarctica's transition to an ice-covered continent.

The onset of Antarctic glaciation has long been linked with the opening of ocean gateways following the tectonic separation of Antarctica from Australia (the Tasman Gateway) and South America (the Drake Passage) (Kennett, 1977; Livermore et al., 2007; Scher and Martin, 2006; Shackleton and Kennett, 1976). It was thought that the opening and subsequent deepening of these gateways represented the final barrier of the development of strong circumpolar ocean flow, the ACC. It is hypothesised the developing ACC led to the thermal isolation of Antarctica from the lower latitudes, cooling the southern high latitude climate and allowing for the formation of the AIS. Testing this hypothesis has focused on several aspects, e.g., dating of gateway opening and deepening, also by reconstructing the evolution of the oceanic frontal systems (Bijl et al., 2018, Evangelinos et al., 2020; Salabarnada et al., 2018; Sangiorgi et al., 2018), and correlating these to ice sheet inception, and using climate and ice sheet models to determine how significant the opening of these gateways was to the climate of Antarctica.

Meanwhile, ice sheet models have been used to test the alternative explanation for the formation of the AIS, that glaciation was caused by a drop in atmospheric CO_2 across the EOT. A key advantage of modelling is that it allows the individual role of these different cooling and glaciation mechanisms to be examined.

Some of the earliest modelling studies produced conflicting results as to the significance of ocean gateway opening. Ocean modelling showed several degrees of surface ocean cooling around Antarctica with the opening of the Drake and Tasman gateways following a reduction in heat convergence to the Southern Ocean, supporting the thermal isolation hypothesis (Nong et al., 2000; Toggweiler and Bjornsson, 2000). This contrasted with a coupled ocean–atmosphere study showing a much smaller effect on sea surface temperatures around Antarctica with the opening of the Tasman Gateway (Huber et al., 2004). More recent modelling work with high-resolution, eddy-permitting ocean simulations and detailed paleobathymetry demonstrate a dramatic surface water cooling along the Antarctic coast (>4°C) by changing the gateways depth from 300 to 600 m (Sauermilch et al., 2021). In contrast, other recent modelling work has shown that the formation of the AIS had a large effect on ocean circulation, suggesting that some of the observed changes in circulation may have been *feedbacks* from glaciation, rather than a *forcing* from the opening of gateways (Goldner et al., 2014; Kennedy-Asser et al., 2015). Although the relative importance of atmospheric CO_2 vs ocean gateways as the primary control on the climate changes across the EOT continues to be debated, it is clear that both had an impact on regional climate and oceanography.

As the direct impact of Southern Ocean gateways on Antarctic climate continues to be debated, focus has shifted to a wider view of Earth system changes around the EOT. The interconnection between ocean basins and processes has led some studies to assess the impact of low latitude and Northern Hemisphere gateways on Southern Ocean and Antarctic climate (Yang et al., 2014). Changes in these distant gateways have been shown to potentially cause Southern Ocean cooling by enhancing North Atlantic overturning at the expense of Southern Ocean overturning (Yang et al., 2014). However, an alternative theory is that overturning shifted from the North Pacific to the North Atlantic due to salt-advection feedbacks, while Southern Ocean overturning continued across the EOT (Hutchinson et al., 2019; McKinley et al., 2019). Other studies have assessed how ocean gateway changes can have impacts on ocean overturning and carbon uptake (Elsworth et al., 2017; Fyke et al., 2015; Mikolajewicz et al., 1993). Although these investigations are somewhat idealised, they emphasise the importance of ocean gateways as potential long-term contributors to the changes associated with the EOT, even if the direct climatic effect associated with gateway changes is limited. In a recent synthesis of EOT modelling, the reduction in atmospheric CO_2, rather than the opening of ocean gateways or formation of the AIS, showed a better match to global proxy reconstructions from before and after the EOT (Hutchinson et al., 2021).

Notable challenges remain in modelling the high latitude climate before, after and across the EOT. Not least this includes issues regarding the higher absolute polar temperatures that are recorded in the proxy observations compared to many model simulations at reasonable CO_2 levels. Some of the most recent modelling studies appear to have made progress in reducing this mismatch, citing improved cloud and water vapour feedbacks associated with higher resolution (Baatsen et al., 2018, 2020) or enhanced climate sensitivity to CO_2 (Hutchinson et al., 2018). A significant component of this warming is also due to changes in land-surface properties, independent of CO_2 forcing. Lunt et al. (2020) found that when moving from pre-industrial to

Eocene boundary conditions in a multi-model ensemble, global mean temperatures increased by 3°C−5°C, due to changes in palaeogeography, vegetation, aerosols and the removal of ice sheets. These factors are all 'prescribed' in Eocene climate models, whereas in reality they may act as feedbacks to a perturbation such as the Antarctic glaciation.

Another challenge is the need for models simulating oceanographic conditions in high, eddy-permitting resolutions (<0.25°). Most paleo simulations run in low resolution (1−3.75°) using coarse paleobathymetries without prominent seafloor roughness and detailed geometric features (Baatsen et al., 2018; Huber et al., 2004) or modified present-day bathymetry grids (Hill et al., 2013), as boundary conditions. Recently, numerous studies in the field of modern physical oceanography have shown that small-scale features in ocean currents and seafloor topography strongly influence large-scale ocean circulation patterns. Mesoscale eddies (<100 km) are key components for ocean circulation and, in large parts of the ocean, eddies are responsible for the majority of the heat transport (Griffies et al., 2015; Jayne and Marotzke, 2002; Morrison and Hogg, 2012; Newsom et al., 2016; Viebahn et al., 2016), and small-scale seafloor features with slope gradients in the order of 10^{-4} to 10^{-5} (>0.05°) significantly influence subsurface velocities and vertical structure of ocean currents (LaCasce, 2017; LaCasce et al., 2019). Using low-resolution ocean model configurations with simplified bathymetries to reconstruct reorganisations of oceanographic conditions during key climate changes, such as the EOT, carries the potential to severely underestimate ocean heat transport processes, as well as the influence of tectonic processes, on changing ocean circulation.

This has been shown in a recent study using high-resolution (0.25°) ocean model configurations with detailed paleobathymetry (Sauermilch et al., 2021). This model demonstrates that the Southern Ocean's tectonic processes across the EOT (including gateway deepening and paleolatitudinal shifts) play a crucial role in reorganising the ocean circulation patterns and temperature distribution around Antarctica (Hochmuth et al., 2020; Xing et al., 2020; Sauermilch et al., 2021). Changing the gateway depths from only 300 to 600 m causes the Pacific and Indo-Atlantic subpolar gyres to weaken and Antarctic surface waters to cool dramatically by >4°C. These results are consistent with existing paleo-sea surface temperatures from proxy records, as well as plankton biogeographic patterns and Neodymium records constraining surface and bottom currents, respectively (Sauermilch et al., 2021).

In addition to climate modelling studies, ice sheet models coupled in various ways to climate models have also tried to determine what caused the rapid onset of Antarctic glaciation across the EOT. The first Antarctic ice sheet study to systematically test the two leading hypotheses of ocean gateway opening vs CO_2 drawdown was carried out by DeConto and Pollard (2003). They used a slab ocean model to mimic the effect of the thermal isolation of Antarctica and compared this to the impact of declining atmospheric CO_2 and changes in Earth's orbit and obliquity. Their results showed a strong sensitivity to CO_2 drawdown, with a critical threshold of 780 ppm identified for ice sheet formation. This value was in remarkable agreement to proxy reconstructions of atmospheric CO_2 concentrations published a few years later (Pagani et al., 2011; Pearson et al., 2009). Although CO_2 was the predominant control on ice sheet volume, reduced ocean heat transport had a smaller impact, increasing the CO_2 threshold by 140 ppm (DeConto and Pollard, 2003). A data-driven (CO_2 and summer insolation) ice sheet model was able to simulate the ice evolution through the EOT and to reproduce the timing and the magnitude of the two steps evident in oxygen isotope records (Ladant et al., 2014). The latter study suggests a substantial impact of the combination of ice feedbacks (height-mass balance and albedo) and insolation, which gives a large-scale glaciation threshold of about 900 ppm more than 100–150 ppm higher than previously thought by DeConto and Pollard (2003) (Fig. 7.12).

Using a range of different GCMs Gasson et al. (2014) showed a larger spread in the glacial CO_2 threshold between ~560 and 920 ppm, although still clustered around the earlier estimate of DeConto and Pollard (2003). One model (HadCM3L) did not produce any glaciation, even with a much lower concentration of atmospheric CO_2 (280 ppm). Part of the cause of the large spread in results was model-dependent. In particular, the topography over the Antarctic continent differed significantly among the models used (Gasson et al., 2014). The results of a recent model inter-comparison project in which GCMs are setup in an identical way (Lunt et al., 2020) will help shed new light on the nature of the model dependency of glaciation threshold. Other important aspects that are not adequately incorporated in ice sheet models are ice–ocean interactions and GIA effects, incorporation of which will further refine simulations of EOT glaciation in the future (Fig. 7.13).

7.5.2 Relative sea-level change around Antarctica

The onset of the first continent-size AIS across the EOT was accompanied by a regression that left a clear mark in ocean-sediment cores and geologic sections in the Northern Hemisphere (Houben et al., 2012; Kominz and Pekar, 2001; Miller et al., 2008).

These far-field sites haves recorded a 60 ± 20 m sea-level drop that, according to the concept of glacioeustasy, and hence under the assumption that the Earth is rigid and non-gravitating, is consistent with a larger Antarctic continent, thus capable of hosting a large ice sheet (Wilson et al., 2012).

However, near-field sections and cores from East Antarctica, such as Wilkes Land (IODP 318, Escutia et al., 2011), Prydz Bay (ODP, Barron et al., 1991; O'Brien et al., 2001) and Cape Roberts (CRP-3, Fielding et al., 2001; Galeotti et al., 2012), show significant local and regional deviations of

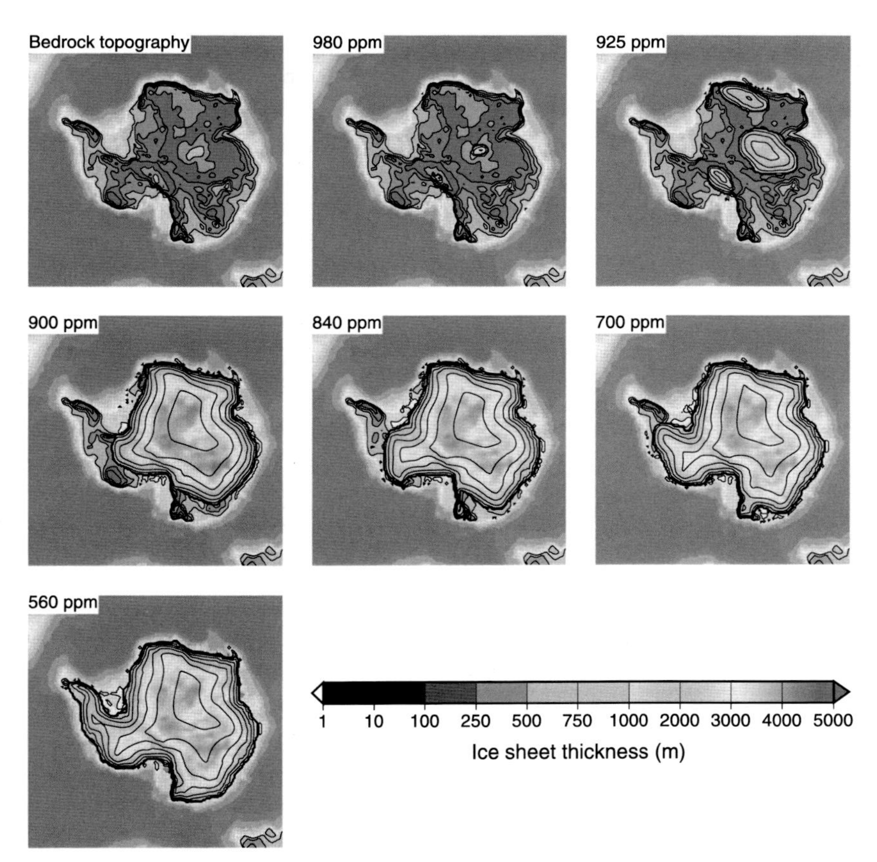

FIGURE 7.12 Numerical simulation of the maximum volume of ice sheet attained with different atmospheric CO_2 concentrations based on a data-driven (CO_2 and summer insolation) ice sheet model. The modelled CO_2 threshold for the onset of a large-scale glaciation is approximately 900–925 ppm. Fully glaciated conditions are reached with atmospheric CO_2 level of approximately 560 ppm. *From Ladant et al. (2014).*

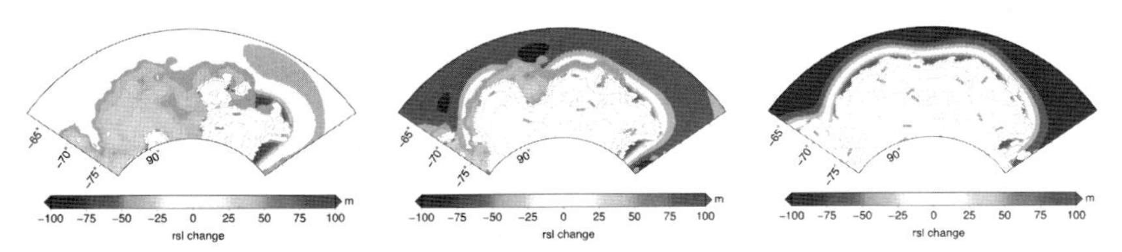

FIGURE 7.13 Seven hundred kyr-long EAIS growth and GIA-driven RSL changes at three model run times, and relative to the pre-glacial state. The ice-sheet extent is shown in white. Left: model run time is 1.5 Myr (since the beginning of the simulation). Centre: model run time is 1.55 Myr, which correlates to the end of the first $\delta^{18}O$ step. Right: model run time is 2.2 Myr.

sea-level change from the Northern Hemisphere trend (Galeotti et al., 2016; Stocchi et al., 2013). Furthermore, the inferred ice-proximal sea-level variations across the EOT vary in space and time, mostly as a function of the distance from the ice sheet margin.

The occurrence of a hiatus in the sediment cores and stratigraphic sections from the offshore sites at Wilkes Land (IODP 318, Escutia et al., 2011) implies that a sea-level drop, which is consistent with the eustatic trend from the NH caused the erosion of late Eocene material as the ice sheet was growing. This is particularly clear because early Oligocene sediments contain reworked late Eocene microfossils (Bijl et al., 2018; Houben et al., 2013), demonstrating profound erosion on the Wilkes Land continental margin, possibly as a consequence of tectonic deformation from the inception of Antarctic glaciation.

At the same time, the occurrence of fining upward Oligocene sediments above the hiatus shows that accommodation space was created on the shelf (Stocchi et al., 2013), which can only be explained with a local sea-level rise. This is obviously at odds with the 60 ± 20 m glacioeustatic sea-level drop as inferred from the Northern Hemisphere geological records.

The change in local sea-level trend in the offshore sites and the continuous increase of bathymetry along the inner margins cannot be explained under the glacioeustatic approximation. By solving the SLE, Stocchi et al. (2013) have shown that when the AIS grows during the EOT, the newly formed excess of Antarctic ice mass causes the solid Earth's surface to subside in order to reestablish an isostatic equilibrium. Because of basin subsidence, the coastal areas experience a progressive relative sea-level rise throughout the EOT.

However, as the ice sheet was growing, the gravitational pull exerted by the ice mass on the surrounding ocean water caused the mean sea surface, which is an equipotential surface of gravity, to rise towards the ice margins. This process, known as self-gravitation, further exacerbated the regional RSL change pattern. It increased the RSL drop over the uplifting forebulges (thus decreasing the column of water) and increased the RSL rise along the coasts, thus facilitating the formation of accommodation space.

The sedimentary cycles observed in the CRP-3 section (Cape Roberts, Ross Sea) are the result of periodic fluctuations in both grounding-line proximity and RSL change (Fielding et al., 2001). The interpretation of the sedimentological evidence implies that glacial maxima and minima locally coincided with times of minimum and maximum RSL, respectively (Fielding et al., 2001). This is in line with the glacioeustatic approximation. However, given the ice marginal position of the CRP-3 site, any proximal ice-thickness variation would have resulted in local RSL change with the opposite sign of eustatic trends and of larger amplitude. Galeotti et al. (2016) suggest that the GIA-induced RSL rise caused by the expansion and grounding of the ice sheet at the CRP-3 site had to be counter-balanced by a strong RSL drop in response to the uplifting forebulge driven by a synchronous thickening of the EAIS. Therefore, the appearance of marine-grounded ice near the CRP-3 site was facilitated by flexural crustal uplift as the EAIS expanded, resulting in a RSL fall (>40 m) in phase with the hypothetical eustatic trend. Accordingly, Galeotti et al. (2016) postpone the timing of maximum, continental-scale ice growth to 32.8 Ma, when CO_2 levels dropped below 600 ppm.

Important outcomes from the numerical modelling of the EOT GIA processes can be summarised as follows:

1. The large RSL rise in the proximity of the Antarctic ice-sheet margins might provide a strong dynamical feedback on ice-sheet stability (Gomez et al., 2010). Accordingly, the near-field processes such as local sea-level change influence the equilibrium state obtained by an ice sheet grounding line. This implies that ice sheets models should be fully coupled to Earth models that capture GIA. It has been shown that self-gravitation operates such as a self-regulating mechanism. As shown by Gomez et al. (2010) and De Boer et al. (2014, 2017), self-gravitation and subsidence cause major local RSL rise at the ice margin, which prevents from further growth over the continental shelf, thus limiting the maximum size that an ice sheet can grow. Obviously, this also works the other way around. When and ice sheet melts, crustal uplift and sea level drop might limit its retreat and actually favour an expansion.
2. The infill of sediments is expected to further alter the local RSL change along the Antarctic margins and should therefore be included in a gravitationally self-consistent manner within the GIA models (Boulton, 1990; Ohneiser et al., 2015; Stocchi et al., 2013; Whitehouse et al., 2019).
3. The uneven local RSL changes might have impacted Southern Ocean surface and deep oceanography by affecting the bathymetry, as shown by Rugenstein et al. (2014). The GIA-induced regional deformations of the sea bottom, although of the order of 50 m, are large enough to affect pressure and density variations, which drive the ocean flow around Antarctica. Throughout the Southern Ocean, frontal patterns are shifted several degrees, velocity changes are regionally more than 100%, and the zonal transport decreases in mean and variability. The model analysis suggests that GIA-induced ocean flow variations alone could impact local nutrient variability, erosion and sedimentation rates, as well as ocean heat transport.

7.6 Summary

The Eocene-Oligocene climatic transition marks the crossing of a threshold in a long-term cooling trend that began following the early Eocene Climatic Optimum ca. 51 Ma. For about 17 Myr global temperatures cooled gradually, although with some more abrupt changes, in concert with a generalised decrease of atmospheric CO_2 levels. A range of data sources from Antarctic drillholes and outcrops outlines a picture of the response to this long-term change and provides crucial information on the environmental conditions that allowed the onset of Antarctica's glaciation as well as its aftermath. The integration of these records with lower latitude data and paleoclimate numerical climate and ice sheet modelling is crucial to test existing hypotheses explaining the transition from a greenhouse to an icehouse climate state.

7.6.1 Early–middle Eocene polar warmth

The early Eocene climate record is associated with high atmospheric CO_2 concentrations and global temperatures much warmer than present day. Evidence from fossil plants, palynology and geochemistry indicates that the early Eocene climate was characterised by warm and wet climates at high latitudes, with the development of tropical forests on the Antarctic margin and temperate forest conditions at high altitudes. The oldest record of some glacial activity, i.e. valley-type tillites, comes from the middle Eocene of King George Island, indicating the presence of alpine glaciers. While the sporadic presence of alpine-type glaciers at high elevation cannot be excluded, floras of middle Eocene age from King George and Seymour islands suggest warm to cool temperate climates, generally moist and probably frost-free on coastal settings. Accordingly, marine isotope records suggest that climates were generally warm until the middle Eocene, although the climate trend was towards cooling.

7.6.2 Late Eocene cooling

Following the middle Eocene climatic optimum at ~40 Ma, stable isotope records capture the continuation of the long-term cooling trend that initiated at ~51 Ma. Marine geochemical proxy records, while uncertain, are indicative of atmospheric CO_2 levels decreasing from ~1000 ppm in the late Eocene to ~700–800 ppm across the EOT.

A generalised cooling trend is evident in geochemical, sedimentary and palaeontological Antarctic records. The period of relative tropicality in southern high latitudes that dominated the early Eocene persisted to some extent into the middle Eocene and ended in the late Eocene. A number of geological data sources suggest that during the early late Eocene, climates cooled but perhaps not to the point of allowing significant ice build-up. The late Eocene sediment record in the Ross Sea region (McMurdo Erratics, magnetic and clay mineral record) and in the Prydz Bay area could be indicative of cold climates, but the coastal/open marine shelf and fluvial-deltaic environments in these two areas, respectively, do not show signs of the presence of persistent ice. Further cooling and a major oceanographic reorganisation with the emplacement of an Antarctic Countercurrent, starting at about 35.7 Ma, is recorded in organic biomarker values. Cooling of surface waters across this time interval is also captured by clumped isotope records.

The late Eocene intensification of the Antarctic Countercurrent and its climatic and environmental feedbacks may have helped set the stage of minor scale, ephemeral Antarctic glaciations prior to the EOT. In the Weddell Sea, there is evidence of iceberg calving from both West and East Antarctica from approximately 36.5 Ma. Evidence for precursor glaciation comes from the Prydz Bay, 0.5 Ma prior to the EOT. A pulse of IRD and a perturbation in the Nd record indicates the presence of a small precursor glaciation at the Kerguelen Plateau. This pattern of transient precursor glaciations leading to major and sudden glacial expansion is in line with the hypothesised role of powerful nonlinear feedbacks in the coupled climate–ice sheet system.

7.6.3 Eocene-Oligocene transition

The widespread occurrence of glaciomarine deposits and/or hiatuses and/or increased physical weathering at a number of sites provide a robust ensemble of evidence for a major glacial expansion over Antarctica across the EOT. Refinement of the biomagnetostratigraphic interpretation (Prydz Bay), astrochronological calibration (Ross Sea) and newly available sedimentary archives from drilling sites (Wilkes Land) allows these Antarctic sedimentary, geochemical and palaeontological records to be correlated with the global record of climate change captured by oxygen isotope values and far-field proxy records in general. The latter indicates a ~500-kyr period characterised by intensified climate instability as the world transitioned from greenhouse to icehouse. The global temperature drop and the build-up of ice across the EOT were followed by a prolonged interval of maximal glacial growth known as the earliest Oligocene Glacial Maximum (EOGM). While near-field and far-field evidence define a picture of major glacial expansion across the EOT, no conclusive agreement on the amount of ice has been reached. A ~70 m sea-level drop together with geochemical evidence suggest that during the EOGM the AIS might have been larger than it is today. While a larger extent of land above sea level could have accommodated a large terrestrial ice sheet, contrasting signals leave the precise size of the EOGM ice sheet in question. While no individual proxy can quantify the amount of ancient Antarctic ice based on near-field sedimentary archives, several lines of independent evidence suggest that glaciation did not reach its maximal extent until after the EOGM. Such data include the persistence of some vegetation, as observed in earliest Oligocene records from marginal settings. Furthermore, the ice sheet did not appear to have reached its maximum seaward extent at some localities.

In summary, the EOT and the associated timing and character of glaciation were much more complex than previously reconstructed from sparse Antarctic evidence and marine oxygen isotope records. A relatively major glacial expansion over East Antarctica did occur across the EOT but it possibly took millions of years for the ice sheet to reach its maximum extent over East Antarctica and parts of West Antarctica. This newly-revised view of Antarctic glacial evolution seems to be more consistent with estimates of ice sheet volume based upon oxygen isotopes (e.g., Bohaty et al., 2012; DeConto et al., 2008; Wilson et al., 2013). Accordingly, the persistence of vegetation at several Antarctic localities provides a clear indication that climate was warmer than present day during the early Oligocene.

Acknowledgements

SG and LL acknowledges funding from the Department of Pure and Applied Sciences. CE acknowledges funding by the Spanish Ministry of Economy, Industry and Competitivity (grants CTM2017-89711-C2-1/2-P), cofunded by the European Union through FEDER funds. IS was supported by the Australian Research Council Discovery Project 180102280. A.T. Kennedy Asser was supported by NERC funding (grant no. NE/L002434/1) Edward Gasson is funded by the Royal Society. EG is funded by the Royal Society. AS thanks the European Research Council for Consolidator Grant #771497 (SPANC).

References

Acosta Hospitaleche, C., 2014. New giant penguin bones from Antarctica: systematic and paleobiological significance. Comptes Rendus Palevol 13 (7), 555–560. Available from: https://doi.org/10.1016/j.crpv.2014.03.008.

Acosta Hospitaleche, C., Reguero, M., 2014. Palaeeudyptes klekowskii, the best-preserved penguin skeleton from the Eocene–Oligocene of Antarctica: taxonomic and evolutionary remarks. Geobios 47 (3), 77–85. Available from: https://doi.org/10.1016/j.geobios.2014.03.003.

Amenábar, C., Montes, M., Nozal, F., Santillana, S., 2020. Dinoflagellate cysts of the La Meseta Formation (middle to late Eocene), Antarctic Peninsula: Implications for biostratigraphy, palaeoceanography and palaeoenvironment. Geological Magazine 157 (3), 351–366. Available from: https://doi.org/10.1017/S0016756819000591.

Anagnostou, E., John, E.H., Babila, T.L., et al., 2020. Proxy evidence for state-dependence of climate sensitivity in the Eocene greenhouse. Nature Communications 11, 4436. Available from: https://doi.org/10.1038/s41467-020-17887-x.

Anagnostou, E., John, E., Edgar, K., et al., 2016. Changing atmospheric CO_2 concentration was the primary driver of early Cenozoic climate. Nature 533, 380–384. Available from: https://doi.org/10.1038/nature17423.

Anderson, J., Warny, S., Askin, R., Wellner, J., Bohaty, S., Kirshner, A., et al., 2011. Progressive Cenozoic cooling and the demise of Antarctica's last refugium. Proceedings of the National Academy of Sciences 108, 11356–11360. Available from: https://doi.org/10.1073/pnas.1014885108.

Askin, R.A., 1992. Late Cretaceous–Early Tertiary Antarctic outcrop evidence for past vegetation and climate. In: Kennett, J.P., Warnke, D.A. (Eds.), The Antarctic Paleoenvironment: A Perspective on Global Change. Antarctic Research Series, vol. 56. American Geophysical Union, Washington, DC, pp. 61–75.

Askin, R.A., 2000. Spores and pollen from the McMurdo sound erratics, Antarctica. In: Stilwell, J.D., Feldmann, R.M. (Eds.), Paleobiology and Paleoenvironments of Eocene Rocks, McMurdo Sound, East Antarctica. Antarctic Research Series, vol. 76. American Geophysical Union, Washington, DC, pp. 161–181.

Askin, R.A., Raine, J.I., 2000. Oligocene and early Miocene terrestrial palynology of the cape roberts drill hole CRP-2/2A, Victoria Land Basin, Antarctica. Terra Antartica 7 (4), 493–501.

Baatsen, M., von der Heydt, A., Huber, M., Kliphuis, M., Bijl, P., Sluijs, A., et al., 2020. The middle-to-late Eocene greenhouse climate, modelled using the CESM 1.0.5. Climate of the Past 16, 2573–2597. Available from: https://doi.org/10.5194/cp-16-2573-2020.

Baatsen, M., van Hinsbergen, D.J.J., von der Heydt, A.S., Dijkstra, H.A., Sluijs, A., Abels, H.A., et al., 2016. Reconstructing geographical boundary conditions for palaeoclimate modelling during the Cenozoic. Climate of the Past 12, 1635–1644. Available from: https://doi.org/10.5194/cp-12-1635-2016.

Baatsen, M.L.J., von der Heydt, A.S., Kliphuis, M., Viebahn, J., Dijkstra, H.A., 2018. Multiple states in the late Eocene ocean circulation. Global and Planetary Change 163 (2018), 18–28. Available from: https://doi.org/10.1016/j.gloplacha.2018.02.009.

Barker, P.F., 2001. Scotia Sea regional tectonic evolution: implications for mantle flow and palaeocirculation. Earth-Science Reviews 55 (1–2), 1–39.

Barker, P.F., Filippelli, G.M., Florindo, F., Martin, E.E., Scher, H.D., 2007. Onset and role of the Antarctic circumpolar current. Deep Sea Research II 54, 2388–2398.

Barker, P.F., Kennett, J.P., O'Connell, S., Berkowitz, S., Bryant, W.P., Burckle, H., et al., 1988. Preliminary-results of ODP Leg-113 of Joides-resolution in the Weddell Sea – history of the Antarctic Glaciation. Comptes Rendus de L'Academie des Sciences Serie II 306, 73–78.

Barrett, P.J. (Ed.), 1989. Antarctic Cenozoic History from the CIROS-1 Drillhole, McMurdo Sound, vol. 245. New Zealand DSIR Bulletin, Wellington, p. 251.

Barrett, P.J., 1996. Antarctic paleoenvironments through Cenozoic time – a review. Terra Antartica 3, 103–119.

Barrett, P.J., Hambrey, M.J., Harwood, D.M., Pyne, A.R., Webb, P.-N., 1989. Synthesis. In: Barrett, P.J. (Ed.), Antarctic Cenozoic History from the CIROS-1 Drillhole, McMurdo Sound, vol. 245. New Zealand DSIR Bulletin, Wellington, pp. 241–251.

Barrett, P.J., Hambrey, M.J., Robinson, P.H., 1991. Cenozoic glacial and tectonic history from CIROS-1, McMurdo Sound. In: Thomson, M.R.A., Crame, A., Thomson, J.W. (Eds.), Geological Evolution of Antarctica. Cambridge University Press, New York, pp. 651–656.

Barron, J., Larsen, B., Shipboard Scientific Party, 1991. In: Proceedings of the Ocean Drilling Program, Scientific Results, Leg 119. Available from: <http://www-odp.tamu.edu/publications/119_SR/119TOC.HTM>.

Beamud, E., Montes, M., Santillana, S., Nozal, F., Marenssi, S., 2015. Magnetostratigraphic dating of Paleogene sediments in the Seymour Island (Antarctic Peninsula): a preliminary chronostratigraphy. American Geophysical Union, Fall Meeting, Abstract GP51B-1331.

Bijl, P., Schouten, S., Sluijs, A., et al., 2009. Early Palaeogene temperature evolution of the southwest Pacific Ocean. Nature 461, 776–779. Available from: https://doi.org/10.1038/nature08399.

Bijl, P.K., Bendle, J.A., Bohaty, S.M., Pross, J., Schouten, S., Tauxe, L., et al., 2013. Eocene cooling linked to early flow across the Tasmanian Gateway. Proceedings of the National Academy of Sciences 110 (24), 9645–9650.

Bijl, P.K., Houben, A.J.P., Hartman, J.D., Pross, J., Salabarnada, A., Escutia, C., et al., 2018. Paleoceanography and ice sheet variability offshore Wilkes Land, Antarctica – part 2: insights from Oligocene–Miocene dinoflagellate cyst assemblages. Climate of the Past 14, 1015–1033. Available from: https://doi.org/10.5194/cp-14-1015-2018.

Bindoff, N.L., Rosenberg, M.A., Warner, M.J., 2000. On the circulation and water masses over the Antarctic continental slope and rise between 80 and 150°E. Deep Sea Research Part II: Topical Studies in Oceanography 47 (12–13), 2299–2326. Available from: https://doi.org/10.1016/S0967-0645(00)00038-2.

Birkenmajer, K., 1980a. A revised lithostratigraphic standard for the tertiary of King George Island, South Shetland Islands (West Antarctica). Bulletin of the Polish Academy of Sciences, Earth Sciences 27 (1–2), 49–57.

Birkenmajer, K., 1980b. Tertiary volcanic–sedimentary succession at Admiralty Bay, King George Island (South Shetland Islands, Antarctica). Studia Geologica Polonica 64, 7–65.

Birkenmajer, K., 1981. Lithostratigraphy of the Point Hennequin Group (Miocene volcanics and sediments) at King George Island (South Shetland Islands, Antarctica). Studia Geologica Polonica 74, 175–197.

Birkenmajer, K., 1989. A guide to tertiary geochronology of King George Island, West Antarctica. Polish Polar Research 10, 555–579.

Birkenmajer, K., 1990. Geochronology and climatostratigraphy of tertiary glacial and interglacial successions on King George Island, South Shetland Islands (West Antarctica). Zentralblatt für Geologie und Paläontologie 1, 141–151.

Birkenmajer, K., 1997. Tertiary glacial/interglacial palaeoenvironments and sea-level changes, King George Island, West Antarctica. An overview. Bulletin of the Polish Academy of Sciences, Earth Sciences 44, 157–181.

Birkenmajer, K., Delitala, M.C., Narebski, W., Nicoletti, M., Petrucciani, C., 1986. Geochronology of tertiary island-arc volcanics and glacigenic deposits, King George Island, South Shetland Islands (West Antarctica). Bulletin of the Polish Academy of Sciences, Earth Sciences 34 (3), 257–273.

Birkenmajer, K., Gazdzicki, A., Krajewski, K.P., Przybycin, A., Solecki, A., Tatur, A., et al., 2005. First Cenozoic glaciers in West Antarctica. Polish Polar Research 26, 3–12.

Birkenmajer, K., Zastawniak, E., 1989a. Late cretaceous–early tertiary floras of King George Island, West Antarctica: their stratigraphic distribution and palaeoclimatic significance, origins and evolution of the Antarctic biota. Geological Society of London Special Publication 147, 227–240.

Birkenmajer, K., Zastawniak, E., 1989b. Late cretaceous–early Neogene vegetation history of the Antarctic Peninsula sector, Gondwana breakup and tertiary glaciations. Bulletin of the Polish Academy of Sciences, Earth Sciences 37, 63–88.

Bo, S., Siegert, M., Mudd, S., Sugden, D., Fujita, S., Xiangbin, C., et al., 2009. The Gamburtsev mountains and the origin and early evolution of the Antarctic Ice Sheet, Nature 459, 690–693. Available from: https://doi.org/10.1038/nature08024.

Bohaty, S.M., Delaney, M.L., Zachos, J.C., 2012. Foraminiferal Mg/Ca and Mn/Ca ratios across the Eocene–Oligocene transition. Earth and Planetary Science Letters 317–318, 251.

Boulton, G.S., 1990. Processes and sediments. In: Dowdeswell, J.A., Scourse, J.D. (Eds.), Glaciomarine Environments, vol. 53. Geological Society of London, Special Publications, pp. 15–52.

Brea, M., 1996. Analisis de los anillos de crecimiento de leños fosiles de coniferas de la Formacion La Meseta, Isla Seymour, Antartida. Congreso Paleogeno de America del Sur. Resumenes, Santa Rosa, p. 28.

Brea, M., 1998. Analisis de los Anillos de Crecimiento en Leños Fosiles de Coniferas de la Formacion La Meseta, Isla Seymour (Marambio), Antartida. In: Casadio, S. (Ed.), Paleogeno de America del Sur y de la Peninsula Antartica. Asociacion Paleontologica Argentina, Buenos Aires, pp. 163–175. Publicacion Especial.

Brinkhuis, H., Munsterman, D.K., Sengers, S., Sluijs, A., Warnaar, J., Williams, G.L., 2003. Late Eocene–quaternary dinoflagellate cysts from ODP site 1168, off Western Tasmania. In: Exon, N.F., Kennett, J.P., Malone, M.J. (Eds.), Proceedings of the Ocean Drilling Program, Scientific Results, 189. Available from: <http://www-odp.tamu.edu/publications/189_SR/105/105.htm>.

Caballero, R., Huber, M., 2013. State-dependent climate sensitivity in past warm climates and its implications for future climate projections. PNAS 110 (35), 14162–14167.

Cande, S.C., Mutter, J.C., 1982. A revised identification of the oldest sea-floor spreading anomalies between Australia and Antarctica. Earth and Planetary Science Letters 58 (2), 151–160.

Cande, S.C., Stock, J.M., 2004. Cenozoic reconstructions of the australia-new zealand-south pacific sector of Antarctica. Geophysical Monograph Series 151, 5–17. Available from: https://doi.org/10.1029/151GM02.

Carter, A., Riley, T.R., Hillenbrand, C.-D., Rittner, M., 2017. Widespread Antarctic glaciation during the Late Eocene. Earth and Planetary Science Letters 458, 49–57. Available from: https://doi.org/10.1016/j.epsl.2016.10.045.

Case, J.A., 1988. Paleogene floras from Seymour Island, Antarctic Peninsula. Memoir of the Geological Society of America 169 (1), 523–539. Available from: https://doi.org/10.1130/MEM169-p523.

Colleoni, F., De Santis, L., Montoli, E., Olivo, E., Sorlien, C.C., Bart, P.J., et al., 2018. Past continental shelf evolution increased Antarctic ice sheet sensitivity to climatic conditions. Scientific Reports 8 (1), 11323. Available from: https://doi.org/10.1038/s41598-018-29718-7.

Contreras, L., Pross, J., Bijl, P.K., Koutsodendris, A., Raine, J.I., van de Schootbrugge, B., et al., 2013. Early to Middle Eocene vegetation dynamics at the Wilkes Land Margin (Antarctica). Review of Palaeobotany and Palynology 197, 119–142. Available from: https://doi.org/10.1016/j.revpalbo.2013.05.009.

Cooper, A.K., O'Brien, P.E., 2004. Leg 188 synthesis: transitions in the glacial history of the Prydz Bay region, East Antarctica, from ODP drilling. In: Cooper, A.K., O'Brien, P.E., Richter, C. (Eds.), Proceedings of the Ocean Drilling Program, Scientific Results, 188. Available from: <http://www-odp.tamu.edu.publications/188_SR/synth/synth.htm>.

Cooper, A.K., O'Brien, P.E., Shipboard Scientific Party, 2004. Prydz Bay – Co-operation Sea, Antarctica: glacial history and paleoceanography sites 1165–1167. In: Proceedings of the Ocean Drilling Program, Scientific Results, p. 188. Available from: <http://www-odp.tamu.edu.publications/188_SR/188TOC.HTM>.

Cooper, A.K., Stagg, H., Geist, E.L., 1991. Seismic stratigraphy and structure of Prydz Bay, Antarctica: implications for Leg 119 drilling. In: Proceedings of the Ocean Drilling Program, Scientific Results, 119. Available from: <http://www-odp.tamu.edu/publications/119_SR/119TOC.HTM>.

Costa, E., Garcés, M., Sáez, A., Cabrera, L., López-Blanco, M., 2011. The age of the "Grande Coupure" mammal turnover: new constraints from the Eocene–Oligocene record of the Eastern Ebro Basin (NE Spain). Palaeogeography, Palaeoclimatology, Palaeoecology 301, 97–107.

Coxall, H.K., Huck, C.E., Huber, M., et al., 2018. Export of nutrient rich northern component water preceded early Oligocene Antarctic glaciation. Nature Geoscience 11, 190–196. Available from: https://doi.org/10.1038/s41561-018-0069-9.

Coxall, H.K., Pearson, P.N., 2007. The Eocene-Oligocene transition. Deep Time Perspectives on Climate Change: Marrying the Signal from Computer Models and Biological Proxies. pp. 351–387.

Coxall, H.K., Wilson, P.A., 2011. Early Oligocene glaciation and productivity in the eastern equatorial Pacific: insights into global carbon cycling. Paleoceanography 26, PA2221.

Coxall, H.K., Wilson, P.A., Palike, H., Lear, C.H., Backman, J., 2005. Rapid stepwise onset of Antarctic glaciation and deeper calcite compensation in the Pacific Ocean. Nature 433, 53–57.

Crame, J.A., Beu, A.G., Ineson, J.R., Francis, J.E., Whittle, R.J., Bowman, V.C., 2014. The early origin of the Antarctic Marine Fauna and its evolutionary implications. PLoS One 9 (12), e114743. Available from: https://doi.org/10.1371/journal.pone.0114743.

Cramwinckel, M.J., Huber, M., Kocken, I.J., et al., 2018. Synchronous tropical and polar temperature evolution in the Eocene. Nature 559, 382–386. Available from: https://doi.org/10.1038/s41586-018-0272-2.

CRP Science Team, 2000. Studies from the Cape Roberts Project, Ross Sea, Antarctica. Initial Report on CRP-3. Terra Antartica 7, 1–209. With Supplement, 305 pp.

Dalziel, I., 2006. On the extent of the active West Antarctic Rift System. Terra Antartica Reports 12, 193–202.

Davis, S.N., Torres, C.R., Musser, G.M., Proffitt, J.V., Crouch, N.M.A., Lundelius, E.L., et al., 2020. New mammalian and avian records from the late Eocene la Meseta and Submeseta formations of Seymour Island, Antarctica. PeerJ 2020 (1), 8268. Available from: https://doi.org/10.7717/peerj.8268.

De Boer, B., Stocchi, P., van de Wal, R.W.S., 2014. A fully coupled 3-D ice- sheet – sea-level model: algorithm and applications. Geoscience Model Development Discussions 7, 2141–2156. Available from: https://doi.org/10.5194/gmd-7-2141-2014.

DeConto, R.M., Pollard, D., 2003. Rapid Cenozoic glaciation of Antarctica induced by declining atmospheric CO_2. Nature 421, 245–249.

DeConto, R.M., Pollard, D., Harwood, D., 2007. Sea ice feedback and Cenozoic evolution of Antarctic climate and ice sheets. Paleoceanography 22, PA3214. Available from: https://doi.org/10.1029/2000PA000567.

DeConto, R.M., Pollard, D., Wilson, P.A., Pälike, H., Lear, C.H., Pagani, M., 2008. Thresholds for Cenozoic bipolar glaciation. Nature 455 (7213), 652–656. Available from: https://doi.org/10.1038/nature07337.

Devereux, I., 1967. Oxygen isotope paleotemperature measurements on New Zealand tertiary fossils. New Zealand Journal of Science 10, 988–1011.

Diester-Haass, L., Zahn, R., 1996. Eocene–Oligocene transition in the Southern Ocean: history of water mass circulation and biological productivity. Geology 24, 163–166.

Diester-Haass, L., Zahn, R., 2001. Paleoproductivity increase at the Eocene–Oligocene climatic transition: ODP/DSDP sites 763 and 592. Palaeogeography, Palaeoclimatology, Palaeoecology 172, 153–170.

Dingle, R., Lavelle, M., 1998. Late Cretaceous–Cenozoic climatic variations of the Northern Antarctic Peninsula: new geochemical evidence and review. Palaeogeography, Palaeoclimatology, Palaeoecology 141 (3), 215–232.

Doktor, M., Gazdzicki, A.J., Jermanska, A., Porebski, S., Zastawaniak, E., 1996. A plant–fish assemblage from the Eocene La Meseta Formation of Seymour Island (Antarctic Peninsula) and its environmental implications. Acta Palaeontologica Polonica 55, 127–146.

Doktor, M., Gazdzicki, A., Marenssi, S., Porebski, S., Santillana, S., Vrba, A., 1988. Argentine–Polish geological investigations on Seymour (Marambio) Island, Antarctica. Polish Polar Research 9, 521–541.

Douglas, P.M.J., Affek, H.P., Ivany, L.C., et al., 2014. Pronounced zonal heterogeneity in Eocene southern high-latitude sea surface temperatures. PNAS 111 (18), 6582–6587.

Dunbar, G.B., Naish, T.R., Barrett, P.J., Fielding, C.R., Powell, R.D., 2008. Constraining the amplitude of late Oligocene bathymetric changes in Western Ross Sea during orbitally-induced oscillations in the East Antarctic Ice Sheet: (1) Implications for glacimarine sequence stratigraphic models, Palaeogeography, Palaeoclimatology, Palaeoecology, 260 (1–2), 50–65. Available from: https://doi.org/10.1016/j.palaeo.2007.08.018.

Duncan, B., McKay, R., Bendle, J., Naish, T., Inglis, G.N., Moossen, H., et al., 2019. Lipid biomarker distributions in Oligocene and Miocene sediments from the Ross Sea region, Antarctica: implications for use of biomarker proxies in glacially-influenced settings. Palaeogeography, Palaeoclimatology, Palaeoecology 516, 71–89. Available from: https://doi.org/10.1016/j.palaeo.2018.11.028.

Dutton, A., Lohmann, K., Zinsmeister, W.J., 2002. Stable isotope and minor element proxies for Eocene climate of Seymour Island, Antarctica. Paleoceanography 17 (2), 1016. Available from: https://doi.org/10.1029/2000PA000593.

Eagles, G., Livermore, R., Morris, P., 2006. Small basins in the Scotia Sea; the Eocene drake passage gateway. Earth and Planetary Science Letters 242 (3–4), 343–353.

Ehrmann, W., 1997. Smectite concentrations and crystallinities: Indications for Eocene age of glaciomarine sediments in the CIROS-1 Drill Hole, McMurdo Sound, Antarctic. In: Ricci, E.A. (Ed.), The Antarctic Region, Geological Evolution and Processes. Museo Nazionale dell'Antartide, Siena, pp. 771–780.

Ehrmann, W.U., Mackensen, A., 1992. Sedimentological evidence for the formation of an East Antarctic Ice Sheet in Eocene/Oligocene time. Palaeogeography, Palaeoclimatology, Palaeoecology 93, 85–112.

Eittreim, S.L., Cooper, A.K., Wannesson, J., 1995. Seismic stratigraphic evidence of ice-sheet advances on the Wilkes Land margin of Antarctica. Sedimentary Geology 96 (1–2), 131–156.

Elliot, D.H., Trautman, T.A., 1982. Lower tertiary strata on Seymour Island, Antarctic Peninsula. In: Craddock, C. (Ed.), Antarctic Geoscience. University of Wisconsin Press, Madison, pp. 287–298.

Elsworth, G., Galbraith, E., Halverson, G., Yang, S., 2017. Enhanced weathering and CO2 drawdown caused by latest Eocene 1445 strengthening of the Atlantic meridional overturning circulation. Nature Geoscience 10 (3), 213–216. Available from: https://doi.org/10.1038/ngeo2888.

Escutia, C., Brinkhuis, H., 2014. From greenhouse to icehouse at the Wilkes Land Antarctic Margin: IODP Expedition 318 synthesis of results. Developments in Marine Geology 7, 295–328.

Escutia, C., Brinkhuis, H., Klaus, A., the Expedition 318 Scientists, 2011. Proceedings of Integrated Ocean Drilling Program 318 (Integrated Ocean Drilling Program Management International).

Evangelinos, D., Escutia, C., Etourneau, J., Hoem, F., Bijl, P., Boterblom, W., et al., 2020. Late Oligocene-Miocene proto-Antarctic circumpolar current dynamics off the Wilkes Land margin, East Antarctica. Global and Planetary Change 191, 103221. Available from: https://doi.org/10.1016/j.gloplacha.2020.103221.

Evans, D., Wade, B.S., Henehan, M., Erez, J., Müller, W., 2016. Revisiting carbonate chemistry controls on planktic foraminifera Mg/Ca: implications for sea surface temperature and hydrology shifts over the Paleocene–Eocene Thermal Maximum and 1455 Eocene–Oligocene transition. Climate of the Past 12 (4), 819–835. Available from: https://doi.org/10.5194/cp-12-819-2016.

Farrell, W.E., Clark, J.A., 1976. On postglacial sea level. Geophysical Journal International 46 (3), 647–667. Available from: https://doi.org/10.1111/j.1365-246X.1976.tb01252.x.

Fielding, C., Naish, T.R., Woolfe, K.J., 2001. Facies architecture of the CRP-3 drillhole, Victoria Land Basin, Antarctica. Terra Antarctica 8 (3), 217–224.

Fielding, C.R., Whittaker, J., Henrys, S.A., Wilson, T.J., Naish, T.R., 2008. Seismic facies and stratigraphy of the Cenozoic succession in McMurdo Sound, Antarctica: Implications for tectonic, climatic and glacial history. Palaeogeography, Palaeoclimatology, Palaeoecology 260 (1–2), 8–29.

Fielding, C.R., Woolfe, K.J., Purdon, R.G., Lavelle, M.A., Howe, J.A., 1997. Sedimentological and stratigraphical re-evaluation of the CIROS-1 Core, McMurdo Sound, Antarctica. Terra Antartica 4, 149–160.

Fioroni, C., Villa, G., Persico, D., Wise, S.W., Pea, L., 2012. Revised middle Eocene-upper Oligocene calcareous nannofossil biozonation for the Southern Ocean. Revue de Micropaléontologie 55, 53–70. Available from: https://doi.org/10.1016/j.revmic.2012.03.001.

Florindo, F., Wilson, G.S., Roberts, A.P., Sagnotti, L., Verosub, K.L., 2005. Magnetostratigraphic chronology of a late Eocene to early Miocene glacimarine succession from the Victoria Land Basin, Ross Sea, Antarctica. Global and Planetary Change 45, 207–236.

Foster, G.L., Royer, D.L., Lunt, D.J., 2017. Future climate forcing potentially without precedent in the last 420 million years. Nature Communications 8, 14845. Available from: https://doi.org/10.1038/ncomms14845.

Francis, J.E., 1991. Palaeoclimatic significance of Cretaceous–Early Tertiary fossil forests of the Antarctic Peninsula. In: Thomson, M.R.A., Crame, A., Thomson, J.W. (Eds.), Geological Evolution of Antarctica. Cambridge University Press, New York, pp. 623–627.

Francis, J.E., 1999. Evidence from fossil plants for Antarctic palaeoclimates over the past 100 million years. In: Barrett, P.J., Orombelli, G. (Eds.), Geological Records of Global and Planetary Changes. Terra Antartica Report 3, Siena, pp. 43–52.

Francis, J.E., 2000. Fossil wood from Eocene high latitude forests, McMurdo Sound, Antarctica. In: Stilwell, J.D., Feldmann, R.M. (Eds.), Paleobiology and Palaeoenvironments of Eocene Rocks, McMurdo Sound, East Antarctica. Antarctic Research Series, vol. 76. American Geophysical Union, Washington, DC, pp. 253–260.

Francis, J.E., Ashworth, A., Cantrill, D.J., Crame, J.A., Howe, J., Stephens, R., et al., 2008. 100 Million Years of Antarctic Climate Evolution: Evidence from Fossil Plants. United States Geological Survey and The National Academies; USGS OFR-2007.

Francis, J.E., Marenssi, S., Levy, R., Hambrey, M., Thorn, V.C., Mohr, B., et al., 2009. Chapter 8: From greenhouse to icehouse – the Eocene/Oligocene in Antarctica. In: Florindo, F., Siegert, M.J. (Eds.), Developments in Earth and Environmental Sciences. Antarctic Climate Evolution, vol. 8. Elsevier, pp. 309–368.

Francis, J.E., Poole, I., 2002. Cretaceous and early tertiary climates of Antarctica: evidence from fossil wood. Palaeogeography, Palaeoclimatology, Palaeoecology 182 (1–2), 47–64.

Francis, J.E., Tosolini, A.-M., Cantrill, D., 2004. Biodiversity and climate change in Antarctic Palaeogene floras. In: VII International Organisation of Palaeobotany Conference, Bariloche, Argentina, pp. 33–34, Abstract Volume.

Fyke, J.G., D'Orgeville, M., Weaver, A.J., 2015. Drake passage and Central American Seaway controls on the distribution of the oceanic carbon reservoir. Global and Planetary Change 128 (0), 72–82. Available from: https://doi.org/10.1016/j.gloplacha.2015.02.011.

Galeotti, S., DeConto, R., Naish, T., Stocchi, P., Florindo, F., Pagani, M., et al., 2016. Antarctic Ice Sheet variability across the Eocene-Oligocene boundary climate transition. Science (New York, N.Y.) 352 (6281), 76–80. Available from: https://doi.org/10.1126/science.aab0669.

Galeotti, S., Lanci, L., Florindo, F., Naish, T.R., Sagnotti, L., Sandroni, S., et al., 2012. Cyclochronology of the Eocene–Oligocene transition from a glacimarine succession off the Victoria Land coast, Cape Roberts Project, Antarctica. Palaeogeography, Palaeoclimatology, Palaeoecology 335–336, 84–94. Available from: https://doi.org/10.1016/j.palaeo.2011.08.011.

Gandolfo, M.A., Hoc, P., Santillana, S., Marenssi, S.A., 1998a. Una Flor Fosil Morfologicamente Afın a las Grossulariaceae (Orden Rosales) de la Formacion La Meseata (Eoceno medio), Isla Marambio, Antartida. In: Casadio, S. (Ed.), Paleógeno de America del Sur y de la Peninsula Antartica. Asociacion Paleontologica Argentina, Buenos Aires, pp. 147–153. Publicación Especial.

Gandolfo, M.A., Marenssi, S.A., Santillana, S.N., 1998b. Flora y Paleoclima de la Formacion La Meseta (Eoceno medio), Isla Marambio (Seymour), Antartida. In: Casadio, S. (Ed.), Paleogeno de America del Sur y de la Peninsula Antartica. Asociacion Paleontologica Argentina, Buenos Aires, pp. 155–162. Publicacion Especial.

Gandolfo, M.A., Marenssi, S.A., Santillana, S.N., 1998c. Flora y Paleoclima de la Formacion La Meseta (Eoceno-Oligoceno Inferior?) Isla Marambio (Seymour), Antartida. In: I Congreso del Paleogeno de America del Sur, La Pampa, Argentina, Actas, pp. 31–32.

Gasson, E.G.W., Keisling, B.A., 2020. The antarctic ice sheet: a paleoclimate modeling perspective. Oceanography 33 (2), 91–100. Available from: https://doi.org/10.5670/oceanog.2020.208.

Gasson, E., Lunt, D.J., Deconto, R., Goldner, A., Heinemann, M., Huber, M., et al., 2014. Uncertainties in the modelled CO_2 threshold for Antarctic glaciation. Climate of the Past 10 (2). Available from: https://doi.org/10.5194/cp-10-451-2014.

Gelfo, J.N., Goin, F.J., Bauzá, N., Reguero, M., 2019. The fossil record of Antarctic land mammals: commented review and hypotheses for future research. Advances in Polar Science 30, 274–292.

Goldner, A., Herold, N., Huber, M., 2014. Antarctic glaciation caused ocean circulation changes at the Eocene-Oligocene transition. Nature 511 (7511), 574–577. Available from: https://doi.org/10.1038/nature13597.

Goldner, A., Huber, M., Caballero, R., 2013. Does antarctic glaciation cool the world? Climate of the Past 9 (1), 173–189. Available from: https://doi.org/10.5194/cp-9-173-2013.

Gomez, N., et al., 2010. A new projection of sea level change in response to collapse of marine sectors of the Antarctic Ice-Sheet. Geophysical Journal International 180, 623–634.

Gothan, W., 1908. Die Fossilen Holzer von der Seymour und Snow Hill Insel. In: Nordenskjold, O. (Ed.), Wissenschaftliche Erbegnesse Schwedischen Sudpolar Expedition 1901–1903. pp. 1–33. Stockholm.

Greenwood, D.R., Wing, S.L., 1995. Eocene continental climates and latitudinal temperature gradients. Geology 23 (11), 1044–1048. Available from: https://doi.org/10.1130/0091-7613.

Griffies, S.M., Winton, M., Anderson, W.G., Benson, R., Delworth, T.L., Dufour, C.O., et al., 2015. Impacts on ocean heat from transient mesoscale eddies in a hierarchy of climate models. Journal of Climate 28 (3), 952–977. Available from: https://doi.org/10.1175/JCLI-D-14-00353.1.

Gulick, S.P.S., Shevenell, A.E., Montelli, A., Fernandez, R., Smith, C., Warny, S., et al., 2017. Initiation and long-term instability of the East Antarctic Ice Sheet. Nature 552 (7684), 225–229. Available from: https://doi.org/10.1038/nature25026.

Hambrey, M.J., Barrett, P.J., 1993. Cenozoic sedimentary and climatic record, Ross Sea region, Antarctica. In: Kennett, J.P., Warnke, D.A. (Eds.), The Antarctic Paleoenvironment: A Perspective on Global Change, Part 2. Antarctic Research Series, vol. 60. American Geophysical Union, Washington, DC, pp. 91–124.

Hambrey, M.J., Barrett, P.J., Ehrmann, E.H., Larsen, B., 1992. Cenozoic sedimentary processes on the Antarctic continental shelf: The record from deep drilling. Zeitschrift für Geomorphologie 86 (Suppl), 73–99.

Hambrey, M.J., Barrett, P.J., Powell, R.D., 2002. Late Oligocene and early Miocene glacimarine sedimentation in the SW Ross Sea, Antarctica: the record from offshore drilling. In: O'Cofaigh, C., Dowdeswell, J.A. (Eds.), Glacier-Influence Sedimentation on High-Latitude Continental Margins, 203. Geological Society of London Special Publication, pp. 105–128.

Hambrey, M.J., Barrett, P.J., Robinson, P.H., 1989. Stratigraphy. In: Barrett, P.J. (Ed.), Antarctic Cenozoic History from the CIROS-1 Drillhole, McMurdo Sound, 245. New Zealand DSIR Bulletin, Wellington, pp. 23–48.

Hambrey, M.J., Ehrmann, E.H.R., Larsen, B., 1991. The Cenozoic glacial record of the Prydz Bay continental shelf, East Antarctica. In: Barron, J., Larsen, B., Shipboard Scientific Party (Eds.), Proceedings of the Ocean Drilling Program, Scientific Results, p. 119. Available from: <http://www-odp.tamu.edu/publications/119_SR/119TOC.HTM>.

Haomin, L., 1994. Early tertiary fossil hill flora from Fildes Peninsula of King George Island, Antarctica. In: Yanbin, S. (Ed.), Stratigraphy and Palaeontology of Fildes Peninsula, King George Island, Antarctica, State Antarctic Committee. Science Press, Beijing, pp. 165–171. Monograph 3.

Hartman, J.D., Sangiorgi, F., Salabarnada, A., Peterse, F., Houben, A.J.P., Schouten, S., et al., 2018. Paleoceanography and ice sheet variability offshore Wilkes Land, Antarctica-Part 3: Insights from Oligocene-Miocene TEX86-based sea surface temperature reconstructions. Climate of the Past 14 (9), 1275–1297. Available from: https://doi.org/10.5194/cp-14-1275-2018.

Haywood, A.M., Smellie, J.L., Ashworth, A.C., Cantrill, D.J., Florindo, F., Hambrey, M.J., et al., 2008. Middle Miocene to Pliocene history of Antarctica and the Southern Ocean 405. In: Florindo, F., Siegert, M. (Eds.), Antarctic Climate Evolution, vol. 8. Elsevier, Amsterdam, pp. 401–463.

Henrys, S.A., Wilson, T., Whittaker, J.M., Fielding, C., Hall, J., Naish, T., 2007. Tectonic history of mid-Miocene to present southern Victoria Land Basin, inferred from seismic stratigraphy in McMurdo Sound, Antarctica. In: Cooper, A.K., Raymond, C.R., et al. (Eds.), Antarctica: A Keystone in a Changing World – Online Proceedings of the 10th International Symposium on Antarctic Earth Sciences, United States Geological Survey Open-File Report 2007-1047, Short Research Paper 049. 4 pp. Available from: https://doi.org/10.3133/of2007-1047.srp049.

Herold, N., Huber, M., Müller, R.D., Seton, M., 2012. Modeling the Miocene climatic optimum: ocean circulation. Paleoceanography 27 (1), PA1209. Available from: https://doi.org/10.1029/2010PA002041.

Hiemstra, J.F., 1999. Microscopic evidence of grounded ice in the sediments of the CIROS-1 Core, McMurdo Sound, Antarctica. Terra Antartica 6, 365–376.

Hill, D.J., Haywood, A.M., Valdes, P.J., Francis, J.E., Lunt, D.J., Wade, B.S., 2013. Paleogeographic controls on the onset of the Antarctic circumpolar current. Geophysical Research Letters 40 (19), 5199–5204.

Hill, R.S., 1989. Palaeontology-fossil leaf. In: Barrett, P.J. (Ed.), Antarctic Cenozoic History from the CIROS-1 Drillhole, McMurdo Sound, 245. New Zealand DSIR Bulletin, Wellington, pp. 143–144.

Hill, P.J., Exon, N.F., 2004. Tectonics and basin development of the offshore Tasmanian area incorporating results from deep ocean drilling. Geophysical Monograph Series 151, 19–42. Available from: https://doi.org/10.1029/151GM03.

Hochmuth, K., et al., 2020. The evolving paleobathymetry of the Circum-Antarctic Southern Ocean since 34 Ma: a key to understanding past cryosphere-ocean developments. Geochemistry, Geophysics, Geosystems 21 (8), e2020GC009122.

Houben, A.J.P., et al., 2012. The Eocene–Oligocene transition: changes in sea level, temperature or both? Palaeogeography, Palaeoclimatology, Palaeoecology 335–336, 75–83.

Houben, A.J.P., 2019. Late Eocene Southern Ocean cooling and invigoration of circulation preconditioned Antarctica for full-scale glaciation. Geochemistry, Geophysics, Geosystems 20 (5), 2214–2234.

Houben, A.J.P., Bijl, P.K., Guerstein, G.R., Sluijs, A., Brinkhuis, H., 2011. Malvinia escutiana, a new biostratigraphically important Oligocene dinoflagellate cyst from the Southern Ocean. Review of Palaeobotany and Palynology 165 (3–4), 175–182. Available from: https://doi.org/10.1016/j.revpalbo.2011.03.002.

Houben, A.J.P., Bijl, P.K., Pross, J., Bohaty, S.M., Stckley, C.E., Passchier, S., et al., 2013. Reorganization of Southern ocean plankton ecosystem at the onset of Antarctic glaciation. Science (New York, N.Y.) 340 (6130), 341–344.

Huang, X., Gohl, K., Jokat, W., 2014. Variability in Cenozoic sedimentation and paleo-water depths of the Weddell Sea basin related to pre-glacial and glacial conditions of Antarctica. Global and Planetary Change 118, 25–41. Available from: https://doi.org/10.1016/j.gloplacha.2014.03.010.

Huber, M., Caballero, R., 2011. The early Eocene equable climate problem revisited. Climate of the Past 7, 603–633. Available from: https://doi.org/10.5194/cp-7-603-2011.

Huber, M., Nof, D., 2006. The ocean circulation in the southern hemisphere and its climatic impacts in the Eocene. Palaeogeography, Palaeoclimatology, Palaeoecology 231, 9–28.

Huber, M., Brinkhuis, H., Stickley, C.E., Döös, K., Sluijs, A., Warnaar, J., et al., 2004. Eocene circulation of the Southern Ocean: Was Antarctica kept warm by subtropical waters? Paleoceanography 19 (4), PA4026. Available from: https://doi.org/10.1029/2004PA001014.

Hunt, R., 2001. Biodiversity and Palaeoecology of Tertiary Fossil Floras in Antarctica. Ph.D. Thesis, University of Leeds, Leeds, United Kingdom.

Hunt, R.J., Poole, I., 2003. Paleogene West Antarctic climate and vegetation history in light of new data from King George Island. In: Wing, S.L., Gingerich, P.D., Schmitz, B., Thomas, E. (Eds.), Causes and Consequences of Globally Warm Climates in the Early Paleogene, vol. 369. Geological Society of America, Boulder, CO, pp. 395–412.

Hutchinson, D.K., Coxall, H.K., Lunt, D.J., Steinthorsdottir, M., de Boer, A.M., Baatsen, M., et al., 2021. The Eocene-Oligocene transition: a review of marine and terrestrial proxy data, models and model-data comparisons. Climate of the Past 17 (1), 269–315. Available from: https://doi.org/10.5194/cp-17-269-2021.

Hutchinson, D.K., Coxall, H.K., O'Regan, M., Nilsson, J., Caballero, R., de Boer, A.M., 2019. Arctic closure as a trigger for Atlantic overturning at the Eocene-Oligocene Transition. Nature Communications 10 (1). Available from: https://doi.org/10.1038/s41467-019-11828-z.

Hutchinson, D.K., de Boer, A.M., Coxall, H.K., Caballero, R., Nilsson, J., Baatsen, M.J.L., 2018. Climate sensitivity and meridional overturning circulation in the late Eocene using GFDL CM2.1. Climate of the Past 14, 789–810. Available from: https://doi.org/10.5194/cp-14-789-2018.

Ivany, L.C., Lohmann, K.C., Hasiuk, F., Blake, D.B., Glass, A., Aronson, R.B., et al., 2008. Eocene climate record of a high southern latitude continental shelf: Seymour Island, Antarctica. Bulletin of the Geological Society of America 120 (5–6), 659–678. Available from: https://doi.org/10.1130/B26269.1.

Jadwiszczak, P., Mörs, T., 2019. First partial skeleton of Delphinornis larseni Wiman, 1905, a slender-footed penguin from the Eocene of Antarctic Peninsula. Paleontologia Electronica 22, 1–31. Available from: https://doi.org/10.26879/933.

Jayne, S.R., Marotzke, J., 2002. The oceanic eddy heat transport. Journal of Physical Oceanography 32 (12), 3328–3345.

Jones, C.M., 2000. The first record of a fossil bird from East Antarctic. In: Stilwell, J.D., Feldmann, R.M. (Eds.), Paleobiology and Palaeoenvironments of Eocene Rocks, McMurdo Sound, East Antarctica. Antarctic Research Series, vol. 76. American Geophysical Union, Washington, DC, pp. 359–364.

Judd, E.J., Ivany, L.C., DeConto, R.M., Halberstadt, A.R.W., Miklus, N.M., Junium, C.K., et al., 2019. Seasonally resolved proxy data from the Antarctic Peninsula support a heterogeneous middle Eocene Southern Ocean. Paleoceanography and Paleoclimatology 34 (5), 787–799. Available from: https://doi.org/10.1029/2019PA003581.

Katz, M.E., Cramer, B.S., Toggweiler, J.R., Esmay, G., Liu, C., Miller, K.G., Rosenthal, Y., Wade, B.S., Wright, J.W., 2011. Impact of Antarctic Circumpolar Current Development on Late Paleogene Ocean Structure. Science 332, 1076–1079. Available from: https://doi.org/10.1126/science.1202122.

Katz, M.E., Miller, K.G., Wright, J.D., Wade, B.S., Browning, J.V., Cramer, B.S., et al., 2008. Stepwise transition from the Eocene greenhouse to the Oligocene icehouse. Nature Geoscience 1, 329. Available from: https://doi.org/10.1038/ngeo179 [online].

Keigwin Jr., L.D., 1980. Palaeoceanographic change in the pacific at the Eocene-Oligocene boundary. Nature 287 (5784), 722–725. Available from: https://doi.org/10.1038/287722a0.

Keigwin, L.D., Corliss, B.H., 1986. Stable isotopes in late middle Eocene to Oligocene foraminifera. Geological Society of America Bulletin 97 (3), 335–345. Available from: https://doi.org/10.1130/0016-7606(1986)97(335:SIILME)2.0.CO;2.

Kennedy-Asser, A.T., Farnsworth, A., Lunt, D.J., Lear, C.H., Markwick, P.J., 2015. Atmospheric and oceanic impacts of Antarctic glaciation across the Eocene-Oligocene transition. Philosophical Transactions of the Royal Society A: Mathematical, Physical and Engineering Sciences 373 (2054). Available from: https://doi.org/10.1098/rsta.2014.0419.

Kennedy-Asser, A.T., Lunt, D.J., Valdes, P.J., Ladant, J.-B., Frieling, J., Lauretano, V., 2020. Changes in the high-latitude Southern Hemisphere through the Eocene-Oligocene transition: a model-data comparison. Climate of the Past 16 (2), 555–573. Available from: https://doi.org/10.5194/cp-16-555-2020.

Kennett, J.P., 1977. Cenozoic evolution of Antarctic glaciation, the circum-Antarctic oceans and their impact on global paleoceanography. Journal of Geophysical Research 82, 3843–3859.

Kennett, J.P., Shackleton, N.J., 1976. Oxygen isotopic evidence for the development of the psychryosphere 38 my ago. Nature 260, 513–515.

Klages, J.P., Salzmann, U., Bickert, T., et al., 2020. Temperate rainforests near the South Pole during peak Cretaceous warmth. Nature 580, 81–86. Available from: https://doi.org/10.1038/s41586-020-2148-5.

Kominz, M.A., Pekar, S.F., 2001. Oligocene eustasy from two-dimensional sequence stratigraphic backstripping. GSA Bulletin 113, 291–304.

Kraatz, B.P., Geisler, J.H., 2010. Eocene–Oligocene transition in Central Asia and its effects on mammalian evolution. Geology 38, 111–114. Available from: https://doi.org/10.1130/G30619.1.

LaCasce, J.H., 2017. The prevalence of oceanic surface modes. Geophysical Research Letters 44 (21), 11–097.

LaCasce, J.H., Escartin, J., Chassignet, E.P., Xu, X., 2019. Jet instability over smooth, corrugated, and realistic bathymetry. Journal of Physical Oceanography 49 (2), 585–605.

Ladant, J.B., Donnadieu, Y., Lefebvre, V., Dumas, C., 2014. The respective role of atmospheric carbon dioxide and orbital parameters on ice sheet evolution at the Eocene-Oligocene transition. Paleoceanography 29 (8), 810–823. Available from: https://doi.org/10.1002/2013PA002593.

Lagabrielle, Yves, et al., 2009. The tectonic history of Drake Passage and its possible impacts on global climate. Earth and Planetary Science Letters 279 (3–4), 197–211.

Langton, S.J., Rabideaux, N.M., Borrelli, C., Katz, M.E., 2016. Southeastern Atlantic deep-water evolution during the late-middle Eocene to earliest Oligocene (ocean drilling program site 1263 and deep sea drilling project site 366). Geosphere 12 (3), 1032–1047. Available from: https://doi.org/10.1130/GES01268.1.

Lear, C.H., Bailey, T.R., Pearson, P.N., Coxall, H.K., Rosenthal, Y., 2008. Cooling and ice growth across the Eocene-Oligocene transition. Geology 36 (3), 251. Available from: https://doi.org/10.1130/G24584A.1.

Levy, R.H., Harwood, D.M., 2000a. Sedimentary lithofacies of the McMurdo sound erratics. In: Stilwell, J.D., Feldmann, R.M. (Eds.), Paleobiology and Paleoenvironments of Eocene Rocks, McMurdo Sound, East Antarctica. Antarctic Research Series, vol. 76. American Geophysical Union, Washington, DC, pp. 39–61.

Levy, R.H., Harwood, D.M., 2000b. Tertiary marine palynomorphs from the McMurdo sound erratics, Antarctica. In: Stilwell, J.D., Feldmann, R.M. (Eds.), Paleobiology and Paleoenvironments of Eocene Rocks, McMurdo Sound, East Antarctica. Antarctic Research Series, vol. 76. American Geophysical Union, Washington, DC, pp. 183–242.

Lewis, A.R., Marchant, D.R., Ashworth, A.C., Hedenäs, L., Hemming, S.R., Johnson, J.V., et al., 2008. Mid-Miocene cooling and the extinction of tundra in continental Antarctica. Proceedings of the National Academy of Sciences 105, 10676–10680.

Li, H.M., 1992. Early Tertiary palaeoclimate of King George Island, Antarctica – evidence from the fossil hill flora. In: Yoshida, Y., Kaminuma, K., Shiraishi, K. (Eds.), Recent Progress in Antarctic Earth Science. Terra Scientific Publishing Company, Tokyo, pp. 371–375.

Liu, Z., Pagani, M., Zinniker, D., DeConto, R., Huber, M., Brinkhuis, H., et al., 2009. Global cooling during the eocene-oligocene climate transition. Science (New York, N.Y.) 323 (5918), 1187–1190. Available from: https://doi.org/10.1126/science.1166368.

Livermore, R., Hillenbrand, C.D., Meredith, M., Eagles, G., 2007. Drake Passage and Cenozoic climate: an open and shut case? Geochemistry, Geophysics, Geosystems 8 (1). Available from: https://doi.org/10.1029/2005GC001224.

Long, D.H., Stilwell, J.D., 2000. Fish remains from the Eocene of Mount Discovery, East Antarctic. In: Stilwell, J.D., Feldmann, R.M. (Eds.), Paleobiology and Palaeoenvironments of Eocene Rocks, McMurdo Sound, East Antarctica. Antarctic Research Series, vol. 76. American Geophysical Union, Washington, DC, pp. 349–354.

López-Quirós, A., Escutia, C., Sánchez-Navas, A., Nieto, F., García-Casco, A., Martín-Algarra, A., et al., 2019. Glaucony authigenesis, maturity and alteration in the Weddell Sea: An indicator of paleoenvironmental conditions before the onset of Antarctic glaciation. Scientific Reports 9 (1), 13580. Available from: https://doi.org/10.1038/s41598-019-50107-1.

Lunt, D.J., Bragg, F., Chan, W.-L., Hutchinson, D.K., Ladant, J.-B., Niezgodzki, I., et al., 2020. DeepMIP: YEAR Model intercomparison of early Eocene climatic optimum (EECO) large-scale climate features and comparison with proxy data. Climate of the Past Discussions. Available from: https://doi.org/10.5194/cp-2019-149.

Lunt, D.J., Dunkley Jones, T., Heinemann, M., Huber, M., LeGrande, A., Winguth, A., et al., 2012. A model– data comparison for a multi-model ensemble of early Eocene atmosphere–ocean simulations: EoMIP. Climate of the Past 8, 1717–1736. Available from: https://doi.org/10.5194/cp-8-1717-2012.

Marenssi, S.A., 2006. Eustatically controlled sedimentation recorded by Eocene strata of the James Ross Basin, Antarctica. Geological Society, London, Special Publications 258, 125–133. Available from: https://doi.org/10.1144/GSL.SP.2006.258.01.09.

Marenssi, S.A., Santillana, S., Rinaldi, C.A., 1998. Stratigraphy of the La Meseta Formation (Eocene), Marambio (Seymour) Island, Antarctica. In: Casadıio, S. (Ed.), Paleogeno de America del Sur y de la Penınsula Antartica, Publicacion Especial 5. Asociacion Paleontologica Argentina, Buenos Aires, pp. 137–146.

Matthews, R.K., Poore, R.Z., 1980. Tertiary delta^{18}O record and glacio-eustatic sea-level fluctuations. Geology 8 (10), 501–504. Available from: https://doi.org/10.1130/0091-7613(1980)8(501:TORAGS)2.0.CO.

McKay, R.M., et al., 2021. Cenozoic History of Antarctic Glaciation and Climate from onshore and offshore studies. In: Florindo, F., et al. (Eds.), Antarctic Climate Evolution, second edition, Elsevier.

McKinley, C.C., Thomas, D.J., Le Vay, L.J., Rolewicz, Z., 2019. Nd isotopic structure of the Pacific Ocean 40–10 Ma, and evidence for the reorganization of deep North Pacific Ocean circulation between 36 and 25 Ma. Earth and Planetary Science Letters 521, 139–149. Available from: https://doi.org/10.1016/j.epsl.2019.06.009.

Mikolajewicz, U., Maier-Reimer, E., Crowley, T.J., Kim, K.-Y., 1993. Effect of Drake and Panamanian gateways on the circulation of an ocean model. Paleoceanography 8 (4), 409–426.

Mildenhall, D.C., 1989. Terrestrial palynology. In: Barrett, P.J. (Ed.), Antarctic Cenozoic History from the CIROS-1 Drillhole, McMurdo Sound, vol. 245. New Zealand DSIR Bulletin, Wellington, pp. 119–127.

Miller, K.G., Browning, J.V., Aubry, M.-P., Wade, B.S., Katz, M.E., Kulpecz, A.A., et al., 2008. Eocene-Oligocene global climate and sea-level changes: St. Stephens Quarry, Alabama. Bulletin of the Geological Society of America 120 (1–2), 34–53. Available from: https://doi.org/10.1130/B26105.1.

Miller, K.G., Curry, W.B., 1982. Eocene to Oligocene benthic foraminiferal isotopic record in the Bay of Biscay. Nature 296 (5855), 347–350. Available from: https://doi.org/10.1038/296347a0.

Miller, K.G., Fairbanks, R.G., Mountain, G.S., 1987. Tertiary oxygen isotope synthesis, sea level history, and continental margin erosion. Paleoceanography 2 (1), 1–19. Available from: https://doi.org/10.1029/PA002i001p00001.

Miller, K.G., Wright, J.D., Fairbanks, R.G., 1991. Unlocking the ice house: Oligocene-Miocene oxygen isotopes, eustasy, and margin erosion. Journal of Geophysical Research 96 (B4), 6829–6848. Available from: https://doi.org/10.1029/90JB02015.

Mohr, B.A.R., 1990. Eocene and Oligocene sporomorphs and dinoflagellate cysts from Leg 113 Drill Sites, Weddell Sea, Antarctica. In: Stewart, N.J. (Ed.). Proceedings of the Ocean Drilling Program, Scientific Results, vol. 113, pp. 595–612.

Morrison, A.K., Hogg, A.M., 2012. On the relationship between Southern Ocean overturning and ACC transport. Journal of Physical Oceanography 43, 140–148. Available from: https://doi.org/10.1175/JPO-D-12-057.1.

Murphy, M.G., Kennett, J.P., 1986. Development of latitudinal thermal gradients during the Oligocene: oxygen-isotope evidence from the southwest Pacific. Initial reports DSDP, Leg 90, Noumea, New Caledonia to Wellington, New Zealand. Part 2, pp. 1347–1360.

Naish, T.R., Woolfe, K.J., Barrett, P.J., et al., 2001. Orbitally induced oscillations in the East Antarctic Ice Sheet at the Oligocene/Miocene boundary. Nature 413, 719–723.

Nawrocki, J., Pańczyk, M., Williams, I.S., 2011. Isotopic ages of selected magmatic rocks from King George Island (West Antarctica) controlled by magnetostratigraphy. Geological Quarterly 55 (4), 301–322.

Newsom, E.R., et al., 2016. Southern Ocean deep circulation and heat uptake in a high-resolution climate model. Journal of Climate 29 (7), 2597–2619.

Nong, G.T., Najjar, R.G., Seidov, D., Peterson, W., 2000. Simulation of ocean temperature change due to the opening of Drake Passage. Geophysical Research Letters 27, 2689–2692.

O'Brien, P.E., Cooper, A.K., Richter, C., et al., 2001. Proceedings of ODP, Init. Repts., 188 [CD-ROM]. Available from: Ocean Drilling Program, Texas A&M University, College Station, TX 77845-9547, United States.

Ohneiser, C., Florindo, F., Stocchi, P., Roberts, A.P., DeConto, R.M., Pollard, P., 2015. Antarctic glacio-eustatic contributions to late Miocene Mediterranean desiccation and reflooding. Nature Communications. Available from: https://doi.org/10.1038/ncomms9765.

Orsi, A.H., Whitworth III, T., Nowlin Jr., W.D., 1995. On the meridional extent and fronts of the Antarctic Circumpolar Current. Deep-Sea Research Part I 42 (5), 641–673. Available from: https://doi.org/10.1016/0967-0637(95)00021-W.

Pagani, M., Huber, M., Liu, Z., Bohaty, S.M., Henderiks, J., Sijp, W., et al., 2011. The role of carbon dioxide during the onset of Antarctic glaciation. Science (New York, N.Y.) 334 (6060), 1261–1264. Available from: https://doi.org/10.1126/science.1203909.

Pagani, M., Zachos, J.C., Freeman, K.H., Tipple, B., Bohaty, S., 2005. Marked decline in atmospheric carbon dioxide concentrations during the Paleogene. Science (New York, N.Y.) 309, 600–603.

Paleobiology and paleoenvironments of Eocene Rocks, McMurdo Sound, East Antarctica. In: Stilwell, J.D., Feldmann, R.M. (Eds.), Antarctic Research Series, vol. 76. American Geophysical Union, Washington, DC, p. 372.

Pälike, H., et al., 2012. A Cenozoic record of the equatorial Pacific carbonate compensation depth. Nature 488 (7413), 609–614. Available from: https://doi.org/10.1038/nature11360.

Passchier, S., Bohaty, S.M., Jiménez-Espejo, F., Pross, J., Röhl, U., Van De Flierdt, T., et al., 2013. Early eocene to middle miocene cooling and aridification of east Antarctica. Geochemistry, Geophysics, Geosystems 14 (5), 1399–1410. Available from: https://doi.org/10.1002/ggge.20106.

Passchier, S., Ciarletta, D.J., Miriagos, T.E., Bijl, P.K., Bohaty, S.M., 2016. An Antarctic stratigraphic record of stepwise ice growth through the Eocene-Oligocene transition. GSA Bulletin . Available from: https://doi.org/10.1130/B31482.1.

Passchier, S., Ciarletta, D.J., Miriagos, T.E., Bijl, P.K., Bohaty, S.M., 2017. An Antarctic stratigraphic record of stepwise ice growth through the Eocene-Oligocene transition. GSA Bulletin 129 (3–4), 318–330. Available from: https://doi.org/10.1130/B31482.1.

Paxman, G.J.G., Gasson, E.G.W., Jamieson, S.S.R., Bentley, M.J., Ferraccioli, F., 2021. Long-term increase in Antarctic Ice Sheet vulnerability driven by bed topography evolution. Geophysical Research Letters 47 (20), e2020GL090003. Available from: https://doi.org/10.1029/2020GL090003.

Paxman, G.J.G., Jamieson, S.S.R., Ferraccioli, F., Bentley, M.J., Ross, N., Armadillo, E., et al., 2018. Bedrock Erosion Surfaces Record Former East Antarctic Ice Sheet Extent. Geophysical Research Letters. Available from: https://doi.org/10.1029/2018GL077268.

Paxman, G.J.G., Jamieson, S.S.R., Hochmuth, K., Gohl, K., Bentley, M.J., Leitchenkov, G., et al., 2019. Reconstructions of Antarctic topography since the Eocene–Oligocene boundary. Palaeogeography, Palaeoclimatology, Palaeoecology 535 (May), 109346. Available from: https://doi.org/10.1016/j.palaeo.2019.109346.

Pearson, P.N., Foster, G.L., Wade, B.S., 2009. Atmospheric carbon dioxide through the Eocene-Oligocene climate transition. Nature 461 (7267), 1110–1113. Available from: https://doi.org/10.1038/nature08447.

Pearson, P.N., McMillan, I.K., Wade, B.S., Jones, T.D., Coxall, H.K., Bown, P.R., Lear, C.H., 2008. Extinction and environmental change across the Eocene-Oligocene boundary in Tanzania. Geology 36 (2), 179–182. Available from: https://doi.org/10.1130/G24308A.1.

Pérez, L.F., Hernández-Molina, F.J., Lodolo, E., Bohoyo, F., Galindo-Zaldívar, J., Maldonado, A., 2019. Oceanographic and climatic consequences of the tectonic evolution of the southern scotia sea basins, Antarctica. Earth-Science Reviews 198. Available from: https://doi.org/10.1016/j.earscirev.2019.102922.

Persico, D., Fioroni, C., Villa, G., 2012. A refined calcareous nannofossil biostratigraphy for the middle Eocene-early Oligocene Southern Ocean ODP sites. Palaeogeography, Palaeoclimatology, Palaeoecology 335–336, 12–23. Available from: https://doi.org/10.1016/j.palaeo.2011.05.017.

Petersen, S.V., Schrag, D.P., 2015. Antarctic ice growth before and after the Eocene-Oligocene transition: new estimates from clumped isotope paleothermometry. Paleoceanography 30, 1305–1317. Available from: https://doi.org/10.1002/2014PA002769.

Pole, M., Hill, B., Harwood, D.M., 2000. In: Stilwell, J.D., Feldmann, R.M. (Eds.), Paleobiology and Paleoenvironments of Eocene Rocks, McMurdo Sound, East Antarctica. Antarctic Research Series, 76. American Geophysical Union, Washington, DC, pp. 243–251.

Poole, I., Cantrill, D.J., Utescher, T., 2005. A multi-proxy approach to determine Antarctic terrestrial palaeoclimate during the Late Cretaceous and Early Tertiary. Palaeogeography, Palaeoclimatology, Palaeoecology 222, 95–121.

Poole, I., Hunt, R., Cantrill, D., 2001. A fossil wood flora from King George Island: ecological implications for an Antarctic Eocene vegetation. Annals of Botany 88 (1), 33–54.

Poore, R.Z., Matthews, R.K., 1984. Oxygen isotope ranking of late Eocene and Oligocene planktonic foraminifers: implications for Oligocene sea-surface temperatures and global ice-volume. Marine Micropaleontology 9 (2), 111–134. Available from: https://doi.org/10.1016/0377-8398(84)90007-0.

Porebski, S.J., 1995. Facies architecture in a tectonically-controlled incised-valley estuary: La Meseta Formation (Eocene) of Seymour Island, Antarctic Peninsula. Studia Geologica Polonica 107, 7–97.

Pound, M.J., Salzmann, U., 2017. Heterogeneity in global vegetation and terrestrial climate change during the late Eocene to early Oligocene transition. Scientific Reports 7, 43386. Available from: https://doi.org/10.1038/srep43386.

Powell, R.D., Krissek, L.A., van der Meer, J.J.M., 2000. Preliminary depositional environmental analysis of CRP-2/2A, Victoria Land Basin, Antarctica: Palaeoglaciological and palaeoclimatic inferences. Terra Antartica 7, 313–322.

Powell, R., Naish, T.R., Fielding, C.R., Krissek, L.A., van der Meer, J.J.M., 2001. Depositional environments for strata cored in CRP-3 (Cape Roberts Project), Victoria Land Basin, Antarctica: Palaeoglaciological and palaeoclimatological inferences. Terra Antartica 8 (3), 207–216.

Prebble, J.G., Raine, J.I., Barrett, P.J., Hannah, M.J., 2006. Vegetation and climate from two Oligocene glacioeustatic sedimentary cycles (31 and 24 Ma) cored by the Cape Roberts Project, Victoria Land Basin, Antarctica. Palaeogeography, Palaeoclimatology, Palaeoecology 231, 41–57.

Prentice, M.L., Matthews, R.K., 1988. Cenozoic ice-volume history: development of a composite oxygen isotope record. Geology 16 (11), 963–966. Available from: https://doi.org/10.1130/0091-7613(1988)016(0963:CIVHDO)2.3.CO;2.

Pross, J., Contreras, L., Bijl, P.K., Greenwood, D.R., Bohaty, S.M., Schouten, S., et al., 2012. Persistent near-tropical warmth on the Antarctic continent during the early Eocene epoch. Nature 488, 73. Available from: https://doi.org/10.1038/nature11300.

Prothero, D.R., 1994. The late Eocene-Oligocene extinctions. Annual Review of Earth and Planetary Sciences 22, 145–165.

Raine, J.I., Askin, R.A., 2001. Terrestrial palynology of Cape Roberts Project Drillhole CRP-3, Victoria Land Basin, Antarctica. Terra Antartica 8 (4), 389–400.

Reguero, M.A., Marenssi, S.A., Santillana, S.N., 2002. Antarctic Peninsula and South America (Patagonia) Paleogene terrestrial environments: biotic and biogeographic relationships. Palaeogeography, Palaeoclimatology, Palaeobiology 179 (3–4), 189–210.

Reguero, M.A., Marenssi, S.A., Santillana, S.N., 2012. Weddellian marine/coastal vertebrates diversity from a basal horizon (Ypresian, Eocene) of the Cucullaea I Allomember, La Meseta formation, Seymour (Marambio) Island, Antarctica. Revista Peruana de Biología 19, 275–284.

Roberts, A.P., Wilson, G.S., Harwood, D.M., Verosub, K.L., 2003. Glaciation across the Oligocene–Miocene boundary in southern McMurdo Sound, Antarctica: new chronology from the CIROS-1 Drill-Hole. Palaeogeography, Palaeoclimatology, Palaeoecology 198 (1), 113–130.

Rose, K.C., Ferraccioli, F., Jamieson, S.S.R., Bell, R.E., Corr, H., Creyts, T.T., et al., 2013. Early East Antarctic Ice Sheet growth recorded in the landscape of the Gamburtsev Subglacial Mountains. Earth and Planetary Science Letters 375, 1–12. Available from: https://doi.org/10.1016/j.epsl.2013.03.05.

Rugenstein, M., Stocchi, P., von der Heydt, A., Dijkstra, H., Brinkhuis, H., 2014. Emplacement of Antarctic ice sheet mass affects circumpolar ocean flow. Global and Planetary Change . Available from: https://doi.org/10.1016/j.gloplacha.2014.03.011.

Sadler, P., 1988. Geometry and stratification of uppermost Cretaceous and Paleogene units on Seymour Island, Northern Antarctic Peninsula. In: Feldmann, R.M., Woodburne, M.O. (Eds.), Geology and Paleontology of Seymour Island, Antarctic Peninsula, vol. 169. Geological Society of America Memoir, pp. 303–320.

Sagnotti, L., Florindo, F., Verosub, K.L., Wilson, G.S., Roberts, A.P., 1998. Environmental magnetic record of Antarctic palaeoclimate from Eocene/Oligocene glaciomarine sediments, Victoria Land Basin. Geophysical Journal International 134, 653–662.

Salabarnada, A., Escutia, C., Röhl, U., Nelson, C.H., McKay, R., Jiménez-Espejo, F.J., et al., 2018. Paleoceanography and ice sheet variability offshore Wilkes Land, Antarctica – Part 1: Insights from late Oligocene astronomically paced contourite sedimentation. Climate of the Past 14, 991–1014. Available from: https://doi.org/10.5194/cp-14-991-2018.

Salamy, K.A., Zachos, J.C., 1999. Latest Eocene-early Oligocene climate change and Southern Ocean fertility: inferences from sediment accumulation and stable isotope data. Palaeogeography, Palaeoclimatology, Palaeoecology 145 (1–3), 61–77. Available from: https://doi.org/10.1016/S0031-0182(98)00093-5.

Sandroni, S., Talarico, F., 2001. Petrography and provenance of basement clasts and clast variability in CRP-3 drillcore (Victoria Land Basin, Antarctica). Terra Antarctica 8 (4), 449–467.

Sangiorgi, F., Bijl, P.K., Passchier, S., et al., 2018. Southern Ocean warming and Wilkes Land ice sheet retreat during the mid-Miocene. Nature Communications 9, 317. Available from: https://doi.org/10.1038/s41467-017-02609-7.

Sauermilch, I., Whittaker, J. M., Klocker, A., Munday, D. R., Hochmuth, K., LaCasce, J. H., Bijl, P., 2021. Gateway-driven weakening of ocean gyres leads to Southern Ocean cooling. Nature Communications, in press (manuscript number NCOMMS-20-23968B).

Savin, S.M., 1977. The history of the Earth's surface temperature during the past 100 million years. Annual Review of Earth and Planetary Sciences 5, 319–355.

Savin, S.M., Douglas, R.G., Stehli, F.G., 1975. Tertiary marine paleotemperatures. Bulletin of the Geological Society of America 86 (11), 1499–1510. Available from: https://doi.org/10.1130/0016-7606(1975)86(1499:TMP)2.0.CO;2.

Scher, H.D., Martin, E.E., 2006. Timing and climatic consequences of the opening of Drake Passage. Science (New York, N.Y.) 312, 428–430.

Scher, H.D., Martin, E.E., 2008. Oligocene deep water export from the north Atlantic and the development of the Antarctic circumpolar current examined with neodymium isotopes. Paleoceanography 23 (1). Available from: https://doi.org/10.1029/2006PA001400.

Scher, H.D., Bohaty, S.M., Smith, B.W., Munn, G.H., 2014. Isotopic interrogation of a suspected late Eocene glaciation. Paleoceanography 29 (6), 628–644. Available from: https://doi.org/10.1002/2014PA002648.

Scher, H.D., Bohaty, S.M., Zachos, J.C., Delaney, M.L., 2011. Two-stepping into the icehouse: East Antarctic weathering during progressive ice-sheet expansion at the Eocene-Oligocene transition. Geology 39 (4), 383–386. Available from: https://doi.org/10.1130/G31726.1.

Scher, H.D., Whittaker, J.M., Williams, S.E., Latimer, J.C., Kordesch, W.E.C., Delaney, M.L., 2015. Onset of Antarctic circumpolar current 30 million years ago as Tasmanian Gateway aligned with westerlies. Nature 523 (7562), 580.

Scotese, C.R., Wright, N., 2018. PALEOMAP Paleodigital Elevation Models (PaleoDEMS) for the Phanerozoic PALEOMAP Project. Available from: <https://www.earthbyte.org/paleo-dem-resource-scotese-and-wright-2018/>.

Shackleton, N., Kennett, J., 1976. Oxygen isotopic evidence for the development of the psychrosphere 38 Myr ago. Nature 260, 513–515. Available from: https://doi.org/10.1038/260513a0.

Shackleton, N.J., Backman, J., Zimmerman, H., Kent, D.V., Hall, M.A., Roberts, D.G., et al., 1984. Oxygen isotope calibration of the onset of ice-rafting and history of glaciation in the North Atlantic region. Nature 307 (5952), 620–623. Available from: https://doi.org/10.1038/307620a0.

Shackleton, N.J., Kennett, J.P., 1975. Paleotemperature history of the Cenozoic and the initiation of Antarctic glaciation: oxygen and carbon isotope analyses in DSDP sites 277, 279 and 281, Initial Reports Deep Sea Drilling Projects, vol. 29, 743.

Sheldon, N.D., Retallack, G.J., Tanaka, S., 2002. Geochemical climofunction from North American soils and application to paleosols across the Eocene-Oligocene boundary in Oregon. The Journal of Geology 110, 687–696.

Shen, Y., 1994. Subdivision and correlation of Cretaceous to Paleogene volcano-sedimentary sequence from Fildes Peninsula, King George Island, Antarctica. In: Shen, Y. (Ed.), Stratigraphy and Palaeontology of Fildes Peninsula, King George Island, Antarctica, State Antarctic Committee. Science Press, Beijing, pp. 1–36. Monograph 3.

Shipboard Scientific Party, 2001a. Leg 188 Summary: Prydz Bay-Co-Operation Sea, Antarctica. In: O'Brien, P. E., Cooper, A. K., Richter, C., et al. (Eds.), Proceedings of the Ocean Drilling Program, Initial Reports, vol. 188. Available from: <http://www-odp.tamu.edu/publications/188_IR/188TOC.HTM>.

Shipboard Scientific Party, 2001b. Leg summary. In: Exon, N.F., Kennett, J.P., Malone, M.J., et al. (Eds.), Proceedings of the Ocean Drilling Program, Initial Reports, vol. 189. Available from: <http://www-odp.tamu.edu/publications/189_IR/chap_01/chap_01.htm>.

Shipboard Scientific Party, 2001c. Site 1170. In: Exon, N.F., Kennett, J.P., Malone, M.J., et al. (Eds.), Proceedings of the Ocean Drilling Program, Initial Reports, vol. 189. Available from: <http://www-odp.tamu.edu/publications/189_IR/chap_05/chap_05.htm>.

Shipboard Scientific Party, 2001d. Site 1172. In: Exon, N.F., Kennett, J.P., Malone, M.J., et al. (Eds.), Proceedings of the Ocean Drilling Program, Initial Reports, vol. 189. Available from: <http://www-odp.tamu.edu/publications/189_IR/chap_06/chap_06.htm>.

Sijp, W.P., England, M.H., Huber, M., 2011. Effect of the deepening of the Tasman Gateway on the global ocean. Paleoceanography 26 (4), 1–18. Available from: https://doi.org/10.1029/2011PA002143.

Sloan, L.C., Rea, D.K., 1995. Atmospheric carbon dioxide and early Eocene climate: a general circulation modeling sensitivity study. Palaeogeography, Palaeoclimatology, Palaeoecology 119, 275–292.

Smellie, J.L., Pankhurst, R.J., Thomson, M.R.A., Davies, R.E.S., 1984. The Geology of the South Shetland Islands: VI. Stratigraphy, Geochemistry and Evolution. British Antarctic Survey, pp. 1–85, Scientific Report 87.

Spada, G., Stocchi, P., 2007. SELEN: a Fortran 90 program for solving the "sea level equation.". Computers & Geosciences 33. Available from: https://doi.org/10.1016/j.cageo.2006.08.006.

Spicer, R.A., Ahlberg, A., Herfort, A.B., Hofmann, C.-C., Raikevich, M., Valdes, P.J., et al., 2008. The Late Cretaceous continental interior of Siberia: A challenge for climate models. Earth and Planetary Science Letters 267, 228–235. Available from: https://doi.org/10.1016/j.epsl.2007.11.049.

Stickley, C.E., Brinkhuis, H., Schellenberg, S.A., Sluijs, A., Röhl, U., Fuller, M., et al., 2004. Timing and nature of the deepening of the Tasmanian Gateway. Paleoceanography 19 (4), PA4027. Available from: https://doi.org/10.1029/2004PA001022.

Stilwell, J., Zinsmeister, W., 1992. Molluscan systematics and biostratigraphy, lower tertiary La Meseta formation, Seymour Island, Antarctic Peninsula, Antarctic Research Series, vol. 55. American Geophysical Union, Washington, DC, p. 192.

Stocchi, P., Escutia, C., Houben, A.J.P., Vermeersen, B.L.A., Bijl, P.K., Brinkhuis, H., et al., 2013. Relative sea-level rise around East Antarctica during Oligocene glaciation. Nature Geoscience 6, 380. Available from: https://doi.org/10.1038/ngeo1783.

Strand, K., Passchier, S., Näsi, J., 2003. Implications of quartz grain microtextures for onset Eocene/Oligocene glaciation in Prydz Bay, ODP Site 1166, Antarctica. Palaeogeography, Palaeoclimatology, Palaeoecology 198 (1–2), 101–111.

Strother, S.L., Salzmann, U., Sangiorgi, F., Bijl, P.K., Pross, J., Escutia, C., et al., 2017. A new quatitative approach to identify reworking in Eocene to Miocene pollen records from offshore Antarctica using red fluorescence and digital imaging. Biogeosciences 14 (8), 2089–2100. Available from: https://doi.org/10.5194/bg-14-2089-2017.

Tambussi, C.P., Acosta Hospitaleche, C.I., Reguero, M.A., Marenssi, S.A., 2006. Late Eocene penguins from West Antarctica; systematics and biostratigraphy. In: Francis, J.E., Pirrie, D., Crame, J.A. (Eds.), Cretaceous-Tertiary High-Latitude Palaeoenvironments; James Ross Basin, Antarctica, 258. Geological Society Special Publications, pp. 145–161.

Tauxe, L., Stickley, C.E., Sugisaki, S., Bijl, P.K., Bohaty, S.M., Brinkhuis, H., et al., 2012. Chronostratigraphic framework for the IODP expedition 318 cores from the Wilkes Land margin: constraints for paleoceanographic reconstruction. Paleoceanography 27, PA2214. Available from: https://doi.org/10.1029/2012PA002308.

Tochilin, C.J., Reiners, P.W., Thomson, S.N., Gehrels, G.E., Hemming, S.R., Pierce, E.L., 2012. Erosional history of the Prydz Bay sector of East Antarctica from detrital apatite and zircon geo- and thermochronology multidating. Geochemistry, Geophysics, Geosystems 13 (11), Q1101.

Toggweiler, J.R., Bjornsson, H., 2000. Drake Passage and paleoclimate. Journal of Quaternary Science 15, 319–328. Available from: https://doi.org/10.1029/2012GC004364.

Torres, T., Marenssi, S.A., Santillana, S., 1994. Maderas Fósiles de la isla Seymour, Formación La Meseta, Antártica. Serie Científica del INACH, Santiago de Chile 44, 17–38.

Troedson, A., Smellie, J., 2002. The Polonez Cove formation of King George Island, Antarctica: stratigraphy, facies and implications for mid-Cenozoic cryosphere development. Sedimentology 49, 277–301.

Truswell, E.M., 1997. Palynomorph assemblages from marine Eocene sediments on the West Tasmanian continental margin and the South Tasman rise. Australian Journal of Earth Sciences 4, 633–654.

Vandenberghe, N., Hilgen, F.J., Speijer, R.P., Ogg, J.G., Gradstein, F.M., Hammer, O., et al., 2012. The Paleogene period. In: Gradstein, F.M., Ogg, J.G., Schmitz, M.D., Ogg, G.M., (Eds.), The Geological Time Scale 2012, Amsterdam, Netherlands, Elsevier, pp. 855–921.

Van Hinsbergen, D.J.J., De Groot, L.V., Van Schaik, S.J., Spakman, W., Bijl, P.K., Sluijs, A., et al., 2015. A paleolatitude calculator for paleoclimate studies. PLoS One 10 (6). Available from: https://doi.org/10.1371/journal.pone.0126946.

Viebahn, J.P., von der Heydt, Le Bars, D., Dijkstra, H.A., 2016. Effects of Drake Passage on a strongly eddying global ocean. Paleoceanography 31 (5), 564–581. Available from: https://doi.org/10.1002/2015PA002888.

Villa, G., Fioroni, C., Persico, D., Roberts, A.P., Florindo, F., 2013. Middle Eocene to late Oligocene Antarctic glaciation/deglaciation and Southern Ocean productivity. Paleoceanography 29, 223–237. Available from: https://doi.org/10.1002/2013PA002518.

Warny, S., Kymes, C.M., Askin, R., Krajewski, K.P., Tatur, A., 2018. Terrestrial and marine floral response to latest Eocene and Oligocene events on the Antarctic Peninsula. Palynology 43 (1), 4–21. Available from: https://doi.org/10.1080/01916122.2017.1418444.

Wei, W., Thierstein, H.R., 1991. Upper Cretaceous and Cenozoic calcareous nannofossils of the Kerguelen Plateau (southern Indian Ocean) and Prydz Bay (East Antarctica). In: Proceedings of the ODP, Scientific Results, J. Barron et al., (Eds.) (Ocean Drilling Program, College Station, TX, 1991), vol. 119, pp. 467–493.

Westerhold, T., Marwan, N., Drury, A.J., Liebrand, D., Agnini, C., Anagnostou, E., et al., 2020. An astronomically dated record of earth's climate and its predictability over the last 66 million years. Science (New York, N.Y.) 369 (6509), 1383–1387. Available from: https://doi.org/10.1126/science.aba6853.

Whitehead, J.M., Quilty, P.G., McKelvey, B.C., O'Brien, P.E., 2006. A review of the Cenozoic stratigraphy and glacial history of the Lambert Graben – Prydz Bay Region, East Antarctica. Antarctic Science 18, 83–99.

Whitehouse, P.L., Gomez, N., King, M.A., Wiens, D.A., 2019. Solid Earth change and the evolution of the Antarctic Ice Sheet. Nature Communications 10 (1), 503. Available from: https://doi.org/10.1038/s41467-018-08068-y.

Wilch, T.I., McIntosh, W.C., 2000. Eocene and Oligocene volcanism at Mount Petras, Marie Byrd land: implications for middle Cenozoic ice sheet reconstructions in West Antarctica. Antarctic Science 12, 477–491.

Williams, S.E., Whittaker, J.M., Halpin, J.A., Müller, R.D., 2019. Australian-antarctic breakup and seafloor spreading: balancing geological and geophysical constraints. Earth-Science Reviews 188, 41–58. Available from: https://doi.org/10.1016/j.earscirev.2018.10.011.

Willis, P.M.A., Stilwell, J.D., 2000. A possible piscivorous crocodile from Eocene deposits of McMurdo Sound, Antarctic. In: Stilwell, J.D., Feldmann, R.M. (Eds.), Paleobiology and

Palaeoenvironments of Eocene Rocks, McMurdo Sound, East Antarctica. Antarctic Research Series, vol. 76. American Geophysical Union, Washington, DC, pp. 355–358.

Wilson, D.S., Jamieson, S.S., Barrett, P.J., Leitchenkov, G., Gohl, K., Larter, R.D., 2012. Antarctic topography at the Eocene-Oligocene boundary. Palaeogeography, Palaeoclimatology, Palaeoecology 335–336, 24–34. Available from: https://doi.org/10.1016/j.palaeo.2011.05.028.

Wilson, D.S., Luyendyk, B.P., 2009. West antarctic paleotopography estimated at the Eocene-Oligocene climate transition. Geophysical Research Letters 36, L16302. Available from: https://doi.org/10.1029/2009GL039297.

Wilson, D.S., Pollard, D., Deconto, R.M., Jamieson, S.S.R., Luyendyk, B.P., 2013. Initiation of the West Antarctic Ice Sheet and estimates of total Antarctic ice volume in the earliest Oligocene. Geophysical Research Letters 40 (16), 4305–4309. Available from: https://doi.org/10.1002/grl.50797.

Wilson, G.S., 2000. Glacial geology and origin of fossiliferous-erratic-bearing moraines, southern McMurdo Sound, Antarctica – an alternative ice sheet hypothesis. In: Stilwell, J.D., Feldmann, R.M. (Eds.), Paleobiology and Paleoenvironments of Eocene Rocks, McMurdo Sound, East Antarctica. Antarctic Research Series, vol. 76. American Geophysical Union, Washington, DC, pp. 19–37.

Wilson, G.S., Naish, T.R., Aitken, A.R.A., Johnston, L.J., Damaske, D., Timms, C.J., et al., 2006. Basin development beneath the southern McMurdo Sound ice shelf (SMIS) – combined geophysical and glacial geological evidence for a potential Paleogene drilling target for ANDRILL. In: Open Science Conference, XXIX SCAR/COMNAP XVII, Hobart, Australia, Abstract.

Wilson, G.S., Roberts, A.P., Verosub, K.L., Florindo, F., Sagnotti, L., 1998. Magnetobiostratigraphic chronology of the Eocene–Oligocene transition in the CIROS-1 core, Victoria Land Margin, Antarctica: implications for Antarctic glacial history. Geological Society of America Bulletin 110, 35–47.

Wilson, T.J., 1999. Cenozoic structural segmentation of the Transantarctic Mountains rift flank in southern Victoria Land, Antarctica. In: Van der Wateren, F.M., Cloetingh, S. (Eds.), Lithosphere dynamics and environmental change of the Cenozoic West Antarctic rift system Global and Planetary Change 23, 105–127.

Yang, S., Galbraith, E., Palter, J., 2014. Coupled climate impacts of the Drake Passage and the Panama Seaway. Climate Dynamics 43, 37–52. Available from: https://doi.org/10.1007/s00382-013-1809-6.

Zachos, J., Pagani, M., Sloan, L., Thomas, E., Billups, K., 2001. Trends, rhythms, and aberrations in global climate 65 Ma to present. Science (New York, N.Y.) 292, 686–693.

Zachos, J.C., Breza, J.R., Wise, S.W., 1992. Early Oligocene ice-sheet expansion on Antarctica – stable isotope and sedimentological evidence from Kerguelen Plateau, Southern Indian Ocean. Geology 20, 569–573.

Zachos, J.C., Quinn, T.M., Salamy, K.A., 1996. High-resolution (104 years) deep-sea foraminiferal stable isotope records of the Eocene–Oligocene climate transition. Paleoceanography 11, 251–266.

Zachos, J.C., Stott, L.D., Lohmann, K.C., 1994. Evolution of early Cenozoic marine temperatures. Paleoceanography 9 (2), 353–387. Available from: https://doi.org/10.1029/93PA03266.

Zastawniak, E., Wrona, R., Gazdzicki, A.J., Birkenmajer, K., 1985. Plant remains from the top part of the Point Hennequin Group (Upper Oligocene), King George Island (South Shetland Islands, Antarctica). Studia Geologica Polonica 81, 143–164.

Zhang, Y.G., Pagani, M., Liu, Z., Bohaty, S.M., DeConto, R., 2013. A 40-million-year history of atmospheric CO2. Phil Trans R Soc A 371, 20130096. Available from: https://doi.org/10.1098/rsta.2013.0096.

Chapter 8

Antarctic Ice Sheet dynamics during the Late Oligocene and Early Miocene: climatic conundrums revisited

Tim R. Naish[1], Bella Duncan[1], Richard Levy[1,2], Robert M. McKay[1], Carlota Escutia[3], Laura De Santis[4], Florence Colleoni[4], Edward G.W. Gasson[5], Robert M. DeConto[6] and Gary Wilson[2]

[1]Antarctic Research Centre, Victoria University of Wellington, Wellington, New Zealand, [2]GNS Science, Lower Hutt, New Zealand, [3]Andalusian Institute of Earth Sciences, CSIC and Universidad de Granada, Armilla, Spain, [4]National Institute of Oceanography and Applied Geophysics – OGS, Sgonico, Italy, [5]School of Geographical Sciences, University of Bristol, Bristol, United Kingdom, [6]Department of Geosciences, University of Massachusetts, Amherst, Amherst, MA, United States

8.1 Introduction

Investigating the drivers of ice volume change in the geological past has been a primary goal of the Scientific Committee on Antarctic Research (SCAR) Past Antarctic Ice Sheet (PAIS) Dynamics strategic research programme. These drivers include: (1) millennial-scale variations in local insolation, atmosphere and ocean forcing caused by quasi-periodic changes in Earth's orbital parameters ('Croll-Milankovitch cycles'), (2) longer-term, million-year trends controlled by more gradual changes in atmospheric greenhouse gas concentrations (e.g., Masson-Delmotte et al., 2013; Zhang et al., 2013), and (3) slowly-evolving plate tectonic changes modulating ocean currents and the flow of heat around Antarctica (Bijl et al., 2013; Kennett, 1977; Kennett et al., 1974; Sijp et al., 2014), as well as impacting the topography and elevation of the continent (Paxman et al., 2019, 2020; Wilson and Luyendyk, 2009).

The Late Oligocene to Early Miocene (~26–22 Ma) is an important period of Earth history during which the Antarctic Ice Sheet (AIS) was

Antarctic Climate Evolution. DOI: https://doi.org/10.1016/B978-0-12-819109-5.00013-X

highly dynamic. Here, we use it to examine the relative influences of plate tectonics, atmospheric carbon dioxide and orbital forcing on southern high-latitude regional climate and the evolution of the ice sheet. We provide a synthesis of the latest research as an update on the first edition of this chapter, which focussed primarily on the role of orbital forcing on ice sheet variability (Wilson et al., 2009). Our approach summarises the latest and most complete reconstructions of geography and bathymetry, Southern Ocean and atmospheric temperature, atmospheric carbon dioxide, and Antarctic ice volume and its influence on global sea level, using geological proxies from locations in the near and far-field of the AIS (Fig. 8.1).

This chapter presents some major advances over the last 10 years, and also highlights some remaining challenges that still hamper reconstructions, and a better understanding of Earth's climate system using geological proxies. These include:

1. Cenozoic atmospheric CO_2 compilations reconstructed from alkenone and boron have limited resolution (Badger et al., 2013; Pagani et al., 2005, 2010, 2011; Seki et al., 2010; Super et al., 2018; Zhang et al., 2013) and boron (Bartoli et al., 2011; Foster et al., 2012; Greenop et al., 2014, 2019; Honisch et al., 2009; Martinez-Boti et al., 2015) proxies. This is especially true during the Oligocene and Miocene, where the relationship between longer-term secular variation on millions of year timescales and shorter-term millennial-scale fluctuations in the carbon cycle are difficult to reconcile (e.g., Greenop et al., 2019).
2. Deep-ocean benthic foraminiferal $\delta^{18}O$ records provide the most detailed proxy for long-term and glacial-interglacial climate variability during the Cenozoic (e.g., Cramer et al., 2009; De Vleeschouwer et al., 2017; Westerhold et al., 2020; Zachos et al., 2001a). However, deconvolving the relative influence of ocean temperature and ice volume on $\delta^{18}O$ composition is not straight forward. It is generally accepted that prior to the development of continental scale ice sheets across Northern Hemisphere continents, which began $\sim$3 Ma, the Cenozoic benthic $\delta^{18}O$ record reflects predominantly Antarctic ice volume and bottom-water temperature. Although these two properties typically co-vary, they appear to decouple in the Late Oligocene (e.g., Greenop et al., 2019; Hartman et al., 2018; Hauptvogel et al., 2017; Pekar et al., 2006).
3. To address this, several studies have used paired Mg/Ca paleothermometry, on the same foraminiferal calcite used in benthic $\delta^{18}O$ records, to extract the temperature component (e.g., Cramer et al., 2009; Lear et al., 2015; Miller et al., 2020; Shevenell et al., 2008). Unfortunately, most of these records are restricted to mid- and low-latitude sites, because of the poor-preservation of calcitic-microfossils in the cold CO_2 saturated waters of the Southern Ocean. Moreover, this method contains significant uncertainties outlined in detail by Raymo et al. (2018). Recently, upper Southern Ocean

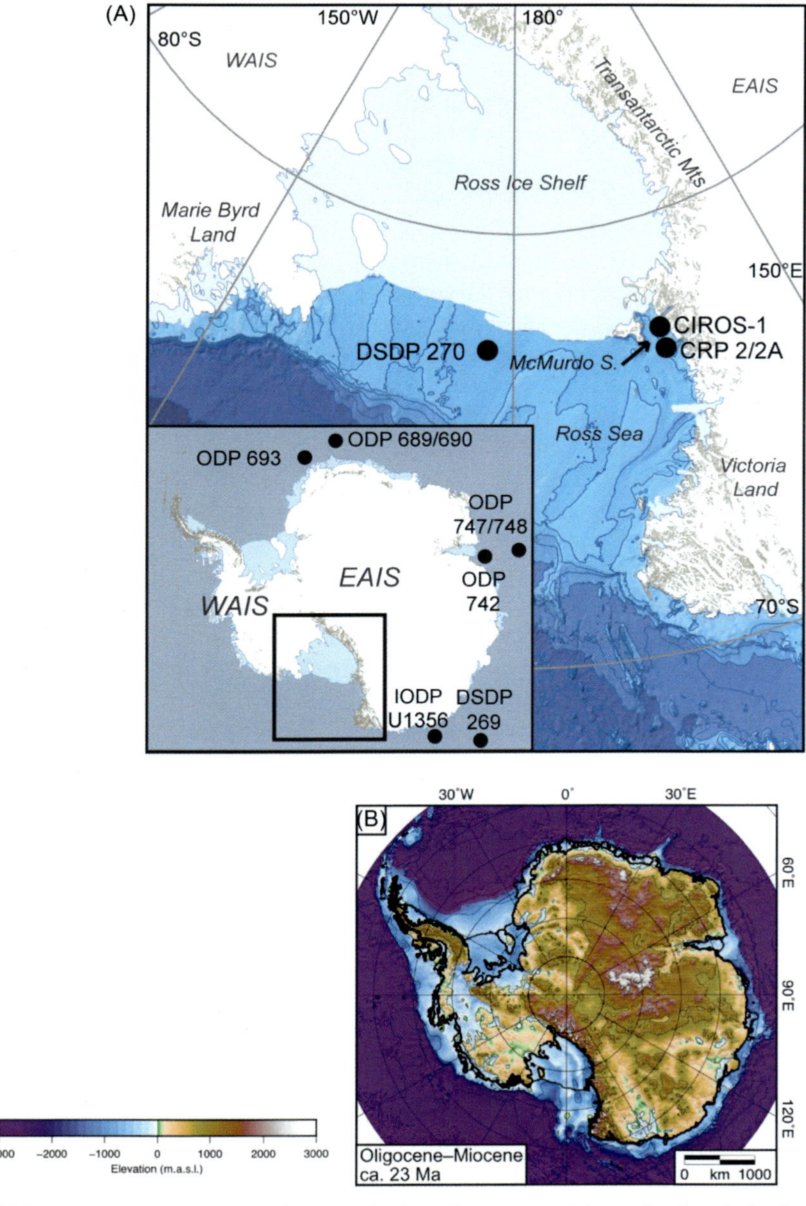

FIGURE 8.1 (A) Location map of present day Ross Sea sector and Antarctica (inset) showing drill sites and locations referred to in the text. (B) Reconstructed topography for the Oligocene-Miocene Transition. The elevations are given relative to present-day mean sea level, and contoured at 1 km intervals. Note that elevations are for fully isostatically-relaxed ice-free conditions. *Reproduced from Paxman et al. (2019).*

temperatures have been derived from sediment cores on the Antarctic continental margin using glycerol dialkyl glycerol tetraethers (GDGTs), membrane lipids formed by archaea and some bacteria (e.g., Hartman et al., 2018; Levy et al., 2016; Sangiorgi et al., 2018). Significant advances in assessing the drivers of long-term AIS volume change from the benthic $\delta^{18}O$ record are now being made through the development of independent records of high-southern latitude ocean temperature through the use of lipid biomarkers (e.g., Hartman et al., 2018; Sangiorgi et al., 2018).

4. Alternatively, ice volume calibrations of the benthic $\delta^{18}O$ record have been constrained using sea-level reconstructions from back-stripped continental margins (shallow marine sedimentary sequences that have tectonic, isostatic and compaction effects removed; e.g., Cramer et al., 2011; Grant et al., 2019; Miller et al., 2005, 2020; Kominz et al., 2008; Pekar and Kominz, 2001), but the accuracy is often hampered by the limited precision of paleodepth indicators and the influence of sediment erosion during sea-level lowstands, leading to underestimation of the full amplitude of sea-level change.
5. Antarctic paleogeographic reconstructions show that the morphology of the bedrock surface has evolved substantially over the past 34 million years. At the Eocene-Oligocene Transition (see Galeotti et al., 2016), most of the West and East AISs formed on a land surface above sea level (De Santis et al., 1999; Hochmuth et al., 2020; Paxman et al., 2019, 2020; Wilson and Luyendyk, 2009). With time, tectonic subsidence, sediment erosion, and ice and water loading have resulted deepening of many previously emergent sectors of the Antarctic margin (and that are situated below sea-level today). These paleogeographic reconstructions, together with the timing of subsidence, have significant implications for understanding the evolution of the AIS. It is much easier to grow an ice sheet on land than on a submarine bed. For example, continental-scale dynamic ice sheet modelling studies, that use restored Antarctic paleogeography for the Oligocene and Miocene, show West Antarctica accommodated a much larger reservoir of ice than today, even when ocean temperatures in the Ross Sea were relatively warm (e.g., Colleoni et al., 2018; Gasson et al., 2016; Paxman et al., 2020; Wilson et al., 2013). Marine-based ice is inherently more sensitive to ocean warming than atmospheric heat, and retreat is exacerbated by non-linear processes such as marine ice sheet instability on reverse-slope bed topography and/or marine ice cliff instability if ice shelves disintegrate (DeConto and Pollard, 2016; Gasson et al., 2016; Pollard et al., 2015), due primarily to incursions of warm water toward and at ice sheet grounding lines (e.g., Levy et al., 2019).

8.2 Oligocene-Miocene Transition in Antarctic geological records and its climatic significance

Early definition of the Oligocene-Miocene Transition (OMT) relied on the identification of the last occurrence of the calcareous nannofossil

Dictyococcites bisectus (23.7 Ma; Berggren et al., 1985). However, the use of this stratigraphic marker proved problematic in the colder waters, coarse sediments and hiatus prone strata of the Antarctic and Southern Ocean. The reassignment of the Epoch boundary by Cande and Kent (1992, 1995) to the slightly older base of Magnetic Polarity subchron C6Cn.2n (23.8 Ma; Fig. 8.2) made its identification more straightforward in Antarctica and the Southern Ocean, but only in relatively complete and continuous stratigraphic successions (e.g., Kulhanek et al., 2019; Roberts et al., 2003; Wilson et al., 2002). More recently, the recognition of astronomically-influenced cyclical physical properties, and $\delta^{18}O$ records in continuously deposited deep successions, has enabled astronomical calibration of Late Oligocene through Early Miocene time. The astronomical calibration suggested that, while still coincident with the base of subchron C6Cn.2n, the boundary was in fact nearly a million years younger (22.9 $\pm$ 0.1 Ma; Pälike et al., 2004; Shackleton et al., 2000; 23.03 Ma; Billups et al., 2004; Gradstein et al., 2004; Fig. 8.2).

The climatic significance of this was outlined by Zachos et al. (1997, 2001a,b) and Pälike et al. (2006) who recognised the coincidence of the OMT with the culmination of a $\sim$300 kyr-duration, up to $\sim +1‰$, $\delta^{18}O$ excursion (Mi-1 event; Miller et al., 1991) with a $\sim$1.2 myr low-amplitude variability in the eccentricity and obliquity of the Earth's orbit (Fig. 8.2). An attempt to constrain the ice volume component of the Mi-1, $\delta^{18}O$ excursion using the global sea-level estimates from the New Jersey continental margin (Pekar et al., 2006), implied the AIS may have grown to 120% of its present-day size, which should be regarded as a maximum as these estimates are uncorrected for glacio-isostatic adjustment (GIA) and contain large uncertainties (discussed above). Notwithstanding this, the New Jersey records appear to be consistent with estimates from numerical ice sheet simulations using reconstructed paleotopographies, $\sim$300 ppm atmospheric CO_2, and cold orbits run to equilibrium (Gasson et al., 2016; Wilson et al., 2002).

Levy et al. (2019), using an analysis of the most recent $\delta^{18}O$ compilation for the Cenozoic (De Vleeschouwer et al., 2017), identified a progressive increase in the obliquity sensitivity parameter (S_{obl}) between 24.5 and 23.5 Ma and associated this with a significant period of AIS growth. Indeed, using proximal geological evidence, Levy et al. (2019) showed that the Mi-1 glaciation was the first major expansion of ice sheets across the Ross Sea continental shelf since the Eocene-Oligocene Transition and the onset of Antarctic glaciation (e.g., Galeotti et al., 2016). This was based on the first occurrence of ice proximal glacimarine sediments in the DSDP Site 270 (Kulhanek et al., 2019) and disconformities in Cape Roberts Project (CRP) site CRP-2/2A (Naish et al., 2001). Further evidence for continent-wide glacial expansion and climatic deterioration comes from the occurrence of mass transport deposits (MTDs) on the continental rise at International Ocean Discovery Program (IODP) Site U1356 in the Wilkes Land region (Escutia et al., 2011, 2014), alongside a large turnover in Southern Ocean phytoplankton (Crampton et al., 2016) and Antarctic

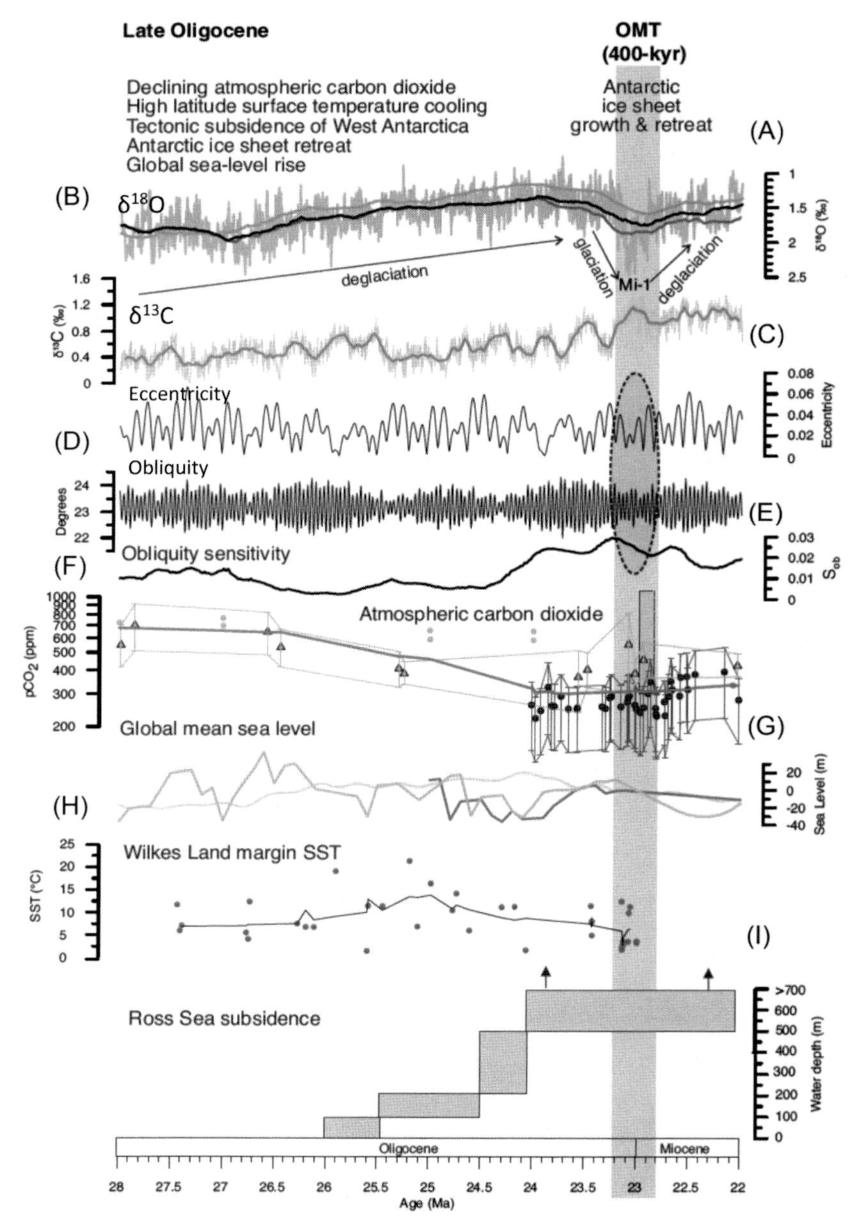

FIGURE 8.2 Late Oligocene and Oligocene-Miocene climate Transition (OMT). (A) Smoothed 500-kyr moving averages of benthic foraminiferal $\delta^{18}O$ stacks from Cramer et al. (2009), De Vleeschouwer et al. (2017) and Miller et al. (2005), superimposed on the high resolution benthic $\delta^{18}O$ record of Liebrand et al. (2011). (B) Benthic foraminiferal $\delta^{13}C$ record from Liebrand et al. (2011). (C) Eccentricity and (D) obliquity forcing (Laskar et al., 2004). (E) Obliquity sensitivity in

(*Continued*)

shelf-wide unconformities in seismic data (De Santis et al., 1995; Donda et al., 2008; Escutia et al., 1997, 2005; Mckay et al., 2021; Sorlien et al., 2007) (Fig. 8.1).

A reassessment of the chronostratigraphy of Deep Sea Drilling Project (DSDP) Site 270 in the Ross Sea (Kulhanek et al., 2019) provides the most complete ice proximal stratigraphic record of the Oligocene-Miocene boundary in Antarctica. It is correlated with an interval of mudstone containing ice-berg rafted debris between 149 and 146 m below sea-floor (mbsf), but may potentially fall within an ~100-kyr-duration unconformity at 149 mbsf. The boundary occurs ~70 m above grounding-line proximal diamictite, sandstone and laminated mudstone, raising questions about the additional role of hydro-glacio-isostatic adjustment (GIA) on near-field sea-level change at this time. In the even more ice-proximal CRP-2A core adjacent to the Victoria Land coast the OMT and the majority of the preceding Mi-1, $\delta^{18}O$ excursion is missing within a ~3 Myr-duration unconformity at the base of Sequence 8 (Naish et al., 2008). Roberts et al. (2003) placed the OMT at 274 m in the CIROS-1 core, ~35 km south of CRP-2A (Fig. 8.1), within an interval of ice-proximal glacimarine diamictite. However, following the relocation of the original position of OMT (e.g., Wilson et al., 2002) by Naish et al. (2008), it is also missing in the CIROS-1 core and most likely correlates with an unconformity at 92 m (Wilson et al., 2009).

Ice-distal, Late Oligocene-Early Miocene strata reported from the East Antarctic margin include Maud Rise (ODP Leg 113 Sites 689 and 690; Barker et al., 1999), the Weddell Sea margin (ODP Leg 113 Site 693; Barker et al., 1999) and Kerguelen Plateau (ODP Leg 120 sites 747 and 748; Fraass et al., 2019; Schlich and Wise, 1992; Fig. 8.1). The record at Maud Rise is relatively thin and comprises exclusively siliceous and carbonate ooze, although rare glacial dropstones are reported in strata from Site 689 (Barker et al., 1999). In the Weddell Sea margin, Oligocene-Miocene sediments are also fine-grained and include diatom mud, clay and ooze (Barker et al., 1988). OMT sediments at Kerguelen Plateau are also carbonate ooze (Schlich and Wise, 1992). Foraminifera preservation and age resolution were good enough at ODP Site 747 to yield a benthic oxygen isotope stratigraphy across the boundary at that site (Billups and Schrag, 2002; Wright and Miller, 1992), although a recent re-

---

the benthic $\delta^{18}O$ stack of De Vleeschouwer et al. (2017), calculated by Levy et al. (2019) and considered to represent the relative presence of marine-based AISs. (F) Atmospheric CO_2 (red line is median) reconstructed from alkenones (orange triangles; Pagani et al., 2011; Zhang et al., 2013), boron (black dots; Sosdian et al., 2018; Greenop et al., 2019), phytane (pink dots, Witkowski et al., 2018), and stomatal gas exchange method (green box; Reichgelt et al., 2016). (G). Global mean sea-level reconstructed from sequence stratigraphic back-stripping of the New Jersey continental margin (Cramer et al., 2011; Kominz et al., 2008; Miller et al., 2020). (H) Near surface ocean temperature off Wilkes Land, Antarctica (Hartman et al., 2018). (I) Water depth estimates from DSDP Site 270, Ross Sea, Antarctica (Leckie and Webb, 1983; Kulhanek et al., 2019).

evaluation of the age model and new benthic oxygen isotope stratigraphy suggests that the OMT occurs within a ~200 kyr unconformity (Fraass et al., 2019). The amplitude of the Mi-1 event was much reduced compared to equatorial values, with a $\delta^{18}O$ shift of only 0.3‰ across the OMT.

Drilling on the East Antarctic margin in Prydz Bay (ODP Legs 119 and 188; Cooper and O'Brien, 2004; Hambrey et al., 1991) did not yield any Late Oligocene-Miocene age strata and Hambrey et al. (1991) concluded that this was due to erosion beneath an expanded middle Miocene ice sheet. Seismic stratigraphy on the inner continental shelf across the Prydz Channel to the shallow outer shelf, that intersects with ODP drill sites, shows widespread erosion that correlates with a probable Late Oligocene-Early Miocene stratigraphic hiatus at the drill Site ODP 742 (Mckay et al., 2021). While, this does not provide unequivocal evidence of large-scale grounding of a marine-based ice sheet after that time, it does suggest ice advanced into fjords and across the shallow continental shelf during, or after, the OMT (Barron and Larsen, 1989; Hambrey et al., 1991; Mckay et al., 2021). Although the OMT interval was not directly sampled, marine geological and geophysical data from the continental shelf seaward of the Aurora subglacial basin Gulick et al. (2017) infer that grounded ice advanced across, and retreated from, the Sabrina Coast continental shelf at least 11 times during the Oligocene and Miocene epochs.

At Site U1356, located in ~4 km of water depth adjacent to the East Antarctic Wilkes Land margin, the OMT overlies a poorly recovered interval of alternations of laminated and bioturbated diatomaceous contourites and sediment gravity flow deposits, punctuated by Mass Transport Deposits (MTDs) and numerous hiatuses (Escutia et al., 2011; 2014). The OMT is associated with a dramatic change in depositional style on the Wilkes Land margin from MTD-dominated to turbidite/contourite dominated deposition (Escutia et al., 2011; 2014; Tauxe et al., 2012). TEX_{86} records show cooling from 25 to 23 Ma with lowest temperatures coincident with the OMT (Hartman et al., 2018; Fig. 8.2). Sedimentological and geochemical indicators, dinocyst associations and SST variations point to a dynamic oceanic frontal system paced by glacial-interglacial climate variability across the OMT (Bjil et al., 2018; Hartman et al., 2018; Salabarnada et al., 2018).

A recent reassessment of the chronostratigraphy of DSDP Leg 28, Site 269, located on the abyssal plain ~280 km seaward from Site U1356, revealed sediments dated ~24.2 to ~23 Ma (Evangelinos et al., 2020). Although the site was not continuously cored, the recovered sediments contain the record that was interrupted by MTD deposits at Site U1356, thus providing snapshots into the oceanic conditions that prevailed in this sector of the Southern Ocean when ice sheets were expanding repeatedly onto the Wilkes Land continental shelf during the latest Oligocene. Evangelinos et al. (2020) imply that overall cooling and ice sheet advance was punctuated by warm interglacials associated with enhanced upwelling of Circumpolar Deep Water (CDW) as the polar front episodically migrated southward.

8.3 Conundrums revisited

8.3.1 What caused major transient glaciation of Antarctica across the OMT?

The OMT is characterised by up to +1‰ cooling event that occurs as a two-step increase in benthic foraminiferal $\delta^{18}O$ over 200–300 kyr (Mi-1 glaciation of Miller et al., 1991), followed by a dramatic two-step rebound of −1.2‰ within 100 kyr (e.g., Liebrand et al., 2011; 2016; 2017; Fig. 8.2). Although it is not possible to discount a Northern Hemisphere ice contribution (e.g., DeConto et al., 2008), the OMT has been widely attributed to expansion and collapse of the AIS, driving changes in global mean sea-level of between 30 and 90 m (Greenop et al., 2019; Kominz et al., 2008; Liebrand et al., 2011; Levy et al., 2019; Mawbey and Lear, 2013; Miller et al., 1991, 2005, 2020; Naish et al., 2001; Pälike et al., 2006; Paul et al., 2000; Pekar et al., 2002).

The entire OMT spans a 400-kyr eccentricity cycle. Numerous studies have noted the coincidence of the glacial phase (Mi-1 glaciation) with a 1.2-Myr-paced minimum in the modulation of the Earth's axial tilt (an obliquity node), as well as a minimum in the 400-kyr-long eccentricity cycle (a very circular orbit), which together favour lower mean annual insolation and lower seasonality, conducive to ice sheet growth (Coxall et al., 2005; Pälike et al., 2006; Zachos et al., 2001a) (Fig. 8.2). While obliquity variance was low, the AIS appeared to be highly responsive to ice volume variability in the obliquity band (Naish et al., 2001; Levy et al., 2019). This heightened 'obliquity sensitivity' (S_{obl}) (Fig. 8.2) was attributed to the first major expansion of the AIS into the ocean, as near-surface temperatures cooled through a threshold, whereby marine-based ice could be sustained on the continental shelf. Both Naish et al. (2009) and Levy et al. (2019) have argued that the mass balance of marine-based ice sheets is modulated by obliquity via its influence on the pole-equator temperature gradient controlling the upwelling of CDW (e.g., Toggweiler et al., 2007), with consequences for melt rates at marine grounding lines.

However, Greenop et al. (2019) observed that the obliquity nodes and eccentricity minima occur regularly throughout the Late Oligocene (Laskar et al., 2004; Pälike et al., 2006) and the amplitude of the preceding node at 24.4 Ma is more extreme than the one associated with the OMT (Fig. 8.2). This is also consistent with strong orbital modulation of proximal glacimarine sedimentation in the Ross Sea (Naish et al., 2001), and SSTs and ocean dynamics on the Wilkes Land margin prior to the OMT, as discussed above (Bjil et al., 2018; Evangelinos et al., 2020; Hartman et al., 2018; Salabarnada et al., 2018).

On this basis, Greenop et al. (2019) proposed the additional role of a long-term decline in atmospheric CO_2, which dipped below 400 ppm for the first time in the Cenozoic (Fig. 8.2), cooling the climate to a state whereby an extreme orbital configuration could trigger widespread Antarctic glaciation and the development of the first marine-based ice margins. A similar CO_2 threshold of ~400 ppm for marine-based ice sheet development had been proposed by Foster and Rohling (2013) based on the empirical relationship between global ice volume and atmospheric CO_2 in proxy records, and by Naish et al. (2009) and Levy et al. (2016; 2019) based on direct geological evidence from Antarctic margin drill cores.

The AIS expansion had a dramatic effect on Southern Ocean sea surface temperatures and sea ice extent via the influence of albedo on regional temperatures and its elevation on low-level winds. Numerical climate and ice sheet simulations clearly show that a growing Mi-1 ice sheet cooled Southern Ocean sea surface temperatures by several degrees, pushing the 0°C isotherm equatorward and increasing the area, thickness and concentration of seasonal and perennial sea ice cover (DeConto et al., 2007). Furthermore, those simulations suggest that as the katabatic wind field increased in intensity it enhanced polar easterlies, and the expansion of sea-ice may have enhanced westerlies increasing ocean frontal divergence and upwelling, with possible implications for the marine carbon cycle and CO_2 drawdown observed in proxy records (Greenop et al., 2019; Masson-Delmotte et al., 2013; Pagani et al., 2011; Super et al., 2018; Zhang et al., 2013). Moreover, the long-term increase of 0.8‰ in carbon isotopes from 24 to 22.9 Ma (Fig. 8.2) has been attributed to an increase in global organic carbon burial and the associated reduction in atmospheric CO_2 (Mawbey and Lear, 2013; Paul et al., 2000; Stewart et al., 2017; Zachos et al., 1997).

Arguably even more remarkable than the Mi-1 glaciation, was its transient nature and the rapid deglaciation that followed within less than 100 kyrs (Greenop et al., 2019). The global ice volume loss may have been equivalent to as much as ~80 m of global sea-level rise implying continent-wide deglaciation of Antarctica according to some estimates (e.g. Miller et al., 2020) (Fig. 8.2), albeit the proxy sea-level estimates have significant uncertainties and are uncorrected for GIA. Atmospheric CO_2, reconstructed using the boron isotope method, appears to have rebounded by ~ 65ppm during the deglaciation from a low of ~265 ppm at the maximum of Mi-1 excursion, apparently in near-phase with a benthic $\delta^{18}O$ increase of +1.2‰ in the Atlantic Ocean equatorial ODP Site 926 record (Greenop et al., 2019) (Fig. 8.3), suggesting a direct coupling between the carbon cycle and orbitally-paced, ice-volume variability. This was in agreement with direct geological evidence for orbitally-driven fluctuations of the East Antarctic Ice Sheet (EAIS) from the dating of glacimarine sedimentary cycles in the Cape Roberts Project cores in the western Ross Sea (Naish et al., 2001). However, the magnitude of the implied OMT deglaciation is at odds with an oxygen-isotope enabled, ice-sheet

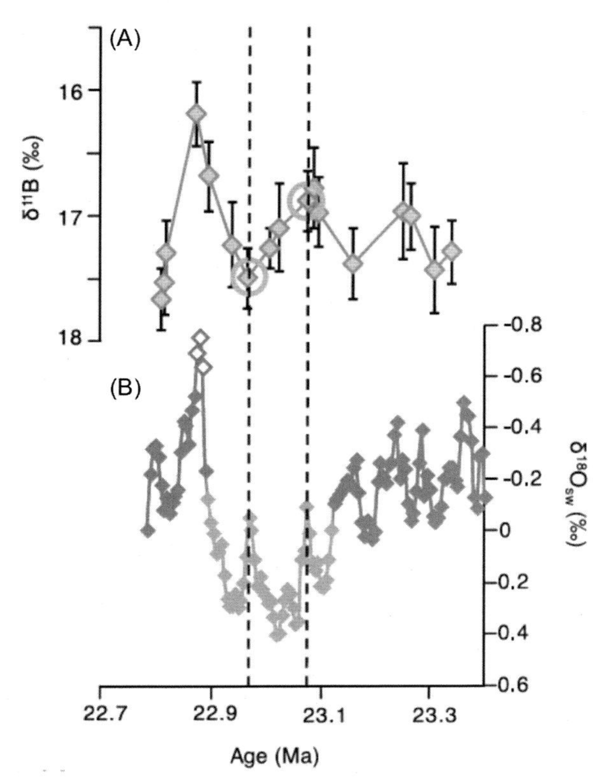

FIGURE 8.3 The relationship between $\delta^{11}B$ (Boron isotope atmospheric CO_2 proxy) and $\delta^{18}O_{sw}$ (global ice volume proxy) across the OMT. (A) The $\delta^{11}B$ record from Site 926 in the western equatorial Atlantic Ocean focused on 22.7–23.4 Ma from Greenop et al. (2019). The pink circles highlight $\delta^{11}B$ samples that fall within "peak glaciation conditions." Note the axis is reversed. (B) Relative $\delta^{18}O_{sw}$ change colour-coded for peak glacial (blue) and preglacial/postglacial conditions (red; Mawbey and Lear, 2013). Open circles are $\delta^{18}O_{sw}$ estimates within a dissolution event and therefore bias toward negative values. The dashed black lines show the coincident timing of the two $\delta^{11}B$ data points that sit on the pre/post-OMT/Mi-1 glaciation event lines.

modelling study by Gasson et al. (2016), that requires an extreme warm orbital configuration and atmospheric CO_2 concentration of more than 800 ppm to effect widespread retreat of terrestrial ice sheet. For a range of atmospheric CO_2 concentrations between 280 and 500 ppm, the $\delta^{18}O$ change was 0.52‰–0.66‰, or a sea level equivalent change of 30–36 m. Moreover, a recent transient ice sheet modelling study for deglaciations during the middle Miocene climatic optimum reveals significantly lower Antarctic ice volume variations (Stap et al., 2019), implying a relatively greater contribution by bottom water temperature to the $\delta^{18}O$ excursion. This modelling is consistent with a recent reconstruction of Middle-Late Miocene bottom water temperatures using Mg/Ca paleothermometry (Bradshaw et al., 2021). Intriguingly, while Early Miocene CO_2 concentrations derived from marine proxies generally stay below ~500 ppm, two new quantitative CO_2 estimates using the stomatal leaf proxy method (Steinthorsdottir et al., 2019) and a leaf gas-exchange model (Reichgelt et al., 2016) from lake sediments in New Zealand suggest the CO_2 concentration may have been considerably higher, up to 1000 ppm (Steinthorsdottir et al., 2021). The well-dated, high-resolution varved record of the OMT from Foulden Maar implies atmospheric CO_2 more than doubled from ~500 ppm at the Mi-1 glacial maximum to ~1100 ppm within the first 20 kyr of the deglaciation phase (Reichgelt et al., 2016) (Fig. 8.2). While there are large uncertainties associated with the CO_2 proxy reconstructions, and the precise amplitude of the atmospheric CO_2 perturbation may be debatable, it appears to be orbitally-paced. Exactly how exchange between ocean and atmospheric reservoirs occurs at this magnitude and timescale remains an open question.

8.3.2 Apparent decoupling of Late Oligocene climate and ice volume?

Preceding the Mi-1 glaciation, a 3-Myr trend towards lower values (decrease up to ~1‰) in Late Oligocene (~27–24 Ma) composite benthic $\delta^{18}O$ records has been widely interpreted as a significant global warming event (e.g., Cramer et al., 2009; De Vleeschouwer et al., 2017; Liebrand et al., 2011, 2016, 2017; Westerhold et al., 2020; Zachos et al., 2001a) (Fig. 8.2). Bottom water temperature estimates using Mg/Ca paleothermometry on the same foraminiferal calcite as the $\delta^{18}O$ measurements suggest the mid-low latitude ocean warmed by up to 2°C (Billups and Schrag, 2002; Cramer et al., 2011). However, a $\delta^{18}O$ record from the high southern latitude Site ODP 690 on Maud Rise, displays a more muted decrease (Hauptvogel et al., 2017).

Paleoenvironmental information from stratigraphic drill cores in the Ross Sea region (Fig. 8.1) appears to preclude significant warming and, in fact, is more consistent with cooling during this time. Nannofossil, foraminiferal and marine macrofossil assemblages observed in DSDP Site 270 and CRP Ross Sea cores imply temperatures did not warm in regions proximal to the continent (Leckie and Webb, 1983; Taviani and Beu, 2003; Watkins and Villa, 2000). Moreover, pollen assemblages, clay minerals and chemical weathering indices in CRP cores record a cooling trend during the Late Oligocene (Ehrmann et al., 2005; Prebble et al., 2006a, 2006b; summarised in Barrett, 2007; Passchier and Krissek, 2008). In addition, orbitally-paced glacial advances recorded in shallow-marine sediments from western Ross Sea drill cores, showed the climate was periodically cold enough for EAIS trunk glaciers to down cut through the Transantarctic Mountains and ground on the inner continental shelf during the Late Oligocene (Hambrey et al., 1989; Naish et al., 2001; Roberts et al., 2003).

Late Oligocene glacio-eustatic sea-levels, reconstructed from sequence stratigraphic backstripping of shallow-marine continental margin records from sites in the far-field of the AIS, imply a long term lowering of global mean sea-level as ice grew on Antarctica from 26.5 Ma, with superimposed, orbital-scale sea-level fluctuations of up to $\pm$ ~50 m (Miller et al., 2005, 2020; Kominz et al., 2008), although this magnitude of Antarctic ice volume variability remains difficult to reconcile in equilibrium or transient numerical ice sheet simulations (Gasson et al., 2016; Stap et al., 2019) (Figs 8.2, 8.4 and 8.5).

In contrast, cooling on the Wilkes Land margin of East Antarctica appears to have come later. A TEX_{86}-based sea-surface temperature (SST) record from IODP Site U1356 reveals a warming between 28–25 Ma followed by cooling through to ~23 Ma, culminating with the OMT (Hartman et al., 2018). A continuous Late Oligocene succession of contourites between 26 and 25 Ma at this site, records changes in bottom water strength and oxygenation interpreted to result from considerable fluctuations in oceanic

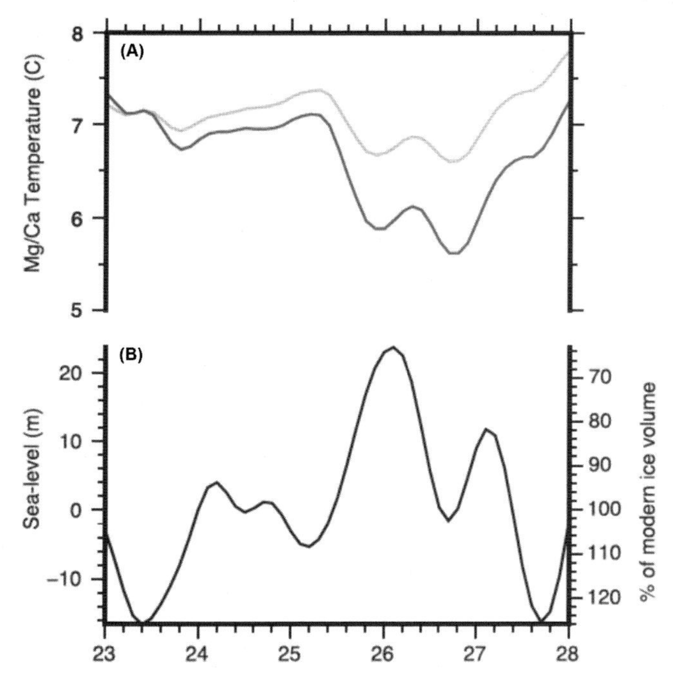

FIGURE 8.4 (A) Mg/Ca-based Pacific Ocean bottom water temperature estimates and (B) sea level with the per cent of modern ice volume estimates from Cramer et al. (2011) based on the sea level record of Kominz et al. (2008) and updated to the Gradstein et al. (2012) time scale. The blue line represents equation 7a and orange for equation 7b that Cramer et al. (2011) used to calculate bottom water temperature. Sea level is normalised to the modern ice volume equivalent of ~64 m. *Reproduced from Hauptvogel et al. (2017).*

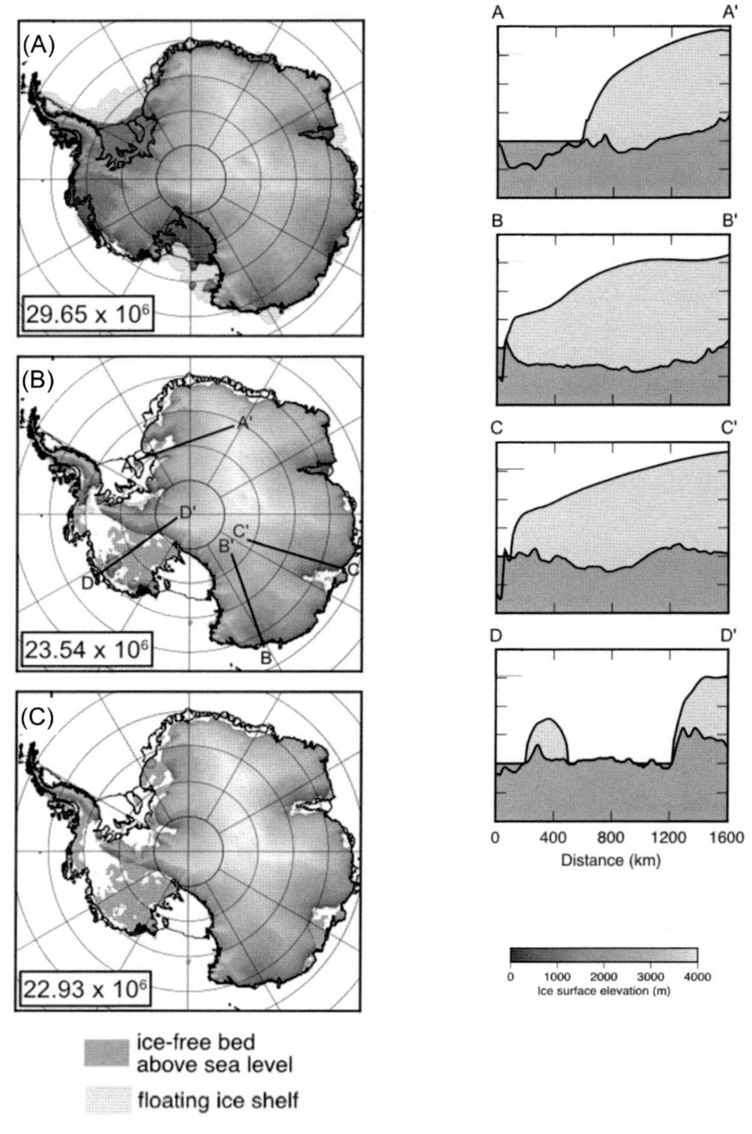

FIGURE 8.5 Ice sheet sensitivity to atmospheric CO_2 reconstructed and simulated for the Oligocene Miocene transition (OMT) after Paxman et al. (2020). (A) Shows an ice sheet grown under a colder climate ($pCO_2 = 280$ ppm) on the OMT median modern topography. (B) Shows the ice sheet elevation on the same topography following climate and ocean warming ($pCO_2 = 500$ ppm; 5°C ocean temperature rise). Profiles A-A′, B-B′, C-C′, and D-D′ are shown on the right. (C) Shows the ice sheet elevation following climate and ocean warming with a larger increase in pCO_2 to 840 ppm. Modelled total ice sheet volumes (in km^3) are given in each panel.

frontal systems in response to glacial-interglacial cycles paced by obliquity (Escutia et al., 2011; Salabarnada et al., 2018; Tauxe et al., 2012). Temperate surface waters between 27 and 25 Ma are also supported by the abundance of nannofossils preserved in the sediments (Escutia et al., 2011; Salabarnada et al., 2018), the lack of sea-ice-related dinocyst species and the relative abundance of oligotrophic, temperate dinocyst taxa (Bjil et al., 2018). Additionally, the absence of iceberg rafted debris (IRD) and reconstructions of cool-temperate vegetation (Strother et al., 2017) indicate that glaciers/ice caps were predominantly terrestrial occupying the topographic highs and lowlands in the now over-deepened Wilkes Subglacial Basin (Salabarnada et al., 2018). Cooling and development of marine-based ice on the Wilkes Land continental shelf occurred between ~25 and 23 Ma, and is expressed in Site U1356 cores as MTDs on the continental rise, associated with repeated AIS expansions to the shelf break during the latest Oligocene (Escutia et al., 2011; 2014; Evangelinos et al., 2020). However, episodic occurrence of carbonate cements, nannofossils and temperate dinocysts implies interglacials remained warm (e.g., Bjil et al., 2018; Escutia et al., 2011; Salabarnada et al., 2018), perhaps associated with orbitally-paced ocean front migration and CDW-upwelling (Evangelinos et al., 2020).

The disconnect between apparent cooling in some sectors of Antarctica and the warming implied by $\delta^{18}O$ records has been a source of ongoing investigation and speculation since a seminal paper by Barrett (2007) first pointed this out. One explanation invokes the development of poleward temperature gradient, whereby drill sites north of the Antarctic Circumpolar Current (ACC) were bathed with a warmer deep-water mass of Northern Hemisphere provenance (Pekar et al., 2006), while some sites south of the ACC were thermally-isolated (Scher et al., 2014), and under the influence of a locally produced cooler bottom waters (Hauptvogel et al., 2017). Furthermore, Hauptvogel et al. (2017) have argued that this may have masked AIS expansion in $\delta^{18}O$ records.

Notwithstanding this, the plot appears to thicken. While deep-water masses may explain the gradient in $\delta^{18}O$ between high and low latitudes, a trend towards lower $\delta^{18}O$ beginning after glacial Oi2b at 26.7 Ma and extending through the Late Oligocene (Fig. 8.2), is apparent in all benthic $\delta^{18}O$ records. Thus, unless temperature-dependent fractionation of foraminiferal calcite dominates the overall $\delta^{18}O$ signature during this time (i.e., the opposite of Quaternary $\delta^{18}O$ records), then AIS volume must have decreased despite the far-field sea-level records implying a global mean sea-level fall.

Continental-scale ice sheet models using restored Antarctic paleogeography for time slices in the Oligocene and Miocene (e.g., Paxman et al., 2019; Wilson and Luyendyk, 2009) are able to store more ice on west Antarctica than today (Colleoni et al., 2018, 2021; Gasson et al., 2016; Paxman et al., 2020; Wilson et al., 2002, 2013). This is because much of the region was subaerial providing more land area for a large terrestrial ice sheet to develop (De Santis et al., 1995, 1999; Paxman et al., 2019; Wilson and Luyendyk, 2009), Based on evidence from seismic data (De Santis et al., 1995), and sediments in the central Ross Sea from DSDP Site 270 (e.g. Kulhanek et al., 2019, Leckie and Webb, 1983) (Fig. 8.2), a number of authors have proposed that a marine transgression across West Antarctica began at ~26 Ma when it tectonically-subsided, and this may have caused ocean-induced melt of a significant part of the terrestrial West Antarctica Ice Sheet (WAIS) (Greenop et al., 2019; Levy et al., 2016; Liebrand et al., 2017). This resulted in a gradual and progressive decrease in AIS extent and volume through the Late Oligocene, and may account for the overall decrease in average $\delta^{18}O$ values in the benthic isotope stacks (e.g., Cramer et al.; 2009; De Vleeschouwer et al., 2017; Zachos et al., 2001a).

While this potentially provides a tantalising explanation for the apparent decoupling between atmospheric CO_2 and ice volume, as expressed in the benthic $\delta^{18}O$ record during the Late Oligocene, the evidence for circum-Antarctic surface ocean cooling is ambiguous. More high-quality sea-surface temperature and climate records will be needed to confirm this hypothesis.

A further wrinkle in this reconciliation of global and Antarctic Late Oligocene climate and ice volume proxy records, is the apparent fall in global sea-level between ~26 and 23 Ma implied by far-field sea-level records from the New Jersey continental margin (Hauptvogel et al., 2017; Kominz et al., 2008; Miller et al., 2020; Fig. 8.4). We suggest, that ongoing efforts should focus on improving the precision of the age models and reducing uncertainties for far-field continental margin sea-level estimates and their correlation with benthic $\delta^{18}O$ records.

8.4 Concluding remarks

Considerable progress has been made over the last 10 years in our understanding of AIS dynamics across the OMT from analysis of both geological records and the improvement of numerical ice sheet models. When the first edition of this chapter was written (Wilson et al., 2009), numerical ice sheet models displayed strong hysteresis due to surface elevation−mass balance feedback as a result of the atmospheric lapse rate and surface albedo, possibly strengthened by the cooling effect of the ice sheet on the surrounding Southern Ocean (DeConto et al., 2007; Ladant et al., 2014; Pollard and DeConto 2005). Since then, four developments by the modelling community have resulted in a much better comparison between models and data: (1) A climate−ice sheet coupling method that uses a high-resolution atmospheric model to account for ice sheet−climate feedbacks (e.g., Gasson et al., 2016); (2) The ability to model instability mechanisms that act only on marine-based ice, particularly where the ice sits on reverse-sloped beds (e.g., Pollard and DeConto, 2009); (3) Some ice sheet models now include mechanisms for retreat into deep subglacial basins caused by ice-cliff failure and ice-shelf hydrofracture (e.g., DeConto and

Pollard, 2016; Pollard et al., 2015); and (4) Models now have the ability to account for changes in the oxygen isotopic composition of the ice sheet by using isotope-enabled climate and ice sheet models (Gasson et al., 2016).

Notwithstanding this progress, challenges still remain in model development and parameterisation, the quality, precision, and stratigraphic completeness of proxy data and our ability to develop high-resolution chronologies for Antarctic margin and Southern Ocean records, as outlined in the Introduction section (see also Siegert and Golledge, 2021). To address this, more ice-proximal to ice-distal marine sediment cores are needed to understand and reconcile regional differences (e.g., Wilkes Land vs Ross Sea), which result from differing tectonic subsidence histories and sub-glacial basin topographies affecting ice drainage and ocean-ice connections.

In this chapter we have summarised the latest proxy datasets for ice volume, atmospheric CO_2, ocean carbon cycle dynamics and ocean temperature, to build a data-driven picture of a highly dynamic AIS across the OMT. The AIS initially evolved from a terrestrial ice sheet into a marine-based ice sheet for the first time, which was perhaps 120% larger than today, and then retreated with ice potentially disappearing from the Antarctic continental interior implying a very rapid, Late Quaternary Northern Hemisphere-style, orbitally-modulated, CO_2-driven continental-scale collapse and deglaciation.

The OMT provides insights into the potential of threshold behaviour, irreversible on human timescales, if certain CO_2 concentrations are crossed (e.g., 400 ppm) (e.g. Masson-Delmotte et al., 2013). In this chapter, we have highlighted the relative roles of CO_2 and orbital forcing, agreeing with model predictions (e.g., DeConto et al., 2008; Gasson et al., 2016; Pollard and DeConto, 2009) and strengthening geological evidence (e.g., Escutia et al., 2011, 2014; Levy et al., 2016, 2019; Naish et al., 2009) that the AIS is highly dynamic within a relatively small range of atmospheric CO_2 concentrations. Proxy data and model simulations suggest thresholds for glaciation (e.g. LGM, see Siegert et al., 2021) and loss of marine-based ice sheets may be around 300 and 400 ppm, respectively (e.g., DeConto et al., 2008; Foster and Rohling, 2013; Levy et al., 2019). Under sustained current atmospheric CO_2 levels of 400 ppm, the Intergovernmental Panel on Climate Change 5th Assessment Report (Masson-Delmotte et al., 2013), states orbital forcing alone cannot cause a future LGM-style glaciation. This report also implies that we may be close to initiating irreversible loss of Antarctica's marine based ice sheets, potentially taking our world back to the Pliocene when sea-levels were ~20 m higher than today (e.g., Grant and Naish, 2021; Grant et al., 2019).

Acknowledgements

T.N., B.D., R.L., and R.M. were funded by the NZ MBIE Antarctic Science Platform (ANTA1801). R.M. was also funded by the Royal Society of New Zealand Te Apārangi Marsden Fund (18-VUW-089). L.D.S. and F.D. were funded by the Programma Nazionale delle Ricerche in Antartide (PNRA16_00016 project and PNRA 14_00119). C.E.

acknowledges funding by the Spanish Ministry of Science and Innovation (grant CTM2017-89711-C2-1/2-P), co-funded by the European Union through FEDER funds.

References

Badger, M., Lear, C.H., Pancost, R.D., Foster, G.L., Bailey, T., et al., 2013. CO_2 drawdown following the middle Miocene expansion of the Antarctic Ice Sheet. Paleoceanography 28, 42–53. Available from: https://doi.org/10.1002/palo.20015.

Barker, P.F., Carmerlenghi, A., Acton, G.D., et al., 1999. Proceedings of the Ocean Drilling Program, Initial reports, v. 178. Ocean Drilling Program, College Station, TX [online].

Barker, P.F., Kennett, J.P.Scientific Party, 1988. Weddell Sea Palaeoceanography: preliminary results of ODP Leg 113. Palaeogeography, Palaeoclimatology, Palaeoecology 67, 75–102.

Barrett, P.J., 2007. Cenozoic climate and sea level history from glacimarine strata off the Victoria Land coast, Cape Roberts Project, Antarctica. In: Hambrey, M.J., Christoffersen, P., Glasser, N.F., Hubbard, B. (Eds.), Glacial Sedimentary Processes and Products, 39. International Association of Sedimentologists, pp. 259–287.

Barron, J., Larsen, B., et al., 1989. College Station, TX (Ocean Drilling Program). doi:10.2973/odp.proc.ir.119.1989. Proc. ODP, Init. Repts. 119. Available from: https://doi.org/10.2973/odp.proc.ir.119.1989.

Bartoli, G., Hönisch, B., Zeebe, R.E., 2011. Atmospheric CO_2 decline during the Pliocene intensification of Northern Hemisphere glaciations. Paleoceanography 26 (4). Available from: https://doi.org/10.1029/2010PA002055.

Berggren, W.A., Kent, D.V., Flynn, J., Van Couvering, J.A., 1985. Cenozoic geochronology. Geological Society of America Bulletin 96, 1407–1418.

Bijl, P.K., Bendle, J.A., Bohaty, S.M., Pross, J., Schouten, S., Tauxe, L., et al.,Expedition 318 Scientists 2013. Eocene cooling linked to early flow across the Tasmanian Gateway. Proceedings of the National Academy of Sciences of the United States of America 110 (24), 9645–9650. Available from: https://doi.org/10.1073/pnas.1220872110.

Billups, K., Pälike, H., Channell, J.E.T., Zachos, J.C., Shackleton, N.J., 2004. Astronomic calibration of the late Oligocene through early Miocene geomagnetic polarity time scale. Earth and Planetary Science Letters 224, 33–44.

Billups, K., Schrag, D.P., 2002. Paleotemperatures and ice volume of the past 27 Myr revisited with paired Mg/Ca and $^{18}O/^{16}O$ measurements on benthic foraminifera. Paleoceanography 17. Available from: https://doi.org/10.1029/2000PA000567.

Bjil, P.K., Houben, A.J.P., Hartman, J.D., Pross, J., Salabarnada, A., Escutia, C., et al., 2018. Paleoceanography and ice sheet variability off-shore Wilkes Land, Antarctica – Part 2: Insights from Oligocene–Miocene dinoflagellate cyst assemblages. Climate of the Past 14, 1015–1033. Available from: https://doi.org/10.5194/cp-14-1015-2018.

Bradshaw, C.D., Langebroek, P.M., Lear, C.H., et al., 2021. Hydrological impact of Middle Miocene Antarctic ice-free areas coupled to deep ocean temperatures. Nature Geoscience. Available from: https://doi.org/10.1038/s41561-021-00745-w.

Cande, S.C., Kent, D.V., 1992. A new geomagnetic polarity timescale for the Late Cretaceous and Cenozoic. Journal of Geophysical Research 97, 13917–13951.

Cande, S.C., Kent, D.V., 1995. Revised calibration of the geomagnetic polarity timescale for the Late Cretaceous and Cenozoic. Journal of Geophysical Research 100, 6093–6095.

Colleoni, F., De Santis, L., Montoli, E., Olivo, E., Sorlien, C.C., Bart, P.J., et al., 2018. Past continental shelf evolution increased Antarctic ice sheet sensitivity to climatic conditions. Scientific Reports 8, 11323. Available from: https://doi.org/10.1038/s41598-018-29718-7.

Colleoni, F., et al., PAIS community 2021. Past Antarctic ice sheet dynamics (PAIS) and implications for future sea-level change. In: Florindo, F., et al., (Eds.), Antarctic Climate Evolution, Second ed. Elsevier.

Cooper, A.K., O'Brien, P.E., 2004. Leg 188 synthesis: transitions in the glacial history of the Prydz Bay Region, East Antarctica, from ODP Drilling. Proceedings of the Ocean Drilling Programme, Scientific Results 188. Available from: https://doi.org/10.2973/odp.proc.sr.188.001.2004.

Coxall, H.K., Wilson, P.A., Pälike, H., Lear, C.H., Backman, J., 2005. Rapid stepwise onset of Antarctic glaciation and deeper calcite compensation in the Pacific Ocean. Nature 433 (7021), 53–57.

Cramer, B.S., Miller, K.G., Barrett, P.J., Wright, J.D., 2011. Late Cretaceous–Neogene trends in deep ocean temperature and continental ice volume: Reconciling records of benthic foraminiferal geochemistry ($\delta^{18}O$ and Mg/Ca) with sea level history. Journal of Geophysical Research 116, C12023. Available from: https://doi.org/10.1029/2011JC007255.

Cramer, B.S., Toggweiler, J.R., Wright, J.D., Katz, M.E., Miller, K.G., 2009. Ocean overturning since the Late Cretaceous: inferences from a new benthic foraminiferal isotope compilation. Paleoceanography 24, PA4216. Available from: https://doi.org/10.1029/2008PA001683.

Crampton, J.S., Cody, R.D., Levy, R., Harwood, D., McKay, R., Naish, T.R., 2016. Southern Ocean phytoplankton turnover in response to stepwise Antarctic cooling over the past 15 million years. Proceedings of the National Academy of Sciences 113 (25), 6868–6873.

De Santis, L., Anderson, J.B., Brancolini, G., Zayatz, I., 1995. In: Cooper, A.K., et al., (Eds.), Antarctic Research Series, vol. 71. Springer International, pp. 235–260.

De Vleeschouwer, D., Vahlenkamp, M., Crucifix, M., Pälike, H., 2017. Alternating Southern and Northern Hemisphere climate response to astronomical forcing during the past 35 m.y. Geology 45 (4), 375–378. Available from: https://doi.org/10.1130/G38663.1.

DeConto, R., Pollard, D., Wilson, P., et al., 2008. Thresholds for Cenozoic bipolar glaciation. Nature 455, 652–656. Available from: https://doi.org/10.1038/nature07337.

DeConto, R.M., Pollard, D., 2016. Contribution of Antarctica to past and future sea level rise. Nature 531, 591–597. Available from: https://doi.org/10.1038/nature17145.

DeConto, R.M., Pollard, D., Harwood, D., 2007. Sea ice feedback and Cenozoic evolution of Antarctic climate and ice sheets. Paleoceanography 22, PA3214. Available from: https://doi.org/10.1029/2006PA001350.

Donda, F., O'Brien, P.E., De Santis, L., Rebesco, M., Brancolini, G., 2008. Mass wasting processes in the Western Wilkes Land margin: Possible implications for East Antarctic glacial history. Palaeogeography, Palaeoclimatology, Palaeoecology 260, 77–91.

Ehrmann, W., Setti, M., Marinoni, L., 2005. Clay minerals in Cenozoic sediments off Cape Roberts (McMurdo Sound, Antarctica) reveal palaeoclimatic history. Palaeogeography, Palaeoclimatology, Palaeoecology 229, 187–211.

Escutia, C., De Santis, L., Donda, F., Dunbar, R.B., Brancolini, G., Eittreim, S.L., et al., 2005. Cenozoic ice sheet history from east Antarctic Wilkes Land continental margin sediments. Global and Planetary Change 45 (1–3), 51–81.

Escutia, C., Brinkhuis, H.and the Expedition 318 Science Party, 2014. From greenhouse to icehouse at the Wilkes Land Antarctic margin: IODP 318 synthesis of results. Developments in Marine Geology 7, 295–328. Available from: https://doi.org/10.1016/B978-0-444-62617-2.00012-8.

Escutia, C., Brinkhuis, H., Klaus, A., and the Expedition 318 Scientists, 2011. Proceedings of the IODP, vol. 318, Tokyo (Integrated Ocean Drilling Program Management International, Inc.). doi:10.2204/iodp.proc.318.2011.

Escutia, C., Eittreim, S., Cooper, A.K., 1997. Cenozoic sedimentation on the Wilkes Land continental rise, Antarctica. In: Ricci, C.A. (Ed.), The Antarctic Region: Geological Evolution and Processes. Terra Antartica, pp. 791–795.

Evangelinos, D., Escutia, C., Etourneau, J., Hoem, F., Bijl, P., Boterblom, W., et al., 2020. A. Late Oligocene-Miocene proto-Antarctic circum polar current dynamics off the Wilkes Land margin, east Antarctica. Global and Planetary Change 191, 103221. Available from: https://doi.org/10.1016/j.gloplacha.2020.103221.

Foster, G.L., Lear, C.H., Rae, J., W., B., 2012. The evolution of pCO_2, ice volume and climate during the middle Miocene. Earth and Planetary Science Letters 341–344, 243–254. Available from: https://doi.org/10.1016/j.epsl.2012.06.007.

Foster, G., Rohling, E., 2013. Relationship between sea level and climate forcing by CO_2 on geological timescales. PNAS. 110. Available from: https://doi.org/10.1073/pnas.1216073110.

Fraass, A., Leckie, M., DeConto, R., McQuaid, C., Burns, S., Zachos, J., 2019. Reappraisal of Oligocene–Miocene chronostratigraphy and the Mi-1 event: ocean drilling program site 744, Kerguelen Plateau, southern Indian Ocean. Stratigraphy 15 (4), 265–278.

Galeotti, S., DeConto, R.M., Naish, T.R., Stocchi, P., Florindo, F., Pagani, M., et al., 2016. Antarctic ice sheet variability across the Eocene–Oligocene boundary climate transition. Science 352, 76–80. Available from: https://doi.org/10.1126/science.aab0669.

Gasson, E., DeConto, R.M., Pollard, D., Levy, R., 2016. Dynamic Antarctic ice sheet during the early to mid-Miocene. Proceedings of the National Academy of Sciences of the United States of America 113 (13), 3,459–3,464. Available from: https://doi.org/10.1073/pnas.1516130113.

Gradstein, F.M., Ogg, J.G., Smith, A.G. (Eds.), 2004. A Geological Time Scale. Cambridge University Press, Cambridge, p. 589.

Gradstein, F.M., Ogg, J.G., Schmitz, M.D., Ogg, G.M., 2012. The geological time scale 2012. Elsevier B.V., Oxford, U.K.

Grant, G.R., Naish, T.R., Dunbar, G.B., Stocchi, P., Kominz, M.A., Kamp, P.J., et al., 2019. The amplitude and origin of sea-level variability during the Pliocene epoch. Nature 574 (7777), 237–241. Available from: https://doi.org/10.1038/s41586-019-1619-z. doi.org/10.22498/pages.29.1.4.

Grant, G., Naish, T., 2021. Pliocene sea-level revisited: is there more than meets the eye? PAGES Magazine 29, doi.org/10.22498/pages.29.1.4.

Greenop, R., Foster, G.L., Wilson, P.A., Lear, C.H., 2014. Middle Miocene climate instability associated with high-amplitude CO_2 variability. Paleoceanography 29. Available from: https://doi.org/10.1002/2014PA002653.

Greenop, R., Sosdian, S.M., Henehan, M.J., Wilson, P.A., Lear, C.H., Foster, G.L., 2019. Orbital forcing, ice volume, and CO_2 Across the Oligocene-Miocene Transition. Paleoceanography and Paleoclimatology 34 (3), 316–328. Available from: https://doi.org/10.1029/2018PA003420.

Gulick, S., Shevenell, A., Montelli, A., et al., 2017. Initiation and long-term instability of the East Antarctic Ice Sheet. Nature 552, 225–229. Available from: https://doi.org/10.1038/nature25026.

Hambrey, M.J., Barrett, P.J., Robinson, P.H., 1989. Stratigraphy. In Barrett, P.J. (Ed.), Antarctic Cenozoic History from the CIROS-1 Drillhole, McMurdo Sound. DSIR Publishing, Wellington, New Zealand, pp. 23–48.

Hambrey, M.J., Ehrmann, W.U., Larsen, B., 1991. Cenozoic glacial record of the Prydz Bay Continental Shelf, East Antarctica. Proceedings of the Ocean Drilling Programme, Scientific Results 120, 77–132.

Hartman, J.D., Sangiorgi, F., Salabarnada, A., Peterse, F., Houben, A.J., Schouten, S., et al., 2018. Paleoceanography and ice sheet variability offshore Wilkes Land, Antarctica-Part 3: insights from Oligocene-Miocene TEX86-based sea surface temperature reconstructions. Climate of the Past 14 (9), 1275–1297. Available from: https://doi.org/10.5194/cp-14-1275-2018.

Hauptvogel, D.W., Pekar, S.F., Pincay, V., 2017. Evidence for a heavily glaciated Antarctica during the late Oligocene "warming" (27.8–24.5 Ma): Stable isotope records from ODP Site 690. Paleoceanography 32 (4), 384–396.

Hochmuth, K., Gohl, K., Leitchenkov, G., Sauermilch, I., Whittaker, J.M., Uenzelmann-Neben, G., et al., 2020. The evolving paleobathymetry of the circum-Antarctic Southern Ocean since 34 Ma – a key to understanding past cryosphere-ocean developments. Geochemistry, Geophysics, Geosystems e2020GC009122. Available from: https://doi.org/10.1029/2020GC009122.

Honisch, B., Hemming, G., Archer, D., Siddal, M., McManus, J., 2009. Atmospheric carbon dioxide concentration across the Mid-Pleistocene Transition. Science 324, 1551–1554. Available from: https://doi.org/10.1126/science.1171477.

Kennett, J.P., 1977. Cenozoic evolution of Antarctic glaciation, the circum-Antarctic Ocean, and their impact on global paleoceanography. Journal of Geophysical Research 82, 3843–3860.

Kennett, J.P., Houtz, R.E., et al., 1974. Initial Reports of the Deep Sea Drilling Project, vol. 29. U.S. Government Printing Office, Washington, DC, p. 1197.

Kominz, M.A., Browning, J.V., Miller, K.G., Sugarman, P.J., Mizintseva, S., Scotese, C.R., 2008. Late cretaceous to Miocene sea-level estimates from the New Jersey and Delaware coastal plain coreholes: an error analysis. Basin Research 20 (2), 211–226. Available from: https://doi.org/10.1111/j.1365-2117.2008.00354.x.

Kulhanek, D.K., Levy, R.H., Clowes, C.D., Prebble, J.G., Rodelli, D., Jovane, L., et al., 2019. Revised chronostratigraphy of DSDP Site 270 and late Oligocene to early Miocene paleoecology of the Ross Sea sector of Antarctica. Global and Planetary Change 178, 46–64.

Ladant, J.B., Donnadieu, Y., Lefebvre, V., Dumas, C., 2014. The respective role of atmospheric carbon dioxide and orbital parameters on ice sheet evolution at the Eocene–Oligocene transition. Paleoceanography 29 (8), 810–823.

Laskar, J., Robutel, P., Joutel, F., Gastineau, M., Correia, A.C.M., Levrard, B., 2004. A long-term numerical solution for the insolation quantities of the Earth. Astronomy and Astrophysics 428 (1), 261–285. Available from: https://doi.org/10.1051/0004-6361:20041335.

Lear, C.H., Coxall, H.K., Foster, G.L., Lunt, D.J., Mawbey, E.M., Rosenthal, Y., et al., 2015. Neogene ice volume and ocean temperatures: Insights from infaunal foraminiferal Mg/Ca paleothermometry. Paleoceanography 30, 1437–1454. Available from: https://doi.org/10.1002/2015PA002833.

Leckie, R.M., Webb, P.-N., 1983. Late Oligocene–early Miocene glacial record of the Ross Sea, Antarctica: evidence from DSDP Site 270. Geology 11, 578–582.

Levy, R., Harwood, D., Florindo, F., Sangiorgi, F., Tripati, R., von Eynatten, H., et al., 2016. SMS science team. 2016. Early to mid-Miocene Antarctic ice sheet dynamics. Proceedings of the National Academy of Sciences 113 (13), 3453–3458. Available from: https://doi.org/10.1073/pnas.1516030113.

Levy, R.H., Meyers, S.R., Naish, T.R., Golledge, N.R., McKay, R.M., Crampton, J.S., et al., 2019. Antarctic ice-sheet sensitivity to obliquity forcing enhanced through ocean connections. Nature Geoscience 12, 132–137. Available from: https://doi.org/10.1038/s41561-018-0284-4.

Liebrand, D., Beddow, H.M., Lourens, L.J., et al., 2016. Cyclostratigraphy and eccentricity tuning of the early Oligocene through early Miocene (30.1–17.1 Ma): cibicides mundulus stable oxygen and carbon isotope records from Walvis Ridge Site 1264. Earth and Planetary Science Letters 450, 392–405.

Liebrand, D., de Bakker, A., Beddow, H., Wilson, P., Bohaty, S., Ruessink, G., et al., 2017. Evolution of the early Antarctic ice ages. Proceedings of the National Academy of Sciences 114 (15), 3867–3872. Available from: https://doi.org/10.1073/pnas.1615440114.

Liebrand, D., Lourens, L.J., Hodell, D.A., de Boer, B., van de Wal, R.S.W., Palike, H., 2011. Antarctic ice sheet and oceanographic response to eccentricity forcing during the early Miocene. Climate of the Past 7 (3), 869–880. Available from: https://doi.org/10.5194/cp-7-869-2011.

Martinez-Boti, M.A., Foster, G.L., Chalk, T.B., Rohling, E.J., Sexton, P.F., et al., 2015. Plio-Pleistocene climate sensitivity evaluated using high-resolution CO_2 records. Nature 518, 49–54. Available from: https://doi.org/10.1038/nature14145.

Masson-Delmotte, V., Schulz, M., Abe-Ouchi, A., Beer, J., Ganopolski, A., Gonzáles Rouco, J.F., et al., 2013. Information from paleoclimate archives. In: Stocker, T.F., Qin, D., Plattner, G.K., Tignor, M., Allen, S.K., Boschung, J., et al.,Climate Change 2013: The Physical Science Basis. Contribution of Working Group I to the Fifth Assessment Report of the Intergovernmental Panel on Climate Change. Cambridge University Press, Cambridge, UK and New York.

Mawbey, E.M., Lear, C.H., 2013. Carbon cycle feedbacks during the Oligocene–Miocene transient glaciation. Geology 41 (9), 963–966. Available from: https://doi.org/10.1130/G34422.1.

McKay, R., et al., 2021. Cenozoic History of Antarctic Glaciation and Climate from Onshore and Offshore Studies. In: Florindo, F., et al., (Eds.), Antarctic Climate Evolution, Second ed. Elsevier.

Miller, K.G., Browning, J.V., Schmelz, W.J., Kopp, R.E., Mountain, G.S., Wright, J.D., 2020. Cenozoic sea-level and cryospheric evolution from deep-sea geochemical and continental margin records. Science Advances 6 (20), eaaz1346. Available from: https://doi.org/10.1126/sciadv.aaz1346.

Miller, K.G., Kominz, M.A., Browning, J.V., Wright, J.D., Mountain, G.S., Katz, M.E., et al., 2005. The phanerozoic record of global sea-level change. Science 310 (5752), 1293–1298. Available from: https://doi.org/10.1126/science.1116412.

Miller, K.G., Wright, J.D., Fairbanks, R.G., 1991. Unlocking the ice house: Oligocene–Miocene oxygen isotopes, eustasy, and margin erosion. Journal of Geophysical Research: Solid Earth 96 (B4), 6829–6848. Available from: https://doi.org/10.1029/90JB02015.

Naish, T., Powell, R., Levy, R., Wilson, G., Scherer, R., Talarico, F., et al., 2009. Obliquity-paced Pliocene West Antarctic ice sheet oscillations. Nature 458, 322–328. Available from: https://doi.org/10.1038/nature07867.

Naish, T.R., Wilson, G.S., Dunbar, G., Barrett, P.J., 2008. Constraining the amplitude of late Oligocene bathymetric changes in western Ross Sea during orbitally-induced oscillations in the East Antarctic ice sheet: (2) Implications for global sea–level changes. Palaeogeography, Palaeoclimatology, Palaeoecology 260, 55–65.

Naish, T.R., Woolfe, K.J., Barrett, P.J., Wilson, G.S., Atkins, C., Bohaty, S., et al., 2001. Orbitally induced oscillations in the East Antarctic ice sheet at the Oligocene/Miocene boundary. Nature 413, 719–723.

Pagani, M., Liu, Z., LaRiviere, J., Ravelo, A.C., 2010. High Earth-system climate sensitivity determined from Pliocene carbon dioxide concentrations. Nature Geoscience 3, 27–30. Available from: https://doi.org/10.1038/ngeo724.

Pagani, M., Zachos, J.C., Freeman, K.H., Tipple, B., Bohaty, S., 2005. Carbon dioxide concentrations during the Paleogene. Science 309, 600–603. Available from: https://doi.org/10.1126/science.1110063.

Pagani, M., Huber, M., Liu, Z., Bohaty, S., Henderiks, J., et al., 2011. The role of carbon dioxide during the onset of Antarctic glaciation. Science 334, 1261–1264. Available from: https://doi.org/10.1126/science.1203909.

Pälike, H., Laskar, J., Shackleton, N.J., 2004. Geologic constraints on the chaotic diffusion of the solar system. Geology 32, 929–932.

Pälike, H., Norris, R.D., Herrle, J.O., Wilson, P.A., Coxall, H.K., Lear, C.H., et al., 2006. The heartbeat of the Oligocene climate system. Science 314, 1894–1898.

Passchier, S., Krissek, L.A., 2008. Oligocene-Miocene Antarctic continental weathering record and paleoclimatic implications, Cape Roberts Drilling Project, Ross Sea, Antarctica. Palaeogeography, Palaeoclimatology, Palaeoecology 260, 30–40.

Paul, H., Zachos, J.C., Flower, B., Tripati, A., 2000. Orbitally induced climate and geochemical variability across the Oligocene/Miocene boundary. Paleoceanography 15 (5), 471–485. Available from: https://doi.org/10.1029/1999PA000443.

Paxman, G.J.G., Gasson, E.G.W., Jamieson, S.S.R., Bentley, M.J., Ferraccioli, F., 2020. Long-term increase in Antarctic ice sheet vulnerability driven by bed topography evolution. Geophysical Research Letters 47, e2020GL090003. Available from: https://doi.org/10.1029/2020GL090003.

Paxman, G.J., Jamieson, S.S., Hochmuth, K., Gohl, K., Bentley, M.J., Leitchenkov, G., et al., 2019. Reconstructions of Antarctic topography since the Eocene–Oligocene boundary. Palaeogeography, Palaeoclimatology, Palaeoecology 535, 109346. Available from: https://doi.org/10.1016/j.palaeo.2019.109346.

Pekar, S.F., DeConto, R.M., Harwood, D.M., 2006. Resolving a late Oligocene conundrum: deep-sea warming and Antarctic glaciation. Palaeogeography, Palaeoclimatology, Palaeoecology 231 (1–2), 29–40.

Pekar, S.F., Christie-Blick, N., Kominz, M.A., Miller, K.G., 2002. Calibration between eustatic estimates from backstripping and oxygen isotopic records for the Oligocene. Geology 30 (10), 903–906. Available from: https://doi.org/10.1130/0091-7613.

Pekar, S.F., Kominz, M.A., 2001. Two-dimensional paleoslope modeling: a new methods for estimating water depths for benthic foraminiferal biofacies and paleo shelf margins. Journal of Sedimentary Research 71, 608–620.

Pollard, D., DeConto, R.M., 2005. Hysteresis in Cenozoic Antarctic ice-sheet variations. Global and Planetary Change 45 (1–3), 9–21. Available from: https://doi.org/10.1016/j.gloplacha.2004.09.011.

Pollard, D., DeConto, R.M., 2009. Modelling West Antarctic ice sheet growth and collapse through the past five million years. Nature 458 (7236), 329–332. Available from: https://doi.org/10.1038/nature07809.

Pollard, D., DeConto, R.M., Alley, R.B., 2015. Potential Antarctic ice sheet retreat driven by hydrofracturing and ice cliff failure. Earth and Planetary Science Letters 412, 112–121. Available from: https://doi.org/10.1016/j.epsl.2014.12.035.

Prebble, J.G., Hannah, M.J., Barrett, P.J., 2006a. Changing Oligocene climate recorded by palynomorphs from two glacio-eustatic sedimentary cycles, Cape Roberts Project, Victoria Land Basin, Antarctica. Palaeogeography, Palaeoclimatology, Palaeoecology 231 (1–2), 58–70.

Prebble, J.G., Raine, J.I., Barrett, P.J., Hannah, M.J., 2006b. Vegetation and climate from two Oligocene glacioeustatic sedimentary cycles (31 and 24 Ma) cored by the Cape Roberts Project, Victoria Land Basin, Antarctica. Palaeogeography, Palaeoclimatology, Palaeoecology 231 (1–2), 41–57.

Raymo, M.E., Kozdon, R., Evans, D., Lisiecki, L., Ford, H., 2018. The accuracy of mid-Pliocene $\delta^{18}O$-based ice volume and sea level reconstructions. Earth-Science Reviews 177, 291–302. Available from: https://doi.org/10.1016/j.earscirev.2017.11.022.

Reichgelt, T., D'Andrea, W.J., Fox, B.R., 2016. Abrupt plant physiological changes in southern New Zealand at the termination of the Mi-1 event reflect shifts in hydroclimate and pCO_2. Earth and Planetary Science Letters 455, 115–124.

Roberts, A.P., Wilson, G.S., Harwood, D.M., Verosub, K.L., 2003. Glaciation across the Oligocene-Miocene boundary in southern McMurdo Sound, Antarctica: new chronology from the CIROS-1 drill hole. Palaeogeography, Palaeoclimatology, Palaeoecology 198, 113–130.

Salabarnada, A., Escutia, C., Röhl, U., Nelson, C.H., McKay, R., Jiménez-Espejo, F.J., et al., 2018. Paleoceanography and ice sheet variability offshore Wilkes Land, Antarctica – part 1: insights from late Oligocene astronomically paced contourite sedimentation. Climates of the Past 14, 991–1014. Available from: https://doi.org/10.5194/cp-14-991-2018.

Sangiorgi, F., Bijl, P.K., Passchier, S., Salzmann, U., Schouten, S., McKay, R., et al., 2018. Southern Ocean warming and Wilkes Land ice sheet retreat during the mid-Miocene. Nature Communications 9 (1), 1–11. Available from: https://doi.org/10.1038/s41467-017-02609-7.

Scher, H.D., Bohaty, S.M., Smith, B.W., Munn, G.H., 2014. Isotopic interrogation of a suspected late Eocene glaciation. Paleoceanography 29 (6), 628–644.

Schlich, R., Wise Jr, S.W., 1992. The geologic and tectonic evolution of the Kerguelen Plateau: an introduction to the scientific results of Leg 120. Proceedings of the Ocean Drilling Programme, Scientific results 120, 5–30.

Seki, O., Foster, G.L., Schmidt, D.N., Mackensen, A., Kawamura, K., Pancost, R.D., 2010. Alkenone and boron based Plio-Pleistocene pCO_2 records. Earth and Planetary Science Letters 292, 201–211. Available from: https://doi.org/10.1016/j.epsl.2010.01.037.

Shackleton, N.J., Hall, M.A., Raffi, I., Tauxe, L., Zachos, J., 2000. Astronomical calibration for the Oligocene–Miocene boundary. Geology 28, 447–450.

Shevenell, A.E., Kennett, J.P., Lea, D.W., 2008. Middle Miocene ice sheet dynamics, deep-sea temperatures, and carbon cycling: a Southern Ocean perspective. Geochemistry, Geophysics, Geosystems 9, Q02006. Available from: https://doi.org/10.1029/2007GC001736.

Siegert, M.J., Golledge, N.R., 2021. Advances in numerical modelling of the Antarctic ice sheet. In: Florindo, F., et al., (Eds.), Antarctic Climate Evolution, second edition. Elsevier (this volume).

Siegert, M.J., Hein, A.S., White, D.A., Gore, D.B., De Santis, L., Hillenbrand, C.D., 2021. Antarctic ice sheet changes since the Last Glacial Maximum. In: Florindo, F., et al. (Eds.), Antarctic Climate Evolution, second edition. Elsevier (this volume).

Sijp, W.P., Anna, S., Dijkstra, H.A., Flögel, S., Douglas, P.M., Bijl, P.K., 2014. The role of ocean gateways on cooling climate on long time scales. Global and Planetary Change 119, 1–22.

Sorlien, C., Luyendyk, B., Wilson, D., Decesari, R., Bartek, L., Diebold, J., 2007. Oligocene development of the West Antarctic ice sheet recorded in eastern Ross Sea strata. Geology 35 (5), 467–470. Available from: https://doi.org/10.1130/G23387A.1.

Stap, L.B., Sutter, J., Knorr, G., Stärz, M., Lohmann, G., 2019. Transient variability of the Miocene Antarctic ice sheet smaller than equilibrium differences. Geophysical Research Letters 46, 4288–4298. Available from: https://doi.org/10.1029/2019GL082163.

Steinthorsdottir, M., Coxall, H.K., de Boer, A.M., Huber, M., Barbolini, N., Bradshaw, C.D., et al., 2021. The Miocene: The future of the past. Paleoceanography and Paleoclimatology 36, e2020PA004037. Available from: https://doi.org/10.1029/2020PA004037.

Steinthorsdottir, M., Vajda, V., Pole, M., 2019. Significant transient pCO_2 perturbation at the New Zealand Oligocene–Miocene transition recorded by fossil plant stomata. Palaeogeography, Palaeoclimatology, Palaeoecology 515, 152–161. Available from: https://doi.org/10.1016/j.palaeo.2018.01.039.

Stewart, J.A., James, R.H., Anand, P., Wilson, P.A., 2017. Silicate weathering and carbon cycle controls on the Oligocene–Miocene transition glaciation. Paleoceanography 32, 1070–1085. Available from: https://doi.org/10.1002/2017PA003115.

Strother, S.L., Salzmann, U., Sangiorgi, F., Bijl, P.K., Pross, J., Escutia, C., 2017. A new quantitative approach to identify reworking in Eocene to Miocene pollen records from offshore Antarctica using red fluorescence and digital imaging. Biogeosciences 14, 2089–2100. Available from: https://doi.org/10.5194/bg-14-2089-2017.

Super, J.R., Thomas, E., Pagani, M., Huber, M., O'Brien, C., Hull, P.M., 2018. North Atlantic temperature and pCO_2 coupling in the early-middle Miocene. Geology 46 (6), 519–522. Available from: https://doi.org/10.1130/G40228.1.

Tauxe, L., Stickley, C.E., Sugisaki, S., Bijl, P.K., Bohaty, S.M., Brinkhuis, H., et al., 2012. Chronostratigraphic framework for the IODP Expedition 318 Wilkes land margin: constraints for paleoceanographic reconstruction. Paleoceanography 27 (2). Available from: https://doi.org/10.1029/2012PA002308 PA2214.

Taviani, M., Beu, A.G., 2003. The palaeoclimatic significance of Cenozoic marine macrofossil assemblages from Cape Roberts Project drillholes, McMurdo Sound, Victoria Land Basin, East Antarctica. Palaeogeography, Palaeoclimatology, Palaeoecology 198 (1–2), 131–143.

Toggweiler, J.R., Russell, J.L., Carson, S.R., 2007. Mid-latitude westerlies, atmospheric CO_2, and climate change. Paleoceanography 21. Available from: https://doi.org/10.1029/2005PA001154(2007).

Watkins, D.K., Villa, G., 2000. Palaeogene calcareous nannofossils from CRP-2/2A, Victoria Land Basin, Antarctica. Terra Antarctica 7 (4), 443–452.

Westerhold, T., Marwan, N., Drury, A.J., Liebrand, D., Agnini, C., Anagnostou, E., et al., 2020. An astronomically dated record of Earth's climate and its predictability over the last 66 million years. Science 369 (6509), 1383–1387. Available from: http://doi.org/10.1126/science.aba6853.

Wilson, D.S., Luyendyk, B.P., 2009. West Antarctic paleotopography estimated at the Eocene–Oligocene climate transition. Geophysical Research Letters 36 (16).

Wilson, G.S., Lavelle, M., McIntosh, W.C., Roberts, A.P., Harwood, D.M., Watkins, D.K., et al., 2002. Integrated chronostratigraphic calibration of the Oligocene–Miocene boundary at 24.0 ± 0.1 Ma from the CRP-2A drill core, Ross Sea, Antarctica. Geology 30, 1043–1046.

Wilson, G.S., Pekar, S.F., Naish, T.R., Passchier, S., DeConto, R., 2009. The Oligocene-Miocene Boundary—Antarctic Climate Response to Orbital Forcing. In: Florindo, F., Siegert, M. (Eds.), Developments in Earth and Environmental Sciences, Vol. 8: Antarctic Climate Evolution. Elsevier, Amsterdam, pp. 369–400.

Wilson, D.S., Pollard, D., DeConto, R., Jamieson, S.S.R., Luyendyk, B.P., 2013. Initiation of the West Antarctic ice sheet and estimates of total Antarctic ice volume in the earliest Oligocene. Geophysical Research Letters 40 (16), 4305–4309.

Witkowski, C.R., Weijers, J.W., Blais, B., Schouten, S., Sinninghe Damsté, J.S., 2018. Molecular fossils from phytoplankton reveal secular pCO_2 trend over the Phanerozoic. Science Advances 4 (11), eaat4556.

Zachos, J.C., Flower, B.P., Paul, H., 1997. Orbitally paced climate oscillations across the Oligocene/Miocene boundary. Nature 388 (6642), 567–570. Available from: https://doi.org/10.1038/41528.

Zachos, J., Pagani, M., Sloan, L., Thomas, E., Billups, K., 2001a. Trends, rhythms, and aberrations in global climate 65 Ma to present. Science 292 (5517), 686–693. Available from: https://doi.org/10.1126/science.1059412.

Zachos, J.C., Shackleton, N.J., Revenaugh, J., Pälike, H., Flower, B.P., 2001b. Climate response to orbital forcing across the Oligocene–Miocene boundary. Science 292, 274–278.

Zhang, Y.G., Pagani, M., Liu, Z., Bohaty, S.M., DeConto, R.M., 2013. A 40-million-year history of atmospheric CO_2. Philosophical Transactions of the Royal Society A 371, 20130096. Available from: https://doi.org/10.1098/rsta.2013.0096.

Chapter 9

Antarctic environmental change and ice sheet evolution through the Miocene to Pliocene — a perspective from the Ross Sea and George V to Wilkes Land Coasts

Richard H. Levy[1,2], Aisling M. Dolan[3], Carlota Escutia[4], Edward G.W. Gasson[5], Robert M. McKay[2], Tim Naish[2], Molly O. Patterson[6], Lara F. Pérez[7], Amelia E. Shevenell[8], Tina van de Flierdt[9], Warren Dickinson[2], Douglas E. Kowalewski[10], Stephen R. Meyers[11], Christian Ohneiser[12], Francesca Sangiorgi[13], Trevor Williams[14], Hannah K. Chorley[2], Laura De Santis[15], Fabio Florindo[16], Nicholas R. Golledge[2], Georgia R. Grant[1], Anna Ruth W. Halberstadt[17], David M. Harwood[18], Adam R. Lewis[19], Ross Powell[20] and Marjolaine Verret[2]

[1]GNS Science, Lower Hutt, New Zealand, [2]Antarctic Research Centre, Victoria University of Wellington, Wellington, New Zealand, [3]School of Earth and Environment, University of Leeds, Leeds, United Kingdom, [4]Andalusian Institute of Earth Sciences, CSIC and Universidad de Granada, Armilla, Spain, [5]Centre for Geography and Environmental Science, University of Exeter, Cornwall Campus, Penryn, United Kingdom, [6]Department of Geological Sciences and Environmental Studies, Binghamton University, Binghamton, NY, United States, [7]British Antarctic Survey, Cambridge, United Kingdom, [8]College of Marine Science, University of South Florida, St. Petersburg, FL, United States, [9]Department of Earth Science and Engineering, Imperial College London, London, United Kingdom, [10]Department of Earth, Environment, and Physics, Worcester State University, Worcester, MA, United States, [11]Department of Geoscience, University of Wisconsin-Madison, Madison, WI, United States, [12]Department of Geology, University of Otago, Dunedin, New Zealand, [13]Laboratory of Palaeobotany and Palynology, Department of Earth Sciences, Marine Palynology and Paleoceanography, Utrecht University, Utrecht, The Netherlands, [14]International Ocean Discovery Program, Texas A&M University, College Station, TX, United States, [15]National Institute of Oceanography and Applied Geophysics — OGS, Sgonico, Italy, [16]National Institute of Geophysics and Volcanology, Rome, Italy, [17]Climate System Research Center, University of Massachusetts, Amherst, MA, United States,

Antarctic Climate Evolution. DOI: https://doi.org/10.1016/B978-0-12-819109-5.00014-1

[18]Department of Earth and Atmospheric Sciences, University of Nebraska, Lincoln, NE, United States, [19]Department of Geosciences, North Dakota State University, Fargo, ND, United States, [20]Department of Geology and Environmental Geosciences, Northern Illinois University, DeKalb, IL, United States

9.1 Introduction

9.1.1 Overview and relevance

The Miocene to Pliocene interval of Earth's history occurred between 23.04 and 2.58 million years ago (Ma) (Gradstein and Ogg, 2020; Raffi et al., 2020). During this time, Earth's climate transitioned from relatively warm conditions in the Oligocene, where ice sheets in Antarctica were largely land-based, to the onset of bi-polar glaciation. The long-term cooling trend was punctuated by several, global-scale climatic events and transitions (De Vleeschouwer et al., 2017; Westerhold et al., 2020; Zachos et al., 2001a) including the Miocene Climatic Optimum (MCO) (Flower, 1999; Flower and Kennett, 1994; Steinthorsdottir et al., 2021), Middle Miocene Climate Transition (MMCT) (Flower and Kennett, 1994; Shevenell et al., 2004; Tian et al., 2013), Tortonian Thermal Maximum (Holbourn et al., 2013; Westerhold et al., 2020), Late Miocene Cooling (Herbert et al., 2016; Holbourn et al., 2018; Loutit and Kennett, 1979), and the Pliocene Warm Period (Beu, 1974; Dowsett et al., 1996; Haywood et al., 2016; Poore and Sloan, 1996). These events are captured in regional paleoenvironmental proxy records and their global nature is reflected in the deep-sea benthic foraminifer oxygen and carbon isotope records (Fig. 9.1). Importantly, variations in the oxygen isotopic composition of sea water recorded in the tests of benthic foraminifers reflect changes in bottom water temperature and ice volume, which are used to infer the evolution of high latitude climate and ice sheets (De Vleeschouwer et al., 2017; Kennett, 1977; Miller et al., 1991a; Westerhold et al., 2020; Zachos et al., 2001a, 2008). Sea level reconstructions from geological archives (Carlson et al., 2019; Grant et al., 2019; John et al., 2011; Kominz et al., 2008, 2016; Miller et al., 2005; Miller et al., 2005, 2012, 2020) provide independent records of ice volume variability but can be contradictory due to uncertainty in age control and non-climatic processes (e.g., glacial isostatic adjustment, dynamic topography and tectonics) that affect local sea level. In this chapter, we set the global climate scene and then focus on Antarctica's climate and ice sheet evolution through the Miocene and Pliocene. We review and summarise paleoclimate observations from Antarctica during these global climatic episodes and discuss modelling studies. We take a regional approach and focus our discussion towards geological data from the Ross Sea, George V Coast and Wilkes Land (Fig. 9.2). However, we acknowledge that data and insights into Miocene and Pliocene climate change and ice dynamics are available from other regions around Antarctica (Haywood et al., 2008; McKay et al., 2021).

By the end of this century, atmospheric carbon dioxide (CO_2) is projected to reach concentrations between 430 and 1000 parts per million in volume (ppmv), depending upon future emission scenarios (IPCC, 2013; Meinshausen et al., 2011) (Fig. 9.1). Proxy-based atmospheric CO_2 reconstructions suggest that the Pliocene was the last time when atmospheric CO_2 levels exceeded 400 ppm (Badger et al., 2013a; Bartoli et al., 2011; Martinez-Boti et al., 2015; Pagani et al., 2010; Seki et al., 2010) and global mean temperatures were 2°C–3°C warmer than pre-industrial, making this time period an important target for understanding environmental conditions in light of ongoing climate change (Burke et al., 2018; Dolan et al., 2012; Dowsett et al., 2012; Haywood et al., 2013, 2016; Lawrence et al., 2009; Lunt et al., 2012; Masson-Delmotte et al., 2013; Raymo, 1994). The long-term trend of the global benthic marine oxygen isotope stack implies that during the early to mid Pliocene global ice volume was reduced, compared to today (Lisiecki and Raymo, 2005; Miller et al., 2012), with estimates of global mean sea level (GMSL) ranging between 6 and 25 metres higher than present (Dumitru et al., 2019; Dutton et al., 2015; Grant et al., 2019; Miller et al., 2012; Raymo et al., 2011; Rovere et al., 2020). Proxy-based reconstructions suggest CO_2 concentrations were even higher during the MCO, at times exceeding 600 ppm, although the range of concentration estimates is high (Greenop et al., 2014; Sosdian et al., 2018; Steinthorsdottir et al., 2021). Global mean surface temperatures during this interval were between 3°C and 8°C warmer than today (Shevenell et al., 2004; Steinthorsdottir et al., 2021; You, 2010; You et al., 2009) and sea level records suggest Antarctica's ice sheets were at times much smaller than now (John et al., 2011; Kominz et al., 2008, 2016; Miller et al., 2005, 2020). Significant research effort is focused on the Miocene as it is recognised as an important target for evaluating the Earth system response to future warming (Steinthorsdottir et al., 2020), particularly if efforts to mitigate greenhouse emissions fall short.

Warm climate intervals from the Miocene and Pliocene help us understand the sensitivity of Antarctica's ice sheets and the high southern latitudes to climate change; information that can inform our planet's possible future. Presently, a disconnect exists between paleoceanographic interpretations of ice volume change derived from the deep-sea records and ice-proximal records of environmental variability and ice sheet dynamics from the margins of Antarctica. An example is the long-lasting debate over Antarctic ice sheet stability (Clapperton and Sugden, 1990; Sugden et al., 1993) vs dynamism (Webb and Harwood, 1991; Wilson, 1995) since the middle Miocene. The stability hypothesis is based, in part, on the deep-sea oxygen-isotope record, which shows a stepwise and sustained increase in $\delta^{18}O$ values beginning at ~14 Ma ago (Flower and Kennett, 1993a; Kennett et al., 1975) and

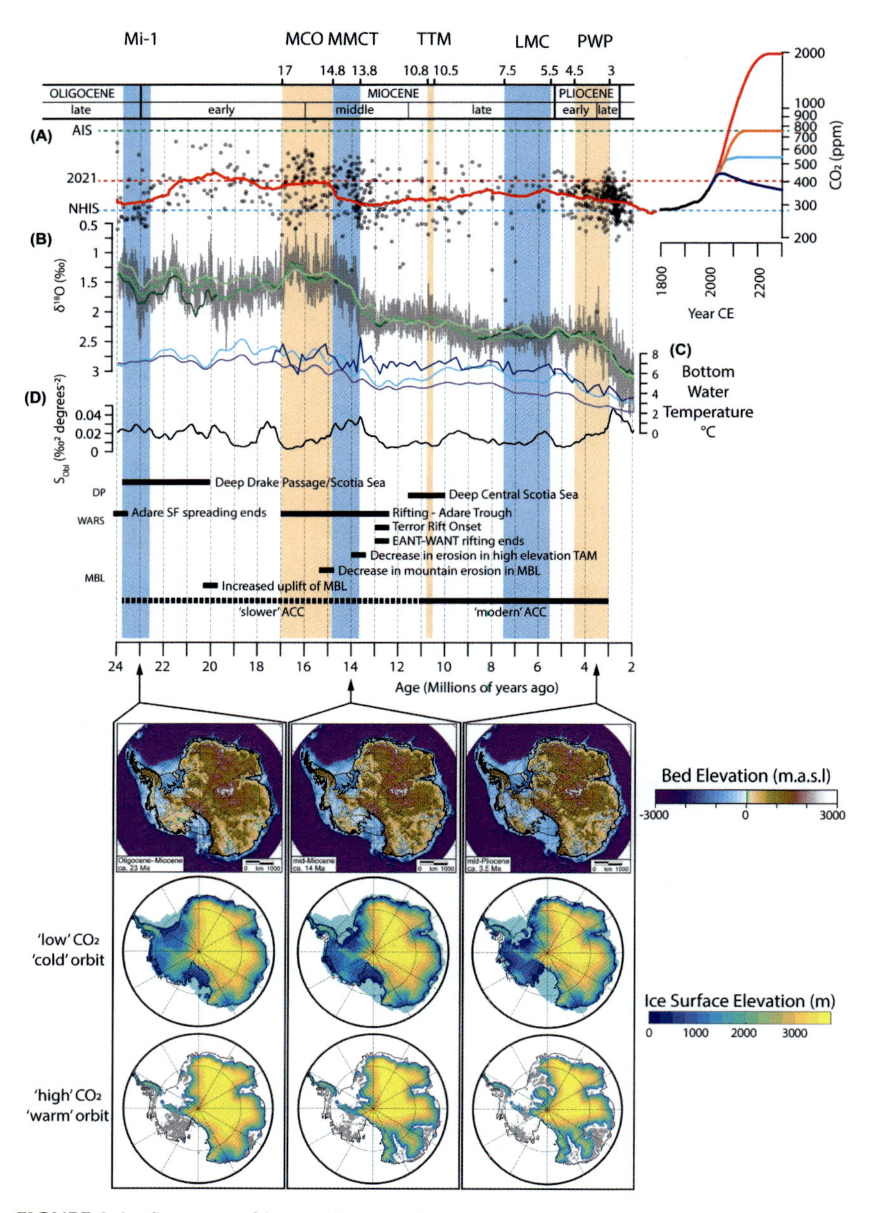

FIGURE 9.1 Summary of key environmental records and tectonic events that have influenced Antarctic climate and ice sheet evolution through the Miocene to Pliocene. Major climatic transitions include the Oligocene/Miocene Boundary (Mi-1) event, the Miocene Climatic Optimum (MCO), Middle Miocene Climate Transition (MMCT), Tortonian Thermal Maximum (TTM), Late Miocene Cooling (LMC), and Pliocene Warm Period (PWP). (A) Atmospheric CO_2 compilation comprises data from a range of proxies including boron isotopes (Bartoli et al., 2011;

(Continued)

has been interpreted to represent ice build-up and stabilisation of the East Antarctic Ice Sheet (EAIS). However, as: (1) the geographic distribution of geological studies around the Antarctic margin has increased and the fidelity of the records has improved (Bertram et al., 2018; Cook et al., 2013, 2017; Escutia et al., 2019; Gulick et al., 2017; Levy et al., 2019; McKay et al., 2012; Patterson et al., 2014; Sangiorgi et al., 2018; Wilson et al., 2018), (2) new sea level reconstructions have been produced (Dumitru et al., 2019; Grant et al., 2019; Miller et al., 2012; Rovere et al., 2020), and (3) modelling approaches have improved (Berends et al., 2019; de Boer et al., 2014; DeConto and Pollard, 2016; DeConto et al., 2012; Dolan et al., 2012; Gasson et al., 2016b; Haywood et al., 2016; Pollard and DeConto, 2009; Pollard et al., 2015), it is now clear that Antarctica's ice sheets behaved dynamically throughout the Neogene and into the Quaternary.

Here we outline recent advances in knowledge of Antarctic Ice Sheet (AIS) behaviour and climate evolution through the Neogene. We focus our review on ice proximal geological records from the Ross Sea and offshore George V Land and Wilkes Land (Fig. 9.2); three regions where important data sets have become available over the past 15 years. We also present

---

◀ Foster et al., 2012; Greenop et al., 2014, 2019; Martinez-Boti et al., 2015; Raitzsch et al., 2021; Seki et al., 2010; Sosdian et al., 2018), B/Ca (Badger et al., 2013b), phytoplankton/alkenones (Badger et al., 2013b, 2019; Bolton et al., 2016; Pagani et al., 2011; Seki et al., 2010; Super et al., 2018; Zhang et al., 2013), leaf stomata (Beerling et al., 2009; Kürschner et al., 1996, 2008; Retallack, 2009a; Van der Burgh et al., 1993), paleosols (Cerling, 1992; Ekart et al., 1999; Ji et al., 2018; Retallack, 2009b), phytanes (Witkowski et al., 2018) and diatoms (Mejía et al., 2017). Solid red line displays a two million year moving average. CO_2 concentrations from 1765 to 2300 are shown for representative concentration pathways (RCPs) 2.6 (dark blue), 4.5 (light blue), 6 (orange) and 8.5 (red) (Meinshausen et al., 2011). Atmospheric CO_2 concentration thresholds for Antarctic ice sheet formation (green dashed line) and Northern Hemisphere (bipolar) glaciation (blue dashed line) are based on modelling results (DeConto and Pollard, 2003; DeConto et al., 2008). Present day average atmospheric CO_2 concentration shown by the red dashed line and is like those that characterised much of the early to middle Miocene. (B) Splice of deep sea benthic foraminifera $\delta^{18}O$ data (light grey) reflect changes in ice volume and deep/bottom water temperature (De Vleeschouwer et al., 2017). A comparison between three different records/splices is shown by the three green lines that display a 500 kyr moving average through the splice of Miller et al. (2020) (dark green), De Vleeschouwer et al. (2017) (medium green), and Cramer et al. (2009) (light green). (C) Bottom water temperature (BWT) records are derived from Ocean Drilling Program (ODP) Site 806 (dark blue) (Lear et al., 2015) Equation 7a (light blue) and 7b (purple) from the compilation of Cramer et al. (2011). (D) Obliquity sensitivity (S_{obl}) offers a proxy for ice sheet extent where higher values reflect times during which ice sheets expanded across Antarctica's continental shelves (Levy et al., 2019). Timing and duration of significant tectonic and oceanic events in the region around the Drake Passage (DP), West Antarctic Rift System (WARS), and Marie Byrd Land (MBL) are indicated by black horizontal bars (see text for details). Reconstructed topography for Antarctica is shown for three key time slices (23, 14, and 3.5 Ma) with associated ice sheet model simulations under low (280 ppm) CO_2 forcing and cold astronomical configuration and high (840 ppm) CO_2 forcing and warm astronomical configuration (Paxman et al., 2020).

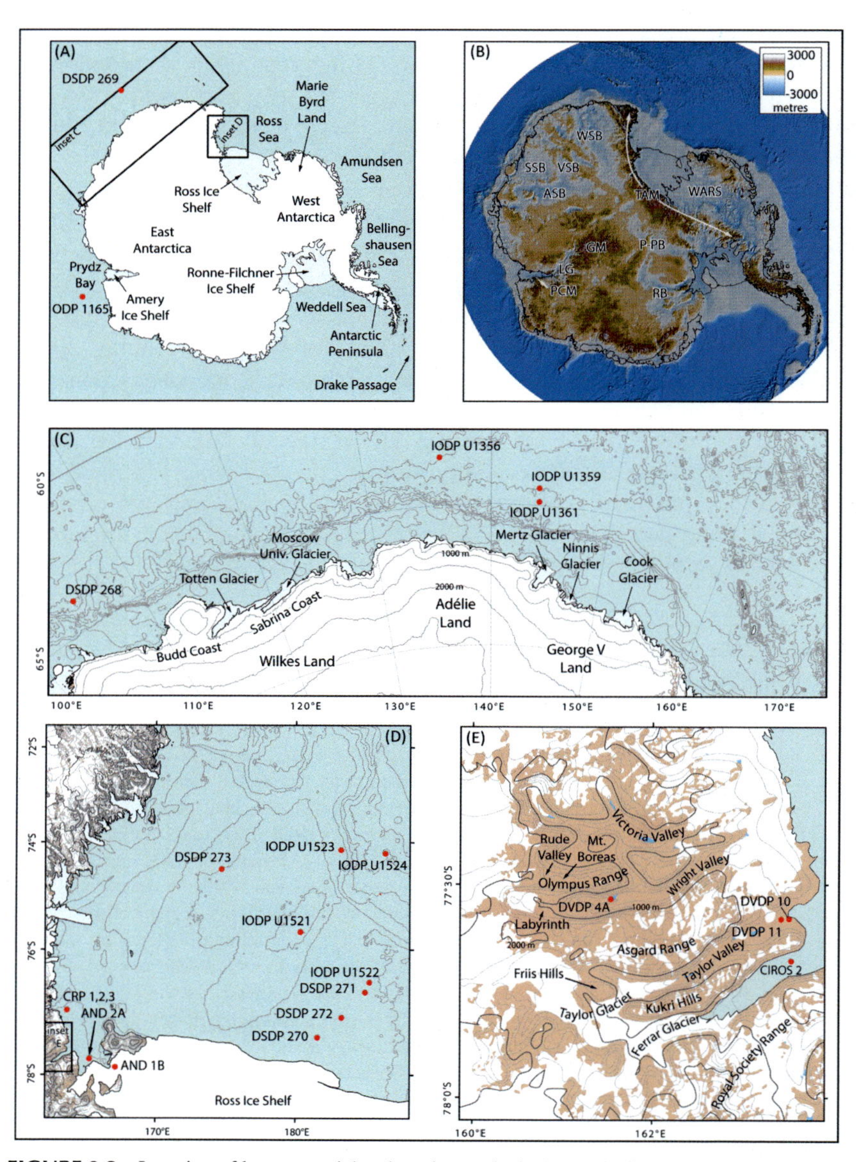

FIGURE 9.2 Location of key terrestrial and marine geological records discussed in the text. Deep Sea Drilling Project (DSDP), Ocean Drilling Program (ODP), Integrated Ocean Discovery Program (IODP), Dry Valleys Drilling Project (DVDP), Cenozoic Investigation of the Ross Sea Core (CIROS-1), Cape Roberts Project (CRP), and ANDRILL (AND) drill sites are indicated by red dots. Blue regions = ocean, white = ice, brown = exposed rock and sediment. Ice sheet elevation and ocean bathymetric contours are every 500 m. (A) Antarctica. (B) Subglacial geography of Antarctica derived from Bedmap2 (Fretwell et al., 2013) with major subglacial basins: *WSB*, Wilkes Subglacial Basin; *ASB*, Aurora Subglacial Bain; *SSB*, Sabrina Subglacial Basin; *VSB*, (*Continued*)

and discuss recent modelling studies that investigate climate–ice sheet interactions in the Miocene and Pliocene. For a broader review of other data sets from around the Antarctic, particularly offshore Prydz Bay and the Antarctic Peninsula, we refer the reader to the previous version of this chapter (Haywood et al., 2008) and other review papers and documents (Barrett, 2013; Bart and De Santis, 2012; Escutia et al., 2019; McKay et al., 2021).

9.1.2 Far-field records of climate and ice sheet variability

9.1.2.1 *The Early Miocene*

The onset of the Miocene is marked by a major transition in global climate from relative warmth of the late Oligocene to the generally cooler climates that persisted through much of the Neogene (Naish et al., 2021). Atmospheric CO_2 declined in the latest Oligocene (Zhang et al., 2013) reaching concentrations below 400 ppm across the Oligocene-Miocene Transition (OMT) (Fig. 9.1) (Greenop et al., 2019; Super et al., 2018). A large ($\sim$1‰) transient positive oxygen isotope excursion, broadly known as the Mi-1 event (Miller et al., 1991a; Zachos et al., 2001a; Zachos et al., 2001c) coincides with globally-distributed evidence for sea level lowstands of 20 to 40 m (Miller et al., 2020; Miller et al., 2005; Śliwińska et al., 2014; Zelilidis et al., 2002). Early Miocene cooling coincides with the onset of modern monsoon-like climate and widespread loess deposition in northern China (Guo et al., 2008; Zhang et al., 2018), $\sim$2°C cooling in NW Europe (Śliwińska et al., 2014), and cooler sea surface temperature (SST) in the north Atlantic Ocean (Egger et al., 2018; Super et al., 2020; Super et al., 2018). High resolution ice-distal benthic foraminifer stable isotope records suggest that the AIS expanded across the OMT and reached a volume comparable to present (Liebrand et al., 2017; Liebrand et al., 2011). Ice volume was potentially up to 25% larger than today if assumptions are made about the stability of the $\delta^{18}O$ composition of Antarctica's ice sheets (Pekar et al., 2006). Far field records suggest that the AIS decreased in size over the one million years after the OMT, in response to climatic warming and a hypothesised subsidence-induced reduction in terrestrial Antarctic land mass (Liebrand et al., 2017). Earliest Miocene proxy CO_2 concentrations are variable, with post-OMT estimates as high as 1000 ppm (Reichgelt et al.,

◀ Vincennes Subglacial Basin; *P-PB*, Pensacola-Pole Basin; *RB*, Recovery Basin; *WARS*, West Antarctic Rift System; *LG*, Lambert Graben; *TAM*, Transantarctic Mountains, *GM*, Gamburtsev Mountains; *PCM*, Prince Charles Mountains. (C) George V Land, Adélie Land, and Wilkes Land region of East Antarctica. (D) Ross Sea Region, in the panel D a square indicating the location of panel E. (E) McMurdo Dry Valleys. Land contours in (D) and (E) are every 200 m.

2016). However, a majority of the proxy data indicate values between 350 and 450 ppm from 23 to 19.5 Ma (Greenop et al., 2019; Londoño et al., 2018; Reichgelt et al., 2020; Super et al., 2018; Zhang et al., 2013). Glacial-interglacial variability through much of the early Miocene was paced by variations in short and long-period eccentricity (Holbourn et al., 2005; Liebrand et al., 2017) and likely saw fluctuations between 50% and 125% of the modern EAIS (Pekar and DeConto, 2006). Bottom water temperatures (BWT) likely varied up to 3°C on million year time scales (Cramer et al., 2011).

9.1.2.2 The mid-Miocene

Following the relatively cool climate of the earliest Miocene, environmental proxies indicate that the Planet entered an interval of relative warmth. The Miocene Climatic Optimum (MCO; ~17 to ~14.8 Ma) is characterised by low average deep sea benthic foraminifera $\delta^{18}O$ values (De Vleeschouwer et al., 2017; Flower and Kennett, 1993a; Holbourn et al., 2015; Westerhold et al., 2020; Woodruff and Savin, 1991) (Figs 9.1 and 9.3). Fossil floral and faunal evidence indicate that this was the warmest interval of the past 35 million years (Böhme, 2003; Hornibrook, 1992; Mosbrugger et al., 2005; Prebble et al., 2017; Utescher et al., 2011). Average global surface temperatures were likely 3°C to 4°C warmer than present (You, 2010), with intervals of peak warmth during which average surface temperatures were ~7°C to 8°C warmer than today (Goldner et al., 2014; Steinthorsdottir et al., 2020, 2021). Near surface and surface water temperature (upper 200 m) in the Southern Ocean were at times between 7°C and ~16°C warmer than present (Sangiorgi et al., 2018; Shevenell et al., 2004) and global deep sea water temperatures were between 5°C and 9°C warmer than today (Billups and Schrag, 2003; Lear et al., 2010, 2015; Modestou et al., 2020; Shevenell et al., 2008). Causes of warming and cooling through the MCO have largely been attributed to changes in carbon cycling (e.g., greenhouse gas concentrations) (Foster and Rohling, 2013; Foster et al., 2012; Vincent and Berger, 1985; Woodruff and Savin, 1989). There is a wide range of estimates of atmospheric CO_2 concentrations during the MCO of between <400 and >1000 ppm (Cui et al., 2020; Greenop et al., 2014; Ji et al., 2018; Sosdian et al., 2018; Steinthorsdottir et al., 2021). The inferred increase in CO_2 has been linked to rapid eruption of the Columbia River Flood basalts between 16.7 and 15.9 Ma (Foster et al., 2012; Kasbohm and Schoene, 2018). Statistical analysis of available proxy data suggests that average values through the MCO were ~400 ppm (Figs 9.1 and 9.3) but estimates based on boron proxy data from Ocean Drilling Program (ODP) Site 761 off NW Australia suggest eccentricity-paced (100 kyr) variations between 200 and 600 ppm occurred (Greenop et al., 2014). Mechanisms and processes driving these relatively large high frequency changes in the carbon cycle require further exploration. However, there is little doubt they support other climate proxies that indicate a dynamic astronomically-paced climate during the MCO (Holbourn et al., 2015) (Fig. 9.3) and that climate sensitivity may have been higher than present.

A more detailed examination of benthic foraminifera $\delta^{18}O$ data suggests that, like most intervals in Earth's history, the MCO cannot be characterised by a single 'time slice' reconstruction. The onset of the MCO is marked by a large (~1‰), relatively rapid shift in mean $\delta^{18}O$ values (Fig. 9.1) and is followed by low to moderate variability on eccentricity-paced time scales (Holbourn et al., 2015) (Figs 9.1 and 9.3). These data suggest subdued environmental variability during the early MCO (~17 to 16 Ma) and reflect warmer ocean bottom waters and/or reduced ice volume relative to the early Miocene (Figs 9.1 and 9.3). This period of relative warmth and stability was followed by an interval from 16 to 15 Ma characterised by high amplitude $\delta^{18}O$ variability (Flower and Kennett, 1994) with shifts up to 1.5‰ on eccentricity-paced glacial/interglacial timescales (Holbourn et al., 2015) (Fig. 9.3B). MCO bottom water temperatures (BWT) fluctuated by as much as 3°C in the Southern Ocean on hundred thousand year timescales (Shevenell et al., 2008) and up to 6°C at lower latitude sites across million year periods (Lear et al., 2015) (Fig. 9.3F). These relatively large BWT changes may explain a significant portion of the $\delta^{18}O$ variability between 16 and 15 Ma, although global sea level reconstructions suggest that large ice volume variations during this time cannot be ruled out (Miller et al., 2020). These large $\delta^{18}O$ variations also coincide with eccentricity-paced (~400 kyr) changes in benthic foraminifera $\delta^{13}C$ data. The astronomically-paced excursions are superimposed on the ~2 million year-long global ~1‰ increase in $\delta^{13}C$ referred to as the Monterrey Carbon Isotopic Excursion (Vincent and Berger, 1985) (Fig. 9.3D). The Monterey Excursion has nine 400 kyr-paced $\delta^{13}C$ maxima (Holbourn et al., 2007; Pagani et al., 1999a; Woodruff and Savin, 1989) that reflect changes in global carbon cycling, an interpretation further supported by observations of global Carbonate Compensation Depth variability (Holbourn et al., 2015) and intervals of enhanced organic carbon deposition and preservation on the circum-Pacific continental shelves (Sosdian et al., 2020; Vincent and Berger, 1985).

Sedimentary records from the New Jersey margin (Kominz et al., 2008, 2016; Miller et al., 2005) and the Marion Plateau, south of Australia (John et al., 2011), indicate sea level rose and fell up to ~35 m during the MCO (Fig. 9.3), requiring fluctuations of up to 60% of the present AIS. SSTs in high northern latitudes were up to 15°C warmer than present during the MCO but were not significantly higher than the early Miocene (Super et al., 2018) and there is little evidence for ice-rafting in the Arctic from 17.5 to 16 Ma (St. John, 2008). These data suggest that the fluctuating

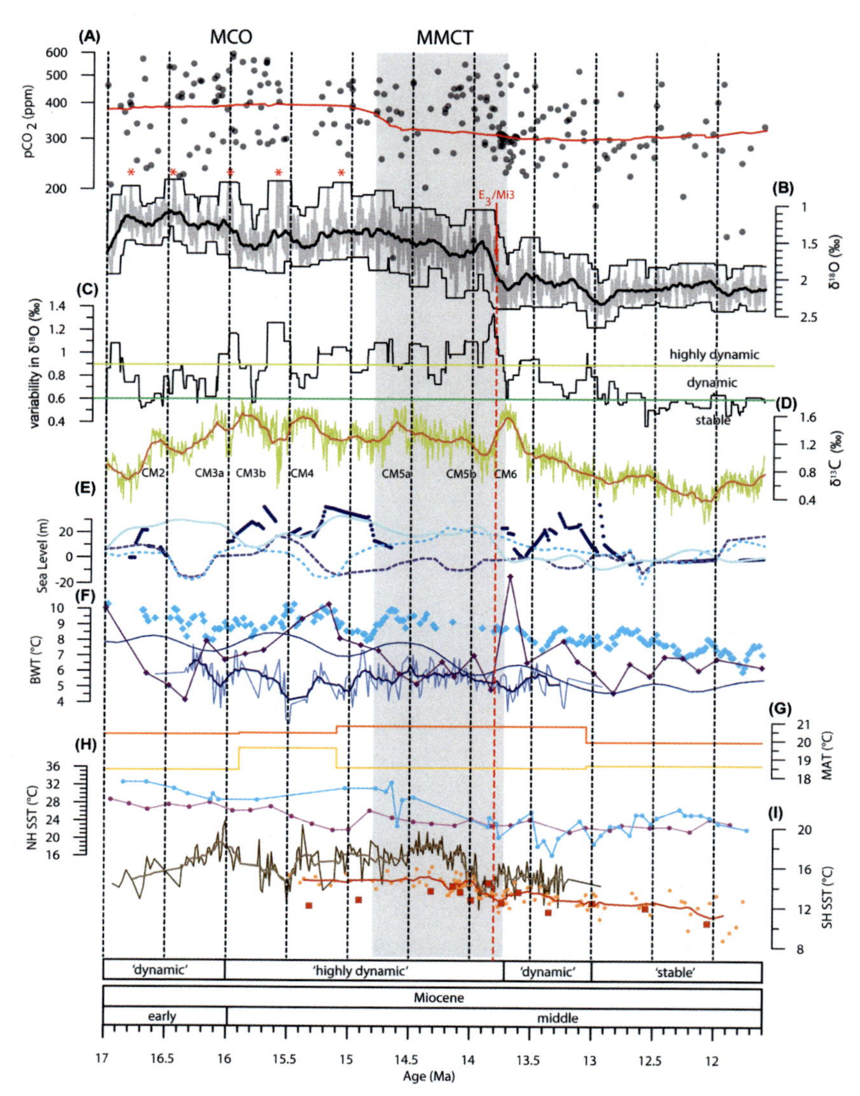

FIGURE 9.3 Environmental data from the late early Miocene through middle Miocene (17 to 11.6 Ma). (A) Atmospheric CO_2 compilation comprises data from a range of proxies outlined in Fig. 9.1. Solid red line displays a two million year moving average. (B) Splice of deep sea benthic foraminifera $\delta^{18}O$ data (light grey) reflect changes in ice volume and deep/bottom water temperature (De Vleeschouwer et al., 2017). Solid black line displays a 150 kyr moving average. Maximum and minimum values are determined within each 150 kyr window and create the envelope (black lines) that bound the $\delta^{18}O$ data. Red asterisks indicate intervals of peak warmth and maximum carbonate dissolution in the eastern equatorial Pacific (Holbourn et al., 2015). E_3/Mi-3 is the large stepwise $\delta^{18}O$ increase (Flower and Kennett, 1993a; Miller et al., 1991b; Woodruff and Savin, 1991) that marks the end of the MMCT. (C) A $\delta^{18}O$ variability 'index',

(Continued)

glaciers and ice sheets that drove changes in sea level during the MCO were in Antarctica. However, questions remain regarding the minimum extent of Antarctic ice during this time. Did the AIS completely melt during the MCO or did high elevation and inland areas remain ice covered during the warmest interglacials? Recent studies using far-field records suggest that ice completely disappeared from Antarctica during intervals of peak warmth in the MCO (Miller et al., 2020). However, equilibrium ice sheet simulations (Gasson et al., 2016b; Halberstadt et al., 2021; Paxman et al., 2020) and proxy data (Levy et al., 2016; Shevenell et al., 2008) suggest these 'total deglaciation' events were rare or did not occur. Whereas data presented throughout this review shed some light regarding AIS behaviour in the MCO, new ice-proximal sedimentary records are required to determine the sensitivity of the AIS to middle Miocene warmth (McKay et al., 2021).

Termination of the MCO and the onset of cooling and glacial expansion through the Middle Miocene Climate Transition (MMCT) is indicated by an increase in $\delta^{18}O$ maxima starting at ~14.8 Ma, which was followed by a gradual increase of ~0.3‰ in average $\delta^{18}O$ values during the subsequent

---

◀ interpreted as a measure of glacial/interglacial variability, is determined by subtracting the maximum values in (B) from the minimum value, for each 150 kyr window. We indicate arbitrary 'thresholds' in these data (green and yellow solid lines) and suggest that values below 0.6‰ reflect a relatively 'stable' high latitude environment and values that exceed 0.9‰ indicate a highly dynamic environment with large changes in BWT and/or ice volume over glacial–interglacial time scales. (D) Yellow line shows high resolution deep sea benthic foraminifera $\delta^{13}C$ splice (Westerhold et al., 2020). Solid deep orange line displays a 150 kyr moving average and carbon maxima events (Vincent and Berger, 1985; Woodruff and Savin, 1991) are labelled (CM2–6). (E) Sea level curves are from Kominz et al. (2016) (blue dots), Kominz et al. (2008) (pale blue dashed line), Miller et al. (2005) (blue dashed line), and Miller et al. (2020) (pale blue line). (F) BWT data from ODP Site 761 (east Indian Ocean: 16.738°S, 115.535°E) (Lear et al., 2010) (light blue), ODP Site 806 (western equatorial Pacific Ocean: 0.319°N, 159.361°E) (Lear et al., 2015) (purple), ODP Site 1171 (southwest Tasman Sea: 48.4999°S, 146.1115°E) (Shevenell et al., 2008) (blue), and derived from the compilation of Cramer et al. (2011) using Equation 7a (dark blue). Solid line through data from ODP Site 1171 displays a 9-pt running average. (G) Pollen-based Mean Annual Temperature estimates for New Zealand display 'envelope' where upper bound (dark orange) = mean of warmest 20% of samples and lower bound (light orange) = mean of coldest 20% of samples (Prebble et al., 2017). (H) SST data for the Northern Hemisphere from DSDP Site 608 (northwest Atlantic Ocean: 42.8367°N, 23.0875°W) (Super et al., 2018) (light blue) and ODP Site 982 (northwest Atlantic Ocean: 57.516°N, 15.866°W) (Super et al., 2020) (pink). (I) SST data for the Southern Hemisphere from ODP Site 1171 using the Mg/Ca proxy (Shevenell et al., 2004) (dark brown) and the TEX_{86} proxy (orange) (Leutert et al., 2020) using the calibration of Ho and Laepple (2016). Solid lines through the Mg/Ca and TEX_{86} data display a 9-pt running average. Red squares display temperature estimates derived from clumped isotopes (Δ_{47}) (Leutert et al., 2020). Grey vertical band highlights the Middle Miocene Climate Transition and vertical red dashed line indicates likely threshold in the climate system, across which high latitudes cooled and sea ice and Antarctic marine-based ice sheets became a more frequent and persistent feature.

~800 kyrs (Holbourn et al., 2014) (Fig. 9.3B). By the end of the MMCT (at ~13.8 Ma), global $\delta^{18}O$ had increased by 1‰ to 1.5‰ (Holbourn et al., 2005, 2014; Kennett, 1977; Shackleton and Kennett, 1975a). A ~100 ppm decrease (400 to 300 ppm) in average CO_2 concentrations occurred at the onset of the MMCT (Fig. 9.3A), although proxy data indicate a high degree of variability continued following the initial drop, with values exceeding 500 ppm at times and dropping below 200 ppm at others (Fig. 9.3A). In contrast to the MCO, BWT variability during the MMCT was low, generally fluctuating by ~1°C on 100 and 400 kyr time scales in the Southern Ocean (Shevenell et al., 2008). BWT generally decreased during the MMCT by ~2°C (Cramer et al., 2011; Modestou et al., 2020; Shevenell et al., 2008). SSTs records suggest mid southern latitudes cooled by 3°C to 4°C between 14.1 and 13.8 Ma (Leutert et al., 2020; Shevenell et al., 2004, 2008).

Sea level records through much of the MMCT are somewhat ambiguous (Fig. 9.3E). Records from the New Jersey margin indicate sea level amplitudes varied by at least 10 m (Kominz et al., 2016), but different data sets produce significantly different maximum and minimum magnitudes (Fig. 9.3E). Furthermore, data from the Marion Plateau indicate sea level rose and fell by ~35 m between 14.7 and 14 Ma (John et al., 2011). Large amplitude (1‰) high frequency variations in the $\delta^{18}O$ record occur through the interval (Holbourn et al., 2014; Shevenell et al., 2008). Given that BWTs remained relatively stable, these large excursions suggest glacial–interglacial changes in ice volume were significant. A gradual increase in ice-rafted debris (IRD) from 15.6 to 14 Ma in the North Atlantic suggests sea ice presence and/or ice growth and glacial advance to coastlines around the margins of the Greenland Sea (St. John, 2008). While a modelling study suggests northern hemisphere ice sheets may have grown when CO_2 dropped below 280 ppm (DeConto et al., 2008), it is still unclear if significant growth of ice occurred in the northern hemisphere at this time. This implies that the AIS was highly dynamic and advanced and retreated throughout the MMCT. However, an overall shift to more positive $\delta^{18}O$ values from the MCO to MMCT suggests the AIS grew larger during glacials and retreated less during interglacials between 14.8 and 13.9 Ma (Holbourn et al., 2014; Shevenell et al., 2008).

The gradual increase in average $\delta^{18}O$ values across the MMCT culminated between 13.9 and 13.8 Ma with a global ~1.2‰ increase, the most striking feature in Neogene oxygen isotope records (Figs 9.1 and 9.3). Prior to 13.8 Ma, $\delta^{18}O$ values generally fluctuated by 1‰ on glacial–interglacial timescales. However, at ~13.8 Ma it appears that an environmental tipping point was reached (Kennett, 1977). Rather than 'rebounding' to prior interglacial values, an additional rapid increase in $\delta^{18}O$ occurred, driving a ~0.6‰ stepwise increase in average deep sea $\delta^{18}O$ values. This pronounced increase was first recorded in low resolution records from Deep Sea Drilling Project (DSDP) Sites 289 (0.4987°S, 158.5115°E) and 291 (12.8072°N,

127.8308°E) in the western Pacific Ocean (Shackleton and Kennett, 1975b). Moderate resolution benthic and planktic foraminifera isotopic data from DSDP Site 590 (31.167°S, 163.3585°E) first showed a two-step increase occurred during this interval (Kennett et al., 1986). As the resolution of the deep sea benthic isotope records improved, this interval of maximum $\delta^{18}O$ values was formally identified as oxygen isotope zone Mi3 (Miller et al., 1991a) and $\delta^{18}O$ maximum E_3 (Woodruff and Savin, 1991). High-resolution data from DSDP Site 588 (26.1117°S, 161.2267°E) in the north Tasman Sea resolved three distinct increases through Mi3/event E, that were identified as E_1, E_2, and E_3 (Flower and Kennett, 1993a). Most recently, well-dated high-resolution records from the mid to low latitudes (Holbourn et al., 2007, 2014) and astronomically-tuned $\delta^{18}O$ splices (De Vleeschouwer et al., 2017; Miller et al., 2020; Westerhold et al., 2020) place the final step of the MMCT (Mi3/E_3) between 13.9 and 13.8 Ma.

Approximately 70% of the observed ~1.2‰ shift between 13.9 and 13.8 Ma is attributed to AIS growth (Shevenell et al., 2004; Wright et al., 1992), which implies ~76 to 100 m of sea level fall if the Pleistocene calibration of 0.08‰–0.11‰ per 10 metres of sea level is applied (Adkins et al., 2002; Fairbanks and Matthews, 1978). However, sea level records through this final MMCT step are ambiguous and somewhat contradictory. A major sequence boundary characterises the entire MMCT at the shallow continental shelf sites offshore New Jersey (Kominz et al., 2016). However, it is impossible to identify whether a series of sea level rise and fall events occurred across the MMCT or that a single sea level fall at ~13.8 Ma eroded older sediments. A sea level fall of ~30 m between 13.9 and 13.7 Ma is inferred from sea level calibrations applied to $\delta^{18}O$ records (Miller et al., 2020) and a sea level fall of 59 ± 6 m at 13.8 Ma is inferred from stratigraphic sequences on the Marion Plateau (John et al., 2011). The Marion Plateau data suggest global ice volume grew by ~26 m sea level equivalent (s.l.e.) more at ~13.8 Ma than during previous glacial maxima within the MMCT (e.g., at 14.7 Ma). This 'additional' ice likely filled Antarctica's marine basins and expanded to the edge of Antarctica's prograding and advancing continental shelves (Colleoni et al., 2018; De Santis et al., 1995, 1999; Levy et al., 2016, 2019; McKay et al., 2019; Pérez et al., 2021a) but may have also formed in the northern hemisphere (DeConto et al., 2008). Whereas there is no definitive evidence for ice sheet growth in the northern hemisphere during the MMCT, a pronounced cooling of 6°C in the north Atlantic Ocean occurred between 14.5 Ma and 13.8 Ma (Super et al., 2018). Furthermore, a major change in foraminifera assemblage (Kender and Kaminski, 2013) suggests that perennial sea ice persisted in the Arctic Ocean since 14 Ma. Clearly, the Earth's high latitudes cooled at this time.

Understanding the drivers of environmental change recorded by increasing $\delta^{18}O$ across the MMCT, and the large rapid stepwise increase at its end has been a focus of many studies (Flower and Kennett, 1993a, 1994; Kennett,

1977; Shevenell et al., 2004). Changes in ocean circulation (Shevenell et al., 2004) and the draw down in atmospheric CO_2 concentration (Holbourn et al., 2005, 2013) are often cited as likely catalysts of the MMCT. Warm surface ocean temperatures at ODP Site 1171 (48.4999°S 149.1115°E) suggest that changes in the hydrological cycle and an increase in precipitation over Antarctica beginning at ~15.4 Ma may have contributed to initial ice sheet expansion (Shevenell et al., 2004). Constriction of the Tethys seaway (Hamon et al., 2013; Hsü and Bernoulli, 1978; Woodruff and Savin, 1989), ongoing deepening of the Drake Passage, and enhanced circulation of the Antarctic Circumpolar Current (Dalziel et al., 2013) may have slowly reduced meridional heat and vapour transport to Antarctica (Lewis et al., 2006; Shevenell et al., 2004, 2008). Subsequent Southern Ocean surface cooling and intensified climatic response to changes in Earth's orbital eccentricity increased thermal isolation of Antarctica and drove maximum ice sheet expansion recorded by the E_3 $\delta^{18}O$ isotope event (Holbourn et al., 2005; Shevenell et al., 2004) (Fig. 9.3E). SSTs at high northern latitudes also cooled between 2°C and 6°C (Super et al., 2020). Terrestrial records in central Europe also indicate a ~17°C decline in soil temperature occurred over ~350 kyrs between 14.5 and 14 Ma (Methner et al., 2020).

Astronomical variations in insolation likely also played a role in driving high latitude cooling and ice growth at this time. Records from ODP Site 1171 in the Southern Ocean suggest that climate was paced by long-period eccentricity variations (~400 kyr) between 15.4 and 13.5 Ma (Shevenell et al., 2004). Higher resolution records from IODP Site U1338 (2.5078°N 117.9693°W) in the eastern Pacific indicate that while prominent 400 and 100 kyr eccentricity cycles dominated the warm MCO, a switch to obliquity-paced climate variability occurred at 14.7 Ma, with an increase in response to eccentricity (100 kyr) forcing occurring after 14.1 Ma (Holbourn et al., 2014). The major shift in $\delta^{18}O$ and inferred advance of the AIS coincides with an obliquity node between 14.2 and 13.8 Ma (Holbourn et al., 2005).

The role of carbon cycling across the MMCT remains unclear (Shevenell et al., 2008), in part due to the range of atmospheric CO_2 estimates from the proxy data and the cessation of the Monterey $\delta^{13}C$ Excursion (Fig. 9.3D). Whereas average CO_2 values remained relatively constant across the E_3 event (Fig. 9.3), high resolution data from the Ras il-Pellegrin section in Malta suggest atmospheric CO_2 concentration dropped and remained below 300 ppm for at least 200 kyrs following the event (Badger et al., 2013b). This drop in atmospheric CO_2 occurred as $\delta^{13}C$ values increased during the last major excursion of the Monterrey event (CM6) (Woodruff and Savin, 1991) (Fig. 9.3E). Large increases in opal accumulation in the eastern equatorial Pacific at 14 and 13.8 Ma and uplifted marginal marine sedimentary sequences around the Pacific Rim suggest high silica-based productivity contributed to atmospheric CO_2 drawdown (Flower and Kennett, 1993a, 1993b;

Holbourn et al., 2014; Vincent and Berger, 1985). These observations are consistent with hypotheses that an increase in organic carbon burial drove a decrease in atmospheric CO_2 and contributed to global cooling (Vincent and Berger, 1985).

The interval of climatic cooling and ice expansion reflected by the Mi3/E_3 isotope shift was a transient event. Sea surface and bottom water temperatures warmed by 2°C to 3°C between 13.7 and 13.5 Ma (Lear et al., 2010; Modestou et al., 2020; Shevenell et al., 2004, 2008). Notably, the large variations in $\delta^{13}C$ that characterised the MCO and MMCT ceased by ~13.5 Ma and began a gradual decline of ~0.5‰ in average values through to the end of the middle Miocene (Fig. 9.3E) (Vincent and Berger, 1985). This drop in carbon isotope values and coincident warming suggests that feedbacks associated with ice expansion may have influenced global carbon cycling (Shevenell et al., 2004, 2008) and is supported by modelling experiments (Knorr and Lohmann, 2014). Deep sea $\delta^{18}O$ records indicate the AIS remained variable at this time and $\delta^{18}O$ minima of <1.5‰ (Fig. 9.3B) suggest the AIS retreated to, and possibly inland of, the terrestrial continental margin during interglacial intervals.

Surface temperatures at mid latitudes in the Southern Ocean began to cool again at ~13.5 Ma and gradually declined by 2°C to 3°C through the late middle Miocene (Leutert et al., 2020; Shevenell et al., 2004). Mean annual terrestrial temperatures in New Zealand also cooled by ~1°C (Prebble et al., 2017). However, cooling was not ubiquitous, as indicated by data from the North Atlantic, which show SSTs warmed by approximately 5°C (Super et al., 2018, 2020). An interval during which average deep sea $\delta^{18}O$ values remained relatively low and glacial–interglacial amplitude variability was relatively small (<0.6‰), occurred between 13.5 and 13 Ma. Relatively large ice sheets may have persisted during this interval. However, variability in BWT was also muted (~1°C) during this interval (Cramer et al., 2011) and may account for ~0.22‰ of the 0.6‰ $\delta^{18}O$ shift. It follows that the remaining ~0.38‰ of $\delta^{18}O$ not accounted for by temperature may reflect ice volume changes equivalent to between 47 and 35 metres sea level equivalent if the Pleistocene calibration of 0.08–0.11‰ per 10 metres of sea level is applied (Adkins et al., 2002; Fairbanks and Matthews, 1978). This magnitude of sea level change requires relatively large-scale growth and retreat of ice on Antarctica. However, if ice sheets occupied the northern hemisphere, then these bi-polar ice masses could have each varied by smaller amounts. Data from northern hemisphere IODP Site 302 (87.8666°N, 136.1774°E) hints that glacier ice may have reached the coast around the Arctic at this time (St. John, 2008). Proxy data indicate atmospheric CO_2 concentration dropped below 300 ppm (Badger et al., 2013b) (Fig. 9.3A) and an idealised modelling study suggests ice may have grown in the northern hemisphere under these climatic conditions (DeConto et al., 2008). New geological data are required to confirm or refute whether significant volumes of ice grew in high northern latitudes following the MMCT.

9.1.2.3 The Late Miocene

The late Miocene was recently dubbed the 'cool house' because it comprises the bridge from the middle Miocene 'warm house' to the modern 'icehouse' (Westerhold et al., 2020). However, global SST reconstructions indicate ocean surface temperatures remained significantly warmer than present even after the final 'cooling step' in $\delta^{18}O$ records at the end of the MMCT (Herbert et al., 2016). Furthermore, whereas bottom water temperatures cooled across the MMCT (Cramer et al., 2011; Lear et al., 2015; Modestou et al., 2020; Shevenell et al., 2008), they were still 6°C to 9°C warmer than present in the Indian Ocean in the late Miocene (Modestou et al., 2020). SSTs at high northern latitudes were also significantly warmer (up to 17°C) relative to modern conditions in the late Miocene and latitudinal temperature gradients remained low (Super et al., 2020).

Sea level reconstructions from the New Jersey margin record an episode of major sea level fall between ~11 and 10.5 Ma (Kominz et al., 2008; Miller et al., 2005) (Fig. 9.4E). This drop is one of the largest Cenozoic lowstand events recorded in classic sequence stratigraphic studies (Haq et al., 1987) but is less obvious in recent reconstructions based on analysis of the $\delta^{18}O$ record (Miller et al., 2020, 2005). A dramatic shoaling of the carbon compensation depth at ~10.5 Ma marks the start of the late Miocene 'carbonate crash' (Diester-Haass et al., 2004; Lyle et al., 1995; Pälike et al., 2012; Peterson et al., 1992; Preiss-Daimler et al., 2021), which has been attributed to changes in the intensity of chemical weathering and riverine input of calcium and carbonate ions to the ocean (Lübbers et al., 2019) and/or changes in ocean circulation related to the restriction of the central American seaway (Newkirk and Martin, 2009). The Benguela upwelling system off southern Africa is thought to have initiated at ~10 Ma (Diester-Haass et al., 2004; Rommerskirchen et al., 2011; Siesser, 1980). Today this major upwelling system delivers cold and nutrient-rich Antarctic sourced waters to support vast populations of phytoplankton and its inception may reflect change at high southern latitudes in the late Miocene. Pollen records from New Zealand indicate a shift to cooler and wetter climate in the mid southern latitudes occurred at this time (Prebble et al., 2017). An increase in the altitude of the Tibetan Plateau is thought to have occurred between 10 and 8 Ma and may have caused enhanced aridity in the Asian interior and contributed to the onset of the Indian and east Asian monsoons (Raymo and Ruddiman, 1992; Zhisheng et al., 2001). However, questions remain around the timing of uplift in the Neogene, the orographic influence on regional climate dynamics, and the synchroneity of paleoclimatic indicators in the region (Molnar et al., 2010).

Whereas textural analysis of sand grains from ODP Site 918 (63.0928°N 38.6389°W) suggests glaciers were present on Greenland ~11 Ma (Helland and Holmes, 1997), drop stones in glacial marine records from the same site suggest glaciers first reached the coastline of Greenland approximately 7 Ma (Larsen et al., 1994; St John and Krissek, 2002). In addition, analysis of beryllium and aluminium isotopes in sand grains from marine sediments show that the East Greenland Ice Sheet existed over the past 7.5 million years (Bierman et al., 2016). These data and analysis of regional seismic data around the Greenland margin (Pérez et al., 2018) offer compelling evidence for continental-scale ice sheets in both polar regions by the late Miocene. A major shift in marine $\delta^{13}C$ occurs between 8 and 7 Ma (Diester-Haass et al., 2002; Farrell et al., 1995; Holbourn et al., 2018; Loutit and Kennett, 1979) (Fig. 9.4D) and is coincident with the onset of global cooling and inferred bi-polar glaciation. This shift has been attributed to changes in the Southern Hemisphere due to either reduced ventilation of Southern Component Water (SCW) following expansion of the West Antarctic Ice Sheet (WAIS) (Kennett and Barker, 1990), or decreased input of Northern Component Water (NCW) to the Southern Ocean and/or decreased exchange of CO_2 between the atmosphere and surface water in Antarctic source areas (Hodell and Venz-Curtis, 2006). Expansion of C_4 grasslands at the expense of C_3 vegetation occurred at this time (Osborne and Beerling, 2006; Pagani et al., 1999b; Tipple and Pagani, 2007) and coincides with major changes in large terrestrial fauna to a dominance by browsers (Badgley et al., 2008). Rare but important specimens assigned to *Sahelanthropus tchadensis* suggest the hominin lineage first appeared at 7 Ma (Brunet, 2010; Brunet et al., 2002; Vignaud et al., 2002), although the taxonomy of these fossil taxa is still developing (Haile-Selassie et al., 2004) and their linkages to modern humans are debated (Wood and Harrison, 2011).

Modest cooling of global SSTs that characterised the early-late Miocene accelerated in the Messinian (Herbert et al., 2016). SSTs decreased by 2°C to 3°C in the Southern Ocean and by up to 5°C at high northern latitudes (Herbert et al., 2016) (Fig. 9.4H). This Late Miocene Cooling (LMC) episode coincides with an increase in zonal and meridional temperature gradients in the Pacific at ~7 Ma (Zhang et al., 2014), aridification in the subtropical regions (e.g., south-central Andes) (Amidon et al., 2017), and establishment of the Sahara Desert (Schuster et al., 2006). Mid-latitude SSTs reached near modern values at ~6 Ma (Fig. 9.4H) (Herbert et al., 2016), coincident with pronounced cooling and an increase in precipitation in New Zealand (Prebble et al., 2017). The Messinian Salinity Crisis (MSC) (6.2 to 5.5 Ma) was perhaps the most dramatic oceanographic event in the latest Miocene (Hsü et al., 1973). Sea level fall and subsequent desiccation of the Mediterranean is recorded by >1000 m of evaporite deposits in the Mediterranean Basin

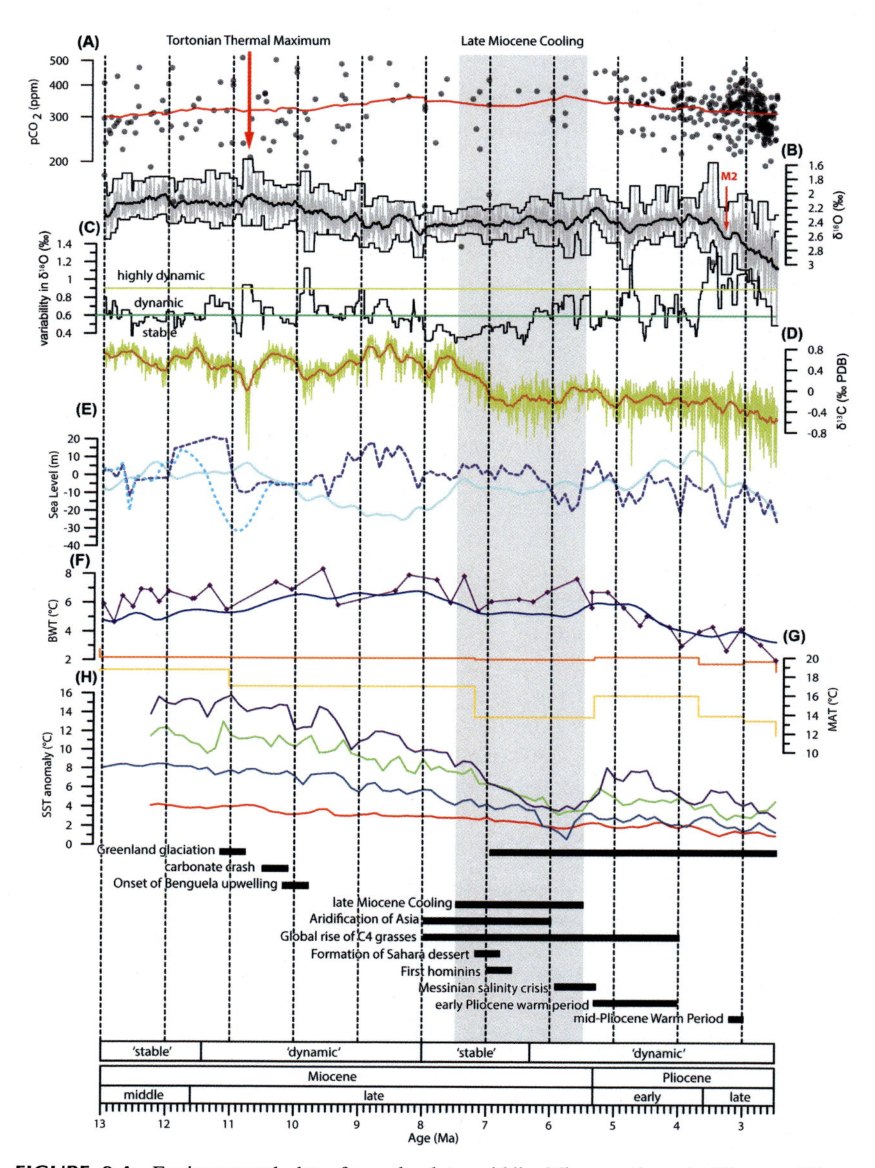

FIGURE 9.4 Environmental data from the late middle Miocene through Pliocene (13 to 2.58 Ma). (A) Atmospheric CO_2 compilation comprises data from a range of proxies outlined in Fig. 9.1. Solid red line displays a two million year moving average. (B) Splice of deep sea benthic foraminifera $\delta^{18}O$ data (light grey) reflect changes in ice volume and deep/bottom water temperature (De Vleeschouwer et al., 2017). Solid black line displays a 150 kyr moving average. Maximum and minimum values are determined within each 150 kyr window and create the envelope (black lines) that bound the $\delta^{18}O$ data. (C) A $\delta^{18}O$ variability 'index', interpreted as a measure of glacial/interglacial variability, is determined by subtracting the maximum values in

(Continued)

(Barber, 1981; Hodell et al., 1986). Debate on the causes of the MSC persists, with tectonic restriction between the Mediterranean Sea and the Atlantic Ocean (Garcia-Castellanos and Villaseñor, 2011), and glacial eustatic controls (Zhang and Scott, 1996) or an interplay between the two being the most likely scenario (Ohneiser et al., 2015).

The episode of Late Miocene Cooling culminated in the MSC and terminated with relatively rapid warming of ~3°C to 4°C in the mid-latitudes into the early Pliocene (Herbert et al., 2016). This climatic 'rebound' at the end of the Miocene is also reflected in Southern Hemisphere mid-latitude records of terrestrial palynomorphs (Prebble et al., 2017), marine molluscs (Beu, 1990) and shallow water foraminifera (Hornibrook, 1992) from New Zealand.

9.1.2.4 *The Pliocene*

Geological proxies indicate that the warm Early to mid-Pliocene (5.3 to 3.0 Ma) interval is the last time in Earth's history that atmospheric CO_2 concentrations (ca. 400 ppm) were similar to present day and global mean surface temperatures were 2°C–3°C warmer than modern (Martinez-Boti et al., 2015; Masson-Delmotte et al., 2013). Other prominent features during this warm interval include a perennially ice-free Arctic Ocean (Ballantyne et al., 2010, 2013; Dowsett et al., 2012), Arctic summer temperatures approximately 8°C–19°C warmer than modern (Brigham-Grette et al., 2013; Salzmann et al., 2011), reduced meridional and zonal SST gradients (Brierley et al., 2015; Fedorov et al., 2013; Tierney et al., 2019), and enhanced production and export of Northern Component Water (NCW) – the ancient correlative to North Atlantic Deep Water (NADW) (Billups et al., 1997; Frank et al., 2002; Kwiek and Ravelo, 1999; Ravelo and Andreasen, 2000). Vegetation

---

◀ (B) from the minimum value, for each 150 kyr window. We indicate arbitrary 'thresholds' in these data (green and yellow solid lines) and suggest that values below 0.6‰ reflect a relatively 'stable' high latitude environment and values that exceed 0.9‰ indicate a highly dynamic environment with large changes in BWT and/or ice volume over glacial–interglacial time scales. (D) Yellow line shows high resolution deep sea benthic foraminifera $\delta^{13}C$ splice (Westerhold et al., 2020). Solid deep orange line displays a 150 kyr moving average and carbon maxima events (Vincent and Berger, 1985; Woodruff and Savin, 1991) are labelled (CM2–6). (E) Sea level curves are from Kominz et al. (2008) (pale blue dashed line), Miller et al., (2005) (blue dashed line), and Miller et al. (2020) (pale blue line). (F) BWT data from ODP Site 926 (western Atlantic Ocean: 3.7191°N, 42.9081°W) (Lear et al., 2003, 2020) and derived from the compilation of Cramer et al. (2011) using Equation 7a (dark blue). (G) Pollen-based Mean Annual Temperature estimates for New Zealand display 'envelope' where upper bound (dark orange) = mean of warmest 20% of samples and lower bound (light orange) = mean of coldest 20% of samples (Prebble et al., 2017). (H) Stacked SST curves from alkenone based reconstructions for the SST anomalies for the Northern Hemisphere >50°N (purple line) and 30 to 50°N (green line), Tropics (red line), and Southern Hemisphere 30 to 50°S (blue line) (Herbert et al., 2016). Grey vertical band highlights Late Miocene Cooling. Timing and duration of key global events through indicated by horizontal black bars (see text for details).

reconstructions (Salzmann et al., 2008) imply that the global extent of arid deserts decreased and boreal forests replaced tundra. Atmosphere-Ocean Global Circulation Models (AOGCMs) predict an enhanced hydrological cycle, but with large inter-model spread (Haywood et al., 2013, 2020). The East Asian Summer Monsoon, as well as other monsoon systems, may have been enhanced at this time (Wan et al., 2010). Furthermore, high southern latitudes were characterised by episodic retreat and collapse of the marine-based WAIS (Naish et al., 2009; Pollard and DeConto, 2009) and the marine margins of the EAIS (Bertram et al., 2018; Cook et al., 2013; Patterson et al., 2014; Reinardy et al., 2015), with reduced coastal sea ice (relative to modern) in the Ross Sea, Prydz Bay and Antarctic Peninsula regions (Escutia et al., 2009; McKay et al., 2012; Scherer et al., 2016; Whitehead and Bohaty, 2003; Whitehead et al., 2005; Winter et al., 2010a).

SSTs reconstructions and diatom assemblage data from the Ross Sea, imply contraction or breakdown of the Antarctic Polar Front allowed sub-Antarctic diatom flora to migrate across the Antarctic continental shelf (McKay et al., 2012). Mean annual near surface temperatures in the Ross Sea were up to 6°C and inhibited growth of sea ice (McKay et al., 2012). Reduced sea ice extent in the Southern Ocean between ~3.6 to 2.75 Ma has been associated with enhanced air-sea gas exchange with the deep ocean resulting from increased ventilation of water masses in the South Atlantic sector of the Southern Ocean (Hodell and Venz-Curtis, 2006; Waddell et al., 2009; Woodard et al., 2014). In terms of areal extent, the majority of marine-based ice sheet loss would have occurred in the Pacific sectors of the WAIS and EAIS (e.g., Ross Sea and Wilkes Land), with implications for the overturning ocean circulation and marine biogeochemistry (Bertram et al., 2018; Cook et al., 2013; DeConto and Pollard, 2016; McKay et al., 2012; Naish et al., 2009; Pollard et al., 2015; Taylor-Silva and Riesselman, 2018). Pan-Antarctic ice sheet simulations for the warm early to mid-Pliocene yield ice volume loss equivalent to between 8.5 and 16 m of sea-level from these marine-based sectors (de Boer et al., 2015; DeConto and Pollard, 2016; Golledge et al., 2017b; Pollard and DeConto, 2009). The climatic implications of the increase in oceanic area as water occupied regions previously occupied by ice is currently underrepresented in climate and ice sheet modelling studies (Woodard et al., 2014). However, one recent study suggests that changes in ice sheet extent may have affected the rate of the Pacific overturning circulation (Hill et al., 2017).

Pliocene sea-level changes have been reconstructed using a variety of geological techniques including: (1) marine benthic oxygen-isotope ($\delta^{18}O$) records paired with Mg/Ca paleothermometry (Miller et al., 2012), (2) an algorithm incorporating sill-depth, salinity and the $\delta^{18}O$ record from the Mediterranean and Red Seas (Rohling et al., 2014), (2) uplifted paleo-shorelines (Miller et al., 2012; Rovere et al., 2014), (4) submerged speleothems (Dumitru et al., 2019), and (5) backstripped continental margins (Grant et al., 2019; Miller et al., 2012; Naish and Wilson, 2009). An assessment of the suite of far-field estimates (Dutton et al., 2015) suggests a range for highest GMSL during the Pliocene of between 5–40 m above present. PlioSeaNZ (Grant et al., 2019) is a continuous floating sea-level record for the Mid-Pliocene Warm Period (3.3–3 Ma) derived from Whanganui, New Zealand and is independent of the global benthic $\delta^{18}O$ record. Sea level estimates are based on a theoretical relationship between sediment transport by waves and water depth that is applied to a grain size record from a well-dated, continuous, shallow marine sequence. If all the glacial–inter-glacial variability in the PlioSeaNZ record was above present-day sea level, then GMSL during the warmest mid-Pliocene interglacial was at least +4.1 m and no more than +20.7 m, with a median of +10.7 m and likely (66%) range between 6.2 m (16th percentile) and 16.7 m (84th percentile) (Grant and Naish, 2021).

9.1.3 Southern Ocean Paleogeography and Paleoceanography

The dominant oceanographic feature of the Southern Ocean circulation is the Antarctic Circumpolar Current (ACC) (Carter et al., 2021). The ACC is the largest global ocean current. It flows clockwise around Antarctica with an average transport volume of ~130–160 Sv and is strongly constrained by seafloor topography (Barker and Thomas, 2004; Olbers et al., 2004; Orsi et al., 1995; Rintoul, 1991; Rintoul et al., 2001). Most of the circumpolar flow takes place along the Polar Front (PF) and Subantarctic Front (SAF), which extend from the surface to the seafloor (Sokolov and Rintoul, 2007). The deep layers of the ACC comprise relatively warm Circumpolar Deep Water (CDW), which reaches the surface along steeply rising isopycnal (line of equal density) surfaces to the south of the PF. Northward advection of nutrient-rich upwelled waters feed subduction of Antarctic Intermediate Water, and southward advection feeds the subduction of Antarctic Bottom Waters. Due to the absence of land barriers, the Southern Ocean circulation connects the three main ocean basins, making it a critical feature of the modern global overturning circulation, which distributes heat, carbon, and nutrients around the globe, compensating the sinking of deep waters in the North Atlantic Ocean (Marshall and Speer, 2012; Rintoul, 2018). Today, incursions of warm CDW onto the Antarctic continental shelves have been shown to cause melting and thinning of Antarctic ice shelves through basal melting (Paolo et al., 2015; Pritchard et al., 2012; Rintoul et al., 2016; Thoma et al., 2008).

The transport and storage of heat, carbon dioxide, and fresh water, by the ACC have a significant influence on global and regional climate (Rintoul and da Silva, 2019). Opening of a path for flow between Australia and Antarctica and South America and Antarctica was required to allow circum-Antarctic flow of the ACC. Initial formation and subsequent variability in the strength of the ACC may have played a key role in modulating climate at high southern latitudes, with implications for ice sheet dynamics (Kennett, 1977; Lyle et al., 2007; Sijp and England, 2004). However, development of the ACC may have played a secondary role to changes in greenhouse gas concentrations in driving initial growth of the AIS (DeConto and Pollard, 2003; Huber and Nof, 2006).

The time of opening and deepening of the Tasman Strait to deep water flow is widely accepted at ~37 Ma (Exon et al., 2001, 2004; Lawver and Gahagan, 2003; Stickley et al., 2004). In addition, neodymium isotopes from records on opposite sides of Tasmania suggest that an eastward flowing deep-water current has been present since 30 Ma (Scher et al., 2015). Analysis of grain size at ODP and DSDP sites across the Tasmanian Gateway, as well as hiatuses, indicate water current speed increased and water masses became more homogeneous at 23.95 Ma and suggest the ACC was well established at this time (Barron and Keller, 1982; Pfuhl and McCave, 2005). Furthermore, a late Oligocene-early Miocene (~25–23 Ma) onset of the ACC is also inferred from evidence for current activity in the South Pacific along the path of the ACC (Lyle et al., 2007). Numerical modelling studies also show that, while an open circum-Antarctic gateway existed since the late Eocene, throughflow of the ACC was still limited during the Oligocene (Hill et al., 2013) because Australia and South America were substantially closer to Antarctica than today (Markwick, 2007). Recent analysis of sediment cores from conjugate margins of the modern Southern Ocean between Australia and Antarctica (Bijl et al., 2018b; Evangelinos et al., 2020; Hartman et al., 2018; Salabarnada et al., 2018; Sangiorgi et al., 2018) suggest a strong, 'near-modern' ACC did not form until the late Miocene (~11 Ma).

Onset of a 'modern and strong' ACC is likely linked to the establishment of deep-water circulation through the Drake Passage and Scotia Sea. However, the timing of opening, widening, and deepening of this key ocean gateway is still widely debated (Barker et al., 2007; Eagles and Jokat, 2014; Hodel et al., 2021; Maldonado et al., 2014; van de Lagemaat et al., 2021). Tectonic reconstructions based on age constraints from oceanic spreading magnetic anomalies and heat flow suggest continental blocks began to separate in the middle Eocene ~45 to 41 Ma (Eagles and Jokat, 2014; Livermore et al., 2007). Paleoceanographic reconstructions, based on neodymium isotopic ratios, propose shallow flows across the Drake Passage started in the late Eocene (Scher and Martin, 2006). However, stronger through flow of a proto-ACC may not have been possible until the late Oligocene to early Miocene (Hill et al., 2013; Hodel et al., 2021; Martos et al., 2013). Tectonic studies suggest that broad (100–300 km wide) and deep (>2.5 km) to intermediate depth (2.5 to 1 km) oceanic pathways were well developed in the Scotia Sea by 20 Ma (Barker et al.,

2007; Eagles and Jokat, 2014) allowing circulation of deep water that may have enhanced formation of year-round sea-ice and increased cooling of surface air temperature (Sijp and England, 2004). Sparse contourite deposits in the Scotia Sea suggest through flow of the CDW began in the early/middle Miocene (Maldonado et al., 2003; Pérez et al., 2019) but strong and widespread deep water currents associated with Southern Component Water production first occurred in the region during the middle to late Miocene (13.8 to 8.4 Ma) (Maldonado et al., 2003; Pérez et al., 2021b). Unobstructed deep through flow of the ACC that characterises the region today may not have occurred until the late middle to late Miocene (Carter et al., 2014; Dalziel et al., 2013; Pérez et al., 2019, 2021b). Clearly more information is required to improve our knowledge regarding the evolution of Southern Ocean circulation and dynamics and its influence on climate and the cryosphere.

9.1.4 Land elevation change and influences on Antarctic Ice Sheet evolution

Tectonic changes have played a significant role in the evolution of water mass circulation in the Southern Ocean and around the Antarctic margin (Carter et al., 2021). But tectonic uplift and subsidence across the Antarctic continental shelves and interior regions has also influenced the growth and extent of the AIS through the Cenozoic (Halberstadt et al., 2021; Hochmuth et al., 2020; Kerr et al., 1999, 2000; Paxman et al., 2020; Sorlien et al., 2012; Wilson et al., 2009, 2012a, 2013). Furthermore, physical processes, including erosion, transport, and deposition of sediment, modify the landscape and can form over deepened basins that inhibit ice sheet growth or create sedimentary platforms across which ice can expand (Pollard and DeConto, 2007, 2020). The interaction between climate, tectonics, ice sheets, and sedimentary processes has been a focus of the Scientific Committee on Antarctic Research (SCAR) ANTscape (Barrett et al., 2009) coordinated initiative within the ACE program, that developped in the Circum-Antarctic Stratigraphy and Paleobathymetry subcommitee activity within the PAIS program for the past two decades. Key results from these efforts include time slice reconstructions for the Eocene-Oligocene (~34 Ma), Oligocene-Miocene (~23 Ma), Middle Miocene (~14 Ma), and Pliocene (~3.5 Ma) (Hochmuth et al., 2020; Paxman et al., 2019b) (Fig. 9.1).

These reconstructions highlight the significant topographic changes that occurred across West Antarctica as it evolved from a subaerial landmass in the late Eocene (Wilson et al., 2009) to a region characterised by deep subglacial basins today (Drewry and Jordan, 1983; Fretwell et al., 2013; Morlighem et al., 2020). Formation of this submarine topography occurred throughout the Miocene and Pliocene as rifting and thermal subsidence caused the region to deepen (Fielding et al., 2005, 2007; Henrys et al., 2007; Kulhanek et al.,

2019; McKay et al., 2021; Sorlien et al., 2012) and glacial erosion scoured deep basins across the continental shelves (Bart and De Santis, 2012; Bart and Owolana, 2012; Levy et al., 2019; McKay et al., 2019). Similar processes likely deepened interior basins in East Antarctica including the Wilkes, Aurora, Pensacola-Pole and Recovery subglacial basins, which occupy large regions of the continental interior (Drewry and Jordan, 1983; Fretwell et al., 2013; Morlighem et al., 2020; Paxman et al., 2019a, 2019b). Today, these areas contain ice that is grounded well below sea level and is vulnerable to warming ocean temperatures with the potential to raise sea level by $\sim$19 m if it were to melt (Fretwell et al., 2013).

9.2 Records of Miocene to Pliocene climate and ice sheet variability from the Antarctic margin

9.2.1 Introduction to stratigraphic records

Because deep-sea $\delta^{18}O$ records provide insight into both bottom water temperature and ice volume change (Emiliani, 1955, 1966; Shackleton, 1967; Shackleton and Kennett, 1975b), ice-proximal records of climate change and ice volume fluctuations from Antarctica's continental margins are required to augment the continuous but distal data from the deep sea. Geological records of past environmental change from the Antarctic continental margin are relatively rare, as much of the sedimentary stratigraphic archive is covered by ocean and ice (Escutia et al., 2019). However, geophysical surveys and geological mapping in Antarctica have uncovered stratigraphic sequences preserved on, and along the margins of, Antarctica's continental shelves and at the mouths of outlet glaciers around the continent (McKay et al., 2021). Over the past 50 years, the Deep Sea Drilling Project (DSDP), Ocean Drilling Program (ODP), and Integrated Ocean Drilling Program (IODP) and International Ocean Discovery Program (IODP) have undertaken seventeen legs and expeditions in sub-Antarctic and Antarctic waters to recover rock and sediment cores from the ocean basins and continental shelves around the continent (Escutia et al., 2019; Gohl et al., 2019; Lamy et al., 2019; McKay et al., 2019; Weber et al., 2019). In addition to drilling in the deep ocean basins and on the outer continental shelf, records from locations proximal to the modern Antarctic coast in the Ross Sea have also been recovered by international drilling efforts including the Dry Valley Drilling Project (DVDP) (McGinnis, 1981), Cenozoic Investigations in the Western Ross Sea (CIROS) Project (Barrett, 1989), Cape Roberts Project (CRP) (Barrett, 2007, 2009), and Antarctic Drilling (ANDRILL) Program (Harwood et al., 2008, 2009; Naish et al., 2007). Here we take a regional approach and summarise the available geological data from Neogene outcrop and drill cores from the Ross Sea region and offshore George V Land (Wilkes subglacial basin) (Fig. 9.5). We acknowledge that significant information from the Neogene is also available from onshore and offshore Prydz Bay (Barron et al., 1991; Cooper et al., 2004) and the Antarctic Peninsula (Barker et al., 2002) and that the majority of new information from recent drilling expeditions is forthcoming (Gohl et al., 2019; Lamy et al., 2019; McKay et al., 2019; Weber et al., 2019).

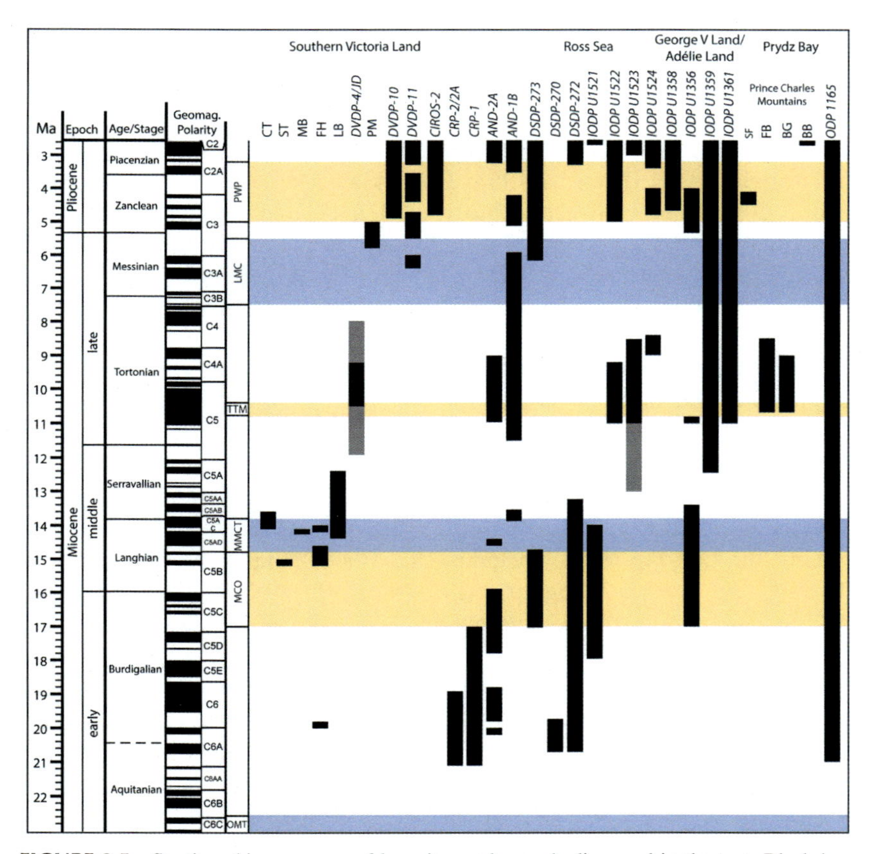

FIGURE 9.5 Stratigraphic summary of key sites and records discussed in the text. Black bars indicate approximate interval of geologic time preserved in each record. Grey bars indicate intervals for which ages are less well constrained. Italics = drill core records, regular type face = outcrop. *CT*, Cirque till; *ST*, Sessruminir till; *MB*, Mount Boreas; *FH*, Friis Hills; *LB*, Labyrinth; *DVDP*, Dry Valleys Drilling Project; *JD*, Jason diamicton; *PM*, Prospect Mesa Gravels; *CIROS*, Cenozoic Investigations in the Ross Sea; *CRP*, Cape Roberts Project; *AND*, ANDRILL; *DSDP*, Deep Sea Drilling Project; *SF*, Sørsdal Formation; *FB*, Fischer Bench Formation; *BG*, Battye Glacier Formation; *BB*, Bardin Bluffs Formation. Significant climate intervals and transitions include the *OMT*, Oligocene/Miocene Transition; *MCO*, Miocene Climatic Optimum; *MMCT*, Middle Miocene Climate Transition; *TTM*, Tortonian Thermal Maximum; *LMC*, Late Miocene Cooling; *PWP*, Pliocene Warm Period.

9.2.2 George V Land to Wilkes Land Margin

9.2.2.1 Geological setting

The segment of the East Antarctic continental margin between the northwestern Ross Sea (164°E) and Prydz Bay (90°E) is characterised by two major subglacial basins, the Wilkes subglacial basin (WSB) and Aurora subglacial basin (ASB) (Fig. 9.1), two areas where the EAIS is largely grounded below sea level (Fretwell et al., 2013; Morlighem et al., 2020). The WSB is the biggest of the EAIS and its marine-based portion alone contains an ice volume of 3–4 m sea level equivalent (Golledge et al., 2017a; Mengel and Levermann, 2014; Pollard et al., 2015). The WSB is 1400 km long and 200 to 600 km wide and is located east of the Mertz Shear Zone and west of the Transantarctic Mountains, inland of the George V Coast (Fig. 9.2) (Drewry, 1983; Drewry and Jordan, 1983; Ferraccioli et al., 2009; Fretwell et al., 2013; Mengel and Levermann, 2014; Morlighem et al., 2020). Below a thick ice cover, the northern WSB presents plateau-like surfaces, which are laterally continuous over tens to hundreds of kilometres (Paxman et al., 2019a). The flat surfaces are separated by a complex network of sub-basins up to 80 km wide, wherein the ice sheet bed lies up to 2.1 km below sea level (Ferraccioli et al., 2009). The elevations of the flat plateau-like surfaces are broadly uniform across the basin, with a modal elevation of 560 m below sea level. Subglacial topography exerts a fundamental influence on the dynamics of the AIS (Austermann et al., 2015; Colleoni et al., 2018; Gasson et al., 2015; Golledge et al., 2017a; Paxman et al., 2020; Wilson et al., 2013). Glaciers draining through the WSB, including the Cook, Ninnis and Mertz, have inland-sloping bedrock topography (Fretwell et al., 2013; Morlighem et al., 2020), which makes them more vulnerable to rapid ice sheet retreat in response to ocean and climate warming (DeConto and Pollard, 2016; Golledge et al., 2015, 2017a, 2017b; Mengel and Levermann, 2014; Pollard et al., 2015). Therefore, it is necessary to reconstruct past bedrock topography for particular time slices in order to accurately simulate AIS dynamics during those times in the past (Paxman et al., 2019b, 2020).

In front of the WSB, the continental shelf has an average width of 125 km and an average water depth of 450–500 m. The shelf exhibits an overdeepened and landward-sloping bathymetric profile that is caused by glacial erosion and sediment loading (Ten Brink and Cooper, 1992; Vanney and Johnson, 1979a, 1979b). The inner and outer shelf topography is irregular, with troughs that contain inner shelf deep basins (>1000 m) that shoal towards the shelf break and are bound by shallow flat shelf banks (the Adélie and Mertz Banks). The continental shelf troughs were eroded by ice streams during times of glacial advances while the banks illustrate where grounded ice was slow moving (Beaman et al., 2010; De Santis et al., 2003; Eittreim et al., 1995; Escutia et al., 2005). The continental slope, which extends from the shelf break to about 2000–2500 m water depth, is steep, narrow and incised by submarine canyons (Escutia et al., 2000; Porter-Smith et al., 2003). Deposition dominates the eastern flank of the channel, with the asymmetry between the levees, the result of the Coriolis effect (Donda et al., 2003; Escutia et al., 2000). Seaward of the slope, the continental rise is also relatively steep and rugged because of (1) the presence of a complex network of tributary-like channels that continue from the slope canyons, (2) the very high-relief levee systems associated to the channels, and (3) a system of sediment ridges (Escutia et al., 2000; Escutia et al., 2002).

The ASB extends up to 1000 km inland from the Antarctic coast from 106°E to 122°E, and it is composed of smaller subglacial basins, including the Vincennes and Sabrina subglacial basins (Aitken et al., 2016; Drewry, 1976; Roberts et al., 2011). The Sabrina basin, closest to the coast, is separated by a discontinuous ridge from the inland Aurora basin on the west side and Vestfold basin inland on the east. The main ice drainage is through the Totten Glacier, with Vanderford Glacier to the west and Sabrina Coast glaciers (including the Moscow University Glacier) to the east. Several overdeepened basins (Aitken et al., 2016; Young et al., 2011) and an active subglacial hydrology (Wright et al., 2021) characterise the catchment and ultimately imply it may be susceptible to coupled atmospheric and oceanic forcings with a potential ice volume contribution of ~4 m of sea level rise equivalent (SLE) (Aitken et al., 2016).

In front of the ASB, the continental shelf width increases from 130 km at the eastern end of the Moscow University ice shelf terminus to 180 km at the Totten Glacier terminus (Fernandez et al., 2018), but narrows to 40 km in the Budd Coast. The continental shelf exhibits an irregular overdeepened morphology with a cross-shelf trough offshore the Moscow University with depths ranging from over 1000 m water depth close to the coast to 450–550 m at the shelf edge (Fernandez et al., 2018; O'Brien et al., 2020). The continental slope is relatively gently dipping (~2°) and is incised by northeast-southwest trending canyons that are separated by topographic ridges (O'Brien et al., 2020). Two distinct areas are defined by their geomorphology. To the east of the Totten Glacier Canyons, sediment ridges between canyons are fed by fine sediments from turbidity currents and hemipelagic deposition being entrained by westward flowing currents. To the west, the ridges form by accretion of suspended sediment moving along the slope as a broad plume.

9.2.2.2 Oceanography of the Adélie coast

Today, the suite of sites drilled in the Georges V Land and the eastern Wilkes Land margin by the DSDP Leg 28 and IODP Expedition 318 lie south of the Antarctic PF, between the Southern Boundary of the ACC, near the Antarctic Divergence at ~63°S (i.e., IODP Site U1356) and north of the

Southern Antarctic Counter Current Front (i.e., DSDP Site 269) (Bindoff et al., 2000; Orsi et al., 1995).

The Adélie Coast continental shelf is one of the locations where the Antarctic Bottom Waters (AABW), the densest water masses of the world ocean fuelling the global ocean circulation, are produced by brine rejection during sea ice formation and the heat loss to the atmosphere (Gordon and Tchernia, 1978; Orsi et al., 1999). The Adélie Land Bottom Water (ALBW) fills the bottom of the Australian sector of the Southern Ocean (Aoki et al., 2005) and contributes ~25% of the AABW volume (Campagne et al., 2015). ALBW is produced when warm modified Circumpolar Deep Water (mCDW) flow southward across the shelf break into the Adélie Depression. Brine rejection in the Mertz Glacier Polynya system produces High-Salinity Shelf Water (HSSW), which circulates across the depression and melts the base of the Mertz Glacier Tongue to produce Ice Shelf Water (ISW). ISW ascends and supplies freshwater to the upper shelf waters. Mixing of the ISW and HSSW with other shelf waters produces Dense Shelf Water (DSW) that is exported through the Adélie sill (Williams and Bindoff, 2003; Williams et al., 2008, 2010). As these waters flow north of the continental shelf, they descend as dense cold water plumes and gravity currents into the complex channel-levee system along the Wilkes Land margin (Caburlotto et al., 2010; Williams et al., 2010). This water is funnelled through the canyons and produces ALBW, some of which is incorporated into the Antarctic Slope Current where it mixes with Ross Sea Bottom Water (Williams et al., 2008). A large polynya occurs to the west of the Mertz Glacier and is an area of high biological productivity.

9.2.2.3 Seismic stratigraphy off the George V Land to Wilkes Land Margin

The Georges V Land and Adélie Coast margins (Fig. 9.2) have been a target for unravelling the Cenozoic glacial history of Antarctica through acquisition of seismic stratigraphic and sedimentological data over many decades of national and international expeditions. A review of the main regional unconformities, seismic units, and their seismic attributes is provided by (Cooper et al., 2008; Escutia et al., 2005). Sediments recovered along this margin during IODP Expedition 318 (Escutia et al., 2011a) have provided unique chronostratigraphic control for the previously defined seismic units (Fig. 9.6), which is summarised in the chapter by McKay et al. (2021).

Recent studies collate available seismic reflection profiles along the Adélie Coast and Wilkes Land margin (west of 145°E) and across to the conjugate Australian margin (Sauermilch et al., 2019). These authors provide for the first time a consistent seismic stratigraphic framework across the

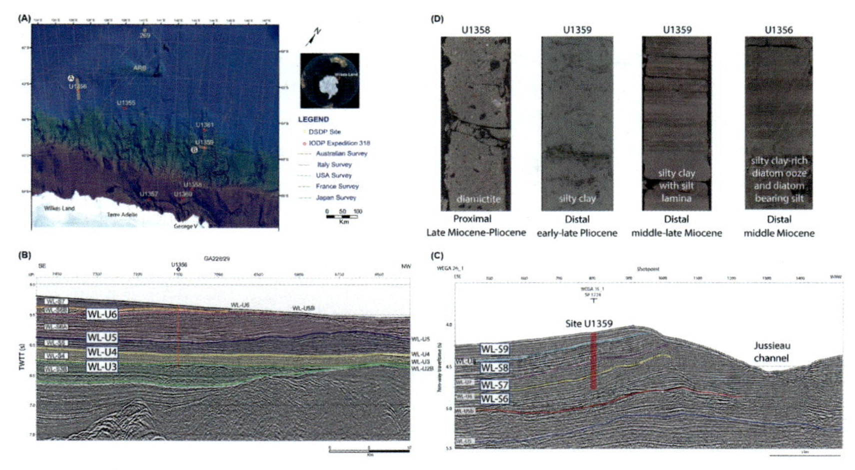

FIGURE 9.6 (A) Location of sites drilled off the George V and Adélie Land continental margin during IODP Expedition 318. Sites U1358, U1359, U1361 and U1356 recovered sediments of Miocene and Pliocene age (Escutia et al., 2011a). (B and C) Seismic reflection profiles across sites U1356 and U1359, respectively, and include seismic units (WL-S) and bounding unconformities (WL-U). WL-S9: Pliocene-Pleistocene; WL-S8: late Miocene-early Pliocene; WL-S7: middle-late Miocene; WL-S6: early-middle Miocene (Escutia et al., 2011a, 2011b). (D) Examples of Miocene and Pliocene facies that characterise the range of depositional environments recovered by drilling at each site.

Australian-Antarctic basin, revising the horizons and seismic units defined by different authors ranging from the Cretaceous to the Miocene.

9.2.2.4 *Drill core records from the George V Land to Wilkes Land Margin*

The first drilling offshore the Wilkes Land margin and Adélie Coast that recovered Pliocene and older sediments took place in 1973 during DSDP Leg 28 at Sites 268 and 269 on the continental rise and abyssal plain, respectively (Hayes et al., 1975a) (Figs 9.2 and 9.5).

DSDP *Site 268* (63.9498°S, 105.1556°E) was drilled at a water depth of ~3500 m and reached a sub-bottom depth of 474.5 m with an average core recovery of 35%. Recovered clays, silty clays and nannofossil oozes with varying amount of Ice Rafted Debris (IRD) were dated Oligocene to Quaternary. Oligocene to Miocene sediments were interpreted to be dominated by contourites (Piper and Brisco, 1975), deposited when the ice sheet first advanced onto the shelf (Hayes et al., 1975c). However, water at the time was warm enough to support calcareous biogenic sedimentation, but ice rafting and contourite deposits provide evidence for nearby ice on East Antarctica and existing bottom water currents. Pliocene and Quaternary deposition was dominated by turbidites, which was interpreted to result from progradation of the margin (Piper and Brisco, 1975).

DSDP *Site 269* (61.6761°S, 140.0701°E) drilled two holes 269 and 269A at water depth of ~4300 m, near the southeastern edge of the South Indian abyssal plain (Hayes et al., 1975a). Drilling penetrated 958 m below seafloor (mbsf) and recovered an average 42% of the cored sediments. The recovered section was assigned a tentative age of late Eocene-early Oligocene based on several reworked specimens of dinoflagellates, to Quaternary (Hayes et al., 1975a). The base of the drilled section has been assigned a late Oligocene age (24.2 Ma) based on magnetostratigraphy constrained by dinocyst and calcareous nannofossil biostratigraphy (Evangelinos et al., 2020). These authors show late Oligocene-earliest Miocene sedimentation to be dominated by persistent reworking by the ACC of sediments sourced by gravity flows and hemipelagic settling, recording fluctuating current intensities driven by the migration of the frontal system in response to climatic changes.

In 2010 IODP Expedition 318 drilled seven sites offshore the Adélie Coast (Fig. 9.6A). Four of the Sites U1356, U1358, U1359, and U1361 recovered sediments from the Miocene and Pliocene (Escutia et al., 2011a) (Fig. 9.5).

IODP *Site U1356* (63.3102°S, 135.999°E) was drilled at a water depth of 3997 m, ~350 km offshore the Adélie Coast, just south of the Antarctic Divergence, an upwelling area south of the Antarctic PF, where mean annual ocean temperatures are ~0°C. The site was drilled to a depth of 1006.4 mbsf with an average recovery of 35% of the cored section (Escutia et al., 2011). Miocene sediments were recovered in Hole 1356A between 97 and 431.5 mbsf (Cores 11R to 46R, lithological Units I, II and III, recovery ~29%) (Escutia et al., 2011a; Tauxe et al., 2012). Sediments from 97 to 400 mbsf span the interval from ~10.8 to 17 Ma, with a hiatus from ~11 to 13.4 Ma (Sangiorgi et al., 2018; Tauxe et al., 2012) (Fig. 9.5). Sediments at 431.5 mbsf were dated earliest Miocene (Tauxe et al., 2012). The record from Site U1356A thus includes the MCO and the MMCT. Miocene sediments mostly consist of clay-rich diatom ooze, diatom-rich and diatom-bearing silty clays, with different degrees of lamination and bioturbation (Escutia et al., 2011a; Salabarnada et al., 2018; Sangiorgi et al., 2018).

IODP *Site U1359* (64.904°S, 143.9603°E) drilled four holes at a water depth 3014 m (Escutia et al., 2011a). The site is located on the eastern levee of the Jussieu submarine channel: one of the many channel systems along the margin that act as conduits for dense water masses and sediment transport from the continental shelf to the deep ocean (Escutia et al., 2000). This site was positioned in an upper fan environment where the levee relief is ~400 m (Escutia et al., 2000, 2011a). Pliocene sediments were recovered from Holes U1359A-C (lithologic Units IIa and IIb) and middle to upper Miocene sediments from ~6 to 12.5 Ma from Hole U1359D (Cores 4R to

46R, lithologic Units IIb, IIc and III) (Escutia et al., 2011a; Tauxe et al., 2012) (Fig. 9.5).

Based on shipboard visual core descriptions (Escutia et al., 2011a), middle to late Miocene sediments mostly consist of laminated clays and silty clays with a higher clast abundance in the late Miocene sediments (Unit IIc), and an upper Miocene nannofossil-bearing mudstone. Although delivery of sediment to the continental rise may have been derived from turbidity currents, the laminated clays in this interval are characterised by persistent sub-mm to mm silt laminae with sharp top and bottom contacts, consistent with redeposition by contour currents in a poorly ventilated benthic environment. However, upper Miocene to upper Pliocene sediments are bioturbated diatom-bearing to diatom-rich silty clays interbedded with massive and laminated silty clays. In this interval, the laminated silty clays are characterised by graded silt laminae fining upwards into massive silty clays, consistent with deposition by low-density overbank turbidity currents in a levee environment (Escutia et al., 2011a). The diatom content in the sediments is suggestive of increased surface water productivity or reduction in terrigenous input. As the continental shelf margin continued to prograde and the channels became more proximal to the sediment source during the late Miocene and Pliocene, low-density muddy turbidity currents resulted in the deposition of well-defined packages of silt laminations on the levee. In general, the depositional environment at Site U1359 during the Miocene was characterised by open, well ventilated waters during interglacial periods, and enhanced sediment delivery and reduced ventilation during glacials.

IODP *Site U1361* (64.4095°S, 143.8867°E) drilled two holes on the lower continental rise (Escutia et al., 2011a), ~280 km north of the Adélie Coast and ~200 km south of the southern boundary of the ACC. Hole U1361A penetrated to 386 mbsf with a recovery of 87%. It was drilled in the east (right) levee bank of the Jussieu submarine channel. Site U1361 recovered more than 200 m of upper middle to upper Miocene (12.5 to 5.33 Ma) sediments (Lithologic Unit IIb. and lower Unit IIa; Cores 14H to 41X, ~134–386 mbsf) (Escutia et al., 2011a; Tauxe et al., 2012) (Fig. 9.5).

This site was cored on the same levee system as Site U1359. The facies associations are essentially the same and are characterised by alternating beds of bioturbated diatom-rich mudstones and laminated silty clays (with a shift in laminae style between the Late Miocene and Pliocene). However, for the late Miocene, post-cruise analysis of Cores 33X and 37X (~302–350 mbsf) focused on an interval with anomalous bioturbated mudstones beds deposited between 11.7 to 10.8 Ma, whereby diatom content was either replaced or diluted by calcareous nannofossils. This nannofossil ooze was initially (and tentatively) interpreted to have formed in response to an increase in surface freshwater as the EAIS melted at this time. An increase in freshwater run off may have enhanced surface water stratification and nutrient delivery via turbid meltwater plumes (Escutia et al., 2011a).

Site U1361 also recovered a substantial Pliocene section from 34.9 to 134 mbsf (Unit IIa). Pliocene sediments are characterised by repetitive interbedded laminated olive-grey silty clays and bioturbated greenish grey diatom-rich silty clay, but with persistent ice rafted debris throughout. As with Site U1359, it is interpreted as being deposited by low-density overbank turbidity currents in a levee environment, superimposed by ice rafting processes, and with enhanced pelagic sedimentation (or reduced turbidite input) and bottom current reworking during interglacials (Patterson et al., 2014). Sediments contain high abundances of reworked sporomorphs suggesting continuous strong erosion in the hinterland (Escutia et al., 2011a).

9.2.2.5 Neogene history of the George V Land to Wilkes Land margin

Multiproxy palynological, geochemical and sedimentological analyses of IODP Expedition 318 sediments (Bijl et al., 2018a; Hartman et al., 2018; Sangiorgi et al., 2018) indicate warm-temperate ice-free ocean conditions characterised the Adelie Coast during much of the MCO. TEX_{86}-based paleothermometry indicates upper ocean temperatures ranged between 11.2 and 16.6°C ± 2.8°C (Sangiorgi et al., 2018). These warm upper ocean conditions favoured ice melt and sustained ice-free conditions at the continental margin (Halberstadt et al., 2021; Levy et al., 2019; Sangiorgi et al., 2018) where temperate vegetation grew in-situ under mild atmospheric temperatures (Passchier et al., 2013b; Strother et al., 2017). Temperature gradients between southern high and mid latitudes (South Tasmanian Sea) were significantly weaker than today (Sangiorgi et al., 2018), whereas the gradient between the Adélie Coast and Ross Sea was significantly higher than today (Levy et al., 2016; Sangiorgi et al., 2018). Mean Annual Precipitation (MAP) reconstructions based on the Chemical Index of Alteration of sediments indicate elevated values (500–800 mm/year) compared to the present-day (150–400 mm/year) (Passchier et al., 2013b). Sediments deposited during the MCO contain no sea-ice indicators (Bijl et al., 2018a) or ice rafted debris, which suggests ice sheets terminated landward of the coastal margin (Halberstadt et al., 2021; Levy et al., 2019; Sangiorgi et al., 2018).

Proxy data at Site U1356 indicate a major environmental transition occurred after the onset of the MMCT at 14.8 Ma. TEX_{86}-based paleothermometry indicate ocean temperatures off the Adélie Coast cooled by ~6°C between ~16.5 and 14.5 Ma and dinocyst assemblages suggest sea-ice became a more persistent feature by ~14.2 Ma (Sangiorgi et al., 2018). An increase in sedimentary clasts at ~14.6 Ma indicates nearby glaciers extended into ocean and that IRD-bearing icebergs transported debris across the drill site. Clast occurrence and abundance peaked between 14 and 13.8 Ma and dropped again between 13.8 and 13.5 Ma (Sangiorgi et al., 2018). Clast provenance studies suggest the clasts were derived from an inland source area that extended along the eastern part of the Adélie Craton (the western margin of the WSB) (Pierce et al., 2017). These data suggest a dynamic coastal environment in which an ice sheet repeatedly advanced and retreated from interior regions across the WSB shelf throughout the MMCT (Pierce et al., 2017) and that the ice sheet and glaciers reached their maximum extent between 14 and 13.8 Ma.

Sediments and seismic sequences that overly a major hiatus (WL-U5b) between 13.4 and ~11 Ma at Site U1356 (Fig. 9.6), formed as channel levee systems migrated across the site when the margin of the expanded EAIS advanced and retreated over the continental shelf (Escutia et al., 2011a; Sangiorgi et al., 2018). The abundance of soil organic matter in these sediments is lower than in sediments deposited during the MCO, consistent with a reduction in soil formation and extent due to colder climatic conditions and extensive regional ice cover (Sangiorgi et al., 2018). Vegetation composition (from pollen) remained broadly unchanged between the MCO and late Miocene, although an increase in the relative abundance of southern beech (*Nothofagus* spp.) suggest temperatures were colder (Sangiorgi et al., 2018). Geochemical proxies indicate mean annual terrestrial temperatures between 6°C–8°C at 10.8 Ma compared to temperatures of 7°C–12°C indicated through the MCO (Passchier et al., 2013a; Sangiorgi et al., 2018). TEX_{86}-based temperature estimates show upper ocean temperatures were between 4°C and 10°C (±4°C). A decrease in the relative abundance of protoperidinioid dinocysts suggests a decrease in sea-ice at this time. Nannofossil-rich sediments appear in a discrete sequence at Site U1361 (Pretty, 2019) and indicate that between 11.7 and 11 Ma, temperatures were either warm enough to support coccolithophore growth (Balch et al., 2016) or the local CCD deepened enough to allow carbonate deposition. Together proxy environmental data from Sites U1356 and U1361 indicate warming of water masses that outcrop close to the continent along the George V and Adélie Coasts in the late Miocene (DeCesare and Pekar, 2016; Evangelinos et al., 2020), well after the end of the MMCT. However, the increase in diatomaceous sediments after 11 Ma suggests this late Miocene warming was relatively short-lived, at least offshore of the Adélie Coast (Pretty, 2019; Sangiorgi et al., 2018).

Younger sediments recovered from the continental rise at Sites U1359 and U1361 indicate the EAIS margin remained dynamic into and through the Pliocene. Sedimentological data including lithofacies characterisation, grain size, and major and minor trace element ratios from continental shelf sediments helped identify oscillating periods of open-marine conditions and glacial advances to the outer shelf during the Pliocene (Orejola and Passchier, 2014; Patterson et al., 2014; Reinardy et al., 2015). The geochemical provenance of fine-grained detrital sediments recovered at continental rise Site U1361 suggests a dynamic early Pliocene (5.3 to 3.3 Ma) ice sheet margin, with repeated ice sheet advance and retreat well into the WSB (Cook et al., 2013), as indicated by the relative contribution of inland lithologies of the Ferrar and Beacon supergroup lithologies to the detrital sediment signature offshore. Analyses of a variety of geochemical and mineralogical provenance proxies in the Pliocene sediments (clay mineralogy, fine-grained Sr and Nd isotope composition, ice-rafted hornblende $^{40}Ar/^{39}Ar$ ages) highlight that ice sheet reconstructions are more robust when multiple methods are deployed. For example, the provenance of ice-rafted hornblende grains points to far-travelled sources, indicating ice sheet instability in West Antarctica at the same time ice retreat occurred into the WSB (Cook et al., 2017).

Geochemical proxies, including Mn/Al and biogenic barite, have been used to determine oceanic redox conditions at Site U1359 and point to possible incursions of CDW onto the continental shelf, which may have promoted glacial retreat in the Pliocene (Hansen and Passchier, 2017). Fossil diatom and silicoflagellate assemblage data furthermore suggest that the warmest interglacials coincided with a poleward shift of Southern Ocean frontal systems. Such periods were characterised by prolonged ice-free open water conditions with higher productivity and strong surface ocean stratification (Armbrecht et al., 2018; Taylor-Silva and Riesselman, 2018).

Pliocene fluctuations in ice sheet volume and extent, as well as surface ocean conditions in the vicinity of the WSB, appear to be paced by changes in incoming solar radiation through astronomical (orbital) cycles (Armbrecht et al., 2018; Hansen et al., 2015; Patterson et al., 2014; Taylor-Silva and Riesselman, 2018). Data from offshore cores show that during the warm Early Pliocene marine based margins of the EAIS responded to ~40-kyr changes in mean annual southern high latitude insolation paced by changes in Earth's axial tilt (obliquity). Between 3.5 and 3.3 Ma significant shift in EAIS dynamics occurred in response to Southern Ocean cooling and development of a perennial sea-ice cover which limited the role of oceanic forcing on ice sheet extent (Patterson et al., 2014). After ~3.3 Ma, substantial retreat of the ice sheet margin occurred only during austral summer insolation maxima that are controlled by 20-kyr precession cycles modulated by changes in short (100 kyr) eccentricity cycles.

High-resolution far-field sea level records recovered from the Whanganui Basin of New Zealand support these findings and suggest that the average amplitude of sea-level changes during the mid-Pliocene (3.3–2.7 Ma) was ~7 to 18 m and oscillated in response to Antarctic ice volume changes that were driven by 20-kyr changes in local insolation (Grant et al., 2019). These data contradict estimates of ice volume and temperature derived from a global stack of benthic $\delta^{18}O$ records (Lisiecki and Raymo, 2005), which suggest 40-kyr changes in mean annual insolation drove ice volume change at this time (de Boer et al., 2015; Pollard and DeConto, 2009). Explanations for the apparent dominance of obliquity at this time have been proposed (Huybers and Tziperman, 2008; Raymo and Nisancioglu, 2003; Raymo et al., 2006) but these new sea level data suggest that further investigation is required. The Wanganui sea level record also suggests the magnitude of glacial retreat was much larger during the KM3 (3.158 Ma) oxygen isotope event than during the M1 (3.251 Ma) and KM5 (3.198 Ma) events. This observation is supported by diatom assemblage data from Site U1361 that imply a major poleward migration of the Antarctic PF during this interval of warmth, which destabilised the ice sheet margins (Taylor-Silva and Riesselman, 2018). Oxygen isotope event KM3 is an important interval to examine Antarctic ice sheet sensitivity to climate warming. The 18 m amplitude change in sea level observed during this interglacial is best explained by mass loss from Antarctica's marine-based ice sheet margins with limited contribution from the Greenland Ice Sheet (Grant et al., 2019; Shakun et al., 2018). If the contribution to sea level from Greenland was small, then the observed magnitude of sea level change requires complete melt of the WAIS and major retreat of the EAIS across East Antarctica's large subglacial basins, including the WSB and ASB (Fig. 9.2).

High resolution (suborbital) geochemical provenance data from Site U1361 (Figs 9.2 and 9.6), alongside XRF records of oceanic productivity and records of IRD mass accumulation during selected Pliocene warm intervals indicate a strong coupling of changes in provenance (e.g., ice sheet margin retreat) and marine biological productivity (e.g., sea ice melting). Increases in iceberg calving appear to precede ice sheet retreat in this sector and may have contributed to fertilising the Southern Ocean (Bertram et al., 2018). These high-resolution data indicate that the time between deglacial onset and maximum grounding line retreat within the WSB was a few thousand years (Bertram et al., 2018), a conclusion that is consistent with modelling results (Golledge et al., 2017a, 2017b; Pollard et al., 2015).

Satellite-based analyses of modern ice sheet and sea ice dynamics suggest that the Wilkes Land margin is more vulnerable to warming temperatures than the George V margin (Miles et al., 2013, 2016). However, glaciers that drain the ASB (including the Totten and Moscow University Glaciers) sit on beds with geomorphological features and slopes that promote glacial stability (Morlighem et al., 2020) and will have to retreat several km inland before they reach a destabilising retrograde bed. Whether or not the margin of the EAIS along the Wilkes Land coastal region was highly dynamic during the warm Pliocene is unclear. Evidence for glacial destabilisation and ice margin retreat within in the ASB during the Pliocene comes from sedimentologic evidence for diatom oozes (Gulick et al., 2017) and geochemical studies of ice rafted debris in cores from offshore the ASB and Prydz Bay (Cook et al., 2014; Williams et al., 2010). Geochemical data suggest iceberg armadas entered the Southern Ocean from the Wilkes Land margin, and possibly the adjacent low-lying ASB, during the early Pliocene (Cook et al., 2014; Williams et al., 2010). These icebergs may have carried sedimentary detritus with a characteristic provenance fingerprint of the rocks from the Wilkes Land region all the way to Prydz Bay, despite elevated SSTs. However, multichannel seismic data from offshore the ASB catchment suggest a muted glacial response to Pliocene warmth (Gulick et al., 2017). Modelling studies also produce contrasting results; some suggest little retreat across the ASB during peak Pliocene warmth (de Boer et al., 2015; Golledge et al., 2017b), while others indicate significant grounding zone retreat can occur across the region (de Boer et al., 2015; Pollard et al., 2015). Further insight into past glacial dynamics in this potentially sensitive region remains elusive as no continuous marine sedimentary record currently exists from the Sabrina Coast (Gulick et al., 2017). However, current efforts are underway to develop a drilling program to recover new records (Gulick et al., 2017; McKay et al., 2021; Montelli et al., 2019).

9.2.3 The Ross Sea Embayment and Southern Victoria Land

9.2.3.1 Geological setting

The Ross Sea Embayment (RSE) is bordered by the 3500 km long Transantarctic Mountain (TAM) chain to the west and Marie Byrd Land (MBL) and Siple Coast to the east (Fig. 9.2). The modern WAIS is grounded within the lower topographic relief of the West Antarctic Rift System (WARS), one of Earth's major continental extension zones extending across Antarctica from the western RSE to the Antarctic Peninsula (AP) and separating West and East Antarctica (Jordan et al., 2020). Most of the WAIS sits on continental crust that lies well below sea level, in places reaching depths well over two kilometres (Fretwell et al., 2013; Morlighem et al., 2020). Clastic and biogenic sediments have accumulated in the sedimentary basins that formed as the WARS evolved through the Cretaceous and Cenozoic (Decesari et al., 2005, 2007; Jordan et al., 2020; Luyendyk et al., 2001; Wilson et al., 2009). Clastic sediments that comprise the Miocene and Pliocene sequences reviewed herein are sourced from pre-Quaternary rocks from the TAM and beneath West Antarctica. Sediment provenance studies have become important tools for reconstructing past ice dynamics across the area so here we provide a brief overview of the regional geology. See Talarico et al. (2021) for a more comprehensive review.

Basement rocks exposed along the TAM primarily comprise early Paleozoic igneous intrusives of the Ross Orogeny (Allibone et al., 1993; Cox et al., 2012; Goodge et al., 2012; Gunn, 1963; Gunn and Warren, 1962; Paulsen et al., 2013; Stump, 1995). Post Ross Orogeny exhumation and erosion of the granitic and metamorphic basement provided Kukri Peneplain upon which the clastic Beacon Supergroup was deposited (Barrett, 1971;

Barrett et al., 1972, 1986; McKelvey et al., 1977). This sequence was intruded by Ferrar Dolerite at c.180 Ma (Compston et al., 1968; Ferrar, 1907; Harrington, 1958; Licht et al., 2014). Little is known of the basement rocks inland of the TAM because these are covered by the EAIS (Palmer et al., 2012). Late Cenozoic volcanics in the western Ross Sea region comprise the McMurdo Volcanic Group, part of the Erebus Volcanic Province (Kyle, 1990a, 1990b), which range in age from ~19 Ma to present. Drillcore evidence extends the volcanic record in the western Ross Sea back to 26 Ma (Di Vincenzo et al., 2009; Gamble et al., 1986; McIntosh, 1998; 2000).

The best exposed basement rocks in West Antarctica are in Marie Byrd Land and comprise the Neoproterozoic to Cambrian Swanson Formation and Devonian to mid Cretaceous magmatic rocks (Korhonen et al., 2010; Pankhurst et al., 1998; Siddoway, 2008; Siddoway and Fanning, 2009; Tingey, 1991). Starting in the Oligocene, the region was subjected to intense alkaline volcanism and uplift of the Marie Byrd Land Dome (Hole and LeMasurier, 1994; Winberry and Anandakrishnan, 2004), although other studies suggest this uplift event began in the early Miocene ~10 million years later (Spiegel et al., 2016). Today the Marie Byrd Land Dome extends to an altitude of 2700 m and is a feature that likely shed sediment into the WARS over the Miocene and Pliocene. Eighteen major volcanoes and many smaller centres across MBL are composed of felsic alkaline lavas (Panter et al., 2000). Other volcanic centres likely exist under the WAIS and have been imaged via aerogeophyscial surveys (Behrendt et al., 1996, 2002, 2004). Some of the MBL volcanoes have been active during the Pleistocene-Holocene, potentially affecting geothermal heat flow and basal WAIS dynamics (de Vries et al., 2017).

Basins on the mid to outer continental shelf in the RSE contain sedimentary rocks dating to the late Eocene (Barrett, 1989, 2009; Galeotti et al., 2012; Harwood and Levy, 2000; Levy and Harwood, 2000a, 2000b). However, reworked palynomorphs in seabed surface samples collected across the Ross Sea indicate terrestrial sedimentary rocks dating to the Cretaceous were deposited across West Antarctica (Truswell and Drewry, 1984), but in-situ sediments have yet to be recovered. Reworked marine microfossils are also common components in glacial sediments recovered from beneath the WAIS and offer insight into the likely in-situ sequences that occur beneath the WAIS. Siliceous microfossils from discrete time intervals in the early Miocene, middle Miocene and late Miocene are preserved in sediment clasts and matrix from the Ross Ice Shelf Project (RISP) J-9 core (Harwood et al., 1989) and indicate in situ Neogene sediments occur upstream of the site. Microfossil analyses also show that reworked Eocene, Miocene, Pliocene, and Pleistocene diatoms are widespread beneath the Whillans and Kamb ice streams (Coenen et al., 2020; Scherer et al., 1998). Whereas these data provide tantalising evidence for the occurrence of Neogene strata beneath the WAIS and inner Ross Ice Shelf, thus far no in-situ pre-LGM samples have been collected. These strata are a target of future drilling efforts (McKay et al., 2021). In contrast, sedimentary basins located north of the Ross Ice Shelf have been the target of multiple drilling campaigns and offer the most complete and comprehensive stratigraphy's through the Miocene and Pliocene and are discussed in detail below.

9.2.3.2 *Oceanography and climate in the Ross Sea Region*

The Ross Sea Embayment covers an area of approximately 1,137,000 km^2 and includes an extensive ocean cavity beneath the world's largest ice shelf. Relatively warm CDW enters the cyclonic flow of the Ross Gyre at its eastern limb (Orsi and Wiederwohl, 2009; Whitworth et al., 1998). At the continental shelf break, CDW locally flows onto the continental shelf and mixes to become mCDW with temperatures of 1°C to 1.5°C (Budillon and Spezie, 2000; Dinniman et al., 2003, 2007). Sea-ice formation in the Ross Sea converts AASW and/or shoaling mCDW, into cold and dense High Salinity Shelf Water (HSSW). Cyclonic circulation of HSSW beneath the Ross Ice Shelf is inferred to occur within the major troughs that connect the continental slope with the grounding line of the Ross Ice Shelf (RIS) (Dinniman and Klinck, 2002). Ice shelf water (ISW) is created via contact between water and the Ross Ice Shelf at depth and is characterised by temperatures below the surface freezing point. Supercooled ISW emerges from beneath the Ross Ice Shelf in the west central region of the continental shelf (Dinniman et al., 2003).

HSSW is exported from the Ross Sea continental shelf where it mixes with CDW and contributes to AABW. At present, most of the abyssal layers of the world's oceans are filled with water that is influenced by AABW that primarily originates from the Weddell and Ross Seas. Thus, changes in ice shelf extent, water temperature, and/or meltwater input to the Ross Sea could significantly disrupt present-day global meridional overturning circulation (MOC) (Jacobs et al., 2002; Orsi et al., 2002; Purkey and Johnson, 2010). Over the past 40 years, Ross Sea-derived AABW have freshened as a result of increased meltwater input to the Amundsen and Bellingshausen Seas from melting ice shelves/glacial systems upstream from the Ross Sea (Jacobs and Giulivi, 2010; Jacobs et al., 2002; Silvano et al., 2018). Recent hydrographic observations in the late 2010s have revealed a reversal of the freshening trend hinting at the existence of decadal variations or oscillations in meltwater contribution from the Ross Sea region (Aoki et al., 2020; Castagno et al., 2019).

At present, the strong westward-flowing Antarctic Slope Current (ASC) and its sharp subsurface front (Antarctic Slope Front – ASF), separates AASW from CDW on the lower continental slope. The ASF serves as a dynamical barrier that limits the transfer of CDW onto the Ross Sea continental shelf (Ainley and Jacobs, 1981; Smith et al., 2012; Thompson et al., 2018). However, a decrease in wind shear stress across the ASF may cause shoaling of isopycnals and enhanced eddy-driven transport of CDW across the continental shelf margin (Stewart and Thompson, 2014). Furthermore, modelling indicates that flow of CDW onto the Ross Sea continental shelf is directly influenced by the volume of dense waters that descend off the continental shelf to form AABW (Morrison et al., 2020). Incursion of warm CDW is sensitive to the morphology of the continental margin, vigour of the ASC (and associated eddy activity), and production of AASW and AABW (Thompson et al., 2018). Changes in atmospheric and ocean dynamics at the continental shelf margin clearly influence the incursion of warm water onto the Ross Sea Continental Shelf and will likely increase ice shelf and ice sheet melt. Positive feedbacks associated with increasing melt and freshwater flux will enhance intrusion of CDW and potentially accelerate ice sheet retreat.

Terrestrial rock outcrops span the western and eastern margins of the Ross Sea Region (Fig. 9.2). The McMurdo Dry Valleys (MDV) are the best studied and monitored and are the focus of the terrestrial environmental records discussed in this review. The MDV represent one of the largest ice-free regions in Antarctica, covering approximately 4000 km^2 in the central Transantarctic Mountains. Today, the Dry Valleys feature a hyperarid, cold-desert climate. Mean annual temperature at Lake Bonney, on the floor of Taylor Valley is −17.9°C (Doran et al., 2002). Climate at higher elevations in the MDV is difficult to constrain due to a lack of long-term monitoring data. However, a meteorological station at 1581 m above sea level in the Friis Hills recorded weather conditions from 2005 to 2010 (Bliss et al., 2011). During this period, mean summer temperature was −13.2°C with an average wind speed of 4.7 m/s. The winter mean was −29.7°C with an average wind speed of 4.2 m/s (Bliss et al., 2011). Based on measured regional lapse rate of 6.4°C/km (Bliss et al., 2011), the range in mean annual temperature in the Friis Hills, Asgard and Olympus ranges would approach −24°C to −28°C. Additional weather station data collected from the Friis Hills between 2011 to 2018 recorded an average mean annual air temperature of −22.7°C ± 1.3°C (Doran and Fountain, 2019). Climate in the Friis Hills and other high elevation regions of the MDV (Fig. 9.2) is classified as a severe polar desert with subfreezing temperatures and strong winds occurring throughout the year (Lewis and Ashworth, 2015). Precipitation measured at Lake Bonney is less than 150 mm/year, which falls as snow (Fountain et al., 1999). Precipitation at locations farther inland and higher is likely less (Doran et al., 2002).

9.2.3.3 Seismic stratigraphic records in the Ross Sea

The Ross Sea Embayment contains a dense network of seismic profiles (McKay et al., 2021). Regional interpretation of the major stratigraphic discontinuities allows for a well-defined seismic-stratigraphic framework, constrained by deep-drilling sites from across the Ross Sea (McKay et al., 2021) (Fig. 9.7). The broad seismic stratigraphy of the Ross Sea was first resolved by the SCAR Antarctic Offshore Stratigraphy project, ANTOSTRAT (Brancolini et al., 1995a; Cooper and Davey, 1987). Above the seismic basement, the underlying sequence holds syn-rift sediments of Mesozoic to early Eocene age (Cooper and Davey, 1987). The overlying sedimentary sequence contains

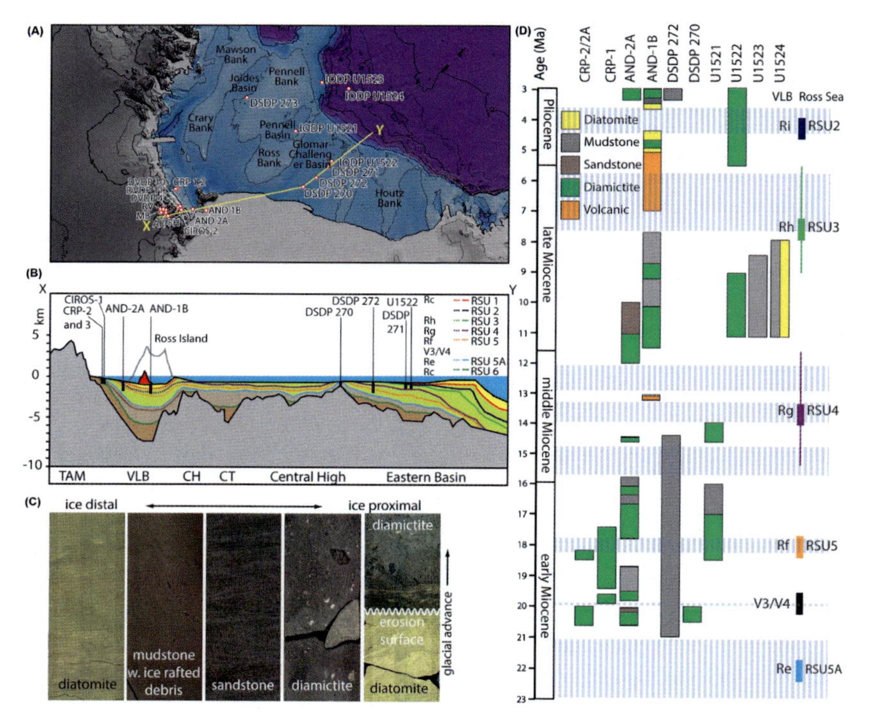

FIGURE 9.7 (A) Ross Sea region with location of key drill cores and outcrop discussed in the text: *DVD*, Dry Valley Drilling Project; *MB*, Mount Boreas; *AR*, Asgard Range; *RV*, Rude Valley; *FH*, Friis Hills; *CRP*, Cape Roberts Project; *CIROS*, Cenozoic Investigations in the Ross Sea, *AND*, ANDRILL; *DSDP*, Deep Sea Drilling Project; *IODP*, Integrated Ocean Discovery Program. Line X–Y indicates approximate location of cross section shown in (B). (B) Generalised cross section from the Transantarctic Mountains (TAM) to the continental shelf margin. Solid coloured lines indicate major Ross Sea seismic unconformities (RSUs). Correlative surfaces in the Victoria Land Basin (VLB) are dashed as direct correlations are ambiguous due to the occurrence of basement highs and faults at the VLB margin. (C) Representative examples of sediment/facies types and transitions observed in Ross Sea drillcores. (D) Chronostratigraphic framework and generalised lithofacies for drill cores discussed in this text. Dashed blue horizontal lines highlight episodes where drill core unconformities coincide with seismic unconformities, which are indicated by coloured lines at right of diagram (thin vertical lines indicate age uncertainty). Drill core chronostratigraphy is based on published age models (Florindo et al., 2005; Levy et al., 2016; McKay et al., 2019; Naish et al., 2009; Wilson et al., 2012b) and/or has been adjusted to fit up-to-date ages for key biostratigraphic datums (Crampton et al., 2016; Florindo et al., 2013).

seven major unconformities referred to as Ross Sea Unconformities (RSU's) that bound eight intervening seismic sequences named Ross Sea Sequences (RSS's) (McKay et al., 2021). The RSS are largely defined by studies in the Eastern and Central basins of the Ross Sea, whereas different nomenclatures exist in the active rift zones of the Western Ross Sea (Davey et al., 2000; Fielding, 2018; Fielding et al., 2008; Henrys et al., 1998, 2001; Levy et al., 2016). Here we describe briefly the seismic stratigraphic sequences and unconformities through the Miocene and Pliocene.

Seismic-stratigraphic sequences RSS2 to RSS6 (oldest to youngest) are Miocene (Brancolini et al., 1995a). RSS2 sits above RSU6 and grades upwards into well-stratified reflections that are laterally continuous. RSU6 has never been drilled but an age of >27 Ma is generally accepted (Busetti and Cooper, 1994). This unconformity is restricted to the deepest sedimentary basins, where it forms the top of stratified sediments infilling channels and structural valleys and onlaps structural highs (De Santis et al., 1995). RSS2 is thickest along the inner continental shelf and is thin to absent along a SE-NW band over Glomar Challenger Basin and the Ross and Iselin banks (Brancolini et al., 1995a; De Santis et al., 1995). RSU5 (~18 Ma) forms the top of RSS2 and is expressed regionally as a high amplitude reflection that is locally erosive on the eastern and inner part of the continental shelf. RSS3 overlies RSU5 and includes thick prograding wedges that occupy the main sedimentary basins of the Ross Sea continental shelf. Locally, strata within the upper part of the unit exhibit aggradational features (Pérez et al., 2021a). RSU4a (~17 to 16.4 Ma) defines the top of RSS3 and is characterised by a regionally extensive lateral reflection that locally truncates underlying sediments. RSS4 occurs above RSU4a and includes seismic facies with regionally variable amplitudes and laterally discontinuous reflectors. The sequence is thick and extends across the central basins. RSS4 is truncated by RSU4 (~14.6 to 12 Ma), an erosional surface that extends across most of the Ross Sea continental shelf. RSS5 and RSS6 are relatively discrete units that occur locally in the central and eastern Ross Sea, where they thicken towards and along the outer shelf. These sequences typically comprise prograding wedges that are bounded by RSU3 (~8 to 6 Ma), which is generally considered an erosional surface across much of the inner to mid Ross Sea continental shelf (De Santis et al., 1995). RSU2 generally deepens landwards (De Santis et al., 1999) and has an estimated age between 4.0 and 2.8 Ma in the western Ross Sea (Brancolini et al., 1995b; Granot et al., 2010).

The origin of the RSUs has been linked to a range of processes including tectonic events, glacial erosion, and sea level oscillations (Alonso et al., 1992; Anderson and Bartek, 1992; Bartek et al., 1991; Busetti et al., 1993; Brancolini et al., 1995a; Cooper et al., 1995; De Santis et al., 1995; Hinz and Block, 1984). The aggradational pattern in the upper part of RSS2 is inferred to reflect ice distal glaciomarine conditions in the Ross Sea that preceded deposition of ice proximal progradational wedges within RSS3. These prograding strata likely represent a period of ice sheet advance into marine environments across the Ross Sea continental shelf (De Santis et al., 1995, 1999; Pérez et al., 2021a). The regionally variable but erosive signature of RSU4a suggests expansion of ice caps from local highs at locations across the Ross Sea formed this feature. The relatively uniform stratified pattern that characterises RSS4 likely reflects a period of open marine conditions and hemipelagic sedimentation with little evidence of ice advance and retreat across the Ross Sea during the MCO (De Santis et al., 1995; McKay et al., 2019; Pérez et al., 2021a). The regional erosional signature of RSU4 reflects an episode of extensive ice sheet advance across the continental shelf of the Ross Sea (Bart and De Santis, 2012; De Santis et al., 1995, 1999; McKay et al., 2019; Pérez et al., 2021a). Prograding wedges that form RSS5 and RSS6 were deposited during multiple episodes of marine ice sheet advance and retreat during late Miocene, which delivered glacial marine sediments to the outer continental shelf. The landward deepening surface of RSU2 formed as a thick and extensive marine ice sheet grounded across the continental shelf during the late Pliocene and early Pleistocene (De Santis et al., 1999; Lindeque et al., 2016).

9.2.3.4 Stratigraphic records from drill cores in the Ross Sea

The history of geological drilling in the Ross Sea is relatively long and has contributed significant discoveries that are fundamental to our understanding of Antarctic ice sheets and environmental evolution. In December 1973, Leg 28 of the Deep Sea Drilling Project (DSDP) (Hayes et al., 1975b) recovered upper Oligocene to lower Miocene glacial sediments at Site 270, instantly extending our known history of Cenozoic glaciation back some 25 million years (Barrett, 1975; Hayes et al., 1975d). Miocene and Pliocene sediments were recovered at Sites 272 and 273, providing insight into the history of glacial advance and retreat across the Ross Sea continental shelf. In 2018, IODP Expedition 374 drilled five holes and recovered Miocene to Pleistocene sediments that add to our expanding knowledge of marine ice sheet behaviour in the central Ross Sea through the Neogene (Figs 9.5 and 9.6) (McKay et al., 2019).

Site U1521 (75.6839°S, 179.6718°W) recovered a 648 m sequence of Miocene to Recent sediments that is divided into seven lithostratigraphic units (I–VII) (McKay et al., 2019). Contacts between the units range from sharp to gradational. Mudstone and diamictite account for ~90% of the recovered core, and minor lithofacies include chert and conglomerate. The assemblage of facies reflects open-marine to ice-proximal depositional environments at this location since at least the early Miocene (~18 Ma). Site U1521 drilled through a thick seismic sequence of progradational foresets above RSU5 and recovered a ~300 m interval (Units VII and VI) of diamictites interbedded with thin mudstone layers that suggest input of a large volume of glacially eroded material by marine-terminating glaciers or ice sheets. Unit V is a 40 m interval of poorly recovered chert nodules and silica cemented mudstone. Unit IV is a ~70 m of dark grey massive to stratified diatom-bearing clast-poor sandy diamictite. Unit III includes an ~120 m-thick interval of diatom-rich muds and diatom oozes provisionally assigned an age spanning 16.7 to 15.8 Ma (McKay et al., 2019). These predominantly open water sediments were deposited during an interval of ice sheet retreat spanning the earliest part of the MCO. Unit II includes an ~80 m thick sequence of dark grey massive clast-poor muddy diamictite interbedded with bioturbated diatom-bearing/rich mudstone with dispersed clasts. The lower ~40 m of the unit consists of diamictite and includes a disconformity that correlates with RSU4 and preliminary age constraints date this surface at ~14.6 Ma (Pérez et al., 2021a). Facies in Lithostratigraphic Unit II indicate highly truncated cycles of subglacial, glaciomarine and open marine sedimentation (McKay et al., 2019) that reflect repetitive cycles of marine-based ice sheet advance and retreat across the Ross Sea continental shelf since the middle Miocene (Pérez et al., 2021a).

Site U1522 (76.5538°S, 174.7578°W) penetrated to 701 m below the sea floor and recovered 279.57 m of core (40%). The 696 m thick succession of upper Miocene to Recent sediments is divided into four lithostratigraphic units (McKay et al., 2019). Lithostratigraphic Unit II spans the Pliocene and includes consists of ~195 m of massive interbedded diatom-bearing/rich sandy/muddy diamictite with mudstone laminae. Units III and IV recovered upper Miocene sediments spanning the interval from ~11 to 9 Ma. Unit III includes 250 m of muddy diatomite and diatom-bearing/rich diamictite and is divided into three subunits based on the style of interbedding and presence of lithologic accessories. Unit IV consists of ~50 m of interbedded diatom-bearing sandy diamictite and muddy diatomite. RSU3 intersects the core between 400 and 500 mbsf, which constrains the age of this surface to <9 and >5.5 Ma.

Site U1523 (74.1503°S, 176.7951°W) was cored on the outermost continental shelf, at a site under the influence of the ASC. A range of coring systems were deployed at Site U1523 as the operations crew attempted to recover material from the location in challenging coring conditions. A total of sixty-four cores were collected from five holes. These cores provide an ~220 m-thick composite sequence spanning the (?)middle Miocene to Pleistocene. The sequence is divided into three lithostratigraphic units and (?)middle Miocene and Pliocene sediments occur in unit II and III and are characterised by sand/gravel-rich beds alternating with diatom-bearing/rich mud – interpreted to reflect shifting ASC current strength through time. Due to the unconsolidated nature of the facies in this interval, and the rotary coring methods, core recovery of Unit III was extremely poor and, while the sequence may extend back to 13 Ma, lithological information is restricted to strata younger than 11 Ma, although downhole logs measurement were conducted throughout this interval of Site U1523, and will continuous records of the downhole variability/cyclicity of facies. Approximately 50 m of upper Miocene sediments within Unit III include massive to laminated diatom-bearing/rich mud interbedded at the decimetre scale with diatom-bearing diamict (McKay et al., 2019).

Site U1524 (74.21738°S, 173.6336°W) is located on the continental rise, and drilled an expanded levee of the Hillary Canyon, one of the largest conduits of AABW feeding the abyssal ocean from the Ross Sea continental rise and includes three holes, the deepest of which was cored to 441.9 m bsf. A composite sequence recovered in three holes includes upper Miocene to Pleistocene sediment that is divided into three lithostratigraphic units (I–III) (McKay et al., 2019). Contacts between units and subunits are mostly gradational and are distinguished by gradual changes in diatom content. Unit II is mid Pliocene and consists of ~117 m of massive to laminated muddy diatom ooze interbedded with bioturbated to laminated diatom ooze and diatom-rich sandy mud. Unit III is 120 m thick and consists of massive to laminated mud interbedded with massive diatom-rich sandy mud with dispersed clasts of muddy diamict and muddy diatom ooze. A major disconformity at ~320 mbsf spans the interval from 8.4 and 4.8 Ma and separates Unit III and II. The fine-grained, and graded nature of the laminae is interpreted as representing a largely continuous record of turbidity current activity associated with AABW flow down the Hillary Canyon, with an ultra-expanded mid- to Late Pliocene section (McKay et al., 2019).

A series of onshore to nearshore geological drilling projects between 1976 and 2008 that have revealed much about Antarctic Cenozoic climate history is the southwest corner of the Ross Sea (Fig. 9.7). These drilling campaigns recovered sediments from discrete intervals through the past 36 million years (McKay et al., 2021). Neogene strata were recovered from onshore drilling within the Dry Valley Drilling Project (McGinnis, 1981; McKelvey, 1975) and from offshore locations during the McMurdo Sound Sediment and Tectonic Studies drilling Project (Barrett, 1986), Cenozoic Investigations in the Ross Sea Project (Barrett, 1989), Cape Roberts Project (Barrett et al., 1998; Barrett and Ricci, 2000, 2001; Davey et al., 2001), and the ANDRILL McMurdo Ice Shelf (Naish et al., 2007; Naish et al., 2008) and Southern McMurdo Sound Projects (Harwood et al., 2008, 2009) (Fig. 9.5). These drill cores provide detailed records of environmental variability and reveal the evolution of glacial marine conditions in Southern Victoria Land through the Neogene and are described in more detail below.

The *Cape Roberts Project* drilled at three locations recovering a composite stratigraphy comprising stratigraphic snap shots from the late Eocene to Pleistocene (Figs 9.2D, 9.5, 9.7). Lower Miocene sediments were recovered between 23 and 184 mbsf in CRP-2/2A (77.006°S, 163.719°E) and 50 and

150 mbsf in CRP-1 (77.008°S, 163.755°E) (Florindo et al., 2005). The Miocene sediments consist of six recurrent lithofacies: diamictite, rhythmically interlaminated fine-grained sandstone and siltstone, well-stratified sandstone, poorly-stratified muddy sandstone, coarse-grained siltstone and fine-grained siltstone (Cape-Roberts-Science-Team, 1998). These data indicate multiple phases of advance and retreat of the McKay glacier occurred through the early Miocene.

The *ANDRILL-2A* drill core (77.76°S, 165.28°E), a 1,138-m-long stratigraphic archive of climate and ice sheet variability from the McMurdo Sound sector of the western Ross Sea (Figs 9.2D, 9.7), was recovered in 2008 by drilling from an ~8.5-m-thick floating sea ice platform in 380 m of water, located ~30 km off the coast of Southern Victoria Land. The core offers an ice-proximal stratigraphic archive over the period c.20.2–14.2 Ma, with a more fragmentary record of Late Miocene and Pliocene time (Figs 9.5, 9.7). The sedimentary sequence is divided into 74 high-frequency glacimarine cycles recording repeated advances and retreats of glaciers into and out of the Victoria Land Basin (Fielding et al., 2011; Jovane et al., 2019; Levy et al., 2016; Passchier et al., 2011).

The *ANDRILL-1B* hole (77.89°S, 167.09°E) (Figs 9.2D, 9.7) was cored to 1284.85 m bsf and was completed on 26 December 2007 (Naish et al., 2007). Rock and sediments below ~150 mbsf are upper Miocene to Pliocene (Naish et al., 2007, 2009; Wilson et al., 2007) (Figs 9.5, 9.7). Sedimentary facies in the core are highly cyclic, comprising stratified and massive diamictite, breccia, conglomerate, sandstone, siltstone, mudstone and diatomite (Krissek et al., 2007b), and three distinct facies associations (termed Motifs) were identified in the cores highlighting major shifts in the glacial mass balance controls, as well as strong orbital pacing of ice sheet volume, over the past ~13.6 Ma (McKay et al., 2009; Naish et al., 2009; Rosenblume and Powell, 2019).

CIROS-2 (77.6833°S, 163.5333°E) was drilled in 1984 from a sea-ice platform near the middle of Ferrar Fjord, western McMurdo Sound in 211 m of water (Figs 9.2E, 9.7). The hole penetrated 166.47 m of Plio-Pleistocene sediments, terminating in basement gneiss. The core includes 13 lithologic units which alternate between massive and stratified diamictite, faintly stratified mudstone, and fine and very fine sand and sandstone (Barrett and Hambrey, 1992). The lowermost sequence (162 to 100 mbsf) covers most of the Pliocene (Levy et al., 2012) and is dominated by diamictite but is interbedded with thin (dm to m-scale) marine mudstone beds.

The *Dry Valley Drilling Project (DVDP)*, run by the United States Antarctic Program in the early to mid-1970s, recovered geological data from onshore sites at Ross Island and in the MDV (Figs 9.2E, 9.7). DVDP holes-10 (77.5833°S, 163.5166°E) and -11 (77.5833°S, 163.4167°E) were drilled during 1974/75 at the eastern end of Taylor Valley (McKelvey, 1975, 1981; Mudrey and McGinnis, 1975; Treves and McKelvey, 1975). DVDP-10 penetrated to 185.47 m and recovered a sequence dating back to the early Pliocene (~5.0 Ma) (Levy et al., 2012; Ohneiser and Wilson, 2012). DVDP 11 penetrated to 328 m and recovered a sequence dating back to ~6 Ma (Levy et al., 2012; Ohneiser and Wilson, 2012). DVDP-10 contains five lithostratigraphic units that include sandstone, pebbly sandstone, conglomerate, diamictite, breccia and laminated sandy mudstone (McKelvey, 1975, 1981). Units 4 and 5 are Pliocene. DVDP-11 was divided into eight lithostratigraphic units (McKelvey, 1975, 1981). The lowermost units (9–13) are Pliocene to upper Miocene and comprise an ~130-metre-thick sequence of sediments dominated by diamictite and minor sandy mudstone with some conglomerate (Levy et al., 2012; McKelvey, 1981).

9.2.3.5 Terrestrial records from Southern Victoria Land

Geological and glaciological research on the terrestrial records in Southern Victoria Land has a long history, beginning with observations by members of Scott and Shackleton's expeditions (David and Priestley, 1914; Ferrar, 1907; Wright and Priestley, 1922). While many of the subsequent early studies centred on mapping extensive exposures of Paleozoic and Mesozoic rocks in the mountains and valleys (Gunn and Warren, 1962), efforts were also made to decipher the glacial history preserved in the regional geomorphology and more recent sedimentary record (Bull et al., 1962; Calkin, 1974; Denton et al., 1971; Selby, 1971; Selby and Wilson, 1971). Since this early work, there have been extensive studies on the geomorphological and climatic evolution of the Dry Valleys region since the Miocene (Denton et al., 1984; Hall and Denton, 2005; Kowalewski et al., 2006; Lewis et al., 2006, 2007, 2008; Marchant et al., 1993; Prentice et al., 1993; Prentice and Krusic, 2005; Sugden and Denton, 2004; Sugden et al., 1991, 1995). Importantly, many of these deposits and landforms are mantled by, or interbedded with, volcanic ash and tephra, which provide critical age constraints (Lewis and Ashworth, 2015; Lewis et al., 2006; Marchant et al., 1993, 1996; Schiller et al., 2019).

Terrestrial sedimentary deposits of Miocene and Pliocene age in the Ross Embayment are primarily located in the MDV but are relatively few and most post-date the MCO (Table 9.1). These records are composed of tills, fluvial-lacustrine sequences and colluvium. Many tills consist of unconsolidated units that are often metres thick with characteristics including striated clasts, underlying striated bedrock and cross bedding suggesting fluvial processes likely from supraglacial melting. Some of the oldest well dated terrestrial glacial deposits observed in the Ross Embayment occur in the Friis Hills (Lewis and Ashworth, 2015), Western Olympus Range (Lewis et al., 2007) and Asgard Range (Marchant et al., 1993) (Fig. 9.2D).

TABLE 9.1 Summary of key terrestrial deposits in the McMurdo Dry Valleys listed in chronostratigraphic order.

Feature	Location	Age	Ref
Alpine II	East-Central Wright Valley	<2.8 ± 0.2 Ma	Hall et al. (1997)
Onyx drift	Lower Wright Valley	<3.3 ± 0.3 Ma	Hall and Denton (2005)
Wright drift	Lower Wright Valley	<3.4 ± 0.2 Ma	Hall and Denton (2005)
Valkyrie drift	Lower Wright Valley		
Loop drift	Lower Wright Valley	<3.3 ± 0.3 Ma	Hall and Denton (2005)
Alpine III	East-Central Wright Valley	<3.3 ± 0.2 Ma	Hall et al. (1997)
Till (lateral moraines (WUIII)	Upper Wright Valley	Overlies Peleus <3.8 Ma (<3.5 Ma)*	Bockheim and McLeod (2006)
Alpine IV	Lower Wright Valley	Overlies Peleus	
Peleus Till/ WU IV	Wright Valley	>3.8<5.5 (<3.5 to >3)*	Bockheim (2010); Prentice (1998); Summerfield et al. (1999)
Pecten Gravels (Prospect Mesa)	Wright Valley	Stratigraphically below Peleus (>3.5)*	
Jason Glaciomarine Diamicton	Wright Valley	9 ± 1.5 Ma	Prentice et al. (1993)
Rock Glaciers	Western Asgard Range	<12.5 Ma	Marchant et al. (1993)
Jotunheim Till	Western Asgard Range	>10.5 Ma	Marchant et al. (1993)
Dido drift 1	Western Olympus Range	>12.44 Ma	Lewis et al. (2007)
Dido till 2a	Western Olympus Range	<13.62>12.44 Ma	Lewis et al. (2007)
Dido till 2b	Western Olympus Range	<13.72>13.62 Ma	Lewis et al. (2007)

(*Continued*)

TABLE 9.1 (Continued)

Feature	Location	Age	Ref
Nibelungun Till	Western Asgard Range	<15.2**>13.6** Ma	Marchant et al. (1993)
Channels and Potholes in Asgard Range (Sessrumir Valley)	Western Asgard Range	Inferred to post-date Asgard Till	Sugden et al. (1991)
Asgard Till	Western Asgard Range	<15.2**>13.6 Ma	Marchant et al. (1993), Sugden et al. (1991)
Dido till 2c	Western Olympus Range	>13.85 Ma	Lewis et al. (2007)
Dido drift 2	Western Olympus Range	>13.94 Ma	Lewis et al. (2007)
Electra Colluvium	Western Olympus Range	>13.94 Ma**	Lewis et al. (2007)
Circe Till	Western Olympus Range	>13.94 Ma	Lewis et al. (2007)
Mount Boreas	Western Olympus Range	14.07 ± 0.05 Ma**	Lewis et al. (2008)
Friis II drift (upper)	Eastern and Central Friis Hills	~14.2 to 13.9 Ma	Chorley (2021)
Friis II drift (lower)	Eastern and Central Friis Hills	~15.1 to 14.4 Ma	Chorley (2021)
Koenig Colluvium	Western Asgard Range	>15.2 Ma	Marchant et al. (1993)
Inland Forts Till	Western Asgard Range	>13.5 Ma	Marchant et al. (1993)
Sessrumnir Till	Western Asgard Range	>15.2 to 15 Ma	Marchant et al. (1993)
Friis I Drift	Eastern Friis Hills	19.76 ± 0.11 Ma	Lewis and Ashworth (2015)

The Friis Hills (Figs 9.2, 9.8) are located at the head of the Taylor Valley in the MDV and, at an average elevation of 1325 metres (m) above sea level, constitute an isolated inselberg with peaks up to 1500 m above sea level. Their unique position has allowed for Miocene terrestrial glacio-fluvial-

FIGURE 9.8 Images from the Friis Hills, upper Taylor Valley, Southern Victoria Land, Antarctica. (A) View from the northern peak in the Friis Hills looking across the broad south-eastern facing Friis central valley that contains glacial lacustrine sediments of the Friis II drift (Lewis and Ashworth, 2015). (B) Drill rig used to recover core shown in (C). (C) Core showing diamictite, lacustrine sediments and tephra from middle Miocene Friis II upper sediments.

lacustrine sediments to be preserved in high-elevation basins that have been protected from subsequent EAIS ice sheet expansions (Lewis and Ashworth, 2015). The Friis I drift sits on striated granitic basement within a small paleovalley in the easternmost region of the Friis Hills and includes an upper and lower diamicton interpreted as subglacial (lodgement/traction) tills up to 5.5 m-thick, separated stratigraphically by an interval of interbedded fluvio-lacustrine muds, sands, gravels and a volcanic ash with an $^{40}Ar/^{39}Ar$ age of 19.76 ± 0.11 Ma (Lewis and Ashworth, 2015). This relatively thin deposit is the only well-dated glacial sequence in the Transantarctic Mountains that pre-dates 15.2 Ma.

The stratigraphically younger Friis II drift occurs above Friis I drift within the paleovalley in the eastern Friis Hills but is most extensive across the central basin where it is deposited on top of basement rocks (Fig. 9.8A). The Friis II sequence includes at least sixteen sedimentary cycles comprising diamicts and interbeds of well-sorted, fine grained interglacial fluvial and lacustrine deposits (Chorley, 2021; Lewis and Ashworth, 2015; Verret et al., 2020). Friis II lower units were deposited during advance and retreat of wet-based temperate style glaciers flowing to the southeast. The Friis II upper sequence includes diamict that likely records a transition to more regionally expansive eastward directed glacial advance (Lewis and Ashworth, 2015). Fossiliferous beds in the Friis II drift record a *Nothofagus* shrub tundra with mean summer temperatures of ~6°C–7°C (Lewis and Ashworth, 2015). The Friis II drift was cored in 2017 and tephra recovered in the cored sequence and a paleomagnetic stratigraphy indicate an age between ~15.1 and 13.8 Ma for the Friis II upper sequence (Chorley, 2021; Verret et al., 2020). These age constraints are consistent with meteoric beryllium analyses that suggest the Friis Hills sequence is older than ~14 Ma (Valletta et al., 2015).

Miocene sedimentary deposits in the western Asgard Range include the Sessrumnir, Inland Forts, Asgard, Nibelungan and Jotenheim tills and the Koenig colluvium (Marchant, 1993). These sequences contain tephra that constrain the age of the glacial sediments to between ~15.2 and 10.5 Ma (Table 9.1). Nibelungan and Jotenheim tills are restricted in areal extent, whereas the other deposits are more extensive and are discussed in more detail here. Sessrumnir till is an poorly sorted massive and unconsolidated sandy diamict that overlies striated and moulded Ferrar Dolerite and granite bedrock and was likely deposited by alpine glaciers that extended from local alpine ice caps at the end of the MCO (Marchant, 1993). Inland Forts till is a silt-rich and unconsolidated diamict that mantles the striated sandstone bedrock floor (Fig. 9.9) and is inferred to have formed beneath a southward flowing wet-based glacier that occupied Inland Forts, similar to nearby Sessrumnir till (Ackert, 1990; Marchant, 1993). Asgard till is a silt-rich and

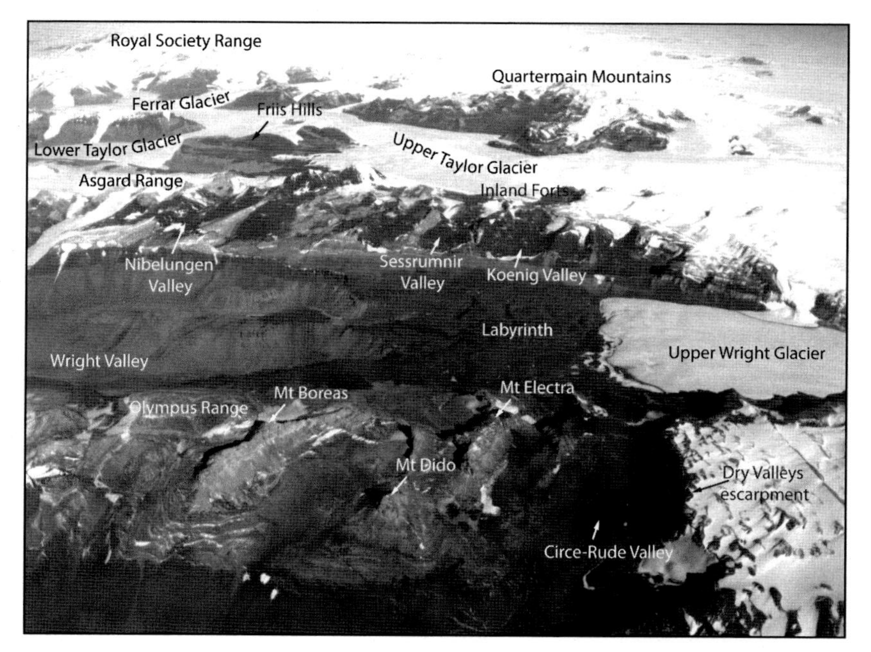

FIGURE 9.9 Oblique aerial photograph (USN TMA2299 0063) looking southwest towards the Olympus Range (foreground), Asgard Range (mid image), Quartermain Mountains and Royal Society Range (top of image). Features and locations containing Miocene terrestrial deposits discussed in text are labelled.

unconsolidated diamict containing granite erratics and striated sandstone and dolerite clasts. It is the most widespread surface deposit in the region and generally occurs in lobate map patterns at the mouths of hanging valleys in the Asgard Range and Quartermain Mountains. Radioisotopic dates on sanidine crystals in ash from deposits beneath the till indicate a maximum age of 14.8 Ma and from ash in sand wedges that cut the till indicate a minimum age of 13.6 Ma (Marchant, 1993). The Asgard till post-dates the MCO and was deposited during the MMCT (14.8 to 13.6 Ma) (Table 9.1). The sandstone and granite clasts and erratics and the morphological features of the Asgard till indicate that depositing ice flowed from the east, down major trunk valleys and spilled southwards into hanging valleys and likely reflects the influence of larger regional scale ice sheets (Marchant, 1993). However, thermal conditions at the base of the ice were likely highly variable with warm-wet conditions in the linear valleys and cold-frozen conditions across the plateaus with minimal disruption to underlying deposits.

Koenig colluvium is a bright orange (highly oxidised) and poorly sorted diamicton, composed of well-developed ventifacts within a sandy matrix that is stratigraphically between the underlying Sessrumnir till and overlying Asgard till. There are no striated, moulded or glacially polished clasts in Koenig colluvium and the deposit is inferred to reflect minor hillslope degradation and subaerial weathering in a desert climate. The age of Koenig colluvium is constrained by phonolitic ash that rests on buried ventifact pavement that covers the colluvium. Radioisotopic dates from sanidine in the ash indicate Koenig colluvium is older than 14.8 Ma and was deposited prior to the onset of the MCT (Table 9.1). This ash also provides a maximum age for the Asgard till which overlies the colluvium (Marchant, 1993).

Miocene sedimentary deposits in the Olympus Range include Circe till, Electra colluvium, Dido drift (Lewis et al., 2007) and fossil-rich moraine-damned glacial-lacustrine deposits (Lewis et al., 2008). Circe till is a remnant of regional alpine glaciation originating from Circe-Rude Valley in the upper western boundary of Wright Valley (Fig. 9.9). Circe till varies from 1–4 m in thickness, sits directly on moulded and striated bedrock, and includes a lower mud-rich diamicton and upper stratified diamicton with sand and gravel lenses. Clasts are commonly striated and cross bedding is apparent in the sandy lenses. The lower unit is interpreted as a lodgement till deposited beneath wet-based ice and the upper unit an ablation till and likely represent a single cycle of glacial advance and retreat (Lewis et al., 2007). Circe till and Electra colluvium are possible equivalents to Sessrumnir/Inland Forts till and Koenig colluvium in the Asgard Range. Dido drift includes disparate patches of till comprising grain-supported sandy diamicton. These drift sediments were most likely deposited in the absence of water by sublimation of cold-based alpine glaciers (Lewis et al., 2007). The stratigraphically youngest unit (Dido Ib drift) is the only till unit in the valley that is not dissected. This feature suggests the till has not been modified due to post-depositional glacial overriding. Age constraints for the Circe-Rude valley deposits come from volcanic ash that occurs as infill within relict sand-wedge troughs, laterally extensive interbeds, and small pods that infill voids in the Dido drift. Radioisotopic ($^{40}Ar/^{39}Ar$) dates on sanidine crystals from an ash in a wedge that dissects the lowermost Dido drift constrains the minimum age of Circe till, and the oldest episode of wet-based glaciation in the region, to $>13.94 \pm 0.75$ Ma (Lewis et al., 2007). Dates on ash in voids in the Dido Ib till suggest that regional glacial overriding at high elevations (>1200 m) has not occurred since 12.44 ± 0.15 Ma.

Middle Miocene glacial lacustrine sediments with exceptionally well-preserved fossil plants and freshwater aquatic organisms occur in a small moraine-dammed basin near Mount Boreas in the western Olympus Range (Fig. 9.9) (Lewis et al., 2008). This site is one of several that occur within north-facing valleys and contain moraine-damned glacial-lacustrine deposits in the western Olympus Range. The lacustrine beds at the Mount Boreas site are near horizontal, unconsolidated and are comprised of glaciolacustrine silt and sand beds beneath a sequence of diatomaceous mud. The lake sediments are overlain by fluvial sands and debris flow deposits. The extent and elevation of the beds behind the moraine suggest the lake was as much as 8 m deep (Lewis et al., 2008). Exquisite examples of freeze-dried moss are abundant in the mud rich layers. Fossil diatom assemblages and well-preserved ostracods indicate the moraine-dammed lake was ice free for at least 2 to 3 months each year. The age of the fluvial lacustrine deposits is 14.07 ± 0.05 Ma and is based on correlation to a tephra in another lake deposit in Fritsen Valley to the east (Lewis et al., 2008).

The oldest known marine sedimentary deposit on the floor of the Wright Valley is the Jason Diamicton (Prentice et al., 1993). The diamicton is a pebbly muddy sand to sandy mud that crops out extensively along the north shore of Lake Vanda in Wright Valley and was recovered at the base of DVDP-4A (Prentice et al., 1993). DVDP-4A was drilled near the centre of Lake Vanda (77°32′S 161°32′E) at an elevation of 83.6 metres above sea level and recovered sediment between the lake floor at 68 m and granitic basement, which was intersected at 80.6 metres below the lake surface (Cartwright et al., 1974). The lowermost section between 5.74 and 11.2 m in DVDP-4A comprises siltstone with minor amounts of sandstone and conglomerate and contains a late Miocene marine diatom assemblage (Brady, 1979). The absence of diamictite in the marine sequence indicates the Wright Valley fjord remained ice free during deposition of the sequence. The diatom assemblage indicates the Wright Valley sediments in DVDP-4A are older than sediments at the base of DVDP-10 and 11. Whole and fragmented marine diatoms are abundant in the Jason diamicton and include *Denticulopsis dimorpha*, *D. lauta*, *Actinocyclus ingens* and *Thalassiosira nitzchiodes* (Brady, 1979, 1982; Prentice et al., 1993). This assemblage indicates a depositional age between 13.66 and 9.23 Ma based on a revised biostratigraphic age presented herein (Fig. 9.10)

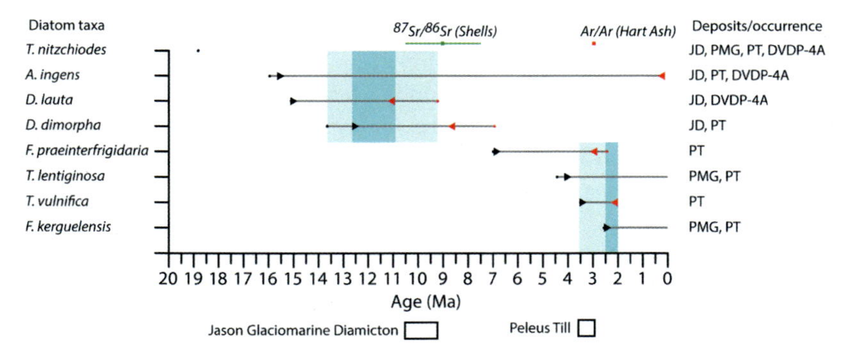

FIGURE 9.10 Diatom range chart for species reported in Dry Valley Drilling Project (DVDP) – 4A, the Jason Diamicton (JD), Peleus Till (PT) and Prospect Mesa Gravels (PMG). Ages for first appearance and last appearance datums for taxa are from the Constrained Optimisation (CONOP) hybrid range model (arrows) and total range model (dots) of Florindo et al. (2013). Deposits in which the diatoms are reported are shown at right (Prentice et al., 1993). $^{87}Sr/^{86}Sr$ date on shell from the Jason Diamicton from Prentice et al. (1993) and $^{40}Ar/^{39}Ar$ date on the Hart Ash from Schiller et al. (2019). Pale blue boxes indicate maximum likely age range of each deposit based on diatom biostratigraphy (darker blue boxes indicate minimum likely age range). Likely age of the Jason Diamicton and Peleus Till based on the combination of available chronostratigraphic data is indicated by boxes at bottom of diagram.

using total range model data for the diatom datums (Florindo et al., 2013). This age is compatible with estimates derived from $^{87}Sr/^{86}Sr$ analyses on shell fragments recovered from the diamicton, which suggests a minimum age of 9 ± 1.5 Ma (Prentice et al., 1993), indicating that the floor of the Wright Valley was a marine fjord during the late Miocene.

Pliocene sedimentary deposits in the MDV region consist of drift sheets, glacimarine sediments and volcaniclastic sequences associated with local cinder cones. Most of these deposits occur at lower elevations in the Taylor and Wright Valleys. However, at least one Pliocene glacial drift deposit occurs at relatively high elevation (1400 to 1500 m) in Vernier Valley in the Wilkniss Mountains. This drift is one of a series of four with a terminating moraine ranging from 1 to 4 metres in height, that were deposited by ice flowing from the Ferrar Glacier and into Vernier Valley. A minimum age of 3.39 ± 0.19 Ma was determined via cosmogenic ^{3}He and ^{21}Ne analysis from pyroxenes in surface boulders on the terminal moraine that extends furthest into the valley and is located at the highest elevation (Staiger et al., 2006). These data indicate that the margin of the Ferrar Glacier was at least 120 m higher than today during at least one glacial advance in the late Pliocene.

Four distinct surficial sedimentary deposits in the upper Wright Valley (WU IV through I) record major advances of the Wright Upper Glacier (Calkin and Bull, 1972), which is fed from Taylor Dome. The lowermost unit is now recognised as the Peleus Till (Prentice et al., 1993). Peleus Till is silty, contains striated clasts and crops out discontinuously across Wright Valley from 180 m to 1150 m elevation (Hall et al., 1997; Prentice et al., 1993). The till lacks moraines and exhibits feather edges and is interpreted to be a subglacial till deposited during multiple glaciations from the west, east, or local alpine areas (Prentice et al., 1993). The age of the Peleus Till is somewhat contentious. The Peleus till has been correlated with the Asgard till (Hall et al., 1993a), which implies a depositional age of between 15 and 13 Ma and requires that the Wright Valley had been cut to its near present morphology prior to the middle Miocene. The absence of basaltic clasts in the till has been used to suggest the till predates ~3.8 Ma, the age of oldest clasts in overlying drifts (Hall et al., 1993a). However, the till is well exposed in a section at Prospect Mesa where a 6-m thick sequence overlies the pecten gravels (Prentice et al., 1993), which requires the till post-dates the late Miocene/early Pliocene gravels. However stratigraphic integrity at the location was questioned by Hall et al. (1997), who suggested original deposition of the till occurred before the Prospect Mesa Gravels were deposited and that the Peleus Till was remobilised down slope and emplaced on top of the Pecten Gravels. Regardless, diatoms in the Peleus Till at Prospect Mesa include *Thalassiosira vulnifica, T. lentiginosa, T. insigna (reported as Cosmiodiscus insignus), and Fragilariopsis kerguelensis (reported as Nitzschia kerguelensis)*, which reflect a Pliocene age (Prentice et al., 1993). Currently accepted ages for the first and last appearances of these taxa (Florindo et al., 2013) (Fig. 9.10) indicate a depositional age between ~3.5 and 2 Ma. The Hart Ash, which has been recovered at several locations in the lower Wright Valley (Schiller et al., 2019), provides additional age constraint for the Peleus Till. Whereas no direct stratigraphic relationship between the till and ash has been observed, the ash would not survive wet-based glacial advances and therefore the ash places a minimum age on the Peleus Till. The Hart Ash was originally dated to 3.9 ± 0.3 Ma (Hall, 1992) but recent work suggests a younger age of 2.97 ± 0.02 Ma (Schiller et al., 2019). Based on available constraints presented in this review, we conclude that the Peleus Till was deposited between 3.5 and 3 Ma.

Pliocene surficial glacial sedimentary deposits in central Wright Valley include Alpine IV, III and II drifts (Hall et al., 1993a). These drifts are highly oxidised and contain weathered and fractured boulders. Alpine IV drift overlies Peleus Till and lateral moraines associated with the drift underlie moraines comprised of Alpine III and Alpine II drifts (Hall et al., 1993b). A chronology for the drifts was established through $^{40}Ar/^{39}Ar$ dating of reworked volcanic clasts contained within the drifts, which provide a maximum age for the deposits (i.e. the drifts cannot be older than the youngest dated clast). Alpine III and IV drifts contain clasts that range between 3.8 and 3.3 Ma, which indicates the drifts are younger than late Pliocene.

Relatively extensive glacial deposits in lower Wright Valley include the Valkyrie, Wright and Onyx drifts (Hall and Denton, 2005). Wright drift overlies Alpine III and IV drifts and is dissected by Alpine II Drift. Onyx drift is a massive coarse-sand diamicton that overlies Alpine III and IV drifts but is cut by Alpine II drift. Loop drift is an unconsolidated, massive, coarse-sand diamicton with common stained and ventifacted clasts that are occasionally striated. Loop drift is inferred to be older than Valkyrie, Wright and Onyx drifts based on the degree of weathering (Hall and Denton, 2005). Basaltic clasts in the Loop, Wright and Onyx drifts constrain their maximum age to 3.4 to 3.3 Ma (Hall and Denton, 2005) (Table 9.1). Basaltic cones are not known from the eastern Wright Valley so the likely source of basalt in the drifts is McMurdo Sound, which indicates the drift sediments were deposited by ice that extended into the Wright Valley from the Ross Sea. Drift geomorphology and extent suggests they were deposited by ice that was 400 to 500 metres thicker than the present-day Wilson Piedmont Glacier (Hall and Denton, 2005).

9.2.3.6 Neogene history in the Ross Sea Region

The Ross Sea mid-continental shelf began to subside in the late Oligocene (~26 Ma) and increasingly large regions of West Antarctica became inundated by the sea (Kulhanek et al., 2019; Paxman et al., 2019b, 2020). Ice sheets and glaciers advanced and retreated across regions of the Ross Sea throughout the late Oligocene, periodically expanding from ice caps on Marie Byrd Land, topographic highs in the central and eastern Ross Sea (De Santis et al., 1995, 1999; Sorlien et al., 2007a, 2007b), and outlet glaciers through the TAM (Fielding et al., 2000). Water depths in the region around DSDP 270 reached 500 m by the early Miocene (Kulhanek et al., 2019; Leckie and Webb, 1983, 1986).

A significant transient glacial advance across the Oligocene/Miocene boundary is captured in DSDP 270 and the Cape Roberts Project cores by major disconformities that overly glacimarine glacial-interglacial sedimentary cycles leading up to 23 Ma (Florindo et al., 2005; Kulhanek et al., 2019; Naish et al., 2001b, 2021). Depositional packages that immediately post-date the disconformities indicate ice retreated from these sites in the early Miocene. Whereas the composition of the pollen assemblage in DSDP Site 270 suggests terrestrial environmental conditions did not change across the O/M boundary, an increase in relative abundance of *Nothofagidites* suggests moderate climatic cooling occurred in the early Miocene (Kulhanek et al., 2019).

Environmental conditions were highly variable in the Ross Sea through much of the early and middle Miocene (Levy et al., 2016). SSTs ranged from cold sub-zero values similar to those recorded today to short intervals of warmth during which temperatures peaked at ~10°C (Levy et al., 2016; Warny et al., 2009). Whereas the AIS was generally restricted to terrestrial environments, at least four episodes of extensive marine ice sheet advances are reflected by drill core and seismic data (Levy et al., 2016). Two of these episodes of marine ice sheet advance (MISA) occurred in the early Miocene. The first is captured by a major disconformity in AND-2A which correlates in time with RSU 5A and records advance of ice across regions of the Ross Sea continental shelf at ~19.8 Ma. This disconformity occurs above a sequence of strata at the base of AND-2A assigned to stratigraphic Motifs 2 and 5 (Fielding et al., 2011; Passchier et al., 2011), which are interpreted to record a broad environmental range from a sub-polar glacial regime with significant meltwater influence to minimal (distal) glacial influence (Fielding et al., 2011). Pollen in AND-2A (Griener et al., 2015; Warny et al., 2009) and sequences in the Friis Hills (Lewis and Ashworth, 2015) indicate tundra occupied regions spanning the coast to at least 60 km inland during interglacial conditions before and after the MISA at 19.8 Ma. These data reflect a highly variable climate in which the ice sheet margin advanced and retreated across the coastal margin of the Ross Sea, at times extending into marine environments.

An unusual period of cold and relatively stable climate is indicated by proxies in a prominent thick sequence of fine-grained sediments in AND-2A (Levy et al., 2016). This conspicuous stratigraphic interval is characterised by very low amounts of pollen and spores and persistently low surface water temperature (SWT = upper 200 m of water column) (−1.3°C to 2.6°C) and is thought to have accumulated in a dark environment beneath semipermanent sea ice or an ice shelf. This episode of apparent environmental stability coincides with an interval characterised by low variability in sea level (Kominz et al., 2008) and orbital eccentricity (Laskar et al., 2004, 2011; Levy et al., 2016) (Fig. 9.11). During this time, the ice sheet margin retreated from the coast and the grounding zone remained distal to the AND-2A site through many glacial–interglacial cycles (~700 kyrs). Variability in the $\delta^{18}O$ record decreased during this interval but glacial–interglacial amplitudes still ranged between 0.6‰ and 1‰ (Fig. 9.11). Given the apparent relative stability of the adjacent ice sheet, much of the variability in the $\delta^{18}O$ record must have been driven by changes in bottom water temperature. Today, significant volumes of bottom water are generated in the Ross Sea (Carter et al., 2021; Orsi et al., 1999). If we assume similar quantities of Earth's bottom water was generated in the Ross Sea during the early Miocene, then the 3°C to 4°C range of SWTs recorded in AND-2A through this interval may have driven changes in BWT of similar magnitude. Changes of 3°C to 4°C in BWT can explain between 90% and 100% of the total glacial–interglacial variation in the $\delta^{18}O$ records between 19.4 and 18.6 Ma (up to 0.88‰ if we apply 0.22‰ per 1°C of bottom water temperature change). It appears possible that terrestrial portions of the AIS remained relatively stable during intervals of low variability in Earth's eccentricity while ocean temperature at high latitudes fluctuated up to 4°C. These observations highlight the potential for distinct sensitivities of the cryosphere and ocean to climate and raise questions regarding the fundamental mechanisms and processes that drive change in Antarctica through time.

A second episode of early Miocene marine ice sheet advance across the Ross Sea is captured in both the AND-2A and IODP Site U1521 cores and by RSU5 (Pérez et al., 2021a) (Fig. 9.7D). Disconformity U2 removed sediments equivalent to ~900 kyrs of record from AND-2A and is overlain by a 100 m-thick diamictite dominated sequence that is sheared and deformed (Fielding et al., 2011; Passchier et al., 2011). The AND-2A U2 disconformity coincides with RSU5, which marks the base of RSS-3 and correlates to the bottom of Lithostratigraphic Unit VI at 566 mbsf in Site U1521, which is a thick progradational package (McKay et al., 2019; Pérez et al., 2021a). RSU5 is not interpreted as an erosive surface at Site U1521, but the overlying interbedded massive to stratified diamictite and mudstones in Lithostratigraphic Unit VI contain shell fragments and calcium carbonate nodules and concretions that reflect advance and retreat of glaciers south of the site in the central Ross Sea. These regional disconformities and diamictite dominated sequences along the western margin and across the central Ross Sea, suggest highly erosive ice sheets and glaciers extended well beyond the Antarctic coast into marine environments between 17.8 and 17.3 Ma (Levy et al., 2016, 2019; McKay et al., 2019). This marine-based ice sheet advance approximately coincides with a prominent sea level lowstand at the New Jersey margin (Kominz et al., 2008) (Fig. 9.11E), which supports significant ice sheet growth at this time. Whereas there is no obvious excursion in recent high resolution $\delta^{18}O$ compilations (De Vleeschouwer et al., 2017; Westerhold et al., 2020), the episode does coincide with the Mi1b isotope event (Miller et al., 1991a). Furthermore, this distinct episode of marine-based ice sheet advance occurs during a prominent peak in obliquity sensitivity and diatom turnover (Fig. 9.11I and J), which is inferred to reflect AIS expansion and enhanced connection between marine-based ice sheets and the Southern Ocean (Crampton et al., 2016; Levy et al., 2019). These observations show the AIS extended well beyond the TAM and periodically occupied large regions of the Ross Sea prior to the MMCT.

Retreat of the AIS from the Ross Sea Embayment during the MCO is captured by the 124 m-thick sequence of bioturbated diatomaceous mudstone with rare clasts that constitutes Unit III in IODP Site U1521 (McKay et al., 2019). The diatomaceous unit was deposited sometime between 17 and 15 Ma (McKay et al., 2019), is bounded by sequences that include diamictites and lacks evidence for proximal glacial influence. Sequences spanning the MCO in AND-2A contain diverse lithologies, lack massive diamictites (Fielding et al., 2011; Passchier et al., 2011) and indicate temperate wet-based glaciers calved at the coastline through glacial–interglacial cycles. Several units comprise sedimentary motifs and facies associations that indicate the grounding zone of the temperate glaciers that fed the basin were distal to the drill site. However, only three short interglacial intervals contain multiproxy data that suggest peak warmth and maximum ice margin retreat.

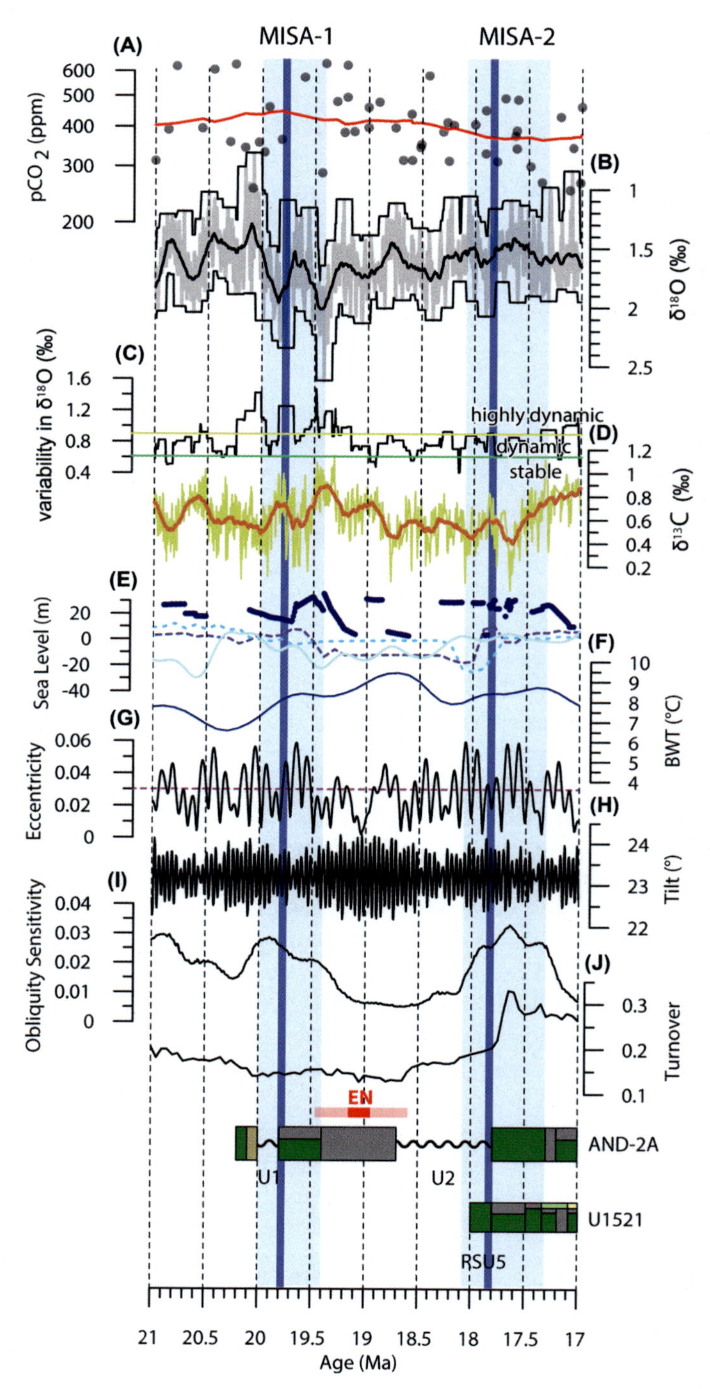

FIGURE 9.11 Environmental data from the early Miocene (21 to 17 Ma). (A) Atmospheric CO_2 compilation comprises data from a range of proxies outlined in Fig. 9.1. Solid red line displays a two million year moving average. (B) Splice of deep sea benthic foraminifera $\delta^{18}O$ data (light grey) reflect changes in ice volume and deep/bottom water temperature (De Vleeschouwer

(*Continued*)

During the warmest of these events, SWT reached 10°C (Levy et al., 2016), diatoms and dinocyst abundance peaked, and podocarp and pollen numbers suggest tundra vegetation expanded (Griener et al., 2015; Warny et al., 2009) and perhaps reflect hydrologically active intervals during the MCO (Feakins et al., 2012). Two of these short intervals of maximum warmth captured in AND-2A (at ~16.4 and 16 Ma) coincide with oxygen and carbon isotope minima that mark intervals of unique peak warmth that align with 400 kyr eccentricity maxima (Holbourn et al., 2015).

Terrestrial environmental records through most of the MCO have not yet been recovered from the Ross Sea region. A 'gap' in the ash stratigraphy in the TAM occurs between the 19.76 Ma ash recovered from the Friis Hills (Lewis and Ashworth, 2015) and the oldest ash in the Asgard Range that dates to 15.15 ± 0.2 Ma (Marchant et al., 1993, 1996). Analyses of volcanic material in the AND-2A core indicate eruptions from Mount Morning continued throughout the MCO (Di Vincenzo et al., 2009; Nyland et al., 2013) so it is unclear why ash deposits from the mid Miocene are not present in the TAM. Perhaps the climate was highly variable and glaciers too dynamic and erosive through this interval to preserve sediments in the TAM? Two tephra layers have been recovered in drill cores from the central Friis Hills and indicate the sequence post-dates ~15 Ma (Chorley et al., 2021; Verret et al., 2020). Interglacial deposits contain a pollen assemblage characteristic of tundra vegetation that characterised the region since the late Oligocene. The search for Miocene terrestrial deposits should continue so that we can more

---

◀ et al., 2017). Solid black line displays a 150 kyr moving average. Maximum and minimum values are determined within each 150 kyr window and create the envelope (black lines) that bound the $\delta^{18}O$ data. (C) A $\delta^{18}O$ variability 'index', interpreted as a measure of glacial/interglacial variability, is determined by subtracting the maximum values in (B) from the minimum value, for each 150 kyr window. We indicate arbitrary 'thresholds' in these data (green and yellow solid lines) and suggest that values below 0.6% reflect a relatively 'stable' high latitude environment and values that exceed 0.9‰ indicate a highly dynamic environment with large changes in BWT and/or ice volume over glacial-interglacial time scales. (D) Yellow line shows high resolution deep sea benthic foraminifera $\delta^{13}C$ splice (Westerhold et al., 2020). Solid deep orange line displays a 150 kyr moving average. (E) Sea level curves are from Kominz et al. (2008) (pale blue dashed line), Kominz et al. (2016) (blue dots), Miller et al. (2005) (blue dashed line), and Miller et al. (2020) (pale blue line). (F) BWT data are derived from the compilation of Cramer et al. (2011) using Equation 7a (dark blue). (G) Astronomical eccentricity and (H) axial tilt (obliquity) from Laskar et al. (2011). (I) obliquity sensitivity after Levy et al. (2019). (J) Rate of diatom species extinction and speciation (turnover) after Crampton et al. (2016). Lithologic logs from the AND-2A and U1521 drill cores highlight disconformities (U1, U2, RSU5) and diamictite dominated intervals that are inferred to indicate marine ice sheet advances (MISA events) (Levy et al., 2016; McKay et al., 2019). Note that peak values in obliquity sensitivity and diatom turnover coincide with the MISA event between ~18 and 17.4 Ma. Red bars highlight an interval of low eccentricity that coincides with a thick mudstone unit in AND-2A and reflects a period of low variability in AIS grounding line variability and 'stable' glacial dynamics.

fully assess ice sheet and environmental sensitivity during peak warm intervals in the MCO.

In contrast to the MCO, terrestrial deposits that capture the transition into and through the MMCT in the Ross Sea Region are relatively common. The thickest, most continuous record occurs in the Friis Hills (Chorley, 2021; Lewis and Ashworth, 2015; Verret et al., 2020). Here, up to 80 m of strata described from outcrop (Lewis and Ashworth, 2015) and drillcore (Chorley, 2021; Verret et al., 2020) include approximately 15 glacial—interglacial cycles comprising diamictite, glacifluvial, and glacilacustrine sediments. Most of the sediments assigned to the Friis Hills II lower sequence were deposited during the last few hundred thousand years of the MCO (~15 Ma). Diamictites in this sequence are muddy and were deposited beneath wet-based alpine glaciers. Interglacial units contain *Nothofagus* leaves, pollen and spores from a tundra vegetation, and freshwater diatoms. Friis II upper sediments are younger than 14.8 Ma (Verret et al., 2020) and contain diamictites with sedimentary features that suggest deposition under thicker ice and colder, drier climatic conditions (Lewis and Ashworth, 2015). Interglacial units still contain abundant evidence for liquid water across the region and fossil remains indicate tundra vegetation persisted through the MMCT to ~14 Ma (Valletta et al., 2015). Fossil-rich glacial-lacustrine deposits at Mount Boreas in the Olympus Range also show relatively warm and wet conditions persisted in the TAM through to at least 14.07 ± 0.05 Ma (Lewis et al., 2008). By 13.85 ± 0.03 Ma the climate had cooled by at least 8°C, muddy glacial diamictites were superseded by sandy sublimation tills indicative of drier conditions, and evidence for tundra vegetation had disappeared (Lewis et al., 2008; Lewis et al., 2007). Sedimentary deposits and geomorphological features in the Asgard Range generally support observations from sequences in the Friis Hills and Olympus Range. Wet based sediments deposited by alpine glaciers, including the Sessrumnir and Inland Forts tills, predate 15 and 13.5 Ma, respectively (Table 9.1). The laterally extensive Asgard Till is silt-rich, contains striated clasts, was deposited sometime between 13.6 and 15.2 Ma, and records expansion of a large glacier down the paleo Wright Valley (Marchant et al., 1993). Together, these terrestrial data indicate climate became colder and drier, and glaciers thicker and more extensive, during glacial episodes leading up to the final step of the MMCT between 13.9 and 13.8 Ma. But interglacial episodes remained relatively warm. These observations are reflected by an increase in glacial/higher values (colder/more ice) in the deep-sea $\delta^{18}O$ record between ~14.8 and 13.9 Ma while low interglacial/warm values (warmer/less ice) persisted until 13.9 Ma (Holbourn et al., 2014; Shevenell et al., 2004, 2008).

Marine sedimentary records support terrestrial evidence of progressive cooling and ice expansion during glacial episodes beginning at ~14.8 Ma. A major disconformity at 264 mbsf in the AND-2A core (U3) spans 14.6 to

15.8 Ma and coincides with the appearance of massive diamictites and associated facies that reflect a cold, subpolar/polar glacial regime with minor meltwater involvement (Fielding et al., 2011; Levy et al., 2016; Passchier et al., 2011). Another major disconformity at 214.13 mbsf separates sediments deposited at ~14.4 Ma from overlying sediments that are most likely younger than 11 Ma (Acton et al., 2008–2009). These two major unconformities are equivalent to RSU4, which suggests several shelf-wide advances of grounded marine-based ice originating from East Antarctica started between 14.8 and 14.6 Ma and continued through to ~11 Ma (Fig. 9.7D). This regionally extensive unconformity represents the first unequivocal seismic evidence for a glacially carved trough in the central Ross Sea (Anderson, 1999; Colleoni et al., 2018; De Santis et al., 1995; Ten Brink and Schneider, 1995). Approximately 250 m of till foreset and aggrading bottomset strata occur above RSU4 (De Santis et al., 1999) and RSU4 overlies outwash channels (tunnel valley features), suggesting the region was influenced by large volumes of erosive sediment-laden meltwater during glacial advance associated with the MMCT beginning at ~14.6 Ma (Anderson and Bartek, 1992; Chow and Bart, 2003; McKay et al., 2019; Pérez et al., 2021a).

Dissected drift sheets, potholes and subglacial melt channels at modern elevations >1000 m in the TAM provide evidence for glacial overriding event(s) in the middle Miocene (Sugden and Denton, 2004). The age of these features suggests ice flowing through the mountains reached its maximum thickness within 200 kyrs after the Mi3/E_3 isotope event and termination of the MMCT. For example, glacial plucking of bedrock and sedimentary deposits including the Circe and Electra tills in Western Olympus Range occurred between 13.62 and 12.44 Ma (Lewis et al., 2007). Formation of potholes and in the Asgard Range occurred after the Asgard Till was deposited between 15 and 13.6 Ma (Marchant, 1993; Sugden et al., 1995). Subglacial meltwater channels that characterise the Labyrinth formed beneath thick ice that flowed into the Wright Valley sometime between 14.4 and 12.4 Ma (Lewis et al., 2006). We note that evidence for subsequent overriding has not been recovered or reported, but that overriding events in the early and middle Miocene cannot be excluded.

The Cavendish Drift forms a series of stacked cycles of diamicton and fine-grained interbeds that blanket the southeastern margin of the Friis Hills (Lewis and Ashworth, 2015). Climbing ripples and crossbeds are common in sandy interbeds and rare dropstones are scattered throughout most beds. The sequence was deposited by wet-based ice but all beds are non-fossiliferous, which suggests tundra was no longer present in the region (Lewis and Ashworth, 2015). However, the climate was clearly warm enough during interglacial episodes to allow glacier surface melting and water to flow along the ice margin. The age of the drift is difficult to constrain; however, it truncates the Friis Hills II sequence so must be younger than 14 Ma (Lewis and

Ashworth, 2015; Valletta et al., 2015; Verret et al., 2020) and likely post-dates the episode of erosion and moulding of bedrock that occurred during glacial overriding between 13.6 and 12.4 Ma (Lewis and Ashworth, 2015; Lewis et al., 2007). A suite of $^{40}Ar/^{39}Ar$ dates indicate glacial drift was deposited on the Rhone Platform at the margin of a paleo Taylor Glacier between 10.76 and 10.39 Ma (Hartman, 1998). It is plausible that the most recent phase of downcutting of the Taylor Valley to its present depth and deposition of the Cavendish Drift occurred between ~11 and 10 Ma.

Geological data from neighbouring Wright Valley support a late Miocene age for the formation of modern valley topography in the MDV. Fossil marine diatoms in the Jason Diamicton deposited on the floor of the Wright Valley (Brady, 1982) indicate glacial expansion and erosion during and after the MMCT had cut the valley to its present depth sometime between 13.66 and 9.23 Ma. Furthermore, strontium isotope dates on fossil shells suggest the marine incursion that formed the deposit within the Wright Valley marine fjord occurred after 10.5 Ma (Prentice et al., 1993, 1999). This date aligns with the postulated age of deposition of the Cavendish Drift (Lewis and Ashworth, 2015). Furthermore, it coincides with the transition from diamictite-dominated sequences of Motif 1 below 1060 mbsf in AND-1B to deposition of mudstone-dominated Motif 3 (McKay et al., 2009; Wilson et al., 2012b). This distinct change in facies in AND-1B reflects an environmental shift in the western Ross Sea at ~10 Ma from a polar glacial regime to a subpolar meltwater dominated system that was similar to Spitsbergen today (McKay et al., 2009; Rosenblume and Powell, 2019; Wilson et al., 2012b). Sequence 67 occurs between 101.03 and 142.34 mbsf in the AND-2A core and includes an ~25 m-thick stratified sandstone that reflects a high-latitude temperate glacial regime with wet-based glaciers (Fielding et al., 2011; Passchier et al., 2011). Two reworked lava clasts within the sequence are dated at 11.43 ± 0.46 and 11.363 ± 0.072 Ma and provide a maximum age for the sediments (Di Vincenzo et al., 2009). The first appearance of the planktic foraminifera, *Neogloboquadrina pachyderma* at 83.80 mbsf in AND-2A indicates sediments above this depth must be younger than 11.04 Ma (Acton et al., 2008–2009; Patterson and Ishman, 2012). Together these chronostratigraphic data suggest that sequence 67 is ~11 Ma or younger. Evidence for ice free marine fjords in the Dry Valleys and subpolar and temperate conditions in drillcores from McMurdo Sound indicate that climate warmed and ice retreated inland between ~11 and 8 Ma, well after the end of the MMCT.

Sedimentary deposits from the latest Miocene (8 to 5.5 Ma) are relatively rare in the Ross Sea region (Figs 9.5 and 9.7). Preliminary age analysis of cores from IODP Expedition 374 indicate a late Miocene hiatus from ~9 to 5.5 Ma (McKay et al., 2019). This disconformity ties with RSU3, which defines major cross-shelf paleotroughs and is associated with enhanced progradation of the continental shelf into the Eastern Ross Sea (Bart and De

Santis, 2012; Chow and Bart, 2003; De Santis et al., 1995). RSU3 reflects expansion of marine based WAIS in the eastern Ross Sea, that likely coalesced with ice derived from the EAIS to drive continental shelf wide advances of marine-based ice sheets across the Ross Sea (De Santis et al., 1995; 1999). The latest Miocene record in the AND-1B drill core comprises an ~175 m-thick volcanic sequence (Di Roberto et al., 2010; Krissek et al., 2007a; McKay et al., 2009). The sequence is subdivided into a 70 m-thick succession of volcanic-rich mudstone and sandstone and a 105 m-thick section of interbedded tuff, lapilli tuff and volcanic diamictite that were deposited in an open water ice free environment. $^{40}Ar/^{39}Ar$ dates on a 2.81-m-thick lava flow in the middle of the sequence indicates the sequence was deposited around 6.48 Ma. Seismic and drill core data from the central Ross Sea show marine-based ice advanced across large areas of the Ross Sea in the late Miocene, but evidence from AND-1B also shows the ice sheet also episodically retreated to the terrestrial continental margin.

Importantly, the 'missing record' spanning ~8 to 5.5 Ma and inferred frequent expansion and retreat of a marine-based AIS associated with RSU3, approximately coincides with the episode of late Miocene cooling observed across our planet (Herbert et al., 2016; Holbourn et al., 2018). Furthermore, cosmogenic nuclides in the AND-1B core suggest major retreat at the terrestrial margins of the AIS along the TAM front has been minimal since ~8 Ma (Shakun et al., 2018). Cosmogenic nuclide studies have also been used to constrain the onset of persistent aridity in the MDV and East Antarctica. Specifically, meteoric ^{10}Be ($^{10}Be_{met}$) is a tracer of water infiltration and its migration in permafrost soils (Dickinson et al., 2012). $^{10}Be_{met}$ is formed in the upper atmosphere by cosmic rays and is transferred to Earth's surface by precipitation of deposition of dust. $^{10}Be_{met}$ nuclides can move into soils by infiltration and clay illuviation, making it a suitable environmental tracer of water movement through soil profiles over million-year timescales. Analysis of meteoric beryllium ($^{10}Be_{met}$) in the MDV suggests water infiltrated stable upland soils until ~6.6 Ma but has not penetrated since (Dickinson et al., 2012). It appears that the dry and arid polar climate that characterises Victoria Land and the MDV today, first became a persistent feature of the Antarctic climate during the peak cold of the late Miocene.

SSTs data from several ocean basins and across a range of latitudes indicate global climate warmed again in the early Pliocene (Fig. 9.4H) (Herbert et al., 2016). Diatomites in AND-1B indicate frequent collapse of the WAIS through the Pliocene. Diatom assemblages and geochemical paleothermometry indicate ocean temperatures were as much as 4°C warmer than present during the warmest interglacial episodes (McKay et al., 2012; Naish et al., 2009). Contemporaneous paleothermometry data from the nearby DVDP-11 drill core indicates Taylor Fjord surface waters were up to 5°C warmer than today (Ohneiser et al., 2020). Despite climatic warming, the hyper arid conditions that were established along the Victoria Land Coast at ~6 Ma appear to have persisted into the Pliocene. Fine grained sedimentary lithofacies in AND-1B are dominated by pelagic diatom ooze, indicating that the offshore environment was starved of terrigenous sediment supply and meltwater discharge was reduced during Pliocene interglacials. These facies are a stark contrast to the thick (10s of metres) glaciomarine mudstone-rich facies deposited during late Miocene interglacials that indicate large volumes of turbid outwash from marine and/or terrestrial-terminating glacial systems (McKay et al., 2009; Rosenblume and Powell, 2019).

A relatively thick diatomite sequence in the AND-1B core, deposited between 3.6 and 3.4 Ma, indicates warmer-than-present conditions persisted in the western Ross Sea into the early late Pliocene. However, a cooling trend is reflected in diatom assemblage data from the diatomite (McKay et al., 2012; Winter et al., 2010a, 2010b). A thick diamictite overlies the diatomite and indicates readvance of grounded ice from West Antarctica beginning at 3.4 Ma. Subsequent diatomite units contain diatom assemblages that indicate colder conditions and more persistent winter sea ice (McKay et al., 2012). The last major expansion of a wet-based glacier from the EAIS through the Wright Valley deposited the Peleus Till sometime between 3.5 and 3 Ma (Table 9.1) (Levy et al., 2012). Sand-rich alpine drift deposits were deposited between 3.8 and 3.3 Ma (Table 9.1) (Levy et al., 2012). Plant biomarkers in the DVDP-11 indicate vegetation disappeared from coastal regions of the Taylor Valley at some time between 3 and 4 Ma. This transition from silt- to sand-rich glacial deposits along valley floors in the MDV indicates a shift from wet- to cold-based glacial conditions occurred between 3.5 and 3 Ma. This observed readvance of marine ice sheets and transition from wet to dry glacial conditions in the lower MDV coincides with the major M2 glaciation recorded in $\delta^{18}O$ records (Lisiecki and Raymo, 2005). By ~3.0 Ma, relatively thick packages of diamictite and gravel dominated sediment accumulated in DVDP-10 and -11 at the mouths of the fjord valleys, indicating the drill sites were often covered by grounded ice, thick multiyear sea ice or an ice shelf (Levy et al., 2012).

Paleobathymetric reconstructions suggest that RSU2 formed as the continental shelf in the Ross Sea deepened inland of the continental shelf margin (De Santis et al., 1999). Overdeepening developed as marine-based ice sheets expanded and retreated across the Ross Sea continental shelf eroding middle and late Miocene sediments below RSU3. Most of the eroded sediment was transported to the continental shelf break and redeposited on the upper slope to form large trough-mouth fans (Colleoni et al., 2018; Cooper et al., 2008; De Santis et al., 1999). While the age of RSU2 is poorly constrained, we hypothesise the initial erosional event that formed this basin-wide unconformity coincided with the M2 isotope event and the transition to 'near modern' polar conditions in the Ross Sea between 3.4 and 3 Ma (Levy et al., 2012; McKay et al., 2012).

9.3 Numerical modelling

9.3.1 Miocene

Although there have been relatively few modelling studies of the Miocene AIS, efforts are increasingly shifting to examination of this important interval in Earth's climate history (Steinthorsdottir et al., 2020). Until recently, most ice sheet model (ISM) studies that have covered the Miocene have been long-duration simulations spanning multiple epochs or idealised studies representative of a Miocene-like Antarctic (De Boer et al., 2010; Huybrechts et al., 1993; Langebroek et al., 2009, 2010; Pollard and DeConto, 2005, 2020). Only a few recent modelling studies have specifically targeted the Miocene AIS (Colleoni et al., 2018; Gasson et al., 2016b; Halberstadt et al., 2021; Stap et al., 2019) and only recently with an Antarctic topography appropriate for the Miocene (Paxman et al., 2020). In contrast to the Pliocene (covered in the next section), there are some unique modelling challenges presented when simulating the Miocene. Intervals of the Miocene were warmer and with higher concentrations of atmospheric CO_2 (Foster et al., 2012; Greenop et al., 2014). There was greater retreat of the AIS causing greater variability in sea level (Lear et al., 2010; Miller et al., 2020). In the mid-Pliocene, ice sheet retreat was centred on the large subglacial basins of Antarctica; however, during the mid-Miocene, there was additional retreat of the terrestrial ice sheet, grounded above sea level (Gasson and Keisling, 2020). At times during the Miocene, it is likely that surface temperatures were warm enough to generate widespread surface melting of the ice sheet, although the extent of this retreat is difficult to constrain (Fielding et al., 2011; Levy et al., 2016; Passchier et al., 2011; Sangiorgi et al., 2018). Simulating this magnitude of retreat with ISMs has proven difficult (Foster and Rohling, 2013).

The first simulations of a Miocene-like AIS were transient simulations run from the onset of Antarctic glaciation across the EOT (Pollard and DeConto, 2005). These experiments showed that to reproduce the large ice volume changes inferred for the Miocene from oxygen isotope and sea level records at the time (Kominz et al., 2008; Miller et al., 2005; Zachos et al., 2001b) required a large climate forcing. This is needed to overcome the strong cooling caused by the elevation feedback during the ice sheet growth (Oerlemans, 2002). This feedback is further strengthened by albedo feedback, meaning early simulations required an Antarctic warming of $\sim 15°C$ (Huybrechts et al., 1993) or an increase of atmospheric CO_2 to $\sim 8x$ pre-industrial concentrations (Pollard and DeConto, 2005) to produce a negative surface mass balance and retreat of the Miocene Antarctic Ice Sheet. This is much higher than proxy records of atmospheric CO_2 for the mid-Miocene and therefore presented an enigma known as the Antarctic Ice Sheet hysteresis problem (Foster et al., 2012).

As this effect is also seen in long-term future simulations of the AIS (Garbe et al., 2020), the Miocene is an important test-bed for this effect.

Another way to simulate relatively large magnitude of AIS retreat with lower atmospheric CO_2 requires greater polar amplification than is currently simulated by coupled ocean-atmosphere climate models (Langebroek et al., 2009). Langebroek et al. (2009, 2010) used a reduced complexity climate box model and a 2-d ISM, which allowed them to increase or decrease Antarctic polar amplification. This study showed that when polar amplification increases, AIS hysteresis is reduced, and ice sheet retreat can be initiated at a lower concentration of atmospheric CO_2 than in the study of Pollard and DeConto (2005). However, the cause of this inferred increase in polar amplification in the Miocene has yet to be explored or explained. In a follow-up study, Langebroek et al. (2010) added oxygen isotope tracing to their model and showed that changes in the oxygen isotope composition of the ice sheet could have caused part (15%) of the shift in the oxygen isotope composition of sea water across the middle-Miocene, with the remaining 85% of the shift caused by a change in ice volume.

Gasson et al. (2016b) simulated a dynamic Miocene AIS (~35 m sea level equivalent changes) that was consistent with oxygen isotope records (Lear et al., 2010) and atmospheric CO_2 reconstructions (Foster et al., 2012; Greenop et al., 2014). However, they showed smaller changes in sea level than some recent reconstructions (Miller et al., 2020). In their simulations, increased retreat of the ice sheet was achieved with a high resolution asynchronously coupled ice sheet and climate model. Use of a high-resolution regional climate model allowed processes in the narrow Antarctic ablation zone to be captured. As a result, feedbacks from retreating ice on the climate were simulated, which caused a positive feedback that increased surface melting. Gasson et al. (2016b) also simulated changes in the oxygen isotope composition of the Antarctic Ice Sheet, following Langebroek et al. (2010) but with 3-d isotope tracing. Modelled changes in the oxygen isotopic composition of the ice sheet could explain ~45% of the change in the oxygen isotope composition of seawater, with the rest caused by ice mass. Importantly, this feature is something that is not accounted for in the sea level reconstruction of Miller et al. (2020), which may partially explain the discrepancy between those studies.

Recent transient simulations of the Miocene ice sheet suggests a highly dynamic AIS occupied Antarctica during the Miocene (Stap et al., 2019). Stap et al. (2019) used the COSMOS climate model, which shows a particularly strong sensitivity to a change to Miocene boundary conditions, with an increase in the global mean temperature of 2.4°C with pre-industrial CO_2 relative to a control simulation (Stap et al., 2017, 2019; Stärz et al., 2017). They highlight that transient simulations show a smaller change in volume (~66%) when compared to equilibrium simulations. This difference is largely because, in the transient simulations (and reality), the ice sheet does not reach equilibrium with the surface climate because of a continually changing orbit. The strength of this effect will depend on how quickly the ice sheet responds to the climate forcing and highlights the possibility that equilibrium simulations may overestimate ice sheet response to climate forcing (Stap et al., 2019).

Other transient ice sheet simulations have been run across several geological epochs, including the Miocene (De Boer et al., 2010), using an inverse routine (Van de Wal et al., 2005) to force a one-dimensional ISM using the benthic $\delta^{18}O$ record (De Boer et al., 2010). This simple model produced a solution for surface temperature, ice volume and benthic $\delta^{18}O$ (De Boer et al., 2010) and atmospheric CO_2 (Stap et al., 2017). The major strength of this approach is that it calculated a self-consistent high-resolution record of Antarctic ice volume (De Boer et al., 2010), but it lacks a predictive capability as it is forced with proxy ice volume reconstructions. Therefore, the paleo data cannot be used to evaluate the model, nor can the model be used to make inferences about the future response of the AIS to climate changes. This second issue has been addressed by including atmospheric CO_2 in the inversion scheme (Stap et al., 2017). This study explored two components of the AIS hysteresis – height mass balance feedback and albedo feedback. Results showed that, for cooler climate states, albedo is the stronger feedback on the AIS, whereas in a warmer Miocene-like state the two feedbacks are balanced in strength (Stap et al., 2017).

Bed topography is one of the most important boundary conditions for an ISM and, until recently, an Antarctic topography for the Miocene has not been available. Studies have variously used the modern-day bed topography or created idealised 'Miocene-like' topographies by scaling between modern-day and an earlier reconstruction for the ice-free Antarctic of the late Eocene (Colleoni et al., 2018; Gasson et al., 2016b; Wilson et al., 2012a). Such studies used an idealised topography to show how the evolution for the continental shelf around Antarctica changed the ice sheet sensitivity to climate forcing, with the ice sheet effectively increasing its own sensitivity to ocean forcing. On a smaller scale, Halberstadt et al. (2021) investigated how the elevation and uplift of the TAM affected sensitivity to climate forcing in that sector. The publication of reconstructed Miocene Antarctic topography (Paxman et al., 2019b) and bathymetry (Hochmuth et al., 2020) allowed these processes to be explored in more detail and with greater confidence. Paxman et al. (2020) quantified how changing Antarctic bed has increased ice sheet sensitivity. Using an equivalent climate forcing, they found that the modern-day AIS is ~28% more sensitive to climate forcing than the mid-Miocene AIS. This result is sobering given the magnitude of sea level swings observed in the Miocene.

The climate forcing used in the studies discussed above has been ad-hoc, without a consistent set of boundary conditions (e.g., paleogeography) adopted in the various climate models used (Burls et al., 2021). A new initiative, MIOMIP (the Miocene model intercomparison project) is seeking to change this. This will allow a clearer examination of what impact the climate forcing and the climate model used has on simulations of the Miocene AIS and how polar amplification varies across models. This is something that has been successfully achieved for the Pliocene (Dolan et al., 2018) (see below).

9.3.2 Pliocene

Over the last decade, there has been a rapid increase in efforts to understand both the broader climate of the Pliocene from a modelling perspective (for example the Pliocene Model Intercomparison Projects [PlioMIP] 1 and 2) (Haywood et al., 2013, 2020) and to also provide well-constrained geological data to assess global modelling results (Dowsett et al., 2019; McClymont et al., 2020; Tierney et al., 2019). There have also been several different and evolving approaches to understand the stability of the AIS using computer models. Building on efforts from the geological proxy data community to understand in detail specific regions or glacial drainage basins (Cook et al., 2013; Naish et al., 2009; Passchier, 2011; Patterson et al., 2014; Williams et al., 2010), and reconstructions of Pliocene sea level (Dumitru et al., 2019; Dutton et al., 2015; Grant et al., 2019; Miller et al., 2012), ice-sheet modellers have primarily focussed on reconstructing the broad-scale extent of Antarctica during this warm interval and, in that, understanding potential areas of ice sheet instability for the past and future AIS and also the impact of changes on global mean sea level (de Boer et al., 2015; DeConto and Pollard, 2016; DeConto et al., 2021; Dolan et al., 2018; Gasson et al., 2016a; Golledge et al., 2017b).

Some of the first ISM simulations that focused on the mid-Pliocene Warm Period (mPWP; 3.264–3.025 Ma) identified the WSB and ASB as areas that were most susceptible to increased temperatures (Hill et al., 2007). These studies used climatological forcing fields (temperature and precipitation) from a coupled atmosphere-ocean general circulation model (GCM; HadCM3), which had been set up with Pliocene boundary conditions, to force an offline shallow-ice approximation (SIA) ISM over East Antarctica (Hill et al., 2007). Hill et al. (2007) simulated retreat which was broadly in line with Pliocene sea-level high-stand estimates at that time (Dowsett and Cronin, 1990). Using a similar modelling framework, Dolan et al. (2011) tested the impact of Pliocene-specific extremes in orbital forcing on reconstructions of the EAIS ice sheet size, showing a possible global sea level high stand of $\sim +21$ m under high Southern Hemisphere (SH) summer insolation scenarios. The area of EAIS retreat was again restricted to the WSB and ASB, but the overall approach was limited by the more simplistic modelling framework employed.

Simulations of the WAIS and marine-basins of the EAIS require models with the enhanced capability to simulate grounding line and ice shelf dynamics (termed shallow-shelf approximation dynamics), in addition to shallow ice approximation dynamics (SIA-SSA ISMs). Such a model, which included these dynamics and a parameterisation of the grounding line movement, was employed by Pollard and DeConto (2009) to simulate the evolution of the WAIS over the last 5 Ma. For the Pliocene portion of the record, the modelling results showed good agreement with the ANDRILL-1B records (Naish et al., 2009) and indicated rapid transitions with the WAIS collapsing periodically to leave only small isolated ice caps on the West Antarctic islands. However, for these simulations, only marginal ice sheet retreat in East Antarctica was demonstrated. This differs from other modelling results mentioned previously, using solely shallow ice approximation dynamics, with this difference explained by a stronger climate model forcing arising from the PRISM3 boundary conditions in the studies of Hill et al. (2007) and Dolan et al. (2011).

Inconsistencies between continental scale retrodictions of AIS extent led to efforts for a more structured assessment comparing modelled predictions of ice sheets for the Pliocene (e.g., the Pliocene Ice Sheet Modelling Intercomparison Project; PLISMIP) (de Boer et al., 2015; Dolan et al., 2018, 2012; Yan et al., 2016). de Boer et al. (2015) coordinated simulations from six SIA-SSA ISMs to simulate the complete Antarctic domain (including grounded and floating ice). HadCM3 climate fields from simulations comparable to those used in PlioMIP1 experiment (Bragg et al., 2012) were used to force the ISMs offline. de Boer et al. (2015) tested multiple scenarios within the modelling framework including the impact of different ISM initialisation methods (e.g., starting with a present-day AIS or with a reduced Pliocene AIS) and also the impact of different underlying bedrock topographies including Bedmap1 (Lythe et al., 2001) and Bedmap2 (Fretwell et al., 2013). One key result from this study was that none of the models using Bedmap1 predicted a retreat of the EAIS across the WSB and ASB as has been suggested by studies of marine sediments (Cook et al., 2013) and previous simulations. Our understanding of the vulnerability of different sectors of the ice sheet continues to evolve as improvements to bedrock topography reconstructions of Antarctica (Morlighem et al., 2020) and the identification of new subglacial basins (e.g., Recovery Basin). Reconstructions of the Antarctic topography during the geological past by accounting for the effects of glacial erosion and tectonics are also highlighting how the ice sheet sensitivity to basal conditions has changed through time (Colleoni et al., 2018; Paxman et al., 2020). Importantly, these reconstructions show minimal differences in Antarctic topography between the Pliocene and present-day, making the Pliocene a more suitable analogue for future warming than geological periods further back in time (Paxman et al., 2019b, 2020). However, Austermann et al. (2015) additionally show that factors such as dynamic topography related changes in bed elevation (not accounted for in the reconstructions of Paxman et al., 2019b) during the Pliocene can also have a significant effect on the predicted stability of ice in the marine-based WSB.

Following a similar experimental design to de Boer et al. (2015), Dolan et al. (2018) used results from PlioMIP1 (Haywood et al., 2011, 2013) to test the dependency of reconstructions of the mid-Pliocene AIS on the climate model used to provide the climate forcing fields to an ISM. The outputs of seven fully coupled atmosphere-ocean climate models were used to predict the whole AIS using three ISMs (of differing levels of complexity). Dolan et al. (2018) showed mean sea level changes relative to the pre-industrial control simulations of $+7.8 \pm 4.1$ m for simulations starting from a reduced (PRISM3) ice sheet, and $+2.4 \pm 3.5$ m for those starting from a present-day AIS configuration within the ISMs. It was demonstrated that there is a high level of model dependency on predictions of AIS extent and volume, and that this is exacerbated by imposed initial conditions (e.g., the PRISM3 AIS) that are a necessary requirement of the coupled atmosphere-ocean GCM modelling framework (e.g., choice of AIS configuration in the underlying climate model, the impact of bedrock topography and the initial conditions within the ISM). Despite the broad range in climate forcing from the different GCMs used, none of these simulations showed retreat of the EAIS grounding line.

Despite uncertainties in geological reconstructions of Pliocene sea level, the Pliocene has also been used as a calibration target for ISMs in order to predict future sea-level rise. ISM experiments that are set-up to perform well in simulating Last Interglacial (LIG) and Pliocene sea level show that Antarctica has the potential to contribute more than a metre of sea-level rise by 2100 in the future if emissions of greenhouse gases continue unabated (DeConto and Pollard, 2016). Simulations set up for the Pliocene that use climatological inputs from a Regional Climate Model (RCM) configured with a warm austral summer orbit, atmospheric CO_2 at 400 ppmv, and 2°C of ocean warming imposed within the model, suggest that Antarctica can contribute up to 11.3 m of sea-level rise (DeConto and Pollard, 2016). Significant ice sheet retreat in these simulations occurred within marine basins of WAIS and EAIS (particularly around the present-day Ninnis, Mertz, Totten and Recovery glaciers in East Antarctica). One key difference between this work and previous modelling was the implementation of new processes, based on theoretical work (Bassis and Walker, 2012), which link atmospheric warming with hydrofracturing of buttressing ice shelves and the structural collapse of marine terminating ice cliffs (termed the Marine Ice Cliff instability or MICI) (DeConto and Pollard, 2016; Pollard et al., 2015). However, observational evidence for MICI is minimal, largely because the conditions that would trigger it have not yet been met in Antarctica. There is also considerable debate about the theoretical basis of this process, in particular whether ice shelves can be removed rapidly enough to leave behind vulnerable ice cliffs (Robel and Banwell, 2019), whether surface meltwater can be efficiently removed from the ice shelf surface (Bell et al., 2017), and what the critical failure height of sheer ice cliffs is and how this depends of the extent of crevassing (Clerc et al., 2019). Marine-terminating glaciers of the Greenland Ice Sheet such as Jackobshavn Isbrae provide better insight into whether this process is plausible (Parizek et al., 2019). However, the relatively narrow channels of these glaciers, which have a strong seasonal cycle in calving which is arrested by ice melange during winter (Joughin et al., 2020), mean that they are not directly comparable to calving fronts of Antarctic glaciers, which can be tens of kilometres wide. There is also debate about whether these additional mechanisms are needed to explain Pliocene sea levels (Edwards et al., 2019; Gasson et al., 2016a; Gasson and Keisling, 2020). Low end estimates of Pliocene sea level (Miller et al., 2019; Raymo et al., 2018) can be explained with existing ISM physics (Edwards et al., 2019). Although there is continued debate around MICI, more recent modelling by DeConto et al. (2021) has reiterated the importance of ice-cliff calving in order to improve geological data-model comparisons and to constrain ISM physics in climate states that are warmer than seen in the satellite era. DeConto et al. (2021) show a maximum GMSL contribution of 20.85 m from their Pliocene Antarctic simulation, which is consistent with recent geological observations (Dumitru et al., 2019; Grant et al., 2019). In this study, sea level estimates from the Pliocene and LIG were used alongside observed ice loss between 1992 and 2017 (Shepherd et al., 2018) to predict future scenarios of ice sheet retreat for the Antarctic ice sheet. A median value of 34 cm by 2100 is predicted for the AIS contribution to GMSL for global mean warming limits of +3°C (DeConto et al., 2021). Improved model physics and revised atmospheric forcing in this study relative to DeConto and Pollard (2016) explain the discrepancy in future estimates of change at 2100, although the contribution to GMSL does reach 1 m by 2125. Without the Pliocene sea level constraint, these future projections have larger uncertainties (DeConto et al., 2021).

Golledge et al. (2017b) have also investigated Antarctic climate and ice sheet configuration during the early Pliocene (c.4.23 Ma). Using a regional climate model (RegCM3) coupled offline to the PISM (Parallel Ice Sheet Model) three-dimensional, thermodynamic SIA-SSA ISM, alongside a new synthesis of high-latitude paleoenvironmental proxy data to define a probable climatic envelope, they simulate an AIS contribution of 8.6 ± 2.8 m to GMSL. In the simulations presented by Golledge et al. (2017b), substantial grounding-line retreat is evident in the WSB, but the Aurora and Recovery basins are less affected by early Pliocene conditions (Dolan et al., 2011). The ISM used in this study includes a sub-grid grounding line melt mechanism, which is used as a way of compensating for the low spatial resolution of the model. However, there is debate about this parameterisation, and experiments with higher resolution models have shown that it may overestimate ocean melting at the grounding line and generate numerical errors (Cornford et al., 2020; Seroussi and Morlighem, 2018). In their simulations the WAIS is deglaciated, with only some ice remaining on the West Antarctic islands. These results are corroborated by evidence for vegetation at the margins of a fjord in the Taylor Valley, Southern Victoria Land during the early Pliocene (Ohneiser et al., 2020). Model-based predictions for vegetation in this area are most compatible with simulations that show a complete collapse of the WAIS and significant retreat of the EAIS from the subglacial basins (including Wilkes) (Ohneiser et al., 2020).

When the modelling and the data communities work in combination to tackle a challenge there is potential for significant progress in our understanding of the nature of specific characteristics of the Pliocene AIS over the coming years. For example, cosmogenic nuclide exposure ages alongside ice sheet modelling have been used to suggest that the EAIS continental interior could have been up to 600 m higher than present around the mid-Pliocene (Yamane et al., 2015). Sediment provenance records derived from $^{40}Ar/^{39}Ar$ ages of hornblende grains form ODP Site 1165 near Prydz Bay have been combined with iceberg trajectory modelling to understand the record of IRD during the Pliocene (Cook et al., 2014). This demonstrated, that declining SSTs over the period of the mid-Pliocene allowed Wilkes Land icebergs to travel further before melting.

Oxygen isotope data (Miller et al., 2012; Winnick and Caves, 2015) combined with simulations of the oxygen isotope composition of the AIS for a range of configurations have also added to our understanding of the impact of changes to the Pliocene AIS on sea level (Gasson et al., 2016a). Gasson et al. (2016a) were able to identify ice sheet configurations that are consistent with the oxygen isotope record and conclude that a maximum contribution of ~13 m to Pliocene sea level highstands should be expected from Antarctica. However, there are additional uncertainties in this method that hinder accurate results (Raymo et al., 2018), determining the temperature contribution to changes in the oxygen isotope composition of benthic foraminifera is particularly difficult. There is no current consensus on the Pliocene sea level high stand using this approach.

Finally, there has been a recent focus on reconstructing the transient nature of Pliocene ice sheets (Berends et al., 2019; de Boer et al., 2017). Berends et al. (2019) used a coupled hybrid ice-sheet-climate model and a matrix method using fields from snapshot climate model simulations to reconstruct ice sheet geometry and sea level over the Late Pliocene (3.6 to 2.58 Ma). Combined simultaneous simulations of the Northern Hemisphere ice sheets and AIS, suggested a sea level drop of ~16 m at MIS M2, associated with a significant expansion of the AIS and surrounding ice shelves.

9.4 Synthesis/summary of key climate episodes and transitions in Antarctica through the Miocene and Pliocene

9.4.1 Early to mid-Miocene

While there is evidence that Antarctica's ice margins periodically extended into marine settings prior to the Neogene (Barrett, 1989; Carter et al., 2017; Galeotti et al., 2016; Gulick et al., 2017; Levy et al., 2000; Wilson et al., 1997), geological data suggest the first major expansion of marine ice sheets in Antarctica occurred at the Oligocene/Miocene transition (see Naish et al., 2021). Sedimentary facies analysis from drill cores (Naish et al., 2001a, 2001b), seismic data (Sorlien et al., 2007b) and geochemical proxies (Duncan et al., 2019) show that large regions of the AIS advanced into marine environments in the Ross Sea during the latest Oligocene and formed a major disconformity at ~23 Ma (Florindo et al., 2005; Kulhanek et al., 2019). Furthermore, an increase in mass transport deposits and ice rafted debris in the latest Oligocene a major regional unconformity (WL-U5) corresponding to a major hiatus within lower Miocene rocks at Site U1356 (Bijl et al., 2018a, 2018b; Escutia et al., 2011a; Hartman et al., 2018) suggests significant environmental change occurred across the O/M at the George V and Adélie Coasts and supports a shift to a glacial regime in which marine-ice sheet advance became more frequent. Together these data suggest that the O/M transition and Mi1 oxygen isotope event record the first extensive advance of Antarctica's ice sheets and ice caps into marine environments (Bart and De Santis, 2012; De Santis et al., 1995; Escutia and Brinkhuis, 2014; Escutia et al., 2011a; Levy et al., 2019; Sorlien et al., 2007b).

Climatic warming in the earliest Miocene ended the transient glacial advance recorded by the Mi1 isotope event (Naish et al., 2021), although we note that pollen data suggest climatic conditions in the Ross Sea embayment in the early Miocene were similar to those that characterised the late Oligocene (Kulhanek et al., 2019; Prebble et al., 2006). Proxy data suggest atmospheric CO_2 increased in the early Miocene, from values <300 ppm that characterised the Mi1 oxygen isotope event, to average concentrations around 400 ppm (Figs 9.1 and 9.13). These data support other studies that suggest 400 ppm is a threshold for marine glaciation, below which climatic conditions in the high southern latitudes are conducive to the advance of marine-based ice sheets (Gasson et al., 2016b; Halberstadt et al., 2021; Levy et al., 2019, 2016; Naish et al., 2009). Evidence for episodic advance of the AIS margin well beyond the present coastline and into glacial marine settings during the early and middle Miocene is captured by U-shaped valley features within lower to middle Miocene strata (RSS2 and RSS3) offshore of the outlet glaciers along the Southern Victoria Land coast (Brancolini et al., 1995). Thick prograding wedges characterise strata within RSS3 at many locations across the Ross Sea (De Santis et al., 1999; Pérez et al., 2021a) and in seismic sections through the continental shelves offshore George V Land and the Adélie Coast (Eittreim et al., 1995, 2011a) and suggest glacimarine and subglacial depositional environments across the continental shelves during the early to early middle Miocene.

Whereas the age of Sirius Group rocks at locations across the TAM is mostly poorly constrained (Barrett, 2013; Scherer et al., 2016; Webb et al., 1984), the location of diamictite-rich units at Table Mountain inland of, and higher than, middle Miocene deposits in the Asgard and Olympus Ranges and Friis Hills, suggest they were potentially deposited during the early Miocene (Lewis and Ashworth, 2015). We hypothesise that these glacial sediments were deposited during episodes of ice sheet expansion that coincide with disconformities and diamictite-dominated units at Sites AND-2A (Levy et al., 2016; Passchier et al., 2011) and U1521 (McKay et al., 2019). One of these episodes of major marine ice sheet expansion occurred between 18 and 17 Ma and coincides with a significant fall in eustatic sea level and an increase in Earth system sensitivity to obliquity (Levy et al., 2019). Glacial sediments deposited at the AND-2A site during this time interval reflect a polar climatic and glacial regime (Passchier et al., 2011) and show that 'cold-based' glacial events and potentially arid climatic conditions occurred in Antarctica well before the MMCT (Fig. 9.13E).

Evidence from offshore Prydz Bay (Cooper and O'Brien, 2004; Haywood et al., 2008; Williams and Handwerger, 2005) and the ASB (Gulick et al., 2017) also suggest the EAIS margin was dynamic during the early Miocene. At times the ice sheet advanced across the continental shelf and others it retreated inland of the terrestrial coastal margin (Williams and Handwerger, 2005). Drill core data from the Ross Sea Embayment indicate environmental conditions were highly variable throughout the early Miocene and show that sediments were deposited under a range of glacial regimes at different times to include terrestrial, subpolar, and polar climates (Fielding, 2018; Fielding et al., 2011; Levy et al., 2016; Passchier et al., 2011) (Fig. 9.13). Large areas of West Antarctica's land surface were either subaerial and/or significantly shallower than today in the early Miocene and this topography may have enabled ice to more easily grow and expand across the Ross Sea continental shelf (Colleoni et al., 2018; Paxman et al., 2020; Pollard and DeConto, 2020). However, climatic conditions must have occasionally been cold enough to allow thick ice margins to advance into relatively deep marine environments (Gasson et al., 2016b; Kulhanek et al., 2019; Levy et al., 2019; McKay et al., 2019). TEX_{86}-based data indicate surface water temperatures were at times similar to Holocene values and coastal environments may have supported multi-season sea ice and/or small ice shelves (Levy et al., 2016). During warmer intervals surface water temperatures reached 5°C to 6°C ± 2.8°C and tundra vegetation occupied regions from the coast (Griener et al., 2015; Kulhanek et al., 2019; Levy et al., 2016) to at least 60 km inland within the MDV (Lewis and Ashworth, 2015).

9.4.2 Miocene Climate Optimum

Beginning at ~17 Ma, the onset of the MCO is reflected in the deep sea benthic foraminifera records by a major decrease in $\delta^{18}O$ values and increase in $\delta^{13}C$ values (De Vleeschouwer et al., 2017; Westerhold et al., 2020; Woodruff and Savin, 1991). While the deep-sea data and a range of paleoclimate proxies from across the globe indicate a shift to warmer conditions at this time, environmental variability through the MCO remained high. This variability is most clearly reflected by large excursions (up to 1.5‰) in the deep sea $\delta^{18}O$ data that occur on glacial–interglacial timescales and require either large changes in ice volume (Miller et al., 2020) and/or bottom water temperature (Lear et al., 2015; Modestou et al., 2020; Shevenell et al., 2008). So, while the average climate state was warm during the MCO, environmental changes through this key interval were large and frequent.

Antarctic records of environmental conditions throughout the MCO primarily come from three key drill cores: AND-2A and IODP Site U1521 from the Ross Sea, and IODP Site U1356 from offshore the WSB. Interestingly, terrestrial records through the MCO have yet to be recovered, although snapshots from the end of the optimum occur in the Friis Hills (Chorley, 2021; Verret et al., 2020) and Asgard Range (Marchant, 1993) in the MDV. The onset of the MCO occurs at ~550 mbsf in AND-2A and is somewhat unremarkable as there are no obvious changes in sedimentary facies or other environmental proxies (Fielding et al., 2011; Levy et al., 2016; Passchier et al., 2011). However, sedimentary facies and sequences deposited at AND-2A during the MCO are lithologically diverse and generally mud-rich (Fielding et al., 2011; Passchier et al., 2011). Importantly, there is no evidence that glaciers extended beyond the drill site during the MCO (Fielding et al., 2011; Levy et al., 2016; Passchier et al., 2011). Environmental data through the interval between ~15.8 and ~14.6 Ma are missing in a disconformity in the AND-2A core (Levy et al., 2016). The warmth of the MCO is reflected more clearly at Site U1521 in the central Ross Sea where a distinctive change in sedimentary facies occurs at ~210 mbsf. This depth marks the boundary between lithostratigraphic Units IV that comprises dark grey massive diatom-bearing clast-poor sandy diamictites and lithostratigraphic Unit III, a 124 m-thick sequence of diatom-bearing/rich mudstones (McKay et al., 2019). The paucity of coarse sediments in this interval suggests the ice sheet margin was restricted to the terrestrial coastal region across the Ross Embayment between 17 and 16 Ma at least. The environmental record between ~16 and ~14.6 Ma at Site U1521 is missing in a disconformity (McKay et al., 2019). The stratigraphic interval spanning the MCO at Site U1356 is characterised by diatomaceous and cherty mudstones lacking outsized clasts (Escutia et al., 2011a; Sangiorgi et al., 2018). These sediments were deposited within a distal channel levee setting with subsequent minor reworking by bottom currents of variable strength and bioturbation. There is little evidence for ice rafting over the drill site during the MCO (Sangiorgi et al., 2018), which suggests the coastal marine margins across the WSB remained ice free during the MCO. However, we note that core recovery through the MCO was relatively low and inferences regarding ice sheet extent and variability at this site must consider that clast-rich intervals may not have been recovered.

Sedimentological data from three drill sites offer evidence that the AIS did not extend far beyond the terrestrial coastal margin between at least 17 and 16 Ma. However, diamictites at the base of each sedimentary cycle throughout the section spanning the MCO in AND-2A indicate ice reached the Southern Victoria Land coast during glacial episodes. Coupled ice sheet and climate modelling experiments using Miocene boundary conditions show the AIS remained land-based during cold (glacial) astronomical configurations when CO_2 forcing was set at 460 ppm or greater (Halberstadt et al., 2021). However, the simulated AIS could extend across Antarctica's continental shelves in 'cold orbit' simulations where lower CO_2 concentrations are used (Gasson et al., 2016b; Halberstadt et al., 2021; Paxman et al., 2020). These results suggest atmospheric CO_2 concentrations were unlikely to have dropped below 400 ppm during the MCO, as suggested by proxy studies (Greenop et al., 2014).

In contrast to sedimentological data that inform maximum glacial extent, direct geological constraints on minimum AIS extent and volume during the MCO are few to absent. Geochemical proxy data (BIT index) from Site U1356 suggest that soil was able to form on extensive ice-free regions along the near-coastal lowlands at the margins of the WSB during intervals of ice retreat (Sangiorgi et al., 2018). Pollen data indicate these regions were covered by woody vegetation dominated by southern beech and *Podocarpaceae* conifers growing in a temperate climate with MATs between 5.8°C and 13°C and mean summer temperatures >10°C (Sangiorgi et al., 2018). Offshore surface-water temperature reconstructions based on TEX_{86} suggest temperatures of 11.2°C–16.7°C ± 2.8°C. Intervals that indicate peak warmth and more extensive ice margin retreat from the coast in AND-2A are rare. One such interval contains thick-shelled costate scallops and venerid clams that indicate that water temperatures were at least 5°C warmer than in the Ross Sea today (Beu and Taviani, 2013). These observations are supported by TEX_{86} and Δ_{47} data, which indicate surface water temperatures reached a maximum between 7.0°C ± 2.8°C and 10.4°C ± 2.5°C (Levy et al., 2016). Another interval contains abundant diatoms and dinoflagellates and minimal gravel and sand clasts, which suggests the glacier grounding line retreated inland from the coast as ice did not calve into the marine environment. Pollen and spores are abundant in this interval and indicate a coastal vegetation of mossy tundra with shrub podocarps and southern beech and suggest a cool terrestrial climate (up to 10°C January mean air temperature) (Levy et al., 2016; Warny et al., 2009).

While these glacial marine data indicate regional temperature increases and coastal ice margin retreat during peak warm intervals of the MCO, they do not provide constraints on the extent of inland ice margin retreat. However, coupled numerical climate and ice sheet modelling experiments offer a means to examine ice sheet response to maximum warming during the MCO. Results show significant volumes of ice remained on Antarctica under warm astronomical configurations and 'extreme' greenhouse gas forcing (>840 ppm) (Gasson et al., 2016b; Gasson and Keisling, 2020; Halberstadt et al., 2021; Paxman et al., 2020). It appears that the terrestrial regions of the AIS are sensitive to warming under CO_2 conditions that exceed 540 ppm but they do not completely melt even under high CO_2 (>840 ppm) and warm orbits (Gasson et al., 2016b; Halberstadt et al., 2021). Ice that sits on West Antarctica and the WSB and ASB is particularly susceptible to warming. The minimum simulated ice volume produced on reconstructed Miocene topographies under high (840 ppm) atmospheric CO_2 concentration and a warm (interglacial) astronomical configuration ranges between 14.1×10^6 km^3 (Gasson et al., 2016b) and 18.94×10^6 km^3 (Paxman et al., 2020). These volumes are equivalent to ~53% and ~71% of the modern AIS (26.54×10^6 km^3 excluding ice shelves) (Fretwell et al., 2013). If we assume that maximum ice volume during glacial episodes in the MCO was no more than modern (which is generally consistent with drill core data), then the likely maximum glacial–interglacial ice volume change during the MCO ranges between 27 and 17 m sea level equivalent. These outputs are consistent with records from the Marion Plateau that indicate sea level amplitude changes of 27 ± 1 m at 16.5 and 15.4 Ma (John et al., 2011), and the New Jersey margin with maximum amplitudes of between 20 and 30 m through the MCO (Kominz et al., 2008, 2016) (Fig. 9.3E). These results contrast with a recent study that suggests complete collapse of the AIS occurred during the MCO (Miller et al., 2020).

The moderate changes in ice volume indicated by ice sheet modelling studies and sea level records require significant changes in bottom water temperature to explain the large (up to 1.5‰) excursions in deep sea $\delta^{18}O$ records. Paleotemperature records from ODP Site 1172 in the Southern Ocean indicate bottom water temperature variations up to 3°C occurred on astronomical time scales through the MCO (Shevenell et al., 2008) (Fig. 9.3E). Relatively low resolution records derived from ODP Site 806 (western equatorial Pacific) using similar Mg/Ca-based proxy techniques indicate bottom water temperatures varied by up to 6°C over million year time scales (Lear et al., 2015). These temperature estimates are also supported by Δ_{47} (clumped isotope) data from ODP Site 761 which indicate bottom water temperatures in the Indian Ocean were up to 9°C warmer than today during the MCO and varied by up to 4°C through the optimum (Modestou et al., 2020). These bottom waters were likely produced at high southern latitudes including in the Ross Sea Embayment and offshore George V and Adélie Land. Data from AND-2A indicate surface water temperatures ranged between −1.8°C and 10°C in the Ross Sea during the MCO. It seems reasonable to infer that the temperature of bottom water derived from these regions varied by 3°C to 6°C on interglacial time scales. This magnitude of change in BWT can explain between 0.63‰ and 1.26‰ of the change in deep sea $\delta^{18}O$ records and suggests that estimates of ice volume changes of 27 m sea level equivalent through the MCO are reasonable (e.g. Bradshaw et al., 2021).

9.4.3 Miocene Climate Transition

The onset of the middle Miocene Climate Transition is reflected in deep sea $\delta^{18}O$ records by a relatively subtle increase in glacial values and a trend towards higher average values beginning at ~14.8 Ma (Holbourn et al., 2014). The MMCT lasted for approximately one million years and culminated in a large (~0.4‰) stepped increase in average $\delta^{18}O$ values at 13.8 Ma that coincides with the final major increase in deep sea $\delta^{13}C$ events (CM6) at the end of the Monterrey carbon excursion (Vincent and Berger, 1985). This last major step in the MMCT coincides with a drop in atmospheric CO_2 concentration below 300 ppm (Badger et al., 2013b) (Fig. 9.12A). It has long been inferred that the MMCT records the expansion and stabilisation of the EAIS and the cooling and aridification of Antarctica's terrestrial environments (Clapperton and Sugden, 1990; Sugden and Denton, 2004; Sugden et al., 1993). This inference is primarily based on extensive evidence from deposits at high elevation (>1000 m) in the TAM (Sugden et al., 1993) but also relies on the deep sea $\delta^{18}O$ data. Of particular importance to the 'stabilist' perspective are 15 million year old ash deposits in the Asgard Range of the MDV that suggest little to no landscape modification has occurred at these high elevation locations since the end of the MCO (Marchant, 1993). This inference is supported by surface exposure studies that utilise cosmogenic nuclides, which indicate little erosion has occurred at high elevations in the TAM over the past 15 million years (Spector and Balco, 2021; Valletta et al., 2015). Arguments for expansion of the EAIS during the MMCT are based on geomorphological evidence for glacial overriding including subglacial potholes in the Asgard Range (Sugden et al., 1991) and channels that form the Labyrinth in upper Wright Valley (Lewis et al., 2006). However, these results are debated, with another study suggesting that cosmogenic nuclide data indicate relatively high erosion rates at high elevations in the MDV since the middle Miocene (Middleton et al., 2012). These results also require a much younger or more complex history for the formation of the channelised landforms in the Asgard Range and upper Wright Valley (Middleton et al., 2012).

Here we summarise environmental events in the Ross Sea and offshore George V Land that occurred through the MMCT and show that while climate cooled and the AIS advanced to the continental margin, the ice sheet expansion was a transient event (Figs 9.12 and 9.13). Disconformities in the AND-2A and Site U1521 cores at ~14.8 and 14.6 Ma, respectively, indicate ice grounded and expanded beyond the terrestrial coastal margin and across the Ross Sea continental shelf at this time. These disconformities correlate with Ross Sea Unconformity 4 (RSU4), which is the oldest spatially extensive erosional surface preserved within Neogene-aged seismic sequences across the Ross Sea (De Santis et al., 1995, 1999; Pérez et al., 2021a). The transition to a polar glacial regime at this time is reflected by a shift to facies sequences dominated by mud-poor massive diamictites above the disconformity in the AND-2A core (Passchier et al., 2011). Increases in deep sea $\delta^{18}O$ data align with the drill core and seismic evidence for more extensive advances of grounded ice into marine settings during glacial episodes. Furthermore, the Marion Plateau sea level record indicates maximum sea level fall of 33 ± 5 m at 14.7 Ma, which is slightly more than the maximum recorded amplitude of sea level fall during the MCO (−27 ± 1 m) and supports the near field Antarctic records and deep sea $\delta^{18}O$ data. However, it is critical to recognise that the deep sea $\delta^{18}O$ data also show ice volume decreased again during interglacial episodes through the MMCT. Importantly, flora and fauna from fossil-rich glacial–fluvial and glacial–lacustrine deposits in the Friis Hills indicate interglacial climate remained relatively warm and wet until at least 14 Ma (Chorley, 2021; Lewis and Ashworth, 2015; Verret et al., 2020) and suggest ice margins retreated from the coast. These observations indicate the AIS was highly variable throughout much of the million-year-long MMCT.

Major disconformities in AND-2A and Site U1531 post-date ~14.4 and ~14 Ma, respectively, and form major stratigraphic breaks between two and three million years in duration (Figs 9.7 and 9.13D). These major disconformities correlate with RSU4, which likely formed as AIS advance during glacial episodes became progressively more extensive during the later stages of the MMCT. By 14.2 Ma maximum values of 2.2‰ in deep sea $\delta^{18}O$ data indicate the AIS was growing to its greatest extent and volume since the onset of the MMCT. However, the AIS still underwent significant retreat during interglacial events. A large decrease in deep sea $\delta^{18}O$ records at ~13.9 Ma may capture the 'last gasp' of relative warmth and significant inland retreat of terrestrial ice sheets during warm interglacial episodes in the middle Miocene (Fig. 9.12B). Fossil flora and fauna preserved in lacustrine deposits at Friis Hills and Mt Boreas indicate climate remained warm

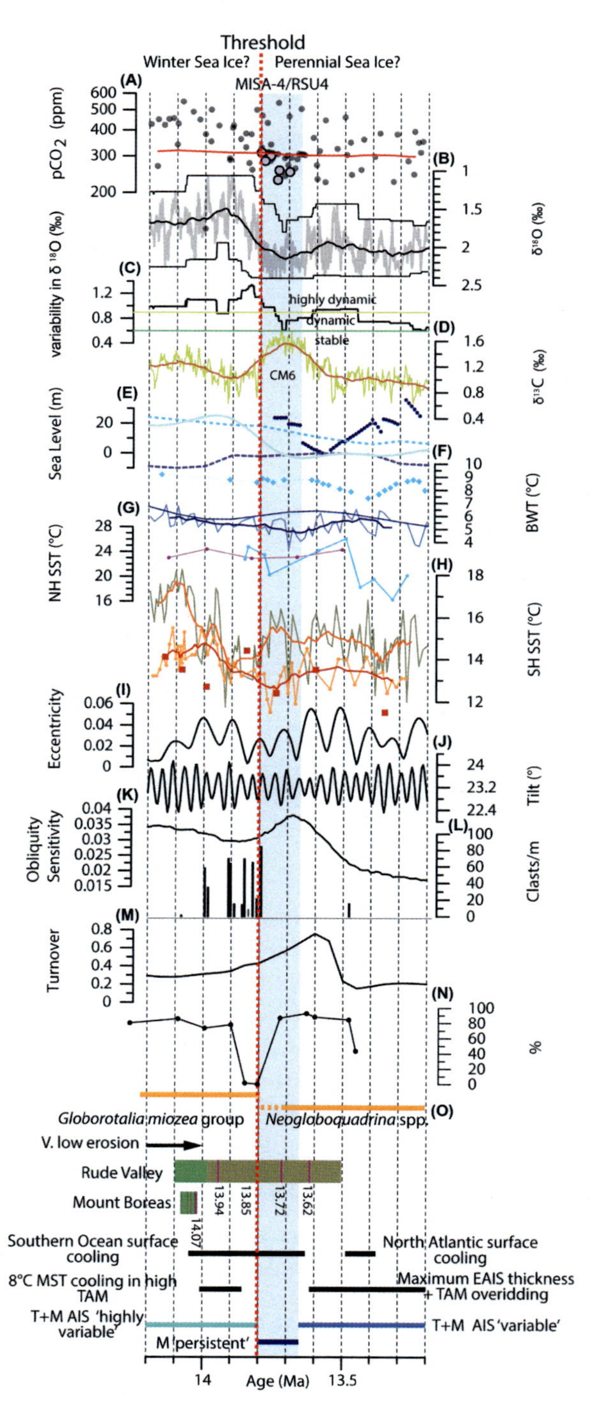

FIGURE 9.12 Environmental data from the middle Miocene (14.2 to 13.2 Ma). (A) Atmospheric CO_2 compilation comprises data from a range of proxies outlined in Fig. 9.1. Solid red line displays a two million year moving average. Purple dots highlight boron proxy data from Malta (Badger et al., 2013b). (B) Splice of deep sea benthic foraminifera $\delta^{18}O$ data (light

(Continued)

enough at this time to support tundra vegetation at relatively high inland sites (Lewis and Ashworth, 2015; Lewis et al., 2008; Verret et al., 2020). Significant increases in IRD are recorded at IODP Site U1356 and ODP Site 1165 between 14.1 and 13.8 Ma (O'Brien et al. 2001; Sangiorgi et al., 2018). Geochemical provenance analysis of clasts in these cores indicates the EAIS margin

---

◀ grey) reflect changes in ice volume and deep/bottom water temperature (De Vleeschouwer et al., 2017). Solid black line displays a 150 kyr moving average. Maximum and minimum values are determined within each 150 kyr window and create the envelope (black lines) that bound the $\delta^{18}O$ data. (C) A $\delta^{18}O$ variability 'index', interpreted as a measure of glacial/interglacial variability, is determined by subtracting the maximum values in (B) from the minimum value, for each 150 kyr window. We indicate arbitrary 'thresholds' in these data (green and yellow solid lines) and suggest that values below 0.6‰ reflect a relatively 'stable' high latitude environment and values that exceed 0.9‰ indicate a highly dynamic environment with large changes in BWT and/or ice volume over glacial-interglacial time scales. (D) Yellow line shows high resolution deep sea benthic foraminifera $\delta^{13}C$ splice (Westerhold et al., 2020) and carbon maxima event CM6 (Vincent and Berger, 1985; Woodruff and Savin, 1991). Solid deep orange line displays a 150 kyr moving average. (E) Sea level curves are from Kominz et al. (2008) (pale blue dashed line), (Kominz et al., 2016) (blue dots), Miller et al. (2005) (blue dashed line) and Miller et al. (2020) (pale blue line). (F) BWT data from ODP Site 761 (Lear et al., 2010) (light blue), ODP Site 1171 (Shevenell et al., 2008) (blue), and derived from the compilation of Cramer et al. (2011) using Equation 7a (dark blue). Solid line through data from ODP Site 1171 displays a 9-pt running average. (G) SST data for the Northern Hemisphere from ODP Site 608 (Super et al., 2018) (light blue) and ODP Site 902 (Super et al., 2020) (pink). (H) SST data for the Southern Hemisphere from ODP Site 1171 using the Mg/Ca proxy (Shevenell et al., 2004) (dark brown) and the TEX_{86} proxy (orange) (Leutert et al., 2020), using the calibration of Ho and Laepple (2016). Solid lines through the Mg/Ca and TEX_{86} data display a 9-pt running average. Red squares display temperature estimates derived from clumped isotopes (Δ_{47}) (Leutert et al., 2020). (I) Astronomical eccentricity and (J) axial tilt (obliquity) from (Laskar et al., 2011). (K) obliquity sensitivity after Levy et al. (2019). (L) Ice rafted debris from IODP Site U1356 (Sangiorgi et al., 2018). (M) Rate of diatom species extinction and speciation (turnover) after Crampton et al. (2016). (N) Percentage protoperidinioid dinoflagellate cysts at IODP Site U1356 (Sangiorgi et al., 2018). (O) Orange bars indicate significant changes in foraminifera assemblage composition in the Southern Ocean (Verducci et al., 2009). The coincident changes in diatom (M), dinocyst (N), and foraminifera (O) assemblages suggests an increase in sea ice persistence and extent occurred after 13.8 Ma. A decrease in erosion rates in the TAM after 15 Ma is suggested by cosmogenic nuclide studies (Spector and Balco, 2021). Generalised stratigraphic data from the Olympus Range in the McMurdo Dry Valleys indicate a transition from wet-based, fossiliferous, glacial-lacustrine deposits (green), to dry-based glacial drifts (brown) occurred between 14 and 13.9 Ma based on tephra chronology (pink layers) (Lewis et al., 2008; Lewis et al., 2007). Summary of climatic events (black bars) are based on environmental proxies within this figure [interval of maximum EAIS thickness inferred from regional geomorphology in the MDV (Lewis et al., 2006; Sugden et al., 1991)]. Blue vertical band indicates interval characterised by major unconformities in Ross Sea drill cores and the occurrence of erosional seismic surface RSU4. Antarctic Ice Sheet variability and extent (*T*, terrestrial; *M*, Marine) inferred from drill core data (Levy et al., 2016; McKay et al., 2019; Sangiorgi et al., 2018), modelling studies (Gasson et al., 2016b; Halberstadt et al., 2021), obliquity sensitivity (K), and high resolution $\delta^{18}O$ records (B). Red dashed vertical line highlights likely threshold in the climate system across which atmospheric cooling enhanced sea ice and marine ice sheet expansion (see text for discussion).

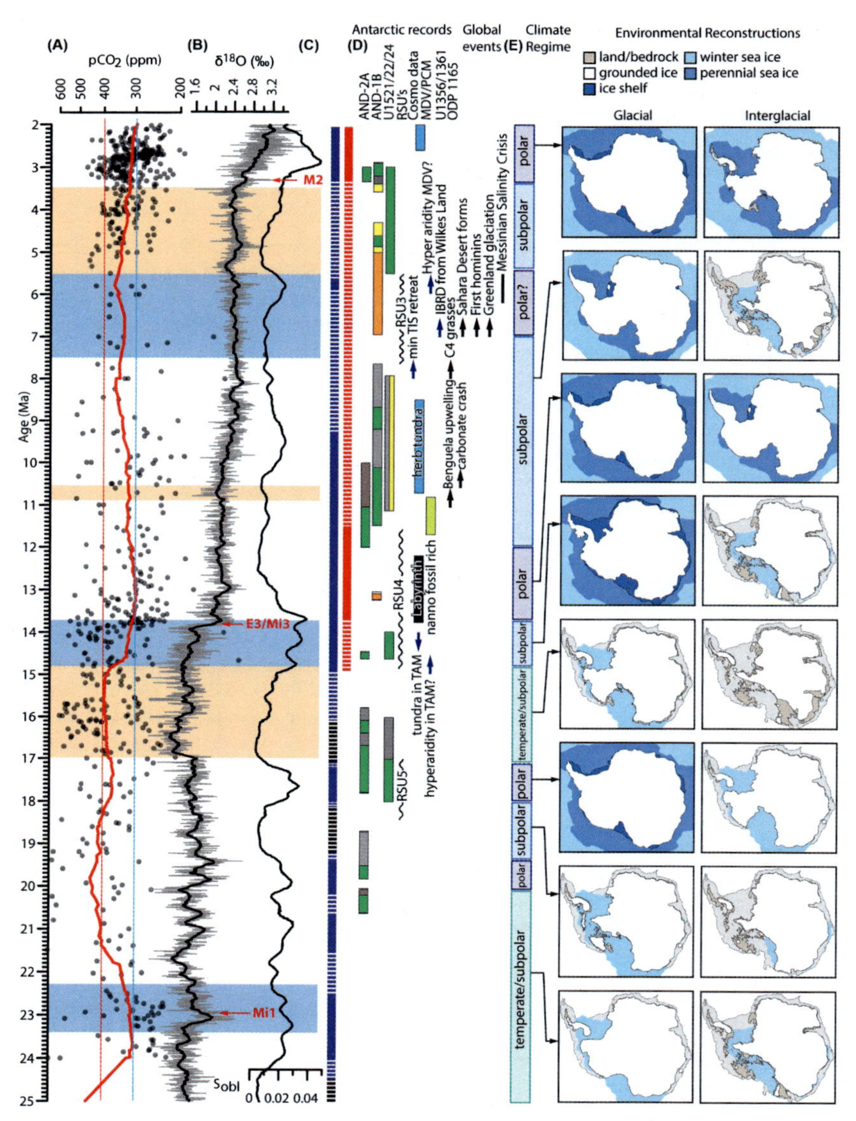

FIGURE 9.13 Summary of Antarctic ice sheet and high southern latitude climate change through the Miocene and Pliocene. (A) Atmospheric CO_2 compilation comprises data from a range of proxies outlined in Fig. 9.1. Solid red line displays a two million year moving average. (B) Splice of deep sea benthic foraminifera $\delta^{18}O$ data (light grey) reflect changes in ice volume and deep/BWT (De Vleeschouwer et al., 2017). Solid black line displays a 150 kyr moving average. (C) Obliquity sensitivity (S_{obl}) calculated from the obliquity forcing variance and the $\delta^{18}O$ record and is interpreted as a proxy for ocean-ice sheet connectivity (Levy et al., 2019). Coloured bars reflect the ice-sheet extent: T, terrestrial (black); T + M, terrestrial and marine (blue), PSI, perennial sea ice (red); dashed lines, transitional. (D) Graphical summary of key

(*Continued*)

repeatedly retreated inland across the WSB and Lambert Graben and advanced again into marine environments delivering IRD-rich ice to the drill sites (Pierce et al., 2017). SSTs in the Southern Ocean dropped by 2°C to 6°C between ~14.1 and 13.8 Ma (Leutert et al., 2020; Shevenell et al., 2004).

Sea level fell by 59 ± 6 m at ~13.9 Ma (John et al., 2011), which suggests the AIS expanded to the edge of the continental shelf at this time. Coupled ice sheet and climate modelling experiments show that ice volume in the mid Miocene could have increased by $\sim 18.3 \times 10^6$ km^3 between warm interglacial climates under moderate CO_2 (500 ppm) and cold glacial climates under low CO_2 (280 ppm) (Gasson et al., 2016b). This change in volume would cause a 36 m fall in sea level (Gasson et al., 2016b). These results suggest that either the observed magnitude of sea level fall at the Marion Plateau is larger than eustatic (i.e. influenced by local geodynamic effects) (Austermann et al., 2015, 2017), minimum (interglacial) AIS volumes were less than modelled and/or glacial expansion was greater (climate may have been colder than simulated), or that ice sheets may have also grown in the northern hemisphere (DeConto et al., 2008).

Deep sea $\delta^{18}O$ records and geological, paleoceanographic and paleoecological data from the Antarctic margin and Southern Ocean show a major environmental threshold was crossed at ~13.8 Ma (Fig. 9.12). Summer mean air temperature in the TAM cooled by 8°C between 14.07 ± 0.05 and 13.85 ± 0.03 Ma, tundra vegetation disappeared from high elevations and glacial systems became dry-based (Lewis et al., 2007, 2008). Bottom water temperature decreased by ~2°C in the Southern Ocean (Shevenell et al., 2008) between ~14 and 13.7 Ma and by 2.9°C ± 2.5°C in the Indian Ocean by 13.5 Ma (Modestou et al., 2020). Immediately after 13.8 Ma, an episode of low variability in deep sea $\delta^{18}O$ data (Fig. 9.12B and C) suggests the AIS 'stabilised' at the continental shelf margin and persisted at this position for ~150 kyrs. This period of maximum marine based ice extent coincides with the CM6 $\delta^{13}C$ excursion and a peak in obliquity sensitivity (Levy et al., 2019) (Fig. 9.12D and K). A major turnover pulse in diatom species (an episode of coincident species extinction and appearance) also occurred during this interval (Fig. 9.12M) and may have happened, in part, because marine based ice sheets covered Antarctica's continental shelves for an extended period (Crampton et al., 2016).

Climate and ice sheet modelling studies show that extensive sea ice forms around Antarctica when atmospheric CO_2 concentrations drop, orbits are particularly cold, and the AIS grows large (DeConto et al., 2007; Halberstadt

◀ Antarctic records and events and major global events discussed in text. (E) Summary of changes in climate regime (temperate, subpolar, and polar) and associated reconstructions of glacial-interglacial ice sheet, ice shelf, and sea ice extent (maps at right) is based on proxy evidence discussed in text and on climate and ice sheet modelling experiments (Gasson et al., 2016b; Golledge et al., 2017b; Halberstadt et al., 2021).

et al., 2021). These results suggest expansion and persistence of perennial sea ice across the Southern Ocean also occurred between 13.8 and 13.6 Ma, as: (1) the interval coincides with a node in obliquity and low eccentricity; (2) proxy CO_2 data show values dropped below 300 ppm (Badger et al., 2013b); and (3) geological data show the AIS expanded to the continental margin (Figs 9.12 and 9.13E). The onset of persistent and extensive perennial sea ice after 13.8 Ma is also supported by the occurrence of abundant protoperidinioid dinocyst species at Site U1356 (Sangiorgi et al., 2018) and a prominent change in formaminiferal fauna at ODP Hole 747A (Kerguelen Plateau, southern Indian Ocean). Here taxa with a warm water affinity (e.g., *Globorotalia miozea* group) were replaced by a fauna with typical polar characters and dominated by neogloboquadrinids (Verducci et al., 2009). This expansion of sea ice and transition to a 'polar' climate in the Southern Ocean likely contributed to the major perturbation in Southern Ocean phytoplankton at 13.6 Ma (Crampton et al., 2016). However, it is important to note that this climatic cooling in the Southern Ocean was transient and that after ~12.9 Ma warm water foraminifera species became more abundant once again (Majewski, 2010; Verducci et al., 2009).

The timeline of events and range of paleoenvironmental data outlined above is compelling and shows that climate gradually cooled and the AIS progressively expanded during glacial episodes through the MMCT but was highly dynamic until ~13.8 Ma. An environmental threshold was crossed at ~13.8 Ma, which may have been driven by a combination of favourable 'cold' astronomical configurations and a decrease in atmospheric CO_2 concentration below 300 ppm (Badger et al., 2013b). The AIS expanded across marine basins and continental shelves at this time and remained relatively stable for ~150 kyrs. Persistent perennial sea ice occupied the circum-Antarctic ocean and 'polar' climatic conditions characterised the high southern latitudes (Fig. 9.13E). However, these cold polar conditions were transient and as the Antarctic climate warmed the AIS again became highly dynamic after 13.6 Ma, which highlights an ongoing puzzle regarding the timing of the onset, and subsequent persistence, of hyper arid climate in the TAM and the transition to a modern polar climate.

Cosmogenic exposure age data suggest erosion rates have been minimal since 15 to 14 Ma and are used to infer an arid climate since this time (Spector and Balco, 2021; Valletta et al., 2015). However, this inference is incongruous with data described above that indicate relatively warm interglacial climates persisted at high elevations until at least 13.9 Ma (Lewis and Ashworth, 2015; Lewis et al., 2008; Verret et al., 2020). Furthermore, mud-rich facies in AND-1B offshore the MDV indicate a subpolar glacial regime delivered plumes of terrestrial material during the warm late Miocene (McKay et al., 2009). Climate and ice sheet modelling also suggests climatic conditions that were warm and wet enough to support tundra vegetation, which persisted at the coast even when atmospheric CO_2 concentrations were low (<460 ppm) and ice occupied most of Antarctica's terrestrial regions (Halberstadt et al., 2021). Finally, meteoric beryllium data suggest water percolated into the surface at high elevation locations into the late Neogene (Dickinson et al., 2012).

9.4.4 Late Miocene

Climate in the late Miocene was cooler than during peak warm episodes of the MCO, but proxy data show surface temperatures remained significantly warmer than today (Herbert et al., 2016; Prebble et al., 2017; Super et al., 2020). Northern high latitude temperature anomalies ranged between 10°C and 15°C higher than today between 11.5 and 8 Ma and mid-latitude temperatures in the southern hemisphere were between 5°C and 8°C warmer (Herbert et al., 2016). Whereas bottom water temperatures cooled by ~2°C across the MMCT (Modestou et al., 2020; Shevenell et al., 2008), they were still at least 6°C warmer than present in the Indian Ocean during the late Miocene (Modestou et al., 2020). Sedimentological (Passchier et al., 2011) and paleoecological (Verducci et al., 2009) data show that polar climatic conditions characterised southern high latitudes by the end of the MMCT, but a return to a subpolar glacial regime around Antarctica by the late Miocene is reflected in geological records from Wilkes Land (Pretty, 2019; Sangiorgi et al., 2018), the Ross Sea (McKay et al., 2009) and Prydz Bay (Hambrey and McKelvey, 2000; Whitehead et al., 2006a, 2006b). Geological data and observations that document this return to warm subpolar conditions at the Antarctic margin are summarised below.

The appearance of nannofossil-rich mudstones at IODP Site U1361 indicates a return to warm oceanic conditions, enhanced meltwater production and ice sheet retreat along the Wilkes Land Coast between 11.7 and 11 Ma (Pretty, 2019). TEX_{86}-based paleotemperature reconstruction from the site show surface waters was 6°C to 10°C at this time (Sangiorgi et al., 2018). These data suggest temperatures were ~8°C to 12°C above sub-zero temperatures that characterise the modern environment and are similar to the anomalies recorded at high northern latitudes and indicate a 2- to 3-times polar amplification. Sedimentary facies changes in AND-1B reflect a shift from a polar to subpolar glacial regime at ~10 Ma in the Ross Sea region (McKay et al., 2009; Rosenblume and Powell, 2019; Wilson et al., 2012b). Massive diamicts and waterlain tills deposited at ODP Site 739 indicate the Lambert ice stream was highly dynamic with multiple episodes of advance and retreat across the continental shelf in Prydz Bay during the late Miocene (Barron and Larsen, 1989). In-situ marine diatoms in the Battye Glacier and Fisher Bench Formations show the Lambert Glacier margin retreated well inland between 10 and 8.5 Ma (Hambrey and McKelvey, 2000; Whitehead et al., 2006b). Sediments deposited during this marine incursion contain palynological assemblages that reflect a herb tundra vegetation similar in form to that of the present-day cool to cold sub-Antarctic regions (Wei et al., 2013). Pollen assemblages in AND-2A contain some elements that are similar to the Prydz Bay assemblages (Taviani et al., 2008–2009), which suggests climatic conditions that support modern sub-Antarctic flora also persisted in the Ross Sea region into the early late Miocene. The first major marine incursions up the 'modern' Wright Valley also occurred at this approximate time and represent relatively warm ice-free conditions at low elevations in the MDV (Brady, 1979; Prentice et al., 1993). Up to 675 m of strata have been imaged offshore the ASB (package Ms-II) that contain additional erosive surfaces that truncate reflectors and exhibit rough morphology and channels indicative of glacial erosion in a meltwater-rich environment (Gulick et al., 2017). While age constraints on this unit are few, it predates the late Miocene (~8 Ma) (Gulick et al., 2017) and may include sediments deposited during the late middle to early late Miocene. These environmental indicators show warming occurred across a range of latitudes and longitudes during the late Miocene and indicate Antarctica transitioned out of a polar regime that characterised environmental conditions for several million years through and immediately after the MMCT (Fig. 9.13).

Glacial–interglacial environmental variability reflected in deep sea $\delta^{18}O$ records was low though the late middle Miocene and suggests a relatively stable climate system persisted between ~13 and 11.5 Ma, after which these deep sea data capture a shift to a more dynamic state (De Vleeschouwer et al., 2017; Holbourn et al., 2013; Westerhold et al., 2020) (Fig. 9.4B and C). A major feature in the $\delta^{18}O$ record at ODP Site 1146 (South China Sea) is an abrupt warming event ($\delta^{18}O$ drop of ~1‰) at ~10.8 Ma that is associated with a large decline in $\delta^{13}C$ of ~1‰ (Holbourn et al., 2013). Several subsequent abrupt warming events ($\delta^{18}O$ drops) occur in the Site 1146 record between 9 and 10 Ma. This interval has been identified as the Tortonian thermal maximum (Westerhold et al., 2020). The 'hyperthermal' at 10.8 Ma approximately coincides with an abrupt but transient (2°C to 4°C) increase in bottom water temperature recorded in a relatively low resolution Mg/Ca record from ODP Site 747 (Billups and Schrag, 2002). These warm events coincide with evidence for warm surface water temperatures (Sangiorgi et al., 2018) and glacial retreat at the Antarctic margin (Whitehead et al., 2006a, 2006b), including the carbonate rich unit from offshore Wilkes Land (Escutia et al., 2011a; Pretty, 2019), and suggest the AIS was sensitive to global climate change and remained highly dynamic following the end of the MMCT. This interval of time certainly warrants more attention.

Environmental records through the interval from ~8 to 5.5 Ma are generally missing from drill cores in the Ross Sea region (Figs 9.7 and 9.13)

(Acton et al., 2008–2009; McKay et al., 2019). This major disconformity correlates with RSU3 at Site U1522 and suggests another phase of marine based ice sheet advance and retreat across the continental shelf began at the onset of the LMC. This episode of cooling and AIS advance is also reflected in beryllium data from AND-1B, which indicate that terrestrial outlet glaciers along the Victoria Land coast have not retreated much further inland of their current position since ~8 Ma (Shakun et al., 2018). Iceberg rafted debris derived from the Wilkes and Adélie Land margin of the EAIS first appears at ODP Site 1165 at ~7 Ma (Williams et al., 2010). These observations suggest that only after 7 Ma were SSTs cold enough to allow ice bergs to transit the large distance from offshore the WSB and ASB to Site 1165 without melting (Williams et al., 2010). Glacial expansion of both the EAIS and WAIS is also inferred from mass accumulation rates of terrigenous matter and iron that intensified between ~7.2 and 6.6 Ma offshore Prydz Bay and the Antarctic Peninsula (Grützner et al., 2005). Furthermore, seismic data and short cores collected offshore of the ASB reveal a 0–110-m-thick package of sub-horizontal to landward-dipping strata that are most likely younger than 6.9 Ma (Gulick et al., 2017). These strata contain no visible channels, which suggests reduced meltwater influence, and erosional surfaces in these late Miocene strata record advance and retreat of an expanded EAIS offshore the ASB. A lack of accumulation and preservation of open marine sediments suggests limited regional ice retreat or shorter interglacials since the late Miocene (Gulick et al., 2017).

Most of the evidence discussed above suggests the AIS became larger, drier and less dynamic at ~7 Ma. Meteoric beryllium data from sites in the MDV suggest modern hyper arid conditions were established at these high southern latitudes at ~6 Ma (Dickinson et al., 2012; Schiller et al., 2009). At the same time, Southern Hemisphere ocean surface water temperatures reached modern values for a brief period (Herbert et al., 2016). This late Miocene increase in Antarctic ice volume and areal extent likely peaked between 5.96 and 5.6 Ma coincident with the Messinian salinity event that impacted the Mediterranean (Garcia-Castellanos and Villaseñor, 2011; Hsü et al., 1973; Ohneiser et al., 2015). However, this interval of peak cold conditions and maximum ice volume ended as climate warmed into the early Pliocene.

9.4.5 Pliocene

The Pliocene is an important period to examine as it offers insight into the equilibrium response of the different sectors of the AIS to climate change under current atmospheric CO_2 concentration. Pliocene research also has the potential to constrain the amplitude of ice volume and sea-level change that may arise from a future deglaciation (DeConto et al., 2021). The last 10 years has seen a significant increase in the coverage and quality of proxy data from paleoclimate archives, which in turn has driven innovation and increased performance and skill of numerical ice sheet simulations, such that the modelled ice volume changes are broadly consistent with paleoclimate constraints for the warm interglacials of the Pliocene.

ANDRILL provided geological evidence that the WAIS had grown and collapsed numerous times between 5 and 2.6 Ma in response to astronomically-paced climate cycles when atmospheric CO_2 was last ~400 ppm (McKay et al., 2012; Naish et al., 2009; Pollard and DeConto, 2009). However, results from IODP Expedition 318 showed that East Antarctic ice within the catchment of the WSB was also highly dynamic during the Pliocene (Bertram et al., 2018; Cook et al., 2013, 2017; Patterson et al., 2014; Reinardy et al., 2015) and that advance and retreat of both the EAIS and WAIS drove global sea-level changes of up to 20 m (Dumitru et al., 2019; Grant et al., 2019; Miller et al., 2012). These discoveries challenged some of the early Pliocene ISM reconstructions which displayed strong hysteresis and could only release 7–9 m of sea-level equivalent ice volume (Pollard and DeConto, 2009). Subsequent work to improve the ISMs has improved alignment between model results and geological constraints. These improvements are primarily due to: (1) better climate–ice sheet coupling to account for ice sheet–climate feedbacks (DeConto and Pollard, 2016; DeConto et al., 2021; Golledge et al., 2015, 2017b, 2019); (2) better characterisation of uncertainty associated with different modelling frameworks (de Boer et al., 2015; Dolan et al., 2018); (3) the ability to simulate instability mechanisms that act on marine-based ice, particularly the ice that sits on reverse-sloped beds; and (4) numerical representation of processes that enhance ice sheet retreat including ice-cliff failure and ice-shelf hydrofracture (DeConto and Pollard, 2016; DeConto et al., 2021; Gasson et al., 2016a; Pollard et al., 2015), although the need to employ these processes and ice sheet physics is still debated (Gasson and Keisling, 2020).

Proxy data from IODP Expedition 318 and ANDRILL imply that the marine-based sectors of both West and East Antarctica were sensitive to obliquity forcing, especially under a warmer mean climate state when CO_2 was between 350 and 400 ppm prior to ~3.3 Ma and the M2 glaciation (Naish et al., 2009; Patterson et al., 2014). Several studies have argued that this was due to the influence of obliquity on modulating the temperature gradient of the Southern Hemisphere (Raymo and Nisancioglu, 2003), promoting wind-driven upwelling of CDW with consequences for melt rates at grounding lines and in ice shelf cavities (Hansen and Passchier, 2017; Levy et al., 2019; Naish et al., 2009). While the duration of both glacimarine cycles in AND-1B and IRD mass accumulation in Site U1316 cores was dominated by obliquity prior to regional cooling associated with the M2 glaciation in benthic foraminifera

$\delta^{18}O$ records (Patterson et al., 2014), the East Antarctic WSB displays a strong eccentricity/precession-paced ice volume variability in IRD mass accumulation between 3.3 and 2.5 Ma (Fig. 9.14).

Ice sheet, SST and sea ice reconstructions from sediment cores in the Ross Sea and eastern Wilkes Land margins provide evidence for expansion of the AIS that began at ~3.3 Ma (Levy et al., 2012; McKay et al., 2012), essentially terminating Pliocene warmth in the southern high-latitudes. This was associated with coastal SST cooling of ~2.5°C, a stepwise expansion of sea ice and polynya formation in the Ross Sea between 3.3 and 2.5 Ma (Fig. 9.14). The intensification of Antarctic cooling resulted in strengthened westerly winds and invigorated ocean circulation. The associated northward migration of Southern Ocean fronts has been linked with reduced Atlantic Meridional Overturning Circulation by restricting surface water connectivity between the ocean basins, with implications for heat transport to the high latitudes of the North Atlantic. McKay et al. (2012) imply this may in turn have preconditioned the northern hemisphere for continental-scale glaciation when atmospheric CO_2 concentration fell below 300 ppm between ~2.9 and 2.7 Ma (Masson-Delmotte et al., 2013).

Peak sea-level estimates during the warmest Pliocene interglacials from far-field sea-level records are as high as +40 m above present. Based on the present global ice-sheet budget the sea-level equivalent of marine-based sectors of the AIS can account for 22.7 m, the Greenland Ice Sheet contains 7.3 m, and an additional 35.6 m is available from terrestrial sectors of the AIS. Amplitudes > + 30 m can only be explained by melting the terrestrial sectors of the AIS, and/or by having more ice on the Northern Hemisphere continents during glacial periods than can be explained by the available geological data (Thiede et al., 2011). Larger than Holocene Antarctic glacial ice volumes cannot be excluded by proximal geological data for glacials during the mPWP (Naish et al., 2009), but retreat of the EAIS inland of its terrestrial margins since ~8 Ma appears to be precluded by a recent study that found extremely low concentrations of cosmogenic ^{10}Be and ^{26}Al isotopes in the ANDRILL-1B marine sediment core (Shakun et al., 2018). Therefore, the maximum contribution to GMSL from the Antarctic ice sheets during the mPWP is the volume of the marine-based sectors of the ice sheet (modern day = ~23 m sea-level equivalent). A recent reassessment of the far-field estimates for the warmest mid-Pliocene interglacials shows GMSL was at least +4.1 m and no more than +20.7 m, with a median of +10.7 m (Grant and Naish, 2021). This median value is consistent with ISM simulations constrained by ice proximal geological data (DeConto and Pollard, 2016; Golledge et al., 2017b; Pollard et al., 2015).

This range also implies an equilibrium polar ice-sheet sensitivity of 2–8 m of sea-level change for every degree of temperature change, with a mean value of 4 m/AIS (Grant and Naish, 2021). However, the

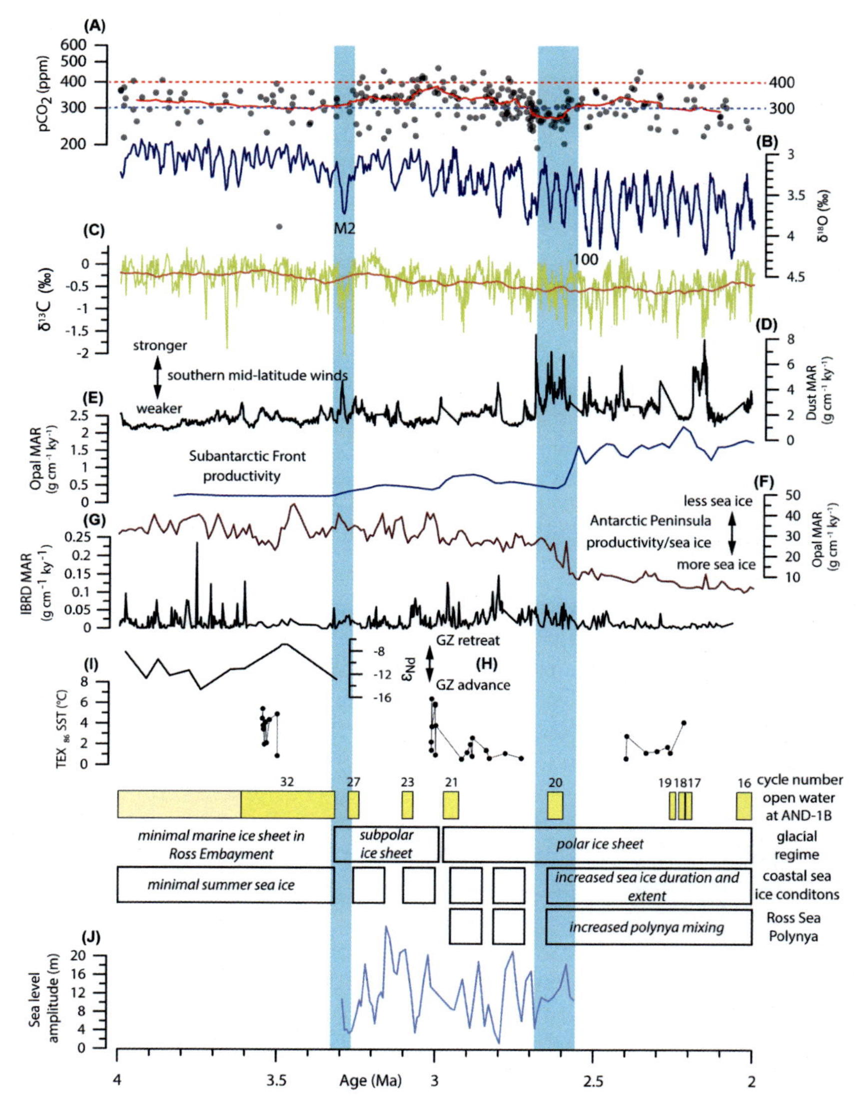

FIGURE 9.14 Environmental data from the Pliocene to early Pleistocene (4 to 2 Ma) highlight Southern Hemisphere climate system feedbacks and responses and their relationship to Antarctic climate evolution. Cooling steps in Pliocene climate at ~3.3 and ~2.6 Ma are highlighted by blue dashed lines. (A) Atmospheric CO_2 compilation comprises data from a range of proxies outlined in Fig. 9.1. Solid red line displays a 17-point running average. (B) Benthic foraminifera $\delta^{18}O$ proxy for ice volume and temperature (Lisiecki and Raymo, 2005). (C) High resolution deep sea benthic foraminifera $\delta^{13}C$ splice (Westerhold et al., 2020). Solid deep orange line displays a 150 kyr moving average. (D) Atmospheric circulation (relative westerly wind strength) from dust mass accumulation rates for ODP Site 1090 (Martinez-Garcia et al., 2011). (E) Southern Ocean primary productivity based on biogenic opal mass accumulation rates at ODP Site 1091 shows a sharp increase coincident with increased windiness and nutrient supply by Fe-rich dust at ~2.6 Ma (Cortese et al., 2004; Hillenbrand and Cortese, 2006). (F) Onset of Antarctic sea ice at ~2.6 Ma marked by a decline in primary productivity recorded in biogenic

(Continued)

empirical estimate does not consider ice sheet dynamics, such as a potential stability threshold in the AIS, caused by the loss of ice shelves, which may be crossed at 1.5°C–2°C of global warming, after which ongoing mass loss may be rapid and non-linear (DeConto et al., 2021; Golledge et al., 2015). Given the analogous nature of the warm Pliocene to future projections (Burke et al., 2018), the paleoclimate reconstructions presented in this chapter offer a stark warning about the potential future of the AIS if warming continues at its current rate. If a threshold is exceeded, AIS instabilities would likely be irreversible on multi-century timescales (DeConto et al., 2021; Golledge et al., 2015).

9.5 Next steps

In this chapter, we have summarised efforts over the past 50 years to obtain proxy environmental data from geological records of the Miocene and Pliocene, that are informed by modelling outputs to offer insight into AIS behaviour during key episodes and transitions through the Neogene. Despite the progress that has been made, there are many outstanding questions regarding AIS response to past climate variability that require future focus (Colleoni et al., 2021; McKay et al., 2021). There remains significant uncertainty over AIS sensitivity to past intervals of warmth when Earth's average surface temperatures were similar to those forecast for our future, particularly if we fail to achieve emissions targets to limit global warming to well below 2°C (Meinshausen et al., 2011; Rogelj et al., 2016; UNFCCC, 2015). In particular, there is a need to document and constrain the extent of ice margin retreat and amount of ice volume loss during the Miocene and Pliocene, when Earths average surface temperatures exceeded 2°C. These data offer important benchmarks for testing ice sheet and climate models that are used for future forecasts and projections. These paleo-targets allow us to identify potential thresholds, and the resultant implications for the AIS, when key tipping points in the Earth System are crossed (DeConto et al., 2021).

Whereas sediment cores and seismic records from the margin provide information on maximum AIS extent, we still rely on sea level records and

---

◀ opal mass accumulation at Antarctic Peninsula ODP Site 1096 (Hillenbrand and Cortese, 2006). (G) Ice berg rafted debris mass accumulate rate from IODP Site U1361 (Patterson et al., 2014). (H) Detrital Nd isotope composition (ε_{Nd}) for Pliocene marine sediments from IODP Site U1361 (Cook et al., 2013, 2017). Lower values indicate grounding zone (GZ) retreat. (I) Summary of ocean, sea ice and ice sheet evolution in Ross Embayment based on the AND-1B record (McKay et al., 2012; Naish et al., 2009). Note the cooling in SST and the return of periodic grounded ice sheets to western Ross Embayment occurs at ~3.3 Ma, ending a ~1.2 Ma period of relatively warm, ice free, open ocean conditions, and the appearance of sea ice and the development Ross Sea polynya between 3.2 and 2.6 Ma. (J) Sea level record (PlioSeaNZ) from the Whanganui Basin, New Zealand, shows maximum glacial-interglacial amplitude change over 20 m (Grant et al., 2019).

deep sea benthic $\delta^{18}O$ records for estimates of minimum ice volume. Unfortunately, far field sea level records with high temporal precision and well constrained estimates of amplitude are rare (Dumitru et al., 2019; Grant and Naish, 2021; Grant et al., 2019). Furthermore, estimates of ice volume change from deep sea oxygen isotope records require independent estimates of bottom water temperature. These temperature records are difficult to obtain, and proxies are still being developed, tested and improved. Large (~1‰–1.5‰) changes in $\delta^{18}O$ over glacial–interglacial timescales (100 kyrs) during the MCO (Holbourn et al., 2015) are particularly difficult to explain and highlight the challenge we still face in our efforts to understand Earth's climate system. While some sea level studies propose that the AIS completely melted during peak warm episodes in the MCO (Miller et al., 2020), modelling experiments suggest large areas of inland Antarctica likely remained ice covered throughout the Miocene (Gasson et al., 2016b; Halberstadt et al., 2021; Paxman et al., 2020). Furthermore, a recent climate modelling study suggests that large spatial variability of the AIS in the Miocene was more rapid than previously thought, and in turn enabled large BWT swings (up to 4°C) (Bradshaw et al., 2021). All these studies suffer from a lack of direct evidence to constrain the maximum extent of AIS retreat during intervals of peak warmth and $\delta^{18}O$ minima.

This 'challenge' to better document and constrain maximum ice sheet retreat is also true for the Pliocene. Advance and retreat of the AIS across regions that were grounded below sea level likely occurred throughout the Pliocene and drove sea level amplitude variations of up to 20 m (Dumitru et al., 2019; Grant et al., 2019; Miller et al., 2012). While data from offshore of the WSB (Cook et al., 2013; Patterson et al., 2014) and in the Ross Sea (McKay et al., 2012; Naish et al., 2009) show that margins of both the WAIS and EAIS retreated during warm episodes in the Pliocene, direct observations from inland portions of these marine basins have yet to be recovered. To better investigate and explain AIS dynamics during the Neogene, we need sedimentary records from the continental interior of West and East Antarctica (McKay et al., 2016). Constraints on bed topography are also critical given the sensitivity of ISMs (and by inference the ice sheet) to this boundary condition (Paxman et al., 2020). These paleotopographic constraints are generally limited to DSDP Site 270 in the Ross Sea (Kulhanek et al., 2019; Leckie and Webb, 1983) and in-situ bedrock samples from the East Antarctic craton do not exist. Clearly, more data are needed. Efforts are underway to acquire new Miocene and Pliocene records from near the grounding zone of the WAIS at the Siple Coast and from beneath the Crary Ice Rise (Levy et al., 2020). Site survey data from beneath the EAIS are required to identify potential drilling targets but are difficult to obtain. Recent advances in overice seismic data acquisition using a vibroseis source and snow streamer recording technology show promise and offer the opportunity to more rapidly acquire high quality data sets (Eisen et al., 2015;

Hofstede et al., 2013). Drilling systems that can penetrate the thick ice to acquire rock and sediment samples from the beneath the ice are also being developed (Goodge and Severinghaus, 2016; Hodgson et al., 2016; Timoney et al., 2020) but have yet to be fully tested in remote locations.

While the Antarctic community endeavours to obtain new ice proximal records of ice sheet and environmental variability, new high-fidelity surface and bottom water temperature records at high temporal resolution across latitudinal transects also need to be established (Koppers and Coggon, 2020). In addition, new multi-proxy CO_2 records at high temporal resolution across key transitions need to be developed. These paleotemperature and paleo-CO_2 data will most likely be generated from deep sea sedimentary records. Legacy cores housed in repositories can be targeted for new studies at the same time new targets could be identified for future drilling through the next phase of the IODP and the International Continental Drilling Program. Finally, the PlioSEANZ record from the Wanganui Basin in New Zealand (Grant et al., 2019) highlights the value of highly resolved records of sea level amplitude change as a constraint on past ice sheet behaviour that is independent of the deep sea $\delta^{18}O$ records. Similar records through key episodes in the Neogene are highly desirable. The MCO and MMCT are obvious targets for new proxy records but the Tortonian Thermal Maxiumum and late Miocene cooling are also intervals of major change for which we have few records and relatively sparse data.

Finally, next generation atmosphere-ocean models that can be fully coupled with state-of-the-art ice sheet and GIA models are highly desirable. Development of computing capacity may provide the means to run these increasingly complex models over longer (geological) timescales and will help address outstanding questions regarding the connectivity of the AIS to the broader Earth system. Outputs from these modelling experiments will be integrated with existing and new observations to improve our understanding of key processes that influence AIS dynamics and provide targets to advance our ability to reduce uncertainty and forecast change as the world warms over the coming decades and centuries. These major objectives are the focus of the new Scientific Community on Antarctic Research INStabilities and Thresholds in ANTarctica (INSTANT) Scientific Research Program.

Acknowledgements

We wish to acknowledge the support of National Antarctic Programmes and the International Scientific Drilling Programmes and Projects that have allowed our community to acquire the critical records of environmental change that have been discussed in this review. We thank Jenny Black, GNS Science, for her assistance with Fig. 9.2. R.L., T.N., R.M., C.O. and N.G. acknowledge funding support from the New Zealand Ministry of Business and Innovation and Employment through the Antarctic Science Platform contract (ANTA1801) Antarctic Ice Dynamics Project (ASP-021-01). C.E. acknowledges funding by the Spanish Ministry of Economy, Industry and Competitivity (grant CTM2017-89711-C2-1/2-P), co-funded by the European Union through FEDER funds. L.F.P. was funded through the European Union's Horizon 2020 research and innovation program under the Marie Sklodowska-Curie grant agreement number 792773 for the West Antarctic Margin Signatures of Ice Sheet Evolution (WAMSISE) Project.

References

Ackert Jr., R.P., 1990. Surficial geology and geomorphology of the western Asgard Range, Antarctica: implications for late Tertiary glacial history. University of Maine, pp. 147.

Acton, G., Crampton, J., Di Vincenzo, G., Fielding, C.R., Florindo, F., Hannah, M., et al., and ANDRILL SMS Science Team 2008–2009. Preliminary integrated chronostratigraphy of the AND-2A Core, ANDRILL Southern McMurdo Sound Project Antarctica. In: Harwood, D., Florindo, F., Talarico, F., Levy, R.H. (Eds.), Studies from the ANDRILL Southern McMurdo Sound Project, Antarctica – Initial Science Report on AND-2A. Terra Antartica, pp. 211–220.

Adkins, J.F., McIntyre, K., Schrag, D.P., 2002. The salinity, temperature, and $\delta^{18}O$ of the glacial deep ocean. Science 298 (5599), 1769–1773.

Ainley, D.G., Jacobs, S.S., 1981. Sea-bird affinities for ocean and ice boundaries in the Antarctic. Deep Sea Research Part A. Oceanographic Research Papers 28 (10), 1173–1185.

Aitken, A., Roberts, J., Van Ommen, T., Young, D., Golledge, N., Greenbaum, J., et al., 2016. Repeated large-scale retreat and advance of Totten Glacier indicated by inland bed erosion. Nature 533 (7603), 385–389.

Allibone, A.H., Cox, S.C., Smillie, R.W., 1993. Granitoids of the Dry Valleys area, southern Victoria Land: geochemistry and evolution along the early Paleozoic Antarctic Craton margin. New Zealand Journal of Geology and Geophysics 36, 299–316.

Alonso, B., Anderson, J.B., Diaz, J.I., Bartek, L.R., Elliot, D.H., 1992. Pliocene-Pleistocene seismic stratigraphy of the Ross Sea; evidence for multiple ice sheet grounding episodes Contributions to Antarctic research III. Antarctic Research Series 57, 93–103.

Amidon, W.H., Fisher, G.B., Burbank, D.W., Ciccioli, P.L., Alonso, R.N., Gorin, A.L., et al., 2017. Mio-Pliocene aridity in the south-central Andes associated with Southern Hemisphere cold periods. Proceedings of the National Academy of Sciences 114 (25), 6474–6479.

Anderson, J.B., Bartek, L.R., 1992. Cenozoic glacial history of the Ross Sea revealed by intermediate resolution seismic reflection data combined with drill site information. In: Kennett, J.P., Warnke, D.A. (Eds.), The Antarctic Paleoenvironment: A perspective on global change. American GeophysicalUnion, Washington, D.C., pp. 231–263. Antarctic Research Series.

Anderson, J.B., 1999. Antarctic Marine Geology. Cambridge University Press, pp. 289.

Aoki, S., Rintoul, S.R., Ushio, S., Watanabe, S., Bindoff, N.L., 2005. Freshening of the Adélie Land Bottom water near 140 E. Geophysical Research Letters 32 (23).

Aoki, S., Yamazaki, K., Hirano, D., Katsumata, K., Shimada, K., Kitade, Y., et al., 2020. Reversal of freshening trend of Antarctic Bottom Water in the Australian-Antarctic Basin during 2010s. Scientific Reports 10 (1), 1–7.

Armbrecht, L.H., Lowe, V., Escutia, C., Iwai, M., McKay, R., Armand, L.K., 2018. Variability in diatom and silicoflagellate assemblages during mid-Pliocene glacial-interglacial cycles determined in Hole U1361A of IODP Expedition 318, Antarctic Wilkes Land Margin. Marine Micropaleontology 139, 28–41.

Austermann, J., Mitrovica, J.X., Huybers, P., Rovere, A., 2017. Detection of a dynamic topography signal in last interglacial sea-level records. Science Advances 3 (7).

Austermann, J., Pollard, D., Mitrovica, J.X., Moucha, R., Forte, A.M., DeConto, R.M., et al., 2015. The impact of dynamic topography change on Antarctic ice sheet stability during the mid-Pliocene warm period. Geology 43 (10), 927–930.

Badger, M.P., Chalk, T.B., Foster, G.L., Bown, P.R., Gibbs, S.J., Sexton, P.F., et al., 2019. Insensitivity of alkenone carbon isotopes to atmospheric CO 2 at low to moderate CO 2 levels. Climate of the Past 15 (2), 539–554.

Badger, M.P., Schmidt, D.N., Mackensen, A., Pancost, R.D., 2013a. High-resolution alkenone palaeobarometry indicates relatively stable p CO_2 during the Pliocene (3.3–2.8 Ma). Philosophical Transactions of the Royal Society A: Mathematical, Physical and Engineering Sciences 371 (2001), 20130094.

Badger, M.P.S., Lear, C.H., Pancost, R.D., Foster, G.L., Bailey, T.R., Leng, M.J., et al., 2013b. CO_2 drawdown following the middle Miocene expansion of the Antarctic Ice Sheet. Paleoceanography 28 (1), 42–53.

Badgley, C., Barry, J.C., Morgan, M.E., Nelson, S.V., Behrensmeyer, A.K., Cerling, T.E. et al., 2008. Ecological changes in Miocene mammalian record show impact of prolonged climatic forcing. Proceedings of the National Academy of Sciences 105 (34), 12145–12149.

Balch, W.M., Bates, N.R., Lam, P.J., Twining, B.S., Rosengard, S.Z., Bowler, B.C., et al., 2016. Factors regulating the Great Calcite Belt in the Southern Ocean and its biogeochemical significance. Global Biogeochemical Cycles 30 (8), 1124–1144.

Ballantyne, A., Greenwood, D., Sinninghe Damsté, J., Csank, A., Eberle, J., Rybczynski, N., 2010. Significantly warmer Arctic surface temperatures during the Pliocene indicated by multiple independent proxies. Geology 38 (7), 603–606.

Ballantyne, A.P., Axford, Y., Miller, G.H., Otto-Bliesner, B.L., Rosenbloom, N., White, J. W., 2013. The amplification of Arctic terrestrial surface temperatures by reduced sea-ice extent during the Pliocene. Palaeogeography, Palaeoclimatology, Palaeoecology 386, 59–67.

Barber, P.M., 1981. Messinian subaerial erosion of the proto-Nile Delta. Marine Geology 44 (3–4), 253–272.

Barker, P., Thomas, E., 2004. Origin, signature and palaeoclimatic influence of the Antarctic Circumpolar Current. Earth-Science Reviews 66 (1–2), 143–162.

Barker, P.F., Camerlenghi, A., Baker, P.F.e., Camerlenghi, A.e., Acton, G.D.e., Brachfeld, S.A., et al., 2002. Glacial history of the Antarctic Peninsula from Pacific margin sediments. In: Proceedings of the Ocean Drilling Program, Scientific Results, Antarctic Glacial History and Sea-Level Change; covering Leg 178 of the cruises of the drilling vessel JOIDES Resolution; Punta Arenas, Chile, to Cape Town, South Africa; sites 1095–1103; 5 February–9 April 1998. Proceedings of the Ocean Drilling Program, Scientific Results (CD-ROM), 178: 40.

Barker, P.F., Filippelli, G.M., Florindo, F., Martin, E.E., Scher, H.D., 2007. Onset and role of the Antarctic Circumpolar Current. Deep Sea Research Part II: Topical Studies in Oceanography 54 (21), 2388–2398.

Barrett, P., Francis, J., Gohl, K., Haywood, A., Siddoway, C., Wilson, D., 2009. ANTscape: antarctic Paleotopographic maps for the last 100 million years, AGU Fall Meeting Abstracts, p. PP43A-1561.

Barrett, P.J. (Ed.), 1989. Antarctic Cenozoic history from the CIROS-1 drillhole, McMurdo Sound, Antarctica. NZ DSIR Bulletin, p. 254.

Barrett, P.J., Ricci, C.A., 2000. Studies from the Cape Roberts Project, Ross Sea, Antarctica; scientific report of CPR-2/2A. Terra Antartica 7 (3–4), 211–654.

Barrett, P.J., Ricci, C.A., 2001. Studies from the Cape Roberts Project, Ross Sea, Antarctica; scientific report of CRP-3. Terra Antartica 8 (3–4), 121–620.

Barrett, P.J., 1971. Stratigraphy and paleogeography of the Beacon Supergroup in the Transantarctic Mountains. Antarctica. 2nd Gondwana Symp., Proc. Pap. 249–256.

Barrett, P.J., 1975. Textural characteristics of Cenozoic preglacial and glacial sediments at Site 270, Ross Sea, Antarctica. Initial Reports of the Deep Sea Drilling Project 28, 757–766.

Barrett, P.J., 1986. Antarctic Cenozoic history from the MSSTS-1 drillhole, McMurdo Sound. NZ DSIR Bulletin 237, 172.

Barrett, P.J., 2007. Cenozoic climate and sea level history from glacimarine strata off the Victoria Land coast, Cape Roberts Project, Antarctica. In: Hambrey, M.J., Christoffersen, P., Glasser, N.F., Hubbart, B. (Eds.), Glacial Sedimentary Processes and Products. Blackwell: International Association of Sedimentologists, pp. 259–287.

Barrett, P.J., 2009. Cenozoic climate and sea level history from Glacimarine Strata off the Victoria Land Coast, Cape Roberts Project, Antarctica. Glacial Sedimentary Processes and Products. Blackwell Publishing Ltd., pp. 259–287.

Barrett, P.J., 2013. Resolving views on Antarctic Neogene glacial history—the Sirius debate. Earth and Environmental Science Transactions of the Royal Society of Edinburgh 104 (1), 31–53.

Barrett, P.J., Hambrey, M.J., 1992. Plio-Pleistocene sedimentation in Ferrar Fjord, Antarctica. Sedimentology 39, 109–123.

Barrett, P.J., Elliot, D.H., Lindsay, J.F., 1986. The Beacon Supergroup (Devonian-Triassic) and Ferrar Group (Jurassic) in the Beardmore Glacier area, Antarctica. Antarctic Research Series 36 (14), 339–428.

Barrett, P.J., Fielding, C.R.A., Wise, S. (Eds.), 1998. Initial Report on CRP-1, Cape Roberts Project, Antarctica, 187. Terra Antartica, Siena.

Barrett, P.J., Grindley, G.W., Webb, P.N., 1972. The Beacon Supergroup of East Antarctica. Antarctic geology and geophysics. International Union of Geological Sciences. Series B 1, 319–332.

Barron, J.A., Keller, G., 1982. Widespread Miocene deep-sea hiatuses: coincidence with periods of global cooling. Geology 10 (11), 577–581.

Barron, J., Larsen, B., et al., 1989. *Proc. ODP, Init. Repts.*, 119. College Station, TX (Ocean Drilling Program). Available from: https://doi.org/10.2973/odp.proc.ir.119.1989.

Barron, J., Larsen, B., et al., 1991. *Proc. ODP, Sci. Results*, 119. College Station, TX (Ocean Drilling Program). Available from: https://doi.org/10.2973/odp.proc.sr.119.1991.

Bart, P.J., De Santis, L., 2012. Glacial intensification during the Neogene: a review of seismic stratigraphic evidence from the Ross Sea. Antarctica, continental shelf. Oceanography 25 (3), 166–183.

Bart, P.J., Owolana, B., 2012. On the duration of West Antarctic Ice Sheet grounding events in Ross Sea during the Quaternary. Quaternary Science Reviews 47, 101–115.

Bartek, L.R., Vail, P.R., Anderson, J.B., Emmett, P.A., Wu, S., 1991. The effect of Cenozoic ice sheet fluctuations in Antarctica on the stratigraphic signature of the Neogene. Journal of Geophysical Research 96B, 6753–6778.

Bartoli, G., Hönisch, B., Zeebe, R.E., 2011. Atmospheric CO_2 decline during the Pliocene intensification of Northern Hemisphere glaciations. Paleoceanography 26 (4).

Bassis, J.N., Walker, C., 2012. Upper and lower limits on the stability of calving glaciers from the yield strength envelope of ice. Proceedings of the Royal Society A: Mathematical, Physical and Engineering Sciences 468 (2140), 913–931.

Beaman, R.J., O'Brien, P.E., Post, A.L., De Santis, L., 2010. A new high-resolution bathymetry model for the Terre Adélie and George V continental margin, East Antarctica. Antarctic Science 23 (1), 95.

Beerling, D.J., Fox, A., Anderson, C.W., 2009. Quantitative uncertainty analyses of ancient atmospheric CO_2 estimates from fossil leaves. American Journal of Science 309 (9), 775–787.

Behrendt, J.C., Blankenship, D.D., Morse, D.L., Bell, R.E., James, T.Se, Jacka, T.He, et al., 2004. Shallow-source aeromagnetic anomalies observed over the West Antarctic ice sheet compared with coincident bed topography from radar ice sounding; new evidence for glacial "removal" of subglacially erupted late Cenozoic rift-related volcanic edifices. Ice sheets and neotectonics. Global and Planetary Change 42 (1–4), 177–193.

Behrendt, J.C., Blankenship, D.D., Morse, D.L., Finn, C.A., Bell, R.E., Smellie, J.Le, et al., 2002. Subglacial volcanic features beneath the West Antarctic ice sheet interpreted from aeromagnetic and radar ice sounding. Volcano-ice interaction on Earth and Mars. Geological Society Special Publications 202, 337–355.

Behrendt, J.C., Saltus, R., Damaske, D., McCafferty, A.E., Finn, C.A., Blankenship, D., et al., 1996. Patterns of late Cenozoic volcanic and tectonic activity in the West Antarctic rift system revealed by aeromagnetic surveys. Tectonics 15 (3), 660–676.

Bell, R.E., Chu, W., Kingslake, J., Das, I., Tedesco, M., Tinto, K.J., et al., 2017. Antarctic ice shelf potentially stabilized by export of meltwater in surface river. Nature 544 (7650), 344–348.

Berends, C.J., de Boer, B., Dolan, A.M., Hill, D.J., van de Wal, R.S.W., 2019. Modelling ice sheet evolution and atmospheric CO_2 during the Late Pliocene. Climate of the Past 15 (4), 1603–1619.

Bertram, R.A., Wilson, D.J., van de Flierdt, T., McKay, R.M., Patterson, M.O., Jimenez-Espejo, F.J., et al., 2018. Pliocene deglacial event timelines and the biogeochemical response offshore Wilkes Subglacial Basin, East Antarctica. Earth and Planetary Science Letters 494, 109–116.

Beu, A., Taviani, M., 2013. Early Miocene Mollusca from McMurdo Sound, Antarctica (ANDRILL 2A drill core), with a review of Antarctic Oligocene and Neogene Pectinidae (Bivalvia). Palaeontology 57 (2), 299–342.

Beu, A., 1990. Molluscan generic diversity of New Zealand Neogene stages: extinction and biostratigraphic events. Palaeogeography, Palaeoclimatology, Palaeoecology 77 (3–4), 279–288.

Beu, A.G., 1974. Molluscan evidence of warm sea temperatures in New Zealand during Kapitean (late Miocene) and Waipipian (middle Pliocene) time. New Zealand Journal of Geology and Geophysics 17 (2), 465–479.

Bierman, P.R., Shakun, J.D., Corbett, L.B., Zimmerman, S.R., Rood, D.H., 2016. A persistent and dynamic East Greenland Ice Sheet over the past 7.5 million years. Nature 540 (7632), 256–260.

Bijl, P.K., Houben, A.J., Bruls, A., Pross, J., Sangiorgi, F., 2018a. Stratigraphic calibration of Oligocene–Miocene organic-walled dinoflagellate cysts from offshore Wilkes Land, East Antarctica, and a zonation proposal. Journal of Micropalaeontology 37 (1), 105–138.

Bijl, P.K., Houben, A.J., Hartman, J.D., Pross, J., Salabarnada, A., Escutia, C., et al., 2018b. Paleoceanography and ice sheet variability offshore Wilkes Land, Antarctica–part 2: insights from Oligocene–Miocene dinoflagellate cyst assemblages. Climate of the Past 14 (7), 1015–1033.

Billups, K., Ravelo, A.C., Zachos, J.C., 1997. Early Pliocene deep-water circulation: stable isotope evidence for enhanced Northern Component Deep Water. Proc. Ocean Drill. Program, Scientific Results 154, 319–330.

Billups, K., Schrag, D.P., 2002. Paleotemperatures and ice volume of the past 27 Myr revisited with paired Mg/Ca and $^{18}O/^{16}O$ measurements on benthic foraminifera. Paleoceanography 17 (1), 1003.

Billups, K., Schrag, D.P., 2003. Application of benthic foraminiferal Mg/Ca ratios to questions of Cenozoic climate change. Earth and Planetary Science Letters 209, 181–195.

Bindoff, N.L., Rosenberg, M.A., Warner, M.J., 2000. On the circulation and water masses over the Antarctic continental slope and rise between 80 and 150 E. Deep Sea Research Part II: Topical Studies in Oceanography 47 (12–13), 2299–2326.

Bliss, A.K., Cuffey, K.M., Kavanaugh, J.L., 2011. Sublimation and surface energy budget of Taylor Glacier, Antarctica. Journal of Glaciology 57 (204), 684–696.

Bockheim, J.G., 2010. Soil preservation and ventifact recycling from dry-based glaciers in Antarctica. Antarctic Science 22, 409–417.

Bockheim, J.G., McLeod, M., 2006. Soil formation in Wright Valley, Antarctica since the late Neogene. Geoderma 137, 109–116.

Böhme, M., 2003. The Miocene climatic optimum: evidence from ectothermic vertebrates of Central Europe. Palaeogeography, Palaeoclimatology, Palaeoecology 195 (3–4), 389–401.

Bolton, C.T., Hernández-Sánchez, M.T., Fuertes, M.-Á., González-Lemos, S., Abrevaya, L., Mendez-Vicente, A., et al., 2016. Decrease in coccolithophore calcification and CO_2 since the middle Miocene. Nature Communications 7 (1), 10284.

Bradshaw, C.D., Langebroek, P.M., Lear, C.H., Lunt, D.J., Coxall, H.K., Sosdian, S.M., et al., 2021. Hydrological impact of Middle Miocene Antarctic ice-free areas coupled to deep ocean temperatures. Nature Geoscience 14 (6), 429–436.

Brady, H.T., 1979. A diatom report on DVDP cores 3, 4a, 12, 14, 15 and other related surface sections. In: Nagata, T. (Ed.), Proceedings of the Seminar III on Dry Valley Drilling Project, 1978. Memoirs of National Institute of Polar Research, Special Issue. pp. 150–163.

Brady, H.T., 1982. Late Cenozoic history of Taylor and Wright valleys and McMurdo Sound inferred from diatoms in Dry Valley Drilling Project cores. In: Craddock, C. (Ed.), Antarctic Geoscience. University of Wisconsin Press, Madison, pp. 1123–1131.

Bragg, F., Lunt, D.J., Haywood, A., 2012. Mid-Pliocene climate modelled using the UK hadley centre model: PlioMIP experiments 1 and 2. Geoscientific Model Development 5 (5), 1109–1125.

Brancolini, G., Busetti, M., Marchetti, A., De Santis, L., Zanolla, C., Cooper, A.K., et al., 1995a. Descriptive text for the seismic stratigraphic atlas of the Ross Sea, Antarctica. Geology and seismic stratigraphy of the Antarctic margin. Antarctic Research Series 68, A271–A286.

Brancolini, G., Cooper, A.K., Coren, F., Cooper, A.Ke, Barker, P.Fe, Brancolini, Ge, 1995b. Seismic facies and glacial history in the western Ross Sea (Antarctica). Geology and seismic stratigraphy of the Antarctic margin. Antarctic Research Series 68, 209–233.

Brancolini, G., Cooper, A.K., Coren, F., 1995. Seismic facies and glacial history in the western Ross Sea (Antarctica). Geology and Seismic Stratigraphy of the Antarctic Margin, 68. pp. 209–233.

Brierley, C., Burls, N., Ravelo, C., Fedorov, A., 2015. Pliocene warmth and gradients. Nature Geoscience 8 (6), 419–420.

Brigham-Grette, J., Melles, M., Minyuk, P., Andreev, A., Tarasov, P., DeConto, R., et al., 2013. Pliocene warmth, polar amplification, and stepped Pleistocene cooling recorded in NE Arctic Russia. Science 340 (6139), 1421–1427.

Brunet, M., 2010. Two new Mio-Pliocene Chadian hominids enlighten Charles Darwin's 1871 prediction. Philosophical Transactions of the Royal Society B: Biological Sciences 365 (1556), 3315–3321.

Brunet, M., Guy, F., Pilbeam, D., Mackaye, H.T., Likius, A., Ahounta, D., et al., 2002. A new hominid from the Upper Miocene of Chad, Central Africa. Nature 418 (6894), 145–151.

Budillon, G., Spezie, G., 2000. Thermohaline structure and variability in the Terra Nova Bay polynya, Ross Sea. Antarctic Science 12 (4), 493–508.

Bull, C., McKelvey, B.C., Webb, P.N., 1962. Quaternary glaciations in southern Victoria Land, Antarctica. Journal of Glaciology 4 (31), 63–78.

Burke, K., Williams, J., Chandler, M., Haywood, A., Lunt, D., Otto-Bliesner, B., 2018. Pliocene and Eocene provide best analogs for near-future climates. Proceedings of the National Academy of Sciences 115(52), 13288–13293.

Burls, N.J., Bradshaw, C., De Boer, A.M., Herold, N., Huber, M., Pound, M., et al., 2021. Simulating Miocene warmth: insights from an opportunistic multi-model ensemble (MioMIP1). Paleoceanography and Paleoclimatology 36, e2020PA004054.

Busetti, M., Camerlenghi, A., Carta, A., Lodolo, E., Sauli, C., 1993. The Ross Sea (Antarctica): a review of the geological and geophysical exploration. Bollettino di Geofisica Teorica ed Applicata 35, 245–263.

Busetti, M., Cooper, A.K., 1994. Possible ages and origins of unconformity U6 in the Ross Sea, Antarctica. Terra Antartica 1 (Special Issue), 341–343.

Caburlotto, A., Lucchi, R., De Santis, L., Macrì, P., Tolotti, R., 2010. Sedimentary processes on the Wilkes Land continental rise reflect changes in glacial dynamic and bottom water flow. International Journal of Earth Sciences 99 (4), 909–926.

Calkin, P.E., 1974. Subglacial geomorphology surrounding the ice-free valleys of Southern Victoria Land, Antarctica. Journal of Glaciology 13, 415–429.

Calkin, P.E., Bull, C., 1972. Interaction of the East Antarctic Ice Sheet, alpine glaciations and sea-level in the Wright Valley area, southern Victoria Land. In: Adie, R.J. (Ed.), Antarctic Geology and Geophysics. Universitetsforlaget, Oslo, pp. 435–440.

Campagne, P., Crosta, X., Houssais, M.-N., Swingedouw, D., Schmidt, S., Martin, A., et al., 2015. Glacial ice and atmospheric forcing on the Mertz Glacier Polynya over the past 250 years. Nature Communications 6 (1), 1–9.

Cape Roberts Science Team, 1998. Miocene strata in CRP-1, Cape Roberts Project, Antarctica. Terra Antartica 5 (1), 63–124.

Carlson, A.E., Dutton, A., Long, A.J., Milne, G.A., 2019. PALeo constraints on SEA level rise (PALSEA): ice-sheet and sea-level responses to past climate warming. Quaternary Science Reviews 212, 28–32.

Carter, A., Curtis, M., Schwanethal, J., 2014. Cenozoic tectonic history of the South Georgia microcontinent and potential as a barrier to Pacific-Atlantic through flow. Geology 42 (4), 299–302.

Carter, L., Bostock-Lyman, H., Bowen, M., 2021. Water masses, circulation and change in the modern Southern Ocean. In: Florindo, F., et al. (Eds.), Antarctic Climate Evolution, second ed. Elsevier (this volume).

Carter, A., Riley, T.R., Hillenbrand, C.-D., Rittner, M., 2017. Widespread Antarctic glaciation during the Late Eocene. Earth and Planetary Science Letters 458, 49–57.

Cartwright, K., Treves, S.B., Torii, T., 1974. Geology of DVDP 4, Lake Vanda, Wright Valley, Antarctica. Dry Valley Drilling Project Bulletin 3.

Castagno, P., Capozzi, V., DiTullio, G.R., Falco, P., Fusco, G., Rintoul, S.R., et al., 2019. Rebound of shelf water salinity in the Ross Sea. Nature Communications 10 (1), 1–6.

Cerling, T.E., 1992. Use of carbon isotopes in paleosols as an indicator of the P (CO2) of the paleoatmosphere. Global Biogeochemical Cycles 6 (3), 307–314.

Chorley, H., 2021. Antarctic ice sheet and climate evolution during the mid-Miocene. Open Access Victoria University of Wellington | Te Herenga Waka. Thesis. Available from: https://doi.org/10.26686/wgtn.14150105.v1.

Chow, J.M., Bart, P.J., 2003. West Antarctic ice sheet grounding events on the Ross Sea outer continental shelf during the middle Miocene. Palaeogeography, Palaeoclimatology, Palaeoecology 198, 169–186.

Clapperton, C.M., Sugden, D.E., 1990. Late Cenozoic glacial history of the Ross embayment, Antarctica. Quaternary Science Reviews 9 (2–3), 253–272.

Clerc, F., Minchew, B.M., Behn, M.D., 2019. Marine ice cliff instability mitigated by slow removal of ice shelves. Geophysical Research Letters 46 (21), 12108–12116.

Coenen, J., Scherer, R., Baudoin, P., Warny, S., Castañeda, I., Askin, R., 2020. Paleogene marine and terrestrial development of the West Antarctic Rift System. Geophysical Research Letters 47 (3), e2019GL085281.

Colleoni, F., Santis, L. De, Naish, T., DeConto, R.M., Escutia, C., Stocchi, P., et al., 2021. Past Antarctic ice sheet dynamics and implications for future sea-level change. In: Florindo, F., et al. (Eds.), Antarctic Climate Evolution, second ed. Elsevier (this volume).

Colleoni, F., Santis, De, Siddoway, L., Bergamasco, C.S., Golledge, A., Lohmann, N.R., et al., 2018. Spatio-temporal variability of processes across Antarctic ice-bed-ocean interfaces. Nature Communications 9 (1), 2289.

Compston, W., McDougall, I., Heier, K., 1968. Geochemical comparison of the mesozoic basaltic rocks of Antarctica, South Africa, South America and Tasmania. Geochimica et Cosmochimica Acta 32 (2), 129–149.

Cook, C.P., Hemming, S.R., van de Flierdt, T., Davis, E.L.P., Williams, T., Galindo, A.L., et al., 2017. Glacial erosion of East Antarctica in the Pliocene: a comparative study of multiple marine sediment provenance tracers. Chemical Geology 466, 199–218.

Cook, C.P., Hill, D.J., van de Flierdt, T., Williams, T., Hemming, S.R., Dolan, A.M., et al., 2014. Sea surface temperature control on the distribution of far-traveled Southern Ocean ice-rafted detritus during the Pliocene. Paleoceanography 29 (6), 533–548.

Cook, C.P., van de Flierdt, T., Williams, T., Hemming, S.R., Iwai, M., Kobayashi, M., et al., 2013. Dynamic behaviour of the East Antarctic ice sheet during Pliocene warmth. Nature Geoscience 6, 765–769.

Cooper, A.K., Barker, P.F., Brancolini, G., 1995. Geology and Seismic Stratigraphy of the Antarctic Margin, Atlas, CD-ROMs, Antarctic Research Series. American Geophysical Union, Washington, DC, p. 301.

Cooper, A.K., Brancolini, G., Escutia, C., Kristoffersen, Y., Larter, R., Leitchenkov, G., et al., 2008. Cenozoic climate history from seismic reflection and drilling studies on the Antarctic continental margin. Developments in Earth and Environmental Sciences 8, 115–234.

Cooper, A.K., O'Brien, P.E., 2004. Leg 188 synthesis: transitions in the glacial history of the Prydz Bay region, East Antarctica, from ODP drilling. In: Proceedings of the Ocean Drilling Program: Scientific Results, pp. 1–42.

Cooper, A.K., O'Brien, P.E., Richter, C., 2004. Scientific results. Proceedings of the Ocean Drilling Programme, p. 188.

Cooper, A.K., Davey, F.J., 1987. The Antarctic Continental Margin; Geology and Geophysics of the Western Ross Sea. Circum-Pacific Council for Energy and Mineral Resources, Earth Science Series, 5B, 253 pp.

Cornford, S.L., Seroussi, H., Asay-Davis, X.S., Gudmundsson, G.H., Arthern, R., Borstad, C., et al., 2020. Results of the third Marine Ice Sheet Model Intercomparison Project (MISMIP+). The Cryosphere 14 (7), 2283–2301.

Cortese, G., Gersonde, R., Hillenbrand, C.-D., Kuhn, G., 2004. Opal sedimentation shifts in the World Ocean over the last 15 Myr. Earth and Planetary Science Letters 224 (3–4), 509–527.

Cox, S.C., Turnbull, I., Isaac, M., Townsend, D.B., Lyttle, B.S., 2012. Geology of Southern Victoria Land, Antarctica.

Cramer, B., Toggweiler, J., Wright, J., Katz, M., Miller, K., 2009. Ocean overturning since the Late Cretaceous: Inferences from a new benthic foraminiferal isotope compilation. Paleoceanography 24 (4).

Cramer, B.S., Miller, K.G., Barrett, P.J., Wright, J.D., 2011. Late Cretaceous–Neogene trends in deep ocean temperature and continental ice volume: reconciling records of benthic foraminiferal geochemistry ($\delta^{18}O$ and Mg/Ca) with sea level history. Journal of Geophysical Research: Oceans 116 (C12), C12023.

Crampton, J.S., Cody, R.D., Levy, R., Harwood, D., McKay, R., Naish, T.R., 2016. Southern Ocean phytoplankton turnover in response to stepwise Antarctic cooling over the past 15 million years. Proceedings of the National Academy of Sciences 113 (25), 6868–6873.

Cui, Y., Schubert, B.A., Jahren, A.H., 2020. A 23 my record of low atmospheric CO_2. Geology.

Dalziel, I.W.D., Lawver, L.A., Pearce, J.A., Barker, P.F., Hastie, A.R., Barfod, D.N., et al., 2013. A potential barrier to deep Antarctic circumpolar flow until the late Miocene? Geology 41 (9), 947–950.

Davey, F., Cita, M.B., Vander Meer, J.J.M., Tessohnson, F., Thomson, M.R., Barrett, P.J., et al., 2001. Drilling for Antarctic Cenozoic Climate and Tectonic History ate Cape Roberts, Southwestern Ross Sea. EOS Transactions of the American Geophysical Union 82 (48), 585–590.

Davey, F.J., Brancolini, G., Hamilton, R.J., Henrys, S.A., Sorlien, C.C., Bartek, L.R., 2000. A revised correlation of seismic stratigraphy at the Cape Roberts drillsites with the seismic stratigraphy of the Victoria Land basin. In: Davey, F.J., Jarrard, R.D. (Eds.), Studies from the Cape Roberts Project, Ross Sea, Antarctica; Scientific Results of CRP-2/2A; Part I, Geophysics and Physical Properties Studies for CRP-2/2A. Terra Antartica, pp. 215–220.

David, T.W.E., Priestley, R.E., 1914. Glaciology, physiography and tectonic geology of south Victoria Land: British Antarctic Expedition, 1907–09, reports of scientific investigations. Geology 1, 1–319.

de Boer, B., Dolan, A.M., Bernales, J., Gasson, E., Goelzer, H., Golledge, N.R., et al., 2015. Simulating the Antarctic ice sheet in the late-Pliocene warm period: PLISMIP-ANT, an ice-sheet model intercomparison project. The Cryosphere 9 (3), 881–903.

de Boer, B., Dolan, A.M., Bernales, J., Gasson, E., Goelzer, H., Golledge, N.R., et al., 2014. Simulating the Antarctic ice sheet in the Late-Pliocene warm period: PLISMIP-ANT, an ice-sheet model intercomparison project. The Cryosphere Discuss 8 (6), 5539–5588.

de Boer, B., Stocchi, P., Whitehouse, P.L., van de Wal, R.S.W., 2017. Current state and future perspectives on coupled ice-sheet—sea-level modelling. Quaternary Science Reviews 169, 13–28.

De Boer, B., Van de Wal, R., Bintanja, R., Lourens, L., Tuenter, E., 2010. Cenozoic global ice-volume and temperature simulations with 1-D ice-sheet models forced by benthic $\delta^{18}O$ records. Annals of Glaciology 51 (55), 23–33.
De Santis, L., Anderson, J.B., Brancolini, G., Zayatz, I., 1995. Seismic record of Late Oligocene through Miocene glaciation on the central and eastern Continental shelf of the Ross Sea, Geology and Seismic Stratigraphy of the Antarctic Margin. Antarctic Research Series. AGU, Washington, D.C., pp. 235–260.
De Santis, L., Brancolini, G., Donda, F., Harris, P.Te, Brancolini, Ge, Bindoff, N.L.E., et al., 2003. Seismo-stratigraphic analysis of the Wilkes Land continental margin (East Antarctica); influence of glacially driven processes on the Cenozoic deposition. Recent investigations of the Mertz Polynya and George Vth Land continental margin, East Antarctica. Deep-Sea Research. Part II: Topical Studies in Oceanography 50 (8–9), 1563–1594.
De Santis, L., Prato, S., Brancolini, G., Lovo, M., Torelli, L., 1999. The Eastern Ross Sea continental shelf during the Cenozoic: implications for the West Antarctic ice sheet development. Global and Planetary Change 23 (1–4), 173–196.
De Vleeschouwer, D., Vahlenkamp, M., Crucifix, M., Pälike, H., 2017. Alternating Southern and Northern Hemisphere climate response to astronomical forcing during the past 35 m.y. Geology 45 (4), 375–378.
de Vries, M.v.W., Bingham, R.G., Hein, A., 2017. A new volcanic province: an inventory of subglacial volcanoes in West Antarctica. In: Siegert, M.J., Jamieson, S.S.R., White, D.A. (Eds.), Exploration of Subsurface Antarctica: Uncovering Past Changes and Modern Processes. Geological Society, London, Special Publications 461.
DeCesare, M., Pekar, S., 2016. Data report: foraminiferal stable isotope and percent calcium carbonate analysis from IODP Expedition 318 Hole U1361A. In: Proceedings of IODP, 2 pp.
Decesari, R.C., Wilson, D.S., Faulkner, M., Luyendyk, B.P., Sorlien, C.C., 2005. A Model for Cretaceous and Tertiary Extension of the Ross Sea, Antarctica. EOS (American Geophysical Union Transactions), 86(52): Fall Meet. Suppl., Abstract T53E-08.
Decesari, R.C., Wilson, D.S., Luyendyk, B.P., Faulkner, M., 2007. Cretaceous and Tertiary extension throughout the Ross Sea, Antarctica. In: Cooper, A.K., Raymond, C.R., et al. (Eds.), Antarctica: A Keystone in a Changing World–Online Proceedings of the 10th ISAES X. Available from: https://doi.org/10.3133/of2007-1047.srp3098.
DeConto, R.M., Pollard, D., 2003. Rapid Cenozoic glaciation of Antarctica induced by declining atmospheric CO_2. Nature (London) 421 (6920), 245–249.
DeConto, R.M., Pollard, D., 2016. Contribution of Antarctica to past and future sea-level rise. Nature 531 (7596), 591–597.
DeConto, R.M., Pollard, D., Harwood, D., 2007. Sea ice feedback and Cenozoic evolution of Antarctic climate and ice sheets. Paleoceanography 22, PA3214. Available from: https://doi.org/10.1029/2006PA001350.
DeConto, R.M., Pollard, D., Kowalewski, D., 2012. Modeling Antarctic ice sheet and climate variations during Marine Isotope Stage 31. Global and Planetary Change 88–89, 45–52.
DeConto, R.M., Pollard, D., Alley, R.B., Velicogna, I., Gasson, E., Gomez, N., et al., 2021. The Paris Climate Agreement and future sea-level rise from Antarctica. Nature 593 (7857), 83–89.
DeConto, R.M., Pollard, D., Wilson, P.A., Pälike, H., Lear, C., Pagani, M., 2008. Thresholds for Cenozoic Bipolar Glaciation. Nature 455, 652–657.
Denton, G.H., Armstrong, R.L., Stuiver, M., 1971. The Late Cenozoic Glacial History of Antarctica: the Late Cenozoic glacial ages. Yale University Press, New Haven, CT, pp. 267–306.
Denton, G.H., Prentice, M.L., Kellogg, D.E., Kellogg, T.B., 1984. Late Tertiary history of the Antarctic ice sheet; evidence form the dry valleys. Geology (Boulder) 12 (5), 263–267.

Di Roberto, A., Pompilio, M., Wilch, T.I., 2010. Late Miocene submarine volcanism in ANDRILL AND-1B drill core, Ross Embayment, Antarctica. Geosphere 6 (5), 524–536.

Di Vincenzo, G., Bracciali, L., Del Carlo, P., Panter, K., Rocchi, S., 2009. ^{40}Ar-^{39}Ar dating of volcanogenic products from the AND-2A core (ANDRILL Southern McMurdo Sound Project, Antarctica): correlations with the Erebus Volcanic Province and implications for the age model of the core. Bulletin of Volcanology 72, 487–505.

Dickinson, W.W., Schiller, M., Ditchburn, B.G., Graham, I.J., Zondervan, A., 2012. Meteoric Be-10 from Sirius Group suggests high elevation McMurdo Dry Valleys permanently frozen since 6 Ma. Earth and Planetary Science Letters 355–356, 13–19.

Diester-Haass, L., Meyers, P.A., Bickert, T., 2004. Carbonate crash and biogenic bloom in the late Miocene: Evidence from ODP Sites 1085, 1086, and 1087 in the Cape Basin, southeast Atlantic Ocean. Paleoceanography 19 (1).

Diester-Haass, L., Meyers, P.A., Vidal, L., Christensen, B.A.E., Giraudeau, J.E., 2002. The late Miocene onset of high productivity in the Benguela Current upwelling system as part of a global pattern. Marine Geology 180 (1–4), 87–103.

Dinniman, M.S., Klinck, J.M., 2002. The influence of open vs periodic alongshore boundaries on circulation near submarine canyons. Journal of Atmospheric and Oceanic Technology 19 (10), 1722–1737.

Dinniman, M.S., Klinck, J.M., Smith Jr, W.O., 2003. Cross-shelf exchange in a model of the Ross Sea circulation and biogeochemistry. Deep Sea Research Part II: Topical Studies in Oceanography 50 (22–26), 3103–3120.

Dinniman, M.S., Klinck, J.M., Smith Jr., W.O., 2007. Influence of sea ice cover and icebergs on circulation and water mass formation in a numerical circulation model of the Ross Sea, Antarctica. Journal of Geophysical Research: Oceans 112 (C11).

Dolan, A.M., Haywood, A.M., Hill, D.J., Dowsett, H.J., Hunter, S.J., Lunt, D.J., Pickering, S.J., 2011. Sensitivity of Pliocene ice sheets to orbital forcing. Palaeogeography, Palaeoclimatology, Palaeoecology 309, 98–110.

Dolan, A.M., De Boer, B., Bernales, J., Hill, D.J., Haywood, A.M., 2018. High climate model dependency of Pliocene Antarctic ice-sheet predictions. Nature Communications 9 (1), 1–12.

Dolan, A.M., Koenig, S.J., Hill, D.J., Haywood, A.M., DeConto, R.M., 2012. Pliocene Ice Sheet Modelling Intercomparison Project (PLISMIP)—experimental design. Geoscientific Model Development 5 (4), 963–974.

Donda, F., Brancolini, G., De Santis, L., Trincardi, F., Harris, P.T.E., Brancolini, G.E., et al., 2003. Seismic facies and sedimentary processes on the continental rise off Wilkes Land (East Antarctica); evidence of bottom current activity. Recent investigations of the Mertz Polynya and George Vth Land continental margin, East Antarctica. Deep-Sea Research. Part II: Topical Studies in Oceanography 50 (8–9), 1509–1527.

Doran, P., Fountain, A., 2019. McMurdo Dry Valleys LTER - High frequency measurements from Friis Hills Meteorological Station (FRSM) - Taylor Valley, Antarctica - 1993 to present ver 11. Environmental Data Initiative. Available from: https://doi.org/10.6073/pasta/9ff39ef4762d45d4ff35576b79d63a2b.

Doran, P.T., McKay, C.P., Clow, G.D., Dana, G.L., Fountain, A.G., Nylen, T., et al., 2002. Valley floor climate observations from the McMurdo dry valleys, Antarctica, 1986–2000. Journal of Geophysical Research: Atmospheres 107 (D24), ACL 13-11–ACL 13-12.

Dowsett, H.J., Cronin, T.M., 1990. High eustatic sea level during the middle Pliocene; evidence from the Southeastern United States Atlantic Coastal Plain; with Suppl. Data 90–13. Geology (Boulder) 18 (5), 435–438.

Dowsett, H.J., Barron, J.A., Poore, R.Z., Poore, R.Ze, Sloan, L.C.E., 1996. Middle Pliocene sea surface temperatures: a global reconstruction. Climates and climate variability of the Pliocene. Marine Micropaleontology 27 (1–4), 13–25.

Dowsett, H.J., Robinson, M.M., Foley, K.M., Herbert, T.D., Otto-Bliesner, B.L., Spivey, W., 2019. Mid-piacenzian of the North Atlantic Ocean. Stratigraphy 16 (3), 119–144.

Dowsett, H.J., Robinson, M.M., Haywood, A.M., Hill, D.J., Dolan, A.M., Stoll, D.K., et al., 2012. Assessing confidence in Pliocene sea surface temperatures to evaluate predictive models. Nature Climate Change 2 (5), 365–371.

Drewry, D.J., 1976. Sedimentary basins of the East Antarctic craton from geophysical evidence. Tectonophysics 36, 301–314.

Drewry, D.J. (Ed.), 1983. Antarctica: Glaciological and Geophysical Folio. Scott Polar Research Institute, Cambridge, 8 maps pp.

Drewry, D.J., Jordan, S.R., 1983. The Bedrock Surface of Antarctica. Scott Polar Research Institute, Cambridge.

Dumitru, O.A., Austermann, J., Polyak, V.J., Fornós, J.J., Asmerom, Y., Ginés, J., et al., 2019. Constraints on global mean sea level during Pliocene warmth. Nature 574 (7777), 233–236.

Duncan, B., McKay, R., Bendle, J., Naish, T., Inglis, G.N., Moossen, H., et al., 2019. Lipid biomarker distributions in Oligocene and Miocene sediments from the Ross Sea region, Antarctica: implications for use of biomarker proxies in glacially-influenced settings. Palaeogeography, Palaeoclimatology, Palaeoecology 516, 71–89.

Dutton, A., Carlson, A.E., Long, A.J., Milne, G.A., Clark, P.U., DeConto, R., et al., 2015. Sea-level rise due to polar ice-sheet mass loss during past warm periods. Science 349 (6244).

Eagles, G., Jokat, W., 2014. Tectonic reconstructions for paleobathymetry in Drake Passage. Tectonophysics 611, 28–50.

Edwards, T.L., Brandon, M.A., Durand, G., Edwards, N.R., Golledge, N.R., Holden, P.B., et al., 2019. Revisiting Antarctic ice loss due to marine ice-cliff instability. Nature 566 (7742), 58–64.

Egger, L.M., Bahr, A., Friedrich, O., Wilson, P.A., Norris, R.D., van Peer, T.E., et al., 2018. Sea-level and surface-water change in the western North Atlantic across the Oligocene–Miocene Transition: a palynological perspective from IODP Site U1406 (Newfoundland margin). Marine Micropaleontology 139, 57–71.

Eisen, O., Hofstede, C., Diez, A., Kristoffersen, Y., Lambrecht, A., Mayer, C., et al., 2015. On-ice vibroseis and snowstreamer systems for geoscientific research. Polar Science 9 (1), 51–65.

Eittreim, S.L., Cooper, A.K., Wannesson, J., Davies, T.Ae, Coffin, M.Fe, Wise, S.We, 1995. Seismic stratigraphic evidence of ice-sheet advances on the Wilkes Land margin of Antarctica. Selected topics relating to the Indian Ocean basins and margins. Sedimentary Geology 96 (1–2), 131–156.

Ekart, D.D., Cerling, T.E., Montanez, I.P., Tabor, N.J., 1999. A 400 million year carbon isotope record of pedogenic carbonate; implications for paleoatomospheric carbon dioxide. American Journal of Science 299 (10), 805–827.

Emiliani, C., 1955. Pleistocene temperatures. Journal of Geology 63 (6), 538–578.

Emiliani, C., 1966. Isotopic paleotemperatures. Science 154, 851–857.

Escutia, C., Brinkhuis, H., 2014. Chapter 3.3—From greenhouse to icehouse at the Wilkes Land Antarctic Margin: IODP expedition 318 synthesis of results. In: Stein, R., Blackman, D.K., Inagaki, F., Larsen, H.-C. (Eds.), Developments in Marine Geology. Elsevier, pp. 295–328.

Escutia, C., Bárcena, M.A., Lucchi, R.G., Romero, O., Ballegeer, A.M., Gonzalez, J.J., et al., 2009. Circum-Antarctic warming events between 4 and 3.5Ma recorded in marine sediments from the Prydz Bay (ODP Leg 188) and the Antarctic Peninsula (ODP Leg 178) margins. Global and Planetary Change 69 (3), 170–184.

Escutia, C., Brinkhuis, H., Klaus, A.and Expedition 318 Scientists (Eds.), 2011a. Proceedings of the Integrated Ocean Drilling Program. Integrated Ocean Drilling Program Management International, Inc., Tokyo.

Escutia, C., Brinkhuis, H., Klaus, A. and the I.E.S., 2011b. IODP Expedition 318: From Greenhouse to Icehouse at the Wilkes Land Antarctic Margin. Scientific Drilling 12, 15–23.

Escutia, C., De Santis, L., Donda, F., Dunbar, R., Cooper, A., Brancolini, G., et al., 2005. Cenozoic ice sheet history from East Antarctic Wilkes Land continental margin sediments. Global and Planetary Change 45 (1–3), 51–81.

Escutia, C., DeConto, R.M., Dunbar, R., Santis, L.D., Shevenell, A., Naish, T., 2019. Keeping an eye on Antarctic Ice Sheet stability. Oceanography 32 (1), 32–46.

Escutia, C., Eittreim, S.L., Cooper, A.K., Nelson, C.H., 2000. Morphology and acoustic character of the antarctic Wilkes Land turbidite systems; ice-sheet-sourced vs river-sourced fans. Journal of Sedimentary Research, Section A: Sedimentary Petrology and Processes 70 (1), 84–93.

Escutia, C., Nelson, C.H., Acton, G.D., Eittreim, S.L., Cooper, A.K., Warnke, D.A., et al., 2002. Current controlled deposition on the Wilkes Land continental rise, Antarctica. Deep-water contourite systems; modern drifts and ancient series, seismic and sedimentary characteristics. Memoir – Geological Society of London 22, 373–384.

Evangelinos, D., Escutia, C., Etourneau, J., Hoem, F., Bijl, P., Boterblom, W., et al., 2020. Late Oligocene-Miocene proto-Antarctic Circumpolar Current dynamics off the Wilkes Land margin, East Antarctica. Global and Planetary Change 191, 103221.

Exon, N.F., Kennett, J.P., Malone, M.J., 2004. Leg 189 Synthesis: Cretaceous-Holocene History of the Tasmanian Gateway. In: Exon, N.F., Kennett J.P., Malone M.J. (Eds.), Proceedings of the Ocean Drilling Program.

Exon, N.F., Kennett, J.P., Malone, M.J., Brinkhuis, H., Chaproniere, G.C.H., Ennyu, A., et al., 2001. Proceedings of the Ocean Drilling Program, initial reports, the Tasmanian Gateway, Cenozoic climatic and oceanographic development; covering Leg 189 of the cruises of the drilling vessel JOIDES Resolution; Hobart, Tasmania, to Sydney, Australia; sites 1168–1172, 11 March–6 May 2000. Proceedings of the Ocean Drilling Program, Part A: Initial Reports, 189: (variously paginated).

Fairbanks, R.G., Matthews, R.K., 1978. The marine oxygen isotope record in Pleistocene coral, Barbados, West Indies. Quaternary Research (New York) 10 (2), 181–196.

Farrell, J.W., Raffi, I., Janecek, T.R., Murray, D.W., Levitan, M., Dadey, K.A., et al., 1995. Late Neogene sedimentation patterns in the eastern Equatorial Pacific Ocean. Proceedings of the Ocean Drilling Program; scientific results; eastern Equatorial Pacific, covering Leg 138 of the cruises of the drilling vessel JOIDES Resolution, Balboa, Panama, to San Diego, California, Sites 844–854, 1 May–4 July 1991. Proceedings of the Ocean Drilling Program, Scientific Results, 138: 717–756.

Feakins, S.J., Warny, S., Lee, J.-E., 2012. Hydrologic cycling over Antarctica during the middle Miocene warming. Nature Geoscience 5 (8), 557–560.

Fedorov, A.V., Brierley, C.M., Lawrence, K.T., Liu, Z., Dekens, P.S., Ravelo, A.C., 2013. Patterns and mechanisms of early Pliocene warmth. Nature 496 (7443), 43–49.

Fernandez, R., Gulick, S., Domack, E., Montelli, A., Leventer, A., Shevenell, A., et al., 2018. Past ice stream and ice sheet changes on the continental shelf off the Sabrina Coast. East Antarctica. Geomorphology 317, 10–22.

Ferraccioli, F., Armadillo, E., Jordan, T., Bozzo, E., Corr, H., 2009. Aeromagnetic exploration over the East Antarctic Ice Sheet: A new view of the Wilkes Subglacial Basin. Tectonophysics 478 (1–2), 62–77.

Ferrar, H.T., 1907. Report on the field geology of the region explored during the "Discovery" Antarctic Expedition 1901-4. National Antarctic Expedition 1, 1–100 (Geology).

Fielding, C., Whittaker, J., Henrys, S., Wilson, T., Naish, T., 2007. Seismic facies and stratigraphy of the Cenozoic succession in McMurdo Sound, Antarctica: Implications for tectonic, climatic, and glacial history, Short Research Paper 090. In: Cooper A.K., Raymond C. (Eds.), Antarctica: A Keystone in a Changing World—Online Proceedings of the 10th ISAES. USGS Open-File Report 2007-1047, Santa Barbara, United States, 4 p.

Fielding, C.R., 2018. Stratigraphic architecture of the Cenozoic succession in the McMurdo Sound region, Antarctica: an archive of polar palaeoenvironmental change in a failed rift setting. Sedimentology 65 (1), 1–61.

Fielding, C.R., Browne, G.H., Field, B., Florindo, F., Harwood, D.M., Krissek, L.A., et al., 2011. Sequence stratigraphy of the ANDRILL AND-2A drillcore, Antarctica: a long-term, ice-proximal record of Early to Mid-Miocene climate, sea-level and glacial dynamism. Palaeogeography, Palaeoclimatology, Palaeoecology 305 (1–4), 337–351.

Fielding, C.R., Henrys, S.A., Wilson, T.J., 2005. Rift history of the Western Victoria Land Basin: a new perspective based on integration of cores with seismic reflection data. In: Fütterer, D.K., Damaske, D., Kleinschmidt, G., Miller, H., Tessensohn, F. (Eds.), Antarctica: Contributions to Global Earth Sciences. Springer-Verlag, Berlin, Heidelberg, New York, pp. 307–316.

Fielding, C.R., Naish, T., Woolfe, K.J., Lavelle, M.J., 2000. Facies analysis and sequence stratigraphy of CRP-2/2A, Victoria Land Basin, Antarctica. Terra Antartica 7, 323–338.

Fielding, C.R., Whittaker, J., Henrys, S.A., Wilson, T.J., Naish, T.R., 2008. Seismic facies and stratigraphy of the Cenozoic succession in McMurdo Sound, Antarctica: implications for tectonic, climatic and glacial history. Palaegeography, Palaeoclimatology, Palaeoecology 260, 8–29.

Florindo, F., Farmer, R.K., Harwood, D.M., Cody, R.D., Levy, R., Bohaty, S.M., et al., 2013. Paleomagnetism and biostratigraphy of sediments from Southern Ocean ODP Site 744 (southern Kerguelen Plateau): implications for early-to-middle Miocene climate in Antarctica. Global and Planetary Change 110 (Part C), 434–454.

Florindo, F., Roberts, A.P., 2005. Eocene-Oligocene magnetobiochronology of ODP Sites 689 and 690, Maud Rise, Weddell Sea, Antarctica. Geological Society of America Bulletin 117 (1–2), 46–66.

Florindo, F., Wilson, G.S., Roberts, A.P., Sagnotti, L., Verosub, K.L., 2005. Magnetostratigraphic chronology of a late Eocene to early Miocene glacimarine succession from the Victoria Land Basin, Ross Sea, Antarctica. Global and Planetary Change 45 (1–3), 207–236.

Flower, B.P., 1999. Warming without high CO_2? Nature (London) 399 (6734), 313–314.

Flower, B.P., Kennett, J.P., 1993a. Middle Miocene ocean-climate transition: high-resolution oxygen and carbon isotopic records from Deep Sea Drilling Project Site 588A, southwest Pacific. Paleoceanography 8 (6), 811–843.

Flower, B.P., Kennett, J.P., 1993b. Relations between Monterey Formation deposition and middle Miocene global cooling; Naples Beach section, California. Geology (Boulder) 21 (10), 877–880.

Flower, B.P., Kennett, J.P., 1994. The middle Miocene climatic transition: East Antarctic ice sheet development, deep ocean circulation and global carbon cycling. Palaeogeography, Palaeoclimatology, Palaeoecology 108 (3–4), 537–555.

Foster, G.L., Rohling, E.J., 2013. Relationship between sea level and climate forcing by CO_2 on geological timescales. Proceedings of the National Academy of Sciences 110 (4), 1209–1214.

Foster, G.L., Lear, C.H., Rae, J.W.B., 2012. The evolution of pCO_2, ice volume and climate during the middle Miocene. Earth and Planetary Science Letters 341–344, 243–254.

Fountain, A.G., Lyons, W.B., Burkins, M.B., Dana, G.L., Doran, P.T., Lewis, K.J., et al., 1999. Physical controls on the Taylor Valley ecosystem, Antarctica. Bioscience 49 (12), 961–971.

Frank, M., Whiteley, N., Kasten, S., Hein, J.R., O'Nions, K., 2002. North Atlantic Deep Water export to the Southern Ocean over the past 14 Myr: evidence from Nd and Pb isotopes in ferromanganese crusts. Paleoceanography 17 (2), 12-11–12-19.

Fretwell, P., Pritchard, H.D., Vaughan, D.G., Bamber, J.L., Barrand, N.E., Bell, R., et al., 2013. Bedmap2: improved ice bed, surface and thickness datasets for Antarctica. The Cryosphere 7 (1), 375–393.

Galeotti, S., DeConto, R., Naish, T., Stocchi, P., Florindo, F., Pagani, M., et al., 2016. Antarctic Ice Sheet variability across the Eocene-Oligocene boundary climate transition. Science 352 (6281), 76–80.

Galeotti, S., Lanci, L., Florindo, F., Naish, T.R., Sagnotti, L., Sandroni, S., et al., 2012. Cyclochronology of the Eocene–Oligocene transition from the Cape Roberts Project-3 core, Victoria Land basin, Antarctica. Palaeogeography, Palaeoclimatology, Palaeoecology 335–336, 84–94.

Gamble, J.A., Barrett, P.J., Adams, C.J., 1986. Basaltic clasts from Unit 8. In: Barrett, P.J. (Ed.), Antarctic Cenozoic History from the MSSTS-1 Drillhole, McMurdo Sound, DSIR Bulletin. Science Information Publishing Centre, Wellington, pp. 145–152.

Garbe, J., Albrecht, T., Levermann, A., Donges, J.F., Winkelmann, R., 2020. The hysteresis of the Antarctic Ice Sheet. Nature 585 (7826), 538–544.

Garcia-Castellanos, D., Villaseñor, A., 2011. Messinian salinity crisis regulated by competing tectonics and erosion at the Gibraltar arc. Nature 480 (7377), 359–363.

Gasson, E., DeConto, R., Pollard, D., 2015. Antarctic bedrock topography uncertainty and ice sheet stability. Geophysical Research Letters 42 (13), 5372–5377.

Gasson, E., DeConto, R.M., Pollard, D., 2016a. Modeling the oxygen isotope composition of the Antarctic ice sheet and its significance to Pliocene sea level. Geology 44 (10), 827–830.

Gasson, E., DeConto, R.M., Pollard, D., Levy, R.H., 2016b. Dynamic Antarctic ice sheet during the early to mid-Miocene. Proceedings of the National Academy of Sciences 113 (13), 3459–3464.

Gasson, E.G., Keisling, B.A., 2020. The Antarctic Ice Sheet. Oceanography 33 (2), 90–100.

Gohl, K., Wellner, J.S., Klaus, A., and the Expedition 379 Scientists, 2019. Expedition 379 Preliminary Report: Amundsen Sea West Antarctic Ice Sheet History. International Ocean Discovery Program. Available from: https://doi.org/10.14379/iodp.pr.379.2019.

Goldner, A., Herold, N., Huber, M., 2014. The challenge of simulating the warmth of the mid-Miocene climatic optimum in CESM1. Climate of the Past 10 (2), 523–536.

Golledge, N.R., Keller, E.D., Gomez, N., Naughten, K.A., Bernales, J., Trusel, L.D., et al., 2019. Global environmental consequences of twenty-first-century ice-sheet melt. Nature 566 (7742), 65–72.

Golledge, N.R., Kowalewski, D.E., Naish, T.R., Levy, R.H., Fogwill, C.J., Gasson, E.G.W., 2015. The multi-millennial Antarctic commitment to future sea-level rise. Nature 526 (7573), 421–425.

Golledge, N.R., Levy, R.H., McKay, R.M., Naish, T.R., 2017a. East Antarctic ice sheet most vulnerable to Weddell Sea warming. Geophysical Research Letters 44 (5), 2343–2351.

Golledge, N.R., Thomas, Z.A., Levy, R.H., Gasson, E.G.W., Naish, T.R., McKay, R.M., et al., 2017b. Antarctic climate and ice-sheet configuration during the early Pliocene interglacial at 4.23 Ma. Climate of the Past 13 (7).

Goodge, J.W., Severinghaus, J.P., 2016. Rapid Access Ice Drill: a new tool for exploration of the deep Antarctic ice sheets and subglacial geology. Journal of Glaciology 62 (236), 1049–1064.

Goodge, J.W., Fanning, C.M., Norman, M.D., Bennett, V.C., 2012. Temporal, isotopic and spatial relations of early Paleozoic Gondwana-margin arc magmatism, central Transantarctic Mountains, Antarctica. Journal of Petrology 53 (10), 2027–2065.

Gordon, A.L., Tchernia, P.L., 1978. Waters of the continental margin off Adélie Coast, Antarctica. Antarctica Oceanology II: The Australian–New Zealand Sector 19, 59–69.

Gradstein, F.M., Ogg, J.G., 2020. Chapter 2—The chronostratigraphic scale. In: Gradstein, F.M., Ogg, J.G., Schmitz, M.D., Ogg, G.M. (Eds.), Geologic Time Scale 2020. Elsevier, pp. 21–32.

Granot, R., Cande, S.C., Stock, J.M., Davey, F.J., Clayton, R.W., 2010. Postspreading rifting in the Adare Basin, Antarctica: regional tectonic consequences. Geochemistry, Geophysics, Geosystems 11 (Q08005), 29.

Grant, G., Naish, T., 2021. Pliocene sea-level revisited: is there more than meets the eye? PAGES Magazine 29.

Grant, G.R., Naish, T.R., Dunbar, G.B., Stocchi, P., Kominz, M.A., Kamp, P.J.J., et al., 2019. The amplitude and origin of sea-level variability during the Pliocene epoch. Nature 574 (7777), 237–241.

Greenop, R., Foster, G.L., Wilson, P.A., Lear, C.H., 2014. Middle Miocene climate instability associated with high-amplitude CO_2 variability. Paleoceanography 29 (9), 2014PA002653.

Greenop, R., Sosdian, S.M., Henehan, M.J., Wilson, P.A., Lear, C.H., Foster, G.L., 2019. Orbital forcing, ice vVolume, and CO_2 across the Oligocene-Miocene Transition. Paleoceanography and Paleoclimatology 34 (3), 316–328.

Griener, K.W., Warny, S., Askin, R., Acton, G., 2015. Early to middle Miocene vegetation history of Antarctica supports eccentricity-paced warming intervals during the Antarctic icehouse phase. Global and Planetary Change 127, 67–78.

Grützner, J., Hillenbrand, C.-D., Rebesco, M., 2005. Terrigenous flux and biogenic silica deposition at the Antarctic continental rise during the late Miocene to early Pliocene: implications for ice sheet stability and sea ice coverage. Global and Planetary Change 45 (1–3), 131–149.

Gulick, S.P.S., Shevenell, A.E., Montelli, A., Fernandez, R., Smith, C., Warny, S., et al., 2017. Initiation and long-term instability of the East Antarctic Ice Sheet. Nature 552, 225.

Gunn, B.M., 1963. Geological structure and stratigraphic correlation in Antarctica. New Zealand Journal of Geology and Geophysics 6 (3), 423–443.

Gunn, B.M., Warren, G., 1962. Geology of Victoria Land between the Mawson and Mulock Glaciers, Antarctica. New Zealand Geological Survey Bulletin 100 (71), 157.

Guo, Z.T., Sun, B., Zhang, Z.S., Peng, S.Z., Xiao, G.Q., Ge, J.Y., et al., 2008. A major reorganization of Asian climate by the early Miocene. Climate of the Past 4 (3), 153–174.

Haile-Selassie, Y., Suwa, G., White, T.D., 2004. Late Miocene teeth from Middle Awash, Ethiopia, and early hominid dental evolution. Science 303 (5663), 1503–1505.

Halberstadt, A.R.W., Chorley, H., Levy, R.H., Naish, T., DeConto, R.M., Gasson, E., et al., 2021. CO_2 and tectonic controls on Antarctic climate and ice-sheet evolution in the mid-Miocene. Earth and Planetary Science Letters 564, 116908.

Hall, B.L., Denton, G.H., 2005. Surficial geology and geomorphology of eastern and central Wright Valley, Antarctica. Geomorphology 64 (1–2), 25–65.

Hall, B.L., 1992. Surficial geology and geomorphology of eastern Wright Valley, Antarctica: Implications for Plio-Pleistocene ice-sheet dynamics. Master of Science thesis, University of Maine.

Hall, B.L., Denton, G.H., Lux, D.R., Bockheim, J.G., 1993a. Late Tertiary Antarctic Paleoclimate and Ice-Sheet Dynamics inferred from surficial deposits in Wright Valley. Geografiska Annaler. Series A, Physical Geography 75 (4), 239–267.

Hall, B.L., Denton, G.H., Lux, D.R., Schluechter, C., 1997. Pliocene paleoenvironment and Antarctic ice sheet behavior; evidence from Wright Valley. Journal of Geology 105 (3), 285–294.

Hall, B.L., Denton, G.H., Lux, D.R., Bockheim, J.G., Sugden, D.E.e., Marchant, D.R.e. et al., 1993b. Late Tertiary Antarctic paleoclimate and ice-sheet dynamics inferred from surficial deposits in Wright Valley. The case for a stable East Antarctica ice sheet. Proceedings. Geografiska Annaler. Series A: Physical Geography 75 (4), 239–267.

Hambrey, M.J., McKelvey, B.C., 2000. Neogene fjordal sedimentation on the western margin of the Lambert Graben, East Antarctica. Sedimentology 47 (3), 577–607.

Hamon, N., Sepulchre, P., Lefebvre, V., Ramstein, G., 2013. The role of eastern Tethys seaway closure in the Middle Miocene Climatic Transition (ca. 14 Ma). Climate of the Past 9 (6), 2687–2702.

Hansen, M.A., Passchier, S., 2017. Oceanic circulation changes during early Pliocene marine ice-sheet instability in Wilkes Land, East Antarctica. Geo-Marine Letters 37 (3), 207–213.

Hansen, M.A., Passchier, S., Khim, B.-K., Song, B., Williams, T., 2015. Threshold behavior of a marine-based sector of the East Antarctic Ice Sheet in response to early Pliocene ocean warming. Paleoceanography 30 (6), 789–801.

Haq, B.U., Hardenbol, J., Vail, P.R., 1987. Chronology of fluctuating sea levels since the Triassic. Science 235 (4793), 1156–1167.

Harrington, H.J., 1958. Nomenclature of Rock Units in the Ross Sea Region, Antarctica. Nature 182 (4631), 290.

Hartman, B.N., 1998. Miocene Paleoclimate and Ice-Sheet Dynamics as Recorded in Central Taylor Valley, Antarctica, Boston University. Boston, Massachusetts 179.

Hartman, J.D., Sangiorgi, F., Salabarnada, A., Peterse, F., Houben, A.J.P., Schouten, S., et al., 2018. Paleoceanography and ice sheet variability offshore Wilkes Land, Antarctica—part 3: insights from Oligocene–Miocene TEX86-based sea surface temperature reconstructions. Climate of the Past 14 (9), 1275–1297.

Harwood, D., Florindo, F., Talarico, F., Levy, R.H. (Eds.), 2008–2009. Studies from the ANDRILL Southern McMurdo Sound Project, Antarctica – Initial Science Report on AND-2A, 235 pp.

Harwood, D., Florindo, F., Talarico, F., Levy, R., Kuhn, G., Naish, T., et al., 2009. Antarctic Drilling recovers stratigraphic records from the continental margin. EOS, Transactions of the American Geophysical Union 90 (11), 90–91.

Harwood, D.M., Levy, R.H., 2000. The McMurdo erratics: introduction and overview. In: Stilwell, J.D., Feldmann, R.M. (Eds.), Paleobiology and Paleoenvironments of Eocene Rocks, McMurdo Sound, East Antarctica: Antarctic Research Series. American Geophysical Union, Washington, D.C., pp. 1–18.

Harwood, D.M., Scherer, R.P., Webb, P.-N., 1989. Multiple Miocene marine productivity events in West Antarctica as recorded in upper Miocene sediments beneath the Ross Ice Shelf (Site J-9). Marine Micropaleontology 15 (1), 91–115.

Hayes, D.E., Frakes, L.A. and Shipboard Scientific Party, 1975d. Sites 270, 271, 272. In: Hayes, D.E., Frakes, L.A. (Eds.), Initial Reports of the Deep Sea Drilling Project, Leg 28. pp. 211–334.

Hayes, D.E., Frakes, L.A., Barrett, P.J., Burns, D.A., Chen, P.-H., Ford, A.B., et al., 1975a. Freemantle, Australia, to Christchurch, New Zealand; December 1972–February 1973. Initial Reports of the Deep Sea Drilling Project 28, 1017.

Hayes, D.E., Frakes, L.A., Barrett, P.J., Burns, D.A., Chen, P.-H., Ford, A.B., et al., 1975b. Introduction. Initial Reports of the Deep Sea Drilling Project 28, 5–18.

Hayes, D.E., Frakes, L.A., Barrett, P.J., Burns, D.A., Chen, P.-H., Ford, A.B., et al., 1975c. Site 268. Initial Reports of the Deep Sea Drilling Project 28, 153–178.

Haywood, A.M., Dowsett, H.J., Dolan, A.M., Rowley, D., Abe-Ouchi, A., Otto-Bliesner, B., et al., 2016. The Pliocene Model Intercomparison Project (PlioMIP) phase 2: scientific objectives and experimental design. Climate of the Past 12 (3), 663–675.

Haywood, A.M., Hill, D.J., Dolan, A.M., Otto-Bliesner, B.L., Bragg, F., Chan, W.L., et al., 2013. Large-scale features of Pliocene climate: results from the Pliocene Model Intercomparison Project. Climate of the Past 9 (1), 191–209.

Haywood, A.M., Smellie, J.L., Ashworth, A.C., Cantrill, D.J., Florindo, F., Hambrey, M.J., et al., 2008. Middle Miocene to Pliocene history of Antarctica and the southern ocean. Developments in Earth and Environmental Sciences 8, 401–463.

Haywood, A.M., Tindall, J.C., Dowsett, H.J., Dolan, A.M., Foley, K.M., Hunter, S.J., et al., 2020. The Pliocene Model Intercomparison Project Phase 2: large-scale climate features and climate sensitivity. Climate of the Past 16 (6), 2095–2123.

Helland, P., Holmes, M.A., 1997. Surface textural analysis of quartz sand grains from ODP Site 918 off the southeast coast of Greenland suggests glaciation of southern Greenland at 11 Ma. Palaeogeography, Palaeoclimatology, Palaeoecology 135 (1–4), 109–121.

Henrys, S.A., Bartek, L.R., Brancolini, G., Luyendyk, B.P., Hamilton, R.J., Sorlien, C.C., et al., 1998. Seismic stratigraphy of the pre-Quaternary strata off Cape Roberts and their correlation with strata cored in the CIROS-1 drillhole, McMurdo Sound Studies from the Cape Roberts Project; Ross Sea, Antarctica; scientific report of CRP-1. Terra Antartica 5 (3), 273–279.

Henrys, S.A., Buecker, C.J., Niessen, F., Bartek, L.R., Buecker, Ce, Davey, F.Je, 2001. Correlation of seismic reflectors with the CRP-3 drillhole, Victoria Land Basin, Antarctica Studies from the Cape Roberts Project; Ross Sea, Antarctica; scientific report of CRP-3; Part 1, Geophysics and tectonic studies for CRP-3. Terra Antartica 8 (3), 127–136.

Henrys, S.A., Wilson, T.J., Whittaker, J.M., Fielding, C.R., Hall, J., Naish, T.R., 2007. Tectonic history of mid-Miocene to present southern Victoria Land Basin, inferred from seismic stratigraphy in McMurdo Sound, Antarctica. USGS and National Academies, USGS OF-2007-1047, Short Research Paper 049.

Herbert, T.D., Lawrence, K.T., Tzanova, A., Peterson, L.C., Caballero-Gill, R., Kelly, C.S., 2016. Late Miocene global cooling and the rise of modern ecosystems. Nature Geoscience 9 (11), 843–847.

Hill, D.J., Haywood, A.M., Hindmarsh, R.C.A., Valdes, P.J., 2007. Characterizing ice sheets during the Pliocene: evidence from data and models. In: Williams, M., Haywood, A.M., Gregory, F.J., Schmidt, D.N., (Eds.) Deep-time perspectives on climate change: marrying the signal from computer models and biological proxies, 517–538. London, Geological Society of London. (Micropalaeontological Society special publications).

Hill, D.J., Bolton, K.P., Haywood, A.M., 2017. Modelled ocean changes at the Plio-Pleistocene transition driven by Antarctic ice advance. Nature Communications 8, 14376.

Hill, D.J., Haywood, A.M., Valdes, P.J., Francis, J.E., Lunt, D.J., Wade, B.S., et al., 2013. Paleogeographic controls on the onset of the Antarctic circumpolar current. Geophysical Research Letters 40 (19), 2013GL057439.

Hillenbrand, C.-D., Cortese, G., 2006. Polar stratification: A critical view from the Southern Ocean. Palaeogeography, Palaeoclimatology, Palaeoecology 242 (3–4), 240–252.

Hinz, K., Block, M., 1984. Results of geophysical investigations in the Weddell Sea and in the Ross Sea, Antarctica, Proceedings 11th World Petroleum Congress, London. Wiley, New York, pp. 279-291.

Ho, S.L., Laepple, T., 2016. Flat meridional temperature gradient in the early Eocene in the subsurface rather than surface ocean. Nature Geoscience 9 (8), 606–610.

Hochmuth, K., Gohl, K., Leitchenkov, G., Sauermilch, I., Whittaker, J.M., Uenzelmann-Neben, G., et al., 2020. The evolving paleobathymetry of the circum-Antarctic Southern Ocean since 34 Ma: a key to understanding past cryosphere-ocean developments. Geochemistry, Geophysics, Geosystems 21 (8), e2020GC009122.

Hodel, F., Grespan, R., de Rafélis, M., Dera, G., Lezin, C., Nardin, E., et al., 2021. Drake Passage gateway opening and Antarctic Circumpolar Current onset 31 Ma ago: the message of foraminifera and reconsideration of the Neodymium isotope record. Chemical Geology 570, 120171.

Hodell, D.A., Venz-Curtis, K.A., 2006. Late Neogene history of deepwater ventilation in the Southern Ocean. Geochemistry, Geophysics, Geosystems 7 (9).

Hodell, D.A., Elmstrom, K.M., Kennett, J.P., 1986. Latest Miocene benthic ^{18}O changes, global ice volume, sea level and the "Messinian salinity crisis". Nature (London) 320 (6061), 411–414.

Hodgson, D.A., Bentley, M.J., Smith, J.A., Klepacki, J., Makinson, K., Smith, A.M., et al., 2016. Technologies for retrieving sediment cores in Antarctic subglacial settings. Philosophical Transactions of the Royal Society A: Mathematical, Physical and Engineering Sciences 374 (2059), 20150056.

Hofstede, C., Eisen, O., Diez, A., Jansen, D., Kristoffersen, Y., Lambrecht, A., et al., 2013. Investigating englacial reflections with vibro- and explosive-seismic surveys at Halvfarryggen ice dome, Antarctica. Annals of Glaciology 54 (64), 189–200.

Holbourn, A., Kuhnt, W., Clemens, S., Prell, W., Andersen, N., 2013. Middle to late Miocene stepwise climate cooling: evidence from a high-resolution deep water isotope curve spanning 8 million years. Paleoceanography 28 (4), 688–699.

Holbourn, A., Kuhnt, W., Kochhann, K.G.D., Andersen, N., Sebastian Meier, K.J., 2015. Global perturbation of the carbon cycle at the onset of the Miocene Climatic Optimum. Geology 43 (2), 123–126.

Holbourn, A., Kuhnt, W., Lyle, M., Schneider, L., Romero, O., Andersen, N., 2014. Middle Miocene climate cooling linked to intensification of eastern equatorial Pacific upwelling. Geology 42 (1), 19–22.

Holbourn, A., Kuhnt, W., Schulz, M., Erlenkeuser, H., 2005. Impacts of orbital forcing and atmospheric carbon dioxide on Miocene Ice Sheet expansion. Nature 438, 483–487.

Holbourn, A., Kuhnt, W., Schulz, M., Flores, J.-A., Andersen, N., 2007. Orbitally-paced climate evolution during the middle Miocene "Monterey" carbon-isotope excursion. Earth and Planetary Science Letters 261 (3–4), 534–550.

Holbourn, A.E., Kuhnt, W., Clemens, S.C., Kochhann, K.G.D., Jöhnck, J., Lübbers, J., et al., 2018. Late Miocene climate cooling and intensification of southeast Asian winter monsoon. Nature Communications 9 (1), 1584.

Hole, M., LeMasurier, W., 1994. Tectonic controls on the geochemical composition of Cenozoic, mafic alkaline volcanic rocks from West Antarctica. Contributions to Mineralogy and Petrology 117 (2), 187–202.

Hornibrook, Nd.B., 1992. New Zealand Cenozoic Marine Paleoclimates: A Review Based on the Distribution of Some Shallow Water and Terrestrial Biota. Pacific Neogene: Environment, Evolution, and Events. University of Tokyo Press, Tokyo, pp. 83–106.

Hsü, K.J., Bernoulli, D., 1978. Genesis of the Tethys and the Mediterranean. Available from: https://doi.org/10.2973/dsdp.proc.42-1.149.1978.

Hsü, K.J., Ryan, W.B., Cita, M.B., 1973. Late Miocene desiccation of the Mediterranean. Nature 242 (5395), 240–244.

Huber, M., Nof, D., 2006. The ocean circulation in the southern hemisphere and its climatic impacts in the Eocene. Palaeogeography, Palaeoclimatology, Palaeoecology 231 (1–2), 9–28.

Huybers, P., Tziperman, E., 2008. Integrated summer insolation forcing and 40,000-year glacial cycles: the perspective from an ice-sheet/energy-balance model. Paleoceanography 23 (1).

Huybrechts, P., Sugden, D.E.e., Marchant, D.R.e., Denton, G.H.E., 1993. Glaciological modelling of the late Cenozoic East Antarctic ice sheet; stability or dynamism? The case for a stable East Antarctica ice sheet; proceedings. Geografiska Annaler. Series A: Physical Geography 75 (4), 221–238.

IPCC, 2013. Climate change 2013: the physical science basis. In: Stocker, T.F., Qin, D., Plattner, G.-K., Tignor, M., Allen, S.K., Boschung, J., Nauels, A., Xia, Y., Bex, V., Midgley, P.M. (Eds.), Contribution of Working Group I to the Fifth Assessment Report of the Intergovernmental Panel on Climate Change. Cambridge University Press, Cambridge, UK and New York, NY, p. 1535.

Jacobs, S.S., Giulivi, C.F., 2010. Large multidecadal salinity trends near the Pacific–Antarctic continental margin. Journal of Climate 23 (17), 4508–4524.

Jacobs, S.S., Giulivi, C.F., Mele, P.A., 2002. Freshening of the Ross Sea during the late 20th century. Science 297 (5580), 386–389.

Ji, S., Nie, J., Lechler, A., Huntington, K.W., Heitmann, E.O., Breecker, D.O., 2018. A symmetrical CO_2 peak and asymmetrical climate change during the middle Miocene. Earth and Planetary Science Letters 499, 134–144.

John, C.M., Karner, G.D., Browning, E., Leckie, R.M., Mateo, Z., Carson, B., et al., 2011. Timing and magnitude of Miocene eustasy derived from the mixed siliciclastic-carbonate stratigraphic record of the northeastern Australian margin. Earth and Planetary Science Letters 304 (3–4), 455–467.

Jordan, T.A., Riley, T.R., Siddoway, C.S., 2020. The geological history and evolution of West Antarctica. Nature Reviews Earth & Environment 1 (2), 117–133.

Joughin, I., Shean, D.E., Smith, B.E., Floricioiu, D., 2020. A decade of variability on Jakobshavn Isbræ: ocean temperatures pace speed through influence on mélange rigidity. The cryosphere 14 (1), 211–227.

Jovane, L., Florindo, F., Acton, G., Ohneiser, C., Sagnotti, L., Strada, E., et al. (2019). Miocene glacial dynamics recorded by variations in magnetic properties in the ANDRILL-2A drillcore. Journal of Geophysical Research: Solid Earth 124, 2297–2312. Available from: https://doi.org/10.1029/2018JB016865.

Kasbohm, J., Schoene, B., 2018. Rapid eruption of the Columbia River flood basalt and correlation with the mid-Miocene climate optimum. Science Advances 4 (9), eaat8223.

Kender, S., Kaminski, M.A., 2013. Arctic Ocean benthic foraminiferal faunae change associated with the onset of perennial sea ice in the Middle Miocene. Journal of Foraminiferal Research 43 (1), 99–109.

Kennett, J.P., 1977. Cenozoic evolution of Antarctic Glaciation, the Circum-Antarctic Ocean, and their impact on global paleoceanography. Journal of Geophysical Research 82 (27), 3843–3859.

Kennett, J.P., Barker, P.F., 1990. Latest Cretaceous to Cenozoic climate and oceanographic developments in the Weddell Sea, Antarctica: an ocean-drilling perspective. In: Proceedings of the Ocean Drilling Program. Scientific Results, 113, pp. 937–960.

Kennett, J.P., Houtz, R.E., Andrews, P.B., Edwards, A.R., Gostin, V.A., Hajos, M., et al., 1975. Cenozoic paleoceanography in the Southwest Pacific Ocean, Antarctic glaciation, and the development of the Circum-Antarctic Current. Initial Reports of the Deep Sea Drilling Project 29, 1155–1169.

Kennett, J.P., von der Borch, C.C., Baker, P.A., Barton, C.E., Boersma, A., Caulet, J.P., et al., 1986. Miocene to early Pliocene oxygen and carbon isotope stratigraphy in the Southwest Pacific, Deep Sea Drilling Project Leg 90. Initial reports of the Deep Sea Drilling Project covering Leg 90 of the cruises of the drilling vessel Glomar Challenger, Noumea, New Caledonia, to Wellington, New Zealand, December 1982–January 1983. Initial Reports of the Deep Sea Drilling Project 90 (Part 2), 1383–1411.

Kerr, A., Huybrechts, P., van der Wateren, F.Me, Cloetingh, S.A.P.L.E., 1999. The response of the East Antarctic ice-sheet to the evolving tectonic configuration of the Transantarctic Mountains. Lithosphere dynamics and environmental change of the Cenozoic West Antarctic Rift system. Global and Planetary Change 23 (1–4), 213–229.

Kerr, A., Sugden, D.E., Summerfield, M.A., 2000. Linking tectonics and landscape development in a passive margin setting: the Transantarctic Mountains. In: Summerfield, M.A. (Ed.), Geomorphology and Global Tectonics. John Wiley & Sons, pp. 303–319.

Knorr, G., Lohmann, G., 2014. Climate warming during Antarctic ice sheet expansion at the Middle Miocene transition. Nature Geoscience 7 (5), 376–381.

Kominz, M.A., Browning, J.V., Miller, K.G., Sugarman, P.J., Mizintseva, S., Scotese, C.R., 2008. Late Cretaceous to Miocene sea-level estimates from the New Jersey and Delaware coastal plaoin coreholes: an error analysis. Basin Research 20, 211–226.

Kominz, M.A., Miller, K.G., Browning, J.V., Katz, M.E., Mountain, G.S., 2016. Miocene relative sea level on the New Jersey shallow continental shelf and coastal plain derived from one-dimensional backstripping: A case for both eustasy and epeirogeny. Geosphere 12 (5), 1437–1456. Available from: https://doi.org/10.1130/GES01241.1.

Koppers, A.A.P., Coggon, R. (Eds.), 2020. Exploring Earth by Scientific Ocean Drilling: 2050 Science Framework. 124 pp. Available from: https://doi.org/10.6075/J0W66J9H.

Korhonen, F., Saito, S., Brown, M., Siddoway, C., Day, J., 2010. Multiple generations of granite in the Fosdick Mountains, Marie Byrd Land, West Antarctica: implications for polyphase intracrustal differentiation in a continental margin setting. Journal of Petrology 51 (3), 627–670.

Kowalewski, D.E., Marchant, D.R., Levy, J., Head, J., 2006. Quantifying low rates of summertime sublimation for buried glacier ice in Beacon Valley, Antarctica. Antarctic Science 18 (3), 421–428.

Krissek, L., Browne, G., Carter, L., Cowan, E., Dunbar, G., McKay, R., et al., and ANDRILL MIS Science Team 2007a. Sedimentology and Stratigraphy of the AND-1B core, ANDRILL McMurdo Ice Shelf Project, Antarctica. In: Naish, T., Powell, R., Levy, R. (Eds.), Studies from the ANDRILL, McMurdo Ice Shelf Project, Antarctica—Initial Science Report on AND-1B. Terra Antartica, pp. 185–222.

Krissek, L., Browne, G., Carter, L., Cowan, E.A., Dunbar, G., McKay, R., et al., 2007b. Sedimentology and stratigraphy of the AND-1B core, ANDRILL McMurdo Ice Shelf Project, Antarctica. Terra Antartica 14 (3), 185–222.

Kulhanek, D.K., Levy, R.H., Clowes, C.D., Prebble, J.G., Rodelli, D., Jovane, L., et al., 2019. Revised chronostratigraphy of DSDP Site 270 and late Oligocene to early Miocene paleoecology of the Ross Sea sector of Antarctica. Global and Planetary Change 178, 46–64.

Kürschner, W.M., Kvaček, Z., Dilcher, D.L., 2008. The impact of Miocene atmospheric carbon dioxide fluctuations on climate and the evolution of terrestrial ecosystems. Proceedings of the National Academy of Sciences 105 (2), 449–453.

Kurschner, W.M., Van der Burgh, J., Visscher, H., Dilcher, D.L., 1996. Oak leaves as biosensors of late Neogene and early Pleistocene paleoatmospheric CO_2 concentrations. Marine Micropaleontology 27 (1–4), 299–312.

Kwiek, P.B., Ravelo, A.C., 1999. Pacific Ocean intermediate and deep water circulation during the Pliocene. Palaeogeography, Palaeoclimatology, Palaeoecology 154, 191–217.

Kyle, P.R., 1990a. Erebus Volcanic Province. In: LeMasurier, W.E., Thomson, J.W. (Eds.), Volcanoes of the Antarctic Plate and Southern Ocean, Antarctic Research Series. American Geophysical Union, Washington, D. C, pp. 81–145.

Kyle, P.R., 1990b. McMurdo Volcanic Group, western Ross embayment. In: LeMasurier, W.E., Thomson, J.W. (Eds.), Volcanoes of the Antarctic Plate and Southern Oceans, Antarctic Research Series. American Geophysical Union, Washington, D. C, pp. 19–25.

Lamy, F., Winckler, G., Alvarez Zarikian, C.A., and the Expedition 383 Scientists, 2019. Expedition 383 Preliminary Report: Dynamics of the Pacific Antarctic Circumpolar Current. International Ocean Discovery Program. Available from: https://doi.org/10.14379/iodp.pr.383.2019.

Langebroek, P.M., Paul, A., Schulz, M., 2009. Antarctic ice-sheet response to atmospheric CO_2 and insolation in the Middle Miocene. Climate of the Past 5 (4), 633–646.

Langebroek, P.M., Paul, A., Schulz, M., 2010. Simulating the sea level imprint on marine oxygen isotope records during the middle Miocene using an ice sheet–climate model. Paleoceanography 25 (4).

Larsen, H.C., Saunders, A.D., Clift, P.D., Beget, J., Wei, W., Spezzaferri, S., 1994. Seven Million Years of Glaciation in Greenland. Science 264 (5161), 952–955.

Laskar, J., Fienga, A., Gastineau, M., Manche, H., 2011. La2010: a new orbital solution for the long-term motion of the Earth. Astronomy & Astrophysics 532, 15.

Laskar, J., Robutel, P., Gastineau, M., Correia, A.C.M., Levrard, B., 2004. A long term numerical solution for the insolation quantities of the Earth. Astronomy and Astrophysics 428, 261–285.

Lawrence, K.T., Herbert, T.D., Brown, C.M., Raymo, M.E., Haywood, A.M., 2009. High-amplitude variations in North Atlantic sea surface temperature during the early Pliocene warm period. Paleoceanography 24 (2).

Lawver, L.A., Gahagan, L.M., 2003. Evolution of Cenozoic seaways in the circum-Antarctic region. Palaeogeogr. Palaeoclimatol. Palaeoecol. Available from: https://doi.org/10.1016/S0031-0182(03)00392-4.

Lear, C.H., Coxall, H.K., Foster, G.L., Lunt, D.J., Mawbey, E.M., Rosenthal, Y., et al., 2015. Neogene ice volume and ocean temperatures: Insights from infaunal foraminiferal Mg/Ca paleothermometry. Paleoceanography 30 (11), 1437−1454.

Lear, C.H., Mawbey, E.M., Rosenthal, Y., 2010. Cenozoic benthic foraminiferal Mg/Ca and Li/Ca records: toward unlocking temperatures and saturation states. Paleoceanography 25 (4), PA4215.

Lear, C.H., Rosenthal, Y., Wright, J.D., 2003. The closing of a seaway: ocean water masses and global climate change. Earth and Planetary Science Letters 210 (3−4), 425−436.

Lear, C.H., Rosenthal, Y., Wright, J.D., 2020. Compiled bottom water temperatures of ODP site 154−926. In: Lear, C.H., et al., (Eds.), Atlantic and Pacific Benthic Mg/Ca Temperatures 0−12 Ma. PANGAEA. Available from: https://doi.org/10.1594/PANGAEA.913906.

Leckie, R.M., Webb, P.-N., 1983. Late Oligocene-early Miocene glacial record of the Ross Sea, Antarctica; evidence from DSDP Site 270. Geology (Boulder) 11 (10), 578−582.

Leckie, R.M., Webb, P.-N., 1986. Late Paleogene and early Neogene foraminifers of Deep Sea Drilling Project Site 270, Ross Sea, Antarctica. In: Kennett, J.P., van der Borch, C.C. (Eds.), Initial Reports DSDP, Leg 90, Noumea, New Caledonia to Wellington, New Zealand, Part 2. United States Government Printing Office, Washington, D.C., pp. 1,093-091,142.

Leutert, T.J., Auderset, A., Martínez-García, A., Modestou, S., Meckler, A.N., 2020. Coupled Southern Ocean cooling and Antarctic ice sheet expansion during the middle Miocene. Nature Geoscience 13 (9), 634−639.

Levy, R., Harwood, D., Florindo, F., Sangiorgi, F., Tripati, R., von Eynatten, H., et al., 2016. Antarctic ice sheet sensitivity to atmospheric CO_2 variations in the early to mid-Miocene. Proceedings of the National Academy of Sciences 113 (13), 3453−3458.

Levy, R., Patterson, M., van de Flierdt, T., Jiménez Espejo, F.J., Stocchi, P., Klages, J., et al., 2020. The SWAIS 2C Project-sensitivity of the West Antarctic ice sheet in a warmer world.

Levy, R.H., Harwood, D.M., 2000a. Sedimentary lithofacies of the McMurdo Sound Erratics. In: Stilwell, J.D., Feldmann, R.M. (Eds.), Paleobiology and Paleoenvironments of Eocene Rocks, McMurdo Sound, East Antarctica: Antarctic Research Series. American Geophysical Union, Washington, D.C, pp. 39−60.

Levy, R.H., Harwood, D.M., 2000b. Tertiary marine palynomorphs from the McMurdo Sound Erratics. In: Stilwell, J.D., Feldmann, R.M. (Eds.), Paleobiology and Paleoenvironments of Eocene Rocks, McMurdo Sound, East Antarctica: Antarctic Research Series. American Geophysical Union, Washington, D.C, pp. 183−242.

Levy, R.H., Cody, R.D., Crampton, J., Fielding, C., Golledge, N., Harwood, D.M., et al., 2012. Late Neogene climate and glacial history of the Southern Victoria Land coast from integrated drill core, seismic and outcrop data. Global and Planetary Change 80−81 (Special Issue), 61−84.

Levy, R.H., Harwood, D.M., Stilwell, J.De, Feldmann, R.Me, 2000. Sedimentary lithofacies of the McMurdo Sound erratics. Paleobiology and paleoenvironments of Eocene rocks, McMurdo Sound, East Antarctica. Antarctic Research Series 76, 39−61.

Levy, R.H., Meyers, S.R., Naish, T.R., Golledge, N.R., McKay, R.M., Crampton, J.S., et al., 2019. Antarctic ice-sheet sensitivity to obliquity forcing enhanced through ocean connections. Nature Geoscience 12 (2), 132−137.

Lewis, A.R., Ashworth, A.C., 2015. An early to middle Miocene record of ice-sheet and landscape evolution from the Friis Hills, Antarctica. Geological Society of America Bulletin.

Lewis, A.R., Marchant, D.R., Ashworth, A.C., Hedenäs, L., Hemming, S.R., Johnson, J.V., et al., 2008. Mid-Miocene cooling and the extinction of tundra in continental Antarctica. Proceedings of the National Academy of Sciences 105 (31), 10676−10680.

Lewis, A.R., Marchant, D.R., Ashworth, A.C., Hemming, S.R., Machlus, M.L., 2007. Major middle Miocene global climate change: evidence from East Antarctica and the Transantarctic Mountains. Geological Society of America Bulletin 119 (11–12), 1449–1461.

Lewis, A.R., Marchant, D.R., Kowalewski, D.E., Baldwin, S.L., Webb, L.E., 2006. The age and origin of the Labyrinth, western Dry Valleys, Antarctica: evidence for extensive middle Miocene subglacial floods and freshwater discharge to the Southern Ocean. Geology 34 (7), 513–516.

Licht, K.J., Hennessy, A.J., Welke, B.M., 2014. The U-Pb detrital zircon signature of West Antarctic ice stream tills in the Ross embayment, with implications for Last Glacial Maximum ice flow reconstructions. Antarctic Science 26 (6), 687–697.

Liebrand, D., de Bakker, A.T.M., Beddow, H.M., Wilson, P.A., Bohaty, S.M., Ruessink, G., et al., 2017. Evolution of the early Antarctic ice ages. Proceedings of the National Academy of Sciences 114 (15), 3867–3872.

Liebrand, D., Lourens, L.J., Hodell, D.A., de Boer, B., van de Wal, R.S.W., Pälike, H., 2011. Antarctic ice sheet and oceanographic response to eccentricity forcing during the early Miocene. Climate of the Past 7 (3), 869–880.

Lindeque, A., Gohl, K., Wobbe, F., Uenzelmann-Neben, G., Henrys, S., Davy, B., 2016. Pre-glacial to full glacial sedimentation along the Pacific margin of West Antarctica: Record of shifting pattern of supply and transport, SCAR Open Science Conference, Kuala Lumpur, Malaysia, 10 August - 20 August 2016.

Lisiecki, L.E., Raymo, M.E., 2005. A Pliocene-Pleistocene stack of 57 globally distributed benthic ^{18}O records. Paleoceanography 20 (PA 1003), 1–17.

Livermore, R., Hillenbrand, C.D., Meredith, M., Eagles, G., 2007. Drake Passage and Cenozoic climate: an open and shut case? Geochemistry, Geophysics, Geosystems 8 (1).

Londoño, L., Royer, D.L., Jaramillo, C., Escobar, J., Foster, D.A., Cárdenas-Rozo, A.L., et al., 2018. Early Miocene CO2 estimates from a Neotropical fossil leaf assemblage exceed 400 ppm. American Journal of Botany 105 (11), 1929–1937.

Loutit, T.S., Kennett, J., 1979. Application of Carbon Isotope Stratigraphy to Late Miocene Shallow Marine Sediments, New Zealand. Science 204 (4398), 1196–1199.

Lübbers, J., Kuhnt, W., Holbourn, A.E., Bolton, C.T., Gray, E., Usui, Y., et al., 2019. The middle to late Miocene "Carbonate Crash" in the equatorial Indian Ocean. Paleoceanography and Paleoclimatology 34 (5), 813–832.

Lunt, D.J., Haywood, A.M., Schmidt, G.A., Salzmann, U., Valdes, P.J., Dowsett, H.J., et al., 2012. On the causes of mid-Pliocene warmth and polar amplification. Earth and Planetary Science Letters 321–322, 128–138.

Luyendyk, B.P., Sorlien, C.C., Wilson, D.S., Bartek, L.R., Siddoway, C.S., 2001. Structural and tectonic evolution of the Ross Sea Rift in the Cape Colbeck region, eastern Ross Sea, Antarctica. Tectonics 20 (6), 933–958.

Lyle, M., Dadey, K.A., Farrell, J.W., 1995. The Late Miocene (11–8 Ma) Eastern Pacific Carbonate Crash: evidence for reorganization of deep-water Circulation by the closure of the Panama Gateway. In: 1995 Proceedings of the Ocean Drilling Program, Scientific Results, 138.

Lyle, M., Gibbs, S., Moore, T., Rea, D., 2007. Late Oligocene initiation of the Antarctic Circumpolar Current: evidence from the South Pacific. Geology 35 (8), 691–694.

Lythe, M.B., Vaughan, D.G., Lambrecht, A., Miller, H., Nixdorf, U., Oerter, H., et al., 2001. BEDMAP; a new ice thickness and subglacial topographic model of Antarctica. Journal of Geophysical Research, B, Solid Earth and Planets 106 (6), 11335–11,351.

Majewski, W., 2010. Planktonic foraminiferal response to Middle Miocene cooling in the Southern Ocean (ODP Site 747, Kerguelen Plateau). Acta Palaeontologica Polonica 55 (3), 541–560. 520.

Maldonado, A., Barnolas, A., Bohoyo, F., Galindo-Zaldivar, J., Hernandez-Molina, J., Lobo, F., et al., 2003. Contourite deposits in the central Scotia Sea; the importance of the Antarctic Circumpolar Current and the Weddell Gyre flows Antarctic Cenozoic palaeoenvironments; geologic record and models. ANTOSTRAT Symposium; The Geologic Record of the Antarctic Ice Sheet from Drilling, Coring and Seismic Studies 198 (1−2), 187−221.

Maldonado, A., Bohoyo, F., Galindo-Zaldívar, J., Hernández-Molina, F.J., Lobo, F.J., Lodolo, E., et al., 2014. A model of oceanic development by ridge jumping: opening of the Scotia Sea. Global and Planetary Change 123, 152−173.

Marchant, D.R., 1993. Miocene glacial stratigraphy and landscape evolution of the western Asgard Range, Antarctica, Geografiska Annaler. Series A: Physical Geography 303−330.

Marchant, D.R., Denton, G.H., Sugden, D.E., Swisher III, C.C., 1993. Miocene glacial stratigraphy and landscape evolution of the western Asgard Range, Antarctica. Geografiska Annaler: Series A, Physical Geography 75 (4), 303−330.

Marchant, D.R., Denton, G.H., Swisher III, C.C., Potter Jr., N., 1996. Late Cenozoic Antarctic paleoclimate reconstructed from volcanic ashes in the dry valleys region of southern Victoria Land. Geological Society of America Bulletin 108 (2), 181−194.

Markwick, P.J., 2007. The palaeogeographic and palaeoclimatic significance of climate proxies for data-model comparisons. In: Williams, M., Haywood, A.M., Gregory, F.J., Schmidt, D. N. (Eds.), Deep-Time Perspectives on Climate Change: Marrying the Signal from Computer Models and Biological Proxies. The Micropalaeontological Society, Special Publications. The Geological Society, London, 251−312.

Marshall, J., Speer, K., 2012. Closure of the meridional overturning circulation through Southern Ocean upwelling. Nature Geoscience 5 (3), 171−180.

Martinez-Boti, M.A., Foster, G.L., Chalk, T.B., Rohling, E.J., Sexton, P.F., Lunt, D.J., et al., 2015. Plio-Pleistocene climate sensitivity evaluated using high-resolution CO2 records. Nature 518 (7537), 49−54.

Martinez-Garcia, A., Rosell-Mele, A., Jaccard, S.L., Geibert, W., Sigman, D.M., Haug, G.H., 2011. Southern Ocean dust-climate coupling over the past four million years. Nature 476 (7360), 312−315.

Martos, Y.M., Maldonado, A., Lobo, F.J., Hernández-Molina, F.J., Pérez, L.F., 2013. Tectonics and palaeoceanographic evolution recorded by contourite features in southern Drake Passage (Antarctica). Marine Geology 343, 76−91.

Masson-Delmotte, V., M. Schulz, A. Abe-Ouchi, J. Beer, A. Ganopolski, J.F. González Rouco, E. Jansen, K. Lambeck, J. Luterbacher, T. Naish, T. Osborn, B. Otto-Bliesner, T. Quinn, R. Ramesh, M. Rojas, X. Shao and A. Timmermann, 2013: Information from Paleoclimate Archives. In: Climate Change 2013: The Physical Science Basis. Contribution of Working Group I to the Fifth Assessment Report of the Intergovernmental Panel on Climate Change [Stocker, T.F., D. Qin, G.-K. Plattner, M. Tignor, S.K. Allen, J. Boschung, A. Nauels, Y. Xia, V. Bex and P.M. Midgley (eds.)]. Cambridge University Press, Cambridge, United Kingdom and New York, NY, USA.

McClymont, E.L., Ford, H.L., Ho, S.L., Tindall, J.C., Haywood, A.M., Alonso-Garcia, M., et al., 2020. Lessons from a high-CO_2 world: an ocean view from $\sim$3 million years ago. Climate of the Past 16 (4), 1599−1615.

McGinnis, L.D., 1981. Dry Valley Drilling Project. American Geophysical Union, Washington, D.C, p. 465.

McIntosh, W.C., 1998. 40Ar/39Ar geochronology of volcanic clasts and pumice in CRP-1 Core, Cape Roberts, Antarctica. Studies from the Cape Roberts Project; Ross Sea, Antarctica; scientific report of CRP-1. Terra Antartica 5 (3), 683−690.

McIntosh, W.C., 2000. $^{40}Ar/^{39}Ar$ geochronology of tephra and volcanic clasts in CRP-2A, Victoria Land Basin, Antarctica. Studies from the Cape Roberts Project, Ross Sea, Antarctica; scientific report of CRP-2/2A; part II, chronology and chronostratigraphy for CRP-2/2A. Terra Antartica 7 (4–5), 621–630.

McKay, R., Browne, G., Carter, L., Cowan, E., Dunbar, G., Krissek, L., et al., 2009. The stratigraphic signature of the late Cenozoic Antarctic Ice Sheets in the Ross Embayment. Geological Society of America Bulletin 121 (11–12), 1537–1561.

McKay, R., Escutia, C., Santis, L. De., Donda, F., Duncan, B., Gohl, K., 2021. A history of Antarctic Cenozoic Glaciation. In: Florindo, F., et al. (Eds.), Antarctic Climate Evolution, second ed. Elsevier (this volume).

McKay, R., Naish, T., Carter, L., Riesselman, C., Dunbar, R., Sjunneskog, C., et al., 2012. Antarctic and Southern Ocean influences on Late Pliocene global cooling. Proceedings of the National Academy of Sciences 109 (17), 6423–6428.

McKay, R., De Santis, L., Kulhanek, D., Expedition 374 Scientists, 2019. Ross Sea West Antarctic Ice Sheet History, Proceedings of the International Ocean Discovery Program. International Ocean Discovery Program, College Station.

McKay, R.M., Barrett, P.J., Levy, R.S., Naish, T.R., Golledge, N.R., Pyne, A., 2016. Antarctic Cenozoic climate history from sedimentary records: ANDRILL and beyond. Philosophical Transactions of the Royal Society A: Mathematical, Physical and Engineering Sciences 374 (2059).

McKelvey, B., Webb, P., Kohn, B., 1977. Stratigraphy of the Taylor and lower Victoria Groups (Beacon Supergroup) between the Mackay Glacier and Boomerang Range, Antarctica. New Zealand Journal of Geology and Geophysics 20 (5), 813–863.

McKelvey, B.C., 1975. Preliminary site reports, DVDP sites 10 and 11, Taylor Valley. In: Mudrey Jr., M.G., McGinnis, L.D. (Eds.), Dry Valley Drilling Project (DVDP) Bulletin. pp. 16–60.

McKelvey, B.C., 1981. The lithologic logs of DVDP cores 10 and 11, Eastern taylor Valley. In: McGinnis, L.D. (Ed.), Dry Valley Drilling Project. Antarctic Research Series. American Geophysical Union, pp. 63–94.

Meinshausen, M., Smith, S.J., Calvin, K., Daniel, J.S., Kainuma, M.L.T., Lamarque, J.F., et al., 2011. The RCP greenhouse gas concentrations and their extensions from 1765 to 2300. Climatic Change 109 (1–2), 213–241.

Mejía, L.M., Méndez-Vicente, A., Abrevaya, L., Lawrence, K.T., Ladlow, C., Bolton, C., et al., 2017. A diatom record of CO_2 decline since the late Miocene. Earth and Planetary Science Letters 479, 18–33.

Mengel, M., Levermann, A., 2014. Ice plug prevents irreversible discharge from East Antarctica. Nature Climate Change 4 (6), 451–455.

Methner, K., Campani, M., Fiebig, J., Löffler, N., Kempf, O., Mulch, A., 2020. Middle Miocene long-term continental temperature change in and out of pace with marine climate records. Scientific Reports 10 (1).

Middleton, J.L., Ackert Jr, R.P., Mukhopadhyay, S., 2012. Pothole and channel system formation in the McMurdo Dry Valleys of Antarctica: new insights from cosmogenic nuclides. Earth and Planetary Science Letters 355, 341–350.

Miles, B.W.J., Stokes, C.R., Jamieson, S.S.R., 2016. Pan–ice-sheet glacier terminus change in East Antarctica reveals sensitivity of Wilkes Land to sea-ice changes. Science Advances 2 (5).

Miles, B.W.J., Stokes, C.R., Vieli, A., Cox, N.J., 2013. Rapid, climate-driven changes in outlet glaciers on the Pacific coast of East Antarctica. Nature 500 (7464), 563–566.

Miller, K.G., Browning, J.V., Schmelz, W.J., Kopp, R.E., Mountain, G.S., Wright, J.D., 2020. Cenozoic sea-level and cryospheric evolution from deep-sea geochemical and continental margin records. Science Advances 6 (20), eaaz1346. Available from: https://doi.org/10.1126/sciadv.aaz1346.

Miller, K.G., Kominz, M.A., Browning, J.V., Wright, J.D., Mountain, G.S., Katz, M.E., et al., 2005. The Phanerozoic record of global sea-level Change. Science 310 (5752), 1293–1298.

Miller, K.G., Raymo, M., Browning, J., Rosenthal, Y., Wright, J., 2019. Peak sea level during the warm Pliocene: errors, limitations, and constraints. PAGES Magazine 27 (1), 1–2.

Miller, K.G., Wright, J.D., Faribanks, R.G., 1991b. Unlocking the ice house: Oligocene-Miocene oxygen isotopes, eustasy, and margin erosion. Journal of Geophysical Research 96 (B4), 6829–6848.

Miller, K.G., Wright, J.D., Browning, J.V., Kulpecz, A., Kominz, M., Naish, T.R., et al., 2012. High tide of the warm Pliocene: implications of global sea level for Antarctic deglaciation. Geology 40 (5), 407–410.

Miller, K.G., Wright, J.D., Fairbanks, R.G., Cloetingh, S.E., 1991a. Unlocking the ice house; Oligocene-Miocene oxygen isotopes, eustasy, and margin erosion. Special section on long-term sea level changes. AGU Chapman Conference on Causes and consequences of Long-Term Sea Level Changes 96 (4), 6829–6848.

Modestou, S.E., Leutert, T.J., Fernandez, A., Lear, C.H., Meckler, A.N., 2020. Warm middle Miocene Indian Ocean bottom water temperatures: comparison of clumped isotope and Mg/Ca-based estimates. Paleoceanography and Paleoclimatology 35 (11), e2020PA003927.

Molnar, P., Boos, W.R., Battisti, D.S., 2010. Orographic controls on climate and paleoclimate of Asia: thermal and mechanical roles for the Tibetan Plateau. Annual Review of Earth and Planetary Sciences 38 (1), 77–102.

Montelli, A., Gulick, S.P.S., Fernandez, R., Frederick, B.C., Shevenell, A.E., Leventer, A., et al., 2019. Seismic stratigraphy of the Sabrina Coast shelf, East Antarctica: early history of dynamic meltwater-rich glaciations. GSA Bulletin 132 (3–4), 545–561.

Morlighem, M., Rignot, E., Binder, T., Blankenship, D., Drews, R., Eagles, G., et al., 2020. Deep glacial troughs and stabilizing ridges unveiled beneath the margins of the Antarctic ice sheet. Nature Geoscience 13, 132–137.

Morrison, A., Hogg, A.M., England, M.H., Spence, P., 2020. Warm Circumpolar Deep Water transport toward Antarctica driven by local dense water export in canyons. Science Advances 6 (18), eaav2516. Available from: https://doi.org/10.1126/sciadv.aav2516.

Mosbrugger, V., Utescher, T., Dilcher, D.L., 2005. Cenozoic continental climatic evolution of Central Europe. Proceedings of the National Academy of Sciences of the United States of America 102 (42), 14964–14969.

Mudrey, M.G., Jr., McGinnis, L.D., 1975. Dry Valley Drilling Project (DVDP) Bulletin, 280 pp.

Naish, T., Duncan, B., Levy, R., McKay, R., Escutia, C., Santis, L. De., et al., 2021. Antarctic Ice Sheet dynamics during the Late Oligocene and Early Miocene: climatic conundrums revisited. In: Florindo, F., et al. (Eds.), Antarctic Climate Evolution, second ed. Elsevier (this volume).

Naish, T., Powell, R., Levy, R. (Eds.), 2007. Studies from the ANDRILL, McMurdo Ice Shelf Project, Antarctica—Initial Science Report on AND-1B, pp. 121–327.

Naish, T., Powell, R., Levy, R., Wilson, G., Scherer, R., Talarico, F., et al., 2009. Obliquity-paced Pliocene West Antarctic ice sheet oscillations. Nature 458 (7236), 322–328.

Naish, T.R., Wilson, G.S., 2009. Constraints on the amplitude of Mid-Pliocene (3.6–2.4 Ma) eustatic sea-level fluctuations from the New Zealand shallow-marine sediment record. Philosophical Transactions of the Royal Society A: Mathematical, Physical and Engineering Sciences 367 (1886), 169–187.

Naish, T.R., Barrett, P.J., Dunbar, G.B., Woolfe, K.J., Dunn, A.G., Henrys, S.A., et al., 2001a. Sedimentary cyclicity in CRP drillcore, Victoria Land Basin, Antarctica. Studies from the Cape Roberts Project, Ross Sea, Antarctica; scientific report of CRP-3; part I, sedimentary environments for CRP-3. Terra Antartica 8 (3), 225–244.

Naish, T.R., Powell, R.D., Barrett, P.J., Levy, R.H., Henrys, S., Wilson, G.S., et al. and ANDRILL MIS Science Team, 2008. Late Cenozoic Climate History of the Ross Embayment from the AND-1B Drill Hole: Culmination of three decades of Antarctic Margin Drilling. In: Cooper, A.K., Barrett, P. J., Stagg, H., Storey, B., Stump, E., Wise, W., and the 10th ISAES Editorial Team (Eds.), Antarctica: A Keystone in a Changing World. Proceedings of the 10th International Symposium on Antarctic Earth Sciences. The National Academies Press, Washington, D.C., pp. 71–82. Available from: https://doi.org/10.3133/of2007-1047.kp3107.

Naish, T.R., Woolfe, K.J., Barrett, P.J., Wilson, G.S.E.A., 2001b. Orbitally induced oscillations in the East Antarctic ice sheet at the Oligocene/Miocene boundary. Nature 413, 719–723.

Newkirk, D.R., Martin, E.E., 2009. Circulation through the Central American Seaway during the Miocene carbonate crash. Geology 37 (1), 87–90.

Nyland, R.E., Panter, K.S., Rocchi, S., Di Vincenzo, G., Del Carlo, P., Tiepolo, M., et al., 2013. Volcanic activity and its link to glaciation cycles: Single-grain age and geochemistry of Early to Middle Miocene volcanic glass from ANDRILL AND-2A core, Antarctica. Journal of Volcanology and Geothermal Research 250, 106–128.

O'Brien, P., Post, A., Edwards, S., Martin, T., Caburlotto, A., Donda, F., et al., 2020. Continental slope and rise geomorphology seaward of the Totten Glacier, East Antarctica (112 E-122 E). Marine Geology 427, 106221.

O'Brien, P.E., Cooper, A.K., Richter, C., et al., 2001. Proc. ODP, Init. Repts., 188: College Station, TX (Ocean Drilling Program). Available from: https://doi.org/10.2973/odp.proc.ir.188.2001.

Oerlemans, J., 2002. On glacial inception and orography. Quaternary International 95–96, 5–10.

Ohneiser, C., Florindo, F., Stocchi, P., Roberts, A.P., DeConto, R.M., Pollard, D., 2015. Antarctic glacio-eustatic contributions to late Miocene Mediterranean desiccation and reflooding. Nature Communications 6 (1), 1–10.

Ohneiser, C., Wilson, G.S., 2012. Revised magnetostratigraphic chronologies for New Harbour Drill cores, southern Victoria Land, Antarctica. Global and Planetary Change 82–83 (Special Issue), 12–24.

Ohneiser, C., Wilson, G.S., Beltran, C., Dolan, A.M., Hill, D.J., Prebble, J.G., 2020. Warm fjords and vegetated landscapes in early Pliocene East Antarctica. Earth and Planetary Science Letters 534, 116045.

Olbers, D., Borowski, D., Völker, C., Wolff, J.-O., 2004. The dynamical balance, transport and circulation of the Antarctic Circumpolar Current. Antarctic Science 16 (4), 439–470.

Orejola, N., Passchier, S., 2014. Sedimentology of lower Pliocene to upper Pleistocene diamictons from IODP site U1358, Wilkes Land margin, and implications for East Antarctic Ice Sheet dynamics. Antarctic Science 26 (2), 183.

Orsi, A.H., Johnson, G.C., Bullister, J.L., 1999. Circulation, mixing, and production of Antarctic Bottom Water. Progress in Oceanography 43 (1), 55–109.

Orsi, A.H., Smethie Jr, W.M., Bullister, J.L., 2002. On the total input of Antarctic waters to the deep ocean: A preliminary estimate from chlorofluorocarbon measurements. Journal of Geophysical Research: Oceans 107 (C8), 31-1–31-14.

Orsi, A.H., Whitworth, T., Nowlin, W.D., 1995. On the meridional extent and fronts of the Antarctic Circumpolar Current. Deep Sea Research Part I: Oceanographic Research Papers 42 (5), 641–673.

Orsi, A.H., Wiederwohl, C.L., 2009. A recount of Ross Sea waters. Deep Sea Research Part II: Topical Studies in Oceanography 56 (13–14), 778–795.

Osborne, C.P., Beerling, D.J., 2006. Nature's green revolution: the remarkable evolutionary rise of C4 plants. Philosophical Transactions of the Royal Society B: Biological Sciences 361 (1465), 173–194.

Pagani, M., Arthur, M.A., Freeman, K.H., 1999a. Miocene evolution of atmospheric carbon dioxide. Paleoceanography 14 (3), 273–292.

Pagani, M., Freeman, K.H., Arthur, M.A., 1999b. Late Miocene atmospheric CO_2 concentrations and the expansion of C_4 grasses. Science 285 (5429), 876–879.

Pagani, M., Huber, M., Liu, Z., Bohaty, S.M., Henderiks, J., Sijp, W., et al., 2011. The role of carbon dioxide during the onset of Antarctic Glaciation. Science 334 (6060), 1261–1264.

Pagani, M., Liu, Z., LaRiviere, J., Ravelo, A.C., 2010. High Earth-system climate sensitivity determined from Pliocene carbon dioxide concentrations. Nature Geoscience 3 (1), 27–30.

Pälike, H., Lyle, M.W., Nishi, H., Raffi, I., Ridgwell, A., Gamage, K., et al., 2012. A Cenozoic record of the equatorial Pacific carbonate compensation depth. Nature 488 (7413), 609–614.

Palmer, E.F., Licht, K.J., Swope, R.J., Hemming, S.R., 2012. Nunatak moraines as a repository of what lies beneath the East Antarctic ice sheet. Geological Society of America Special Paper 487, 97–104.

Pankhurst, R.J., Weaver, S.D., Bradshaw, J.D., Storey, B.C., Ireland, T.R., 1998. Geochronology and geochemistry of pre-Jurassic superterranes in Marie Byrd Land, Antarctica. Journal of Geophysical Research 1033 (B2), 2529–2547.

Panter, K.S., Hart, S.R., Kyle, P.R., Blusztanjn, J., Wilch, T.I., 2000. Geochemistry of late Cenozoic basalts from the Crary Mountains: characterization of mantle sources in eastern Marie Byrd Land, Antarctica. Chemical Geology 165, 215–241.

Paolo, F.S., Fricker, H.A., Padman, L., 2015. Volume loss from Antarctic ice shelves is accelerating. Science 348 (6232), 327–331.

Parizek, B.R., Christianson, K., Alley, R.B., Voytenko, D., Vaňková, I., Dixon, T.H., et al., 2019. Ice-cliff failure via retrogressive slumping. Geology 47 (5), 449–452.

Passchier, S., 2011. Linkages between East Antarctic Ice Sheet extent and Southern Ocean temperatures based on a Pliocene high-resolution record of ice-rafted debris off Prydz Bay, East Antarctica. Paleoceanography 26 (4).

Passchier, S., Bohaty, S.M., Jiménez-Espejo, F., Pross, J., Röhl, U., van de Flierdt, T., et al., 2013a. Early Eocene to middle Miocene cooling and aridification of East Antarctica. Geochemistry, Geophysics, Geosystems 14 (5), 1399–1410.

Passchier, S., Browne, G., Field, B., Fielding, C.R., Krissek, L.A., Panter, K., et al.and Science Team, A.-S., 2011. Early and middle Miocene Antarctic glacial history from the sedimentary facies distribution in the AND-2A drill hole, Ross Sea, Antarctica. Geological Society of America Bulletin, B30334.30331.

Passchier, S., Falk, C.J., Florindo, F., 2013b. Orbitally paced shifts in the particle size of Antarctic continental shelf sediments in response to ice dynamics during the Miocene climatic optimum. Geosphere 9 (1), 54–62.

Patterson, M.O., Ishman, S.E., 2012. Neogene benthic foraminiferal assemblages and paleoenvironmental record for McMurdo Sound, Antarctica. Geosphere 8 (6), 1331–1341.

Patterson, M.O., McKay, R., Naish, T., Escutia, C., Jimenez-Espejo, F.J., Raymo, M.E., et al., 2014. Orbital forcing of the East Antarctic ice sheet during the Pliocene and Early Pleistocene. Nature Geoscience 7 (11), 841–847.

Paulsen, T., Encarnación, J., Grunow, A., Valencia, V., Pecha, M., Layer, P., et al., 2013. Age and significance of 'outboard' high-grade metamorphics and intrusives of the Ross orogen, Antarctica. Gondwana Research 24 (1), 349–358.

Paxman, G.J., Jamieson, S.S., Ferraccioli, F., Bentley, M.J., Ross, N., Watts, A.B., et al., 2019a. The role of lithospheric flexure in the landscape evolution of the Wilkes Subglacial Basin and Transantarctic Mountains, East Antarctica. Journal of Geophysical Research: Earth Surface 124 (3), 812–829.

Paxman, G.J.G., Gasson, E.G.W., Jamieson, S.S.R., Bentley, M.J., Ferraccioli, F., 2020. Long-term increase in Antarctic Ice Sheet vulnerability driven by bed topography evolution. Geophysical Research Letters 47 (20), e2020GL090003.

Paxman, G.J.G., Jamieson, S.S.R., Hochmuth, K., Gohl, K., Bentley, M.J., Leitchenkov, G., et al., 2019b. Reconstructions of Antarctic topography since the Eocene–Oligocene boundary. Palaeogeography, Palaeoclimatology, Palaeoecology 535, 109346.

Pekar, S., DeConto, R.M., 2006. High-resolution ice-volume estimates for the early Miocene: Evidence for a dynamic ice sheet in Antarctica. Palaeogeography, Palaeoclimatology, Palaeoecology 231, 101–109.

Pekar, S.F., DeConto, R.M., Harwood, D.M., 2006. Resolving a late Oligocene conundrum: deep sea warming and Antarctic glaciation. Palaeogeography, Palaeoclimatology, Palaeoecology 231 (1–2), 29–40.

Pérez, L.F., De Santis, L., McKay, R.M., Larter, R.D., Ash, J., Bart, P.J., et al. and IODP Expedition 374 Scientists, 2021a. Early and middle Miocene ice sheet dynamics in the Ross Sea: results from integrated core-log-seismic interpretation. GSA Bulletin. Available from: https://doi.org/10.1130/B35814.1.

Pérez, L.F., Hernández-Molina, F.J., Lodolo, E., Bohoyo, F., Galindo-Zaldívar, J., Maldonado, A., 2019. Oceanographic and climatic consequences of the tectonic evolution of the southern scotia sea basins, Antarctica. Earth-Science Reviews 198, 102922.

Pérez, L.F., Martos, Y.M., García, M., Weber, M.E., Raymo, M.E., Williams, T., et al., 2021b. Miocene to present oceanographic variability in the Scotia Sea and Antarctic ice sheets dynamics: Insight from revised seismic-stratigraphy following IODP Expedition 382. Earth and Planetary Science Letters 553, 116657.

Pérez, L.F., Nielsen, T., Knutz, P.C., Kuijpers, A., Damm, V., 2018. Large-scale evolution of the central-east Greenland margin: New insights to the North Atlantic glaciation history. Global and Planetary Change 163, 141–157.

Peterson, L., Murray, D., Ehrmann, W., Hempel, P., 1992. Cenozoic carbonate accumulation and compensation depth changes in the Indian Ocean. Synthesis of Results from Scientific Drilling in the Indian Ocean 70, 311–333.

Pfuhl, H.A., McCave, I.N., 2005. Evidence for late Oligocene establishment of the Antarctic Circumpolar Current. Earth and Planetary Science Letters 235 (3), 715–728.

Pierce, E.L., van de Flierdt, T., Williams, T., Hemming, S.R., Cook, C.P., Passchier, S., 2017. Evidence for a dynamic East Antarctic ice sheet during the mid-Miocene climate transition. Earth and Planetary Science Letters 478 (Supplement C), 1–13.

Piper, D., Brisco, C., 1975. Deep-water continental-margin sedimentation, DSDP Leg 28, Antarctica. Initial Reports of the Deep Sea Drilling Project 28, 727–755.

Pollard, D., DeConto, R.M., 2005. Hysteresis in Cenozoic Antarctic ice-sheet variations. Global and Planetary Change 45, 9–21.

Pollard, D., DeConto, R.M., 2007. A coupled ice-sheet/ice-shelf/sediment model applied to a marine-margin flowline: Forced and unforced variations. In: Hambrey, M., Christoffersen, P., Glasser, N.F., Hubbard, B. (Eds.), Glacial Sedimentary Processes and Products. Blackwell, pp. 37–52.

Pollard, D., DeConto, R.M., 2009. Modelling West Antarctic ice sheet growth and collapse through the past five million years. Nature 458, 329–332.

Pollard, D., DeConto, R.M., 2020. Continuous simulations over the last 40 million years with a coupled Antarctic ice sheet-sediment model. Palaeogeography, Palaeoclimatology, Palaeoecology 537, 109374.

Pollard, D., DeConto, R.M., Alley, R.B., 2015. Potential Antarctic Ice Sheet retreat driven by hydrofracturing and ice cliff failure. Earth and Planetary Science Letters 412, 112–121.

Poore, R.Ze, Sloan, L.Ce, 1996. Climates and climate variability of the Pliocene. Marine Micropaleontology 27 (1–4), 326.

Porter-Smith, R., Harris, P.T.E., Brancolini, G.E., Bindoff, N.L.E., De Santis, L.E., 2003. Bathymetry of the George Vth land shelf and slope. Recent investigations of the Mertz Polynya and George Vth Land continental margin, East Antarctica. Deep-Sea Research. Part II: Topical Studies in Oceanography 50 (8–9), 1337–1341.

Prebble, J.G., Raine, J.I., Barrett, P.J., Hannah, M.J., 2006. Vegetation and climate from two Oligocene glacioeustatic sedimentary cycles (31 and 24 Ma) cored by the Cape Roberts Project, Victoria Land Basin, Antarctica. Palaeogeography, Palaeoclimatology, Palaeoecology 231 (1), 41–57.

Prebble, J.G., Reichgelt, T., Mildenhall, D.C., Greenwood, D.R., Raine, J.I., Kennedy, E.M., et al., 2017. Terrestrial climate evolution in the Southwest Pacific over the past 30 million years. Earth and Planetary Science Letters 459, 136–144.

Preiss-Daimler, I., Zarkogiannis, S.D., Kontakiotis, G., Henrich, R., Antonarakou, A., 2021. Paleoceanographic perturbations and the marine carbonate System during the Middle to Late Miocene Carbonate Crash—a critical review. Geosciences 11 (2), 94.

Prentice, M.L., Bockheim, J.G., Wilson, S.C., Burckle, L.H., Hodell, D.A., Schluchter, C., et al., 1993. Late Neogene Antarctic glacial history: evidence from central Wright Valley. In: Kennett, J.P., Warnke, D.A. (Eds.), The Antarctic Paleoenvironment: A Perspective on Global Change, Part Two: Antarctic Research Series. American Geophysical Union, Washington, D.C, pp. 207–250.

Prentice, M.L., Kleman, J.L., Stroeven, A.P., 1998. The composite glacial erosional landscape of the northern McMurdo Dry Valleys: Implications for Antarctic Tertiary glacial history. Ecosystem Dynamics in a Polar Desert: the Mcmurdo Dry Valleys. Antarctica 72, 1–38.

Prentice, M.L., Ishman, S.E., Clemens, S.C., McIntosh, W.C., Clarke, K., 1999. Late Neogene stable isotope records from fjords within the McMurdo Dry Valleys, Antarctica. In: Skinner, D.N.B. (Ed.), 8th International Symposium on Antarctic Earth Sciences. Victoria University of Wellington, New Zealand, p. 249.

Prentice, M.L., Krusic, A.G., 2005. Early Pliocene alpine glaciation in Antarctica: terrestrial vs tidewater glaciers in Wright Valley. Geografiska Annaler, Series A 87, 87–109.

Pretty, R., 2019. Ice dynamics and ocean productivity during the Late Miocene, offshore Wilkes Land, East Antarctica. Victoria University of Wellington, Wellington, p. 148.

Pritchard, H.D., Ligtenberg, S.R.M., Fricker, H.A., Vaughan, D.G., van den Broeke, M.R., Padman, L., 2012. Antarctic ice-sheet loss driven by basal melting of ice shelves. Nature 484 (7395), 502–505.

Purkey, S.G., Johnson, G.C., 2010. Warming of global abyssal and deep Southern Ocean waters between the 1990s and 2000s: contributions to global heat and sea level rise budgets. Journal of Climate 23 (23), 6336–6351.

Raffi, I., Wade, B.S., Pälike, H., Beu, A.G., Cooper, R., Crundwell, M.P., et al., 2020. Chapter 29 – The Neogene period. In: Gradstein, F.M., Ogg, J.G., Schmitz, M.D., Ogg, G. M. (Eds.), Geologic Time Scale 2020. Elsevier, pp. 1141–1215.

Raitzsch, M., Bijma, J., Bickert, T., Schulz, M., Holbourn, A., Kučera, M., 2021. Atmospheric carbon dioxide variations across the middle Miocene climate transition. Climate of the Past 17 (2), 703–719.

Ravelo, A.C., Andreasen, D.H., 2000. Enhanced circulation during a warm period. Geophysical Research Letters 27 (7), 1001–1004.

Raymo, M.E., Nisancioglu, K.H., 2003. The 41 kyr world; Milankovitch's other unsolved mystery. Paleoceanography 18 (1), 11.11–11.16.

Raymo, M.E., Ruddiman, W.F., 1992. Tectonic forcing of late Cenozoic climate. Nature (London) 359 (6391), 117–122.

Raymo, M.E., 1994. The initiation of Northern Hemisphere glaciation. Annual Review of Earth and Planetary Sciences 22, 353–383.

Raymo, M.E., Kozdon, R., Evans, D., Lisiecki, L., Ford, H.L., 2018. The accuracy of mid-Pliocene δ^{18}O-based ice volume and sea level reconstructions. Earth-Science Reviews 177, 291–302.

Raymo, M.E., Lisiecki, L., Nisancioglu, K.H., 2006. Plio-Pleistocene ice volume, Antarctic climate, and the global δ^{18}O record. Science 313 (5786), 492–495.

Raymo, M.E., Mitrovica, J.X., O/'Leary, M.J., DeConto, R.M., Hearty, P.J., 2011. Departures from eustasy in Pliocene sea-level records. Nature Geoscience 4 (5), 328–332.

Reichgelt, T., D'Andrea, W.J., Fox, B.R.S., 2016. Abrupt plant physiological changes in southern New Zealand at the termination of the Mi-1 event reflect shifts in hydroclimate and pCO_2. Earth and Planetary Science Letters 455, 115–124.

Reichgelt, T., D'Andrea, W.J., Valdivia-McCarthy, A.C., Fox, B.R.S., Bannister, J.M., Conran, J.G., et al., 2020. Elevated CO2, increased leaf-level productivity, and water-use efficiency during the early Miocene, Clim. Past 16, 1509–1521. Available from: https://doi.org/10.5194/cp-16-1509-2020.

Reinardy, B., Escutia, C., Iwai, M., Jimenez-Espejo, F., Cook, C., van de Flierdt, T., et al., 2015. Repeated advance and retreat of the East Antarctic Ice Sheet on the continental shelf during the early Pliocene warm period. Palaeogeography, Palaeoclimatology, Palaeoecology 422, 65–84.

Retallack, G.J., 2009a. Greenhouse crises of the past 300 million years. Geological Society of America Bulletin 121 (9–10), 1441–1455.

Retallack, G.J., 2009b. Refining a pedogenic-carbonate CO_2 paleobarometer to quantify a middle Miocene greenhouse spike. Palaeogeography, Palaeoclimatology, Palaeoecology 281 (1), 57–65.

Rintoul, S., Silvano, A., Pena-Molino, B., van Wijk, E., Rosenberg, M., Greenbaum, J. et al., 2016. Ocean heat drives rapid basal melt of the Totten Ice Shelf. Science Advances.

Rintoul, S.R., da Silva, C.E., 2019. Antarctic Circumpolar Current☆. In: Cochran, J.K., Bokuniewicz, H.J., Yager, P.L. (Eds.), Encyclopedia of Ocean Sciences, third ed. Academic Press, Oxford, pp. 248–261.

Rintoul, S.R., 1991. South Atlantic interbasin exchange. Journal of Geophysical Research 96, 5493–5550.

Rintoul, S.R., 2018. The global influence of localized dynamics in the Southern Ocean. Nature 558 (7709), 209–218.

Rintoul, S.R., Hughes, C.W., Olbers, D., 2001. The Antarctic Circumpolar Current System, International Geophysics. Elsevier, pp. 271–XXXVI.

Robel, A.A., Banwell, A.F., 2019. A speed limit on ice shelf collapse through hydrofracture. Geophysical Research Letters 46 (21), 12092–12100.

Roberts, J.L., Warner, R.C., Young, D., Wright, A., van Ommen, T.D., Blankenship, D.D., et al., 2011. Refined broad-scale sub-glacial morphology of Aurora Subglacial Basin, East Antarctica derived by an ice-dynamics-based interpolation scheme. Cryosphere 5 (3), 551–560.

Rogelj, J., den Elzen, M., Höhne, N., Fransen, T., Fekete, H., Winkler, H., et al., 2016. Paris Agreement climate proposals need a boost to keep warming well below 2°C. Nature 534 (7609), 631–639.

Rohling, E., Foster, G.L., Grant, K., Marino, G., Roberts, A., Tamisiea, M.E., et al., 2014. Sea-level and deep-sea-temperature variability over the past 5.3 million years. Nature 508 (7497), 477–482.

Rommerskirchen, F., Condon, T., Mollenhauer, G., Dupont, L., Schefuss, E., 2011. Miocene to Pliocene development of surface and subsurface temperatures in the Benguela Current system. Paleoceanography 26 (3).

Rosenblume, J.A., Powell, R.D., 2019. Glacial sequence stratigraphy of ANDRILL-1B core reveals a dynamic subpolar Antarctic Ice Sheet in Ross Sea during the late Miocene. Sedimentology 66 (6), 2072–2097.

Rovere, A., Pappalardo, M., Richiano, S., Aguirre, M., Sandstrom, M.R., Hearty, P.J., et al., 2020. Higher than present global mean sea level recorded by an Early Pliocene intertidal unit in Patagonia (Argentina). Communications Earth & Environment 1 (1), 1–10.

Rovere, A., Raymo, M.E., Mitrovica, J., Hearty, P.J., O'Leary, M., Inglis, J., 2014. The Mid-Pliocene sea-level conundrum: glacial isostasy, eustasy and dynamic topography. Earth and Planetary Science Letters 387, 27–33.

Salabarnada, A., Escutia, C., Röhl, U., Nelson, C.H., McKay, R., Jiménez-Espejo, F.J., et al., 2018. Paleoceanography and ice sheet variability offshore Wilkes Land, Antarctica–Part 1: insights from late Oligocene astronomically paced contourite sedimentation. Climate of the Past 14 (7), 991–1014.

Salzmann, U., Haywood, A., Lunt, D., Valdes, P., Hill, D., 2008. A new global biome reconstruction and data-model comparison for the middle Pliocene. Global Ecology and Biogeography 17 (3), 432–447.

Salzmann, U., Riding, J.B., Nelson, A.E., Smellie, J.L., 2011. How likely was a green Antarctic Peninsula during warm Pliocene interglacials? A critical reassessment based on new palynofloras from James Ross Island. Palaeogeography, Palaeoclimatology, Palaeoecology 309 (1–2), 73–82.

Sangiorgi, F., Bijl, P.K., Passchier, S., Salzmann, U., Schouten, S., McKay, R., et al., 2018. Southern Ocean warming and Wilkes Land ice sheet retreat during the mid-Miocene. Nature Communications 9 (1), 317.

Sauermilch, I., Whittaker, J.M., Bijl, P.K., Totterdell, J., Jokat, W., 2019. Tectonic, oceanographic, and climatic controls on the Cretaceous-Cenozoic sedimentary record of the Australian-Antarctic Basin. Journal of Geophysical Research: Solid Earth 124 (8), 7699–7724.

Scher, H.D., Martin, E.E., 2006. Timing and climatic consequences of the opening of Drake passage. Science 312 (5772), 428–430.

Scher, H.D., Whittaker, J.M., Williams, S.E., Latimer, J.C., Kordesch, W.E.C., Delaney, M.L., 2015. Onset of Antarctic Circumpolar Current 30 million years ago as Tasmanian Gateway aligned with westerlies. Nature 523 (7562), 580–583.

Scherer, R.P., Aldahan, A.A., Tulaczyk, S., Possnert, G., Engelhardt, H., Kamb, B., 1998. Pleistocene collapse of the West Antarctic ice sheet. Science 281 (5373), 82–85.

Scherer, R.P., DeConto, R.M., Pollard, D., Alley, R.B., 2016. Windblown Pliocene diatoms and East antarctic ice sheet retreat. Nature Communications 7 (1), 1–9.

Schiller, M., Dickinson, W., Ditchburn, R.G., Graham, I.J., Zondervan, A., 2009. Atmospheric 10Be in an Antarctic soil: Implications for climate change. Journal of Geophysical Research 114 (F1), F01033.

Schiller, M., Dickinson, W.W., Iverson, N.A., Baker, J.A., 2019. A re-evaluation of the Hart Ash, an important stratigraphic marker: Wright Valley, Antarctica. Antarctic Science 31 (3), 139–149.

Schuster, M., Duringer, P., Ghienne, J.-F., Vignaud, P., Mackaye, H.T., Likius, A., et al., 2006. The age of the Sahara Desert. Science 311 (5762), 821.

Seki, O., Foster, G.L., Schmidt, D.N., Mackensen, A., Kawamura, K., Pancost, R.D., 2010. Alkenone and boron-based Pliocene pCO_2 records. Earth and Planetary Science Letters 292 (1–2), 201–211.

Selby, M.J., 1971. Slopes and their development in an ice-free, arid area of Antarctica. Geografiska Annaler: Series A, Physical Geography 53 (3–4), 235–245.

Selby, M.J., Wilson, A.T., 1971. The Origin of the Labyrinth, Wright Valley, Antarctica. GSA Bulletin 82 (2), 471–476.

Seroussi, H., Morlighem, M., 2018. Representation of basal melting at the grounding line in ice flow models. The Cryosphere 12 (10), 3085–3096.

Shackleton, N., 1967. Oxygen isotope analyses and pleistocene temperatures re-assessed. Nature (London) 215 (5096), 15–17.

Shackleton, N.J., Kennett, J.P., 1975a. Late Ceonozic oxygen and carbon isotope changes at DSDP Site 284: Implications for glacial history of the northern hemisphere and Antarctica. Initial Report Deep Sea Drilling Project 29, 801–807.

Shackleton, N.J., Kennett, J.P., 1975b. Paleotemperature history of the Cenozoic and the initiation of Antarctic glaciation; oxygen and carbon isotope analyses in DSDP sites 277, 279, and 281. Initial Reports of the Deep Sea Drilling Project 29, 743–755.

Shakun, J.D., Corbett, L.B., Bierman, P.R., Underwood, K., Rizzo, D.M., Zimmerman, S.R., et al., 2018. Minimal East Antarctic Ice Sheet retreat onto land during the past eight million years. Nature 558 (7709), 284–287.

Shepherd, A., Ivins, E., Rignot, E., Smith, B., van den Broeke, M., Velicogna, I., et al., 2018. Mass balance of the Antarctic Ice Sheet from 1992 to 2017. Nature 558 (7709), 219–222.

Shevenell, A.E., Kennett, J.P., Lea, D.W., 2004. Middle Miocene Southern Ocean Cooling and Antarctic Cryosphere Expansion. Science 305 (5691), 1766–1770.

Shevenell, A.E., Kennett, J.P., Lea, D.W., 2008. Middle Miocene ice sheet dynamics, deep-sea temperatures, and carbon cycling: a Southern Ocean perspective. Geochemistry Geophysics, Geosystems 9 (2), Q02006.

Siddoway, C.S., Fanning, C.M., 2009. Paleozoic tectonism on the East Gondwana margin: Evidence from SHRIMP U–Pb zircon geochronology of a migmatite–granite complex in West Antarctica. Tectonophysics 477 (3–4), 262–277.

Siddoway, C.S., 2008. Tectonics of the West Antarctic Rift System: new light on the history and dynamics of distributed intracontinental extension. In: Cooper, A.K., Barrett, P.J., Stagg, H., Storey, B., Stump, E., Wise, W., The 10th ISAES Editorial Team (Eds.), Antarctica: A Keystone in a Changing World. Proceedings of the 10th International Symposium on Antarctic Earth Sciences. The National Academies Press, Washington, D.C., pp. 91–114. Available from: http://doi.org/10.3133/of2007-1047.kp3109.

Siesser, W.G., 1980. Late Miocene origin of the Benguela upswelling system off northern Namibia. Science 208 (4441), 283–285.

Sijp, W.P., England, M.H., 2004. Effect of the Drake passage throughflow on global climate. Journal of Physical Oceanography 34 (5), 1254–1266.

Silvano, A., Rintoul, S.R., Peña-Molino, B., Hobbs, W.R., van Wijk, E., Aoki, S., et al., 2018. Freshening by glacial meltwater enhances melting of ice shelves and reduces formation of Antarctic Bottom Water. Science Advances 4 (4), eaap9467.

Śliwińska, K.K., Dybkjær, K., Schoon, P.L., Beyer, C., King, C., Schouten, S., et al., 2014. Paleoclimatic and paleoenvironmental records of the Oligocene–Miocene transition, central Jylland, Denmark. Marine Geology 350, 1–15.

Smith Jr, W.O., Sedwick, P.N., Arrigo, K.R., Ainley, D.G., Orsi, A.H., 2012. The Ross Sea in a sea of change. Oceanography 25 (3), 90–103.

Sokolov, S., Rintoul, S.R., 2007. On the relationship between fronts of the Antarctic Circumpolar Current and surface chlorophyll concentrations in the Southern Ocean. Journal of Geophysical Research: Oceans 112 (C7).

Sorlien, C.C., Luyendyk, B.P., Wilson, D.S., Decesari, R.C., Bartek, L.R., Diebold, J.B., 2007a. Buried Oligocene glacial topography beneath a smooth middle Miocene unconformity in the southeast Ross Sea: From temperate to glacial West Antarctica? In: Cooper, A. K., Raymond C.R., et al. (Eds.), Antarctica: A Keystone in a Changing World—Online Proceedings of the 10th ISAES X. United States Geological Survey and The National Academies, pp. Extended Abstract 099; 091-094. <http://pubs.usgs.gov/of/2007/1047/ea/of2007-1047ea2099.pdf>.

Sorlien, C.C., Luyendyk, B.P., Wilson, D.S., Decesari, R.C., Bartek, L.R., Diebold, J.B., 2007b. Oligocene development of the West Antarctic Ice Sheet recorded in eastern Ross Sea strata. Geology 35 (5), 467–470.

Sorlien, C.C., Wilson, D.S., Luyendyk, B.P., Wardell, N., Santis, L.D., Bart, P.J.C.S., et al., 2012. Subsidence and tilting of pre-25Ma wave-cut platform in Ross Sea, Antarctica. In: SCAR Open Science Conference, Portland, Oregon.

Sosdian, S.M., Babila, T.L., Greenop, R., Foster, G.L., Lear, C.H., 2020. Ocean Carbon Storage across the middle Miocene: a new interpretation for the Monterey Event. Nature Communications 11 (1), 134.

Sosdian, S.M., Greenop, R., Hain, M.P., Foster, G.L., Pearson, P.N., Lear, C.H., 2018. Constraining the evolution of Neogene ocean carbonate chemistry using the boron isotope pH proxy. Earth and Planetary Science Letters 498, 362–376.

Spector, P., Balco, G., 2021. Exposure-age data from across Antarctica reveal mid-Miocene establishment of polar desert climate. Geology 49 (1), 91–95.

Spiegel, C., Lindow, J., Kamp, P.J.J., Meisel, O., Mukasa, S., Lisker, F., et al., 2016. Tectonomorphic evolution of Marie Byrd Land—implications for Cenozoic rifting activity and onset of West Antarctic glaciation. Global and Planetary Change 145, 98–115.

St John, K.E., Krissek, L.A., 2002. The late Miocene to Pleistocene ice-rafting history of southeast Greenland. Boreas 31 (1), 28–35.

St. John, K., 2008. Cenozoic ice-rafting history of the central Arctic Ocean: terrigenous sands on the Lomonosov Ridge. Paleoceanography 23 (1).

Staiger, J.W., Marchant, D.R., Schaefer, J.M., Oberholzer, P., Johnson, J.V., Lewis, A.R., et al., 2006. Plio-Pleistocene history of Ferrar Glacier, Antarctica: implications for climate and ice sheet stability. Earth and Planetary Science Letters 243, 489–503.

Stap, L.B., Sutter, J., Knorr, G., Stärz, M., Lohmann, G., 2019. Transient variability of the Miocene Antarctic Ice Sheet smaller than equilibrium differences. Geophysical Research Letters 46 (8), 4288–4298.

Stap, L.B., Van De Wal, R.S., Boer, Bd, Bintanja, R., Lourens, L.J., 2017. The influence of ice sheets on temperature during the past 38 million years inferred from a one-dimensional ice sheet–climate model. Climate of the Past 13 (9), 1243–1257.

Stärz, M., Jokat, W., Knorr, G., Lohmann, G., 2017. Threshold in North Atlantic-Arctic Ocean circulation controlled by the subsidence of the Greenland-Scotland Ridge. Nature Communications 8 (1), 1–13.

Steinthorsdottir, M., Coxall, H.K., de Boer, A.M., Huber, M., Barbolini, N., Bradshaw, C.D., et al., 2020. The Miocene: the future of the past. Paleoceanography and Paleoclimatology 35, e2020PA004037.

Steinthorsdottir, M., Jardine, P.E., Rember, W.C., 2021. Near-Future pCO_2 During the Hot Miocene Climatic Optimum. Paleoceanography and Paleoclimatology 36 (1), e2020PA003900.

Stewart, A.L., Thompson, A.F., 2014. Eddy-mediated transport of warm circumpolar deep water across the Antarctic Shelf Break. Geophysical Research Letters 2014, GL062281.

Stickley, C.E., Brinkhuis, H., McGonigal, K., Chaproniere, G.C.H., Fuller, M., Kelly, D.C., et al., 2004. Late Cretaceous-Quaternary biomagnetostratigraphy of ODP Sites 1168, 1170, 1171, and 1172, Tasmanian Gateway. In: Exon, N.F., Kennett, J.P., and Malone, M.J. (Eds.), Proceedings of the Ocean Drilling Program, Scientific Results Volume 189, pp.1–57.

Strother, S.L., Salzmann, U., Sangiorgi, F., Bijl, P.K., Pross, J., Escutia, C., et al., 2017. A new quantitative approach to identify reworking in Eocene to Miocene pollen records from offshore Antarctica using red fluorescence and digital imaging. Biogeosciences 14 (8), 2089–2100.

Stump, E., 1995. The Ross Orogen of the Transantarctic Mountains. Cambridge University Press.

Sugden, D., Denton, G., 2004. Cenozoic landscape evolution of the Convoy Range to Mackay Glacier area, Transantarctic Mountains; onshore to offshore synthesis. Geological Society of America Bulletin 116 (7–8), 840–857.

Sugden, D.E., Denton, G.H., Marchant, D.R., 1991. Subglacial meltwater channel systems and ice sheet overriding, Asgard Range, Antarctica. Geografiska Annaler. Series A: Physical Geography 73 (2), 109–121.

Sugden, D.E., Denton, G.H., Marchant, D.R., 1995. Landscape evolution of the dry valleys, transantarctic mountains: tectonic implications. Journal of Geophysical Research: Solid Earth 100 (B6), 9949–9967.

Sugden, D.E., Marchant, D.R., Denton, G.H., 1993. The case for a stable East Antarctic Ice Sheet. Geografiska Annaler 75A, 151–351.

Summerfield, M.A., Sugden, D.E., Denton, G.H., Marchant, D.R., Cockburn, H.A.P., Stuart, F. M., 1999. Cosmogenic isotope data support previous evidence of extremely low rates of denudation in the Dry Valleys region, 162. Geological Society, London, Special Publications, Southern Victoria Land, Antarctica, pp. 255–267.

Super, J.R., Thomas, E., Pagani, M., Huber, M., O'Brien, C., Hull, P.M., 2018. North Atlantic temperature and pCO_2 coupling in the early-middle Miocene. Geology 46 (6), 519–522.

Super, J.R., Thomas, E., Pagani, M., Huber, M., O'Brien, C.L., Hull, P.M., 2020. Miocene evolution of North Atlantic sea surface temperature. Paleoceanography and Paleoclimatology 35 (5), e2019PA003748.

Talarico, F., Ghezzo, C., Kleinschmidt, G., 2021. The Antarctic Continent in Gondwana: a perspective from the Ross Embayment and Potential Research Targets for Future

Investigations. In: Florindo, F., et al. (Eds.), Antarctic Climate Evolution, second ed. Elsevier (this volume).

Tauxe, L., Stickley, C.E., Sugisaki, S., Bijl, P.K., Bohaty, S.M., Brinkhuis, H., et al., 2012. Chronostratigraphic framework for the IODP Expedition 318 cores from the Wilkes Land Margin: Constraints for paleoceanographic reconstruction. Paleoceanography 27 (2) . Available from: https://doi.org/10.1029/2012PA002308.

Taviani, M., Hannah, M., Harwood, D.M., Ishman, S.E., Johnson, K., Olney, M., et al., and ANDRILL SMS Science Team 2008–2009. Palaentological characterization an analysis of the AND-2A Core, ANDRILL Southern McMurdo Sound Project Antarctica. In: Harwood, D., Florindo, F., Talarico, F., Levy, R.H. (Eds.), Studies from the ANDRILL Southern McMurdo Sound Project, Antarctica—Initial Science Report on AND-2A. Terra Antartica, pp. 113–146.

Taylor-Silva, B., Riesselman, C., 2018. Polar frontal migration in the warm late Pliocene: diatom evidence from the Wilkes Land margin, East Antarctica. Paleoceanography and Paleoclimatology 33 (1), 76–92.

Ten Brink, U., Cooper, A., 1992. Modeling the bathymetry of Antarctic continental margins. Recent Progress in Antarctic Earth Science. Terra Scientific Publishing Co, Tokyo, pp. 763–772.

Ten Brink, U.S., Schneider, C., 1995. Glacial morphology and depositional sequences of the Antarctic continental shelf. Geology (Boulder) 23 (7), 580–584.

Thiede, Jr, Jessen, C., Knutz, P., Kuijpers, A., Mikkelsen, N., Spielhagen, R.F., 2011. Millions of years of Greenland Ice Sheet history recorded in ocean sediments. Polarforschung 80 (3), 141–159.

Thoma, M., Jenkins, A., Holland, D., Jacobs, S., 2008. Modelling circumpolar deep water intrusions on the Amundsen Sea continental shelf, Antarctica. Geophysical Research Letters 35 (18), L18602.

Thompson, A.F., Stewart, A.L., Spence, P., Heywood, K.J., 2018. The Antarctic slope current in a changing climate. Reviews of Geophysics 56 (4), 741–770. Available from: https://doi.org/10.1029/2018RG000624.

Tian, J., Yang, M., Lyle, M.W., Wilkens, R., Shackford, J.K., 2013. Obliquity and long eccentricity pacing of the Middle Miocene climate transition. Geochemistry, Geophysics, Geosystems 14 (6), 1740–1755.

Tierney, J.E., Haywood, A.M., Feng, R., Bhattacharya, T., Otto-Bliesner, B.L., 2019. Pliocene warmth consistent with greenhouse gas forcing. Geophysical Research Letters 46 (15), 9136–9144.

Timoney, R., Worrall, K., Firstbrook, D., Harkness, P., Rix, J., Ashurst, D., et al., 2020. A low resource subglacial bedrock sampler: the percussive rapid access isotope drill (P-RAID). Cold Regions Science and Technology 177, 103113.

Tingey, R.J., 1991. The regional geology of Archaean and Proterozoic rocks in Antarctica. In: Tingey, R.J. (Ed.), The Geology of Antarctica, Oxford Monographs on Geology and Geophysics. Clarendon Press, Oxford, pp. 1–73.

Tipple, B.J., Pagani, M., 2007. The early origins of terrestrial C4 photosynthesis. Annual Review of Earth and Planetary Sciences 35, 435–461.

Treves, S.B., McKelvey, B.C., 1975. Drilling in Antarctica, September–December, 1974/1975. In: Mudrey Jr., M.G., McGinnis, L.D. (Eds.), Dry Valley Drilling Project (DVDP) Bulletin. pp. 5–10.

Truswell, E.M., Drewry, D.J., 1984. Distribution and provenance of recycled palynomorphs in surficial sediments of the Ross Sea, Antarctica. Marine Geology 59 (1–4), 187–214.

UNFCCC, 2015. Adoption of the Paris agreement. FCCC/CP/2015/L.9/Rev.1. United Nations Framework Convention on Climate Change. Available from: http://unfccc.int/paris_agreement/items/9485.php.

Utescher, T., Bruch, A.A., Micheels, A., Mosbrugger, V., Popova, S., 2011. Cenozoic climate gradients in Eurasia—a palaeo-perspective on future climate change? Palaeogeography, Palaeoclimatology, Palaeoecology 304 (3–4), 351–358.

Valletta, R.D., Willenbring, J.K., Lewis, A.R., Ashworth, A.C., Caffee, M., 2015. Extreme decay of meteoric beryllium-10 as a proxy for persistent aridity. Scientific Reports 5, 17813.

van de Lagemaat, S.H.A., Swart, M.L.A., Vaes, B., Kosters, M.E., Boschman, L.M., Burton-Johnson, A., et al., 2021. Subduction initiation in the Scotia Sea region and opening of the Drake Passage: when and why? Earth-Science Reviews 215, 103551.

Van de Wal, R., Greuell, W., van den Broeke, M.R., Reijmer, C., Oerlemans, J., 2005. Surface mass-balance observations and automatic weather station data along a transect near Kangerlussuaq, West Greenland. Annals of Glaciology 42, 311–316.

Van der Burgh, J., Visscher, H., Dilcher, D.L., Kuerschner, W.M., 1993. Paleoatmospheric signatures in Neogene fossil leaves. Science 260 (5115), 1788–1790.

Vanney, J.-R., Johnson, G., 1979a. Wilkes land continental margin physiography, East Antarctica. Polarforschung 49 (1), 20–29.

Vanney, J.-R., Johnson, G.L., 1979b. The sea floor morphology seaward of Terre Adélie (Antarctica). Deutsche Hydrografische Zeitschrift 32 (2), 77–87.

Verducci, M., Foresi, L.M., Scott, G.H., Sprovieri, M., Lirer, F., Pelosi, N., 2009. The Middle Miocene climatic transition in the Southern Ocean: evidence of paleoclimatic and hydrographic changes at Kerguelen plateau from planktonic foraminifers and stable isotopes. Palaeogeography, Palaeoclimatology, Palaeoecology 280 (3), 371–386.

Verret, M., Dickinson, W., Lacelle, D., Fisher, D., Norton, K., Chorley, H., et al., 2020. Cryostratigraphy of mid-Miocene permafrost at Friis Hills, McMurdo Dry Valleys of Antarctica. Antarctic Science 1–15.

Vignaud, P., Duringer, P., Mackaye, H.T., Likius, A., Blondel, C., Boisserie, J.-R., et al., 2002. Geology and palaeontology of the Upper Miocene Toros-Menalla hominid locality, Chad. Nature 418 (6894), 152–155.

Vincent, E., Berger, W.H., 1985. Carbon dioxide and polar cooling in the Miocene; the Monterey Hypothesis. In: Sundquist, E.T., Broecker, W.S. (Eds.), The Carbon Cycle and Atmospheric CO_2: Natural Variations Archean to Present. pp. 455–468. Geophysical Monograph 32.

Waddell, L.M., Hendy, I.L., Moore, T.C., Lyle, M.W., 2009. Ventilation of the abyssal Southern Ocean during the late Neogene: a new perspective from the subantarctic Pacific. Paleoceanography 24 (3).

Wan, S., Tian, J., Steinke, S., Li, A., Li, T., 2010. Evolution and variability of the East Asian summer monsoon during the Pliocene: evidence from clay mineral records of the South China Sea. Palaeogeography, Palaeoclimatology, Palaeoecology 293 (1–2), 237–247.

Warny, S., Askin, R.A., Hannah, M.J., Mohr, B.A.R., Raine, J.I., Harwood, D.M., et al., and the SMS Science Team 2009. Palynomorphs from a sediment core reveal a sudden remarkably warm Antarctica during the middle Miocene. Geology 37 (10), 955–958.

Webb, P.-N., Harwood, D.M., 1991. Late Cenozoic glacial history of the Ross Embayment, Antarctica. Quaternary Science Reviews 10 (2–3), 215–223.

Webb, P.-N., Harwood, D.M., McKelvey, B.C., Mercer, J.H., Stott, L.D., 1984. Cenozoic marine sedimentation and ice-volume variation on the East Antarctic craton. Geology 12 (5), 287–291.

Weber, M.E., Raymo, M.E., Peck, V.L., Williams, T., and the Expedition 382 Scientists, 2019. Expedition 382 Preliminary Report: Iceberg Alley and Subantarctic Ice and Ocean Dynamics. International Ocean Discovery Program. Available from: https://doi.org/10.14379/iodp.pr.382.2019.

Wei, L.J., Raine, J.I., Liu, X.H., 2013. Terrestrial palynomorphs of the Cenozoic Pagodroma Group, northern Prince Charles Mountains, East Antarctica. Antarctic Science 26, 69–79.

Westerhold, T., Marwan, N., Drury, A.J., Liebrand, D., Agnini, C., Anagnostou, E., et al., 2020. An astronomically dated record of Earth's climate and its predictability over the last 66 million years. Science 369 (6509), 1383–1387.

Whitehead, J.M., Bohaty, S.M., 2003. Pliocene summer sea surface temperature reconstruction using silicoflagellates from Southern Ocean ODP Site 1165. Paleoceanography 18 (3), 1075.

Whitehead, J.M., Ehrmann, W., Harwood, D.M., Hillenbrand, C.-D., Quilty, P.G., Hart, C., et al., 2006a. Late Miocene paleoenvironment of the Lambert Graben embayment, East Antarctica, evident from: mollusc paleontology, sedimentology and geochemistry. Global and Planetary Change 50 (3–4), 127–147.

Whitehead, J.M., Quilty, P.G., Mckelvey, B.C., O'Brien, P.E., 2006b. A review of the Cenozoic stratigraphy and glacial history of the Lambert Graben—Prydz Bay region, East Antarctica. Antarctic Science 18 (1), 83–99.

Whitehead, J.M., Wotherspoon, S., Bohaty, S.M., 2005. Minimal Antarctic sea ice during the Pliocene. Geology 33 (2), 137–140.

Whitworth III, T., Orsi, A., Kim, S., Nowlin, W., Locarmini, R., 1998. Ocean, ice, and atmosphere: interactions at the Antarctic Continental Margin. In: Jacobs, S.S., Weiss, R.F. (Eds.), Antarctic Research Series, vol. 75. pp. 1–28.

Williams, G., Bindoff, N., 2003. Wintertime oceanography of the Adélie Depression. Deep Sea Research Part II: Topical Studies in Oceanography 50 (8–9), 1373–1392.

Williams, G., Aoki, S., Jacobs, S., Rintoul, S., Tamura, T., Bindoff, N., 2010. Antarctic bottom water from the Adélie and George V Land coast, East Antarctica (140–149 E). Journal of Geophysical Research: Oceans 115 (C4).

Williams, G., Bindoff, N., Marsland, S., Rintoul, S., 2008. Formation and export of dense shelf water from the Adélie Depression, East Antarctica. Journal of Geophysical Research: Oceans 113 (C4).

Williams, T., Handwerger, D., 2005. A high-resolution record of early Miocene Antarctic glacial history from ODP Site 1165, Prydz Bay. Paleoceanography 20 (2).

Williams, T., Flierdt, Tvd, Hemming, S.R., Chunga, E., Roya, M., Goldstein, S.L., 2010. Evidence for iceberg armadas from East Antarctica in the Southern Ocean during the late Miocene and early Pliocene. Earth and Planetary Science Letters 290, 351–361.

Wilson, D.J., Bertram, R.A., Needham, E.F., van de Flierdt, T., Welsh, K.J., McKay, R.M., et al., 2018. Ice loss from the East Antarctic Ice Sheet during late Pleistocene interglacials. Nature 561 (7723), 383–386.

Wilson, D.S., Jamieson, S.S.R., Barrett, P.J., Leitchenkov, G., Gohl, K., Larter, R.D., 2012a. Antarctic topography at the Eocene-Oligocene Boundary. Palaeogeography, Palaeoclimatology, Palaeoecology 335–336, 24–34.

Wilson, D.S., Luyendyk, B., Pollard, D., DeConto, R., 2009. Antarctic paleotopography estimates at the Eocene-Oligocene climate transition and their implications for Ice Sheet Growth (invited). In: Escutia Dotti, C. (Ed.), Program: First Antarctic Climate Evolution Symposium. SCAR, Granada, p. 30.

Wilson, D.S., Pollard, D., DeConto, R.M., Jamieson, S.S.R., Luyendyk, B.P., 2013. Initiation of the West Antarctic Ice Sheet and estimates of total Antarctic ice volume in the earliest Oligocene. Geophysical Research Letters 40, 1–5.

Wilson, G.S., 1995. The Neogene East Antarctic ice sheet: a dynamic or stable feature? Quaternary Science Reviews 14 (2), 101–123.

Wilson, G.S., Levy, R.H., Naish, T.R., Powell, R.D., Florindo, F., Ohneiser, C., et al., 2012b. Neogene tectonic and climatic evolution of the Western Ross Sea, Antarctica—Chronology of events from the AND-1B drill hole. Global and Planetary Change 96–97, 189–203.

Wilson, G.S., Roberts, A.P., Verosub, K.L., Florindo, F., Sagnotti, L., 1997. Magnetobiostratigraphic chronology of the Eocene-Oligocene transition in the CIROS-1 core, Victoria Land margin, Antarctica: implications for Antarctic glacial history. Geological Society of America Bulletin 109 (12), 2–14.

Wilson, G.S., Levy, R., Browne, G., Cody, R., Dunbar, N., Florindo, F., et al., 2007. Preliminary Integrated Chronostratigraphy of the AND-1B Core, ANDRILL McMurdo Ice Shelf Project, Antarctica. In: Naish, T., Powell, R., Levy, R. (Eds.), Studies from the ANDRILL, McMurdo Ice Shelf Project, Antarctica - Initial Science Report on AND-1B. Terra Antartica. pp. 297–316.

Winberry, J.P., Anandakrishnan, S., 2004. Crustal structure of the West Antarctic rift system and Marie Byrd Land hotspot. Geology 32 (11), 977–980.

Winnick, M.J., Caves, J.K., 2015. Oxygen isotope mass-balance constraints on Pliocene sea level and East Antarctic Ice Sheet stability. Geology 43 (10), 879–882.

Winter, D.M., Sjunesskog, C., Harwood, D., 2010a. Early to mid-Pliocene environmentally constrained diatom assemblages from the AND-1B drillcore, McMurdo Sound, Antarctica. Stratigraphy 7 (2–3), 207–227.

Winter, D.M., Sjunneskog, C., Scherer, R.P., Maffioli, P., Riesselman, C., Harwood, D.M., 2010b. Pliocene-Pleistocene diatom biostratigraphy of nearshore Antarctica from the AND-1B drillcore, McMurdo Sound. Global and Planetary Change. (2), 96–97. Available from: https://doi.org/10.1016/j.gloplacha.2010.04.004.

Witkowski, C.R., Weijers, J.W.H., Blais, B., Schouten, S., Sinninghe Damsté, J.S., 2018. Molecular fossils from phytoplankton reveal secular Pco_2 trend over the Phanerozoic. Science Advances 4 (11), eaat 4556.

Wood, B., Harrison, T., 2011. The evolutionary context of the first hominins. Nature 470 (7334), 347–352.

Woodard, S.C., Rosenthal, Y., Miller, K.G., Wright, J.D., Chiu, B.K., Lawrence, K.T., 2014. Antarctic role in Northern Hemisphere glaciation. Science 346 (6211), 847–851.

Woodruff, F., Savin, S., 1991. Mid-Miocene Isotope stratigraphy in the Deep Sea: high-resolution correlations, paleoclimatic cycles, and sediment preservation. Paleoceanography 6 (6), 755–806.

Woodruff, F., Savin, S.M., 1989. Miocene deepwater oceanography. Paleoceanography 4 (1), 87–140.

Wright, A.P., Young, D.A., Roberts, J.L., Dowdeswell, J.A., Bamber, J.L., Young, N., LeBrocq, A.M., Warner, R.C., Payne, A.J., Blankenship, D.D., van Ommen, T. and Siegert, M.J., 2021. Evidence for a hydrological connection between the ice divide and ice sheet margin in the Aurora Subglacial Basin sector of East Antarctica. Journal of Geophysical Research, Earth Surface 117, F01033, 15pp. Available from: https://doi.org/10.1029/2011JF002066.

Wright, C., Priestley, R., 1922. Glaciology, British Antarctic (Terra Nova) Expedition 1910–13. Harrison & Sons, Ltd., London, p. 581.

Wright, J.D., Miller, K.G., Fairbanks, R., 1992. Early and middle Miocene stable isotopes: implications for deepwater circulation and climate. Paleoceanography 7, 357–389.

Yamane, M., Yokoyama, Y., Abe-Ouchi, A., Obrochta, S., Saito, F., Moriwaki, K., et al., 2015. Exposure age and ice-sheet model constraints on Pliocene East Antarctic ice sheet dynamics. Nature Communications 6 (1), 1–8.

Yan, Q., Wei, T., Korty, R.L., Kossin, J.P., Zhang, Z., Wang, H., 2016. Enhanced intensity of global tropical cyclones during the mid-Pliocene warm period. Proceedings of the National Academy of Sciences 113 (46), 12963–12967.

You, Y., 2010. Climate-model evaluation of the contribution of sea-surface temperature and carbon dioxide to the Middle Miocene Climate Optimum as a possible analogue of future climate change. Australian Journal of Earth Sciences: An International Geoscience Journal of the Geological Society of Australia 57 (2), 207–219.

You, Y., Huber, M., Müller, R.D., Poulsen, C.J., Ribbe, J., 2009. Simulation of the Middle Miocene Climate Optimum. Geophysical Research Letters 36, 1–5.

Young, D.A., Wright, A.P., Roberts, J.L., Warner, R.C., Young, N.W., Greenbaum, J.S., et al., 2011. A dynamic early East Antarctic Ice Sheet suggested by ice-covered fjord landscapes. Nature 474, 72–75.

Zachos, J., Pagani, M., Sloan, L., Thomas, E., Billups, K., 2001a. Trends, rhythms, and aberrations in global climate 65Ma to present. Science 292 (5517), 686–693.

Zachos, J.C., Dickens, G.R., Zeebe, R.E., 2008. An early Cenozoic perspective on greenhouse warming and carbon-cycle dynamics. Nature 451, 279–283.

Zachos, J.C., Pagani, M., Sloan, L., Thomas, E., Billups, K., 2001b. Trends, rhythms, and aberrations in global climate 65 Ma to present. Science 292, 686–693.

Zachos, J.C., Shackleton, N.J., Revenaugh, J.S., Palike, H., Flower, B.P., 2001c. Climate response to orbital forcing across the Oligocene-Miocene boundary. Science 292 (5515), 274–278.

Zelilidis, A., Piper, D.J.W., Kontopoulos, N., 2002. Sedimentation and basin evolution of the Oligocene-Miocene Mesohellenic Basin, Greece. AAPG Bulletin 86 (1), 161–182.

Zhang, J., Scott, D.B., 1996. Messinian deep-water turbidites and glacioeustatic sea-level changes in the North Atlantic: linkage to the Mediterranean salinity crisis. Paleoceanography 11 (3), 277–297.

Zhang, R., Zhang, Z., Jiang, D., 2018. Global cooling contributed to the establishment of a modern-like East Asian Monsoon Climate by the Early Miocene. Geophysical Research Letters 45 (21), 11,941–911,948.

Zhang, Y.G., Pagani, M., Liu, Z., 2014. A 12-million-year temperature history of the Tropical Pacific Ocean. Science 344 (6179), 84–87.

Zhang, Y.G., Pagani, M., Liu, Z., Bohaty, S.M., DeConto, R., 2013. A 40-million-year history of atmospheric CO_2. Philosophical Transactions of the Royal Society A: Mathematical, Physical and Engineering Sciences 371 (2001).

Zhisheng, A., Kutzbach, J.E., Prell, W.L., Porter, S.C., 2001. Evolution of Asian monsoons and phased uplift of the Himalaya-Tibetan Plateau since late Miocene times. Nature (London) 411 (6833), 62–66.

Chapter 10

Pleistocene Antarctic climate variability: ice sheet, ocean and climate interactions

David J. Wilson[1], Tina van de Flierdt[2], Robert M. McKay[3] and Tim R. Naish[3]
[1]Department of Earth Sciences, University College London, London, United Kingdom, [2]Department of Earth Science and Engineering, Imperial College London, London, United Kingdom, [3]Antarctic Research Centre, Victoria University of Wellington, Wellington, New Zealand

10.1 Background and motivation

10.1.1 Introduction

Following the warm mid-Pliocene interval, Earth's climate cooled through the Late Pliocene and passed a threshold causing intensification of glaciation in the Northern Hemisphere, with periodic expansion and retreat of ice sheets in Eurasia and North America since ~2.6–2.7 Ma (Bailey et al., 2013; DeConto et al., 2008; Lawrence et al., 2006; Lisiecki and Raymo, 2005; Raymo, 1994). This event marked the onset of the Pleistocene epoch (0–2.6 Ma) and, for the first time in the Cenozoic, Earth's climate system was characterised by bipolar glaciation (Zachos et al., 2001). Pleistocene glacial–interglacial cycles were paced by orbital forcing on ~21, ~41 and ~100-kyr timescales (Hays et al., 1976; Imbrie et al., 1992, 1993), leading to high-amplitude fluctuations in global temperature, atmospheric composition, ice sheet extent and sea level (Chappell and Shackleton, 1986; Elderfield et al., 2012; Naish et al., 1998; Lisiecki and Raymo, 2005; Petit et al., 1999). The prevailing view is that ice volume variance over the Pleistocene has been dominated by ice sheets expanding to cover large parts of North America and Eurasia during glacial periods, and mostly disappearing during interglacials, with only Greenland remaining partially glaciated during all but the warmest interglacials. For the past 800,000 years, these changes in climate were consistently accompanied by atmospheric carbon dioxide (CO_2) variability, which ranged between ~180 ppm during glacials and ~280 ppm during the warmest interglacials (Fig. 10.1).

Antarctic Climate Evolution. DOI: https://doi.org/10.1016/B978-0-12-819109-5.00001-3

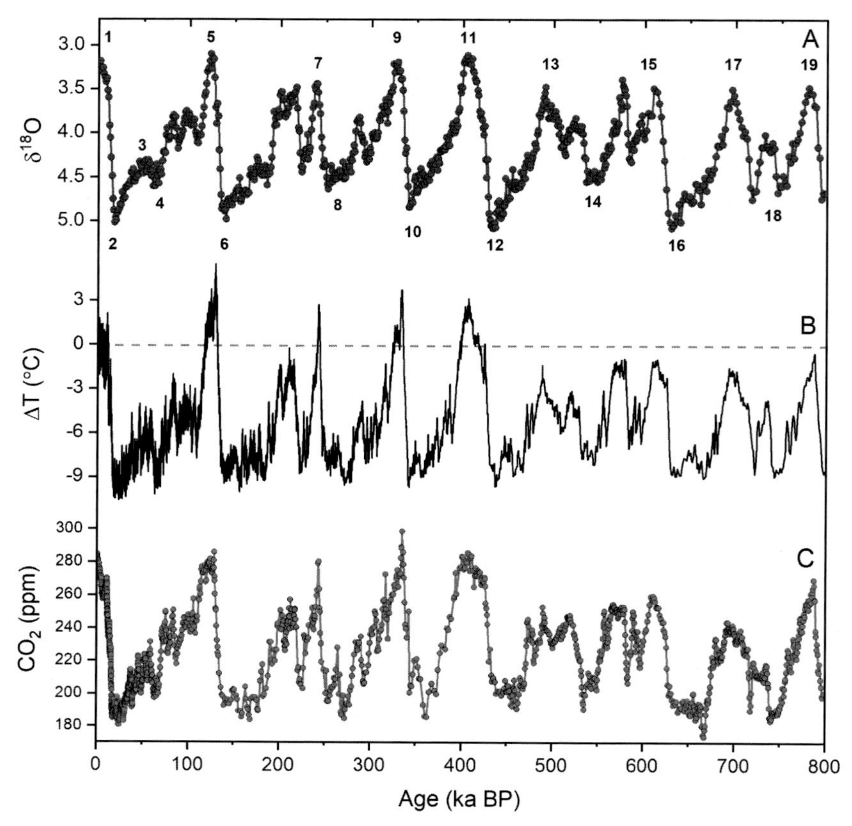

FIGURE 10.1 Marine and ice core records of late Pleistocene climate variability. (A) 'LR04' benthic foraminiferal oxygen isotope ($\delta^{18}O$) stack (Lisiecki and Raymo, 2005). (B) Antarctic ice core temperature reconstruction (ΔT, difference from the mean of the last millennium) based on deuterium isotopes (δD_{ice}) in EPICA Dome C (Jouzel et al., 2007). (C) Atmospheric carbon dioxide (CO_2) concentrations from Antarctic ice cores (Bereiter et al., 2015). Marine Isotope Stage (MIS) numbers are shown in (A). Note (1) the strong connection between Antarctic temperatures, benthic foraminiferal $\delta^{18}O$ values and atmospheric CO_2; (2) the dominance of quasi-100-kyr cycles for the last 800 kyr; and (3) the enhanced warmth and elevated atmospheric CO_2 of Antarctic interglacials since MIS 11 (i.e. the Mid-Brunhes Event at ~430 ka).

The Earth is currently in the Holocene (0–11.7 ka BP), an interglacial period with a remarkably stable climate state. However, human interference in the Earth system, in particular through greenhouse gas emissions, has significantly perturbed this stability and has already resulted in atmospheric CO_2 levels that exceed 410 ppm. Such atmospheric conditions are not thought to have existed since the Pliocene, leading to the proposal that we have entered a new epoch known as the Anthropocene (Crutzen and Stoermer, 2000). Assessments by the Intergovernmental Panel on Climate Change (IPCC) indicate that it is virtually certain that these elevated CO_2

levels will prevent an orbitally-induced inception of glaciation before the end of the next millennium, even under reduced emission scenarios (Masson-Delmotte et al., 2013). It is clear that, due to anthropogenic activity, we are entering into a non-analogue state of the Earth system, and one in which changes in Antarctica may play a relatively more important role than they did during the glacial–interglacial cycles of the Pleistocene.

Exploring paleoclimate and paleoenvironmental change during the Pleistocene has proven fundamental to our understanding of the Earth's climate system. Important advances have included testing the orbital theory of ice ages (Hays et al., 1976; Milankovitch, 1941), assessing oceanic mechanisms for centennial scale climate perturbations (Chen et al., 2015; Marcott et al., 2014), and quantifying sea level rise from melting polar ice sheets (DeConto and Pollard, 2016; Dutton et al., 2015). These three examples highlight, respectively, the critical role of Earth system feedbacks for climate, the potential for rapid non-linear changes to occur in the Earth system, and the value of geological evidence for predicting future changes in response to our evolving anthropogenic experiment. Much is known, but many important questions remain unanswered, particularly in the Southern Ocean and Antarctica, and it is against this backdrop that the present chapter is set.

Perhaps the most pressing of these questions is how, and how fast, the polar ice sheets will respond to a future warming climate, which has significant implications for global sea level rise (IPCC, 2019). The Antarctic Ice Sheet contains a potential contribution to global sea levels of ~58 m (Fretwell et al., 2013; Morlighem et al., 2020), comprising ~4 m of marine-based ice in West Antarctica and ~19 m of marine-based ice and ~34 m of terrestrial-based ice in East Antarctica (Fretwell et al., 2013). Whereas the terrestrial portions of the Antarctic Ice Sheet are expected to be relatively stable, the idea that the marine-based sectors could be susceptible to retreat or collapse under only moderate climate warming was recognised some time ago (Mercer, 1978). Theoretical calculations and modern observations suggest that two key factors affecting ice sheet mass balance could be critical for such retreat. First, where marine-based ice sheets rest on a reverse sloping bed that deepens inland, they are susceptible to marine ice sheet instability, a positive feedback process governing ice flow that enables rapid retreat following an initial perturbation (Schoof, 2007; Weertman, 1974). Second, most ice streams are significantly supported by offshore buttressing ice shelves, whose thinning or removal would reduce the back-stress on the ice and accelerate upstream ice sheet flow (Dupont and Alley, 2005; Fürst et al., 2016; Rignot et al., 2004). Recent observations indicate that marine ice sheet instability is potentially already underway in parts of West Antarctica (Joughin et al., 2014; Rignot et al., 2014; Wouters et al., 2015), while the West Antarctic ice shelves are currently losing mass (Paolo et al., 2015). Despite recognising the importance of these processes, the levels of atmospheric or ocean warming required to generate retreat of the marine-based portions of the Antarctic Ice Sheet are not well constrained, making it essential to assess their past behaviour under different climatic regimes. Since reverse sloping beds of Antarctica's continental shelves are thought to have become more widespread in the late Pliocene (Bart and Iwai, 2012; Bart et al., 1999; De Santis et al., 1999; McKay et al., 2019), and the ice shelves became more persistent and extensive with Pleistocene cooling (McKay et al., 2009), the Pleistocene interval provides the closest analogue for the modern and future Antarctic Ice Sheet (Colleoni et al., 2018).

10.1.2 Orbital cyclicity and climate

Building on earlier work by James Croll, who had proposed that an orbital influence on seasonality and winter snowfall led to the development of past ice ages (Croll, 1864), Milutin Milankovitch took the critical step of linking high-latitude summer insolation in the Northern Hemisphere to the growth of continental scale ice sheets during the Pleistocene (Milankovitch, 1941). In the Milankovitch model, intervals with low summer insolation are predicted to produce fewer positive degree days and a positive ice mass balance over multiple seasons, thereby enabling ice sheet expansion. Seasonal and spatial variations in insolation are themselves determined by the Earth's orbital configuration, with significant periodic variations on timescales of ~21 kyr (precession), ~41 kyr (obliquity, or axial tilt) and ~100 kyr (orbital eccentricity). In support of the Milankovitch theory, Earth's climate during the early Pleistocene (~2.6 Ma to ~900 ka) was largely characterised by ~41 kyr cycles forced by obliquity (Imbrie et al., 1992; Lisiecki and Raymo, 2005), with intervals of low obliquity leading to reduced seasonality, cooler summers and ice sheet growth. Whereas the early Pleistocene glacial and interglacial periods had similar durations and climate responded fairly linearly to insolation changes, the climate cyclicity changed significantly during the Mid Pleistocene Transition (MPT), which occurred gradually over a few hundred thousand years at ~900 ka (Berends et al., 2021; Ruddiman et al., 1989). Glacial–interglacial cycles since the MPT have been strongly asymmetric, with long glacial periods and shorter interglacials recurring on a quasi-100-kyr timescale (Fig. 10.1A) (Lisiecki and Raymo, 2005), indicating a more complex non-linear response of the Earth system to orbital forcing (Imbrie et al., 1993).

In detail, the exact manner by which orbital insolation has controlled the Pleistocene glacial–interglacial cycles remains a matter of debate, with several questions remaining to be fully answered. These include

1. the detailed timing of glacial–interglacial transitions in relation to orbital forcing (Drysdale et al., 2009; Kawamura et al., 2007; Thomas et al., 2009);
2. the importance of peak summer insolation values (Milankovitch model) vs summer duration, seasonally-integrated insolation, or latitudinal insolation gradients (Huybers, 2006; Huybers and Denton, 2008) in forcing Earth's climate;

3. the cause of temporal shifts in the expression of different orbital frequencies in Earth's ice volume and climate records, as seen at the MPT (Maslin and Brierley, 2015; Ruddiman et al., 1989);
4. the extent to which the late Pleistocene 100-kyr cycles reflect a nonlinear Earth system response to eccentricity forcing (Imbrie et al., 1993); forcing by multiple obliquity cycles (Huybers and Wunsch, 2005), by combined obliquity and precession cycles (Huybers, 2011; Ruddiman, 2006), or by multiple precession cycles (Cheng et al., 2016); or internal oscillations of the climate system (Berger, 1999; Rial et al., 2013);
5. the necessity for orbital and millennial scale processes to combine to cause glaciation and deglaciation (Anderson et al., 2009; Barker et al., 2011; Cheng et al., 2009); and
6. the relative importance of Northern Hemisphere vs Southern Hemisphere forcing, and in particular the role of local insolation in governing the ice sheet mass balance of terrestrial and marine-based ice sheets (He et al., 2013; Huybers, 2006; Huybers and Denton, 2008; Knorr and Lohmann, 2003; Laepple et al., 2011; Raymo et al., 2006).

10.1.3 Antarctic feedbacks in the global climate system

Although orbital forcing is recognised as the pacemaker of Pleistocene glacial cycles, it is important to emphasise that the globally-integrated annual insolation budget is insensitive to obliquity and precession, and only weakly sensitive to eccentricity. Therefore, the direct effect of orbital forcing on Earth's energy balance is inadequate to explain glacial–interglacial cycles. Instead, a series of internal Earth system processes and feedbacks are required to translate and amplify subtle regional and seasonal changes driven by orbital forcing into the large-scale Pleistocene climate cycles (Imbrie et al., 1993; Ruddiman, 2006; Shackleton, 2000). These feedbacks arise from interactions between different components of the climate system, including pole-to-equator temperature gradients (which alter the atmospheric and ocean circulation), the biosphere, the cryosphere and the carbon cycle. The close link between a global stack of benthic foraminiferal oxygen isotope records ($\delta^{18}O$), indicating Earth's climate state (i.e. a combination of ice volume and deep ocean temperature), and the ice core records of atmospheric CO_2 concentrations exemplifies these critical feedbacks (Fig. 10.1A and C). Exploring and quantifying how these feedbacks operated in the past represents a key step in understanding the Earth's climate system and informing us about possible future changes that will arise from anthropogenic forcing.

Historically, explanations for Pleistocene climate change focused on the Northern Hemisphere ice sheets, which made the major contribution to sea level variability, and on Atlantic Ocean processes, which can convey climate signals into the Southern Hemisphere and globally. However, ice core records indicate a striking correlation between local Antarctic temperatures and atmospheric CO_2 concentrations (Fig. 10.1B and C). As such, research is increasingly recognising the critical importance of Southern Ocean processes, such as sea ice formation, ocean stratification, upwelling, deep water formation and biological productivity in the Pleistocene climate system (Ferrari et al., 2014; Kohfeld and Chase, 2017; Martinez-Garcia et al., 2011; Sigman and Boyle, 2000). Furthermore, evidence is emerging for a more dynamic behaviour of the Antarctic Ice Sheet in response to Pleistocene climate cycles than had previously been recognised (Blackburn et al., 2020; Dutton et al., 2015; McKay et al., 2012b; Rohling et al., 2019; Scherer et al., 2008; Turney et al., 2020; Wilson et al., 2018). Therefore, the Antarctic Ice Sheet may be both an active player in the interconnected climate system (Fogwill et al., 2015; Schloesser et al., 2019; Weaver et al., 2003) and a highly sensitive component that could respond dramatically to future anthropogenic warming (DeConto and Pollard, 2016; Golledge et al., 2017; Rintoul et al., 2018).

While this chapter will touch on some of the above debates concerning the operation of the Pleistocene climate system, including mechanisms for the MPT, we do not seek to provide an authoritative review of Pleistocene glacial–interglacial cycles or their link with orbital forcing at a global scale. Such a topic would be deserving of an entire book in itself. Instead, we focus on changes in Antarctic climate and the Antarctic Ice Sheet during this interval, and the role of Southern Ocean and Antarctic processes in the interconnected global Earth system. We make a particular effort to discuss some of the key interactions between the Antarctic Ice Sheet, ocean circulation and global carbon cycle changes, across orbital and millennial timescales.

10.1.4 Strengths of Pleistocene research on Antarctica

Although we now appear to be entering into a world in which large Northern Hemisphere ice sheets will no longer exert a dominant influence on global climate variability, an understanding of the Earth system behaviour during the Pleistocene epoch is highly relevant for constraining ice sheet–ocean–climate interactions on the modern and future Earth. Below we highlight some of the key strengths of research on this interval:

1. Boundary conditions: geological boundary conditions during the Pleistocene were close to those of the modern day, in terms of the latitudinal distribution and topography of the continents, the bathymetry and geometry of ocean gateways, and the subglacial topography beneath ice sheets. These are important factors that determine atmospheric and ocean circulation, global heat transport, and ice sheet stability and dynamics. The Pleistocene behaviour is therefore expected to provide a useful guide to how a wide range of climate feedbacks could operate and interact in future, with the similarities in boundary conditions beneficial for climate modelling.
2. Temporal resolution: when compared to earlier intervals of the Cenozoic, Pleistocene paleoclimate records from archives such as ocean sediments,

lake sediments and speleothems generally benefit from better temporal resolution and a wider range of precise dating methods. This geological evidence provides the opportunity to constrain processes operating in the climate system from orbital to sub-centennial timescales, with the potential in some cases to determine rates of change, although this latter goal remains challenging beyond the Last Glacial Maximum (LGM, see Siegert et al., Chapter 11, this volume).

3. Ice core records: extremely well-resolved records of late Pleistocene polar temperatures and atmospheric compositions are contained in ice cores, which span the last ~100 kyr in Greenland and ~800 kyr in Antarctica (Fig. 10.1). These archives provide critical evidence on the local climate forcing acting on the ice sheets at an annual to centennial resolution, as well as recording global carbon cycle changes and interhemispheric patterns of climate variability.
4. Spatial integration: Pleistocene sediment records are readily obtained from the seafloor and lake beds using shallow coring techniques, as well as extended piston coring, leading to a wide spatial coverage of climate archives. Combining such records reveals interhemispheric patterns of climate variability and connections between different components of the climate system. As such, and in combination with climate modelling, it is possible to constrain how the critical processes and feedbacks that amplify climate responses are operating.
5. Cold climates: The glacial periods of the late Pleistocene represent a good example of Earth system behaviour in a significantly colder climate state than the modern day, which provides a robust test for our understanding of the climate system and, in particular, for the processes and parameterisations employed in ocean, ice sheet and climate models. Critically, we are also able to explore Earth system responses to natural warming events that had a similar magnitude to projected scenarios for the future, such as the 5°C–6°C of global warming during the last glacial termination.
6. Future analogues: The warmest interglacials of the Pleistocene were globally ~1°C–2°C warmer than the pre-industrial Holocene and, potentially, warmer than this for short intervals in Antarctica due to polar amplification (Fig. 10.1B). Therefore, records from these periods provide insight into Earth system behaviour under levels of modest warming that are relevant to climate projections for the coming decades.

10.2 Archives of Pleistocene Antarctic climate and climate-relevant processes

10.2.1 Polar ice cores

10.2.1.1 Background and characteristics of ice core records

Continuous deep ice cores provide highly-resolved records of Antarctic temperatures and atmospheric carbon dioxide (CO_2) concentrations over the last eight glacial–interglacial cycles (Fig. 10.1B and C). Past temperatures can be derived from the isotopic composition of oxygen ($\delta^{18}O_{ice}$) or deuterium (δD_{ice}) in the ice, while past concentrations of atmospheric CO_2 and other greenhouse gases are recorded in the trapped gas bubbles. Recent discussions of Antarctic ice core science, including the historical background, theoretical and practical considerations, and interpretations of the records can be found in reviews by Alley (2010), Brook and Buizert (2018) and Jouzel (2013), and only a short summary is given below.

The earliest deep drilling was carried out at Byrd Station in central West Antarctica (Gow et al., 1968), but it was several decades before the first continuous record covering multiple glacial–interglacial cycles was recovered, at Vostok Station in East Antarctica (Petit et al., 1999) (Fig. 10.2). The Vostok record extends through four glacial cycles to ~420 ka, while the European Programme for Ice Coring in Antarctica (EPICA) project subsequently recovered an even longer core from Dome C (EDC) (Fig. 10.2) from the present day to ~800 ka (EPICA Community Members, 2004; Jouzel et al., 2007). To achieve such long records drilling must be carried out on the East Antarctic Plateau, where low snow accumulation rates and slow ice flow mean that old ice can be reliably obtained from deep parts of the ice sheet. However, low accumulation rates also limit the temporal resolution of the records and restrict the ability to date the ice by layer-counting, so dating in these settings requires a combination of glaciological ice flow/snow accumulation modelling and orbital tuning.

At lower-lying settings, including East Antarctic coastal regions and in West Antarctica, snow accumulation rates are higher and annual layers are thicker. This scenario enables dating by layer-counting (at least in the upper portions of cores where there has been less compaction) and the acquisition of highly-resolved climate records for the last glacial cycle. Recent examples include the deglacial record from the West Antarctic Ice Sheet (WAIS) Divide ice core (WDC) (Marcott et al., 2014; WAIS Divide Project Members, 2013) and a Holocene record from the Roosevelt Island Climate Evolution (RICE) core (Bertler et al., 2018) (Fig. 10.2). Since these settings experience a stronger marine influence than the East Antarctic Plateau, they may also provide more sensitive records of changes in ocean circulation and sea ice in the Southern Ocean (Bertler et al., 2018; WAIS Divide Project Members, 2013), making them highly complementary to the interior records. In addition, comparisons between records from the Atlantic, Indian and Pacific sectors of the continent can provide evidence on the spatial patterns of climate variability and sea ice extent (Buizert et al., 2018; Holloway et al., 2017; Stenni et al., 2011). By comparing records between multiple ice cores, there is also potential to recover past changes in local ice sheet elevation, which may in turn provide evidence on ice sheet mass balance, ice flow and glacio-isostatic adjustment (Bradley et al., 2012; Sutter et al., 2020).

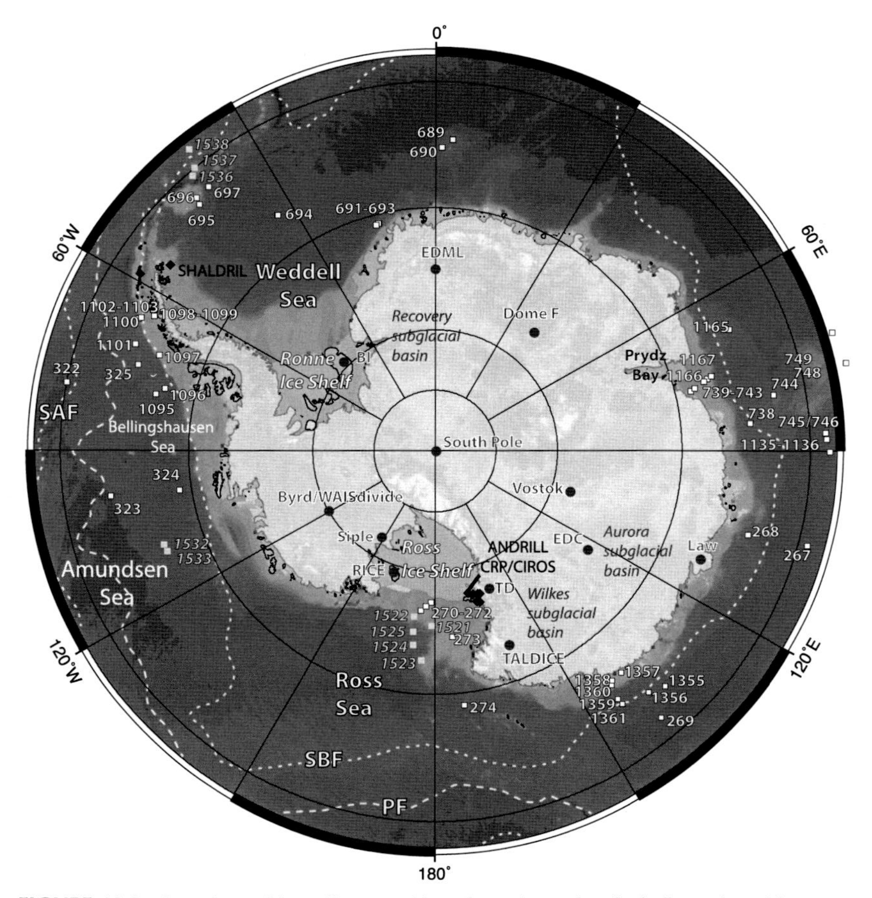

FIGURE 10.2 Locations of key climate archives from Antarctica, including selected ice cores (blue circles) and marine sediment cores (white squares, yellow squares, black diamonds), and subglacial topography of Antarctica (Bedmap2; Fretwell et al., 2013). Sediment core sites from DSDP/ODP/IODP expeditions (white squares) include DSDP Leg 28 (Ross Sea), DSDP Leg 35 (Bellingshausen Sea), ODP Leg 113 (Weddell Sea), ODP Legs 119 and 188 (Prydz Bay), ODP Leg 178 (Antarctic Peninsula), and IODP Expedition 318 (Wilkes Land). Sites from recent IODP drilling legs (yellow squares) include IODP Expedition 374 (Ross Sea), IODP Expedition 379 (Amundsen Sea), and IODP Expedition 382 (Iceberg Alley). Nearshore drilling sites (black diamonds) include Cape Roberts Project (CRP), Cenozoic Investigations in the Western Ross Sea (CIROS), Antarctic Drilling Project (ANDRILL) in the Ross Sea, and SHALDRILL on the Antarctic Peninsula. Abbreviations for ice cores: *BI*, Berkner Island; *EDC*, EPICA Dome C; *EDML*, EPICA Dronning Maud Land; *RICE*, Roosevelt Island Climate Evolution; *TALDICE*, Talos Dome Ice Core; *TD*, Taylor Dome; *WAIS Divide*, West Antarctic Ice Sheet Divide. Abbreviations for ocean fronts (white dashed lines): *SBF*, Southern Boundary Front; *PF*, Polar Front; *SAF*, Subantarctic Front.

For the last glacial cycle, records of Northern Hemisphere climate have also been recovered using ice cores in Greenland, thereby enabling inter-hemispheric comparisons. In particular, the ice cores recovered by the European Greenland Ice-core Project (GRIP) and the Greenland Ice Sheet Project 2 (GISP2) were instrumental in providing high-resolution records of millennial and centennial climate change in the North Atlantic region during the last glacial period and the last deglaciation (Dansgaard et al., 1993; Grootes et al., 1993; Johnsen et al., 1992). More recently, the North Greenland Ice-core Project (NGRIP) recovered a core from a more northerly site, providing a record that extends further back into the last interglacial period (NGRIP Members, 2004). Chronologies in Greenland are generally based on layer counting in the upper portions of cores, with age uncertainties of less than ~1 kyr back to 60 ka in the GICC05 chronology (Svensson et al., 2008), but approaches such as glaciological flow modelling, orbital tuning and correlations based on volcanic ash layers or methane content are required at deeper depths where the layers are more compressed.

10.2.1.2 *Ice core climate proxies*

Past Antarctic temperatures can be derived from measurements of $\delta^{18}O_{ice}$ or δD_{ice} in the ice, using an approach based on the relatively well-understood process of Rayleigh distillation. As water vapour is transported through the atmosphere, colder temperatures lead to more removal as rain or snow, which preferentially takes out the heavy isotopes (^{18}O, D) and leads to a lighter isotopic composition (i.e. more ^{16}O, H) in the remaining vapour, which is ultimately recorded in the ice core record (Jouzel et al., 2007). This relationship appears relatively simple in the modern day, but additional considerations are required when applying it to the past, such as changes in the nature and location of the moisture source, changes in seasonality and changes in ice sheet elevation (Bradley et al., 2012; Jouzel et al., 2003). Evidence of past climate change can also be preserved more directly in the temperature of the ice itself, enabling independent estimates of temperature changes during the last deglaciation to be obtained by modelling of borehole thermometry data (Cuffey et al., 2016). For times of rapid climate change, modelling of the isotopic composition of nitrogen ($\delta^{15}N$) trapped in gas bubbles can also be used to reconstruct temperature changes (Severinghaus and Brook, 1999), although this approach has been more widely used in Greenland than in Antarctica (Cuffey et al., 2016). Overall, the application of these alternative methods has provided temperature estimates that add confidence to the high-resolution reconstructions based on $\delta^{18}O_{ice}$ or δD_{ice}.

Atmospheric compositions are recorded in trapped bubbles of gas, enabling reconstructions of past changes in greenhouse gas concentrations, such as CO_2, methane and nitrous oxides (Petit et al., 1999; Spahni et al., 2005). However, gas bubbles continuously exchange with the atmosphere and only become trapped (or 'locked-in') when snow is converted to ice, which typically occurs at ~50–120 m depth depending on local site conditions (Parrenin et al., 2013). Therefore, in any given sample, the gas compositions are always younger than the surrounding ice, with the gas-ice age offset (Δ_{age}) determined by the lock-in depth and the accumulation rate. For Antarctic coastal sites with rapid accumulation rates, Δ_{age} is typically only ~200–1500 years during the last glacial period, whereas it can be as high as ~6–8 kyr where accumulation rates are low on the East Antarctic Plateau (Marcott et al., 2014).

When comparing records of gas composition (e.g., atmospheric CO_2) to records obtained on the ice itself (e.g., $\delta^{18}O_{ice}$), a correction is routinely applied for Δ_{age}. However, this correction introduces an uncertainty which limits the ability to resolve the detailed phasing between atmospheric composition and climate. Suggestions that deglacial warming in Antarctica preceded the atmospheric CO_2 increase by a few hundred years at the start of the last deglaciation (Fischer et al., 1999; Monnin et al., 2001) may have underestimated the uncertainty in Δ_{age}, whereas a more recent analysis with improved age constraints indicates that the onset of deglacial CO_2 and temperature change was synchronous within error (Parrenin et al., 2013). Most recently, using data from the WAIS Divide Core (WDC), which benefits from high snow accumulation rates and a small Δ_{age} (~200–300 years), it has been suggested that atmospheric CO_2 changes may actually have slightly led the temperature changes for much of the deglaciation (Marcott et al., 2014).

Measurements of methane gas concentrations serve two additional purposes, beyond quantifying its role as a greenhouse gas. First, because changes in atmospheric methane concentrations are virtually synchronous across the globe, comparison of the methane signal allows for synchronisation of ice core age models between Greenland and Antarctica, which is crucial for evaluating the interhemispheric phasing of climate (Blunier and Brook, 2001; EPICA Community Members, 2006; WAIS Divide Project Members, 2015). In principle, a similar approach would be possible using CO_2 records, but the Greenland ice cores do not provide reliable CO_2 reconstructions due to in situ production from chemical impurities. Second, since the atmospheric methane budget is believed to be dominated by Northern Hemisphere processes, methane records may provide markers within ice cores for rapid Northern Hemisphere climate changes (e.g., Dansgaard–Oeschger warming events). An interesting consequence is that measurements of methane and CO_2 on the same gas phase in an Antarctic ice core can be used to make a direct comparison (independent of uncertainty in Δ_{age}) between the timing of Northern Hemisphere processes (recorded by methane concentrations) and changes in the global carbon cycle (recorded by CO_2 concentrations) (Marcott et al., 2014).

Many other chemical constituents can be measured in ice cores to provide proxy evidence on regional or global climate forcings and responses. Examples include past sea ice extent (recorded by sea-salt sodium fluxes, ssNa), dust deposition rates (recorded by non-sea-salt calcium fluxes, nssCa), and inputs of biological aerosols (e.g., methanesulphonate or sulphate) (Wolff et al., 2006). While providing highly valuable records, source and transport effects typically need to be considered for a full understanding of those proxies (Levine et al., 2014), and the relative role of these processes may differ between ice core locations and through time. To complement the dust records, measurements of radiogenic isotopes (e.g., neodymium, strontium and lead) can be used to trace the dust source regions (Aciego et al., 2009; Grousset and Biscaye, 2005; Vallelonga et al., 2010) and therefore to evaluate changes in dust emission and/or atmospheric circulation patterns in the past. Similarly, records of volcanic eruptions can be obtained from tephra layers in ice, providing evidence on the volcanic forcing of climate (Sigl et al., 2015), as well as being a useful chronological tool for dating and correlating records (Hillenbrand et al., 2008; Narcisi et al., 2005; Parrenin et al., 2013; Turney et al., 2020).

10.2.1.3 Recent advances in ice core proxies and attempts to obtain ice older than one million years

Recent pioneering advances in generating climate records from Antarctic ice cores include the development of analytically challenging measurements on the gas bubbles and the recovery of ice older than ~800 ka. Isotopic measurements have previously been made on the oxygen gas ($\delta^{18}O_{air}$) to monitor changes in the atmospheric oxygen cycle, which is linked to variability in the global monsoon (Petit et al., 1999; Severinghaus et al., 2009). More recently, methods have been developed to measure carbon isotopes of the trace constituent CO_2 ($\delta^{13}C_{CO2}$), providing novel constraints on the sources and sinks of atmospheric CO_2 and new insights into the global carbon cycle (Bauska et al., 2016). Another pioneering development is the measurement of the krypton/nitrogen (Kr/N_2) ratio, which serves as a proxy for global mean ocean temperature because the solubility of each gas has a different temperature dependence (Bereiter et al., 2018).

Whereas the existing East Antarctic ice cores provide a continuous climate record back to ~800 ka, a major long-term goal is to obtain older records, in particular to resolve the nature and causes of the MPT at ~900 ka (Jouzel and Masson-Delmotte, 2010). One recent approach has involved drilling horizontal blue-ice sections from marginal settings of the ice sheet where old ice upwells towards the surface. This method has obtained discontinuous snapshots of older ice from short intervals at ~1, ~1.5 and ~2 Ma (Higgins et al., 2015; Yan et al., 2019), providing new constraints on the carbon cycle–climate relationship before the MPT. However, those data remain challenging to interpret because they represent short time intervals and are subject to large dating uncertainties. To circumvent those issues, and to provide a continuous record through the MPT, a number of efforts are underway to obtain a deep Antarctic ice core that extends to $\sim$1.5 Ma, which modelling studies suggest is a feasible goal (Fischer et al., 2013; Sutter et al., 2019). The 'Beyond EPICA' project proposed that a suitable site exists at Little Dome C near Concordia Station on the East Antarctic Plateau and drilling operations were set to begin in the 2020–2021 field season.

10.2.2 Deep-sea paleoceanographic records

Deep-sea sediment cores have been collected from throughout the global oceans and provide relatively continuous archives of past changes in ocean properties. Analyses on such cores have enabled reconstructions of many aspects of ocean chemistry, biology and physics, from which changes in regional and global climate states and oceanic and atmospheric processes can be inferred. The general approach is well demonstrated by the series of international programmes that have used riserless drilling to recover long sediment cores from the seafloor, namely, the Deep Sea Drilling Project (DSDP, 1968–1983), Ocean Drilling Program (ODP, 1983–2003), Integrated Ocean Drilling Program (IODP, 2003–2013), and International Ocean Discovery Program (IODP, 2013–2023). These programmes have been running for more than half a century and have facilitated many of the major advances made in the Earth sciences over this interval, in particular in paleoceanography and paleoclimate. Since Pleistocene sediments are generally accessible to a wider range of drilling and coring approaches than sequences from older epochs, a large number of cores have also been collected during expeditions led by particular countries or institutes. These endeavours have typically provided records from the last deglaciation, but in some cases longer records have been recovered extending through multiple glacial–interglacial cycles.

10.2.2.1 Proxies for climate and ocean–atmosphere–ice sheet processes

By analysing the fossil and mineral contents, grain sizes, magnetic properties and chemical and isotopic compositions of various fractions of the sediment, it is possible to recover both qualitative and quantitative evidence on a multitude of climate-relevant parameters. These parameters include global ice volume, deep-ocean temperatures, sea surface temperatures (SSTs), salinity, water mass sourcing, ocean current strength, biological productivity, nutrient utilisation, pH, carbonate ion concentration, nutrient content, iceberg rafting, sediment provenance, dust input and sea ice extent. In general, observations and measurements on sediment cores provide 'proxies' for the parameter of interest, rather than enabling a direct reconstruction. Therefore, such proxies must be tested and calibrated using a combination of theoretical calculations, laboratory experiments, core-top calibrations and numerical ocean modelling.

General reviews of paleoceanographic approaches and proxies can be found in the literature (Henderson, 2002; Katz et al., 2010; Robinson and Siddall, 2012), and details of methodology and proxy developments are beyond the scope of this contribution. Here we restrict ourselves to listing a range of the most widely applied proxies, together with notes highlighting some important aspects of their application, and selected references to guide the reader towards pioneering studies and recent instructive updates (Table 10.1).

10.2.2.2 Pleistocene age models

In order to apply such proxy methods in the past, and to make comparisons between marine sediment cores and ice cores, it is necessary to generate age models that convert sediment core depths to calendar ages. For sediments of Pleistocene age, a wide range of approaches are possible and these techniques are often used in combination to achieve the most robust and/or highest resolution age model:

1. Radiocarbon ($\Delta^{14}C$) dating of carbonate fossils or organic matter: This method provides absolute ages for individual sediment layers, subject to consideration of surface reservoir ages (Skinner et al., 2019) and conversion from radiocarbon years to calendar years (Reimer et al., 2013). Due to the short half-life of radiocarbon, this method is only applicable back to $\sim$40 ka. In addition, bioturbation or reworking of microfossils and inputs of aged terrestrial organic carbon can present challenges, particularly on the Antarctic continental shelf (Anderson et al., 2014).
2. Correlation of carbonate $\delta^{18}O$ records to a benthic $\delta^{18}O$ stack: This method is routinely used where foraminiferal carbonate is available in a sediment core, but carbonate is sparse and typically absent from Southern Ocean and Antarctic margin sediments. In this region, other types of proxy records can be used for correlation with a benthic $\delta^{18}O$ stack, but this approach leads to greater uncertainties when assessing leads and lags with far-field records. Some caution is also needed because deep ocean $\delta^{18}O$ changes are not synchronous on millennial timescales (Skinner and Shackleton, 2005), while $\delta^{18}O$ stacks are typically constructed using orbital tuning with an assumed phase lag between insolation changes and the climate response (e.g., SPECMAP, Martinson et al., 1987; LR04, Lisiecki and Raymo, 2005).
3. Biostratigraphy using carbonate or siliceous microfossils and nannofossils: Although it usually provides lower resolution age constraints than the above two methods, diatom biostratigraphy is widely and effectively used in the Southern Ocean where carbonate is often absent (Cody et al., 2012; Gersonde and Barcena, 1998). However, it is limited by the need to assume linear sedimentation rates between datums that are often widely separated in time, and by the reworking processes that affect many depositional settings on the Antarctic shelf (McKay et al., 2019). A specific challenge for

TABLE 10.1 Key paleoceanographic parameters and their reconstruction from proxy measurements.

Parameter	Proxy	Notes	References
Ice volume	Benthic foraminiferal oxygen isotopes ($\delta^{18}O$)	Need benthic Mg/Ca data or climate modelling to remove temperature effect	Bintanja et al. (2005), Elderfield et al. (2012), Shackleton (2000)
Deep ocean temperature	Benthic foraminiferal oxygen isotopes ($\delta^{18}O$)	Need to remove ice volume effect; also sensitive to salinity	Marchitto et al. (2014), Zachos et al. (2001)
	Benthic foraminiferal Mg/Ca ratios	Potential additional carbonate ion influence	Elderfield et al. (2012), Lear et al. (2000)
	Clumped isotopes of oxygen and carbon (Δ_{47}) in carbonates	Independent of fluid composition, but need to consider species-dependent vital effects in corals	Ghosh et al. (2006), Spooner et al. (2016)
Sea surface temperature (SST)	Planktonic foraminiferal oxygen isotopes ($\delta^{18}O$)	Need to remove ice volume effect; also sensitive to salinity	Emiliani (1955), Pearson (2012)
	Planktonic foraminiferal Mg/Ca ratios	Need species- and/or location-specific calibrations; foraminifera cleaning method is important	Barker et al. (2005), Elderfield and Ganssen (2000), Nürnberg et al. (1996), Vázquez Riveiros et al. (2016)
	Unsaturation index in alkenones derived from coccoliths ($U^{K'}_{37}$ and variants)	Calibrations for different regions and temperature ranges; coccoliths can be absent near the poles; need to consider seasonality effects	Conte et al. (2006), Sikes and Volkman (1993)
	Tetraether index in organic membrane lipids (TEX_{86} and variants)	Versions of the index exist for different latitudes; regional calibrations may be useful	Ho et al. (2014), Schouten et al. (2002)
	Long Chain Diol Index (LDI)	Wide distribution including high latitudes; calibration extends to cold temperatures; research needed on source of diols, mechanism and seasonality effect	Lopes dos Santos et al. (2013), Rampen et al. (2012)
	Heterocyst diol and triol indices (HDIs and HTIs)	Developed in lakes; application in the ocean is in its infancy	Bauersachs et al. (2015)

(*Continued*)

TABLE 10.1 (Continued)

Parameter	Proxy	Notes	References
Salinity	Foraminiferal oxygen isotopes ($\delta^{18}O$)	Need to couple with other proxies to remove temperature and ice volume effects; salinity effects of sea ice are not readily resolved	Rohling (2000), Schmidt et al. (2004)
Meltwater input	Diatom oxygen isotopes ($\delta^{18}O$)	Promising tracer in polar regions where foraminifera are absent; need to consider species, size fraction, vital effects and diagenesis	Shemesh et al. (1994), Swann and Leng (2009)
Water mass sourcing	Authigenic neodymium isotopes (ε_{Nd})	Need to constrain water mass endmembers; additional influences from weathering changes and boundary exchange	Goldstein and Hemming (2003), van de Flierdt et al. (2016)
	Benthic foraminiferal carbon isotopes ($\delta^{13}C$)	Need to constrain water mass endmembers; also influenced by nutrient regeneration (linked to surface productivity and deep ocean ventilation rate)	Curry and Oppo (2005), Lynch-Stieglitz and Fairbanks (1994), Mackensen and Schmiedl (2019)
	Benthic foraminiferal Cd/Ca ratios	Need to constrain water mass endmembers; also influenced by nutrient regeneration (linked to surface productivity and deep ocean ventilation rate)	Boyle (1988a), Marchitto and Broecker (2006)
Deep ocean ventilation	Radiocarbon in benthic foraminifera or deep-sea corals ($\Delta^{14}C$)	Need to compare measured $\Delta^{14}C$ to $\Delta^{14}C$ of contemporaneous surface ocean or atmosphere	Adkins et al. (1998), Cook and Keigwin (2015), Shackleton et al. (1988)
Deep water export rates	Protactinium/thorium ($^{231}Pa/^{230}Th$)	Significant controls from productivity and particle scavenging	Bradtmiller et al. (2014), Chase et al. (2002)
Local current strength	Mean grain size of sortable silt fraction (SS)	Need to test that the sediment is current-transported; require local calibrations for quantitative reconstructions	McCave and Hall (2006), McCave et al. (1995), McCave et al. (2017)

(*Continued*)

TABLE 10.1 (Continued)

Parameter	Proxy	Notes	References
Export productivity	Organic carbon % or mass accumulation rate (can be based on specific organic molecules)	Strongly influenced by preservation (linked to factors such as sedimentation rate, temperature, and oxygenation)	Martinez-Garcia et al. (2009), Sachs and Anderson (2005)
	Opal content	Influenced by preservation (linked to sedimentation rate)	Mortlock et al. (1991), Ragueneau et al. (2000)
	Barite content, barium concentrations, or Ba/Al ratios	May be less sensitive to preservation than other productivity proxies	Francois et al. (1995), Paytan and Griffith (2007)
Nutrient utilisation	Stable nitrogen isotopes in bulk or microfossil-bound organic matter ($\delta^{15}N$)	Microfossil-bound data are less susceptible to transport and diagenesis than bulk data; consider diatom species and seasonality	Francois et al. (1997), Studer et al. (2015)
	Stable silica isotopes in diatoms ($\delta^{30}Si$)	Need to ensure separation from clay; isotopic fractionation is size- and species-dependent	de la Rocha et al. (1998), Egan et al. (2012)
Surface pH or atmospheric CO_2	Planktonic foraminiferal or coral boron isotopes ($\delta^{11}B$)	Calibrations are species-dependent; conversion to atmospheric CO_2 requires ocean-atmosphere equilibrium	Foster and Rae (2016), Hönisch et al. (2004), Sanyal et al. (1997)
Deep ocean pH	Benthic foraminiferal or deep-sea coral boron isotopes ($\delta^{11}B$)	Calibrations are species-dependent; need to understand internal pH modification of calcifying fluid (particularly for corals)	Foster and Rae (2016), Rae et al.(2011), Sanyal et al. (1997)
Deep ocean carbonate ion	Benthic B/Ca ratios	Calibrations are species-dependent	Yu and Elderfield (2007), Yu et al. (2014)
	Carbonate percentage, mass accumulation rate, Ca/Ti ratios or preservation indices	Also influenced by surface ocean productivity and porewater dissolution; method is most sensitive near lysocline depth	Anderson and Archer (2002), Broecker and Clark (2001), Gottschalk et al. (2018)
Deep ocean nutrient content	Benthic foraminiferal Cd/Ca ratios	Also influenced by water mass mixing	Boyle and Keigwin (1982), Marchitto and Broecker (2006)
	Benthic foraminiferal carbon isotopes ($\delta^{13}C$)	Also influenced by changes in water mass endmembers and mixing	Gebbie (2014), Lynch-Stieglitz and Fairbanks (1994), Mackensen and Schmiedl (2019)

(Continued)

TABLE 10.1 (Continued)

Parameter	Proxy	Notes	References
Deep ocean oxygen content	$\Delta\delta^{13}C$ gradient between epifaunal and deep infaunal benthic foraminifera	Requires suitable foraminifera species to coexist; calibration appears best at low to moderate oxygen levels	Hoogakker et al. (2015), McCorkle and Emerson (1988)
	Alkenone preservation	Need independent tracer for surface productivity	Anderson et al. (2019)
	Authigenic U, Mo, Mn or other redox-sensitive trace metals	Controlled by porewater chemistry so need independent tracer for productivity to infer deep water oxygenation	Jaccard et al. (2009)
Iceberg rafting (i.e. iceberg-rafted debris, IRD)	Percentage or accumulation rate of coarse grain size fraction (e.g., >125 μm)	Other potential controls from ocean current transport, iceberg survival (linked to ocean temperature) and debris content of icebergs	Ruddiman (1977), Patterson et al. (2014)
Provenance of iceberg rafted debris	U-Pb, K-Ar, Ar-Ar ages on specific minerals in coarse grain size fraction	Potential for biases due to nature of source lithology, grain size and transport processes	Hemming et al. (1998), Licht and Hemming (2017)
Provenance of fine-grained detrital sediment	Nd, Pb, Sr isotopes in fine-grained or bulk fraction	Need to remove authigenic fraction; need to consider transport, grain size, and weathering effects (particularly for Sr)	Farmer et al. (2006), Licht and Hemming (2017)
Dust input fluxes	Fe concentrations or mass accumulation rates (or Fe/Al, Fe/Ti ratios by scanning-XRF)	Sediment redistribution can be addressed using method of ^{230}Th normalisation	Kumar et al. (1995), Martinez-Garcia et al. (2009)
	^{232}Th mass accumulation rate	^{232}Th is a lithogenic tracer but grain size and provenance can be additional influences	McGee et al. (2016), Winckler et al. (2008)
Sea ice extent	Diatom and radiolarian assemblages	Most effective for winter sea ice extent; consider seasonality; location of cores is critical	Benz et al. (2016), Gersonde et al. (2005)
	Biomarkers (e.g., highly branched isoprenoids, HBIs)	New approach still in development; potential to additionally resolve summer sea ice extent and seasonality	Collins et al. (2013), Vorrath et al. (2019)

the late Pleistocene is the lack of species turnover events, with only nine diatom events spanning the last 800 kyr (Cody et al., 2012), only one of which is a first appearance datum (FAD), whereas the majority are last appearance datums (LADs) that are susceptible to reworking.

4. Tephrostratigraphy: Discrete tephra layers in sediment cores can be identified from their mineralogy and geochemistry, providing a valuable tool for regional correlation and dating that is often used in combination with other approaches. In Antarctica, this approach is most promising in regions of active volcanism associated with the West Antarctic Rift system. For example, $^{40}Ar/^{39}Ar$ dating of tephra deposits provides absolute constraints for the development of Neogene chronologies in the Ross Sea (Naish et al., 2009). In addition, the recognition of distinct tephra or crypto-tephra layers can also enable correlation between Antarctic marine sediments and ice core records (Di Roberto et al., 2019, 2021; Hillenbrand et al., 2008; Narcisi et al., 2005; Parrenin et al., 2013; Turney et al., 2020).
5. Magnetostratigraphy: Magnetic measurements are routinely conducted on cores during IODP expeditions, enabling precise stratigraphic alignment at major reversals (e.g., Brunhes-Matayama boundary; Bassinot et al., 1994) and coarse-scale interpolation in between, often assuming constant sedimentation rates. Additional approaches such as relative paleointensity are also being developed to provide finer-resolution age models away from magnetic reversals (Laj et al., 2004).
6. Correlation to ice core records: In the North Atlantic region, high-resolution age models for glacial periods have been generated based on correlation of North Atlantic SST proxies to Greenland ice core climate records (Shackleton et al., 2000) or synthetic tuning targets (Barker et al., 2011), based on the assumption that rapid millennial climate events are effectively synchronous across the region. Although Pleistocene climate changes in the Southern Ocean were generally less abrupt than in Greenland, SST records in South Atlantic and Southern Ocean cores typically show a strong similarity to Antarctic ice core temperatures on glacial–interglacial and millennial timescales, enabling the inter-comparison of marine and ice core records (Mortyn et al., 2003). The validity of transferring Antarctic ice core age scales to sediment cores in this way has recently been demonstrated (Hoffman et al., 2017). However, because of the assumption of synchronicity between these signals, caution is required when considering the relative phasing of climate events between marine and ice core records. In certain Southern Ocean settings, distinct glacial–interglacial and millennial timescale changes in dust inputs have also been inferred from proxies in marine sediment cores (e.g., iron content, leaf waxes, or magnetic susceptibility), enabling correlation to dust records in Antarctic ice cores (Martinez-Garcia et al., 2011; Pugh et al., 2009; Weber et al., 2012).

7. Correlations using sediment core properties: It is often possible to correlate between records from the same region where there is systematic climate-related variability in sediment properties such as carbonate percentage, colour index or magnetic susceptibility. This correlation can allow an age model to be transferred from a core with a well-established chronology and depositional model to another core that may have fewer age constraints.

10.2.2.3 Bioturbation and resolution

Unlike ice cores, sediment cores are generally subject to bioturbation, in which burrowing organisms living in oxic or sub-oxic conditions mix sediment components vertically over depths of up to tens of centimetres. Ocean sediments can also experience periods of erosion or non-deposition related to ocean current strength, sea level change, submarine channel activity or iceberg scouring. In combination, bioturbation, hiatuses and low or variable sedimentation rates limit the age resolution that is achievable in proxy records to typically a few hundred to a few thousand years. However, higher resolution paleoceanographic records can be obtained in settings with high accumulation rates, such as sediment drifts where sediment is locally focused, or highly productive regions with high biogenic particle fluxes. Nearshore sites with significant terrestrial inputs can also generate high-resolution records, but much of the Antarctic continental shelf has been characterised by polar glacial regimes during the late Pleistocene that are sediment-starved due to limited discharge of turbid meltwater (McKay et al., 2009). An additional challenge in the cold, corrosive waters of the Southern Ocean arises from the poor preservation of carbonate microfossils, which can further limit sediment accumulation rates, age model resolution and the range of proxy approaches available.

10.2.2.4 Deep-sea coral archives

Since they are not subject to bioturbation or highly variable sedimentation rates, deep-sea scleractinian corals have emerged as a promising archive that is suitable for absolute dating and the application of multiple geochemical tracers (Robinson et al., 2014). Measurements on fossil corals are particularly valuable in the Southern Ocean because it circumvents the challenges of carbonate dissolution and age model resolution that affect foraminifera-based proxies. As such, deep-sea corals have enabled glacial and deglacial reconstructions of Southern Ocean ventilation (radiocarbon; Burke and Robinson, 2012; Hines et al., 2015), water mass sourcing (neodymium isotopes; Robinson and van de Flierdt, 2009; Wilson et al., 2020), pH (boron isotopes; Rae et al., 2018) and surface nutrient utilisation (nitrogen isotopes; Wang et al., 2017). Future research in the Southern Ocean will be significantly enhanced by an increasing spatial and temporal coverage of deep-sea coral measurements and an expansion in the range of proxy measurements, as well as by integrating evidence from deep-sea corals with sediment and ice core records.

10.2.3 Ice-proximal sedimentary records

Globally-distributed sediment cores provide evidence on large-scale changes in Earth's climate system, with reconstructions of global deep water chemistry being particularly informative on processes occurring in the polar regions where deep waters form (Gebbie and Huybers, 2011; Lisiecki and Raymo, 2005). However, a full understanding of Antarctic changes requires evidence from sites that are more proximal to the Antarctic Ice Sheet, for a number of reasons. First, because of the polar amplification of climate, knowledge of globally-averaged temperatures is not sufficient for understanding the local climate forcing acting on the Antarctic Ice Sheet. Given the strong oceanic influences on ice sheet mass balance for Antarctica's marine-based ice sheets, records from both the surface and subsurface of the high latitude Southern Ocean are required. Second, changes in sea ice extent, biological productivity and sub-surface water mass properties can vary spatially over short distances, such as across Southern Ocean fronts, which necessitates an array of well-distributed local records. Third, and crucially for reconstructing past changes in ice sheet behaviour, local records of iceberg rafted debris (IRD), sediment provenance and sedimentary signatures of glacial grounding can be obtained. Since each catchment of the Antarctic Ice Sheet likely has a different sensitivity to climate and ocean forcing (Golledge et al., 2017), reconstructions of past ice sheet behaviour and proximal ocean dynamics need to be made on a sector-by-sector basis.

Proximal sites include the Antarctic continental shelves, the nearby continental rise and the deep ocean. Records from shelf sites provide the only way to obtain direct evidence on past changes in ice sheet extent, based on subglacial vs open marine sedimentation (McKay et al., 2012b; Smith et al., 2019) or the erosion and deformation signatures produced by ice grounding (Domack et al., 1999; McKay et al., 2012b; Reinardy et al., 2015). However, these settings are often characterised by discontinuous records with low sedimentation rates, poor core recovery and significant erosional hiatuses due to glacial over-riding, making them difficult to date. Conventional dating tools using foraminiferal $\delta^{18}O$ or SST reconstructions are rarely available, and dating is generally achieved at a coarse resolution using magnetostratigraphy, biostratigraphy and tephrostratigraphy, as well as radiocarbon where possible.

Where foraminifera are present at shelf sites, radiocarbon-based approaches are viable to generate age models covering the last $\sim$40 kyr, but such records are rare (e.g., Bart et al., 2018; Hillenbrand et al., 2017; Mackintosh et al., 2011; McKay et al., 2016). Many studies rely on radiocarbon dating of bulk organic matter or its acid-insoluble fraction, but in shelf settings characterised by low sedimentation rates and extensive glacial reworking of pre-LGM deposits, such measurements are usually compromised by the presence of reworked fossil carbon leading to anomalously old ages (Anderson et al., 2014). However, novel approaches are being developed to overcome the influence of such contamination, including ramped pyrolysis (Rosenheim et al., 2013) and compound-specific analyses (Ohkouchi et al., 2003), which attempt to extract the youngest populations of organic carbon present in the sediment. In the Ross Sea, recent studies have observed in-situ calcareous foraminifera in certain sub-ice-shelf and grounding-line-proximal settings (Bart et al., 2016), which has provided better radiocarbon constraints for records of the last deglaciation (Bart et al., 2018), as well as enabling more robust radiocarbon age models to be developed based on a combination of bulk, compound-specific and foraminiferal analyses (Prothro et al., 2020).

On the continental rise and in the deep ocean, depositional settings with more continuous sedimentation are more common, allowing the recovery of longer records with fewer (if any) erosional hiatuses that are more suitable for dating. While the general lack of carbonate sediments restricts conventional $\delta^{18}O$-based approaches in much of the Southern Ocean, methods including magnetostratigraphy, biostratigraphy and cyclostratigraphy are generally applicable. In specific settings, these methods may be supplemented by correlating proxy records for dust with Antarctic ice core records (Pugh et al., 2009), or by assuming a link between biological productivity and glacial–interglacial climate state (Bonn et al., 1998; Hillenbrand et al., 2009; Wu et al., 2017). Although the continental rise is typically located hundreds or thousands of kilometres from ice sheet margins (Fig. 10.2), sediment sequences recovered from levees on the banks of canyons influenced by turbidity currents have proven to be a valuable archive of orbital-scale continental margin processes (Escutia et al., 2011; Wilson et al., 2018). Aspects of past ice sheet behaviour are recorded by proxies including IRD fluxes (Patterson et al., 2014; Weber et al., 2014; Wilson et al., 2018; Wu et al., 2021) and mineralogical or geochemical provenance signatures of the sediments (Farmer et al., 2006; Hillenbrand et al., 2009; Licht and Hemming, 2017; Roy et al., 2007; Wilson et al., 2018; Wu et al., 2021). Records of IRD indicate the presence of ice grounded at sea level and can be used to infer dynamic ice margin behaviour. However, it is important to emphasise that IRD accumulation rates are not simply related to ice volume (or to changes in ice volume), since they are also influenced by the original sediment content of the ice, variability in the loss of basal sediment during ice shelf and iceberg transport (Clark and Pisias, 2000; Smith et al., 2019), and ocean temperatures and ocean currents which determine the locus of deposition (Cook et al., 2014; Licht and Hemming, 2017). Provenance of the fine-grained or bulk sediment provides information about the source rocks undergoing subglacial erosion on the proximal continent (Roy et al., 2007) and can therefore be used to infer changes in erosional sources linked to past ice sheet extent (Wilson et al., 2018) or to indicate sediment transport pathways (Simões Pereira et al., 2018). The other advantage of records from such offshore settings is that they are usually well-located to provide complementary evidence (often in the same samples) on paleoceanographic and climate changes, enabling records of ice sheet behaviour to be integrated with regional and global climate forcings and effects.

10.3 Records of global and Southern Ocean climate during the Pleistocene

In this section, we present a compilation of key records of global and Southern Ocean climate, derived from sediment cores and ice cores over the last 800 kyr (Fig. 10.3) and from sediment cores over the last 3 Myr (Fig. 10.4). Lisiecki and Raymo (2005) combined benthic foraminiferal $\delta^{18}O$ records from 57 globally-distributed sites to generate a Plio-Pleistocene reference curve (the 'LR04 stack') that reflects a combination of deep ocean temperature and ice volume. The LR04 stack indicates glacial–interglacial cycles in global climate superimposed on a progressive cooling during the Pleistocene, which was mostly expressed in more extreme glacial conditions (Fig. 10.4G). The stack also demonstrates two major climate shifts (Lisiecki and Raymo, 2005): a switch in cyclicity from dominantly ~41 to ~100-kyr cycles at the MPT (~900 ka, yellow bar in Fig. 10.4), and an increase in interglacial intensity since the Mid Brunhes Event (MBE, ~430 ka, yellow bar in Fig. 10.3) (Jansen et al., 1986).

10.3.1 Global sea level

While the LR04 stack provides an excellent stratigraphic tool for the Pleistocene (Lisiecki and Raymo, 2005), it is influenced by changes in both global ice volume and deep ocean temperatures (Shackleton, 2000), and their relative contributions to that record have varied through time (Elderfield et al., 2012). By combining records of benthic foraminiferal $\delta^{18}O$ and Mg/Ca ratios in the same sediment core, the $\delta^{18}O$ signal can be deconvolved into separate deep ocean temperature and 'ice volume' components (Elderfield et al., 2012; Sosdian and Rosenthal, 2009) (Fig. 10.3B, light blue line). However, converting the ice volume component in an individual record into a global ice volume or sea level record is complicated by local hydrographic (salinity) changes and also requires assumptions about the $\delta^{18}O$ compositions of past ice sheets (Jakob et al., 2020). A similar deconvolution approach has been proposed using a global compilation of planktonic $\delta^{18}O$ records and SST estimates (Shakun et al., 2015), while other methods have included scaling benthic $\delta^{18}O$ records to coral-derived sea levels for the last deglaciation

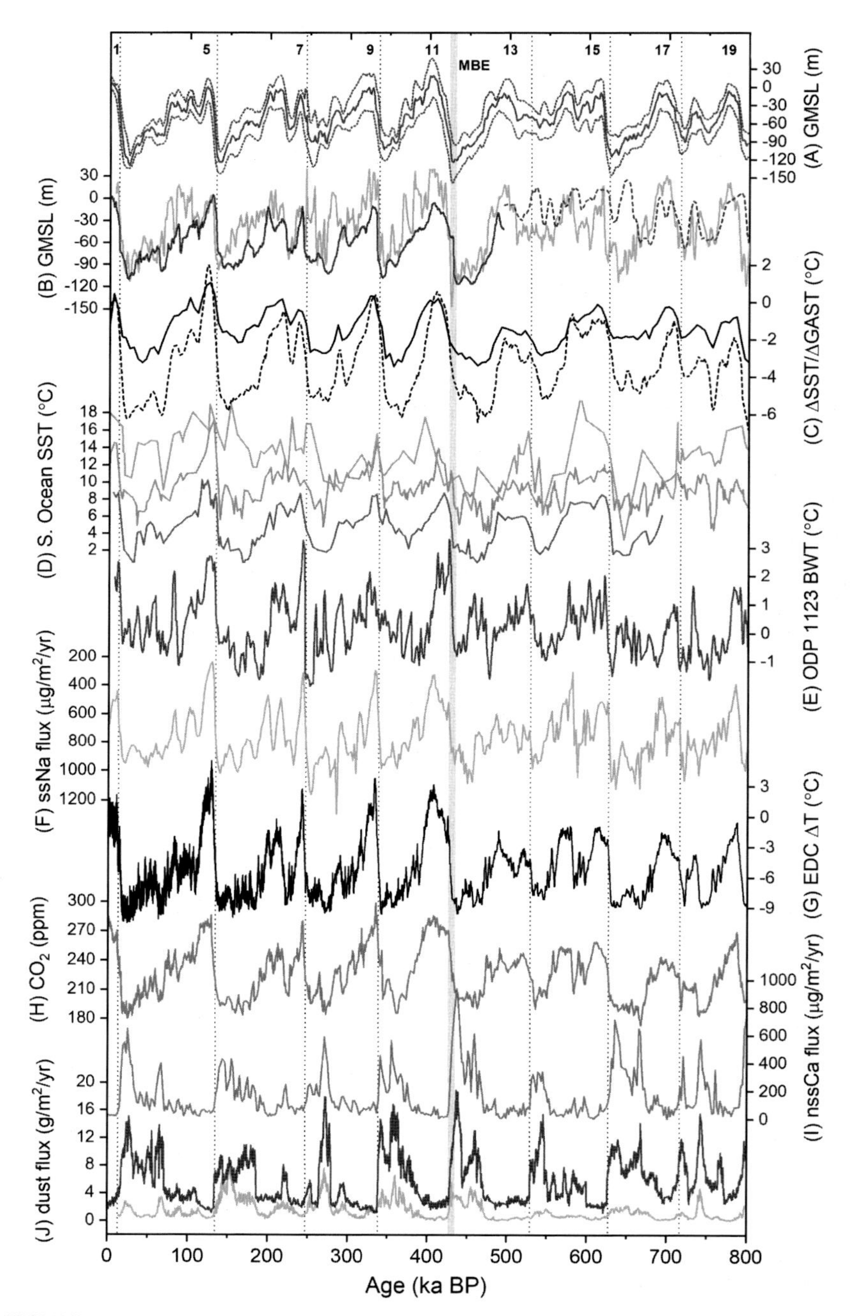

FIGURE 10.3 Global and Antarctic climate from 0 to 800 ka based on ice core and marine records. (A) Global mean sea level (GMSL) stack, with dashed lines showing 95% confidence

(Continued)

(Waelbroeck et al., 2002) or combining benthic $\delta^{18}O$ data with ice sheet modelling (Bintanja et al., 2005). An alternative approach for generating continuous sea level reconstructions has exploited planktonic foraminiferal $\delta^{18}O$ records from the Red Sea (Grant et al., 2014; Rohling et al., 2004) and the Mediterranean Sea (Rohling et al., 2014), by modelling the hydrographic response of these restricted basins to eustatic sea level change and local isostatic adjustments. The Red Sea record provides a sea level reconstruction back to ~500 ka (Grant et al., 2014) (Fig. 10.3B, solid blue line), while the Mediterranean Sea record extends back into the Pliocene (Fig. 10.3B and Fig. 10.4A, dashed blue line), although with larger uncertainties for the earlier section (Rohling et al., 2014). A recent statistical approach has combined a number of the above records with other similar ones to provide a late Pleistocene sea level stack for the last ~800 kyr (Spratt and Lisiecki, 2016) (Fig. 10.3A).

The above methods largely agree on the scale of the late Pleistocene sea level changes (Fig. 10.3A and B), indicating glacial–interglacial fluctuations of up to ~130 m since the MBE, with smaller changes both before the MBE and during the Marine Isotope Stage (MIS) 10–9 and MIS 8–7 transitions. However, the individual records are subject to quite large uncertainties (e.g., compare the two records in Fig. 10.3B), and a statistical treatment indicates uncertainties of 9–12 m (1σ) for the sea level stack (Spratt and Lisiecki, 2016) (Fig. 10.3A). Therefore, while these approaches provide valuable evidence on global sea level change through the late Pleistocene on orbital timescales, they are not well suited for resolving changes in Antarctic ice volume, which are expected to represent only ~10% of the global sea level signal during this interval (Tigchelaar et al., 2018). A particular issue here is

◀ interval (Spratt and Lisiecki, 2016). (B) Global mean sea level (GMSL) based on combined benthic $\delta^{18}O$ and Mg/Ca in ODP Site 1123 (light blue; Elderfield et al., 2012), and based on modelling of planktonic $\delta^{18}O$ in the Red Sea (solid blue, 0–492 ka; Grant et al., 2014) and the Mediterranean Sea (dashed blue, 492–800 ka; Rohling et al., 2014). (C) Change in global sea surface temperature (ΔSST) from an alkenone-based stack (solid; Martínez-Botí et al., 2015a) and change in global average surface temperature (ΔGAST) from a combined data-modelling approach (dashed; Snyder, 2016). (D) Southern Ocean SSTs based on alkenones from the subantarctic Southeast Pacific (purple, PS75/34-2; Ho et al., 2012), subantarctic South Atlantic (pink, PS2489-2/ODP Site 1090; Martinez-Garcia et al., 2010), and subtropical Tasman Sea (orange, DSDP Site 593; McClymont et al., 2016). (E) Southwest Pacific bottom water temperatures (BWT) from benthic foraminiferal Mg/Ca (ODP Site 1123; Elderfield et al., 2012). (F) Sea-salt sodium (ssNa) flux as a sea ice proxy in EDC ice core (Wolff et al., 2006). (G) Antarctic temperature difference (ΔT) from δD_{ice} in EDC ice core (Jouzel et al., 2007). (H) Atmospheric CO_2 concentration from an ice core compilation (Bereiter et al., 2015). (I) Non-sea-salt calcium (nssCa) fluxes as a dust proxy in EDC ice core (Wolff et al., 2006). (J) Dust fluxes to Southern Ocean marine cores in the Atlantic (brown, ODP 1090; Martinez-Garcia et al., 2011) and Pacific (orange, PS75/076-2; Lamy et al., 2014) sectors. Marine Isotope Stage numbers for interglacials are shown at the top. Vertical dotted lines indicate glacial terminations I–VIII. Yellow bar indicates the approximate timing of the Mid Brunhes Event (MBE). For clarity, uncertainties are not plotted on individual records.

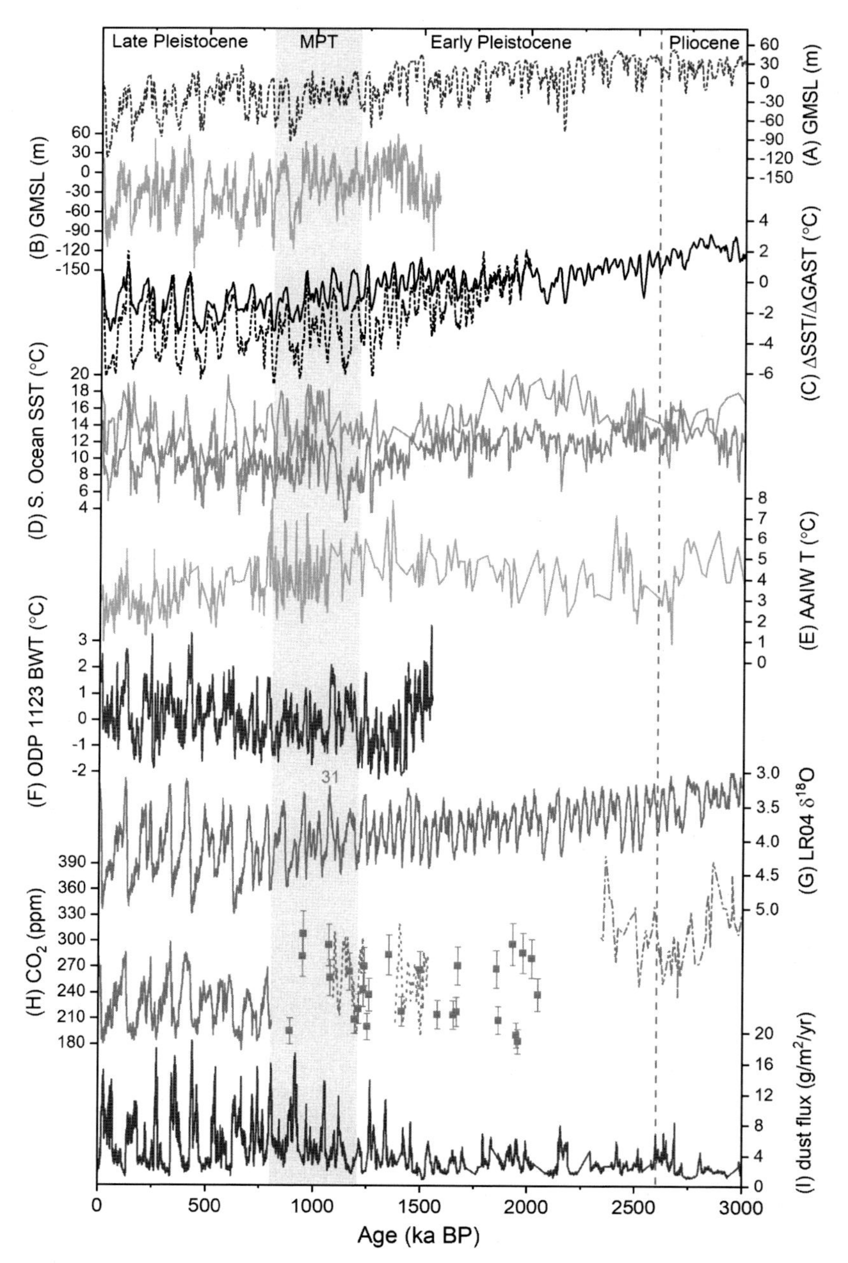

FIGURE 10.4 Global and Antarctic climate from 0 to 3 Ma based on marine records. (A) Global mean sea level (GMSL) based on modelling of planktonic $\delta^{18}O$ in the Mediterranean Sea (Rohling et al., 2014). (B) Global mean sea level (GMSL) based on combined benthic $\delta^{18}O$ and Mg/Ca in ODP Site 1123 (Elderfield et al., 2012). (C) Change in global sea surface temperature

(*Continued*)

that coral-based reconstructions indicate peak sea levels higher than the Holocene for certain recent interglacials (e.g., 6–9 m during MIS 5e (~116–129 ka) and 6–13 m during MIS 11; Dutton et al., 2015), which would implicate ice loss from the Antarctic Ice Sheet, but this magnitude of past sea level variability is not resolvable in the marine proxy records given their uncertainties. It is also important to note that the sea level estimates from sediment cores provide a globally-integrated measure of ice volume and cannot identify the locations where ice volume changes occurred, or the phasing of ice sheet advance and retreat between hemispheres. Therefore, while Elderfield et al. (2012) suggested that an expansion of global ice volume at ~900 ka during the MPT (Fig. 10.4B) may have included a significant Antarctic contribution, direct evidence is needed to test the location of this ice growth.

10.3.2 Sea surface temperatures

In terms of Earth's surface climate, the best global coverage of past changes is derived from proxy-based SST reconstructions in widely-distributed marine cores. The construction of a stack from these records minimises the effects of local changes in ocean currents or frontal positions, as well as proxy uncertainties, and enhances the signal/noise ratio to produce a meaningful global reconstruction. A global alkenone-based SST stack based on 10 high-resolution Pleistocene records (Martínez-Botí et al., 2015a) indicates glacial–interglacial changes of ~3°C–4°C in the late Pleistocene since the MBE (Fig. 10.3C, black line), ~2°C–3°C before the MBE and ~1°C–2°C in the early Pleistocene (Fig. 10.4C, black line). In addition, a multi-proxy compilation of over 20,000 SST data points (including alkenones, Mg/Ca

◀ (ΔSST) from an alkenone-based stack (solid; Martínez-Botí et al., 2015a) and change in global average surface temperature (ΔGAST) from a combined data-modelling approach (dashed; Snyder, 2016). (D) Southern Ocean SSTs based on alkenones from the subantarctic South Atlantic (pink, PS2489-2/ODP Site 1090; Martinez-Garcia et al., 2010) and the subtropical Tasman Sea (orange, DSDP Site 593; McClymont et al., 2016). (E) Antarctic Intermediate Water (AAIW) temperatures from the subtropical Tasman Sea (DSDP Site 593; McClymont et al., 2016). (F) Southwest Pacific bottom water temperatures (BWT) from benthic foraminiferal Mg/Ca (ODP Site 1123; Elderfield et al., 2012). (G) LR04 benthic $\delta^{18}O$ stack (Lisiecki and Raymo, 2005). (H) Atmospheric CO_2 concentration from ice cores (solid line; Bereiter et al., 2015) and from marine boron isotope records: low-resolution Pleistocene data (symbols and error bars; Hönisch et al., 2009); early MPT (dashed line; Chalk et al., 2017); early Pleistocene (dotted line; Dyez et al., 2018); late Pliocene-Pleistocene (dash-dot line; Martínez-Botí et al., 2015a; Sosdian et al., 2018). Note that uncertainties on those marine records are typically ~30–40 ppm in the Pleistocene and ~100 ppm in the Pliocene. (I) Dust fluxes to a Southern Ocean marine core in the Atlantic sector (ODP 1090; Martinez-Garcia et al., 2011). Vertical dashed grey line indicates Plio-Pleistocene boundary. Yellow bar indicates approximate timing of the Mid Pleistocene Transition (MPT). Super-interglacial MIS 31 is indicated on panel (G). For clarity, uncertainties are not plotted on individual records.

and foraminiferal and radiolarian transfer functions) from 59 marine cores was used to generate a 2 Myr long record of global average surface temperature (GAST) by spatially weighting the SST data in latitudinal bands and converting to GAST using a climate model (Snyder, 2016). This reconstruction indicates glacial–interglacial GAST changes of ~5°C–6°C in the late Pleistocene since the MBE (Fig. 10.3C, dashed black line), compared to ~4°C–5°C in the interval preceding the MBE, and ~2°C–3°C in the earlier Pleistocene (1.2–2.0 Ma) (Fig. 10.4C, dashed black line). Note that the larger changes in GAST than in SST reconstructions arise because larger temperature changes occur over land than over the ocean.

We plot three individual alkenone-based SST records to demonstrate changes in different regions of the Southern Ocean. Two records are from the Subantarctic Zone, in the South Atlantic (PS2489-2/ODP Site 1090; Martinez-Garcia et al., 2009, 2010) and the Southeast Pacific (PS75/34-2; Ho et al., 2012). The third record is from the Subtropical Zone, just north of the modern Subtropical Front, in the Tasman Sea (DSDP Site 593; McClymont et al., 2016). These records all demonstrate glacial cooling, resulting from a combination of global cooling, polar amplification and shifts in Southern Ocean fronts. The magnitude of glacial–interglacial temperature changes in the subantarctic South Atlantic and Southeast Pacific Ocean were very similar, with both recording glacial cooling of ~6°C–8°C since the MBE in comparison to glacial cooling of ~4°C–6°C in the interval immediately before it (Ho et al., 2012; Martinez-Garcia et al., 2009) (Fig. 10.3D). These results are also in good agreement with subantarctic South Atlantic estimates from foraminiferal transfer functions over the last 550 kyr (Becquey and Gersonde, 2003). For the Tasman Sea, which today is influenced by subtropical waters, the late Pleistocene glacial–interglacial changes were slightly larger at ~8°C–10°C, suggesting northward glacial shifts of the Subtropical Front (McClymont et al., 2016), with no distinct change at the MBE (Fig. 10.3D). In both subantarctic and subtropical settings, the glacial–interglacial variability during the early Pleistocene was muted compared to the late Pleistocene (Fig. 10.4D), being ~2°C–4°C in the subantarctic South Atlantic (Martinez-Garcia et al., 2010) and ~4°C–5°C in the Tasman Sea (McClymont et al., 2016).

10.3.3 Intermediate and deep ocean temperatures

Antarctic Intermediate Water (AAIW) forms today in the vicinity of the Subantarctic Front and propagates northwards at intermediate depths into each ocean basin. Therefore, a record of its changing properties reflects changes in surface conditions in the Subantarctic Zone of the Southern Ocean and/or changes in the location or mode of AAIW formation. Benthic foraminiferal Mg/Ca measurements at DSDP Site 593 in the Tasman Sea have been used to establish AAIW temperatures through the Pleistocene

(McClymont et al., 2016). Glacial–interglacial variability in AAIW temperature was ~3°C–4°C since the MPT and ~3°C during the early Pleistocene, superimposed on a long-term cooling trend since ~1.3 Ma (Fig. 10.4E). Note that the trends in the Pliocene and early Pleistocene intervals of that record are sensitive to poorly constrained seawater Mg/Ca ratios for that time, but the mid to late Pleistocene cooling is a robust feature (McClymont et al., 2016).

A large proportion of the global deep ocean is filled by deep waters that form in the Antarctic Zone of the high latitude Southern Ocean, i.e. Antarctic Bottom Water (AABW) and Lower Circumpolar Deep Water (LCDW) (Gebbie and Huybers, 2011). Therefore, past deep water temperatures reflect conditions in the regions of deep water formation near Antarctica. At ODP Site 1123, benthic foraminiferal Mg/Ca ratios provide a record of past LCDW temperatures in the pathway of the largest inflow of Antarctic-sourced deep waters into the Pacific Ocean. At this site, glacial temperatures have persistently been −1°C to −2°C (i.e. close to the freezing point of seawater) for the past ~1.5 Myr (Fig. 10.4F) (Elderfield et al., 2012). Before the MBE, interglacial temperatures were ~1°C–2°C and glacial–interglacial variability was ~2°C–3°C, whereas since the MBE there have been slightly higher interglacial peak temperatures (~2°C–3°C) and enhanced glacial–interglacial variability (~3°C–4°C) (Fig. 10.3E).

10.3.4 Antarctic temperatures and atmospheric CO_2

Antarctic air temperature reconstructions for the last 800 kyr based on ice core δD_{ice} indicate a distinct ~100-kyr cyclicity and a sawtooth pattern of slow stepped cooling and rapid warming that mirrors the LR04 benthic $\delta^{18}O$ stack (Fig. 10.1). Late Pleistocene glacial–interglacial variations were ~10°C–12°C since the MBE, compared to ~7°C–8°C before the MBE (Fig. 10.3G). This difference arose mostly as a result of the significantly cooler 'lukewarm' interglacials that preceded the MBE, whereas glacial temperatures changed by only ~1°C across the MBE (Jouzel et al., 2007). In comparison with estimates of GAST, Antarctica experienced polar amplification of around 1.6 times (2σ range of 1.2 to 2.3 times) in the late Pleistocene (Snyder, 2016). It is also important to note that peak Antarctic temperatures during the last four interglacials (MIS 5e, MIS 7, MIS 9 and MIS 11) were warmer than the Holocene by 2°C–4°C for a few thousand years (Jouzel et al., 2007) (Fig. 10.3G).

Atmospheric CO_2 concentrations from a compilation of Antarctic ice core records (Bereiter et al., 2015) demonstrate a very close link to Antarctic temperature variations. Since the MBE, glacial–interglacial variability in atmospheric CO_2 concentrations has averaged ~90 ppm, fluctuating between ~180 and 190 ppm during glacial periods and ~270 and 290 ppm during interglacials (Fig. 10.3H). Before the MBE, the glacial–interglacial variability was smaller (~60–70 ppm), with glacial CO_2 levels similar to the more recent glacials (~170–190 ppm) while interglacial CO_2 levels were lower (peaks of ~240–260 ppm) (Fig. 10.3H). Beyond the ice core record, atmospheric CO_2 reconstructions for the early Pleistocene are based on foraminiferal boron isotope records (Fig. 10.4H), which have larger uncertainties but appear to indicate slightly higher glacial CO_2 concentrations before the MPT (Chalk et al., 2017), as well as a decline in CO_2 concentrations since the late Pliocene and earliest Pleistocene (Martínez-Botí et al., 2015a).

10.3.5 Sea ice extent and dust supply

Other processes in the Southern Ocean region also varied on a ~100-kyr timescale, although sometimes with different magnitudes or timing within a glacial cycle. Reconstructions of Southern Ocean sea ice extent based on ice-core ssNa content (Wolff et al., 2006) indicate initial expansions in sea ice at the onset of glaciation (e.g., MIS 5e to MIS 5d transition), with further enhancements towards maximum sea ice extent during full glacials (e.g., MIS 4) and glacial maxima (e.g., MIS 2) (Fig. 10.3F). Although quantitative interpretation of the ssNa proxy faces some challenges, ssNa fluxes vary by up to a factor of four and show a striking similarity to Antarctic temperature records (Fig. 10.3F cf. Fig. 10.3G). Interglacials prior to the MBE experienced higher ssNa than more recent interglacials (Fig. 10.3F), suggesting that interglacial sea ice retreat was more muted during the lukewarm interglacials before the MBE.

Atmospheric dust supply to the Antarctic region is recorded in ice cores by non sea-salt calcium fluxes (nssCa fluxes) (Wolff et al., 2006), which vary by a factor of ~20 through glacial cycles (Fig. 10.3I) in a similar pattern to iron fluxes. Glacial periods were characterised by an increased dust supply that was particularly pronounced during glacial maxima (Fig. 10.3I). While changes in atmospheric transport to Antarctica may influence dust fluxes, the temporal patterns are well reproduced in dust flux records from Southern Ocean marine cores, such as ODP Site 1090 in the Subantarctic Zone of the South Atlantic (Martinez-Garcia et al., 2011) (Fig. 10.3J, brown) and PS75/076-2 in the South Pacific (Fig. 10.3J, orange), albeit with smaller relative glacial–interglacial changes. In combination with grain-size evidence for minimal variations in the wind-driven Antarctic Circumpolar Current during the LGM and Holocene (McCave et al., 2013), the agreement among these dust records suggests that a major part of the signal was due to changes in continental aridity, local wind strength or shelf exposure in the Southern Hemisphere dust source regions (e.g., Patagonia). Whereas highly-resolved sea ice reconstructions are lacking beyond the oldest ice core record at ~800 ka, high-resolution marine records provide dust reconstructions for the entire Pleistocene and indicate an approximate doubling of glacial dust fluxes to the Southern Ocean at the MPT (Martinez-Garcia et al., 2011) (Fig. 10.4I).

10.4 Late Pleistocene carbon cycle and climate dynamics

10.4.1 Controls on glacial–interglacial atmospheric CO_2

Changes in the carbon cycle were a fundamental feature of late Pleistocene glacial–interglacial cycles (Fig. 10.3) and were crucial for translating orbital forcing into changes in Earth's energy balance and mean global temperature. While colder ocean temperatures during glacial periods (Fig. 10.3C–E) would have directly enhanced oceanic uptake of CO_2 due to its greater solubility in cold water, thereby creating a positive feedback loop, the effect of these changes was modest in comparison to the full magnitude of glacial–interglacial atmospheric CO_2 changes (Sigman and Boyle, 2000). Furthermore, this temperature effect was almost cancelled out by opposing effects from reduced carbon storage in a saltier glacial ocean and the reduced size of the terrestrial biosphere (Sigman and Boyle, 2000). Therefore, more active mechanisms of carbon storage must have been operating to explain the consistent ~90 ppm decrease in atmospheric CO_2 during each glacial period since the MBE (Fig. 10.3H).

Because the deep ocean contains approximately 60 times more carbon than the atmosphere, it has been recognised for some time that carbon storage in the deep ocean was probably key to explaining these atmospheric CO_2 variations. Various mechanisms were proposed involving changes in deep ocean circulation, ocean nutrient inventories, marine productivity and alkalinity (Boyle, 1988b; Broecker and Denton, 1989; Martin, 1990; Rickaby et al., 2010). Much of the early focus was on global ocean nutrient chemistry and North Atlantic Deep Water (NADW) formation, whereas more recently there has been a shift towards Southern Ocean carbon cycle mechanisms (Sigman and Boyle, 2000; Sigman et al., 2010). A significant role for processes in this region is supported by the close correlation between Antarctic temperatures and atmospheric CO_2 during the last deglaciation (Monnin et al., 2001; Parrenin et al., 2013) and throughout the last 800 kyr (Bereiter et al., 2015; Jouzel et al., 2007; Luthi et al., 2008) (Fig. 10.1). Since the temporal evolution of Antarctic temperature differed from Northern Hemisphere temperatures, this powerful observation has driven a search for theories and data to constrain the role of Antarctic and Southern Ocean processes in the global carbon cycle.

10.4.2 Southern Ocean mechanisms based on sea ice, ocean circulation and deep stratification

An important role for Southern Ocean circulation in the glacial–interglacial carbon cycle was proposed by Toggweiler (1999), who suggested that changes in the strength and/or position of the Southern Hemisphere westerly winds could determine the structure of the overturning circulation and hence its capacity to store carbon (Menviel et al., 2018; Toggweiler et al., 2006). In simple terms, equatorward shifts in the westerly winds during glacial periods could weaken the wind forcing over the Antarctic Circumpolar Current, leading to reduced Southern Ocean upwelling and weaker overturning. This scenario would decrease CO_2 release to the atmosphere through the Southern Ocean surface (i.e. reduced upwelling), and simultaneously increase carbon storage in the deep ocean through the combination of the low-latitude biological pump and a longer water residence time in the deep ocean (i.e. reduced overturning). The glacial expansion of sea ice in response to high-latitude cooling was proposed as an additional driver of deep ocean carbon storage, since it would restrict gas exchange between upwelling Southern Ocean deep waters and the atmosphere (Elderfield and Rickaby, 2000; Keeling and Stephens, 2001).

More recently, the magnitude of the wind-driven mechanism proposed by Toggweiler (1999) has been challenged because eddy compensation in the Antarctic Circumpolar Current weakens the link between wind strength and Southern Ocean upwelling (Farneti et al., 2010). However, while our detailed understanding has evolved since the original hypothesis, the general concept remains influential, and a related idea has emerged that combines aspects of both the ventilation and sea ice mechanisms. Specifically, it has been proposed that changes in sea ice extent in the Southern Ocean control the surface ocean buoyancy forcing and hence the global deep ocean circulation structure (Adkins, 2013; Ferrari et al., 2014; Watson et al., 2015). The glacial expansion of sea ice would shoal the boundary between northern-sourced (NADW) and underlying southern-sourced (AABW) water masses, effectively increasing the volume of the deep overturning cell (Fig. 10.5). In addition, a shoaled boundary would occupy a water depth where there is less rough seafloor bathymetry and significantly weaker vertical mixing, leading to reduced mixing between the two cells and increased deep stratification (Fig. 10.5). This scenario would significantly enhance the capacity for glacial carbon storage in the deep overturning cell (Brovkin et al., 2007; Ferrari et al., 2014; Lund et al., 2011; Marzocchi and Jansen, 2019; Stein et al., 2020; Watson et al., 2015).

Observations consistent with those ideas include the enhanced salinity of glacial southern-sourced deep waters based on pore water chlorinity profiles (Adkins et al., 2002), increased glacial deep ocean stratification inferred from benthic foraminiferal carbon isotopes ($\delta^{13}C$) (Curry and Oppo, 2005; Hodell et al., 2003), increased glacial density stratification derived from $\delta^{18}O$ gradients (Lund et al., 2011) and restriction of NADW from the lower overturning cell inferred from neodymium isotopes (Wilson et al., 2020). There is also strong evidence from sediment cores and deep-sea corals to support enhanced carbon storage in a more isolated glacial deep ocean and rapid carbon release during the deglaciation, which is derived from radiocarbon (Burke and Robinson, 2012; Skinner et al., 2010, 2017),

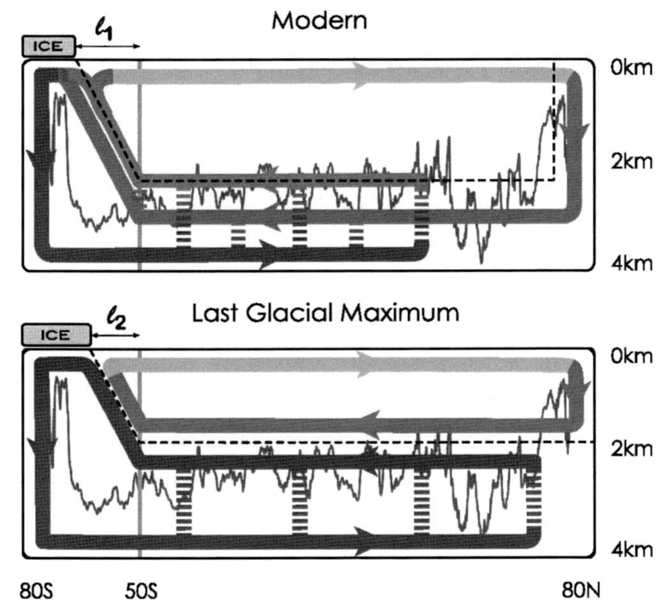

FIGURE 10.5 Schematic representation of ocean circulation structure during (A) the modern interglacial state and (B) the Last Glacial Maximum. Ribbons represent water masses: green, NADW; blue, AABW; red, Pacific and Indian deep waters; and yellow, AAIW. During glacial periods, extended Southern Ocean sea ice is proposed to shift the boundary between buoyancy loss and buoyancy gain northwards, reducing the width of the channel between the sea ice margin and 50°S (l_2 vs l_1). Because interior isopycnal slopes in the channel are largely unchanged, the effect is to shoal the interface (dashed line) between the upper northern-sourced cell (NADW) and the lower southern-sourced cell (AABW). Shoaling of that water mass boundary places it above the depth of most rough seafloor bathymetry (black line), leading to reduced vertical mixing (vertical dashed lines) between the northern and southern cells, increased deep stratification and an enhanced capacity for carbon storage in the lower cell. *Reproduced from Ferrari, R., Jansen, M.F., Adkins, J.F., Burke, A., Stewart, A.L., Thompson, A.F., 2014. Antarctic sea ice control on ocean circulation in present and glacial climates. Proceedings of National Academy of Sciences of the United States of America 111, 8753–8758 [p. 8755].*

boron isotopes (Martínez-Botí et al., 2015b; Rae et al., 2018) and oxygenation proxies (Anderson et al., 2019; Jaccard et al., 2016). These changes in ocean carbon storage and the associated changes in tracer distributions are also supported by recent ocean modelling studies (Mariotti et al., 2016; Menviel et al., 2018).

Perhaps the biggest challenge in testing the finer details of these sea ice-based mechanisms arises from the difficulty in obtaining direct evidence on past sea ice processes and areal extent. Estimates of circum-Antarctic sea ice extent for the LGM are predominantly based on diatom and radiolarian transfer functions (Gersonde et al., 2005), which indicate a doubling of the modern area of winter sea ice and a northward shift of

~5–10 degrees in the winter sea ice edge in the Indian and Atlantic sectors. Recent studies have improved constraints on the Pacific sector at the LGM, supporting a northward shift of ~5 degrees latitude in the winter sea ice extent (Benz et al., 2016), and provide a clearer view on deglacial changes in the Atlantic sector (Xiao et al., 2016). However, past sea ice extent in some regions, such as the Pacific entrance to the Drake Passage, remains poorly constrained. Beyond reconstructing winter sea ice extent, it is challenging to obtain a more holistic view of past sea ice behaviour, although the distributions of organic biomarkers such as highly branched isoprenoids may provide complementary evidence on summer sea ice extent and seasonality (Collins et al., 2013), which appears to have been enhanced during the LGM (Green et al., 2020). Such evidence will be essential for distinguishing between different sea ice-based processes that control the deep ocean structure, because some mechanisms are based on changes in the summer sea ice position (Ferrari et al., 2014) whereas others are sensitive to sea ice formation rates (Nadeau et al., 2019).

For a full understanding of ocean–atmosphere–climate interactions in the Southern Ocean, it is also important to establish the exact timing of past sea ice changes, particularly during transitions such as the last deglaciation. Challenges arise from age model uncertainties and the limited latitudinal distribution of sediment cores, although sea ice records have been recovered from all the major ocean basins, including the Atlantic (Allen et al., 2005; Collins et al., 2012; Shemesh et al., 2002; Xiao et al., 2016), Indian (Crosta et al., 2004; Xiao et al., 2016) and Pacific oceans (Ferry et al., 2015). A complimentary approach, with potential to provide highly-resolved records of past changes, is based on sea ice proxies such as ssNa in ice cores (Wolff et al., 2006). While atmospheric transport is a major control on ssNa input to Antarctic ice core sites on interannual timescales, ssNa levels over geological timescales appear to be linked to sea ice extent because the sea ice surface is the major source of ssNa (Levine et al., 2014). Despite uncertainties arising from transport processes, records of ssNa and methane sulphonic acid (MSA) (Abram et al., 2013) appear to be useful semiquantitative proxies for past sea ice extent and are particularly useful for recording the timing of rapid sea ice changes. Future observational and modelling studies will be important for translating ice core ssNa and MSA records into a more quantitative understanding of past sea ice behaviour. Better sea ice implementation in ocean and climate models is also required, since the existing models give widely different results and struggle to reproduce a sea ice field for the LGM that is consistent with data constraints (Roche et al., 2012), although that situation is improving (Green et al., 2020). A related challenge arises from modelling bottom water formation around Antarctica, specifically the contribution of dense shelf water production to AABW (Snow et al., 2016), because this process is not included in most climate models at present.

10.4.3 Southern Ocean mechanisms based on dust supply, productivity and nutrient utilisation

The other major way in which Southern Ocean processes could influence the global carbon cycle is through changes in surface ocean productivity. The high-latitude Southern Ocean is presently a high-nutrient low-chlorophyll area, where iron and possibly other micronutrients limit biological productivity (Martin, 1990). In the modern (interglacial) ocean, nutrients in this region are not fully utilised, which leads to a leak in the biological pump. Upwelling deep waters release regenerated CO_2 to the atmosphere, while the subduction of water masses with high preformed nutrient levels reduces the efficiency of carbon sequestration for a given global ocean nutrient inventory (and for a given ocean circulation structure and strength) (Sigman et al., 2010). Recognising this scenario for the modern day led to the 'iron hypothesis', in which it was proposed that increased glacial dust supply to this region could have provided the missing dissolved iron, thereby enhancing Southern Ocean productivity and nutrient utilisation, and increasing glacial carbon storage in the deep ocean (Martin, 1990).

Antarctic ice core records provide a test for the iron hypothesis because dust fluxes can be inferred from nssCa fluxes (Lambert et al., 2008; Wolff et al., 2006). Those records support a major increase in dust supply from Patagonia during glacial periods (Fig. 10.3I), although with the caveat that dust transport to Antarctic ice cores may have differed from dust transport to the Southern Ocean. Evidence that this signal is broadly representative of dust input to the Southern Ocean is found in dust records from marine cores (Fig. 10.3J), which indicate glacial increases in dust fluxes by a factor of 5–10 at ODP Site 1090 in the Subantarctic Zone of the South Atlantic Ocean (Martinez-Garcia et al., 2011), and by a factor of ~3 at a number of sites in the South Pacific sector of the Southern Ocean (Lamy et al., 2014). Complementary studies have confirmed the carbon cycle impact of this mechanism, by demonstrating that increased dust fluxes led to both increased productivity (from alkenone fluxes) and increased nutrient utilisation (from foraminiferal nitrogen isotopes) in the Subantarctic Zone (Martinez-Garcia et al., 2014). A recent compilation of dust records indicates that the dust signal was circumpolar in extent but also highlights the potential for spatial differences in carbon sequestration linked to geographical variations in dust fluxes and the effect of frontal shifts on nutrient supply (Thöle et al., 2019).

The Subantarctic Zone was probably the key region where increased dust fluxes could have enhanced glacial carbon sequestration, because increased sea ice coverage, reduced upwelling and increased near-surface stratification in the Antarctic Zone probably reduced the importance of this latter region for carbon exchange during glacial periods. Indeed, glacial increases in Antarctic Zone nutrient utilisation (diatom nitrogen isotopes) coincided with reduced productivity (opal and barium proxies) for at least the last two glacial cycles (Francois et al., 1997; Studer et al., 2015), which can only be reconciled by a reduction in nutrient supply from below. However, a full mechanistic understanding of the changes in upwelling, near-surface stratification, nutrient supply, iron fertilisation and export productivity across all sectors of the Antarctic and Subantarctic Zones is yet to be established. The challenge in establishing a full Southern Ocean carbon budget is complex because of the interactions between each of these regions, but modelling studies generally agree that subantarctic iron fertilisation could cause a decline in glacial atmospheric CO_2 of $\sim$30–40 ppm (Brovkin et al., 2007; Hain et al., 2010).

In marine records, it is possible to take a further step in assessing the influence of iron fertilisation on the carbon cycle because multiple proxies can be used to simultaneously trace not only dust input and marine export productivity but also the effect on bottom water (or pore water) carbon chemistry and oxygen content (Gottschalk et al., 2016; Jaccard et al., 2016). As a result, it may be possible to separate changes in deep ocean carbon storage that occurred due to productivity variations (predominantly in the Subantarctic Zone) from changes that were linked to stratification and deep ocean ventilation (predominantly in the Antarctic Zone). Evidently, both mechanisms are required to explain CO_2 changes through glacial–interglacial cycles (Jaccard et al., 2013, 2016), and they likely also played a role in carbon cycle dynamics on millennial timescales (Gottschalk et al., 2016; Jaccard et al., 2016).

10.4.4 Sequence of changes through the last glacial cycle

Explaining glacial–interglacial cycles in terms of changes in the atmospheric CO_2 budget between glacial and interglacial states is not sufficient for a full understanding of how the Earth's climate system operated. In addition, transient events must be studied to address questions such as what triggered glaciation or deglaciation, what feedbacks were operating, and how fast these feedbacks acted. Ongoing efforts are seeking to refine proxies and age models, to generate high-resolution paleo-records, and to integrate them with modelling approaches to delve into the mechanisms.

Here we summarise such efforts to constrain the sequence and timing of changes that operated in the progression towards full glaciation, drawing heavily on a recent compilation of regional SST records and other proxy reconstructions for the last glacial cycle (Kohfeld and Chase, 2017) (Fig. 10.6). Early in the glacial cycle, at the MIS 5e to MIS 5d transition, both hemispheres experienced SST changes that were largest at high latitudes and probably linked to a major decline in summer insolation forcing (Kohfeld and Chase, 2017). The direct effect of cooling would only have been a modest CO_2 drawdown due to the solubility effect, but the indirect effects were larger. Antarctic ice core ssNa records indicate an expansion of

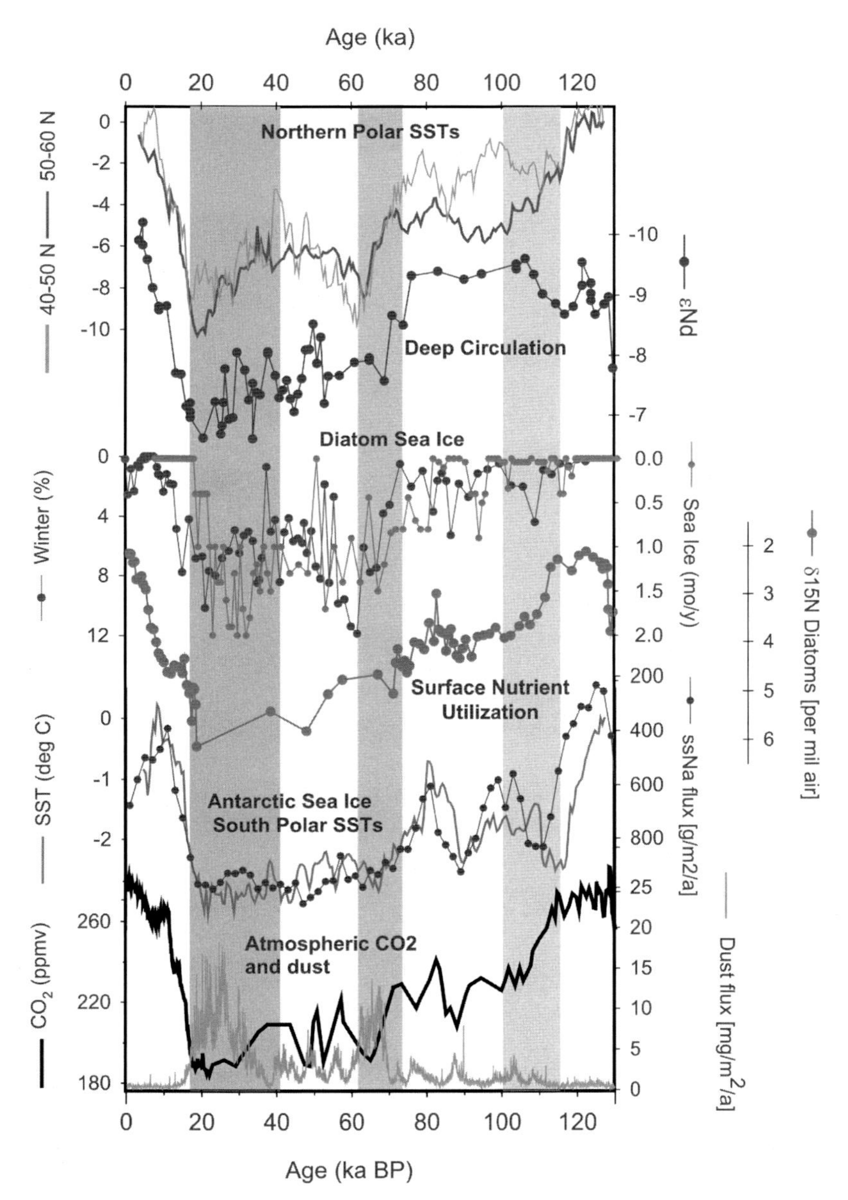

FIGURE 10.6 Controls on the carbon cycle and climate through the last glacial cycle from 0 to 130 ka. Coloured bars indicate three intervals of CO_2 drawdown: MIS 5e to 5d transition (blue), MIS 5 to MIS 4 transition (red), LGM (grey). Records are (top to bottom): Northern Hemisphere SST stacks plotted as relative changes (Kohfeld and Chase, 2017); deep Indian Ocean circulation from neodymium isotopes (Wilson et al., 2015); sea ice proxies in the Southeast Atlantic Ocean (winter sea ice extent from % *F. curta*; Gersonde and Zielinski, 2000) and Southeast Indian Ocean

(*Continued*)

winter sea ice at this time (Wolff et al., 2006), although marine diatom records indicate that the sea ice expansion was moderate and spatially variable (Fig. 10.6). These changes were accompanied by increased nutrient utilisation in the Antarctic Zone (Studer et al., 2015) (Fig. 10.6), probably because near-surface stratification caused by seasonal melting of sea ice led to a restricted supply of nutrients from below. Together, the cooling, sea ice expansion and increased surface stratification would have restricted carbon release to the atmosphere and increased deep ocean carbon storage (Keeling and Stephens, 2001; Sigman et al., 2010), helping to explain the ~35 ppm drop in atmospheric CO_2.

At the MIS 5 to MIS 4 transition, there was further global cooling, leading to average SST ~3°C cooler than during interglacials, with the most pronounced changes at high latitudes (Kohfeld and Chase, 2017). It was accompanied by Antarctic sea ice expansion (Fig. 10.6) and major changes in ocean circulation, as recorded by benthic $\delta^{13}C$ (Oliver et al., 2010) and neodymium isotope records (Piotrowski et al., 2005; Wilson et al., 2015) (Fig. 10.6). These circulation changes could have arisen directly from polar cooling and sea ice expansion (Ferrari et al., 2014) (Fig. 10.5), with an additional potential influence from AABW formation by brine rejection (Adkins, 2013; Bouttes et al., 2010). Dust fluxes increased significantly at the MIS 5 to MIS 4 transition, as recorded in ice cores (Lambert et al., 2008; Wolff et al., 2006) (Fig. 10.6) and Southern Ocean sediment cores (Martinez-Garcia et al., 2009, 2014) (Fig. 10.3J), indicating that increased nutrient utilisation by iron fertilisation of the Subantarctic Zone was restricted to this later step. Hence, a combination of sea ice expansion, ocean circulation changes and iron fertilisation could explain the ~40 ppm drop in atmospheric CO_2. Modelling studies also suggest that the switch in deep ocean structure could have magnified the CO_2 drawdown effect of the polar surface ocean stratification and sea ice changes that emerged in MIS 5d (Hain et al., 2010), leading to additional carbon storage at this time.

After increases in atmospheric CO_2 of ~20–30 ppm during MIS 3, the lowest glacial CO_2 values were reached during MIS 2. Many of the proxies that changed during the MIS 5 to MIS 4 transition reached their most extreme values at the LGM (Fig. 10.6), indicating maximum contributions to carbon drawdown from high latitude temperatures, ocean circulation and iron fertilisation of the Subantarctic Zone.

◀ (sea ice duration; Crosta et al., 2004); surface nutrient utilisation from nitrogen isotopes in the Antarctic Zone of the Pacific Ocean (Studer et al., 2015); Antarctic SST stack from 50°S to 60°S plotted as relative change (Kohfeld and Chase, 2017); sea ice proxy from ice core ssNa flux (Wolff et al., 2006); dust flux based on nssCa in EDC ice core (Lambert et al., 2008); atmospheric CO_2 from EDC ice core (EPICA Community Members, 2004). *Reproduced from Kohfeld, K.E., Chase, Z., 2017. Temporal evolution of mechanisms controlling ocean carbon uptake during the last glacial cycle. Earth and Planetary Science Letters 472, 206–215 [p. 210].*

10.4.5 Millennial climate variability and the bipolar seesaw

Glacial climates were also highly variable on sub-orbital timescales, and here we discuss some of the records of those changes and the mechanisms responsible. Early evidence for millennial and sub-millennial climate change in the North Atlantic region during the last glacial period and deglaciation was provided by Greenland ice core $\delta^{18}O_{ice}$ records (Dansgaard et al., 1993). Rapid fluctuations of more than 10°C between cold stadial and warm interstadial intervals were termed Dansgaard–Oeschger events (Figs 10.7 and 10.8A), which were also observed in reconstructions of regional North Atlantic climate (Fig. 10.8B) and IRD input to the oceans (Bond et al., 1997; McManus et al., 1999). Specific incidences of partial ice sheet collapse from the North American (Laurentide) ice sheet, known as Heinrich events, were recorded by distinct IRD layers in marine sediment cores

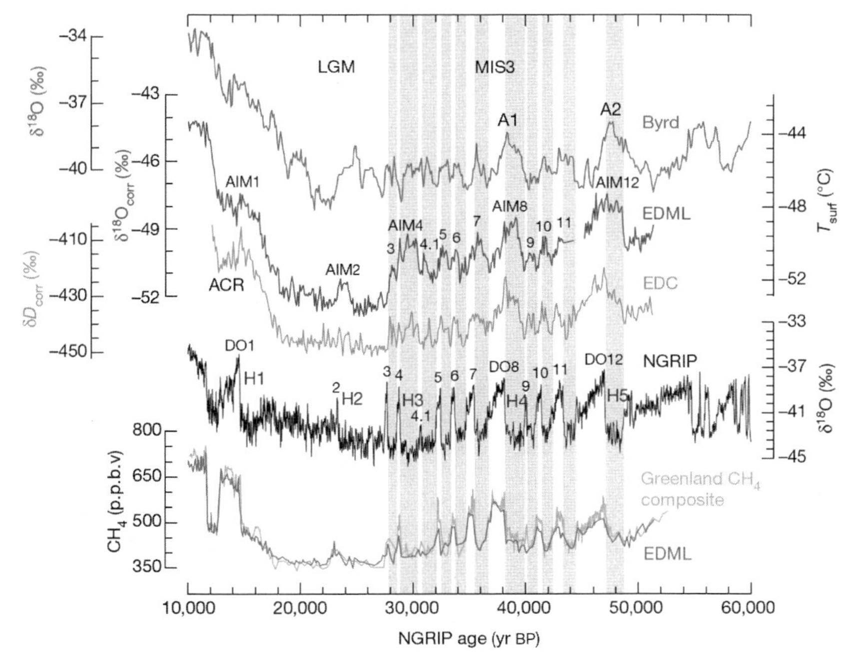

FIGURE 10.7 Record of the bipolar seesaw in ice core climate records. Upper curves: Antarctic δD_{ice} or $\delta^{18}O_{ice}$ in different ice cores (Byrd, pink; EDML, purple; EDC, blue), with temperature scale on the right corresponding to the EDML record. Middle curve: Greenland $\delta^{18}O_{ice}$ (NGRIP, black). Lower curves: methane records from Antarctica (EDML, blue) and Greenland (composite, orange) used to synchronise the age scales. Yellow bars indicate warming events in Antarctica coinciding with Northern Hemisphere stadials. *ACR*, Antarctic Cold Reversal; *AIMx*, Antarctic Isotope Maximum events; *Ax*, Antarctic warm events; *DOx*, Dansgaard–Oeschger events; *Hx*, Heinrich events; *LGM*; Last Glacial; *MIS3*, Marine Isotope Stage 3. Maximum. *Reproduced from EPICA Community Members, 2006. One-to-one coupling of glacial climate variability in Greenland and Antarctica. Nature 444, 195–198 [p. 196].*

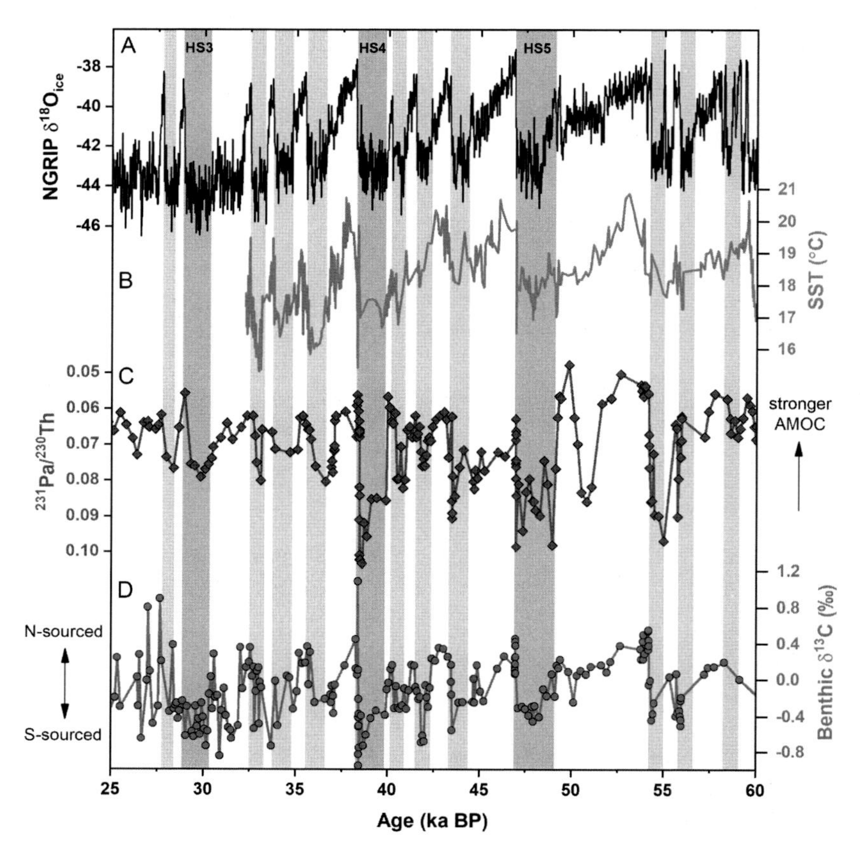

FIGURE 10.8 Records of surface climate and deep ocean circulation from the North Atlantic region during the last glacial period. (A) Greenland $\delta^{18}O_{ice}$ in NGRIP (NGRIP Members, 2004; Svensson et al., 2008). (B) North Atlantic SST at Bermuda Rise (Sachs and Lehman, 1999). (C) Proxy for AMOC strength based on protactinium/thorium ($^{231}Pa/^{230}Th$) ratios at Bermuda Rise (Henry et al., 2016). (D) Proxy for water mass mixing based on benthic foraminiferal $\delta^{13}C$ at Bermuda Rise (Henry et al., 2016). Vertical blue bars indicate stadials and purple bars indicate Heinrich Stadials (labelled HSx).

during some of the longest stadials (Alley and MacAyeal, 1994; Bond et al., 1992; Hemming, 2004).

In comparison to the relative stability of Holocene climate, the greater variability of regional and global climate during late Pleistocene glacial and deglacial periods has been attributed to climatic and oceanographic boundary conditions that led to a bi-stable ocean circulation mode in the Atlantic Ocean (Broecker et al., 1985, 1990; Ganopolski and Rahmstorf, 2001). Comparing ice core climate records from Greenland and Antarctica reveals that this signal was region-specific, with the Southern Ocean warming during cold Greenland stadials (yellow bars in Fig. 10.7), and then starting to cool following the rapid onset of Greenland interstadials (Blunier and Brook, 2001; EPICA Community Members, 2006) (Fig. 10.7). This antiphased bipolar pattern is pervasive and has also been recognised in SST reconstructions from North and South Atlantic sediment cores (Barker et al., 2009). Hence, the concept of a bipolar seesaw emerged (Broecker, 1998), in which a strong Atlantic meridional overturning circulation (AMOC) would lead to warming in the North Atlantic and cooling in the Southern Ocean, whereas a weak AMOC with reduced NADW formation would lead to cooling in the North Atlantic and the accumulation of heat in the South Atlantic and Southern Ocean (Broecker, 1998; Ganopolski and Rahmstorf, 2001; Stocker and Johnsen, 2003). The more gradual changes in Antarctica than Greenland (Fig. 10.7) were initially linked to the large volume of the Southern Ocean which could act as a heat capacitor (Stocker and Johnsen, 2003), supported by the observation that longer stadial events in Greenland coincided with greater warming in Antarctica (EPICA Community Members, 2006).

The concept of a bipolar seesaw has often been used to refer to both the observed antiphasing in polar temperatures and the ocean circulation mechanism that may be behind it, even though the forcing mechanisms responsible have yet to be fully established. For example, it is not clear whether the forcing originates from processes controlling deep water formation in the North Atlantic (Ganopolski and Rahmstorf, 2001) or the Southern Ocean (Buizert and Schmittner, 2015), or some combination. It has also not been established whether oceanic processes (WAIS Divide Project Members, 2015) or atmospheric teleconnections (Hogg et al., 2016) play the dominant role in transferring the climate signal. Regardless of the mechanism, there are many examples of the signal of Dansgaard–Oeschger or Heinrich events being transmitted globally and almost instantaneously, leading to changes in multiple components of the Earth system. For stadials and/or Heinrich events, the effects include weakening of the Asian monsoon system (Wang et al., 2001), a southward shift of the intertropical convergence zone (Peterson et al., 2000; Mulitza et al., 2017; Wang et al., 2004), South Atlantic warming (Barker et al., 2009) and a strengthening and/or southward shift of the southern westerly winds (Anderson et al., 2009; Lamy et al., 2007; Whittaker et al., 2011). With highly-resolved and absolute-dated reconstructions, it may be possible to distinguish specific roles for oceanic and atmospheric processes in transmitting such signals (WAIS Divide Project Members, 2015). Records from the WDC ice core indicate roles for both the atmosphere and the ocean, with an instantaneous but spatially-variable atmospheric response in the southern westerly winds being followed a few hundred years later by a spatially-uniform oceanic response (Buizert et al., 2018).

There is also an increasing wealth of evidence indicating deep ocean circulation changes on millennial timescales during the last glacial cycle, including records from the Bermuda Rise in the deep Northwest Atlantic (Henry et al., 2016) (Fig. 10.8C and D), the Portuguese margin of the

Northeast Atlantic (Martrat et al., 2007) and the Cape Basin of the deep Southeast Atlantic (Gottschalk et al., 2018; Piotrowski et al., 2005). What remains less clear is the exact nature of the link between deep ocean circulation changes and these millennial events. While Heinrich events could force ocean circulation changes by freshening the North Atlantic surface ocean in regions of deep water formation (Ganopolski and Rahmstorf, 2001), iceberg discharge may also simply provide a positive feedback that lengthens or deepens a stadial following an initial perturbation (McManus et al., 1999). Recent evidence supports this latter view, because SST changes occurred before IRD input during Heinrich events in North Atlantic records covering the last ~400 kyr (Barker et al., 2015). The observation that similar circulation changes occurred during most stadials of the last glacial period, both those with and without Heinrich events (Fig. 10.8C and D), would also support that view (Henry et al., 2016). However, such observations are not inconsistent with an ocean circulation origin for some abrupt climate events, such as the Younger Dryas (Muschitiello et al., 2019). Regardless of the actual trigger, which could differ between events, it is possible to effectively model spatial and temporal patterns of millennial climate variability during Heinrich stadials, and also many of the related changes in atmospheric and Southern Ocean processes, using North Atlantic freshwater hosing experiments (Ganopolski and Rahmstorf, 2001; Menviel et al., 2018).

Other components of the Southern Ocean atmosphere–ocean system also record sub-orbital scale variability. For example, millennial variability is seen in records of the Agulhas leakage (Marino et al., 2013), which is the 'warm water return route' that advects salt from the Indian Ocean via the southern tip of Africa to the Atlantic Ocean. Increases in the Agulhas leakage during North Atlantic stadials may have influenced the AMOC by resupplying salt to the Atlantic Ocean, leading to the potential for a dynamic feedback through the influence of the southern westerly winds and Southern Ocean fronts on the Agulhas leakage (Marino et al., 2013). Changes in the Agulhas leakage could also have influenced late Pleistocene glacial terminations by preconditioning the interglacial resumptions of AMOC (Peeters et al., 2004). The strength of the subantarctic portion of the Antarctic Circumpolar Current in the northern Drake Passage also appears to have fluctuated on millennial timescales during the last glacial period and deglaciation, with enhanced flow during Antarctic warm periods (North Atlantic stadials), possibly linked to southward shifts of the southern westerly winds to better align with the Drake Passage (Lamy et al., 2015). Similar to the effects of the Agulhas leakage on the warm water return route, these changes could have influenced the 'cold water return route' to the North Atlantic Ocean on both millennial (Lamy et al., 2015) and orbital or longer timescales (Toyos et al., 2020), although their origin and dynamics are yet to be fully understood.

Understanding millennial climate variability is important because it may have played a crucial role in determining the timing and sequence of events during late Pleistocene glacial terminations (Anderson et al., 2009; Barker et al., 2011; Cheng et al., 2009; Denton et al., 2010; Menviel et al., 2018; Roberts et al., 2010; Shakun et al., 2012). As summarised by Denton et al. (2010), North Atlantic freshwater forcing during Heinrich Stadial 1 and the associated AMOC weakening could have led to Southern Ocean warming, sea ice retreat and a southward shift of the southern westerly winds. In combination, these changes would have acted to enhance near-surface upwelling and reduce deep ocean stratification, thereby allowing carbon stored in the deep ocean to escape to the atmosphere (Anderson et al., 2009; Martínez-Botí et al., 2015b; Rae et al., 2018). The ubiquitous association between such bipolar seesaw oscillations and late Pleistocene terminations (Barker et al., 2011) suggests that a particularly large or sustained millennial event, in combination with appropriate orbital forcing, could enable the climate system to cross a critical threshold that allows further positive feedbacks to move the system into a full interglacial period. We also note that the reinvigoration of AMOC at the start of the Bolling–Allerod warm period has been identified as the likely origin of a centennial-scale CO_2 release from the ocean to the atmosphere (Chen et al., 2015; Marcott et al., 2014), highlighting just how fast the ocean–atmosphere system can act.

It is also important to understand the role of AMOC variability in heat transport, because the bipolar seesaw can cause significant local warming of the ocean and atmosphere in the vicinity of either Greenland or Antarctica that is above the expected 'background' levels for a given climate state. Such warming could be crucial for both atmospheric and ocean-driven mechanisms that influence ice sheet stability (Clark et al., 2020; Golledge et al., 2014; Weber et al., 2014). In this regard, we note that the operation of a bipolar seesaw is not restricted to glacial periods of the late Pleistocene, since it has also been observed during the early Pleistocene (Birner et al., 2016), with evidence also emerging for AMOC instability during recent interglacial periods (Galaasen et al., 2020). Therefore, when considering Antarctic climate records and the response of the Antarctic Ice Sheet, the operation of a bipolar seesaw should be considered as a pervasive (although likely variable) factor over at least the duration of the Pleistocene. Changes in the bipolar seesaw and southern westerly winds were probably also important earlier in the Neogene, with the influence of Antarctic cooling on the Agulhas leakage potentially contributing to the onset of Northern Hemisphere Glaciation (McKay et al., 2012a).

10.5 Antarctic Ice Sheet dynamics in the late Pleistocene

10.5.1 Climate context

The predominantly marine-based WAIS (~4 m sea level equivalent) and marine-based sectors of the East Antarctic Ice Sheet (EAIS) (~19 m sea level equivalent) are susceptible to retreat through a combination of reduced buttressing through ice shelf thinning and marine ice sheet instability, as outlined earlier. To constrain the levels of atmospheric or ocean warming required to generate ice sheet retreat, it is essential to assess past ice sheet behaviour in response to different climatic and oceanographic regimes, in particular during times that were warmer than today. Recent Pleistocene interglacials provide a good target, because both Antarctic ice cores and other globally-distributed marine and terrestrial paleoclimate reconstructions indicate that certain interglacials have been warmer than the pre-industrial Holocene, in particular MIS 5e, MIS 9 and MIS 11 (Capron et al., 2014; Holloway et al., 2017; Jouzel et al., 2007; Lang and Wolff, 2011; Snyder, 2016; Yin and Berger, 2015) (Fig. 10.3).

In terms of average global SSTs, MIS 5e was probably warmer than pre-industrial conditions by only around 0.5°C–1°C (Hoffman et al., 2017; McKay et al., 2011) (Fig. 10.3C), although larger changes occurred at high latitudes (Capron et al., 2014; Hoffman et al., 2017). In Antarctica, peak temperatures during MIS 5e, MIS 9 and MIS 11 reached 2°C–4°C warmer than the Holocene (Jouzel et al., 2007) (Fig. 10.3G). Given similar greenhouse gas forcing to the Holocene (Fig. 10.3H), the Antarctic warmth of these previous interglacials likely reflects differences in both insolation forcing (Yin and Berger, 2015) and interhemispheric heat transport by the AMOC (Holden et al., 2010). It has also been suggested that some component of the peak interglacial warming of Antarctica and the Southern Ocean could have arisen as a local consequence of feedbacks from WAIS collapse (Holden et al., 2010), although evidence for it during the late Pleistocene remains inconclusive. Changes in sea ice extent and atmospheric transport could also have contributed to changes in ice core records during these intervals of peak warmth (Holloway et al., 2016).

10.5.2 Global evidence on the Antarctic Ice Sheet

Global sea level reconstructions based on benthic foraminiferal $\delta^{18}O$ records provide constraints on the magnitude and timing of past changes in globally-integrated ice volumes (Figs 10.3A and B and 10.4A and B). However, as discussed earlier, a number of caveats and uncertainties limit the precision of these methods, severely restricting their ability to quantitatively reconstruct past peak interglacial sea levels or the timing of Antarctic changes. Furthermore, such approaches cannot resolve where a given sea level change originated from, which is a critical factor when thinking about specific contributions from Antarctica through time as well as the local fingerprints of future sea level change on a warming planet (Hay et al., 2014).

Instead, the most robust evidence on global sea levels during late Pleistocene interglacials is based on reconstructions from absolute-dated coral terraces, once corrections are made for tectonic uplift and glacio-isostatic adjustments (GIA) that affect local (relative) sea levels (Dutton et al., 2015;

Raymo and Mitrovica, 2012). Such studies indicate elevated global mean sea levels during warm late Pleistocene interglacials, with estimates of ~6–9 m during MIS 5e (Dutton and Lambeck, 2012; Kopp et al., 2009) and ~6–13 m during MIS 11 (Raymo and Mitrovica, 2012). An additional source of error in those estimates comes from uncertainty in the GIA corrections that arise from differences in the size and distribution of the Northern Hemisphere ice sheets at the preceding glacial maxima. In light of the possibility that a greater proportion of Northern Hemisphere ice was in Eurasia rather than North America at the penultimate glacial maximum compared to the LGM, it has been suggested that global mean sea level during MIS 5e may have been another ~2 m higher than those estimates (Rohling et al., 2017).

Since a significant amount of ice remained on Greenland during MIS 5e, a reduction in the Greenland ice sheet likely contributed no more than ~2 m to global sea level rise above Holocene levels (Clark et al., 2020; Colville et al., 2011; Dahl-Jensen et al., 2013; Yau et al., 2016a). Noble gas measurements in ice cores indicate a short-lived global mean ocean warming of ~1.1°C ± 0.3°C above Holocene levels at ~129 ka for around 2 kyr, leading to a sea level contribution from thermal expansion of 0.7 ± 0.3 m, followed by a return to stable temperatures similar to the Holocene for the remainder of MIS 5e (Shackleton et al., 2020). Given the above constraints, the global sea level estimate of ~6–9 m for MIS 5e implies significant ice loss from Antarctica (Clark et al., 2020; Dutton et al., 2015). In addition, ice loss in Greenland and Antarctica may not have occurred synchronously (Capron et al., 2014; Clark et al., 2020; Rohling et al., 2019; Yau et al., 2016b), and therefore simply subtracting the Greenland contribution from the total sea level change during MIS 5e (i.e. the 'mass balance' approach) could underestimate Antarctic contributions to sea level change. Ice loss from Antarctica during MIS 11 is subject to more uncertainty, because of the wider range in global mean sea level estimates (~6–13 m), as well as evidence for a greater sea level contribution from Greenland ice sheet collapse (Irvalı et al., 2020; Reyes et al., 2014). Overall, the sea level evidence points to a reduced ice volume in Antarctica during both MIS 5e and MIS 11, presumably from retreat of some combination of the WAIS and the marine-based sectors of the EAIS, but the location of ice loss is not resolved by the sea level records.

Compared to absolute sea level estimates, the magnitude and rates of sea level variability within interglacials remain less well constrained and are strongly debated. Significant sea level fluctuations of several metres were proposed to have occurred within MIS 5e (Kopp et al., 2009), whereas recent studies indicate relatively stable sea levels (Barlow et al., 2018; Polyak et al., 2018). For example, Polyak et al. (2018) reported phreatic overgrowths on coastal speleothems from Mallorca that indicate a stable local sea level during MIS 5e, which (after GIA adjustments) is consistent with an early MIS 5e peak in global mean sea level followed by a gradual decline. In terms of rates of change, estimates based on coral terraces suggest maximum rates of sea level rise within MIS 5e of ~3–7 mm/yr (Barlow et al., 2018; Kopp et al., 2013). In contrast, while subject to large uncertainties, the sea level reconstruction from the Red Sea (Fig. 10.3B) suggests potentially much higher rates of sea level rise of up to ~9–35 mm/yr for the most rapid millennial event within MIS 5e, which has been attributed mostly to changes in the Antarctic Ice Sheet volume (Rohling et al., 2019).

10.5.3 Regional studies of Antarctic Ice Sheet behaviour before the LGM

During the LGM, the Antarctic Ice Sheet extended across the continental shelves, reaching the shelf break along many but not all margins (RAISED Consortium, 2014). Hence, the glacial Antarctic Ice Sheet was characterised by a different geometry than today, with thickening near the present-day ice margins and an overall increase in ice volume (see Chapter 11 for details, Siegert et al., 2021). Such LGM and deglacial reconstructions draw heavily on radiocarbon-based dating of shelf sediments and terrestrial deposits (e.g., lake sediments, raised beaches and moraines), as well as using cosmogenic nuclides (e.g., ^{10}Be, ^{14}C, ^{26}Al) to constrain past rock exposure and ice sheet elevation. Our understanding of the LGM provides a useful context for assessing the likely geometry and volume of the ice sheet through previous glacial–interglacial cycles, but such reconstructions are more challenging due to the effects of LGM erosion and the difficulty in dating events in shelf and terrestrial records beyond the radiocarbon dating window. Recent advances in understanding the LGM and Holocene ice sheet have particularly benefited from the use of in-situ ^{14}C exposure dating (e.g., Nichols et al., 2019), which is more sensitive to recent changes in the ice sheet and less susceptible to inheritance than cosmogenic ^{10}Be or ^{26}Al methods. For previous glacial cycles, ^{14}C exposure dating is unavailable, while ^{10}Be and ^{26}Al data are more suitable for understanding the long-term evolution of the ice sheet than for detailed reconstructions of variability within individual glacial or interglacial periods. The likelihood of multi-stage exposure and burial histories means that interpretations are typically non-unique (e.g., Jones et al., 2017 and references therein), while the relatively long half-lives of ^{10}Be and ^{26}Al preclude their use to constrain short climate events. For the above reasons, most of our knowledge of the earlier Pleistocene behaviour of the Antarctic Ice Sheet comes from offshore marine sediment cores and ice sheet modelling, which therefore form our main focus below.

10.5.4 Regional evidence on the West Antarctic Ice Sheet

In light of the sea level contributions that appear to be required from Antarctica during recent warm interglacials, and the sensitivity of West

Antarctic catchments such as the Pine Island/Thwaites Glacier system (Amundsen Sea Embayment) and the Siple Coast (Ross Sea) (Fig. 10.2) to ocean warming in models (Golledge et al., 2017), a partial or full collapse of the WAIS might be suspected for MIS 5e and/or MIS 11. However, geological evidence either supporting or refuting this suggestion has been hard to obtain. There is direct evidence for WAIS collapse during at least one late Pleistocene interglacial within the past 1.1 Myr, which comes from the discovery of Pleistocene open-ocean diatoms in tills beneath the modern Whillans Ice Stream at the Siple Coast (Scherer et al., 1998). Such a late Pleistocene collapse is also supported by faunal evidence from bryozoans for a trans-Antarctic seaway (Barnes and Hillenbrand, 2010). However, while Scherer et al. (1998) provide unequivocal evidence for at least one collapse event, neither study was able to determine which interglacial such collapse occurred in, how long it lasted or whether collapse occurred multiple times. Future studies based on the molecular genetics of benthic marine species may provide better constraints on past gateway openings (Strugnell et al., 2018), but the detailed timing of any such Pleistocene events is yet to be resolved by this method (e.g., Collins et al., 2020).

Offshore sedimentary evidence from the late Pleistocene has been obtained from sediment cores on the Ross Sea shelf during the Antarctic Drilling Project (ANDRILL; see also Chapter 9) (Fig. 10.2). An extensive sequence of Pliocene sediments was recovered in core AND-1B and indicates significant Pliocene variability in the WAIS, including events of ice sheet advance across the Ross Sea shelf (recorded by subglacial diamictites) as well as extended intervals of retreat (recorded by diatomite) (Naish et al., 2009). The AND-1B record extends into the Pleistocene, where sediments indicative of ice retreat become less prominent, but open ocean deposition is indicated for some Pleistocene intervals (Naish et al., 2009). The Pleistocene record was analysed in detail by McKay et al. (2012b), who suggested that the late Pleistocene sedimentary environment generally fluctuated between two states: subglacial or grounding zone sedimentation during glacial periods, and deposition beneath a floating ice shelf during interglacials. Critically, loss of the Ross Ice Shelf at some point in the last ~250 kyr is indicated by a thick layer of reworked volcanic glass sourced from Mt Erebus, which suggests that WAIS deglaciation occurred during either MIS 5e or MIS 7 (McKay et al., 2012b). The ANDRILL records can be challenging to interpret because of the combined influences of EAIS and WAIS dynamics on ice flow in the Ross Sea (McKay et al., 2012b), but the above interpretation is consistent with models that indicate a strong link between removal of the Ross Ice Shelf and WAIS deglaciation (Martin et al., 2019). Other limitations in acquiring a detailed late Pleistocene reconstruction from AND-1B arise from the presence of major erosional hiatuses and a lack of dating constraints (McKay et al., 2012b), indicating the need for further research in this region. The broader context of the Plio-Pleistocene AND-1B record, including a comparison to more proximal records from the Southern Victoria Land margin and geological sequences in the Transantarctic Mountains, can be found in Levy et al. (2012).

Pleistocene evidence from the vicinity of the Amundsen Sea Embayment (Fig. 10.2) is rather limited, but sequences from the continental rise have been used to address the possibility of a late Pleistocene WAIS collapse. In an early study, no collapse was inferred to have occurred over the past ~1.8 Ma, based on the absence of distinct sedimentological changes that would be expected to have characterised such an event (Hillenbrand et al., 2002). However, subsequent research revealed an extended interval of MIS 13–15 that was characterised by elevated productivity and distinctive sediment inputs from the hinterland that hinted at WAIS collapse during this time (Hillenbrand et al., 2009). Collapse or retreat during this lukewarm late Pleistocene interglacial may have been aided by the extended duration of warmth (Fig. 10.3G) or by shifts in the Amundsen Sea low-pressure system that led to increased incursions of warm Circumpolar Deep Water (CDW) onto the shelf since the MPT (Konfirst et al., 2012). However, if the WAIS collapsed during MIS 13–15, it might seem surprising to find no evidence here for more recent collapse events, given the warmer interglacial temperatures since the MBE, both globally and in Antarctica (Fig. 10.3). In light of modelling outputs indicating that WAIS deglaciation in the Amundsen Sea Embayment could potentially occur without the removal of the Ross Ice Shelf (Clark et al., 2020), it will be important to obtain better constraints on the late Pleistocene ice sheet history in this region.

Sedimentary records providing evidence on the Antarctic Peninsula Ice Sheet were recovered from the continental rise of the Bellingshausen Sea during ODP Leg 178 (Sites 1095, 1096, 1101; Fig. 10.2). Clay mineral constraints (Hillenbrand and Ehrmann, 2005) and sedimentological changes (Cowan et al., 2008) both indicate Pleistocene advance and retreat of the Antarctic Peninsula Ice Sheet across the shelf, but a more detailed understanding is limited by the low resolution of those records. Since the Antarctic Peninsula Ice Sheet may survive even while the WAIS retreats significantly (Sutter et al., 2016), records from this region may not provide strong constraints on past or future behaviour of the WAIS or on global sea level contributions from West Antarctica.

Overall, the offshore sedimentary records provide support for late Pleistocene variability of the WAIS, including retreat or collapse in certain interglacials, but the record is fragmentary and does not provide a clear picture of a consistent response between sectors during each of the recent warm interglacials. While ongoing marine geological studies are seeking to significantly improve this picture, alternative approaches are exploring how an ice sheet change such as WAIS collapse would be imprinted on Antarctic ice core records. For example, ice core $\delta^{18}O_{ice}$ or δD_{ice} records could be affected directly by changes in ice sheet elevation and glacio-isostatic adjustment, or indirectly through the impacts of ice sheet melting on ocean and atmospheric temperatures, sea ice extent and atmospheric circulation (Bradley et al., 2012; Steig et al., 2015). Unfortunately, the direct effect of ice sheet elevation changes linked to WAIS collapse may not have had a significant impact at the location of existing East Antarctic ice cores (Bradley et al., 2012) (Fig. 10.2), thereby limiting the ability to resolve such a change, which points to a need for ice cores in targeted locations. Modelling studies have also suggested that peaks in ice core $\delta^{18}O_{ice}$ or δD_{ice} that occurred early in MIS 5e could record the effect of significant reductions in sea ice linked to warm SSTs, without requiring WAIS collapse (Holloway et al., 2016).

The latest evidence indicating changes to the WAIS during MIS 5e is emerging from blue-ice sites. Both δD_{ice} and other geochemical tracers measured on MIS 5e ice at the Mount Moulton blue-ice site in Marie Byrd Land of West Antarctica are consistent with the regional climate changes that would be expected to result from WAIS collapse (Korotkikh et al., 2011; Steig et al., 2015). In addition, a blue-ice record from the Patriot Hills Blue Ice Area, located ~50 km inland from the modern grounding line of the Filchner-Ronne Ice Shelf, provides indirect constraints on ice sheet changes in the Weddell Sea Embayment (Turney et al., 2020). Measurements of δD_{ice} in a short section of ice corresponding to Heinrich Stadial 11 suggest that temperatures were elevated above Holocene levels during the penultimate deglaciation. Critically, a subsequent hiatus between 130 and 80 ka appears to indicate a retreated grounding line in the Weddell Sea Embayment which, in combination with ice sheet modelling, would support significant mass loss from both the WAIS and the Recovery Basin of the EAIS during the last interglacial (Turney et al., 2020). Future studies of horizontal blue-ice cores can be expected to provide additional new insights, but research is also required to test the relationship between ice sheet grounding line positions and upstream flow patterns, as well as the origin of unconformities in blue-ice records (e.g., Winter et al., 2016).

10.5.5 Regional evidence on the East Antarctic Ice Sheet

Until recently, evidence constraining the behaviour of the EAIS through late Pleistocene glacial–interglacial cycles (and during many other time periods) had been limited by the lack of suitable sedimentary archives recovered from this challenging and remote region (Fig. 10.2). Early views on the EAIS generally favoured a history of stable behaviour through the Neogene because of its predominantly terrestrial-based geometry (Sugden et al., 1993), although counter-arguments for more dynamic behaviour were put forward (Webb and Harwood, 1991). It is now clear that nearly one-third of the ice in East Antarctica (~19 m sea level equivalent; Fretwell et al., 2013) is contained within marine-based catchments underlain by deep subglacial basins, notably the Wilkes, Aurora and Recovery Subglacial Basins (Fig. 10.2). For these regions, ice sheet modelling suggests that the dynamics were probably quite different from the continental-based EAIS interior (DeConto and Pollard, 2016; Fogwill et al., 2014; Golledge et al., 2017; Mengel and Levermann, 2014).

Recent studies from the continental shelf offshore of the Aurora Basin (Fig. 10.2) provide support for a dynamic EAIS. Specifically, marine geophysical data indicative of a surface-meltwater rich subpolar glacial system supports retreat of this marine margin in the Oligocene and Miocene (Gulick et al., 2017), with a transition to a polar environment occurring in the late Miocene. Marine sequences from the Prydz Bay area, recovered from the uplifted terrestrial margin and from offshore sites during ODP Leg 188 (Sites 1165–1167; Fig. 10.2), also clearly indicate dynamic behaviour of the EAIS during the Pliocene (Cook et al., 2014; Hambrey and McKelvey, 2000; Quilty et al., 2000). For example, characteristic $^{40}Ar/^{39}Ar$ ages indicate that far-travelled IRD from the vicinity of the Aurora Subglacial Basin reached ODP Site 1165 during the Pliocene (Cook et al., 2014). At the same time, cosmogenic isotope data from the Ross Sea (core AND-1B) (Fig. 10.2) support the idea that the terrestrial portion of the EAIS has remained largely intact and has not retreated significantly over at least the last 8 Myr (Shakun et al., 2018). Based on the above studies, it is clear that attention should be focused on the marine basins when assessing possible EAIS variability during the Pleistocene, with the caveat that the specific subglacial bedrock topography of each basin may result in significant spatial variability in ice sheet behaviour between sectors (Golledge et al., 2017; Morlighem et al., 2020). Furthermore, both the onshore and offshore bedrock topography of Antarctica has evolved over geological timescales, in response to multiple tectonic, erosional and sedimentary processes, with implications for past ice sheet size and stability (e.g., Austermann et al., 2015; Colleoni et al., 2018; Jamieson et al., 2010; Paxman et al., 2019, 2020). Only recently have continent-scale topographic reconstructions for multiple time slices been attempted (e.g., Paxman et al., 2019), and these will require ongoing validation and refinement using geological records.

Evidence constraining the late Pleistocene behaviour of the EAIS has come from cosmogenic isotopes, distal Southern Ocean IRD records and more proximal studies of sedimentology and geochemical provenance. Cosmogenic isotope analyses adjacent to the Skelton Glacier, an outlet glacier that drains a portion of the EAIS into the Ross Sea, indicate that it had surface elevations at least $\sim$200 m higher than today for a significant portion of the Pleistocene, presumably representing past glacial periods with an expanded ice sheet (Jones et al., 2017). Since cosmogenic isotope studies do not indicate thickening of the EAIS interior during Pleistocene glacial periods (e.g., Lilly et al., 2010; Suganuma et al., 2014), elevation changes in the transition regions between the interior and the marine margins may have been driven by ice shelf and marine processes (Jones et al., 2017). While the Dronning Maud Land region of East Antarctica experienced very limited elevation changes during the last deglaciation, more than 500 m of long-term thinning is indicated for the EAIS in this region during the Pleistocene (Suganuma et al., 2014). Thinning of the EAIS interior since the Pliocene is also reported in other regions (Yamane et al., 2015) and may have been linked to global cooling, sea ice expansion and reduced atmospheric moisture transport, leading to reduced accumulation rates (Suganuma et al., 2014; Yamane et al., 2015). Overall, while the cosmogenic isotope evidence points to spatially-variable Pleistocene elevation changes of the EAIS, these data do not constrain the lateral extent of the ice sheet and are not well-suited for constraining the possibility of past short-lived retreat events leading to ice surface elevations similar to or lower than today.

Teitler et al. (2010) presented distal IRD records from ODP Leg 177 Site 1090 in the Agulhas Basin of the Southeast Atlantic covering multiple glacial cycles over the past ~500 kyr. They observed almost continuous IRD deposition over that interval, but significantly more IRD during glacial periods, which was attributed to an influence of cold SSTs on iceberg survivability into lower latitudes. The mineralogical provenance of the IRD indicates a dominantly EAIS source, such that the continuous presence of IRD was used to infer persistent marine margins of the EAIS across this interval (Teitler et al., 2010). That inference is consistent with sea level reconstructions during late Pleistocene interglacials (Dutton et al., 2015) and the stability of the interior EAIS (Shakun et al., 2018), because even sea level contributions from Antarctica of up to ~6–13 m (Dutton et al., 2015) would not imply retreat of the ice sheet margins onto land. However, the strong control of iceberg survivability makes their record highly sensitive to climate variability, limiting the ability to infer changes in iceberg production and to draw conclusions on the dynamic behaviour of EAIS margins (Teitler et al., 2010).

For the Lambert Glacier-Amery Ice Shelf system, which drains a large catchment of the EAIS into Prydz Bay (Fig. 10.2), data from ODP Legs 119 and 188 provide information on its Plio-Pleistocene behaviour. Focusing on ODP Site 1166 on the continental shelf and ODP Site 1167 on the continental slope (Fig. 10.2), Passchier et al. (2003) presented sedimentological evidence indicating a switch in behaviour between the early and late Pleistocene. Alternations between glacially-derived debrites and fine-grained pelagic interbeds at ODP Site 1167 during the early Pleistocene were attributed to advance and retreat of the grounding line, implying that the Lambert Glacier was a highly dynamic ice stream (Passchier et al., 2003). In contrast, since the MPT, grounding-line debrites were replaced by glacio-marine deposition, from which it was inferred that the glacial advances of the Lambert Glacier became less extreme and no longer reached the shelf edge, suggesting a less dynamic ice sheet (Passchier et al., 2003). Nevertheless, clay mineralogy and IRD records from the Prydz Bay continental rise covering the last ~500 kyr do indicate late Pleistocene variability in the Lambert Glacier-Amery Ice Shelf system (Wu et al., 2021). As well as providing evidence for glacial advances across the shelf and interglacial retreat, differences in clay mineralogy between individual glacial and interglacial periods suggest a sensitive response of the ice sheet in this region to subtle differences in oceanic and atmospheric forcing (Wu et al., 2021).

The most prominent developments in understanding the past behaviour of the EAIS have come from IODP Expedition 318, which retrieved sediment cores from the continental shelf and rise in the vicinity of the Wilkes Subglacial Basin (Escutia et al., 2011) (Sites U1355–1361; Fig. 10.2). While many new insights from that expedition concerned earlier periods of glacial history, Pleistocene sections were recovered from the continental rise sites, including Site U1361, which provides important constraints on the late Pleistocene variability of the EAIS in this region. A recent study used neodymium and strontium isotope measurements on the bulk and fine-grained sediment fractions to trace changes in sediment provenance during the late Pleistocene (Fig. 10.9E), from which variability of the ice sheet margin was inferred (Wilson et al., 2018). These authors proposed that the ice margin retreated in the vicinity of the Wilkes Subglacial Basin during MIS 5e, MIS 9 and MIS 11, based on provenance signatures that differed from the Holocene sediments but which could be explained by a greater contribution from Ferrar Large Igneous Province and associated Beacon lithologies that are found inland in the Wilkes Subglacial Basin (Ferraccioli et al., 2009). Differing responses for those warm interglacials compared to the Holocene and MIS 7 indicate that retreat occurred when Antarctic air temperatures were at least 2°C warmer than pre-industrial temperatures for 2500 years or more (Jouzel et al., 2007) (Fig. 10.9A), suggesting a role for extended warmth (rather than the absolute magnitude of peak interglacial temperatures) in forcing the ice sheet behaviour (Wilson et al., 2018).

Overall, the Site U1361 record fingerprints a contribution to late Pleistocene interglacial sea levels from ice loss at the EAIS margin (Wilson et al., 2018), but those data are not able to quantify the extent of retreat or the total sea level contribution. However, the striking similarity between this record (Fig. 10.9) and changes during the Pliocene (Bertram et al., 2018; Cook et al., 2013) is an important observation. In light of global climate cooling since the Pliocene (Fig. 10.4C), the occurrence of similar ice sheet behaviour suggests that the late Pleistocene EAIS in the vicinity of the Wilkes Subglacial Basin has been more sensitive to glacial–interglacial changes than might have been anticipated. Because of the lower atmospheric CO_2 levels of Pleistocene interglacials compared to the Pliocene (Fig. 10.4H), and the slightly cooler global average surface temperatures, it appears that other factors may have compensated to maintain a sensitive ice sheet. One key factor may have been the operation of a stronger bipolar seesaw in ocean heat transport during the late Pleistocene (Barker et al., 2011; Stocker and Johnsen, 2003), which could have played a major role in generating peak interglacial Antarctic temperatures that were ~2°C–4°C warmer

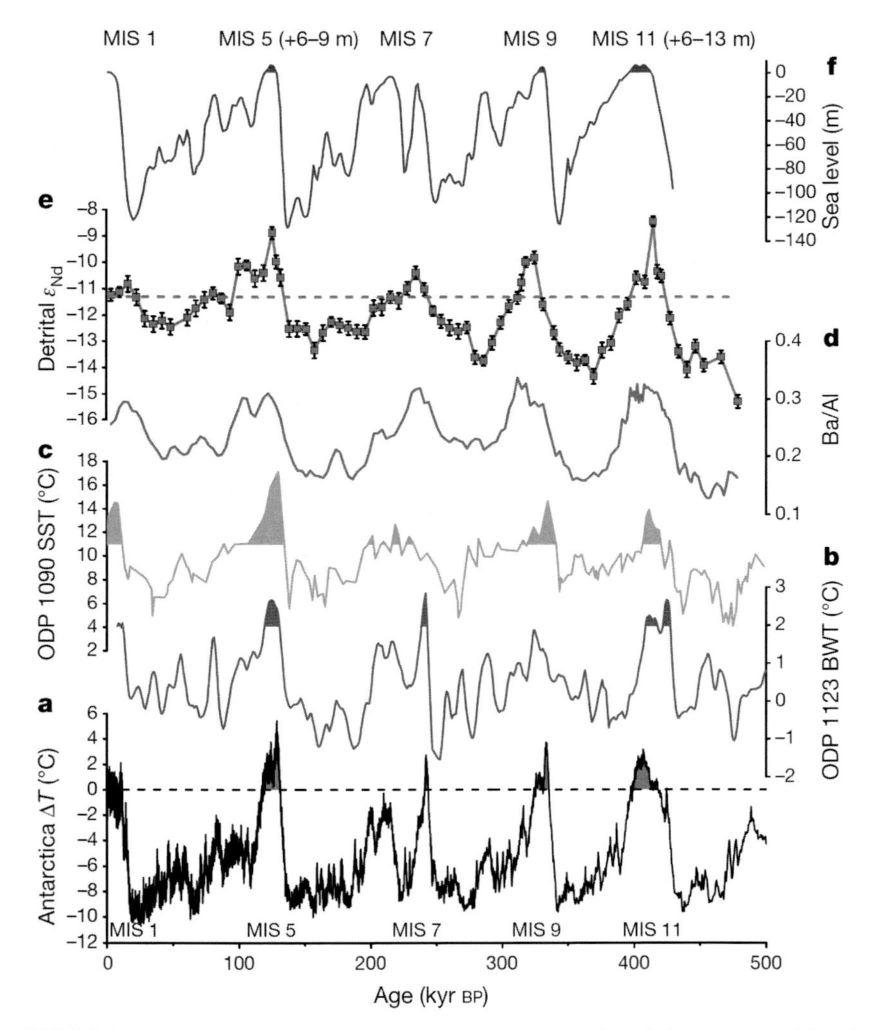

FIGURE 10.9 Late Pleistocene variability of the EAIS in the Wilkes Subglacial Basin inferred from geochemical proxies at Site U1361 in relation to global sea level and climate reconstructions. (A) Antarctic temperature change (ΔT, difference from mean values of the last millennium) from δD in EDC ice core (Jouzel et al., 2007). (B) Southern Ocean bottom water temperature (BWT) from benthic foraminiferal Mg/Ca at ODP Site 1123 (Elderfield et al., 2012). (C) Southern Ocean SST from alkenones at ODP Site 1090 (Martinez-Garcia et al., 2009). (D) Ba/Al ratios in U1361A (X-ray fluorescence [XRF]-scanner counts, three-point smoothed; Wilson et al., 2018), with higher values indicating higher marine productivity. (E) Bulk detrital sediment Nd isotopes in U1361A (Wilson et al., 2018), as an indicator of inland (more radiogenic, up) vs coastal (less radiogenic, down) erosion. (F) Global sea level proxy from benthic $\delta^{18}O$ (Waelbroeck et al., 2002), labelled with MIS numbers and sea level estimates from MIS 5e and MIS 11 (Dutton et al., 2015). Shading in (A–C and F) represents intervals with values above modern or late Holocene; red dashed line in (E) indicates the core top value. *Reproduced from Wilson, D.J., Bertram, R.A., Needham, E.F., van de Flierdt, T., Welsh, K.J., McKay, R.M., Mazumder, A., Riesselman, C.R., Jimenez-Espejo, F.J., Escutia, C., 2018. Ice loss from the East Antarctic Ice Sheet during late Pleistocene interglacials. Nature 561, 383–386 [p. 385].*

than the Holocene (Holden et al., 2010; Jouzel et al., 2007) (Fig. 10.9A). Sea ice feedbacks may also have contributed to localised Southern Ocean warming (Holloway et al., 2016). Regardless of the combination of mechanisms responsible for the local warmth, it appears critical to consider local rather than global temperatures in forcing the ice sheet behaviour (Clark et al., 2020; Rohling et al., 2019). An additional contributing factor may have been the deepening of bed topography in the Wilkes Subglacial Basin through time (Colleoni et al., 2018), which could have enhanced the sensitivity of this catchment to ocean forcing, although only modest changes have been reconstructed between the Pliocene and late Pleistocene (Colleoni et al., 2018; Paxman et al., 2020).

Two complementary studies have recently emerged, providing apparently contrasting views on the interglacial stability of ice in the Wilkes Subglacial Basin. Blackburn et al. (2020) showed that subglacial opal and calcite precipitates in the Wilkes Subglacial Basin record a level of ^{234}U enrichment that is consistent with a major ice margin retreat during MIS 11 followed by the isolation of a marine-derived fluid reservoir beneath the ice sheet since this time (Blackburn et al., 2020). To first order, those data agree with the offshore marine record, which also recorded the most extreme provenance changes during MIS 11 (Fig. 10.9E; Wilson et al., 2018), together providing support for a globally significant sea level contribution from the Wilkes Subglacial Basin of up to 3–4 m at that time. In contrast, no such major retreat is inferred for MIS 5e (Blackburn et al., 2020), which may indicate that the MIS 5e retreat indicated by the marine record (Fig. 10.9E) was more limited in extent. Evidence on the local ice sheet elevation during MIS 5e from $\delta^{18}O_{ice}$ values in the Talos Dome (TALDICE) ice core (Fig. 10.2) also appears to rule out a collapse on the scale inferred for MIS 11, which restricts MIS 5e sea level contributions from the Wilkes Subglacial Basin to a maximum of ~0.4–0.8 m (Sutter et al., 2020). Therefore, while Wilson et al. (2018) highlighted the value in comparing recent super-interglacials to the slightly cooler Holocene or MIS 7 intervals, the response of the ice sheet to subtle differences between super-interglacials should also be explored.

A major question to be addressed with future research is whether the EAIS in the vicinity of the Wilkes Subglacial Basin could even have been more sensitive to environmental change than the WAIS. If so, it would challenge a long-standing assumption that WAIS collapse would precede EAIS collapse, opening up the possibility that feedbacks from retreat of EAIS margins could contribute to WAIS collapse (Fogwill et al., 2015; Phipps et al., 2016). It will also be important to obtain similar constraints on the late Pleistocene behaviour of the EAIS margin in other regions, which are likely to have differing sensitivities to oceanic and atmospheric warming (Golledge et al., 2017). Late Pleistocene paleoenvironmental work is underway at the Sabrina Coast offshore of the Aurora Subglacial Basin (Holder et al., 2020; Tooze et al., 2020), and in Prydz Bay offshore of the Lambert

Glacier-Amery Ice Shelf system (Wu et al., 2021), but much more extensive sediment provenance data will be required from these regions. In addition, the integration of ice sheet models with both erosion models and bedrock geology maps holds significant potential to improve interpretations of detrital provenance records in terms of past ice sheet dynamics (Aitken and Urosevic, 2021).

10.5.6 Mechanisms of Antarctic Ice Sheet retreat and insights from ice sheet modelling

The above evidence from geological reconstructions is important for constraining ice sheet models. In particular, the requirement for ice margin retreat in the Wilkes Subglacial Basin during Pleistocene super-interglacials (Blackburn et al., 2020; Wilson et al., 2018) (Fig. 10.9E), similar to the Pliocene (Bertram et al., 2018; Cook et al., 2013), may implicate dynamic mechanisms such as ice shelf hydrofracture and ice cliff failure (in addition to marine ice sheet instability) in increasing the sensitivity of the Antarctic Ice Sheet to warm ocean temperatures (DeConto and Pollard, 2016; Pollard et al., 2015). With those processes incorporated, the output from the Penn State University ice sheet model (PSU-ISM) for MIS 5e (Fig. 10.10C) appears to be in agreement with the geological data (Fig. 10.9) and suggests that Antarctica could contribute ~6–7 m to global mean sea level during this time (DeConto and Pollard, 2016). That scenario involves local ice margin retreat into the subglacial basins of East Antarctica, and also the loss of large parts of the WAIS (Fig. 10.10C), consistent with the observations from Turney et al. (2020). In contrast, running the same model without implementing hydrofracture or ice cliff failure led to no collapse of the WAIS and virtually no changes in East Antarctica (Fig. 10.10D). The initial boundary conditions can also affect estimates of sea level change from ice sheet models, although they have a modest effect compared to differences in model physics. For example, the choice of glacial vs modern initial conditions leads to a difference of ~1.5 m in the predicted peak MIS 5e sea level rise (Fig. 10.10B cf. Fig. 10.10C), because the deeper bed elevations resulting from a glacial starting point act to enhance ice margin retreat (DeConto and Pollard, 2016).

From the above data-model comparison, it could be inferred that models lacking the ice cliff failure mechanism, such as the Parallel Ice Sheet Model (PISM) used by Golledge et al. (2017), or other statistical approaches (Ritz et al., 2015), may underestimate the sensitivity of the Antarctic Ice Sheet to ocean warming. However, significant caution is warranted here, because ice sheet collapse in the Wilkes Subglacial Basin (as required to explain the geological data from past super-interglacials) has been simulated using the PISM model (Mengel and Levermann, 2014) under a similar warming scenario to that used by DeConto and Pollard (2016). Indeed, a recent statistical

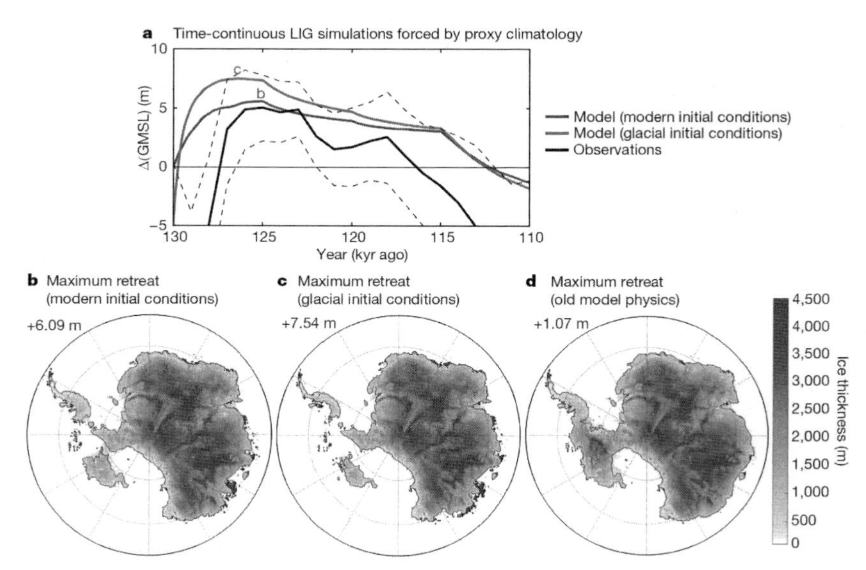

FIGURE 10.10 Ice sheet modelling results for MIS 5e based on the Penn State University ice sheet model (DeConto and Pollard, 2016). (A) Time series evolution of the Antarctic contribution to global mean sea level (GMSL) during MIS 5e, based on the outputs from two different models starting at 130 ka (panels B and C) and a probabilistic data-based assessment with uncertainties (16th and 84th percentiles) (Kopp et al., 2009). (B–D) Maps showing modelled ice thickness (with numbers indicating Antarctic contributions to global mean sea level) for three different model scenarios. The models are all driven by the same time-evolving proxy-based atmosphere and ocean climatologies, and the results are plotted for the maximum MIS 5e retreat state. (B) New model incorporating ice shelf hydrofracture and cliff failure processes, based on modern initial conditions. (C) New model as for (B), but based on glacial initial conditions. (D) Old model physics, based on modern initial conditions. Comparing (B) with (C) indicates the modest effect of initiating the model run from glacial rather than interglacial initial conditions (i.e. deeper bed elevations for glacial initial conditions). Comparing (B) with (D) indicates the major effect of incorporating new processes (i.e. ice shelf hydrofracture and cliff failure) in the models. Brown areas are ice-free land surfaces. *LIG*, Last Interglacial. *Reproduced from DeConto, R.M., Pollard, D., 2016. Contribution of Antarctica to past and future sea-level rise. Nature 531, 591–597 [p. 594].*

analysis suggests that MIS 5e retreat could be achieved in the DeConto and Pollard (2016) model even without applying cliff failure mechanisms (Edwards et al., 2019). Most recently, using a revised version of the PSU-ISM model, it was proposed that marine ice cliff failure cannot be excluded as a major ice loss mechanism unless the Antarctic Ice Sheet contribution to peak MIS 5e sea levels was less than 3.5 m, while Antarctic contributions of over 6 m would require the operation of an ice cliff failure mechanism (Gilford et al., 2020). With the present observational constraints from MIS 5e, it is unclear whether such dynamic mechanisms are required, but constraining the role of ice cliff instability during past retreat events would represent a major step for accurately modelling the rates of change associated with Antarctic contributions to past interglacial sea levels. Furthermore, evidence on whether these processes were at play in the past would provide essential information to constrain the high-end predictions of future sea level change and to project near-future rates of sea level change (Bassis et al., 2021; Edwards et al., 2019; Gilford et al., 2020).

While atmospheric forcing for an ice sheet model can reasonably be approximated based on Antarctic ice core temperature reconstructions (Blasco et al., 2019; DeConto and Pollard, 2016; Jouzel et al., 2007) or using local insolation and atmospheric CO_2 forcing (Tigchelaar et al., 2018), past ocean forcing of the Antarctic Ice Sheet is harder to implement. Challenges arise from the lack of suitable paleoceanographic archives from the Southern Ocean, in particular from sites proximal to the ice sheet where the forcing acts, and a lack of understanding of local ocean dynamics under ice shelves and near ice sheet grounding lines. Therefore, our present understanding of the mechanisms that may have influenced ocean forcing and contributed to melting of the Antarctic Ice Sheet is largely based on modern observations and modelling. As described above, changes in ocean heat transport linked to the bipolar seesaw (Holden et al., 2010; Stocker and Johnsen, 2003) may have been a major factor behind the elevated Southern Ocean warmth affecting surface, intermediate and deep waters during deglacial or interglacial periods (Fig. 10.9B and C). Alternatively, enhanced melting of the Antarctic Ice Sheet during MIS 5e could have been linked to warming of CDW arising from changes in the temperature of NADW or its fractional contribution to CDW (Duplessy et al., 2007; Rohling et al., 2019). Regardless of the exact controlling mechanisms, CDW temperatures are important because where this water mass is able to cross the Antarctic continental shelves (such as in deep channels or where the shelves are narrow), its warmth can melt the undersides of ice shelves or the deep portions of an ice sheet near its grounding line (Alley et al., 2016; Nakayama et al., 2019; Pritchard et al., 2012; Silvano et al., 2016; Turner et al., 2017). If CDW was no more than $\sim 1°C$ warmer during MIS 5e than during the pre-industrial Holocene (Duplessy et al., 2007; Shackleton et al., 2020), these changes may have been insufficient on their own to cause the magnitude of retreat proposed to have occurred (Fig. 10.10C). However, increases in the upwelling of CDW and/or more efficient transport of CDW across the Antarctic shelves could also have been important for increasing the heat supply to the ice sheet margins (Fogwill et al., 2014; Hillenbrand et al., 2017; Turner et al., 2017). Hence, the possibility arises for forcing by atmospheric processes because of their control on CDW upwelling and transport (Anderson et al., 2009; Fogwill et al., 2014; Turner et al., 2017).

In the absence of a full understanding of the above processes, ocean forcing has often been applied in a fairly simplistic way in ice sheet models, for example using a spatially uniform warming (DeConto and Pollard, 2016; Golledge et al., 2017; Mengel and Levermann, 2014; Turney et al., 2020). Where an ocean climatology derived from an atmosphere–ocean general circulation model was applied to an ice sheet model, both the climate forcing and the modelled ice sheet response were found to be inconsistent with paleoclimate records (Sutter et al., 2016), such that the application of a uniform warming was recommended. Regardless of the detailed approach, the majority of models applying an ocean forcing have suggested that collapse of the WAIS during MIS 5e was feasible in light of paleoclimate constraints (Capron et al., 2014; Hoffman et al., 2017). Specifically, modelled collapses occurred under regional ocean warming of ~0.5°C (Golledge et al., 2017), ~1°C (Turney et al., 2020), ~2°C (DeConto and Pollard, 2016) and ~2°C–3°C (Sutter et al., 2016), and also in a model with transient variations in ocean forcing derived from a general circulation model that was forced to replicate reconstructed AMOC variability (Clark et al., 2020). In addition, a high-resolution modelling study using the BISICLES model also indicates the sensitivity of WAIS to collapse on centennial to millennial timescales in response to CDW incursions and ice shelf removal from any of its major ice shelves (Amundsen Sea, Ronne, or Ross) (Martin et al., 2019). While not specifically targeted at MIS 5e, and probably somewhat extreme in the forcing assumed, that study indicates one mechanism through which the WAIS could respond to ocean warming and reduced ice shelf buttressing during MIS 5e. In contrast, one recent modelling study found that WAIS collapse during late Pleistocene interglacials only occurred when forced by subsurface warming of ~4°C (Tigchelaar et al., 2018), which the authors considered to be unrealistically high, but their suggestion that both the WAIS and EAIS were relatively stable creates a problem in explaining the global sea level data from MIS 5e and MIS 11. Marine Isotope Stage 11 has received less targeted attention from modelling, but a recent study simulated WAIS collapse during this interval with an intermediate ocean warming of only ~0.4°C when sustained for ~4 kyr (Mas e Braga et al., 2021), which represents warming that is around an order of magnitude lower than in Tigchelaar et al. (2018). It will be important to resolve the causes of such model differences, which will require higher resolution ice sheet models, the inclusion of both sea ice processes and cross-shelf CDW transport in coupled ocean-ice sheet models (Nakayama et al., 2019), and the incorporation of dynamical ice sheet–ocean–climate feedbacks, as well as better high latitude paleotemperature constraints and verification with geological records.

Finally, it is important to emphasise that the relative contributions of different processes to ice loss are likely to differ between sectors of Antarctica and to vary with the climate state (Golledge et al., 2017; Morlighem et al., 2020). Although late Pleistocene and near-future changes are likely to occur in marine basins, variable contributions can be expected from atmospheric and ocean warming, with WAIS catchments most sensitive to ocean warming and EAIS catchments sensitive to a combination of atmospheric and ocean warming (Golledge et al., 2017). Given the likelihood of differing sensitivities between regions, the existing late Pleistocene constraints on the Wilkes Subglacial Basin

(Blackburn et al., 2020; Sutter et al., 2020; Wilson et al., 2018) are not sufficient on their own to validate or tune an ice sheet model. It will therefore be important to obtain further geological evidence from other sectors of the EAIS and WAIS in order to test the sensitivity of specific Antarctic catchments to atmospheric and/or ocean warming. Another major target for the future will be data constraining the possibility of WAIS collapse in each of the recent late Pleistocene interglacials, because at present the evidence is sparse and equivocal. Nevertheless, whereas earlier model studies tended to implicate MIS 7 as the late Pleistocene interglacial with the greatest potential for Antarctic Ice Sheet retreat due to the high insolation peak at this time (e.g., de Boer et al., 2014; Pollard and DeConto, 2009), recent models have indicated greatest retreat during MIS 5e and MIS 11 (e.g., Sutter et al., 2019), in better agreement with constraints from the Wilkes Subglacial Basin (Blackburn et al., 2020; Wilson et al., 2018) and with global sea level reconstructions (Dutton et al., 2015; Spratt and Lisiecki, 2016).

10.5.7 Millennial variability and ice sheet–ocean–climate feedbacks

A recent assessment of global sea level variability during MIS 5e, based on the Red Sea $\delta^{18}O$ record in combination with a salinity reconstruction from the surface ocean near Greenland, has implicated asynchronous ice volume changes between Greenland and Antarctica during this interval (Rohling et al., 2019). These authors suggested that an early MIS 5e sea level high stand at ~129–125 ka was caused by Antarctic melting during and following Heinrich Stadial 11, and that these changes preceded significant melting in Greenland. Given the coincidence of Antarctic melting with a cool interval in the Northern Hemisphere, and the asynchronicity with Greenland retreat, this millennial scale response of the Antarctic Ice Sheet has been attributed to changes in the supply of warm water to the Antarctic shelves linked to the bipolar seesaw (Clark et al., 2020; Holden et al., 2010; Marino et al., 2015; Turney et al., 2020). In addition to warming of the interior ocean and Southern Ocean during Heinrich Stadial 11 (Duplessy et al., 2007; Shackleton et al., 2020), southward shifts in the southern westerly winds likely occurred as a bipolar seesaw response to weakening of the AMOC (Menviel et al., 2018) and could have enhanced incursions of warm CDW onto the shelves (Fogwill et al., 2014).

Millennial changes in the Antarctic Ice Sheet have also been proposed for the last glacial period. Sea level reconstructions from the Red Sea indicate sea level rise of up to ~30 m at ~20 mm/yr during the Antarctic warm events of MIS 3, which was originally attributed to approximately equal contributions from Antarctic and Northern Hemisphere ice sheets (Rohling et al., 2004). However, an Antarctic contribution of ~15 m would appear to be an overestimate, since it would match or exceed the deglacial changes in Antarctica between the LGM and the Holocene (RAISED Consortium, 2014; Tigchelaar et al., 2018). A recent ice sheet model supports ice loss from the

Antarctic Ice Sheet in response to millennial scale climate variability during the last glacial period, but of a smaller magnitude, with ocean warming contributing to sea level rise of up to ~6 m from a combination of both WAIS and EAIS margins (Blasco et al., 2019). Evidence for rapid drops in global sea level during the last glacial period has also been reported, including a sea level fall of ~17 m towards the end of the LGM (Yokoyama et al., 2018), although a large proportion of that ice growth was presumably in the Northern Hemisphere. Regardless, these data suggest that ice sheets may also gain volume rapidly, so the origin of such ice growth events must be explored for a better understanding of ice sheet dynamics.

Whereas salinity reconstructions and IRD records have provided extensive evidence on millennial variability of the North American and Greenland ice sheets during the late Pleistocene (Hemming, 2004; Rohling et al., 2019), comparable marine geological evidence on the Antarctic Ice Sheet has been more limited. In an early pioneering study, IRD records from the high latitude Southeast Atlantic Subtropical Zone (Cape Basin core TNO57-21) and Antarctic Zone (ODP Site 1094) were used to infer millennial variability in iceberg discharge from the Weddell Sea sector of the WAIS during MIS 3 (Kanfoush et al., 2000). These authors suggested that such changes may have been linked to rapid sea level variability or to the enhanced warmth of CDW due to increased NADW contributions during Northern Hemisphere interstadials (Fig. 10.8). However, such a mechanism would seem inconsistent with the Red Sea and modelling studies described above, in which Antarctic retreat was linked to Northern Hemisphere stadials, and it might be suspected that complications from iceberg survivability in this setting make it challenging to infer the exact timing of discharge events from the Antarctic Ice Sheet.

A more recent study used IRD records from cores in Iceberg Alley in the Scotia Sea to explore millennial variability in the Antarctic Ice Sheet during the last deglaciation (Weber et al., 2014). That study revealed centennial to millennial scale fluctuations in IRD content during the deglaciation, potentially indicating variability in ice sheet processes on very short timescales. Pronounced deglacial IRD peaks occurred during early Heinrich Stadial 1 (~17–16 ka) and at the onset of the Antarctic Cold Reversal (~15–14 ka), with the latter consistent with an Antarctic contribution to meltwater pulse (MWP) 1A (Weber et al., 2014). Although it is difficult to quantify the contribution of local icebergs from the Antarctic Peninsula (or the Weddell Sea region more generally) vs far-travelled icebergs from East Antarctica in those records, future geochemical and mineralogical provenance studies may provide this information. Unfortunately, the effects of changes in basal sediment content and its survival, both in ice shelves as they flow to the calving line (Smith et al., 2019) and in icebergs in the open ocean (Clark and Pisias, 2000), may challenge a simple interpretation of ice dynamics, but it is interesting to note that ice sheet models have reproduced a very similar temporal pattern in Antarctic ice discharge through the deglaciation (Golledge et al.,

2014). The debate concerning the origin of MWP-1A is active and ongoing, with the case also being made for a dominantly Northern Hemisphere meltwater source (Mackintosh et al., 2011), while a recent data-model comparison using planktonic $\delta^{18}O$ records is consistent with significant contributions from both hemispheres (Yeung et al., 2019).

The Antarctic Ice Sheet may not only respond to changes in AMOC and interhemispheric heat transport on millennial timescales, but it could also melt through local positive feedbacks following an initial perturbation. In particular, meltwater input and surface stratification could generate further local subsurface warming in the Southern Ocean, thereby enhancing ice shelf and ice sheet melting and accelerating ice sheet retreat (Bronselaer et al., 2018; Fogwill et al., 2015; Golledge et al., 2014; Schloesser et al., 2019). Such a positive feedback mechanism has been proposed to explain Antarctic melting during MWP-1A during the last deglaciation (Golledge et al., 2014; Weber et al., 2014), as well as during early MIS 5e (Rohling et al., 2019). However, it is also worth noting that local processes near an ice margin could lead to negative feedbacks as well as positive ones. For example, the expansion of large ice shelf cavities in response to ice sheet retreat from the continental shelves could trigger extensive supercooling of near-surface waters and sea ice growth, thereby reducing ice shelf melting and slowing retreat (e.g., Ashley et al., 2021). As well as causing local changes, Antarctic meltwater inputs could also affect the global ocean circulation, because stratification of the near-surface ocean would restrict AABW formation, as suggested for MWP-1A (Golledge et al., 2014) and early MIS 5e (Hayes et al., 2014). Furthermore, since meltwater inputs would be likely to enhance sea ice formation and generate surface cooling, they could cause northward shifts in ocean fronts and atmospheric circulation patterns, with the effects extending even to the intertropical convergence zone (Bronselaer et al., 2018). Therefore, in a case where Antarctic Ice Sheet melting arose from Southern Ocean warming and southward-shifted westerly winds linked to an AMOC reduction, these meltwater-induced changes could potentially provide a negative feedback on the original global-scale ocean and atmospheric circulation changes. With millennial variability of the Antarctic Ice Sheet only now starting to be addressed in ice sheet models (Blasco et al., 2019; Clark et al., 2020; Golledge et al., 2014; Martin et al., 2019), these kinds of dynamic ice sheet–ocean–climate feedbacks (both local and global) will need to be incorporated for these models to be effective.

10.6 Antarctica during earlier Pleistocene climate states

10.6.1 Lukewarm interglacials

Before the MBE at ~430 ka, interglacial periods in Antarctica were cooler than the more recent ones and are often described as 'lukewarm' interglacials

(Jouzel et al., 2007) (Fig. 10.3G). Interglacial atmospheric CO_2 concentrations were also lower during this interval (Fig. 10.3H), consistent with a greenhouse gas influence on global temperatures, but a complete explanation for the changes at the MBE is lacking. While oceanic processes play a major role in setting atmospheric CO_2 levels, there is currently little evidence to suggest significant differences in deep water source regions or ocean circulation across the MBE. Benthic $\delta^{13}C$ and neodymium isotope data from the deep North Atlantic Ocean (Site U1313) during the early Pleistocene indicate that similar glacial and interglacial modes of deep ocean circulation have operated since the onset of significant glacial Southern Ocean stratification and sea ice expansion at the start of the Pleistocene (Lang et al., 2016). Similarly, high-resolution neodymium isotope data spanning the last $\sim$800 kyr from the mid-depth equatorial Atlantic (ODP Site 929) record no evidence of interglacial changes in Atlantic Ocean circulation across the MBE (Howe and Piotrowski, 2017), indicating a clear decoupling from atmospheric CO_2 and Antarctic temperatures. Since sea ice proxies in ice cores record a shift across the MBE that resembles Antarctic temperature changes (Wolff et al., 2006) (Fig. 10.3F), Southern Ocean processes are a strong candidate to explain interglacial changes across the MBE. This suggestion is also supported by increases in interglacial Antarctic Zone productivity and deep water ventilation at ODP Site 1094 after the MBE (Jaccard et al., 2013).

Here we highlight this interval because constraints on the Antarctic Ice Sheet response to the lukewarm interglacials could complement the evidence on more recent super-interglacials such as MIS 5e and MIS 11. However, there is currently a dearth of evidence from Antarctica during this interval. The ANDRILL cores from the Ross Sea are challenging to interpret during the lukewarm interglacials because clearly defined ice shelf sediments are less frequent than during the more recent interglacials, although there may have been an episode of ice shelf retreat and/or open ocean conditions during one of these interglacials (McKay et al., 2012b). As described earlier, the possibility of retreat or collapse of the WAIS during MIS 13–15 was put forward based on elevated productivity and distinctive sediment inputs in an Amundsen Sea core (Hillenbrand et al., 2009). If correct, this result would imply either the importance of extended warmth rather than peak warmth in triggering WAIS retreat, or a role for local ocean forcing at this time (Hillenbrand et al., 2009), but more evidence from other settings is required to confirm this event.

For the EAIS, a clay minerology record from Prydz Bay also hints at an enhanced retreat of the Lambert Glacier-Amery Ice Shelf system during MIS 13 (Wu et al., 2021), but that possibility needs to be tested with additional proxies and a longer record. At Site U1361 offshore of the Wilkes Subglacial Basin, it was not possible to conduct provenance work on the lukewarm interglacials due to core disturbance (Wilson et al., 2018). However, recent studies using other piston cores from this East Antarctic margin have provided paleoenvironmental records that extend back into interglacials prior to the MBE (Jimenez-Espejo et al., 2020; Tolotti et al., 2018). On the basis of elemental chemistry and X-ray fluorescence (XRF) core-scanning records, Jimenez-Espejo et al. (2020) suggested that the lukewarm interglacials were characterised by a reduced variability in ocean conditions and ice sheet size, while Tolotti et al. (2018) indicated changes in diatom assemblages at the MBE. Future work using isotopic provenance and IRD records in this region could therefore provide further constraints on the behaviour of the ice sheet in the Wilkes Subglacial Basin during lukewarm interglacials.

10.6.2 Super-interglacial MIS 31

Similar to MIS 5e and MIS 11, certain early Pleistocene interglacials also experienced warmer than usual interglacial conditions, in both Antarctica and globally. A distinctive example is MIS 31 (1.08–1.06 Ma), when the high latitude Southern Ocean warmed by several degrees above modern-day values (Beltran et al., 2020; Scherer et al., 2008; Villa et al., 2008), accompanied by deep ocean warming and/or global ice volume decrease (Lisiecki and Raymo, 2005; Raymo et al., 2006) (Fig. 10.4G). This interval provides an additional test for the Antarctic Ice Sheet response to high latitude warming above present-day temperatures, similar to the Pliocene but under more comparable ice sheet boundary conditions to the modern day. Notably, recent data from MIS 31 include SST reconstructions that indicate warming of ~5°C at the Antarctic margin, based on organic biomarkers ($U^{k'}_{37}$; Long Chain Diol Index, LDI) at Site U1361 offshore of the Wilkes Subglacial Basin and at ODP Site 1101 offshore of the Antarctic Peninsula (Beltran et al., 2020) (Fig. 10.2). This excess warmth has been linked to a specific alignment of orbital parameters (i.e. high obliquity and eccentricity) that led to an unusually high summer insolation anomaly at high latitudes (Berger and Loutre, 1991). However, such a magnitude of warming cannot be explained by insolation forcing alone, leading to the suggestion that a minor initial warming triggered ice sheet collapse, with subsequent dynamic ice sheet–ocean–climate feedbacks leading to the major warming event (Beltran et al., 2020).

The best constraints on the Antarctic Ice Sheet behaviour during MIS 31 come from the Ross Sea, in the form of an unusual carbonate unit at Cape Roberts Project site CRP-1 (Scherer et al., 2008) and an interval of extended sedimentation of diatomite and volcanic sands at AND-1B (McKay et al., 2012b; Naish et al., 2009) (Fig. 10.2). Both observations indicate sea ice retreat and open ocean deposition and have been interpreted to indicate the retreat of the Ross Ice Shelf and the WAIS as a whole. This scenario can also be reproduced in ice sheet models forced by warm ocean waters, with model outputs suggesting ~6–8 m sea level rise, with ~4 m from WAIS collapse and the remainder from small-scale retreat or ice thinning in the subglacial basins of East Antarctica (Beltran et al., 2020; DeConto et al., 2012).

These changes in the WAIS may also have had more widespread global consequences. As discussed earlier, freshwater input linked to ice sheet retreat could generate a series of feedbacks in the climate system, through reduced AABW formation, increased sea ice formation and shifts in atmospheric circulation (Bronselaer et al., 2018). Retreat of the WAIS during MIS 31 (McKay et al., 2012b) may provide such an example, because it is proposed to have played a role in reducing AABW formation and inflow to the deep Southwest Pacific (Hall et al., 2001) and to have influenced the remarkably warm interglacial temperatures at Lake El'gygytgyn in Siberia through feedbacks involving ocean circulation and/or sea level change (Melles et al., 2012).

Unfortunately, geological constraints on the EAIS during this interval are rather poor, making it difficult to test ice sheet model outputs that typically suggest only minor changes (Beltran et al., 2020; DeConto et al., 2012). Teitler et al. (2015) compared IRD records through MIS 31 from a proximal site on the Prydz Bay continental rise (ODP Site 1165) (Fig. 10.2) and a distal South Atlantic site (ODP Site 1090). At the distal ODP Site 1090, there was a significantly reduced IRD content during this warm interval, similar to MIS 5e and MIS 11, which was likely controlled by SST warming. The proximal ODP Site 1165 in Prydz Bay recorded increased sedimentation rates and increased IRD supply during MIS 31, coupled to evidence for ocean warming from the elevated foraminiferal carbonate content. Therefore, these data do not support loss of the marine ice margin, consistent with model studies (Beltran et al., 2020), but do suggest some local instability and/or ice margin retreat. However, detailed stratigraphy is challenging through this period, in both ODP Site 1165 (Teitler et al., 2015) and the Ross Sea records (McKay et al., 2012b), and future research will be required to provide better constraints on Antarctic Ice Sheet dynamics during MIS 31.

10.6.3 Mid-Pleistocene Transition

At the MPT (~900 ka), there was an intensification of glacial conditions and a major switch in the predominant cyclicity of global climate records (Fig. 10.4). The ~41-kyr cycles of the early Pleistocene were replaced by ~100-kyr cycles in the late Pleistocene (Berends et al., 2021; Lisiecki and Raymo, 2005; Ruddiman et al., 1989), but the cause of this event remains unexplained. Early ideas include the crossing of a CO_2 threshold that made the Northern Hemisphere ice sheets more resistant to melting (Berger et al., 1999) or which influenced their sensitivity to integrated summer insolation forcing (Huybers, 2006); a change in the dynamics of those ice sheets linked to regolith removal from their base (Clark and Pollard, 1998); a transition from predominantly terrestrial-based ice sheets to expanded marine-based ice sheets in Antarctica (Raymo et al., 2006); and a change in bottom water formation around Antarctica that enhanced glacial ocean carbon storage

(Paillard and Parrenin, 2004). While the Northern Hemisphere mechanisms continue to be explored (Willeit et al., 2019), here we focus on the two mechanisms linked to Antarctica and discuss the new constraints that have emerged in recent years to test them.

10.6.3.1 Role of Antarctic Ice Sheet dynamics

Raymo et al. (2006) suggested that the MPT may have resulted from a change in Antarctic Ice Sheet dynamics, specifically the development of marine-based ice sheets in Antarctica. In this hypothesis, the dominant 41-kyr signal in global climate records before the MPT is explained as an in-phase response of the Antarctic and Northern Hemisphere ice sheets to obliquity forcing. Because the influence of precession forcing is out of phase between the hemispheres, it was proposed that the precession component of ice sheet variance was also out of phase, such that the coincidence of Northern Hemisphere ice sheet advance with Antarctic Ice Sheet retreat led to an approximate cancelling out of the precession signal in globally-integrated $\delta^{18}O$ records. According to this hypothesis, the expansion of marine-based ice sheets in Antarctica at the MPT would have resulted in an increased sensitivity of the Antarctic Ice Sheet to Northern Hemisphere climate and ice volume variability, propagated through changes in ocean circulation and sea level. Hence, more synchronous ice sheet behaviour would emerge at this time, leading to greater overall ice volumes and stronger $\sim$21 and $\sim$100-kyr signals (Raymo et al., 2006). Evidence from a benthic foraminiferal $\delta^{18}O$ record in the deep Southwest Pacific Ocean was taken to support an increase in Antarctic ice volume at the MPT (Elderfield et al., 2012) (Fig. 10.4B), but this scenario is equivocal because that record could have been influenced by regional changes in deep ocean circulation (Ford and Raymo, 2020), while the location of ice build-up cannot be determined by this method.

The idea of a switch from a terrestrial to a marine-based EAIS (Raymo et al., 2006) can now be addressed with more direct evidence, in particular from Site U1361 which constrains changes in the vicinity of the Wilkes Subglacial Basin (Fig. 10.2). The key observation here, based on a range of sedimentological and geochemical evidence, is that a marine-based ice sheet existed in East Antarctica from the Early Pliocene to the early Pleistocene (Escutia et al., 2011; Patterson et al., 2014), periodically extending to the continental shelf edge (Reinardy et al., 2015), and also in the late Pleistocene (Wilson et al., 2018) and presumably in between (Bertram, 2018). Therefore, it is hard to support the idea that the MPT originated from a switch from a smaller terrestrial EAIS to an expanded marine-based EAIS. For the WAIS, there is also strong evidence from the Ross Sea (AND-1B) (Fig. 10.2) for marine margins in both the Pliocene and Pleistocene (Naish et al., 2009), and for a control on its dynamics by obliquity rather than precession cycles in the Pliocene and early Pleistocene (Naish et al., 2009). In contrast to those observations, and depending on the climate scenario, some ice sheet model runs have suggested that a switch to a marine-based WAIS could have occurred around the MPT (Sutter et al., 2019). A ~200-kyr long unconformity at around this time in the ANDRILL record would be consistent with the expansion of a marine-based ice sheet in the Ross Sea Embayment at the MPT and is interpreted as representing a cooling climate in the high southern latitudes that allowed ice shelves to persist into interglacial periods (McKay et al., 2009, 2012b). However, even if the WAIS expanded at this time, it would not be possible to explain the majority of the MPT ice volume signal in this way.

While the above evidence does not support a transition between a terrestrial and a marine Antarctic Ice Sheet at the MPT, it is interesting that temporal shifts in the orbital cyclicity of IRD records from the EAIS have been recorded within the Plio-Pleistocene. For example, Patterson et al. (2014) demonstrate a much earlier switch in the orbital behaviour of the EAIS in the Wilkes Subglacial Basin, with IRD records from Site U1361 indicating ~41-kyr cyclicity during the warm mid-Pliocene followed by a shift to ~21 and ~100-kyr cyclicity around the onset of the Pleistocene. Hence, the idea that local forcing could create ~21-kyr cyclicity in Antarctic Ice Sheet dynamics during the early Pleistocene (Raymo et al., 2006) appears to be supported, and such changes could conceivably have been cancelled out in marine $\delta^{18}O$ records by similar opposing changes in the Northern Hemisphere. However, the significant IRD content indicates the presence of a marine-terminating margin during the early Pleistocene, so this Antarctic precession signal was clearly not the result of a terrestrial EAIS. Instead, Patterson et al. (2014) suggested that an increased sea ice extent linked to Plio-Pleistocene cooling could have shielded the early Pleistocene ice sheet from marine forcing and left it relatively more sensitive to local precession forcing. The growth of Northern Hemisphere ice sheets and an accompanying decrease in global sea level has also been proposed to play a possible role in increasing the stability of the marine sectors of the EAIS within the early Pleistocene (Jakob et al., 2020).

The late Pleistocene IRD record at Site U1361 was generated from material deposited under lower sedimentation rates than during the Pliocene, making it difficult to characterise orbital cyclicity (Wilson et al., 2018). Nevertheless, the observation of IRD pulses during terminations of the last five glacial periods appears consistent with Antarctic Ice Sheet variability that was synchronised to changes in the Northern Hemisphere ice sheets. Modelling studies suggest that synchronisation in the late Pleistocene could arise through a combination of global sea level and CO_2 changes (Gomez et al., 2020; Tigchelaar et al., 2018), and the deglacial sea level forcing acting on the marine margins of the Antarctic Ice Sheet would certainly have increased with the development of larger Northern Hemisphere ice sheets at the MPT. In addition, larger Northern Hemisphere ice sheets since the MPT may have displayed more dynamic behaviour (Hodell et al., 2008), potentially leading to increased millennial variability in the Atlantic deep ocean circulation and a stronger bipolar seesaw (Stocker and Johnsen, 2003). If early interglacial retreat of the Antarctic Ice Sheet occurred in response to heat transport changes during terminal Heinrich Stadial events (Clark et al., 2020; Marino et al., 2015; Rohling et al., 2019), such a scenario could effectively synchronise the ice sheets on orbital timescales, even while the behaviour is asynchronous on millennial timescales. Overall, while changes in the magnitude and length of glacial cycles at the MPT are likely to have affected the behaviour of the Antarctic Ice Sheet, there is a lack of evidence at present to suggest that a change in Antarctic Ice Sheet dynamics played the significant role in this transition.

10.6.3.2 Role of the Southern Ocean carbon cycle

In light of the dominant control of Southern Ocean processes on the late Pleistocene carbon cycle (Fig. 10.6), this region is a strong candidate to have played a role in the declining glacial atmospheric CO_2 levels at the MPT (Fig. 10.4H). As proposed by Paillard and Parrenin (2004), changes in SST or sea ice extent could impact on polar ocean stratification, deep ocean circulation and carbon storage, leading to extended glacial periods that are more resistant to deglaciation. A recent study has tested this idea using paired planktonic and benthic $\delta^{18}O$ records from ODP Site 1094 to reconstruct the surface stratification of the Southern Ocean Antarctic Zone across the MPT (Hasenfratz et al., 2019). These authors demonstrated that a stronger halocline has formed during glacial periods since the MPT, in association with reduced communication between the deep ocean and the Antarctic surface, with these two factors potentially combining to form a positive feedback loop. In this scenario, glacial carbon storage could be enhanced and a stronger orbital forcing would be required to initiate deglacial carbon release, allowing glacial periods to be extended and larger Northern Hemisphere ice sheets to grow (Hasenfratz et al., 2019). This mechanism is supported by atmospheric CO_2 reconstructions from marine boron isotope records (Chalk et al., 2017; Hönisch et al., 2009) (Fig. 10.4H) and from Antarctic blue-ice cores (Higgins et al., 2015; Yan et al., 2019), which indicate a decline in glacial CO_2 levels of ~30 ppm at the MPT. A change of this magnitude is readily achievable through changes in the Antarctic Zone (Hain et al., 2010; Watson et al., 2015), but the CO_2 records do not uniquely constrain the mechanism of carbon drawdown. A further complexity is that while a period of global cooling preceding the MPT may have pre-conditioned the Earth for such a mechanism (Snyder, 2016) (Fig. 10.4C), the stalling of global average temperatures at ~1.2 Ma suggests that another trigger was ultimately required to generate the ice volume increase at ~800–900 ka (Fig. 10.4A and B).

Evidence for changes in deep ocean circulation and chemistry in the Atlantic Ocean is recorded across the MPT, including a weakening or shoaling of the glacial AMOC (neodymium isotopes; Pena and Goldstein, 2014) that coincided with enhanced deep ocean carbon storage (benthic B/Ca and Cd/Ca; Farmer et al., 2019). Those records are consistent with the Southern Ocean circulation changes expected from the Antarctic Zone mechanism (Ferrari et al., 2014; Hasenfratz et al., 2019). Furthermore, the distinct circulation switch during MIS 22–24 (Pena and Goldstein, 2014) appears to have coincided with the glacial ice volume expansion at ~900 ka (Elderfield et al., 2012), which suggests that a sensitive response of the ocean circulation to a combination of forcings may have ultimately determined the timing for the emergence of the 100-kyr cycles. Low orbital eccentricity at ~900 ka has been proposed to play a role, although the extent to which orbital forcing acted as a Northern Hemisphere (Pena and Goldstein, 2014) or Southern Hemisphere (Elderfield et al., 2012) driver remains to be determined. Further work will be required to understand the mechanistic links between near-surface stratification and deep water formation in the Southern Ocean, as well as the feedbacks between Southern Ocean processes, Atlantic Ocean circulation and the Northern Hemisphere ice sheets during this interval.

Evidence for Subantarctic Zone changes has also been obtained over the MPT, with planktonic foraminiferal Mg/Ca and $\delta^{18}O$ reconstructions at ODP Site 1090 indicating an early cooling and freshening of the surface ocean during glacial periods at ~1250 ka (Rodríguez-Sanz et al., 2012). Those results could indicate either a sea ice expansion or a latitudinal migration of the westerly wind belt driving shifts in the Southern Ocean fronts (Rodríguez-Sanz et al., 2012), which may have helped precondition the Earth system for the changes at the MPT. In addition, glacial dust fluxes in the Subantarctic Zone of the Atlantic sector approximately doubled at the MPT (Martinez-Garcia et al., 2011) (Fig. 10.4I), which suggests that iron fertilisation of surface ocean productivity could have contributed to the glacial carbon drawdown at the MPT and to longer-lasting glacial periods thereafter (Chalk et al., 2017; Martinez-Garcia et al., 2011). Although fertilisation of the Subantarctic Zone in the late Pleistocene is responsible for ~30 ppm glacial CO_2 drawdown (Hain et al., 2010), the dust changes at the MPT might be expected to have arisen as a result of increased global ice volume and reduced sea level, in which case this process may have acted as a positive feedback rather than as the trigger for the MPT, similar to its role during more recent glacial cycles (Fig. 10.6).

At present, Antarctic ice core records do not extend across the MPT (Bereiter et al., 2015) and CO_2 reconstructions from marine proxies have lower resolution and larger uncertainties (Chalk et al., 2017; Hönisch et al., 2009) (Fig. 10.4H). Therefore, obtaining continuous ice core records across this interval is an important target (Fischer et al., 2013; Sutter et al., 2019). High-resolution records of Antarctic climate, sea ice, dust and atmospheric

CO_2, all measured in the same archive, will be invaluable for determining which combination of these mechanisms and feedbacks was responsible for the MPT.

10.7 Future research on Antarctica in the Pleistocene

10.7.1 Motivation and outlook

The boundary conditions and climatic forcing affecting each sector of the Antarctic Ice Sheet, and indeed each ice stream, are different (e.g., subglacial bed topography, basal lithology, shelf width and geometry, proximity to CDW inflow, ice shelf extent). Hence, a spatial perspective is critical for understanding the past and future responses of the ice sheet to climate change. While this chapter has highlighted the existing reconstructions that constrain the dynamic behaviour of the WAIS and EAIS in the Pleistocene, their value is limited by the small number of observations and their restricted spatial distribution, as well as in some cases by poor temporal resolution.

With the challenges of drilling in the Southern Ocean and on the Antarctic margin, there have been relatively few expeditions of the IODP (and its precursors) to this region, with only seven expeditions in its first 50 years from 1968 to 2017 (Fig. 10.2; white squares). Following the success of IODP Expedition 318 to the George V Land margin of East Antarctica in 2010, a concerted community effort led to four expeditions to the Antarctic margin and/or high latitude Southern Ocean in 2018 and 2019 (Fig. 10.2; yellow squares). It is anticipated that samples from these expeditions will significantly improve upon the existing spatial and temporal resolution of records of Antarctic and Southern Ocean variability in the Pleistocene, particularly along the WAIS margin. However, increased efforts to obtain Pleistocene records of EAIS variability are required, particularly from the margins of the marine-based basins.

As well as records of ice sheet behaviour, improved spatial coverage in records of changing Southern Ocean processes and properties is required, including evidence on sea ice formation, upwelling, deep convection, dust supply, productivity, surface fronts and oceanic temperatures on the shelves and near the ice margins. In particular, there is a strong need for better records of oceanic temperatures at intermediate water depths near the continental shelf breaks, which models indicate to be a key control on ice shelf thinning that can trigger marine-based ice sheet retreat. Another important target is to constrain changes in Southern Ocean sea ice formation, which influences the locations and strengths of deep convection during the Pleistocene. Both oceanic temperatures and sea ice show complex spatial variability, and neither are well reproduced in the current generation of ocean circulation or climate models. New paleo-data will therefore be important for indicating system behaviour and for ground-truthing new and improved modelling approaches. For sea ice, the Cycles of Sea-Ice Dynamics in the Earth System (C-SIDE) working group of PAGES has been addressing some of these issues, for example by compiling a database of proxy data for the last glacial cycle, identifying data gaps, and conducting data-model comparisons (Chadwick et al., 2019), with the seasonality of sea ice emerging as an important factor to consider (Green et al., 2020). In combination, better records of sea ice extent, ocean temperatures and deep water formation will guide a better understanding of the various dynamic feedbacks between the ice sheets and the ocean. More broadly, evidence on ice sheet–ocean–climate interactions in the Southern Ocean is crucial for understanding the role of Southern Hemisphere processes in heat and carbon transport, over timescales ranging from glacial–interglacial cycles through to millennial and sub-centennial climate variability.

By way of introducing some of this exciting future research, we end this chapter with a brief overview of the sediment sequences recovered in four recent IODP expeditions (Fig. 10.2). The discussion below is limited to preliminary findings from immediate shipboard observations, whereas the science emanating from those expeditions, both individually and when synthesised together, will fuel at least the next decade of research on the Pleistocene history of Antarctica. It is anticipated that future analysis of these cores will allow us to address many crucial questions that remain about Antarctica's role in the global climate system.

10.7.2 IODP Expedition 374: Ross Sea West Antarctic Ice Sheet History

In 2018 IODP Expedition 374 drilled a latitudinal and depth transect of five sites (U1521–1525) from the shelf to the rise in the Ross Sea (Fig. 10.2). The aim was to evaluate the Neogene to Quaternary response of the WAIS to ocean and atmospheric forcing, orbital forcing and sea level forcing under a range of climatic boundary conditions and to establish its contribution to far-field sea level change (McKay et al., 2019). In total 1293 m of high-quality sediment cores from five sites were recovered, spanning the early Miocene to late Pleistocene. Because the shelf sites were located on the outer shelf, they will provide complementary evidence on climate forcing and WAIS behaviour to the existing records from the more ice-proximal ANDRILL sites in the western Ross Sea.

Coring at Sites U1521 and U1522 on the shelf recovered some Pleistocene sections that will help constrain the shelf environment, but recovery was very poor in the Pleistocene intervals of these cores. At the shelf-edge, Site U1523 is well placed to evaluate Plio-Pleistocene changes in the Antarctic Slope Current, which could exert an important control on ice sheet stability because it influences the ability of modified CDW to penetrate onto the Ross Sea shelf. At deeper depths on the continental slope, Site U1525 will provide a complimentary view on the Antarctic Slope Current from early to mid-Pleistocene strata. Continental rise Site U1524 is the deepest site, located on the southeastern levee of the Hillary Canyon, one of the largest conduits for newly-formed Ross Sea Bottom Water (a type of AABW). Drilling at this site recovered a mostly continuous ~120 m long Pleistocene record of AABW outflow. This expanded record is particularly exciting because it opens up the prospect of coupling evidence on the ice margin behaviour, based on turbidite activity and IRD/fine-grained sediment provenance, to a highly resolved record of changes in climate, meltwater input and AABW production (McKay et al., 2019).

10.7.3 IODP Expedition 379: Amundsen Sea West Antarctic Ice Sheet History

The primary goal of IODP Expedition 379 in 2019 was to constrain the past behaviour of the WAIS in the Amundsen Sea Embayment (Gohl et al., 2019) (Fig. 10.2), a region that is currently experiencing significant and rapid ice loss (Shepherd et al., 2018) and in which ice sheet retreat by marine ice sheet instability may already be underway (Joughin et al., 2014; Rignot et al., 2014). Unlike the Ross Sea or Weddell Sea, this setting benefits from the ability to record changes in WAIS dynamics with no direct influence from the EAIS.

Continuous Plio-Pleistocene records were obtained from the continental rise at Sites U1532 and U1533 (Fig. 10.2) and are characterised by strong lithological cyclicity. These cycles are inferred to be related to glacial–interglacial changes in glacially-derived sediment inputs to the Southern Ocean, and to oceanographic variability linked to the wind-driven currents of the Antarctic Circumpolar Current. Analyses of IRD accumulation rates and provenance hold the potential to constrain potential retreat or collapse events, and in particular to assess changes during super-interglacials of the late Pleistocene (e.g., MIS 11) and earlier Pleistocene (e.g., MIS 31).

10.7.4 IODP Expedition 382: Iceberg Alley and Subantarctic Ice and Ocean Dynamics

IODP Expedition 382 in 2019 targeted the first deep drilling in the Scotia Sea, with the goal of constraining the Plio-Pleistocene dynamics of the Antarctic Ice Sheet and its role in sea level variations, as well as exploring ice sheet–ocean–atmosphere interactions in the wider region (Weber et al., 2019). The South Scotia Sea lies in the pathway of 'Iceberg Alley', where icebergs escape the coastal ocean into the Antarctic Circumpolar Current, making it an excellent location to trace iceberg calving (IRD fluxes) and sources (geochemical provenance tracing). South Falkland Slope sites were also drilled, in a sensitive location to trace movements of the Subantarctic

Front and AAIW properties, as well as westerly wind changes and iron fertilisation of surface ocean productivity.

The Scotia Sea sites (U1536, U1537, U1538; Fig. 10.2) recovered diatom oozes and silty clays that form a continuous Plio-Pleistocene record. The Pleistocene sediments show clear glacial–interglacial variability in physical properties, colour and biogenic opal content and will provide crucial evidence on Antarctic Ice Sheet variability, Drake Passage throughflow and shifts in ocean fronts and sea ice over this interval. These continuous high-resolution records will allow the effects of Northern Hemisphere Glaciation, the MPT and the MBE to be explored, with preliminary shipboard data indicating the potential for a high fidelity orbital signal to be present in these records. The South Falkland Slope sites (U1534, U1535; outside the map area of Fig. 10.2) recovered late Pliocene and Pleistocene foraminifera-bearing silts and clays, including a continuous late Pleistocene section back to ~700 ka that should allow millennial variability in frontal positions and AAIW properties to be reconstructed back to at least MIS 17, as well as providing some earlier Pleistocene and late Pliocene snapshots.

10.7.5 IODP Expedition 383: Dynamics of Pacific Antarctic Circumpolar Current

IODP Expedition 383 to the South and Southeast Pacific in 2019 did not directly target Antarctic Ice Sheet dynamics but will provide important evidence on Miocene to Holocene atmosphere–ocean–climate dynamics in the Southern Ocean region (Lamy et al., 2019). Specifically, it fills a significant spatial gap in records of the Antarctic Circumpolar Current, which was previously mostly constrained in the Atlantic and Indian sectors. Its main focus was on recovering changes in the past vertical structure of the Antarctic Circumpolar Current on glacial–interglacial and millennial timescales, as well as exploring changes during super-interglacials and the MPT. The results from this expedition will be highly complementary for interpreting the more proximal records of Antarctic Ice Sheet variability from other recent expeditions, both by providing information on the ocean conditions influencing the ice sheet and by revealing how the ice sheet has influenced ocean circulation and climate.

Sites in the central South Pacific (U1539, U1540, U1541) and eastern South Pacific (U1543) recovered diatom and carbonate oozes from the northern margin of the Antarctic Circumpolar Current at 3.6–4.1 km depth. Site U1539 will provide a particularly valuable Pleistocene record of LCDW paleoceanography because of its high late Pleistocene sedimentation rates and its continuation through the MPT. Nearby Site U1540 has a lower sedimentation rate but extends through much of the Plio-Pleistocene, while sites U1541 and U1543 also contain mostly continuous records that extend back into the Miocene. The Chilean margin sites (U1542, U1544) are located at shallower water depths of 1.1 and 2.1 km and will provide high-resolution late Pleistocene records. Despite the dominance of siliciclastic sediments, these two sites will be suitable for monitoring late Pleistocene changes in AAIW and CDW properties and flow, providing insight into interconnected changes in the Southern Hemisphere ice sheet–ocean–climate system.

Acknowledgements

We thank the reviewers for their positive and helpful comments on this chapter. D.J.W. and T.v.d.F acknowledge the Natural Environment Research Council, the Leverhulme Trust and the Kristian Gerhard Jebsen Foundation for supporting their Southern Ocean and Antarctic research. D.J.W. is supported by a NERC independent research fellowship (NE/T011440/1). R.M.M was funded by the NZ Marsden Fund (18-VUW-089) R.M.M. and T.R.N. acknowledge support from MBIE Antarctic Science Platform contract ANTA1801.

References

Abram, N.J., Wolff, E.W., Curran, M.A.J., 2013. A review of sea ice proxy information from polar ice cores. Quaternary Science Reviews. 79, 168–183.

Aciego, S.M., Bourdon, B., Lupker, M., Rickli, J., 2009. A new procedure for separating and measuring radiogenic isotopes (U, Th, Pa, Ra, Sr, Nd, Hf) in ice cores. Chemical Geology 266, 194–204.

Adkins, J.F., 2013. The role of deep ocean circulation in setting glacial climates. Paleoceanography 28, 539–561.

Adkins, J.F., Cheng, H., Boyle, E.A., Druffel, E.R.M., Edwards, R.L., 1998. Deep-sea coral evidence for rapid change in ventilation of the deep North Atlantic 15,400 years ago. Science 280, 725–728.

Adkins, J.F., McIntyre, K., Schrag, D.P., 2002. The salinity, temperature, and $\delta^{18}O$ of the glacial deep ocean. Science 298, 1769–1773.

Aitken, A.R.A., Urosevic, L., 2021. A probabilistic and model-based approach to the assessment of glacial detritus from ice sheet change. Palaeogeography, Palaeoclimatology, Palaeoecology 561, 110053. Available from: https://doi.org/10.1016/j.palaeo.2020.110053.

Allen, C.S., Pike, J., Pudsey, C.J., Leventer, A., 2005. Submillennial variations in ocean conditions during deglaciation based on diatom assemblages from the southwest Atlantic. Paleoceanography 20, PA2012. Available from: https://doi.org/10.1029/2004PA001055.

Alley, R.B., 2010. Reliability of ice-core science: historical insights. Journal of Glaciology 56, 1095–1103.

Alley, R.B., MacAyeal, D.R., 1994. Ice-rafted debris associated with binge/purge oscillations of the Laurentide Ice Sheet. Paleoceanography 9, 503–511.

Alley, K.E., Scambos, T.A., Siegfried, M.R., Fricker, H.A., 2016. Impacts of warm water on Antarctic ice shelf stability through basal channel formation. Nature Geoscience 9, 290–293.

Anderson, D.M., Archer, D., 2002. Glacial–interglacial stability of ocean pH inferred from foraminifer dissolution rates. Nature 416, 70–73.

Anderson, J.B., Conway, H., Bart, P.J., Witus, A.E., Greenwood, S.L., McKay, R.M., et al., 2014. Ross Sea paleo-ice sheet drainage and deglacial history during and since the LGM. Quaternary Science Reviews 100, 31–54.

Anderson, R.F., Ali, S., Bradtmiller, L.I., Nielsen, S.H.H., Fleisher, M.Q., Anderson, B.E., et al., 2009. Wind-driven upwelling in the Southern Ocean and the deglacial rise in atmospheric CO_2. Science 323, 1443–1448.

Anderson, R.F., Sachs, J.P., Fleisher, M.Q., Allen, K.A., Yu, J., Koutavas, A., et al., 2019. Deep-sea oxygen depletion and ocean carbon sequestration during the last ice age. Global Biogeochemical Cycle 33, 301–317.

Ashley, K.E., McKay, R., Etourneau, J., Jimenez-Espejo, F.J., Condron, A., Albot, A., et al., 2021. Mid-Holocene Antarctic sea-ice increase driven by marine ice sheet retreat. Climate of the Past 17, 1–19.

Austermann, J., Pollard, D., Mitrovica, J.X., Moucha, R., Forte, A.M., DeConto, R.M., et al., 2015. The impact of dynamic topography change on Antarctic ice sheet stability during the mid-Pliocene warm period. Geology 43, 927–930.

Bailey, I., Hole, G.M., Foster, G.L., Wilson, P.A., Storey, C.D., Trueman, C.N., et al., 2013. An alternative suggestion for the Pliocene onset of major northern hemisphere glaciation based on the geochemical provenance of North Atlantic Ocean ice-rafted debris. Quaternary Science Reviews. 75, 181–194.

Barker, S., Cacho, I., Benway, H., Tachikawa, K., 2005. Planktonic foraminiferal Mg/Ca as a proxy for past oceanic temperatures: a methodological overview and data compilation for the Last Glacial Maximum. Quaternary Science Reviews 24, 821–834.

Barker, S., Diz, P., Vautravers, M.J., Pike, J., Knorr, G., Hall, I.R., et al., 2009. Interhemispheric Atlantic seesaw response during the last deglaciation. Nature 457, 1097–1102.

Barker, S., Chen, J., Gong, X., Jonkers, L., Knorr, G., Thornalley, D., 2015. Icebergs not the trigger for North Atlantic cold events. Nature 520, 333–336.

Barker, S., Knorr, G., Edwards, R.L., Parrenin, F., Putnam, A.E., Skinner, L.C., et al., 2011. 800,000 years of abrupt climate variability. Science 334, 347–351.

Barlow, N.L.M., McClymont, E.L., Whitehouse, P.L., Stokes, C.R., Jamieson, S.S.R., Woodroffe, S.A., et al., 2018. Lack of evidence for a substantial sea-level fluctuation within the Last Interglacial. Nature Geoscience 11, 627–634.

Barnes, D.K.A., Hillenbrand, C.D., 2010. Faunal evidence for a late Quaternary trans-Antarctic seaway. Global Change Biology 16, 3297–3303.

Bart, P.J., Coquereau, L., Warny, S., Majewski, W., 2016. In situ foraminifera in grounding zone diamict: a working hypothesis. Antarctic Science 28, 313–321.

Bart, P.J., De Batist, M., Jokat, W., 1999. Interglacial collapse of Crary Trough-mouth fan, Weddell Sea, Antarctica; implications for Antarctic glacial history. Journal of Sedimentary Research 69, 1276–1289.

Bart, P.J., DeCesare, M., Rosenheim, B.E., Majewski, W., McGlannan, A., 2018. A centuries-long delay between a paleo-ice-shelf collapse and grounding-line retreat in the Whales Deep Basin, eastern Ross Sea, Antarctica. Scientific Reports 8, 12392. Available from: https://doi.org/10.1038/s41598-018-29911-8.

Bart, P.J., Iwai, M., 2012. The overdeepening hypothesis: how erosional modification of the marine-scape during the early Pliocene altered glacial dynamics on the Antarctic Peninsula's Pacific margin. Palaeogeography, Palaeoclimatology, Palaeoecology 335, 42–51.

Bassinot, F.C., Labeyrie, L.D., Vincent, E., Quidelleur, X., Shackleton, N.J., Lancelot, Y., 1994. The astronomical theory of climate and the age of the Brunhes-Matuyama magnetic reversal. Earth and Planetary Science Letters 126, 91–108.

Bassis, J.N., Berg, B., Crawford, A.J., Benn, D.I., 2021. Transition to marine ice cliff instability controlled by ice thickness gradients and velocity. Science 372, 1342–1344.

Bauersachs, T., Rochelmeier, J., Schwark, L., 2015. Seasonal lake surface water temperature trends reflected by heterocyst glycolipid-based molecular thermometers. Biogeosciences 12, 3741–3751.

Bauska, T.K., Baggenstos, D., Brook, E.J., Mix, A.C., Marcott, S.A., Petrenko, V.V., et al., 2016. Carbon isotopes characterize rapid changes in atmospheric carbon dioxide during the last deglaciation. Proceedings of the National Academy of Sciences 113, 3465–3470.

Becquey, S., Gersonde, R., 2003. A 0.55-Ma paleotemperature record from the Subantarctic zone: implications for Antarctic Circumpolar Current development. Paleoceanography 18, 1014. Available from: https://doi.org/10.1029/2000PA000576.

Beltran, C., Golledge, N.R., Ohneiser, C., Kowalewski, D.E., Sicre, M.-A., Hageman, K.J., et al., 2020. Southern Ocean temperature records and ice-sheet models demonstrate rapid Antarctic ice sheet retreat under low atmospheric CO_2 during Marine Isotope Stage 31. Quaternary Science Reviews 228, 106069. Available from: https://doi.org/10.1016/j.quascirev.2019.106069.

Benz, V., Esper, O., Gersonde, R., Lamy, F., Tiedemann, R., 2016. Last Glacial Maximum sea surface temperature and sea-ice extent in the Pacific sector of the Southern Ocean. Quaternary Science Reviews 146, 216–237.

Bereiter, B., Eggleston, S., Schmitt, J., Nehrbass-Ahles, C., Stocker, T.F., Fischer, H., et al., 2015. Revision of the EPICA Dome C CO_2 record from 800 to 600 kyr before present. Geophysical Research Letters 42, 542–549.

Bereiter, B., Shackleton, S., Baggenstos, D., Kawamura, K., Severinghaus, J., 2018. Mean global ocean temperatures during the last glacial transition. Nature 553, 39–44.

Berends, C.J., Köhler, P., Lourens, L.J., van de Wal, R.S.W., 2021. On the cause of the mid-Pleistocene transition. Reviews of Geophysics 59, e2020RG000727. Available from: https://doi.org/10.1029/2020RG000727.

Berger, W.H., 1999. The 100-kyr ice-age cycle: internal oscillation or inclinational forcing? International Journal of Earth Sciences 88, 305–316.

Berger, A., Li, X.S., Loutre, M.-F., 1999. Modelling northern hemisphere ice volume over the last 3 Ma. Quaternary Science Reviews 18, 1–11.

Berger, A., Loutre, M.F., 1991. Insolation values for the climate of the last 10 million years. Quaternary Science Reviews 10, 297–317.

Bertler, N.A.N., Conway, H., Dahl-Jensen, D., Emanuelsson, D.B., Winstrup, M., Vallelonga, P.T., et al., 2018. The Ross Sea Dipole-temperature, snow accumulation and sea ice variability in the Ross Sea region, Antarctica, over the past 2700 years. Climate of the Past 14, 193–214.

Bertram, R.A., 2018. Reconstructing the East Antarctic Ice Sheet During the Plio-Pleistocene Using Geochemical Provenance Analysis. Ph.D. thesis, Imperial College London.

Bertram, R.A., Wilson, D.J., van de Flierdt, T., McKay, R.M., Patterson, M.O., Jimenez-Espejo, F. J., et al., 2018. Pliocene deglacial event timelines and the biogeochemical response offshore Wilkes Subglacial Basin, East Antarctica. Earth and Planetary Science Letters 494, 109–116.

Bintanja, R., van de Wal, R.S.W., Oerlemans, J., 2005. Modelled atmospheric temperatures and global sea levels over the past million years. Nature 437, 125–128.

Birner, B., Hodell, D.A., Tzedakis, P.C., Skinner, L.C., 2016. Similar millennial climate variability on the Iberian margin during two early Pleistocene glacials and MIS 3. Paleoceanography 31, 203–217.

Blackburn, T., Edwards, G.H., Tulaczyk, S., Scudder, M., Piccione, G., Hallet, B., et al., 2020. Ice retreat in Wilkes Basin of East Antarctica during a warm interglacial. Nature 583, 554–559.

Blasco, J., Tabone, I., Alvarez-Solas, J., Robinson, A., Montoya, M., 2019. The Antarctic Ice Sheet response to glacial millennial-scale variability. Climate of the Past 15, 121–133.

Blunier, T., Brook, E.J., 2001. Timing of millennial-scale climate change in Antarctica and Greenland during the last glacial period. Science 291, 109–112.

Bond, G., Heinrich, H., Broecker, W., Labeyrie, L., McManus, J., Andrews, J., et al., 1992. Evidence for massive discharges of icebergs into the North Atlantic Ocean during the last glacial period. Nature 360, 245–249.

Bond, G., Showers, W., Cheseby, M., Lotti, R., Almasi, P., deMenocal, P., et al., 1997. A pervasive millennial-scale cycle in North Atlantic Holocene and glacial climates. Science 278, 1257–1266.

Bonn, W.J., Gingele, F.X., Grobe, H., Mackensen, A., Futterer, D.K., 1998. Palaeoproductivity at the Antarctic continental margin: opal and barium records for the last 400 ka. Paleogeography, Paleoclimatology, Paleoecology 139, 195–211.

Bouttes, N., Paillard, D., Roche, D.M., 2010. Impact of brine-induced stratification on the glacial carbon cycle. Climate of the Past 6, 575–589.

Boyle, E.A., 1988a. Cadmium: chemical tracer of deepwater paleoceanography. Paleoceanography 3, 471–489.

Boyle, E.A., 1988b. Vertical oceanic nutrient fractionation and glacial/interglacial CO_2 cycles. Nature 331, 55–56.

Boyle, E.A., Keigwin, L.D., 1982. Deep circulation of the North Atlantic over the last 200,000 years: geochemical evidence. Science 218, 784–787.

Bradley, S.L., Siddall, M., Milne, G.A., Masson-Delmotte, V., Wolff, E., 2012. Where might we find evidence of a Last Interglacial West Antarctic Ice Sheet collapse in Antarctic ice core records? Global and Planetary Change 88, 64–75.

Bradtmiller, L.I., McManus, J.F., Robinson, L.F., 2014. $^{231}Pa/^{230}Th$ evidence for a weakened but persistent Atlantic meridional overturning circulation during Heinrich Stadial 1. Nature Communications 5, 5817. Available from: https://doi.org/10.1038/ncomms6817.

Broecker, W., Clark, E., 2001. An evaluation of Lohmann's foraminifera weight dissolution index. Paleoceanography 16, 531–534.

Broecker, W.S., 1998. Paleocean circulation during the last deglaciation: a bipolar seesaw? Paleoceanography 13, 119–121.

Broecker, W.S., Bond, G., Klas, M., Bonani, G., Wolfli, W., 1990. A salt oscillator in the glacial Atlantic? 1. The concept. Paleoceanography 5, 469–477.

Broecker, W.S., Denton, G.H., 1989. The role of ocean-atmosphere reorganizations in glacial cycles. Geochimica et Cosmochimica Acta 53, 2465–2501.

Broecker, W.S., Peteet, D.M., Rind, D., 1985. Does the ocean-atmosphere system have more than one stable mode of operation? Nature 315, 21–26.

Bronselaer, B., Winton, M., Griffies, S.M., Hurlin, W.J., Rodgers, K.B., Sergienko, O.V., et al., 2018. Change in future climate due to Antarctic meltwater. Nature 564, 53–58.

Brook, E.J., Buizert, C., 2018. Antarctic and global climate history viewed from ice cores. Nature 558, 200–208.

Brovkin, V., Ganopolski, A., Archer, D., Rahmstorf, S., 2007. Lowering of glacial atmospheric CO_2 in response to changes in oceanic circulation and marine biogeochemistry. Paleoceanography 22, PA4202. Available from: https://doi.org/10.1029/2006PA001380.

Buizert, C., Schmittner, A., 2015. Southern Ocean control of glacial AMOC stability and Dansgaard-Oeschger interstadial duration. Paleoceanography 30, 1595–1612.

Buizert, C., Sigl, M., Severi, M., Markle, B.R., Wettstein, J.J., McConnell, J.R., et al., 2018. Abrupt ice-age shifts in southern westerly winds and Antarctic climate forced from the north. Nature 563, 681–685.

Burke, A., Robinson, L.F., 2012. The Southern Ocean's role in carbon exchange during the last deglaciation. Science 335, 557–561.

Capron, E., Govin, A., Stone, E.J., Masson-Delmotte, V., Mulitza, S., Otto-Bliesner, B., et al., 2014. Temporal and spatial structure of multi-millennial temperature changes at high latitudes during the Last Interglacial. Quaternary Science Reviews 103, 116–133.

Chadwick, M., Jones, J., Lawler, K., Prebble, J., Kohfeld, K., Crosta, X., 2019. Understanding glacial-interglacial changes in Southern Ocean sea ice. PAGES Magazine 27, 86.

Chalk, T.B., Hain, M.P., Foster, G.L., Rohling, E.J., Sexton, P.F., Badger, M.P.S., et al., 2017. Causes of ice age intensification across the Mid-Pleistocene Transition. Proceedings of the National Academy of Sciences 114, 13114–13119.

Chappell, J., Shackleton, N.J., 1986. Oxygen isotopes and sea level. Nature 324, 137–140.

Chase, Z., Anderson, R.F., Fleisher, M.Q., Kubik, P.W., 2002. The influence of particle composition and particle flux on scavenging of Th, Pa and Be in the ocean. Earth and Planetary Science Letters 204, 215–229.

Chen, T.Y., Robinson, L.F., Burke, A., Southon, J., Spooner, P., Morris, P.J., et al., 2015. Synchronous centennial abrupt events in the ocean and atmosphere during the last deglaciation. Science 349, 1537–1541.

Cheng, H., Edwards, R.L., Broecker, W.S., Denton, G.H., Kong, X.G., Wang, Y.J., et al., 2009. Ice Age terminations. Science 326, 248–252.

Cheng, H., Edwards, R.L., Sinha, A., Spötl, C., Yi, L., Chen, S., et al., 2016. The Asian monsoon over the past 640,000 years and ice age terminations. Nature 534, 640–646.

Clark, P.U., He, F., Golledge, N.R., Mitrovica, J.X., Dutton, A., Hoffman, J.S., et al., 2020. Oceanic forcing of penultimate deglacial and last interglacial sea-level rise. Nature 577, 660–664.

Clark, P.U., Pisias, N.G., 2000. Interpreting iceberg deposits in the deep sea. Science 290, 51.

Clark, P.U., Pollard, D., 1998. Origin of the middle Pleistocene transition by ice sheet erosion of regolith. Paleoceanography 13, 1–9.

Cody, R., Levy, R., Crampton, J., Naish, T., Wilson, G., Harwood, D., 2012. Selection and stability of quantitative stratigraphic age models: Plio-Pleistocene glaciomarine sediments in the ANDRILL 1B drillcore, McMurdo Ice Shelf. Global and Planetary Change 96–97, 143–156.

Colleoni, F., De Santis, L., Montoli, E., Olivo, E., Sorlien, C.C., Bart, P.J., et al., 2018. Past continental shelf evolution increased Antarctic ice sheet sensitivity to climatic conditions. Scientific Reports 8, 11323. Available from: https://doi.org/10.1038/s41598-018-29718-7.

Collins, L.G., Allen, C.S., Pike, J., Hodgson, D.A., Weckström, K., Massé, G., 2013. Evaluating highly branched isoprenoid (HBI) biomarkers as a novel Antarctic sea-ice proxy in deep ocean glacial age sediments. Quaternary Science Reviews 79, 87–98.

Collins, L.G., Pike, J., Allen, C.S., Hodgson, D.A., 2012. High-resolution reconstruction of southwest Atlantic sea-ice and its role in the carbon cycle during marine isotope stages 3 and 2. Paleoceanography 27, PA3217. Available from: https://doi.org/10.1029/2011PA002264.

Collins, G.E., Hogg, I.D., Convey, P., Sancho, L.G., Cowan, D.A., Lyons, W.B., et al., 2020. Genetic diversity of soil invertebrates corroborates timing estimates for past collapses of the West Antarctic Ice Sheet. Proceedings of the National Academy of Sciences 117, 22293–22302.

Colville, E.J., Carlson, A.E., Beard, B.L., Hatfield, R.G., Stoner, J.S., Reyes, A.V., et al., 2011. Sr-Nd-Pb isotope evidence for ice-sheet presence on southern Greenland during the Last Interglacial. Science 333, 620–623.

Conte, M.H., Sicre, M.A., Rühlemann, C., Weber, J.C., Schulte, S., Schulz-Bull, D., et al., 2006. Global temperature calibration of the alkenone unsaturation index ($U^{K'37}$) in surface waters and comparison with surface sediments. Geochemistry, Geophysics, Geosystems 7, Q02005. Available from: https://doi.org/10.1029/2005GC001054.

Cook, M.S., Keigwin, L.D., 2015. Radiocarbon profiles of the NW Pacific from the LGM and deglaciation: evaluating ventilation metrics and the effect of uncertain surface reservoir ages. Paleoceanography 30, 174–195.

Cook, C.P., van de Flierdt, T., Williams, T., Hemming, S.R., Iwai, M., Kobayashi, M., et al., 2013. Dynamic behaviour of the East Antarctic ice sheet during Pliocene warmthIODP Expedition 318 Scientists Nature Geoscience 6, 765–769.

Cook, C.P., Hill, D.J., van de Flierdt, T., Williams, T., Hemming, S.R., Dolan, A.M., et al., 2014. Sea surface temperature control on the distribution of far-traveled Southern Ocean ice-rafted detritus during the Pliocene. Paleoceanography 29, 533–548.

Cowan, E.A., Hillenbrand, C.-D., Hassler, L.E., Ake, M.T., 2008. Coarse-grained terrigenous sediment deposition on continental rise drifts: a record of Plio-Pleistocene glaciation on the Antarctic Peninsula. Palaeogeography, Palaeoclimatology, Palaeoecology 265, 275–291.

Croll, J., 1864. XIII. On the physical cause of the change of climate during geological epochs. The London, Edinburgh, and Dublin Philosophical Magazine and Journal of Science 28, 121–137.

Crosta, X., Sturm, A., Armand, L., Pichon, J.-J., 2004. Late Quaternary sea ice history in the Indian sector of the Southern Ocean as recorded by diatom assemblages. Marine Micropaleontology 50, 209–223.

Crutzen, P.J., Stoermer, E.F., 2000. The "Anthropocene.". Global Change Newsletter 41, 17–18.

Cuffey, K.M., Clow, G.D., Steig, E.J., Buizert, C., Fudge, T., Koutnik, M., et al., 2016. Deglacial temperature history of West Antarctica. Proceedings of the National Academy of Sciences 113, 14249–14254.

Curry, W.B., Oppo, D.W., 2005. Glacial water mass geometry and the distribution of $\delta^{13}C$ of ΣCO_2 in the western Atlantic Ocean. Paleoceanography 20, PA1017. Available from: https://doi.org/10.1029/2004pa001021.

Dahl-Jensen, D., Albert, M.R., Aldahan, A., Azuma, N., Balslev-Clausen, D., Baumgartner, M., et al., 2013. Eemian interglacial reconstructed from a Greenland folded ice core. Nature 493, 489–494.

Dansgaard, W., Johnsen, S.J., Clausen, H.B., Dahljensen, D., Gundestrup, N.S., Hammer, C.U., et al., 1993. Evidence for general instability of past climate from a 250-kyr ice-core record. Nature 364, 218–220.

de Boer, B., Lourens, L.J., van de Wal, R.S.W., 2014. Persistent 400,000-year variability of Antarctic ice volume and the carbon cycle is revealed throughout the Plio-Pleistocene. Nature Communications 5, 2999. Available from: https://doi.org/10.1038/ncomms3999.

de la Rocha, C.L., Brzezinski, M.A., DeNiro, M.J., Shemesh, A., 1998. Silicon-isotope composition of diatoms as an indicator of past oceanic change. Nature 395, 680–683.

De Santis, L., Prato, S., Brancolini, G., Lovo, M., Torelli, L., 1999. The Eastern Ross Sea continental shelf during the Cenozoic: implications for the West Antarctic ice sheet development. Global and Planetary Change 23, 173–196.

DeConto, R.M., Pollard, D., 2016. Contribution of Antarctica to past and future sea-level rise. Nature 531, 591–597.

DeConto, R.M., Pollard, D., Wilson, P.A., Pälike, H., Lear, C.H., Pagani, M., 2008. Thresholds for Cenozoic bipolar glaciation. Nature 455, 652–656.

DeConto, R.M., Pollard, D., Kowalewski, D., 2012. Modeling Antarctic ice sheet and climate variations during Marine Isotope Stage 31. Global and Planetary Change 96, 181–188.

Denton, G.H., Anderson, R.F., Toggweiler, J.R., Edwards, R.L., Schaefer, J.M., Putnam, A.E., 2010. The last glacial termination. Science 328, 1652–1656.

Di Roberto, A., Colizza, E., Del Carlo, P., Petrelli, M., Finocchiaro, F., Kuhn, G., 2019. First marine cryptotephra in Antarctica found in sediments of the western Ross Sea correlates with englacial tephras and climate records. Scientific Reports 9, 10628. Available from: https://doi.org/10.1038/s41598-019-47188-3.

Di Roberto, A.D., Scateni, B., Vincenzo, G.D., Petrelli, M., Fisauli, G., Barker, S.J., et al., 2021. Tephrochronology and provenance of an early Pleistocene (Calabrian) tephra from IODP Expedition 374 Site U1524, Ross Sea (Antarctica). Geochemistry, Geophysics, Geosystems, e2021GC009739. Available from: https://doi.org/10.1029/2021GC009739.

Domack, E.W., Jacobson, E.A., Shipp, S., Anderson, J.B., 1999. Late Pleistocene–Holocene retreat of the West Antarctic Ice-Sheet system in the Ross Sea: part 2-sedimentologic and stratigraphic signature. Geological Society of America Bulletin 111, 1517–1536.

Drysdale, R.N., Hellstrom, J.C., Zanchetta, G., Fallick, A.E., Goni, M.F.S., Couchoud, I., et al., 2009. Evidence for obliquity forcing of Glacial Termination II. Science 325, 1527–1531.

Duplessy, J.C., Roche, D.M., Kageyama, M., 2007. The deep ocean during the last interglacial period. Science 316, 89–91.

Dupont, T.K., Alley, R.B., 2005. Assessment of the importance of ice-shelf buttressing to ice-sheet flow. Geophysical Research Letters 32, L04503. Available from: https://doi.org/10.1029/2004GL022024.

Dutton, A., Lambeck, K., 2012. Ice volume and sea level during the last interglacial. Science 337, 216–219.

Dutton, A., Carlson, A.E., Long, A.J., Milne, G.A., Clark, P.U., DeConto, R., et al., 2015. Sea-level rise due to polar ice-sheet mass loss during past warm periods. Science 349, aaa4019. Available from: https://doi.org/10.1126/science.aaa4019.

Dyez, K.A., Hönisch, B., Schmidt, G.A., 2018. Early Pleistocene obliquity-scale pCO_2 variability at ~1.5 million years ago. Paleoceanography and Paleoclimatology 33, 1270–1291.

Edwards, T.L., Brandon, M.A., Durand, G., Edwards, N.R., Golledge, N.R., Holden, P.B., et al., 2019. Revisiting Antarctic ice loss due to marine ice-cliff instability. Nature 566, 58–64.

Egan, K.E., Rickaby, R.E.M., Leng, M.J., Hendry, K.R., Hermoso, M., Sloane, H.J., et al., 2012. Diatom silicon isotopes as a proxy for silicic acid utilisation: a Southern Ocean core top calibration. Geochimica et Cosmochimica Acta 96, 174–192.

Elderfield, H., Ganssen, G., 2000. Past temperature and $\delta^{18}O$ of surface ocean waters inferred from foraminiferal Mg/Ca ratios. Nature 405, 442–445.

Elderfield, H., Ferretti, P., Greaves, M., Crowhurst, S., McCave, I.N., Hodell, D., et al., 2012. Evolution of ocean temperature and ice volume through the Mid-Pleistocene Climate Transition. Science 337, 704–709.

Elderfield, H., Rickaby, R.E.M., 2000. Oceanic Cd/P ratio and nutrient utilization in the glacial Southern Ocean. Nature 405, 305–310.

Emiliani, C., 1955. Pleistocene temperatures. The Journal of Geology 63, 538–578.

EPICA Community Members, 2004. Eight glacial cycles from an Antarctic ice core. Nature 429, 623–628.

EPICA Community Members, 2006. One-to-one coupling of glacial climate variability in Greenland and Antarctica. Nature 444, 195–198.

Escutia, C., Brinkhuis, H., Klaus, A., Expedition 318 Scientists, 2011. Proceedings of IODP, 318, Wilkes Land glacial history. Integrated Ocean Drilling Program Management International, Tokyo.

Farmer, G.L., Licht, K., Swope, R.J., Andrews, J., 2006. Isotopic constraints on the provenance of fine-grained sediment in LGM tills from the Ross Embayment, Antarctica. Earth and Planetary Science Letters 249, 90–107.

Farmer, J.R., Hönisch, B., Haynes, L.L., Kroon, D., Jung, S., Ford, H.L., et al., 2019. Deep Atlantic Ocean carbon storage and the rise of 100,000-year glacial cycles. Nature Geoscience 12, 355–360.

Farneti, R., Delworth, T.L., Rosati, A.J., Griffies, S.M., Zeng, F., 2010. The role of mesoscale eddies in the rectification of the Southern Ocean response to climate change. Journal of Physical Oceanography 40, 1539–1557.

Ferraccioli, F., Armadillo, E., Jordan, T., Bozzo, E., Corr, H., 2009. Aeromagnetic exploration over the East Antarctic Ice Sheet: a new view of the Wilkes Subglacial Basin. Tectonophysics 478, 62–77.

Ferrari, R., Jansen, M.F., Adkins, J.F., Burke, A., Stewart, A.L., Thompson, A.F., 2014. Antarctic sea ice control on ocean circulation in present and glacial climates. Proceedings of the National Academy of Sciences of the United States of America 111, 8753–8758.

Ferry, A.J., Crosta, X., Quilty, P.G., Fink, D., Howard, W., Armand, L.K., 2015. First records of winter sea ice concentration in the southwest Pacific sector of the Southern Ocean. Paleoceanography 30, 1525–1539.

Fischer, H., Severinghaus, J., Brook, E., Wolff, E., Albert, M., 2013. Where to find 1.5 million yr old ice for the IPICS "Oldest Ice" ice core. Climate of the Past. 9, 2489–2505.

Fischer, H., Wahlen, M., Smith, J., Mastroianni, D., Deck, B., 1999. Ice core records of atmospheric CO_2 around the last three glacial terminations. Science 283, 1712–1714.

Fogwill, C.J., Phipps, S.J., Turney, C.S.M., Golledge, N.R., 2015. Sensitivity of the Southern Ocean to enhanced regional Antarctic ice sheet meltwater input. Earth's Future 3, 317–329.

Fogwill, C.J., Turney, C.S.M., Meissner, K.J., Golledge, N.R., Spence, P., Roberts, J.L., et al., 2014. Testing the sensitivity of the East Antarctic Ice Sheet to Southern Ocean dynamics: past changes and future implications. Journal of Quaternary Science 29, 91–98.

Ford, H.L., Raymo, M.E., 2020. Regional and global signals in seawater $\delta^{18}O$ records across the mid-Pleistocene transition. Geology 48, 113–117.

Foster, G.L., Rae, J.W.B., 2016. Reconstructing ocean pH with boron isotopes in foraminifera. Annual Review of Earth and Planetary Sciences 44, 207–237.

Francois, R., Altabet, M.A., Yu, E.-F., Sigman, D.M., Bacon, M.P., Frank, M., et al., 1997. Contribution of Southern Ocean surface-water stratification to low atmospheric CO_2 concentrations during the last glacial period. Nature 389, 929–935.

Francois, R., Honjo, S., Manganini, S.J., Ravizza, G.E., 1995. Biogenic barium fluxes to the deep sea: implications for paleoproductivity reconstruction. Global Biogeochemical Cycle 9, 289–303.

Fretwell, P., Pritchard, H.D., Vaughan, D.G., Bamber, J.L., Barrand, N.E., Bell, R., et al., 2013. Bedmap2: improved ice bed, surface and thickness datasets for Antarctica. Cryosphere 7, 375–393.

Fürst, J.J., Durand, G., Gillet-Chaulet, F., Tavard, L., Rankl, M., Braun, M., et al., 2016. The safety band of Antarctic ice shelves. Nature Climate Change 6, 479–482.

Galaasen, E.V., Ninnemann, U.S., Kessler, A., Irvalı, N., Rosenthal, Y., Tjiputra, J., et al., 2020. Interglacial instability of North Atlantic Deep Water ventilation. Science 367, 1485–1489.

Ganopolski, A., Rahmstorf, S., 2001. Rapid changes of glacial climate simulated in a coupled climate model. Nature 409, 153–158.

Gebbie, G., 2014. How much did Glacial North Atlantic Water shoal? Paleoceanography 29, 190–209.

Gebbie, G., Huybers, P., 2011. How is the ocean filled? Geophysical Research Letters 38, L06604. Available from: https://doi.org/10.1029/2011GL046769.

Gersonde, R., Barcena, M.A., 1998. Revision of the upper Pliocene-Pleistocene diatom biostratigraphy for the northern belt of the Southern Ocean. Micropaleontology 44, 84–98.

Gersonde, R., Crosta, X., Abelmann, A., Armand, L., 2005. Sea-surface temperature and sea ice distribution of the Southern Ocean at the EPILOG Last Glacial Maximum—a circum-Antarctic view based on siliceous microfossil records. Quaternary Science Reviews 24, 869–896.

Gersonde, R., Zielinski, U., 2000. The reconstruction of late Quaternary Antarctic sea-ice distribution—the use of diatoms as a proxy for sea-ice. Palaeogeography, Palaeoclimatology, Palaeoecology 162, 263–286.

Ghosh, P., Adkins, J., Affek, H., Balta, B., Guo, W., Schauble, E.A., et al., 2006. $^{13}C-^{18}O$ bonds in carbonate minerals: a new kind of paleothermometer. Geochimica et Cosmochimica Acta 70, 1439–1456.

Gilford, D.M., Ashe, E.L., DeConto, R.M., Kopp, R.E., Pollard, D., Rovere, A., 2020. Could the Last Interglacial constrain projections of future Antarctic ice mass loss and sea-level rise? Journal of Geophysical Research: Earth Surface 125. Available from: https://doi.org/10.1029/2019JF005418.

Gohl, K., Wellner, J.S., Klaus, A., Expedition 379 Scientists, 2019. Expedition 379 Preliminary Report: Amundsen Sea West Antarctic Ice Sheet History. International Ocean Discovery Program. <https://doi.org/10.14379/iodp.pr.379.2019>.

Goldstein, S.L., Hemming, S.R., 2003. Long-lived isotopic tracers in oceanography, paleoceanography and ice sheet dynamics. In: Elderfield, H. (Ed.), The Oceans and Marine Geochemistry. Elsevier-Pergamon, Oxford, pp. 453–489.

Golledge, N.R., Levy, R.H., McKay, R.M., Naish, T.R., 2017. East Antarctic ice sheet most vulnerable to Weddell Sea warming. Geophysical Research Letters 44, 2343–2351.

Golledge, N.R., Menviel, L., Carter, L., Fogwill, C.J., England, M.H., Cortese, G., et al., 2014. Antarctic contribution to meltwater pulse 1A from reduced Southern Ocean overturning. Nature Communications 5, 5107. Available from: https://doi.org/10.1038/ncomms6107.

Gomez, N., Weber, M.E., Clark, P.U., Mitrovica, J.X., Han, H.K., 2020. Antarctic ice dynamics amplified by Northern Hemisphere sea-level forcing. Nature 587, 600–604.

Gottschalk, J., Hodell, D.A., Skinner, L.C., Crowhurst, S.J., Jaccard, S.L., Charles, C., 2018. Past carbonate preservation events in the deep Southeast Atlantic Ocean (Cape Basin) and their implications for Atlantic overturning dynamics and marine carbon cycling. Paleoceanography and Paleoclimatology 33, 643–663.

Gottschalk, J., Skinner, L.C., Lippold, J., Vogel, H., Frank, N., Jaccard, S.L., et al., 2016. Biological and physical controls in the Southern Ocean on past millennial-scale atmospheric CO_2 changes. Nature Communications 7, 11539. Available from: https://doi.org/10.1038/ncomms11539.

Gow, A.J., Ueda, H.T., Garfield, D.E., 1968. Antarctic ice sheet: preliminary results of first core hole to bedrock. Science 161, 1011–1013.

Grant, K.M., Rohling, E.J., Ramsey, C.B., Cheng, H., Edwards, R.L., Florindo, F., et al., 2014. Sea-level variability over five glacial cycles. Nature Communications 5, 5076. Available from: https://doi.org/10.1038/ncomms6076.

Green, R.A., Menviel, L., Meissner, K.J., Crosta, X., 2020. Evaluating seasonal sea-ice cover over the Southern Ocean from the Last Glacial Maximum. Climate of the Past Discussions 1–23. Available from: https://doi.org/10.5194/cp-2020-155.

Grootes, P.M., Stuiver, M., White, J.W.C., Johnsen, S., Jouzel, J., 1993. Comparison of oxygen isotope records from the GISP2 and GRIP Greenland ice cores. Nature 366, 552–554.

Grousset, F.E., Biscaye, P.E., 2005. Tracing dust sources and transport patterns using Sr, Nd and Pb isotopes. Chemical Geology 222, 149–167.

Gulick, S.P.S., Shevenell, A.E., Montelli, A., Fernandez, R., Smith, C., Warny, S., et al., 2017. Initiation and long-term instability of the East Antarctic Ice Sheet. Nature 552, 225–229.

Hain, M.P., Sigman, D.M., Haug, G.H., 2010. Carbon dioxide effects of Antarctic stratification, North Atlantic Intermediate Water formation, and subantarctic nutrient drawdown during the last ice age: diagnosis and synthesis in a geochemical box model. Global Biogeochemical Cycle 24, GB4023. Available from: https://doi.org/10.1029/2010GB003790.

Hall, I.R., McCave, I.N., Shackleton, N.J., Weedon, G.P., Harris, S.E., 2001. Intensified deep Pacific inflow and ventilation in Pleistocene glacial times. Nature 412, 809–812.

Hambrey, M.J., McKelvey, B., 2000. Neogene fjordal sedimentation on the western margin of the Lambert Graben, East Antarctica. Sedimentology 47, 577–607.

Hasenfratz, A.P., Jaccard, S.L., Martínez-García, A., Sigman, D.M., Hodell, D.A., Vance, D., et al., 2019. The residence time of Southern Ocean surface waters and the 100,000-year ice age cycle. Science 363, 1080–1084.

Hay, C., Mitrovica, J.X., Gomez, N., Creveling, J.R., Austermann, J., Kopp, R.E., 2014. The sea-level fingerprints of ice-sheet collapse during interglacial periods. Quaternary Science Reviews 87, 60–69.

Hayes, C.T., Martínez-García, A., Hasenfratz, A.P., Jaccard, S.L., Hodell, D.A., Sigman, D.M., et al., 2014. A stagnation event in the deep South Atlantic during the last interglacial period. Science 346, 1514–1517.

Hays, J.D., Imbrie, J., Shackleton, N.J., 1976. Variations in the Earth's orbit: pacemaker of the ice ages. Science 194, 1121–1132.

He, F., Shakun, J.D., Clark, P.U., Carlson, A.E., Liu, Z., Otto-Bliesner, B.L., et al., 2013. Northern Hemisphere forcing of Southern Hemisphere climate during the last deglaciation. Nature 494, 81–85.

Hemming, S.R., 2004. Heinrich events: massive late Pleistocene detritus layers of the North Atlantic and their global climate imprint. Reviews of Geophysics 42, RG1005. Available from: https://doi.org/10.1029/2003rg000128.

Hemming, S.R., Broecker, W.S., Sharp, W.D., Bond, G.C., Gwiazda, R.H., McManus, J.F., et al., 1998. Provenance of Heinrich layers in core V28–82, northeastern Atlantic: $^{40}Ar/^{39}Ar$ ages of ice-rafted hornblende, Pb isotopes in feldspar grains, and Nd–Sr–Pb isotopes in the fine sediment fraction. Earth and Planetary Science Letters 164, 317–333.

Henderson, G.M., 2002. New oceanic proxies for paleoclimate. Earth and Planetary Science Letters 203, 1–13.

Henry, L.G., McManus, J.F., Curry, W.B., Roberts, N.L., Piotrowski, A.M., Keigwin, L.D., 2016. North Atlantic Ocean circulation and abrupt climate change during the last glaciation. Science 353, 470–474.

Higgins, J.A., Kurbatov, A.V., Spaulding, N.E., Brook, E., Introne, D.S., Chimiak, L.M., et al., 2015. Atmospheric composition 1 million years ago from blue ice in the Allan Hills, Antarctica. Proceedings of the National Academy of Sciences 112, 6887–6891.

Hillenbrand, C.-D., Ehrmann, W., 2005. Late Neogene to Quaternary environmental changes in the Antarctic Peninsula region: evidence from drift sediments. Global and Planetary Change 45, 165–191.

Hillenbrand, C.-D., Fütterer, D.K., Grobe, H., Frederichs, T., 2002. No evidence for a Pleistocene collapse of the West Antarctic Ice Sheet from continental margin sediments recovered in the Amundsen Sea. Geo-Marine Letters 22, 51–59.

Hillenbrand, C.-D., Moreton, S.G., Caburlotto, A., Pudsey, C.J., Lucchi, R.G., Smellie, J.L., et al., 2008. Volcanic time-markers for Marine Isotopic Stages 6 and 5 in Southern Ocean sediments and Antarctic ice cores: implications for tephra correlations between palaeoclimatic records. Quaternary Science Reviews 27, 518–540.

Hillenbrand, C.-D., Kuhn, G., Frederichs, T., 2009. Record of a Mid-Pleistocene depositional anomaly in West Antarctic continental margin sediments: an indicator for ice-sheet collapse? Quaternary Science Reviews 28, 1147–1159.

Hillenbrand, C.-D., Smith, J.A., Hodell, D.A., Greaves, M., Poole, C.R., Kender, S., et al., 2017. West Antarctic Ice Sheet retreat driven by Holocene warm water incursions. Nature 547, 43–48.

Hines, S.K.V., Southon, J.R., Adkins, J.F., 2015. A high-resolution record of Southern Ocean intermediate water radiocarbon over the past 30,000 years. Earth and Planetary Science Letters 432, 46–58.

Ho, S.L., Mollenhauer, G., Fietz, S., Martínez-Garcia, A., Lamy, F., Rueda, G., et al., 2014. Appraisal of TEX_{86} and TEX_{86}^{L} thermometries in subpolar and polar regions. Geochimica et Cosmochimica Acta 131, 213–226.

Ho, S.L., Mollenhauer, G., Lamy, F., Martínez-Garcia, A., Mohtadi, M., Gersonde, R., et al., 2012. Sea surface temperature variability in the Pacific sector of the Southern Ocean over the past 700 kyr. Paleoceanography 27, PA4202. Available from: https://doi.org/10.1029/2012PA002317.

Hodell, D.A., Channell, J.E.T., Curtis, J.H., Romero, O.E., Röhl, U., 2008. Onset of "Hudson Strait" Heinrich events in the eastern North Atlantic at the end of the middle Pleistocene transition ($\sim$640 ka)? Paleoceanography 23, PA4218. Available from: https://doi.org/10.1029/2008PA001591.

Hodell, D.A., Venz, K.A., Charles, C.D., Ninnemann, U.S., 2003. Pleistocene vertical carbon isotope and carbonate gradients in the South Atlantic sector of the Southern Ocean. Geochemistry, Geophysics, Geosystems 4, 1004. Available from: https://doi.org/10.1029/2002gc000367.

Hoffman, J.S., Clark, P.U., Parnell, A.C., He, F., 2017. Regional and global sea-surface temperatures during the last interglaciation. Science 355, 276–279.

Hogg, A., Southon, J., Turney, C., Palmer, J., Ramsey, C.B., Fenwick, P., et al., 2016. Punctuated shutdown of Atlantic meridional overturning circulation during Greenland Stadial 1. Scientific Reports 6, 25902. Available from: https://doi.org/10.1038/srep25902.

Holden, P.B., Edwards, N.R., Wolff, E.W., Lang, N.J., Singarayer, J.S., Valdes, P.J., et al., 2010. Interhemispheric coupling, the West Antarctic Ice Sheet and warm Antarctic interglacials. Climate of the Past 6, 431–443.

Holder, L., Duffy, M., Opdyke, B., Leventer, A., Post, A., O'Brien, P., et al., 2020. Controls since the Mid-Pleistocene Transition on sedimentation and primary productivity downslope of Totten Glacier, East Antarctica. Paleoceanography and Paleoclimatology 35. Available from: https://doi.org/10.1029/2020PA003981.

Holloway, M.D., Sime, L.C., Allen, C.S., Hillenbrand, C.D., Bunch, P., Wolff, E., et al., 2017. The spatial structure of the 128 ka Antarctic sea ice minimum. Geophysical Research Letters 44, 11129–11139.

Holloway, M.D., Sime, L.C., Singarayer, J.S., Tindall, J.C., Bunch, P., Valdes, P.J., 2016. Antarctic last interglacial isotope peak in response to sea ice retreat not ice-sheet collapse. Nature Communications 7, 12293. Available from: https://doi.org/10.1038/ncomms12293.

Hönisch, B., Hemming, N.G., Archer, D., Siddall, M., McManus, J.F., 2009. Atmospheric carbon dioxide concentration across the mid-Pleistocene transition. Science 324, 1551–1554.

Hönisch, B., Hemming, N.G., Grottoli, A.G., Amat, A., Hanson, G.N., Bijma, J., 2004. Assessing scleractinian corals as recorders for paleo-pH: empirical calibration and vital effects. Geochimica et Cosmochimica Acta 68, 3675–3685.

Hoogakker, B.A.A., Elderfield, H., Schmiedl, G., McCave, I.N., Rickaby, R.E.M., 2015. Glacial–interglacial changes in bottom-water oxygen content on the Portuguese margin. Nature Geoscience 8, 40–43.

Howe, J.N.W., Piotrowski, A.M., 2017. Atlantic deep water provenance decoupled from atmospheric CO_2 concentration during the lukewarm interglacials. Nature Communications 8, 2003. Available from: https://doi.org/10.1038/s41467-017-01939-w.

Huybers, P., 2006. Early Pleistocene glacial cycles and the integrated summer insolation forcing. Science 313, 508–511.

Huybers, P., 2011. Combined obliquity and precession pacing of late Pleistocene deglaciations. Nature 480, 229–232.

Huybers, P., Denton, G., 2008. Antarctic temperature at orbital timescales controlled by local summer duration. Nature Geoscience 1, 787–792.

Huybers, P., Wunsch, C., 2005. Obliquity pacing of the late Pleistocene glacial terminations. Nature 434, 491–494.

Imbrie, J., Berger, A., Boyle, E.A., Clemens, S.C., Duffy, A., Howard, W.R., et al., 1993. On the structure and origin of major glaciation cycles 2. The 100,000-year cycle. Paleoceanography 8, 699–735.

Imbrie, J., Boyle, E.A., Clemens, S.C., Duffy, A., Howard, W.R., Kukla, G., et al., 1992. On the structure and origin of major glaciation cycles 1. Linear responses to Milankovitch forcing. Paleoceanography 7, 701–738.

IPCC, 2019. Special report on the Ocean and Cryosphere in a changing climate [H.-O. Pörtner, D.C. Roberts, V. Masson-Delmotte, P. Zhai, M. Tignor, E. Poloczanska, K. Mintenbeck, A. Alegría, M. Nicolai, A. Okem, J. Petzold, B. Rama, N.M. Weyer (eds.)]. IPCC, Geneva.

Irvalı, N., Galaasen, E.V., Ninnemann, U.S., Rosenthal, Y., Born, A., Kleiven, H.K.F., 2020. A low climate threshold for south Greenland Ice Sheet demise during the Late Pleistocene. Proceedings of the National Academy of Sciences 117, 190–195.

Jaccard, S.L., Galbraith, E.D., Martínez-García, A., Anderson, R.F., 2016. Covariation of deep Southern Ocean oxygenation and atmospheric CO_2 through the last ice age. Nature 530, 207–210.

Jaccard, S.L., Galbraith, E.D., Sigman, D.M., Haug, G.H., Francois, R., Pedersen, T.F., et al., 2009. Subarctic Pacific evidence for a glacial deepening of the oceanic respired carbon pool. Earth and Planetary Science Letters 277, 156–165.

Jaccard, S.L., Hayes, C.T., Martínez-García, A., Hodell, D.A., Anderson, R.F., Sigman, D.M., et al., 2013. Two modes of change in Southern Ocean productivity over the past million years. Science 339, 1419–1423.

Jakob, K.A., Wilson, P.A., Pross, J., Ezard, T.H.G., Fiebig, J., Repschläger, J., et al., 2020. A new sea-level record for the Neogene/Quaternary boundary reveals transition to a more stable East Antarctic Ice Sheet. Proceedings of the National Academy of Sciences 117, 30980–30987.

Jamieson, S.S.R., Sugden, D.E., Hulton, N.R.J., 2010. The evolution of the subglacial landscape of Antarctica. Earth and Planetary Science Letters 293, 1–27.

Jansen, J.H.F., Kuijpers, A., Troelstra, S.R., 1986. A mid-Brunhes climatic event: long-term changes in global atmosphere and ocean circulation. Science 232, 619–622.

Jimenez-Espejo, F.J., Presti, M., Kuhn, G., Mckay, R., Crosta, X., Escutia, C., et al., 2020. Late Pleistocene oceanographic and depositional variations along the Wilkes Land margin (East Antarctica) reconstructed with geochemical proxies in deep-sea sediments. Global and Planetary Change 184, 103045. Available from: https://doi.org/10.1016/j.gloplacha.2019.103045.

Johnsen, S.J., Clausen, H.B., Dansgaard, W., Fuhrer, K., Gundestrup, N., Hammer, C.U., et al., 1992. Irregular glacial interstadials recorded in a new Greenland ice core. Nature 359, 311–313.

Jones, R.S., Norton, K.P., Mackintosh, A.N., Anderson, J.T.H., Kubik, P., Vockenhuber, C., et al., 2017. Cosmogenic nuclides constrain surface fluctuations of an East Antarctic outlet glacier since the Pliocene. Earth and Planetary Science Letters 480, 75–86.

Joughin, I., Smith, B.E., Medley, B., 2014. Marine ice sheet collapse potentially under way for the Thwaites Glacier Basin, West Antarctica. Science 344, 735–738.

Jouzel, J., 2013. A brief history of ice core science over the last 50 yr. Climate of the Past 9, 2525–2547.

Jouzel, J., Masson-Delmotte, V., 2010. Deep ice cores: the need for going back in time. Quaternary Science Reviews 29, 3683–3689.

Jouzel, J., Vimeux, F., Caillon, N., Delaygue, G., Hoffmann, G., Masson-Delmotte, V., et al., 2003. Magnitude of isotope/temperature scaling for interpretation of central Antarctic ice cores. Journal of Geophysical Research: Atmospheres 108, 4361. Available from: https://doi.org/10.1029/2002JD002677.

Jouzel, J., Masson-Delmotte, V., Cattani, O., Dreyfus, G., Falourd, S., Hoffmann, G., et al., 2007. Orbital and millennial Antarctic climate variability over the past 800,000 years. Science 317, 793–796.

Kanfoush, S.L., Hodell, D.A., Charles, C.D., Guilderson, T.P., Mortyn, P.G., Ninnemann, U.S., 2000. Millennial-scale instability of the Antarctic ice sheet during the last glaciation. Science 288, 1815–1818.

Katz, M.E., Cramer, B.S., Franzese, A., Honisch, B., Miller, K.G., Rosenthal, Y., et al., 2010. Traditional and emerging geochemical proxies in foraminifera. Journal of Foraminiferal Research 40, 165–192.

Kawamura, K., Parrenin, F., Lisiecki, L., Uemura, R., Vimeux, F., Severinghaus, J.P., et al., 2007. Northern Hemisphere forcing of climatic cycles in Antarctica over the past 360,000 years. Nature 448, 912–916.

Keeling, R.F., Stephens, B.B., 2001. Antarctic sea ice and the control of Pleistocene climate instability. Paleoceanography 16, 112–131.

Knorr, G., Lohmann, G., 2003. Southern Ocean origin for the resumption of Atlantic thermohaline circulation during deglaciation. Nature 424, 532–536.

Kohfeld, K.E., Chase, Z., 2017. Temporal evolution of mechanisms controlling ocean carbon uptake during the last glacial cycle. Earth and Planetary Science Letters 472, 206–215.

Konfirst, M.A., Scherer, R.P., Hillenbrand, C.-D., Kuhn, G., 2012. A marine diatom record from the Amundsen Sea - Insights into oceanographic and climatic response to the Mid-Pleistocene Transition in the West Antarctic sector of the Southern Ocean. Marine Micropaleontology 92, 40–51.

Kopp, R.E., Simons, F.J., Mitrovica, J.X., Maloof, A.C., Oppenheimer, M., 2009. Probabilistic assessment of sea level during the last interglacial stage. Nature 462, 863–867.

Kopp, R.E., Simons, F.J., Mitrovica, J.X., Maloof, A.C., Oppenheimer, M., 2013. A probabilistic assessment of sea level variations within the last interglacial stage. Geophysical Journal International 193, 711–716.

Korotkikh, E.V., Mayewski, P.A., Handley, M.J., Sneed, S.B., Introne, D.S., Kurbatov, A.V., et al., 2011. The last interglacial as represented in the glaciochemical record from Mount Moulton Blue Ice Area, West Antarctica. Quaternary Science Reviews 30, 1940–1947.

Kumar, N., Anderson, R.F., Mortlock, R.A., Froelich, P.N., Kubik, P., Dittrich-Hannen, B., et al., 1995. Increased biological productivity and export production in the glacial Southern Ocean. Nature 378, 675–680.

Laepple, T., Werner, M., Lohmann, G., 2011. Synchronicity of Antarctic temperatures and local solar insolation on orbital timescales. Nature 471, 91–94.

Laj, C., Kissel, C., Beer, J., 2004. High resolution global paleointensity stack since 75 kyr (GLOPIS-75) calibrated to absolute values, Timescales of the Paleomagnetic Field, Vol. 145. American Geophysical Union Geophysical Monograph Series, Washington, DC, pp. 255–265.

Lambert, F., Delmonte, B., Petit, J.R., Bigler, M., Kaufmann, P.R., Hutterli, M.A., et al., 2008. Dust-climate couplings over the past 800,000 years from the EPICA Dome C ice core. Nature 452, 616–619.

Lamy, F., Arz, H.W., Kilianc, R., Lange, C.B., Lembke-Jene, L., Wengler, M., et al., 2015. Glacial reduction and millennial-scale variations in Drake Passage throughflow. Proceedings of the National Academy of Sciences of the United States of America 112, 13496–13501.

Lamy, F., Gersonde, R., Winckler, G., Esper, O., Jaeschke, A., Kuhn, G., et al., 2014. Increased dust deposition in the Pacific Southern Ocean during glacial periods. Science 343, 403–407.

Lamy, F., Kaiser, J., Arz, H.W., Hebbeln, D., Ninnemann, U., Timm, O., et al., 2007. Modulation of the bipolar seesaw in the Southeast Pacific during Termination 1. Earth and Planetary Science Letters 259, 400–413.

Lamy, F., Winckler, G., Alvarez Zarikian, C.A., Expedition 383 Scientists, 2019. Expedition 383 Preliminary Report: Dynamics of the Pacific Antarctic Circumpolar Current. International Ocean Discovery Program. <https://doi.org/10.14379/iodp.pr.383.2019>.

Lang, D.C., Bailey, I., Wilson, P.A., Chalk, T.B., Foster, G.L., Gutjahr, M., 2016. Incursions of southern-sourced water into the deep North Atlantic during late Pliocene glacial intensification. Nature Geoscience 9, 375–379.

Lang, N., Wolff, E.W., 2011. Interglacial and glacial variability from the last 800 ka in marine, ice and terrestrial archives. Climate of the Past 7, 361–380.

Lawrence, K.T., Liu, Z., Herbert, T.D., 2006. Evolution of the eastern tropical Pacific through Plio-Pleistocene glaciation. Science 312, 79–83.

Lear, C.H., Elderfield, H., Wilson, P.A., 2000. Cenozoic deep-sea temperatures and global ice volumes from Mg/Ca in benthic foraminiferal calcite. Science 287, 269–272.

Levine, J.G., Yang, X., Jones, A.E., Wolff, E.W., 2014. Sea salt as an ice core proxy for past sea ice extent: a process-based model study. Journal of Geophysical Research: Atmospheres 119, 5737–5756.

Levy, R., Cody, R., Crampton, J., Fielding, C., Golledge, N., Harwood, D., et al., 2012. Late Neogene climate and glacial history of the Southern Victoria Land coast from integrated drill core, seismic and outcrop data. Global and Planetary Change 96–97, 157–180.

Licht, K.J., Hemming, S.R., 2017. Analysis of Antarctic glacigenic sediment provenance through geochemical and petrologic applications. Quaternary Science Reviews 164, 1–24.

Lilly, K., Fink, D., Fabel, D., Lambeck, K., 2010. Pleistocene dynamics of the interior East Antarctic ice sheet. Geology 38, 703–706.

Lisiecki, L.E., Raymo, M.E., 2005. A Pliocene-Pleistocene stack of 57 globally distributed benthic $\delta^{18}O$ records. Paleoceanography 20, PA1003. Available from: https://doi.org/10.1029/2004PA001071.

Lopes dos Santos, R.A., Spooner, M.I., Barrows, T.T., De Deckker, P., Sinninghe Damsté, J.S., Schouten, S., 2013. Comparison of organic ($U^{K'}_{37}$, TEX^{H}_{86}, LDI) and faunal proxies (foraminiferal assemblages) for reconstruction of late Quaternary sea surface temperature variability from offshore southeastern Australia. Paleoceanography 28, 377–387.

Lund, D.C., Adkins, J.F., Ferrari, R., 2011. Abyssal Atlantic circulation during the Last Glacial Maximum: constraining the ratio between transport and vertical mixing. Paleoceanography 26, PA1213. Available from: https://doi.org/10.1029/2010PA001938.

Luthi, D., Le Floch, M., Bereiter, B., Blunier, T., Barnola, J.M., Siegenthaler, U., et al., 2008. High-resolution carbon dioxide concentration record 650,000–800,000 years before present. Nature 453, 379–382.

Lynch-Stieglitz, J., Fairbanks, R.G., 1994. A conservative tracer for glacial ocean circulation from carbon isotope and palaeo-nutrient measurements in benthic foraminifera. Nature 369, 308–310.

Mackensen, A., Schmiedl, G., 2019. Stable carbon isotopes in paleoceanography: atmosphere, oceans, and sediments. Earth-Science Reviews 197, 102893. Available from: https://doi.org/10.1016/j.earscirev.2019.102893.

Mackintosh, A., Golledge, N., Domack, E., Dunbar, R., Leventer, A., White, D., et al., 2011. Retreat of the East Antarctic ice sheet during the last glacial termination. Nature Geoscience 4, 195–202.

Marchitto, T.M., Broecker, W.S., 2006. Deep water mass geometry in the glacial Atlantic Ocean: a review of constraints from the paleonutrient proxy Cd/Ca. Geochemistry, Geophysics, Geosystems 7, Q12003. Available from: https://doi.org/10.1029/2006GC001323.

Marchitto, T.M., Curry, W.B., Lynch-Stieglitz, J., Bryan, S.P., Cobb, K.M., Lund, D.C., 2014. Improved oxygen isotope temperature calibrations for cosmopolitan benthic foraminifera. Geochimica et Cosmochimica Acta 130, 1–11.

Marcott, S.A., Bauska, T.K., Buizert, C., Steig, E.J., Rosen, J.L., Cuffey, K.M., et al., 2014. Centennial-scale changes in the global carbon cycle during the last deglaciation. Nature 514, 616–619.

Marino, G., Rohling, E.J., Rodríguez-Sanz, L., Grant, K.M., Heslop, D., Roberts, A.P., et al., 2015. Bipolar seesaw control on last interglacial sea level. Nature 522, 197–201.

Marino, G., Zahn, R., Ziegler, M., Purcell, C., Knorr, G., Hall, I.R., et al., 2013. Agulhas salt-leakage oscillations during abrupt climate changes of the Late Pleistocene. Paleoceanography 28, 599–606.

Mariotti, V., Paillard, D., Bopp, L., Roche, D.M., Bouttes, N., 2016. A coupled model for carbon and radiocarbon evolution during the last deglaciation. Geophysical Research Letters 43, 1306–1313.

Martin, D.F., Cornford, S.L., Payne, A.J., 2019. Millennial-Scale Vulnerability of the Antarctic Ice Sheet to Regional Ice Shelf Collapse. Geophysical Research Letters 46, 1467–1475.

Martin, J.H., 1990. Glacial-interglacial CO_2 change: the iron hypothesis. Paleoceanography 5, 1–13.

Martínez-Botí, M.A., Foster, G.L., Chalk, T.B., Rohling, E.J., Sexton, P.F., Lunt, D.J., et al., 2015a. Plio-Pleistocene climate sensitivity evaluated using high-resolution CO_2 records. Nature 518, 49–54.

Martínez-Botí, M.A., Marino, G., Foster, G.L., Ziveri, P., Henehan, M.J., Rae, J.W.B., et al., 2015b. Boron isotope evidence for oceanic carbon dioxide leakage during the last deglaciation. Nature 518, 219–222.

Martinez-Garcia, A., Rosell-Mele, A., Geibert, W., Gersonde, R., Masque, P., Gaspari, V., et al., 2009. Links between iron supply, marine productivity, sea surface temperature, and CO_2 over the last 1.1 Ma. Paleoceanography 24, PA1207. Available from: https://doi.org/10.1029/2008PA001657.

Martinez-Garcia, A., Rosell-Melé, A., McClymont, E.L., Gersonde, R., Haug, G.H., 2010. Subpolar link to the emergence of the modern equatorial Pacific cold tongue. Science 328, 1550–1553.

Martinez-Garcia, A., Rosell-Mele, A., Jaccard, S.L., Geibert, W., Sigman, D.M., Haug, G.H., 2011. Southern Ocean dust-climate coupling over the past four million years. Nature 476, 312–315.

Martinez-Garcia, A., Sigman, D.M., Ren, H., Anderson, R.F., Straub, M., Hodell, D.A., et al., 2014. Iron fertilization of the Subantarctic Ocean during the last ice age. Science 343, 1347–1350.

Martinson, D.G., Pisias, N.G., Hays, J.D., Imbrie, J., Moore, T.C., Shackleton, N.J., 1987. Age dating and the orbital theory of the Ice Ages: development of a high-resolution 0 to 300,000-year chronostratigraphy. Quaternary Research 27, 1–29.

Martrat, B., Grimalt, J.O., Shackleton, N.J., de Abreu, L., Hutterli, M.A., Stocker, T.F., 2007. Four climate cycles of recurring deep and surface water destabilizations on the Iberian margin. Science 317, 502–507.

Marzocchi, A., Jansen, M.F., 2019. Global cooling linked to increased glacial carbon storage via changes in Antarctic sea ice. Nature Geoscience 12. Available from: https://doi.org/10.1038/s41561-019-0466-8.

Mas e Braga, M., Bernales, J., Prange, M., Stroeven, A.P., Rogozhina, I., 2021. Sensitivity of the Antarctic ice sheets to the warming of marine isotope substage 11c. The Cryosphere 15, 459–478.

Maslin, M.A., Brierley, C.M., 2015. The role of orbital forcing in the Early Middle Pleistocene Transition. Quaternary International 389, 47–55.

Masson-Delmotte, V., Schulz, M., Abe-Ouchi, A., Beer, J., Ganopolski, A., González Rouco, J., et al., 2013. Information from paleoclimate archives, Climate Change 2013: the physical science basis. In: Stocker, T.F., Qin, D., Plattner, G.-K., Tignor, M., Allen, S.K., Boschung, J., Nauels, A., Xia, Y., Bex, V., Midgley, P.M. (Eds.), Contribution of Working Group I to the Fifth Assessment Report of the Intergovernmental Panel on Climate Change. Cambridge University Press, Cambridge and New York.

McCave, I.N., Crowhurst, S.J., Kuhn, G., Hillenbrand, C.D., Meredith, M.P., 2013. Minimal change in Antarctic Circumpolar Current flow speed between the last glacial and Holocene. Nature Geoscience 7, 113–116.

McCave, I.N., Hall, I.R., 2006. Size sorting in marine muds: processes, pitfalls, and prospects for paleoflow-speed proxies. Geochemistry, Geophysics, Geosystems. 7, Q10N05. Available from: https://doi.org/10.1029/2006GC001284.

McCave, I.N., Manighetti, B., Robinson, S.G., 1995. Sortable silt and fine sediment size composition slicing – parameters for paleocurrent speed and paleoceanography. Paleoceanography 10, 593–610.

McCave, I.N., Thornalley, D.J.R., Hall, I.R., 2017. Relation of sortable silt grain-size to deep-sea current speeds: calibration of the 'Mud Current Meter'. Deep Sea Research Part I: Oceanographic Research Papers 127, 1–12.

McClymont, E.L., Elmore, A.C., Kender, S., Leng, M.J., Greaves, M., Elderfield, H., 2016. Pliocene-Pleistocene evolution of sea surface and intermediate water temperatures from the southwest Pacific. Paleoceanography 31, 895–913.

McCorkle, D.C., Emerson, S.R., 1988. The relationship between pore water carbon isotopic composition and bottom water oxygen concentration. Geochimica et Cosmochimica Acta 52, 1169–1178.

McGee, D., Winckler, G., Borunda, A., Serno, S., Anderson, R.F., Recasens, C., et al., 2016. Tracking eolian dust with helium and thorium: impacts of grain size and provenance. Geochimica et Cosmochimica Acta 175, 47–67.

McKay, N.P., Overpeck, J.T., Otto-Bliesner, B.L., 2011. The role of ocean thermal expansion in Last Interglacial sea level rise. Geophysical Research Letters 38, L14605. Available from: https://doi.org/10.1029/2011GL048280.

McKay, R., Browne, G., Carter, L., Cowan, E., Dunbar, G., Krissek, L., et al., 2009. The stratigraphic signature of the late Cenozoic Antarctic Ice Sheets in the Ross Embayment. Geological Society of America Bulletin 121, 1537–1561.

McKay, R., Golledge, N.R., Maas, S., Naish, T., Levy, R., Dunbar, G., et al., 2016. Antarctic marine ice-sheet retreat in the Ross Sea during the early Holocene. Geology 44, 7–10.

McKay, R., Naish, T., Carter, L., Riesselman, C., Dunbar, R., Sjunneskog, C., et al., 2012a. Antarctic and Southern Ocean influences on Late Pliocene global cooling. Proceedings of the National Academy of Sciences 109, 6423–6428.

McKay, R., Naish, T., Powell, R., Barrett, P., Scherer, R., Talarico, F., et al., 2012b. Pleistocene variability of Antarctic Ice Sheet extent in the Ross Embayment. Quaternary Science Reviews 34, 93–112.

McKay, R.M., De Santis, L., Kulhanek, D.K., Expedition 374 Scientists, 2019. Ross Sea West Antarctic Ice Sheet History. In Proceedings of the International Ocean Discovery Program, vol. 374. International Ocean Discovery Program, College Station, TX. doi: 10.14379/iodp.proc.374.2019.

McManus, J.F., Oppo, D.W., Cullen, J.L., 1999. A 0.5-million-year record of millennial-scale climate variability in the North Atlantic. Science 283, 971–975.

Melles, M., Brigham-Grette, J., Minyuk, P.S., Nowaczyk, N.R., Wennrich, V., DeConto, R.M., et al., 2012. 2.8 million years of Arctic climate change from Lake El'gygytgyn, NE Russia. Science 337, 315–320.

Mengel, M., Levermann, A., 2014. Ice plug prevents irreversible discharge from East Antarctica. Nature Climate Change 4, 451–455.

Menviel, L., Spence, P., Yu, J., Chamberlain, M.A., Matear, R.J., Meissner, K.J., et al., 2018. Southern Hemisphere westerlies as a driver of the early deglacial atmospheric CO_2 rise. Nature Communications 9, 2503. Available from: https://doi.org/10.1038/s41467-018-04876-4.

Mercer, J.H., 1978. West Antarctic ice sheet and CO_2 greenhouse effect: a threat of disaster. Nature 271, 321–325.

Milankovitch, M.M., 1941. Canon of insolation and the ice age problem. Royal Serbian Academy Special Publication 132.

Monnin, E., Indermuhle, A., Dallenbach, A., Fluckiger, J., Stauffer, B., Stocker, T.F., et al., 2001. Atmospheric CO_2 concentrations over the last glacial termination. Science 291, 112–114.

Morlighem, M., Rignot, E., Binder, T., Blankenship, D., Drews, R., Eagles, G., et al., 2020. Deep glacial troughs and stabilizing ridges unveiled beneath the margins of the Antarctic ice sheet. Nature Geoscience 13, 132–137.

Mortlock, R.A., Charles, C.D., Froelich, P.N., Zibello, M.A., Saltzman, J., Hays, J.D., et al., 1991. Evidence for lower productivity in the Antarctic Ocean during the last glaciation. Nature 351, 220–223.

Mortyn, P.G., Charles, C.D., Ninnemann, U.S., Ludwig, K., Hodell, D.A., 2003. Deep sea sedimentary analogs for the Vostok ice core. Geochemistry, Geophysics, Geosystems 4, 8405. Available from: https://doi.org/10.1029/2002GC000475.

Mulitza, S., Chiessi, C.M., Schefuß, E., Lippold, J., Wichmann, D., Antz, B., et al., 2017. Synchronous and proportional deglacial changes in Atlantic meridional overturning and northeast Brazilian precipitation. Paleoceanography 32, 622–633.

Muschitiello, F., D'Andrea, W.J., Schmittner, A., Heaton, T.J., Balascio, N.L., DeRoberts, N., et al., 2019. Deep-water circulation changes lead North Atlantic climate during deglaciation. Nature Communications 10, 1272. Available from: https://doi.org/10.1038/s41467-019-09237-3.

Nadeau, L.-P., Ferrari, R., Jansen, M.F., 2019. Antarctic sea ice control on the depth of North Atlantic Deep Water. Journal of Climate 32, 2537–2551.

Naish, T.R., Abbott, S.T., Alloway, V., Beu, A.G., Carter, R.M., Edwards, A.R., et al., 1998. Astronomical calibration of a southern hemisphere Plio-Pleistocene reference section, Wanganui Basin, New Zealand. Quaternary Science Reviews 17, 695–710. Available from: https://doi.org/10.1016/S0277-3791(97)00075-9.

Naish, T., Powell, R., Levy, R., Wilson, G., Scherer, R., Talarico, F., et al., 2009. Obliquity-paced Pliocene West Antarctic ice sheet oscillations. Nature 458, 322–328.

Nakayama, Y., Manucharyan, G., Zhang, H., Dutrieux, P., Torres, H.S., Klein, P., et al., 2019. Pathways of ocean heat towards Pine Island and Thwaites grounding lines. Scientific Reports 9, 16649. Available from: https://doi.org/10.1038/s41598-019-53190-6.

Narcisi, B., Petit, J.-R., Delmonte, B., Basile-Doelsch, I., Maggi, V., 2005. Characteristics and sources of tephra layers in the EPICA-Dome C ice record (East Antarctica): implications for past atmospheric circulation and ice core stratigraphic correlations. Earth and Planetary Science Letters 239, 253–265.

NGRIP Members, 2004. High-resolution record of Northern Hemisphere climate extending into the last interglacial period. Nature 431, 147–151.

Nichols, K.A., Goehring, B.M., Balco, G., Johnson, J.S., Hein, A.S., Todd, C., 2019. New Last Glacial Maximum ice thickness constraints for the Weddell Sea Embayment, Antarctica. The Cryosphere 13, 2935–2951.

Nürnberg, D., Bijma, J., Hemleben, C., 1996. Assessing the reliability of magnesium in foraminiferal calcite as a proxy for water mass temperatures. Geochimica et Cosmochimica Acta 60, 803–814.

Ohkouchi, N., Eglinton, T.I., Hayes, J.M., 2003. Radiocarbon dating of individual fatty acids as a tool for refining Antarctic margin sediment chronologies. Radiocarbon 45, 17–24.

Oliver, K.I.C., Hoogakker, B.A.A., Crowhurst, S., Henderson, G.M., Rickaby, R.E.M., Edwards, N.R., et al., 2010. A synthesis of marine sediment core $\delta^{13}C$ data over the last 150 000 years. Climate of the Past 6, 645–673.

Paillard, D., Parrenin, F., 2004. The Antarctic ice sheet and the triggering of deglaciations. Earth and Planetary Science Letters 227, 263–271.

Paolo, F.S., Fricker, H.A., Padman, L., 2015. Volume loss from Antarctic ice shelves is accelerating. Science 348, 327–331.

Parrenin, F., Masson-Delmotte, V., Köhler, P., Raynaud, D., Paillard, D., Schwander, J., et al., 2013. Synchronous change of atmospheric CO_2 and Antarctic temperature during the last deglacial warming. Science 339, 1060–1063.

Passchier, S., O'Brien, P.E., Damuth, J.E., Januszczak, N., Handwerger, D.A., Whitehead, J.M., 2003. Pliocene–Pleistocene glaciomarine sedimentation in eastern Prydz Bay and development of the Prydz trough-mouth fan, ODP Sites 1166 and 1167, East Antarctica. Marine Geology 199, 279–305.

Patterson, M.O., McKay, R., Naish, T., Escutia, C., Jimenez-Espejo, F.J., Raymo, M.E., et al., 2014. Orbital forcing of the East Antarctic ice sheet during the Pliocene and Early PleistoceneIODP Expedition 318 Scientists Nature Geoscience 7, 841–847.

Paxman, G.J.G., Gasson, E.G.W., Jamieson, S.S.R., Bentley, M.J., Ferraccioli, F., 2020. Long-term increase in Antarctic Ice Sheet vulnerability driven by bed topography evolution. Geophysical Research Letters 47 Available from: https://doi.org/10.1029/2020GL090003.

Paxman, G.J.G., Jamieson, S.S.R., Hochmuth, K., Gohl, K., Bentley, M.J., Leitchenkov, G., et al., 2019. Reconstructions of Antarctic topography since the Eocene–Oligocene boundary. Paleogeography, Paleoclimatology, Paleoecology 535, 109346. Available from: https://doi.org/10.1016/j.palaeo.2019.109346.

Paytan, A., Griffith, E.M., 2007. Marine barite: recorder of variations in ocean export productivity. Deep Sea Research Part II: Topical Studies in Oceanography 54, 687–705.

Pearson, P.N., 2012. Oxygen isotopes in foraminifera: overview and historical review. The Paleontological Society Papers 18, 1–38.

Peeters, F.J.C., Acheson, R., Brummer, G.J.A., de Ruijter, W.P.M., Schneider, R.R., Ganssen, G. M., et al., 2004. Vigorous exchange between the Indian and Atlantic oceans at the end of the past five glacial periods. Nature 430, 661–665.

Pena, L.D., Goldstein, S.L., 2014. Thermohaline circulation crisis and impacts during the mid-Pleistocene transition. Science 345, 318–322.

Peterson, L.C., Haug, G.H., Hughen, K.A., Röhl, U., 2000. Rapid changes in the hydrologic cycle of the tropical Atlantic during the last glacial. Science 290, 1947–1951.

Petit, J.R., Jouzel, J., Raynaud, D., Barkov, N.I., Barnola, J.M., Basile, I., et al., 1999. Climate and atmospheric history of the past 420,000 years from the Vostok ice core, Antarctica. Nature 399, 429–436.

Phipps, S.J., Fogwill, C.J., Turney, C.S.M., 2016. Impacts of marine instability across the East Antarctic Ice Sheet on Southern Ocean dynamics. The Cryosphere 10, 2317–2328.

Piotrowski, A.M., Goldstein, S.L., Hemming, S.R., Fairbanks, R.G., 2005. Temporal relationships of carbon cycling and ocean circulation at glacial boundaries. Science 307, 1933–1938.

Pollard, D., DeConto, R.M., 2009. Modelling West Antarctic ice sheet growth and collapse through the past five million years. Nature 458, 329–332.

Pollard, D., DeConto, R.M., Alley, R.B., 2015. Potential Antarctic Ice Sheet retreat driven by hydrofracturing and ice cliff failure. Earth and Planetary Science Letters 412, 112–121.

Polyak, V.J., Onac, B.P., Fornós, J.J., Hay, C., Asmerom, Y., Dorale, J.A., et al., 2018. A highly resolved record of relative sea level in the western Mediterranean Sea during the last interglacial period. Nature Geoscience 11, 860–864.

Pritchard, H.D., Ligtenberg, S.R.M., Fricker, H.A., Vaughan, D.G., van den Broeke, M.R., Padman, L., 2012. Antarctic ice-sheet loss driven by basal melting of ice shelves. Nature 484, 502–505.

Prothro, L.O., Majewski, W., Yokoyama, Y., Simkins, L.M., Anderson, J.B., Yamane, M., et al., 2020. Timing and pathways of East Antarctic Ice Sheet retreat. Quaternary Science Reviews 230, 106166. Available from: https://doi.org/10.1016/j.quascirev.2020.106166.

Pugh, R.S., McCave, I.N., Hillenbrand, C.-D., Kuhn, G., 2009. Circum-Antarctic age modelling of Quaternary marine cores under the Antarctic Circumpolar Current: ice-core dust–magnetic correlation. Earth and Planetary Science Letters 284, 113–123.

Quilty, P.G., Lirio, J.M., Jillett, D., 2000. Stratigraphy of the Pliocene Sørsdal Formation, Marine Plain, Vestfold Hills, East Antarctica. Antarctic Science 12, 205–216.

Rae, J.W.B., Burke, A., Robinson, L.F., Adkins, J.F., Chen, T., Cole, C., et al., 2018. CO_2 storage and release in the deep Southern Ocean on millennial to centennial timescales. Nature 562, 569–573.

Rae, J.W.B., Foster, G.L., Schmidt, D.N., Elliott, T., 2011. Boron isotopes and B/Ca in benthic foraminifera: proxies for the deep ocean carbonate system. Earth and Planetary Science Letters 302, 403–413.

Ragueneau, O., Tréguer, P., Leynaert, A., Anderson, R.F., Brzezinski, M.A., DeMaster, D.J., et al., 2000. A review of the Si cycle in the modern ocean: recent progress and missing gaps in the application of biogenic opal as a paleoproductivity proxy. Global and Planetary Change 26, 317–365.

RAISED Consortium, 2014. A community-based geological reconstruction of Antarctic Ice Sheet deglaciation since the Last Glacial Maximum. Quaternary Science Reviews 100, 1–9.

Rampen, S.W., Willmott, V., Kim, J.-H., Uliana, E., Mollenhauer, G., Schefuß, E., et al., 2012. Long chain 1, 13-and 1, 15-diols as a potential proxy for palaeotemperature reconstruction. Geochimica et Cosmochimica Acta 84, 204–216.

Raymo, M.E., 1994. The initiation of Northern Hemisphere glaciation. Annual Review of Earth and Planetary Sciences 22, 353–383.

Raymo, M.E., Lisiecki, L.E., Nisancioglu, K.H., 2006. Plio-Pleistocene ice volume, Antarctic climate, and the global δ^{18} O record. Science 313, 492–495.

Raymo, M.E., Mitrovica, J.X., 2012. Collapse of polar ice sheets during the stage 11 interglacial. Nature 483, 453–456.

Reimer, P.J., Bard, E., Bayliss, A., Beck, J.W., Blackwell, P.G., Ramsey, C.B., et al., 2013. IntCal13 and Marine13 radiocarbon age calibration curves 0–50,000 years cal BP. Radiocarbon 55, 1869–1887.

Reinardy, B.T.I., Escutia, C., Iwai, M., Jimenez-Espejo, F.J., Cook, C., de Flierdt, T.V., et al., 2015. Repeated advance and retreat of the East Antarctic Ice Sheet on the continental shelf during the early Pliocene warm period. Paleogeography, Paleoclimatology, Paleoecology 422, 65–84.

Reyes, A.V., Carlson, A.E., Beard, B.L., Hatfield, R.G., Stoner, J.S., Winsor, K., et al., 2014. South Greenland ice-sheet collapse during marine isotope stage 11. Nature 510, 525–528.

Rial, J.A., Oh, J., Reischmann, E., 2013. Synchronization of the climate system to eccentricity forcing and the 100,000-year problem. Nature Geoscience 6, 289–293.

Rickaby, R.E.M., Elderfield, H., Roberts, N., Hillenbrand, C.D., Mackensen, A., 2010. Evidence for elevated alkalinity in the glacial Southern Ocean. Paleoceanography 25, PA1209. Available from: https://doi.org/10.1029/2009PA001762.

Rignot, E., Casassa, G., Gogineni, P., Krabill, W., Rivera, A.U., Thomas, R., 2004. Accelerated ice discharge from the Antarctic Peninsula following the collapse of Larsen B ice shelf. Geophysical Research Letters 31, L18401. Available from: https://doi.org/10.1029/2004GL020697.

Rignot, E., Mouginot, J., Morlighem, M., Seroussi, H., Scheuchl, B., 2014. Widespread, rapid grounding line retreat of Pine Island, Thwaites, Smith, and Kohler glaciers, West Antarctica, from 1992 to 2011. Geophysical Research Letters 41, 3502–3509.

Rintoul, S.R., Chown, S.L., DeConto, R.M., England, M.H., Fricker, H.A., Masson-Delmotte, V., et al., 2018. Choosing the future of Antarctica. Nature 558, 233–241.

Ritz, C., Edwards, T.L., Durand, G., Payne, A.J., Peyaud, V., Hindmarsh, R.C.A., 2015. Potential sea-level rise from Antarctic ice-sheet instability constrained by observations. Nature 528, 115–118.

Roberts, N.L., Piotrowski, A.M., McManus, J.F., Keigwin, L.D., 2010. Synchronous deglacial overturning and water mass source changes. Science 327, 75–78.

Robinson, L.F., Adkins, J.F., Frank, N., Gagnon, A.C., Prouty, N.G., Roark, E.B., et al., 2014. The geochemistry of deep-sea coral skeletons: a review of vital effects and applications for palaeoceanography. Deep Sea Research Part II: Topical Studies in Oceanography 99, 184–198.

Robinson, L.F., Siddall, M., 2012. Palaeoceanography: motivations and challenges for the future. Philosophical Transactions of the Royal Society A: Mathematical, Physical and Engineering Sciences 370, 5540–5566.

Robinson, L.F., van de Flierdt, T., 2009. Southern Ocean evidence for reduced export of North Atlantic Deep Water during Heinrich event 1. Geology 37, 195–198.

Roche, D.M., Crosta, X., Renssen, H., 2012. Evaluating Southern Ocean sea-ice for the Last Glacial Maximum and pre-industrial climates: PMIP-2 models and data evidence. Quaternary Science Reviews 56, 99–106.

Rodríguez-Sanz, L., Mortyn, P.G., Martínez-Garcia, A., Rosell-Mele, A., Hall, I.R., 2012. Glacial Southern Ocean freshening at the onset of the Middle Pleistocene climate transition. Earth and Planetary Science Letters 345, 194–202.

Rohling, E.J., 2000. Paleosalinity: confidence limits and future applications. Marine Geology 163, 1–11.

Rohling, E.J., Foster, G.L., Grant, K.M., Marino, G., Roberts, A.P., Tamisiea, M.E., et al., 2014. Sea-level and deep-sea-temperature variability over the past 5.3 million years. Nature 508, 477–482.

Rohling, E.J., Hibbert, F.D., Williams, F.H., Grant, K.M., Marino, G., Foster, G.L., et al., 2017. Differences between the last two glacial maxima and implications for ice-sheet, $\delta^{18}O$, and sea-level reconstructions. Quaternary Science Reviews 176, 1–28.

Rohling, E.J., Hibbert, F.D., Grant, K.M., Galaasen, E.V., Irvalı, N., Kleiven, H.F., et al., 2019. Asynchronous Antarctic and Greenland ice-volume contributions to the last interglacial sea-level highstand. Nature Communications 10, 5040. Available from: https://doi.org/10.1038/s41467-019-12874-3.

Rohling, E.J., Marsh, R., Wells, N.C., Siddall, M., Edwards, N.R., 2004. Similar meltwater contributions to glacial sea level changes from Antarctic and northern ice sheets. Nature 430, 1016–1021.

Rosenheim, B.E., Santoro, J.A., Gunter, M., Domack, E.W., 2013. Improving Antarctic sediment ^{14}C dating using ramped pyrolysis: an example from the Hugo Island Trough. Radiocarbon 55, 115–126.

Roy, M., van de Flierdt, T., Hemming, S.R., Goldstein, S.L., 2007. $^{40}Ar/^{39}Ar$ ages of hornblende grains and bulk Sm/Nd isotopes of circum-Antarctic glacio-marine sediments: implications for sediment provenance in the southern ocean. Chemical Geology 244, 507–519.

Ruddiman, W.F., 1977. Late Quaternary deposition of ice-rafted sand in the subpolar North Atlantic (lat 40° to 65°N). Geological Society of America Bulletin 88, 1813–1827.

Ruddiman, W.F., 2006. Orbital changes and climate. Quaternary Science Reviews 25, 3092–3112.

Ruddiman, W.F., Raymo, M.E., Martinson, D.G., Clement, B.M., Backman, J., 1989. Pleistocene evolution: Northern Hemisphere ice sheets and North Atlantic Ocean. Paleoceanography 4, 353–412.

Sachs, J.P., Anderson, R.F., 2005. Increased productivity in the subantarctic ocean during Heinrich events. Nature 434, 1118–1121.

Sachs, J.P., Lehman, S.J., 1999. Subtropical North Atlantic temperatures 60,000 to 30,000 years ago. Science 286, 756–759.

Sanyal, A., Hemming, N.G., Broecker, W.S., Hanson, G.N., 1997. Changes in pH in the eastern equatorial Pacific across stage 5–6 boundary based on boron isotopes in foraminifera. Global Biogeochemical Cycle 11, 125–133.

Scherer, R.P., Aldahan, A., Tulaczyk, S., Possnert, G., Engelhardt, H., Kamb, B., 1998. Pleistocene collapse of the West Antarctic ice sheet. Science 281, 82–85.

Scherer, R.P., Bohaty, S.M., Dunbar, R.B., Esper, O., Flores, J.A., Gersonde, R., et al., 2008. Antarctic records of precession-paced insolation-driven warming during early Pleistocene Marine Isotope Stage 31. Geophysical Research Letters 35, L03505. Available from: https://doi.org/10.1029/2007GL032254.

Schloesser, F., Friedrich, T., Timmermann, A., DeConto, R.M., Pollard, D., 2019. Antarctic iceberg impacts on future Southern Hemisphere climate. Nature Climate Change 9, 672–677.

Schmidt, M.W., Spero, H.J., Lea, D.W., 2004. Links between salinity variation in the Caribbean and North Atlantic thermohaline circulation. Nature 428, 160–163.

Schoof, C., 2007. Ice sheet grounding line dynamics: steady states, stability, and hysteresis. Journal of Geophysical Research: Earth Surface 112, F03S28. Available from: https://doi.org/10.1029/2006JF000664.

Schouten, S., Hopmans, E.C., Schefuß, E., Damste, J.S.S., 2002. Distributional variations in marine crenarchaeotal membrane lipids: a new tool for reconstructing ancient sea water temperatures? Earth and Planetary Science Letters 204, 265–274.

Severinghaus, J.P., Beaudette, R., Headly, M.A., Taylor, K., Brook, E.J., 2009. Oxygen-18 of O_2 records the impact of abrupt climate change on the terrestrial biosphere. Science 324, 1431–1434.

Severinghaus, J.P., Brook, E.J., 1999. Abrupt climate change at the end of the last glacial period inferred from trapped air in polar ice. Science 286, 930–934.

Shackleton, N.J., 2000. The 100,000-year ice-age cycle identified and found to lag temperature, carbon dioxide, and orbital eccentricity. Science 289, 1897–1902.

Shackleton, N.J., Duplessy, J.-C., Arnold, M., Maurice, P., Hall, M.A., Cartlidge, J., 1988. Radiocarbon age of last glacial Pacific deep water. Nature 335, 708–711.

Shackleton, N.J., Hall, M.A., Vincent, E., 2000. Phase relationships between millennial-scale events 64,000–24,000 years ago. Paleoceanography 15, 565–569.

Shackleton, S., Baggenstos, D., Menking, J., Dyonisius, M., Bereiter, B., Bauska, T., et al., 2020. Global ocean heat content in the Last Interglacial. Nature Geoscience 13, 77–81.

Shakun, J.D., Clark, P.U., He, F., Marcott, S.A., Mix, A.C., Liu, Z.Y., et al., 2012. Global warming preceded by increasing carbon dioxide concentrations during the last deglaciation. Nature 484, 49–54.

Shakun, J.D., Lea, D.W., Lisiecki, L.E., Raymo, M.E., 2015. An 800-kyr record of global surface ocean $\delta^{18}O$ and implications for ice volume-temperature coupling. Earth and Planetary Science Letters 426, 58–68.

Shakun, J.D., Corbett, L.B., Bierman, P.R., Underwood, K., Rizzo, D.M., Zimmerman, S.R., et al., 2018. Minimal East Antarctic Ice Sheet retreat onto land during the past eight million years. Nature 558, 284–287.

Shemesh, A., Burckle, L.H., Hays, J.D., 1994. Meltwater input to the Southern Ocean during the last glacial maximum. Science 266, 1542–1544.

Shemesh, A., Hodell, D., Crosta, X., Kanfoush, S., Charles, C., Guilderson, T., 2002. Sequence of events during the last deglaciation in Southern Ocean sediments and Antarctic ice cores. Paleoceanography 17, 1056. Available from: https://doi.org/10.1029/2000PA000599.

Shepherd, A., Fricker, H.A., Farrell, S.L., 2018. Trends and connections across the Antarctic cryosphere. Nature 558, 223–232.

Siegert, M.J., Hein, A.S., White, D.A., Gore, D.B., De Santis, L., Hillenbrand, C.D., 2021. Antarctic ice sheet changes since the Last Glacial Maximum. In: Florindo, F., et al. (Eds.), Antarctic Climate Evolution, second edition. Elsevier (this volume).

Sigl, M., Winstrup, M., McConnell, J.R., Welten, K.C., Plunkett, G., Ludlow, F., et al., 2015. Timing and climate forcing of volcanic eruptions for the past 2,500 years. Nature 523, 543–549.

Sigman, D.M., Boyle, E.A., 2000. Glacial/interglacial variations in atmospheric carbon dioxide. Nature 407, 859–869.

Sigman, D.M., Hain, M.P., Haug, G.H., 2010. The polar ocean and glacial cycles in atmospheric CO_2 concentration. Nature 466, 47–55.

Sikes, E.L., Volkman, J.K., 1993. Calibration of alkenone unsaturation ratios ($U^{k'}_{37}$) for paleotemperature estimation in cold polar waters. Geochimica et Cosmochimica Acta 57, 1883–1889.

Silvano, A., Rintoul, S.R., Herraiz-Borreguero, L., 2016. Ocean-ice shelf interaction in East Antarctica. Oceanography 29, 130–143.

Simões Pereira, P., van de Flierdt, T., Hemming, S.R., Hammond, S.J., Kuhn, G., Brachfeld, S., et al., 2018. Geochemical fingerprints of glacially eroded bedrock from West Antarctica: detrital thermochronology, radiogenic isotope systematics and trace element geochemistry in Late Holocene glacial-marine sediments. Earth-Science Reviews 182, 204–232.

Skinner, L.C., Fallon, S., Waelbroeck, C., Michel, E., Barker, S., 2010. Ventilation of the deep Southern Ocean and deglacial CO_2 rise. Science 328, 1147–1151.

Skinner, L.C., Muschitiello, F., Scrivner, A.E., 2019. Marine reservoir age variability over the last deglaciation: implications for marine carbon cycling and prospects for regional radiocarbon calibrations. Paleoceanography and Paleoclimatology 34, 1807–1815.

Skinner, L.C., Primeau, F., Freeman, E., de la Fuente, M., Goodwin, P.A., Gottschalk, J., et al., 2017. Radiocarbon constraints on the glacial ocean circulation and its impact on atmospheric CO_2. Nature Communications 8, 16010. Available from: https://doi.org/10.1038/ncomms16010.

Skinner, L.C., Shackleton, N.J., 2005. An Atlantic lead over Pacific deep-water change across Termination I: implications for the application of the marine isotope stage stratigraphy. Quaternary Science Reviews 24, 571–580.

Smith, J.A., Graham, A.G.C., Post, A.L., Hillenbrand, C.-D., Bart, P.J., Powell, R.D., 2019. The marine geological imprint of Antarctic ice shelves. Nature Communications 10, 5635. Available from: https://doi.org/10.1038/s41467-019-13496-5.

Snow, K., Hogg, A.M., Sloyan, B.M., Downes, S.M., 2016. Sensitivity of Antarctic bottom water to changes in surface buoyancy fluxes. Journal of Climate 29, 313–330.

Snyder, C.W., 2016. Evolution of global temperature over the past two million years. Nature 538, 226–228.

Sosdian, S., Rosenthal, Y., 2009. Deep-sea temperature and ice volume changes across the Pliocene-Pleistocene climate transitions. Science 325, 306–310.

Sosdian, S.M., Greenop, R., Hain, M., Foster, G.L., Pearson, P.N., Lear, C.H., 2018. Constraining the evolution of Neogene ocean carbonate chemistry using the boron isotope pH proxy. Earth and Planetary Science Letters 498, 362–376.

Spahni, R., Chappellaz, J., Stocker, T.F., Loulergue, L., Hausammann, G., Kawamura, K., et al., 2005. Atmospheric methane and nitrous oxide of the late Pleistocene from Antarctic ice cores. Science 310, 1317–1321.

Spooner, P.T., Guo, W., Robinson, L.F., Thiagarajan, N., Hendry, K.R., Rosenheim, B.E., et al., 2016. Clumped isotope composition of cold-water corals: a role for vital effects? Geochimica et Cosmochimica Acta 179, 123–141.

Spratt, R.M., Lisiecki, L.E., 2016. A Late Pleistocene sea level stack. Climate of the Past 12, 1079–1092.

Steig, E.J., Huybers, K., Singh, H.A., Steiger, N.J., Ding, Q., Frierson, D.M., et al., 2015. Influence of West Antarctic ice sheet collapse on Antarctic surface climate. Geophysical Research Letters 42, 4862–4868.

Stein, K., Timmermann, A., Kwon, E.Y., Friedrich, T., 2020. Timing and magnitude of Southern Ocean sea ice/carbon cycle feedbacks. Proceedings of the National Academy of Sciences 117, 4498–4504.

Stenni, B., Buiron, D., Frezzotti, M., Albani, S., Barbante, C., Bard, E., et al., 2011. Expression of the bipolar see-saw in Antarctic climate records during the last deglaciation. Nature Geoscience 4, 46–49.

Stocker, T.F., Johnsen, S.J., 2003. A minimum thermodynamic model for the bipolar seesaw. Paleoceanography 18, 1087. Available from: https://doi.org/10.1029/2003pa000920.

Strugnell, J.M., Pedro, J.B., Wilson, N.G., 2018. Dating Antarctic ice sheet collapse: proposing a molecular genetic approach. Quaternary Science Reviews 179, 153–157.

Studer, A.S., Sigman, D.M., Martínez-García, A., Benz, V., Winckler, G., Kuhn, G., et al., 2015. Antarctic Zone nutrient conditions during the last two glacial cycles. Paleoceanography 30, 845–862.

Suganuma, Y., Miura, H., Zondervan, A., Okuno, Ji, 2014. East Antarctic deglaciation and the link to global cooling during the Quaternary: evidence from glacial geomorphology and 10Be surface exposure dating of the Sør Rondane Mountains, Dronning Maud Land. Quaternary Science Reviews 97, 102–120.

Sugden, D.E., Marchant, D.R., Denton, G.H., 1993. The case for a stable East Antarctic ice sheet: the background. Geografiska Annaler: Series A, Physical Geography 75, 151–154.

Sutter, J., Eisen, O., Werner, M., Grosfeld, K., Kleiner, T., Fischer, H., 2020. Limited retreat of the Wilkes Basin ice sheet during the Last Interglacial. Geophysical Research Letters 47, e2020GL088131. Available from: https://doi.org/10.1029/2020GL088131.

Sutter, J., Fischer, H., Grosfeld, K., Karlsson, N.B., Kleiner, T., Van Liefferinge, B., et al., 2019. Modelling the Antarctic Ice Sheet across the mid-Pleistocene transition–implications for Oldest Ice. The Cryosphere 13, 2023–2041.

Sutter, J., Gierz, P., Grosfeld, K., Thoma, M., Lohmann, G., 2016. Ocean temperature thresholds for Last Interglacial West Antarctic Ice Sheet collapse. Geophysical Research Letters 43, 2675–2682.

Svensson, A., Andersen, K.K., Bigler, M., Clausen, H.B., Dahl-Jensen, D., Davies, S.M., et al., 2008. A 60 000 year Greenland stratigraphic ice core chronology. Climate of the Past 4, 47–57.

Swann, G.E.A., Leng, M.J., 2009. A review of diatom $\delta^{18}O$ in palaeoceanography. Quaternary Science Reviews 28, 384–398.

Teitler, L., Florindo, F., Warnke, D.A., Filippelli, G.M., Kupp, G., Taylor, B., 2015. Antarctic Ice Sheet response to a long warm interval across Marine Isotope Stage 31: a cross-latitudinal study of iceberg-rafted debris. Earth and Planetary Science Letters 409, 109–119.

Teitler, L., Warnke, D.A., Venz, K.A., Hodell, D.A., Becquey, S., Gersonde, R., et al., 2010. Determination of Antarctic Ice Sheet stability over the last ~500 ka through a study of iceberg-rafted debris. Paleoceanography 25, PA1202. Available from: https://doi.org/10.1029/2008PA001691.

Thöle, L.M., Amsler, H.E., Moretti, S., Auderset, A., Gilgannon, J., Lippold, J., et al., 2019. Glacial-interglacial dust and export production records from the Southern Indian Ocean. Earth and Planetary Science Letters 525, 115716. Available from: https://doi.org/10.1016/j.epsl.2019.115716.

Thomas, A.L., Henderson, G.M., Deschamps, P., Yokoyama, Y., Mason, A.J., Bard, E., et al., 2009. Penultimate deglacial sea-level timing from uranium/thorium dating of Tahitian corals. Science 324, 1186–1189.

Tigchelaar, M., Timmermann, A., Pollard, D., Friedrich, T., Heinemann, M., 2018. Local insolation changes enhance Antarctic interglacials: insights from an 800,000-year ice sheet simulation with transient climate forcing. Earth and Planetary Science Letters 495, 69–78.

Toggweiler, J.R., 1999. Variation of atmospheric CO_2 by ventilation of the ocean's deepest water. Paleoceanography 14, 571–588.

Toggweiler, J.R., Russell, J.L., Carson, S.R., 2006. Midlatitude westerlies, atmospheric CO_2, and climate change during the ice ages. Paleoceanography 21, PA2005. Available from: https://doi.org/10.1029/2005PA001154.

Tolotti, R., Bárcena, M.A., Macrì, P., Caburlotto, A., Bonci, M.C., De Santis, L., et al., 2018. Wilkes Land Late Pleistocene diatom age model: from bio-events to quantitative biostratigraphy. Revue de Micropaléontologie 61, 81–96.

Tooze, S., Halpin, J.A., Noble, T.L., Chase, Z., O'Brien, P.E., Armand, L., 2020. Scratching the surface: a marine sediment provenance record from the continental slope of central Wilkes Land, East Antarctica. Geochemistry, Geophysics, Geosystems 21. Available from: https://doi.org/10.1029/2020GC009156.

Toyos, M.H., Lamy, F., Lange, C.B., Lembke-Jene, L., Saavedra-Pellitero, M., Esper, O., et al., 2020. Antarctic Circumpolar Current dynamics at the Pacific entrance to the Drake Passage over the past 1.3 million years. Paleoceanography and Paleoclimatology 35. Available from: https://doi.org/10.1029/2019PA003773.

Turner, J., Orr, A., Gudmundsson, G.H., Jenkins, A., Bingham, R.G., Hillenbrand, C.D., et al., 2017. Atmosphere-ocean-ice interactions in the Amundsen Sea Embayment, West Antarctica. Reviews of Geophysics 55, 235–276.

Turney, C.S.M., Fogwill, C.J., Golledge, N.R., McKay, N.P., van Sebille, E., Jones, R.T., et al., 2020. Early Last Interglacial ocean warming drove substantial ice mass loss from Antarctica. Proceedings of the National Academy of Sciences 117, 3996–4006.

Vallelonga, P., Gabrielli, P., Balliana, E., Wegner, A., Delmonte, B., Turetta, C., et al., 2010. Lead isotopic compositions in the EPICA Dome C ice core and Southern Hemisphere Potential Source Areas. Quaternary Science Reviews 29, 247–255.

van de Flierdt, T., Griffiths, A.M., Lambelet, M., Little, S.H., Stichel, T., Wilson, D.J., 2016. Neodymium in the oceans: a global database, a regional comparison and implications for palaeoceanographic research. Philosophical Transactions of the Royal Society A: Mathematical, Physical and Engineering Sciences 374. Available from: https://doi.org/10.1098/rsta.2015.0293.

Vázquez Riveiros, N., Govin, A., Waelbroeck, C., Mackensen, A., Michel, E., Moreira, S., et al., 2016. Mg/Ca thermometry in planktic foraminifera: improving paleotemperature estimations for *G. bulloides* and *N. pachyderma* left. Geochemistry, Geophysics, Geosystems 17, 1249–1264.

Villa, G., Lupi, C., Cobianchi, M., Florindo, F., Pekar, S.F., 2008. A Pleistocene warming event at 1 Ma in Prydz Bay, East Antarctica: evidence from ODP site 1165. Palaeogeography, Palaeoclimatology, Palaeoecology 260, 230–244.

Vorrath, M.-E., Müller, J., Esper, O., Mollenhauer, G., Haas, C., Schefuß, E., et al., 2019. Highly branched isoprenoids for Southern Ocean sea ice reconstructions: a pilot study from the Western Antarctic Peninsula. Biogeosciences 16, 2961–2981.

Waelbroeck, C., Labeyrie, L., Michel, E., Duplessy, J.C., McManus, J.F., Lambeck, K., et al., 2002. Sea-level and deep water temperature changes derived from benthic foraminifera isotopic records. Quaternary Science Reviews 21, 295–305.

WAIS Divide Project Members, 2013. Onset of deglacial warming in West Antarctica driven by local orbital forcing. Nature 500, 440–444.

WAIS Divide Project Members, 2015. Precise interpolar phasing of abrupt climate change during the last ice age. Nature 520, 661–665.

Wang, X., Auler, A.S., Edwards, R.L., Cheng, H., Cristalli, P.S., Smart, P.L., et al., 2004. Wet periods in northeastern Brazil over the past 210 kyr linked to distant climate anomalies. Nature 432, 740–743.

Wang, X.T., Sigman, D.M., Prokopenko, M.G., Adkins, J.F., Robinson, L.F., Hines, S.K., et al., 2017. Deep-sea coral evidence for lower Southern Ocean surface nitrate concentrations during the last ice age. Proceedings of the National Academy of Sciences 114, 3352–3357.

Wang, Y.J., Cheng, H., Edwards, R.L., An, Z.S., Wu, J.Y., Shen, C.C., et al., 2001. A high-resolution absolute-dated Late Pleistocene monsoon record from Hulu Cave, China. Science 294, 2345–2348.

Watson, A.J., Vallis, G.K., Nikurashin, M., 2015. Southern Ocean buoyancy forcing of ocean ventilation and glacial atmospheric CO_2. Nature Geoscience 8, 861–864.

Weaver, A.J., Saenko, O.A., Clark, P.U., Mitrovica, J.X., 2003. Meltwater pulse 1 A from Antarctica as a trigger of the Bolling-Allerod warm interval. Science 299, 1709–1713.

Webb, P.-N., Harwood, D.M., 1991. Late Cenozoic glacial history of the Ross embayment, Antarctica. Quaternary Science Reviews 10, 215–223.

Weber, M.E., Clark, P.U., Kuhn, G., Timmermann, A., Sprenk, D., Gladstone, R., et al., 2014. Millennial-scale variability in Antarctic ice-sheet discharge during the last deglaciation. Nature 510, 134–138.

Weber, M.E., Kuhn, G., Sprenk, D., Rolf, C., Ohlwein, C., Ricken, W., 2012. Dust transport from Patagonia to Antarctica – a new stratigraphic approach from the Scotia Sea and its implications for the last glacial cycle. Quaternary Science Reviews 36, 177–188.

Weber, M.E., Raymo, M.E., Peck, V.L., Williams, T., Expedition 382 Scientists, 2019. Expedition 382 Preliminary Report: Iceberg Alley and Subantarctic Ice and Ocean Dynamics. International Ocean Discovery Program. <https://doi.org/10.14379/iodp.pr.382.2019>.

Weertman, J., 1974. Stability of the junction of an ice sheet and an ice shelf. Journal of Glaciology 13, 3–11.

Whittaker, T.E., Hendy, C.H., Hellstrom, J.C., 2011. Abrupt millennial-scale changes in intensity of Southern Hemisphere westerly winds during marine isotope stages 2–4. Geology 39, 455–458.

Willeit, M., Ganopolski, A., Calov, R., Brovkin, V., 2019. Mid-Pleistocene transition in glacial cycles explained by declining CO_2 and regolith removal. Science Advances 5, eaav7337. Available from: https://doi.org/10.1126/sciadv.aav7337.

Wilson, D.J., Bertram, R.A., Needham, E.F., van de Flierdt, T., Welsh, K.J., McKay, R.M., et al., 2018. Ice loss from the East Antarctic Ice Sheet during late Pleistocene interglacials. Nature 561, 383–386.

Wilson, D.J., Piotrowski, A.M., Galy, A., Banakar, V.K., 2015. Interhemispheric controls on deep ocean circulation and carbon chemistry during the last two glacial cycles. Paleoceanography 30, 621–641.

Wilson, D.J., Struve, T., van de Flierdt, T., Chen, T., Li, T., Burke, A., et al., 2020. Sea-ice control on deglacial lower cell circulation changes recorded by Drake Passage deep-sea corals. Earth and Planetary Science Letters 544, 116405. Available from: https://doi.org/10.1016/j.epsl.2020.116405.

Winckler, G., Anderson, R.F., Fleisher, M.Q., McGee, D., Mahowald, N., 2008. Covariant glacial-interglacial dust fluxes in the equatorial Pacific and Antarctica. Science 320, 93–96.

Winter, K., Woodward, J., Dunning, S.A., Turney, C.S., Fogwill, C.J., Hein, A.S., et al., 2016. Assessing the continuity of the blue ice climate record at Patriot Hills, Horseshoe Valley, West Antarctica. Geophysical Research Letters 43, 2019–2026.

Wolff, E.W., Fischer, H., Fundel, F., Ruth, U., Twarloh, B., Littot, G.C., et al., 2006. Southern Ocean sea-ice extent, productivity and iron flux over the past eight glacial cycles. Nature 440, 491–496.

Wouters, B., Martin-Español, A., Helm, V., Flament, T., van Wessem, J.M., Ligtenberg, S.R.M., et al., 2015. Dynamic thinning of glaciers on the Southern Antarctic Peninsula. Science 348, 899–903.

Wu, L., Wang, R., Xiao, W., Ge, S., Chen, Z., Krijgsman, W., 2017. Productivity-climate coupling recorded in Pleistocene sediments off Prydz Bay (East Antarctica). Palaeogeography, Palaeoclimatology, Palaeoecology 485, 260–270.

Wu, L., Wilson, D.J., Wang, R., Passchier, S., Krijgsman, W., Yu, X., et al., 2021. Late Quaternary dynamics of the Lambert Glacier-Amery Ice Shelf system, East Antarctica. Quaternary Science Reviews 252, 106738. Available from: https://doi.org/10.1016/j.quascirev.2020.106738.

Xiao, W., Esper, O., Gersonde, R., 2016. Last Glacial-Holocene climate variability in the Atlantic sector of the Southern Ocean. Quaternary Science Reviews 135, 115–137.

Yamane, M., Yokoyama, Y., Abe-Ouchi, A., Obrochta, S., Saito, F., Moriwaki, K., et al., 2015. Exposure age and ice-sheet model constraints on Pliocene East Antarctic ice sheet dynamics. Nature Communications 6, 7016. Available from: https://doi.org/10.1038/ncomms8016.

Yan, Y., Bender, M.L., Brook, E.J., Clifford, H.M., Kemeny, P.C., Kurbatov, A.V., et al., 2019. Two-million-year-old snapshots of atmospheric gases from Antarctic ice. Nature 574, 663–666.

Yau, A.M., Bender, M.L., Blunier, T., Jouzel, J., 2016a. Setting a chronology for the basal ice at Dye-3 and GRIP: implications for the long-term stability of the Greenland Ice Sheet. Earth and Planetary Science Letters 451, 1–9.

Yau, A.M., Bender, M.L., Robinson, A., Brook, E.J., 2016b. Reconstructing the last interglacial at Summit, Greenland: insights from GISP2. Proceedings of the National Academy of Sciences 113, 9710–9715.

Yeung, N.K.H., Menviel, L., Meissner, K.J., Sikes, E., 2019. Assessing the spatial origin of Meltwater Pulse 1A using oxygen-isotope fingerprinting. Paleoceanography and Paleoclimatology 34, 2031–2046.

Yin, Q., Berger, A., 2015. Interglacial analogues of the Holocene and its natural near future. Quaternary Science Reviews 120, 28–46.

Yokoyama, Y., Esat, T.M., Thompson, W.G., Thomas, A.L., Webster, J.M., Miyairi, Y., et al., 2018. Rapid glaciation and a two-step sea level plunge into the Last Glacial Maximum. Nature 559, 603–607.

Yu, J., Anderson, R.F., Rohling, E.J., 2014. Deep ocean carbonate chemistry and glacial-interglacial atmospheric CO_2 changes. Oceanography 27, 16–25.

Yu, J., Elderfield, H., 2007. Benthic foraminiferal B/Ca ratios reflect deep water carbonate saturation state. Earth and Planetary Science Letters 258, 73–86.

Zachos, J., Pagani, M., Sloan, L., Thomas, E., Billups, K., 2001. Trends, rhythms, and aberrations in global climate 65 Ma to present. Science 292, 686–693.

Chapter 11

Antarctic Ice Sheet changes since the Last Glacial Maximum

Martin Siegert[1], Andrew S. Hein[2], Duanne A. White[3], Damian B. Gore[4], Laura De Santis[5] and Claus-Dieter Hillenbrand[6]
[1]Grantham Institute and Department of Earth Science and Engineering, Imperial College London, London, United Kingdom, [2]School of GeoSciences, University of Edinburgh, Edinburgh, United Kingdom, [3]Institute for Applied Ecology, University of Canberra, Canberra, ACT, Australia, [4]Department of Earth and Environmental Sciences, Macquarie University, Sydney, NSW, Australia, [5]National Institute of Oceanography and Applied Geophysics —OGS, Sgonico, Italy, [6]British Antarctic Survey, Cambridge, United Kingdom

11.1 Introduction

Antarctica is comprised of two main grounded ice sheets: the West Antarctic Ice Sheet (WAIS) and the East Antarctic Ice Sheet (EAIS) separated by the Transantarctic Mountains (McMillan, 2018). The marine-terminating margins of these two ice sheets are characterised by direct interaction with the ocean, either at a calving front or through floating ice shelves. Another, smaller ice sheet, the Antarctic Peninsula Ice Sheet (APIS), covers the Antarctic Peninsula in West Antarctica (Fig. 11.1). The present volume of ice in Antarctica is around 27 million km^3 (equivalent to $\sim$58 m of global sea level) of which 91% is within the EAIS (Fretwell et al., 2013). The maximum ice-sheet surface elevation of 4093 m occurs at Dome A, which, along with several other subsidiary ice domes connected by ridges, forms a major ice divide through the centre of East Antarctica. Ice thickness in the central regions of the continent varies commonly between 2000 and 4000 m, mainly as a consequence of bed topography which is known to vary spatially by more than a vertical kilometre over just a few kilometres (Fretwell, 2013; Morlighem, 2020). If the ice were to be removed from East Antarctica, the bed surface would be largely above sea level (apart from notable low-lying areas such as the Wilkes and Aurora basins). However, if the same were to happen over West Antarctica, even accounting for isostatic rebound, most of the bed underlying the WAIS would remain well below the modern sea level (e.g., Jamieson et al., 2014). Thus, the WAIS is often referred to as a 'marine-based' ice sheet (Mercer, 1978).

Antarctic Climate Evolution. DOI: https://doi.org/10.1016/B978-0-12-819109-5.00002-5

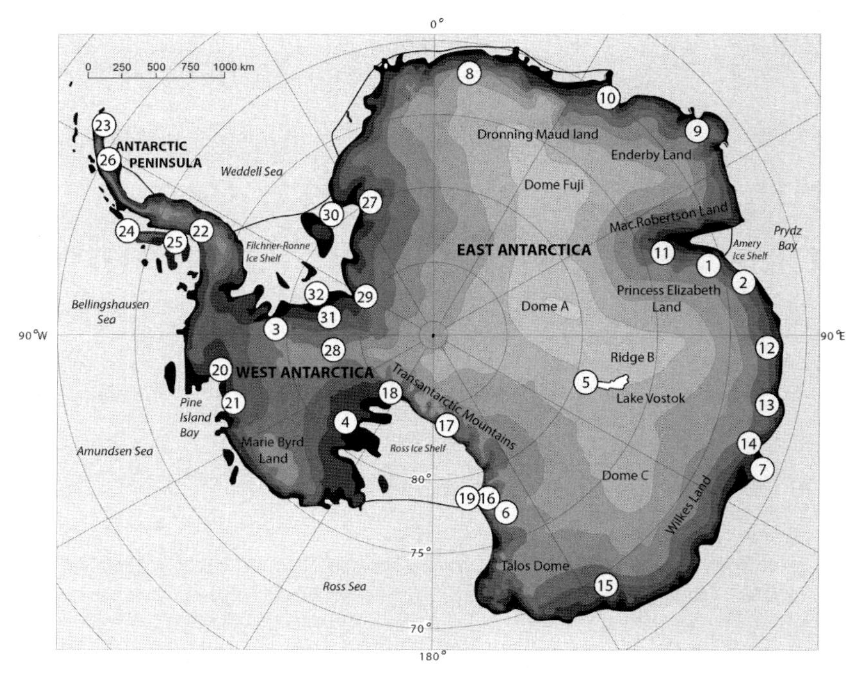

FIGURE 11.1 Surface elevation of Antarctica with placenames mentioned in the text. Contours are at 500 m intervals, with the elevation at Dome A over 4000 m. Numbers refer to locations cited in the text as follows: (1) Larsemann Hills; (2) Vestfold Hills; (3) Ellsworth Mountains; (4) Siple Coast; (5) Vostok Station (with Subglacial Lake Vostok outlined); (6) Taylor Dome; (7) Law Dome; (8) Schirmacher Oasis with Lake Glubokoye; (9) Mount Riiser Larsen; (10) Lützow-Holm Bay; (11) Grove Mountains; (12) Gaussberg; (13) Bunger Hills; (14) Windmill Islands; (15) Mertz and Ninnis Glaciers; (16) McMurdo Dry Valleys; (17) Beardmore Glacier; (18) Reedy Glacier; (19) Ross Island; (20) Pine Island Glacier; (21) Thwaites Glacier; (22) Mount Jackson; (23) James Ross Island; (24) Marguerite Bay; (25) Alexander Island and George VI Ice Shelf; (26) Larsen A & B embayments; (27) Shackleton Range; (28) Whitmore Mountains; (29) Pensacola Mountains; (30) Filchner Trough; (31) Robin Subglacial Basin; (32) Bungenstock Ice Rise.

Ice drains from the interior domes predominantly via fast flowing rivers of ice known as 'ice streams', where the dominant method of flow is by sliding or basal sediment deformation. These contribute grounded ice to numerous floating ice shelves that surround the ice sheets. The Ross and the Filchner-Ronne ice shelves are Antarctica's largest, with areas of 497,000 and 438,000 km^2, respectively (e.g., Rignot et al., 2013). Icebergs, usually of tabular form, calve from the marine margins of ice shelves, where ice thickness is usually 250–300 m. This process represents an important mechanism by which ice is lost from the ice sheet system, accounting for 52%–79% (Rignot et al., 2013) or 60%–70% (Depoorter et al., 2013) of all ice discharged from the Antarctic Ice Sheets. Sub-ice shelf melting and other basal processes cause the remaining ice loss; they vary locally and account for

10%–90% of mass loss from an individual glacial system (e.g., Depoorter et al., 2013; Rignot and Jacobs, 2002; Rignot et al., 2013, 2019; Shepherd et al., 2018). Better knowledge of ice loss processes, especially at ice-sheet grounding lines and beneath ice shelves, represents a major challenge in glaciology as they are critical to projecting future sea level change through ice sheet modelling (Siegert et al., 2020; Siegert and Golledge, 2021, this volume). Evidence of past changes, where the ice sheet grew during cool periods and shrank during warming episodes, is also important in glaciology, as it can help us understand ice sheet sensitivity and vulnerability to external climate and ocean forcing, and help constrain and improve ice sheet models. In this chapter, we concentrate on the geological evidence for ice sheet expansion at the last glacial maximum (LGM, centred at ~20 ka) and for the changes that occurred subsequently. Such knowledge can be used in ice-sheet modelling exercises to understand the processes responsible for changes observed, to help determine where important data gaps remain and to improve the models where ice flow calculations are at irreconcilable odds with observations.

11.2 Response of the ice sheets to glacial climate and late Quaternary ice sheet reconstructions

The climate around Antarctica during 'full glacial' periods, such as the LGM, is likely to have been highly conducive to the presence of expanded ice sheets, and so the 'minimum' reconstruction for the LGM Antarctic ice sheets is similar to the present configuration. However, due to the sea-level reduction of around 120–130 m (e.g., Austermann et al., 2013; Bard et al., 1990; Shackleton, 2000; Siegert, 2001; Simms et al., 2019) and ocean and air temperature reductions that occurred during the LGM, the ice sheets in Antarctica are known to have grown out towards the continental shelf edge in most places. The ice shelves bordering the Antarctic ice sheets thickened, grounded and became part of the ice sheet proper. A 'maximum' reconstruction would therefore be a major expansion of the Antarctic ice sheets (compared with today) to reach the continental shelf break right around the continent (see Siegert, 2001; The RAISED Consortium, 2014, for a broad overview).

Evidence from ice cores and internal radio-echo layering from around Antarctica (Siegert, 2003) reveals a much lower rate of ice accumulation at the LGM compared with today. Towards the East Antarctic Ice Sheet margin, the Taylor Dome ice core also shows that LGM (and post-LGM) wind and storm path directions were different from the modern day. This implies a reorganisation of the climate system that needs to be taken into account when interpreting and correlating paleoclimate data, at least at this part of the ice sheet (Morse et al., 1998), a feature corroborated at nearby Talos Dome (Mezgec et al., 2017; Siegert and Leysinger Vieli, 2007). While reductions in both sea level and air and seawater temperatures at the LGM would have been conducive to ice sheet expansion, decreasing ice accumulation may well have had a restraining effect on ice sheet growth. Resolving how the ice sheets responded to such complicated changes in forcing requires the application of numerical models.

Ice-sheet behaviour in the late Quaternary is coupled with, and sensitive to, connected processes in the cryosphere, ocean, Earth and atmosphere (e.g., Colleoni et al., 2018; Noble et al., 2020; Siegert et al., 2020; Whitehouse, 2018). For example, if the sea ice extent increased due to cooling of the air temperature over the Southern Ocean (Armand, 2000; Gersonde et al., 2005), then the moisture supply to the ice sheets may have decreased and so ice sheet growth may have been impeded. This introduces the possibility that independent mountain glaciers in Antarctica may have responded differently to LGM conditions than areas connected to the ice sheet (as it is known to happen today, see Miles et al., 2013). In other words, ice sheet advance may have been predominantly related to sea level fall while, at the same time, glacier decay may be related to a significant reduction in rates of snow precipitation.

Since the influential CLIMAP (1976) reconstruction of ice age Earth, there have been numerous reconstructions of LGM ice extent in Antarctica based on evidence from sediments and geomorphology, isostatic rebound and ice-flow modelling. These reconstructions provide estimates of ice volume varying considerably, from just 0.5-2.0 m up to 38 m of sea level equivalent (SLE) (e.g., Budd and Smith, 1982; Colhoun et al., 1992; Nakada and Lambeck, 1988). While the techniques used vary, many of the early estimates placed the grounding line at the edge of the continental shelf, with relatively steep ice profiles. As such these are at the upper end of estimates of sea-level lowering. More recent work has advanced our understanding of the dynamics of ice flow (in particular the longitudinal profile of ice streams and their controls), produced sophisticated 3D thermomechanical ice sheet models and provided reliable geomorphic evidence that indicates relatively limited ice expansion in some areas. These advances have rendered the upper estimates implausible. More recent reconstructions indicate a more limited increase of Antarctic ice volume at the LGM. Compilations by Anderson et al. (2002), Bentley (1999) and Simms et al. (2019) show that average estimates for Antarctica's contribution to the LGM sea-level lowstand were 25 m SLE in the literature published until CE (common era) 2000 (e.g., Denton et al., 1991; Nakada et al., 2000), 13 m SLE in studies published between CE 2000 and 2010 (e.g., Bassett et al., 2007; Huybrechts, 2002; Pollard and DeConto, 2009), and just 10 m SLE in papers published after CE 2010 (e.g., Briggs et al., 2014; Golledge et al., 2013, 2014; Whitehouse et al., 2012). Estimates of 10 m SLE are largely in agreement with reconstructions from geological data (e.g., The RAISED Consortium, 2014) and with a recent global reconstruction of LGM ice volume (Gowan et al., 2021). However, far field data on global sea level at the LGM and calculated meltwater contributions from all LGM ice outside Antarctica still suggest a larger Antarctic contribution in some studies (e.g., Clark and Tarasov, 2014; Lambeck et al., 2014; Simms et al., 2019).

11.3 Constraining late Quaternary ice sheet extent, volume and timing

The geometry of the Antarctic ice sheets throughout the last glacial cycle has been compiled using a number of different methods. In terrestrial environments, the areal extent of former ice cover is generally recognised by mapping the lateral extent of ice-marginal landforms such as moraines or pro-glacial lake sediments, or by mapping subglacial landforms such as drumlins and erosional features such as striae. Ice volume can equally be reconstructed by mapping glacial debris, striae, the locations of erratic boulders and the height/extent of mountain trimlines. The relative age of each mapped unit is usually constrained using a measure of exposure age, most often by the degree of weathering (such as soil formation or development of tafoni or iron staining on boulders) and by optically stimulated luminescence (OSL) dating or the measurement of isotopes produced by cosmogenic radiation. Stratigraphic correlations are generally difficult due to a lack of natural exposures.

Similar techniques are also used in marine environments, but rather than investigating the geomorphology of former glacial landscapes using field mapping and aerial photography, the ocean floor is imaged using techniques such as swath bathymetry and side-scan sonar. These techniques commonly identify ice marginal landforms such as moraines, or subglacial landforms such as mega-scale glacial lineations, meltwater channels and drumlins. Stratigraphic techniques are more useful in the marine environment due to the ease of data acquisition via acoustic and seismic surveys, particularly when the bedforms and sedimentary strata are also sampled by coring.

Establishing detailed late Quaternary glacial chronologies in Antarctica can be challenging (e.g., Anderson et al., 2002; Ingólfsson, 2004; see also Wilson, 2021). The most direct dating is usually achieved in present terrestrial environments, through cosmogenic exposure or OSL dating of clasts from moraines or exposed bedrock and through radiocarbon or electron spin resonance dating (e.g., Takada et al., 1998, 2003) of fossil organisms contained within emergent marine sediments or ice-marginal lakes. However, factors such as cosmogenic nuclide production rate uncertainties (e.g., Gosse and Phillips, 2001), recycling of clasts with prior exposure and post-depositional reworking of glacial sediments (e.g., Brook et al., 1995) can reduce the precision of exposure age and OSL chronologies. Where such materials are preserved, the algal mats commonly found in proglacial lake sediments can provide reasonably accurate radiocarbon ages if the water column is well mixed and at least a part of the lake-surface ice cover melts out each summer (Gore, 1997b; Hendy and Hall, 2006). Limiting ages on glacial events in terrestrial environments can also be constrained by radiocarbon dating of the small amount of organic material sometimes found in ice-marginal or postglacial sediments (e.g., Baroni and Orombelli, 1994a; Burgess et al., 1994).

Radiocarbon dates on carbon sourced from marine environments, such as the bodies of seals and penguins, or carbonate shells in former (and emergent) marine environments such as raised beaches, are subject to a marine reservoir effect that varies around the Antarctic continent but is on average about 1.3 ka (e.g., Berkman and Forman, 1996; Harden et al., 1992; but also see Kiernan et al., 2003). On the continental shelf, the majority of published ages for grounded ice retreat after the LGM, and the few dates constraining pre-LGM ice advance, have been obtained by Atomic Mass Spectrometry (AMS) radiocarbon dating of calcareous microfossils or, due to the scarcity of calcareous microfossils in Antarctic shelf sediments, the acid-insoluble fraction of organic material (AIO; mostly derived from diatoms) that is preserved in glacimarine sediments above or below a layer of subglacially-derived sediment (e.g., Heroy and Anderson, 2007; Livingstone et al., 2012). While non-reworked calcareous microfossils frequently occur well above the subglacial-glacimarine transition in sediment cores, several problems remain with AIO dating, mainly due to the influx of subglacially eroded and reworked fossil organic matter. This is evident from the fact that AIO dates from the modern sediment/water interface often provide uncorrected radiocarbon ages of >2 ^{14}C ka before present (BP; referring to CE 1950), and sometimes >6 ^{14}C ka BP in areas of little biological productivity and/or where carbon-rich rocks are eroded (e.g., Andrews et al., 1999; Hemer et al., 2007; Mosola and Anderson, 2006; Pudsey et al., 2006). These ages exceed the marine reservoir effect of $\sim$1.3 ka, which is derived from ^{14}C ages of carbonate shells from living biota at such sites. Usually, down-core AIO ^{14}C dates are corrected by subtracting the core-top AIO ^{14}C age (e.g., Heroy and Anderson, 2007; Pudsey et al., 2006), but this approach assumes that contamination with fossil organic matter was constant through time and that the age of this contaminating material remained the same. In some studies, the problem of AIO ^{14}C dating is overcome by relying on (1) AMS ^{14}C dates of discrete carbonate shells only (e.g., Kirshner et al., 2012), or (2) a combination of AMS ^{14}C dates from AIO and calcareous microfossils (e.g., Domack et al., 2005; Leventer et al., 2006; Prothro et al., 2020). More recent studies applied AMS ^{14}C dating by selective extraction of the most labile portions of the dispersed organic phase through compound specific extraction (e.g., Yamane et al., 2014; Yokoyama et al., 2016), ramped pyrolysis (Rosenheim et al., 2008, 2013; Subt et al., 2016, 2017) or utilising the MICADAS ^{14}C dating technique, which requires only a very small amount of calcareous material (e.g., Arndt et al., 2017; Klages et al., 2014). In addition, sediment cores from the Antarctic continental shelf have been successfully dated by analysing their relative geomagnetic palaeointensity (Brachfeld et al., 2003; Hillenbrand et al., 2010a; Klages et al., 2017; Willmott et al., 2006), and in few cases using tephra layers or cryptotephra as volcanic event markers for age correlation (Di Roberto et al., 2019). Despite the difficulties in dating Antarctic shelf sediments, considerable progress has been made in constraining the timing of post-LGM ice retreat from the shelf in many sectors of the Antarctic continent.

11.4 Last interglacial (Eemian, ~130–116 ka)

Data from sources including dated coral terraces and the oceanic $\delta^{18}O$ record have fixed the timing of the Eemian interglacial at 130–116 ka (Shackleton et al., 2003; Stirling et al., 1998). During this period, there was less ice in the world than at present and global sea level was between 3 and 9 m higher than today (Dutton et al., 2015; Kopp et al., 2009; Overpeck et al., 2006; Stirling et al., 1998; Voosen, 2018). While part of this sea-level rise occurred through decay of the Greenland Ice Sheet (e.g., Cuffey and Marshall, 2000; Tarasov and Peltier, 2003), preservation of ice in northwest Greenland during this time limits the Greenland Ice Sheet contribution to ~2 m SLE (Dahl-Jensen et al., 2013), although a possible contribution of up to 5 m SLE has also been suggested (Yau et al., 2016). Some believe that the Eemian sea-level highstand was caused by a smaller WAIS (e.g., Mercer, 1978; Scherer et al., 1998), which could have contributed up to 3.5 m of sea level rise (Bamber et al., 2009; DeConto and Pollard, 2016; Hein et al., 2016a), and/or the marine portions of the EAIS (Wilson et al., 2018), which could contribute over 10 m in a substantive deglacial event. Evidence to support the reduction in volume of at least a part of the WAIS at the last interglacial comes from observations of marine diatoms of Quaternary age and the measurement of high concentrations of ^{10}Be in subglacial sediments recovered from the UpB drill site beneath the Whillans Ice Stream on the Siple Coast of the Ross Sea embayment (Scherer et al., 1998). ^{10}Be and diatoms are not thought to accumulate under a grounded ice sheet, and any small amounts of ^{10}Be and diatoms that may be introduced by subglacial melt at the base of the ice sheet will be eroded. Scherer et al. (1998) argued that the marine components in the tills retrieved from the UpB site were probably emplaced during the last interglacial, when marine conditions prevailed. However, a recent study on modern subglacial tills from the Siple Coast ice streams, including samples from the UpB site, detected young ^{14}C-dated organic carbon and concluded that the WAIS grounding line had undergone a short-term retreat to the UpB and other drill sites at the beginning of the Holocene (Kingslake et al., 2018). This finding raises the question whether ocean currents also had advected marine diatoms and any adhering ^{10}Be to the UpB site during the last and/or earlier glacial terminations, which would not require a major WAIS collapse. In any case, the current dynamics of the

WAIS is certainly affected by the location of these low shear strength, now subglacial, sediments (Anandakrishnan et al., 1998; Siegert et al., 2016; Studinger et al., 2001).

11.5 Last Glacial Maximum, subsequent deglaciation and the Holocene (~20–0 ka)

Regardless of ice sheet geometry at the last interglacial, there is persuasive evidence from the geological record to indicate that the Antarctic ice sheets were larger than present around the time of the global sea-level lowstand at ~20 ka, although the extent of this expansion is well constrained at only a few sites around the continental margin (The RAISED Consortium, 2014; Whitehouse, 2018). Evidence of former ice surface elevations is particularly sparse, in part due to the lack of sites at which such information can be preserved but also due to the difficulty in accessing remote inland mountain ranges, where this evidence might occur.

Since the publication of the Antarctic Climate Evolution book in 2008 (Florindo and Siegert, 2008), there has been a significant amount of new research to explore the past behaviour of the Antarctic ice sheets over a range of timescales. Much of this research covering the time span back to 20 ka has been synthesised in a Reconstruction of Antarctic Ice Sheet Deglaciation (RAISED) in a coordinated effort by the Antarctic glacial geology community in 2014 (The RAISED Consortium, 2014). The RAISED Consortium developed a synthesis of Antarctic ice-sheet history and created a series of time-slice maps that described ice thickness and extent at 25, 20, 15, 10 and 5 ka based on a comprehensive review of available marine and terrestrial geological and glaciological data (The RAISED Consortium, 2014). In the next sections, we discuss glacial history by dividing Antarctica into seven sectors (see Fig. 11.2, and placename locations in Fig. 11.1). Our summaries rely heavily on the RAISED Consortium review for each sector (The RAISED Consortium, 2014, and related sector papers), and we

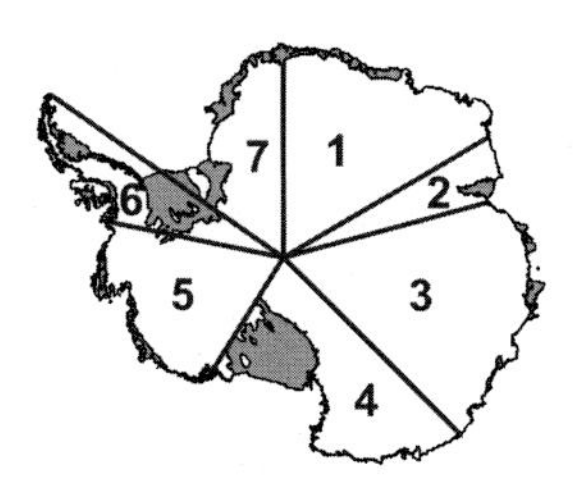

FIGURE 11.2 Broad locations of the sectors discussed in this chapter, including (1) Queen Maud/Enderby Land; (2) Mac.Robertson Land/Lambert Glacier/Amery Ice Shelf/Prydz Bay; (3) Princess Elizabeth Land to Wilkes Land; (4) Ross Sea sector; (5) Amundsen-Bellingshausen Seas; (6) Antarctic Peninsula; and (7) Weddell Sea Embayment. Ice shelves are grey shaded. *Modified from Wright et al. (2008).*

encourage readers to consult these reviews for full details of the original publications that present the data underpinning the reconstructions. We also discuss the most recent advances in our understanding of ice sheet change in each sector, with a focus on the age of maximum extent, and the timing and style of deglaciation.

The changes in ice thickness, and timings of deglaciation across the first three sectors are summarised in Fig. 11.3 (after Mackintosh et al., 2014). Descriptions of notable geological evidence from these East Antarctic sectors are provided, followed by a review of evidence from the other Antarctic sectors.

11.5.1 Queen Maud/Enderby Land

Schirmacher Oasis (11.5°E) is a 34 km^2 large area located at the transition between the Novolazarevskaya Ice Shelf and the EAIS. Infra-red stimulated luminescence dating of lake floor sediments yielded burial ages of ~52 ka in Lake Glubokoye (Krause et al., 1997), which was in part corroborated by AMS ^{14}C ages of 35 ka BP from 2 m above the glacial diamict at the base of the lake sediments (see also Mackintosh et al., 2014). On the northern side of Fimbulheimen, the mountain range to the south of Schirmacher Oasis, deposition of mumijo (proventricular ejecta of the snow petrel, *Pagodroma nivea*) throughout the LGM indicates that ice thickening in this sector of the ice sheet was also limited, with <80 m thickening occurring at Insel Range (72°S, 11°E) during this time (Hiller et al., 1995).

The 46–30 old ka marine shorelines along the eastern shore of Lützow–Holm Bay (39.6°E) indicate that the northern islands may have remained ice-free through the LGM, constraining the expansion of the ice sheet in this area (Igarashi et al., 1995, 1998). Ice was covering the peak of Skarvsnes in the southern Soya Coast and then retreated ca. 10 km landward from 9 to 5 ka (Kawamata et al., 2020). Ice-free areas in the southern shore record at least 350 m of thinning between 10 and 6 ka (Yamane et al., 2011) and correlate with a regional decrease in ice load (Nakada et al., 2000) recorded by Holocene shorelines reaching 15–20 m above present sea level and dating from 8.0 cal. (calibrated) ka BP to the present day (Maemoku et al., 1997; Miura et al., 1998; Yamane et al., 2011; Yoshida and Moriwaki, 1979). Marine sediment cores collected in Lützow–Holm Bay only recovered glacimarine deposits but no subglacial tills. Thus, their oldest AIO dates constrain the last ice advance to sometime before 13.0 cal. ka BP (Igarashi et al., 2001).

Despite an abundance of small ice-free areas that are suitable for Quaternary studies, including Mt Riiser–Larsen (50.7°E), Øygarden Group (57.5°E) and Stillwell Hills (59.3°E), there is limited information regarding the geometry of the Enderby Land sector of the EAIS during or following the LGM. Ice thinning of at least 400 m occurred along lower Rayner

Glacier between 9 and 7 ka (White and Fink, 2014) (Fig. 11.3). At Mt Riiser–Larsen, basal fresh-water sediments from Richardson Lake reveal that the area has been deglaciated since at least 9.9 uncorrected ^{14}C ka BP (Zwartz et al., 1998a). Mega-scale glacial lineations in Edward VIII Gulf (57.9°E) record a likely LGM advance across the continental shelf, but sediment cores from this site did not retrieve subglacial or grounding-line proximal glacimarine diamictons and provide only minimum limiting retreat ages around 8.7 cal. ka BP (Dove et al., 2020).

11.5.2 Mac.Robertson Land/Lambert Glacier-Amery Ice Shelf/Prydz Bay

^{10}Be and ^{26}Al cosmogenic isotope exposure ages on glacial erratics show that the ice sheet thickness at Framnes Mountains (62.5°E) was ~350 m greater than present during the LGM, had begun lowering at 13 ka and had reached the modern ice margin by 6 ka (Mackintosh et al., 2007) (Fig. 11.3). This evidence agrees reasonably well with side-scan sonar/swath bathymetry and sediment core data which indicate that grounded ice had retreated from the mid-outer continental shelf in Nielsen Basin ~80 km to the east of Framnes Mountains by ~14 cal. ka BP (Leventer et al., 2006; Mackintosh et al., 2011) and had reached the inner shelf by 6 cal. ka BP (Harris and O'Brien, 1998). The mid-inner shelf in Burton Basin between Nielsen Basin and Cape Darnley had become free of grounded ice by ~13 cal. ka BP (Borchers et al., 2015).

At Mt Harding (72°88′S, 75°03′E) in Grove Mountains, the EAIS forming the eastern flank of Lambert Glacier did not thicken above the present ice surface altitude throughout the LGM (Lilly, 2008; Lilly et al., 2010). Former ice advances of the Lambert Glacier-Amery Ice Shelf system (70°E) into Prydz Bay are relatively well constrained through geophysical

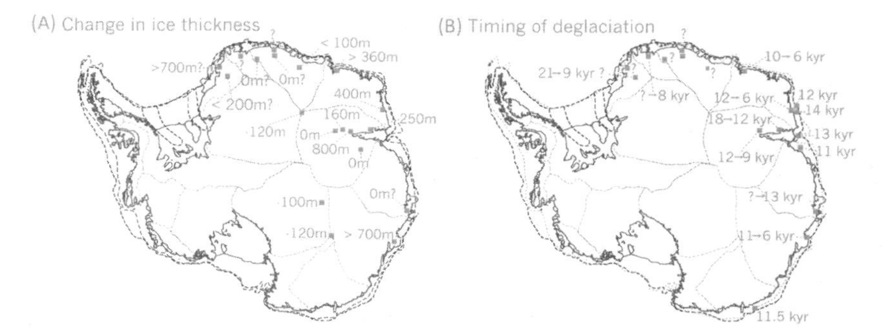

FIGURE 11.3 (A) Change in ice thickness and (B) timing of deglaciation across a range of sites in East Antarctica. *Taken from Mackintosh et al. (2014) and reproduced with permission from Elsevier.*

investigations and marine sediment cores (Domack et al., 1991, 1998; O'Brien et al., 1999; Taylor and McMinn, 2002), and by drilling through the ice shelf (Hemer and Harris, 2003; Hemer et al., 2007). The sedimentary record indicates that ice was grounded on shallow (<500 m water depth) banks in Prydz Bay and across most of the area currently occupied by the Amery Ice Shelf at the LGM. However, the age of the base of an uninterrupted sequence of terrigenous glacimarine and biogenic-bearing seasonally open-marine sediments was constrained by a radiocarbon date on foraminifera to >34 ^{14}C ka BP (uncorrected) within Prydz Channel, a 700 m deep trench that cuts across the continental shelf in Prydz Bay (Domack et al., 1998). This suggests that grounded ice may not have extended all the way to the continental shelf at the LGM.

Investigations into glacial and lake sediments deposited on mountains flanking the major outlet glaciers (Lilly, 2008; Lilly et al., 2010' Wagner et al., 2004; White and Hermichen, 2007) also constrain the LGM thickness of the ice sheet. With the exception of the area around the southern tip of the modern Amery Ice Shelf, glacial sediments dating from the LGM are restricted to <200 m above the modern ice margin. Also, there are no emergent shorelines around epishelf Beaver Lake (Adamson et al., 1997), but there are subaerially deposited postglacial sediments on the lake bottom at 60 m below modern sea level (Wagner et al., 2007). These data support the results from the continental shelf and indicate that both thickening and expansion of ice in this region at the LGM was limited (e.g., Hemer et al., 2007). One reason for this may be the deep basin along which the outlet glaciers flow, which reduces the ability of the ice sheet to advance to the continental shelf edge (Taylor et al., 2004).

Deglaciation from the ice maximum is first recorded by cosmogenic records of ice thinning from 18 to 12 ka at Loewe Massif in the Amery Oasis beside Amery Ice Shelf (White et al., 2011), with thinning proceeding inland, concluding at ~9 ka at the southern part of the modern grounding line at Mt Stinear, and a hundred kilometres further inland at Mt Ruker. Minimum ages for grounding line retreat are available from only a few marine sediment cores in Prydz Bay, but it is clear that retreat was underway from the most seaward grounding zone wedge at Prydz Channel by 11.3 cal. ka BP (Domack et al., 1991), while the northern part of the Amery Ice Shelf has been ungrounded since at least 13.5 cal. ka BP (Hemer and Harris, 2003; Hemer et al., 2007). Grounded ice had retreated from the near-coastal part of Svenner Channel, located just to the east of Prydz Bay, by ~10.1 cal. ka BP (Barbara et al., 2010; Leventer et al., 2006).

11.5.3 Princess Elizabeth Land to Wilkes Land

Larsemann Hills (76.2°E) host lake sediments dated to ^{14}C background (>42 ka BP). Lake sediment stratigraphy (Hodgson et al., 2005, 2009) and cosmogenic ^{10}Be exposure ages (Kiernan et al., 2009) have been used to infer continual exposure from the ice sheet since the last interglacial (see also Mackintosh et al., 2014). An OSL age of 21 ka was obtained from glaciofluvial sediments within 500 m of the ice margin (Hodgson et al., 2001), supporting interpretations of continual exposure through the LGM.

Rauer Islands (77.8°E), on the south side of Sørsdal Glacier, are formed by small peninsulas jutting out from the ice sheet to small islands some 12 km offshore. Ice free conditions were recorded in the islands from before 44 to <32 cal. ka BP. A modest expansion of ice occurred during the LGM, with progressive re-exposure of the outer islands from 16 ka to the present margin by 11 ka (Berg et al., 2010, 2016; White et al., 2009). This modest regional ice expansion created Holocene emergent shorelines of <10 m above sea level (asl).

Vestfold Hills (78°E) have emergent marine shorelines to <10 m asl (Zwartz et al., 1998b). A sediment core from Abraxas Lake in the northeast of the Hills suggests that the LGM ice sheet did not cover that area (Gibson et al., 2009). Ice-free conditions enabled penguin occupation of the northeastern portion of the hills by 14 cal. ka BP (Gao et al., 2018) and a ^{14}C age on a shell demonstrated that the Sørsdal Glacier margin was near its present position by 8.4 cal. ka BP (Adamson and Pickard, 1986a). Near-shore geomorphological features mapped on the seabed by multibeam and sonar data revealed a landscape formed by slow-moving grounded ice that was subsequently modified by glacimarine processes during and after deglaciation (O'Brien et al., 2015). Cosmogenic ^{10}Be analyses indicate that the ice sheet margin had retreated to within 5 km of the present margin by 12.5–9 ka (Fabel et al., 1997) and to within 1 km since 8 ka (Lilly 2008). There has been a minor (<4 km) lateral expansion and retraction of the flank of Sørsdal Glacier during the late Holocene (Adamson and Pickard, 1983; Gore, 1997a).

Gaussberg (89.2°E) is a 370 m high, glacially striated volcano on the coast. Its benched morphology and presence of palagonite encrusted pillow lavas indicate an eruption that occurred at 56 ka in a water filled subglacial vault, and that the ice sheet has since retreated to its present position (Tingey et al., 1983), thereby depositing erratics from the summit to the mountain foot.

Bunger Hills (101°E) have emergent marine shorelines to <11 m asl (Colhoun and Adamson, 1992; Colhoun et al., 1992). OSL ages from glacial lake shorelines and glaciofluvial sediments indicate that deglaciation commenced ~40–30 ka (Gore et al., 2001), with the area largely deglaciated by ~25 ka. Like Vestfold Hills, the oasis attained most of its present form around 11 cal. ka BP and warm conditions from around 9.6 cal. ka BP (Berg et al., 2020).

Windmill Islands (110.3°E) have emergent marine shorelines to 35 m asl (Goodwin, 1993; Goodwin and Zweck, 2000), with deglaciation of the southern islands by 11 cal. ka BP and the northern peninsulas by 8 cal. ka BP (Kirkup et al., 2002). Geomorphological bedforms mapped on the adjacent shelf include lineations and meltwater channels formed at the LGM and potentially during earlier glacial periods as well as mid-late Holocene moraines deposited during episodic grounded ice retreat following ice surging of the Law Dome margin (Carson et al., 2017).

The Sabrina Coast has been recently investigated to seek evidence of the retreat and advance of Totten Glacier. Multibeam bathymetry data from the inner shelf revealed mega-scale glacial lineations and drumlins inside a deep glacial trough and transversal ridges suggesting a step-wise grounded ice retreat possibly after the LGM (Fernandez et al., 2018). Gullies on the adjacent upper continental slope (Post et al., 2020) indicate that ice had expanded to near the shelf edge around this time.

Offshore from the large Mertz and Ninnis glaciers (145–150°E) of George V Land, swath bathymetry identified mega-scale glacial lineations within Mertz Trough (McMullen et al., 2006) and a moraine on Mertz Bank, which flanks this trough to the west on the outer shelf (Beaman and Harris, 2003; Beaman et al., 2011). The bedforms indicate LGM expansion of grounded ice to the outer continental shelf. Successive grounding zone wedge (GZW) deposits, which record pauses in the retreat of the grounded ice from the shelf, were also imaged in Mertz Trough (McMullen et al., 2006). Radiocarbon dating of molluscs and AIO predominantly provided uncorrected ^{14}C ages of 5–6 ^{14}C ka BP, suggesting grounded ice retreat before the Mid-Holocene, but overall the LGM extent of the EAIS on the shelf in this area, as well as the chronology of its retreat, remain poorly constrained (Mackintosh et al., 2014; McMullen et al., 2006).

Further west along the Adélie Land coast (135–145°E) the grounded EAIS had retreated from the mid-shelf in Mertz-Ninnis Trough, which is located just to the west of Mertz Trough on the inner-mid shelf and which is also referred to as George V Basin (Beaman et al., 2011), by ~10.5 cal. ka BP according to ^{14}C dates on calcareous shells and AIO (Leventer et al., 2006; Maddison et al., 2006). Post-LGM deglaciation of the inner shelf in Dumont d'Urville Trough to the west of Adélie Bank, which separates this trough from Mertz-Ninnis Trough, occurred in two phases at 10.6 and 9.0 cal. ka BP (Denis et al., 2009). A rapid glacier readvance happened at 7.7 cal. ka BP but was limited to the inner shelf. The middle to late Holocene was characterised by a series of similar episodes of limited glacier advance and subsequent retreat (Denis et al., 2009).

A new study combining ocean modelling with results from multi-proxy investigations on sediment cores from Dumont d'Urville Trough concluded that the end of cavity expansion under the Ross Ice Shelf caused modification of surface water-mass formation on the Ross Sea and Adélie Land continental shelves at ~4.5 cal. ka BP and contributed to widespread surface water cooling and increased coastal sea ice during the late Holocene (Ashley et al., 2021). Since ~4.5 cal. ka BP ocean-driven glacial ice discharge has increased in the EAIS sector between the Ross Sea and Prydz Bay (Crosta et al., 2018), with the simultaneous sea-ice expansion believed to have slowed down basal ice-shelf melting (Ashley et al., 2021). On decadal timescales, regional changes in sea-ice extent and sea-surface temperatures in this part of the East Antarctic margin were driven by variations of El Niño Southern Oscillations (ENSO) and the Southern Annular Mode (SAM) throughout at least the last 2000 years (Crosta et al., 2021).

11.5.4 Ross Sea sector

Since the 1990s a comprehensive geophysical and sedimentological dataset was compiled across the Ross Sea continental shelf using seismic profiles (see Anderson et al., 2014), swath bathymetry and side-scan sonar imagery of the sea-floor sediments in front of the present-day Ross Ice Shelf (Anderson, 1999; Bart et al., 2017a,b; Domack et al., 1999; Greenwood et al., 2012, 2018; Halberstadt et al., 2016; Lee et al., 2017; Shipp et al., 1999, 2002; Simkins et al., 2017). Recently, even the bathymetry below the ice shelf itself has been modelled (Tinto, 2019). Seaward of the Ross Ice Shelf, a suite of subglacial landforms, such as mega-scale lineations, drumlins and large-scale grooves, document advance of grounded ice streams within bathymetric troughs across the shelf, with GZWs and moraines documenting phases of episodic rapid and slow continuous retreat (e.g., Dowdeswell et al., 2008a). Predominantly on the central and eastern Ross Sea shelf, these subglacial features extend right to the shelf edge, but in the western Ross Sea, they only extend to the mid-outer parts of the shelf (Fig. 11.4). Sediment cores taken from the mega-scale lineations and GZWs retrieved deformation tills and overlying sub-ice shelf and glacimarine sediments of LGM to Holocene age (e.g., Bart et al., 2018; Domack et al., 1999; Licht, 2004; Licht et al., 1996, 1999; McGlannan et al., 2017; McKay et al., 2008, 2016; Mosola and Anderson, 2006; Prothro et al., 2018, 2020), while glacimarine sediments of Pliocene to Pleistocene age were recovered in cores from the outer shelf in the western Ross Sea documenting that grounded ice did not reach the shelf edge there at the LGM (Bart et al., 2011; Licht et al., 1996, 1999; Melis and Salvi, 2020).

Lateral moraines deposited by the many glaciers flowing through the Transantarctic Mountains and into the Ross Sea during the LGM have very similar profiles. These moraines are located high above the present-day ice surface at the feet of these glaciers but merge towards the modern-day surface of the ice sheet when traced upstream. LGM thickening indicated by these moraines therefore increases towards the marine margin, indicative of a thicker, grounded Ross Sea ice sheet, whereas the interior of the ice sheet in East Antarctica appears to have maintained a relatively constant thickness (Broecker and Denton, 1990).

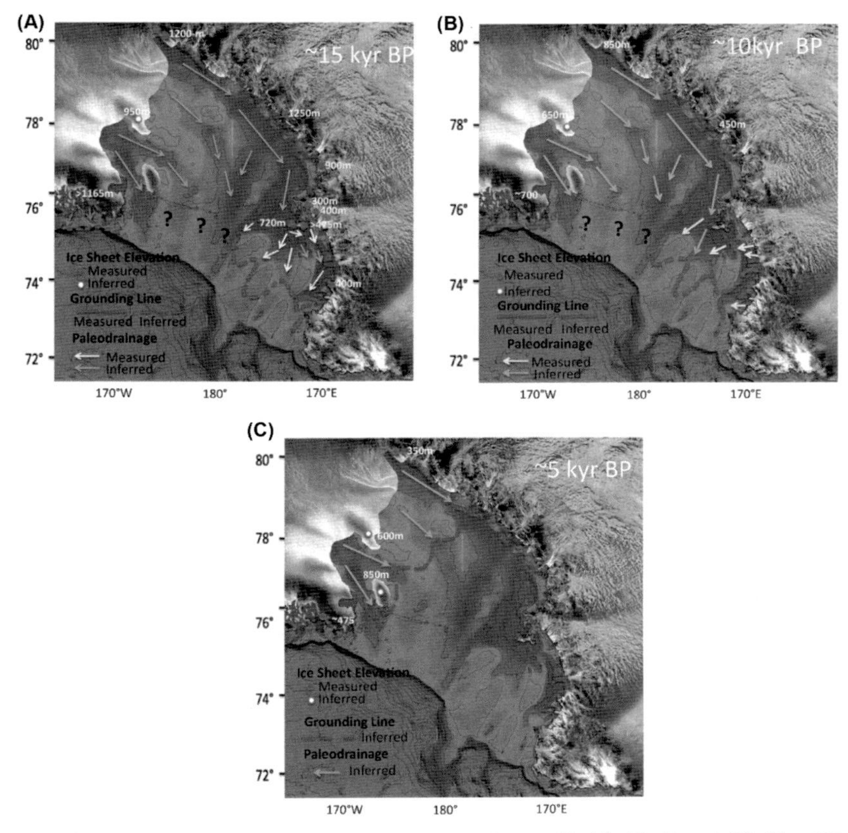

FIGURE 11.4 Ice-sheet drainage maps for the Ross Sea at (A) 15, (B) 10 and (C) 5 kyr BP. *Taken from Anderson et al. (2014) and reproduced with permission from Elsevier.*

The mineralogical, geochemical and geochronological provenance of glacial diamicts, including LGM tills, from the Ross Sea continental shelf is similar to that of source areas in Marie Byrd Land (Perotti et al., 2017) and on the Siple Coast in the eastern Ross Sea, and of sources in the Transantarctic Mountains in the central and western Ross Sea (Farmer et al., 2006; Licht and Palmer, 2013 Licht et al., 2005, 2014). These associations indicate more complicated glacial flow patterns at the LGM (Licht et al., 2005, 2014) than just a simple expansion of the present-day Siple Coast ice streams as previously assumed (e.g., Hughes, 1977) (Fig. 11.5).

Model investigations, which have attempted to reconstruct the late Holocene change in the thickness of Siple Dome from the depth-age relationship of the Siple Dome ice core, and plausible scenarios for the changes in accumulation rate since the LGM, have led to estimates of surface height increase of 200–400 m. For the ice sheet to extend ~1000 km to the

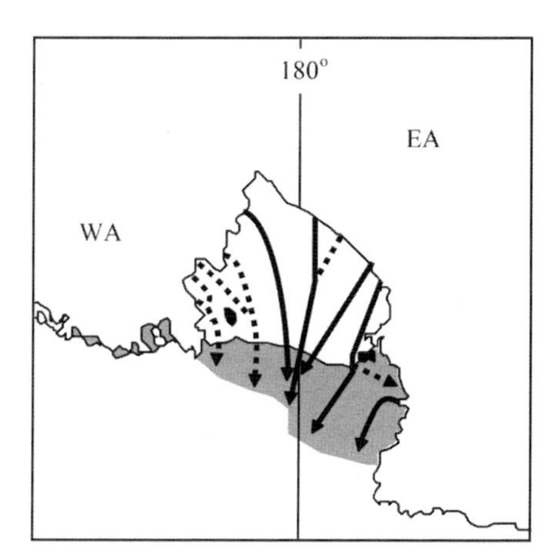

FIGURE 11.5 Flow lines for the Ross Ice Sheet during glacial episodes proposed by Licht et al. (2005) based on Ross Sea, East Antarctic (EA) and West Antarctic (WA) till provenance (cf. Licht et al., 2014). Dashed lines represent flow that could only be inferred due to lack of sample coverage. *Reproduced with permission from Elsevier.*

continental shelf edge, the required ice surface gradient would have been much shallower than that of present-day ice streams, and thus only possible by invoking a very slippery bed (Waddington et al., 2005). This supports the interpretation that ice from the Transantarctic Mountains must have contributed significantly to the Ross Sea ice sheet.

The timing of ice retreat in the Ross Sea sector has received considerable attention and includes some of the best-resolved records in the broader Antarctic region. Records of ice sheet thickness and extent maxima are documented through several different techniques and locations. Fossil algae that grew in ice marginal lakes at the LGM glacial limit on Ross Island, in the McMurdo Dry Valleys and headlands adjacent to western McMurdo Sound, provide strong evidence of ice extent and thickness reaching their maxima between 19.6 and 12.3 cal. ka BP (Christ and Bierman, 2020; Hall et al., 2015). This is supported by cosmogenic ages from erratics at the LGM limit along the Transantarctic Mountains of between ~16 and 19 ka at Reedy Glacier (Todd et al., 2010), 16 and 20 ka at Scott Glacier, and 16 and 23 ka at Beardmore Glacier (Spector et al., 2017), although it is unclear whether the age spread of these erratics represents the true duration of the LGM maxima or simply small amounts of inheritance of cosmogenic nuclides from prior periods of exposure. Radiocarbon dates from algae combined with exposure ages from the valley flanks around proglacial Lake Wellman indicate that Hatherton Glacier slowly thickened from 13.0 to 9.5 cal. ka BP (King et al., 2020). Offshore in the western Ross Sea, AMS ^{14}C dates on (reworked) foraminifera and AIO from diamictons indicate that the grounding line was located at its most advanced position in Pennell Trough between

24.2 and 15.1 cal. ka BP (Prothro et al., 2020), while a date from foraminifera from the central Ross Sea shelf suggests that maximum advance of grounded ice there had occurred by 16.8 cal. ka BP (Licht, 2004).

Multi-proxy data from a sediment core recovered on the upper continental slope offshore from JOIDES Trough document an abrupt warming event at 17.8 cal. ka BP, which was followed by a period of increasing biological productivity (Melis et al., 2021). Grounding line retreat appears to have been underway in several of the shelf basins by ~15 cal. ka BP as it is documented by ^{14}C dates from foraminifera. Ice had started to retreat from the westernmost Ross Sea shelf in JOIDES Trough by ~13 cal. ka BP (Prothro et al., 2020), shortly before ice recession from the headlands adjacent to western McMurdo Sound at 12.8 cal. ka BP (Hall et al., 2015). The Ross Ice Shelf remained extended in outer JOIDES Trough until ca. 9.5 cal. ka BP (Prothro et al., 2020) and its front had retreated to Ross Island by 8.6 cal. ka BP (McKay et al., 2016). Grounded ice retreat started as early as 15.1 cal. ka BP in Pennell Trough (Prothro et al., 2020) and at 14.7 cal. ka BP in Whales Deep Basin in the eastern Ross Sea (Bart et al., 2018). Retreat in these troughs appears to coincide with thinning along the major East Antarctic outlet glaciers draining through the Transantarctic Mountains. Here, ice thinning had begun by ~16 ka in the north at Tucker Glacier (Goehring et al., 2019) and 14.5 ka in the southern Transantarctic Mountains at Beardmore Glacier (Spector et al., 2017). In Marie Byrd Land on the WAIS flank of the Ross Sea sector, exposure age dated moraines and recessional deposits indicate that ice thickness was around 45 m greater than present when thinning began at ~10 ka (Ackert et al., 1999, 2013). In the Ohio Range, located at the divide between the Ross Sea drainage sector of the WAIS and the EAIS, ice had thickened by ~125 m at 11.5 ka (Ackert et al., 2007, 2013).

Timing of ice withdrawal across the shelf basins was complex, and likely influenced by topography. Rapid retreat or thinning has been documented in several glacier systems, for example abrupt thinning of EAIS glaciers draining through the southern Transantarctic Mountains at 8–9 ka (Spector et al., 2017), thinning of Mackay Glacier by ~200 m within a few hundred years at ~7 ka (Jones et al., 2015) and 200 km of ice retreat in Whales Deep Basin by 11.5 cal. ka BP (Bart et al., 2018). Both thinning of Mackay Glacier and retreat in Whales Deep Basin have been linked to grounding zone retreat through over-deepened basins. Ice appears to have reached the modern position at different times. Along the Siple Coast the grounding line apparently had retreated briefly upstream of its present location at the start of the Holocene before it readvanced to its modern position due to isostatic rebound (Kingslake et al., 2018). In the western Ross Sea, ice had largely cleared around Ross Island by 8.6 cal. ka BP (McKay et al., 2016), but retreat is assumed to have continued into the late Holocene near the southern grounding line. Terrestrial ice-free conditions enabled penguin recolonisation progressing southward along the coast from the early to middle Holocene but during the late Holocene conditions for penguin and seal colonies varied (Baroni and Orombelli, 1994b; Emslie et al., 2007; Hall et al., 2006). Cosmogenic records of the cessation of ice retreat are variable. For example, Beardmore Glacier was at or near its modern ice margin by ~8 ka, while other glaciers thinned until the mid-late Holocene – e.g., ~5 ka at Tucker Glacier (Goehring et al., 2019), ~3 ka at Scott Glacier (Spector et al., 2017) and at least ~2.8 ka at Hatherton Glacier (King et al., 2020).

With the exception of its westernmost part, the timing of retreat of the calving front of Ross Ice Shelf is poorly constrained. The ice shelf probably covered considerable portions of the central and eastern Ross Sea shelf until well into the Holocene. A substantive ice shelf occupied outer Whales Deep Basin in the eastern Ross Sea until ~12 cal. ka BP, and the inner part of the basin until ~3 cal. ka BP, several thousand years after grounding line retreat (Bart and Tulaczyk, 2020; Bart et al., 2018). Ice shelf cover of the basins on the central Ross Sea shelf apparently persisted for even longer, with the calving front in the Eastern Basin retreating from the outer shelf to near its modern position from ~5 until 1.5 cal. ka BP (Yokoyama et al., 2016). In the western Ross Sea seasonal open-marine deposition on the outer shelf parts of JOIDES Trough and Pennell Trough began at ~5 cal. ka BP (Yokoyama et al., 2016), while in the westernmost Ross Sea the ice-shelf front had retreated to near Ross Island already by 8.6 cal. ka BP (McKay et al., 2016).

Recent numerical modelling has investigated post-LGM grounding line retreat in the Ross Sea sector (Lowry et al., 2019). The work demonstrates that retreat was forced by two main factors acting in sequence: first, immediately after the glacial maximum, atmospheric conditions moderated the influence of rising ocean temperatures on melting at the grounding line; and second, during the early Holocene, ocean-forced basal melting became dominant, leading to significant retreat of the grounding line. A high-resolution ocean model combined with proxy data from sediment cores retrieved on the Adélie Land margin found that the expansion of the sub-ice shelf cavity under the Ross Ice Shelf was complete by ~5 cal. ka BP (Ashley et al., 2021). This led to modification of surface water masses formation processes on the Ross Sea and Adélie Land shelves and contributed to widespread surface water cooling and increased coastal sea ice during the late Holocene (Ashley et al., 2021). Previously, a study analysing sea-ice proxies in both marine sediment cores and ice cores from the Ross Sea sector found that a combination of wind forcing and an increase in the efficiency of regional latent-heat polynyas had increased coastal sea-ice cover but decreased overall sea-ice extent in the western Ross Sea since 3.6 cal. ka BP (Mezgec et al., 2017).

Today, warm modified Circumpolar Deep Water (CDW) may locally intrude under the Ross Ice Shelf, but it fails to reach the grounding line (Tinto, 2019). However, conditions may have been different when the ice sheet grounding line was positioned on the outer continental shelf at the end of the LGM. Modern and buried gullies, observed at the shelf edge of the Pennell Trough and Glomar Challenger basin, document periods of intense seafloor erosion, likely occurring when ice was grounded near to the shelf edge (Gales et al., 2021). These gullies likely formed at the beginning of the last (and previous) deglacial period, when sediment-laden subglacial meltwater was released at the shelf edge. Further sampling and analysis and dating of sediment cores along the slope are needed to confirm this hypothesis.

11.5.5 Amundsen-Bellingshausen Seas

Exposure dating studies in Marie Byrd Land and around the Amundsen Sea Embayment (ASE) (see also Larter et al., 2014) support the view that ice in this drainage sector of the WAIS was significantly thicker at the LGM (Johnson et al., 2008, 2014; Lindow et al., 2014; Stone et al., 2003; Sugden et al., 2006). Although the exact age for the onset of post-LGM WAIS thinning is still unconstrained in most of these areas, the exposure dates demonstrate that surface lowering began at least at ~10 ka, continued steadily and rapidly throughout the early Holocene until ~6 ka around the ASE, and continued well into the late Holocene in parts of Marie Byrd Land (Johnson et al., 2014, 2017, 2020; Lindow et al., 2014; Stone et al., 2003; Sugden et al., 2006).

Large palaeo-ice stream troughs have been identified offshore from all the major present-day outlet glaciers in the Amundsen Sea sector of West Antarctica (Anderson and Shipp, 2001; Nitsche et al., 2007; Wellner et al., 2001). Subglacial bedforms mapped with swath bathymetry revealed that Pine Island Glacier and Thwaites Glacier had merged in the past and eroded a cross-shelf trough into the eastern ASE shelf (Evans et al., 2006; Graham et al., 2010; Hogan et al., 2020; Jakobsson et al., 2012; Lowe and Anderson, 2002; Nitsche et al., 2013; Fig. 11.6). Gullies incised into the uppermost continental slope seaward of the trough mouth suggest that grounded ice had reached the shelf break at the LGM (Dowdeswell et al., 2006; Gales et al., 2013; Kirshner et al., 2012), with GZWs mapped on the outer and middle shelf parts of the trough documenting stepwise ice-stream retreat after the LGM (Graham et al., 2010; Jakobsson et al., 2012). Grounded ice also reached an outermost shelf position in Dotson-Getz Trough in the western ASE (Graham et al., 2009; Larter et al., 2009). On the westernmost Amundsen Sea shelf, a prominent GZW on the inner shelf part of Hobbs Trough was interpreted to mark a temporary stillstand in post-LGM grounding-line retreat rather than the maximum LGM extent (Klages et al., 2014), while the outer shelf part of Abbot-Cosgrove Trough on the easternmost ASE shelf was only covered by a floating ice shelf during the last glacial period (Klages et al., 2015, 2017). Geomorphological features, such as subglacial meltwater channels and cavities that probably hosted subglacial lakes, as well as sedimentary records, show that a complex subglacial hydrological

(A)

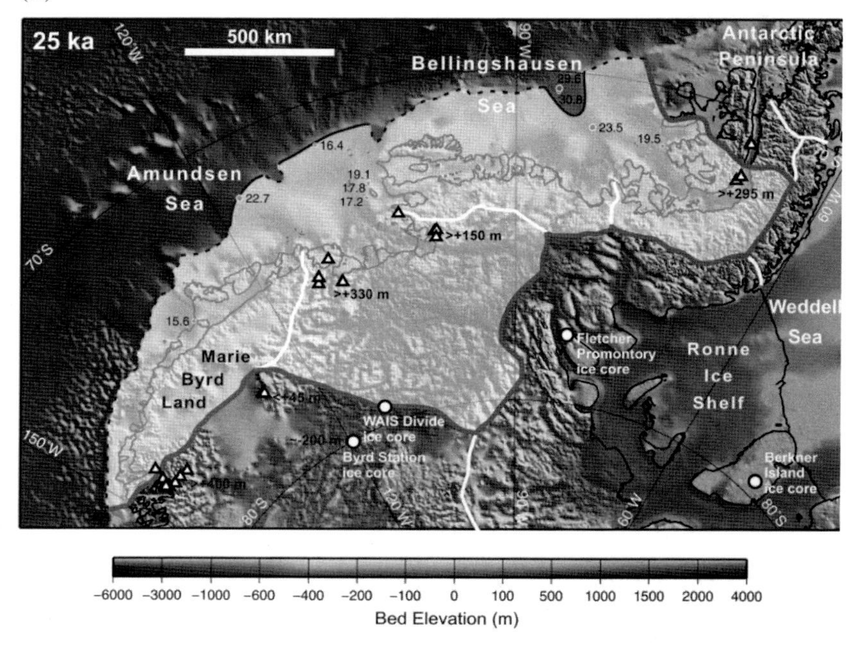

(B)

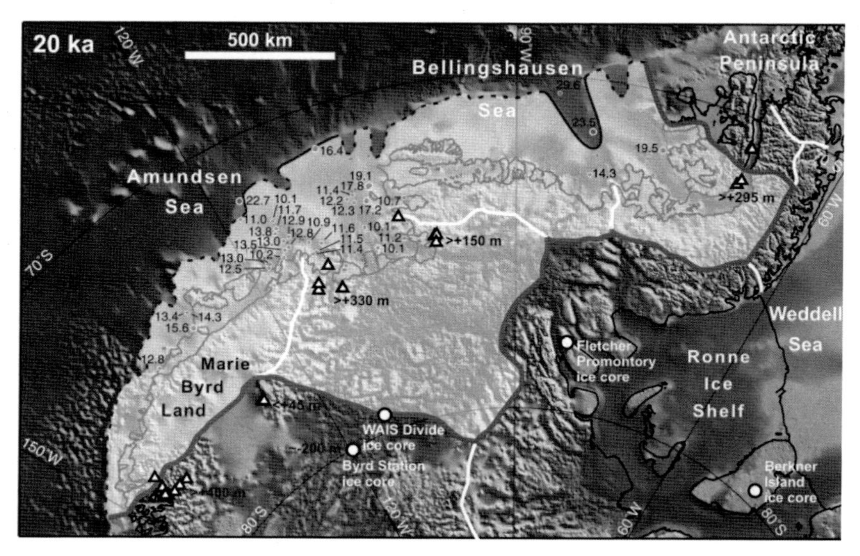

FIGURE 11.6 (*Contined*).

(C)

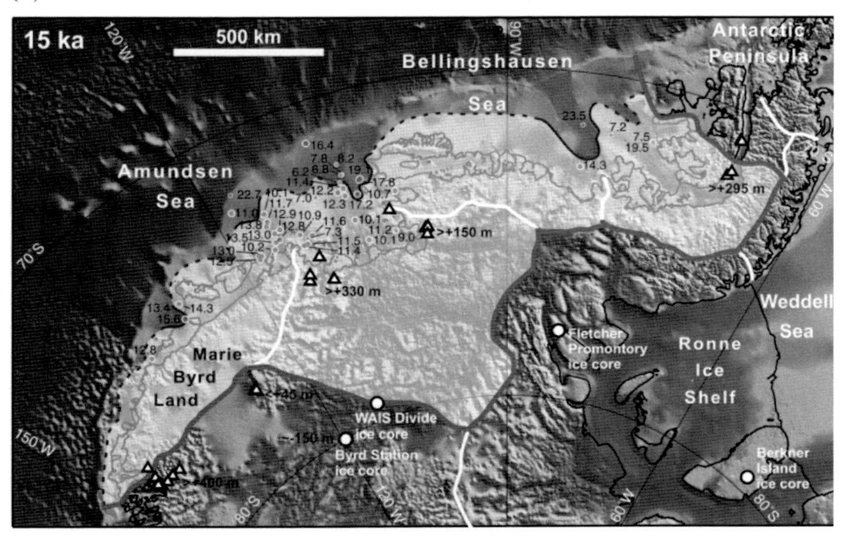

(D)

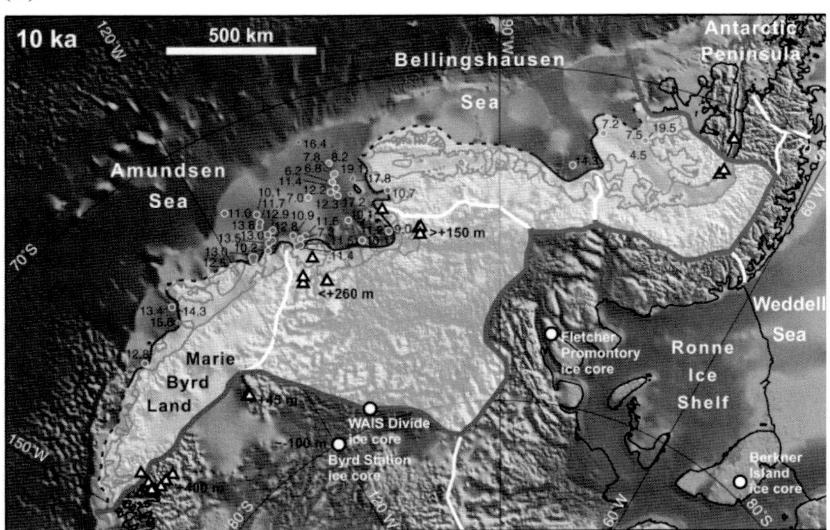

FIGURE 11.6 (*Contined*).

(E)

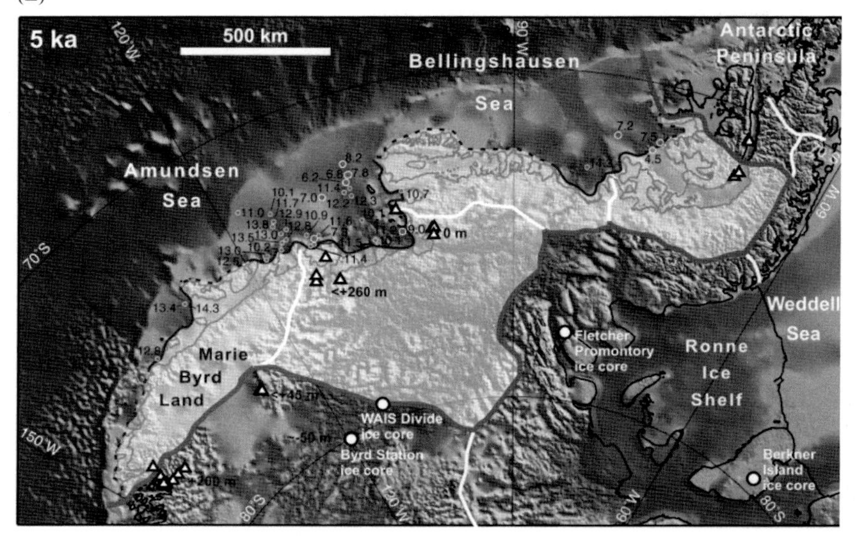

(F)

0 ka
500 km
120°W
90°W
Bellingshausen
Sea
Antarctic
Peninsula
Amundsen
Sea
70°S
Weddell
Sea
Ronne
Ice
Shelf
Marie
Byrd
Land
150°W
80°S
60°W

-1.20 0 0.70
Rate of elevation change, WAIS (m yr^{-1})

-1.5 0.5 0.2 0 0.5 0.5
Rate of elevation change, Antarctic Peninsula and Weddell Sea region (m yr^{-1})

6000 3000 1000 600 400 200 100 0
Sea-floor Depth (m)

FIGURE 11.6 Ice sheet reconstructions for the Amundsen and Bellingshausen Sea sectors at (A) 25 ka, (B) 20 ka, (C) 15 ka, (D) 10 ka, (E) 5 ka and (F) 0 ka. Reconstructions are overlain *(Continued)* system existed at the base of the WAIS on the inner ASE shelf during the LGM and earlier glacial periods (Graham et al., 2009; Kirkham et al., 2019; Kuhn et al., 2017; Lowe and Anderson, 2002, 2003; Nitsche et al., 2013; Smith et al., 2009; Witus et al., 2014).

The Pine Island-Thwaites palaeo-ice stream retreated from the outer shelf before 20.6 cal. ka BP and reached an inner shelf position by 11.7 cal. ka BP (Hillenbrand et al., 2013; Kirshner et al., 2012; Lowe and Anderson, 2002; Smith et al., 2014). Geomorphological and sedimentological evidence from the mid-shelf part of the trough indicates that an ice shelf collapse between 12.3 and 11.4 cal. ka BP had triggered rapid grounding-line retreat into the deep basins on the inner shelf (Jakobsson et al., 2011, 2012; Kirshner et al., 2012), which probably deglaciated by ice-cliff failure (Wise et al., 2017). The palaeo-ice stream occupying Dotson-Getz Trough had cleared the outer shelf by 22.4 cal. ka BP, reached the middle shelf by 13.8 cal. ka BP and the inner shelf near the present ice shelf front between 12.6 and 10.1 cal. ka BP (Smith et al., 2011). An advance of the Getz Ice Shelf front across the inner shelf during the Antarctic Cold Reversal (14.5–12.9 ka) has been concluded from biomarker analyses on a sediment core retrieved from the western tributary of the Dotson-Getz palaeo-ice stream trough (Lamping et al., 2020). Along the entire Amundsen Sea coastline the WAIS grounding line had retreated to within 100 km of, and at most locations much closer to, its modern position between ~20.9 and 11.0 cal. ka BP (Anderson et al., 2002; Hillenbrand et al., 2013; Klages et al., 2014; Minzoni et al., 2017). Ocean forcing, caused by enhanced advection of relatively warm deep water onto the ASE shelf, is believed to have been the main driver of both post-LGM grounding-line retreat (Hillenbrand et al., 2017) and late Holocene ice-shelf retreat at the eastern ASE coast (Minzoni et al., 2017).

Ice-surface elevation upstream of Pine Island Glacier at Mt Moses and Maish Nunatak rapidly decreased by ~142 m from ~8.5 to ~6.0 ka, when it reached its present elevation (Johnson et al., 2014, 2020). This indicates a considerable delay between grounding line retreat of the palaeo-ice stream

---

◀ on Bedmap2 bed topography and bathymetry. Ice sheet extent is indicated by semi-transparent white fill (only shown within the Amundsen-Bellingshausen sector). Thick line is the sector boundary, which follows the main ice-drainage divides. Thick white lines mark other major ice divides. Core sites constraining the minimum time of deglaciation are marked by the circles with yellow outlines (see Larter et al., 2014; for minimum ages of deglaciation). Red fills are ages older than time of reconstruction, blue fills are younger ages; large circles are ages within 5 kyr of time of reconstruction; small circles are ages within 5 and 0 kyr. Cosmogenic surface exposure age sample locations are marked by white-filled triangles, and deep ice core sites by white-filled circles, with surface elevation constraints they provide for time of reconstruction annotated. In the modern configuration (F), contours on the ice sheet (thin grey lines) show surface elevation at 500 m intervals, from Bedmap2. Colours on ice sheet show the rate of change of surface elevation over the period 2003–2007 from Pritchard et al. (2009). *Taken from Larter et al. (2014) and reproduced with permission from Elsevier.*

and rapid thinning in its hinterland, which has been attributed to the collapse of a buttressing ice-shelf covering Pine Island Bay until ~7.5 cal. ka BP (Hillenbrand et al., 2017). Recent thinning, flow acceleration and grounding-line retreat of Pine Island Glacier and other ice streams draining into Pine Island Bay (e.g., Rignot et al., 2019; Turner et al., 2017) are believed to have been triggered by another phase of intensified upwelling of warm deep-water onto the ASE shelf that had started in the 1940s (Hillenbrand et al., 2017; Smith et al., 2017).

On the shelf of the Bellingshausen Sea, only a single but major cross-shelf trough, called 'Belgica Trough', has been mapped with multibeam bathymetry (Graham et al., 2011; Ó Cofaigh et al., 2005b; Wellner et al., 2001). This trough exhibits geomorphological evidence of a palaeo-ice stream, which presumably drained both a large part of the Ellsworth Land sector of the WAIS and the southernmost APIS during the last glacial period (Ó Cofaigh et al., 2005b). Gullies incised into the uppermost continental slope at the mouth of Belgica Trough indicate a shelf-wide advance of the palaeo-ice stream (Dowdeswell et al., 2008b; Gales et al., 2013). As for the glacial troughs on the Amundsen Sea shelf, GZWs documenting stepwise grounding line retreat were also identified in Belgica Trough (Ó Cofaigh et al., 2005b). The reconstructed timing of grounded ice advance and retreat in Belgica Trough is based on AIO ^{14}C dates only and therefore not very well constrained. Available ages suggest that the Belgica palaeo-ice stream had advanced across the shelf after ~40.0 cal. ka BP and had retreated from the outer shelf by ~30.0 cal. ka BP, the middle shelf by ~23.5 cal. ka BP, the inner shelf in Eltanin Bay by ~14.3 cal. ka BP and the inner shelf in Ronne Entrance by ~7.2 cal. ka BP (ages from Hillenbrand et al., 2010b, and calibration from Larter et al., 2014). An early onset of grounded ice retreat may be supported by onshore cosmogenic nuclide data from the eastern side of George VI Ice Shelf, suggesting that thinning of the APIS there had started by 27.2 ka (Bentley et al., 2006).

11.5.6 Antarctic Peninsula

The LGM extent of grounded ice in the Antarctic Peninsula (see also Ó Cofaigh et al., 2014) and the post-LGM glaciological and environmental history in this region have received a significant amount of attention during the past few decades, which is summarised in several review papers (Bentley and Anderson, 1998; Bentley et al., 2009; Davies et al., 2012; Heroy and Anderson, 2005, 2007; Ingólfsson et al., 2003; Johnson et al., 2011; Livingstone et al., 2012;). To date, the RAISED Consortium (Ó Cofaigh et al., 2014) has produced the most comprehensive review of deglaciation of this sector of Antarctica from terrestrial and marine geological and geophysical data (Fig. 11.7). Here, we summarise some of the patterns that emerge from the reconstruction, while also including more recent progress.

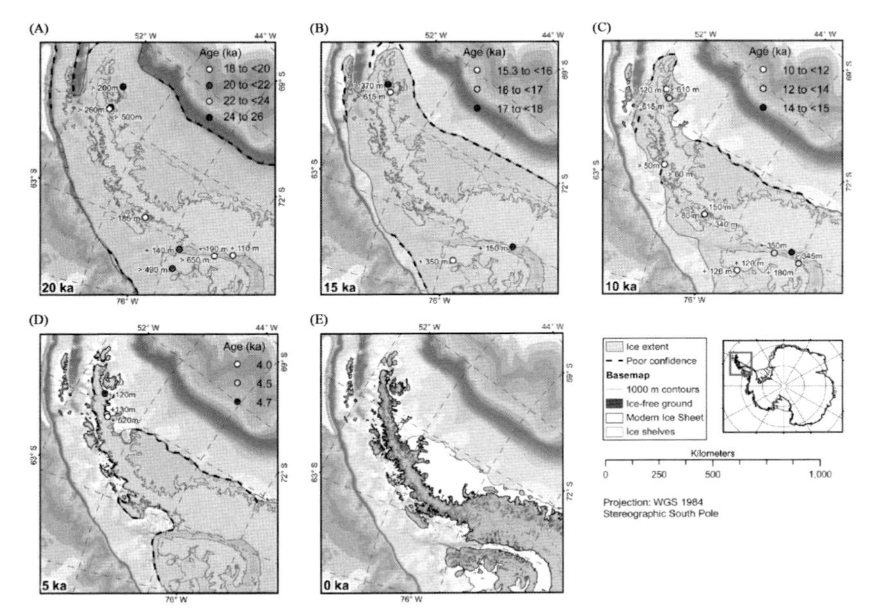

FIGURE 11.7 Ice sheet reconstructions for the Antarctic Peninsula at (A) 20 ka, (B) 15 ka, (C) 10 ka, (D) 5 ka and (E) 0 ka. *Reproduced from Ó Cofaigh et al. (2014) with permission from Elsevier.*

Terrestrial geomorphic evidence including glacial erratics and striations on nunatak summits along the length of the Antarctic Peninsula indicate that the APIS thickened significantly during the LGM. The centre of the ice sheet thickened by at least 500 m in places, reaching a maximum height of 2350 m asl at Mt Jackson (72°S), and there is evidence for at least two distinct ice domes along the spine of the southern part of the peninsula (Bentley et al., 2000, 2006; Ingólfsson, 2004). Striation orientations in the southeastern part of the Antarctic Peninsula also suggest that ice flow was deflected by an expanded WAIS in the Weddell Sea, indicating that ice advance in these two regions was synchronous (Bentley et al., 2006). The timing of maximum LGM ice thickness is not well constrained. Most data derive from cosmogenic nuclide ages from near James Ross Island and Marguerite Bay. At James Ross Island, LGM ice thickness is thought to have exceeded 370 m above present (Davies et al., 2013; Glasser et al., 2014), while to the west of James Ross Island, ice thickness was locally up to 520 m above present sea level (Balco et al., 2013). Ice core data suggest Haddington Ice Cap remained an independent ice dome (Mulvaney et al., 2012), which expanded to cover the island and merge with the APIS (Johnson et al., 2009; Mulvaney et al., 2012; Ó Cofaigh et al., 2014). Further south along the Lassiter Coast exposure dates suggest that the LGM ice surface was at least 385 m above present (Johnson et al., 2019). An LGM ice thickening of

>650 m at 27.2 ka was reconstructed for the western Antarctic Peninsula flanking George VI Ice Shelf, and thickening of >490 m at 21.8 ka is evident from eastern and northwestern Alexander Island (Bentley et al., 2006; Johnson et al., 2012).

Marine geophysical research of the continental shelf around the Antarctic Peninsula using a combination of swath bathymetry, side-scan sonar and seismic surveys, have revealed a comprehensive set of subglacial landforms that indicate the APIS was drained by several topographically-steered ice streams that were grounded at or very near the continental shelf edge during the LGM (e.g., Amblas et al., 2006; Anderson and Oakes-Fretwell, 2008; Banfield and Anderson, 1995; Campo et al., 2017; Canals et al., 2000, 2002; Domack et al., 2006; Dowdeswell et al., 2004; Evans et al., 2004, 2005; Graham and Smith, 2012; Heroy and Anderson, 2005; Larter and Vanneste, 1995; Larter et al., 2019; Lavoie et al., 2015; Livingstone et al., 2013; Ó Cofaigh et al., 2002, 2005a; Pope and Anderson, 1992; Pudsey et al., 1994). Such landforms are observed in cross-shelf bathymetric troughs around the Antarctic Peninsula and include mega-scale glacial lineations and GZWs in the outer and middle shelf sections of the troughs, and ice-moulded bedrock and subglacial meltwater channels in their inner shelf sections. Age control is only available in a few locations, and so the timing and duration of the maximum LGM ice extent remains poorly resolved (Ó Cofaigh et al., 2014). Ice likely remained at or near the shelf-edge in the Vega and Robertson troughs in the northern and northeastern Antarctic Peninsula from 25 to 20 cal. ka BP (Evans et al., 2005; Heroy and Anderson, 2005; Ó Cofaigh et al., 2014).

Retreat of the APIS grounding line from its maximum LGM position at the shelf edge is mainly constrained by a suite of AMS ^{14}C ages obtained from marine sediment cores on the western Antarctic Peninsula shelf, which suggest that its onset progressed from northeast to southwest (Heroy and Anderson, 2007; Livingstone et al., 2012; Ó Cofaigh et al., 2014). The earliest deglaciation ages of 17.5 cal. ka BP come from both Lafond Trough south of Bransfield Basin (Banfield and Anderson, 1995) and Smith Trough, with one older, but less reliable, AIO date of 18.7 cal. ka BP obtained from Biscoe Trough (Heroy and Anderson, 2007). Grounded ice had retreated to the modern coastline in this part of the Antarctic Peninsula between ~12.9 and 10.7 cal. ka BP (Domack et al., 2001, 2006; Subt et al., 2016). Further south, the grounding line of the Marguerite Trough palaeo-ice stream had retreated from the shelf edge before 14.4 cal. ka BP (Heroy and Anderson, 2007; Kilfeather et al., 2011; Pope and Anderson, 1992), the mid-shelf by ~14.1 cal. ka BP and the inner shelf by 9.6 cal. ka BP, and it reached the modern coastline before 9.1 cal. ka BP (Heroy and Anderson, 2007; Peck et al., 2015), shortly after an ice shelf extending to the middle shelf had disintegrated (Kilfeather et al., 2011). Cosmogenic nuclide dating from locations across Marguerite Bay, including Pourquois-Pas Island, Horseshoe

Island, Alexander Island and the Batterby Mountains above George VI Ice Shelf, together with ^{14}C dates on lake sediments and raised beaches show that onshore thinning was underway by 18 ka and that its main phase had ended by 10–9 ka (Bentley et al., 2006, 2011; Çiner et al., 2019; Davies et al., 2017; Hodgson et al., 2013; Johnson et al., 2012). Data from around George VI Ice Shelf also show that its northern part, or even the entire ice shelf, broke up at least once during the early Holocene (Bentley et al., 2005; Davies et al., 2017; Hodgson et al., 2006; Roberts et al., 2009; Smith et al., 2007). While post-LGM deglaciation of the western Antarctic Peninsula did not follow an entirely uniform pattern, its main phase apparently occurred from 15 to 10 cal. ka BP, with modern ice limits having been reached no later than the mid-Holocene (e.g., Heroy and Anderson, 2007; Kim et al., 2018; Ó Cofaigh et al., 2014).

Deglaciation of the northeastern Antarctic Peninsula also progressed from north to south (Ó Cofaigh et al., 2014). Initial retreat had begun by ~18.3 cal. ka BP in Vega Trough (Heroy and Anderson, 2005, 2007), and ice surface elevations had dropped below 370 m at Lachman Crags on James Ross Island at 17.7 ka (Glasser et al., 2014). The APIS grounding line is thought to have reached its modern position by 11–10 cal. ka BP at several locations along the eastern Antarctic Peninsula margin (Campo et al., 2017; Evans et al., 2005; Pudseyet al., 2001), including Prince Gustav Channel and the Larsen A and B embayments, where the transition from grounded to floating ice occurred by 10.7–10.6 cal. ka BP (Domack et al., 2005; Pudsey et al., 2006 Rosenheim et al., 2008). High-resolution multi-beam data suggest that the APIS underwent episodes of very fast grounding-line retreat, possibly >10 km/yr, during the post-LGM deglaciation (Dowdeswell et al., 2020). Cosmogenic nuclide dates from around the Larsen A and B embayments and southern Prince Gustav Channel region indicate that most thinning had occurred by ~9 ka, but that thinning continued until 8–6 ka in the southern Larsen B embayment. A north to south trend in ice sheet/shelf thinning is evident based on the emergence of ice-free areas at 8.0–6.5 ka near Prince Gustav Channel (Balco et al., 2013; Johnson et al., 2011), 6.2–4.7 ka in the Larsen A Embayment (Balco et al., 2013) and 5.1–4.9 ka in the Larsen B Embayment (Jeong et al., 2018). This trend in thinning is also consistent with marine evidence for collapses and subsequent reformations of the Larsen A and Prince Gustav Channel Ice Shelves during the Holocene (Brachfeld et al., 2003; Pudsey and Evans, 2001; Pudsey et al., 2006), which again collapsed in CE 1995 in response to recent regional warming (e.g., Hodgson et al., 2006; Vaughan et al., 2003), whereas the Larsen B Ice Shelf remained a persistent feature throughout the Holocene and did not collapse until CE 2002 (Domack et al., 2005, Rebesco et al., 2014). At the Lassiter Coast to the south, the ice surface dropped by at least 250 m from 7.5 to 6.0 ka according to ^{14}C exposure ages on bedrock (Johnson et al., 2019).

Throughout the Holocene (sub-)surface sea temperatures in Palmer Deep, western Antarctic Peninsula, cooled by either 3°C–4°C (Shevenell et al., 2011) or 1.5°C (Etourneau et al., 2013) and sea-ice cover increased in response to a long-term decline in annual and spring insolation, which influenced the location of the Southern Hemisphere westerlies and upwelling of warm deep water onto the shelf. These long-term trends were superimposed by strong centennial- to millennial-scale variations derived from ENSO variability (Etourneau et al., 2013; Shevenell et al., 2011). Glacial discharge from the APIS in Palmer Deep showed a strong dependence on increasing occurrence of La Niña events and rising levels of summer insolation (Pike et al., 2013).

On the east side of the northern Antarctic Peninsula, heavy sea ice conditions and reduced primary productivity prevailed from 8.8 until 7.4 cal. ka BP, with a southward shift of the westerly winds and the Antarctic Circumpolar Current causing warming and reduced sea ice coverage between 7.4 and 5 cal. ka BP (Barbara et al., 2016). Expansion of the Weddell Gyre and development of a strong oceanic connection between the eastern and western Antarctic Peninsula caused a gradual increase in annual sea-ice duration from 5 cal. ka BP until 1.9 cal. ka BP (Barbara et al., 2016). Glacial discharge from the eastern APIS remained unchanged from 6.3 to 1.6 cal. ka BP but has increased significantly since CE 1700, with a major increase observed since CE 1912 (Dickens et al., 2019). This trend has been attributed to a positive SAM, which drove stronger westerly winds, atmospheric warming and surface ablation on the eastern Antarctic Peninsula (Dickens et al., 2019). Increases in subsurface (50–400 m) ocean temperatures by 0.3°C–1.5°C are believed to have affected the stability and extent of ice shelves at the northeastern Antarctic Peninsula both during the Holocene and since the 1990s (Etourneau et al., 2019). Sea ice and temperature data from both sides of the northern Antarctic Peninsula show ocean warming and sea-ice reduction from CE 1935 to 1950 and notable variability but no clear trend thereafter, suggesting that multi-decadal sea ice variations over the last century were forced by the recent atmospheric warming affecting the Antarctic Peninsula (e.g., Vaughan et al., 2003), whereas the observed annual-to-decadal variability was mainly governed by synoptic and regional wind fields (Barbara et al., 2013).

11.5.7 Weddell Sea Embayment

The Weddell Sea sector may have contributed a similar magnitude of post-LGM sea level rise to that of the Ross Sea sector, yet our knowledge of the extent of the grounding line and the change in thickness of the inland ice is by comparison not as well constrained (see also Hillenbrand et al., 2014) (Fig. 11.8). The great uncertainty in LGM grounded ice thickness and extent has been fully described in the RAISED Consortium review for the Weddell

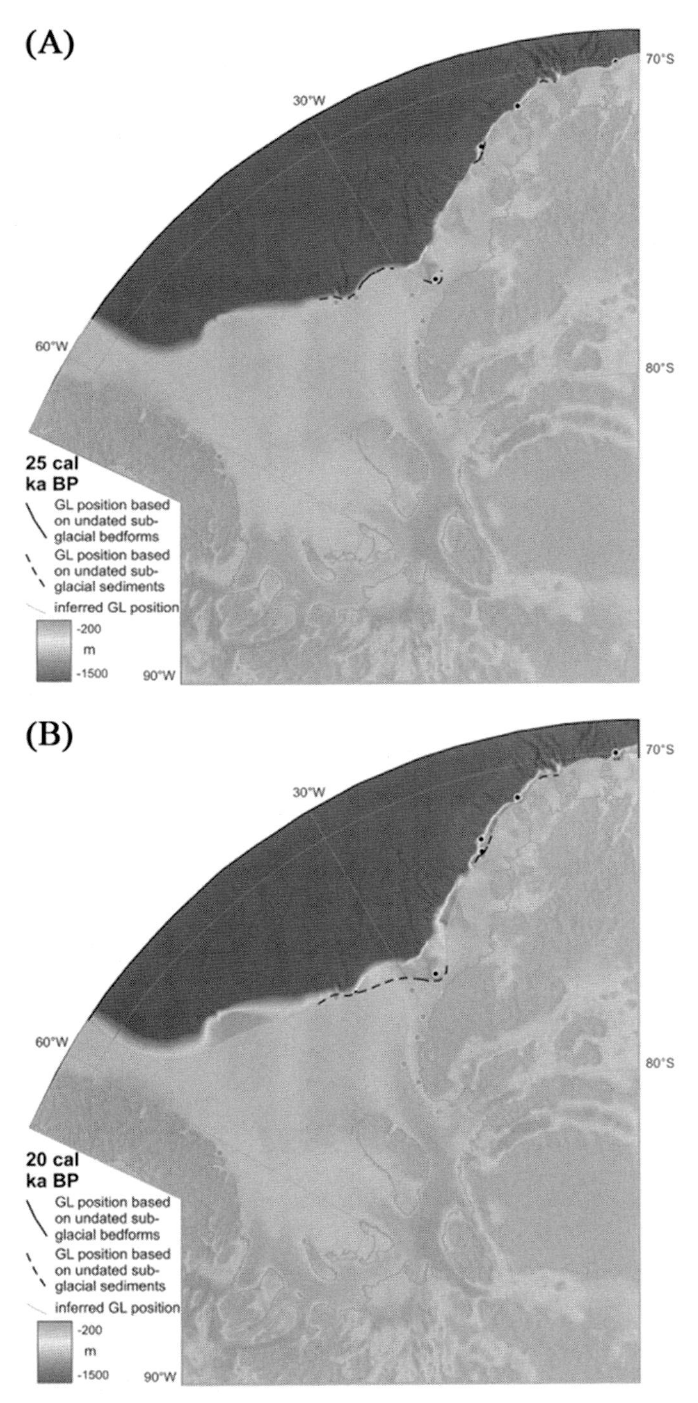

FIGURE 11.8 Ice sheet reconstructions across the Weddell Sea sector of Antarctica, revealing both a 'maximum' LGM extent based mainly on marine data and a 'minimum' LGM extent based mainly on terrestrial geological data. (A) 25 ka maximum, (B) 20 ka maximum, (C) 20 ka minimum, (D) 15 ka maximum, (E) 15 ka minimum, (F) 10 ka maximum, (G) 10 ka minimum, (H) 5 ka maximum, and (I) 5 ka minimum. *Reproduced from Hillenbrand et al. (2014) with permission from Elsevier.*

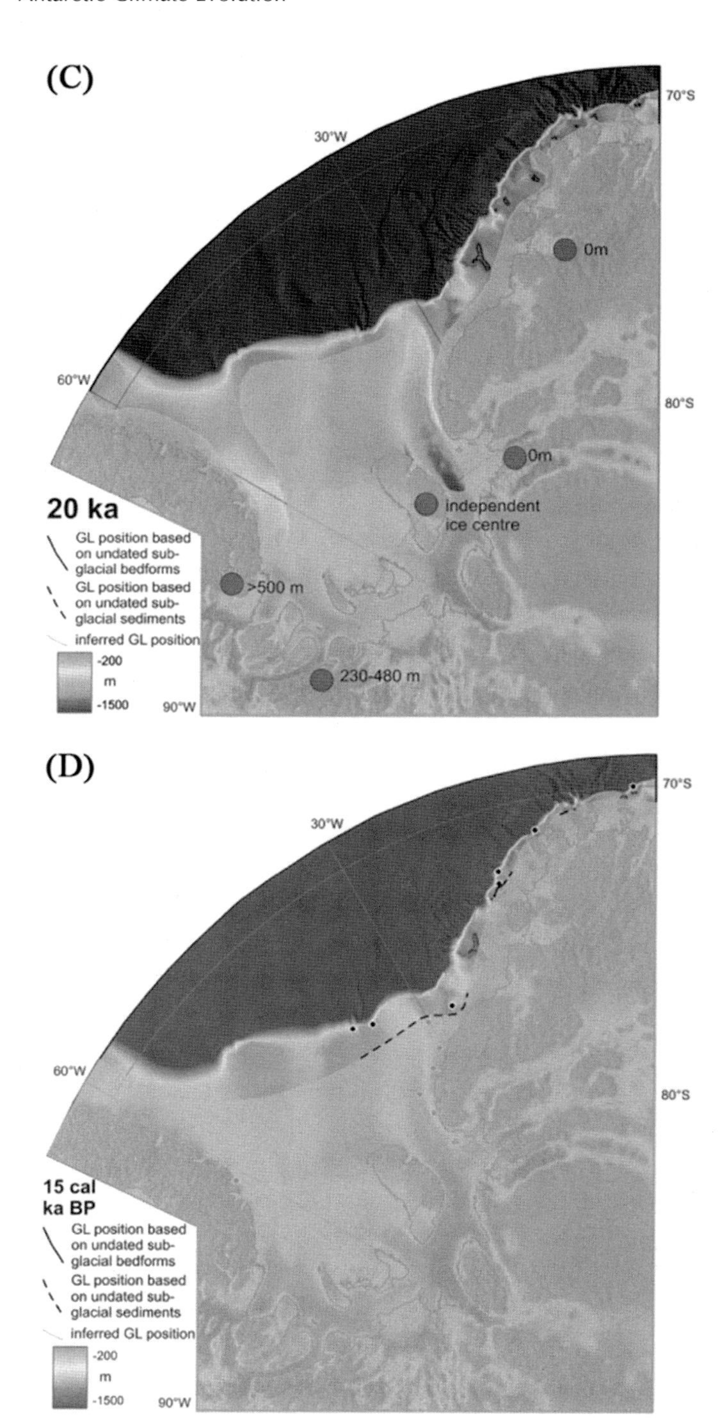

FIGURE 11.8 (*Contined*).

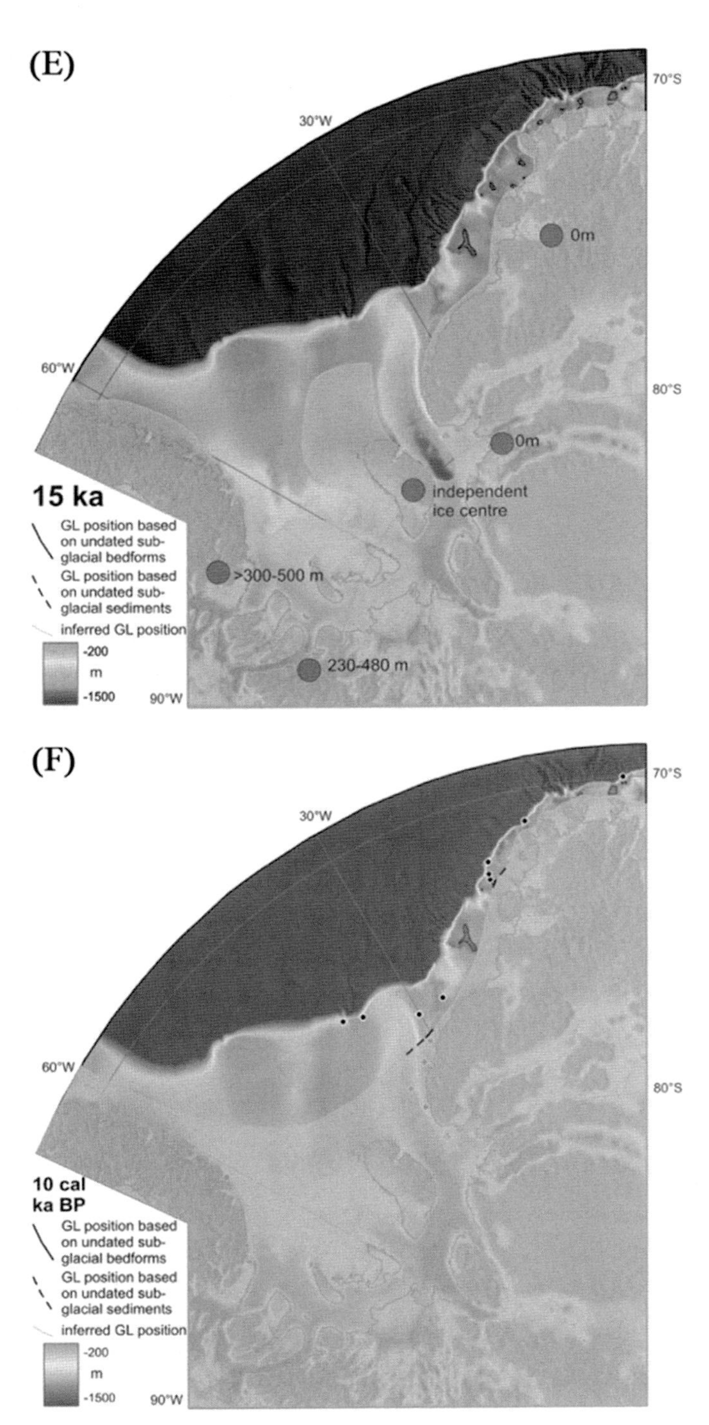

FIGURE 11.8 (*Contined*).

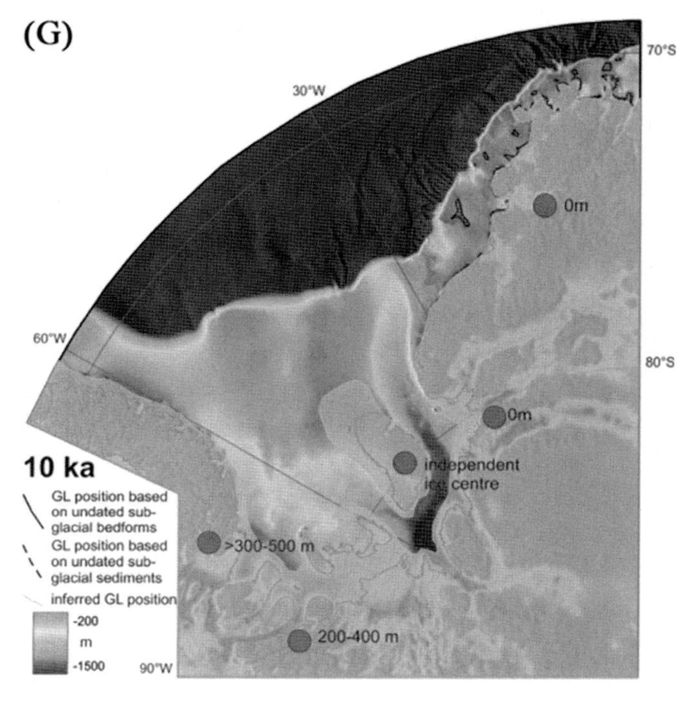

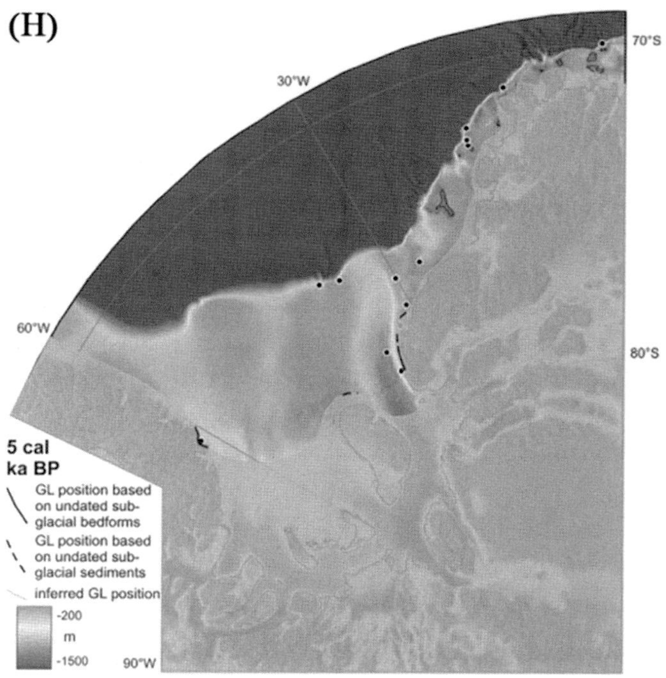

FIGURE 11.8 (*Contined*).

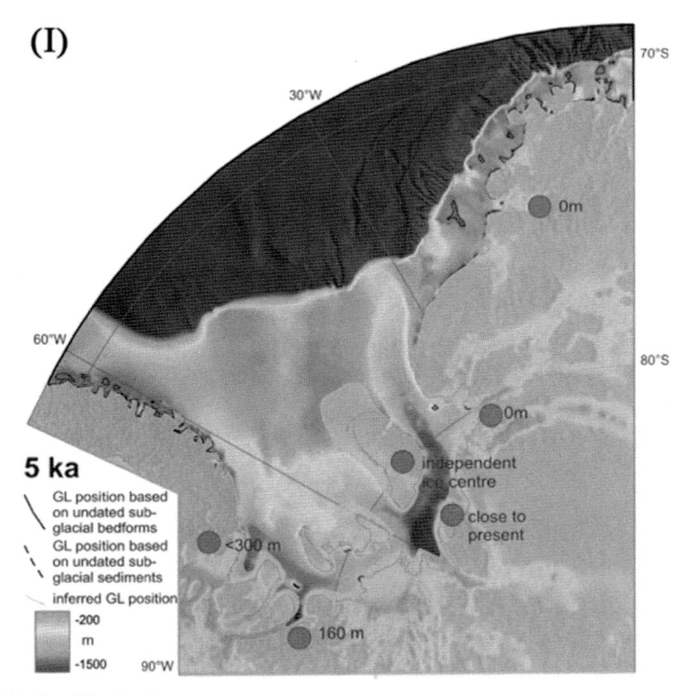

FIGURE 11.8 (*Contined*).

Sea sector (Hillenbrand et al., 2014). In this review, two deglaciation scenarios based on either marine or terrestrial data were presented. The key difference is the extent of grounded ice in the over-deepened troughs that extend beneath Filchner-Ronne Ice Shelf; the marine data favour grounded ice near the shelf edge while the terrestrial data favour more restricted grounded ice within Filchner Trough. Both configurations appear glaciologically plausible based on glacier flowline modelling (Whitehouse et al., 2017). The fact that two incompatible reconstructions of ice extent exist demonstrates the dearth of geochronological and geomorphological data in this sector. The situation is now changing. In the past decade, several new marine and terrestrial studies have provided important information on the thickness and extent of ice in this sector and the timing of its retreat.

In the Shackleton Range (80.2°S, 30.0°W), cosmogenic nuclide exposure-age dating of erratics and surface bedrock has been used to constrain the LGM thickening of the ice near the present-day grounding line of Slessor Glacier, where it joins Filchner Ice Shelf (Fogwill et al., 2004; Hein et al., 2011, 2014; Nichols et al., 2019). Here, ^{10}Be and ^{26}Al exposure dates of LGM age were found only on the modern ice margin, which suggests Slessor Glacier at the LGM was not significantly thicker or more extensive than today (Hein et al., 2011, 2014). Recent advances have made it possible to measure in situ ^{14}C in quartz, and this nuclide is better suited to date recent exposure because of its short half-life, which makes it comparatively insensitive to inheritance. Nichols et al. (2019) measured in situ ^{14}C in some of the same rock samples with ^{10}Be and ^{26}Al data and found clear evidence for thicker ice in the Shackleton Range during the late glacial period and the early Holocene. LGM ice surfaces in the Shackleton Range are now thought to have been between 310 and 650 m above the present Slessor Glacier surface.

In the Dufek Massif, northern Pensacola Mountains (centred at 83.5°S, 57.0°W), mapping of bouldery moraines and cosmogenic surface exposure dates of erratics on the moraines, along with radiocarbon ages around the margins of a pond, had suggested only moderate ice sheet thickening and advance of less than 2.5 km along-valley during the last glacial advance, which was assumed to have occurred at the LGM (Hodgson et al., 2012). Later on, cosmogenic nuclide exposure-age dating of erratics and bedrock was used in the southern Pensacola Mountains to constrain the LGM thickening of the ice near the present-day grounding line of Foundation Ice Stream, where it feeds into Filchner Ice Shelf (reviewed in Nichols et al., 2019; Siegert et al., 2019). Here, recent in situ ^{14}C exposure ages from Schmidt Hills indicate LGM thickening of at least 800 m (Nichols et al., 2019). This result supersedes the results of Balco et al. (2016) and Bentley et al. (2017) that indicated thickening of less than 200 m based on long-lived cosmogenic ^{10}Be data. Further upstream at Williams Hills there is evidence for at least 500 m of thickening of Foundation Ice Stream and Academy Glacier before 11 ka, with this amount of thinning occurring throughout the Holocene from 11 to 2.5 ka (Balco et al., 2016; Bentley et al., 2017). In the Thomas Hills further upstream, LGM thickening of at least 400 m is recorded by both ^{10}Be and in situ ^{14}C ages (Balco et al., 2016; Bentley et al., 2017; Nichols et al., 2019).

In the southern Weddell Sea sector, exposure dates demonstrate that erosional glacial trimlines preserved more than 1000 m above the present ice surface in the Ellsworth Mountains (Denton et al., 1992) are much older than LGM (Sugden et al., 2006). In the southern Heritage Range (80.2°S, 82.0°W), measurements of cosmogenic ^{26}Al, ^{10}Be and ^{21}Ne in a depth-profile through an ice-moulded bedrock surface indicate this erosional trimline pre-dates the Quaternary with an age of first exposure between 3.5 and 5.1 Ma; its true age is likely much older (Sugden and Jamieson, 2018; Sugden et al., 2017).

In the southern Ellsworth Mountains, ice elevations during the LGM were up to 475 m higher than today as evidenced by cosmogenic ^{10}Be exposure ages from elevated, ice marginal blue-ice moraine deposits in Marble Hills (Bentley et al., 2014; Hein et al., 2016a,b). Here, thinning began by 10 ka with a pulse of rapid thinning of ~400 m occurring at 6.5–3.5 ka (Hein et al., 2016b). The present ice margin is the same as it was at 3.5 ka. Thickening of Rutford Ice Stream reached ~900 m at the

LGM with thinning occurring between 14.5 and 6 ka (Fogwill et al., 2014a). Further south in Pirrit Hills, ice elevations reached 320 m above present and thinning occurred from 14 to 4 ka (Spector et al., 2019). In the Whitmore Mountains near the Weddell/Ross ice divide, the maximum ice thickness during the last glacial period were no more than 190 m when compared to the present (Spector et al., 2019). In situ ^{14}C and ^{10}Be data indicate the ice here was thicker than today for no more than 8 thousand years within the last 15 ka. Thickening at this inland site occurred well after the LGM in response to increase in snowfall (Spector et al., 2019), a result that agrees with earlier work in the interior of the WAIS (Ackert et al., 2007, 2013). At the ice divide between the Weddell Sea and Amundsen Sea drainage sectors, Ross et al. (2011) examined internal radio-echo layers, which show that this divide was remarkably stable during Holocene retreat of the ice margin.

In the southwestern Weddell Sea, cosmogenic ^{10}Be and ^{26}Al exposure ages indicate that LGM ice elevations in the Behrendt Mountains (75.5°S, 72.5°W) were at least 300 m higher than today (Bentley et al., 2006). Recent work on the Lassiter Coast (74.5°S, 62.5°W) demonstrates that the elevation of Johnston Glacier exceeded 385 m above present at the LGM (Johnson et al., 2019). Here, a mid-Holocene pulse of thinning of at least 250 m occurred from 7.5 to 6.0 ka according to in situ ^{14}C exposure ages from bedrock surfaces. The timing and magnitude of thinning are similar to those observed in the southern Ellsworth Mountains (Hein et al., 2016b) and Pensacola Mountains (Balco et al., 2016; Bentley et al., 2017).

Terrestrial data indicate greater ice thicknesses at the mouths of major Antarctic ice streams which result in surface gradients that are more compatible with the marine record indicating extensive grounded ice in Filchner Trough at the LGM (Anderson and Andrews, 1999; Anderson et al., 1980; Arndt et al., 2017; Bentley and Anderson, 1998; Elverhøi, 1981; Gales et al., 2013, 2014; Hillenbrand et al., 2012, 2014; Hodgson et al., 2018; Larter et al., 2012; Melles and Kuhn, 1993; Stolldorf et al., 2012). Recent swath bathymetry imagery, acoustic sub-bottom profiles and sediment core data from the outer Filchner Trough (75°50′S) provide evidence for a highly dynamic ice stream until the early Holocene (Arndt et al., 2017). Here, mega-scale glacial lineations, iceberg furrows and a series of GZWs were identified and their formation chronologically constrained. The geomorphology and available AMS ^{14}C dates on foraminifera suggest the grounded ice margin had advanced onto and retreated from the outer shelf before 27.5 cal. ka BP, several thousand years before the global LGM. The Filchner palaeo-ice stream then re-advanced to deposit GZWs at this site at least twice after 11.8 cal. ka BP, with final retreat before 8.7 cal. ka BP (Arndt et al., 2017), when glacimarine conditions established 220 km further upstream (Hillenbrand et al., 2014). This dynamic behaviour of the Filchner palaeo-ice stream may be linked to the reorganisation and re-routing of East and West Antarctic ice stream tributaries during the last glacial period, the post-LGM deglaciation and the Holocene (e.g., Arndt et al., 2017; Fogwill et al., 2014b; Kingslake et al., 2016; Larter et al., 2012; Rosier et al., 2018; Siegert et al., 2013, 2019; Whitehouse et al., 2017; Winter et al., 2015). The timing of early Holocene grounding line retreat in Filchner Trough is consistent with delayed early to mid-Holocene thinning observed at several locations across the southern Weddell Sea sector (Balco et al., 2016; Bentley et al., 2017; Hein et al., 2016a,b; Johnson et al., 2019; Nichols et al., 2019).

The glacial geomorphological footprint to the east of Filchner Trough reveals that the post-LGM retreat of the Filchner-palaeo ice stream resulted in its progressive southwards decoupling from outlet glaciers draining the EAIS along the eastern flank of the Weddell Sea embayment (Hodgson et al., 2018). In the northeastern embayment, a palaeo-ice stream that had formed in the area of the modern Stancomb-Wills Ice Shelf acted as the main conduit for ice drainage from the EAIS at the LGM (Arndt et al., 2020). GZWs document that this ice stream underwent stepwise retreat, with its grounding line having retreated from near the modern ice-shelf front by 10.5 cal. ka BP. Bedforms in front of the modern Brunt Ice Shelf further south hint at slow flow of grounded ice there during the last glacial period, while another glacial trough further west, called 'Halley Trough' (Gales et al., 2014), is assumed to have been covered by floating ice only during the LGM (Arndt et al., 2020).

Most geological records from the Weddell Sea sector indicate that ice margins had reached their present limits by the late Holocene. However, it is also suggested that grounding lines may have retreated significantly further upstream before re-advancing to their present locations as a consequence of glacial isostatic adjustment and ice shelf and ice rise buttressing (Bradley et al., 2015; Kingslake et al., 2016, 2018; Siegert et al., 2019; Wolstencroft et al., 2015). Radar investigations of the subglacial topography and internal ice sheet structures inform us that the Institute Ice Stream's grounding line is positioned on the top of a steep and sizeable reverse bed slope, towards the $>$2 km deep Robin Subglacial Basin (Siegert et al., 2019). However, satellite altimetry reveals little change today, suggesting the ice stream is presently 'stable', despite it being on the threshold of marine ice sheet instability (Ross et al., 2012). Radar stratigraphy reveals that Bungenstock Ice Rise, bordering one side of the ice stream, comprises buckled and broken internal layering beneath horizontal and unbroken layers. The glaciological explanation is that the ice rise has been subject to major changes in ice dynamics, from fast flowing in the past to today's very slow flow (Siegert et al., 2013). The date when this transition happened can be estimated from the age of the stratigraphic transition in the ice, which is around 5 ka. Hence, the region has experienced major ice dynamic change in the mid-Holocene (Siegert et al., 2019).

There are two explanations for this change. The first is that this entire sector of the ice sheet gently relaxed from its LGM position, halting the grounding line at its present location, with alteration in subglacial routing of water dictating ice-stream dynamic change. In this case, the buckled layers in Bungenstock Ice Rise are relics from a former ice stream that switched off ~5000 years ago. An alternative explanation is that, during post-LGM relaxation, the entire Robin Subglacial Basin deglaciated, and the release in the ice loading led to crustal uplift, which in turn caused floating ice across what is now Bungenstock Ice Rise to ground and build up. In this situation, the buckled layers are advected from an ice stream further up flow. At present, there is no conclusive evidence for either explanation (Siegert et al., 2019), though it is planned to undertake ice-core drilling combined with geophysical surveys in the region and utilise their results for constraining numerical models in order to understand how, and under which conditions, postglacial uplift can help to stabilise marine ice-sheet decay.

11.6 Discussion: pattern and timing of post-LGM ice retreat and thinning

The geological record provides convincing evidence that the Antarctic ice sheets were more expanded at the LGM and subsequently decreased in extent and thinned near their margins. The magnitude of ice retreat varies around the continent, ranging from relatively little marginal change at some terrestrial sites, such as the Larsemann Hills, to retreat of the grounding line by hundreds of kilometres in the large embayments currently occupied by the Ross and Filchner-Ronne ice shelves and along the Pacific margin of West Antarctica. At most of the marine sites studied, the evidence implies that the maximum ice extent at the LGM was relatively short-lived, while data from several terrestrial sites (e.g., Bunger Hills, Lützow-Holm Bay) suggest that the LGM was not necessarily the period of greatest ice extent during the last glacial cycle, and that ice sheet margins were more advanced prior to 35 ka.

Similarly, evidence for the timing of ice retreat following the LGM supports different retreat timings at different sites (Anderson, 1999; Livingstone et al., 2012; The RAISED Consortium, 2014). Some areas, such as the Antarctic Peninsula and the Amundsen Sea sector of the WAIS, appear to have responded relatively rapidly to global climate and sea level changes following the LGM. At other locations, including the Ross Sea, the Weddell Sea and the East Antarctic margin at Framnes Mountains, ice retreat appears to have begun relatively late and, in a few places, such as locations in Marie Byrd Land, may have continued to the present day. In the Ross Sea, ice streams fed by ice from West Antarctica interacted with glaciers from East Antarctica, leading to a complex retreat history that has only recently been recognised (Greenwood et al., 2018; Halberstadt et al., 2016; Prothro et al., 2018, 2020). While the evidence for this time-transgressive retreat is not definitive, the pattern may point to the differing sensitivities of each region to external forcing (e.g., Lowry et al., 2019).

Evidence for smaller ice volumes/ice extents during the mid-late Holocene relative to the present day is available at a number of sites. Sediments from the bed of epishelf Moutonée Lake indicate the free movement of icebergs, and therefore the partial or total collapse of the George VI Ice Shelf between Alexander Island and the Antarctic Peninsula, between 9.6 and 8 cal. ka BP (Bentley et al., 2005). Likewise, analyses of sediment cores recovered from the seabed of Prince Gustav Channel on the eastern side of the Antarctic Peninsula have shown that between 5 and 2 ^{14}C ka BP (corrected), clasts were sourced from a range of distal locations, indicating open marine conditions and the collapse of Prince Gustav Channel Ice Shelf (Pudsey and Evans, 2001). The mid-Holocene glacial minimum may not be confined to the Antarctic Peninsula. Goodwin (1996) provides evidence from a number of sources that Law Dome was at least 3–4 km smaller before ~4 ka, while data on penguin colonisation (Emslie et al., 2007) suggest that ice shelves along the western Ross Sea coast were reduced from ~4.0 to ~2.0 cal. ka BP.

Geological evidence suggests that the Larsen B Ice Shelf had been in place since 10.7 cal. ka BP before its collapse in CE 2002 (Domack et al., 2005; Rebesco et al., 2014). In East Antarctica, Verleyen et al. (2005) provide relative sea level (RSL)-based evidence for increased ice load near Larsemann Hills between ~7.1 and 2.5 cal. ka BP. Given the similarity of the RSL curve to that obtained from the Vestfold Hills (Zwartz et al., 1998b), this may have been more than a local feature. This fluctuation may correlate with glacial advances identified through striae patterns and weathering in the Vestfold Hills (Adamson and Pickard, 1986a,b) and the Rauer Group (White et al., 2009).

The idea that ocean forcing, rather than ice surface melting induced by atmospheric warming or global sea-level rise caused by post-LGM decay of Northern Hemisphere ice sheets, triggered rapid thinning and retreat of the Antarctic ice sheets subsequent to the LGM has stimulated a number of studies, which aimed at reconstructing advection of warm deep water onto the continental shelf. Following on from Hillenbrand et al. (2017) and Minzoni et al. (2017), who had reported ocean-forced ice sheet and ice-shelf retreat in the Amundsen Sea Embayment at the end of the LGM and during the Holocene, respectively, evidence from marine sediment records from the northeastern Antarctic Peninsula shelf show that ocean forcing strongly contributed to Holocene and recent ice-shelf collapses in this region (Etourneau et al., 2019). Similarly, the history of ice-sheet thinning and retreat in East Antarctica's Lützow-Holm Bay was also proposed to have resulted from warm deep-water intrusions (Kawamata et al., 2020). Further east in Edward VIII Gulf, however, elevated bed topography landward of deep troughs incised into the shelf apparently helped to stabilise the ice sheet during the Holocene, despite the presence of warm water on the continental shelf (Dove et al., 2020). Because most of the Antarctic continental margin remains poorly investigated and sampled, we are still far from understanding causes and feedbacks of ice-sheet/ocean interactions.

11.7 Summary

There is persuasive marine geological evidence from the relatively well-studied Ross Sea and Amundsen Sea embayments and the western Antarctic Peninsula shelf for an LGM ice sheet grounding line at, or very near to the edge of the continental shelf. A similar situation possibly existed on the eastern side of the Antarctic Peninsula and in the Weddell Sea embayment, although these regions have been less well studied to date. LGM thickening of the WAIS has been identified in several locations with the ice surface in the central regions likely to have been several hundred metres higher than today. Likewise, the APIS was up to 500 m thicker at central sites at that time.

The smaller number of studies undertaken on the EAIS have found wide variations in the extent of LGM ice advance across the continental shelf. It appears that the grounding line in East Antarctica advanced to the continental shelf break in some places, to a mid-shelf position in others, but not significantly at all elsewhere. Furthermore, there are several sites in East Antarctica, including the Bunger Hills, where deglaciation appears to have begun by ~30 ka, well before the LGM.

The timing and rate of deglaciation following the LGM and during the Holocene appears to have varied greatly across the continent. The APIS and the Amundsen Sea sector of the WAIS may have been among the first regions to reach their pre-industrial ice-sheet configurations, while the Ross Sea sector along with many coastal areas in East Antarctica responded less rapidly to climate and sea-level change since the LGM.

At the end of the 20th Century, estimates of the reduction in the total volume of the Antarctic ice sheets since their maximum during the last glacial cycle ranged from 0.5 to 37 m SLE. Analysis of the geological evidence in the late 1990s and early 2000s considered the upper estimates improbable, with a consensus tending towards the upper end of the lower half of this range. Although there was no single agreed value, estimates based on geological evidence at that time (Bentley, 1999, 6.1–13.1 m; Denton and Hughes, 2002, 14 m) started to converge with those based on numerical ice sheet modelling (Ritz et al., 2001, 5.9 m; Huybrechts, 2002, 19.2 m; Philippon et al., 2006, 9.5–17.5 m) and those based on modelling of isostatic postglacial uplift (Nakada et al., 2000, 6.6–16.7 m; Peltier, 2004, 17.3 m) to narrow the likely range to 5.9–19.2 m SLE. The most recent numerical models, published after 2010, indicate an Antarctic contribution of 9.9 ± 1.7 m SLE to the global LGM sea-level lowstand of 120 m (see references in Simms et al., 2019), which is largely in line with constraints from geological data (The RAISED Consortium, 2014) and from a recent global reconstruction of LGM ice volume (Gowan et al., 2021).

Acknowledgements

We thank Andrew Wright for his work on the original version of this paper (Wright et al., 2008), published in the first volume of Antarctic Climate Evolution (Florindo and Siegert, 2008), and the RAISED Consortium for providing figures for this chapter.

References

Ackert Jr., R.P., Barclay, D.J., Borns Jr., H.W., Calkin, P.E., Kurz, M.D., Fastook, J.L., et al., 1999. Measurements of past ice sheet elevations in interior West Antarctica. Science 286, 276–280. Available from: https://doi.org/10.1126/science.286.5438.276.

Ackert, R.P., Mukhopadhyay, S., Parizek, B.R., Borns, H.W., 2007. Ice elevation near the West Antarctic Ice Sheet divide during the last glaciation. Geophysical Research Letters 34, L21506. Available from: https://doi.org/10.1029/2007GL031412.

Ackert, R.P., Putnam, A.E., Mukhopadhyay, S., Pollard, D., DeConto, R.M., Kurz, M.D., et al., 2013. Controls on interior West Antarctic Ice Sheet elevations: inferences from geologic constraints and ice sheet modeling. Quaternary Science Reviews 65, 26–38. Available from: https://doi.org/10.1016/j.quascirev.2012.12.017.

Adamson, D.A., Mabin, M.C.G., Luly, J.G., 1997. Holocene isostasy and late Cenozoic development of landforms including Beaver and Radok Lake basins in the Amery Oasis, Prince Charles Mountains, Antarctica. Antarctic Science 9 (3), 299–306. Available from: https://doi.org/10.1017/S0954102097000382.

Adamson, D.A., Pickard, J., 1983. Late Quaternary ice movement across the Vestfold Hills, East Antarctica. In: Oliver, R.L., James, P.R., Jago, J.B. (Eds.), Antarctic Earth Science. Australian Academy of Science, Canberra, pp. 465–469.

Adamson, D.A., Pickard, J., 1986a. Cainozoic history of the Vestfold Hills. In: Pickard, J. (Ed.), Antarctic oasis. Academic Press, Sydney, pp. 63–97.

Adamson, D.A., Pickard, J., 1986b. Physiography and geomorphology of the Vestfold Hills, Antarctica. In: Pickard, J. (Ed.), Antarctic Oasis. Academic Press, Sydney, pp. 99–139.

Amblas, D., Urgeles, R., Canals, M., Calafat, A.M., Rebesco, M., Camerlenghi, A., et al., 2006. Relationship between continental rise development and palaeo-ice sheet dynamics, Northern Antarctic Peninsula Pacific margin. Quaternary Science Reviews 25, 933–944. Available from: https://doi.org/10.1016/j.quascirev.2005.07.012.

Anandakrishnan, S., Blankenship, D.D., Alley, R.B., Stoffa, P.L., 1998. Influence of subglacial geology on the position of a West Antarctic ice stream from seismic observations. Nature 394, 62–65. Available from: https://doi.org/10.1038/27889.

Anderson, J.B., 1999. Antarctic Marine Geology. Cambridge University Press, p. 289.

Anderson, J.B., Andrews, J.T., 1999. Radiocarbon constraints on ice sheet advance and retreat in the Weddell Sea, Antarctica. Geology 27, 179–182. Available from: https://doi.org/10.1130/0091-7613(1999)027%3C0179:RCOISA%3E2.3.CO;2.

Anderson, J.B., Conway, H., Bart, P.J., Witus, A.E., Greenwood, S.L., McKay, R.M., et al., 2014. Ross Sea paleo-ice sheet drainage and deglacial history during and since the LGM. Quaternary Science Reviews 100, 31–54. Available from: https://doi.org/10.1016/j.quascirev.2013.08.020.

Anderson, J.B., Kurtz, D.D., Domack, E.W., Balshaw, K.M., 1980. Glacial and glacial marine sediments of the Antarctic continental shelf. Journal of Geology 88, 399–414.

Anderson, J.B., Oakes-Fretwell, L., 2008. Geomorphology of the onset area of a palaeo-ice stream, Marguerite Bay, Antarctica Peninsula. Earth Surface Processes and Landforms 33, 503–512.

Anderson, J.B., Shipp, S.S., 2001. Evolution of the West Antarctic Ice Sheet. In: Alley, R.B., Bindschadler, R. (Eds.), The West Antarctic Ice Sheet: Behaviour and Environments, 77. American Geophysical Union, Washington, DC, pp. 45–57.

Anderson, J.B., Shipp, S.S., Lowe, A.L., Wellner, J.S., Mosola, A.B., 2002. The Antarctic Ice Sheet during the last glacial maximum and its subsequent retreat history: a review. Quaternary Science Reviews 21, 49–70. Available from: https://doi.org/10.1016/S0277-3791(01)00083-X.

Andrews, J.T., Domack, E.W., Cunningham, W.L., Leventer, A., Licht, K.J., Jull, A.J.T., et al., 1999. Problems and possible solutions concerning radiocarbon dating of surface marine sediments, Ross Sea, Antarctica. Quaternary Research 52, 206–216. Available from: https://doi.org/10.1006/qres.1999.2047.

Armand, L.K., 2000. An ocean of ice – advances in the estimation of past sea ice in the southern ocean. GSA Today 10, 1–7.

Arndt, J.E., Hillenbrand, C.-D., Grobe, H., Kuhn, G., Wacker, L., 2017. Evidence for a dynamic grounding-line in outer Filchner Trough, Antarctica, until the early Holocene. Geology 45 (11), 1035–1038. Available from: https://doi.org/10.1130/G39398.1.

Arndt, J.E., Larter, R.D., Hillenbrand, C.-D., Sørlie, S.H., Forwick, M., Smith, J.A., et al., 2020. Past ice sheet–seabed interactions in the northeastern Weddell Sea embayment, Antarctica. The Cryosphere 14, 2115–2135. Available from: https://doi.org/10.5194/tc-14-2115-2020.

Ashley, K.E., McKay, R., Etourneau, J., Jimenez-Espejo, F.J., Condron, A., Albot, A., et al., 2021. Mid-Holocene Antarctic sea-ice increase driven by marine ice sheet retreat. Climate of the Past 17, 1–19. Available from: https://doi.org/10.5194/cp-17-1-2021.

Austermann, J., Mitrovica, J.X., Latychev, K., Milne, G.A., 2013. Barbados-based estimate of ice volume at Last Glacial Maximum affected by subducted plate. Nature Geoscience 6, 553–557.

Balco, G., Schaefer, J.M., LARISSA Group, 2013. Exposure-age record of Holocene ice sheet and ice shelf change in the northeast Antarctic Peninsula. Quaternary Science Reviews 59, 101–111. Available from: https://doi.org/10.1016/j.quascirev.2012.10.022.

Balco, G., Todd, C., Huybers, K., Campbell, S., Vermuelen, M., Hegland, M., et al., 2016. Cosmogenic-nuclide exposure ages from the Pensacola Mountains adjacent to the foundation ice stream, Antarctica. American Journal of Science 316, 542–577. Available from: https://doi.org/10.2475/06.2016.02.

Bamber, J.L., Riva, R.E.M., Vermeersen, B.L.A., LeBrocq, A.M., 2009. Reassessment of the potential sea-level rise from a collapse of the West Antarctic Ice Sheet. Science 324, 901–903. Available from: https://doi.org/10.1126/science.1169335.

Banfield, L.A., Anderson, J.B., 1995. Seismic facies investigation of the Late Quarternary glacial history of Bransfield Basin, Antarctica. In: Cooper, A.K., Barker, P.F., Brancolini, G. (Eds.), Geology and Seismic Stratigraphy of the Antarctic Margin. Antarctic Research Series, 68. American Geophysical Union, pp. 123–140. Available from: https://doi.org/10.1029/AR068p0123.

Bard, E., Hamelin, B., Fairbanks, R.G., 1990. U-Th ages obtained by mass spectrometry in corals from Barbados: sea level during the past 130,000 years. Nature 346, 456–458. Available from: https://doi.org/10.1038/346456a0.

Barbara, L., Crosta, X., Leventer, A., Schmidt, S., Etourneau, J., Domack, E., et al., 2016. Environmental responses of the Northeast Antarctic Peninsula to the Holocene climate variability. Paleoceanography 31, 131–147. Available from: https://doi.org/10.1002/2015PA002785.

Barbara, L., Crosta, X., Massé, G., Ther, O., 2010. Deglacial environments in eastern Prydz Bay, East Antarctica. Quaternary Science Reviews 29, 2731–2740. Available from: https://doi.org/10.1016/j.quascirev.2010.06.027.

Barbara, L., Crosta, X., Schmidt, S., Massé, G., 2013. Diatoms and biomarkers evidence for major changes in sea ice conditions prior the instrumental period in Antarctic Peninsula. Quaternary Science Reviews 79, 99–110. Available from: https://doi.org/10.1016/j.quascirev.2013.07.021.

Baroni, C., Orombelli, G., 1994a. Abandoned penguin rookeries as Holocene paleoclimatic indicators in Antarctica. Geology 22, 23–26. Available from: https://doi.org/10.1130/0091-7613(1994)022(0023:APRAHP)2.3.CO;2.

Baroni, C., Orombelli, G., 1994b. Holocene glacier variations in the Terra Nova Bay area (Victoria Land, Antarctica). Antarctic Science 6, 497–505. Available from: https://doi.org/10.1130/0091-7613(1994)022(0023:APRAHP)2.3.CO;2.

Bart, P.J., Anderson, J.B., Nitsche, F., 2017a. Post-LGM grounding-line positions of the Bindschadler Ice Stream in Ross Sea. Journal of Geophysical Research: Earth Surface 122, 1827–1844. Available from: https://doi.org/10.1002/2017JF004259.

Bart, P.J., DeCesare, M., Rosenheim, B.E., Majewski, W., McGlannan, A., 2018. A centuries-long delay between a paleo-ice-shelf collapse and grounding-line retreat in the Whales Deep Basin, eastern Ross Sea, Antarctica. Scientific Reports 8, 12392. Available from: https://doi.org/10.1038/s41598-018-29911-8.

Bart, P.J., Krogmeier, B.J., Bart, M.P., Tulaczyk, S., 2017b. The paradox of a long grounding during West Antarctic Ice Sheet retreat in Ross Sea. Scientific Reports 7, 1262. Available from: https://doi.org/10.1038/s41598-017-01329-8.

Bart, P.J., Sjunneskog, C., Chow, J.M., 2011. Piston-core based biostratigraphic constraints on Pleistocene oscillations of the West Antarctic Ice Sheet in western Ross Sea between North Basin and AND-1B drill site. Marine Geology 289, 86–99. Available from: https://doi.org/10.1016/j.margeo.2011.09.005.

Bart, P., Tulaczyk, S., 2020. A significant acceleration of ice volume discharge preceded a major retreat of a West Antarctic paleo–ice stream. Geology 48, 313–317. Available from: https://doi.org/10.1130/G46916.1.

Bassett, S.E., Milne, G.A., Bentley, M.J., Huybrechts, P., 2007. Modelling Antarctic sea-level data to explore the possibility of a dominant Antarctic contribution to meltwater pulse IA. Quaternary Science Reviews 26, 2113–2127. Available from: https://doi.org/10.1016/j.quascirev.2007.06.011.

Beaman, R.J., Harris, P.T., 2003. Seafloor morphology and acoustic facies of the George V Land shelf. Deep-Sea Research II 50, 1343–1355.

Beaman, R.J., O'Brien, P.E., Post, A.L., De Santis, L., 2011. A new high-resolution bathymetry model for the Terre Adélie and George V continental margin, East Antarctica. Antarctic Science 23 (1), 95–103. Available from: https://doi.org/10.1017/S095410201000074X.

Bentley, M.J., 1999. Volume of Antarctic ice at the Last Glacial Maximum, and its impact on global sea level change. Quaternary Science Reviews 18, 1569–1595. Available from: https://doi.org/10.1016/S0277-3791(98)00118-8.

Bentley, M.J., Anderson, J.B., 1998. Glacial and marine geological evidence for the ice sheet configuration in the Weddell Sea Antarctic Peninsula region during the Last Glacial Maximum. Antarctic Science 10, 309–325. Available from: https://doi.org/10.1017/S0954102098000388.

Bentley, M.J., Fogwill, C.J., Kubik, P.W., Sugden, D.E., 2006. Geomorphological evidence and cosmogenic 10Be/26Al exposure ages for the Last Glacial Maximum and deglaciation of the Antarctic Peninsula Ice Sheet. Geological Society of America Bulletin 118, 1149–1159. Available from: https://doi.org/10.1130/B25735.1.

Bentley, M.J., Hein, A.S., Sugden, D.E., Whitehouse, P.L., Shanks, R., Xu, S., et al., 2017. Deglacial history of the Pensacola Mountains, Antarctica from glacial geomorphology and cosmogenic nuclide surface exposure dating. Quaternary Science Reviews 158, 58–76. Available from: https://doi.org/10.1016/j.quascirev.2016.09.028.

Bentley, M.J., Hodgson, D.A., Smith, J.A., Cofaigh, C.Ó., Domack, E.W., Larter, R.D., et al., 2009. Mechanisms of Holocene palaeoenvironmental change in the Antarctic Peninsula region. The Holocene 19, 51–66. Available from: https://doi.org/10.1177/0959683608096603.

Bentley, M.J., Hodgson, D.A., Sugden, D.E., Roberts, S.J., Smith, J.A., Leng, M.J., et al., 2005. Early Holocene retreat of the George VI Ice Shelf, Antarctic Peninsula. Geology 33, 173–176. Available from: https://doi.org/10.1130/G21203.1.

Bentley, M.J., Kubik, P.W., Sugden, D.E., 2000. Geomorphological evidence and cosmogenic $^{10}Be/^{26}Al$ exposure ages for the Last Glacial Maximum configuration of the Antarctic Peninsula ice sheet. In: Hjort, C. (Ed.), Glaciation of the Weddell Sea Basin. ANTIME Workshop, Abisko.

Bentley, M.J., Johnson, J.S., Hodgson, D.A., Dunai, T., Freeman, S.P.H.T., Cofaigh, C.Ó., 2011. Rapid deglaciation of Marguerite Bay, western Antarctic Peninsula in the Early Holocene. Quaternary Science Reviews 30, 3338–3349. Available from: https://doi.org/10.1016/j.quascirev.2011.09.002.

Bentley, M.J., Cofaigh, C.Ó., Anderson, J.B., Conway, H., et al., 2014. A community-based geological reconstruction of Antarctic Ice Sheet deglaciation since the Last Glacial Maximum. Quaternary Science Reviews 100, 1–9. Available from: https://doi.org/10.1016/j.quascirev.2014.06.025.

Berg, S., Melles, M., Gore, D.B., Verkulich, S., Pushina, Z.V., 2020. Postglacial evolution of marine and lacustrine water bodies in Bunger Hills. Antarctic Science 32, 107–129. Available from: https://doi.org/10.1017/S0954102019000476.

Berg, S., Wagner, B., Cremer, H., Leng, M.J., Melles, M., 2010. Late Quaternary environmental and climate history of Rauer Group, East Antarctica. Palaeogeography, Palaeoclimatology, Palaeoecology 297, 201–213. Available from: https://doi.org/10.1016/j.palaeo.2010.08.002.

Berg, S., White, D.A., Bennike, O., Fülöp, R.-H., Fink, D., 2016. Unglaciated areas in East Antarctica during the Last Glacial (Marine Isotope Stage 3) – new evidence from Rauer Group. Quaternary Science Reviews 153, 1–10. Available from: https://doi.org/10.1016/j.quascirev.2016.08.021.

Berkman, P., Forman, S.L.J., 1996. Pre-bomb radiocarbon and the reservoir correction for calcareous marine species in the Southern Ocean. Geophysical Research Letters 23, 363–366. Available from: https://doi.org/10.1029/96GL00151.

Borchers, S., Dietze, E., Kuhn, G., Esper, O., Voigt, I., Hartmann, K., et al., 2015. Holocene ice dynamics and bottom-water formation associated with Cape Darnley polynya activity recorded in Burton Basin, East Antarctica. Marine Geophysical Research 37, 49–70. Available from: https://doi.org/10.1007/s11001-015-9254-z.

Brachfeld, S., Domack, E.W., Kissel, C., Laj, C., Leventer, A., Ishman, S., et al., 2003. Holocene history of the Larsen-A Ice Shelf constrained by geomagnetic paleointensity dating. Geology 31, 749–752. Available from: https://doi.org/10.1130/G19643.1.

Bradley, S.L., Hindmarsh, R.C., Whitehouse, P.L., Bentley, M.J., King, M.A., 2015. Low post-glacial rebound rates in the Weddell Sea due to late Holocene ice-sheet readvance. Earth and Planetary Science Letters 413, 79–89. Available from: https://doi.org/10.1016/j.epsl.2014.12.039.

Briggs, R.D., Pollard, D., Tarasov, L., 2014. A data-constrained large ensemble analysis of Antarctic evolution since the Eemian. Quaternary Science Reviews 103, 91–115. Available from: https://doi.org/10.1016/j.quascirev.2014.09.003.

Broecker, W.S., Denton, G.H., 1990. The role of ocean-atmosphere reorganizations in glacial cycles. Quaternary Science Reviews 9 (4), 305–341. Available from: https://doi.org/10.1016/0277-3791(90)90026-7.

Brook, E.J., Kurz, M.D., Ackert, R.P., Raisbeck, G.M., Yiou, F., 1995. Cosmogenic nuclide exposure ages and glacial history of late Quaternary Ross Sea drift in McMurdo Sound, Antarctica. Earth and Planetary Science Letters 131 (1–2), 41–56. Available from: https://doi.org/10.1016/0012-821X(95)00006-X.

Budd, W.F., Smith, I.N., 1982. Large scale numerical modelling of the Antarctic Ice Sheet. Annals of Glaciology 3, 42–49.

Burgess, J.S., Spate, A.P., Shevlin, J., 1994. The onset of deglaciation in the Larsemann Hills, Eastern Antarctica. Antarctic Science 6 (4), 491–495.

Campo, J., Wellner, J.S., Lavoie, C., Domack, E., Yoo, K.-C., 2017. Glacial geomorphology of the northwestern Weddell Sea, eastern Antarctic Peninsula continental shelf: shifting ice flow patterns during deglaciation. Geomorphology 280, 89–107. Available from: https://doi.org/10.1016/j.geomorph.2016.11.022.

Canals, M., Casamor, J.L., Urgeles, R., Calafat, A.M., Domack, E.W., Baraza, J., et al., 2002. Seafloor evidence of a subglacial sedimentary system off the northern Antarctic Peninsula. Geology 30, 603–606. Available from: https://doi.org/10.1130/0091-7613(2002)030%3C0603:SEOASS%3E2.0.CO;2.

Canals, M., Urgeles, R., Calafat, A.M., 2000. Deep sea-floor evidence of past ice streams off the Antarctic Peninsula. Geology 28, 31–34. Available from: https://doi.org/10.1130/0091-7613(2000)028%3C0031:DSEOPI%3E2.0.CO;2.

Carson, C.J., Post, A.L., Smith, J., Walker, G., Waring, P., Bartley, R., et al., 2017. The seafloor geomorphology of the Windmill Islands, Wilkes Land, East Antarctica: evidence of Law Dome ice margin dynamics. Geomorphology 292 (2017), 1–15. Available from: https://doi.org/10.1016/j.geomorph.2017.04.031.

Christ, A.J., Bierman, P.R., 2020. The local Last Glacial Maximum in McMurdo Sound, Antarctica: implications for ice-sheet behavior in the Ross Sea Embayment. Geological Society of America Bulletin 132 (1–2), 31–47. Available from: https://doi.org/10.1130/B35139.1.

Çiner, A., Yildirim, C., Sarikaya, M.A., Seong, Y.B., Yu, B.Y., 2019. Cosmogenic ^{10}Be exposure dating of glacial erratics on Horseshoe Island in western Antarctic Peninsula confirms rapid deglaciation in the Early Holocene. Antarctic Science 31 (6), 319–331. Available from: https://doi.org/10.1017/S0954102019000439.

Clark, P.U., Tarasov, L., 2014. Closing the sea level budget at the last glacial maximum. Proceedings of the National Academy of Sciences 111, 15861–15862. Available from: https://doi.org/10.1073/pnas.1418970111.

CLIMAP, 1976. The surface of the Ice Age Earth. Science 191, 1131–1136.

Colhoun, E.A., Adamson, D.A., 1992. Raised beaches of the Bunger Hills. Australian National Antarctic Research Expeditions Report 136, 47.

Colhoun, E.A., Mabin, M.C.G., Adamson, D.A., Kirk, R.M., 1992. Antarctic ice volume and contribution to sea level fall at 20 000 yr BP from raised beaches. Nature 358, 316–319. Available from: https://doi.org/10.1038/358316a0.

Colleoni, F., De Santis, L., Siddoway, C.S., Bergamasco, A., Golledge, N., Lohmann, G., G., et al., 2018. Spatio-temporal variability of processes across Antarctic ice-bed-ocean interfaces. Nature Communications 8, 2289. Available from: https://doi.org/10.1038/s41467-018-04583-0.

Crosta, X., Crespin, J., Swingedouw, D., Marti, O., Masson-Delmotte, V., Etourneau, J., et al., 2018. Ocean as the main driver of Antarctic Ice Sheet retreat during the Holocene. Global and Planetary Change 166, 62–74. Available from: https://doi.org/10.1016/j.gloplacha.2018.04.007.

Crosta, X., Etourneau, J., Orme, L.C., Dalaiden, Q., Campagne, P., Swingedouw, D., et al., 2021. Multi-decadal trends in Antarctic sea-ice extent driven by ENSO–SAM over the last 2,000 years. Nature Geoscience. Available from: https://doi.org/10.1038/s41561-021-00697-1.

Cuffey, K.M., Marshall, S.J., 2000. Substantial contribution to sea-level rise during the last interglacial from the Greenland ice sheet. Nature 404, 591–594. Available from: https://doi.org/10.1038/35007053.

Dahl-Jensen, D., Albert, M., Aldahan, A., et al., 2013. Eemian interglacial reconstructed from a Greenland folded ice core. Nature 493, 489–494. Available from: https://doi.org/10.1038/nature11789.

Davies, B., Hambrey, M.J., Glasser, N.F., Holt, T., Rodes, A., Smellie, J.L., et al., 2017. Ice-dammed lateral lake and epishelf lake insights into Holocene dynamics of Marguerite Trough Ice Stream and George VI Ice Shelf, Alexander Island, Antarctic Peninsula. Quaternary Science Reviews 177, 189–219. Available from: https://doi.org/10.1016/j.quascirev.2017.10.016.

Davies, B.J., Glasser, N.F., Carrivick, J.L., Hambrey, M.J., Smellie, J.L., Nývlt, D., 2013. Antarctic landscape evolution and ice-sheet behaviour in a semi-arid polar environment: James Ross Island, NE Antarctic Peninsula. Palaeoenvironments and Earth Surface Processes. Geological Society of London Special Publication 381, 353–395. Available from: https://doi.org/10.1144/SP381.1.

Davies, B.J., Hambrey, M.J., Smellie, J.L., Carrivick, J.L., Glasser, N.F., 2012. Antarctic Peninsula Ice Sheet evolution during the Cenozoic Era. Quaternary Science Reviews 31, 30–66. Available from: https://doi.org/10.1016/j.quascirev.2014.06.023.

DeConto, R.M., Pollard, D., 2016. Contribution of Antarctica to past and future sea-level rise. Nature 531 (7596), 591–597. Available from: https://doi.org/10.1038/nature17145.

Denis, D., Crosta, X., Schmidt, S., Carson, D.S., Ganeshram, R.S., Renssen, H., et al., 2009. Holocene glacier and deep water dynamics, Adelie Land region, East Antarctica. Quaternary Science Reviews 28, 1291–1303. Available from: https://doi.org/10.1016/j.quascirev.2008.12.024.

Denton, G.H., Bockheim, J.G., Rutford, R.H., Andersen, B.G., 1992. Glacial history of the Ellsworth Mountains, West Antarctica. In: Webers, G.F., Craddock, C., Splettstoesser, J.F. (Eds.), Geology and Palaeontology of the Ellsworth Mountains, West Antarctica. Geological Society of America, Boulder, CO.

Denton, G.H., Hughes, T.P., 2002. Reconstructing the Antarctic Ice Sheet at the Last Glacial Maximum. Quaternary Science Reviews 21, 193–202. Available from: https://doi.org/10.1016/S0277-3791(01)00090-7.

Denton, G.H., Prentice, M.L., Burckle, L.H., 1991. Cainozoic history of the Antarctic Ice Sheet. In: Tingey, R.J. (Ed.), The Geology of Antarctica. Clarendon Press, Oxford, pp. 365–433.

Depoorter, M.A., Bamber, J.L., Griggs, J.A., Lenaerts, J.T.M., Ligtenberg, S.R.M., van den Broeke, M.R., et al., 2013. Calving fluxes and basal melt rates of Antarctic ice shelves. Nature 502, 89–92. Available from: https://doi.org/10.1038/nature12567.

Di Roberto, A., Colizza, E., Del Carlo, P., Petrelli, M., Finocchiaro, F., Kuhn, G., 2019. First marine cryptotephra in Antarctica found in sediments of the western Ross Sea correlates with englacial tephras and climate records. Scientific Reports 9, 10628. Available from: https://doi.org/10.1038/s41598-019-47188-3.

Dickens, W.A., Kuhn, G., Leng, M.J., Graham, A.G.C., Dowdeswell, J.A., Meredith, M.P., et al., 2019. Enhanced glacial discharge from the eastern Antarctic Peninsula since the 1700s associated with a positive Southern Annular Mode. Scientific Reports 9, 14606. Available from: https://doi.org/10.1038/s41598-019-50897-4.

Domack, E.W., Amblas, D., Gilbert, R., Brachfeld, S., Camerlenghi, A., Rebesco, M., et al., 2006. Subglacial morphology and glacial evolution of the Palmer Deep outlet system, Antarctic Peninsula. Geomorphology 75, 125–142. Available from: https://doi.org/10.1016/j.geomorph.2004.06.013.

Domack, E.W., Duran, D., Leventer, A., Ishman, S., Doane, S., McCallum, S., et al., 2005. Stability of the Larsen-B Ice Shelf on the Antarctic Peninsula during the Holocene epoch. Nature 436, 681–685. Available from: https://doi.org/10.1038/nature03908.

Domack, E.W., Jacobsen, E.A., Shipp, S.S., Anderson, J.B., 1999. Late Pleistocene-Holocene retreat of the West Antarctic Ice-Sheet system in the Ross Sea: part 2 – sedimentologic and stratigraphic signature. Geological Society of America Bulletin 111, 1517–1536. Available from: https://doi.org/10.1130/0016-7606(1999)111%3C1517:LPHROT%3E2.3.CO;2.

Domack, E.W., Jull, A.J.T., Nakao, S., 1991. Advance of East Antarctic outlet glaciers during the hypsithermal; implications for the volume state of the Antarctic Ice Sheet under global warming. Geology 19, 1059–1062. Available from: https://doi.org/10.1130/0091-7613(1991)019%3C1059:AOEAOG%3E2.3.CO;2.

Domack, E.W., Leventer, A., Dunbar, R., Taylor, F., Brachfeld, S., Sjunneskog, C., et al., 2001. Chronology of the Palmer Deep site, Antarctic Peninsula: a Holocene palaeoenvironmental reference for the circum-Antarctic. The Holocene 11, 1–9. Available from: https://doi.org/10.1191/095968301673881493.

Domack, E.W., O'Brien, P., Harris, P.T., Taylor, F., Quilty, P.G., De Santis, L., et al., 1998. Late Quaternary sediment facies in Prydz Bay, East Antarctica and their relationship to glacial advance onto the continental shelf. Antarctic Science 10 (3), 236–246. Available from: https://doi.org/10.1017/S0954102098000339.

Dove, I.A., Leventer, A., Metcalf, M.J., Brachfeld, S.A., Dunbar, R.B., Manley, P., et al., 2020. Marine geological investigation of Edward VIII Gulf, Kemp Coast, East Antarctica. Antarctic Science 32, 210–222. Available from: https://doi.org/10.1017/S0954102020000097.

Dowdeswell, J.A., Batchelor, C.L., Montelli, A., Ottesen, D., Christie, F.D.W., Dowdeswell, E. K., et al., 2020. Delicate seafloor landforms reveal past Antarctic grounding-line retreat of kilometers per year. Science 368 (6494), 1020–1024. Available from: https://doi.org/10.1126/science.aaz3059.

Dowdeswell, J.A., Cofaigh, C.Ó., Noormets, R., Larter, R.D., Hillenbrand, C.-D., Benetti, S., et al., 2008b. A major trough-mouth fan on the continental margin of the Bellingshausen Sea, West Antarctica: The Belgica Fan. Marine Geology 252, 129–140. Available from: https://doi.org/10.1016/j.margeo.2008.03.017.

Dowdeswell, J.A., Cofaigh, C.Ó., Pudsey, C.J., 2004. Thickness and extent of the subglacial till layer beneath an Antarctic paleo-ice stream. Geology 32, 13–16. Available from: https://doi.org/10.1130/G19864.1.

Dowdeswell, J.A., Evans, J., Cofaigh, C.Ó., Anderson, J.B., 2006. Morphology and sedimentary processes on the continental slope off Pine Island Bay, Amundsen Sea, West Antarctica. Geological Society of America Bulletin 118 (5/6), 606–619. Available from: https://doi.org/10.1130/B25791.1.

Dowdeswell, J.A., Ottesen, D., Evans, J., Cofaigh, C.Ó., Anderson, J.B., 2008a. Submarine glacial landforms and rates of ice-stream collapse. Geology 36, 819–822. Available from: https://doi.org/10.1130/G24808A.1.

Dutton, A., Carlson, A.E., Long, A.J., Milne, G.A., Clark, P.U., DeConto, R., et al., 2015. Sea-level rise due to polar ice-sheet mass loss during past warm periods. Science 349 (6244), . Available from: https://doi.org/10.1126/science.aaa4019aaa4019-1.

Elverhøi, A., 1981. Evidence for a late Wisconsin glaciation of the Weddell Sea. Nature 293, 641–642. Available from: https://doi.org/10.1038/293641a0.

Emslie, S.D., Coats, L.L., Licht, K.J., 2007. A 45,000 yr record of Adélie penguins and climate change in the Ross Sea, Antarctica. Geology 35 (1), 61–64. Available from: https://doi.org/10.1130/G23011A.1.

Etourneau, J., Collins, L.G., Willmott, V., Kim, J.-H., Barbara, L., Leventer, A., et al., 2013. Holocene climate variations in the western Antarctic Peninsula: evidence for sea ice extent predominantly controlled by changes in insolation and ENSO variability. Climate of the Past 9, 1431–1446. Available from: https://doi.org/10.5194/cp-9-1431-2013.

Etourneau, J., Sgubin, G., Crosta, X., Swingedouw, D., Willmott, V., Barbara, L., et al., 2019. Ocean temperature impact on ice shelf extent in the eastern Antarctic Peninsula. Nature Communications 10, 304. Available from: https://doi.org/10.1038/s41467-018-08195-6.

Evans, J., Dowdeswell, J.A., Cofaigh, C.Ó., 2004. Late Quaternary submarine bedforms and ice-sheet flow in Gerlache Strait and on the adjacent continental shelf, Antarctic Peninsula. Journal of Quaternary Science 19 (4), 397–407. Available from: https://doi.org/10.1002/jqs.831.

Evans, J., Dowdeswell, J.A., Cofaigh, C.Ó., Benham, T.J., Anderson, J.B., 2006. Extent and dynamics of the West Antarctic Ice Sheet on the outer continental shelf of Pine Island Bay during the last glaciation. Marine Geology 230, 53–72. Available from: https://doi.org/10.1016/j.margeo.2006.04.001.

Evans, J., Pudsey, C.J., Cofaigh, C.Ó., Morris, P., Domack, E.W., 2005. Late Quaternary glacial history, flow dynamics and sedimentation along the eastern margin of the Antarctic Peninsula Ice Sheet. Quaternary Science Reviews 24 (5–6), 741–774. Available from: https://doi.org/10.1016/j.quascirev.2004.10.007.

Fabel, D., Stone, J., Fifield, L.K., Cresswell, R.G., 1997. Deglaciation of the Vestfold Hills, East Antarctica: preliminary evidence from exposure dating of three subglacial erratics. In: Ricci, C.A. (Ed.), The Antarctic Region: Geological Evolution and Processes. Terra Antartica Publications, Siena, pp. 829–834.

Farmer, G.L., Licht, K., Swope, R.J., Andrews, J., 2006. Isotopic constraints on the provenance of fine-grained sediment in LGM tills from the Ross Embayment, Antarctica. Earth and Planetary Science Letters 249, 90–107. Available from: https://doi.org/10.1016/j.epsl.2006.06.044.

Fernandez, R., Gulick, S., Domack, E., Montelli, A., Leventer, A., Shevenell, A., et al., 2018. Past ice stream and ice sheet changes on the continental shelf off the Sabrina Coast, East Antarctica. Geomorphology 317, 10–22. Available from: https://doi.org/10.1016/j.geomorph.2018.05.020.

Florindo, F., Siegert, M.J. (Eds.), 2008. Antarctic Climate Evolution. Elsevier, 606 pp. ISBN: 9780444528476.

Fogwill, C.J., Bentley, M.J., Sugden, D.E., Kerr, A.R., Kubik, P.W., 2004. Cosmogenic nuclides ^{10}Be and ^{26}Al imply limited Antarctic Ice Sheet thickening and low erosion in the Shackleton Range for >1 m.y. Geology 32, 265–268. Available from: https://doi.org/10.1130/G19795.1.

Fogwill, C.J., Turney, C.S.M., Golledge, N.R., Rood, D.H., Hippe, K., Wacker, L., et al., 2014b. Drivers of abrupt Holocene shifts in West Antarctic ice stream direction determined from combined ice sheet modelling and geologic signatures. Antarctic Science 26 (6), 674–686. Available from: https://doi.org/10.1017/S0954102014000613.

Fogwill, C.J., Turney, C.S.M., Meissner, K.J., Golledge, N.R., Spence, P., Roberts, J.L., et al., 2014a. Testing the sensitivity of the East Antarctic Ice Sheet to Southern Ocean dynamics: past changes and future implications. Journal of Quaternary Science 29, 91–98. Available from: https://doi.org/10.1002/jqs.2683.

Fretwell, P., et al., 2013. Bedmap2: improved ice bed, surface and thickness datasets for Antarctica. The Cryosphere 7, 375–393. Available from: https://doi.org/10.5194/tc-7-375-2013.

Gales, J.A., Larter, R.D., Mitchell, N.C., Dowdeswell, J.A., 2013. Geomorphic signature of Antarctic submarine gullies: implications for continental slope processes. Marine Geology 337, 112–124. Available from: https://doi.org/10.1016/j.margeo.2013.02.003.

Gales, J.A., Leat, P.T., Larter, R.D., Kuhn, G., Hillenbrand, C.-D., Graham, A.G.C., et al., 2014. Large-scale submarine landslides, channel and gully systems on the southern Weddell Sea margin, Antarctica. Marine Geology 348, 73–87. Available from: https://doi.org/10.1016/j.margeo.2013.12.002.

Gales, J., Rebesco, M., De Santis, L., Bergamasco, A., Colleoni, F., Kim, S., et al. 2021. Role of dense shelf water in the development of Antarctic submarine canyon morphology. Geomorphology 372, 107453. Available from: https://doi.org/10.1016/j.geomorph.2020.107453.

Gao, Y., Yang, L., Wang, J., Xie, Z., Wang, Y., Sun, L., 2018. Penguin colonization following the last glacial-interglacial transition in the Vestfold Hills, East Antarctica. Palaeogeography, Palaeoclimatology, Palaeoecology 490, 629–639. Available from: https://doi.org/10.1016/j.palaeo.2017.11.053.

Gersonde, R., Crosta, X., Abelmann, A., Armand, L., 2005. Sea-surface temperature and sea ice distribution of the Southern Ocean at the EPILOG Last Glacial Maximum – circum-Antarctic view based on siliceous microfossil records. Quaternary Science Reviews 24, 869–896. Available from: https://doi.org/10.1016/j.quascirev.2004.07.015.

Gibson, J.A.E., Paterson, K.S., White, C.A., Swadling, K.M., 2009. Evidence for the continued existence of Abraxas Lake, Vestfold Hills, East Antarctica during the Last Glacial Maximum. Antarctic Science 21 (3), 269–278. Available from: https://doi.org/10.1017/S0954102009001801.

Glasser, N.F., Davies, B.J., Carrivick, J.L., Ròdes, A., Hambrey, M.J., Smellie, J.L., et al., 2014. Ice-stream initiation, duration and thinning on James Ross Island, northern Antarctic Peninsula. Quaternary Science Reviews 86, 78–88. Available from: https://doi.org/10.1016/j.quascirev.2013.11.012.

Goehring, B.M., Balco, G., Todd, C., Moening-Swanson, I., Nichols, K., 2019. Late-glacial grounding line retreat in the northern Ross Sea, Antarctica. Geology 47, 291–294. Available from: https://doi.org/10.1130/G45413.1.

Golledge, N.R., Levy, R.H., McKay, R.M., Fogwill, C.J., White, D.A., Graham, A.G.C., et al., 2013. Glaciology and geological signature of the Last Glacial Maximum Antarctic Ice Sheet. Quaternary Science Reviews 78, 225–247. Available from: https://doi.org/10.1016/j.quascirev.2013.08.011.

Golledge, N.R., Menviel, L., Carter, L., Fogwill, C.J., England, M.H., Cortese, G., et al., 2014. Antarctic contribution to meltwater pulse 1A from reduced Southern Ocean overturning. Nature Communications 5 (5107), 1–10. Available from: https://doi.org/10.1038/ncomms6107.

Goodwin, I.D., 1993. Holocene deglaciation, sea-level change, and the emergence of the Windmill Islands, Budd Coast, Antarctica. Quaternary Research 40, 55–69. Available from: https://doi.org/10.1006/qres.1993.1057.

Goodwin, I.D., 1996. A mid to late Holocene readvance of the Law Dome ice margin, Budd Coast, East Antarctica. Antarctic Science 8, 395–406. Available from: https://doi.org/10.1017/S0954102096000570.

Goodwin, I.D., Zweck, C., 2000. Glacio-isostasy and glacial ice load at Law Dome, Wilkes Land, East Antarctica. Quaternary Research 53, 285–293.

Gore, D.B., 1997a. Last glaciation of Vestfold Hills: extension of the East Antarctic Ice Sheet or lateral expansion of Sørsdal Glacier? Polar Record 33, 5–12.

Gore, D.B., 1997b. Blanketing snow and ice; constraints on radiocarbon dating deglaciation in East Antarctic oases. Antarctic Science 9, 336–348. Available from: https://doi.org/10.1017/S0954102097000412.

Gore, D.B., Rhodes, E.J., Augustinus, P.C., Leishman, M.R., Colhoun, E.A., Rees-Jones, J., 2001. Bunger Hills, East Antarctica: ice free at the Last Glacial Maximum. Geology 29, 1103–1106. Available from: https://doi.org/10.1130/0091-7613(2001)029%3C1103:BHEAIF%3E2.0.CO;2.

Gosse, J.C., Phillips, F.M., 2001. Terrestrial in situ cosmogenic nuclides: theory and application. Quaternary Science Reviews 20, 1475–1560. Available from: https://doi.org/10.1016/S0277-3791(00)00171-2.

Gowan, E.J., Zhang, X., Khosravi, S., Rovere, A., Stocchi, P., Hughes, A.L.C., et al., 2021. A new global ice sheet reconstruction for the past 80 000 years. Nature Communications 12, 1199. Available from: https://doi.org/10.1038/s41467-021-21469-w.

Graham, A.G.C., Larter, R.D., Gohl, K., Dowdeswell, J.A., Hillenbrand, C.-D., Smith, J.A., et al., 2010. Flow and retreat of the Late Quaternary Pine Island-Thwaites palaeo-ice stream, West Antarctica. Journal of Geophysical Research: Earth Surface 115, F03025. Available from: https://doi.org/10.1029/2009JF001482.

Graham, A.G.C., Larter, R.D., Gohl, K., Hillenbrand, C.-D., Smith, J.A., Kuhn, G., 2009. Bedform signature of a West Antarctic palaeo-ice stream reveals a multi-temporal record of flow and substrate control. Quaternary Science Reviews 28, 2774–2793. Available from: https://doi.org/10.1016/j.quascirev.2009.07.003.

Graham, A.G.C., Nitsche, F.O., Larter, R.D., 2011. An improved bathymetry compilation for the Bellingshausen Sea, Antarctica, to inform ice-sheet and ocean models. The Cryosphere 5, 95–106. Available from: https://doi.org/10.5194/tc-5-95-2011.

Graham, A.G.C., Smith, J.A., 2012. Palaeoglaciology of the Alexander Island ice cap, western Antarctic Peninsula, reconstructed from marine geophysical and core data. Quaternary Science Reviews 35, 63–81. Available from: https://doi.org/10.1016/j.quascirev.2012.01.008.

Greenwood, S.L., Gyllencreutz, R., Jakobsson, M., Anderson, J.B., 2012. Ice-flow switching and East/West Antarctic Ice Sheet roles in glaciation of the western Ross Sea. Geological Society of America Bulletin 124, 1736–1749. Available from: https://doi.org/10.1130/B30643.1.

Greenwood, S.L., Simkins, L.M., Halberstadt, A.R.W., Prothro, L.O., Anderson, J.B., 2018. Holocene reconfiguration and readvance of the East Antarctic Ice Sheet. Nature Communications 9, 3176. Available from: https://doi.org/10.1038/s41467-018-05625-3.

Halberstadt, A.R.W., Simkins, L.M., Greenwood, S.L., Anderson, J.B., 2016. Past ice-sheet behaviour: retreat scenarios and changing controls in the Ross Sea, Antarctica. The Cryosphere 10, 1003–1020. Available from: https://doi.org/10.5194/tc-10-1003-2016.

Hall, B.L., Hoelzel, A.R., Baroni, C., Denton, G.H., Le Boeuf, B.J., Overturf, B., et al., 2006. Holocene elephant seal distribution implies warmer-than-present climate in the Ross Sea. Proceedings of the National Academy of Sciences 103 (27), 10213–10217. Available from: https://doi.org/10.1073/pnas.0604002103.

Hall, B., Denton, G., Heath, S., Jackson, M., Koffman, T., 2015. Accumulation and marine forcing of ice dynamics in the western Ross Sea during the last deglaciation. Nature Geoscience 8, 625–628. Available from: https://doi.org/10.1038/ngeo2478.

Harden, S.L., DeMaster, D.J., Nittrouer, C.A., 1992. Developing sediment geochronologies for high-latitude continental shelf deposits: a radiochemical approach. Marine Geology 103, 69–97. Available from: https://doi.org/10.1016/0025-3227(92)90009-7.

Harris, P.T., O'Brien, P., 1998. Bottom currents, sedimentation and ice-sheet retreat facies successions on the Mac.Robertson shelf, East Antarctica. Marine Geology 151, 47–72. Available from: https://doi.org/10.1016/S0025-3227(98)00047-4.

Hein, A., Woodward, J., Marrero, S.M., Dunning, S.A., Steig, E.J., Freeman, S.P.H.T., et al., 2016a. Evidence for the stability of the West Antarctic Ice Sheet divide for 1.4 million years. Nature Communications 7, 10325. Available from: https://doi.org/10.1038/ncomms10325.

Hein, A.S., Fogwill, C.J., Sugden, D.E., Xu, S., 2011. Glacial/interglacial ice-stream stability in the Weddell Sea embayment, Antarctica. Earth and Planetary Science Letters 307, 211–221. Available from: https://doi.org/10.1016/j.epsl.2011.04.037.

Hein, A.S., Fogwill, C.J., Sugden, D.E., Xu, S., 2014. Geological scatter of cosmogenic-nuclide exposure ages in the Shackleton Range, Antarctica: implications for glacial history. Quaternary Geochronology 19, 52–66. Available from: https://doi.org/10.1016/j.quageo.2013.03.008.

Hein, A.S., Marrero, S.M., Woodward, J., Dunning, S.A., Winter, K., Westoby, M.J., et al., 2016b. Mid-Holocene pulse of thinning in the Weddell Sea sector of the West Antarctic Ice Sheet. Nature Communications 7, 12511. Available from: https://doi.org/10.1038/ncomms12511.

Hemer, M., Harris, P., 2003. Sediment core from beneath the Amery Ice Shelf, East Antarctica, suggests mid-Holocene ice-shelf retreat. Geology 31, 127–130. Available from: https://doi.org/10.1130/0091-7613(2003)031%3C0127:SCFBTA%3E2.0.CO;2.

Hemer, M.A., Post, A.L., O'Brien, P.E., Craven, M., Truswell, E.M., Roberts, D., et al., 2007. Sedimentological signatures of the sub-Amery Ice Shelf circulation. Antarctic Science 19, 497–506. Available from: https://doi.org/10.1017/S0954102007000697.

Hendy, C.H., Hall, B.L., 2006. The radiocarbon reservoir effect in proglacial lakes: examples from Antarctica. Earth and Planetary Science Letters 241, 413–421. Available from: https://doi.org/10.1016/j.epsl.2005.11.045.

Heroy, D.C., Anderson, J.B., 2005. Ice-sheet extent of the Antarctic Peninsula region during the Last Glacial Maximum (LGM)-insights from glacial geomorphology. Geological Society of America Bulletin 117 (11/12), 1497–1512. Available from: https://doi.org/10.1130/B25694.1.

Heroy, D.C., Anderson, J.B., 2007. Radiocarbon constraints on Antarctic Peninsula Ice Sheet retreat following the Last Glacial Maximum (LGM). Quaternary Science Reviews 26, 3286–3297. Available from: https://doi.org/10.1016/j.quascirev.2007.07.012.

Hillenbrand, C.-D., Bentley, M.J., Stolldorf, T.D., Hein, A.S., Kuhn, G., Graham, A.G.C., et al., 2014. Reconstruction of changes in the Weddell Sea sector of the Antarctic Ice Sheet since the Last Glacial Maximum. Quaternary Science Reviews 100, 111–136. Available from: https://doi.org/10.1016/j.quascirev.2013.07.020.

Hillenbrand, C.-D., Kuhn, G., Smith, J.A., Gohl, K., Graham, A.G.C., Larter, R.D., et al., 2013. Grounding-line retreat of the West Antarctic Ice Sheet from inner Pine Island Bay. Geology 41, 35–38. Available from: https://doi.org/10.1130/G33469.1.

Hillenbrand, C.-D., Larter, R.D., Dowdeswell, J.A., Ehrmann, W., Cofaigh, C.Ó., Benetti, S., et al., 2010a. The sedimentary legacy of a palaeo-ice stream on the shelf of the southern Bellingshausen Sea: clues to West Antarctic glacial history during the Late Quaternary. Quaternary Science Reviews 29 (19–20), 2741–2763. Available from: https://doi.org/10.1016/j.quascirev.2010.06.028.

Hillenbrand, C.-D., Melles, M., Kuhn, G., Larter, R.D., 2012. Marine geological constraints for the grounding-line position of the Antarctic Ice Sheet on the southern Weddell Sea shelf at the Last Glacial Maximum. Quaternary Science Reviews 32, 25–47. Available from: https://doi.org/10.1016/j.quascirev.2011.11.017.

Hillenbrand, C.-D., Smith, J.A., Hodell, D.A., Greaves, M., Poole, C.R., Kender, S., et al., 2017. West Antarctic Ice Sheet retreat driven by Holocene warm water incursions. Nature 547, 43–48. Available from: https://doi.org/10.1038/nature22995.

Hillenbrand, C.-D., Smith, J.A., Kuhn, G., Esper, O., Gersonde, R., Larter, R.D., et al., 2010b. Age assignment of a diatomaceous ooze deposited in the western Amundsen Sea Embayment after the Last Glacial Maximum. Journal of Quaternary Science 25, 280–295. Available from: https://doi.org/10.1002/jqs.1308.

Hiller, A., Hermichen, W.-D., Wand, U., 1995. Radiocarbon dated subfossil stomach oil deposits from petrel nesting sites: novel paleoenvironmental records from continental Antarctica. Radiocarbon 37, 171–180. Available from: https://doi.org/10.1017/S0033822200030617.

Hodgson, D.A., Bentley, M.J., Schnabel, C., Cziferszky, A., Fretwell, P., Convey, P., et al., 2012. Glacial geomorphology and cosmogenic ^{10}Be and ^{26}Al exposure ages in the northern Dufek Massif, Weddell Sea Embayment, Antarctica. Antarctic Science 24, 377–394. Available from: https://doi.org/10.1017/S0954102012000016.

Hodgson, D.A., Hogan, K., Smith, J.M., Smith, J.A., Hillenbrand, C.-D., Graham, A.G.C., et al., 2018. Deglaciation and future stability of the Coats Land ice margin, Antarctica. The Cryosphere 12, 2383–2399. Available from: https://doi.org/10.5194/tc-12-2383-2018.

Hodgson, D.A., Noon, P.E., Vyverman, W., Bryant, C.L., Gore, D.B., Appleby, P., et al., 2001. Were the Larsemann Hills ice-free through the Last Glacial Maximum? Antarctic Science 13, 440–454. Available from: https://doi.org/10.1017/S0954102001000608.

Hodgson, D.A., Roberts, S.J., Smith, J.A., Verleyen, E., Sterken, M., Labarque, M., et al., 2013. Late Quaternary environmental changes in Marguerite Bay, Antarctic Peninsula, inferred from lake sediments and raised beaches. Quaternary Science Reviews 68, 216–236. Available from: https://doi.org/10.1016/j.quascirev.2011.10.011.

Hodgson, D.A., Verleyen, E., Sabbe, K., Squier, A.H., Keely, B.J., Leng, M.J., et al., 2005. Late Quaternary climate-driven environmental change in the Larsemann Hills, East Antarctica, multi-proxy evidence from a lake sediment core. Quaternary Research 64, 83–99. Available from: https://doi.org/10.1016/j.yqres.2005.04.002.

Hodgson, D.A., Verleyen, E., Squier, A.H., Sabbe, K., Keely, B.J., Saunders, K.M., et al., 2006. Interglacial environments of coastal east Antarctica: comparison of MIS 1 (Holocene) and MIS 5e (Last Interglacial) lake-sediment records. Quaternary Science Reviews 25, 179–197. Available from: https://doi.org/10.1016/j.quascirev.2005.03.004.

Hodgson, D.A., Verleyen, E., Vyverman, W., Sabbe, K., Leng, M.J., Pickering, M., et al., 2009. A geological constraint on relative sea level in Marine Isotope Stage 3 in the Larsemann Hills, Lambert Glacier region, East Antarctica (31 366–33 228 cal yr BP). Quaternary Science Reviews 28, 2689–2696. Available from: https://doi.org/10.1016/j.quascirev.2009.06.006.

Hogan, K.A., Larter, R.D., Graham, A.G.C., Arthern, R., Kirkham, J.D., Minzoni, R.T., et al., 2020. Revealing the former bed of Thwaites Glacier using sea-floor bathymetry: implications for warm-water routing and bed controls on ice flow and buttressing. The Cryosphere 14, 2883–2908. Available from: https://doi.org/10.5194/tc-14-2883-2020.

Hughes, T., 1977. West Antarctic Ice Streams. Reviews of Geophysics and Space Physics 15 (1), 1–45. Available from: https://doi.org/10.1029/RG015i001p00001.

Huybrechts, P., 2002. Sea level changes at the LGM from ice-dynamic reconstructions of the Greenland and Antarctic Ice Sheets during the glacial cycles. Quaternary Science Reviews 21, 203–231. Available from: https://doi.org/10.1016/S0277-3791(01)00082-8.

Igarashi, A., Harada, N., Moriwaki, K., 1995. Marine fossils of 30–40 ka in raised beach deposits, and late Pleistocene glacial history around Lützow-Holm Bay, East Antarctica. Proceedings of the NIPR Symposium, Antarctic Geoscience 8, 219–229.

Igarashi, A., H. Miura, C. Hart, 1998. Amino-acid racemization dates of fossil molluscs from raised beach deposits on East Ongul Island and northern part of Langhovde, Lützow-Holm Bay region, East Antarctica. In: The 18th Symposium on Antarctic Geosciences, Program and Abstracts. NIPR, pp. 59–61.

Igarashi, A., Numanami, H., Tsuchiya, Y., Fukuchi, M., 2001. Bathymetric distribution of fossil foraminifera within marine sediment cores from the eastern part of Lützow-Holm Bay, East Antarctica, and its paleoceanographic implications. Marine Micropaleontology 42, 125–162. Available from: https://doi.org/10.1016/S0377-8398(01)00004-4.

Ingólfsson, O., Hjort, C., Humlum, O., 2003. Glacial and climate history of the Antarctic Peninsula since the Last Glacial Maximum. Arctic Antarctic and Alpine Research 35, 175–186.

Ingólfsson, Ó., 2004. Quaternary glacial and climate history of Antarctica. In: Ehlers, J., Gibbard, P.L. (Eds.), Quaternary Glaciations – Extent and Chronology Part III. Elsevier, pp. 3–43. Available from: https://doi.org/10.1016/S1571-0866(04)80109-X.

Jakobsson, M., Anderson, J.B., Nitsche, F., Dowdeswell, J.A., Gyllencreutz, R., Kirchner, N., et al., 2011. Geological record of ice shelf break-up and grounding line retreat, Pine Island Bay, West Antarctica. Geology 39, 691–694. Available from: https://doi.org/10.1130/G32153.1.

Jakobsson, M., Anderson, J.B., Nitsche, F., Gyllencreutz, R., Kirshner, A., Kirchner, N., et al., 2012. Ice sheet retreat dynamics inferred from glacial morphology of the central Pine Island Bay Trough, West Antarctica. Quaternary Science Reviews 38, 1–10. Available from: https://doi.org/10.1016/j.quascirev.2011.12.017.

Jamieson, S.S.R., Stokes, C.R., Ross, N., Rippin, D.M., Bingham, R.G., Wilson, D.S., et al., 2014. The glacial geomorphology of the Antarctic Ice Sheet bed. Antarctic Science 26, 724–741. Available from: https://doi.org/10.1017/S0954102014000212.

Jeong, A., Lee, J.I., Seong, Y.B., Balco, G., Yoo, K.C., Yoon, H.I., et al., 2018. Late Quaternary deglacial history across the Larsen B embayment, Antarctica. Quaternary Science Reviews 189, 134–148. Available from: https://doi.org/10.1016/j.quascirev.2018.04.011.

Johnson, J.S., Bentley, M.J., Gohl, K., 2008. First exposure ages from the Amundsen Sea Embayment, West Antarctica: the Late Quaternary context for recent thinning of Pine Island, Smith, and Pope Glaciers. Geology 36 (3), 223–226. Available from: https://doi.org/10.1130/G24207A.1.

Johnson, J.S., Bentley, M.J., Roberts, S.J., Binney, S.A., Freeman, S.P., 2011. Holocene deglacial history of the northeast Antarctic Peninsula – a review and new chronological constraints. Quaternary Science Reviews 30, 3791–3802. Available from: https://doi.org/10.1016/j.quascirev.2011.10.011.

Johnson, J.S., Bentley, M.J., Smith, J.A., Finkel, R.C., Rood, D.H., Gohl, K., et al., 2014. Rapid thinning of Pine Island glacier in the Early Holocene. Science 343, 999–1001. Available from: https://doi.org/10.1126/science.1247385.

Johnson, J.S., Everest, J.D., Leat, P.T., Golledge, N.R., Rood, D.H., Stuart, F.M., 2012. The deglacial history of NW Alexander Island, Antarctica, from surface exposure dating. Quaternary Research 77, 273–280. Available from: https://doi.org/10.1016/j.yqres.2011.11.012.

Johnson, J.S., Nichols, K.A., Goehring, B.M., Balco, G., Schaefer, J.M., 2019. Abrupt mid-Holocene ice loss in the western Weddell Sea Embayment of Antarctica. Earth and Planetary Science Letters 518, 127–135. Available from: https://doi.org/10.1016/j.epsl.2019.05.002.

Johnson, J.S., Roberts, S.J., Rood, D.H., Pollard, D., Schaefer, J.M., Whitehouse, P.L., et al., 2020. Deglaciation of Pope Glacier implies widespread early Holocene ice sheet thinning in the Amundsen Sea sector of Antarctica. Earth and Planetary Science Letters 548, 116501. Available from: https://doi.org/10.1016/j.epsl.2020.116501.

Johnson, J.S., Smellie, J.L., Nelson, A.E., Stuart, F.M., 2009. History of the Antarctic Peninsula Ice Sheet since the early Pliocene-Evidence from cosmogenic dating of Pliocene lavas on James Ross Island, Antarctica. Global and Planetary Change 69, 205–213. Available from: https://doi.org/10.1016/j.gloplacha.2009.09.001.

Johnson, J.S., Smith, J.A., Schaefer, J.M., Young, N.E., Goehring, B.M., Hillenbrand, C.-D., et al., 2017. The last glaciation of Bear Peninsula, central Amundsen Sea Embayment of Antarctica: constraints on timing and duration revealed by in situ cosmogenic ^{14}C and ^{10}Be dating. Quaternary Science Reviews 178, 77–88. Available from: https://doi.org/10.1016/j.quascirev.2017.11.003.

Jones, R.S., Mackintosh, A.N., Norton, K.P., Golledge, N.R., Fowill, C.J., Kubik, P.W., et al., 2015. Rapid Holocene thinning of an East Antarctic outlet glacier driven by marine ice sheet instability. Nature Communications 6, 8910. Available from: https://doi.org/10.1038/ncomms9910.

Kawamata, M., Suganuma, Y., Doi, K., Misawa, K., Hirabayashi, M., Hattori, A., et al., 2020. Abrupt Holocene ice-sheet thinning along the southern Soya Coast, Lützow-Holm Bay, East Antarctica, revealed by glacial geomorphology and surface exposure dating. Quaternary Science Reviews 247, 106540. Available from: https://doi.org/10.1016/j.quascirev.2020.106540.

Kiernan, K., Gore, D.B., Fink, D., McConnell, A., Sigurdsson, I.A., White, D.A., 2009. Deglaciation and weathering of Larsemann Hills, East Antarctica. Antarctic Science 21 (4), 373–382. Available from: https://doi.org/10.1017/S0954102009002028.

Kiernan, K., McConnell, A., Colhoun, E., Lawson, E., 2003. Radiocarbon dating of mumiyo from the Vestfold Hills, East Antarctica. Papers & Proceedings of the Royal Society of Tasmania 136, 141–144.

Kilfeather, A.A., Cofaigh, C.Ó., Lloyd, J.M., Dowdeswell, J.A., Xu, S., Moreton, S.G., 2011. Ice-stream retreat and ice-shelf history in Marguerite Trough, Antarctic Peninsula: sedimentological and foraminiferal signatures. Geological Society of America Bulletin 123, 997–1015. Available from: https://doi.org/10.1130/B30282.1.

King, C., Hall, B., Hillebrand, T., Stone, J., 2020. Delayed maximum and recession of an East Antarctic outlet glacier. Geology 48 (6), 630–634. Available from: https://doi.org/10.1130/G47297.1.

Kingslake, J., Martín, C., Arthern, R.J., Corr, H.F.J., King, E.C., 2016. Ice-flow reorganization in West Antarctica 2.5 kyr ago dated using radar-derived englacial flow velocities. Geophysical Research Letters 43, 9103–9112. Available from: https://doi.org/10.1002/2016GL070278.

Kingslake, J., Scherer, R.P., Albrecht, T., Coenen, J., Powell, R.D., Reese, R., et al., 2018. Extensive retreat and re-advance of the West Antarctic Ice Sheet during the Holocene. Nature 558 (7710), 430–434. Available from: https://doi.org/10.1038/s41586-018-0208-x.

Kim, S., Yoo, K.-C., Lee, J.I., Khim, B.-K., Bak, Y.-S., Lee, M.K., et al., 2018. Holocene paleoceanography of Bigo Bay, west Antarctic Peninsula: connections between surface water productivity and nutrient utilization and its implication for surface-deep water mass exchange. Quaternary Science Reviews 192, 59–70. Available from: https://doi.org/10.1016/j.quascirev.2018.05.028.

Kirkham, J.D., Hogan, K.A., Larter, R.D., Arnold, N.S., Nitsche, F.O., Golledge, N.R., et al., 2019. Past water flow beneath Pine Island and Thwaites glaciers, West Antarctica. The Cryosphere 13, 1959–1981. Available from: https://doi.org/10.5194/tc-13-1959-2019.

Kirkup, H., Melles, M., Gore, D.B., 2002. Late Quaternary environment of the southern Windmill Islands, East Antarctica. Antarctic Science 14, 385–394. Available from: https://doi.org/10.1017/S0954102002000202.

Kirshner, A., Anderson, J.B., Jakobsson, M., O'Regan, M., Majewski, W., Nitsche, F., 2012. Post-LGM deglaciation in Pine Island Bay, West Antarctica. Quaternary Science Reviews 38, 11–26. Available from: https://doi.org/10.1016/j.quascirev.2012.01.017.

Klages, J.P., Kuhn, G., Graham, A.G.C., Hillenbrand, C.-D., Smith, J.A., Nitsche, F.O., et al., 2015. Palaeo-ice stream pathways and retreat style in the easternmost Amundsen Sea Embayment, West Antarctica, revealed by combined multibeam bathymetric and seismic data. Geomorphology 245, 207–222. Available from: https://doi.org/10.1016/j.geomorph.2015.05.020.

Klages, J.P., Kuhn, G., Hillenbrand, C.-D., Graham, A.G.C., Smith, J.A., Larter, R.D., et al., 2014. Retreat of the West Antarctic Ice Sheet from the western Amundsen Sea shelf at a pre- or early LGM stage. Quaternary Science Reviews 91, 1–15. Available from: https://doi.org/10.1016/j.quascirev.2014.02.017.

Klages, J.P., Kuhn, G., Hillenbrand, C.-D., Smith, J.A., Graham, A.G.C., Nitsche, F.O., F.O., et al., 2017. Limited grounding-line advance onto the West Antarctic continental shelf in the easternmost Amundsen Sea Embayment during the last glacial period. PLoS One 12 (7), e0181593. Available from: https://doi.org/10.1371/journal.pone.0181593.

Kopp, R., Simons, F., Mitrovica, J., et al., 2009. Probabilistic assessment of sea level during the last interglacial stage. Nature 462, 863–867. Available from: https://doi.org/10.1038/nature08686.

Krause, W.E., Krbetschek, M.R., Stolz, W., 1997. Dating of Quaternary lake sediments from the Schirmacher Oasis (East Antarctica) by infra-red stimulated luminescence (IRSL) detected at the wavelength of 560 nm. Quaternary Science Reviews 16 (3–5), 387–392. Available from: https://doi.org/10.1016/S0277-3791(96)00090-X.

Kuhn, G., Hillenbrand, C.-D., Kasten, S., Smith, J.A., Nitsche, F.O., Frederichs, T., et al., 2017. Evidence for a palaeo-subglacial lake on the Antarctic continental shelf. Nature Communications 8, 15591. Available from: https://doi.org/10.1038/ncomms15591.

Lambeck, K., Rouby, H., Purcell, A., Sun, Y., Sambridge, M., 2014. Sea level and global ice volumes from the Last Glacial Maximum to the Holocene. Proceedings of the National Academy of Sciences 111, 15296–15303. Available from: https://doi.org/10.1073/pnas.1411762111.

Lamping, N., Müller, J., Esper, O., Hillenbrand, C.-D., Smith, J.A., Kuhn, G., 2020. Highly branched isoprenoids reveal onset of deglaciation followed by dynamic sea-ice conditions in the western Amundsen Sea, Antarctica. Quaternary Science Reviews 228, 106103. Available from: https://doi.org/10.1016/j.quascirev.2019.106103.

Larter, R.D., Anderson, J.B., Graham, A.G.C., Gohl, K., Hillenbrand, C.-D., Jakobsson, M., et al., 2014. Reconstruction of changes in the Amundsen Sea and Bellingshausen Sea sector of the West Antarctic Ice Sheet since the Last Glacial Maximum. Quaternary Science Reviews 100, 55–86. Available from: https://doi.org/10.1016/j.quascirev.2013.10.016.

Larter, R.D., Graham, A.G.C., Gohl, K., Kuhn, G., Hillenbrand, C.-D., Smith, J.A., et al., 2009. Subglacial bedforms reveal complex basal regime in a zone of paleo-ice stream convergence, Amundsen Sea embayment, West Antarctica. Geology 37, 411–414. Available from: https://doi.org/10.1130/G25505A.1.

Larter, R.D., Graham, A.G.C., Hillenbrand, C.-D., Smith, J.A., Gales, J.A., 2012. Late Quaternary grounded ice extent in the Filchner Trough, Weddell Sea, Antarctica: new marine geophysical evidence. Quaternary Science Reviews 53, 111–122. Available from: https://doi.org/10.1016/j.quascirev.2012.08.006.

Larter, R.D., Hogan, K.A., Hillenbrand, C.-D., Smith, J.A., Batchelor, C.L., Cartigny, M., et al., 2019. Subglacial hydrological control on flow of an Antarctic Peninsula palaeo-ice stream. The Cryosphere 13, 1583–1596. Available from: https://doi.org/10.5194/tc-13-1583-2019.

Larter, R.D., Vanneste, L.E., 1995. Relict subglacial deltas on the Antarctic Peninsula outer shelf. Geology 23, 33–36. Available from: https://doi.org/10.1130/0091-7613(1995)023%3C0033:RSDOTA%3E2.3.CO;2.

Lavoie, C., Domack, E.W., Pettit, E.C., Scambos, T.A., Larter, R.D., Schenke, H.-W., et al., 2015. Configuration of the Northern Antarctic Peninsula Ice Sheet at LGM based on a new synthesis of seabed imagery. The Cryosphere 9, 613–629. Available from: https://doi.org/10.5194/tc-9-613-2015.

Lee, J.I., McKay, R.M., Golledge, N.R., Yoon, H.I., Yoo, K.-C., Kim, H.J., et al., 2017. Widespread persistence of expanded East Antarctic glaciers in the southwest Ross Sea during the last deglaciation. Geology 45 (5), 403–406. Available from: https://doi.org/10.1130/G38715.1.

Leventer, A., Domack, E.W., Dunbar, R., Pike, J., Stickley, C., Maddison, E., et al., 2006. Marine sediment record from the East Antarctic margin reveals dynamics of ice sheet recession. GSA Today 16, 4–10. Available from: https://www.geosociety.org/gsatoday/archive/16/12/pdf/i1052-5173-16-12-4.pdf.

Licht, K.J., 2004. The Ross Sea's contribution to eustatic sea level during meltwater pulse 1A. Sedimentary Geology 165, 343–353. Available from: https://doi.org/10.1016/j.sedgeo.2003.11.020.

Licht, K.J., Dunbar, N.W., Andrews, J.T., Jennings, A.E., 1999. Distinguishing subglacial till and glacial marine diamictons in the western Ross Sea, Antarctica: implications for a last glacial maximum grounding line. Geological Society of America Bulletin 111 (1), 91–103. Available from: https://doi.org/10.1130/0016-7606(1999)111%3C0091:DSTAGM%3E2.3.CO;2.

Licht, K.J., Hennessy, A.J., Welke, B.M., 2014. The U/Pb detrital zircon signature of West Antarctic ice stream tills in the Ross Embayment, with implications for LGM ice flow reconstructions. Antarctic Science 26, 687–697. Available from: https://doi.org/10.1017/S0954102014000315.

Licht, K.J., Jennings, A.E., Andrews, J.T., Williams, K.M., 1996. Chronology of late Wisconsin ice retreat from the western Ross Sea, Antarctica. Geology 24 (3), 223–226. Available from: https://doi.org/10.1130/0091-7613(1996)024%3C0223:COLWIR%3E2.3.CO;2.

Licht, K.J., Lederer, J.R., Swope, R.J., 2005. Provenance of LGM glacial till (sand fraction) across the Ross embayment, Antarctica. Quaternary Science Reviews 24, 1499–1520. Available from: https://doi.org/10.1016/j.quascirev.2004.10.017.

Licht, K.J., Palmer, E.F., 2013. Erosion and transport by Byrd Glacier, Antarctica during the last glacial maximum. Quaternary Science Reviews 62, 32–48. Available from: https://doi.org/10.1016/j.quascirev.2012.11.017.

Lilly, K., 2008. Three million years of East Antarctic Ice Sheet history from in situ cosmogenic nuclides in the Lambert-Amery Basin. PhD thesis, Australian National University, 183 pp. Available from: <https://openresearch-repository.anu.edu.au/handle/1885/149854>.

Lilly, K., Fink, D., Fabel, D., Lambeck, K., 2010. Pleistocene dynamics of the interior East Antarctic Ice Sheet. Geology 38 (8), 703–706. Available from: https://doi.org/10.1130/G31172x.1.

Lindow, J., Castex, M., Wittmann, H., Johnson, J.S., Lisker, F., Gohl, K., et al., 2014. Glacial retreat in the Amundsen Sea sector, West Antarctica – first cosmogenic evidence from central Pine Island Bay and the Kohler Range. Quaternary Science Reviews 98, 166–173. Available from: https://doi.org/10.1016/j.quascirev.2014.05.010.

Livingstone, S.J., Cofaigh, C.Ó., Stokes, C.R., Hillenbrand, C.-D., Vieli, A., Jamieson, S.S.R., 2012. Antarctic palaeo-ice streams. Earth-Science Reviews 111, 90–128. Available from: https://doi.org/10.1016/j.earscirev.2011.10.003.

Livingstone, S.J., Cofaigh, C.Ó., Stokes, C.R., Hillenbrand, C.-D., Vieli, A., Jamieson, S.S.R., 2013. Glacial geomorphology of Marguerite Bay palaeo-ice stream, western Antarctic Peninsula. Journal of Maps 9, 558–572. Available from: https://doi.org/10.1080/17445647.2013.829411.

Lowe, A.L., Anderson, J.B., 2002. Reconstruction of the West Antarctic Ice Sheet in Pine Island Bay during the Last Glacial Maximum and its subsequent retreat history. Quaternary Science Reviews 21, 1879–1897. Available from: https://doi.org/10.1016/S0277-3791(02)00006-9.

Lowe, A.L., Anderson, J.B., 2003. Evidence for abundant subglacial meltwater beneath the paleo-ice sheet in Pine Island Bay, Antarctica. Journal of Glaciology 49, 125–138. Available from: https://doi.org/10.3189/172756503781830971.

Lowry, D.P., Golledge, N.R., Bertler, N.A.N., Selwyn Jones, R., McKay, R., 2019. Deglacial grounding-line retreat in the Ross Embayment, Antarctica, controlled by ocean and atmosphere forcing. Science Advances 5 (8), eaav8754. Available from: https://doi.org/10.1126/sciadv.aav8754.

Mackintosh, A., Golledge, N., Domack, E., Dunbar, R., Leventer, A., White, D., et al., 2011. Retreat of the East Antarctic Ice Sheet during the last glacial termination. Nature Geoscience 4, 195–202. Available from: https://doi.org/10.3189/172756503781830971.

Mackintosh, A., Verleyan, E., O'Brien, P.E., White, D.A., Jones, R.S., McKay, R., et al., 2014. Retreat history of the East Antarctic Ice Sheet since the Last Glacial Maximum. Quaternary Science Reviews 100, 10–30. Available from: https://doi.org/10.1016/j.quascirev.2013.07.024.

Mackintosh, A., White, D., Fink, D., Gore, D.B., Pickard, J., Fanning, P.C., 2007. Exposure ages from mountain dipsticks in Mac.Robertson Land, East Antarctica, indicate little change in ice sheet thickness since the Last Glacial Maximum. Geology 35, 551–554. Available from: https://doi.org/10.1130/G23503A.1.

Maddison, E., Pike, J., Leventer, A., Dunbar, R., Brachfeld, S., Domack, E., et al., 2006. Post-glacial seasonal diatom record of the Mertz Glacial Polynya, East Antarctica. Marine Micropaleontology 60, 66–88. Available from: https://doi.org/10.1016/j.marmicro.2006.03.001.

Maemoku, H., Miura, H., Saigusa, S., Moriwaki, K., 1997. Stratigraphy of the Late Quaternary raised beach deposits in the northern part of Langhovde, Lützow-Holm Bay, East Antarctica. Proceedings of NIPR Symposium. Antarctic Geoscience 10, 178–186.

McGlannan, A.J., Bart, P.J., Chow, J.M., DeCesare, M., 2017. On the influence of post- LGM ice shelf loss and grounding zone sedimentation on West Antarctic Ice Sheet stability. Marine Geology 392, 151–169. Available from: https://doi.org/10.1016/j.margeo.2017.08.005.

McKay, R.M., Dunbar, G.B., Naish, T.R., Barrett, P.J., Carter, L., Harper, M., 2008. Retreat history of the Ross ice sheet (shelf) since the last glacial maximum from deep-basin sediment cores around Ross Island. Palaeogeography, Palaeoclimatology, Palaeoecology 260 (1), 245–261. Available from: https://doi.org/10.1016/j.palaeo.2007.08.015.

McKay, R., Golledge, N.R., Maas, S., Naish, T., Levy, R., Dunbar, G., et al., 2016. Antarctic marine ice-sheet retreat in the Ross Sea during the early Holocene. Geology 44, 7–10. Available from: https://doi.org/10.1130/G37315.1.

McMillan, M., 2018. The current health of polar ice sheets and implications for sea level. In: Nuttall, M., Christensen, T.R., Siegert, M.J. (Eds.), Routledge Handbook of the Polar Regions. Routledge, pp. 185–197.

McMullen, K., Domack, E.W., Leventer, A., Olson, C., Dunbar, R., Brachfeld, S., 2006. Glacial morphology and sediment formation in the Mertz Trough, East Antarctica. Palaeogeography, Palaeoclimatology, Palaeoecology 231, 169–180. Available from: https://doi.org/10.1016/j.palaeo.2005.08.004.

Melis, R., Capotondi, L., Torricella, F., Ferretti, P., Geniram, A., Hong, J.K., et al., 2021. Last Glacial Maximum to Holocene paleoceanography of the northwestern Ross Sea inferred from sediment core geochemistry and micropaleontology at Hallett Ridge. Journal of Micropalaeontology 40, 15–35. Available from: https://doi.org/10.5194/jm-40-15-2021.

Melles, M., Kuhn, G., 1993. Sub-bottom profiling and sedimentological studies in the southern Weddell Sea, Antarctica: evidence for large-scale erosional/depositional processes. Deep-Sea Research 40 (4), 739–760. Available from: https://doi.org/10.1016/0967-0637(93)90069-F.

Melis, R., Salvi, G., 2020. Foraminifer and Ostracod occurrence in a cool-water carbonate factory of the Cape Adare (Ross Sea, Antarctica): a key lecture for the climatic and oceanographic variations in the last 30,000 years. Geosciences 10, 413. Available from: https://doi.org/10.3390/geosciences10100413.

Mercer, J.H., 1978. West Antarctic Ice Sheet and CO_2 greenhouse effect: a threat of disaster. Nature 271, 321–325. Available from: https://doi.org/10.1038/271321a0.

Mezgec, K., Stenni, B., Crosta, X., Masson-Delmotte, V., Baroni, C., Braida, M., et al., 2017. Holocene sea ice variability driven by wind and polynya efficiency in the Ross Sea. Nature Communications 8, 1334. Available from: https://doi.org/10.1038/s41467-017-01455-x.

Miles, B.W.J., Stokes, C.R., Vieli, A., Cox, N.J., 2013. Rapid, climate-driven changes in outlet glaciers on the Pacific coast of East Antarctica. Nature 500, 563–566. Available from: https://doi.org/10.1038/nature12382.

Minzoni, R.T., Majewski, W., Anderson, J.B., Yokoyama, Y., Fernandez, R., Jakobsson, M., 2017. Oceanographic influences on the stability of the Cosgrove Ice Shelf, Antarctica. The Holocene 27 (11), 1645–1658. Available from: https://doi.org/10.1016/j.quascirev.2004.10.006.

Miura, H., Maemoku, H., Seto, K., Moriwaki, K., 1998. Late Quaternary East Antarctic melting event in the Soya coast region based on stratigraphy and oxygen isotopic ratio of fossil molluscs. Polar Geoscience 11, 260–274.

Morlighem, M., et al., 2020. Deep glacial troughs and stabilizing ridges unveiled beneath the margins of the Antarctic Ice Sheet. Nature Geoscience 13, 132–137. Available from: https://doi.org/10.1038/s41561-019-0510-8.

Morse, D.L., Waddington, E.D., Steig, E.J., 1998. Ice age storm trajectories inferred from radar stratigraphy at Taylor Dome, Antarctica. Geophysical Research Letters 25, 3383–3386. Available from: https://doi.org/10.1029/98GL52486.

Mosola, A.B., Anderson, J.B., 2006. Expansion and rapid retreat of the West Antarctic Ice Sheet in eastern Ross Sea: possible consequence of over-extended ice streams? Quaternary Science Reviews 25, 2177–2196. Available from: https://doi.org/10.1016/j.quascirev.2005.12.013.

Mulvaney, R., Abram, N.J., Hindmarsh, R.C.A., Arrowsmith, C., Fleet, L., Triest, J., et al., 2012. Recent Antarctic Peninsula warming relative to Holocene climate and ice-shelf history. Nature 489, 141–144. Available from: https://doi.org/10.1038/nature11391.

Nakada, M., Kimura, R., Okuno, J., Moriwaki, K., Miura, H., Maemoku, H., 2000. Late Pleistocene and Holocene melting history of the Antarctic Ice Sheet derived from sea level variations. Marine Geology 167, 85–103. Available from: https://doi.org/10.1016/S0025-3227(00)00018-9.

Nakada, M., Lambeck, K., 1988. The melting history of the late Pleistocene Antarctic Ice Sheet. Nature 333, 36–40. Available from: https://doi.org/10.1038/333036a0.

Nichols, K.A., Goehring, B.M., Balco, G., Johnson, J.S., Hein, A.S., Todd, C., 2019. New Last Glacial Maximum ice thickness constraints for the Weddell Sea Embayment, Antarctica. The Cryosphere 13, 2935–2951. Available from: https://doi.org/10.5194/tc-13-2935-2019.

Nitsche, F.O., Gohl, K., Larter, R.D., Hillenbrand, C.-D., Kuhn, G., Smith, J.A., et al., 2013. Paleo-ice flow and subglacial meltwater dynamics in Pine Island Bay, West Antarctica. The Cryosphere 7, 249–262. Available from: https://doi.org/10.5194/tc-7-249-2013.

Nitsche, F.O., Jacobs, S.S., Larter, R.D., Gohl, K., 2007. Bathymetry of the Amundsen Sea continental shelf: implications for geology, oceanography and glaciology. Geochemistry, Geophysics, Geosystems 8, Q10009. Available from: https://doi.org/10.1029/2007GC001694.

Noble, T.L., Rohling, E.J., Aitken, A.R.A., Bostock, H.C., Chase, Z., Gomez, N., N., et al., 2020. The sensitivity of the Antarctic Ice Sheet to a changing climate: past, present, and future. Reviews of Geophysics 58. Available from: https://doi.org/10.1029/2019RG000663.

O'Brien, P.E., De Santis, L., Harris, P.T., Domack, E., Quilty, P.G., 1999. Ice shelf grounding zone features of western Prydz Bay, Antarctica: sedimentary processes from seismic and sidescan images. Antarctic Science 11, 78–91. Available from: https://doi.org/10.1017/S0954102099000115.

O'Brien, P., Smith, J., Stark, J.S., Johnstone, G., Riddle, M., Franklin, D., 2015. Submarine geomorphology and sea floor processes along the coast of Vestfold Hills, East Antarctica, from multibeam bathymetry and video data. Antarctic Science 27, 566–586. Available from: https://doi.org/10.1017/S0954102015000371.

Ó Cofaigh, C., Davies, B.J., Livingstone, S.J., Smith, J.A., Johnson, J.S., Hocking, E.P., et al., 2014. Reconstruction of ice-sheet changes in the Antarctic Peninsula since the Last Glacial Maximum. Quaternary Science Reviews 100, 87–110. Available from: https://doi.org/10.1016/j.quascirev.2014.06.023.

Ó Cofaigh, C., Dowdeswell, J.A., Allen, C.S., Hiemstra, J.F., Pudsey, C.J., Evans, J., et al., 2005a. Flow dynamics and till genesis associated with a marine-based Antarctic palaeo-ice stream. Quaternary Science Reviews 24, 709–740. Available from: https://doi.org/10.1016/j.quascirev.2004.10.006.

Ó Cofaigh, C., Larter, R.D., Dowdeswell, J.A., Hillenbrand, C.-D., Pudsey, C.J., Evans, J., et al., 2005b. Flow of the West Antarctic Ice Sheet on the continental margin of the Bellingshausen Sea at the Last Glacial Maximum. Journal of Geophysical Research 110 (B11103). Available from: https://doi.org/10.1029/2005JB003619.

Ó Cofaigh, C., Pudsey, C.J., Dowdeswell, J.A., Morris, P., 2002. Evolution of subglacial bedforms along a paleo-ice stream, Antarctic Peninsula continental shelf. Geophysical Research Letters 29 (8), 1–4. Available from: https://doi.org/10.1029/2001GL014488.

Overpeck, J.T., Otto-Bliesner, B.L., Miller, G.H., Muhs, D.R., Alley, R.B., Kiehl, J.T., 2006. Paleoclimatic evidence for future ice-sheet instability and rapid sea-level rise. Science 311, 1747–1750. Available from: https://doi.org/10.1126/science.1256697.

Peck, V.L., Allen, C.S., Kender, S., McClymont, E.L., Hodgson, D.A., 2015. Oceanographic variability on the West Antarctic Peninsula during the Holocene and the influence of upper circumpolar deep water. Quaternary Science Reviews 119, 54–65. Available from: https://doi.org/10.1016/j.quascirev.2015.04.002.

Peltier, W.R., 2004. Global glacial isostasy and the surface of the ice-age earth: the ICE-56 (VM2) Model and GRACE. Annual Reviews of Earth and Planetary Science 39, 114–149. Available from: https://doi.org/10.1146/annurev.earth.32.082503.144359.

Perotti, M., Andreucci, B., Talarico, F., Zattin, M., Langone, A., 2017. Multianalytical provenance analysis of Eastern Ross Sea LGM till sediments (Antarctica): petrography, geochronology, and thermochronology detrital data. Geochemistry Geophysics Geosystems 18, 2275–2304. Available from: https://doi.org/10.1002/2016GC006728.

Philippon, G., Ramstein, G., Charbit, S., Kageyama, M., Ritz, C., Dumas, C., 2006. Evolution of the Antarctic ice sheet throughout the last deglaciation: a study with a new coupled climate-north and south hemisphere ice sheet model. Earth and Planetary Science Letters 248, 750–758. Available from: https://doi.org/10.1016/j.epsl.2006.06.017.

Pike, J., Swann, G.E.A., Leng, M.J., Snelling, A.M., 2013. Glacial discharge along the west Antarctic Peninsula during the Holocene. Nature Geoscience 6, 199–202. Available from: https://doi.org/10.1038/ngeo1703.

Pollard, D., DeConto, R.M., 2009. Modelling West Antarctic Ice Sheet growth and collapse through the past five million years. Nature 458, 329–332. Available from: https://doi.org/10.1038/nature07809.

Pope, P.G., Anderson, J.B., 1992. Late Quaternary glacial history of the northern Antarctic Peninsula's western continental shelf: evidence from the marine record. Antarctic Research Series 57, 63–91. Available from: https://doi.org/10.1029/AR057p0063.

Post, A.L., O'Brien, P.E., Edwards, S., Carroll, A.G., Malakoff, K., Armand, L.K., 2020. Upper slope processes and seafloor ecosystems on the Sabrina continental slope, East Antarctica. Marine Geology 422, 106091. Available from: https://doi.org/10.1016/j.margeo.2019.106091.

Pritchard, H., Arthern, R., Vaughan, D., Edwards, L.A., 2009. Extensive dynamic thinning on the margins of the Greenland and Antarctic Ice Sheets. Nature 461, 971–975. Available from: https://doi.org/10.1038/nature08471.

Prothro, L.O., Majewski, W., Yokoyama, Y., Simkins, L.M., Anderson, J.B., Yamane, M., et al., 2020. Timing and pathways of East Antarctic Ice Sheet retreat. Quaternary Science Reviews 230, 106166. Available from: https://doi.org/10.1016/j.quascirev.2020.106166.

Prothro, L.O., Simkins, L.M., Majewski, W., Anderson, J.B., 2018. Glacial retreat patterns and processes determined from integrated sedimentology and geomorphology records. Marine Geology 395, 104–119. Available from: https://doi.org/10.1016/j.margeo.2017.09.012.

Pudsey, C.J., Barker, P.F., Larter, R.D., 1994. Ice sheets retreat from the Antarctic Peninsula shelf. Continental Shelf Research 14, 1647–1675. Available from: https://doi.org/10.1016/0278-4343(94)90041-8.

Pudsey, C.J., Evans, J., 2001. First survey of Antarctic sub-ice shelf sediments reveals Mid-Holocene ice shelf retreat. Geology 29 (9), 787–790. Available from: https://doi.org/10.1130/0091-7613(2001)029%3C0787:FSOASI%3E2.0.CO;2.

Pudsey, C.J., Evans, J., Domack, E.W., Morris, P., Del Valle, R.A., 2001. Bathymetry and acoustic facies beneath the former Larsen-A and Prince Gustav ice shelves, north-west Weddell Sea. Antarctic Science 13, 312–322. Available from: https://doi.org/10.1017/S095410200100044X.

Pudsey, C.J., Murray, J.W., Appleby, P., Evans, J., 2006. Ice shelf history from petrographic and foraminiferal evidence, northeast Antarctic Peninsula. Quaternary Science Reviews 25, 2357–2379. Available from: https://doi.org/10.1016/j.quascirev.2006.01.029.

Rebesco, M., Domack, E., Zgur, F., Lavoie, C., Leventer, A., Brachfeld, S., et al., 2014. Boundary condition of grounding lines prior to collapse, Larsen-B Ice Shelf, Antarctica. Science 345, 1354–1358. Available from: https://doi.org/10.1126/science.1256697.

Rignot, E., Jacobs, S.S., 2002. Rapid bottom melting widespread near Antarctic Ice sheet grounding lines. Science 296 (5575), 2020–2023. Available from: https://doi.org/10.1126/science.1070942.

Rignot, E., Jacobs, S., Mouginot, J., Scheuchl, B., 2013. Ice-shelf melting around Antarctica. Science 341, 266–270. Available from: https://doi.org/10.1126/science.1235798.

Rignot, E., Mouginot, J., Scheuchl, B., van den Broeke, M., van Wessem, M.J., Morlighem, M., 2019. Four decades of Antarctic Ice Sheet mass balance from 1979–2017. Proceedings of the National Academy of Sciences 116, 1095–1103. Available from: https://doi.org/10.1073/pnas.1812883116.

Ritz, C., Rommelaere, V., Dumas, C., 2001. Modelling the evolution of Antarctic Ice Sheet over the last 420,000 years: implications for altitude changes in the Vostok region. Journal of Geophysical Research 106, 31943–31964. Available from: https://doi.org/10.1029/2001JD900232.

Roberts, S.J., Hodgson, D.A., Bentley, M.J., Sanderson, D.C.W., Milne, G., Smith, J.A., et al., 2009. Holocene relative sea-level change and deglaciation on Alexander Island, Antarctic Peninsula, from elevated lake deltas. Geomorphology 112, 122–134. Available from: https://doi.org/10.1016/j.geomorph.2009.05.011.

Rosenheim, B.E., Day, M.B., Domack, E., Schrum, H., Benthien, A., Hayes, J.M., 2008. Antarctic sediment chronology by programmed-temperature pyrolysis: methodology and data treatment. Geochemistry, Geophysics, Geosystems 9 (4), 1–16. Available from: https://doi.org/10.1029/2007GC001816.

Rosenheim, B.E., Santoro, J.A., Gunter, M., Domack, E.W., 2013. Improving Antarctic sediment ^{14}C dating using ramped pyrolysis: an example from the Hugo Island trough. Radiocarbon 55 (1), 115–126. Available from: https://doi.org/10.2458/azu_js_rc.v55i1.16234.

Rosier, S.H.R., Hofstede, C., Brisbourne, A.M., Hattermann, T., Nicholls, K.W., Davis, P.E.D., et al., 2018. A new bathymetry for the southeastern Filchner-Ronne Ice Shelf: implications for modern oceanographic processes and glacial history. Journal of Geophysical Research: Oceans 123, 4610–4623. Available from: https://doi.org/10.1029/2018JC013982.

Ross, N., Siegert, M.J., Woodward, J., Smith, A.M., Corr, H.F.J., Bentley, M.J., et al., 2011. Holocene stability of the Pine Island Glacier-Weddell Sea ice divide, West Antarctica. Geology 39, 935–938. Available from: https://doi.org/10.1130/G31920.1.

Ross, N., Bingham, R.G., Corr, H., Ferraccioli, F., Jordan, T.A., Le Brocq, A., et al. 2012. Steep reverse bed slope at the grounding line of the Weddell Sea sector in West Antarctica. Nature Geoscience 5, 393–396. Available from: https://doi.org/10.1038/ngeo1468.

Scherer, R.P., Aldahan, A., Tulaczyk, S., Possnert, G., Engelhardt, H., Kamb, B., 1998. Pleistocene collapse of the West Antarctic Ice Sheet. Science 281, 82–85. Available from: https://doi.org/10.1126/science.281.5373.82.

Shackleton, N.J., 2000. The 100,000-year ice age cycle identified and found to lag temperature, carbon dioxide, and orbital eccentricity. Science 289 (5486), 1897–1902. Available from: https://doi.org/10.1126/science.289.5486.1897.

Shackleton, N.J., Sanchez-Goni, M.F., Pailler, D., Lancelot, Y., 2003. Marine isotope substage 5e and the Eemian interglacial. Global and Planetary Change 36, 151–155. Available from: https://doi.org/10.1016/S0921-8181(02)00181-9.

Shepherd, A., Ivins, E., Rignot, E., Smith, B., van den Broeke, M., et al., 2018. Mass balance of the Antarctic Ice Sheet from 1992 to 2017. Nature 558, 219–222. Available from: https://doi.org/10.1038/s41586-018-0179-y.

Shevenell, A.E., Ingalls, A.E., Domack, E.W., Kelly, C., 2011. Holocene Southern Ocean surface temperature variability west of the Antarctic Peninsula. Nature 470, 250–254. Available from: https://doi.org/10.1038/nature09751.

Shipp, S.S., Anderson, J.B., Domack, E.W., 1999. Late Pleistocene-Holocene retreat of the West Antarctic Ice Sheet system in the Ross Sea: part 1 – geophysical results. Geological Society of America Bulletin 111, 1486–1516. Available from: https://doi.org/10.1130/0016-7606(1999)111(1486:LPHROT)2.3.CO;2.

Shipp, S.S., Wellner, J.A., Anderson, J.B., 2002. Retreat signature of a polar ice stream: subglacial geomorphic features and sediments from the Ross Sea, Antarctica. In: Dowdeswell, J. A., Cofaigh, C.Ó. (Eds.), Glacier-Influenced Sedimentation on High-Latitude Continental Margins, 203. Geological Society of London Special Publication, pp. 277–304. Available from: https://doi.org/10.1144/GSL.SP.2002.203.01.15.

Siegert, M.J., 2001. Ice Sheets and Late Quaternary Environmental Change. John Wiley, Chichester, p. 231, ISBN: 978-0-471-98570-9.

Siegert, M.J., 2003. Glacial-Interglacial variations in central East Antarctic ice accumulation rates. Quaternary Science Reviews 22, 741–750. Available from: https://doi.org/10.1016/S0277-3791(02)00191-9.

Siegert, M.J., Alley, R.B., Rignot, E., Englander, J., Corell, R., 2020. 21st century sea-level rise could exceed IPCC predictions for strong-warming futures. One Earth 3, 691–703. Available from: https://doi.org/10.1016/j.oneear.2020.11.002.

Siegert, M.J., Kingslake, J., Ross, N., Whitehouse, P.L., Woodward, J., Jamieson, S.S.R., et al., 2019. Late Holocene glacial history of the Weddell Sea sector of the West Antarctic Ice Sheet. Reviews of Geophysics 57, 1197–1223. Available from: https://doi.org/10.1029/2019RG000651.

Siegert, M.J., Leysinger Vieli, G.J.M.C., 2007. Late glacial history of the Ross Sea sector of the West Antarctic Ice Sheet: evidence from englacial layering at Talos Dome, East Antarctica. Journal of Environmental and Engineering Geophysics 12, 63–67. Available from: https://doi.org/10.2113/JEEG12.1.63.

Siegert, M.J., Ross, N., Corr, H., Kingslake, J., Hindmarsh, R., 2013. Late Holocene ice-flow reconfiguration in the Weddell Sea sector of West Antarctica. Quaternary Science Reviews 78, 98–107. Available from: https://doi.org/10.1016/j.quascirev.2013.08.003.

Siegert, M.J., Ross, N., Li, J., Schroeder, D., Rippin, D., Ashmore, D., et al., 2016. Controls on the onset and flow of Institute Ice Stream, West Antarctica. Annals of Glaciology 57, 19–24. Available from: https://doi.org/10.1017/aog.2016.17.

Siegert, M.J., Golledge, N.R., 2021. Advances in numerical modelling of the Antarctic ice sheet. In: Florindo, F., et al. (Eds.), Antarctic Climate Evolution, second edition. Elsevier (this volume).

Simkins, L.M., Anderson, J.B., Greenwood, S.L., Gonnermann, H.M., Prothro, L.O., Halberstadt, A.R.W., et al., 2017. Anatomy of a meltwater drainage system beneath the ancestral East Antarctic Ice Sheet. Nature Geoscience 10 (9), 691–697. Available from: https://doi.org/10.1038/ngeo3012.

Simms, A.R., Lisiecki, L., Gebbie, G., Whitehouse, P.L., Clark, J.F., 2019. Balancing the last glacial maximum (LGM) sea-level budget. Quaternary Science Reviews 205, 143–153. Available from: https://doi.org/10.1016/j.quascirev.2018.12.018.

Smith, J.A., Andersen, T.J., Shortt, M., Gaffney, A.M., Truffer, M., Stanton, T.P., et al., 2017. Sub-ice-shelf sediments record history of twentieth-century retreat of Pine Island Glacier. Nature 541, 77–80. Available from: https://doi.org/10.1038/nature20136.

Smith, J.A., Bentley, M.J., Hodgson, D.A., Roberts, S.J., Leng, M.J., Lloyd, J.M., et al., 2007. Oceanic and atmospheric forcing of early Holocene ice shelf retreat, George VI Ice Shelf, Antarctica Peninsula. Quaternary Science Reviews 26, 500–516. Available from: https://doi.org/10.1016/j.quascirev.2006.05.006.

Smith, J.A., Hillenbrand, C.-D., Kuhn, G., Klages, J.P., Graham, A.G.C., Larter, R.D., et al., 2014. New constraints on the timing of West Antarctic Ice Sheet retreat in the eastern

Amundsen Sea since the Last Glacial Maximum. Global and Planetary Change 122, 224–237. Available from: https://doi.org/10.1016/j.gloplacha.2014.07.015.

Smith, J.A., Hillenbrand, C.-D., Kuhn, G., Larter, R.D., Graham, A.G.C., Ehrmann, W., et al., 2011. Deglacial history of the West Antarctic Ice Sheet in the western Amundsen Sea Embayment. Quaternary Science Reviews 30, 488–505. Available from: https://doi.org/10.1016/j.quascirev.2010.11.020.

Smith, J.A., Hillenbrand, C.-D., Larter, R.D., Graham, A.G.C., Kuhn, G., 2009. The sediment infill of subglacial meltwater channels on the West Antarctic continental shelf. Quaternary Research 71, 190–200. Available from: https://doi.org/10.1016/j.yqres.2008.11.005.

Spector, P., Stone, J., Cowdery, S.G., Hall, B., Conway, H., Bromley, G., 2017. Rapid early-Holocene deglaciation in the Ross Sea, Antarctica. Geophysical Research Letters 44, 7817–7825. Available from: https://doi.org/10.1002/2017GL074216.

Spector, P., Stone, J., Goehring, B., 2019. Thickness of the divide and flank of the West Antarctic Ice Sheet through the last deglaciation. The Cryosphere 13, 3061–3075. Available from: https://doi.org/10.5194/tc-13-3061-2019.

Stirling, C.H., Esat, T.M., Lambeck, K., McCulloch, M.T., 1998. Timing and duration of the last interglacial: evidence for a restricted interval of widespread coral reef growth. Earth and Planetary Science Letters 160, 745–762. Available from: https://doi.org/10.1016/S0012-821X(98)00125-3.

Stolldorf, T., Schenke, H.-W., Anderson, J.B., 2012. LGM ice sheet extent in the Weddell Sea: evidence for diachronous behavior of Antarctic Ice Sheets. Quaternary Science Reviews 48, 20–31. Available from: https://doi.org/10.1016/j.quascirev.2012.05.017.

Stone, J.O., Balco, G.A., Sugden, D.E., Caffee, M.W., Sass, L.C.I., Cowdery, S.G., et al., 2003. Holocene deglaciation of Marie Byrd Land, West Antarctica. Science 299, 99–102. Available from: https://doi.org/10.1126/science.1077998.

Studinger, M., Bell, R.E., Blankenship, D.D., Finn, C.A., Arko, R.A., Morse, D.L., et al., 2001. Subglacial sediments: a regional geological template for ice flow in Antarctica. Geophysical Research Letters 28, 3493–3496. Available from: https://doi.org/10.1029/2000GL011788.

Subt, C., Fangman, K.A., Wellner, J.S., Rosenheim, B.E., 2016. Sediment chronology in Antarctic deglacial sediments: reconciling organic carbon ^{14}C ages to carbonate ^{14}C ages using Ramped PyrOx. The Holocene 26, 265–273. Available from: https://doi.org/10.1177/0959683615608688.

Subt, C., Yoon, H.I., Yoo, K.C., Lee, J.I., Leventer, A., Domack, E.W., et al., 2017. Sub-ice shelf sediment geochronology utilizing novel radiocarbon methodology for highly detrital sediments. Geochemistry, Geophysics, Geosystems 18, 1404–1418. Available from: https://doi.org/10.1002/2016GC006578.

Sugden, D.E., Bentley, M.J., Cofaigh, C.Ó., 2006. Geological and geomorphological insights into Antarctic Ice Sheet evolution. Philosophical Transactions of the Royal Society A 364, 1607–1625. Available from: https://doi.org/10.1098/rsta.2006.1791.

Sugden, D.E., Jamieson, S.S.R., 2018. The pre-glacial landscape of Antarctica. Scottish Geographical Journal 134, 203–223. Available from: https://doi.org/10.1080/14702541.2018.1535090.

Sugden, D.E.A.S., Hein, J., Woodward, S.M., Marrero, Á., Rodés, S.A., Dunning, F.M., et al., 2017. The million-year evolution of the glacial trimline in the southernmost Ellsworth Mountains, Antarctica. Earth and Planetary Science Letters 469, 42–52. Available from: https://doi.org/10.1016/j.epsl.2017.04.006.

Takada, M., Miura, M., Zwartz, D., 1998. Radiocarbon and thermoluminescence ages in the Mt Riiser-Larsen area, Enderby Land, East Antarctica. Polar Geoscience 11, 239–248.

Takada, M., Tani, A., Miura, H., Moriwaki, K., Nagatomo, T., 2003. ESR dating of fossil shells in the Lützow-Holm Bay region, East Antarctica. Quaternary Science Reviews 22 (10–13), 1323–1328. Available from: https://doi.org/10.1016/S0277-3791(03)00040-4.

Tarasov, L., Peltier, W.R., 2003. Greenland glacial hstory, borehole constraints and Eemian extent. Journal of Geophysical Research 108 (B3), 2143. Available from: https://doi.org/10.1029/2001JB001731.

Taylor, F., McMinn, A., 2002. Late Quaternary diatom assemblages from Prydz Bay, Eastern Antarctica. Quaternary Research 57 (1), 151–161. Available from: https://doi.org/10.1006/qres.2001.2279.

Taylor, J., Siegert, M.J., Payne, A.J., Hambrey, M.J., O'Brien, P.E., Cooper, A.K., et al., 2004. Topographic controls on post-Oligocene changes in ice-sheet dynamics, Prydz Bay region, East Antarctica. Geology 32, 197–200. Available from: https://doi.org/10.1130/G20275.1.

The RAISED Consortium, 2014. A community-based geological reconstruction of Antarctic Ice Sheet deglaciation since the Last Glacial Maximum. Quaternary Science Reviews 100, 1–9. Available from: https://doi.org/10.1016/j.quascirev.2014.06.025.

Tingey, R.J., McDougall, I., Gleadow, A.J.W., 1983. The age and mode of formation of Gaussberg. Antarctica. Journal of the Geological Society of Australia 30, 241–246. Available from: https://doi.org/10.1080/00167618308729251.

Tinto, K.J., et al., 2019. Ross Ice Shelf response to climate driven by the tectonic imprint on seafloor bathymetry. Nature Geoscience 12, 441–449. Available from: https://doi.org/10.1038/s41561-019-0370-2.

Todd, C., Stone, J., Conway, H., Hall, B., Bromley, G., 2010. Late Quaternary evolutionof Reedy Glacier, Antarctica. Quaternary Science Reviews 29, 1328–1341. Available from: https://doi.org/10.1016/j.quascirev.2010.02.001.

Turner, J., Orr, A., Gudmundsson, G.H., Jenkins, A., Bingham, R.G., Hillenbrand, C.-D., et al., 2017. Atmosphere-ocean-ice interactions in the Amundsen Sea Embayment, West Antarctica. Reviews of Geophysics 55, 235–276. Available from: https://doi.org/10.1002/2016RG000532.

Vaughan, D.G., Marshall, G., Connolley, W.M., Parkinson, C., Mulvaney, R., Hodgson, D.A., et al., 2003. Recent rapid regional climate warming on the Antarctic Peninsula. Climatic Change 60, 243–274. Available from: https://doi.org/10.1023/A:1026021217991.

Verleyen, E., Hodgson, D.A., Milne, G.A., Sabbe, K., Vyverman, W., 2005. Relative sea-level history from the Lambert Glacier region, East Antarctica, and its relation to deglaciation and Holocene glacier readvance. Quaternary Research 63, 45–52. Available from: https://doi.org/10.1016/j.yqres.2004.09.005.

Voosen, P., 2018. Antarctic ice melt 125,000 years ago offers warning. Science 362, 1339. Available from: https://doi.org/10.1126/science.362.6421.1339.

Waddington, E.D., Conway, H., Steig, E.J., Alley, R.B., Brook, E.J., Taylor, K.C., et al., 2005. Decoding the dipstick: thickness of Siple Dome, West Antarctica, at the Last Glacial Maximum. Geology 33, 281–284. Available from: https://doi.org/10.1130/G21165.1.

Wagner, B., Cremer, H., Hultzsch, N., Gore, D.B., Melles, M., 2004. Late Pleistocene and Holocene history of Lake Terrasovoje, Amery Oasis, East Antarctica, and its climatic and environmental implications. Journal of Paleolimnology 32, 321–339. Available from: https://doi.org/10.1007/s10933-004-0143-8.

Wagner, B., Hultzsch, N., Melles, M., Gore, D.B., 2007. Indications of Holocene sea-level rise in Beaver Lake, East Antarctica. Antarctic Science 19, 125–128. Available from: https://doi.org/10.1017/S095410200700017X.

Wellner, J.S., Lowe, A.L., Shipp, S.S., Anderson, J.B., 2001. Distribution of glacial geomorphic features on the Antarctic continental shelf and correlation with substrate: implications for ice behaviour. Journal of Glaciology 47, 397–411. Available from: https://doi.org/10.3189/172756501781832043.

White, D.A., Bennike, O., Berg, S., Harley, S.L., Fink, D., Kiernan, K., et al., 2009. Geomorphology and glacial history of Rauer Group, East Antarctica. Quaternary Research 72, 80–90. Available from: https://doi.org/10.1016/j.yqres.2009.04.001.

White, D.A., Fink, D., 2014. Late Quaternary glacial history constrains glacio-isostatic rebound in Enderby Land East Antarctica. Journal of Geophysical Research 119, 401–413. Available from: https://doi.org/10.1002/2013JF002870.

White, D.A., Fink, D., Gore, D.B., 2011. Cosmogenic nuclide evidence for enhanced sensitivity of an East Antarctic ice stream to change during the last deglaciation. Geology 39 (1), 23–26. Available from: https://doi.org/10.1130/G31591.1.

White, D.A., Hermichen, W.-D., 2007. Glacial and periglacial history of the southern Prince Charles Mountains, East Antarctica. Terra Antartica 14, 5–12.

Whitehouse, P.L., 2018. Post last glacial maximum processes in the polar regions. In: Nuttall, M., Christensen, T.R., Siegert, M.J. (Eds.), Routledge Handbook of the Polar Regions. Routledge, pp. 209–223. Chapter 16.

Whitehouse, P.L., Bentley, M.J., Le Brocq, A.M., 2012. A deglacial model for Antarctica: geological constraints and glaciological modelling as a basis for a new model of Antarctic glacial isostatic adjustment. Quaternary Science Reviews 32, 1–24. Available from: https://doi.org/10.1016/j.quascirev.2011.11.016.

Whitehouse, P.L., Bentley, M.J., Vieli, A., Jamieson, S.S.R., Hein, A.S., Sugden, D.E., 2017. Controls on Last Glacial Maximum ice extent in the Weddell Sea embayment, Antarctica. Journal of Geophysical Research 122, 371–397. Available from: https://doi.org/10.1002/2016JF004121.

Willmott, V., Domack, E.W., Canals, M., Brachfeld, S., 2006. A high resolution relative paleointensity record from the Gerlache-Boyd paleo-ice stream region, northern Antarctic Peninsula. Quaternary Science Reviews 66, 1–11. Available from: https://doi.org/10.1016/j.yqres.2006.01.006.

Wilson, D.J., Bertram, R.A., Needham, E.F., et al., 2018. Ice loss from the East Antarctic Ice Sheet during late Pleistocene interglacials. Nature 561, 383–386. Available from: https://doi.org/10.1038/s41586-018-0501-8.

Wilson, D.J., et al., 2021. In: Florindo, F. (Ed.), et al., Antarctic Climate Evolution, second ed. Elsevier, (this volume).

Winter, K., Woodward, J., Ross, N., Dunning, S.A., Corr, H.F.J., Bingham, R.G., et al., 2015. Airborne radar evidence for tributary flow switching in the Institute Ice Stream upper catchment, West Antarctica: implications for ice sheet configuration and dynamics. Journal of Geophysical Research 120, 1611–1625. Available from: https://doi.org/10.1002/2015JF003518.

Wise, M.G., Dowdeswell, J.A., Jakobsson, M., Larter, R.D., 2017. Evidence of marine ice-cliff instability in Pine Island Bay from iceberg-keel plough marks. Nature 550 (7677), 506–510. Available from: https://doi.org/10.1038/nature24458.

Witus, A.E., Branecky, C.M., Anderson, J.B., Szczuciński, W., Schroeder, D.M., Blankenship, D.D., et al., 2014. Meltwater intensive retreat in polar environments and investigation of associated sediments: example from Pine Island Bay, West Antarctica. Quaternary Science Reviews 85, 99–118. Available from: https://doi.org/10.1016/j.quascirev.2013.11.021.

Wolstencroft, M., King, M.A., Whitehouse, P.L., Bentley, M.J., Nield, G.A., King, E.C., et al., 2015. Uplift rates from a new high-density GPS network in Palmer Land indicate significant late Holocene ice loss in the southwestern Weddell Sea. Geophysical Journal International 203, 737–754. Available from: https://doi.org/10.1093/gji/ggv327.

Wright, A., White, D., Gore, D., Siegert, M.J., 2008. Antarctica at the LGM. In: Florindo, F., Siegert, M.J. (Eds.), Antarctic Climate Evolution. Developments in Earth & Environmental Sciences, 8. Elsevier, pp. 531–570. Available from: https://doi.org/10.1016/S1571-9197(08)00012-8.

Yamane, M., Yokoyama, Y., Miura, H., Maemoku, H., Iwasaki, S., Matsuzaki, H., 2011. The last deglacial history of Lützow-Holm Bay, East Antarctica. Journal of Quaternary Science 26, 3–6. Available from: https://doi.org/10.1002/jqs.1465.

Yamane, M., Yokoyama, Y., Miyairi, Y., Suga, H., Matsuzaki, H., Dunbar, R.B., et al., 2014. Compound-specific ^{14}C dating of IODP Expedition 318 Core U1357A obtained off the Wilkes Land Coast, Antarctica. Radiocarbon 56 (3), 1009–1017. Available from: https://doi.org/10.2458/56.17773.

Yau, A.M., Bender, M.L., Robinson, A., Brook, E.J., 2016. Reconstructing the last interglacial at Summit, Greenland: insights from GISP2. Proceedings of the National Academy of Sciences 113 (35), 9710–9715. Available from: https://doi.org/10.1073/pnas.1524766113.

Yokoyama, Y., Anderson, J.B., Yamane, M., Simkins, L.M., Miyairi, Y., Yamazaki, T., et al., 2016. Widespread collapse of the Ross Ice Shelf during the late Holocene. Proceedings of the National Academy of Sciences 113, 2354–2359. Available from: https://doi.org/10.1073/pnas.1516908113.

Yoshida, Y., Moriwaki, K., 1979. Some considerations on elevated coastal features and their dates around Syowa Station, Antarctica. Memoirs of the National Institute of Polar Research, Special Issue 13, 220–226.

Zwartz, D., Bird, M., Stone, J., Lambeck, K., 1998b. Holocene sea-level change and ice-sheet history in the Vestfold Hills, East Antarctica. Earth and Planetary Science Letters 155 (1–2), 131–145. Available from: https://doi.org/10.1016/S0012-821X(97)00204-5.

Zwartz, D.P., Miura, H., Takada, M., Moriwaki, K., 1998a. Holocene lake sediments and sea-level change at Mt. Riiser-Larsen. Polar Geoscience 11, 249–259. Available from: https://doi.org/10.1016/S0012-821X(97)00204-5.

Chapter 12

Past Antarctic ice sheet dynamics (PAIS) and implications for future sea-level change

Florence Colleoni[1], Laura De Santis[1], Tim R. Naish[2], Robert M. DeConto[3], Carlota Escutia[4], Paolo Stocchi[5,12], Gabriele Uenzelmann-Neben[6], Katharina Hochmuth[7], Claus-Dieter Hillenbrand[8], Tina van de Flierdt[9], Lara F. Pérez[8], German Leitchenkov[10,11], Francesca Sangiorgi[12], Stewart Jamieson[13], Michael J. Bentley[13], David J. Wilson[14] and the PAIS community*

[1]National Institute of Oceanography and Applied Geophysics – OGS, Sgonico, Italy, [2]Antarctic Research Centre, Victoria University of Wellington, Wellington, New Zealand, [3]Department of Geosciences, University of Massachusetts, Amherst, Amherst, MA, United States, [4]Andalusian Institute of Earth Sciences, CSIC and Universidad de Granada, Armilla, Spain, [5]NIOZ - Royal Netherlands Institute for Sea Research, Coastal Systems Department (TX), and Utrecht University, Den Burg, the Netherlands, [6]Alfred Wegener Institute Helmholtz Center for Polar and Marine Research, Bremerhaven, Germany, [7]School of Geography, Geology and the Environment, University of Leicester, Leicester, United Kingdom, [8]British Antarctic Survey, Cambridge, United Kingdom, [9]Department of Earth Science and Engineering, Imperial College London, London, United Kingdom, [10]Institute for Geology and Mineral Resources of the World Ocean, St. Petersburg, Russia, [11]Institute of Earth Sciences, St. Petersburg State University, St. Petersburg, Russia, [12]Laboratory of Palaeobotany and Palynology, Department of Earth Sciences, Marine Palynology and Paleoceanography, Utrecht University, Utrecht, The Netherlands, [13]Department of Geography, Durham University, Durham, United Kingdom, [14]Institute of Earth and Planetary Sciences, University College London and Birkbeck, University of London, London, United Kingdom

12.1 Research focus of the PAIS programme

Ice sheet and sea-level reconstructions from the past 'warmer-than-present' climates of the last 34 million years provide powerful insights into the long-term response of the polar ice sheets to climate changes projected for the

---

*List of authors is shown before the references.

Antarctic Climate Evolution. DOI: https://doi.org/10.1016/B978-0-12-819109-5.00010-4

21st century and onward. Proximal geological evidence shows the onset of large-scale Antarctic glaciations occurred around the Eocene/Oligocene Transition at approximately 34 Ma (EOT, e.g. Barrett, 1989; Coxall et al., 2005; Escutia et al., 2011; Galeotti et al., 2016, 2021 (this volume); Hambrey et al., 1991; Passchier et al., 2013, 2016). Since then, the Antarctic Ice Sheet (AIS) evolution and variability, recorded in the direct geological archives, have been widely compared with the benthic oxygen isotope proxy records of deep-water temperature and global ice volume from far-field deep ocean locations (e.g. De Vleeschouwer et al., 2017; Zachos et al., 2001). Together, they suggest fluctuations in ice volume and ice sheet extent that can be driven by changes in atmospheric CO_2 level and astronomical forcing (Galeotti et al., 2016; Hansen et al., 2015; Levy et al., 2019; Naish et al., 2001, 2009a; Pälike et al., 2006; Patterson et al., 2014). Looking at the broad picture, the EOT really marked the transition into an ice house world (Miller et al., 1991; 2020b; Westerhold et al., 2020). As the initial cryosphere evolved, only one pole, the South Pole, hosted a continental-size ice sheet. Geological evidence from the Arctic revealed that small ice-sheets or ice caps may have formed in the Northern high latitudes as early as the Eocene (e.g. Eldrett et al 2007; Tripati and Darby, 2018). Then, along with the gradual decrease in atmospheric CO_2 level (Fig. 12.1A), Greenland glaciated from the Late Miocene, rapidly followed by episodic and extensive glaciations of most of the high-latitude Arctic margins, after the onset of Northern Hemisphere glaciations 2.7 Ma (e.g. Thiede et al., 2011).

During the past 34 million years, changes in global climate and polar ice volume have been paced by orbital forcing – the Milankovitch cycles. These regular glacial–interglacial cycles, which are amplified by internal Earth system feedbacks, occur on a background of secular change over millions of years driven by global plate tectonics and the carbon cycle (Fig. 12.1). Rapid stepwise transitions between climate states (e.g. EOT) correspond to thresholds in the Earth system often linked to a combination of tectonic reorganisation, atmospheric greenhouse gas composition and extreme astronomical forcing (Levy et al., 2019; Wilson et al., 2021; Zachos et al., 2001, 2008). Our knowledge of the evolution of the AIS and its influence on global climate is now widely documented by proximal ice and sediment core records (e.g. Barrett, 2007; Bereiter et al., 2015; Escutia et al., 2019; EPICA Community Members, 2006; Levy et al., 2019; McKay et al., 2016).

The circum-Antarctic seismic stratigraphic records of the continental margins reveal erosive unconformities indicative of at least six main periods of massive AIS advances (e.g. Bart and De Santis, 2012; Brancolini et al., 1995a, 1995b; Cooper et al., 1999, 2011; De Santis et al., 1999; Donda et al., 2007; Gohl et al., 2013; Gulick et al., 2017; Kristoffersen and Jokat, 2008; Larter et al., 1997; Lindeque et al., 2016; Steinhauff and Webb, 1987). Not all the unconformities have been dated but Hochmuth et al. (2019, 2020) recently provided the first attempt of pan-Antarctic correlation of the various regional

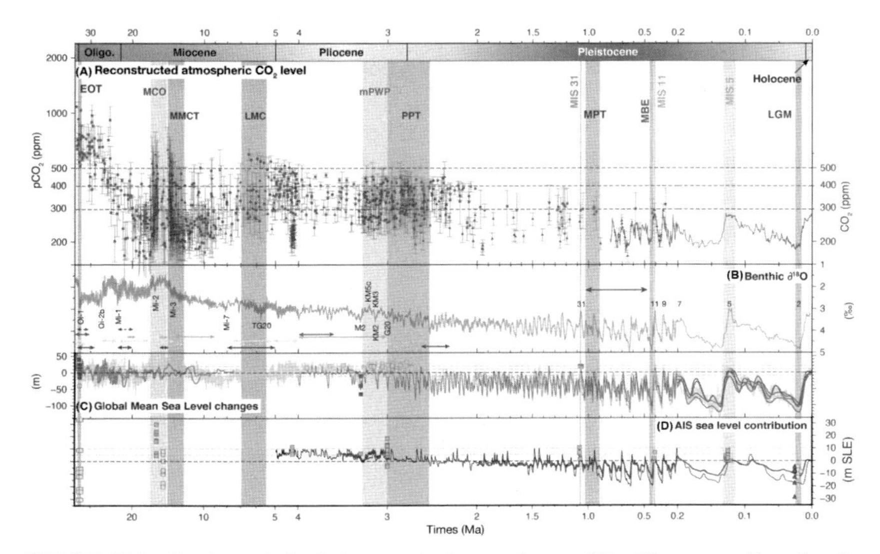

FIGURE 12.1 Proxies and simulations synthesis over the past 34 million years. Note that the time scale is logarithmic. (A) Reconstructed atmospheric p[CO_2] levels are based on alkenone (Badger et al., 2013a, 2013b; Pagani et al., 2005, 2010, 2011; Seki et al., 2010; Super et al., 2018; Zhang et al., 2013) and on boron measurements (Bartoli et al., 2011; Foster et al., 2012; Greenop et al., 2014; Honisch et al., 2009; Martinez-Boti et al., 2015). Late Pleistocene atmospheric CO_2 levels are based on the Antarctic ice core composite record from Bereiter et al. (2015); (B) Deep sea benthic $\delta^{18}O$ record from Zachos et al. (2001a, 2008). Marine isotope stages (glacials and interglacials) discussed or named in the chapter are indicated and based on Miller et al. (1991) for the Oligocene and Miocene, on Haywood et al. (2016) for the Pliocene and on Lisiecki and Raymo (2005) for the Pleistocene. Note that marine isotopes stages EOT-1 (~34.46–33.9 Ma) and EOT-2 (33.7 Ma) are not indicated. Seismic stratigraphic unconformities from different Antarctic sectors are reported with arrows, based on Hochmuth et al. (2020) and references therein (Ross Sea: dark blue; Wilkes Land: green; Weddell Sea: cyan; Amundsen Sea: orange; Cosmonaut Sea: brown; Prydz Bay: purple); (C) Reconstructed (pink, purple) and simulated (blue) global mean sea level changes (GMSL). Proxy-based reconstructions: benthic $\delta^{18}O$ (Miller et al., 2020a) from EOT to present, backstripped sequence stratigraphy from New Jersey from EOT until the Late Miocene (Kominz et al., 2008; Miller et al., 2005). For the EOT (pink solid squares): Bohaty et al. (2012); Hauptvogel and Passchier (2012); Katz et al. (2008), Lear et al. (2008); Miller et al. (2009); Pekar and Christie-Blick (2008); Pekar et al. (2002); Stocchi et al. (2013). For the Pliocene: converted benthic $\delta^{18}O$ record from Dumitru et al. (2019) until the PPT. Pink squares correspond to reconstructed Pliocene highstands (Dumitru et al., 2019; Dwyer and Chandler, 2009; Kulpecz et al., 2009; Miller et al., 2012; Naish and Wilson, 2009b; Sosdian and Rosenthal, 2009; Wardlaw and Quinn, 1991; Winnick and Caves, 2015) and to M2 glaciation (Dwyer and Chandler, 2009; Miller et al., 2005; Naish and Wilson, 2009b). For Pleistocene: sea level reconstructions are taken from Hearty et al. (2020); Sandstrom et al. (2021) for MIS 31 (uncorrected from GIA and dynamic topography) and from Raymo and Mitrovica (2012) and Roberts et al. (2012) for MIS 11. For the last 400 kyrs, reconstructed curves of sea level changes are from Waelbroeck et al. (2002) and from Rohling et al. (2009). For MIS5, palaeoshorelines data are from the compilation in Dutton et al. (2015) and references therein. Simulated GMSL changes are from Bintanja et al. (2005), de Boer et al. (2015) and Stap et al. (2017), and from Raymo et al. (2006) for MIS 31 (blue squares); (D) Simulated

(Continued)

seismic unconformities (Fig. 12.1B). These seismic unconformities likely correspond to: early Antarctic glaciations after the EOT (e.g. Galeotti et al., 2016), a transient glaciation at Oligocene-Miocene boundary ~23–21 Ma (e.g. Naish et al., 2001, 2021, this volume; summarised in Wilson and Luyendyk, 2009), ice sheet re-advance at the end of the Mid-Miocene Climatic Optimum (MCO) about 15.8–14.2 Ma (e.g. Levy et al., 2016), the cooling and expansion of the East Antarctic Ice Sheet (EAIS) at the Middle Miocene Climate Transition (MMCT) ~13.8 Ma (e.g. Levy et al., 2016; Lewis et al., 2007; 2008; Pierce et al., 2017), cooling after a period of warmth known as the Late Miocene Cooling ~8–5 Ma (LMC) (Gulick et al., 2017; Herbert et al., 2016; McKay et al., 2009) and cooling and expansion of marine-based ice during the Plio-Pleistocene Transition (PPT) 3–2.5 Ma (McKay et al., 2012a; Naish et al., 2009a; Patterson et al., 2014) and possibly to the Mid-Pleistocene Transition ~1 Ma (e.g. O'Brien et al., 2007). These periods evidenced by positive excursions in the deep sea benthic $\delta^{18}O$ isotope records (Fig. 12.1B) correspond to episodes of global cooling and ice volume growth periods, generally associated with a decline in atmospheric CO_2 below

Antarctic ice sheet melting contributions (metre Sea Level Equivalent, m SLE) to GMSL changes are from ice sheet simulations (squares and curves) and from Glacio-isostatic-Adjustment simulations (dark red triangles). Note that some of the reported simulated ice volumes do not refer to volumes above floatation. For the EOT: DeConto and Pollard (2003), Gasson et al (2014), Ladant et al. (2014), Liakka et al. (2014), Pollard and DeConto (2005), Wilson et al. (2013). For the Miocene: Colleoni et al. (2018b), Gasson et al. (2016), Langebroek et al. (2009), Stap et al. (2019). For Early Pliocene: Pollard and DeConto (2009, transient and black line), Golledge et al. (2017a, orange squares). For mPWP to Late Pliocene: de Boer et al. (2013, transient blue line), Pollard and DeConto (2009, transient black line). Orange squares come from Tan et al. (2017) for M2 glaciation and remaining symbols are for the mPWP considering Austermann et al. (2015), de Boer et al. (2015), DeConto and Pollard (2016), Dolan et al. (2018), Gasson et al. (2015), Pollard and DeConto (2012), Yan et al. (2016). When simulations were run with averaged mPWP climatic conditions, orange squares are indicatively plotted at 3 Ma. For the entire Pleistocene: blue line – de Boer et al. (2014) and black line – Pollard and DeConto (2009). For MIS 31: Beltran et al. (2020), de Boer et al. (2017a), DeConto et al. (2012). For MIS 11: Mas e Braga et al. (2021), Sutter et al. (2019), Tigchelaar et al. (2018). For MIS 5: Colleoni et al. (2018b), de Boer et al. (2015), DeConto and Pollard (2016), Goelzer et al. (2016), Huybrechts (2002), Pollard and DeConto (2012), Quiquet et al. (2018), Sutter et al. (2016), Sutter et al. (2019), Tigchelaar et al. (2018). For LGM based on ice sheet simulations: Brigg et al. (2014), Colleoni et al. (2018b), de Boer et al. (2013), Golledge et al. (2012), Golledge et al. (2014), Mackintosh et al. (2011), Pollard et al. (2016), Philippon et al. (2006), Quiquet et al. (2018), Sutter et al. (2019). For LGM based on glacio-isostatic adjustment simulations: Argus et al. (2014), Gomez et al. (2013), Ivins and James (2005), Ivins et al. (2013), Lambeck et al. (2014), Peltier (2004), Whitehouse et al. (2012b). Cold periods of interest in this chapter are indicated with blue bars: *EOT*, Eocene-Oligocene Transition; *MMCT*, Mid-Miocene Climatic Transition; *LMC*, Late Miocene Cooling; *PPT*, Plio-Pleistocene Transition; *MPT*, Mid-Pleistocene Transition; *MBE*, mid-Brunhes Event; *LGM*, Last Glacial Maximum. Warm periods mentioned in this chapter are indicated with orange bars: *MCO*, Mid-Miocene Climatic Optimum; *mPWP*, mid-Pliocene Warm Period, MIS 31, MIS 11, MIS 5. Please refer to the online version of this figure for enhanced resolution and magnification.

a threshold (Fig. 12.1A) for triggering the expansion of terrestrial and/or marine-based ice and/or the onset of perennial sea-ice.

Between these cooling periods, multi-proxy global climatic reconstructions and proximal Antarctic geological climate and ice sheet reconstructions provide evidence for intense and/or brief warm periods during which:

- atmospheric CO_2 levels, surface temperatures and global sea level rose well above present-day levels during the MCO (17–15 Ma) and the middle Pliocene Warm Period (mPWP, 3.3–3 Ma) (see Miller et al., 2012 for a review); both periods were characterised by CO_2 concentrations higher than 400 ppm and up to 800 ppm for some specific intervals of the MCO (see Fig. 12.1A and references therein).
- atmospheric CO_2 levels were near pre-industrial levels, i.e. around 300 ppm but were associated with warmer global surface temperatures and higher sea-levels than today during the 'super interglacials' of the Pleistocene. This was likely driven by astronomical forcing. Examples include marine isotope stage (MIS) 31 (1.081–1.062 Ma), and specific warm Late Pleistocene interglacials such as MIS 11 (425–395 ka) and MIS 5e (130–116 ka) (e.g. Dutton et al., 2015; Miller et al., 2020b).

These past warm periods are policy-relevant as they provide accessible examples of how the AIS responded to warmer-than-present global temperatures, comparable to those projected for the coming decades to centuries (IPCC AR5, 2013; IPCC SROCC, 2019). However, using these past warm periods to inform our understanding of the AIS sensitivity under different atmospheric CO_2 levels remains a challenge, in part because Earth system boundary conditions were subtly different than today, and the duration and intensity of these past warm periods was highly variable (e.g. Bracegirdle et al., 2019; Colleoni et al., 2018a; DeConto and Pollard, 2016; Dutton et al., 2015; Noble et al., 2020).

For example, the duration of those past warm periods differs from about 2 million years for the MCO, approximately 300 thousand years for the mPWP to a few millennia for some Pleistocene interglacials. Thus, an approach focusing on specific MCO and mPWP glacial-interglacial cycles and interglacials (with similarities to our present interglacial) is being developed. For example, the Pliocene Model Intercomparison Project community is focussing its ongoing mPWP model-data comparison on the M2-KM5c (3.264–3.205 Ma) and KM5c-KM2 (3.205–3.130 Ma) intervals (Fig. 12.1B) (Haywood et al., 2016). By using more appropriate forcing and boundary conditions for climate model simulations, discrepancies between models and data generally decrease (e.g. Otto-Bliesner et al., 2017).

In most simulations of future ice sheet evolution, model projections typically extend only until the policy horizon of CE (Common Era) 2100. However, some ice sheet models have run projections out as far as CE 2500 (e.g. Clark et al., 2016; DeConto and Pollard, 2016; Golledge et al., 2015).

The recent IPCC special report on 'Ocean and Cryosphere in a Changing Climate' (IPCC ROCC, 2019) utilises these projections and the results of a structured expert judgement approach (Bamber et al., 2019) to present projections to CE 2300, that to some extent account for the long-term thermodynamical response of the Greenland and Antarctic ice sheets and related instabilities (e.g. DeConto and Pollard, 2016; Golledge et al., 2015). Palaeoclimatic changes are often considered at timescales of tens of millennia to millennia and, in a few archives, at sub-millennial timescale. At such timescales, past reconstructions can inform long-term projections over a few millennia (see Golledge et al., 2020 for a review), but some refinements at sub-millennial timescales to investigate some abrupt events of the near past are necessary to reconcile with the projections.

During past warm periods global paleogeography, paleotopography and paleobathymetry can differ substantially from today. Periods prior to the Plio-Pleistocene Transition (3.0–2.5 Ma) were characterised by a very different continental and oceanic configuration that yielded changes in the proportion of emerged lands and their locations. This affected surface elevation, oceanic gateways and bathymetry, which in turn impacted ocean and atmospheric circulation (e.g. Dowsett et al., 2016; Herold et al., 2008; Huang et al., 2017; Kennedy et al., 2015; von der Heydt et al., 2016), on global mean sea level changes (e.g. Miller et al., 2020b) and on heat transport compared to modern conditions. The Antarctic continent and its surface elevation have also evolved throughout the Cenozoic, with important consequences for ice sheet behaviour (e.g. Colleoni et al., 2018b; Paxman et al., 2020). Thus, direct comparison between the past and future AIS sensitivity to high levels of atmospheric greenhouse gases is not straightforward.

Most of the efforts of the PAIS programme, and its predecessor, the Antarctic Climate Evolution (ACE) programme, focused on the past warm periods (e.g. MCO, mPWP, warm interglacials of the Pleisotcene) (Fig. 12.1). More specifically, ice sheet and climate simulations of the MCO emerged during the PAIS programme lifetime. Within its programme, PAIS promoted collaborative work within six specific sub-committees that addressed the following topics for almost all of the warm periods listed above:

- Paleoclimate records from the Antarctic Margin and Southern Ocean (PRAMSO);
- Paleotopographic-Paleobathymetric Reconstructions;
- Subglacial Geophysics;
- Ice Cores and Marine Core Synthesis;
- Recent Ice Sheet Reconstructions;
- Deep-Time Ice Sheet Reconstructions.

Scientific advances related to each of these topics are extensively described in the previous chapters of this book. Many research projects that were initiated within the ACE programme concluded during the PAIS programme. Many of their findings continue to have a significant impact on the community of Antarctic researchers and well beyond. In the following sections, we highlight some of the key findings that have advanced our understanding of the Antarctic Ice Sheet dynamics, instabilities and thresholds during past warm periods. We conclude with a discussion of the PAIS legacy, and highlight emerging issues, knowledge gaps, needs and challenges to be addressed within the next decade by the observational and modelling communities.

12.2 Importance of evolving topography, bathymetry, erosion and pinning points

Ice–sheet–ocean–bedrock interactions are of major importance for understanding the dynamics of the AIS (Colleoni et al., 2018a; Mengel and Levermann, 2014; Paxman et al., 2020; Whitehouse et al., 2019). Surface and basal boundary conditions determine the characteristics and regime of the ice flow. At the base of a terrestrial or marine-based ice sheet, the geothermal heat flux, the morphology as well as the nature of the bed (hard rock or soft sediments), affect the sliding of the ice, generate heat and yield basal meltwater. When the ice sheet advances, it erodes its bed and carries sediment. Eroded material is released into ice shelf cavities and onto the continental shelf at the grounding zone where the ice sheet floats, disconnecting from its bed. Some of this glacigenic detritus finds its way via glacial troughs and via channels across the continental slope and rise and ultimately to the abyssal plain. Some eroded material is also carried by icebergs and deposited offshore as Iceberg Rafted Debris (IBRD). These sediments, preserved in a wide range of marine environments, provide a valuable archive of past ice sheet dynamics and coeval oceanic and atmospheric conditions.

The AIS substantially expanded ~34 Ma and since then has advanced and retreated numerous times (see Galeotti et al., 2016). As a result, the morphology of the bed below the ice sheet evolved considerably. Seismic stratigraphic records from the Antarctic continental margins (e.g. Cooper et al., 1991; De Santis et al.; 1999; Eittreim et al., 1995; Gohl et al., 2013; Huang and Jokat, 2016; Whitehead et al., 2006) clearly show that the shallow continental shelves have been prograding northward through time. Sediment isopach (thickness) reconstructions indicate that much of the sediments have accumulated, and accreted, along the Antarctic continental slope and rise (see references in Hochmuth and Gohl, 2019; Hochmuth et al., 2020 for circum-Antarctic review and reconstructions), implying that a large volume of material has been eroded and removed from inland regions since the onset of continental glaciation (e.g. Hochmuth et al., 2020; Paxman et al., 2019; Wilson et al., 2012).

The most recent circum-Antarctic reconstructions show that the morphology of the bed has changed substantially over the past 34 million years (Fig. 12.2). At the EOT, most of the West and East Antarctic sectors that are currently below sea level were instead above sea level (Paxman et al., 2019; Wilson et al., 2012) (Fig. 12.2). With time, tectonic subsidence and erosion caused those sectors to deepen below sea level. These reconstructions have significant implications for the understanding of the evolution of the AIS and for its ice flow. It is, indeed, much easier to grow an ice sheet on a terrestrial surface than on a submarine bed. On such a restored and emergent topography, simulated Antarctic glaciations at the EOT produce a total ice volume greater than today and similar to that of the Last Glacial Maximum (LGM, ~21 ka) (Ladant et al., 2014; Wilson et al., 2013), even though atmospheric CO_2 levels were much higher than today, ranging from around 780–560 ppm (Fig. 12.1A). At the EOT, the ice sheet did not expand across the continental shelves, because the ocean temperatures were too warm (e.g. Bjill et al., 2018; DeConto et al., 2007). In fact, geological evidence from the Antarctic Peninsula documents a faunal turnover from species adapted to temperate waters (+5°C) to species adapted to cold waters through the EOT (Buono et al., 2019; Kriwet et al., 2016) (see from Galeotti et al., 2021, this volume).

Continental shelf evolution was critical for advances of the AIS across the marine realm after the EOT (e.g. Paxman et al., 2020). The evolution of the shallow continental shelves around Antarctica was connected to the evolution of the topography in the continent's interior. Various reconstructions (Brancolini et al., 1995a, 1995b; Cooper et al., 1991; Eittreim et al., 1995; Hochmuth et al., 2020; Huang and Jokat, 2016; Paxman et al., 2019) suggest that in most of the sectors, the continental shelf edge was located further south than today (Fig. 12.2) and then prograded seaward over time (e.g. Cooper et al., 1991; De Santis et al., 1999; Huang et al., 2014). The stratigraphic records combined with existing Antarctic deep drilling sites suggest that the majority of the continental margin expansion occurred prior to the Pliocene (De Santis et al., 1995, 1999, Hochmuth and Gohl, 2019; McKay et al., 2021), although in some sectors (e.g. Amundsen Sea, Gohl et al., 2013; Prydz Bay, O'Brien et al., 2007; Wilkes Land, Escutia et al., 2011), progradation of the margin was still important throughout the Pliocene (see from McKay et al., 2021, this volume). The Middle to Late Miocene was a period of transition during which the Antarctic ice sheet margin advanced into a cooling ocean, grounding on the continental shelf (Levy et al., 2021, this volume). This is also when prominent marine-based sectors of the AIS developed (e.g., Uenzelmann-Neben, 2019), especially in West Antarctica (Bart, 2003). Numerical ice sheet simulations using new Antarctic Mid-Miocene palaeogeographies, showed that during this period, the AIS became increasingly sensitive to oceanic conditions (Colleoni et al., 2018b), resulting in large glacial-interglacial changes in ice volume (Gasson et al., 2016).

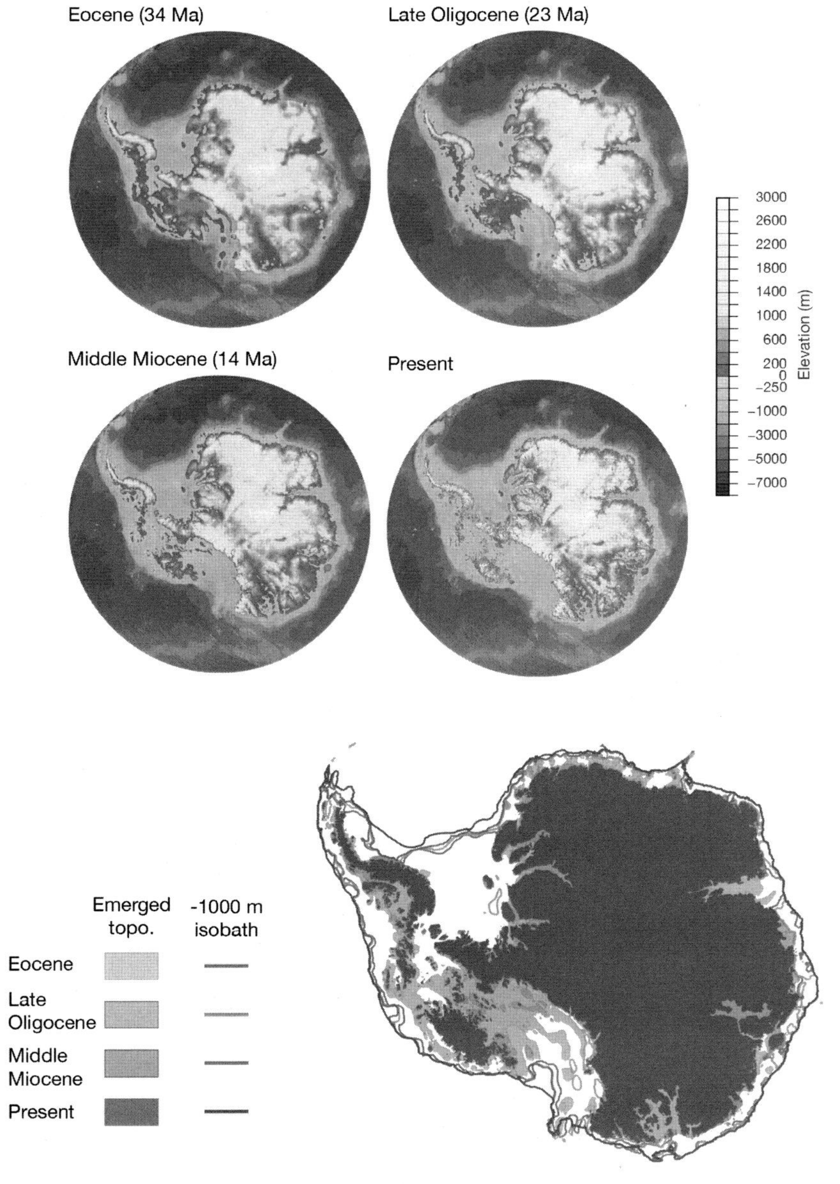

FIGURE 12.2 Top: Pan-Antarctic isostatically-relaxed palaeogeographic reconstructions from Paxman et al. (2019) for the Eocene (34 Ma), the Late Oligocene (23 Ma), the Middle Miocene (14 Ma) and BEDMAP2 for modern pan-Antarctic geography (Fretwell et al., 2013). Bottom: superimposed Eocene (brown), Late Oligocene (orange), Middle Miocene (pink) and modern (blue) emerged topography. Isobath at − 1000 m for each time slices is also indicated.

From the Pliocene onward, the Antarctic continental margin evolved very little. Erosion of the continental interior appears to have been less influential on the ice sheet since the Pliocene than during the earlier Oligocene and Miocene (Naish et al., 2021, this volume), and the terrestrial ice sheet became more stable (Gulick et al., 2017; Kim et al., 2018; McKay et al., 2012a; Passchier et al., 2011). However, fluctuations of marine-based ice in deep subglacial basins still occurred, especially when atmospheric CO_2 was between 400 and 300 ppm, during the early and middle Pliocene between 5 and 3 Ma (Bertram et al., 2018; Cook et al., 2013, 2014; Hansen et al., 2015; Levy et al., 2021, this volume; Naish et al., 2009a; Patterson et al., 2014; Pollard and DeConto, 2009; Reinardy et al., 2015). The relative stability of the terrestrial AIS is further supported by a recent study of cosmogenic nuclide concentrations (e.g. in situ ^{10}Be) in a sediment core from the Ross Sea (ANDRILL Site AND-1B), which suggests minimal retreat of the EAIS onto land during the last 8 million years (Shakun et al., 2018).

Another important aspect of the continental margin evolution is that its orientation or slope gradually changed from seaward dipping until the Early Pliocene, to landward dipping as it is now (Cooper et al., 1991; De Santis et al., 1999) (Fig. 12.3A). This change was caused by the numerous ice sheet advances and retreats and associated erosion and deposition of sediments. In turn, changes in the bed morphology then fed back on the ice sheet dynamics. Thus, since the Pliocene, the bed of the AIS marine-based sectors has generally been characterised by retrograde slopes, which favoured the potential of Marine Ice Sheet Instability (MISI) (Colleoni et al., 2018b; Jamieson et al., 2012; McKay et al., 2016). Prior to the Late Miocene, climatic conditions were generally warm, the continental shelves were less expanded in most of the Antarctic sectors and did not present strong retrograde slope. Consequently, the ice sheet could retreat easily during warm climate episodes with strong surface and oceanic melt (e.g. Gasson et al 2016; Levy et al., 2016). At the end of the Miocene, the climate gradually cooled, which favoured terrestrial EAIS stability, but as the retrograde bed slope began to form instabilities and fast AIS grounding line retreat in the marine-based sectors during phases of prolonged or exceptional warmth became more important (Blackburn et al., 2020; Colleoni et al., 2018b; Cook et al., 2013; DeConto and Pollard, 2016; DeConto et al., 2012; Golledge et al., 2017a; Levy et al., 2019; Naish et al., 2009a; Pollard and DeConto, 2009; Pollard et al., 2015).

Fast retreat of the grounding line can, however, be slowed down or stopped by the occurrence of pinning points at the bed that provide a buttressing backstress that resists seaward ice flow (Mengel and Levermann, 2014). Pinning points or pinning areas can take different forms. Ice rises, for example, form when an ice shelf anchors on a pre-existing bathymetric high, stabilising the flow (Matsuoka et al., 2015) (Fig. 12.3B). They can be tectonic structures or volcanic islands. Today, several ice rises are visible from the

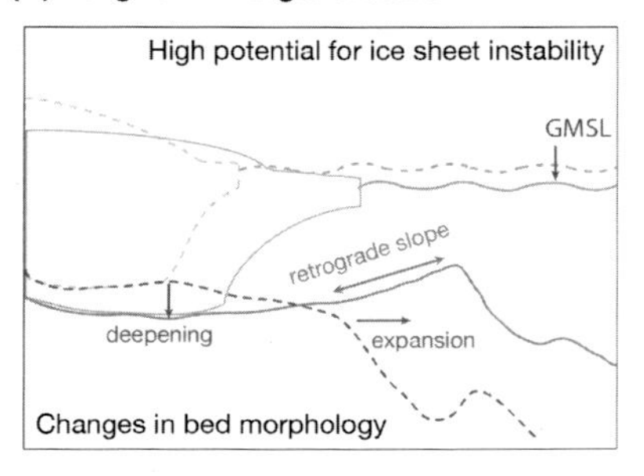

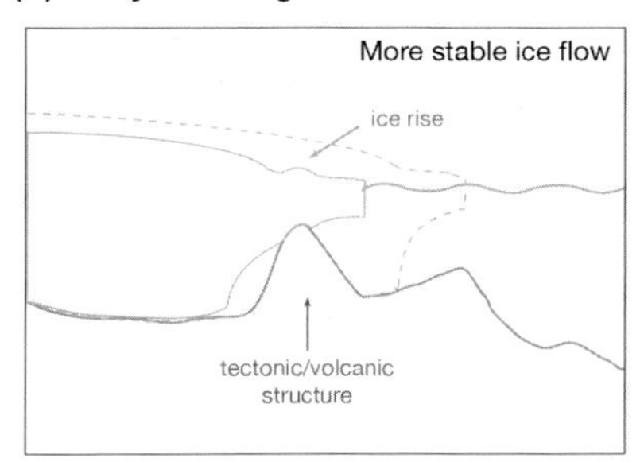

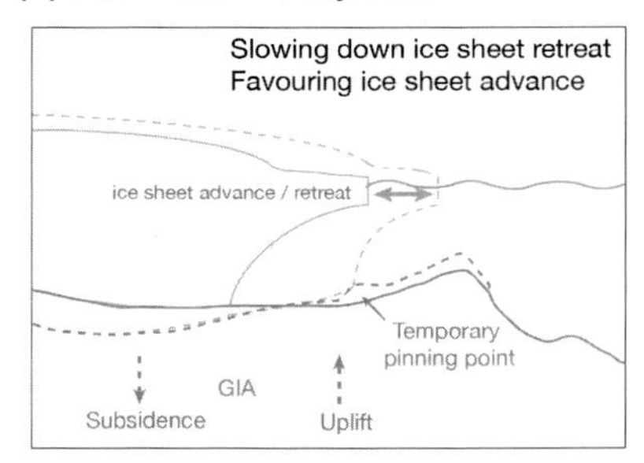

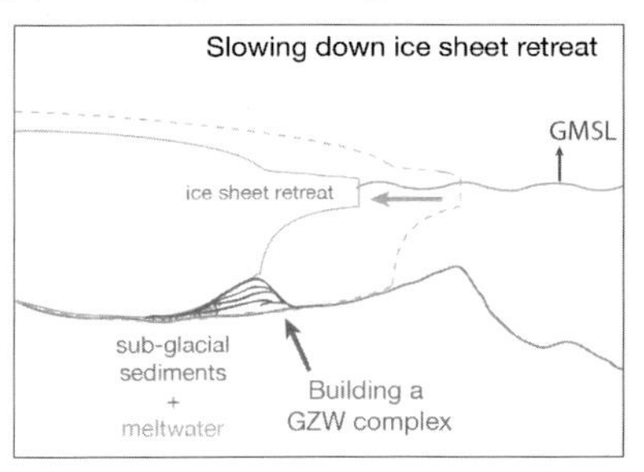

FIGURE 12.3 Schematics of the different ways by which an ice sheet can anchor on the bed. A: Long-term margin evolution; B: bathymetric highs; C: Glacio-isostatic adjustment; D: Grounding Zone wedges. *GMSL*, global mean sea level; *GZW*, grounding zone wedge; *GIA*, glacio-isostatic adjustment.

surface, for example, in the Ross Sea embayment (e.g. Roosevelt Island, Crary Ice Rise, Franklin Islands and Ross Island) and in the Weddell Sea embayment (e.g., Berkner Island). Halberstadt et al. (2016) and Simkins et al. (2018) suggest that the retreat of the grounding line during the last deglaciation was slower in the western Ross Sea than in the eastern Ross Sea, which is characterised by a smoother bed.

Pinning points can also form temporarily due to the uplift of the bed as a result of glacio-isostatic adjustment (GIA) during ice sheet retreat and unloading of the crust (e.g. Whitehouse, 2018; Fig. 12.3C). Numerical simulations show that ignoring GIA during ice sheet advances result in smaller, less extended ice sheets than would occur if GIA was accounted for, because GIA creates pinning opportunities (e.g., Colleoni et al., 2018b). Kingslake et al. (2018) showed that during the early Holocene, the West Antarctic Ice Sheet (WAIS) in both the Weddell Sea and the Ross Sea may have temporarily retreated beyond its present-day grounding line position. It subsequently re-advanced potentially due to uplift of the bed due to GIA (Bradley et al., 2015) and the occurrence of relief (e.g. Bungenstock Ice Rise, Weddell Sea) on which the ice shelf could pin. Similarly, high-resolution bathymetry acquired from a ridge under the Pine Island Glacier Ice Shelf has revealed geomorphological features that may be consistent with a retreat of Pine Island Glacier inland from its present position earlier during the Holocene and a subsequent re-advance to its early 20th Century position (Graham et al., 2013). These pinning points can be subsequently eroded or can simply 'resorb' after glacio-isostatic adjustment of the bed.

Finally, the ice sheet can build its own pinning points by accumulating sediments in grounding zone wedges (GZW) during deglaciations, which slows its retreat (e.g. Alley et al., 2007; Horgan et al., 2013; Fig. 12.3D). An example of this effect was outlined in Bart et al. (2017, 2018), who analysed a complex of GZWs that formed during the last deglaciation in the Whales Deep basin (Eastern Ross Sea). Proxy analyses and dating of sediment cores revealed that the first four GZWs were built during the first ~5000 years of a gradual 75-km southward ice sheet retreat, between ~17 and ~12.3 ka. They were characterised by low sedimentation rates (i.e., the GZWs are very thin) and sediment compositions indicate that the grounding line was pinned on those GZWs, whilst an extensive ice shelf formed during the retreat. The last three GZWs accumulated, with a clear aggradation sequence, in about 800 years between 12.3 and 11.5 ka, implying that the grounding line was not retreating during this brief interval. These GZWs were characterised by very high sedimentation rates and were thus relatively thick compared to the older ones. Sediment compositions seaward of those three GZWs indicate that the ice shelf broke up at the very beginning of this time interval and never reformed, and that the grounding line remained pinned successively on top of those last three GZWs. After building the uppermost GZW, the grounding line stepped back by about 100 km within a few decades, resulting in a very brief, massive ice discharge of about 0.1 mm sea level equivalent (SLE, Bart and Tulaczyk, 2020). Similar mechanisms and sequences have been inferred from the analysis of Pine Island Bay continental shelf multibeam and marine seismic data from the Amundsen Sea Embayment shelf (Jakobsson et al., 2011, 2012; Klages et al., 2015; Uenzelmann-Neben et al., 2007). Data show that the ice stream retreat was paused due to the building of GZWs during the last deglaciation. GZWs can also build as a consequence of ice sheet retreat in a narrow trough in which lateral edges serve as a pinning zone that slows down the retreat. Livingstone et al. (2013) and Jamieson et al. (2012, 2014) mapped a series of GZWs in a paleo-ice stream trough in Marguerite Bay (Antarctic Peninsula). Numerical simulations have shown that ice stream retreat rates slowed as the grounding line passed the laterally narrow parts of the trough. If a constant sedimentation rate was assumed, GZWs could form, thus further slowing the retreat.

The presence or absence of pinning points influences ice sheet dynamics. The estimated rates of sea level change during past periods may have been affected by potential pauses or changes in the rate of AIS (or other ice sheets) advances or retreats, due to ice-bed interactions. These variations may become highly relevant, especially for the interpretation of sea level reconstructions since the LGM, or simulated ice volume changes at the sub-millennial scale (Bart et al., 2018, Kingslake et al., 2018; Klages et al., 2017). However, the erosion of potential paleo-pinning points, during the numerous phases of expansion of the AIS, makes it difficult to know the role of pinning points on past ice sheet variability at sub-millennial time scales in reconstructions older than the LGM. Therefore, pinning points are generally not resolved in paleo-ice sheet simulations, where bed topography cannot be reconstructed with sufficient accuracy or resolution (<5 km). Moreover, shallow ice approximation ice sheet models frequently used for paleo-ice sheet simulations need a smoothing of the bed morphology in order to enable numerical convergence in areas where the morphology is too steep for the applicability of the hydrostatic approximation. For example, the Parallel Ice Sheet Model (PISM, Bueler and Brown, 2009) proposes different levels of smoothing of the bed, and this model is frequently used within the PISM paleo-community (e.g. Albrecht et al., 2020, Golledge et al., 2012). Full-Stokes ice sheet models, in which no hydrostatic approximation is applied, do not require bed smoothing since the physics accounts for both horizontal and vertical shear of the ice flow. However, full-Stokes models are too computationally demanding and are still not usable for most paleoclimate applications (Colleoni et al., 2018a and references therein). Given that most paleogeographic reconstructions (e.g. Hochmuth et al., 2020; Paxman et al., 2019) are very coarse in spatial resolution and highly uncertain in terms of detail, bed smoothing in ice sheet simulations resulting in the loss of local pinning points is generally of lesser importance than the biased controls on ice sheet dynamics induced by uncertain bed morphologies (e.g. Gasson et al., 2015).

12.3 Reconstructions of Southern Ocean sea and air surface temperature gradients

Equator-to-pole surface temperature gradients influence Earth's latitudinal heat distribution. Reconstructions of meridional temperature gradients since the Late Cretaceous clearly show a gradual steepening during the transition from greenhouse to icehouse conditions as the polar regions cooled and ice sheets developed (e.g. Zhang et al., 2019). Reconstructions also show the emergence of oceanic fronts in the sub-tropics and high latitudes, especially in the Northern Hemisphere (e.g. Zhang et al., 2019). The development of the Southern Ocean frontal system is of importance for reconstructing past AISs. The Antarctic Polar Front (APF) is a region marked by elevated current speeds and strong horizontal gradients in seawater density, temperature and salinity. It is currently located at approximately 50°S in the Atlantic and Indian sectors, and around 60°S in the Pacific sector. During warm periods, proxies imply a substantial southward shift of the APF associated with a significant reduction in sea ice extent (e.g. Bjill et al., 2018; Carter et al., 2021, this volume; Chadwick et al., 2020; Evangelinos et al., 2020; Salabarnada et al., 2018; Sangiorgi et al., 2018; Taylor-Silva and Riesselman, 2018). Conversely, during cold periods, the APF shifted northward accompanied by a large expansion of the sea ice cover (Gersonde et al., 2005; Kemp et al., 2010; McKay et al., 2012a).

Fewer sediment cores have been recovered in the Southern Ocean high latitudes than in the Northern Hemisphere high latitudes. South of 50°S, sea surface temperature (SST) proxy reconstructions are rare (Fig. 12.4). The ACE programme and more recently the PAIS programme increased the number of SST and Sea Water Temperature (SWT) measurements at 0–200 m depths from the Southern Ocean's Antarctic margin. SST and SWT records are now available for the MCO from the continental shelf site ANDRILL AND-2A (Western Ross Sea, Levy et al., 2016) and from the continental rise at Integrated Ocean Drilling Program (IODP) Site U1356 (Adélie Land margin, Hartman et al., 2018; Sangiorgi et al., 2018). Mid- to late-Pliocene SST records are available from ANDRILL-1B sediment core (Ross Sea shelf, McKay et al., 2012a) and from other cores on the continental rise and abyssal plains from the Indian sectors (see Dowsett et al., 2013 for an SST compilation). MIS 31 SST records are available from continental rise sites such as Ocean Drilling Program (ODP) Site 1101 (Antarctic Peninsula, Beltran et al., 2020) and IODP Site U1361 (Adélie Land margin, Beltran et al., 2020) and from site ODP Site 1094, (south of APF, South Atlantic sector, Beltran et al., 2020). For MIS 11 and MIS 5e, no Antarctic continental margins SST records are available so far. However, recent International Ocean Discovery Program (IODP) Expedition 374 to the Ross Sea (McKay et al., 2019), IODP Expedition 379 to the Amundsen Sea (Gohl et al., 2019) and expedition INS2017_V01 on the Sabrina Coast (Armand et al., 2018; O'Brien et al., 2020) have recovered highly expanded sedimentary sections from the continental rise. This promises upcoming high-resolution SST records for Pleistocene interglacials, filling a gap where ice proximal SST information have been missing.

A comparison between MCO and mPWP global meridional proxy-based SST gradients highlights the difference between the two periods in the Southern Hemisphere and how much the global climate state has evolved between 17 and 3 Ma (Fig. 12.4A). During the MCO, the air surface temperature gradient strengthened between 30°S to 40°S, as in the Northern Hemisphere, suggesting that the sub-tropical marine frontal system was well developed. The meridional SST and SWT gradients were much weaker than today and a summer warming of 16°C to 22°C (± 5°C) compared to today was observed in geochemical proxies at around 60–65°S on the East Antarctic continental rise (Hartman et al., 2018; Sangiorgi et al., 2018) and a warming of about 2°C to 12°C (± 5°C) compared to today was recorded in

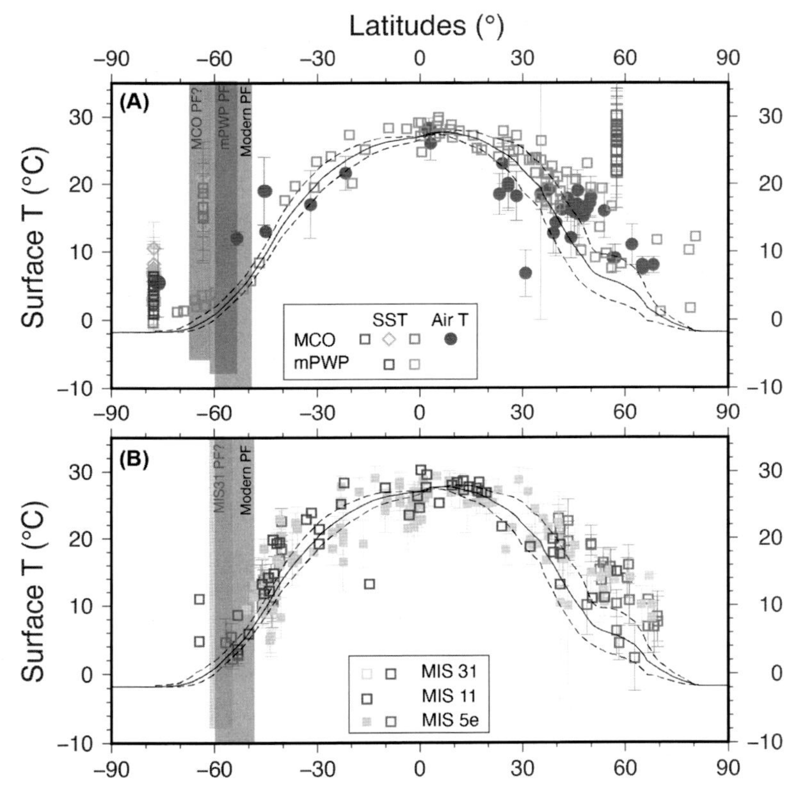

FIGURE 12.4 Proxy compilation of reconstructed meridional sea surface temperatures (SST) for the MCO and the mPWP (A), MIS 31, MIS 11 and MIS 5 (B). Note that vertical colour bars correspond to our interpretation of approximated location of the zonally-averaged circum-Antarctic Polar Front. Due to the paucity of global data for the MCO we include Mean Air Temperature (MAT, brown solid circles) from the Goldner et al. (2014) compilation. SST records for the MCO are from the Super et al. (2020) BAYSPAR calibration of TEX_{86} for North Atlantic ODP Site 982 (brown open squares), from Levy et al. (2016) TEX_{86} Ross Sea SWT 0–200 m depth and Adélie Land margin (Sangiorgi et al., 2018; orange open diamonds) and from Hartman et al. (2018) for TEX_{86} BAYSPAR calibrated Adélie Land margin (red open squares). MAT and SST proxies for the mPWP are from the Dowsett et al. (2012) global compilation (light blue open squares) and the McKay et al. (2012b) TEX_{86} Ross Sea SST record (dark blue open squares). SST proxies for MIS 31 are from the Justino et al. (2017) for global compilation (light green solid squares) and Beltran et al. (2020) for the Antarctic Peninsula, Weddell Sea and Adélie Land margin SST records (dark green open squares). SST proxies for MIS 11 are from Justino et al. (2017). SST proxies for MIS 5 are from Capron et al. (2014) (solid pink squares) and Hoffman et al. (2017) (open purple squares), both at 125 ka. Black continuous (mean annual) and dashed lines (boreal and austral summers) correspond to pre-industrial HadISST reconstruction from Rayner et al. (2003). Note that many of the SST proxies plotted here from the various compilations tend to be more representative of boreal or austral summer conditions rather than of mean annual conditions.

the Ross Sea (Levy et al., 2016, Sangiorgi et al., 2018) indicating a total absence, or rare occurrences, of sea ice during this period. The mPWP was characterised by a meridional SST gradient weaker than today (e.g. Brierley et al., 2009; Haywood et al., 2013) and with a significant Arctic amplification, while the warming anomaly was more subdued in the high southern latitudes (Fig. 12.4A). South of 55°S, East Antarctic continental rise summer SSTs were warmer than today by about 4°C to 6°C on the East Antarctic continental rise and by up to 7°C on the Ross Sea continental shelf (McKay et al., 2012a), indicative of a highly reduced, if not absent, summer sea ice cover in some sectors.

During both the MCO and the mPWP, the APF was probably more contracted towards high latitudes in all sectors around Antarctica (Sangiorgi et al., 2018; Taylor-Silva and Riesselman, 2018). The mPWP presents a meridional SST gradient steeper than during the MCO and it is possible that the APF might not have reached latitudes as poleward as during the MCO. The contrast between the mPWP and MCO meridional SST gradient south of 40°S suggests that the gradient probably steepened during the Late Miocene (Herbert et al., 2016). However, in some sectors of the Antarctic margin, there is strong evidence for warmer conditions during the Early Pliocene compared to the mPWP, i.e. sea-ice reduction and warming west of the Antarctic Peninsula (e.g. Escutia et al., 2009; Hillenbrand and Ehrmann, 2005) and in Prydz Bay (Whitehead et al., 2005) (see also Chapter 9 by Levy et al., 2021, this volume). This was accompanied by a poleward shift of the APF (Bart and Iwai, 2012; Escutia et al., 2009; Whitehead and Bohaty, 2003).

During the Pleistocene, MIS 31, MIS 11 and MIS 5e meridional SST gradients highlight the strong impact of precessional astronomical forcing, with SSTs warmer than today especially in the Northern Hemisphere mid-to-high latitudes (Fig. 12.4B). All three interglacials present SST gradients quite similar to today in the equatorial to sub-tropical latitudinal bands. The discrepancies between them emerge for latitudes poleward of 50°S. For MIS 31, alkenone and long chain diol analysis on sediment cores reveal an SST warming of 4°C to 12°C on the Adélie Land margins and the Antarctic Peninsula (Beltran et al., 2020). Proxies suggest reduced or even absent winter and summer sea ice in the Ross Sea and offshore Adélie Land margin; information is missing for other Antarctic sectors (see references for Fig. 12.6D). The abrupt appearance of foraminiferal oozes and bioclastic limestone in the Ross Sea and coccolith-bearing sediments in Prydz Bay during MIS 31 (Bohaty et al., 1998; Scherer et al., 2003; Villa et al., 2008, 2012) indicates a significant southward migration of the APF. In terms of the SST gradient, MIS 31 shows more similarities with the mPWP in the Southern Hemisphere than with the more recent Late Pleistocene interglacials.

No SST or SWT proxies south of 60°S are available for MIS 11, and thus it is difficult to assess the magnitude of a potential warming closer to the Antarctic margins. Interpretation of diatom and geochemical changes could help to estimate SST or SWT from the Wilkes Subglacial Basin and the Ross Sea margins (see Wilson et al., 2021 for references). North of 40°S, tropical to subtropical MIS 11 SSTs were warmer than modern by about 1°C–4°C, and between 50°S and 60°S only a 1°C–2°C warming above modern is recorded (Kunz-Pirrung et al., 2002). Apart from just a single exception, no SST reconstructions are available for MIS 5e south of 60°S (Capron et al., 2014; Chadwick et al., 2020; Hoffman et al., 2017) and the reconstructed SSTs north of 60°S are similar to those of MIS 11 (Fig. 12.4B). It is hence difficult to assess the magnitude of surface and ice proximal sub-surface ocean warming during MIS 5e. Capron et al. (2014) report warmer than present-day conditions that occurred for a longer time interval in southern high latitudes than in northern high latitudes. They also report an earlier MIS 5e warming in the Southern Ocean starting from 130 ka compared with the Northern high latitudes and synchronous with Antarctic ice core records. Moreover, Chadwick et al. (2020) showed that the sea-ice minima and SST maxima were reached at slightly different times in three Southern Ocean sectors. During both MIS 11 and MIS 5e, the few existing records indicate a seasonally sea ice covered ocean (Chadwick et al., 2020; Escutia et al., 2011; Kunz-Pirrung et al., 2002; Wilson et al., 2018; Wolff et al., 2006). Ice core analyses by Wolff et al. (2006) suggested, on the basis of sea-ice proxies in the EPICA Dome C ice core, that winter sea ice was largely reduced during MIS 5e and MIS 11 in the Indian sector of Antarctica, and that summer sea ice was likely absent. In addition, similar proxy analyses on the EPICA DML ice core from Dronning Maud Land indicate a sea ice reduction in the Atlantic sector during MIS 5e (Schüpbach et al., 2013).

12.4 Extent of major Antarctic glaciations

This section focuses on glaciations that occurred during the Eocene-Oligocene Transition (EOT, ~34 Ma) (Galeotti et al., 2021, this volume; Galeotti et al., 2016), during the Mid-Miocene Climatic Transition (MMCT, 14.5–13.5 Ma) (Levy et al., 2021, this volume), the M2 glaciation (3.312–3.264 Ma) preceding the mPWP, and the Last Glacial Maximum (LGM, ~21 ka) (see Levy et al., 2021; Siegert et al., 2021, this volume).

The EOT was characterised by the development of a continental ice sheet on Antarctica (Barrett, 1989; Hambrey et al., 1991; Wise et al., 1992; Zachos et al., 1992) as atmospheric CO_2 concentrations level fell (e.g. DeConto and Pollard, 2003) and the Southern Ocean cooled (e.g., Bijl et al., 2013) as a result of the opening of ocean gateways (e.g. Kennett, 1977) (Fig. 12.1). Across the EOT, deep-see temperatures cooled by 3°C to 5°C (e.g. Liu et al., 2018) as a consequence of decreasing CO_2 levels (Pagani et al., 2005). Sedimentary cycles from a drill core in the western Ross Sea provided the first direct evidence of orbitally controlled glacial cycles between 34 million and 31 million years ago (Galeotti et al., 2016). Initially, under atmospheric CO_2 levels of ≥ 600 ppm, a smaller AIS, restricted to the terrestrial continent, was highly responsive to local insolation forcing. The establishment of the Antarctic Ice Sheet (AIS) was associated with an approximately +1.5 per mil increase in deep-water marine oxygen isotope values ($\delta^{18}O$) beginning at ~34 million years ago (Ma) and peaking at ~33.6 Ma (Bohaty et al., 2012; Coxall et al., 2005), with two positive $\delta^{18}O$ steps separated by ~200,000 years (Fig. 12.1B). The first positive step in the isotope data primarily reflects a temperature decrease (Lear et al., 2008) (EOT-1, ~34.46–33.9 Ma); the second one has been interpreted as the onset of a prolonged interval of maximum ice extent at ~33.6–33.7 Ma (EOT-2) (Liu et al., 2009). Stratigraphic unconformities identified from the continental margins of Antarctica (Fig. 12.1B, e.g. Cooper et al., 2004; De Santis et al., 1995; Eittreim et al., 1995; Escutia et al., 2005; Gohl et al., 2013; Gulick et al., 2017; Uenzelmann-Neben and Gohl, 2014; Whitehead et al., 2006) presumably correspond to one of these $\delta^{18}O$ excursions.

Galeotti et al. (2016) suggest that a continental-scale AIS with frequent calving at the coastline did not form until ~32.8 million years ago, coincident with the earliest time when atmospheric CO_2 levels fell below ~600 ppm. The atmospheric CO_2 threshold for the onset of large-scale Antarctic glaciations remains, however, uncertain and varies between 900 and 560 ppm in numerical climate and ice sheet simulations (DeConto and Pollard, 2003; Gasson et al., 2014; Ladant et al., 2014, Liakka et al., 2014). Liakka et al. (2014) showed that when accounting for vegetation-albedo feedbacks, large-scale Antarctic glaciations occurred when atmospheric CO_2 dropped between 1120 and 560 ppm. Ladant et al. (2014) simulated a first Antarctic expansion at EOT-1 associated with a first sea level drop of about 10 m (atmospheric CO_2 set to 900 ppm) and a second one coinciding with early Oligocene glaciation Oi-1 (33.4–33.0 Ma, Miller et al.,1991; Zachos and Kump, 2005) of about 63 m (atmospheric CO_2 set to 700 ppm). Sequence boundary and ice volume proxies suggest that the extent of the AIS gradually increased across the EOT and expanded to either near-modern dimensions (Miller et al., 2008; 2020a) or as much as 25% larger than at present day (Katz et al., 2008; Wilson et al., 2013). Numerical ice sheet modelling studies show a large range of ice volumes across the EOT. Simulated glaciations lead to sea level fall clustered around 10 m SLE and 25 m SLE relative to present AIS volume (DeConto and Pollard, 2003; Gasson et al., 2014; Ladant et al., 2014; Liakka et al., 2014; Pollard and DeConto, 2005; Wilson et al., 2013) (Fig. 12.1D), which corresponds to a total simulated AIS volume up to 83 m SLE. This is broadly in agreement with sea level falls up to 70 m estimated from low-latitude shallow-marine sequences (e.g. Cramer et al., 2011) (Fig. 12.1C). Both DeConto and Pollard (2003) and Ladant et al. (2014) simulated large isolated ice caps over East

Antarctic highlands presumably during EOT-1, that ultimately coalesced during Oi-1 (Fig. 12.5). This is supported by the geological evidence of glacimarine deposits in the Wilkes Land continental shelf and rise since 33.6 Ma (Escutia et al., 2005, 2011) and by glacial sediment transport to the continental slope of the Prydz Bay margin since ~35 Ma (O'Brien et al., 2004). No ice grounded on the continental shelf at this time (Fig. 12.5, e.g. Barrett, 1989, 2007), and the continental shelf edge was located further South than present for most of the Antarctic margin (Fig. 12.2). See chapter 7 by Galeotti et al. (2021) for more details about the EOT.

The Mid-Miocene Climatic Transition (MMCT, ~14.8–13.5 Ma) is a period of global cooling following the extreme warmth of the Mid-Miocene Climatic Optimum (MCO). The onset of global climatic cooling at ~14.8 Ma marks the start of the MMCT (Böhme, 2003; Flower and Kennett, 1993; Holbourn et al., 2014; Shevenell et al., 2008). Disconformities in the ANDRILL AND-2A record (Levy et al., 2016) and across the Ross Sea (De Santis et al., 1999), pulsed deposition of ice-rafted debris offshore Prydz Bay and the Adélie Land margin (Pierce et al., 2017) together with an increase in sea ice indicators in

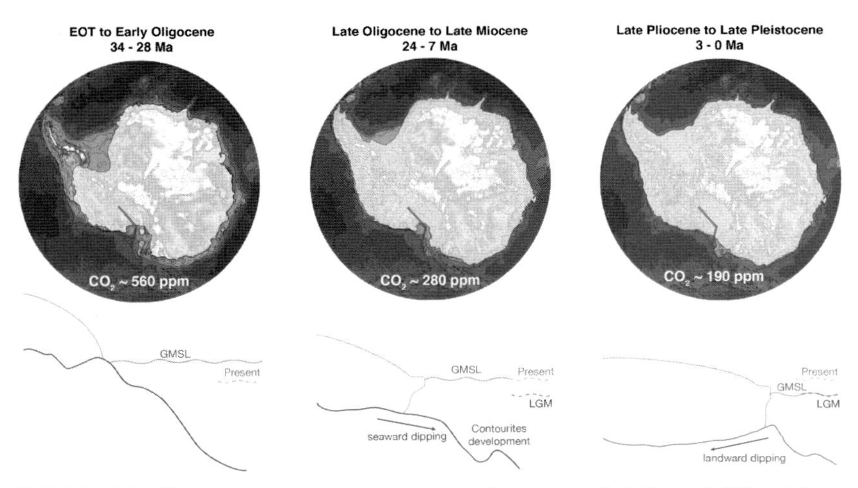

FIGURE 12.5 Simulated Antarctic ice sheet extent during past glaciations of different intervals. EOT to Early Oligocene (34–28 Ma): extent adapted from simulations by Ladant et al. (2014) with prescribed atmospheric CO_2 concentrations of 700–560 ppm. Late Oligocene to Late Miocene (24–7 Ma): extent adapted from simulations by Gasson et al. (2016) and Colleoni et al. (2018b) prescribing with an atmospheric CO_2 level of 280 ppm. Late Pliocene to Late Pleistocene (3 Ma to 0): extent adapted from Colleoni et al. (2018b) with prescribed atmospheric CO_2 values of 190 ppm. Paleotopographies and bathymetries are from Paxman et al. (2019). Note that ice shelves are not represented on the panels. Schematics below each circum-Antarctic view correspond to an idealised transect along the red lines indicated on the Antarctic maps above. Those schematics illustrate the evolution of the continental margin through time, with corresponding global mean sea level variations (GMSL, see Fig. 12.1C) referred to present sea level (dashed orange line) and Last Glacial Maximum sea level (LGM, 21 ka, dashed blue line).

the Ross Sea and off East Antarctica (Levy et al., 2016; Sangiorgi et al., 2018) and major turnover in Southern Ocean diatom species (Crampton et al., 2016) suggest marine ice sheet advances across the Ross Sea during glacial intervals for the first time since the onset of the MCO (for more detail see Levy et al., 2021, this volume) (Fig. 12.5). Ice sheet advance in the Wilkes Land margin is recorded by erosion of older sediments from the shelf (Escutia et al., 2011) and an increase in dinocyst assemblages from the seasonal sea ice zone south of the APF (Sangiorgi et al., 2018). Additionally, less well-dated erosional unconformities in the Weddell Sea (e.g. Huang et al., 2014), Amundsen Sea (e.g. Lindeque et al., 2016; Uenzelmann-Neben and Gohl, 2012), Bellingshausen Sea-Antarctic Peninsula (e.g. Rebesco et al., 2006; Uenzelmann-Neben, 2006), Sabrina Coast (Gulick et al., 2017) and Prydz Bay (e.g. Whitehead et al., 2006), are attributed to MMCT marine-based ice expansion, and together imply that both the EAIS and the WAIS expanded onto the continental shelf at this time (for more detail see Levy et al., 2021). An increase in glacial-interglacial amplitude in the far-field $\delta^{18}O$ data suggests that the AIS expanded further during successively, gradually colder glacial phases. This interval of increased glacial expansion culminated in a major step in the $\delta^{18}O$ record at 13.9 Ma (Fig. 12.1B). During the MMCT, Southern Ocean SSTs cooled by about 6°C (Holbourn et al., 2007; Sangiorgi et al., 2018). Bottom water temperatures generally cooled by 2°C to 3°C (Cramer et al., 2011; Lear et al., 2015; Shevenell et al., 2008) and global sea level may have dropped by as much as 50 m (Miller et al., 2020a), hinting at the possibility of some ice expansion in the Northern Hemisphere (DeConto et al., 2008). Summer temperatures in the Trans-Antarctic Mountains declined by >8°C (Denton and Sugden, 2005) and this cooling has been linked with a shift from temperate climate and wet-based glaciation with a dynamic ice sheet to cold climate and predominantly dry glaciation style with a more stable terrestrial ice sheet under modern-like Antarctic polar climatic conditions (Lewis and Ashworth, 2016; Lewis et al., 2008; Sugden and Denton, 2004).

The largest benthic foraminifera $\delta^{18}O$ shift during this period is of about 1.3‰ (e.g. Holbourn et al., 2013), partly corresponding to an estimated sea level drop of 35–40 m from interglacial to glacial. Interestingly, numerical ice sheet models can only simulate such a large interglacial-to-glacial amplitude in ice volume (Colleoni et al., 2018a; Gasson et al., 2016) when using a reconstructed Mid-Miocene paleogeography (e.g. Paxman et al., 2019). Backstripped sea level data (Kominz et al., 2008; Miller et al., 2005) and calibrated benthic $\delta^{18}O$ sea level changes (Miller et al., 2020a) revealed potential sea level falls up to 10–20 m below the present-day mean sea level during the MMCT (Fig. 12.1C), implying a greatly expanded AIS, perhaps up to 30% larger than today. The compilation of simulated Antarctic ice volume contributions to global mean sea level for this cold period ranges between +10 and −20 m SLE relative to present (Colleoni et al., 2018a; Gasson et al., 2016) (Fig. 12.1D, cyan squares). See Chapter 9 by Levy et al. (2021), this volume for more details about the MMCT.

The Mid-Pliocene M2 glaciation (**~3.312–3.264 Ma**) corresponds to a large transient increase in the deep-sea benthic $\delta^{18}O$ records (Fig. 12.1B) with a cooling of at least 3.5°C preceding the peak of M2 (Karas et al., 2020) and an atmospheric CO_2 level reduction of about 320–343 ppm (de La Vega et al., 2020). A compilation of climate proxy data suggests that during this glaciation, Greenland mountain glaciers expanded (Jansen et al., 2000; Thiede et al., 2011) and that other ice caps also grew in the Northern Hemisphere (De Schepper et al., 2014; Tan et al., 2017). In fact, this glaciation marks the end of global warmth of the early Pliocene (5.5–3.3 Ma) and the beginning of a step-wise transition towards bipolar cooling that culminated in continental-scale Northern Hemisphere glaciations ~2.7 Ma. Numerous sedimentary hiatuses, including the M2 glaciation, are observed in the AND-1B sediment record (western Ross Sea, Naish et al., 2009a) during the Mid to Late Pliocene. Continental margin morphology appears to have allowed the AIS to advance to the shelf edge during the M2 glaciation in the Ross Sea (e.g. Kim et al, 2018; McKay et al., 2012a, 2019), Prydz Bay (O'Brien et al., 2007), Sabrina Coast (Gulick et al., 2017), Wilkes Land (De Santis et al., 2003; Eittreim et al., 1995; Escutia et al., 1997) and the Antarctic Peninsula and Amundsen Sea (Gohl et al., 2013; Rebesco et al., 2006). The glacio-eustatic sea level drop of this period is represented by a major erosional sequence boundary on the New Jersey shelf (Miller et al., 2005) and global sea level fall of about 30 m (Grant et al., 2019; Miller et al., 2012, 2020a; Naish and Wilson, 2009b) (Fig. 12.1C). Based on the review of circum-Antarctic evidence of grounding events and sequence stratigraphy, Bart (2001) suggested an Antarctic ice volume larger than today and almost as large as during the LGM, implying the existence of relatively small Northern Hemisphere ice sheets. In contrast, the compilation of simulated Antarctic ice volume, likely underestimates the ice sheet expansion during this glaciation (Fig. 12.1D) and, instead, yields a global mean sea level rise up to 5 m above present-day mean sea level (De Boer et al., 2017a; Pollard and DeConto, 2009; Tan et al., 2017).

The Last Glacial Maximum (LGM, ~19–23 ka) is the most recent glaciation, and as such, is the best documented. Ice core records have shown that the atmospheric CO_2 levels dropped to about 185 ppm (Lüthi et al., 2008). Global climatic reconstructions revealed that the global mean temperature dropped by about 4°C to 7°C (e.g. Schneider von Deimling et al., 2006; Tierney et al., 2020) compared to present. Marine benthic and planktic foraminifera recorded a clear $\delta^{18}O$ increase (Imbrie et al., 1984; Lisiecki and Raymo, 2005), observable concomitantly to all past cold glacial periods (Imbrie et al., 1984, SPECMAP stack; Lisiecki and Raymo, 2005, LR04 stack; Westerhold et al., 2020, CENOGRID; Zachos et al., 2008, Fig. 12.1B). Calibrated conversions of the $\delta^{18}O$ record or dated paleo coral reefs suggest a global mean sea level drop ranging between 80 and 130 m relative to present (Fig. 12.1C) (e.g. Bard et al., 1990; Shackleton, 2000;

Waelbroeck et al., 2002), with a cluster between 110 and 130 m below present-day mean sea level. Diatoms and radiolarians show that the sea ice cover was larger than today. Winter sea ice cover shifted northward by about 5–10° (e.g. Benz et al., 2016; Gersonde et al., 2005) and the summer sea ice edge, although more uncertain, might have been located around 60.5°S (Green et al., 2020 and references therein). Compilation of climate proxies suggests a potential strengthening and equatorward shift of the Southern Hemisphere westerlies (e.g. Konrad et al., 2014; Lamy et al., 2014; Struve et al., 2020), which remains debated (e.g. Kim et al., 2018; Lamy et al., 2019; Sime et al., 2016).

There is persuasive evidence from the geological record to indicate that the AIS was larger than present around the time of the global sea level low-stand at ~20 ka, although the extent of this expansion is well constrained at only a few sites around the continental margin (Whitehouse, 2018). Both marine and terrestrial geological data indicate that at the LGM, the AIS almost extended to the continental-shelf break in most sectors (Anderson et al., 2002, 2014; Arndt et al., 2017; Bart et al., 2018; Eittreim et al., 1995; Hillenbrand et al., 2012, 2014; Mackintosh et al., 2014; The RAISED Consortium et al., 2014) (Fig. 12.5), as during many previous Pleistocene glaciations (e.g. Escutia et al., 2003). However, the AIS did not advance up to the continental shelf edge in Prydz Bay (Mackintosh et al., 2014; O'Brien et al., 2007; Wu et al., 2021), in the Western Ross Sea (Halberstadt et al., 2016; Prothro et al., 2018) and in parts of the Amundsen Sea (e.g. Klages et al., 2017; Larter et al., 2014). Furthermore, the scenario for ice advance in the Weddell Sea embayment remains uncertain (Nichols et al., 2019; The RAISED Consortium et al., 2014; Whitehouse et al., 2017). Ice sheet expansion during the LGM led to a thickening of the AIS of several hundreds of metres in the margins of almost all sectors, especially around West Antarctica as supported by exposure data (see Siegert et al., 2021, this volume). On the Antarctic plateau, ice core $\delta^{18}O$ isotope records suggest that the elevation increased 270–660 m between the LGM and present-day (Werner et al., 2018). Over West Antarctica, the increase in elevation during the LGM is up to 850–1800 m (Werner et al., 2018).

The relatively small number of proximal geological records on AIS extent and thickness during the LGM prevents an accurate constraint on LGM ice volume. Distal, deep ocean benthic foraminifera $\delta^{18}O$ records may provide overall ice volume estimates but do not allow disentangling contributions from individual ice sheets at the LGM (Simms et al., 2019, and references therein; Clark and Tarasov, 2014). Ice sheet modelling is one of the possible approaches to simulate the volume of the AIS at the LGM. Such modelling suggested an increase in ice volume of 5.9–19.2 m of sea level equivalent (SLE) (Bentley, 1999; Huybrechts, 2002) in the late 1990 and early 2000s. With the improvement of ice sheet models and climate forcing, the range of AIS contributions to sea level change at LGM has narrowed to about −5 to −12 m SLE (e.g. Briggs et al., 2014; Golledge et al., 2012; Gomez et al., 2013; Huybrechts, 2002; Maris et al., 2014; Quiquet et al., 2018, Sutter et al., 2019), with a cluster around −7 to −8 m SLE (Fig. 12.1D). Another approach is to infer AIS thickness by means of glacial-hydro isostatic adjustment (GIA) models, which describe the viscous response of the solid Earth to past changes in surface loading by ice and water (Whitehouse, 2018). This approach has also been used in combination with ice sheet modelling (e.g. Whitehouse et al., 2012b) and/or by making use of constraints on ice thickness from reconstructions based on exposure age dating, as well as satellite observations of current uplift (Argus et al., 2014; Ivins et al., 2013; Whitehouse et al., 2012b). Estimates from GIA modelling for the AIS contribution to global mean sea level amount to −5 to −30 m SLE with most of the contributions smaller than −13 m SLE (Fig. 12.1D). Older studies had estimated large sea level contributions generally above 15 m (e.g. Bassett et al., 2007; Huybrechts, 2002; Nakada et al., 2000; Peltier and Fairbanks, 2006; Philippon et al., 2006), but more recent modelling studies and reconstructions have refined these estimates to below 13.5 m (Argus et al., 2014; Briggs et al., 2014; Gomez et al., 2013; Mackintosh et al., 2011; Whitehouse et al., 2012a) with an average contribution of about −10 m SLE (e.g. Simms et al., 2019 and references therein).

Despite these improvements, AIS contributions to sea level changes at the LGM remain poorly constrained (Simms et al., 2019, and references therein; Clark and Tarasov, 2014) and this has global consequences on the assessment of past, present and future sea level changes. Land ice retreat in both hemispheres during the last deglaciation has produced a residual GIA signal that still affects present-day sea level changes measurements (Martín-Español et al., 2016). This residual signal is estimated from modelled reconstructions of global land ice thickness changes, and spatio-temporal deglaciation history (e.g. ICE-5G to ICE 7G, GLAC-1, ANU; Lambeck and Chappell, 2001; Lambeck et al., 2002; Peltier, 2004; Peltier et al., 2015; Roy and Peltier, 2018; Tarasov and Peltier, 2002, 2003). To date, there is no consensus on AIS volumes at the LGM and through the last deglaciation. A compilation of cosmogenic exposure ages from low-elevation sites shows that the AIS substantially thinned throughout the Holocene, but mainly after the Melt Water Pulse 1A (MWP-1A) (Small et al., 2019). The RAISED Consortium et al. (2014) provided partial pan-Antarctic grounding line positions at ~15, 10 and 5 ka. In the deglaciation scenarios ICE-6G and ICE-7G (Peltier et al., 2015; Roy and Peltier, 2018), the AIS extent at the LGM has been set up to its present-day extent, but with grounded ice filling the embayments currently occupied by the Ross Ice Shelf, by the Ronne-Filchner Ice Shelf and by the Amery Ice Shelf. This surely affects the calculation of GIA and its residual signal and, as such, the assessment of post-glacial and present-day land ice contribution to on-going sea level changes (Martín-Español et al., 2016).

At regional scale, the inclusion of realistic, spatially variable relative sea-level forcing through coupled simulations of 3D ice-sheet and GIA-modulated sea-level change results in a stabilising effect on marine-grounded ice sheet dynamics (Gomez et al., 2010). The grounding line, in fact, advances or retreats in response to the regional ice fluctuation. The latter triggers viscoelastic solid Earth rebound as well as a change of the local geoid height in response to the variation of the gravitational pull (e.g. Stocchi et al., 2013). In particular, the predicted increase in the volume of the WAIS during the last glacial cycle is smaller in the coupled simulations due to negative feedbacks associated with an increase in near-field water depth. The latter stems from the combination of ice-driven solid Earth subsidence and counterintuitive local sea-level rise caused by the gravitational attraction of the growing ice sheet's mass (De Boer et al., 2014b, 2017b; Gomez et al., 2013; Konrad et al., 2014). At global scale, Gomez et al. (2020) showed that the retreat of Northern Hemisphere ice sheets during the last deglaciation and associated sea level rise directly impacts the dynamical behaviour of the AIS and conditions its own retreat. Modelled AIS sensitivity on different paleobathymetries since the Mid-Miocene shows that the position of the AIS advance on the continental shelf depends on glacio-isostatic adjustment (generating pinning points, see Section 12.2) and on the timing and magnitude of global mean sea level changes (Colleoni et al., 2018b; Gomez et al., 2020; Paxman et al., 2020). A similar relationship between the AIS stability and global mean sea level changes has recently been inferred from a North Atlantic deep-ocean benthic $\delta^{18}O$ record of the Plio-Pleistocene Transition and the early Pleistocene (Jakob et al., 2020). Based on a range of Mg/Ca paleothermometer calibrations, the sea level record suggests that the gradual expansion of the Northern Hemisphere ice sheets, and the consequent substantial lowering of global mean sea level, led to an increasing stability of the marine-based sectors of the EAIS (Jakob et al., 2020). Other studies also highlight the sensitivity of the marine-based sectors of the AIS to rapid sea level rise at millennial to sub-millennial time-scales, such as the impact of rapid sea level rise during the various meltwater pulses episodes of past deglaciations (e.g. Golledge et al., 2014; Petrini et al., 2018; Turney et al., 2020; and Siegert et al., 2021, this volume).

12.5 Antarctic ice sheet response to past climate warmings

Assessing the AIS behaviour during past periods warmer than today can inform on the magnitude and timing of past and future sea level changes, as well as on various mechanisms triggering ice sheet retreats that can vary through time (i.e. atmospheric and/or oceanic warming). At millennial to sub-millennial scales, the crossing of tipping points caused by Earth's climate system feedbacks can cause rapid ice sheet retreats. One example is ocean warming triggering MISI. Because the global climatic state has been constantly evolving, the conditions necessary to cross these tipping points have also evolved. To highlight this aspect, several policy-relevant warm periods in the geologic past have been analysed, based on a few climatic and glaciological indicators synthesised in Fig. 12.6. Note that except for the sea ice (Fig. 12.6D), all the other variables are expressed as anomaly relative to their present-day value (20th Century for MAT and SST).

The compilation of global mean sea level changes and simulated Antarctic contributions shown in Fig. 12.1 and synthesised in Fig. 12.6E is exhaustive and illustrates a key research focus of the paleo polar community over recent decades. We do not discard computed estimates of AIS contribution to global mean sea level change that could appear out of the range of data. Instead, we consider such values as part of the uncertainties associated with uncertain physics and boundary conditions. References for all compiled data and simulated ice volumes are provided in the caption of Fig. 12.1.

The Mid-Miocene Climatic Optimum (MCO, 17–14.8 Ma). The MCO presents an interesting analogue for assessment of climate projected for the next decades to centuries (Steinthorsdottir et al., 2020). At that time, Antarctica hosted the only existing continental-size ice sheet. Geological proxy data indicate atmospheric CO_2 concentrations generally varied between 300 and 600 ppm on glacial-interglacial (orbital) time scales during much of the MCO (Foster et al., 2012; Greenop et al., 2014), but it may have reached values as high as 840 ppm (Retallack, 2009) (Fig. 12.1A). The limited existing geological proxies of terrestrial Antarctic temperature, from the Ross Sea (Warny et al., 2009) and off Adélie Land (Sangiorgi et al., 2018), indicate a surface air temperature warming of approximately 14°C–25°C relative to today. Comparison with the global mean annual temperature (MAT) clearly emphasises strong polar amplification occurred during the MCO (Goldner et al., 2014). SWT and SST reconstructions from the Ross Sea (ANDRILL-2A, Levy et al., 2016) and from the Adélie Land margins (Hartman et al., 2018, Sangiorgi et al., 2018) also support this polar amplification (Fig. 12.6C). On the continental rise (paleo latitude 53°S, Sangiorgi et al., 2018), a 5°C–10°C SWT warming (likely summer) was recorded, but on the continental shelf (~77°S) this estimated warming was even larger, reaching 10°C–20°C through the MCO.

Together, far-field data and modelling experiments suggest a highly dynamic ice sheet during the MCO. The AIS was mostly responsive to eccentricity-modulated precession affecting local insolation and leading to widespread inland retreat of the land-terminating ice sheet on glacial-interglacial timescales (e.g. Holbourn et al., 2013). Levy et al. (2019) suggested glacial to interglacial ice volume fluctuations were of about 30–46 m

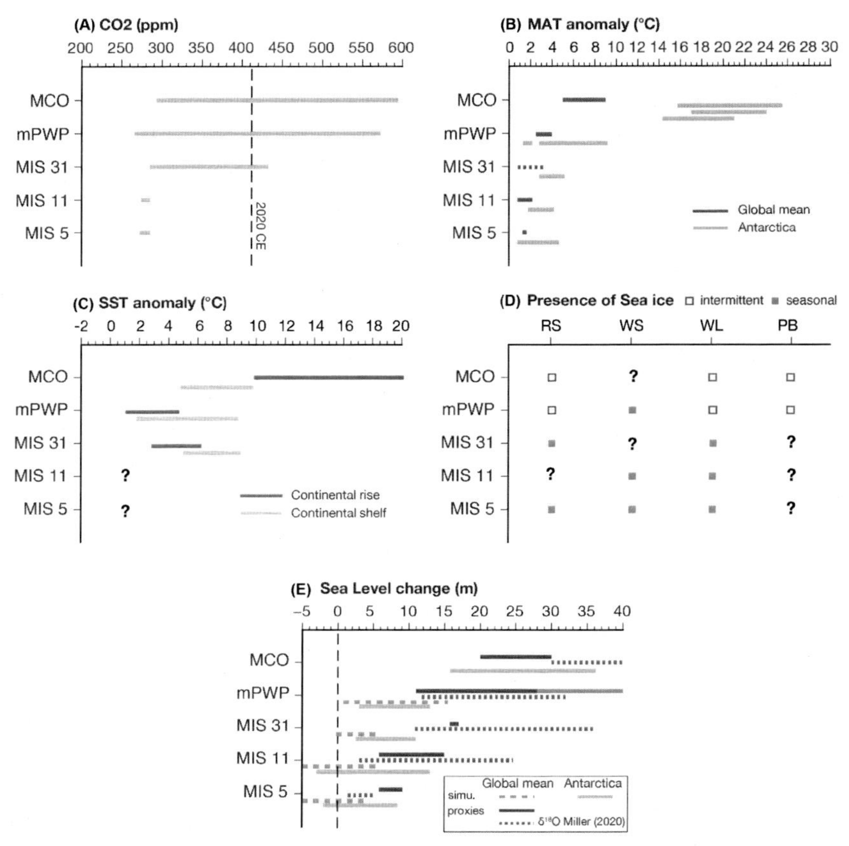

FIGURE 12.6 Main climatic indicators of each warm period considered in the chapter (A–D) and associated simulated range of Antarctic contributions to global mean sea level changes (E). Each panel shows ranges of climatic proxies or simulated quantities at the global scale and for Antarctica relative to their present-day value. Note that for each range (global or Antarctic), minimum and maximum uncertainties of the represented proxies are given. (A) Atmospheric CO_2 levels, see Fig. 12.1A for references. (B) Global mean annual temperature (MAT, °C) anomaly relative to 20th Century average: MCO – Goldner et al. (2014); mPWP – Salzmann et al. (2013) terrestrial proxies (see their Table S3b) and Dowsett et al. (2012) for SST compilation; MIS 31 – Justino et al (2019) averaged SST compilation also accounting for Beltran et al. (2020) Antarctic margin proxy-based SST. Note there are no northern high-latitude MAT or SST reconstructions. Proxies in the northern high latitudes suggest sea ice free conditions (Detlef et al., 2018) as during the mPWP. Given the high similarities with the mPWP SST gradient (Fig. 12.4), MIS 31 global MAT is tentatively extended to mean mPWP MAT (dashed line); MIS 11 – Lang and Wolff (2011) and MIS 5 – Turney and Jones (2010); MAT (°C) anomalies for the Antarctic region (including ice core records) are relative to 1990 at the closest weather station to the sediment cores location: MCO – Warny et al. (2009), mPWP – Haywood et al. (2020), Passchier et al. (2011), MIS 31 – Scherer et al. (2008), MIS 11 – Jouzel et al. (2007), MIS 5 – Jouzel et al. (2007), Lang and Wolff (2011). (C) Sea surface temperature (SST, °C) anomaly relative to present value at each core location: continental shelf values (pink) are from

(Continued)

SLE for a $\delta^{18}O$ shift of about 0.88%, which was successfully simulated by Gasson et al. (2016) and Colleoni et al. (2018a) using idealised (but representative) mid-Miocene boundary conditions. Geological records recovered adjacent to the EAIS suggest it advanced and retreated many times through the TAM during the MCO (Hauptvogel and Passchier, 2012) but did not advance far beyond the coastline during glacial intervals (Levy et al., 2016). The cored interval spanning ~17−15 Ma at Integrated Ocean Discovery Program (IODP) Site U1521 (McKay et al., 2019) consists of diatom-rich mudstone and diatomite, which also indicates ice distal environments in the Ross Sea through the MCO. Sediments collected at IODP Site U1356, off the coast of the Adélie Land margin (East Antarctica), suggest open-water conditions at the site throughout the MCO (Sangiorgi et al., 2018). Modelling studies suggest the Wilkes Subglacial Basin remained free of grounded ice during warm interglacial episodes through the early to mid-Miocene (Colleoni et al., 2018a; Gasson et al., 2016; Paxman et al., 2020) but it is unclear whether grounded ice advanced across the region during glacial intervals prior to the MMCT (Pierce et al., 2017). Mg/Ca calibrated sea level and sequence boundary estimates suggest global mean sea level (GMSL) rise ranging from +20 to +30 m above present (Kominz et al., 2008; Miller et al., 2005) (Fig. 12.6E). The compilation of simulated AIS contributions to GMSL vary between +15 and +35 m SLE (Colleoni et al., 2018b; Gasson et al., 2016; Langebroek et al., 2009; Stap et al., 2019)

Cape Roberts or ANDRILL sites: MCO – Sangiorgi et al. (2018) are sea water temperatures (0−200 m depth), mPWP – McKay et al. (2009), MIS 31 – Scherer et al. (2008). SST records from continental slope and rise (purple) are from: MCO – Sangiorgi et al. (2018) sea water temperature (0−200 m depth), Hartman et al. (2018), mPWP – Dowsett et al. (2012) compilation of proxies below 60°S, MIS 31 – Beltran et al. (2020). Note that many of the SST proxies could be representative of boreal or austral summer conditions rather than annual mean. (D) Proxies for presence/absence of sea ice are shown for different Antarctic sectors, RS (Ross Sea), WS (Weddell Sea), WL (Wilkes Land margin), PB (Prydz Bay). Open blue squares indicate episodic sea ice cover during the period, solid blue squares indicate seasonal sea ice (mostly no sea ice during austral summer) and question marks correspond to the absence of information for the sector. For MCO – Hannah (2006), Levy et al. (2016), Sangiorgi et al. (2018); for mPWP – Burckle et al. (1990), McKay et al. (2012a), Taylor-Silva and Riesselman (2018), Whitehead et al. (2005) for MIS 31 – Bohaty et al. (1998), Beltran et al. (2020), Scherer et al. (2008), Villa et al. (2008); for MIS 11 – Escutia et al. (2011), Kunz-Pirrung et al. (2002), Wilson et al. (2018), Wolff et al. (2006) for MIS 5 – Hartman et al. (2016), Konfirst et al. (2012), Kunz-Pirrung et al. (2002), Presti et al. (2011), Wilson et al. (2018), Wolff et al. (2006). (E) Ranges for global mean sea level changes (GMSL) relative to today from data (dark blue) and models (dotted grey). For mPWP, the proxies indicate a sea level rise up to 40 m above present, indicated by the transparent blue line. For MIS 31, GMSL data are uncorrected from GIA and dynamical topography. Simulated ranges of Antarctic Ice Sheet melting contributions are shown in light blue. A range from recent GMSL reconstructions based on benthic $\delta^{18}O$ records from Miller et al. (2020a) is shown with a dotted purple line. See Fig 12.1C and D and the main text for references.

(Fig. 12.6E). Such a large range mostly results from the use of different ice sheet models, different bed topographies and bathymetries, and different climate forcing in the mid-Miocene experiments. The range of potential Antarctic ice sheet GMSL contribution reduced to +16−17 m in Gasson et al. (2016), and Colleoni et al. (2018b) using an idealised Mid-Miocene paleotopography similar to that of Paxman et al. (2019), and a prescribed atmospheric CO_2 of 500 ppm. See Chapter 9 by Levy et al. (2021) for more details on this period.

The mid-Pliocene Warm Period (mPWP, 3.3−3 Ma) is considered as one of the most geologically accessible and relevant examples of climate change driven by atmospheric CO_2 levels equivalent to present-day (Haywood et al., 2016; Masson-Delmotte et al., 2013; Naish and Zwartz, 2012). Atmospheric CO_2 levels ranged between 300 and 450 ppm and global mean temperature was about 2°C−3°C warmer than present during the warmest interglacials (Masson-Delmotte et al., 2013) (Fig. 12.6A and B). One of the striking characteristics of the mPWP is that the SST proxy compilations reveal a meridional temperature gradient weaker than today (Fig. 12.4) characterised by expansion of tropical to sub-tropical bands, no boreal and reduced austral summer sea-ice and, thus, a strong northern and southern polar amplification (e.g. Haywood et al., 2020; Lunt et al., 2012). Modelling shows that such SST patterns reflected a weaker Hadley circulation than today (Brierley et al., 2009; Haywood et al., 2020). In the Southern high-latitudes, the coastal Antarctic region was up to 6°C warmer than today (e.g. McKay et al., 2012a; compilation in Dowsett et al., 2012) (Fig. 12.6B) mostly due to the fact that summer sea-ice, and the ice sheet, had retreated and the APF had contracted to more southern latitudes (Taylor-Silva and Riesselman, 2018). Evidence documents episodic sea ice in the Ross Sea, and offshore Adélie Land and in Prydz Bay (e.g. McKay et al., 2012a; compilation in Dowsett et al., 2012) (Fig. 12.6D). Seasonal sea ice was likely present in the Weddell Sea (Burckle et al., 1990). Reconstructed SSTs show a pan-Antarctic warming of up to 5°C on the continental slope and rise (Escutia et al., 2009; Whitehead and Bohaty, 2003). SSTs also show a warming up to 6°C in the Ross Sea (McKay et al., 2009) (Fig. 12.6C), which was likely caused by the sea-ice albedo feedback, and decreasing local albedo due to the retreat of coastal land ice.

GMSL reconstructions from (paleo-shorelines and sequence stratigraphy) indicate a sea level rise between about 15 to 28 m (Dumitru et al., 2019; Dwyer and Chandler, 2009; Kulpecz et al., 2009; Miller et al., 2012, 2020a; Naish and Wilson, 2009b; Sosdian and Rosenthal, 2009; Wardlaw and Quinn, 1991; Winnick and Caves, 2015), whereas Mg/Ca paleothermometry calibration of benthic $\delta^{18}O$ records suggest a GMSL up to 40 m above present (Fig. 12.6E). GMSL changes based solely on benthic $\delta^{18}O$ records, however, yield large uncertainties (±15 m) (e.g. Raymo et al., 2018). GMSL change amplitudes larger than 30 m above present can only be explained by melting the terrestrial sectors of the AIS, but retreat of the EAIS in the Ross Sea since 8 Ma appears unlikely because a recent study found extremely low concentrations of cosmogenic ^{10}Be and ^{26}Al isotopes in the ANDRILL AND-1B marine sediment core (Shakun et al., 2018). In addition, many of these peak GMSL estimates (e.g. Hearty et al., 2020; Miller et al., 2012) have not been corrected for regional deviations due to tectonics, glacio-isostatic adjustment and dynamic topography (Dumitru et al., 2019; Raymo et al., 2011; Rovere et al., 2015). A reassessment of Grant et al. (2019) based on far-field data implies GMSL during the warmest mid-Pliocene interglacial was no higher than 21 m (Grant and Naish, 2021, this volume). This new estimate is very close to the average of 20 m above present provided by sea level reconstructions based on sequence stratigraphy and paleo-shore lines.

The compilation of simulated Antarctic ice sheet contributions to GMSL ranges from ~+3 to +15 m SLE (Austermann et al., 2015; De Boer et al., 2015, 2017a, 2017b; DeConto and Pollard, 2016; Dolan et al., 2018; Gasson et al., 2015; Pollard and DeConto, 2009, 2012; Yan et al., 2016) (Fig. 12.6E). Although there is no observational evidence of a potential melting from the Greenland Ice Sheet so far, recent transient numerical simulations suggest that Greenland ice sheet loss could have contributed up to about 6 m SLE to GMSL rise (De Boer et al., 2017a, 2017b). Based on this estimate, the lower bound GMSL rise (+15 m) implies a contribution of the AIS no larger than 9 m SLE. Considering the upper bound of GMSL rise of about +20 to +28 m above present, the maximum contribution of the AIS thus ranges between +15 and +22 m SLE, implying melting of the WAIS and all marine-based sectors of the EAIS (e.g. DeConto and Pollard, 2016; Golledge et al., 2017a, 2017b). ANDRILL AND-1B in the Ross Sea recorded numerous occurrences of open-marine conditions suggesting frequent retreats of the Ross Ice Shelf during the mPWP (Naish et al., 2009a, 2009b). Provenance of fine-grained detritus offshore the Wilkes Subglacial Basin and ice-rafted debris offshore the Aurora Subglacial Basin and Prydz Bay was attributed to the retreat of marine-based sectors of the EAIS (Bertram et al., 2018; Cook et al., 2013, 2014; Whitehead et al., 2006). Similar circum-Antarctic retreat of the marine-based sectors was simulated for one of the Early Pliocene interglacials (Golledge et al., 2017a, 2017b), supported by sedimentological and geological evidence of circum-Antarctic warming events during that period (e.g. Escutia et al 2009; McKay et al., 2012a; Whitehead and Bohaty, 2003). See Chapter 9 by Levy et al. (2021), this volume for more details on this period.

Marine Isotope Stage 31 (MIS 31, 1.081–1.062 Ma) is a prominent mid-Pleistocene interglacial categorised as 'super interglacial' based on the expanded lacustrine sediment record from Lake El'gygytgyn in Siberia (Melles et al., 2012). It corresponded with exceptionally high eccentricity and obliquity values inducing particularly intense high-latitude summers. The level of atmospheric CO_2 is not well known for this interval but ranges from 300 to 420 ppm (Honisch et al., 2009) (Fig. 12.6A). A circum-Antarctic warming has been inferred from sediment core analysis in the Ross Sea (Naish et al., 2009a, 2009b; Scherer et al., 2008), in Prydz Bay (Villa et al., 2008), on the Adélie Land margin and in the Antarctic Peninsula (Beltran et al., 2020). In particular, the presence of diatoms in the Cape Roberts sediment record (Scherer et al., 2008) suggests a 3°C–5°C warming of upper ocean temperatures compared to today, with seasonally open-ocean conditions (no summer sea ice) (Fig. 12.6D). For the Adélie Land margin and the Antarctic Peninsula, Beltran et al. (2020) reconstructed summer SSTs that on average were 3°C to 6°C warmer than today (Fig. 12.6C). Similar warm conditions are recorded in the Ross Sea at ANDRILL AND-1B as indicated by seasonal open-ocean conditions (Fig. 12.6D), suggesting a retreat of the Ross Ice Shelf (McKay et al., 2012b; Naish et al., 2009a, 2009b). GMSL rise during MIS 31 is relatively poorly constrained from far-field records. Sea-level indicators preserved in coastal cliffs of the Northern Cape Province of South Africa and from Cape Range, Western Australia, suggest highstands not higher than 15–16.5 m above present mean sea level (Hearty et al., 2020; Sandstrom et al., 2021). Note that those estimates are not corrected for GIA, dynamic topography and local tectonic. Far-field evidences of four consecutive Middle Pleistocene Transition sea-level highstands between MIS 31 and MIS 35 were identified in a speleothem record from a western Sicily cave (Mediterranean Sea) (Stocchi et al., 2017). The peculiarity of this marine cave is that it was last flooded between MIS 35 and MIS 31 and has been tectonically uplifted to higher elevations afterward. Among several GIA-modulated relative sea level scenarios, only those accounting for a significant AIS retreat up to about 25 m SLE at MIS 31 and 35 are capable to flood the marine Sicilian cave. The compilation of simulated AIS melting contribution to GMSL ranges from 2 to 10 m SLE (Beltran et al., 2020; de Boer et al., 2013; DeConto et al., 2012) (Fig. 12.6E). The upper bound of this range is in agreement with far-field, though uncorrected, sea level changes and indicates a large WAIS retreat, with a modest contribution from East Antarctic marine-based sectors. In fact, mineralogical provenance from IBRD from ODP Site 1090 (South Atlantic) and ODP Site 1165 (Prydz Bay) revealed that the EAIS retreated significantly over MIS 31 and particularly in the Prydz Bay region. However, other sectors of the EAIS were still characterised by active marine margins (Teitler et al., 2015). Beltran et al. (2020) suggested that the AIS retreat was caused by a stronger advection of Circumpolar Deep Water (CDW) resulting from the changes of the westerlies (subpolar jet). Such a process was also inferred from changes in the geochemical composition of Holocene foraminifera shells from the Amundsen Sea and the aeolian dust from a

West Antarctic ice core record. (Hillenbrand et al., 2017). See also chapter 11 by Wilson et al. (2021) for more details on this period.

Marine Isotope Stage 11 (MIS 11, 425–375 ka) occurred close to the Mid-Brunhes Event (Fig. 12.1). It is the Late Pleistocene warm stage considered as one of the closest analogues to our future because astronomical forcing of a few time slices within MIS 11 are very similar to today (Loutre and Berger, 2003). MIS 11 is also the oldest middle Pleistocene interglacial categorised as a 'super interglacial' based on lacustrine sediment records from the Lake El'gygytgyn in Siberia (Melles et al., 2012). Global mean air temperature was 1.5°C–3°C higher compared to modern temperatures (Fig. 12.6B) although atmospheric CO_2 levels were only around 280 ppm (comparable to pre-industrial levels) (Fig. 12.6A). On the Antarctic plateau, the surface air temperature increased was 2°C–3°C greater than today (Jouzel et al., 2007; Uemura et al., 2018). A polar amplification occurred during that period but was reduced compared to MIS 31 or older warm periods. MIS 11 is not really an intense or brief interglacial such as MIS 5e (130–116 ka, see below); its major characteristic is its longer duration of ~50,000 years (Tzedakis et al., 2012), which may have been key to substantial ice sheet melting (Irvali et al., 2020). Reconstructed SSTs were not much warmer than modern temperatures (e.g. Becquey and Gersonde, 2002, 2003a, 2003b; Hodell et al., 2000; King and Howard, 2000). In fact, geological evidence supports the idea that a modest but sustained warming was at the origin of ice sheet retreat in the Wilkes Subglacial Basin during MIS 11 (Blackburn et al., 2020; Wilson et al., 2018). Recent modelling studies show that the WAIS and part of the EAIS retreat could occur with a limited warming of 0.4°C if applied for a duration of 4000 years (Mas e Braga et al., 2021). As with other past intervals, the absence of ice proximal oceanic temperature reconstructions is one of the critical gaps to constrain ice sheet simulations of this interval. Reconstructed GMSL suggest a rise of about 13 m above present sea level (Raymo and Mitrovica, 2012; Roberts et al., 2012) and up to 20 m during MIS 11 (Brigham-Grette, 1999; Hearty et al., 1999; Kindler and Hearty, 2000). Such a range implies the complete melting of both the Greenland Ice Sheet and the WAIS, which would account for about 12 m SLE, leaving about 8 m SLE from the EAIS (e.g. Lythe and Vaughan, 2001; Warrick et al., 1996). Mas e Braga et al. (2021) recently simulated a contribution from the WAIS around 4.3–4.5 m SLE and a contribution from the EAIS ranging from 2.3 to 3.7 m SLE. Sedimentological analyses from Erik Drift, Southeast Greenland reveal that most of South Greenland deglaciated during MIS 11 (Reyes et al., 2014). The compilation of simulated AIS contributions to GMSL ranges from ~−3 to +13 m SLE (Mas e Braga et al., 2021; Sutter et al., 2019; Tigchelaar et al., 2018) (Fig. 12.6E) and in absence of further geological constraints, it is difficult to refine this range. See Chapter 11 by Wilson et al. (2021) for more details on this period.

Marine Isotope Stage 5e or Last Interglacial (LIG, 130–116 ka) was the most recent interglacial with temperatures warmer than today. It has long been considered as an analogue for future climatic changes (Jansen et al., 2007). However, at the peak of the LIG, the astronomical forcing differed too much from the present-day to be a true analogue (Ganopolski and Robinson, 2011). Nevertheless, the LIG presents a very useful time period for understanding the Earth System response (e.g. internal feedbacks in the climate system) to the Paris Agreement temperature targets (e.g. IPCC 1.5°C Special Report). Atmospheric CO_2 concentration was low (Fig. 12.6A), and reconstructed global mean temperature is estimated to have been about 0.5°C–2°C higher than today (Hoffman et al., 2017; Masson-Delmotte et al., 2013) (Fig. 12.6B). The East Antarctic plateau recorded a warming up to 5.5°C at ~128.6 ka followed by a plateau around 2°C (Jouzel et al., 2007; Petit et al., 1999; Watanabe et al, 2003). A polar amplification thus occurred during this period (Capron et al., 2017) and was broadly of the same magnitude as during MIS 11. Antarctic continental margin sediment records imply seasonal sea ice in most of the sensitive marine-based sectors (e.g. Konfirst et al., 2012; Presti et al., 2011) (Fig. 12.6D). GMSL rise is estimated to about 5.9 to 9.3 m above present level from paleo-shorelines (Dutton et al., 2015 and ref. therein) and up to almost 20 m based on calibration of benthic and planktonic $\delta^{18}O$ records (Rohling et al., 2009; Waelbroeck et al., 2002) (Fig. 12.6D), also involving some Greenland Ice Sheet mass loss. However, ice core constraints and modelling studies suggest that the contribution from Greenland was likely only about 2–3 m (Dahl-Jensen et al., 2013), implying a significant meltwater contribution from Antarctica, although also a Greenland Ice Sheet contribution of up to 5.1 m SLE has been suggested (Yan et al., 2016). The compilation of simulated AIS melting contributions to GMSL ranges from about −2 to +8 m SLE (Fig. 12.6E) (e.g. Colleoni et al., 2018b; de Boer et al., 2015; DeConto and Pollard, 2016; Goelzer et al., 2016; Huybrechts, 2002; Pollard and DeConto, 2012; Quiquet et al., 2018; Sutter et al., 2016, 2019; Tigchelaar et al., 2018; compilation in De Boer et al., 2019). Antarctic ice core records of $\delta^{18}O$, considered as a proxy for local ice volume changes, have been analysed in an attempt to better constrain the individual contribution of Antarctica to GMSL. Based on these analyses, numerical climate and ice sheet simulations suggest that part of the $\delta^{18}O$ signal could be explained by sea ice reduction rather than ice sheet retreat (e.g. Holloway et al., 2016). In the absence of ice proximal ocean temperature reconstructions, as for other Late Pleistocene interglacials, it is very difficult to constrain the magnitude and timing of the AIS retreat during this interval. The magnitude of oceanic warming required to trigger a large retreat of the marine-based sectors at that time varies between models from +2°C to +3°C relative to pre-industrial temperature (e.g. DeConto and Pollard, 2016; Sutter et al., 2016; Turney et al., 2020). However, Turney et al. (2020) also showed that with a modest ocean warming of 0.4°C, the major ice shelves could disintegrate within 600 years. While continental margin sediments offshore from the Wilkes Subglacial Basin suggested a reduction of this marine-based sector of the EAIS during MIS 5e (Wilson et al., 2018), geological evidence from the WAIS is contradictory and suggests either that no major ice sheet retreat occurred (e.g. Clark et al., 2020; Hillenbrand et al., 2002, 2009; Spector et al., 2018) or that considerable retreat took place (e.g. Turney et al., 2020). See Chapter 11 by Wilson et al. (2021) for more details on this period.

12.6 Antarctica and global teleconnections: the bipolar seesaw

Inter-hemispheric heat transport, the so-called bipolar see-saw (Stocker and Johnsen, 2003), is another key process affecting AIS evolution. It regulates oceanic and atmospheric temperatures at sub-millennial to millennial time scales. The bipolar see-saw mechanism was hypothesised by Stocker (1998) on the basis of observed asynchronous changes in ice core records between Greenland and Antarctica for some of the Dansgaard/Oeschger events that occurred during the last glacial cycle (Blunier et al., 1998). Stocker and Johnsen (2003) hypothesised that iceberg melting and meltwater discharges close to the North Atlantic convection sites caused a substantial weakening of the Atlantic Meridional Oceanic Circulation (AMOC). Such a slowdown could have induced a gradual heat transfer to the South, with a lag of a few centuries to millenia, thus explaining the asynchronous temperature changes between Greenland and Antarctic ice core records during the last glacial period (EPICA Community Members, 2006, Pedro et al., 2018). Recent findings confirm that cooling of the Antarctic Cold Reversal is synchronous with the Bølling–Allerød warming in the Northern Hemisphere 14,600 years ago (Stenni et al., 2011). The Bølling–Allerød is coincident with the occurrence of meltwater pulse 1A (MWP-1A) that caused a rapid sea level rise of about 9–20 m (e.g. Deschamps et al., 2012; Lambeck et al.,2014; Liu et al., 2016; Peltier et al., 2015) at a rate of 4 m/100 yr (e.g. Carlson and Clark, 2012; Deschamps et al., 2012; Peltier and Fairbanks, 2006). Although some studies have considered the AIS as a potential contributor to MWP-1A (e.g. Clark et al., 1996; Bassett et al., 2007; Golledge et al., 2014; Weaver et al., 2003), most geological and glaciological studies argue against a large Antarctic contribution (e.g. Bentley et al., 2010; Licht, 2004; Spector et al., 2017; The RAISED Consortium et al., 2014). However, IBRD records from 'Iceberg Alley' in the Scotia Sea may record the occurrence of eight events between 20 and 9 ka, including the MWP-1A (e.g. Weber et al., 2014). Etourneau et al. (2019) showed that a 0.3°C to 1.5°C increase in subsurface ocean temperature (50–400 m) in the northeastern Antarctic Peninsula drove a major collapse and recession of the regional ice shelf during both the instrumental period and the last 9000 years. Modelling studies support the idea of responsive marine-based sectors of the AIS at millennial time scales, driven by oceanic melting rather than by atmospheric forcing triggering fast ice sheet instabilities (e.g. Blasco et al., 2019; Golledge et al., 2014; Lowry et al., 2019).

Meltwater sources of such millennial oscillations are poorly constrained, however (e.g. Clark et al., 2002; Liu et al., 2016; Peltier, 2005). This limits our understanding of the causes of the events, i.e. warm water advection to the grounding line (e.g. Golledge et al., 2014) bipolar seesaw caused by melting Northern Hemisphere ice sheets (e.g. Menviel et al., 2011), abrupt global mean sea level rise (e.g. Clark et al., 2002; Golledge et al., 2014; Gomez et al., 2020), atmospheric forcing (e.g. WAIS Divide Project Members, 2015) or feedbacks within the climate system. Past, present and future meltwater-climate feedbacks have been widely studied and there is an extensive literature on modelling the global climate response to freshwater discharge from ice sheets (e.g. Böning et al., 2016; Bronselaer et al., 2018; Golledge et al., 2019; Roche et al., 2014; Sadai et al., 2020; Stammer, 2008). Results from these studies highlight the role of altered inter-hemispheric heat transport on the global climate both in the past and in the future. Different mechanisms respond to the freshwater at different timescales but the overall feedback loop spans the millennial scale (Turney et al., 2020). The sequence of those feedback loops is illustrated in Fig. 12.7 and is based on two recent contributions, i.e., Turney et al. (2020) for the LIG (130–116 ka), and Golledge et al. (2019) for projected climate changes until CE 2100 following RCP 8.5 emission scenario.

Turney et al. (2020) reported evidence of substantial ice discharge across the Weddell Sea sector during the LIG based on a blue-ice core record. Substantiated with climate and ice sheet simulations, they suggest that the ice discharge (and subsequent multi-metre global mean sea level rise) was caused by a millennial-scale oceanic warming following freshwater discharge in the Northern high latitudes (Heinrich event 11 at ~135–130 ka) and a weakening of the AMOC (Böhm et al., 2015). This mechanism corresponds to the bipolar see-saw. Turney et al. (2020) identified a loop of positive ice-sheet-climate feedback that further amplified the warming close to the Antarctic margin. Grant et al. (2014) identified two main meltwater pulses, one at 139 ka pre-dating Heinrich event 11 (135 ± 1 and 130 ± 2 ka) and one occurring at about 133 ka during this Heinrich event. Marino et al. (2015) found that Heinrich event 11 coincided with a rapid sea-level rise mostly explained by Northern hemisphere ice sheet deglaciation. The occurrence of this meltwater pulse supports the positive feedback described in Turney et al. (2020) and potentially explains the delayed timing of the AIS contribution to the GMSL highstand at the LIG. A delay of a few thousand years is supported by the idealised modelling study by Blasco et al. (2019), suggesting

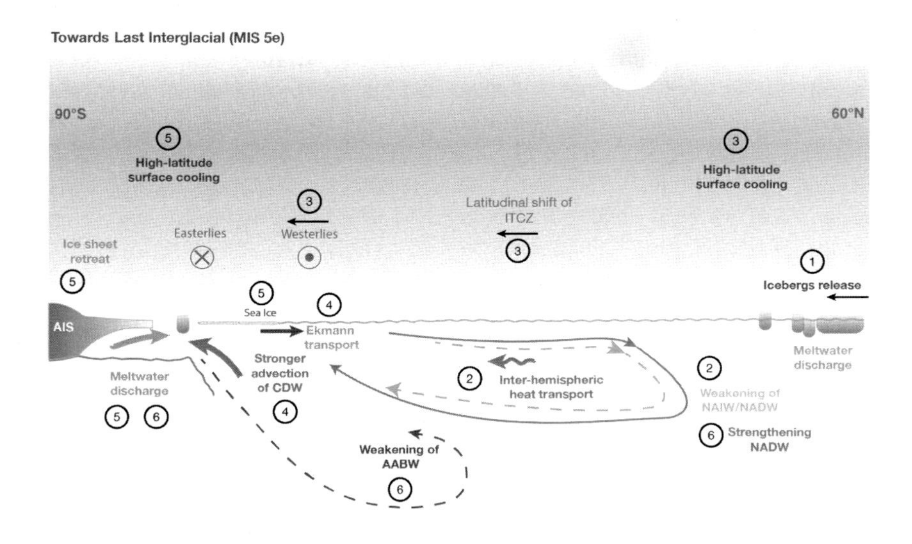

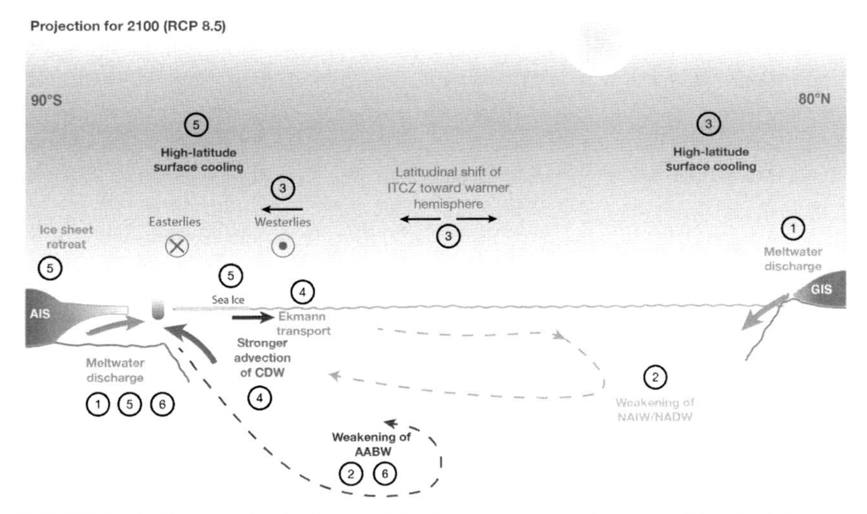

FIGURE 12.7 Cartoons for the impact of freshwater release and associated feedback loops at the global scale and on Antarctica. Top cartoon is based on Turney et al. (2020) and describes the evolution of those feedbacks as a consequence of Heinrich event 11 ($\sim$135 ka) and subsequent evolution until MIS 5e (time frame of few millennia). Numbers indicate the order of the sequence. Bottom cartoon shows similar feedback loops but as projected until 2100 for the RCP 8.5 high-emission scenario (time frame of few decades) based on Golledge et al. (2019). *AABW*, Antarctic Bottom Water; *CDW*, Circumpolar Deep Water; *ITCZ*, Inter-Tropical Convergence Zone; *NAIW*, North Atlantic Intermediate Water; *NADW*, North Atlantic Deep Water.

that the bipolar seesaw accumulated heat in the Southern Hemisphere, enhancing ocean warming on a millennial time scale during the last deglaciation. Similarly, Clark et al. (2020) suggested that the rate of global mean sea level changes during the LIG, as well as spatial sea level variations, could be explained by the responses of the Antarctic and Greenland ice sheets to Heinrich event 11 and associated climate feedbacks. The sequence of feedbacks in Turney et al. (2020) can be applied to other interglacials and is as follows (Fig. 12.7, top):

1. Northern high-latitude freshwater was released during the Heinrich 11 event (~ 135 and 130 ka);
2. Subsequently, a weakening of North Atlantic Deep Water (NADW) flow was observed, and heat was transferred gradually southward.
3. An increased meridional inter-hemispheric thermal gradient due to Northern high latitudes cooling induced a southward shift of the Inter-Tropical Convergence Zone (ITCZ) and of the Southern Hemisphere westerly winds (e.g. Shevenell et al., 2011).
4. The southward shift and strengthening of the westerlies (e.g. Dickens et al., 2019; Etourneau et al., 2019; Lamy et al., 2019; Hillenbrand et al., 2017) drove an enhanced northward Ekman transport and a stronger southward advection of CDW on the continental shelf (e.g. Hillenbrand et al., 2017; Minzoni et al., 2017).
5. Enhanced advection of CDW amplified the melting of the AIS and of the sea ice, triggering the AIS retreat. Northward transport of cool surface waters caused sea ice expansion and local atmospheric cooling.
6. Large freshwater discharge caused a reduction in Antarctic Bottom Water (AABW) formation and a subsequent increase in NADW formation. Increase in NADW formation led to heat transfer towards northern high latitudes, and thus a bipolar see-saw swing towards the north.

Golledge et al. (2019) simulated a similar ice-sheet-climate sequence of feedbacks by considering on-going and projected meltwater discharge from the Greenland and Antarctic ice sheets until CE 2100. Results show that a slow-down of the AMOC occurs in response to Greenland Ice Sheet melting, and that projected meltwater discharge from Antarctica can trap the heat of CDW at intermediate depths on the continental shelf (Bronselaer et al., 2018; Sadai et al., 2020; Silvano et al., 2019), establishing a positive feedback that further enhances AIS melting (Fig. 12.7, bottom), as follows:

1. Projected freshwater release at Northern and Southern high-latitudes;
2. A weakening of NADW formation is observed, and heat is transferred gradually southward. In the South, freshwater stratifies the continental shelf waters.
3. An increased meridional inter-hemispheric thermal gradient induces a weak southward shift of the ITCZ and of the Southern Hemisphere Westerly winds.

4. Southward shift of the westerly winds drives an enhanced northward Ekman transport compensated by a stronger southward advection of CDW on the continental shelves, which amplifies the melting of Antarctic ice shelves and sea ice.
5. Continental shelf water stratification fosters a northward Ekman transport of cool surface waters associated with sea ice expansion and local atmospheric cooling. This mechanism further amplifies the advection of CDW to the AIS grounding line and initiates its retreat.
6. Larger freshwater release further causes a reduction in AABW formation.

Compared to Turney et al. (2020), the sequence of processes and feedbacks in 2100 remains incomplete and stops before all the heat from the North is transferred to South. This suggests that additional decades to centuries are needed for the full effects of the bipolar see-saw on southern high latitudes to be felt.

12.7 The PAIS legacy: bridging the past and the future

12.7.1 The PAIS legacy

The PAIS legacy is clearly one of successfully addressing high-level scientific priorities. Beyond this, it is also the story of a long-lasting network of collaborations, among different nations and researchers, and striving to share scientific infrastructure and capabilities to investigate remote and challenging Antarctic regions. The multidisciplinary concept of the PAIS programme represented the key to its success. Eight years after the start of the programme, PAIS achievements are many, both in terms of field campaigns and in terms of scientific advances concerning Antarctic ice sheet dynamics. Several projects fostered by the PAIS programme, which contributed to major scientific advances in constraining AIS contribution to past sea level changes, are briefly summarised below. This list is far from being exhaustive and interested readers can refer to the other chapters of this book for more detailed descriptions of PAIS research outcomes and other time periods not discussed here.

12.7.1.1 Antarctic ice sheet sensitivity during past high-CO_2 worlds and its contribution to global sea-level change

Geological proxies from the Antarctic continental margin have improved reconstructions of ocean and land temperatures, sea-ice extent, ice sheet extent, subglacial hydrology, carbon cycle feedbacks and paleogeography for past warm climate states. This has provided improved boundary conditions for testing and developing ice sheet and climate models capabilities and performance, as well as evaluating model sensitivity. Significant outcomes include the following:

- Reconciling southern high-latitude meridional temperature gradients and polar amplification between model simulations and data during Greenhouse climates (e.g. DeConto et al., 2012; Pross et al., 2012) and new knowledge of Antarctic margins SSTs and SWTs during the MCO, the mPWP and MIS 31 (Beltran et al, 2020; Hartman et al., 2018; Levy et al., 2016; McKay et al., 2009; Sangiorgi et al., 2018). Polar amplification is much larger during the MCO, mPWP and MIS 31 than during the Pleistocene interglacials. Those findings allow an estimate of Earth's climate sensitivity to high atmospheric CO_2 concentrations.
- Constraining equilibrium and transient ice volumes (e.g. Clark et al., 1996; de Boer et al., 2015; DeConto and Pollard, 2016; DeConto et al., 2012; Dolan et al., 2018; Gasson et al., 2016; Goelzer et al., 2016; Golledge et al., 2017a, 2017b; Pollard et al., 2015; Stap et al., 2019), and the contribution to global sea-level under past 'warmer-than-present' climates (e.g. Dutton et al., 2015; Miller et al., 2012, 2020a, 2020b).
- Recognition of the importance of bed topography and paleobathymetry on past Antarctic ice volume reconstructions (e.g. Gasson et al., 2016, building on Hochmuth and Gohl, 2019; Hochmuth et al., 2020; Paxman et al., 2019, 2020; Wilson and Luyendyk, 2009) and sensitivity to ocean warming (e.g. Colleoni et al., 2018a; Paxman et al., 2020).
- Recognition of the sensitivity of marine-based sectors of the EAIS from models and data (e.g. Aitken et al., 2016; Bertram et al., 2018; Blackburn et al., 2020; Cook et al., 2013, 2014; Gasson et al., 2016; Golledge et al., 2017b; Gulick et al., 2017; Levy et al., 2016; Pierce et al., 2017; Reinardy et al., 2015; Scherer et al., 2016; Simkins et al., 2017; Wilson et al., 2018).
- Insights into the influence of mean climate state (CO_2) on the response of the AIS to orbital forcing (e.g. Dolan et al., 2011; Levy et al., 2019; Patterson et al., 2014; building on concepts in Naish et al., 2009a, 2009b; Stap et al., 2019, 2020; Sutter et al., 2019).

12.7.1.2 Geological evidence of ocean forcing and marine ice sheet instability

The potential for abrupt and non-linear 'runaway' retreat of the marine-based sectors of the AIS due to marine ice sheet instability (MISI) and potentially also marine ice cliff instability (MICI) up until recently had only been mathematically simulated in ice sheet models.

- Geological observations of the last deglaciation and recent observations coupled with models have now identified MISI during the Holocene after atmospheric forcing had weakened in the Ross Sea (e.g. Bart and Tulaczyk, 2020; Jones et al., 2015; McKay et al., 2016; Spector et al.,

2017), and potentially MICI in the Amundsen Sea sector (Wise et al., 2017) and Antarctic Peninsula (Rebesco et al., 2014).
- There are geological and modern oceanographic observations of oceanic warm waters reaching the grounding line of marine-based ice sheets (e.g. Joughin and Alley, 2011; Hansen and Passchier, 2017; Hillenbrand et al., 2017; Rintoul et al., 2016; Schmidtko et al., 2014; Smith et al., 2017).

12.7.1.3 Improved temporal and spatial patterns of AIS retreat and its contribution to global Melt-Water Pulse 1A

Improved geological and bathymetric constraints combined with ice sheet models have shown:

- An improved understanding of the extent and dynamics of the LGM ice sheet and deglaciation into the Holocene (e.g. Anderson et al., 2014; Golledge et al., 2013; Hillenbrand et al., 2014; Hodgson et al., 2014; Johnson et al., 2014; Larter et al., 2014; Lee et al 2017; Mackintosh et al., 2014; McKay et al., 2016; O'Cofaigh et al., 2014; The RAISED Consortium et al., 2014);
- The AIS contributed to melt-water pulse 1A (e.g. Golledge et al., 2014; Weber et al., 2014), though not from all sectors (e.g. Spector et al., 2017) and to other millennial scale fluctuations with estimated contributions to global mean sea level up to 6 m SLE (Blasco et al., 2019; Golledge et al., 2014), at a rate of about 1 m/century in the case of MWP-1A (e.g. Golledge et al., 2014).

12.7.1.4 A better understanding of ice-sheet-ocean interactions

- During the last deglaciation, proxy data and model simulations consistently find that ocean warming drove the ice sheet retreat in different sectors of Antarctica (e.g. Crosta et al., 2018; Hillenbrand et al., 2017; Wilson et al., 2018);
- Ocean warming is also thought to have accelerated the last deglaciation in the Ross Sea during MWP-1A (e.g. Golledge et al., 2014), although this finding is challenged by geological evidence from the Trans-Antarctic Mountains (e.g. Spector et al., 2017). Based on regional ice sheet simulations, atmospheric forcing can enhance, diminish or compensate for oceanic warming during the first half of the deglaciation, while during the second half, ocean warming clearly drove the end of the ice sheet retreat (Buizert et al., 2015; Blasco et al., 2019; Lowry et al., 2019);
- Strengthening of the subpolar jet during deglaciation (e.g. Lamy et al., 2019) enhanced advection of CDW towards the continental margins (Evangelinos et al., 2020; Hillenbrand et al., 2017; Minzoni et al., 2017; Salabarnada et al., 2018);
- Improved understanding of sedimentological facies indicative of sub-ice shelf environments opens up new opportunities for quantifying the

influence of the ocean on the AIS evolution through time (e.g. Smith et al., 2019; Yokoyama et al., 2016);
- Freshwater release from the Northern high latitudes can induce a bipolar seesaw, transferring heat to the Southern Hemisphere and fostering AIS retreat a few thousands of years later (Blasco et al., 2019; Buizert et al., 2015; Clark et al., 2020; Turney et al., 2020). Likewise, freshwater release from Antarctica can stratify the ocean, reduce vertical mixing and the release of heat and gas to the surface, and increase heat transport at the grounding lines of marine-based ice sheets (e.g. Golledge et al., 2019; Silvano et al., 2019).

12.7.1.5 Antarctic ice–Earth interactions and their influence on regional sea-level variability and Antarctic Ice Sheet dynamics

The importance of departures in regional sea-level changes from eustatic sea-level due to rotational, visco-elastic and gravitational changes as water mass is transferred between the ice sheets and the ocean has been identified in the far and near-fields of the AIS from paleo-reconstructions (e.g. Clark et al., 2002; Milne and Mitrovica, 2008; Raymo and Mitrovica, 2012; Raymo et al., 2011; Rovere et al., 2014; Stocchi et al., 2013). This has been established through 1D and 3D glacio-isostatic adjustment models that couple (runtime or asynchronously) ice sheets and solid Earth processes constrained by both near-field and far-field geological reconstructions of sea-level changes. Important outcomes include the following:

- Role of Earth deformation processes (GIA and dynamic topography) on near-field sea-level changes and ice sheet dynamics (e.g. Austermann et al., 2015; Gomez et al., 2015, 2018; Kingslake et al., 2018; Pollard et al., 2017; Stocchi et al., 2013; Whitehouse et al., 2017, 2019);
- Impact of global gravitationally consistent sea level changes induced by Northern Hemisphere ice sheet fluctuations on the retreat of the AIS (e.g. Gomez et al., 2020);
- Impact of long-term global mean sea level change on the stability of EAIS (e.g. Jakob et al., 2020; Shakun et al., 2018).

12.7.1.6 Improved interpretation of subglacial processes from mapping seabed

- Multibeam campaigns in different sectors of Antarctica have mapped the geomorphological footprints of paleo ice streams and their associated paleo-drainage networks (e.g. Kirkham et al., 2019; Larter et al., 2019; Nitsche et al., 2013; Simkins et al., 2017; The RAISED Consortium et al., 2014) as well as other subglacial features (Bart et al., 2018; Dowdeswell et al., 2020; Kuhn et al., 2017; Stokes, 2018).

- Analysis of the characteristics of those geomorphological features inform the long-term mean and potential maximum rates of grounding line retreat (e.g. 1–10 km/yr, Bart et al., 2018; Dowdeswell et al., 2020) but also show that meltwater can enhance ice flow and cause ice surges and meltwater outbursts (e.g. Kuhn et al., 2017; Simkins et al., 2017). These reconstructions provide constraints on the ice flow regime during both advances and retreats and on the mechanics and dynamics of ice stream (Stokes, 2018).

12.7.1.7 Paleo-data calibrated ice sheets models provide revised global sea-level predictions for IPCC scenarios

A new generation of continental scale ice sheet models that simulate MISI and in one case MICI have been developed and tested by reconstructing past AIS volume and extent constrained by paleoclimate and paleo-ice extent data. These models have been used to simulate future Antarctic meltwater contribution to GMSL changes based on the Representative Concentration Pathways (RCPS). Implications include the following:

- The Antarctic contribution to global sea-level rise for the year CE 2100 and beyond may have been underestimated in IPCC AR5 projections especially for high emission scenarios (e.g. DeConto and Pollard, 2016; Edwards et al., 2019; Golledge et al., 2015, 2020; Siegert et al., 2020).
- These paleo-calibrated AIS models show that a threshold for marine ice sheet stability may exist at ~1.5°C–2°C global warming above pre-industrial (e.g. around RCP 2.6, the target of the Paris Agreement) (e.g. Clark et al., 2016; DeConto and Pollard, 2016; Golledge et al., 2015).
- Recent paleo-studies have stressed that a moderate local oceanic warming, lower than the upper bound of 1.5°C–2°C for pan-Antarctic ocean warming can also trigger fast ice sheet retreat if applied for a few centuries: Bakker et al. (2017); Beltran et al. (2020); Golledge et al. (2017a); Turney et al. (2020). This highlights the importance of polar amplification for the fast response of polar areas under past and future global warming conditions.

12.7.2 Challenges for the next programmes

Gaps illustrated above highlight the necessity to assess whether or not the WAIS only partially retreated or totally disintegrated during past warm periods. Records of such massive ice sheet retreats are possibly located below the ice sheet. Locating subglacial drilling sites that could have recorded such extensive retreat represents a high priority challenge worthy of future field campaigns (e.g. Bradley et al., 2012; Spector et al., 2018). Similarly, it is urgent to assess the EAIS marine-based sectors sensitivity to oceanic and atmospheric warming during past warm periods (e.g. Aitken et al., 2016;

Blackburn et al., 2020; Cook et al., 2013, 2014; Gulick et al., 2017; Pierce et al., 2017; Reinardy et al, 2015; Wilson et al., 2018) and their potential contribution to global mean sea level change (e.g. DeConto and Pollard, 2016; Mas e Braga et al., 2021; Paxman et al., 2020) to refine their future contribution to ongoing sea level rise (e.g. Golledge et al., 2017b; Rignot et al., 2019).

Paradoxically, even though the Pleistocene interglacials are more recent and well documented in many places around the world, AIS fluctuations through time have destroyed most of the ice proximal geological evidence of these interglacials on the continental shelf, making direct records of the ice sheet's behaviour difficult to find. Only a few precious SST records are currently available from the Antarctic continental slope and rise and those records are indirect and cannot fill the gap of ice proximal ocean temperature records. This data gap directly impedes the validation of numerical paleoclimate and paleo-ice sheet numerical simulations. The interpretation of sedimentary facies and geomorphological features on the seafloor, however, does allow us to infer the type of sub-glacial environments and, thus, ice flow during past deglaciations to be reconstructed (e.g. Bart et al., 2018; Prothro et al., 2020; Simkins et al., 2017; Smith et al., 2019).

Another observational challenge is to recover records with sub-millennial temporal resolution for the different past warm periods. Such high-resolution archives can be recovered from the continental slope and rise by drilling levee deposits and contourite systems, or from the continental shelf in overdeepened basins and fjords (e.g. ODP Leg 178 The Palmer Deep, Domack et al., 2001; IODP Expedition 318, Ashley et al., 2021; IODP Expedition 374 Ross Sea, McKay et al., 2019; IODP Expedition 379 Amundsen Sea, Gohl et al., 2019; approved IODP proposal 732 Antarctic Peninsula). High-resolution data represent the bridge between the past and the future, in particular for centennial to millennial-scale climate oscillations (e.g. Bakker et al., 2017; Bracegirdle et al., 2019; Golledge et al., 2020; Noble et al., 2020; Weber et al., 2014). High-resolution sedimentary data are also important for correlating marine sediment records with ice core records of the past 8,000,000 years. The on-going Beyond EPICA: Oldest Ice project (e.g. Fischer et al., 2013; Parrenin et al., 2017; https://www.beyondepica.eu/) will allow correlation with expanded sediment records from the Ross Sea (IODP Exp. 374) (McKay et al., 2019) including the MIS-31 super-interglacial event, and from the Amundsen Sea (IODP Exp. 379) (Gohl et al., 2019) across the Mid-Pleistocene Transition and from future expeditions, for example the IODP proposal 732-Full2.

To maintain pace with advances of the observational ice sheet community, the paleoclimate modelling community will need to expand efforts in regional atmospheric and oceanic modelling for different past periods representing both glacial and interglacial contexts. Regional modelling is computationally expensive and also requires highly resolved boundary conditions at high frequency to capture the local variability of processes. Improved large-scale global climate simulations will also be required to support regional modelling. Many on-going data-model comparison initiatives already exist, and some of them focus on the periods described in this chapter, as for example the paleoclimate Model Intercomparison Project (PMIP, now in phase 4) (Kageyama et al., 2018), the Pliocene Model Intercomparison Project (PLIOMIP, now in phase 2) (Haywood et al., 2020), the recently started Miocene Model Intercomparison Project (MIOMIP) (Steinthorsdottir et al., 2020 and related special issue) and the Deep-Time Model Intercomparison Project (DEEPMIP) (Lunt et al., 2017). PMIP focuses on the Late Pleistocene and now also includes transient simulations of entire interglacials using coupled atmosphere-ocean models. DEEPMIP focuses mainly on the EOT and the Eocene warmth. More refined global mean sea level records are also necessary to better assess Antarctic ice volume fluctuations over the past 34 Myrs. Both MCO and mPWP periods are of high interest to assess Earth climate sensitivity to high atmospheric CO_2 concentrations (similar to projected ones) and global mean sea level rise (e.g. Haywood et al., 2016; Steinthorsdottir et al., 2020). Sequence stratigraphy of the continental margins is a powerful approach and the key to fill this gap. However, improvements are needed, especially to correct those records from glacio-hydro-isostasy (e.g. Grant et al., 2019; Grant and Naish, 2021). Thus, coupled ice-sheet-GIA-sediment erosion and transport models are needed (e.g. Pollard and DeConto, 2003, 2020; Whitehouse et al., 2019).

Finally, while the climate and paleoclimate communities are currently putting efforts into the development of fully coupled Earth System Models, such models are too computationally demanding to allow for long-term transient simulations. With upcoming progress in scientific computing, and progress in the computing facilities themselves, using fully coupled Earth System Models now seems an achievable objective for paleo studies.

12.7.3 Long-term projections and role of PAIS and future programs

Future projections of AIS evolution have shown large improvements over the past few years (e.g. DeConto and Pollard, 2016; Edwards et al., 2019; Pattyn et al., 2018). However, related uncertainties remain large, indicating that fundamental knowledge gaps still persist about ice sheet dynamics, and interactions with the atmosphere, ocean and the solid earth (Whitehouse et al., 2019). Morlighem et al. (2020) released an updated subglacial topography map revealing the high-resolution bed morphology of some of the glacial troughs and their potential in causing AIS instability in the case of fast retreat of the grounding line. Many of them are still unexplored, despite their clear importance in understanding past, present and future AIS dynamics. The release of the IPCC Special Report 'The Ocean and Cryosphere in a Changing Climate' (SROCC) in September 2019 (IPCC, 2019) showed that our appreciation of the various contributions to GMSL change has improved since the last IPCC Assessment Report 5 (AR5) in 2013. After the release of AR5, satellite observations revealed that Antarctic ice shelves were thinning faster than previously thought (Paolo et al., 2015), caused by observed warming in the surrounding ocean (Pritchard et al., 2012). Recent re-assessments of 20th century observations confirmed the AIS has been losing mass since the publication of IPCC AR5 and that this mass loss accelerated at the end of the 20th century (e.g. Rignot et al., 2019; Shepherd et al., 2018).

To precisely assess the AIS contributions to GMSL changes, the polar community has increased the monitoring and modelling of AIS evolution. Attention has been focused on ice shelf buttressing and on large partly marine-based drainage basins of the WAIS and EAIS (Fürst et al., 2016) (e.g. Pine Island Glacier, Thwaites glacier, Totten glacier, Recovery ice stream, Foundation ice stream) and ice–ocean interactions around Antarctica. The particularity of most of the marine-based sectors of the AIS is that they are grounded on a bed with retrograde slopes (Joughin and Alley, 2011; Morlighem et al., 2020) and/or that their buttressing ice shelves are pinned on a sill with a retrograde slope and are thus vulnerable to future MISI. New estimates of future GMSL rise from the IPCC SROCC (2019) amount to 0.43 m (0.29–0.59 likely range, RCP 2.6 scenario) and 0.84 m (0.61–1.10, likely range, RCP 8.5) in CE 2100, with the possibility of multi-metre sea level rise by CE 2300 (Clark et al., 2016; Golledge et al., 2015) but with 'deep uncertainty' (IPCC SROCC, 2019). Ice shelf loss is a key pre-requisite for the onset of marine ice shelf instabilities. The large uncertainties in the most recent estimates of sea level rise from Antarctica mostly result from our inability to assess the potentially unstable behaviour of the marine-based sectors of the AIS, and in particular: the sensitivity of ice shelves to sub-ice shelf melting from below and surface warming above. These gaps inevitably lead to model-dependent results, particularly for processes that are parameterised (Asay-Davis et al., 2017; Siegert et al., 2020). This is where the past can close those knowledge gaps and provide necessary observational constraints to model the past and the future evolution of the AIS (e.g. Gasson and Keisling, 2020).

The Earth's past provides a natural laboratory for testing realistic cases of ice-sheet-climate–solid earth interactions at different timescales (Bracegirdle et al., 2019). The research produced within the PAIS programme has shown that the AIS potentially crossed its tipping point for major ice loss many times since the onset of large-scale glaciation (~34 Ma) under climatic conditions warmer than today. Past periods have the potential to identify thresholds for instability (e.g. Cook et al., 2013; Naish et al., 2009a, 2009b; Weber et al., 2014; Wilson et al., 2018) or large retreat/re-advance events (e.g. Golledge et al., 2017a, 2017b; Kingslake et al., 2018; Scherer et al., 2016; Wilson et al., 2018) and thus to provide credibility to future scenarios. Paleo-records also have the potential to reveal new mechanisms, such as for example the marine-ice cliff instability (MICI, DeConto and Pollard, 2016; Pollard et al., 2015). MICI involves the fast disintegration of ice shelves by surface-melt induced hydro-fracturing and rapid calving at thick, marine-terminating ice margins. This mechanism has been inferred to explain rapid major mass loss from the WAIS and EAIS during MIS 5e and the mPWP (DeConto and Pollard, 2016; Pollard et al., 2015) but many open questions about MICI and its possible role in past and future sea level rise remain (e.g., Edwards et al., 2019; Pattyn et al., 2018). Geological and glaciological evidence can also highlight feedbacks in the ice sheet-climate system (Turney et al., 2020) or processes that might not appear policy-relevant but are indeed determinant in understanding the future sensitivity of the AIS and sea level rise to ongoing and projected climate changes (Haywood et al., 2019).

12.8 Coauthors from the PAIS community

Aisling M. Dolan, University of Leeds, Leeds, UK

Alan K. Cooper, U.S. Geological Survey Emeritus, Menlo Park, USA

Alessandra Venuti, Istituto Nazionale di Geofisica e Vulcanologia, Rome, Italy

Amy Leventer, Colgate University, Hamilton, NY, USA

Andrea Bergamasco, C.N.R. (National Research Council) ISMAR, Venice, Italy

Carolina Acosta Hospitaleche, CONICET, División Paleontología Vertebrados, Museo de La Plata (Facultad de Ciencias Naturales y Museo, UNLP) La Plata, Argentina

Carolina Acosta Hospitaleche, CONICET – División Paleontología Vertebrados, Museo de La Plata, Facultad de Ciencias Naturales y Museo, UNLP; La Plata, Argentina

Catalina Gebhardt, Alfred Wegener Institute Helmholtz Centre of Polar and Marine Research, Bremerhaven, Germany

Christine S. Siddoway, Colorado College, Colorado Springs, USA

Christopher C. Sorlien, Earth Research Institute, University of California, Santa Barbara, Santa Barbara, California, USA

David Harwood, University of Nebraska-Lincoln, Lincoln, Nebraska, USA

David Pollard, Pennsylvania State University, University Park, Pennsylvania, USA

David J. Wilson, Department of Earth Sciences, University College London, London, UK

Denise K. Kulhanek, Texas A&M University, College Station, TX, United States

Dominic A. Hodgson, British Antarctic Survey, Cambridge, UK

Edward G.W. Gasson, University of Bristol, UK

Fausto Ferraccioli, NERC/British Antarctic Survey, Cambridge, UK

Fernando Bohoyo, Instituto Geológico y Minero de España, Madrid, Spain

Francesca Battaglia, University of Venice Cá Foscari, Italy

Frank O. Nitsche, Lamont-Doherty Earth Observatory of Columbia University, Palisades, USA

Georgia R. Grant, GNS Science Wellington, New Zealand

Gerhard Kuhn, Alfred-Wegener-Institut Helmholtz-Zentrum für Polar- und Meeresforschung, Bremerhaven, Germany

Guy J.G. Paxman, Lamont-Doherty Earth Observatory, Columbia University, New York, USA

Ian D. Goodwin, Climate Change Research Centre, University of New South Wales, Sydney, Australia

Isabel Sauermilch, University of Tasmania, Institute for Marine and Antarctic Studies, Australia

Jamey Stutz, Antarctic Research Centre at Victoria University of Wellington, New Zealand

Jan Sverre Laberg, Department of Geosciences, UiT The Arctic University of Norway, NO-9037 Tromsø, Norway

Javier N. Gelfo, CONICET – UNLP, División Paleontología Vertebrados, Museo de La Plata, Argentina

Johann P. Klages, Alfred Wegener Institute Helmholtz Center for Polar and Marine Research, Bremerhaven, Germany

Julia S. Wellner, University of Houston, Houston, USA

Karsten Gohl, Alfred Wegener Institute Helmholtz Centre for Polar and Marine Research, Bremerhaven, Germany

Laura Crispini, University of Genova (DISTAV, Genova, Italy)

Leanne K. Armand, Australian National University, Canberra, Australia.

Marcelo A. Reguero, Instituto Antártico Argentino, B1650HMK, San Martín, Buenos Aires, Argentina

Marcelo A. Reguero, Instituto Antártico Argentino, Buenos Aires, Argentina

Marco Taviani, Institute of Marine Sciences (ISMAR), National Research Council (CNR), 40129, Bologna, Italy and Biology Department, Woods Hole Oceanographic Institution, 02543, Woods Hole, USA

Martin J. Siegert, Imperial College London, London, UK

Marvin A. Speece, Montana Technological University, Butte, USA

Mathieu Casado, Alfred Wegener Institute Helmholtz Centre for Polar and Marine Research, Potsdam, Germany

Michele Rebesco, OGS, Trieste, Italy

Mike Weber, University of Bonn, Institute for Geosciences, Department of Geochemistry and Petrology, 53115 Bonn, Germany

Minoru Ikehara, Kochi University, Japan

Nicholas R. Golledge, Antarctic Research Centre Victoria University of Wellington, Wellington 6140, New Zealand

Nigel Wardell, OGS, Trieste, Italy

Paolo Montagna, Institute of Polar Sciences, National Research Council, Bologna, Italy

Peter J. Barrett, Antarctic Research Centre, Victoria University of Wellington, Wellington, New Zealand.

Peter K. Bijl, Utrecht University, Utrecht, The Netherlands

Philip E. O'Brien, Macquarie University, Sydney, Australia

Philip J. Bart, Louisiana State University, Baton Rouge, USA

Raffaella Tolotti, University of Genoa, Genoa, Italy

Reed P. Scherer, Northern Illinois University, DeKalb, IL, USA

Renata G. Lucchi, National Institute of Oceanography and Applied Geophysics (OGS), Sgonico-Trieste, Italy

Riccardo Geletti, National Institute of Oceanography and Applied Geophysics – OGS, Trieste, Italy

Richard C.A. Hindmarsh, British Antarctic Survey & Durham University, Cambridge & Durham, United Kingdom

Richard H. Levy, GNS Science and Victoria University of Wellington, Lower Hutt and Wellington, New Zealand

Robert B. Dunbar, Stanford University, Stanford, California, USA

Robert D. Larter, British Antarctic Survey, Cambridge, UK

Robert M. Mckay, Antarctic Research Centre, Victoria University of Wellington, Wellington, New Zealand

R. Selwyn Jones, Monash University (Melbourne, Australia)

Sandra Passchier, Montclair State University, Montclair, USA

Sean P.S. Gulick, University of Texas at Austin, Austin, Texas

Sidney R. Hemming, Columbia University, New York, USA

Stefanie Brachfeld, Montclair State University, New Jersey, USA

Suzanne OConnell, Wesleyan University, Middletown, CT, USA

Trevor Williams, International Ocean Discovery Program, Texas A&M University, College Station, USA

Ursula Röhl, MARUM, University of Bremen, Bremen, Germany

Yasmina M. Martos, NASA Goddard Space Flight Center, Greenbelt, MD, USA & University of Maryland College Park, MD, USA

Acknowledgements

TN acknowledges support from MBIE Antarctic Science Platform contract ANTA1801.

References

Aitken, A.R.A., Roberts, J.L., Van Ommen, T.D., Young, D.A., Golledge, N.R., Greenbaum, J. S., et al., 2016. Repeated large-scale retreat and advance of Totten Glacier indicated by inland bed erosion. Nature 533 (7603), 385–389.

Albrecht, T., Winkelmann, R., Levermann, A., 2020. Glacial-cycle simulations of the Antarctic Ice Sheet with the Parallel Ice Sheet Model (PISM) – part 1: boundary conditions and climatic forcing. The Cryosphere 14, 599–632. Available from: https://doi.org/10.5194/tc-14-599-2020.

Alley, R.B., Anandakrishnan, S., Dupont, T.K., Parizek, B.R., Pollard, D., 2007. Effect of sedimentation on ice-sheet grounding-line stability. Science 315 (5820), 1838–1841.

Anderson, J.B., Conway, H., Bart, P.J., Witus, A.E., Greenwood, S.L., McKay, R.M., et al., 2014. Ross Sea paleo-ice sheet drainage and deglacial history during and since the LGM. Quaternary Science Reviews 100, 31–54.

Anderson, J.B., Shipp, S.S., Lowe, A.L., Wellner, J.S., Mosola, A.B., 2002. The Antarctic Ice Sheet during the Last Glacial Maximum and its subsequent retreat history: a review. Quaternary Science Reviews 21 (1-3), 49–70.

Argus, D.F., Peltier, W.R., Drummond, R., Moore, A.W., 2014. The Antarctica component of postglacial rebound model ICE-6G_C (VM5a) based on GPS positioning, exposure age dating of ice thicknesses, and relative sea level histories. Geophysical Journal International 198 (1), 537–563.

Armand, L., O'Brien, P., Armbrecht, L., Barker, H., Caburlotto, A., Connell, T., et al., 2018. Interactions of the Totten Glacier with the Southern Ocean through multiple glacial cycles (IN2017-V01): Post-survey report. Australian Antarctic Science Grant Program (AAS# 4333). The grant title is "Interactions of the Totten Glacier with the Southern Ocean through multiple glacial cycles" and the PI's are: Armand, LK O'Brien, P., Post, A., Goodwin, I., Opdyke, B., Leventer, A., Domack, E., Escutia-Dotti, C. DeSantis, L.

Arndt, J.E., Hillenbrand, C.D., Grobe, H., Kuhn, G., Wacker, L., 2017. Evidence for a dynamic grounding line in outer Filchner Trough, Antarctica, until the early Holocene. Geology 45 (11), 1035–1038.

Asay-Davis, X.S., Jourdain, N.C., Nakayama, Y., 2017. Developments in simulating and parameterizing interactions between the Southern Ocean and the Antarctic Ice sheet. Current Climate Change Reports. 3, 316–329. Available from: https://doi.org/10.1007/s40641-017-0071-0.

Ashley, K.E., McKay, R., Etourneau, J., Jimenez-Espejo, F.J., Condron, A., Albot, A., et al., 2021. Mid-Holocene Antarctic sea-ice increase driven by marine ice sheet retreat. Climate of the Past 17 (1), 1–19.

Austermann, J., Pollard, D., Mitrovica, J.X., Moucha, R., Forte, A.M., DeConto, R.M., et al., 2015. The impact of dynamic topography change on Antarctic ice sheet stability during the mid-Pliocene warm period. Geology 43 (10), 927–930. Available from: https://doi.org/10.1130/G36988.1.

Badger, M.P.S., Lear, C.H., Pancost, R.D., Foster, G.L., Bailey, T., et al., 2013a. CO2 drawdown following the middle Miocene expansion of the Antarctic Ice Sheet. Paleoceanography 28, 42–53. Available from: https://doi.org/10.1002/palo.20015.

Badger, M.P.S., Schmidt, D.N., Mackensen, A., Pancost, R.D., 2013b. High resolution alkenone palaeobarometry indicates relatively stable pCO_2 during the Pliocene (3.3 to 2.8 Ma). Philosophical Transactions of the Royal Society A 347, 20130094. Available from: https://doi.org/10.1098/rsta.2013.0094.

Bakker, P., Clark, P.U., Golledge, N.R., Schmittner, A., Weber, M.E., 2017. Centennial-scale Holocene climate variations amplified by Antarctic Ice Sheet discharge. Nature 541 (7635), 72–76.

Bamber, J.L., Oppenheimer, M., Kopp, R.E., Aspinall, W.P., Cooke, R.M., 2019. Ice sheet contributions to future sea-level rise from structured expert judgment. Proceedings of the National Academy of Sciences 116 (23), 11195–11200.

Bard, E., Hamelin, B., Fairbanks, R.G., 1990. U-Th ages obtained by mass spectrometry in corals from Barbados: sea level during the past 130,000 years. Nature 346 (6283), 456–458.

Barrett, P., 1989. Antarctic Cenozoic history from the CIROS-1 drillhole, McMurdo Sound, Antarctica. NZ DSIR Bulletin 245, 254.

Barrett, P.J., 2007. Cenozoic climate and sea level history from glaciomarine strata off the Victoria Land coast, Cape Roberts Project, Antarctica. In: Hambrey, M.J., Christoffersen, P., Glasser, N.F., Hubbard, B., (Eds.), Glacial Sedimentary Processes and Products. International Association of Sedimentologists Special Publications 39, 259–287.

Bart, P.J., 2001. Did the Antarctic ice sheets expand during the early Pliocene? Geology 29 (1), 67–70. Available from: https://doi.org/10.1130/0091-7613(2001)029%3C0067:DTAISE%3E2.0.CO;2.

Bart, P.J., 2003. Were West Antarctic ice sheet grounding events in the Ross Sea a consequence of East Antarctic ice sheet expansion during the middle Miocene? Earth and Planetary Science Letters 216 (1–2), 93–107. Available from: https://doi.org/10.1016/S0012-821X(03)00509-0.

Bart, P.J., De Santis, L., 2012. Glacial intensification during the Neogene: a review of seismic stratigraphic evidence from the Ross Sea, Antarctica, continental shelf. Oceanography 25 (3), 166–183.

Bart, P.J., DeCesare, M., Rosenheim, B.E., Majewski, W., McGlannan, A., 2018. A centuries-long delay between a paleo-ice-shelf collapse and grounding-line retreat in the Whales Deep Basin, eastern Ross Sea, Antarctica. Scientific Reports 8 (1), 1–9. Available from: https://doi.org/10.1038/s41598-018-29911-8.

Bart, P.J., Iwai, M., 2012. The overdeepening hypothesis: how erosional modification of the marine-scape during the early Pliocene altered glacial dynamics on the Antarctic Peninsula's Pacific margin. Palaeogeography, Palaeoclimatology, Palaeoecology 335, 42–51.

Bart, P.J., Krogmeier, B.J., Bart, M.P., Tulaczyk, S., 2017. The paradox of a long grounding during West Antarctic Ice Sheet retreat in Ross Sea. Scientific Reports 7 (1), 1–8. Available from: https://doi.org/10.1038/s41598-017-01329-8.

Bart, P.J., Tulaczyk, S., 2020. A significant acceleration of ice volume discharge preceded a major retreat of a West Antarctic paleo–ice stream. Geology 48 (4), 313–317. Available from: https://doi.org/10.1130/G46916.1.

Bartoli, G., Hönisch, B., Zeebe, R.E., 2011. Atmospheric CO_2 decline during the Pliocene intensification of Northern Hemisphere glaciations. Paleoceanography 26 (4). Available from: https://doi.org/10.1029/2010PA002055.

Bassett, S.E., Milne, G.A., Bentley, M.J., Huybrechts, P., 2007. Modelling Antarctic sea-level data to explore the possibility of a dominant Antarctic contribution to meltwater pulse IA. Quaternary Science Reviews 26 (17–18), 2113–2127.

Becquey, S., Gersonde, R., 2002. Past hydrographic and climatic changes in the Subantarctic Zone of the South Atlantic–The Pleistocene record from ODP Site 1090. Palaeogeography, Palaeoclimatology, Palaeoecology 182 (3–4), 221–239.

Becquey, S., Gersonde, R., 2003a. A 0.55-Ma paleotemperature record from the Subantarctic zone: implications for Antarctic Circumpolar Current development. Paleoceanography 18 (1).

Becquey, S., Gersonde, R., 2003b. 14. Data Report: Early and Mid-Pleistocene (MIS 65-11) Summer Sea-Surface Temperature, Foraminiferal Fragmentation, and Ice-Rafted Debris Records from the Subantarctic (ODP Leg 177 Site 1090). Proceedings of Ocean Drilling Program Science Results 177, 1–23.

Beltran, C., Golledge, N.R., Ohneiser, C., Kowalewski, D.E., Sicre, M.A., Hageman, K.J., et al., 2020. Southern Ocean temperature records and ice-sheet models demonstrate rapid Antarctic ice sheet retreat under low atmospheric CO_2 during Marine Isotope Stage 31. Quaternary Science Reviews 228, 106069. Available from: https://doi.org/10.1016/j.quascirev.2019.106069.

Bentley, M.J., 1999. Volume of Antarctic ice at the Last Glacial Maximum, and its impact on global sea level change. Quaternary Science Reviews 18, 1569–1595.

Bentley, M.J., Fogwill, C.J., Le Brocq, A.M., Hubbard, A.L., Sugden, D.E., Dunai, T.J., et al., 2010. Deglacial history of the West Antarctic Ice Sheet in the Weddell Sea embayment: constraints on past ice volume change. Geology 38 (5), 411–414.

Benz, V., Esper, O., Gersonde, R., Lamy, F., Tiedemann, R., 2016. Last Glacial Maximum sea surface temperature and sea-ice extent in the Pacific sector of the Southern Ocean. Quaternary Science Reviews 146, 216–237.

Bereiter, B., Eggleston, S., Schmitt, J., Nehrbass-Ahles, C., Stocker, T.F., Fischer, H., et al., 2015. Revision of the EPICA Dome C CO_2 record from 800 to 600 kyr before present. Geophysical Research Letters 42 (2), 542–549. Available from: https://doi.org/10.1002/2014GL061957.

Bertram, R.A., Wilson, D.J., van de Flierdt, T., McKay, R.M., Patterson, M.O., Jimenez-Espejo, F.J., et al., 2018. Pliocene deglacial event timelines and the biogeochemical response offshore Wilkes Subglacial Basin, East Antarctica. Earth and Planetary Science Letters 494, 109–116. Available from: https://doi.org/10.1016/j.epsl.2018.04.054.

Bijl, P.K., Bendle, J.A., Bohaty, S.M., Pross, J., Schouten, S., Tauxe, L., et al., 2013. Eocene cooling linked to early flow across the Tasmanian Gateway. Proceedings of the National Academy of Sciences 110 (24), 9645–9650.

Bintanja, R., Van De Wal, R.S., Oerlemans, J., 2005. Modelled atmospheric temperatures and global sea levels over the past million years. Nature 437 (7055), 125–128. Available from: https://doi.org/10.1038/nature03975.

Bjill, P.K., Houben, A.J.P., Hartman, J.D., Pross, J., Salabarnada, A., Escutia, C., et al., 2018. Paleoceanography and ice sheet variability off-shore Wilkes Land, Antarctica – Part 2: Insights from Oligocene–Miocene dinoflagellate cyst assemblages. Climate of the Past 14, 1015–1033. Available from: https://doi.org/10.5194/cp-14-1015-2018.

Blackburn, T., Edwards, G.H., Tulaczyk, S., Scudder, M., Piccione, G., Hallet, B., et al., 2020. Ice retreat in Wilkes Basin of East Antarctica during a warm interglacial. Nature 583 (7817), 554–559. Available from: https://doi.org/10.1038/s41586-020-2484-5.

Blasco, J., Tabone, I., Alvarez-Solas, J., Robinson, A., Montoya, M., 2019. The Antarctic Ice Sheet response to glacial millennial-scale variability. Climate of the Past 15 (1), 121–133.

Blunier, T., Chappellaz, J., Schwander, J., Dällenbach, A., Stauffer, B., Stocker, T.F., et al., 1998. Asynchrony of Antarctic and Greenland climate change during the last glacial period. Nature 394 (6695), 739–743.

Bohaty, S., Scherer, R., Harwood, D.M., 1998. Quaternary diatom biostratigraphy and palaeoenvironments of the CRP-1 drillcore, Ross Sea, Antarctica. Terra Antartica 5 (3), 431–453.

Bohaty, S.M., Zachos, J.C., Delaney, M.L., 2012. Foraminiferal Mg/Ca evidence for southern ocean cooling across the eocene–oligocene transition. Earth and Planetary Science Letters 317, 251–261. Available from: https://doi.org/10.1016/j.epsl.2011.11.037.

Böhme, M., 2003. The Miocene climatic optimum: evidence from ectothermic vertebrates of Central Europe. Palaeogeography, Palaeoclimatology, Palaeoecology 195 (3-4), 389–401.

Böhm, E., Lippold, J., Gutjahr, M., Frank, M., Blaser, P., Antz, B., et al., 2015. Strong and deep Atlantic meridional overturning circulation during the last glacial cycle. Nature 517 (7532), 73–76. Available from: https://doi.org/10.1038/nature14059.

Böning, C.W., Behrens, E., Biastoch, A., Getzlaff, K., Bamber, J.L., 2016. Emerging impact of Greenland meltwater on deepwater formation in the North Atlantic Ocean. Nature Geoscience 9 (7), 523–527. Available from: https://doi.org/10.1038/ngeo2740.

Bracegirdle, T.J., Colleoni, F., Abram, N.J., Bertler, N.A., Dixon, D.A., England, M., et al., 2019. Back to the future: Using long-term observational and paleo-proxy reconstructions to improve model projections of Antarctic climate. Geosciences 9 (6), 255.

Bradley, S.L., Hindmarsh, R.C., Whitehouse, P.L., Bentley, M.J., King, M.A., 2015. Low post-glacial rebound rates in the Weddell Sea due to Late Holocene ice-sheet readvance. Earth and Planetary Science Letters 413, 79–89.

Bradley, S.L., Siddall, M., Milne, G.A., Masson-Delmotte, V., Wolff, E., 2012. Where might we find evidence of a Last Interglacial West Antarctic Ice Sheet collapse in Antarctic ice core records? Global and Planetary Change 88, 64–75. Available from: https://doi.org/10.1016/j.gloplacha.2012.03.004.

Brancolini, G., Busetti, M., Marchetti, A., De Santis, L., Zanolla, C., Cooper, A.K., et al., 1995b. Descriptive text for the Seismic Stratigraphic Atlas of the Ross Sea. In: Cooper, A. K., Barker, P.F., Brancolini, G. (Eds.), Geology and Seismic Stratigraphy of the Antarctic Margin. Antarctic Research Series 68 AGU, Washington, DC, pp. A268–A271.

Brancolini, G., Cooper, A.K., Coren, F., 1995a. Seismic facies and glacial history in the Western Ross Sea Antarctica. In: Cooper, A.K., Barker, P.F., Brancolini, G. (Eds.), Geology and Seismic Stratigraphy of the Antarctic Margin. Antarctic Research Series 68. AGU, Washington, DC, pp. 209–233.

Brierley, C.M., Fedorov, A.V., Liu, Z., Herbert, T.D., Lawrence, K.T., LaRiviere, J.P., 2009. Greatly expanded tropical warm pool and weakened Hadley circulation in the early Pliocene. Science 323 (5922), 1714–1718. Available from: https://doi.org/10.1126/science.1167625.

Briggs, R.D., Pollard, D., Tarasov, L., 2014. A data-constrained large ensemble analysis of Antarctic evolution since the Eemian. Quaternary Science Reviews 103, 91–115.

Brigham-Grette, J., 1999. Marine isotope stage11 high sea level record from northwest Alaska. In: Poore, R.Z., et al., (Eds.), Marine Oxygen Isotope Stage 11 and Associated Terrestrial Records. U.S. Geological Survey, pp. 19–20. Open File Reports, 99–312.

Bronselaer, B., Winton, M., Griffies, S.M., Hurlin, W.J., Rodgers, K.B., Sergienko, O.V., et al., 2018. Change in future climate due to Antarctic meltwater. Nature 564 (7734), 53–58. Available from: https://doi.org/10.1038/s41586-018-0712-z.

Bueler, E., Brown, J., 2009. Shallow shelf approximation as a "sliding law" in a thermomechanically coupled ice sheet model. Journal of Geophysical Research: Earth Surface 114 (F3). Available from: https://doi.org/10.1029/2008JF001179.

Buono, M.R., Fordyce, R.E., Marx, F.G., Fernández, M.S., Reguero, M., 2019. Eocene Antarctica: a window into the earliest history of modern whales. *Advances in Polar*. Science 30 (3), 293–302.

Burckle, L.H., Gersonde, R., Abrams, N., 1990. Late Pliocene-Pleistocene Paleoclimate in the Jane Basin Region ODP Site 697 [Series].

Capron, E., Govin, A., Feng, R., Otto-Bliesner, B.L., Wolff, E.W., 2017. Critical evaluation of climate syntheses to benchmark CMIP6/PMIP4 127 ka Last Interglacial simulations in the high-latitude regions. Quaternary Science Reviews 168, 137–150. Available from: https://doi.org/10.1016/j.quascirev.2017.04.019.

Capron, E., Govin, A., Stone, E.J., Masson-Delmotte, V., Mulitza, S., Otto-Bliesner, B., et al., 2014. Temporal and spatial structure of multi-millennial temperature changes at high latitudes during the Last Interglacial. Quaternary Science Reviews 103, 116–133. Available from: https://doi.org/10.1016/j.quascirev.2014.08.018.

Carlson, A.E., Clark, P.U., 2012. Ice sheet sources of sea level rise and freshwater discharge during the last deglaciation. Reviews of Geophysics 50 (4), Available from: https://doi.org/10.1029/2011RG000371.

Chadwick, M., Allen, C.S., Sime, L.C., Hillenbrand, C.D., 2020. Analysing the timing of peak warming and minimum winter sea-ice extent in the Southern Ocean during MIS 5e. Quaternary Science Reviews 229, 106134.

Clark, P.U., Alley, R.B., Keigwin, L.D., Licciardi, J.M., Johnsen, S.J., Wang, H., 1996. Origin of the first global meltwater pulse following the last glacial maximum. Paleoceanography 11 (5), 563–577.

Clark, P.U., He, F., Golledge, N.R., Mitrovica, J.X., Dutton, A., Hoffman, J.S., et al., 2020. Oceanic forcing of penultimate deglacial and last interglacial sea-level rise. Nature 577 (7792), 660–664. Available from: https://doi.org/10.1038/s41586-020-1931-7.

Clark, P.U., Mitrovica, J.X., Milne, G.A., Tamisiea, M.E., 2002. Sea-level fingerprinting as a direct test for the source of global meltwater pulse IA. Science 295 (5564), 2438–2441.

Clark, P.U., Shakun, J.D., Marcott, S.A., Mix, A.C., Eby, M., Kulp, S., et al., 2016. Consequences of twenty-first-century policy for multi-millennial climate and sea-level change. Nature Climate Change 6 (4), 360–369. Available from: https://doi.org/10.1038/nclimate2923.

Clark, P.U., Tarasov, L., 2014. Closing the sea level budget at the Last Glacial Maximum. Proceedings of the National Academy of Sciences 111 (45), 15861–15862.

Colleoni, F., De Santis, L., Montoli, E., Olivo, E., Sorlien, C.C., Bart, P.J., et al., 2018b. Past continental shelf evolution increased Antarctic ice sheet sensitivity to climatic conditions. Scientific Reports 8 (1), 1–12. Available from: https://doi.org/10.1038/s41598-018-29718-7.

Colleoni, F., De Santis, L., Siddoway, C.S., Bergamasco, A., Golledge, N.R., Lohmann, G., et al., 2018a. Spatio-temporal variability of processes across Antarctic ice-bed–ocean interfaces. Nature Communications 9, 2289. Available from: https://doi.org/10.1038/s41467-018-04583-0.

Cook, C.P., Hill, D.J., van de Flierdt, T., Williams, T., Hemming, S.R., Dolan, A.M., et al., 2014. Sea surface temperature control on the distribution of far-traveled Southern Ocean ice-rafted detritus during the Pliocene. Paleoceanography 29 (6), 533–548. Available from: https://doi.org/10.1002/2014PA002625.

Cook, C.P., Van De Flierdt, T., Williams, T., Hemming, S.R., Iwai, M., Kobayashi, M., et al., 2013. Dynamic behaviour of the East Antarctic ice sheet during Pliocene warmth. Nature Geoscience 6 (9), 765–769. Available from: https://doi.org/10.1038/ngeo1889.

Cooper, A., Barker, P., Barrett, P., Behrendt, J., Brancolini, G., Childs, et al., 2011. The ANTOSTRAT legacy: science collaboration and international transparency in potential marine mineral resource exploitation of Antarctica. Science Diplomacy: Antarctica, Science, and the Governance of International Spaces.

Cooper, A.K., Barrett, P.J., Hinz, K., Traube, V., Letichenkov, G., Stagg, H.M., 1991. Cenozoic prograding sequences of the Antarctic continental margin: a record of glacio-eustatic and tectonic events. Marine Geology 102 (1–4), 175–213.

Cooper, A.K., P.E. O'Brien, C. Richter (Eds.), 2004. Proceedings of the Ocean Drilling Program, Scientific Results, vol. 188. College Station, TX. <https://doi.org/10.2973/odp.proc.sr.188.2004>.

Coxall, H.K., Wilson, P.A., Pälike, H., Lear, C.H., Backman, J., 2005. Rapid stepwise onset of Antarctic glaciation and deeper calcite compensation in the Pacific Ocean. Nature 433 (7021), 53–57.

Cramer, B.S., Miller, K.G., Barrett, P.J., Wright, J.D., 2011. Late Cretaceous–Neogene trends in deep ocean temperature and continental ice volume: Reconciling records of benthic foraminiferal geochemistry (δ18O and Mg/Ca) with sea level history. Journal of Geophysical Research: Oceans 116 (C12).

Crampton, J.S., Cody, R.D., Levy, R., Harwood, D., McKay, R., Naish, T.R., 2016. Southern Ocean phytoplankton turnover in response to stepwise Antarctic cooling over the past 15 million years. Proceedings of the National Academy of Sciences 113 (25), 6868–6873.

Crosta, X., Crespin, J., Swingedouw, D., Marti, O., Masson-Delmotte, V., Etourneau, J., et al., 2018. Ocean as the main driver of Antarctic ice sheet retreat during the Holocene. Global and Planetary Change 166, 62–74. Available from: https://doi.org/10.1016/j.gloplacha.2018.04.007.

Dahl-Jensen, D., Albert, M.R., Aldahan, A., Azuma, N., Balslev-Clausen, D., Baumgartner, M., et al., 2013. Eemian interglacial reconstructed from a Greenland folded ice core. Nature 493 (7433), 489.

de Boer, B., Colleoni, F., Golledge, N.R., DeConto, R.M., 2019. Paleo ice-sheet modeling to constrain past sea level. PAGES Magazine 27 (1), 20.

de Boer, B., Dolan, A.M., Bernales, J., Gasson, E., Golledge, N.R., Sutter, J., et al., 2015. Simulating the Antarctic ice sheet in the late-Pliocene warm period: PLISMIP-ANT, an ice-sheet model intercomparison project. The Cryosphere 9, 881–903. Available from: https://doi.org/10.5194/tc-9-881-2015.

de Boer, B., Haywood, A.M., Dolan, A.M., Hunter, S.J., Prescott, C.L., 2017a. The transient response of ice volume to orbital forcing during the warm late Pliocene. Geophysical Research Letters 44 (20), 10–486. Available from: https://doi.org/10.1002/2017GL073535.

de Boer, B., Haywood, A.M., Dolan, A.M., Hunter, S.J., Prescott, C.L., 2017. The transient response of ice volume to orbital forcing during the warm Late Pliocene. Geophysical Research Letters 44 (20), 10–486.

de Boer, B., Lourens, L.J., Van De Wal, R.S., 2014. Persistent 400,000-year variability of Antarctic ice volume and the carbon cycle is revealed throughout the Plio-Pleistocene. Nature Communications 5, 2999. Available from: https://doi.org/10.1038/ncomms3999.

de Boer, B., Stocchi, P., van de Wal, R.W.S., 2014b. A fully coupled 3-D ice-sheet – sea-level model: algorithm and applications. Geoscience Model Development Discussion 7, 2141–2156. Available from: http://doi.org/10.5194/gmd-7-2141-2014.

de Boer, B., Stocchi, P., Whitehouse, P., Van de Wal, R.S.W., 2017b. Current state and future perspectives on coupled ice-sheet – sea-level modelling. Quaternary Science Reviews 169, 13–28. Available from: https://doi.org/10.1016/j.quascirev.2017.05.013.

de Boer, B., Van de Wal, R.S.W., Lourens, L.J., Bintanja, R., Reerink, T.J., 2013. A continuous simulation of global ice volume over the past 1 million years with 3-D ice-sheet models. Climate Dynamics 41 (5-6), 1365–1384. Available from: https://doi.org/10.1007/s00382-012-1562-2.

DeConto, R.M., Pollard, D., 2003. Rapid Cenozoic glaciation of Antarctica induced by declining atmospheric CO_2. Nature 421 (6920), 245–249. Available from: https://doi.org/10.1038/nature01290.

DeConto, R.M., Pollard, D., Harwood, D., 2007. Sea ice feedback and Cenozoic evolution of Antarctic climate and ice sheets. Paleoceanography 22, PA3214. Available from: https://doi.org/10.1029/2006PA001350.

DeConto, R.M., Pollard, D., Kowalewski, D., 2012. Reprint of: Modeling Antarctic ice sheet and climate variations during Marine Isotope Stage 31. Global and Planetary Change 96, 181–188. Available from: https://doi.org/10.1016/j.gloplacha.2012.05.018.

DeConto, R.M., Pollard, D., Wilson, P.A., Pälike, H., Lear, C., Pagani, M., 2008. Thresholds for Cenozoic bipolar glaciation. Nature 455, 653–656.

DeConto, R.M., Pollard, D., 2016. Contribution of Antarctica to past and future sea level rise. Nature 531, 591–597. Available from: https://doi.org/10.1038/nature17145.

de La Vega, E., Chalk, T.B., Wilson, P.A., Bysani, R.P., Foster, G.L., 2020. Atmospheric CO 2 during the Mid-Piacenzian Warm Period and the M2 glaciation. Scientific reports 10 (1), 1–8.

Denton, G.H., Sugden, D.E., 2005. Meltwater features that suggest Miocene ice-sheet overriding of the Transantarctic Mountains in Victoria Land, Antarctica. Geografiska Annaler: Series A. Physical Geography 87 (1), 67–85.

Deschamps, P., Durand, N., Bard, E., Hamelin, B., Camoin, G., Thomas, A.L., et al., 2012. Ice-sheet collapse and sea-level rise at the Bølling warming 14,600 years ago. Nature 483 (7391), 559–564.

De Santis, L., Anderson, J.B., Brancolini, G., Zayatz, I., 1995. In: Cooper, A.K., et al., (Eds.), In Antarctic Research Series, vol. 71. Springer International.

De Santis, L., Brancolini, G., Donda, F., 2003. Seismo-stratigraphic analysis of the Wilkes Land continental margin (East Antarctica): influence of glacially driven processes on the Cenozoic deposition. Deep Sea Research Part II: Topical Studies in Oceanography 50 (8-9), 1563–1594. Available from: https://doi.org/10.1016/S0967-0645(03)00079-1.

De Santis, L., Prato, S., Brancolini, G., Lovo, M., Torelli, L., 1999. The Eastern Ross Sea continental shelf during the Cenozoic: implications for the West Antarctic ice sheet development. Global and Planetary Change 23 (1–4), 173–196. Available from: https://doi.org/10.1016/S0921-8181(99)00056-9.

De Schepper, S., Gibbard, P.L., Salzmann, U., Ehlers, J., 2014. A global synthesis of the marine and terrestrial evidence for glaciation during the Pliocene Epoch. Earth-Science Reviews 135, 83–102. Available from: https://doi.org/10.1016/j.earscirev.2014.04.003.

Detlef, H., Belt, S.T., Sosdian, S.M., Smik, L., Lear, C.H., Hall, I.R., et al., 2018. Sea ice dynamics across the Mid-Pleistocene transition in the Bering Sea. Nature Communications 9 (1), 1–11.

De Vleeschouwer, D., Vahlenkamp, M., Crucifix, M., Pälike, H., 2017. Alternating Southern and Northern Hemisphere climate response to astronomical forcing during the past 35 m.y. Geology 45 (4), 375–378. Available from: https://doi.org/10.1130/G38663.1.

Dickens, W.A., Kuhn, G., Leng, M.J., Graham, A.G.C., Dowdeswell, J.A., Meredith, M.P., et al., 2019. Enhanced glacial discharge from the eastern Antarctic Peninsula since the 1700s associated with a positive Southern Annular Mode. Scientific Reports 9 (1), 1–11.

Dolan, A.M., De Boer, B., Bernales, J., Hill, D.J., Haywood, A.M., 2018. High climate model dependency of Pliocene Antarctic ice-sheet predictions. Nature Communications 9 (1), 1–12. Available from: https://doi.org/10.1038/s41467-018-05179-4.

Dolan, A.M., Haywood, A.M., Hill, D.J., Dowsett, H.J., Hunter, S.J., Lunt, D.J., Pickering, S.J., 2011. Sensitivity of Pliocene ice sheets to orbital forcing. Palaeogeography, Palaeoclimatology, Palaeoecology 309 (1–2), 98–110.

Domack, E., Leventer, A., Dunbar, R., Taylor, F., Brachfeld, S., Sjunneskog, C., 2001. Chronology of the Palmer Deep site, Antarctic Peninsula: a Holocene palaeoenvironmental reference for the circum-Antarctic. The Holocene 11 (1), 1–9.

Donda, F., Brancolini, G., O'Brien, P.E., De Santis, L., Escutia, C., 2007. Sedimentary processes in the Wilkes Land margin: a record of the Cenozoic East Antarctic Ice Sheet evolution. Journal of the Geological Society of London 164, 243–256.

Dowdeswell, J.A., Batchelor, C.L., Montelli, A., Ottesen, D., Christie, F.D.W., Dowdeswell, E.K., et al., 2020. Delicate seafloor landforms reveal past Antarctic grounding-line retreat of kilometers per year. Science 368 (6494), 1020–1024. https://doi.org/10.1126/science.aaz3059.

Dowsett, H., Dolan, A., Rowley, D., Moucha, R., Forte, A.M., Mitrovica, J.X., et al., 2016. The PRISM4 (mid-Piacenzian) paleoenvironmental reconstruction. Climate of the Past 12 (7), 1519–1538.

Dowsett, H.J., Foley, K.M., Stoll, D.K., Chandler, M.A., Sohl, L.E., Bentsen, M., et al., 2013. Sea surface temperature of the mid-Piacenzian ocean: a data-model comparison. Scientific Reports 3. Available from: https://doi.org/10.1038/srep02013.

Dowsett, H.J., Robinson, M.M., Haywood, A.M., Hill, D.J., Dolan, A.M., Stoll, D.K., et al., 2012. Assessing confidence in Pliocene sea surface temperatures to evaluate predictive models. Nature Climate Change 2 (5), 365–371.

Dumitru, O.A., Austermann, J., Polyak, V.J., Fornós, J.J., Asmerom, Y., Ginés, J., et al., 2019. Constraints on global mean sea level during Pliocene warmth. Nature 574 (7777), 233–236. Available from: https://doi.org/10.1038/s41586-019-1543-2.

Dutton, A., Carlson, A.E., Long, A.J., Milne, G.A., Clark, P., DeConto, R., et al., 2015. Sea-level rise due to polar ice-sheet mass loss during past warm periods. Science 3491, 6244. https://doi.org/10.1126/science.aaa4019.

Dwyer, G.S., Chandler, M.A., 2009. Mid-Pliocene sea level and continental ice volume based on coupled benthic Mg/Ca palaeotemperatures and oxygen isotopes. Philosophical Transactions of the Royal Society A: Mathematical, Physical and Engineering Sciences 367 (1886), 157–168. Available from: https://doi.org/10.1098/rsta.2008.0222.

Edwards, T.L., Brandon, M.A., Durand, G., Edwards, N.R., Golledge, N.R., Holden, P.B., et al., 2019. Revisiting Antarctic ice loss due to marine ice-cliff instability. Nature 566 (7742), 58–64. Available from: https://doi.org/10.1038/s41586-019-0901-4.

Eittreim, S.L., Cooper, A.K., Wannesson, J., 1995. Seismic stratigraphic evidence of ice sheet advances on the Wilkes Land margin of Antarctica. Sedimentary Geology 96 (1–2), 131–156. Available from: https://doi.org/10.1016/0037-0738(94)00130-M.

Eldrett, J.S., Harding, I.C., Wilson, P.A., Butler, E., Roberts, A.P., 2007. Continental ice in Greenland during the Eocene and Oligocene. Nature 446 (7132), 176. Available from: https://doi.org/10.1038/nature05591.

EPICA Community Members, 2006. One-to-one coupling of glacial climate variability in Greenland and Antarctica. Nature 444, 195–198. Available from: https://doi.org/10.1038/nature05301.

Escutia, C., Bárcena, M.A., Lucchi, R.G., Romero, O., Ballegeer, M., Gonzalez, J.J., et al., 2009. Circum-Antarctic warming events between 4 and 3.5 Ma recorded in sediments from the Prydz Bay (ODP Leg 188) and the Antarctic Peninsula (ODP Leg 178) margins. Global and Planetary Change 69, 170–184. Available from: https://doi.org/10.1016/j.gloplacha.2009.09.003.

Escutia, C., H. Brinkhuis, A. Klaus, and the Expedition 318 Scientists, 2011. Wilkes Land Glacial History: Cenozoic East Antarctic Ice Sheet evolution from Wilkes Land margin sediments, in: Proceedings of the Integrated Ocean Drilling Program, vol. 318, Integrated Ocean Drilling Program Management International Inc., Tokyo. <https://doi.org/10.2204/iodp.proc.318.2011>.

Escutia, C., De Santis, L., Donda, F., Dunbar, R.B., Cooper, A.K., Brancolini, G., et al., 2005. Cenozoic ice sheet history from East Antarctic Wilkes Land continental margin sediments. Global and Planetary Change 45 (1–3), 51–81. Available from: https://doi.org/10.1016/j.gloplacha.2004.09.010.

Escutia, C., DeConto, R.M., Dunbar, R., Santis, L.D., Shevenell, A., Naish, T., 2019. Keeping an eye on Antarctic Ice Sheet stability. Oceanography 32 (1), 32–46.

Escutia, C., Eittreim, S.L., Cooper, A.K., 1997. Cenozoic glaciomarine sequences on the Wilkes Land continental rise, Antarctica. Proceedings Volume – VII International Symposium on Antarctic Earth Sciences 791–795.

Escutia, C., Warnke, D., Acton, G.D., Barcena, A., Burckle, L., Canals, M., et al., 2003. Sediment distribution and sedimentary processes across the Antarctic Wilkes Land margin during the Quaternary. Deep Sea Research Part II: Topical Studies in Oceanography 50 (8–9), 1481–1508. Available from: https://doi.org/10.1016/S0967-0645(03)00073-0.

Etourneau, J., Sgubin, G., Crosta, L., Swingedouw, D., Willmott, V., Barbara, L., et al., 2019. Ocean temperature impact on ice shelf extent in the eastern Antarctic Peninsula. Nature Communications 10, 304. Available from: https://doi.org/10.1038/s41467-018-08195-6.

Evangelinos, D., Escutia, C., Etourneau, J., Hoem, F., Bijl, P., Boterblom, W., et al., 2020. Late Oligocene-Miocene proto-Antarctic Circumpolar Current dynamics off the Wilkes Land margin, East Antarctica. Global and Planetary Change 191, 103221. Available from: https://doi.org/10.1016/j.gloplacha.2020.103221.

Fischer, H., Severinghaus, J., Brook, E., Wolff, E., Albert, M., Alemany, O., et al., 2013. Where to find 1.5 million yr old ice for the IPICS "Oldest-Ice" ice core. Climate of the Past 9 (6), 2489–2505.

Flower, B.P., Kennett, J.P., 1993. Middle Miocene ocean-climate transition: High-resolution oxygen and carbon isotopic records from Deep Sea Drilling Project Site 588A, southwest Pacific. Paleoceanography 8 (6), 811–843.

Foster, G.L., Lear, C.H., Rae, J.W.B., 2012. The evolution of pCO_2, ice volume and climate during the middle Miocene. Earth and Planetary Science Letters 341–344, 243–254. Available from: https://doi.org/10.1016/j.epsl.2012.06.007.

Fretwell, P., Pritchard, H.D., Vaughan, D.G., Bamber, J.L., Barrand, N.E., Bell, R., et al., 2013. Bedmap2: improved ice bed, surface and thickness datasets for Antarctica. The Cryosphere 7 (1), 375–393. Available from: https://doi.org/10.5194/tc-7-375-2013.

Fürst, J.J., Durand, G., Gillet-Chaulet, F., Tavard, L., Rankl, M., Braun, M., et al., 2016. The safety band of Antarctic ice shelves. Nature Climate Change 6 (5), 479–482. Available from: https://doi.org/10.1038/nclimate2912.

Galeotti, S., DeConto, R.M., Naish, T.R., Stocchi, P., Florindo, F., Pagani, M., et al., 2016. Antarctic Ice Sheet variability across the Eocene-Oligocene boundary climate transition. Science 352, 76–80. Available from: https://doi.org/10.1126/science.aab0669.

Galeotti, S., et al., 2021. The Eocene-Oligocene boundary climate transition: an Antarctic perspective. In: Florindo, F., et al. (Eds.), Antarctic Climate Evolution, second ed. Elsevier (this volume).

Ganopolski, A., Robinson, A., 2011. The past is not the future. Nature Geoscience 4 (10), 661–663. Available from: https://doi.org/10.1038/ngeo1268.

Gasson, E., DeConto, R., Pollard, D., 2015. Antarctic bedrock topography uncertainty and ice sheet stability. Geophysical Research Letters 42 (13), 5372–5377.

Gasson, E., DeConto, R.M., Pollard, D., 2016. Dynamic Antarctic ice sheet during the early to mid-Miocene. Proceedings of the National Academy of Sciences of the United States of America 113 (13), 3459–3464. Available from: https://doi.org/10.1073/pnas.1516130113.

Gasson, E.G.W., Keisling, B.A., 2020. The Antarctic Ice Sheet: a paleoclimate modeling perspec-tive. Oceanography 33 (2). Available from: https://doi.org/10.5670/oceanog.2020.208.

Gasson, E., Lunt, D.J., DeConto, R., Goldner, A., Heinemann, M., Huber, M., et al., 2014. Uncertainties in the modelled CO_2 threshold for Antarctic glaciation. Climate of the Past. Available from: https://doi.org/10.5194/cp-10-451-2014.

Gersonde, R., Crosta, X., Abelmann, A., Armand, L., 2005. Sea-surface temperature and sea ice distribution of the Southern Ocean at the EPILOG Last Glacial Maximum—a circum-Antarctic view based on siliceous microfossil records. Quaternary Science Reviews 24 (7–9), 869–896.

Goelzer, H., Huybrechts, P., Loutre, M.-F., Fichefet, T., 2016. Last Interglacial climate and sea-level evolution from a coupled ice sheet−climate model. Climate of the Past 12, 2195−2213. Available from: https://doi.org/10.5194/cp-12-2195-2016.

Gohl, K., Uenzelmann-Neben, G., Larter, R.D., Hillenbrand, C.D., Hochmuth, K., Kalberg, T., et al., 2013. Seismic stratigraphic record of the Amundsen Sea Embayment shelf from pre-glacial to recent times: evidence for a dynamic West Antarctic ice sheet. Marine Geology 344, 115−131. Available from: https://doi.org/10.1016/j.margeo.2013.06.011.

Gohl, K., Wellner, J.S., Klaus, A.the Expedition 379 Scientists, 2019. Expedition 379 preliminary report: Amundsen Sea West Antarctic Ice Sheet history. International Ocean Discovery Program . Available from: https://doi.org/10.14379/iodp.pr.379.2019.

Goldner, A., Herold, N., Huber, M., 2014. The challenge of simulating the warmth of the Mid-Miocene Climatic Optimun in CESM1. Climate of the Past . Available from: https://doi.org/10.5194/cp-10-523-2014.

Golledge, N.R., Fogwill, C.J., Mackintosh, A.N., Buckley, K.M., 2012. Dynamics of the last glacial maximum Antarctic ice-sheet and its response to ocean forcing. Proceedings of the National Academy of Sciences of the United States of America 109, 16052−16056. Available from: https://doi.org/10.1073/.02pnas.1205385109.

Golledge, N.R., Keller, E.D., Gomez, N., Naughten, K.A., Bernales, J., Trusel, L.D., et al., 2019. Global environmental consequences of twenty-first-century ice-sheet melt. Nature 566 (7742), 65−72. Available from: https://doi.org/10.1038/s41586-019-0889-9.

Golledge, N.R., Kowalewski, D.E., Naish, T.R., Levy, R.H., Fogwill, C.J., Gasson, E.G., 2015. The multi-millennial Antarctic commitment to future sea-level rise. Nature 526 (7573), 421−425. Available from: https://doi.org/10.1038/nature15706.

Golledge, N.R., Levy, R.H., McKay, R.M., Fogwill, C.J., White, D.A., Graham, A.G., et al., 2013. Glaciology and geological signature of the Last Glacial Maximum Antarctic ice sheet. Quaternary Science Reviews 78, 225−247.

Golledge, N.R., Levy, R.H., McKay, R.M., Naish, T.R., 2017b. East Antarctic ice sheet most vulnerable to Weddell Sea warming. Geophysical Research Letters 44 (5), 2343−2351.

Golledge, N.R., Menviel, L., Carter, L., Fogwill, C.J., England, M.H., Cortese, G., et al., 2014. Antarctic contribution to meltwater pulse 1A from reduced Southern Ocean overturning. Nature Communications 5, 5107. Available from: https://doi.org/10.1038/ncomms6107.

Golledge, N.R., Thomas, Z.A., Levy, R.H., Gasson, E.G., Naish, T.R., McKay, R.M., et al., 2017a. Antarctic climate and ice-sheet configuration during the early Pliocene interglacial at 4.23 Ma. Climate of the Past 13 (7). Available from: https://doi.org/10.5194/cp-13-959-2017.

Golledge, N.R., 2020. Long-term projections of sea-level rise from ice sheets. Wiley Interdisciplinary Reviews: Climate Change 11 (2), e634.

Gomez, N., Latychev, K., Pollard, D., 2018. A coupled ice sheet−sea level model incorporating 3D earth structure: variations in Antarctica during the last deglacial retreat. Journal of Climate 31 (10), 4041−4054.

Gomez, N., Mitrovica, J.X., Huybers, P., Clark, P.U., 2010. Sea level as a stabilizing factor for marine-ice-sheet grounding lines. Nature Geoscience. 3, 850−853.

Gomez, N., Pollard, D., Holland, D., 2015. Sea-level feedback lowers projections of future Antarctic Ice-Sheet mass loss. Nature Communications 6 (1), 1−8. Available from: https://doi.org/10.1038/ncomms9798.

Gomez, N., Pollard, D., Mitrovica, J.X., 2013. A 3-D coupled ice sheet−sea level model applied to Antarctica through the last 40 ky. Earth and Planetary Science Letters 384, 88−99. Available from: https://doi.org/10.1016/j.epsl.2013.09.042.

Gomez, N., Weber, M.E., Clark, P.U., Mitrovica, J.X., Han, H.K., 2020. Antarctic ice dynamics amplified by Northern Hemisphere sea-level forcing. Nature 587 (7835), 600–604.

Graham, A.G., Dutrieux, P., Vaughan, D.G., Nitsche, F.O., Gyllencreutz, R., Greenwood, S.L., et al., 2013. Seabed corrugations beneath an Antarctic ice shelf revealed by autonomous underwater vehicle survey: origin and implications for the history of Pine Island Glacier. Journal of Geophysical Research: Earth Surface 118 (3), 1356–1366.

Grant, G.R., Naish, T.R., Dunbar, G.B., Stocchi, P., Kominz, M.A., Kamp, P.J., et al., 2019. The amplitude and origin of sea-level variability during the Pliocene epoch. Nature 574 (7777), 237–241. Available from: https://doi.org/10.1038/s41586-019-1619-z.

Grant G., Naish T., 2021. Pliocene sea-level revisited: Is there more than meets the eye? PAGES. Available from: https://doi.org/10.22498/pages.29.1.34.

Grant, K.M., Rohling, E.J., Ramsey, C.B., Cheng, H., Edwards, R.L., Florindo, F., et al., 2014. Sea-level variability over five glacial cycles. Nature communications 5 (1), 1–9. Available from: https://doi.org/10.1038/ncomms6076.

Green, R.A., et al., 2020. "Evaluating seasonal sea-ice cover over the Southern Ocean from the Last Glacial Maximum". Climate of the Past Discussions 1–23.

Greenop, R., Foster, G.L., Wilson, P.A., Lear, C.H., 2014. Middle Miocene climate instability associated with high-amplitude CO_2 variability. Paleoceanography 29. Available from: https://doi.org/10.1002/2014PA002653.

Gulick, S.P., Shevenell, A.E., Montelli, A., Fernandez, R., Smith, C., Warny, S., et al., 2017. Initiation and long-term instability of the East Antarctic Ice Sheet. Nature 552 (7684), 225–229. Available from: https://doi.org/10.1038/nature25026.

Halberstadt, A.R.W., Simkins, L.M., Greenwood, S.L., Anderson, J.B., 2016. Past ice-sheet behaviour: retreat scenarios and changing controls in the Ross Sea, Antarctica. The Cryosphere 10, 1003–1020. Available from: https://doi.org/10.5194/tc-10-1003-2016.

Hambrey, M.J., W.U. Ehrmann, B. Larsen, 1991. Cenozoic glacial record of the Prydz Bay continental shelf, East Antarctica. In: Barron, J., Larsen, B., et al. (Eds.), Proceedings of the Ocean Drilling Program Scientific Results, vol. 119. College Station, TX, pp. 77–132. <https://doi.org/10.2973/odp.proc.sr.119.200.1991>.

Hannah, M.J., 2006. The palynology of ODP site 1165, Prydz Bay, East Antarctica: a record of Miocene glacial advance and retreat. Palaeogeography, Palaeoclimatology, Palaeoecology 231 (1–2), 120–133. Available from: https://doi.org/10.1016/j.palaeo.2005.07.029.

Hansen, M.A., Passchier, S., Khim, B.-K., Song, B., Williams, T., 2015. Threshold behavior of a marine-based sector of the East Antarctic Ice Sheet in response to early Pliocene ocean warming. Paleoceanography 30. Available from: https://doi.org/10.1002/2014PA002704.

Hansen, M.A., Passchier, S., 2017. Oceanic circulation changes during early Pliocene marine ice-sheet instability in Wilkes Land, East Antarctica. Geo-Marine Letters. Available from: https://doi.org/10.1007/s00367-016-0489-8.

Hartman, J.D., Sangiorgi, F., Peterse, F., Barcena, M.A., Albertazzi, S., Asioli, A., et al., 2016. Phytoplankton assemblages and lipid biomarkers indicate sea-surface warming and sea-ice decline in the Ross Sea during Marine Isotope sub-Stage 5e. *EGUGA*, EPSC2016-2637.

Hartman, J.D., Sangiorgi, F., Salabarnada, A., Peterse, F., Houben, A.J., Schouten, S., et al., 2018. Paleoceanography and ice sheet variability offshore Wilkes Land, Antarctica-Part 3: insights from Oligocene-Miocene TEX86-based sea surface temperature reconstructions. Climate of the Past 14 (9), 1275–1297. Available from: https://doi.org/10.5194/cp-14-1275-2018.

Hauptvogel, D.W., Passchier, S., 2012. Early–Middle Miocene (17–14 Ma) Antarctic ice dynamics reconstructed from the heavy mineral provenance in the AND-2A drill core, Ross

Sea, Antarctica. Global and Planetary Change 82, 38–50. Available from: https://doi.org/10.1016/j.gloplacha.2011.11.003.

Haywood, A.M., Dowsett, H.J., Dolan, A.M., 2016. Integrating geological archives and climate models for the mid-Pliocene warm period. Nature Communications 7 (1), 1–14. Available from: https://doi.org/10.1038/ncomms10646.

Haywood, A.M., Hill, D.J., Dolan, A.M., Otto-Bliesner, B.L., Bragg, F., Chan, W.L., et al., 2013. Large-scale features of Pliocene climate: results from the Pliocene Model Intercomparison Project. Climate of the Past 9 (1), 191–209.

Haywood, A.M., Tindall, J.C., Dowsett, H.J., Dolan, A.M., Foley, K.M., Hunter, S.J., et al., 2020. A return to large-scale features of Pliocene climate: the Pliocene Model Intercomparison Project Phase 2, Climate of the Past Discussions, in review. <https://doi.org/10.5194/cp-2019-145>.

Haywood, A.M., Valdes, P.J., Aze, T., Barlow, N., Burke, A., Dolan, A.M., et al., 2019. What can Palaeoclimate Modelling do for you? Earth Systems and Environment 3 (1), 1–18. Available from: https://doi.org/10.1007/s41748-019-00093-1.

Hearty, P.J., Kindler, P., Cheng, H., Edwards, R.L., 1999. A +20 m middle Pleistocene sea-level highstand (Bermuda and the Bahamas) due to partial collapse of Antarctic ice. Geology 27 (4), 375–378.

Hearty, P.J., Rovere, A., Sandstrom, M.R., O'Leary, M.J., Roberts, D., Raymo, M.E., 2020. Pliocene-Pleistocene Stratigraphy and Sea-Level Estimates, Republic of South Africa With Implications for a 400 ppmv CO_2 World. Paleoceanography and Paleoclimatology 35 (7), e2019PA003835.

Herbert, T.D., Lawrence, K.T., Tzanova, A., Peterson, L.C., Caballero-Gill, R., Kelly, C.S., 2016. Late Miocene global cooling and the rise of modern ecosystems. Nature Geoscience 9 (11), 843–847. Available from: https://doi.org/10.1038/ngeo2813.

Herold, N., Seton, M., Müller, R.D., You, Y., Huber, M., 2008. Middle Miocene tectonic boundary conditions for use in climate models. Geochemistry, Geophysics, Geosystems 9, Q10009. Available from: https://doi.org/10.1029/2008GC002046.

Hillenbrand, C.-D., Bentley, M.J., Stolldorf, T.D., Hein, A.S., Kuhn, G., Graham, A.G.C., et al., 2014. Reconstruction of changes in the Weddell Sea sector of the Antarctic Ice Sheet since the Last Glacial Maximum. Quaternary Science Review. 100, 111–136. Available from: https://doi.org/10.1016/j.quascirev.2013.07.020.

Hillenbrand, C.D., Ehrmann, W., 2005. Late Neogene to Quaternary environmental changes in the Antarctic Peninsula region: evidence from drift sediments. Global and Planetary Change 45 (1–3), 165–191.

Hillenbrand, C.D., Kuhn, G., Frederichs, T., 2009. Record of a Mid-Pleistocene depositional anomaly in West Antarctic continental margin sediments: an indicator for ice-sheet collapse? Quaternary Science Reviews 28 (13–14), 1147–1159.

Hillenbrand, C.D., Melles, M., Kuhn, G., Larter, R.D., 2012. Marine geological constraints for the grounding-line position of the Antarctic Ice Sheet on the southern Weddell Sea shelf at the Last Glacial Maximum. Quaternary Science Reviews 32, 25–47.

Hillenbrand, C.D., Smith, J.A., Hodell, D.A., Greaves, M., Poole, C.R., Kender, S., et al., 2017. West Antarctic Ice Sheet retreat driven by Holocene warm water incursions. Nature 547 (7661), 43–48. Available from: https://doi.org/10.1038/nature22995.

Hochmuth, K., Gohl, K., 2019. "Seaward growth of Antarctic continental shelves since establishment of a continent-wide ice sheet: Patterns and mechanisms.". Palaeogeography, Palaeoclimatology, Palaeoecology 520, 44–54. Available from: https://doi.org/10.1016/j.palaeo.2019.01.025.

Hochmuth, K., Gohl, K., Leitchenkov, G., Sauermilch, I., Whittaker, J.M., Uenzelmann-Neben, G., et al., 2020. The evolving paleobathymetry of the circum-Antarctic Southern Ocean since 34 Ma – a key to understanding past cryosphere-ocean developments. Geochemistry, Geophysics, Geosystems. Available from: https://doi.org/10.1029/2020GC009122.

Hodell, D.A., Charles, C.D., Ninnemann, U.S., 2000. Comparison of interglacial stages in the South Atlantic sector of the southern ocean for the past 450 kyr: implifications for Marine Isotope Stage (MIS) 11. Global and Planetary Change 24 (1), 7–26.

Hodgson, D.A., Graham, A.G., Roberts, S.J., Bentley, M.J., Cofaigh, C.O., Verleyen, E., et al., 2014. Terrestrial and submarine evidence for the extent and timing of the Last Glacial Maximum and the onset of deglaciation on the maritime-Antarctic and sub-Antarctic islands. Quaternary Science Reviews 100, 137–158.

Hoffman, J.S., Clark, P.U., Parnell, A.C., He, F., 2017. Regional and global sea-surface temperatures during the last interglaciation. Science 355 (6322), 276–279. Available from: https://doi.org/10.1126/science.aai8464.

Holbourn, A., Kuhnt, W., Clemens, S., Prell, W., Andersen, N., 2013. Middle to late Miocene stepwise climate cooling: Evidence from a high-resolution deep water isotope curve spanning 8 million years. Paleoceanography 28 (4), 688–699.

Holbourn, A., Kuhnt, W., Lyle, M., Schneider, L., Romero, O., Andersen, N., 2014. Middle Miocene climate cooling linked to intensification of eastern equatorial Pacific upwelling. Geology 42 (1), 19–22. Available from: https://doi.org/10.1130/G34890.1.

Holbourn, A., Kuhnt, W., Schulz, M., Flores, J.A., Andersen, N., 2007. Orbitally-paced climate evolution during the middle Miocene "Monterey" carbon-isotope excursion. Earth and Planetary Science Letters 261 (3-4), 534–550.

Holloway, M.D., Sime, L.C., Singarayer, J.S., Tindall, J.C., Bunch, P., Valdes, P.J., 2016. Antarctic last interglacial isotope peak in response to sea ice retreat not ice-sheet collapse. Nature Communications 7 (1), 1–9. Available from: https://doi.org/10.1038/ncomms12293.

Honisch, B., Hemming, G., Archer, D., Siddal, M., McManus, J., 2009. Atmospheric carbon dioxide concentration across the Mid-Pleistocene Transition. Science 324, 1551–1554. Available from: https://doi.org/10.1126/science.1171477.

Horgan, H.J., Christianson, K., Jacobel, R.W., Anandakrishnan, S., Alley, R.B., 2013. Sediment deposition at the modern grounding zone of Whillans Ice Stream, West Antarctica. Geophysical Research Letters 40 (15), 3934–3939.

Huang, X., Gohl, K., Jokat, W., 2014. Variability in Cenozoic sedimentation and paleo-water depths of the Weddell Sea basin related to pre-glacial and glacial conditions of Antarctica. Global and Planetary Change 118, 25–41. Available from: https://doi.org/10.1016/j.gloplacha.2014.03.010.

Huang, X., Jokat, W., 2016. Middle Miocene to present sediment transport and deposits in the Southeastern Weddell Sea, Antarctica. Global and Planetary Change 139, 211–225.

Huang, X., Stärz, M., Gohl, K., Knorr, G., Lohmann, G., 2017. Impact of Weddell Sea shelf progradation on Antarctic bottom water formation during the Miocene. Paleoceanography 32 (3), 304–317. Available from: https://doi.org/10.1002/2016PA002987.

Huybrechts, P., 2002. Sea-level changes at the LGM from ice-dynamic reconstructions of the Greenland and Antarctic ice sheets during the glacial cycles. Quaternary Science Review 21, 203–231. Available from: https://doi.org/10.1016/S0277-3791(01)00082-8.

Imbrie J., Hays J.D., Martinson D.G., McIntyre A., Mix A.C., Morely J.J., et al., 1984. The orbital theory of Pleistocene climate: support from a revised chronology of the marined 18O record. In: Berger, A.L. (Ed.), Milankovitch and Climate, vol. 1. D. Reidel, Dordrecht, pp. 269–305.

IPCC A.R.5., 2013. Climate Change (2013): The Physical Science Basis. Contribution of Working Group I to the Fifth Assessment Report of the Intergovernmental Panel on Climate Change [T. F. Stocker, D. Qin, G.-K. Plattner, M. Tignor, S.K. Allen, J. Boschung, A. Nauels, Y. Xia, V. Bex, and P.M. Midgley (Eds.)]. Cambridge University Press, Cambridge, United Kingdom, and New York, pp. 1535. <https://doi.org/10.1017/CBO9781107415324>.

IPCC, S.R.O.C.C., 2019. IPCC Special Report on the Ocean and Cryosphere in a Changing Climate [H.-O. Pörtner, D.C. Roberts, V. Masson-Delmotte, P. Zhai, M. Tignor, E. Poloczanska, K. Mintenbeck, A. Alegría, M. Nicolai, A. Okem, J. Petzold, B. Rama, N.M. Weyer (Eds.)].

Ivins, E.R., James, T.S., 2005. Antarctic glacial isostatic adjustment: a new assessment. Antarctic Science 17 (4), 541. Available from: https://doi.org/10.1017/S0954102005002968.

Ivins, E.R., James, T.S., Wahr, J., O. Schrama, E.J., Landerer, F.W., Simon, K.M., 2013. Antarctic contribution to sea level rise observed by GRACE with improved GIA correction. Journal of Geophysical Research: Solid Earth 118 (6), 3126–3141. Available from: https://doi.org/10.1002/jgrb.50208.

Jakobsson, M., Anderson, J.B., Nitsche, F.O., Dowdeswell, J.A., Gyllencreutz, R., Kirchner, N., et al., 2011. Geological record of ice shelf break-up and grounding line retreat, Pine Island Bay, West Antarctica. Geology 39 (7), 691–694.

Jakobsson, M., Anderson, J.B., Nitsche, F.O., Gyllencreutz, R., Kirshner, A.E., Kirchner, N., et al., 2012. Ice sheet retreat dynamics inferred from glacial morphology of the central Pine Island Bay Trough, West Antarctica. Quaternary Science Reviews 38, 1–10.

Jakob, K.A., Wilson, P.A., Pross, J., Ezard, T.H., Fiebig, J., Repschläger, J., et al., 2020. A new sea-level record for the Neogene/Quaternary boundary reveals transition to a more stable East Antarctic Ice Sheet. Proceedings of the National Academy of Sciences 117 (49), 30980–30987.

Jamieson, S.S., Vieli, A., Cofaigh, C.Ó., Stokes, C.R., Livingstone, S.J., Hillenbrand, C.D., 2014. Understanding controls on rapid ice-stream retreat during the last deglaciation of Marguerite Bay, Antarctica, using a numerical model. Journal of Geophysical Research: Earth Surface 119 (2), 247–263.

Jamieson, S.S., Vieli, A., Livingstone, S.J., Cofaigh, C.Ó., Stokes, C., Hillenbrand, C.D., et al., 2012. Ice-stream stability on a reverse bed slope. Nature Geoscience 5 (11), 799–802. Available from: https://doi.org/10.1038/ngeo1600.

Jansen, E., Fronval, T., Rack, F., Channell, J.E., 2000. Pliocene-Pleistocene ice rafting history and cyclicity in the Nordic Seas during the last 3.5 Myr. Paleoceanography 15 (6), 709–721.

Jansen, E., et al., 2007. In: Solomon, S.D., et al., (Eds.), Climate Change 2007: The Physical Science Basis. Cambridge University Press, pp. 433–497.

Johnson, J.S., Bentley, M.J., Smith, J.A., Finkel, R.C., Rood, D.H., Gohl, K., et al., 2014. Rapid thinning of Pine Island Glacier in the early Holocene. Science 343 (6174), 999–1001.

Jones, R.S., Mackintosh, A.N., Norton, K.P., Golledge, N.R., Fogwill, C.J., Kubik, P.W., et al., 2015. Rapid Holocene thinning of an East Antarctic outlet glacier driven by marine ice sheet instability. Nature Communications 6 (1), 1–9.

Joughin, I., Alley, R.B., 2011. Stability of the West Antarctic ice sheet in a warming world. Nature Geoscience 4 (8), 506–513. Available from: https://doi.org/10.1038/ngeo1194.

Jouzel, J., Masson-Delmotte, V., Cattani, O., Dreyfus, G., Falourd, S., Hoffmann, G., et al., 2007. Orbital and millennial Antarctic climate variability over the past 800,000 years. Science 317 (5839), 793–796. Available from: https://doi.org/10.1126/science.1141038.

Justino, F., Kucharski, F., Lindemann, D., Wilson, A., Stordal, F., 2019. A modified seasonal cycle during MIS31 super-interglacial favors stronger interannual ENSO and monsoon variability. Climate of the Past 15 (2), 735–749. Available from: https://doi.org/10.5194/cp-15-735-2019.

Justino, F., Lindemann, D., Kucharski, F., Wilson, A., Bromwich, D., Stordal, F., 2017. Oceanic response to changes in the WAIS and astronomical forcing during the MIS31 superinterglacial. Climate of the Past 13, 1081–1095. Available from: https://doi.org/10.5194/cp-13-1081-2017.

Kageyama, M., Braconnot, P., Harrison, S.P., Haywood, A.M., Jungclaus, J.H., Otto-Bliesner, B. L., et al., 2018. The PMIP4 contribution to CMIP6 – Part 1: overview and over-arching analysis plan. Geoscientific Model Development 11 (3), 1033–1057.

Karas, C., Khélifi, N., Bahr, A., Naafs, B.D.A., Nürnberg, D., Herrle, J.O., 2020. Did North Atlantic cooling and freshening from 3.65–3.5 Ma precondition Northern Hemisphere ice sheet growth? Global and Planetary Change 185, 103085.

Katz, M.E., Miller, K.G., Wright, J.D., Wade, B.S., Browning, J.V., Cramer, B.S., et al., 2008. Stepwise transition from the Eocene greenhouse to the Oligocene icehouse. Nature Geoscience 1 (5), 329–334. Available from: https://doi.org/10.1038/ngeo179.

Kemp, A.E.S., Grigorov, I., Pearce, R.B., Garabato, A.N., 2010. Migration of the Antarctic Polar Front through the mid-Pleistocene transition: evidence and climatic implications. Quaternary Science Reviews 29 (17–18), 1993–2009. Available from: https://doi.org/10.1016/j.quascirev.2010.04.027.

Kennedy, A.T., Farnsworth, A., Lunt, D.J., Lear, C.H., Markwick, P.J., 2015. Atmospheric and oceanic impacts of Antarctic glaciation across the Eocene–Oligocene transition. Philosophical Transactions of the Royal Society A 373. Available from: https://doi.org/10.1098/rsta.2014.041920140419.

Kennett, J.P., 1977. Cenozoic evolution of Antarctic glaciation, the circum-Antarctic Ocean, and their impact on global paleoceanography. Journal of Geophysical Research 82 (27), 3843–3860.

Kim, S., De Santis, L., Hong, J.K., Cottlerle, D., Petronio, L., Colizza, E., et al., 2018. Seismic stratigraphy of the Central Basin in northwestern Ross Sea slope and rise, Antarctica: clues to the late Cenozoic ice-sheet dynamics and bottom-current activity. Marine Geology 395, 363–379. Available from: https://doi.org/10.1016/j.margeo.2017.10.013.

Kindler, P., Hearty, P.J., 2000. Elevated marine terraces from Eleuthera (Bahamas) and Bermuda: sedimentological, petrographic and geochronological evidence for important deglaciation events during the middle Pleistocene. Global and Planetary Change 24 (1), 41–58.

Kingslake, J., Scherer, R.P., Albrecht, T., Coenen, J., Powell, R.D., Reese, R., et al., 2018. Extensive retreat and re-advance of the West Antarctic Ice Sheet during the Holocene. Nature 558 (7710), 430–434. Available from: https://doi.org/10.1038/s41586-018-0208-x.

King, A.L., Howard, W.R., 2000. Middle Pleistocene sea-surface temperature change in the southwest Pacific Ocean on orbital and suborbital time scales. Geology 28 (7), 659–662.

Kirkham, J.D., Hogan, K.A., Larter, R.D., Arnold, N.S., Nitsche, F.O., Golledge, N.R., et al., 2019. Past water flow beneath Pine Island and Thwaites glaciers, West Antarctica. The Cryosphere 13, 1959–1981. Available from: https://doi.org/10.5194/tc-13-1959-2019.

Klages, J.P., Kuhn, G., Graham, A.G., Hillenbrand, C.D., Smith, J.A., Nitsche, F.O., et al., 2015. Palaeo-ice stream pathways and retreat style in the easternmost Amundsen Sea Embayment, West Antarctica, revealed by combined multibeam bathymetric and seismic data. Geomorphology 245, 207–222.

Klages, J.P., Kuhn, G., Hillenbrand, C.D., Smith, J.A., Graham, A.G., Nitsche, F.O., et al., 2017. Limited grounding-line advance onto the West Antarctic continental shelf in the easternmost Amundsen Sea Embayment during the last glacial period. PLoS One 12 (7), e0181593.

Kominz, M.A., Browning, J.V., Miller, K.G., Sugarman, P.J., Mizintseva, S., Scotese, C.R., 2008. Late Cretaceous to Miocene sea-level estimates from the New Jersey and Delaware coastal plain coreholes: an error analysis. Basin Research 20 (2), 211–226. Available from: https://doi.org/10.1111/j.1365-2117.2008.00354.x.

Konfirst, M.A., Scherer, R.P., Hillenbrand, C.D., Kuhn, G., 2012. A marine diatom record from the Amundsen Sea—insights into oceanographic and climatic response to the Mid-Pleistocene transition in the West Antarctic sector of the Southern Ocean. Marine Micropaleontology 92, 40–51.

Konrad, H., Thoma, M., Sasgen, I., Klemann, V., Grosfeld, K., Barbi, D., et al., 2014. The deformational response of a viscoelastic solid earth model coupled to a thermomechanical ice sheet model. Surveys in Geophysics. 35, 1441–1458. Available from: https://doi.org/10.1007/s10712-013-9257-8.

Kristoffersen, Y., Jokat, W., 2008. The Weddell Sea. In: Cooper et al., Cenozoic climate history from seismic-reflection and drilling studies on the Antarctic continental margin. In: Florindo, F., Siegert, M. (Eds.), Antarctic Climate Evolution, vol. 8. Elsevier, pp. 144–152.

Kriwet, J., Engelbrecht, A., Mörs, T., Reguero, M., Pfaff, C., 2016. Ultimate Eocene (Priabonian) chondrichthyans (Holocephali, Elasmobranchii) of Antarctica. Journal of Vertebrate Paleontology 36 (4), e1160911.

Kuhn, G., Hillenbrand, C.D., Kasten, S., Smith, J.A., Nitsche, F.O., Frederichs, T., et al., 2017. Evidence for a palaeo-subglacial lake on the Antarctic continental shelf. Nature Communications 8 (1), 1–10.

Kulpecz, A.A., Miller, K.G., Browning, J.V., Edwards, L.E., Powars, D.S., McLaughlin, P.P., Jr., et al., 2009. Post-impact deposition in the Chesapeake Bay impact structure: Variations in eustasy, compaction, sediment supply, and passive-aggressive tectonism. In: Gohn, G.S., et al. (Eds.), The ICDP-USGS Deep Drilling Project in the Chesapeake Bay Impact Structure: Results from the Eyreville Core Holes: Geological Society of America Special Paper 458, pp. 811–837. <https://doi.org/10.1130/2009.2458(34)>.

Kunz-Pirrung, M., Gersonde, R., Hodell, D.A., 2002. Mid-Brunhes century-scale diatom sea surface temperature and sea ice records from the Atlantic sector of the Southern Ocean (ODP Leg 177, sites 1093, 1094 and core PS2089-2). Palaeogeography, Palaeoclimatology, Palaeoecology 182 (3–4), 305–328. Available from: https://doi.org/10.1016/S0031-0182(01)00501-6.

Ladant, J.B., Donnadieu, Y., Lefebvre, V., Dumas, C., 2014. The respective role of atmospheric carbon dioxide and orbital parameters on ice sheet evolution at the Eocene-Oligocene transition. Paleoceanography 29 (8), 810–823. Available from: https://doi.org/10.1002/2013PA002593.

Lambeck, K., Chappell, J., 2001. Sea level change through the last glacial cycle. Science 292, 679–686. Available from: https://doi.org/10.1126/science.1059549.

Lambeck, K., Rouby, H., Purcell, A., Sun, Y., Sambridge, M., 2014. Sea level and global ice volumes from the Last Glacial Maximum to the Holocene. Proceedings of the National Academy of Sciences 111 (43), 15296–15303. Available from: https://doi.org/10.1073/pnas.1411762111.

Lambeck, K., Yokoyama, Y., Purcell, A., 2002. Into and out of the Last glacial Maximum sea level change during Oxygen Isotope Stages 3–2. Quaternary Science Reviews 21, 343–360.

Lamy, F., Chiang, J.C.H., Martínez-Méndez, G., Thierens, M., Arz, H.W., Bosmans, J., et al., 2019. Precession modulation of the South Pacific westerly wind belt over the past million years. Proceedings of the National Academy of Sciences 116 (47), 23455–23460. Available from: https://doi.org/10.1073/pnas.1905847116.

Lamy, F., Gersonde, R., Winckler, G., Esper, O., Jaeschke, A., Kuhn, G., et al., 2014. Increased dust deposition in the Pacific Southern Ocean during glacial periods. Science 343 (6169), 403–407.

Langebroek, P.M., Paul, A., Schulz, M., 2009. Antarctic ice-sheet response to atmospheric CO_2 and insolation in the Middle Miocene. Climate of the Past 5 (4). Available from: https://doi.org/10.5194/cp-5-633-2009.

Lang, N., Wolff, E.W., 2011. Interglacial and glacial variability from the last 800 ka in marine, ice and terrestrial archives. Climate of the Past 7 (2), 361–380. Available from: https://doi.org/10.5194/cp-7-361-2011.

Larter, R.D., Anderson, J.B., Graham, A.G., Gohl, K., Hillenbrand, C.D., Jakobsson, M., et al., 2014. Reconstruction of changes in the Amundsen Sea and Bellingshausen sea sector of the West Antarctic ice sheet since the last glacial maximum. Quaternary Science Reviews 100, 55–86.

Larter, R.D., Hogan, K.A., Hillenbrand, C.D., Smith, J.A., Batchelor, C.L., Cartigny, M., et al., 2019. Subglacial hydrological control on flow of an Antarctic Peninsula palaeo-ice stream. Cryosphere 13 (6), 1583–1596. Available from: https://doi.org/10.5194/tc-13-1583-2019.

Larter, R.D., Rebesco, M., Vanneste, L.E., Gambôa, L.A.P., Barker, P.F., 1997. Cenozoic Tectonic, Sedimentary and Glacial History of the Continental Shelf West of Graham Land, Antarctic Peninsula. In: Barker, P.F., Cooper, A.K. (Eds.), Geology and Seismic Stratigraphy of the Antarctic Margin, vol. 2. American Geophysical Union, pp. 1–27.

Lear, C.H., Bailey, T.R., Pearson, P.N., Coxall, H.K., Rosenthal, Y., 2008. Cooling and ice growth across the Eocene-Oligocene transition. Geology 36 (3), 251–254. Available from: https://doi.org/10.1130/g24584a.1.

Lear, C.H., Coxall, H.K., Foster, G.L., Lunt, D.J., Mawbey, E.M., Rosenthal, Y., et al., 2015. Neogene ice volume and ocean temperatures: insights from infaunal foraminiferal Mg/Ca paleothermometry. Paleoceanography 30 (11), 1437–1454.

Lee, J.I., McKay, R.M., Golledge, N.R., Yoon, H.I., Yoo, K.C., Kim, H.J., et al., 2017. Widespread persistence of expanded East Antarctic glaciers in the southwest Ross Sea during the last deglaciation. Geology 45 (5), 403–406.

Levy, R., Harwood, D., Florindo, F., Sangiorgi, F., Tripati, R., von Eynatten, H., et al.,SMS Science Team 2016. Early to mid-Miocene Antarctic Ice Sheet dynamics. Proceedings of the National Academy of Sciences Mar 2016 113 (13), 3453–3458. Available from: https://doi.org/10.1073/pnas.1516030113.

Levy, R.H., Meyers, S.R., Naish, T.R., Golledge, N.R., McKay, R.M., Crampton, J.S., et al., 2019. Antarctic ice-sheet sensitivity to obliquity forcing enhanced through ocean connections. Nature Geoscience 12, 132–137. Available from: https://doi.org/10.1038/s41561-018-0284-4.

Levy, R.H., et al., 2021. Antarctic environmental change and ice sheet evolution through the Miocene to Pliocene - a perspective from the Ross Sea and George V to Wilkes Land Coasts. In: Florindo, F., et al. (Eds.), Antarctic Climate Evolution, second edition. Elsevier (this volume).

Lewis, A.R., Ashworth, A.C., 2016. An early to middle Miocene record of ice-sheet and landscape evolution from the Friis Hills, Antarctica. GSA Bulletin 128 (5–6), 719–738.

Lewis, A.R., Marchant, D.R., Ashworth, A.C., Hedenäs, L., Hemming, S.R., Johnson, J.V., et al., 2008. Mid-Miocene cooling and the extinction of tundra in continental Antarctica. Proceedings of the National Academy of Sciences 105 (31), 10676–10680.

Liakka, J., Colleoni, F., Ahrens, B., Hickler, T., 2014. The impact of climate-vegetation interactions on the onset of the Antarctic ice sheet. Geophysical Research Letters 41 (4), 1269–1276. Available from: https://doi.org/10.1002/2013GL058994.

Licht, K.J., 2004. The Ross Sea's contribution to eustatic sea level during meltwater pulse 1A. Sedimentary Geology 165 (3–4), 343–353.

Lindeque, A., Gohl, K., Henrys, S., Wobbe, F., Davy, B., 2016. Seismic stratigraphy along the Amundsen Sea to Ross Sea continental rise: A cross-regional record of pre-glacial to glacial processes of the West Antarctic margin. Palaeogeography, Palaeoclimatology, Palaeoecology 443, 183–202.

Lisiecki, L.E., Raymo, M.E., 2005. A Pliocene-Pleistocene stack of 57 globally distributed benthic $\delta^{18}O$ records. Paleoceanography 20 (1). Available from: https://doi.org/10.1029/2004PA001071.

Liu, Z., He, Y., Jiang, Y., Wang, H., Liu, W., Bohaty, S.M., et al., 2018. Transient temperature asymmetry between hemispheres in the Palaeogene Atlantic Ocean. Nature Geoscience 11 (9), 656–660. Available from: https://doi.org/10.1038/s41561-018-0182-9.

Liu, J., Milne, G.A., Kopp, R.E., Clark, P.U., Shennan, I., 2016. Sea-level constraints on the amplitude and source distribution of Meltwater Pulse 1A. Nature Geoscience 9 (2), 130–134.

Liu, Z., Pagani, M., Zinniker, D., DeConto, R., Huber, M., Brinkhuis, H., et al., 2009. Global cooling during the Eocene–Oligocene climate transition. Science 323, 1187–1190.

Livingstone, S.J., O'Cofaigh, C., Stokes, C.R., Hillenbrand, C.-D., Vieli, A., Jamieson, S.S.R., 2013. Glacial geomorphology of Marguerite Bay Palaeo-Ice stream, western Antarctic Peninsula. Journal of Maps 9, 558–572.

Loutre, M.F., Berger, A., 2003. Marine Isotope Stage 11 as an analogue for the present interglacial. Global and Planetary Change 36 (3), 209–217. Available from: https://doi.org/10.1016/S0921-8181(02)00186-8.

Lowry, D.P., Golledge, N.R., Bertler, N.A., Jones, R.S., McKay, R., 2019. Deglacial grounding-line retreat in the Ross Embayment, Antarctica, controlled by ocean and atmosphere forcing. Science Advances 5 (8), eaav8754. Available from: https://doi.org/10.1126/sciadv.aav8754.

Lunt, D.J., Haywood, A.M., Schmidt, G.A., Salzmann, U., Valdes, P.J., Dowsett, H.J., et al., 2012. On the causes of mid-Pliocene warmth and polar amplification. Earth and Planetary Science Letters 321, 128–138.

Lunt, D.J., Huber, M., Anagnostou, E., Baatsen, M.L., Caballero, R., DeConto, R., et al., 2017. The DeepMIP contribution to PMIP4: Experimental design for model simulations of the EECO, PETM, and pre-PETM (version 1.0). Geoscientific Model Development 10 (2), 889–901.

Lüthi, D., Le Floch, M., Bereiter, B., Blunier, T., Barnola, J.M., Siegenthaler, U., et al., 2008. High-resolution carbon dioxide concentration record 650,000–800,000 years before present. Nature 453 (7193), 379–382.

Lythe, M.B., Vaughan, D.G., 2001. BEDMAP: A new ice thickness and subglacial topographic model of Antarctica. Journal of Geophysical Research: Solid Earth 106 (B6), 11335–11351.

Mackintosh, A., Golledge, N., Domack, E., Dunbar, R., Leventer, A., White, D., et al., 2011. Retreat of the East Antarctic ice sheet during the last glacial termination. Nature Geoscience 4 (3), 195–202. Available from: https://doi.org/10.1038/ngeo1061.

Mackintosh, A.N., Verleyen, E., O'Brien, P.E., White, D.A., Jones, R.S., McKay, R., et al., 2014. Retreat history of the East Antarctic Ice Sheet since the last glacial maximum. Quaternary Science Reviews 100, 10–30.

Marino, G., Rohling, E.J., Rodríguez-Sanz, L., Grant, K.M., Heslop, D., Roberts, A.P., et al., 2015. Bipolar seesaw control on last interglacial sea level. Nature 522 (7555), 197–201. Available from: https://doi.org/10.1038/nature14499.

Maris, M.N.A., De Boer, B., Ligtenberg, S.R.M., Crucifix, M., Van de Berg, W.J., Oerlemans, J., 2014. Modelling the evolution of the Antarctic ice sheet since the last interglacial. The Cryosphere 8 (4), 1347–1360.

Martín-Español, A., King, M.A., Zammit-Mangion, A., Andrews, S.B., Moore, P., Bamber, J.L., 2016. An assessment of forward and inverse GIA solutions for Antarctica. Journal of Geophysical Research: Solid Earth 121 (9), 6947–6965.

Martinez-Boti, M.A., Foster, G.L., Chalk, T.B., Rohling, E.J., Sexton, P.F., et al., 2015. Plio-Pleistocene climate sensitivity evaluated using high-resolution CO_2 records. Nature 518, 49–54. Available from: https://doi.org/10.1038/nature14145.

Mas e Braga, M., Bernales, J., Prange, M., Stroeven, A.P., Rogozhina, I., 2021. Sensitivity of the Antarctic ice sheets to the peak warming of Marine Isotope Stage 11. The Cryosphere 15, 459–478. Available from: https://doi.org/10.5194/tc-2020-112.

Masson-Delmotte, V., Schulz, M., Abe-Ouchi, A., Beer, J., Ganopolski, A., Gonzáles Rouco, J.F., et al., 2013. Information from paleoclimate archives. In: Stocker, T.F., Qin, D., Plattner, G.K., Tignor, M., Allen, S.K., Boschung, J., Nauels, A., Xia, Y., Bex, V., Midgley, P.M. (Eds.), Climate Change 2013: The Physical Science Basis. Contribution of Working Group I to the Fifth Assessment Report of the Intergovernmental Panel on Climate Change. Cambridge University Press, Cambridge, United Kingdom and New York.

Matsuoka, K., Hindmarsh, R.C., Moholdt, G., Bentley, M.J., Pritchard, H.D., Brown, J., et al., 2015. Antarctic ice rises and rumples: Their properties and significance for ice-sheet dynamics and evolution. Earth-Science Reviews 150, 724–745.

McKay, R., Browne, G., Carter, L., Cowan, E., Dunbar, G., Krissek, L., et al., 2009. The stratigraphic signature of the late Cenozoic Antarctic Ice Sheets in the Ross Embayment. GSA Bulletin 121 (11–12), 1537–1561. Available from: https://doi.org/10.1130/B26540.1.

McKay, R.M., De Santis, L., Kulhanek, D.K.the Expedition 374 Scientists, 2019. Ross Sea West Antarctic Ice Sheet History. Proceedings of the International Ocean Discovery Program, 374:. International Ocean Discovery Program, College Station, TX, https://doi.org/10.14379/iodp.proc.374.2019.

McKay, R., Golledge, N.R., Maas, S., Naish, T., Levy, R., Dunbar, G., et al., 2016. Antarctic marine ice-sheet retreat in the Ross Sea during the early Holocene. Geology 44 (1), 7–10.

McKay, R.M., Naish, T., Carter, L., Riesselman, C., Dunbar, R., Sjunneskog, C., et al., 2012a. Antarctic and Southern Ocean influences on Late Pliocene global cooling. Proceedings of the National Academy of Sciences of the United States of America 109 (17), 6423–6428. Available from: https://doi.org/10.1073/pnas.1112248109.

McKay, R.M., Naish, T., Powell, R., Barrett, P., Talarico, F., Kyle, P., et al., 2012b. Pleistocene variability of Antarctic Ice Sheet extent in the Ross Embayment. Quaternary Science Reviews 34, 93–112. Available from: https://doi.org/10.1016/j.quascirev.2011.12.012.

McKay, R.M., et al., 2021. Cenozoic History of Antarctic Glaciation and Climate from onshore and offshore studies. In: Florindo, F., et al. (Eds.), Antarctic Climate Evolution, second edition. Elsevier (this volume).

Melles, M., Brigham-Grette, J., Minyuk, P.S., Nowaczyk, N.R., Wennrich, V., DeConto, R.M., et al., 2012. 2.8 million years of Arctic climate change from Lake El'gygytgyn, NE Russia. Science 337 (6092), 315–320. Available from: https://doi.org/10.1126/science.1222135.

Mengel, M., Levermann, A., 2014. Ice plug prevents irreversible discharge from East Antarctica. Nature Climate Change 4 (6), 451–455. Available from: https://doi.org/10.1038/nclimate2226.

Menviel, L., Timmermann, A., Timm, O.E., Mouchet, A., 2011. Deconstructing the Last Glacial termination: the role of millennial and orbital-scale forcings. Quaternary Science Reviews 30 (9–10), 1155–1172.

Miller, K.G., Browning, J.V., Schmelz, W.J., Kopp, R.E., Mountain, G.S., Wright, J.D., 2020a. Cenozoic sea-level and cryospheric evolution from deep-sea geochemical and continental margin records. Science Advances, 6 (20), eaaz1346. Available from: https://doi.org/10.1126/sciadv.aaz1346.

Miller, K.G., Kominz, M.A., Browning, J.V., Wright, J.D., Mountain, G.S., Katz, M.E., et al., 2005. The Phanerozoic record of global sea-level change. Science 310 (5752), 1293–1298. Available from: https://doi.org/10.1126/science.1116412.

Miller, K.G., Schmelz, W.J., Browning, J.V., Kopp, R.E., Mountain, G.S., Wright, J.D., 2020b. Ancient sea level as key to the future. Oceanography 33 (2), 32–41. Available from: https://doi.org/10.5670/oceanog.2020.224.

Miller, K.G., Wright, J.D., Browning, J.V., Kulpecz, A., Kominz, M., Naish, T.R., et al., 2012. High tide of the warm Pliocene: Implications of global sea level for Antarctic deglaciation. Geology 40 (5), 407–410. Available from: https://doi.org/10.1130/G32869.1.

Miller, K.G., Wright, J.D., Fairbanks, R.G., 1991. Unlocking the ice house: Oligocene-Miocene oxygen isotopes, eustasy, and margin erosion. Journal of Geophysical Research: Solid Earth 96 (B4), 6829–6848. Available from: https://doi.org/10.1029/90JB02015.

Miller, K.G., Wright, J.D., Katz, M.E., Browning, J.V., Cramer, B.S., Wade, B.S., et al., 2008. A view of Antarctic ice-sheet evolution from sea-level and deep-sea isotope changes during the Late Cretaceous–Cenozoic. Antarctica: A Keystone in a Changing World 55–70.

Miller, K.G., Wright, J.D., Katz, M.E., Wade, B.S., Browning, J.V., Cramer, B.S., et al., 2009. Climate threshold at the Eocene-Oligocene transition: Antarctic ice sheet influence on ocean circulation. Geological Society of America Special Paper 452, 169–178. Available from: https://doi.org/10.1130/2009.2452(11).

Milne, G.A., Mitrovica, J.X., 2008. Searching for eustasy in deglacial sea-level histories. Quaternary Science Reviews 27 (25–26), 2292–2302.

Minzoni, R.T., Majewski, W., Anderson, J.B., Yokoyama, Y., Fernandez, R., Jakobsson, M., 2017. Oceanographic influences on the stability of the Cosgrove Ice Shelf. Antarctica. The Holocene 27 (11), 1645–1658.

Morlighem, M., Rignot, E., Binder, T., Blankenship, D., Drews, R., Eagles, G., et al., 2020. Deep glacial troughs and stabilizing ridges unveiled beneath the margins of the Antarctic ice sheet. Nature Geoscience 13 (2), 132–137. Available from: https://doi.org/10.1038/s41561-019-0510-8.

Naish, T.R., et al., 2001. "Orbitally induced oscillations in the East Antarctic ice sheet at the Oligocene/Miocene boundary.". Nature 413 (6857), 719–723.

Naish, T., Powell, R., Levy, R., Wilson, G., Scherer, R., Talarico, F., et al., 2009a. Obliquity-paced Pliocene West Antarctic Ice Sheet oscillations. Nature 458, 322–328. Available from: https://doi.org/10.1038/nature07867.

Naish, T.R., Wilson, G.S., 2009b. Constraints on the amplitude of Mid-Pliocene (3.6–2.4 Ma) eustatic sea-level fluctuations from the New Zealand shallow-marine sediment record. Philosophical Transactions of the Royal Society A: Mathematical, Physical and Engineering Sciences 367 (1886), 169–187. Available from: https://doi.org/10.1098/rsta.2008.0223.

Naish, T., Zwartz, D., 2012. Looking back to the future. Nature Climate Change 2 (5), 317–318.

Naish, T., Duncan, B., Levy., R., McKay, R., Escutia, C., De Santis, L., et al., 2021. Antarctic Ice Sheet dynamics during the Late Oligocene and Early Miocene: Climatic conundrums revisited. In: Florindo, F., et al. (Eds.), Antarctic Climate Evolution, second edition. Elsevier (this volume).

Nakada, M., Kimura, R., Okuno, J., Moriwaki, K., Miura, H., Maemoku, H., 2000. Late Pleistocene and Holocene melting history of the Antarctic ice sheet derived from sea-level variations. Marine Geology 167 (1–2), 85–103.

Nichols, K.A., Goehring, B.M., Balco, G., Johnson, J.S., Hein, A.A., Todd, C., 2019. New Last Glacial Maximum Ice Thickness constraints for the Weddell Sea sector, Antarctica. The Cryosphere Discussions . Available from: https://doi.org/10.5194/tc-2019-64.

Nitsche, F.O., Gohl, K., Larter, R.D., Hillenbrand, C.D., Kuhn, G., Smith, J.A., et al., 2013. Paleo ice flow and subglacial meltwater dynamics in Pine Island Bay, West Antarctica. The Cryosphere 7, 249–262.

Noble, T.L., Rohling, E.J., Aitken, A.R.A., Bostock, H.C., Chase, Z., Gomez, N., et al., 2020. The sensitivity of the Antarctic Ice Sheet to a changing climate: Past, present and future. Reviews of Geophysics 58 (4). Available from: https://doi.org/10.1029/2019RG000663.

O'Brien, P.E., Cooper, A.K., Florindo, F., Handwerger, D.A., Lavelle, M., Passchier, S., et al., 2004. Prydz channel fan and the history of extreme ice advances in Prydz Bay. Proceedings of the Ocean Drilling Program, Scientific Results 188, 1–32. ISSN 0884-5891.

O'Brien, P.E., Goodwin, I., Forsberg, C.F., Cooper, A.K., Whitehead, J., 2007. Late Neogene ice drainage changes in Prydz Bay, East Antarctica and the interaction of Antarctic ice sheet evolution and climate. Palaeogeography, Palaeoclimatology, Palaeoecology 245 (3–4), 390–410. Available from: https://doi.org/10.1016/j.palaeo.2006.09.002.

O'Brien, P., Opdyke, B., Post, A., Armand, L., 2020. Sabrina Sea Floor Survey (IN2017-V01) Piston Core Images, Visual Logs and Grain Size DATA Summaries IN2017-V01-A005-PC01.

Otto-Bliesner, B.L., Jahn, A., Feng, R., Brady, E.C., Hu, A., Löfverström, M., 2017. Amplified North Atlantic warming in the late Pliocene by changes in Arctic gateways. Geophysical Research Letters 44 (2), 957–964. Available from: https://doi.org/10.1002/2016GL071805.

O'Cofaigh, C., Davies, B.J., Livingstone, S.J., Smith, J.A., Johnson, J.S., Hocking, E.P., et al., 2014. Reconstruction of ice-sheet changes in the Antarctic Peninsula since the Last Glacial Maximum. Quaternary Science Reviews 100, 87–110.

Pagani, M., Huber, M., Liu, Z., Bohaty, S., Henderiks, J., et al., 2011. The role of carbon dioxide during the onset of Antarctic Glaciation. Science 334, 1261–1264. Available from: https://doi.org/10.1126/science.1203909.

Pagani, M., Liu, Z., LaRiviere, J., Ravelo, A.C., 2010. High Earth-system climate sensitivity determined from Pliocene carbon dioxide concentrations. Nature Geoscience 3, 27–30. Available from: https://doi.org/10.1038/ngeo724.

Pagani, M., Zachos, J.C., Freeman, K.H., Tipple, B., Bohaty, S., 2005. Carbon dioxide concentrations during the Paleogene. Science 309, 600–603. Available from: https://doi.org/10.1126/science.1110063.

Pälike, H., Norris, R.D., Herrle, J.O., Wilson, P.A., Coxall, H.K., Lear, C.H., et al., 2006. The heartbeat of the Oligocene climate system. Science 314, 1894–1897.

Paolo, F.S., Fricker, H.A., Padman, L., 2015. Volume loss from Antarctic ice shelves is accelerating. Science 348 (6232), 327–331. Available from: https://doi.org/10.1126/science.aaa0940.

Parrenin, F., Cavitte, M.G., Blankenship, D.D., Chappellaz, J., Fischer, H., Gagliardini, O., et al., 2017. Is there 1.5-million-year-old ice near Dome C, Antarctica? The Cryosphere 11 (6), 2427–2437.

Passchier, S., Bohaty, S.M., Jiménez-Espejo, F., Pross, J., Röhl, U., van de Flierdt, T., et al., 2013. Early Eocene – to – middle Miocene cooling and aridification of East Antarctica. Geochemistry, Geophysics, Geosystems 14 (5), 1399–1410. Available from: https://doi.org/10.1002/ggge.20106.

Passchier, S., Browne, G., Field, B., Fielding, C.R., Krissek, L.A., Panter, K.ANDRILL-SMS Science Team, 2011. Early and middle Miocene Antarctic glacial history from the sedimentary facies distribution in the AND-2A drill hole, Ross Sea, Antarctica. GSA Bulletin 123 (11–12), 2352–2365. Available from: https://doi.org/10.1130/B30334.1.

Passchier, S., Ciarletta, D., Miriagos, T., Bijl, P., Bohaty, S., 2016. An Antarctic stratigraphic record of step-wise ice growth through the Eocene-Oligocene Transition. Geological Society of America Bulletin 129 (3–4), 318–330. Available from: https://doi.org/10.1130/B31482.

Patterson, M.O., McKay, R., Naish, T., Escutia, C., Jimenez-Espejo, F.J., Raymo, M.E., et al., 2014. Orbital forcing of the East Antarctic ice sheet during the Pliocene and Early Pleistocene. Nature Geoscience 7, 841–847. Available from: https://doi.org/10.1038/ngeo2273.

Pattyn, F., Ritz, C., Hanna, E., Asay-Davis, X., DeConto, R., Durand, G., et al., 2018. The Greenland and Antarctic ice sheets under 1.5 C global warming. Nature Climate Change 8 (12), 1053–1061. Available from: https://doi.org/10.1038/s41558-018-0305-8.

Paxman, G.J.G., Gasson, E.G.W., Jamieson, S.S.R., Bentley, M.J., Ferraccioli, F., 2020. Long-term increase in Antarctic Ice Sheet vulnerability driven by bed topography evolution. Geophysical Research Letters 47. Available from: https://doi.org/10.1029/2020GL090003.

Paxman, G.J., Jamieson, S.S., Hochmuth, K., Gohl, K., Bentley, M.J., Leitchenkov, G., et al., 2019. Reconstructions of Antarctic topography since the Eocene–Oligocene boundary. Palaeogeography, Palaeoclimatology, Palaeoecology 535, 109346. Available from: https://doi.org/10.1016/j.palaeo.2019.109346.

Pedro, J.B., Jochum, M., Buizert, C., He, F., Barker, S., Rasmussen, S.O., 2018. Beyond the bipolar seesaw: Toward a process understanding of interhemispheric coupling. Quaternary Science Reviews 192, 27–46. Available from: https://doi.org/10.1016/j.quascirev.2018.05.005.

Pekar, S.F., Christie-Blick, N., Kominz, M.A., Miller, K.G., 2002. Calibration between eustatic estimates from backstripping and oxygen isotopic records for the Oligocene. Geology 30 (10), 903–906. Available from: https://doi.org/10.1130/0091-7613.

Pekar, S.F., Christie-Blick, N., 2008. Resolving apparent conflicts between oceanographic and Antarctic climate records and evidence for a decrease in pCO_2 during the Oligocene through early Miocene (34–16 Ma). Palaeogeography, Palaeoclimatology, Palaeoecology 260 (1–2), 41–49. Available from: https://doi.org/10.1016/j.palaeo.2007.08.019.

Peltier, W.R., Argus, D.F., Drummond, R., 2015. Space geodesy constrains ice age terminal deglaciation: The global ICE-6G_C (VM5a) model. Journal of Geophysical Research: Solid Earth 120 (1), 450–487.

Peltier, W.R., Fairbanks, R.G., 2006. Global glacial ice volume and Last Glacial Maximum duration from an extended Barbados sea level record. Quaternary Science Reviews 25 (23–24), 3322–3337.

Peltier, W.R., 2004. Global glacial isostasy and the surface of the ice-age Earth: the ICE-5G (VM2) model and GRACE. Annual Review of Earth and Planetary Sciences 32, 111–149.

Peltier, W.R., 2005. On the hemispheric origins of meltwater pulse 1a. Quaternary Science Reviews 24 (14–15), 1655–1671.

Petit, J.R., Jouzel, J., Raynaud, D., Barkov, N.I., Barnola, J.M., Basile, I., et al., 1999. Climate and atmospheric history of the past 420,000 years from the Vostok ice core, Antarctica. Nature 399 (6735), 429. Available from: https://doi.org/10.1038/20859.

Petrini, M., Colleoni, F., Kirchner, N., Hughes, A.L., Camerlenghi, A., Rebesco, M., et al., 2018. Interplay of grounding-line dynamics and sub-shelf melting during retreat of the Bjørnøyrenna Ice Stream. Scientific Reports 8 (1), 1--9.

Philippon, G., Ramstein, G., Charbit, S., Kageyama, M., Ritz, C., Dumas, C., 2006. Evolution of the Antarctic ice sheet throughout the last deglaciation: A study with a new coupled climate—north and south hemisphere ice sheet model. Earth and Planetary Science Letters 248 (3–4), 750–758. Available from: https://doi.org/10.1016/j.epsl.2006.06.017.

Pierce, E.L., van de Flierdt, T., Williams, T., Hemming, S.R., Cook, C.P., Passchier, S., 2017. Evidence for a dynamic East Antarctic ice sheet during the mid-Miocene climate transition. Earth and Planetary Science Letters 478, 1–13.

Pollard, D., Chang, W., Haran, M., Applegate, P., DeConto, R., 2016. Large ensemble modeling of the last deglacial retreat of the West Antarctic Ice Sheet: comparison of simple and advanced statistical techniques. Geoscientific Model Development . Available from: https://doi.org/10.5194/gmd-9-1697-2016.

Pollard, D., DeConto, R.M., 2003. Antarctic ice and sediment flux in the Oligocene simulated by a climate–ice sheet–sediment model. Palaeogeography, Palaeoclimatology, Palaeoecology 198 (1–2), 53–67. Available from: https://doi.org/10.1016/S0031-0182(03)00394-8.

Pollard, D., DeConto, R.M., 2005. Hysteresis in Cenozoic Antarctic ice-sheet variations. Global and Planetary Change 45 (1–3), 9–21. Available from: https://doi.org/10.1016/j.gloplacha.2004.09.011.

Pollard, D., DeConto, R.M., 2009. Modelling West Antarctic ice sheet growth and collapse through the past five million years. Nature 458 (7236), 329–332. Available from: https://doi.org/10.1038/nature07809.

Pollard, D., DeConto, R.M., 2012. Description of a hybrid ice sheet-shelf model, and application to Antarctica. Geoscientific Model Development 5 (5), 1273–1295. Available from: https://doi.org/10.5194/gmd-5-1273-2012.

Pollard, D., DeConto, R.M., Alley, R.B., 2015. Potential Antarctic Ice Sheet retreat driven by hydrofracturing and ice cliff failure. Earth and Planetary Science Letters 412, 112–121. Available from: https://doi.org/10.1016/j.epsl.2014.12.035.

Pollard, D., Gomez, N., Deconto, R.M., 2017. Variations of the Antarctic ice sheet in a coupled ice sheet-Earth-sea level model: sensitivity to viscoelastic Earth properties. Journal of Geophysical Research: Earth Surface 122 (11), 2124–2138. Available from: https://doi.org/10.1002/2017JF004371.

Pollard, D., DeConto, R.M., 2020. Continuous simulations over the last 40 million years with a coupled Antarctic ice sheet-sediment model. Palaeogeography, Palaeoclimatology, Palaeoecology 537, 109374. Available from: https://doi.org/10.1016/j.palaeo.2019.109374.

Presti, M., Barbara, L., Denis, D., Schmidt, S., De Santis, L., Crosta, X., 2011. Sediment delivery and depositional patterns off Adélie Land (East Antarctica) in relation to late Quaternary climatic cycles. Marine Geology 284 (1–4), 96–113.

Pritchard, H., Ligtenberg, S.R., Fricker, H.A., Vaughan, D.G., van den Broeke, M.R., Padman, L., 2012. Antarctic ice-sheet loss driven by basal melting of ice shelves. Nature 484 (7395), 502–505. Available from: https://doi.org/10.1038/nature10968.

Pross, J., Contreras, L., Bijl, P.K., Greenwood, D.R., Bohaty, S.M., Schouten, S., et al., 2012. Persistent near-tropical warmth on the Antarctic continent during the early Eocene epoch. Nature 488 (7409), 73–77.

Prothro, L.O., Majewski, W., Yokoyama, Y., Simkins, L.M., Anderson, J.B., Yamane, M., et al., 2020. Timing and pathways of East Antarctic Ice Sheet retreat. Quaternary Science Reviews 230, 106166.

Prothro, L.O., Simkins, L.M., Majewski, W., Anderson, J.B., 2018. Glacial retreat patterns and processes determined from integrated sedimentology and geomorphology records. Marine Geology 395, 104–119.

Quiquet, A., Dumas, C., Ritz, C., Peyaud, V., Roche, D.M., 2018. The GRISLI ice sheet model (version 2.0): calibration and validation for multi-millennial changes of the Antarctic ice sheet. Geoscientific Model Development 11 (12), 5003. Available from: https://doi.org/10.5194/gmd-11-5003-2018.

Raymo, M.E., Kozdon, R., Evans, D., Lisiecki, L., Ford, H.L., 2018. The accuracy of mid-Pliocene $\delta^{18}O$-based ice volume and sea level reconstructions. Earth-Science Reviews 177, 291–302.

Raymo, M.E., Lisiecki, L.E., Nisancioglu, K.H., 2006. Plio-Pleistocene ice volume, Antarctic climate, and the global δ18O record. Science 313 (5786), 492–495. Available from: https://doi.org/10.1126/science.1123296.

Raymo, M.E., Mitrovica, J.X., O'Leary, M.J., DeConto, R.M., Hearty, P.J., 2011. Departures from eustasy in Pliocene sea-level records. Nature Geoscience 4 (5), 328–332.

Raymo, M.E., Mitrovica, J.X., 2012. Collapse of polar ice sheets during the stage 11 interglacial. Nature 483 (7390), 453–456. Available from: https://doi.org/10.1038/nature10891.

Rayner, N., Parker, D.E., Horton, E., Folland, C., Alexander, L., Rowell, D., et al., 2003. Global analyses of sea surface temperature, sea ice, and night marine air temperature since the late nineteenth century. Journal of Geophysical Research 108 (D14), 4407. Available from: https://doi.org/10.1029/2002JD002670.

Rebesco, M., Camerlenghi, A., Geletti, R., Canals, M., 2006. Margin architecture reveals the transition to the modern Antarctic ice sheet ca. 3 Ma. Geology 34 (4), 301–304. Available from: https://doi.org/10.1130/G22000.1.

Rebesco, M., Domack, E., Zgur, F., Lavoie, C., Leventer, A., Brachfeld, S., Pettit, E., 2014. Boundary condition of grounding lines prior to collapse, Larsen-B Ice Shelf, Antarctica. Science 345 (6202), 1354–1358.

Reinardy, B.T.I., Escutia, C., Iwai, M., Jimenez-Espejo, F.J., Cook, C., van de Flierdt, T., et al., 2015. Repeated advance and retreat of the East Antarctic Ice Sheet on the continental shelf during the early Pliocene warm period. Palaeogeography, Palaeoclimatology, Palaeoecology 422, 65–84.

Retallack, G.J., 2009. Refining a pedogenic-carbonate CO_2 paleobarometer to quantify a middle Miocene greenhouse spike. Palaeogeography, Palaeoclimatology, Palaeoecology 281 (1-2), 57–65.

Reyes, A.V., Carlson, A.E., Beard, B.L., Hatfield, R.G., Stoner, J.S., Winsor, K., et al., 2014. South Greenland ice-sheet collapse during marine isotope stage 11. Nature 510 (7506), 525–528. Available from: https://doi.org/10.1038/nature13456.

Rignot, E., Mouginot, J., Scheuchl, B., van den Broeke, M., van Wessem, M.J., Morlighem, M., 2019. Four decades of Antarctic Ice Sheet mass balance from 1979–2017. Proceedings of the National Academy of Sciences 116 (4), 1095–1103. Available from: https://doi.org/10.1073/pnas.1812883116.

Rintoul, S.R., Silvano, A., Pena-Molino, B., van Wijk, E., Rosenberg, M., Greenbaum, J.S., et al., 2016. Ocean heat drives rapid basal melt of the Totten Ice Shelf. Science Advances 2 (12), e1601610.

Roberts, D.L., Karkanas, P., Jacobs, Z., Marean, C.W., Roberts, R.G., 2012. Melting ice sheets 400,000 yr ago raised sea level by 13 m: past analogue for future trends. Earth and Planetary Science Letters 357, 226–237. Available from: https://doi.org/10.1016/j.epsl.2012.09.006.

Roche, D.M., Paillard, D., Caley, T., Waelbroeck, C., 2014. LGM hosing approach to Heinrich Event 1: results and perspectives from data–model integration using water isotopes. Quaternary Science Reviews 106, 247–261. Available from: https://doi.org/10.1016/j.quascirev.2014.07.020.

Rohling, E.J., Grant, K., Bolshaw, M., Roberts, A.P., Siddall, M., Hemleben, C., et al., 2009. Antarctic temperature and global sea level closely coupled over the past five glacial cycles. Nature Geoscience 2 (7), 500–504. Available from: https://doi.org/10.1038/ngeo557.

Rovere, A., Antonioli, F., Bianchi, C.N., 2015. Fixed biological indicators. Handbook of Sea-Level Research. Wiley, pp. 268–280.

Rovere, A., Raymo, M.E., Mitrovica, J.X., Hearty, P.J., O'Leary, M.J., Inglis, J.D., 2014. The Mid-Pliocene sea-level conundrum: glacial isostasy, eustasy and dynamic topography. Earth and Planetary Science Letters 387, 27–33.

Roy, K., Peltier, W., 2018. Relative sea level in the Western Mediterranean Basin: a regional test of the ICE-7G NA (VM7) model and a constraint on Late Holocene Antarctic Deglaciation. Quaternary Science Reviews 183, 76–87.

Sadai, S., Condron, A., DeConto, R., Pollard, D., 2020. Future climate response to Antarctic Ice Sheet melt caused by anthropogenic warming. Science Advances 6 (39), eaaz1169.

Salabarnada, A., Escutia, C., Röhl, U., Nelson, C.H., McKay, R., Jiménez-Espejo, F.J., et al., 2018. Paleoceanography and ice sheet variability offshore Wilkes Land, Antarctica—part 1: insights from late Oligocene astronomically paced contourite sedimentation. Climate of the Past 14 (7), 991–1014. Available from: https://doi.org/10.5194/cp-14-991-2018.

Sandstrom, M.R., O'Leary, M.J., Barham, M., Cai, Y., Rasbury, E.T., Wooton, K.M., et al., 2021. Age constraints on surface deformation recorded by fossil shorelines at Cape Range, Western Australia. GSA Bulletin 133.

Sangiorgi, F., Bijl, P.K., Passchier, S., Salzmann, U., Schouten, S., McKay, R., et al., 2018. Southern Ocean warming and Wilkes Land ice sheet retreat during the mid-Miocene. Nature Communications 9 (1), 1–11. Available from: https://doi.org/10.1038/s41467-017-02609-7.

Scherer, R.P., Bohaty, S.M., Dunbar, R.B., Esper, O., Flores, J.A., Gersonde, R., et al., 2008. Antarctic records of precession-paced insolation-driven warming during early Pleistocene Marine Isotope Stage 31. Geophysical Research Letters 35 (3), Available from: https://doi.org/10.1029/2007GL032254.

Scherer, R., Bohaty, S., Harwood, D., Roberts, A., Taviani, M., 2003. Marine Isotope Stage 31 (1.07 Ma): an extreme interglacial in the Antarctic nearshore zone. Geophysical Research Abstracts 5, 11710.

Scherer, R.P., DeConto, R.M., Pollard, D., Alley, R.B., 2016. Windblown Pliocene diatoms and East Antarctic Ice Sheet retreat. Nature Communications 7 (1), 1–9. Available from: https://doi.org/10.1038/ncomms12957.

Schmidtko, S., Heywood, K.J., Thompson, A.F., Aoki, S., 2014. Multidecadal warming of Antarctic waters. Science 346 (6214), 1227–1231.

Schneider von Deimling, T., Ganopolski, A., Held, H., Rahmstorf, S., 2006. How cold was the last glacial maximum? Geophysical Research Letters 33, 14.

Schüpbach, S., Federer, U., Kaufmann, P., Albani, S., Barbante, C., Stocker, T., et al., 2013. High-resolution mineral dust and sea ice proxy records from the Talos Dome ice core. Climate of the Past 9 (6), 2789–2807. Available from: https://doi.org/10.5194/cp-9-2789-2013.

Seki, O., Foster, G.L., Schmidt, D.N., Mackensen, A., Kawamura, K., Pancost, R.D., 2010. Alkenone and boron based Plio-Pleistocene pCO2 records. Earth and Planetary Science Letters 292, 201–211. Available from: https://doi.org/10.1016/j.epsl.2010.01.037.

Shackleton, N.J., 2000. The 100,000-year Ice-Age cycle identified and found to lag temperature, carbon dioxide, and orbital eccentricity. Science 289, 1897–1902.

Shakun, J.D., Corbett, L.B., Bierman, P.R., Underwood, K., Rizzo, D.M., Zimmerman, S.R., et al., 2018. Minimal East Antarctic Ice Sheet retreat onto land during the past eight million years. Nature 558 (7709), 284–287. Available from: https://doi.org/10.1038/s41586-018-0155-6.

Shepherd, A., Ivins, E., Rignot, E., Smith, B., Van Den Broeke, M., Velicogna, I., et al., 2018. Mass balance of the Antarctic Ice Sheet from 1992 to 2017. Nature 558, 219–222. Available from: https://doi.org/10.1038/s41586-018-0179-y.

Shevenell, A.E., Ingalls, A.E., Domack, E.W., Kelly, C., 2011. Holocene Southern Ocean surface temperature variability west of the Antarctic Peninsula. Nature 470 (7333), 250–254.

Shevenell, A.E., Kennett, J.P., Lea, D.W., 2008. Middle Miocene ice sheet dynamics, deep-sea temperatures, and carbon cycling: a Southern Ocean perspective. Geochemistry, Geophysics, Geosystems 9 (2).

Siegert, M.J., Alley, R.B., Rignot, E., Englander, J., Corell, R., 2020. 21st Century sea-level rise could exceed IPCC predictions for strong-warming futures. One Earth 3, 691–703. Available from: https://doi.org/10.1016/j.oneear.2020.11.002.

Siegert, M.J., Hein, A.S., White, D.A., Gore, D.B., De Santis, L., Hillenbrand, C.D., 2021. Antarctic ice sheet changes since the Last Glacial Maximum. In: Florindo, F., et al. (Eds.), Antarctic Climate Evolution, second edition. Elsevier (this volume).

Silvano, A., Rintoul, S.R., Kusahara, K., Peña-Molino, B., van Wijk, E., Gwyther, D.E., et al., 2019. Seasonality of warm water intrusions onto the continental shelf near the Totten Glacier. Journal of Geophysical Research: Oceans 124 (6), 4272–4289.

Sime, L.C., Hodgson, D., Bracegirdle, T.J., Allen, C., Perren, B., Roberts, S., et al., 2016. Sea ice led to poleward-shifted windsat the Last Glacial Maximum: the influence of state dependency on CMIP5 and PMIP3 models. Climate of the Past 12 (12), 2241–2253. Available from: https://doi.org/10.5194/cp-12-2241-2016.

Simkins, L.M., Greenwood, S.L., Anderson, J.B., 2018. Diagnosing ice sheet grounding line stability from landform morphology. The Cryosphere 12, 2707–2726. Available from: https://doi.org/10.5194/tc-12-2707-2018.

Simkins, L.M., Anderson, J.B., Greenwood, S.L., Gonnermann, H.M., Prothro, L.O., Halberstadt, A.R.W., et al., 2017. Anatomy of a meltwater drainage system beneath the ancestral East Antarctic ice sheet. Nature Geoscience 10 (9), 691–697. Available from: https://doi.org/10.1038/ngeo3012.

Simms, A.R., Lisiecki, L., Gebbie, G., Whitehouse, P.L., Clark, J.F., 2019. Balancing the last glacial maximum (LGM) sea-level budget. Quaternary Science Reviews 205, 143–153.

Small, D., Bentley, M.J., Jones, R.S., Pittard, M.L., Whitehouse, P.L., 2019. Antarctic ice sheet palaeo-thinning rates from vertical transects of cosmogenic exposure ages. Quaternary Science Reviews 206, 65–80.

Smith, J.A., Andersen, T.J., Shortt, M., Gaffney, A.M., Truffer, M., Stanton, T.P., et al., 2017. Sub-ice-shelf sediments record history of twentieth-century retreat of Pine Island Glacier. Nature 541 (7635), 77–80.

Smith, J.A., Graham, A.G., Post, A.L., Hillenbrand, C.D., Bart, P.J., Powell, R.D., 2019. The marine geological imprint of Antarctic ice shelves. Nature Communications 10 (1), 1–16. Available from: https://doi.org/10.1038/s41467-019-13496-5.

Sosdian, S., Rosenthal, Y., 2009. Deep-sea temperature and ice volume changes across the Pliocene-Pleistocene climate transitions. Science 325, 306–310. Available from: https://doi.org/10.1126/science.1169938.

Spector, P., Stone, J., Cowdery, S.G., Hall, B., Conway, H., Bromley, G., 2017. Rapid early-Holocene deglaciation in the Ross Sea, Antarctica. Geophysical Research Letters 44 (15), 7817–7825.

Spector, P., Stone, J., Pollard, D., Hillebrand, T., Lewis, C., Gombiner, J., 2018. West Antarctic sites for subglacial drilling to test for past ice-sheet collapse. The Cryosphere 12 (8), 2741–2757. Available from: https://doi.org/10.5194/tc-12-2741-2018.

Stammer, D., 2008. Response of the global ocean to Greenland and Antarctic ice melting. Journal of Geophysical Research: Oceans 113 (C6). Available from: https://doi.org/10.1029/2006JC004079.

Stap, L.B., Knorr, G., Lohmann, G., 2020. Anti-phased Miocene ice volume and CO_2 changes by transient antarctic ice sheet variability. Paleoceanography and Paleoclimatology 35 (11), e2020PA003971.

Stap, L.B., Sutter, J., Knorr, G., Stärz, M., Lohmann, G., 2019. Transient variability of the Miocene Antarctic ice sheet smaller than equilibrium differences. Geophysical Research Letters 46 (8), 4288–4298.

Stap, L.B., Van De Wal, R.S., De Boer, B., Bintanja, R., Lourens, L.J., 2017. The influence of ice sheets on temperature during the past 38 million years inferred from a one-dimensional ice sheet-climate model. Climate of the Past 13 (9), 1243–1257. Available from: https://doi.org/10.5194/cp-13-1243-2017.

Steinhauff, D.M., Webb, P.-N., 1987. Miocene foraminifera from DSDP site 272, Ross Sea. Geology 11, 578–582.

Steinthorsdottir, M., Coxall, H.K., de Boer, A.M., Huber, M., Barbolini, N., Bradshaw, C.D., et al., 2020. The Miocene: the future of the past. Paleoceanography and Paleoclimatology 35. Available from: https://doi.org/10.1029/2020PA004037e2020PA004037.

Stenni, B., Buiron, D., Frezzotti, M., Albani, S., Barbante, C., Bard, E., et al., 2011. Expression of the bipolar see-saw in Antarctic climate records during the last deglaciation. Nature Geoscience 4 (1), 46–49. Available from: https://doi.org/10.1038/ngeo1026.

Stocchi, P., et al., 2013. Relative sea-level rise around East Antarctica during Oligocene glaciation. Nature Geoscience 6, 380–384. Available from: https://doi.org/10.1038/ngeo1783.

Stocchi, P., Antonioli, F., Montagna, P., Pepe, F., Lo Presti, V., Caruso, A., et al., 2017. A stalactite record of four relative sea-level highstands during the Middle Pleistocene Transition. Quaternary Science Reviews 173, 92–100.

Stocker, T.F., 1998. The seesaw effect. Science 282 (5386), 61–62.

Stocker, T.F., Johnsen, S.J., 2003. A minimum thermodynamic model for the bipolar seesaw. Paleoceanography 18, 1087. Available from: https://doi.org/10.1029/2003PA000920.

Stokes, C.R., 2018. Geomorphology under ice streams: Moving from form to process. Earth Surface Processes and Landforms 43 (1), 85–123. Available from: https://doi.org/10.1002/esp.4259.

Struve, T., Pahnke, K., Lamy, F., Wengler, M., Böning, P., Winckler, G., 2020. A circumpolar dust conveyor in the glacial Southern Ocean. Nature Communications 11 (1), 1–11.

Sugden, D., Denton, G., 2004. Cenozoic landscape evolution of the Convoy Range to Mackay Glacier area, Transantarctic Mountains: onshore to offshore synthesis. Geological Society of America Bulletin 116 (7–8), 840–857.

Super, J.R., Thomas, E., Pagani, M., Huber, M., O'Brien, C., Hull, P.M., 2018. North Atlantic temperature and pCO_2 coupling in the early-middle Miocene. Geology 46 (6), 519–522. Available from: https://doi.org/10.1130/G40228.1.

Sutter, J., Fischer, H., Grosfeld, K., Karlsson, N.B., Kleiner, T., Van Liefferinge, B., et al., 2019. Modelling the Antarctic Ice Sheet across the mid-Pleistocene transition–implications for

Oldest Ice. The Cryosphere 13 (7), 2023–2041. Available from: https://doi.org/10.5194/tc-13-2023-2019.

Sutter, J., Gierz, P., Grosfeld, K., Thoma, M., Lohmann, G., 2016. Ocean temperature thresholds for last interglacial West Antarctic Ice Sheet collapse. Geophysical Research Letters 43 (6), 2675–2682. Available from: https://doi.org/10.1002/2016GL067818.

Tan, N., Ramstein, G., Dumas, C., Contoux, C., Ladant, J.B., Sepulchre, P., et al., 2017. Exploring the MIS M2 glaciation occurring during a warm and high atmospheric CO2 Pliocene background climate. Earth and Planetary Science Letters 472, 266–276. Available from: https://doi.org/10.1016/j.epsl.2017.04.050.

Tarasov, L., Peltier, W.R., 2002. Greenland glacial history and local geodynamic consequences. Geophysical Journal International 150, 198–229. Available from: https://doi.org/10.1046/j.1365-246X.2002.01702.x.

Tarasov, L., Peltier, W.R., 2003. Greenland glacial history, borehole constraints, and Eemian extent. Journal of Geophysical Research. 108, 1–20. Available from: https://doi.org/10.1029/2001JB001731.

Taylor-Silva, B.I., Riesselman, C.R., 2018. Polar frontal migration in the warm late Pliocene: Diatom evidence from the Wilkes Land margin, East Antarctica. Paleoceanography and Paleoclimatology 33 (1), 76–92. Available from: https://doi.org/10.1002/2017PA003225.

Teitler, L., Florindo, F., Warnke, D.A., Filippelli, G.M., Kupp, G., Taylor, B., 2015. Antarctic Ice Sheet response to a long warm interval across Marine Isotope Stage 31: a cross-latitudinal study of iceberg-rafted debris. Earth and Planetary Science Letters 409, 109–119.

The RAISED ConsortiumBentley, M.J., Cofaigh, C.O., Anderson, J.B., Conway, H., Davies, B., Graham, A.G., et al., 2014. A community-based geological reconstruction of Antarctic Ice Sheet deglaciation since the Last Glacial Maximum. Quaternary Science Reviews 100, 1–9. Available from: https://doi.org/10.1016/j.quascirev.2014.06.025.

Thiede, J., Jessen, C., Knutz, P., Kuijpers, A., Mikkelsen, N., Spielhagen, R.F., 2011. Millions of years of Greenland Ice Sheet history recorded in ocean sediments. Polarforschung 80 (3), 141–159.

Tierney, J.E., Zhu, J., King, J., Malevich, S.B., Hakim, G.J., Poulsen, C.J., 2020. Glacial cooling and climate sensitivity revisited. Nature 584, 569–573.

Tigchelaar, M., Timmermann, A., Pollard, D., Friedrich, T., Heinemann, M., 2018. Local insolation changes enhance Antarctic interglacials: Insights from an 800,000-year ice sheet simulation with transient climate forcing. Earth and Planetary Science Letters 495, 69–78. Available from: https://doi.org/10.1016/j.epsl.2018.05.004.

Tripati, A., Darby, D., 2018. Evidence for ephemeral middle Eocene to early Oligocene Greenland glacial ice and pan-Arctic sea ice. Nature Communications 9 (1), 1–11.

Turney, C.S., Fogwill, C.J., Golledge, N.R., McKay, N.P., van Sebille, E., Jones, R.T., et al., 2020. Early Last Interglacial ocean warming drove substantial ice mass loss from Antarctica. Proceedings of the National Academy of Sciences 117 (8), 3996–4006.

Turney, C.S., Jones, R.T., 2010. Does the Agulhas Current amplify global temperatures during super-interglacials? Journal of Quaternary Science 25 (6), 839–843. Available from: https://doi.org/10.1002/jqs.1423.

Tzedakis, P.C., Wolff, E.W., Skinner, L.C., Brovkin, V., Hodell, D.A., McManus, J.F., et al., 2012. Can we predict the duration of an interglacial? Climate of the Past 8, 1473–1485. Available from: https://doi.org/10.5194/cp-8-1473-2012.

Uemura, R., Motoyama, H., Masson-Delmotte, V., et al., 2018. Asynchrony between Antarctic temperature and CO_2 associated with obliquity over the past 720,000 years. Nature Communications 9, 96. Available from: https://doi.org/10.1038/s41467-018-03328-3.

Uenzelmann-Neben, G., 2006. Depositional patterns at Drift 7, Antarctic Peninsula: Along-slope versus down-slope sediment transport as indicators for oceanic currents and climatic conditions. Marine Geology 233 (1–4), 49–62.

Uenzelmann-Neben, G., Gohl, K., Larter, R., Schlüter, P., 2007. Differences in ice retreat across Pine Island Bay, West Antarctica, since the Last Glacial Maximum: Indications from multichannel seismic reflection data. US Geological Survey Open-File Report, 2007, srp084. <http://pubs.usgs.gov/of/2007/1047/srp/srp084/>. https://doi.org/10.3133/of2007-1047.srp084.

Uenzelmann-Neben, G., Gohl, K., 2012. Amundsen Sea sediment drifts: archives of modifications in oceanographic and climatic conditions. Marine Geology 299, 51–62.

Uenzelmann-Neben, G., Gohl, K., 2014. Early glaciation already during the Early Miocene in the Amundsen Sea, Southern Pacific: indications from the distribution of sedimentary sequences. Global and Planetary Change 120, 92–104.

Uenzelmann-Neben, G., 2019. Variations in ice-sheet dynamics along the Amundsen Sea and Bellingshausen Sea West Antarctic Ice Sheet margin. GSA Bulletin 131 (3–4), 479–498.

Villa, G., Lupi, C., Cobianchi, M., Florindo, F., Pekar, S.F., 2008. A Pleistocene warming event at 1 Ma in Prydz Bay, East Antarctica: evidence from ODP site 1165. Palaeogeography, Palaeoclimatology, Palaeoecology 260 (1–2), 230–244. Available from: https://doi.org/10.1016/j.palaeo.2007.08.017.

Villa, G., Persico, D., Wise, S.W., Gadaleta, A., 2012. Calcareous nanofossil evidence for Marine Isotope Stage 31 (1 Ma) in core AND-1B, ANDRILL McMurdo ice shelf project (Antarctica). Global and Planetary Change 96, 75–86.

von der Heydt, A.S., Dijkstra, H.A., van de Wal, R.S., Caballero, R., Crucifix, M., Foster, G.L., et al., 2016. Lessons on climate sensitivity from past climate changes. Current Climate Change Reports 2 (4), 148–158. Available from: https://doi.org/10.1007/s40641-016-0049-3.

Waelbroeck, C., Labeyrie, L., Michel, E., Duplessy, J.C., McManus, J.F., Lambeck, K., et al., 2002. Sea-level and deep water temperature changes derived from benthic foraminifera isotopic records. Quaternary Science Reviews 21 (1–3), 295–305. Available from: https://doi.org/10.1016/S0277-3791(01)00101-9.

WAIS Divide Project Members, 2015. Precise interpolar phasing of abrupt climate change during the last ice age. Nature 520 (7549), 661–665.

Wardlaw, B.R., Quinn, T.M., 1991. The record of Pliocene sea-level change at Enewetak atoll. Quaternary Science Reviews 10, 247–258. Available from: https://doi.org/10.1016/0277-3791(91)90023-N.

Warny, S., Askin, R.A., Hannah, M.J., Mohr, B.A., Raine, J.I., Harwood, D.M.SMS Science Team, 2009. Palynomorphs from a sediment core reveal a sudden remarkably warm Antarctica during the middle Miocene. Geology 37 (10), 955–958. Available from: https://doi.org/10.1130/G30139A.1.

Warrick, R.A., Le Provost, C., Meier, M.F., Oerlemans, J., Woodworth, P.L., 1996. In: Houghton, J.T., et al., (Eds.), Changes in sea level, in Climate Change1995: the science of climate change. Cambridge University Press, New York, pp. 361–405.

Watanabe, O., Jouzel, J., Johnsen, S., Parrenin, F., Shoji, H., Yoshida, N., 2003. Homogeneous climate variability across East Antarctica over the past three glacial cycles. Nature 422 (6931), 509–512. Available from: https://doi.org/10.1038/nature01525.

Weaver, A.J., Saenko, O.A., Clark, P.U., Mitrovica, J.X., 2003. Meltwater pulse 1A from Antarctica as a trigger of the Bølling-Allerød warm interval. Science 299 (5613), 1709–1713.

Weber, M.E., Clark, P.U., Kuhn, G., Timmermann, A., Sprenk, D., Gladstone, R., et al., 2014. Millennial-scale variability in Antarctic ice-sheet discharge during the last deglaciation. Nature 510 (7503), 134–138. Available from: https://doi.org/10.1038/nature13397.

Werner, M., Jouzel, J., Masson-Delmotte, V., Lohmann, G., 2018. Reconciling glacial Antarctic water stable isotopes with ice sheet topography and the isotopic paleothermometer. Nature Communications 9 (1), 1–10.

Westerhold, T., Marwan, N., Drury, A.J., Liebrand, D., Agnini, C., Anagnostou, E., et al., 2020. An astronomically dated record of Earth's climate and its predictability over the last 66 million years. Science 369 (6509), 1383–1387. Available from: https://doi.org/10.1126/science.aba6853.

Whitehead, J.M., Bohaty, S.M., 2003. Pliocene summer sea surface temperature reconstruction using silicoflagellates from Southern Ocean ODP Site 1165. Paleoceanography 18 (3).

Whitehead, J.M., Quilty, P.G., McKelvey, B.C., O'brien, P.E., 2006. A review of the Cenozoic stratigraphy and glacial history of the Lambert Graben-Prydz Bay region, East Antarctica. Antarctic Science . Available from: https://doi.org/10.1017/S0954102006000083.

Whitehead, J.M., Wotherspoon, S., Bohaty, S.M., 2005. Minimal Antarctic sea ice during the Pliocene. Geology 33 (2), 137–140. Available from: https://doi.org/10.1130/G21013.1.

Whitehouse, P.L., Bentley, M.J., Le Brocq, A.M., 2012b. A deglacial model for Antarctica: geological constraints and glaciological modelling as a basis for a new model of Antarctic glacial isostatic adjustment. Quaternary Science Reviews 32, 1–24. Available from: https://doi.org/10.1016/j.quascirev.2011.11.016.

Whitehouse, P.L., Bentley, M.J., Milne, G.A., King, M.A., Thomas, I.D., 2012a. A new glacial isostatic adjustment model for Antarctica: calibrated and tested using observations of relative sea-level change and present-day uplift rates. Geophysical Journal International 190 (3), 1464–1482.

Whitehouse, P.L., Bentley, M.J., Vieli, A., Jamieson, S.S., Hein, A.S., Sugden, D.E., 2017. Controls on last glacial maximum ice extent in the Weddell Sea embayment, Antarctica. Journal of Geophysical Research: Earth Surface 122 (1), 371–397. Available from: https://doi.org/10.1002/2016JF004121.

Whitehouse, P.L., 2018. Glacial isostatic adjustment modelling: historical perspectives, recent advances, and future directions. Earth Surface Dynamics 6 (2), 401–429.

Whitehouse, P.L., Gomez, N., King, M.A., Wiens, D.A., 2019. Solid Earth change and the evolution of the Antarctic Ice Sheet. Nature Communications 10 (1), 1–14. Available from: https://doi.org/10.1038/s41467-018-08068-y.

Wilson, D.J., Bertram, R.A., Needham, E.F., van de Flierdt, T., Welsh, K.J., McKay, R.M., et al., 2018. Ice loss from the East Antarctic Ice Sheet during late Pleistocene interglacials. Nature 561 (7723), 383–386. Available from: https://doi.org/10.1038/s41586-018-0501-8.

Wilson, D.J., et al., 2021. Pleistocene Antarctic climate variability: ice sheet–ocean–climate interactions. In: Florindo, F., et al. (Eds.), Antarctic Climate Evolution, second edition, Elsevier (this volume).

Wilson, D.S., Jamieson, S.S., Barrett, P.J., Leitchenkov, G., Gohl, K., Larter, R.D., 2012. Antarctic topography at the Eocene–Oligocene boundary. Palaeogeography, Palaeoclimatology, Palaeoecology 335, 24–34. Available from: https://doi.org/10.1016/j.palaeo.2011.05.028.

Wilson, D.S., Luyendyk, B.P., 2009. West Antarctic paleotopography estimated at the Eocene-Oligocene climate transition. Geophysical Research Letters 36, 16.

Wilson, D.S., Pollard, D., DeConto, R.M., Jamieson, S.S., Luyendyk, B.P., 2013. Initiation of the West Antarctic Ice Sheet and estimates of total Antarctic ice volume in the earliest Oligocene. Geophysical Research Letters 40 (16), 4305–4309. Available from: https://doi.org/10.1002/grl.50797.

Winnick, M.J., Caves, J.K., 2015. Oxygen isotope mass-balance constraints on Pliocene sea level and East Antarctic Ice Sheet stability. Geology 43 (10), 879–882. Available from: https://doi.org/10.1130/G36999.1.

Wise, M.G., Dowdeswell, J.A., Jakobsson, M., Larter, R.D., 2017. Evidence of marine ice-cliff instability in Pine Island Bay from iceberg-keel plough marks. Nature 550 (7677), 506–510.

Wise Jr., S.W., Schlich, R., et al., 1992. Proceedings of ODP, Science Results, Part 2, vol. 120. Ocean Drilling Program, College Station, TX, pp. 451–1155.

Wolff, E.W., Fischer, H., Fundel, F., Ruth, U., Twarloh, B., Littot, G.C., et al., 2006. Southern Ocean sea-ice extent, productivity and iron flux over the past eight glacial cycles. Nature 440 (7083), 491–496. Available from: https://doi.org/10.1038/nature04614.

Wu, L., Wilson, D.J., Wang, R., Passchier, S., Krijgsman, W., Yu, X., et al., 2021. Late Quaternary dynamics of the Lambert Glacier-Amery Ice Shelf system, East Antarctica. Quaternary Science Reviews 252, 106738.

Yan, Q., Zhang, Z., Wang, H., 2016. Investigating uncertainty in the simulation of the Antarctic ice sheet during the mid-Piacenzian. Journal of Geophysical Research: Atmospheres 121 (4), 1559–1574. Available from: https://doi.org/10.1002/2015JD023900.

Yokoyama, Y., Anderson, J.B., Yamane, M., Simkins, L.M., Miyairi, Y., Yamazaki, T., et al., 2016. Widespread collapse of the Ross Ice Shelf during the late Holocene. Proceedings of the National Academy of Sciences 113 (9), 2354–2359.

Zachos, J.C., Breza, J.R., Wise, S.W., 1992. Early Oligocene ice-sheet expansion on Antarctica: Stable isotope and sedimentological evidence from Kerguelen Plateau, southern Indian Ocean. Geology 20 (6), 569–573.

Zachos, J.C., Dickens, G.R., Zeebe, R.E., 2008. An early Cenozoic perspective on greenhouse warming and carbon-cycle dynamics. Nature 451 (7176), 279–283. Available from: https://doi.org/10.1038/nature06588.

Zachos, J.C., Kump, L.R., 2005. Carbon cycle feedbacks and the initiation of Antarctic glaciation in the earliest Oligocene. Global and Planetary Change 47 (1), 51–66.

Zachos, J., Pagani, M., Sloan, L., Thomas, E., Billups, K., 2001. Trends, rhythms, and aberrations in global climate 65 Ma to present. Science 292 (5517), 686–693. Available from: https://doi.org/10.1126/science.1059412.

Zhang, L., Hay, W.W., Wang, C., Gu, X., 2019. The evolution of latitudinal temperature gradients from the latest Cretaceous through the Present. Earth-Science Reviews 189, 147–158. Available from: https://doi.org/10.1016/j.earscirev.2019.01.025.

Zhang, Y.G., Pagani, M., Liu, Z., Bohaty, S.M., DeConto, R.M., 2013. A 40-million-year history of atmospheric CO_2. Philosophical Transactions of the Royal Society. A 371. Available from: https://doi.org/10.1098/rsta.2013.0096.

Further reading

Briggs, R.D., Tarasov, L., 2013. How to evaluate model-derived deglaciation chronologies: a case study using Antarctica. Quaternary Science Reviews 63, 109–127. Available from: https://doi.org/10.1016/j.quascirev.2012.11.021.

Holden, P.B., Edwards, N.R., Wolff, E.W., Valdes, P.J., Singarayer, J.S., 2011. The Mid-Brunhes event and West Antarctic ice sheet stability. Journal of Quaternary Science 26 (5), 474–477. Available from: https://doi.org/10.1002/jqs.1525.

Irvalı, N., Galaasen, E.V., Ninnemann, U.S., Rosenthal, Y., Born, A., Kleiven, H.F., 2020. A low climate threshold for south Greenland Ice Sheet demise during the Late Pleistocene. Proceedings of the National Academy of Sciences 117, 190–195.

Jacobs, S.S., Hellmer, H.H., Doake, C.S.M., Jenkins, A., Frolich, R.M., 1992. Melting of ice shelves and the mass balance of Antarctica. Journal of Glaciology 38 (130), 375–387. Available from: https://doi.org/10.3189/S0022143000002252.

Kageyama, M., Paul, A., Roche, D.M., Van Meerbeeck, C.J., 2010. Modelling glacial climatic millennial-scale variability related to changes in the Atlantic meridional overturning circulation: a review. Quaternary Science Reviews 29 (21–22), 2931–2956. Available from: https://doi.org/10.1016/j.quascirev.2010.05.029.

Kohfeld, K.E., Graham, R.M., De Boer, A.M., Sime, L.C., Wolff, E.W., Le Quéré, C., et al., 2013. Southern Hemisphere westerly wind changes during the Last Glacial Maximum: paleo-data synthesis. Quaternary Science Reviews 68, 76–95.

Masson-Delmotte, V., Buiron, D., Ekaykin, A., Frezzotti, M., Gallée, H., Jouzel, J., et al., 2011. A comparison of the present and last interglacial periods in six Antarctic ice cores. Climate of the Past 7, 397–423. Available from: https://doi.org/10.5194/cp-7-397-2011.

Milker, Y., Rachmayani, R., Weinkauf, M., Prange, M., Raitzsch, M., Schulz, M., et al., 2013. Global and regional sea surface temperature trends during Marine Isotope Stage 11. Climate of the Past 9 (5), 2231–2252. Available from: https://doi.org/10.5194/cp-9-2231-2013.

Miller, K.G., Mountain, G.S., Wright, J.D., Browning, J.V., 2011. A 180-million-year record of sea level and ice volume variations from continental margin and deep-sea isotopic records. Oceanography 24 (2), 40–53. Available from: https://www.jstor.org/stable/24861267.

Miller, K.G., Fairbanks, R.G., Mountain, G.S., 1987. Tertiary oxygen isotope synthesis, sea level history, and continental margin erosion. Paleoceanography 2 (1), 1–19. Available from: https://doi.org/10.1029/PA002i001p00001.

Passchier, S., 2011. Linkages between East Antarctic Ice Sheet extent and Southern Ocean temperatures based on a Pliocene high-resolution record of ice-rafted debris off Prydz Bay, East Antarctica. Paleoceanography 26 (4). Available from: https://doi.org/10.1029/2010PA002061.

Passchier, S., Ciarletta, D.J., Miriagos, T.E., Bijl, P.K., Bohaty, S.M., 2017. An Antarctic stratigraphic record of stepwise ice growth through the Eocene-Oligocene transition. GSA Bulletin 129 (3–4), 318–330.

Peltier, W.R., 2004. Global glacial isostasy and the surface of the ice-age Earth: the ICE-5G (VM2) model and GRACE. Annual Review of Earth and Planetary Sciences 32, 111–149. Available from: https://doi.org/10.1146/annurev.earth.32.082503.144359.

Rohling, E.J., Hibbert, F.D., Grant, K.M., Galaasen, E.V., Irvalı, N., Kleiven, H.F., et al., 2019. Asynchronous Antarctic and Greenland ice-volume contributions to the last interglacial sea-level highstand. Nature Communications 10 (1), 1–9. Available from: https://doi.org/10.1038/s41467-019-12874-3.

Shevenell, A.E., Kennett, J.P., Lea, D.W., 2004. Middle Miocene southern ocean cooling and Antarctic cryosphere expansion. Science 305 (5691), 1766–1770.

Smith, B., Fricker, H.A., Gardner, A.S., Medley, B., Nilsson, J., Paolo, F.S., et al., 2020. Pervasive ice sheet mass loss reflects competing ocean and atmosphere processes. Science 368 (6496), 1239–1242. Available from: https://doi.org/10.1126/science.aaz5845.

Stap, L.B., de Boer, B., Ziegler, M., Bintanja, R., Lourens, L.J., van de Wal, R.S., 2016. CO_2 over the past 5 million years: continuous simulation and new δ11B-based proxy data. Earth and Planetary Science Letters 439, 1–10. Available from: https://doi.org/10.1016/j.epsl.2016.01.022.

Strugnell, J.M., Pedro, J.B., Wilson, N.G., 2018. Dating Antarctic ice sheet collapse: Proposing a molecular genetic approach. Quaternary Science Reviews 179, 153–157. Available from: https://doi.org/10.1016/j.quascirev.2017.11.014.

Taviani, M., Reid, D.E., Anderson, J.B., 1993. Skeletal and isotopic composition and paleoclimatic significance of Late Pleistocene carbonates, Ross Sea, Antarctica. Journal of Sedimentary Research 63 (1), 84–90. Available from: https://doi.org/10.1306/D4267A96-2B26-11D7-8648000102C1865D.

Turney, C.S., Jones, R.T., Phipps, S.J., Thomas, Z., Hogg, A., Kershaw, A.P., et al., 2017. Rapid global ocean-atmosphere response to Southern Ocean freshening during the last glacial. Nature Communications 8 (1), 1–9. Available from: https://doi.org/10.1038/s41467-017-00577-6.

Vautravers, M.J., Hillenbrand, C.D., 2008. Deposition of planktonic forminifera on the Pacific margin of the Antarctic Peninsula (ODP Site 1101) during the last 1 Myr. Geophysical Research Abstracts 10.

Zecchin, M., Catuneanu, O., Rebesco, M., 2015. High-resolution sequence stratigraphy of clastic shelves IV: high-latitude settings. Marine and Petroleum Geology 68, 427–437. Available from: https://doi.org/10.1016/j.marpetgeo.2015.09.004.

Chapter 13

The future evolution of Antarctic climate: conclusions and upcoming programmes

Martin Siegert[1], Fabio Florindo[2], Laura De Santis[3] and Tim R. Naish[4]
[1]Grantham Institute and Department of Earth Science and Engineering, Imperial College London, London, United Kingdom, [2]National Institute of Geophysics and Volcanology, Rome, Italy, [3]National Institute of Oceanography and Applied Geophysics – OGS, Sgonico, Italy, [4]Antarctic Research Centre, Victoria University of Wellington, Wellington, New Zealand

13.1 Introduction: the past is key to our future

There is no doubt that the Antarctic ice sheet is experiencing significant mass loss through anthropogenic warming and is contributing to global sea-level rise. The satellite altimetric record of ice-sheet surface-elevation changes reveals a sixfold increase in the rate of loss over the last 30 years (Rignot et al., 2019; Shepherd et al., 2018). Since around 1850 the concentration of atmospheric CO_2 has risen from ~280 to over 415 parts per million (ppm), resulting in a global mean temperature rise of ~0.9°C–1.2°C (IPCC, 2018, 2021; Schurer et al., 2017). While thermal expansion of the ocean has been the dominant process forcing just under half of the ~20 cm of sea level rise since 1850, the contribution of glaciers and ice sheets now exceeds thermal expansion and it is likely to become a growing and controlling component with further warming (IPCC, 2019).

While global warming is undoubtedly the underlying cause of ice-sheet changes measured, the processes by which greenhouse-gas emissions lead to ice-sheet decay are complex, yet critical to evaluating how the ice sheet may change in future. The physical relationships involved are important because they will dictate how the ice sheet is likely to change under further warming, be it constrained to 1.5°C above pre-industrial levels by reducing emissions to net zero by mid Century or driven higher by unabated emissions (Rogelj et al., 2018).

The greatest Antarctic ice loss is being observed within deep-marine terminating margins of West Antarctica (around the Amundsen Sea

Antarctic Climate Evolution. DOI: https://doi.org/10.1016/B978-0-12-819109-5.00005-0

Embayment) and also in parts of East Antarctica (e.g., Totten Glacier; 67°S, 116°E). The main driver for such change is warm ocean water and high-salinity current intrusions that can advect across continental shelves and under cavities below floating ice, up to the ice-sheet grounding zone where the ice-sheet margin is melted (IPCC, 2019). This melting leads to ice-shelf and ice-sheet thinning, and grounding zone retreat. In regions where the bed slopes down towards the ice sheet interior, a runaway process called 'marine ice sheet instability' (MISI) can occur, which is particularly relevant in West Antarctica where the bulk of the ice sheet rests on a bed >1 km below sea level, and some parts of East Antarctica. Indeed, some have argued that MISI is already underway in West Antarctica (Joughin et al., 2014; Rignot et al., 2014).

Although we appreciate the role that ocean warmth is playing on ice-sheet mass balance, the details and processes are not known well enough to allow numerical models to describe observations confidently, or to predict with certainty how the situation will evolve this century (Siegert et al., 2020). Modelling is hampered by insufficient knowledge of (1) the bathymetry around the ice sheet margin, despite it controlling how warm and saline water flows towards the ice sheet; (2) coarseness of resolution, disallowing eddies (despite them being key to the transfer of heat from the ocean to the ice sheet) and surface winds to be accounted for adequately; (3) ocean observations along the ice sheet margin, which could provide valuable information with which to constrain models; and (4) sufficiently detailed records of past change around the Antarctic continental margin explaining how, when and why the Antarctic ice sheet has reacted to periods of global warming. Given the potential consequences for rapid global sea level rise if the West Antarctic Ice Sheet and the marine-based sectors of the East Antarctic Ice Sheet were to become unstable, better definition and measurement of past and modern processes, and their incorporation into models, is critical.

Floating ice shelves are key to changes to the ice sheet upstream, as they offer a 'backforce' against the flow of ice that restricts how much ice can enter the ocean and affect sea-level rise irrespective of whether it melts or not. They are able to receive warmth from both the ocean and the atmosphere, and so may suffer loss of ice from both their upper and lower boundaries. In the Antarctic Peninsula, ice-sheet disintegration has occurred suddenly (Banwell et al., 2013), followed by acceleration of the flow of grounded ice. Under further warming this century (especially in the upper-range of possibilities), the stability of ice shelves becomes questionable. If a large ice shelf suddenly disappears, it is possible that the ice-sheet grounding line becomes a vertical wall of ice. If that wall is unable to support itself mechanically it will fail by calving, leading to migration of the wall, and grounding line, upstream. This is the marine ice cliff instability (MICI) hypothesis and has the potential to cause ice loss much more quickly than through MISI (DeConto and Pollard, 2016; Pattyn et al., 2017).

The drivers for rapid ice-sheet change, and sea-level outcomes, are now clearer. While we must improve our ability to model the processes responsible (Siegert et al., 2020), we also need to look into the palaeo record to observe whether, how and under what conditions such changes have occurred in the past and, especially, if irreversible thresholds in conditions and ice-sheet behaviour have led to inevitable and substantial ice-loss episodes.

Under Hutton's principle of 'uniformitarianism' (Hutton, 1788), past changes can be explained by processes witnessed today. That is to say, past changes – even those in deep time – are caused by physical processes that we can quantify now. Under this principle, physical processes behave regularly and if we understand them well enough they can be used to explain the geological past and what may happen in the future. Hence, understanding of physical processes responsible for ice-sheet events in the geological record is essential for predictions of ice-sheet change this century and beyond.

Irrespective of whether we can constrain greenhouse gas emissions, there are likely to be examples in the palaeo record of what to expect from Antarctica in future. Warming, even under the 1.5°C scenario (Siegert et al., 2019), is inevitable, and substantial change under high emissions remains possible. Resolving which pathway we are on, and the consequences for global sea level, is one of the greatest scientific challenges of the modern age.

Models are needed to make quantitative predictions (Siegert and Golledge, 2021, this volume), and while improvements in their design are necessary so too is better resolution and spatially widespread information of past changes, to a level where physical processes can be deciphered, both locally and regionally (see Colleoni et al., 2021, this volume, for further details). If we can achieve this, models can be better 'trained' against the palaeo record and used with more confidence to make predictions, especially regarding how rates of change are represented in the past and can be replicated plausibly in numerical models.

13.2 Upcoming plans and projects

There is great reason to be optimistic that the next ten years will see knowledge of past changes in Antarctica improve to the extent that it can help predict how physical processes will influence our future. Clearly the global sea-level problem is the central focus of this challenge, but there are also consequences for future ocean conditions and marine life that the palaeo record can inform.

The contribution of the Scientific Committee on Antarctic Research (SCAR) to this challenge, in its promotion of both international collaboration and continental-wide appreciation, should not be underestimated. SCAR's role in forming a scientific Horizon Scan (Kennicutt et al., 2015), where the most important 80 scientific questions that need answers within 20 years were defined, has allowed research proposals to focus on shared and essential research ambitions. This, coupled with the response to the Horizon Scan led by the Council of Managers of National Antarctic Programs (COMNAP), where logistics and facilities planning is aligned against scientific drivers (Kennicutt et al., 2016), means that the scientific community is better connected and set-up to deliver the research outcomes necessary to understand past changes in Antarctica and predict its future. The SCAR Research Program 'Past Antarctic Ice Sheet dynamics' (PAIS) has been addressing one of the six priorities of the Horizon Scan: 'Antarctic ice sheets and sea-level – Understand how, where and why ice sheets lose mass'. Furthermore, PAIS, as well as its predecessors 'Antarctic Climate Evolution' (ACE) and the 'ANTarctic Offshore STRATigraphy project' (ANTOSTRAT), has made considerable efforts to coordinate and stimulate international, multidisciplinary research in line with SCAR's ambition for scientists to work collaboratively on the most important problems. Five years on, great progress has been made in many areas (Kennicutt et al., 2019), but because of the long lead times for deep-field projects, the best work lies ahead in the coming decade.

In terms of projects that are upcoming, and relevant to past Antarctic changes, we can immediately point to INSTANT (INStabilities and Thresholds in ANTarctica), SCAR's research programme that developed from PAIS. Importantly, INSTANT aims to "quantify the Antarctic ice sheet contribution to past and future global sea-level change, from improved understanding of climate, ocean and solid Earth interactions and feedbacks with the ice, so that decision-makers can better anticipate and assess the risk in order to manage and adapt to sea-level rise and evaluate mitigation pathways". It will do this by promoting multidisciplinarity and collaboration, and will focus on research programmes that can make a tangible positive contribution to the aim.

SCAR action and expert groups that are underway, and within the INSTANT remit, include:

- PRAMSO (Paleoclimate Records from the Antarctic Margin and Southern Ocean), which will initiate, promote and coordinate scientific research drilling around the Antarctic margin and the Southern Ocean to obtain past records of ice sheet dynamics and ice sheet ocean interactions that are critical for improving the accuracy and precision of predictions of future changes in global and regional temperatures and sea level rise;
- ADMAP (Antarctic Digital Magnetic Anomaly Project) that will compile and integrate all existing Antarctic near-surface and satellite magnetic anomaly data into a digital database and lead to a better appreciation of tectonic structures and history;
- Bedmap3 (Ice thickness and subglacial topographic model of Antarctica) that will compile ice thickness measurements (that now cover the entire continent without major gaps; Cui et al., 2020) to deliver an updated

account of the bed topography in Antarctica, which is a fundamental boundary condition to numerical ice-sheet models; and

- Antarchitecture (internal structure of ice sheets) – providing the first continental-wide assessment of ice sheet stratigraphy that could provide a novel and important way in which ice flow can be constrained.

  Antarctic RINGS - providing more accurate and complete reference of bed topography and bathymetry of the cavity beneath ice shelves around Antarctica. RINGS will clarify current knowledge gaps at the ice-sheet margin and develop protocols to prioritise and systematically collect, analyse and share comprehensive airborne geophysical measurements from the most vulnerable and data-sparse regions. The aim is constraining the ice sheet subglacial hydrology and geology, ice basal mass balance and discharge from all around Antarctica.

Other international programmes that will likely provide answers to why and how fast Antarctica has changed in the past include the International Thwaites Glacier Collaboration, which will focus on understanding modern processes and past changes in the region that has experienced most ice sheet loss and, indeed, is most vulnerable to both MISI and MICI, potentially this century. To improve our knowledge of ice sheet mass balance, SCAR is supporting collaboration with the World Climate Research Programme (WCRP) named 'ISMASS' (Ice Sheet Mass Balance and Sea Level). SCAR, via INSTANT, is also working in partnership with a range of external programmes on this issue, including WCRP's Climate and Cryosphere (CLiC) project, and the numerical model intercomparison projects ISMIP (for ice sheet models) and MISOMIP (specially for marine-based ice sheets).

In addition, there are numerous drilling programmes that are aiming to acquire records from key positions at the margins especially from the East Antarctic sectors, some of which are still completely unknown. A new generation of challenging, shallow (ice-breaker vessel based) and deep (from floating ice) drilling projects have been proposed during the life span of PAIS to be achieved under the coordination of INSTANT, following the successful ANDRILL, SHALDRILL and MeBo projects (see McKay et al., 2021, this volume for further details). They will allow new geological records to be obtained from other ice proximal and coastal areas, crucially needed to fill the knowledge gap between deep-sea marine and ice-core records. Drilling programs are also planned from the ice shelf, near the present-day grounding line and from the centre of today's ice sheet, including within subglacial lakes at the WAIS ice divide, where ancient sedimentary records are expected from geophysical measurements (Smith et al., 2018). Critically, marine sediments from such locations would prove conclusively the absence of ice and, if they can be dated, will determine the last time when the WAIS decayed.

Large efforts will also need to be dedicated to sea-bed mapping and sub-sea bed geophysical surveys, combined with observations of present-day oceanography and ecosystems, because large areas of the Antarctic margin are still very poorly known, despite their importance for reconstructing the ice sheet evolution.

13.3 Conclusions

Predicting 21st Century sea-level change, with a level of precision useful to decision-makers, is a key scientific challenge of our time. Ice-sheet modelling can be used to calculate the rate of ice loss, but improvements are needed in two main areas. The first is by improvement to the models themselves, through better inclusion of physical processes by glaciological theory, laboratory experiments and measurements of ice, ocean and atmospheric interactions. The second is to 'train' models by matching their performance against records of well-known past changes. While great progress has been made in better defining the main stages of the Antarctic ice sheet's evolution, essential work is now needed to provide ice-sheet models with the well-documented case studies they require.

References

Banwell, A.F., MacAyeal, D.R., Sergienko, O.V., 2013. Breakup of the Larsen B Ice Shelf triggered by chain reaction drainage of supraglacial lakes. Geophysical Research Letters 40, 5872–5876. Available from: https://doi.org/10.1002/2013GL057694.

Colleoni, F., et al., 2021. Past Antarctic ice sheet dynamics (PAIS) and implications for future sea-level change. In: Florindo, F., et al., (Eds.), Antarctic Climate Evolution, second ed. Elsevier (this volume).

Cui, X., Jeofry, H., Greenbaum, J.S., Guo, J., Li, L., Lindzey, L.E., et al., 2020. Bed topography of Princess Elizabeth Land in East Antarctica. Earth System Science Data 12, 2765–2774. Available from: https://doi.org/10.5194/essd-12-2765-2020.

DeConto, R.M., Pollard, D.M., 2016. Contribution of Antarctica to past and future sea-level rise. Nature 531, 591–597. Available from: https://doi.org/10.1038/nature17145.

Hutton, J., 1788. Theory of the Earth; or an investigation of the laws observable in the composition, dissolution, and restoration of land upon the globe. Earth and Environmental Science Transactions of the Royal Society of Edinburgh 1 (2), 209–304. Available from: https://doi.org/10.1017/S0080456800029227.

IPCC, 2018. Summary for policymakers. In: Masson-Delmotte, V., et al., (Eds.), Global Warming of 1.5°C. An IPCC Special Report on the Impacts of Global Warming of 1.5°C above Pre-industrial Levels and Related Global Greenhouse Gas Emission Pathways, in the Context of Strengthening the Global Response to the Threat of Climate Change, Sustainable Development, and Efforts to Eradicate Poverty. World Meteorological Organization, p. 32.

IPCC, 2019. Summary for policymakers. In: Pörtner, H.-O., Roberts, D.C., Masson-Delmotte, V., Zhai, P., Tignor, M., Poloczanska, E., et al. (Eds.), IPCC Special Report on the Ocean and Cryosphere in a Changing Climate. Available from: <https://www.ipcc.ch/srocc/chapter/summary-for-policymakers/>.

IPCC, 2021. CLIMATE CHANGE 2021: The Physical Science Basis. Summary for Policymakers. Working Group I Contribution to the Sixth Assessment Report of the Intergovernmental Panel on Climate Change, August 2021.

Joughin, I., Smith, B.E., Medley, B., 2014. Marine ice sheet collapse potentially under way for the Thwaites Glacier basin, West Antarctica. Science (New York, N.Y.) 344, 735–738. Available from: https://doi.org/10.1126/science.1249055.

Kennicutt, M.C., et al., 2015. A roadmap for Antarctic and Southern Ocean science for the next two decades and beyond. Antarctic Science 27, 3–18. Available from: https://doi.org/10.1017/S0954102014000674.

Kennicutt, M.C., et al., 2016. Delivering 21st century Antarctic and Southern Ocean science. Antarctic Science 28, 407–423. Available from: https://doi.org/10.1017/S0954102016000481.

Kennicutt, M.C., et al., 2019. Sustained Antarctic research—a 21st century imperative. One Earth 1, 95–113. Available from: https://doi.org/10.1016/j.oneear.2019.08.014.

McKay, R., et al., 2021. In: Florindo, F., et al., (Eds.), Antarctic Climate Evolution, second ed. Elsevier (this volume).

Pattyn, F., Favier, L., Sun, S., Durand, G., 2017. Progress in numerical modeling of Antarctic ice-sheet dynamics. Current Climate Change Reports 3, 174–184. Available from: https://doi.org/10.1007/s40641-017-0069-7.

Rignot, E., Mouginot, J., Morlighem, M., Seroussi, H., Scheuchl, B., 2014. Widespread, rapid grounding line retreat of pine Island, Thwaites, Smith, and Kohler glaciers, West Antarctica, from 1992 to 2011. Geophysical Research Letters 41, 3502–3509. Available from: https://doi.org/10.1002/2014GL060140.

Rignot, E., Mouginot, J., Scheuchl, B., van den Broeke, M., van Wessem, M.J., Morlighem, M., 2019. Four decades of Antarctic Ice Sheet mass balance from 1979–2017. Proceedings of the National Academy of Sciences 116 (4), 1095. Available from: https://doi.org/10.1073/pnas.1812883116.

Rogelj, J., Popp, A., Calvin, K.V., Luderer, G., Emmerling, J., Gernaat, D., Fujimori, S., Strefler, J., Hasegawa, T., Marangoni, G. et al. 2018. Scenarios towards limiting global mean temperature increase below 1.5°C. Nature Climate Change 8, 325–332. Available from: https://doi.org/10.1038/s41558-018-0091-3.

Schurer, A.P., Mann, M.E., Hawkins, E., Tett, S.F.B., Hegerl, G.C., 2017. Importance of the pre-industrial baseline for likelihood of exceeding Paris goals. Nature Climate Change 7, 563–567. Available from: https://doi.org/10.1038/nclimate3345.

Shepherd, A., Ivins, E., Rignot, E., Smith, B., van den Broeke, M., Velicogna, I., et al., 2018. Mass balance of the Antarctic ice sheet from 1992 to 2017. Nature 558, 219–222. Available from: https://doi.org/10.1038/s41586-018-0179-y.

Siegert, M.J., Alley, R.B., Rignot, E., Englander, J., Corell, R., 2020. 21st Century sea-level rise could exceed IPCC predictions for strong-warming futures. One Earth 3, 691–703. Available from: https://doi.org/10.1016/j.oneear.2020.11.002.

Siegert, M.J., Atkinson, A., Banwell, A., Brandon, M., Convey, P., Davies, B., et al., 2019. The Antarctic Peninsula under a 1.5°C global warming scenario. Frontiers in Environmental Science 7, 102. Available from: https://doi.org/10.3389/fenvs2019.00102.

Siegert, M.J., Golledge, N.R., 2021. Advances in numerical modelling of the Antarctic ice sheet. In: Florindo, F., et al., (Eds.), Antarctic Climate Evolution, second edition Elsevier (this volume).

Smith, A.M., Woodward, J., Ross, N., Bentley, M.J., Hodgson, D.A., Siegert, M.J., et al., 2018. Evidence for the long-term sedimentary environment in an Antarctic subglacial lake. Earth and Planetary Science Letters 504, 139–151. Available from: https://doi.org/10.1016/j.epsl.2018.10.011.

Index

Note: Page numbers followed by "*f*" and "*t*" refer to figures and tables, respectively.

B

C

D

E

K

L

M

N

O

P

Q

R

Printed in the United States
by Baker & Taylor Publisher Services